Sonoran Desert Plants

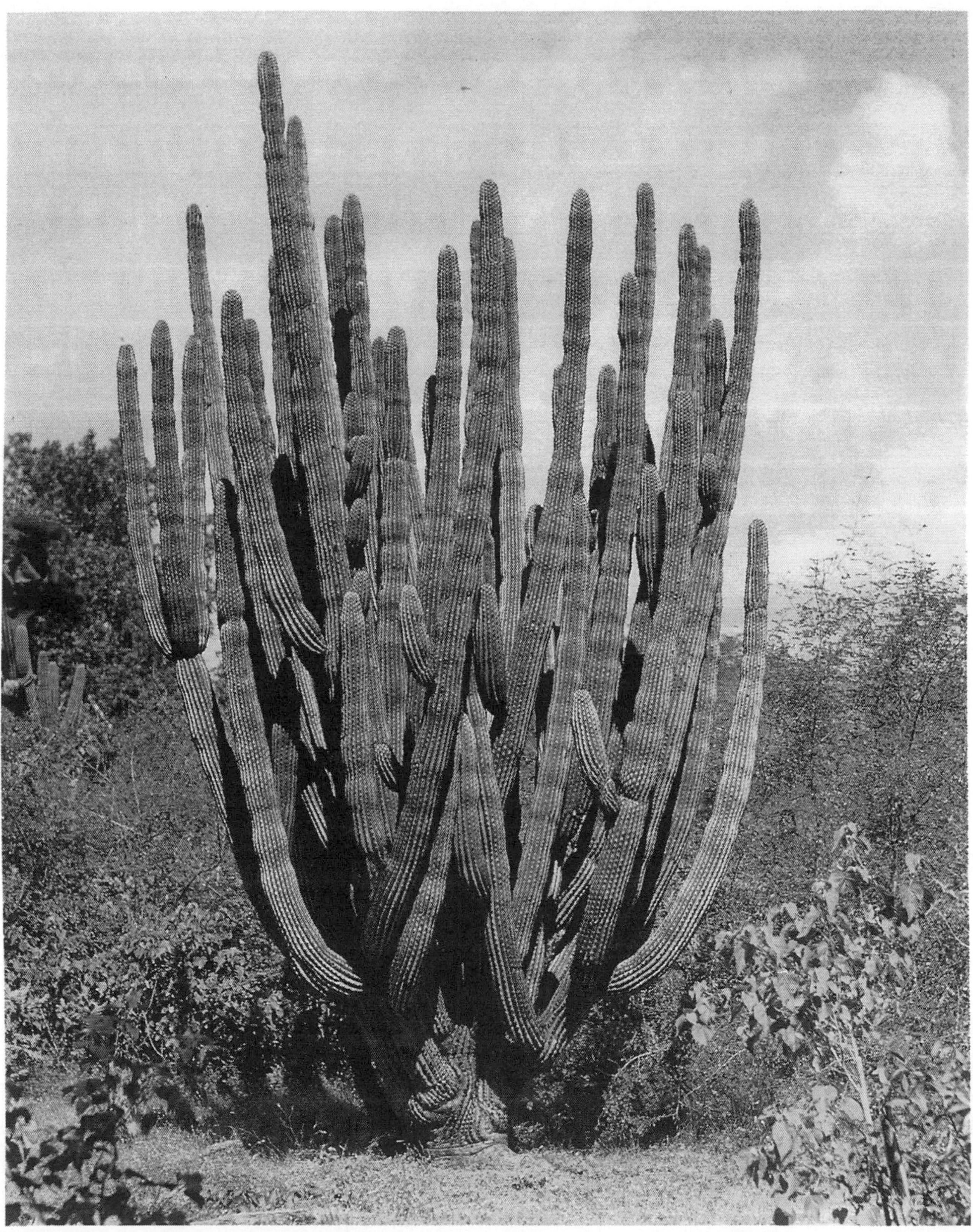

Sonoran Desert Plants

AN ECOLOGICAL ATLAS

Raymond M. Turner

Janice E. Bowers

Tony L. Burgess

THE UNIVERSITY OF ARIZONA PRESS
Tucson

First paperbound printing 2005
The University of Arizona Press

♾ This book is printed on acid-free, archival-quality paper.
Manufactured in the United States of America

10 09 08 07 06 05 7 6 5 4 3 2

Library of Congress Cataloging in Publication Data
Turner, R. M. (Raymond M.)
Sonoran Desert plants: an ecological atlas / Raymond M. Turner, Janice E. Bowers, Tony L. Burgess.
p. cm.
Rev. ed. of: An atlas of some plant distributions in the Sonoran Desert / James Rodney Hastings.
Includes bibliographical references and index.
ISBN-13: 978-0-8165-1532-5 (alk. paper)
ISBN-10: 0-8165-1532-8 (alk. paper)
ISBN-13: 978-0-8165-2519-5 (pbk: alk. paper)
ISBN-10: 0-8165-2519-6 (pbk: alk. paper)
1. Desert plants—Sonoran Desert—Maps. 2. Phytogeography—Southwest, New—Maps. 3. Phytogeography—Mexico—Baja California—Maps. I. Bowers, Janice Emily. II. Burgess, Tony L., 1949– . III. Hastings, James Rodney. Atlas of some plant distributions in the Sonoran Desert. IV. Title.
G1107.S6D2T8 1995 <G&M>
582.90954'0979'022—c20 94-18723

Frontispiece: *Stenocereus thurberi* near Miraflores, Baja California Sur. Photograph by J.R. Hastings.

To Rod Hastings (1923–1974)

Contents

Figures

Distribution Maps

Acknowledgments

To the extent that the success of our enterprise can be gauged by any measure, a large share of the credit is due Rod Hastings, who was present from the conception of the original atlas to its publication in 1972. Rod not only traveled innumerable miles in the field, he also initiated the onerous task of collecting herbarium data. Our debt to him is great.

Among the active contributors to this atlas, one stands out above all the others for his meticulous record keeping. By providing geographic coordinates in addition to verbal descriptions of site locations, Reid Moran saved us many hours of tedious labor. For this, and for his hospitality and his many helpful comments on maps and species accounts, we are deeply indebted. Others who also maintained an interest in this project over the quarter-century during its revival and who unfailingly answered our requests for assistance are Annetta Carter, Howard Scott Gentry, and Charles T. Mason, Jr. Many botanists, biologists, and ecologists, particularly Thomas R. Van Devender, Mary Butterwick, Hyrum Johnson, Arthur M. Phillips III, Wesley E. Niles, Frank Reichenbacher, Peter Warren, and Susan Anderson, went out of their way to provide us with extensive site records from many Sonoran Desert stations. Their efforts have significantly improved our entries for Arizona and California. Among those who cheerfully and promptly resolved numerous distributional and taxonomic problems are Thomas Daniel, Richard Felger, Wendy Hodgson, Geoffrey Levin, Howard Gentry, Melissa Luckow, Donald Pinkava, Rupert Barneby, Andrew Sanders, Steve Boyd, Stephanie Meyer, Mary Sangrey, David Seigler, Guy Nesom, and José Luis León de la Luz. For their courtesy, we thank the curators or directors of the herbaria we visited. Geoffrey Levin and James Ehleringer undertook the heroic task of reading and commenting on the entire manuscript. Many of those named above also reviewed selected species accounts and maps. Other reviewers include Matthew Finn, Steven McLaughlin, Paul Martin, and Matthew Johnson. Peter Van Metre and Douglas Wellington contributed considerable and crucial programming expertise. Carlos Mendoza patiently generated the maps on computer, some of them many times.

We thank them all and regret that their conscientious efforts could not prevent us from making an imperfect product.

Financial support for this work during the 23 years since the original atlas appeared has come from the U.S. Geological Survey.

Introduction

Almost a quarter of a century has passed since the first Sonoran Desert plant atlas appeared (Hastings et al. 1972:iii) with a promise to produce "a more complete atlas showing the distribution of the major perennial plants of the Sonoran Desert." The difficulty of compiling this volume was abruptly magnified by the death of Rod Hastings in 1974. Because Douglas Warren, the third member of the original team, had already left the desert, the full burden of preparing an enlarged volume rested with Raymond Turner alone, and for a while, it seemed the project might founder. A new team emerged when Tony Burgess (in 1978), then Janice Bowers (in 1982), joined Turner, and work on the atlas began again with renewed vigor. The design of the expanded atlas emerged over a period of several years as we added species and decided to provide a narrative, or species account, for each. The pattern for the expanded version was actually set by Forrest Shreve, who published 27 distribution maps and species accounts for 26 common Sonoran Desert plants in his classic *Vegetation of the Sonoran Desert* (1951, 1964).

Scope of the Revised Atlas

The first Sonoran Desert atlas presented maps for 238 species. Most were so abundant or generally distributed that any botanist would have included them. Others—certain minor species in the genera *Ambrosia, Bursera,* and *Randia,* for example—reflected the special interests of the authors. With a few exceptions, all were woody or succulent. This new edition has retained all but 3 of the species presented in the original atlas and has added 104 more, greatly increasing representation in characteristic Sonoran Desert families such as Cactaceae and Agavaceae. Moreover, as a result of our interest in legumes (Turner and Busman 1984), we have brought in most of the woody legumes of the Sonoran Desert. Users familiar with the earlier edition will find here a new emphasis on thornscrub plants, not inappropriate in view of recent redefinition of the desert and thornscrub communities (Brown 1982a; Turner and Brown 1982). Many Mojave Desert species are found in the Sonoran Desert and are included here. The vegetation of the two deserts is similar (Turner 1982), whereas their floras are distinct (McLaughlin 1989, 1992). We have not included any Mojave Desert endemics.

Data Collection

Collection of field data for the atlas began in 1963 when Hastings and Turner first traveled the length of Baja California and began assembling plant distribution data.

This initial trip was followed by others throughout the peninsula and the state of Sonora. On these trips, Hastings and Turner drove both principal highways and lightly used tracks, systematically logging plant occurrences at intervals of 8–16 km. In recent years, we continued to collect field data during our trips to various parts of the desert. For this edition of the atlas, we relied even more heavily than before on unpublished data collected by agencies—the Bureau of Land Management, for example—and individual collectors, most notably Richard Felger and Thomas R. Van Devender. We obtained additional plant localities from herbarium specimens deposited at the Desert Botanical Garden in Phoenix, the San Diego Museum of Natural History, Arizona State University, the University of California at Berkeley, the University of Texas, and the University of Arizona.

Figure 1 shows the Sonoran Desert in its regional setting. The mapped distribution of our study species within this area is not uniform. Data collection was concentrated along highways and better secondary roads (resulting in the linear distribution patterns seen on certain maps). Use of local floras (for example, Simmons 1966; Zabriskie 1979; Bowers 1980) greatly increased the number of observations (and species) for some areas. Examples include the one-degree grid cells to the south and east of 29°N, 110°W; 32°N, 111°W; 29°N, 114°W; and 34°N, 115°W (figures 2a, 2b). Our data base does not substitute for the intensive collecting required to compile local floras, however; as figure 3 suggests, the number of species in the one-degree grid cells is highly dependent on the number of observations in each cell ($r^2=0.686$).

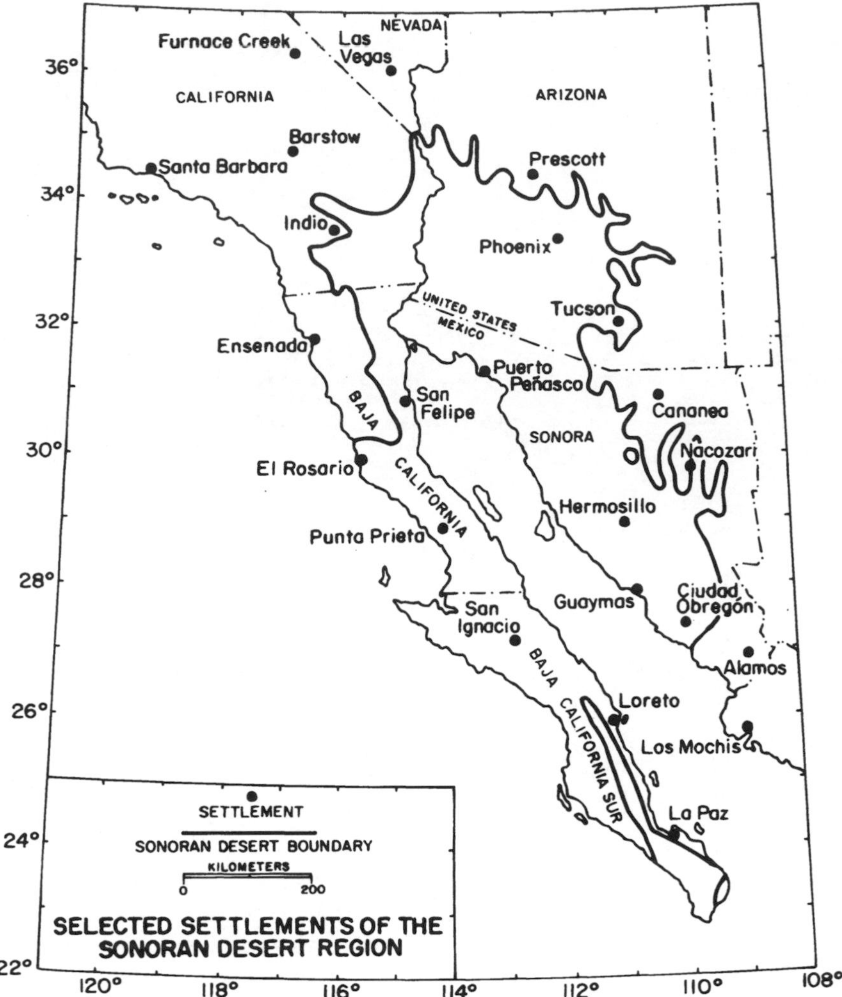

Figure 1. Location map showing Sonoran Desert region and other features of interest.

Our computer data files now contain at least some information for about 3,000 species. The data base comprises more than 100,000 separate entries. As in the original atlas, the coded information for each entry includes latitude and longitude, data source, plant collector or observer, month and year of observation, and phenologic stage. Where known, elevation, soil and terrain characteristics, species abundance, community coverages, and biotic community have also been encoded. Of necessity, only a small part of that information appears here.

Map Preparation

Botanists have made numerous efforts to map plant distributions. These treatments are usually limited to specific taxonomic groups, morphologic groups, or, as in the present effort, specific geographic areas (Perring and Walters 1976; Argus and White 1982; Wetten and Ruben-Sutter 1982; Albee et al. 1988). The emergence of computer technology, with its ability to amass information and display it in a variety of ways, has proven a boon to map makers (for example, Cadbury et al. 1971; Walker and Cocks 1991). The monumental effort to map the trees of the United States using unsophisticated techniques of hand plotting (Little 1971, 1976; Viereck and Little 1975) must be regarded as the last of its kind. As holdings of major herbaria are recorded in computer files, the potential for quickly and easily mapping plant ranges is increasing, provided that written collection locations can be converted to geographic coordinates. When herbarium labels contain too little information, potential data is irretrievably lost.

In the present atlas, each map shows distribution within the Sonoran and Mojave deserts and their periphery (from 108°W to 121°W and 22°N to 37°N). The symbol ∘ indicates a location vouchered by a herbarium specimen and the symbol × an unvouchered location. Because locality

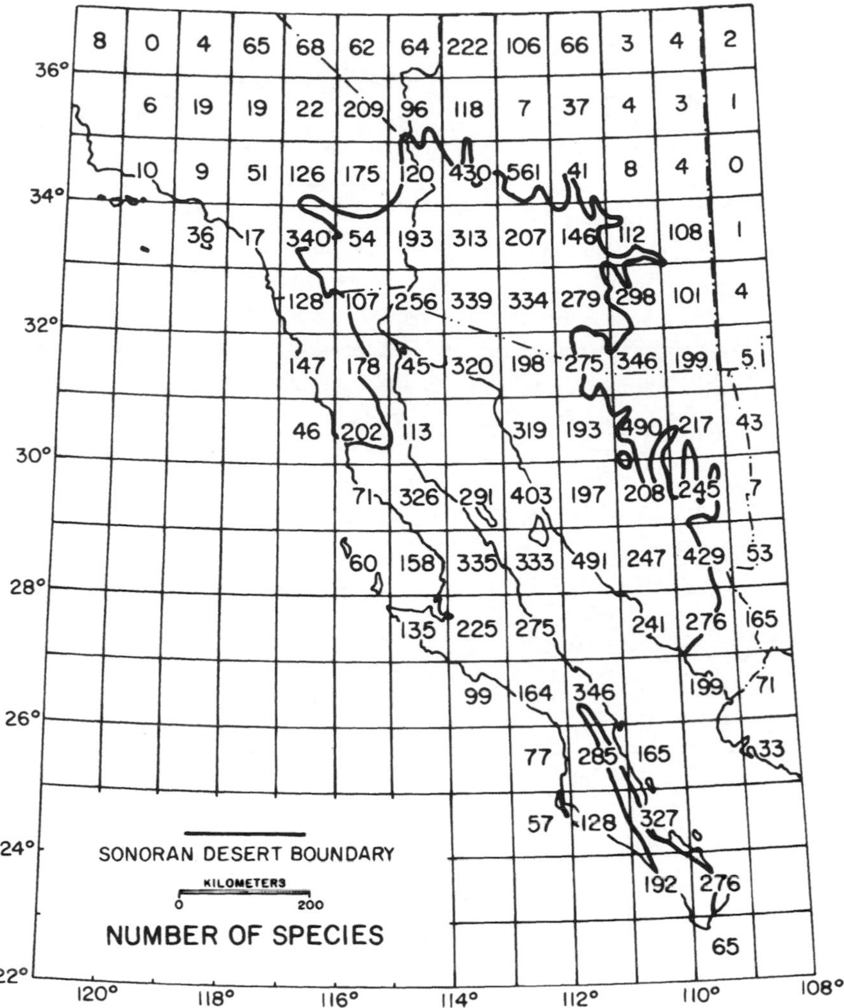

Figure 2a. Number of plant species in one-degree grid cells in the Sonoran Desert region.

resolution is limited to the nearest 0.1°, insular points may appear off-center from an island, and coastal points may look as if they were at sea. An inset of North America south of Canada depicts the range over a larger area.

Species Accounts

The species accounts are arranged alphabetically by genus. Each contains the following elements insofar as possible:

- scientific name and authority
- selected common names
- description
- diagnostic characters of similar species
- taxonomic problems
- habitat
- distributional patterns in the maps and profiles
- biogeography
- phenology
- physiology
- reproductive ecology and pollination
- seedling establishment, growth rate, and life span
- horticulture
- ethnobotany and economic botany

Given the paucity of written information for most of the species in question, our treatment is uneven. We have relied often on our own observations. The species descriptions have been adapted from regional floras (mainly Gentry 1942; Vines 1960; Wiggins 1964; Correll and Johnston 1970; Munz 1974; Benson and Darrow 1981; and Benson 1982). These descriptions—and the brief comparisons with similar species—are intended as an aid to field workers. They are not meant to substitute for the necessary keys and more detailed descriptions available in regional manuals. We have followed David E. Brown's (1982b) classification system in naming biotic communities. Common names (gleaned from many sources including Standley 1920, 1922, 1923, 1924, 1926; Kearney and Peebles 1960; and Martinez 1979) represent those names widely used within the region mapped in this atlas.

Biogeography and Paleoclimate

The species accounts and maps point up fascinating biogeographical problems. Among these we might mention the marked concentration of extralimital extensions in the vicinity of Guaymas and Puerto Libertad, Sonora; the rich concentration of tropical and subtropical species in the Guaymas area; and the remarkable disjunctions shown by *Acacia neovernicosa, Calliandra eriophylla, Mimosa aculeaticarpa,* and other species. As we have noted in the appropriate species accounts, a number of plant distributions are split between the Sonoran mainland and the Baja California peninsula—for example, *Ambrosia camphorata, Ambrosia magdalenae, Bourreria sonorae, Cordia curassavica, Euphorbia ceroderma, Fouquieria columnaris,* and *Viguiera laciniata.* These distributions may predate gulf-floor spreading, or they may represent long-distance dispersal. The disjunct occurrence of Sinaloan thornscrub species in the Cape Region, e.g., *Cercidium praecox, Erythrina flabelliformis, Haematoxylon brasiletto, Lysiloma microphyllum, Senna atomaria,* and others, presents a similar dilemma. In some cases, we have speculated on the origin and meaning of anomalous or interesting distributions. Readers

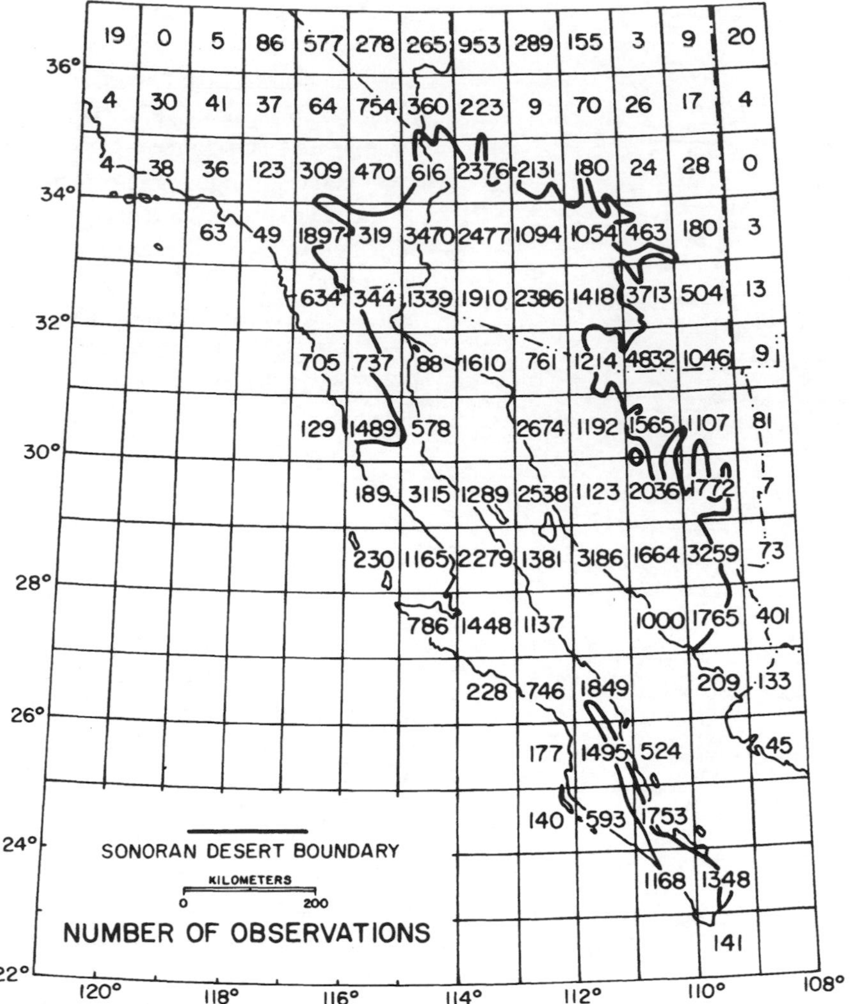

Figure 2b. Number of observations (plant collections and sightings) in one-degree grid cells in the Sonoran Desert region.

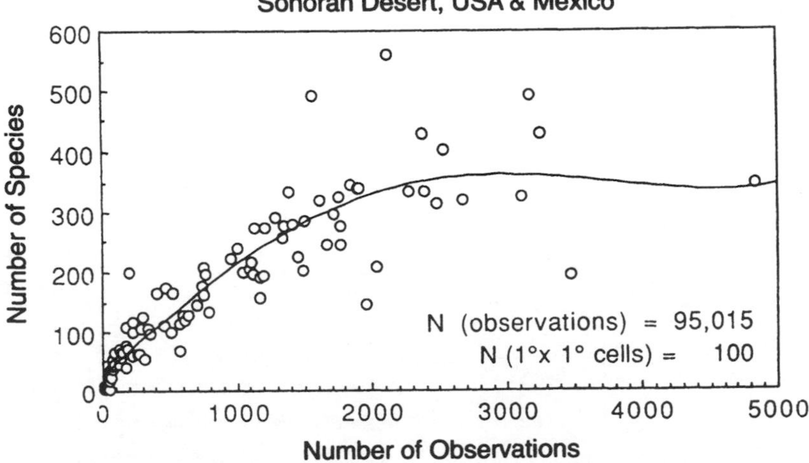

Figure 3. Number of observations (plant collections and sightings) versus number of species in one-degree grid cells in the Sonoran Desert region.

should keep in mind, however, that the evidence is admittedly scanty and our interpretations are provisional, at best.

Our interest in plant distribution is the outgrowth of our fascination with vegetation dynamics. To fully understand a plant's role in a given biotic community, one should know its broad-scale and local ranges (Brown 1984). Does the site in question occur near the periphery or the center of the range? Is the plant common throughout its distributional area or is it sporadic on sites with special characteristics? What are its elevational limits? These and other distributional features may reveal something of the past or even the destiny of a plant species.

The presence of both geographic coordinates and altitudes for many of the plant localities in our data set makes possible graphical display of these range descriptors. For this publication, we have plotted altitude versus latitude, one of many possible combinations. (We refer to these in the species accounts as *elevational profiles.*) Because climatic variables (such as temperature and rainfall seasonality) change especially sharply along gradients defined by altitude and latitude, we felt that maximum insight into a plant's distribution could be gained by their graphical display. As expected, most species show a definite cant upward from low-elevation/low-latitude combinations toward high-elevation/high-latitude pairings. (Two of many examples are ***Carnegiea gigantea*** and ***Cercidium microphyllum.***) This configuration probably reflects a close association with summer rainfall or merely a need for more moisture than is supplied at low-elevation stations at higher latitudes. At their southern limits, where they have evidently been excluded from the thornscrub and short-tree forest communities typical of moderate altitudes, many of these plants are restricted to relatively arid coastal habitats.

Other species occur throughout a wide elevational range over much of their latitudinal ranges. Two of many possible examples are *Trixis californica* and *Viguiera laciniata*. The distribution of *V. laciniata* reflects the pervasive influence of Pacific maritime air throughout its latitudinal range in Baja California. In contrast, *T. californica* is able to thrive in a variety of temperature and rainfall regimes. Each species demonstrates a unique pattern with regard to altitude and latitude, and readers of this atlas will draw many other environmental inferences from these graphical displays. (See, for example, Wells and Berger 1967; Neilson 1987.)

A noteworthy feature of the mapped region is the variation in species richness of woody legumes (figure 4). Because of our special interest in this group, our records for woody legumes are particularly detailed; the number of woody legumes in each one-degree cell is, therefore, relatively independent of the number of observations for that cell ($r^2=0.320$). Figure 4 shows that the species richness of woody legumes peaks at the southern and eastern borders of the area mapped in this atlas and declines to the northwest. This pattern suggests that the amount of warm-season rainfall and winter cold may be major determinants in the distribution of this group. Of further interest is the finding that species richness is greater in southeastern continental grid cells than in southern peninsular ones. This finding suggests that peninsular isolation is associated with a decrease in species richness, which conforms with the island biogeography theory of Robert MacArthur and Edward Wilson (1967).

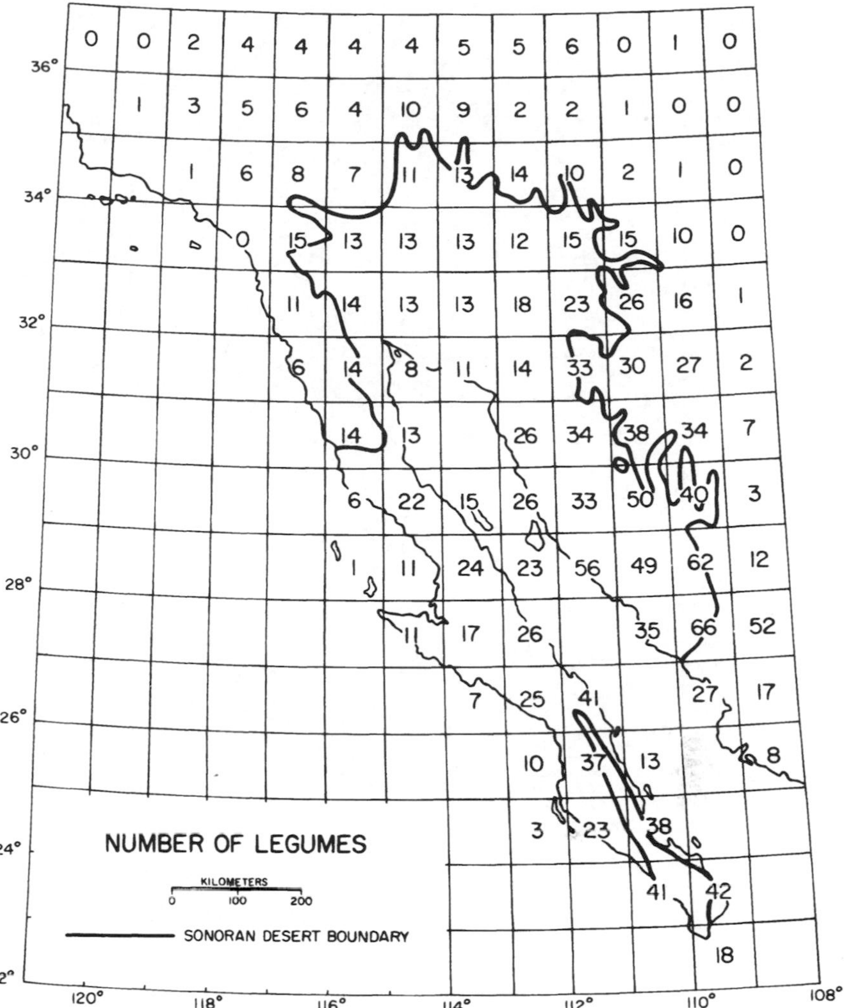

Figure 4. Number of woody legume species in one-degree grid cells in the Sonoran Desert region.

Climates of the Sonoran Desert Region

Our mapped area incorporates small parts of the Chihuahuan and Great Basin deserts, most of the Mojave Desert, and all of the Sonoran Desert. Included are the hottest and driest areas on the North American continent, districts with predominantly winter rainfall and those with mainly summer precipitation. Certain areas are never exposed to freezing temperatures, while others are routinely visited by subzero events in winter. Clearly, our mapped species encounter a broad range of climatic conditions.

Figure 5 shows sea level and the 500 m and 1,300 m contours for the Sonoran Desert region. Almost without exception, our mapped species are confined to elevations below 1,300 m. Indeed, the northern and upper elevational limit of the Sonoran Desert occurs at roughly 900 m on level terrain. This upper threshold gradually decreases toward the south. On south-facing slopes, desert plants often ascend a few hundred meters above their normal upper elevational limits. The 1,300 m limit is therefore a conservative datum for judging affinity for desert conditions. Readers will find this map useful in interpreting the elevational profile that accompanies each distribution map. For example, the profiles for many different species show a rise in elevation from south to north. Using the topographic map, readers will readily discern that this rise occurs as species retreat from low elevations that become increasingly hot and dry toward the north.

The region circumscribed by our maps spans 14 degrees of latitude and 13 degrees of longitude and traverses elevations ranging from 80 m below sea level to more than 3,000 m above. These features alone would engender considerable climatic complexity. Added to them are the meteorological effects that are a result of topographic location. Situated on the western side of the North American con-

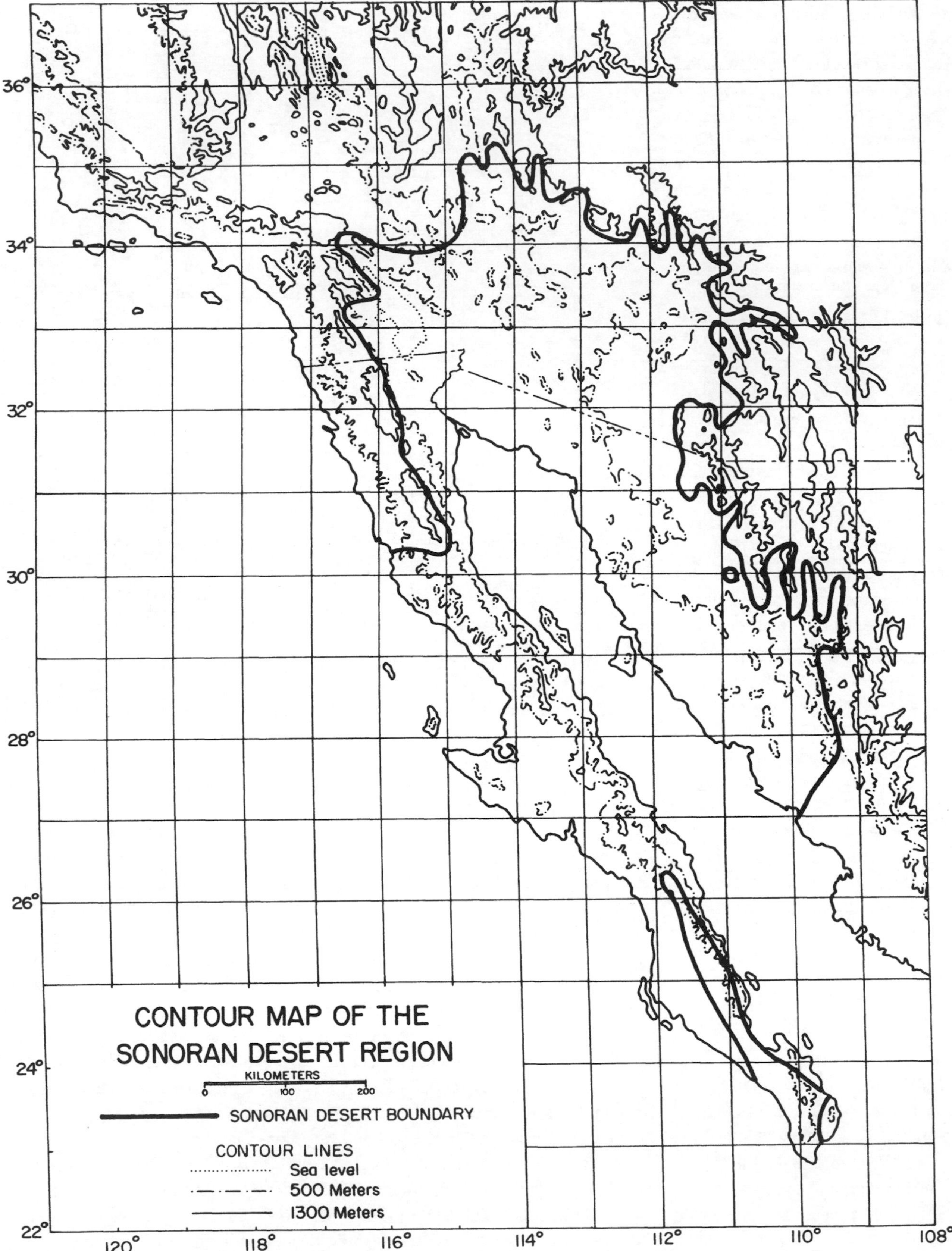

Figure 5. Contour map showing the Sonoran Desert region boundary (after Shreve 1942).

tinent, the Sonoran Desert region occupies a coastal to strongly continental position and does so at latitudes characterized by well-defined, seasonally changing weather systems.

Rainfall seasonality, especially its occurrence with respect to the growing season, is a potent element in determining plant distribution. The Sonoran Desert region has a complex topography dominated by north-south-trending mountain ranges and basins, the result of tectonic interactions along the western edge of the North American crustal plate. Low basins have formed an arid trough in the rain shadows of adjacent mountains. The Sonoran Desert has become organized around this arid corridor. The north-south-trending mountains and basins have skewed the expected polar-equatorial gradient of rainfall seasonality into a northwest-southeast orientation.

The region mapped in this atlas can be viewed as having a core area where the tendency for biseasonality is no stronger than that for a single annual peak. This area of indistinct seasonal tendency lies roughly along the Arizona–New Mexico border, then curves westerly through northern Sonora, and then along a north-south line lying just west of the Colorado River (Bryson 1957). From the U-shaped region defined by this line, the tendency for biseasonality strengthens toward the north through Arizona, and the tendency for single season precipitation becomes dominant southward and eastward (where summer rainfall dominates) and westward (where winter rainfall dominates). A second region of biseasonal rainfall occurs in the central part of the Baja California peninsula. The plants of this hot region are thereby presented with habitats having a wide range of seasonal rainfall possibilities.

Temperature conditions of the region are strongly influenced by elevation, latitude, and proximity and exposure to the ocean. The dubious distinction of being the hottest locality in North America is accorded two regions in the area mapped in this atlas—the head of the Gulf of California and Death Valley (Schmidt 1989). Both spots are surrounded by continental deserts and are leeward of mountain barriers. The coolest growing-season temperatures in our region are found along the West Coast where air temperatures are controlled by the thermal properties of cool offshore Pacific Ocean waters. Continentality, a measure of this control, is given by the difference between the average summer and the average winter temperatures of a location. The least continental climates in the Sonoran Desert region are found along the Pacific coast of Baja California south of 30°N (continentality index=8.6°C). At the other extreme is a continentality index of 20.2°C, which occurs where the states of California, Nevada, and Arizona come together (Turner and Brown 1982). To fully characterize the climate of this highly variable region is beyond the scope of this effort. (Readers interested in more detailed treatments should refer to Vivó and Gomez 1946; Hastings and Turner 1965a; Markham 1972; Huning 1978; Turner and Brown 1982; Schmidt 1983, 1989; Burgess 1985; and Ezcurra and Rodrigues 1986.)

An important objective of this effort has been to discover the fundamental climatic regimes of characteristic Sonoran Desert species. If it were possible to define the seasonal limits of temperature and rainfall at all the known localities for each species, one could roughly characterize the climate under which that species grows. The application to paleoclimatic studies is compelling. Although fossil packrat middens have increasingly gained importance as a means for studying paleoclimates, researchers presently have no objective method for characterizing the climate at the time of fossil packrat midden deposition. As many as 30–40 different plant species can be accurately identified in a single fossil midden assemblage. Knowledge of the present-day climatic controls of those species is the key to knowing the climate under which they grew in the past. No two species have exactly the same range or respond to exactly the same climate. By determining which seasonal climatic regimes coincide for the group of associated midden species, we should be able to characterize the seasonal climate at the time of midden formation. We recognize that our findings will necessarily be general, inasmuch as not all range boundaries are determined by climatic parameters and past climates may have few or no modern analogs. In spite of these possible limitations, the technique promises to have application in the fields of bioclimatology, biogeography, and horticulture.

This is the reference we wished for when we began working in the Sonoran Desert. We hope it will encourage field workers to augment our efforts and avoid our mistakes. Our coverage of some regions and topics is inevitably incomplete; for instance, we have not attempted an analysis of landforms and soils preferred by each species. This revised atlas points up the need for additional ecological information on all Sonoran Desert plants. An unprecedented influx of people into the region has resulted in substantial changes in many biotic communities. Basic knowledge of native species will be an essential tool for creative, sustainable human habitation of the Sonoran Desert.

Species Accounts

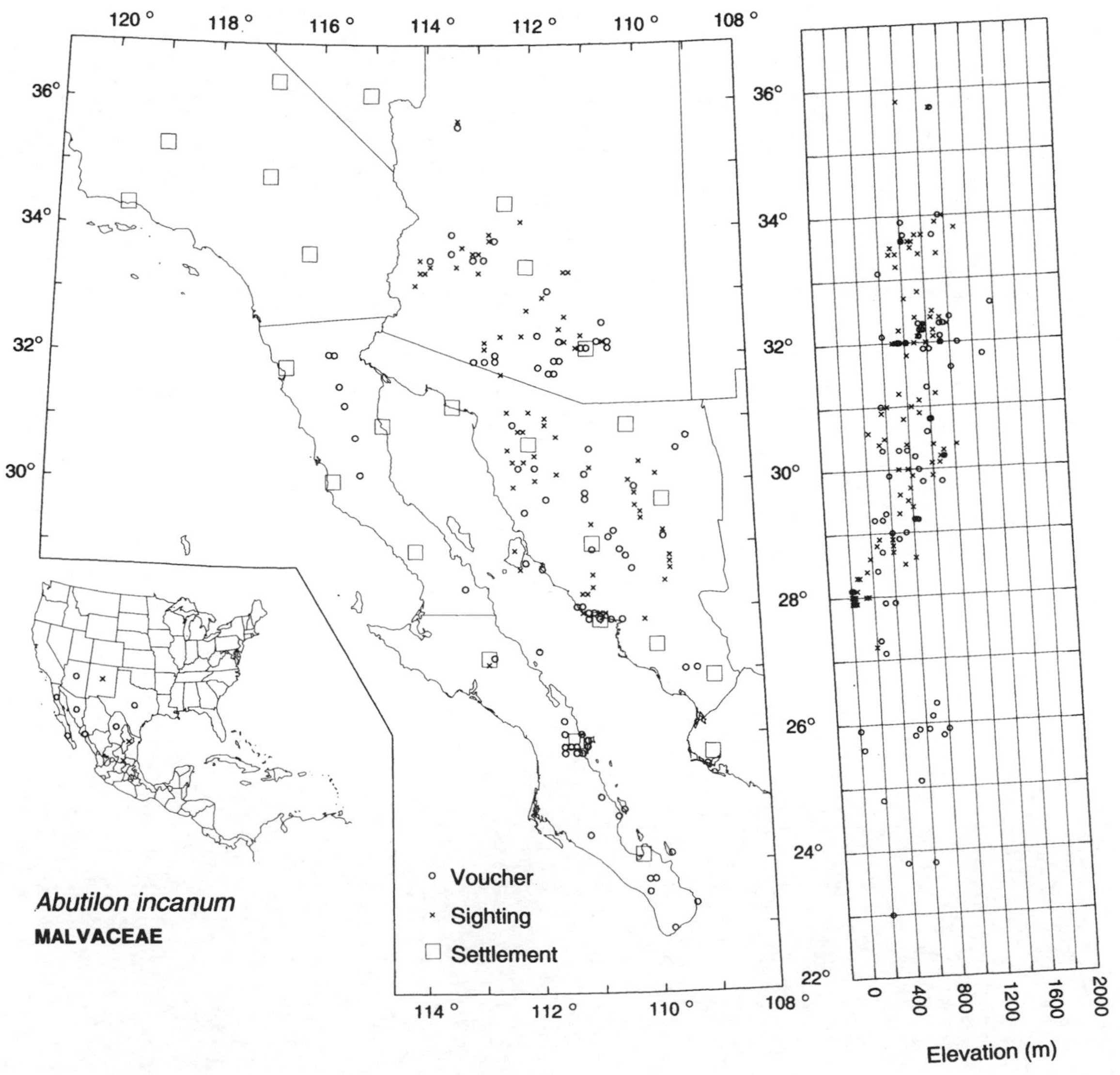

Abutilon incanum
MALVACEAE

KEY TO SPECIES ACCOUNTS

1. Latin name.
2. Authority. The name (or names) following the Latin name identifies the person (or persons) who described the species. Should the original species name be revised, the name of the original authority is shown in parentheses followed by the name of the specialist who supplied the existing combination of genus and species names. The names of these authorities are often abbreviated.
3. Synonym or synonyms. Latin names, with authorities, that have been used in the past but are no longer considered valid.
4. Common name or names.
5. Compendium. Description of the plant and information such as distribution, ecology, life cycle, and human uses.

Not all elements appear in each species account.

Abutilon incanum

(Link) Sweet.

[=*A. pringlei* Hochr., *A. incanum* subsp. *pringlei* (Hochr.) Felger & Lowe]

Pelotazo, pelotazo chico, tronadora

A slender shrub 0.5–2 m tall, ***Abutilon incanum*** has velvety stellate-tomentulose foliage. The ovate to lance-ovate leaves have cordate bases and crenate margins and are 0.5–3 cm wide and 1.5–6 cm long. The 5-petaled flowers are solitary in the leaf axils near the branch tips. Petals are orange-yellow and 6–10 mm long in subsp. ***incanum,*** deep orange spotted with maroon and 4–6 mm long in subsp. ***pringlei.*** Fruits are 4- to 6-celled capsules 5–8 mm high. The haploid number of chromosomes is 7 (Bates and Blanchard 1970).

Abutilon incanum is best separated from other Malvaceae on the basis of the floral and fruit characters provided by Wiggins (1964) or Kearney and Peebles (1960).

Felger and Lowe (1970) considered *A. **incanum*** to be a species complex comprising subsp. ***incanum*** and subsp. ***pringlei.*** Our treatment follows Fryxell (1983).

Growing on rocky slopes and gravelly plains and along arroyos, *A. **incanum*** is widely distributed in the arid, warm regions of North America. It is rare to common in Sonoran desertscrub, Sinaloan thornscrub, and semidesert grassland (Gentry 1942; Benson and Darrow 1981), and it reaches 1,370 m above sea level. The distribution reflects a need for warm-season rain and relatively mild winters. At its southern limits in Sonora, the species is restricted to relatively arid, open vegetation near the coast, perhaps because of competitive exclusion from more mesic Sinaloan thornscrub and Madrean evergreen woodland.

The plants may flower at any time of year. Leaves are virtually evergreen.

Acacia angustissima

(P. Mill.) Kuntze

Cantemó, guajillo, palo de pulque, barbas de chivo, prairie acacia, fern acacia

A thornless shrub or small tree, ***Acacia angustissima*** has striate-ridged branches, bipinnate leaves 10–25 cm long, and 15–30 (or more) pairs of leaflets 4–6.5 mm long on 9–25 pairs of pinnae. Numerous small white flowers are arranged in spherical heads (15 mm diameter), each head with 50–100 elongate stamens. The fruits are 3–6 cm long and 6–9 mm wide.

An exceedingly variable species, *A. angustissima* has several varieties in our area (Wiggins 1942; Benson 1943; Isely 1973). Our map shows *A. angustissima* in its broad sense. One of the varieties has a haploid chromosome number of 13 (Turner and Fearing 1960).

This species ranges northward to Missouri and Kansas and southward to Central America. It is absent from California and the Baja California peninsula. Two closely allied species, *A. mcmurphyi* and *A. brandegeana,* grow on the peninsula. The distribution of *A. angustissima* suggests dependence on ample warm-season rainfall. South of 29°N, the species grows at both high and low elevations, showing a wide range of ecological tolerances consistent with high infraspecific variability. The elevational profiles suggest that *A. pennatula* may competitively displace *A. angustissima.* Competitive displacement may also occur between *A. angustissima* and *A. cochliacantha.*

Acacia angustissima flowers from May to November. The fruits dehisce in late summer or fall. Some seeds often remain on the plants well into winter.

The plants are used to a minor extent as ornamentals.

Acacia brandegeana

I. M. Johnston

A shrub or small tree up to 7 m tall, *Acacia brandegeana* has a single pair of pinnae per leaf and paired, straight, stipular spines up to 3 cm long. The light yellow flowers are in sessile spikes 2 cm long. The linear, curved pods are 4–15 cm long.

Acacia brandegeana belongs to the same species complex as *A. pringlei* subsp. *californica* but is clearly separable by its small, acute, oblong leaflets and its densely lanate legumes. It is also the only species of the group with 4–7 pairs of leaflets per pinna (Lee et al. 1989).

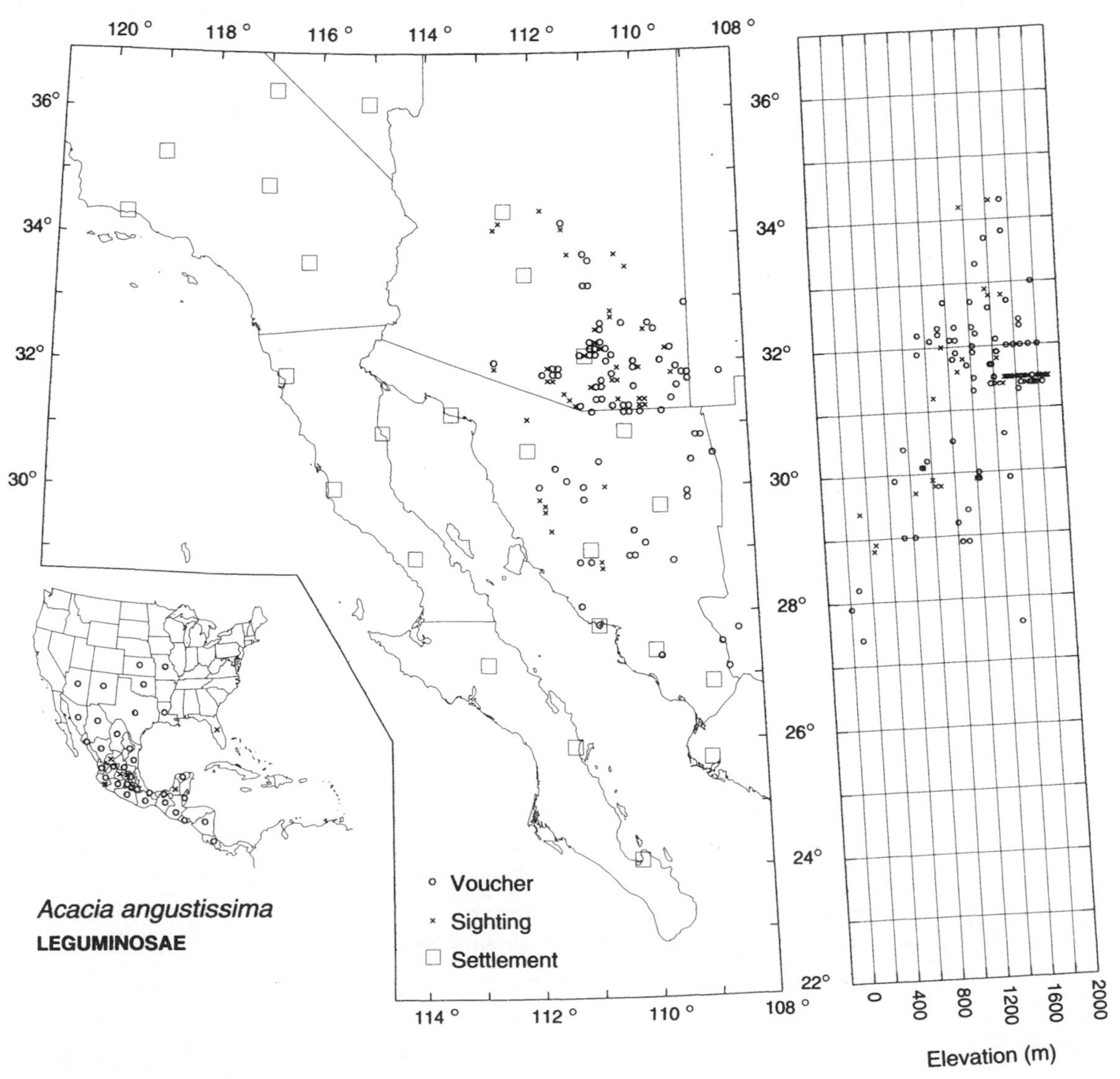

Acacia angustissima
LEGUMINOSAE

A plant of open, disturbed sites, *A. brandegeana* grows only in Baja California Sur from the latitude of Bahía Concepción southward. It reaches both the Pacific and Gulf coasts. The elevational profile shows an abrupt northern limit over a rather wide elevational range. A zone of irregular biseasonal rainfall centered on the Vizcaíno Plain perhaps shapes this northern limit. The elevational profiles suggest strong ecological interactions between *A. brandegeana* and *A. goldmanii* and perhaps also between *A. brandegeana* and *A. kelloggiana*.

Flowers appear in spring, mature fruits during summer rains.

Acacia cochliacantha

Willd.

[=*A. cymbispina* Sprague & Riley]

Boat-thorn acacia, spoon-thorn acacia, cucharillo, cubata, quisache costeño, vinolo, huinole, espino

A shrub or small tree up to 8 m tall, *Acacia cochliacantha* (figure 6) has light gray to almost black bark. On nonflowering branches, the spines typically enlarge, becoming boat shaped and up to 6 cm long and 4 cm wide. The terete, unenlarged spines of the flower-producing branches are only 3–4 mm long. The twice-pinnate leaves are 10–25 cm long. The 9–25 pairs of pinnae each have 30–60 pairs of linear, ciliate leaflets 4–6.5 mm long. The yellow flowers are in globose heads 6–7 mm in diameter. Fruits are broadly cuneate pods 4–7 cm long and 8–12 mm broad.

No other acacia in the Sonoran Desert has boat-shaped spines.

Wiggins (1964) treated this as *A. cymbispina*. In Mexico, however, it is best referred to as *A. cochliacantha* (Rudd 1966). Two forms, separated on the basis of twig and petiole pubescence, have been recognized; only one, forma *cochliacantha*, occurs in the Sonoran Desert Region (Seigler and Ebinger 1988).

This acacia is a plant of rocky hillsides, sandy plains, and clay flats. It frequently occurs in disturbed areas such as road-

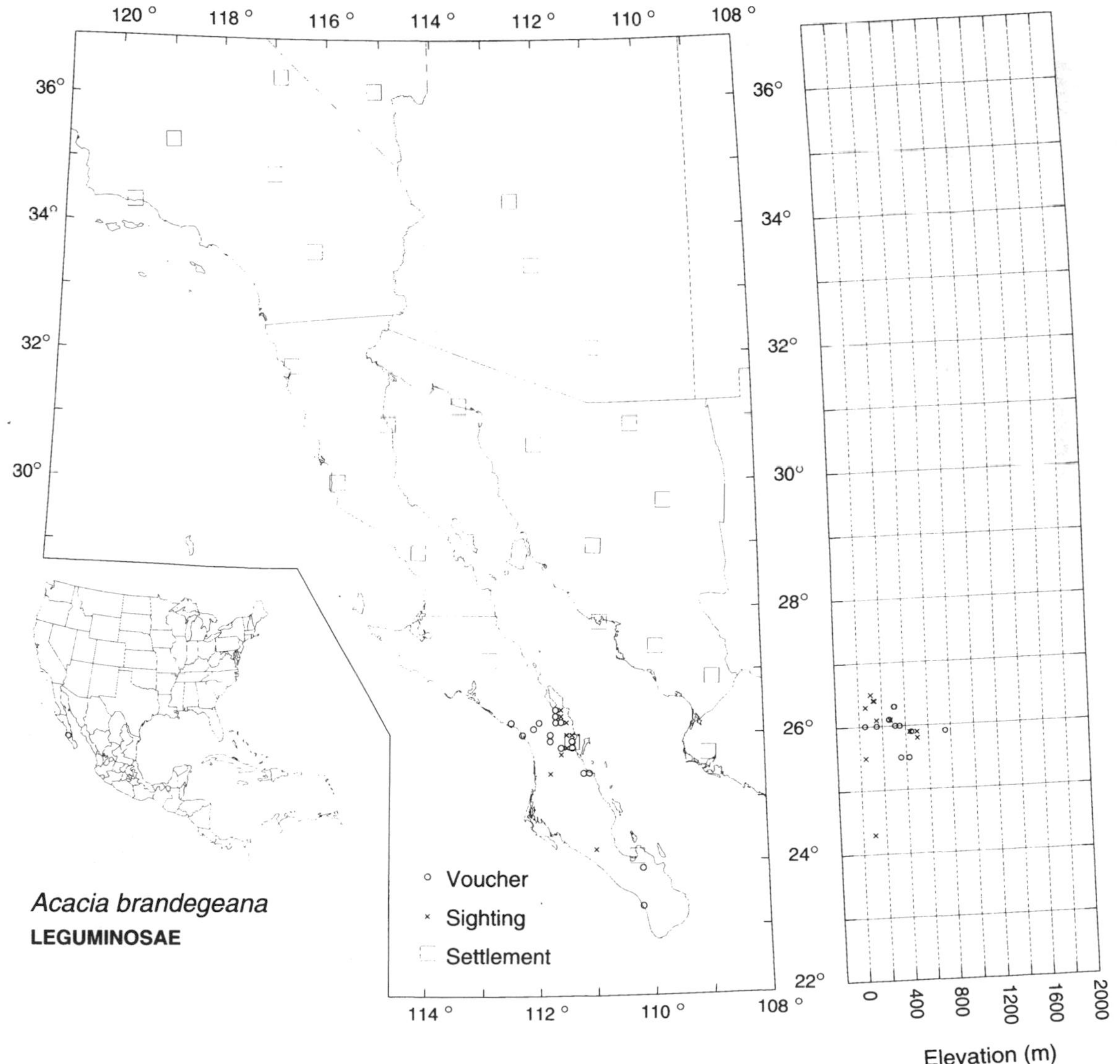

Acacia brandegeana
LEGUMINOSAE

side borrow pits. The northern disjunct (30.5°N) represents a collection by Thomas R. Van Devender (1976) from 22.5 km south-southeast of Magdalena on the road to Cucurpe. The two peninsular disjuncts at 23.8°N, collected near Triumfo by Reid Moran in 1959 and Raymond M. Turner in 1964, may have been recent introductions. An aggressive, colonizing species (Gentry 1942), *A. cochliacantha* can be expected to spread into clearings and along roadsides throughout the Cape Region.

Our map illustrates a species of the dry tropics at its northern and arid limits. Low temperatures probably determine the northern boundary: a low of –4.4°C top-killed cultivated plants in south-central Arizona in December 1978 (Kinnison 1979). Interspecific interactions with *A. constricta* (and perhaps also with *A. greggii* and *A. millefolia*) may affect the northern limit, too. Decreasing summer precipitation may help determine the rather abrupt western limit. South of 29.8°N, the low-elevation limits are shaped by aridity. North of 29.8°N, the elevational profile is more likely shaped by episodes of extreme cold. In the southern part of our area, the plants reach relatively high elevations (1,200–1,500 m), perhaps in response to longer growing seasons. It is hardy to –4°C (Johnson 1993).

Flowers appear in spring (March–May) and again in summer (July–November) (Wiggins 1964). Leaves are cold- and drought-deciduous. Farther south, as in the tropical deciduous forest near Alamos, Sonora, new leaves appear in early July and drop with the onset of spring drought, usually in March (Krizman 1972).

Of the twelve Central American *Acacia* species with boat-shaped spines, all but one, *A. ruddiae,* are obligate ant-acacias (Janzen 1974). Colonies of ants of the genus *Pseudomyrmex* live in the hollow thorns. Adult ants obtain food for their larvae from so-called Beltian bodies on the leaflets and visit petiolar nectaries for their own sustenance. They protect the plant by attacking other insects and removing foreign plant material (Janzen 1974). *Acacia ruddiae* may represent

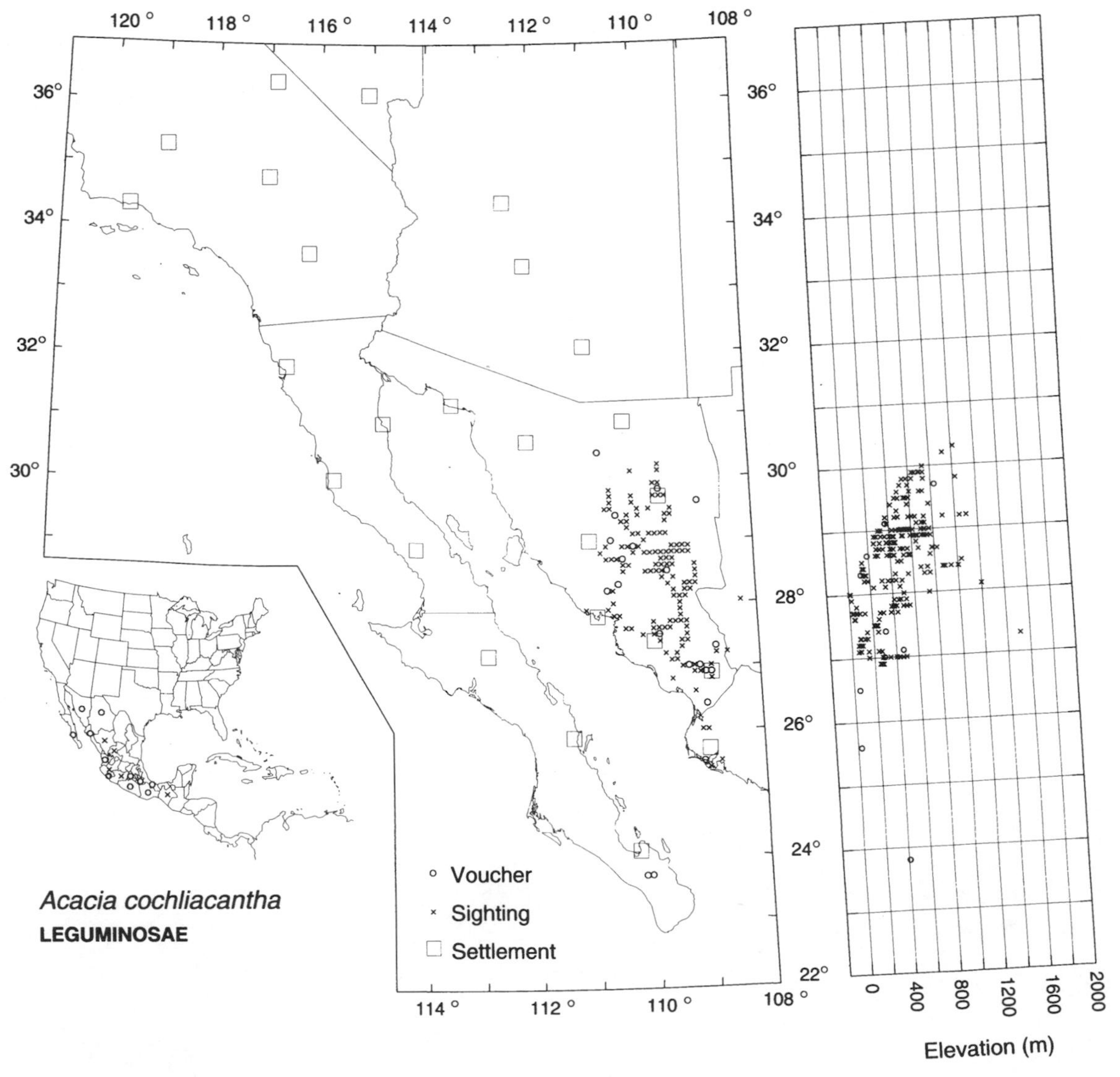

Acacia cochliacantha
LEGUMINOSAE

Figure 6. Acacia cochliacantha, *showing the boat-shaped spines of nonflowering branches. The stem with unenlarged spines is capable of flowering. (Photograph by J. R. Hastings.)*

an intermediate step in the evolution of swollen-thorn acacias (Janzen 1974). *Acacia cochliacantha* likewise lacks a mutualistic relationship with ants, but researchers have not yet determined whether mutualism has gradually disappeared or whether it has yet to develop.

A decoction of *A. cochliacantha* is reported to be used in Sonora for bladder ailments (Standley 1922).

Acacia constricta

Benth.

White thorn, mescat acacia, huisache, gigantillo, vara prieta, chaparro prieto, largoncillo, vinorama

Acacia constricta, a shrub up to 2 (occasionally 3) m high, has light gray or mahogany-colored bark and paired, straight, white spines 0.5–2 cm long. The small, even-pinnate leaves, fascicled on older stems, are 2.5–4 cm long. Each of the 3–9 pairs of pinnae has 4–16 pairs of leaflets up to 3.5 mm long and 1 mm broad. The fragrant yellow flowers are clustered in globose heads about 1 cm across. The pods, 5–12 cm long and 3–6 mm wide, are constricted between the seeds and become twisted before they dehisce. The diploid chromosome number is 52 (Turner 1959).

Several small-leaved leguminous shrubs resemble this species. *Acacia neovernicosa* has markedly resinous foliage and fewer pairs of pinnae (1–2, occasionally 3). *Acacia farnesiana* and *A. minuta* are small trees up to 9 m tall with thick, odoriferous pods. *Acacia cochliacantha* has characteristic boat-shaped spines. *Acacia brandegeana* typically has only a single pair of pinnae. *Acacia pacensis* bears bractlets at the peduncle apex instead of at the midpoint and has indehiscent pods. *Mimosa aculeaticarpa* has curved spines, as does *A. greggii. Calliandra eriophylla* is unarmed and seldom exceeds 1 m in height.

A variety with few or no spines and less glandular leaves has been described (*A. constricta* var. *paucispina,* Wooton and Standley). Isely (1969) rejected this taxon because it is not clearly marked: spininess is variable and is not noticeably correlated with the glandular pubescence of the foliage. Certain authors (Benson and Darrow 1981, for example) make *A. neovernicosa* a variety of *A. constricta.* Isely (1969) considered this association untenable since the former is diploid and the latter tetraploid. In the United States, the tetraploid *A. constricta* is primarily a plant of the Sonoran Desert, and the haploid *A. neovernicosa* is more characteristic of the Chihuahuan Desert. Isely (1969) suggested that *A. neovernicosa* might have given rise to *A. constricta.*

In the more arid portions of its range, *A. constricta* grows along arroyos and washes, where runoff compensates for low rainfall. At somewhat higher elevations, the plants can be common on slopes and ridges. South of 30°N, *A. constricta* is restricted to elevations below 900 m. Its upper limits in this region are roughly contiguous with the northern or lower limits of *A. cochliacantha, A. farnesiana, A. pennatula,* and *A. occidentalis,* perhaps showing that *A. constricta* is competitively excluded from more mesic communities. North of 28°N, aridity controls the lower elevational limit. North of 31°N, cold probably determines the upper elevational limit, although exclusion from denser vegetation may also be involved. The truncation near 34°N shows the combined blocking actions of the steep southern edge of the Colorado Plateau and the westward decrease in warm-season rainfall. The elevational profiles show that in Baja California Sur, *A. constricta* occupies an apparent gap in the distribution of *A. farnesiana.* This suggests competitive displacement. The same may be true also of *A. constricta* and *A. peninsularis.*

The peninsular populations are widely disjunct from the main range. Those in northern Baja California (represented in herbaria by Reid Moran's collections from the Sierra Juárez) are near disjunct populations of *Calliandra eriophylla* and *Mimosa aculeaticarpa.* In Baja California Sur, disjunct *A. constricta* grows on the sandy and calcareous soils of the Magdalena

Plain with various species of columnar cacti, *Jatropha* and *Cercidium.*

Plants bloom in late spring (April–May, occasionally June) and again in summer (July–October). Flowering requires a rain trigger of at least 11 mm. Thereafter, degree-days above the base temperature (15°C) must reach about 522 for the plants to bloom (Bowers and Dimmitt 1994). Leaves are drought- and cold-deciduous. All may drop as the soil dries out, or, more typically, some may remain for most of the year. Leaves are renewed in spring if winter rains have been adequate; otherwise, plants may remain dormant until summer rains are underway (Shreve 1964).

Despite their fragrance, the flowers attract few visitors, probably because they offer no nectar and negligible pollen (Simpson 1977). The waxy seed coat delays germination for at least one year (Shreve 1964). Rodents are important seed predators (Shreve 1964).

At Tumamoc Hill, Tucson, Arizona, seedling establishment is apparently a rare event. During 18 months, we found only two seedlings on a plot 557 m^2. This low level of recruitment, typical of long-lived desert trees and shrubs, is to be expected for *A. constricta,* which lives at least 72 years in the Tucson area (Goldberg and Turner 1986).

Occasionally grown in desert gardens, *A. constricta* can be closely planted to form a barrier or can be trained into a tree (Duffield and Jones 1981). Once the plant is established, it requires little supplemental irrigation except where annual rainfall is less than 25 cm (Duffield and Jones 1981). The plants are frost hardy to −12°C (Johnson 1993).

Seri Indians made a tea from the mashed seeds and leaves to relieve diarrhea or upset stomachs (Felger and Moser 1985). Powdered, dried pods and leaves have been used to treat skin rashes; medicinal tea can be made from the roots (Moore 1989).

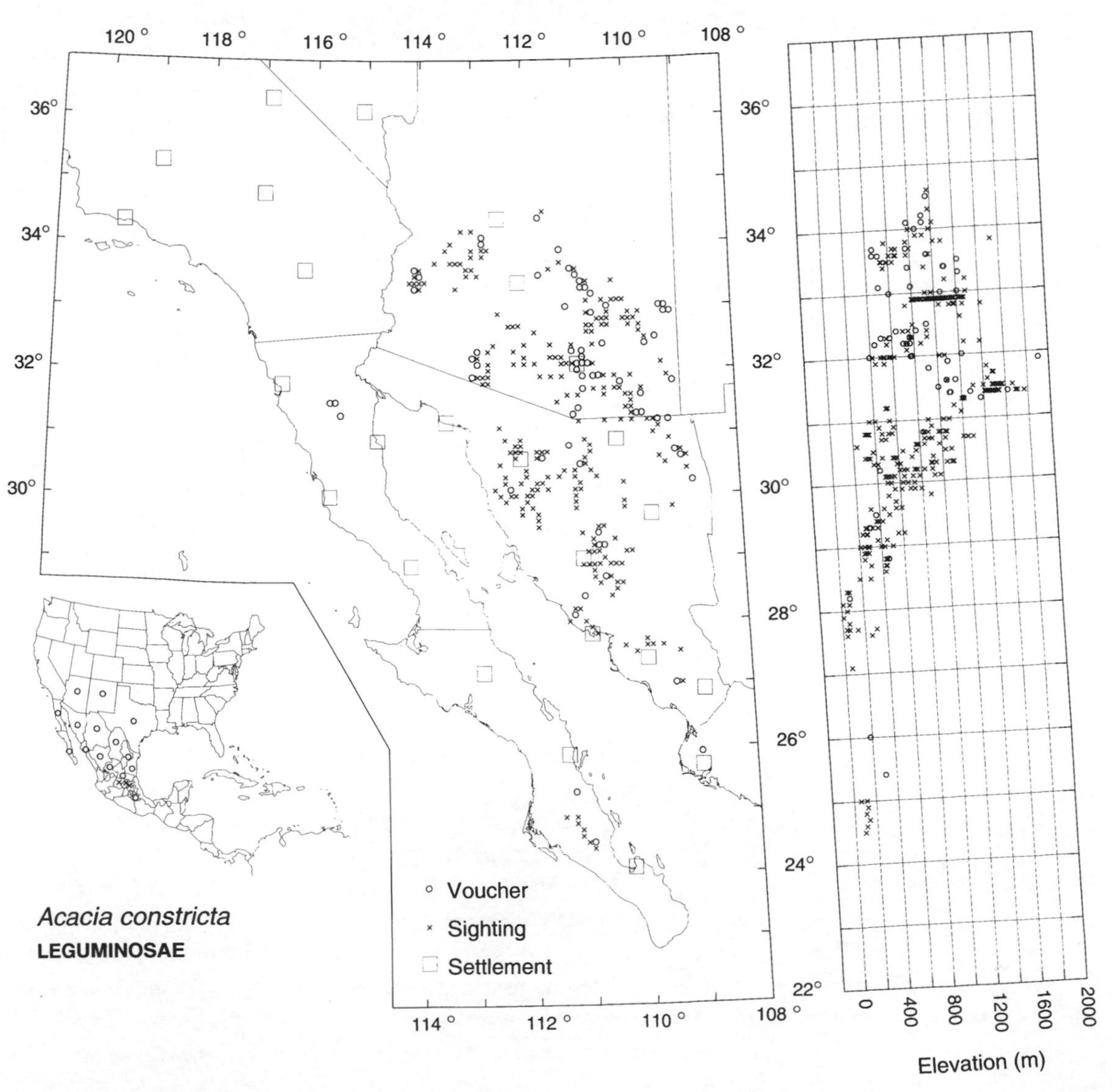

Acacia constricta
LEGUMINOSAE

Acacia coulteri

Benth.

Tepeguaje, palo de arco, huajillo, palo blanco

Acacia coulteri is a slender, unarmed tree or shrub with an open crown. The intricately divided leaves have 10–30 pairs of leaflets 4–6 mm long. The slender flower spikes are 2–10 cm long. Fruits are thin, veiny, oblong pods 7–18 cm long and 15–25 mm wide.

Acacia coulteri bears a striking resemblance to *Lysiloma microphyllum* but has somewhat longer petioles (2.5–4.5 cm in *A. coulteri,* 0.5–2.0 cm in *L. microphyllum*).

From Ures, Sonora, southward to Sinaloa and eastward to Hidalgo, San Luis Potosí, and Tamaulipas, *A. coulteri* extends along mesic corridors into the southern Sonoran and Chihuahuan deserts. The elevational profile is typical of a species that has lower limits determined by aridity. Periodic freezes probably truncate the northern limit. Competitive displacement by *A. millefolia* may also have shaped the northern limit. Interactions with *A. pennatula* appear to influence the upper elevational limit.

The plants flower between April and October.

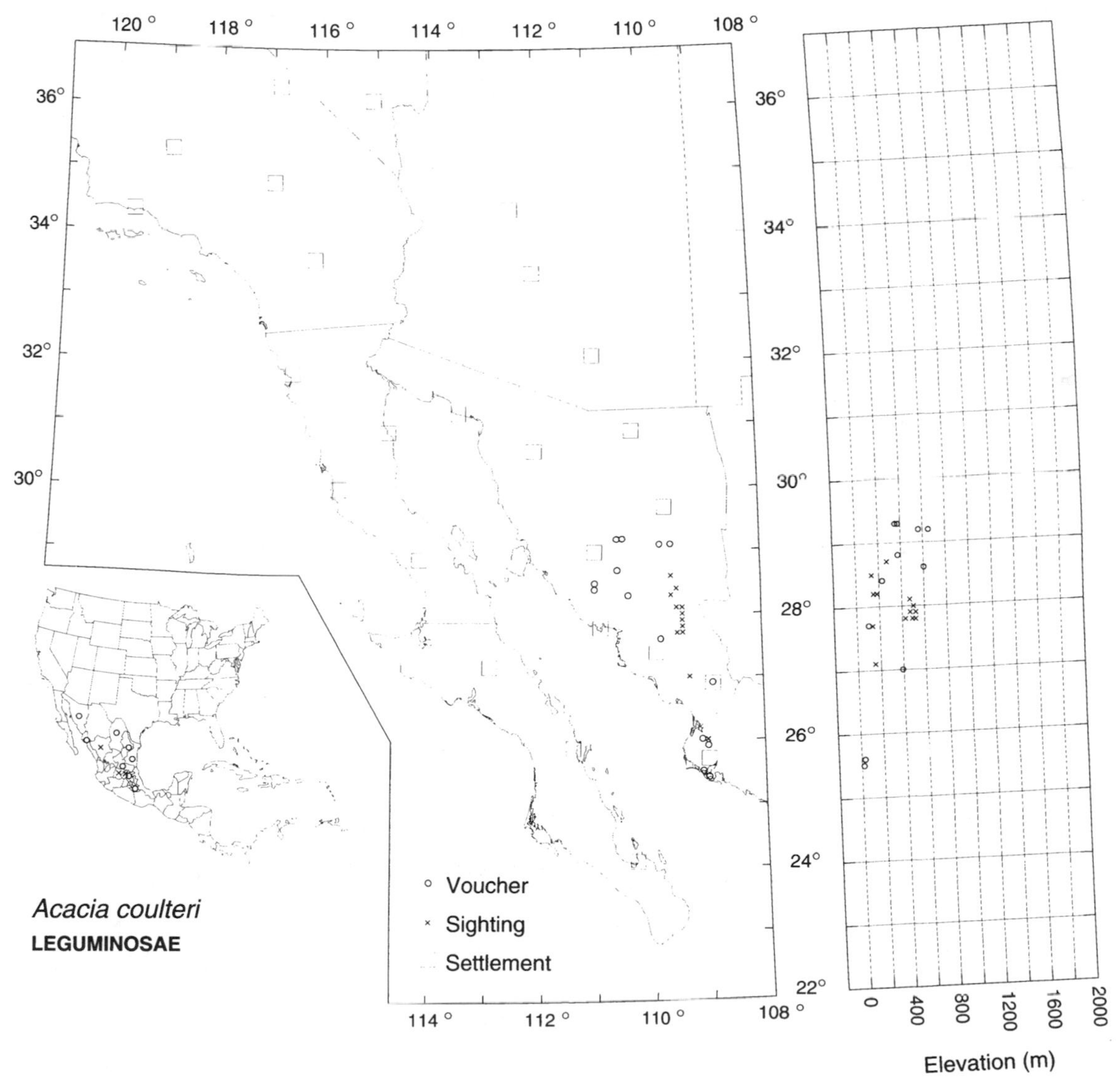

Acacia crinita

Brandegee

La nudita, gato

Acacia crinita, a shrub 4–5 m tall, bears twice-pinnate leaves with a rather open array of leaflets 1.0–1.5 cm long. The pinnae, usually 2–3 pairs per leaf, are 4–10 cm long. They are swollen at the base and bear a pair of bracts on the swelling. Coarse, yellow hairs up to 5 mm long grow along the stems. The pale yellow flowers are in globose heads. The flattened and oblong pods are 5–7 cm long.

Acacia crinita superficially resembles several species of *Caesalpinia, Senna,* and *Calliandra,* but the coarse, yellow stem hairs are distinctive, as are the swollen pinnae bases with their characteristic bracts.

This rarely collected species grows through much of eastern Sonora from 30.3°N southward into Sinaloa and Durango, often in scattered, open colonies. The plants sprout quickly after fire (Gentry 1942).

Flowering specimens have been collected in summer (July–September) and spring (February–March).

Acacia farnesiana

(L.) Willd.

Acacia minuta

(M. E. Jones) Beauchamp

Huisache, vinorama, aroma amarilla, sweet acacia, honey-ball, popinac

A tree up to 10 m tall or a many-stemmed shrub, *Acacia farnesiana* in its broad sense has conspicuous paired, straight, stipular spines up to 5 cm long. Its twice-pinnate leaves are 2–10 cm long with 2–8 pairs of pinnae and 8–25 pairs of leaflets 2–6 mm long. The bright yellow flowers are in conspicuous spherical heads about

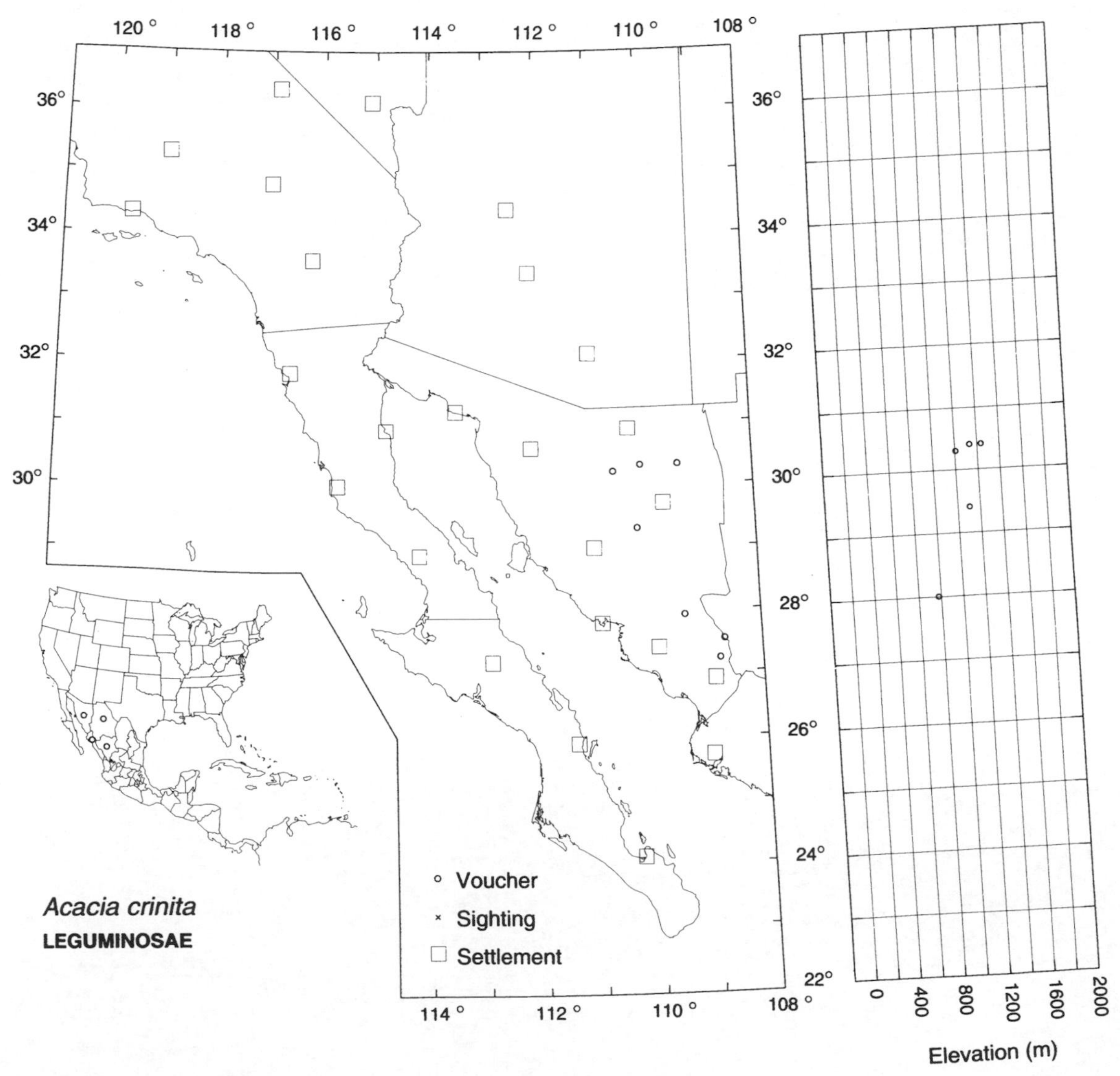

15 mm in diameter. The turgid, blackish pods are 4–7 cm long. A haploid chromosome number of 26 has been reported for *A. farnesiana* (Ghimpu 1929).

Acacia constricta, a widespread species, and *A. pacensis,* a narrow endemic, resemble *A. farnesiana* but have shorter leaves and pinnae and are smaller plants.

Acacia farnesiana has a complex taxonomic history in our region. The Arizona and California populations were referred to as *A. smallii* Isely (Isely 1973), and those of Baja California, California, and Arizona as *A. minuta* (Beauchamp 1980). Other workers concluded that the disjunct Arizona population (Baboquivari Mountains, Pima County) should probably be assigned a new combination under *A. farnesiana* (Clarke et al. 1989). The Sonoran populations may be referable to *A. farnesiana,* but detailed studies are lacking. We have made no attempt to separate this species complex. Many of our records are probably for *A. farnesiana* in the narrow sense, which is separated from *A. minuta* by somewhat subtle inflorescence, fruit, and leaf characters (Isely 1973).

The elevational profile illustrates a widespread, adaptable taxon. On the mainland, the northern limit is apparently determined by a combination of aridity at lower elevations and cold (or exclusion from more mesic communities) at higher elevations. On the Baja California peninsula, the species retreats to lower elevations north of 28°N, perhaps as a result of cold. Along the gradient from lower to higher latitudes, rainfall seasonality shifts from the warm to the cool season. The more northerly plants grow in riparian habitats where seasonal variation in soil moisture is less pronounced. Competition may exclude more southerly plants from riparian communities.

On the Baja California peninsula, *A. farnesiana* overlaps with *A. peninsularis, A. brandegeana,* and the southern populations of *A. greggii.* It is almost contiguous with *A. pacensis* and approximately so with the disjunct range of *A. constricta.*

The extremely fragrant flowers, which appear in the spring when the plants are usually leafless, are rich in volatile compounds (Flath et al. 1983).

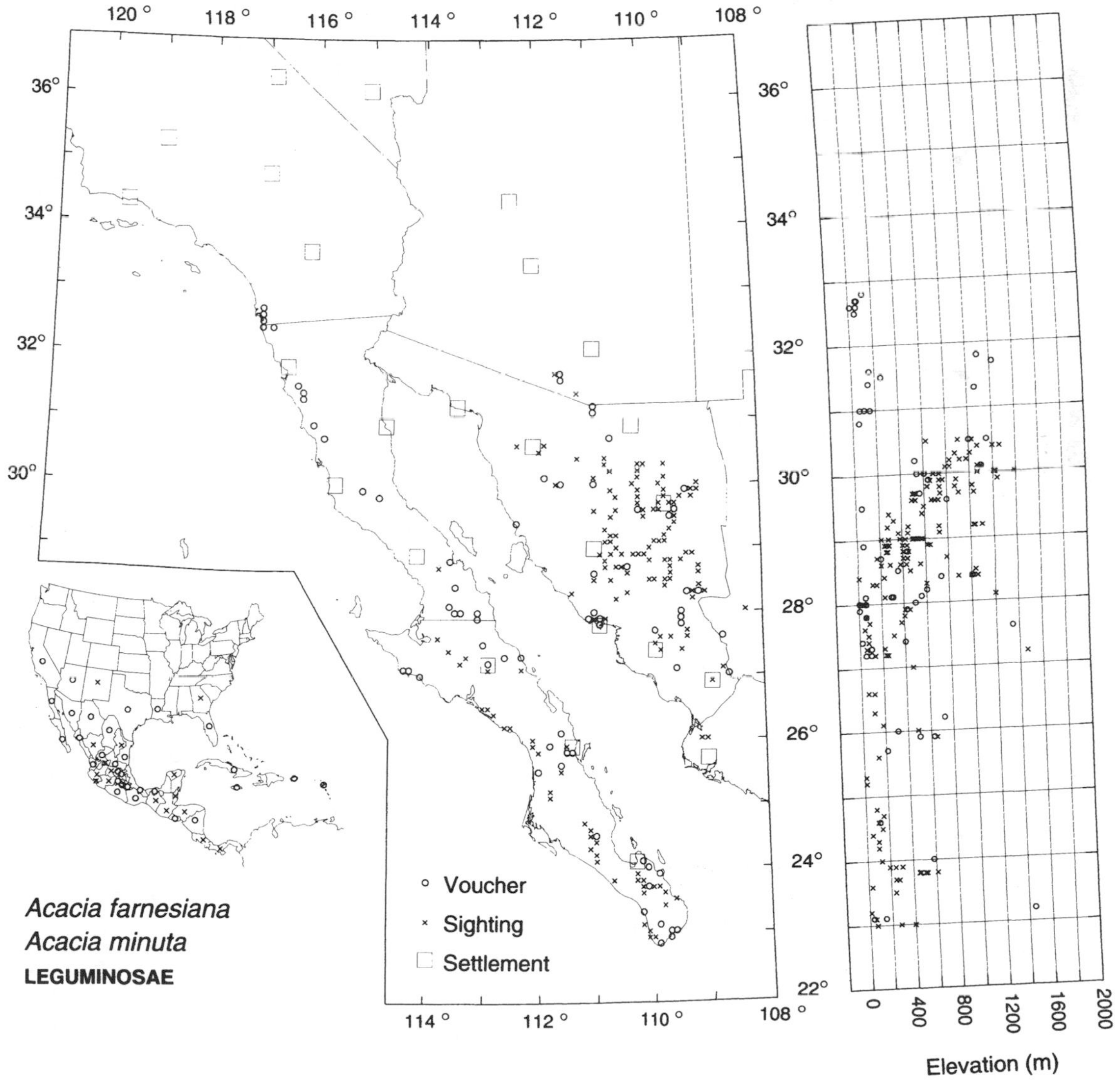

In Texas and eastern Mexico, *A. minuta* is a common weedy roadside plant and an aggressive colonizer of disturbed sites and rangeland (Smith and Rechenthin 1964; Van Auken and Bush 1985; Bush and Van Auken 1986). Attempts to control or eradicate it have included fire (Rasmussen et al. 1983), herbicides (Bovey et al. 1970; Meyer 1982), mowing (Powell et al. 1972), and grubbing (Bontrager et al. 1979).

A high light requirement probably excludes this species from mature forests (Bush and Van Auken 1986). A low nitrogen requirement (Van Auken et al. 1985) lets it prosper in nitrogen-poor soils. Growth is relatively unresponsive to photoperiod and highly responsive to warmth (Peacock and McMillan 1968).

A showy plant in flower, *A. farnesiana* is widely cultivated as an ornamental in tropical and subtropical areas. The plants are well suited to full sun and hot microclimates (Duffield and Jones 1981). The seeds require scarification before they will germinate (Scifres 1974; Vora 1989). In arid regions, *A. farnesiana* grows fastest with moderate or ample water and, once established, can survive without supplemental irrigation where annual rainfall is 25 cm or greater (Duffield and Jones 1981). New growth is damaged at –7°C, and severe frost damage occurs at –10°C (Duffield and Jones 1981; Johnson 1993).

In southern Europe the plants are grown for the flowers from which perfume is manufactured. The bark and fruit are rich in tannin (Seigler et al. 1986). Lac from the pods is used as glue in mending pottery.

Acacia goldmanii

(Britton & Rose) Wiggins

Acacia goldmanii is a shrub 2–4 m tall with slender branches and an open crown. The feathery leaves, up to 15 cm long, have 15–30 pairs of leaflets on 5–8 pairs of pinnae. The cream-colored flowers are in clustered, globose heads 1.5–2 cm in diameter. The oblong pods are 3–5 cm long and

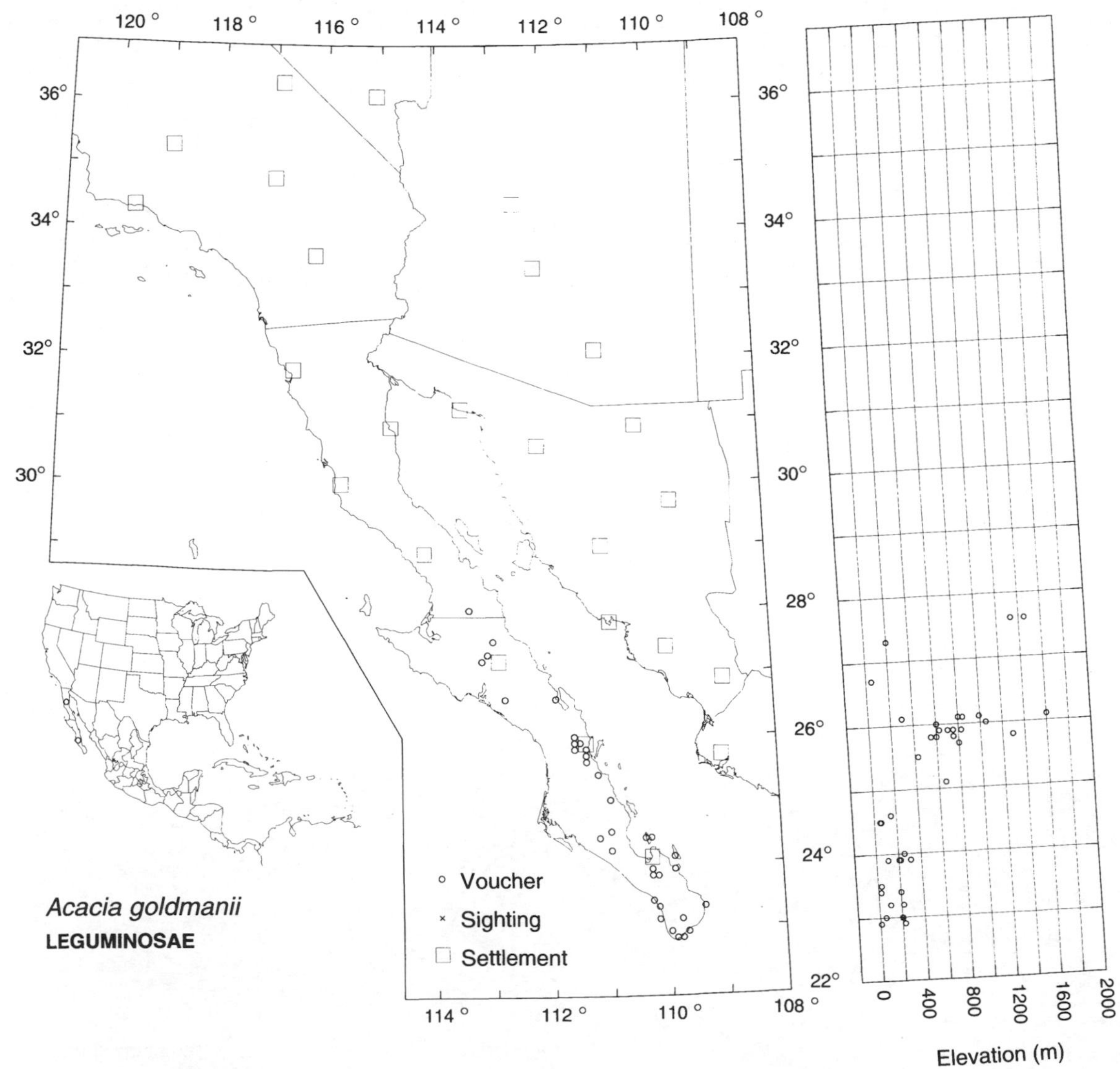

Acacia goldmanii
LEGUMINOSAE

10–12 mm wide.

This peninsular endemic is closely related to *A. angustissima,* which is absent from the peninsula, and to *A. mcmurphyi,* which is limited to the Sierra Giganta region of Baja California Sur. Although *A. mcmurphyi* is reported to have pilose branchlets and rachises compared to glabrous ones in *A. goldmanii* (Wiggins 1940), extensive field collections show that these characters do not segregate geographically (Annetta Carter, personal communication 1986). *Acacia goldmanii* sometimes has 1 or 2 rows of minute glands on the abaxial surface of the petioles, whereas *A. mcmurphyi* does not (Wiggins 1940). Further work is needed on this group.

Acacia goldmanii is confined to the southern half of the peninsula. At its northern limits, *A. goldmanii* grows from near sea level to 1,600 m; at its southern limits, from sea level to about 300 m. The narrowing elevational range toward the south probably reflects exclusion from increasingly dense vegetation; indirect evidence is that the upper limit of *A. goldmanii* corresponds to the lower limit of *Lysiloma microphyllum.* Too little warm-season rain may set the northern limit. Interspecific interactions with *A. brandegeana* and *A. greggii* (and perhaps with overlapping *A. peninsularis*) may also influence the northern limit of *A. goldmanii.*

The plants flower from September through March.

Acacia greggii

A. Gray

Catclaw, devil's claw, algarroba, uña de gato, gatuño, tésota, tepame

A shrub or small tree up to 7 m tall, *Acacia greggii* bears small, bipinnate leaves with 4–6 pairs of leaflets on each of 2–3 pairs of pinnae. The pinnae are 1–1.5 cm long. Strong, sharp, recurved prickles are scattered along the branches. Flowers are in dense, cream-colored spikes 1–4 cm long. Fruits are somewhat twisted, ribbonlike pods 6–13 cm long and 1.0–1.5 cm wide. The roots are not nodulated (Eskew and Ting 1978; Felker and Clark 1981).

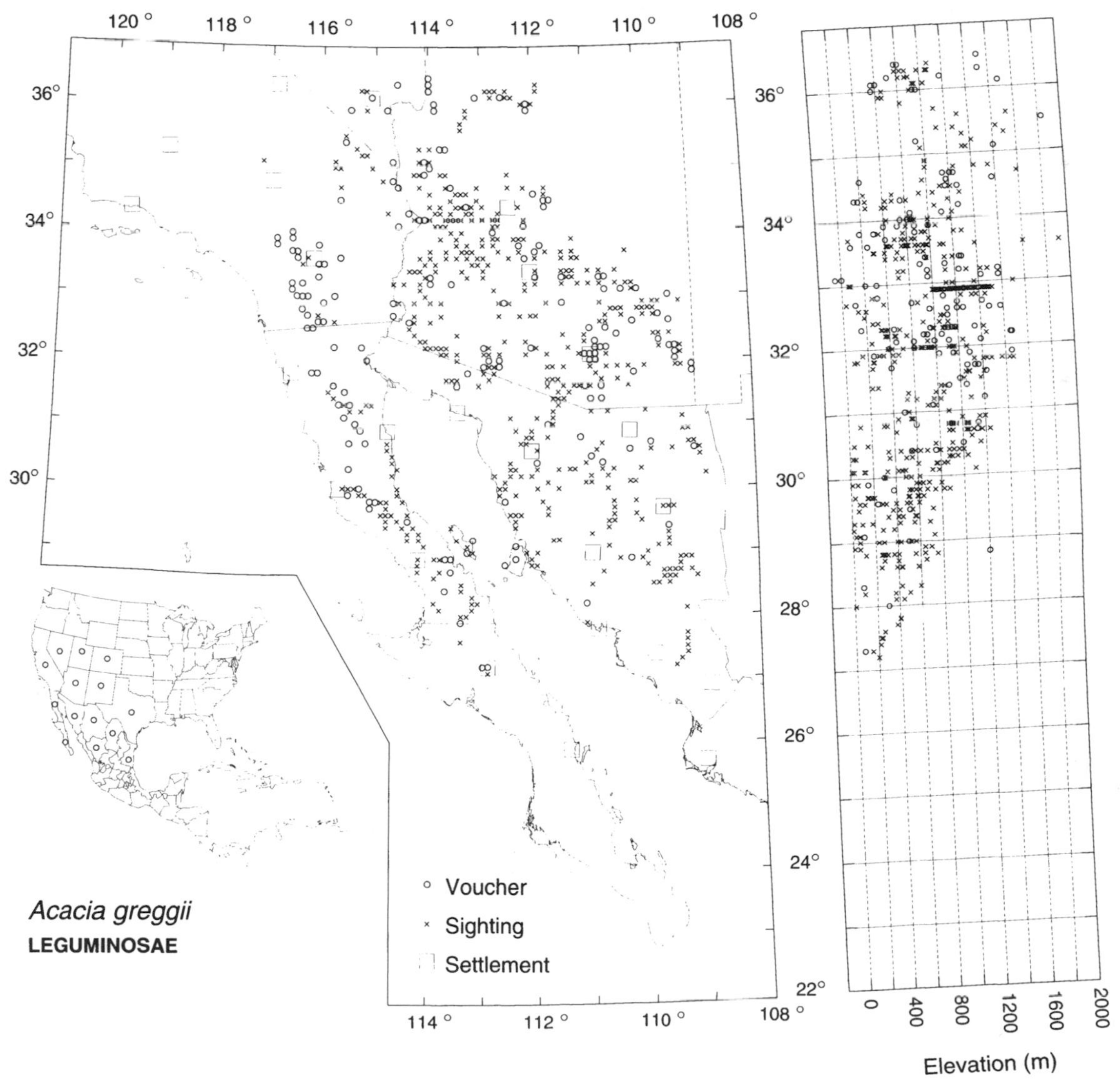

Acacia greggii
LEGUMINOSAE

We combine distributions for *A. greggii* and *A. wrightii* Benth., although Wiggins (1964) recognized them both, because the peninsular collections we have examined could be assigned with equal confidence to either species.

Several leguminous trees look like *A. greggii. Pithecellobium mexicanum* has paired stipular spines instead of internodal prickles. *Acacia greggii* and *A. occidentalis* differ in placement of the petiolar gland: it is at midpetiole in *A. occidentalis* and between the lowest pair of pinnae in *A. greggii* (Turner and Busman 1984). Also, *A. occidentalis* has a capitate inflorescence, *A. greggii* a spicate one. *Olneya tesota* has once-pinnate leaves.

Acacia greggii, the most xerophytic of Sonoran Desert acacias, grows in washes at lower elevations and on slopes at the upper limit of its range. The species has been present in the Sonoran Desert region since 10,500 years B.P., when it was common on rocky slopes in the Puerto Blanco Mountains, Arizona (Van Devender 1987). Hyrum B. Johnson reported the disjunct population at 35.2°N, 117.2°W. The species is nearly absent from the arid corridor lying just west of the Colorado River and north of the Gulf of California (about 34.5 to 31.5°N). The northern limit of the species (slightly north of 37°N; Benson and Darrow 1981) lies outside our map and elevational profile.

Because *A. greggii* occupies riparian habitats in the most arid portions of its range, the mapped distribution does not reflect its true climatic tolerance. A need for occasional warm-season rain (presumably for seedling establishment) seems to shape the northern boundary, while episodes of extreme cold determine the upper elevational limits. North of 33°N, the increasing lower elevational limit is partly a result of rising base level along the Colorado River. At low latitudes, competitive exclusion by *A. constricta, A. coulteri, A. occidentalis,* and others may restrict *A. greggii* to low elevations. *Acacia greggii* reaches its southern limits at the same latitude on both sides of the gulf, surprising in view of the greater aridity on the peninsula.

Acacia greggii is winter deciduous or partly so. Leaves are present from late spring to fall in the coldest places or year-round in the warmer ones (Shreve 1964). Flowering is at the beginning of the arid foresummer (April–May in most areas) and occasionally again during the summer rains.

Like many desert shrubs, *A. greggii* shows its largest photosynthetic rates at higher plant water potentials. In one study, the maximum rate of carbon dioxide assimilation occurred at a dawn plant water potential of –0.5 MPa. The lower limit for carbon dioxide assimilation was about –4 MPa (Szarek and Woodhouse 1978a). Seasonal patterns of dawn plant water potential tend to parallel changes in soil water status, and the seasonal pattern of leaf photosynthesis in turn follows seasonal variation in plant water potential (Szarek and Woodhouse 1978b). During a period of temperature and water stress in May, June, and July, gross photosynthesis was maintained at a relatively low level (2 mg $CO_2/dm^2/hr$). A dramatic increase in photosynthetic rates occurred in October (after a wet September), when mean photosynthesis rose to a maximum (16 mg $CO_2/dm^2/hr$). Photosynthesis declined throughout the winter. In another study, *A. greggii* also showed seasonal photosynthetic adaptation to changing temperatures. Maximum net photosynthesis (about 14 mg CO_2/g dry weight/hr) occurred around noon in January and around 7:00 A.M. in July (Strain 1969). During the summer, leaf water potentials can fall below –4.0 MPa (Nilsen et al. 1984). Plants respond by adjusting osmotic potential and maintaining high conductance in the morning hours (Nilsen et al. 1984).

The blossoms attract a variety of unspecialized pollinators, including bees, wasps, lepidoptera, and beeflies (Simpson 1977; Simpson and Neff 1987). Over a 24-hour period, the heads produced 2.06 mg total sugars (Simpson 1977). (The leaf rachis also bears a nectary; see Pemberton 1988.) Fruits ripen and fall with the onset of summer rains. Seeds undergo delayed germination of at least one year due to the impermeable seed coat (Shreve 1964). Tree-ring chronologies from the Grand Canyon show that the plants live longer than 130 years (Alex McCord, personal communication 1993).

This species can be grown from mechanically scarified seed. Once established, the plants require no supplemental irrigation in areas where annual rainfall exceeds 25 cm, but they grow fastest with occasional deep soaking (Duffield and Jones 1981). The plants survive freezing temperatures as low as –18°C (Johnson 1993).

The flowers are an excellent source of nectar for honeybees. Arizona Indians ground the pods for use in mush and cakes. The wood is strong and has been used locally in various carpentry projects (Kearney and Peebles 1960). A gum similar to gum arabic exudes from the trunks; this substance is used locally in Mexico (Standley 1922). Pods, leaves, and roots can be used to make medicinal teas (Moore 1989).

Acacia kelloggiana

Carter & Rudd

Carabatilla de espina negra

This shrub or small tree grows up to 7 m tall. The leaves are twice-pinnate with 9–23 pairs of leaflets 5–8 mm long on each of 1–13 pairs of pinnae. The spine tips curve downward from broad, laterally compressed bases. The whitish flowers are in globose heads up to 2 cm in diameter. Fruits are flattened, somewhat constricted pods 8–15 cm long and 1.5–2 cm wide. They usually dehisce along one side only.

Acacia kelloggiana differs from the closely related *A. peninsularis* in having paired, stipular spines instead of unpaired, internodal prickles (Turner and Busman 1984).

The species is narrowly confined within Baja California Sur to the northern Sierra de la Giganta and the next range to the north, Sierra de las Palmas (Carter and Rudd 1981).

Leafless at the height of the dry season, *A. kelloggiana* quickly leafs out when summer rains start. Leaves drop at the end of summer. Fruits mature in fall, and by November or December only dehisced fruits remain on the plants (Carter and Rudd 1981).

Acacia mcmurphyi

Wiggins

This unarmed shrub grows up to 5 m tall and has feathery, twice-compound leaves. There are 15–23 pairs of leaflets 4–8 mm long on 3–9 pairs of pinnae. The cream-colored to pink flowers are in globose heads. The reddish brown pods are 4–7 cm long and 8–12 mm wide.

Acacia mcmurphyi differs from *A. goldmanii* in lacking 1 or 2 rows of minute petiolar glands. The degree of pilose pubescence is not useful in telling them apart (Annetta Carter, personal communication 1986).

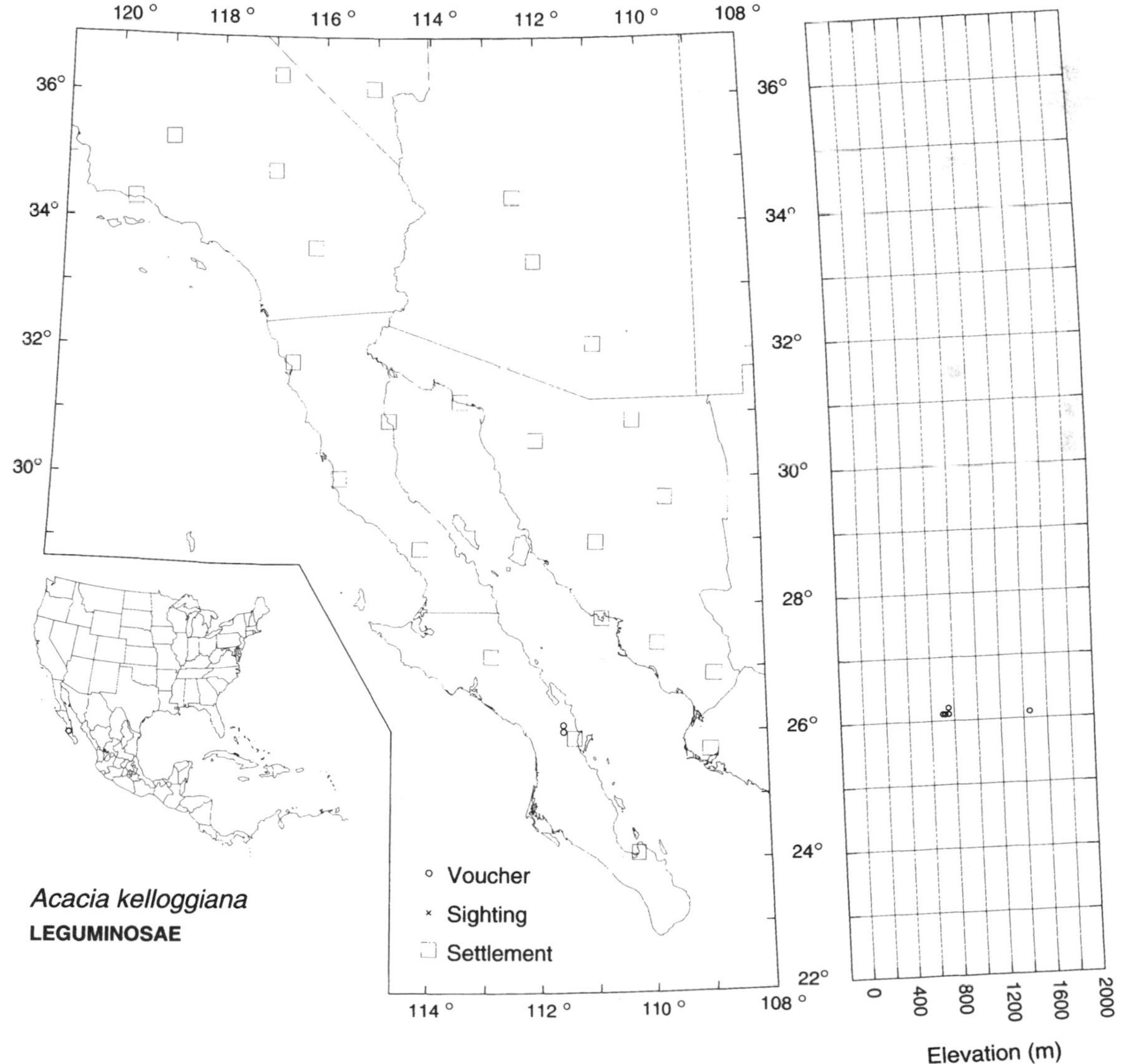

Acacia kelloggiana
LEGUMINOSAE

Competitive interactions with *A. brandegeana, A. farnesiana, A. goldmanii, A. kelloggiana,* and *A. peninsularis* may help define the limits of this narrow endemic.

The plants flower in October and perhaps at other times.

Acacia millefolia
S. Watson

Tepemesquite blanco

Reaching 3 m tall, *Acacia millefolia* has weak stipular spines 2–3 mm long, twice-pinnate leaves, 5–10 pairs of pinnae and 20–30 pairs of small leaflets (1.5–3 mm long). The petioles are U-shaped in cross section. The cream-colored flowers are in spikes 2.5–5 cm long. The flat, veiny pods are 7–15 cm long and 1–1.8 cm wide.

Acacia millefolia superficially resembles *Lysiloma microphyllum* and *L. watsoni,* both unarmed.

Acacia millefolia occurs in a narrow band in southeastern Arizona, southwestern New Mexico, eastern Sonora, northern Sinaloa, and southwestern Chihuahua. The Chihuahuan record, located outside our map, is Edward Palmer's collection (Watson 1886) from Hacienda San José at 450 m elevation (26.8°N, 107.7°W). The Sinaloan record is Arthur Gibson's from Topolobampo.

The elevational profile suggests that *A. millefolia* is squeezed between aridity at its lower limits and dense vegetation at its upper limits. The northern end is somewhat truncated. Interactions with *A. coulteri, A. pringlei,* and perhaps *A. cochliacantha* may have shaped the southern limit.

The plants flower from July to Septem-

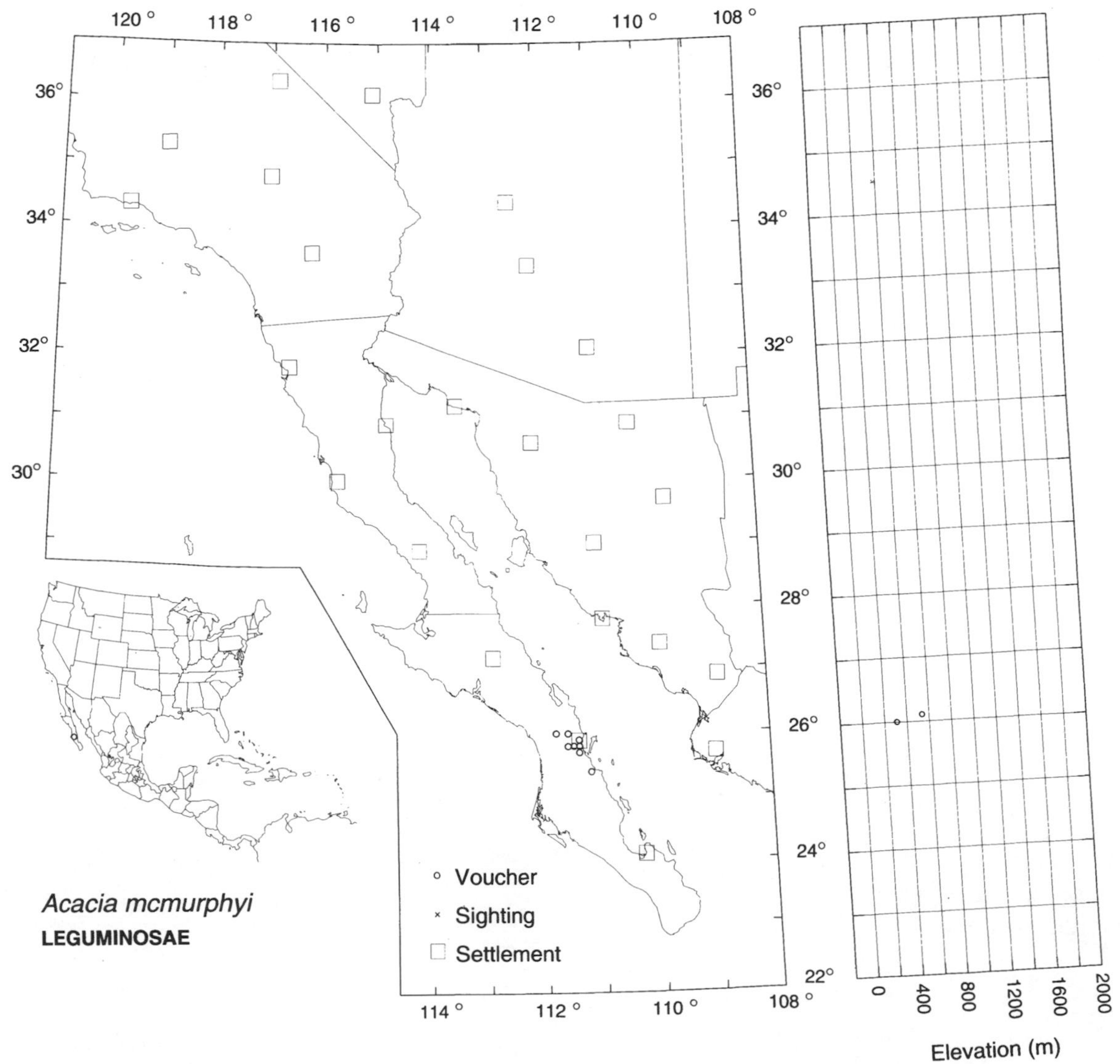

Acacia mcmurphyi
LEGUMINOSAE

ber. In Arizona they are cold deciduous. They survive freezing temperatures as low as –6.5°C (Johnson 1993).

Acacia neovernicosa

Isely

[=*A. vernicosa* Standley, *A. constricta* var. *vernicosa* (Standley) L. Benson]

Chihuahuan white thorn, vinorama, viscid acacia

Acacia neovernicosa, a shrub 1–2 m high, has paired, straight, white spines. Branching is open with several long, straggling stems from the base. Leaves are twice-pinnate. Each of the 1–2 (occasionally 3) pairs of pinnae have 7–9 pairs of crowded leaflets up to 2 mm long. Dotlike glands on the leaves secrete resin, making the foliage somewhat sticky. Flowers are in fragrant, yellow heads about 1 cm in diameter. The pods, 4–7 cm long and 3–5 mm broad, are markedly constricted between the seeds. The diploid chromosome number is 26 (Turner 1959).

This species resembles *A. constricta* in its paired white spines and small, twice-pinnate leaves, but resinous foliage and fewer pinnae help observers distinguish between the two. *Mimosa aculeaticarpa* and *Acacia greggii* have curved spines, and *Calliandra eriophylla* is unarmed.

The diploid *A. neovernicosa* is closely related to the tetraploid *A. constricta* and may be ancestral to it (Isely 1969). Benson (1943) reduced *A. neovernicosa* (as *A. vernicosa*) to a variety of *A. constricta.* Isely (1969), however, noting the lack of intermediates between the two and their different ploidy levels, kept it as a species. Where the two overlap, hybrids are rare, possibly because flowering times overlap little if at all (Correll and Johnston 1970). The difference in chromosome number is surely an even more effective barrier to hybridization (Isely 1969).

Acacia neovernicosa, often common on rocky slopes and plains, can form nearly impenetrable thickets. It is essentially a

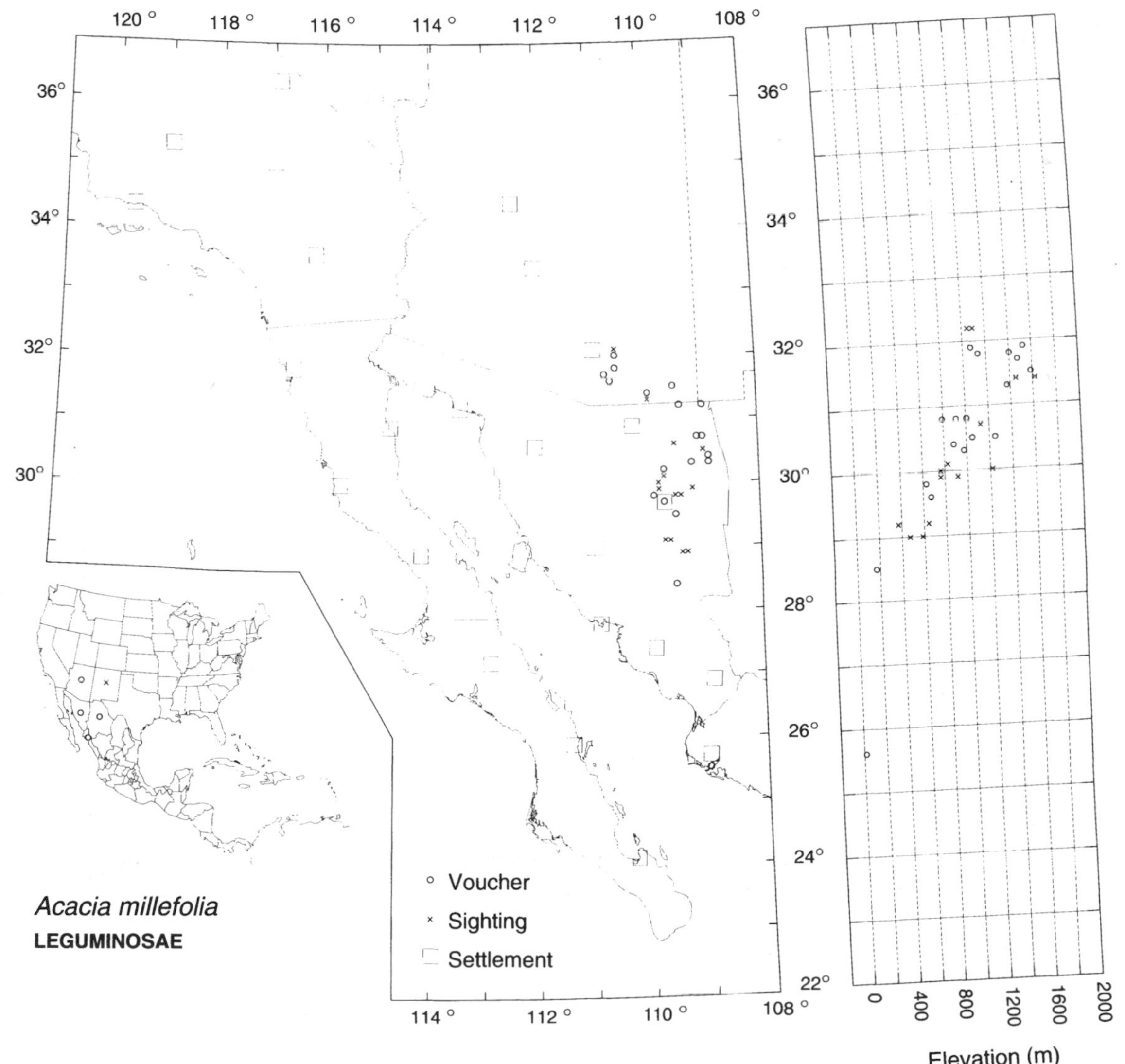

Acacia millefolia
LEGUMINOSAE

shrub of the Chihuahuan Desert, and in southeastern Arizona it often grows on limestone with other Chihuahuan Desert shrubs such as *Flourensia cernua* DC., *Rhus microphylla,* and *Mortonia sempervirens.* Much of its Arizona range was grassland a century ago (Hastings and Turner 1965b). It enters the Sonoran Desert proper in west-central Sonora (30.1 to 30.5°N) on limestone. The sightings in this area are Raymond Turner's from 1969, 1983, and 1987. Wiggins (1964) apparently overlooked these small, scattered populations. Other Chihuahuan disjuncts on limestone in Sonora include *Senna wislizeni* and *Echinocactus horizonthalonius.*

The northwestern limit may occur where warm-season rainfall drops below some necessary threshold. A combination of factors, among them noncalcareous substrates and interspecific competition (especially with *A. constricta* and perhaps also with *A. millefolia* and *A. farnesiana*), have likely shaped the Sonoran distribution to a greater degree than have the direct effects of climate.

The plants flower from June to August. The leaves appear with the first flowers and drop after the first hard frost. The plants survive freezing temperatures as low as –18°C (Johnson 1993).

Acacia occidentalis

Rose

Tree catclaw, tésota, teso, uña de gato

Acacia occidentalis is a small tree or shrub up to 12 m tall with 5–15 pairs of leaflets 3–5 mm long on 2–4 pairs of pinnae. The white or cream-colored flower heads are subglobose. The curved, constricted pods are 4–8 cm long and 2 cm wide.

This plant looks much like *A. greggii, Pithecellobium sonorae,* and *P. mexicanum.* All have small leaflets and small, twice-compound leaves. They differ in inflorescence (capitate in *A. occidentalis,* elongate in *A. greggii*); spine placement (unpaired

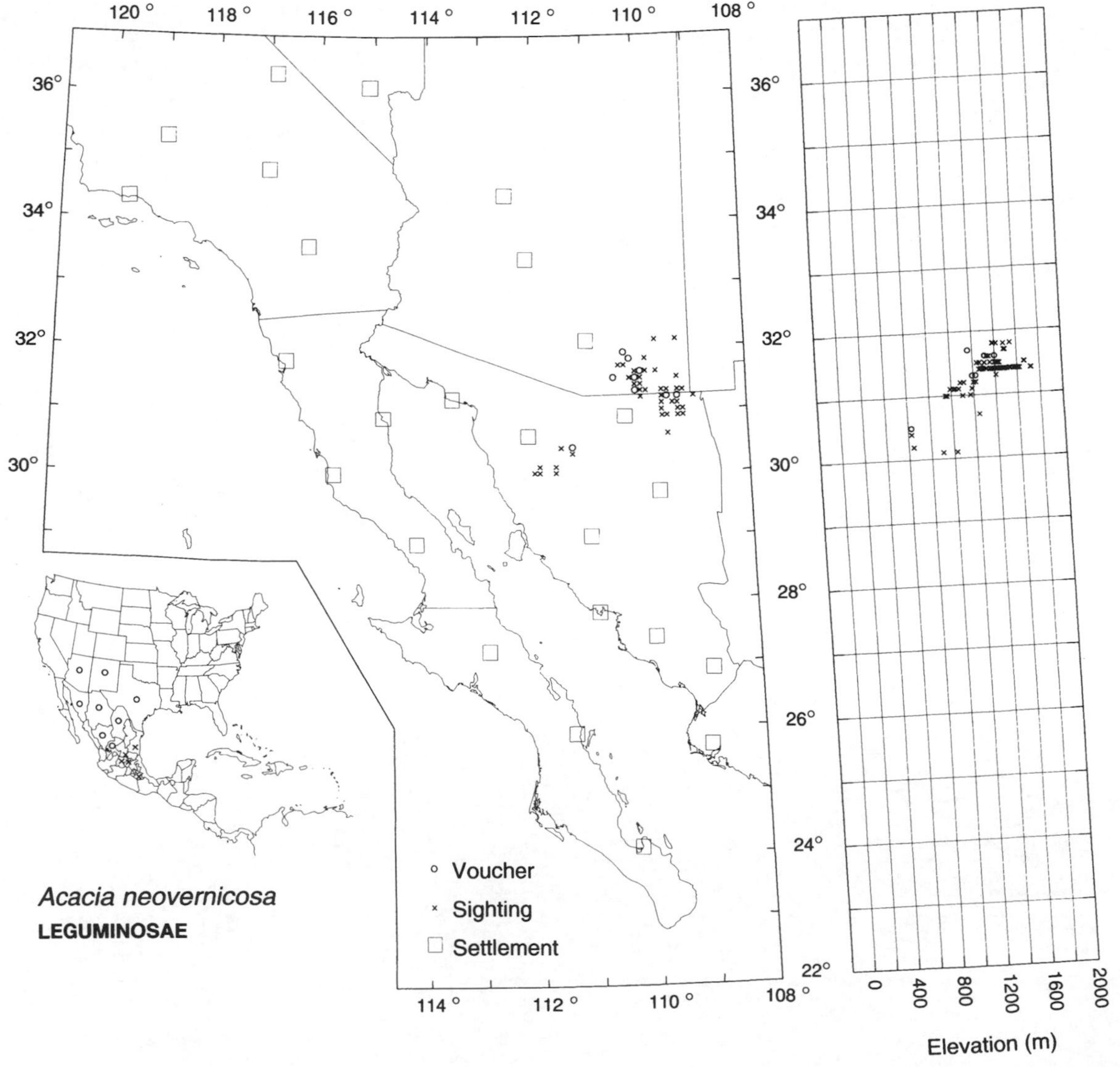

Acacia neovernicosa
LEGUMINOSAE

and internodal in *A. occidentalis,* paired and stipular in *P. mexicanum* and *P. sonorae*); and position of the petiolar gland (near the middle of the petiole in *A. occidentalis,* between the lower pair of pinnae in *A. greggii*).

Acacia occidentalis grows mainly in Sonora east of 111°W as a common tree of arroyo margins. The collection from Isla Tiburón was made in 1890 by W. J. McGee. According to Wiggins (1980), *A. occidentalis* is found on the Baja California peninsula mainly in the vicinity of "Bahía Escondido," which we take to mean Puerto Escondido. We have seen no peninsular vouchers.

The elevational profile suggests adaptation to a rather narrow ecocline (van der Maarel 1990) between coastal arid and montane mesic climates. Nursery-grown plants withstand low temperatures of –9.5°C (Johnson 1991, 1993), so it seems unlikely that freezing truncates the northern limit. On our map, *A. occidentalis* seems squeezed by several congeners, among them *A. angustissima, A. pringlei, A. millefolia,* and *A. coulteri* at the upper, mesic limits and *A. constricta* and *A. greggii* at the northern, arid limits. Juxtaposition of elevational profiles suggests that competition with *P. sonorae* may shape the southern border.

The plants keep their leaves throughout most of the dry season. They flower prolifically in March, perfuming the air so generously that they can be smelled before they are seen (Gentry 1942). Fruits ripen by the summer rainy period.

This species is well worth cultivating as an ornamental in suitable areas (Johnson 1991).

Acacia pacensis

Rudd & Carter

Acacia pacensis, a shrub up to 2.5 m tall, has 2–6 pairs of pinnae on leaves 3–4 cm long. There are 4–10 pairs of leaflets 1–3 mm long. The yellow flowers are in small heads 6–8 mm in diameter. Fruits are linear pods, 6–15 cm long, that are bright red when young and blackish in age.

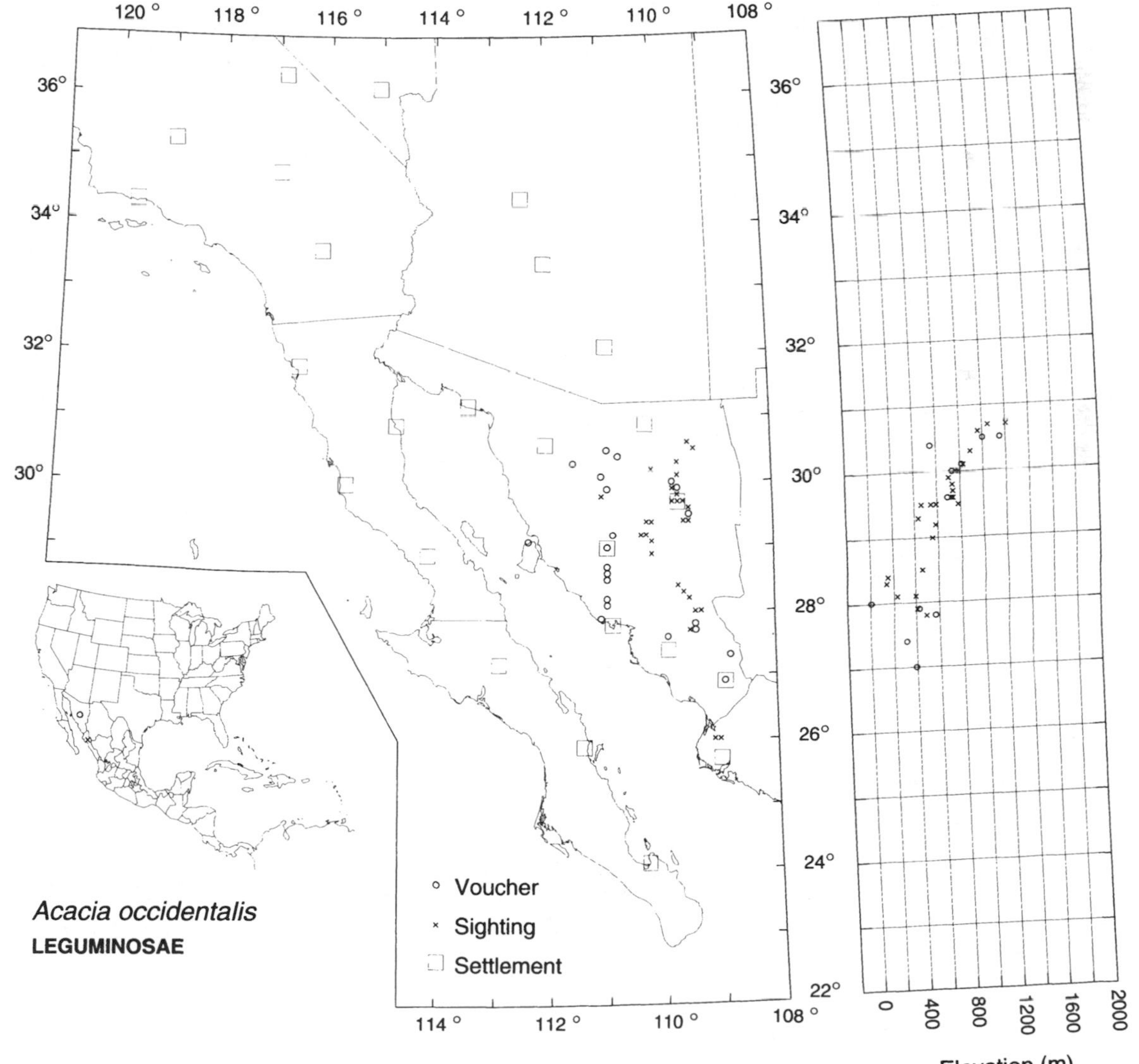

Acacia occidentalis
LEGUMINOSAE

First collected a century ago (Brandegee 1891), *A. pacensis* closely resembles *A. constricta,* the cause of many incorrect identifications. Only recently have its true identity and affinities with other *Acacia* species (e.g., *A. farnesiana, A. cochliacantha*) been noted (Rudd and Carter 1983). In both *A. pacensis* and *A. constricta,* the fruits are long and narrow; however, they are indehiscent with seeds separated by septa in *A. pacensis,* dehiscent and aseptate in *A. constricta.*

This narrowly endemic species grows near La Paz, Baja California Sur, and on two offshore islands, generally on the more arid sites throughout its small range. *Acacia constricta* comes within about 75 km of La Paz, but nowhere does it grow with *A. pacensis.* An inability to invade more mesic communities may have kept *A. pacensis* from spreading, especially during wetter periods of the Pleistocene. Competitive interactions with *A. farnesiana* and possibly also with *A. brandegeana* may have shaped the southern limit.

Flowering is in summer.

Acacia peninsularis
(Britton & Rose) Standley

A shrub or small tree up to 10 m tall, *Acacia peninsularis* has a dense, rounded crown and dark brown bark. The spatulate leaflets, 5–10 mm long and 2.5–6 mm wide, are in 3–8 pairs on 1–2 pairs of pinnae. The greenish white flowers are in dense, globose heads. The flat, oblong pods are 5–10 cm long.

Acacia peninsularis has relatively few and large leaflets compared to other acacias in its range.

This Baja California Sur endemic grows mainly on the east side of the peninsula in a narrow ecotone between tropical

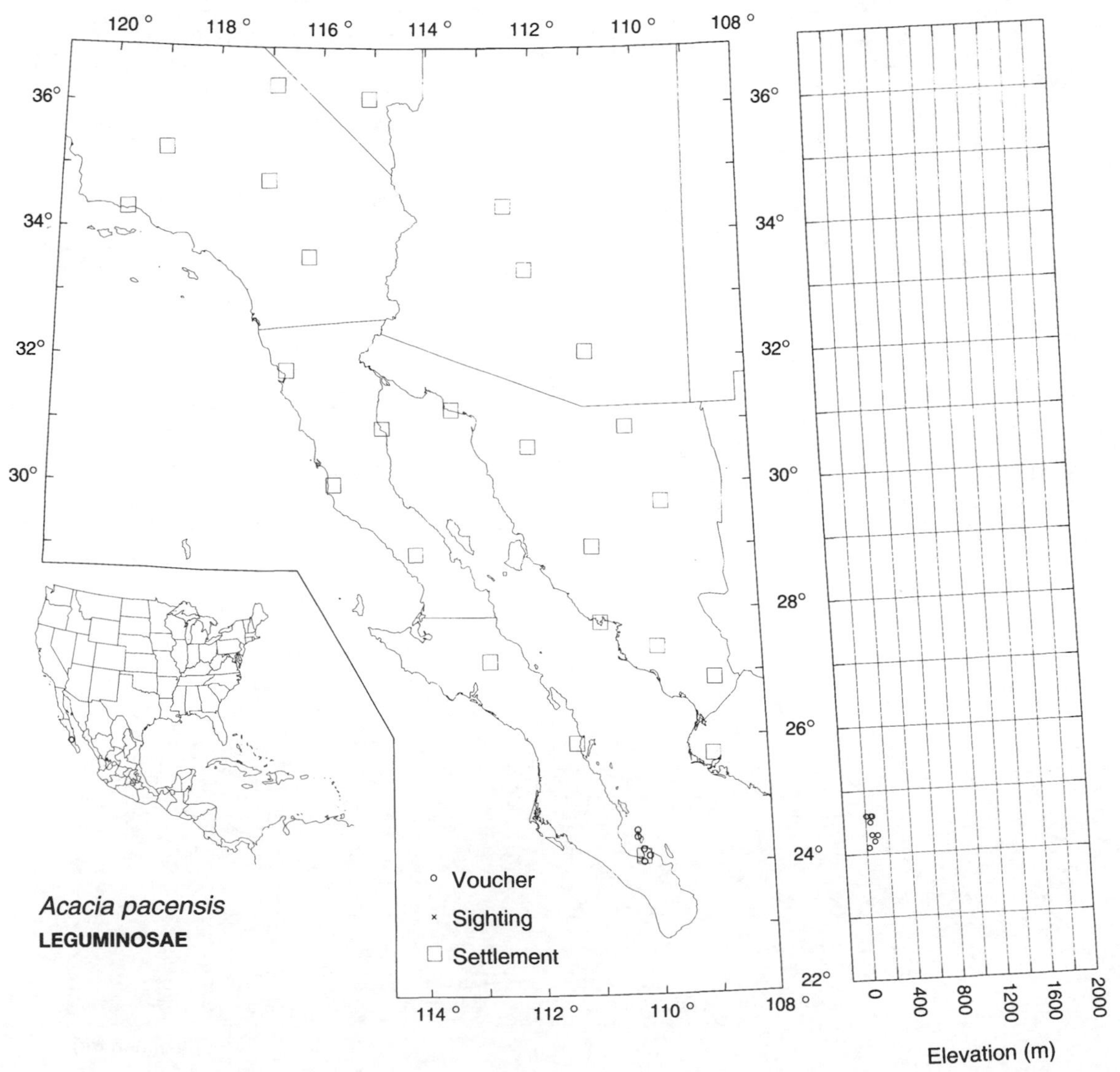

arid and mesic climates. In the north, it seems restricted to riparian habitats. Aridity probably causes the disjunction between 23.8 and 25.2°N. The mapped range of *A. peninsularis* looks much like that of *Lysiloma microphyllum*, but their contiguous elevational profiles suggest that *Lysiloma* competitively replaces *Acacia* along a moisture gradient. The same may be true of *A. peninsularis* and *A. goldmanii*.

The plants flower in the spring.

Acacia pennatula

(Cham. & Schlecht.) Benth.

Algarroba, espino, huizache, huizache tepamo, garrobo, feather acacia

Acacia pennatula is a spiny, copiously pubescent shrub or tree up to 10 m tall. Each leaf (5–10 mm long) has 50–60 pairs of pinnae and 40–70 pairs of tiny, overlapping leaflets. The fragrant yellow flowers are in dense, globose heads. The oblong, flattened fruits, 6–15 cm long and 1.5–2 (occasionally 2.5) cm wide, are tardily dehiscent. Outside, they are woody and black; inside, they are fleshy.

Acacia pennatula has the smallest leaflet area and the greatest leaflet number of any Sonoran Desert acacia (Cody et al. 1983), making it readily recognizable.

Two subspecific taxa have been recognized on the basis of inflorescence length; only subsp. *pennatula* occurs in our region (Seigler and Ebinger 1988).

Widespread in the semiarid tropics, this species reaches its frigid and arid limits along the eastern edge of the Sonoran Desert from the vicinity of Cumpas, Sonora, southward. Reaching 1,525 m, *A. pennatula* is more common in Sinaloan thornscrub and Madrean evergreen woodland than in Sonoran desertscrub. The plants aggressively colonize disturbed areas (Cházaro 1977). Incursions of freezing air across a wide range of

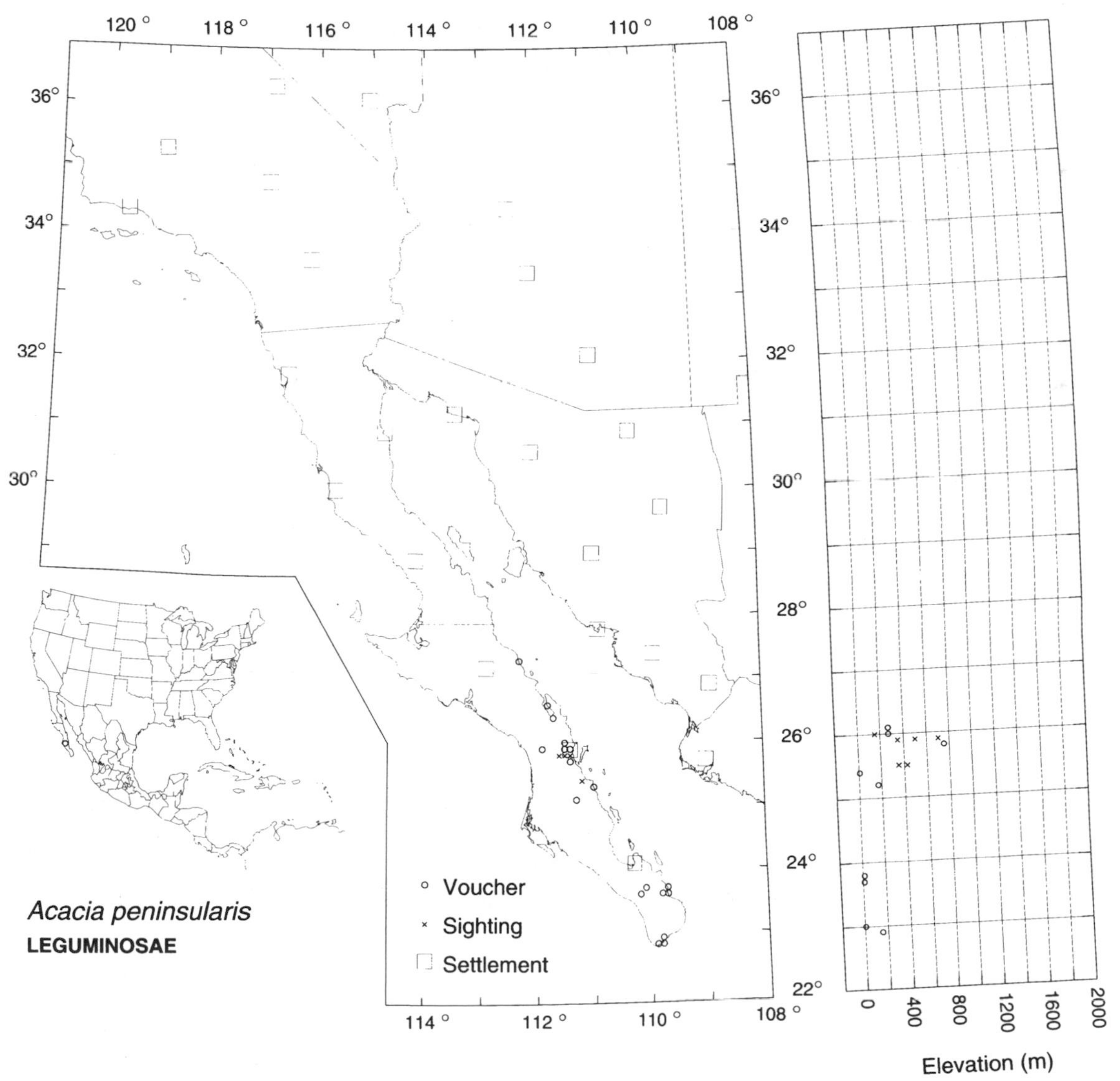

Acacia peninsularis
LEGUMINOSAE

elevations may determine the truncated northern limit seen on our elevational profile. Interactions with *A. angustissima, A. coulteri, A. farnesiana,* and *A. millefolia* may help shape the northern boundary and elevational profile seen on our map.

Flowers appear mainly during April and May. The fruits begin to develop in June but do not reach maturity until the following dry period. They are slow to dehisce and often fall to the ground whole. Plants may flower at four years of age (Cházaro 1977). Approximately 100,000 seeds are produced by a single plant (Cházaro 1977). This species fixes nitrogen (Van Kessel et al. 1983).

Use as an ornamental is limited by sensitivity to freezing temperatures below –4°C (Johnson 1993). Cattle seek the fruits (Cházaro 1977). The bark has been used as a remedy for indigestion (Standley 1922), and the seeds are roasted, ground, and eaten in the form of atole (Gentry 1942).

Acacia pringlei

subsp. *californica*

(Brandegee) Lee, Seigler & Ebinger

[=*A. californica* Brandegee]

Guamuchilillo, guamúchil

Acacia pringlei subsp. *californica* is a tree that typically reaches 5–8 m but may grow as tall as 12 m. Borne in 2–3 pairs on a single pair of pinnae, the leaves are distinctively dimorphic, with terminal leaflets 11–25 mm long, up to twice as long as the others. The paired stipular spines are usually less than 2 cm long. The flowers are in

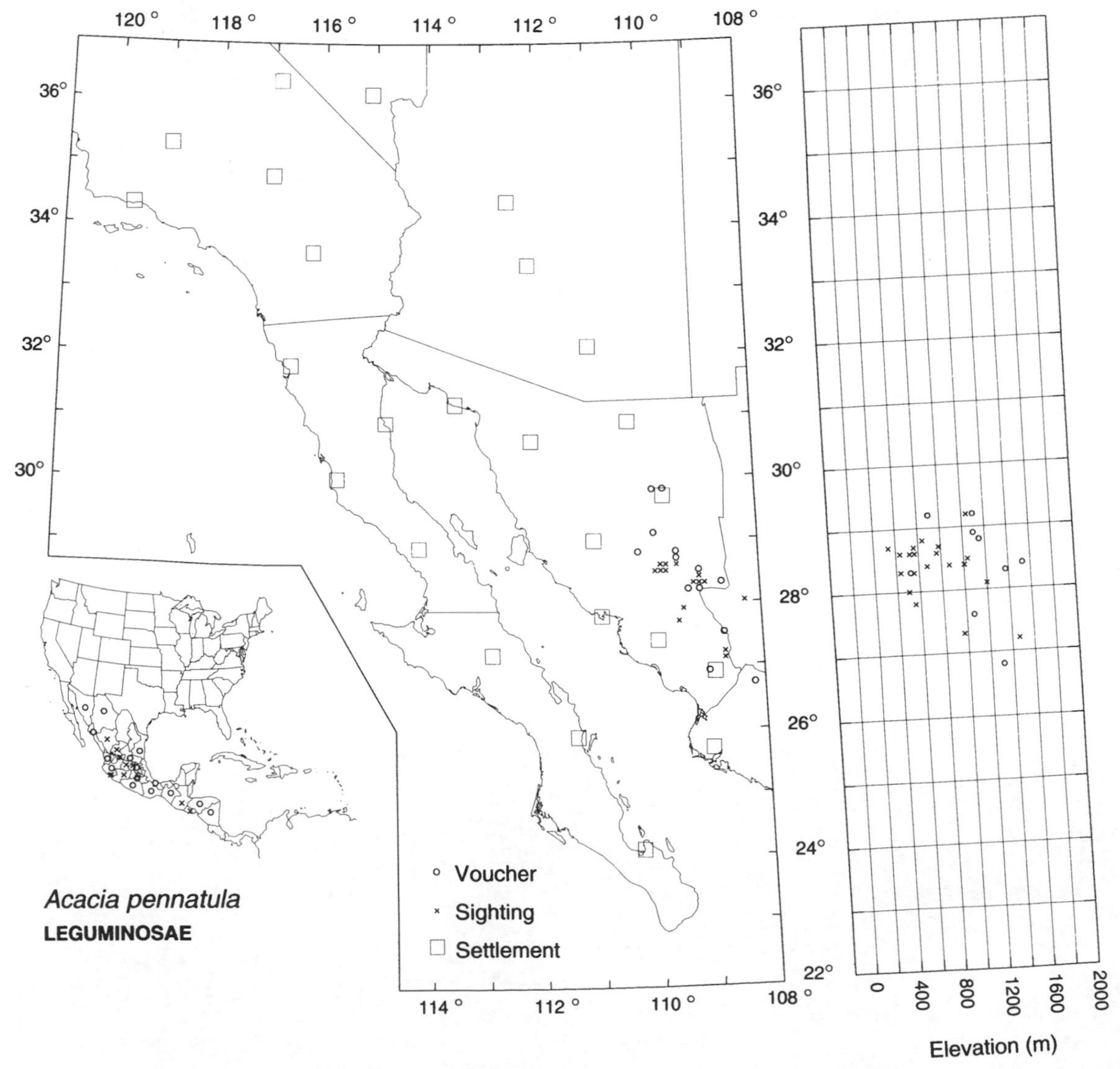

Acacia pennatula
LEGUMINOSAE

elongated white to cream-colored spikes 3–8 cm long. The dark red-brown pods, 10–20 cm long, are linear and striate-veined.

Its closest relative, *A. pringlei* subsp. *pringlei*, differs mainly in leaf pubescence and size of terminal leaflets (Lee et al. 1989) and occurs from Yucatán through Oaxaca and Veracruz, north to Tamaulipas.

Acacia pringlei subsp. *californica* grows in valley bottoms. Like the two subspecies, which are widely disjunct, populations of subsp. *californica* are markedly disjunct between the Cape Region of Baja California Sur and the mountains of Sonora south of 29°N. The pattern of disjunctions suggests that the taxon is old and its range has been split several times. *Acacia pringei* subsp. *californica* grows at lower elevations at the southern end of its range, typical of taxa displaced from more humid tropical communities. The abrupt northern limit at 29°N is unusual for species in the Sonoran mountains. Replacement by *A. millefolia* may be responsible, perhaps along with low-temperature thresholds.

May through June is the flowering season.

Acacia willardiana

Rose

Palo blanco, palo liso

A slender, unarmed tree 3–9 m tall, *Acacia willardiana* (figure 7) has white or cream-colored exfoliating bark. The petiole is a flattened phyllode 1.5–4 mm wide and 6–30 cm long. The 2–8 pinnae have 4–15 pairs of very small, linear leaflets. The pale yellow flowers are in dense spikes 3–6 cm long. Pods are broadly linear, 1–1.8 cm wide and 10–20 cm long.

Locally common on rocky slopes and hill crests, this species is apparently endemic to the state of Sonora. (Standley

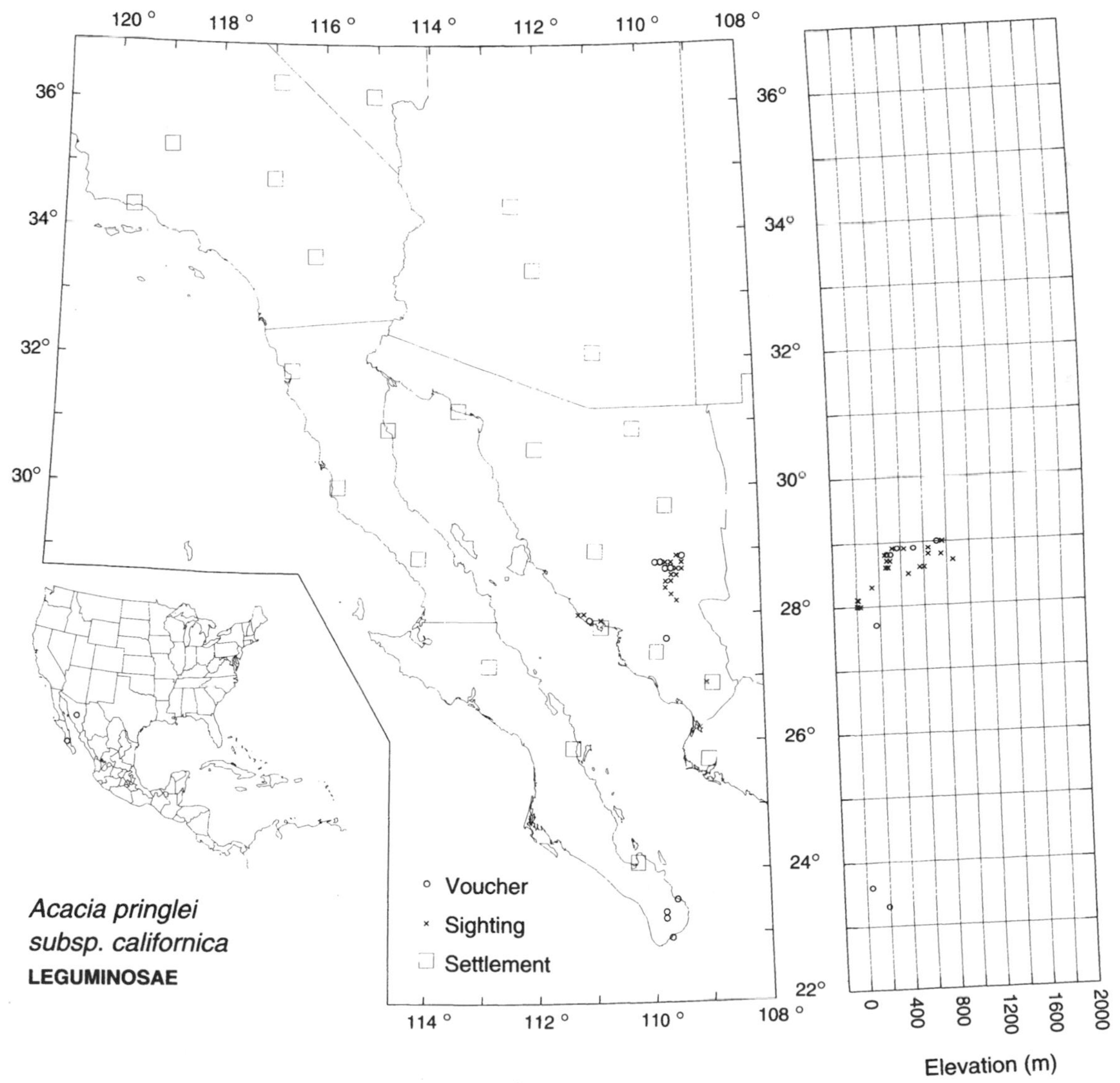

Acacia pringlei subsp. californica
LEGUMINOSAE

[1922] reported it from the Baja California peninsula, but we have seen no specimens.) It reaches its best development in southwestern Sonora (Shreve 1964) and also grows in Sinaloan thornscrub near Alamos and on the Río Yaqui near Soyopa, where reported by Thomas R. Van Devender in 1983 (28.8°N, 109.6°W). The northernmost disjunct is Raymond M. Turner's collection. The nearby sighting is Howard Scott Gentry's (personal communication 1972). The species occurs on the Sonoran islands of Tiburón and San Pedro Nolasco (Moran 1983b).

Acacia willardiana occupies an ecocline (van der Maarel 1990) between arid and semiarid climates where rain falls mainly in the warm season. A shortage of suitable habitat (rocky hills) may keep this species from the semiarid coastal areas near the deltas of the Mayo and Fuerte rivers. Periodic incursions of freezing air probably shape the truncated northern limit.

After the short-lived leaflets drop, the flattened and expanded petioles carry on most of the photosynthesis (Shreve 1964). Flowers appear from February to July, although summer flowering may not occur in a dry season.

This is the only Sonoran Desert plant and the only American acacia with phyllodic petioles (Shreve 1964; Vassal and Guinet 1972). In similar Australian acacias, the phyllode is sclerophyllous and does not produce leaflets (Vassal and Guinet 1972). In contrast, the phyllode of *A. willardiana* is leaflet-bearing and not noticeably sclerophyllous. Vassal and Guinet (1972) assigned *A. willardiana* to a rather broadly defined "Australian group" in subgenus *Heterophyllum*. Pedley (1975) placed it in subgenus *Aculeiferum* with other New World acacias. Willardiine, a rare amino acid first discovered in *A. willardiana*, occurs in two Australian acacias of subgenus *Heterophyllum* (Bell 1971; Guinet and Vassal 1978).

This striking tree is occasionally used for landscaping in areas where winter temperatures do not fall below the range of –6.5 to –4°C (Johnson 1993).

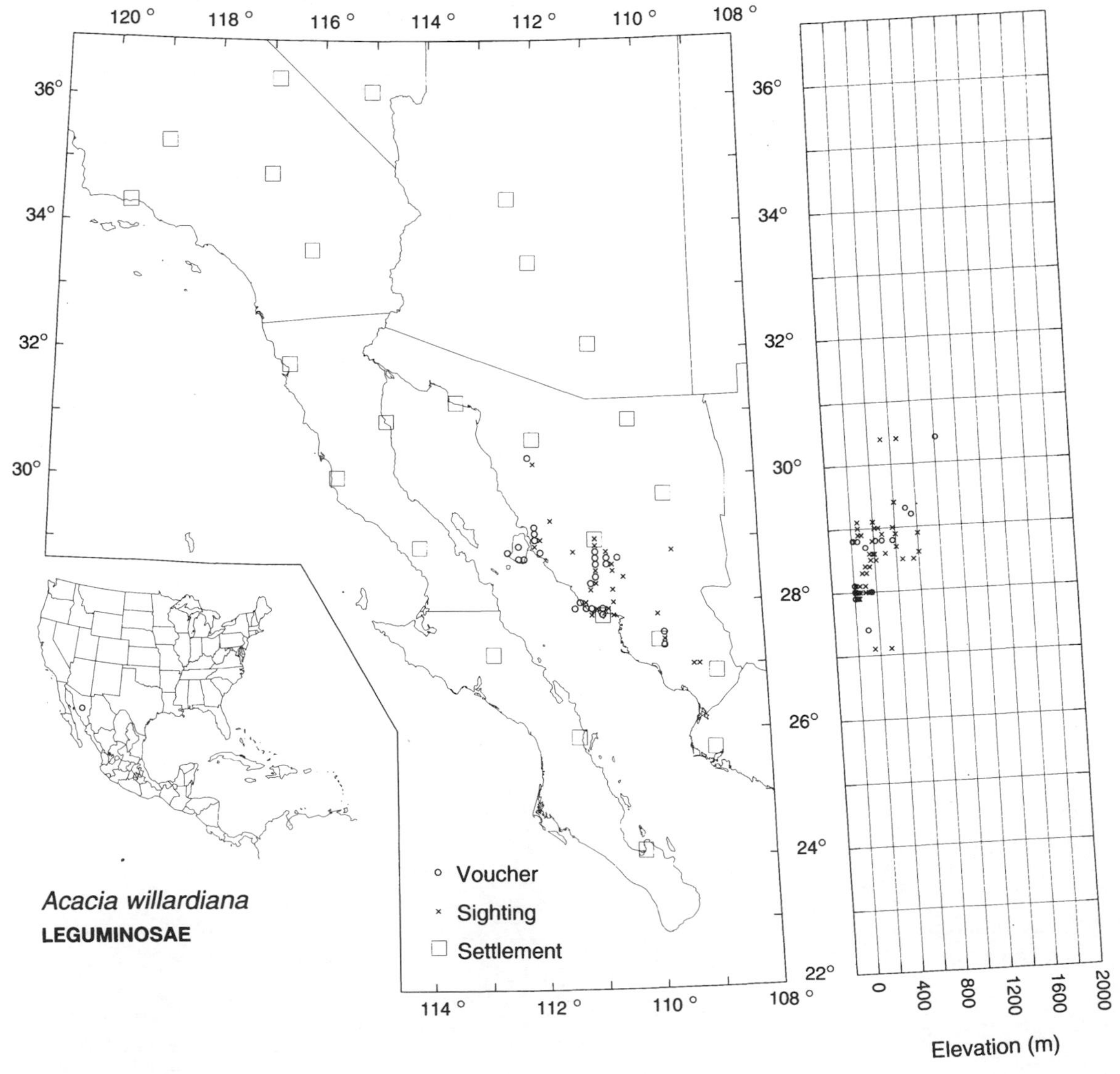

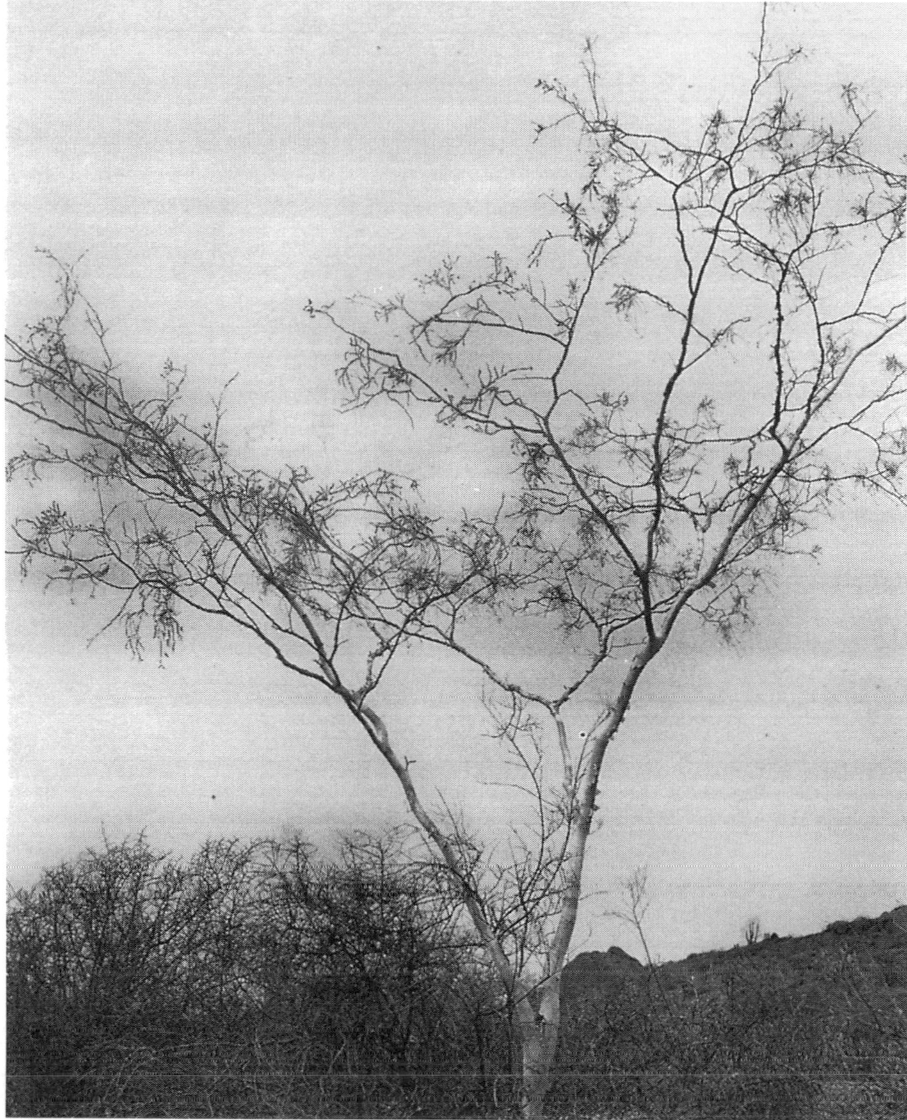

Figure 7. Acacia willardiana *north of Bahía Kino, Sonora, showing the open crown of highly specialized phyllodic leaves. (Photograph by J. R. Hastings.)*

Acanthogilia gloriosa

(Brandegee) Day & Moran

[=*Ipomopsis gloriosa* (Brandegee) A. Grant in V. Grant]

A stiff, spiny shrub 1–3 m tall, *Acanthogilia gloriosa* has dimorphic leaves. Primary leaves are 1–3 cm long and pinnately 3- to 5-lobed with linear, spinose divisions. The linear secondary leaves, fasciculate in the axils, are 7–12 mm long and also spinose. The flowers (3–4.5 cm long) are white, yellow, and orange-red. Fruits are capsules 8–10 mm long. The haploid chromosome number is 9 (Day and Moran 1986). Carlquist and coworkers (1984) have studied the wood anatomy.

Leptodactylon pungens (Torr.) Rydb, a sprawling shrub less than a meter high, also has spinose leaves, but the primary ones are palmately rather than pinnately lobed.

This species has been placed in the genera *Gilia, Ipomopsis, Leptodactylon,* and *Loeselia,* but pollen, leaf, corolla, and seed characters suggest that it is best regarded as a separate genus (Day and Moran 1986). Its closest relative appears to be the Andean genus *Cantua* (Day and Moran 1986).

Acanthogilia gloriosa is locally common on rocky and gravelly benches and along washes, its preferred habitat. Endemic to Baja California, *A. gloriosa* is largely restricted to the Pacific slope mainly below 525 m. The northernmost population (30.7°N) was vouchered in 1976 by Reid Moran and in 1977 by Betty Robinson.

The distribution of this species is typical of a group of endemics, including *Prosopidastrum mexicanum* and *Xylonagra arborea,* found in arid maritime climates with relatively cool, dry summers. Farther inland and along the Gulf of California, the warmer, less humid summers may be too dry, thus determining the eastern limits of this group of endemics. The southern limit occurs near the northern edge of the Vizcaíno Plain where greater aridity and fewer arroyos might combine to stop southward movement. No obvious climatic barrier keeps this species from moving farther north near the Pacific coast. In the transition to Californian coastalscrub, dense vegetation may exclude it from its preferred riparian habitat.

The woody-spinose primary leaves persist two or more years on the stems. The secondary leaves appear after cool-season rains and are shed during summer drought. The flowers appear from January to July and again in October. They last 4–5 days and nights (Day and Moran 1986). Although the white opening of the flowers suggests that moths are the primary pollinators, their lack of scent and their orange-red tubes suggest hummingbird pollination, and hummingbirds do visit them (Day and Moran 1986).

This attractive shrub is well worth cultivating. Even relatively large plants can be transplanted successfully during summer dormancy. In the wild, the lower stems root where they touch the ground (Day and Moran 1986), so propagation by cuttings is probably quite feasible, as well.

Figure 8. The remote, divergent stems and fasciculate leaves of Adelia obovata *represent a common growth habit in the Sonoran Desert. (Photograph by J. R. Hastings.)*

120° 118° 116° 114° 112° 110° 108°

36° 34° 32° 30° 28° 26° 24° 22°

114° 112° 110° 108°

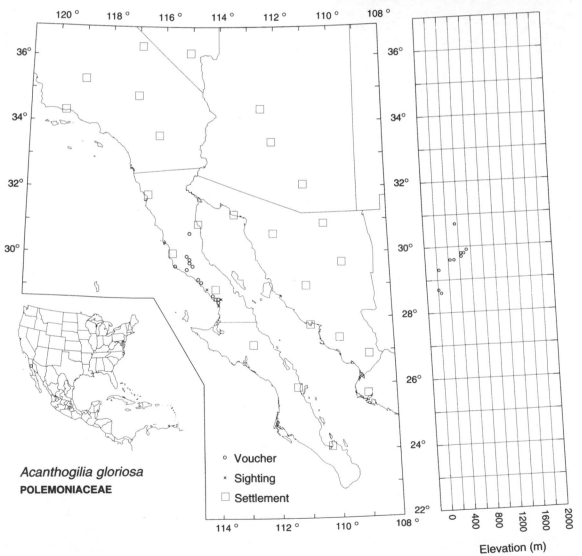

Acanthogilia gloriosa
POLEMONIACEAE

○ Voucher
× Sighting
□ Settlement

0 400 800 1200 1600 2000
Elevation (m)

Adelia obovata
Wiggins & Rollins

Adelia obovata (figure 8), a straggling shrub 1–2 m tall, has fasciculate leaves on rigid branches. The branchlets are not markedly spinescent. The obovate, entire leaves are 8–24 mm long and 5–12 mm wide and are densely pilose on both sides. The inconspicuous flowers are solitary in the leaf axils. Capsules are about 6 mm tall and 5 mm across.

The closely related *Adelia virgata* differs in its spinescent branchlets and glabrous or sparsely pubescent leaves. The two can be difficult to tell apart where they overlap. The sightings between 28.3 and 28.9°N occur in a region of overlap, and some might be *A. virgata*. Critical examination of wild plants and additional collection of herbarium specimens would help in delimiting the two.

A rare to common shrub on rocky hillsides, *A. obovata* is narrowly endemic to the Sonoran Desert. Its distribution reflects speciation at the arid limits of the genus. Apparently, *A. obovata* interacts along its southern boundary with *A. virgata* and other species of coastal thornscrub. The mapped range shows pronounced sensitivity to freezing and a need for summer rain. The elevational profile suggests a northern limit squeezed between coastal aridity and montane cold.

The drought-deciduous leaves appear in response to rain at any time of year. Flowers appear in August and September (Wiggins 1964).

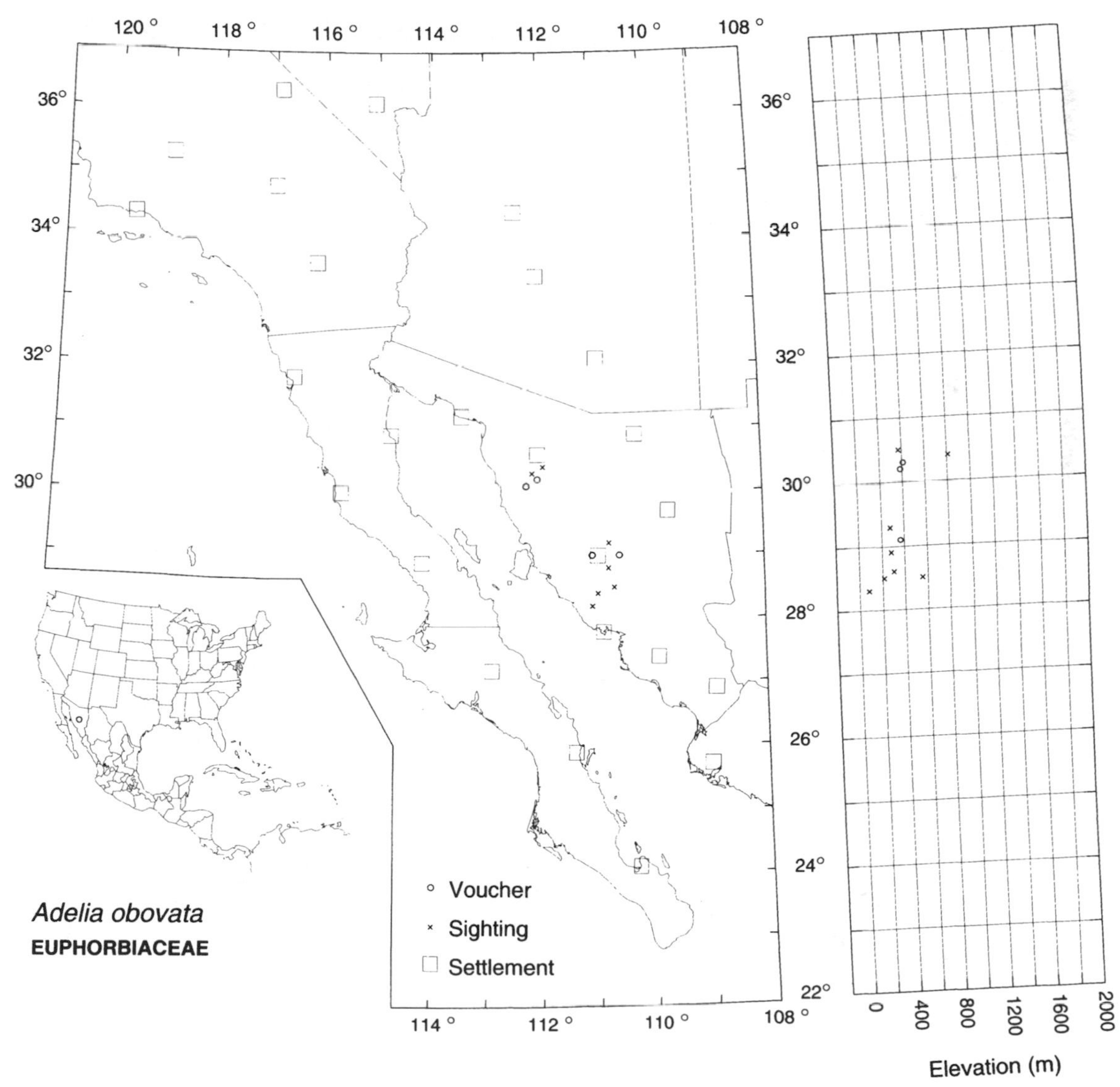

Adelia virgata
Brandegee

This shrub is 2–3 m tall and has long, erect branches and spinescent branchlets. The crowded oblanceolate or oblong leaves are 1.5–3 cm long and 8–12 mm wide. They are glabrous or sparsely pubescent, are usually shallowly emarginate, and are borne on short shoots. The inconspicuous flowers are on spur branches. Fruits are capsules 8–9 mm long and 7 mm across.

The closely related *A. obovata* has densely pilose leaves and usually lacks spinescent branchlets.

Growing on rocky hillsides and along washes, *A. virgata* is rare to occasional throughout its peninsular range. The existence of mainland populations was apparently unknown until 1969, when Raymond M. Turner collected a sterile specimen near Estación Esperanza (27.6°N). Since then, additional collections and sightings have confirmed and extended the mainland range of this species. Like *Fouquieria columnaris, Lysiloma candidum,* and several others, it occupies a wide range of elevations on the peninsula but is restricted to coastal lowlands on the mainland.

The mapped range shows a need for summer rain and winter warmth. On the elevational profile, the populations of Baja California Sur shift upward toward the northern limit, likely a result of aridity at lower elevations. The northern peninsular limit approximately coincides with those of *Cercidium praecox* and *Lysiloma candidum* in an area of biseasonal rainfall. To the north, warm-season moisture declines steeply. Competitive exclusion from more mesic communities restricts *A. virgata* to lower elevations in the Cape Region and on the Sonoran coast.

The drought-deciduous leaves appear in response to rain. Plants bloom in late summer (August–October) (Wiggins 1964).

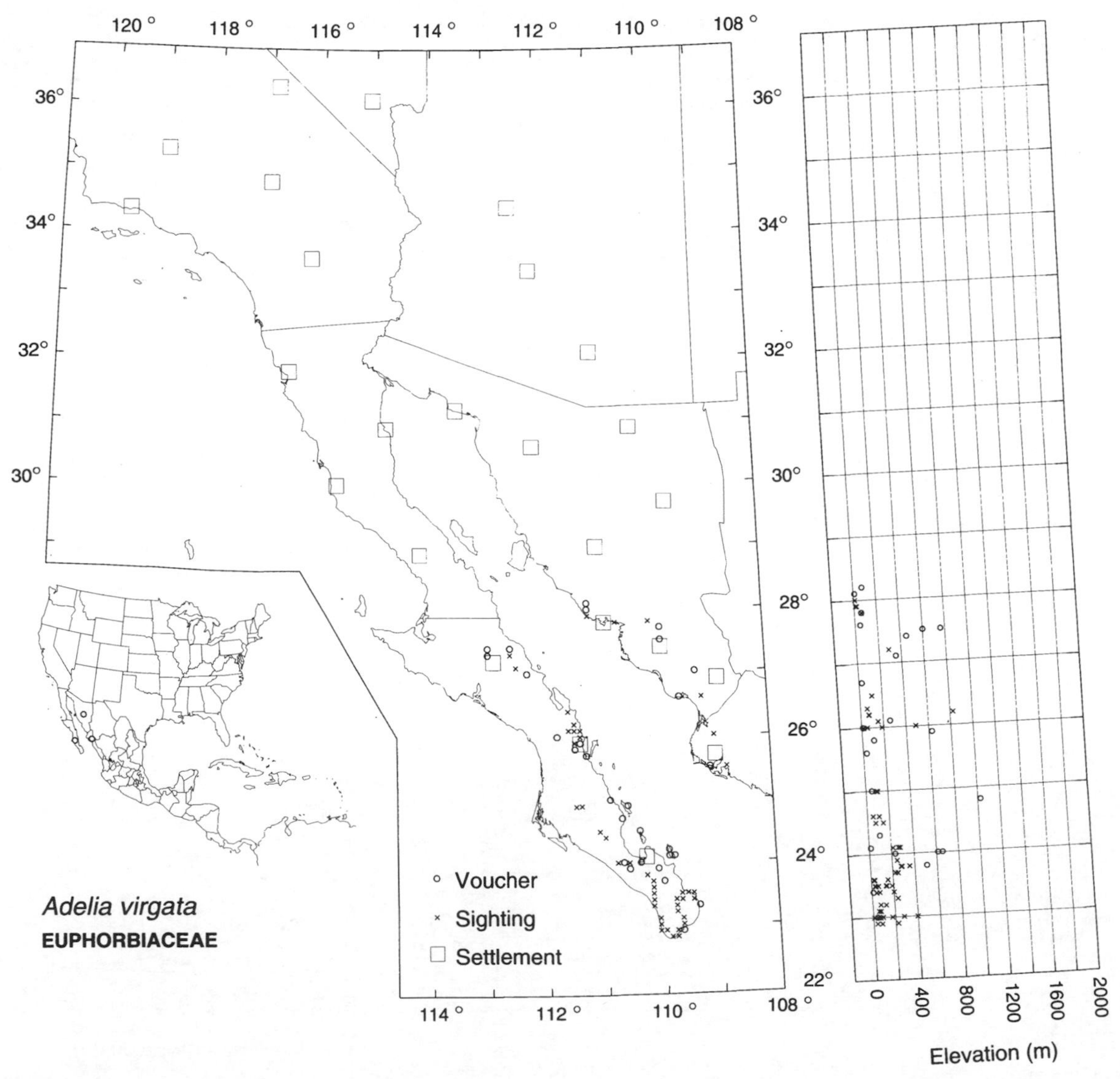

Aeschynomene fascicularis
Schlecht. & Cham.

Popotillo

Aeschynomene fascicularis is a slender-branched shrub 1–2 m tall with once-pinnate leaves and 40–50 leaflets. Glabrous above and appressed pubescent beneath, especially along the midvein and margins, the leaflets are 10–20 mm long and 2–5 mm wide. The yellow flowers, chocolate brown in age, are 8–15 mm long. The fruit breaks transversely into several 1-seeded segments.

Rudd (1955, 1975) treated the taxonomy of *Aeschynomene.*

This species is widespread throughout much of Mexico and Central America. In the area covered by this atlas, it grows beyond the edge of the desert in Sinaloan thornscrub.

The plants flower intermittently from June to November and fruit in autumn.

Aeschynomene nivea
Brandegee

Aeschynomene nivea is a shrub 1–3 m tall with silvery-sericeous foliage and once-pinnate leaves up to 8 cm long. The 30–60 leaflets are 1–2 mm wide and 4–10 mm long. The midvein is strongly off-center. Yellow when fresh, the flowers (10 mm long) turn chocolate brown as they age. The fruit breaks transversely into 1- to 3-seeded segments.

Rudd (1955, 1975) treated the taxonomy of the genus.

The silvery-sericeous leaflets with their markedly eccentric midvein are distinctive.

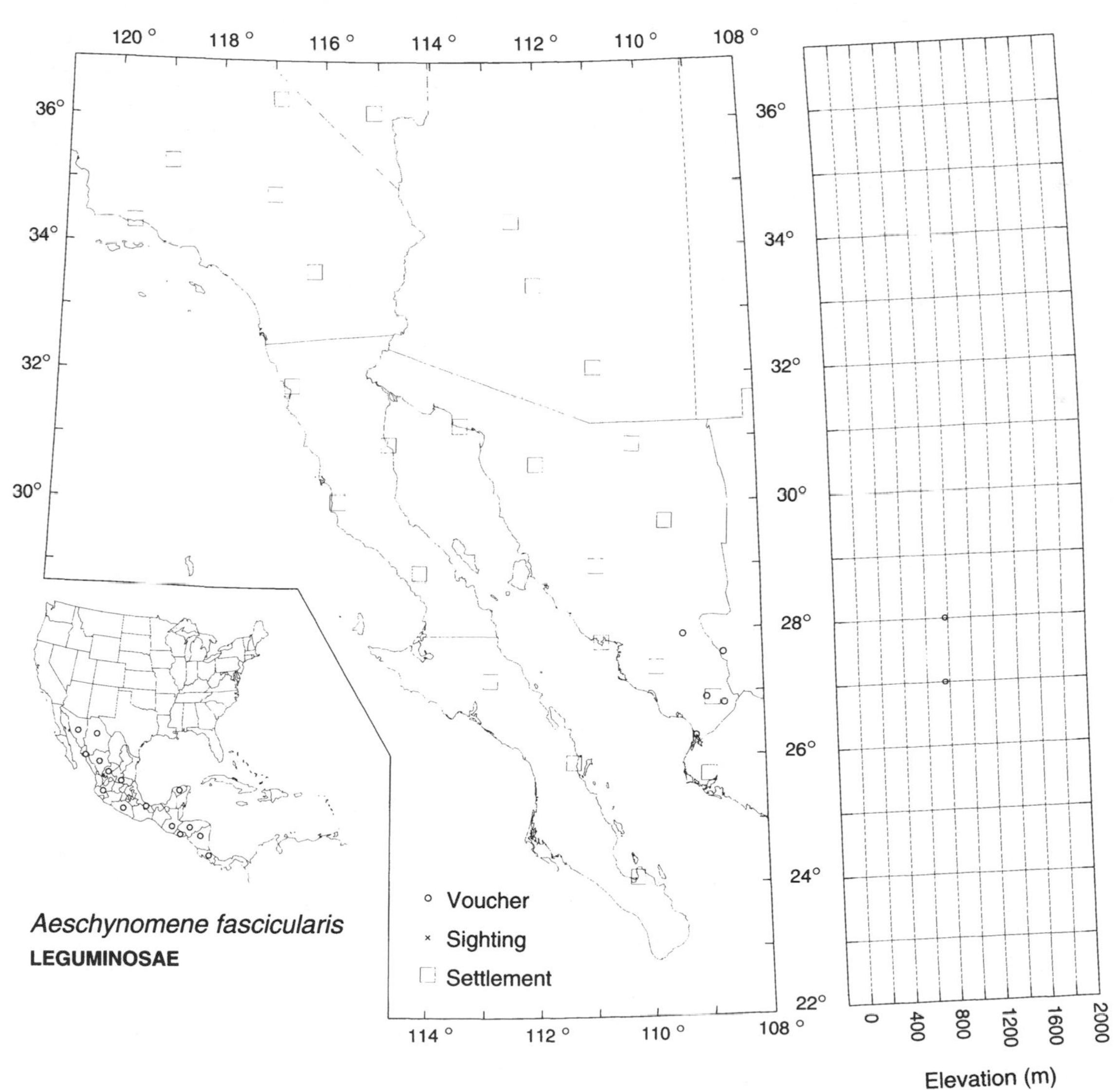

Aeschynomene fascicularis
LEGUMINOSAE

Endemic to Baja California Sur, *Ae. nivea* grows from San Ignacio to Todos Santos and on most of the Gulf of California islands from Isla Coronados (26.1°N) southward (Moran 1983b). Although largely confined to the Gulf Coast, the species extends into some Pacific watersheds, a pattern that suggests insular speciation followed by dispersal to the peninsula. *Aeschynomene nivea* apparently needs regular summer rain and relatively warm winters. It gives way to *Ae. vigil* in the Cape Region.

The flowering season is October to March.

This showy plant is worthy of cultivation.

Aeschynomene vigil
Brandegee

A shrub 1–3 m tall, *Aeschynomene vigil* has white-sericeous stems that turn glabrate and gray-barked with age. The once-pinnate leaves have 8–14 leaflets, each 8–15 mm long and 4–8 mm wide. The leaflet midveins are nearly central. The yellow flowers are 8–10 mm long and discolor to a dark purplish brown. The fruit breaks transversely into a few 1-seeded segments.

Where their ranges overlap, *Ae. vigil* can be readily told apart from *Ae. nivea* by its broader leaflets and nearly central midveins.

This species is narrowly endemic to the southern tip of Baja California. Although it has been found in mountains, it is apparently more frequent in coastal thornscrub.

The plants bloom between October and February and perhaps at other times of the year.

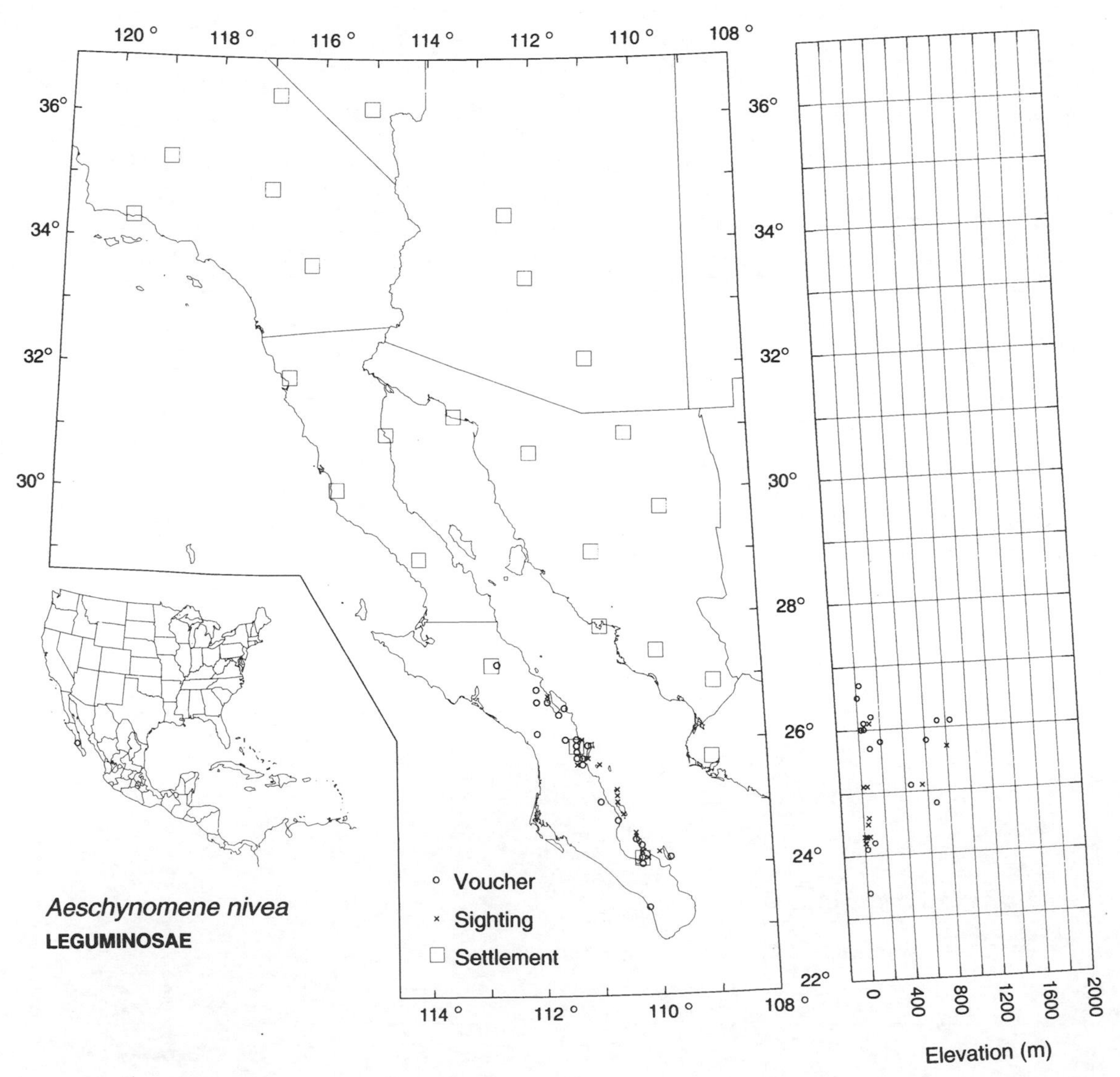

Aeschynomene nivea
LEGUMINOSAE

Aesculus parryi

A. Gray

Buckeye

A shrub or small tree up to 6 m tall, *Aesculus parryi* (figure 9) has palmately compound leaves, each with 5–7 elliptic to ovate leathery leaflets 3–12 cm long and 2–4 cm wide. Creamy white above, the flowers are orange or maroon at the base and 8–18 mm long. They are in narrow panicles 5–20 cm long. Perfect and functionally staminate flowers are produced on the same plant. The obovate to spherical fruits are 2–3 cm in diameter (Hardin 1957b).

Aesculus parryi grows on coastal dunes, rocky slopes, and arroyos. Restricted to the western part of Baja California, it is geographically isolated from all other species of *Aesculus* (Hardin 1957b). On our map, the apparent gap west of 115°W and 30°N is an artifact of our sampling pattern. The distribution seems limited to zones dominated by Pacific maritime air where rain falls exclusively in winter and summers are relatively cool. In the northern part of its range, *Ae. parryi* grows in chaparral up to 900 m. Toward the southern limit, populations occur in Sonoran desertscrub near sea level. Moisture from fog might compensate for low rainfall in this region. Shreve (1936) considered *Ae. parryi* to be endemic to the transition zone between Sonoran desertscrub and California coastalscrub.

According to Hardin (1957a), the genus originated in Central or South America. During its northward migration, *Ae. parryi* and the closely related *Aesculus californica* (Spach) Nutt. became isolated on the Pacific coast of California and Baja California.

The period of flower and fruit development in *Ae. parryi* is quite long (Hardin 1957b). Flowers are present from March through July, fruits from September through February. In contrast, leafing is brief. Leaves appear in November or December, then turn yellow and drop between February and May as the flowers appear.

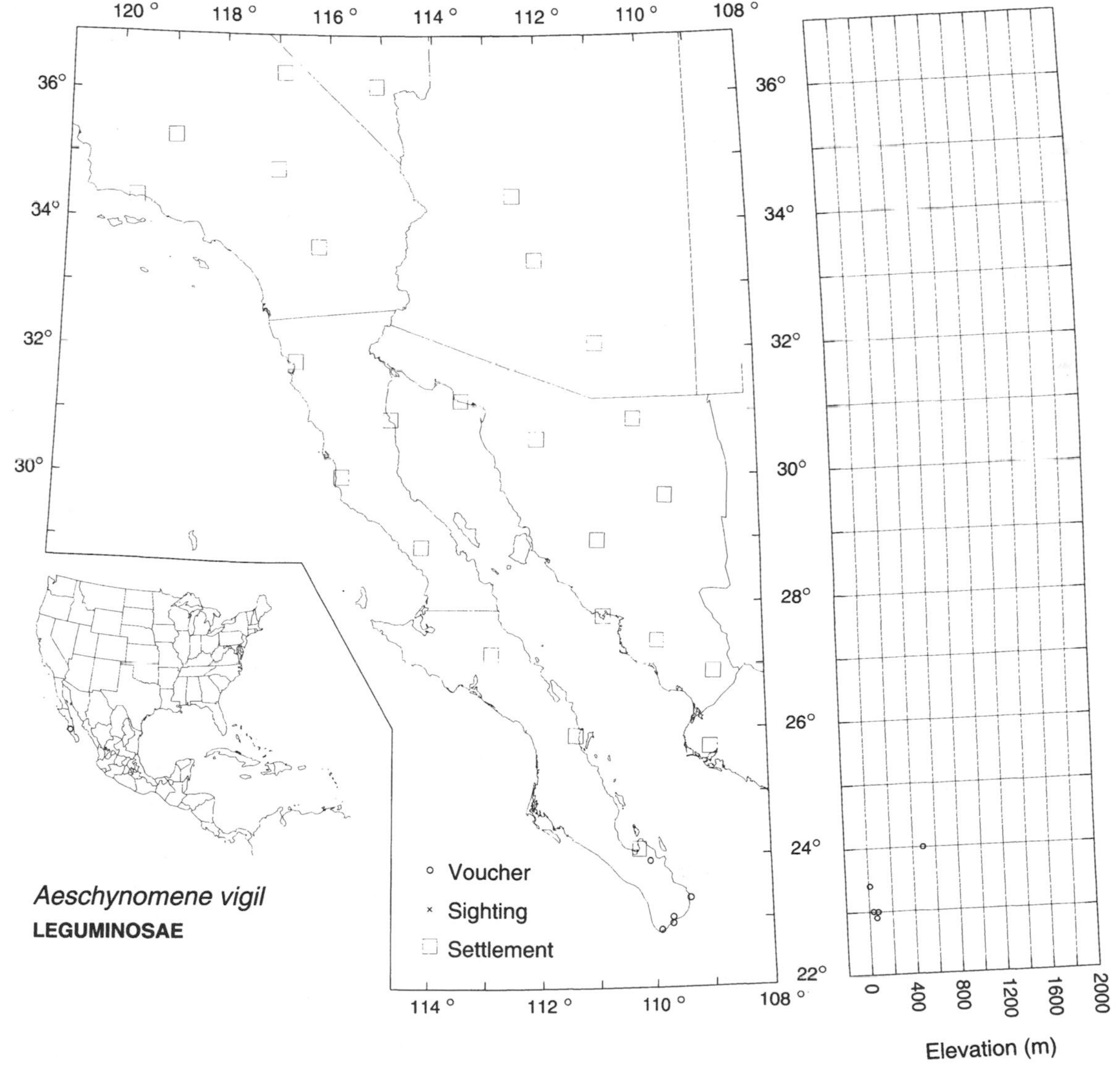

More than half (56%) of the flower visitors to *Ae. californica* are lepidoptera. Butterflies are the main pollinators; the remaining visitors, mostly hymenoptera, are too small to pollinate the flowers (Benseler 1975). As butterflies gather nectar, their wings brush against the anthers and styles, thus transferring pollen (Benseler 1975). Flowers of *Ae. parryi* are probably pollinated in the same fashion. The pollen and nectar of *Ae. californica*, and (one may presume) of *Ae. parryi*, are poisonous to the honeybee (*Apis mellifera*), an exotic species, but not to native bees (Benseler 1975).

Both *Ae. parryi* and *Ae. californica* apparently cannot marshal enough resources to produce a fruit for every flower (Mooney and Bartholomew 1974; Benseler 1975). For *Ae. parryi*, the limiting factor is soil moisture, since two consecutive years of high rainfall are required for sexual reproduction to be successful (Mooney and Bartholomew 1974). The first year of above-average rainfall allows seed set, and the second year fosters seedling establishment. Despite low fertility, *Aesculus* populations persist by virtue of the long life span of individuals and by vigorous root-sprouting.

The short duration of leafiness in *Ae. parryi* would lead one to expect efficient photosynthesis, which is indeed the case, as its photosynthetic rates equal those measured for any tree. In November, when new leaves appear, photosynthesis was relatively low—9.13 mg CO_2/dm/hr. The highest rate, 22.76 mg CO_2/dm/hr, was measured in February (Mooney and Bartholomew 1974). The optimum temperature for photosynthesis was 20°C.

Because flowers and fruits develop in the absence of leaves, much of the growth of *Ae. parryi* occurs at the expense of food reserves (Mooney and Bartholomew 1974). Not surprisingly, a high proportion of carbon is allocated to the roots: during the first growing season, plants in the greenhouse allocated 83% of their biomass (by dry weight) to tap roots, 11.6% to leaves, 4.4% to stems, and 1% to secondary roots (Mooney and Bartholomew 1974). Even juveniles with shoots less than 5 cm tall have relatively large, tuberous tap-

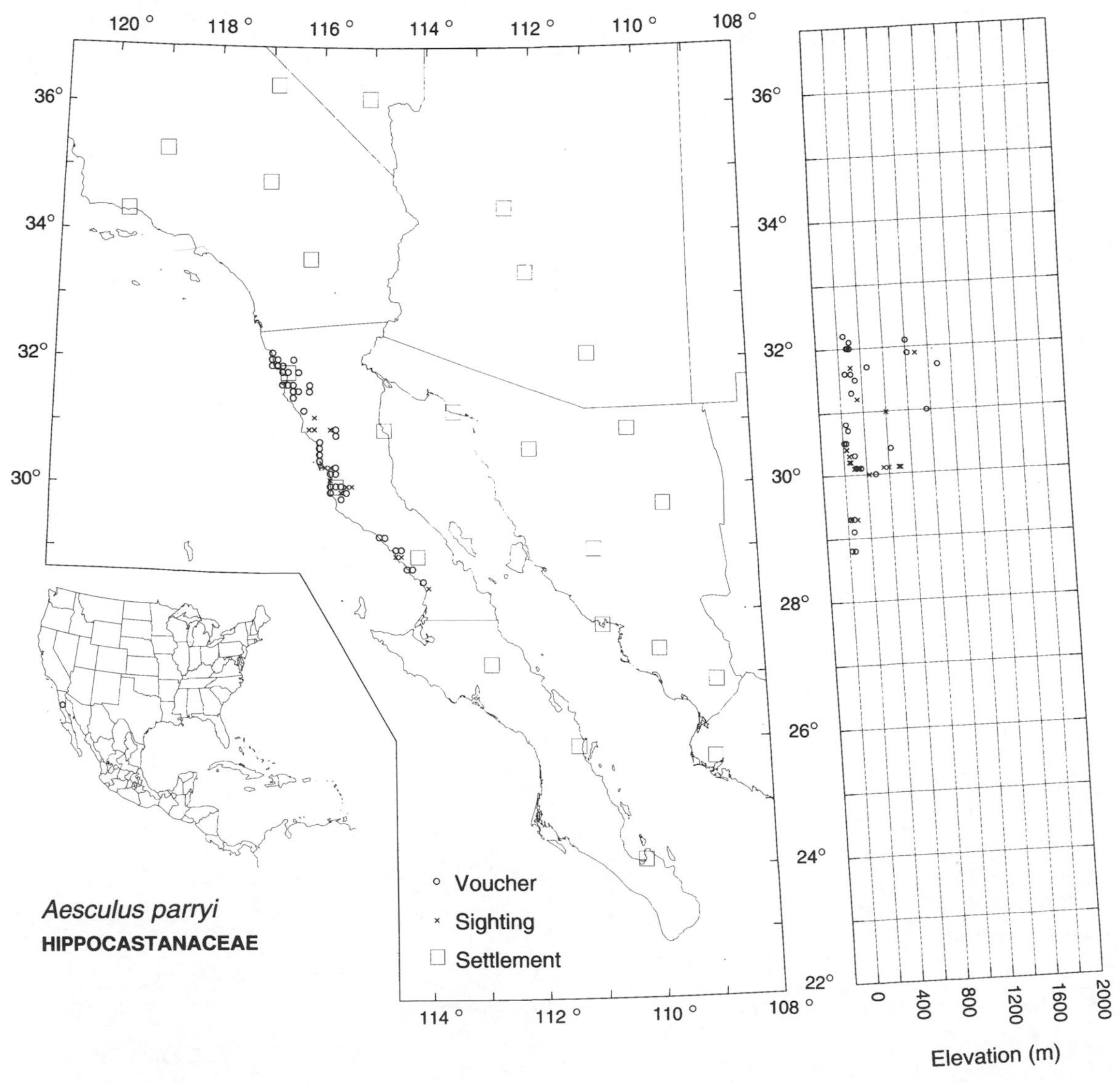

roots. Storage of carbon in roots serves two purposes. In severe drought years, the food reserves can be used to redevelop the canopy once the drought is broken, and in any year and especially after fire, they permit vigorous vegetative reproduction (Mooney and Bartholomew 1974).

Figure 9. Leafless Aesculus parryi *on a north-facing slope along an arroyo east of El Rosario, Baja California. The tree is draped with epiphytic lichens. (Photograph by J. R. Hastings.)*

Agave angustifolia

Haw.

[=*A. owenii* I. M. Johnston, *A. pacifica* Trel., *A. yaquiana* Trel.]

The light green, spreading leaves of *Agave angustifolia* (figure 10) form an open rosette and reach 60–120 cm in length and 3.5–10 cm in width. Linear to lanceolate in shape, they are generally flat to concave above and rigid throughout. The small, reddish brown or dark brown teeth may be closely spaced or remote. Reaching 3–5 m in height, the panicle has 10–20 umbels of green to yellow flowers. Bulbils are occasionally produced in the inflorescence. The diploid number of chromosomes is 60 (Banerjee and Sharma 1988).

Agave angustifolia has proportionally narrower leaves than most other paniculate agaves in the Sonoran Desert. One exception is the closely related *A. aktites* Gentry, which is smaller, more bluish green, and restricted to coastal dunes from Yavaros south to Topolobampo. *Agave angustifolia* also closely resembles *A. datylio,* which is somewhat smaller, more yellowish green and restricted to Baja California Sur. The ranges of *A. angustifolia* and *A. rhodacantha* Trel. overlap in southern Sonora. Gentry (1982) distinguished them by the larger leaves (usually longer than 1.5 m) and stipitate capsules of *A. rhodacantha;* he also noted that hybrids between the two may be difficult to identify. The remaining narrow-leaved agaves in our region have spicate inflorescences and toothless leaf margins.

This highly variable complex has been

Figure 10. Agave angustifolia *near La Colorada, Sonora. (Photograph by J. R. Hastings.)*

treated under nineteen different binomials (Gentry 1982). The forms do not correlate well with geographic distribution, nor are they consistent within a population (Gentry 1982). Six varieties, some known only from cultivation, have been recognized (Gentry 1982).

The most widespread agave in North America, *A. angustifolia* extends from Costa Rica to the Sonoran Desert and grows in Madrean evergreen woodland, Sonoran savanna grassland, Sinaloan thornscrub, and Sonoran desertscrub. On the Yucatán Peninsula, it is a plant of stable coastal dunes (Espejel 1986; Moreno-Casasola and Espejel 1986). *Agave angustifolia* generally grows in open habitats, although it does tolerate light shade. The outlier at 30.4°N, 111.3°W represents a collection by Howard Scott Gentry from the vicinity of Rancho Tinaja. The northwest outposts at 29.6°N and 29.7°N were cited by Felger and Moser (1985). The species has naturalized in West Africa (Leuenberger 1979).

Agave angustifolia reaches its frigid and arid limits at the southeastern margin of the Sonoran Desert. Northern ecotypes can withstand light frost but southern forms cannot (Gentry 1982). Severe leaf tissue damage occurs at –9.4°C, and the plants show limited ability to acclimate to cold (Nobel and Smith 1983). In addition, the winter-flowering habit limits success where freezes are likely and pollinators are torpid. Not surprisingly, therefore, the northern edge of the elevational profile shows a steep slope shaped by incursions of freezing air. Interactions with *A. palmeri* and *A. shrevei* Gentry may also help determine the northern and upper limits. This species acclimates well to heat. The upper limit of its tissue tolerance is about 63°C, approximately the same as for the more arid-adapted *A. deserti* (Nobel and Smith 1983).

Across the range of the species, the protracted flowering period lasts from June to April. In the Sonoran Desert, the plants apparently bloom from November to March. Reproduction is by seed, basal offshoots, and bulbils (Gentry 1982). It is likely that the winter-blooming, tropical *A.*

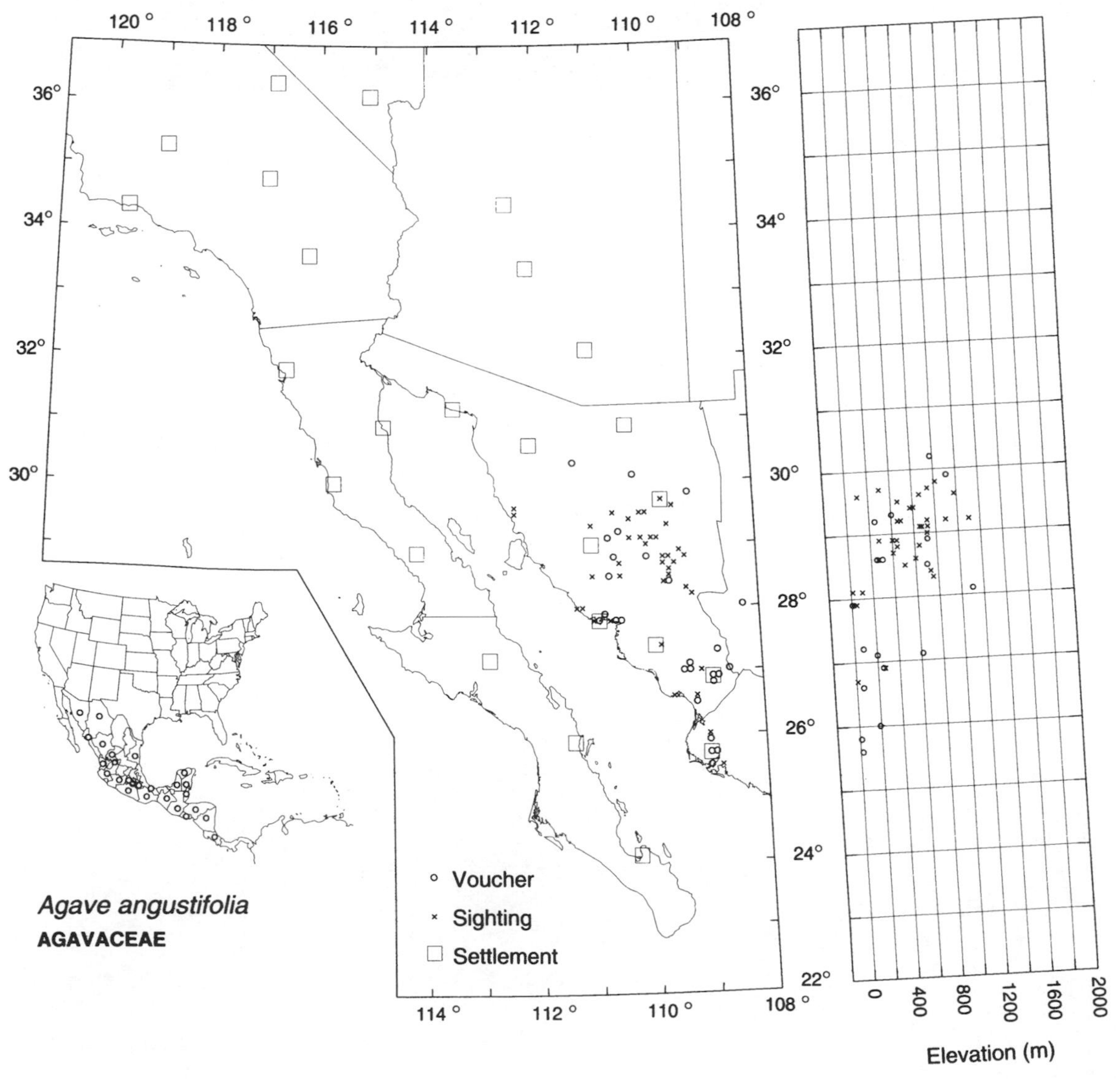

Agave angustifolia
AGAVACEAE

angustifolia shares migratory bat pollinators with the summer- and fall-blooming *A. palmeri* and *A. shrevei.*

Certain varieties are grown for fiber or ornament (Gentry 1982). In Tucson, cultivated plants are prone to freeze damage during most winters. The Seri cooked and ate the stems (Felger and Moser 1985), as did the Tarahumara of Chihuahua (Bye et al. 1975). The Tarahumara also ate the flowers, made a fermented beverage from the hearts and leaf bases, and used the leaf fibers in making rope and thread (Bye et al. 1975).

Agave aurea

Brandegee

Lechuguilla mezcal

Growing 10–12 dm tall, the open rosettes of *Agave aurea* are of widely arching, lanceolate leaves 63–110 cm long and 7–12 cm broad. Light to dark brown teeth 4–7 mm long are spaced at intervals of 1 to 2 cm along the straight to undulate leaf margins. Red to purple in bud, the flowers are yellow to orange when open. They are borne in broad, congested umbels on the upper half of the panicle, which is 2.5 to 5 m tall.

Agave sobria, which is partially sympatric with *A. aurea,* has shorter leaves (25–80 cm) and narrower panicles. *Agave datylio* has narrower leaves (3–4 cm). At its southern limit, *A. aurea* gives way to the related *A. capensis* Gentry, which has smaller leaves (0.3–0.6 m) and a clustering habit, and to *A. promontorii* Trel., which has wider leaves (11–17 cm).

Locally common on lava fields and granitic slopes, *A. aurea* is endemic to Baja California Sur. Gentry (1978) suggested that this and other species in the Campaniflorae evolved in isolation when the southern peninsula was an island. During the Pleistocene, changes in the sea level connected this island to the Cape Region, allowing *A. aurea* to filter southward and exchange genes with the closely related

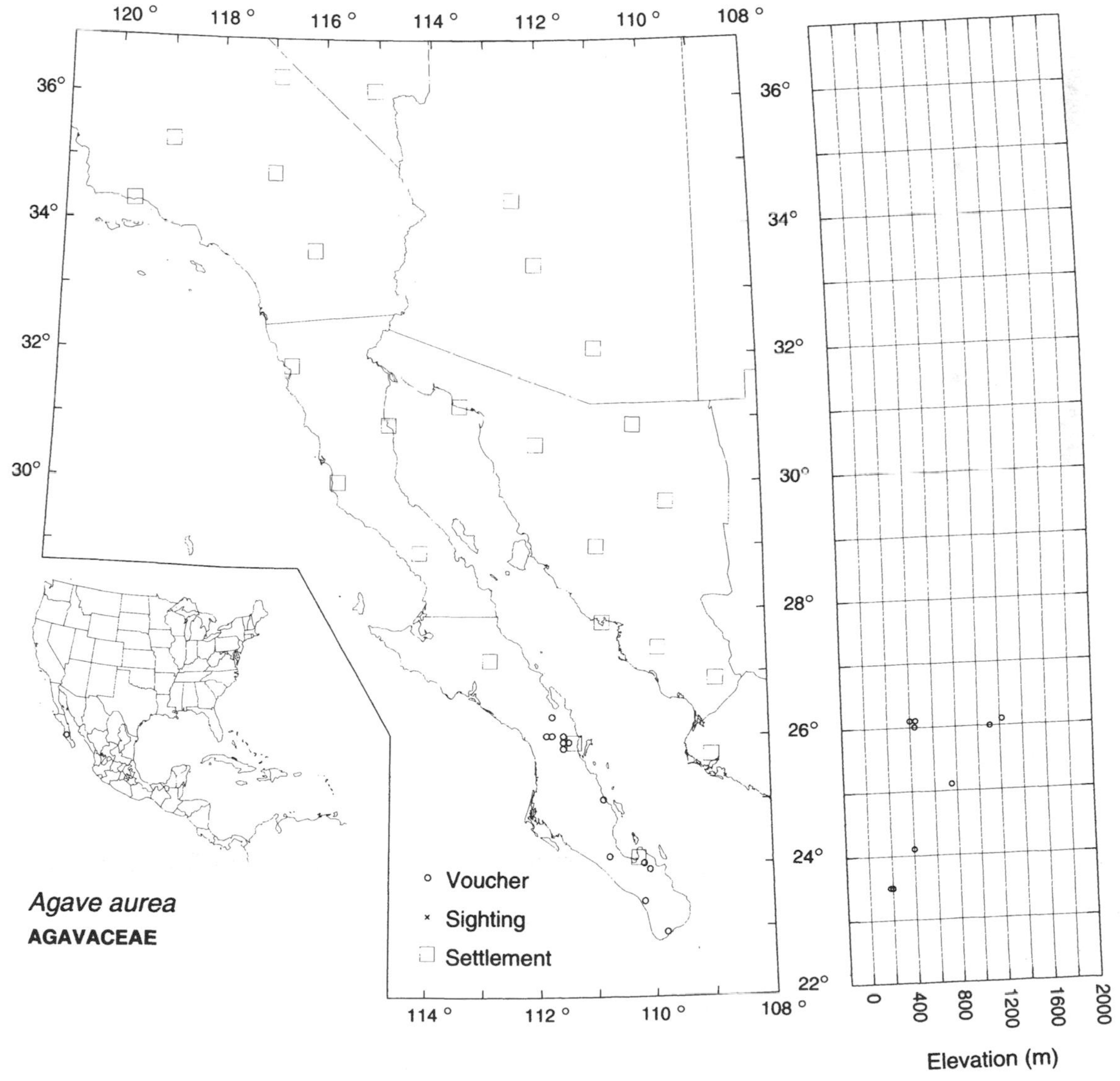

Agave aurea
AGAVACEAE

A. promontorii and *A. capensis* (Gentry 1978).

The northern limits of *A. aurea* are probably shaped by lack of warm-season rain. Climatic effects on the distribution may be mediated through *A. sobria* and *A. gigantensis* Gentry. South of 23.5°N, *A. promontorii* replaces *A. aurea* above 900 m (Gentry 1978). Along the coast, *A. aurea* grades into *A. capensis.*

The flowering season is February to April.

Leaves of wild *A. aurea* have been extensively harvested for fiber, but commercialization seems unlikely in view of the low fiber yield per plant (Gentry 1982).

Agave cerulata
Trel.

This is a small to medium agave with light gray, yellowish, or pale green rosettes 25–70 cm tall (figures 11, 12). The linear to lanceolate leaves bear small (1–5 mm), irregularly shaped, weakly attached teeth and are 25–70 cm long and 4–8 cm wide. The pale yellow flowers are in 5–20 small, lateral umbels on panicles 2–4 m tall.

Agave deserti could be confused with *A. cerulata* subsp. *cerulata* but is more robust, with leaves 4–7 times longer than broad (rather than 5–12 times longer than broad as in subsp. *cerulata*) (Gentry 1982). Also, narrow rings of brown tissue around the bases of the marginal teeth are characteristic of *A. cerulata* but usually not of *A. deserti.* Leaves of *A. shawii* are much wider (8–20 cm), those of *A. moranii* much longer (70–120 cm). *Agave sobria* can generally be recognized by its longer marginal teeth (5–10 mm) and more open rosettes with leaves 8–9 times longer than wide, but some overlap with *A. cerulata* does occur.

Gentry (1982) recognized four taxa in this polymorphic species. Our map combines them. Subspecies *cerulata* has narrow panicles and yellowish, long-acuminate, entire to slightly undulate leaves that are 25–40 cm long, 6–12 times as long as broad, and broadest at or

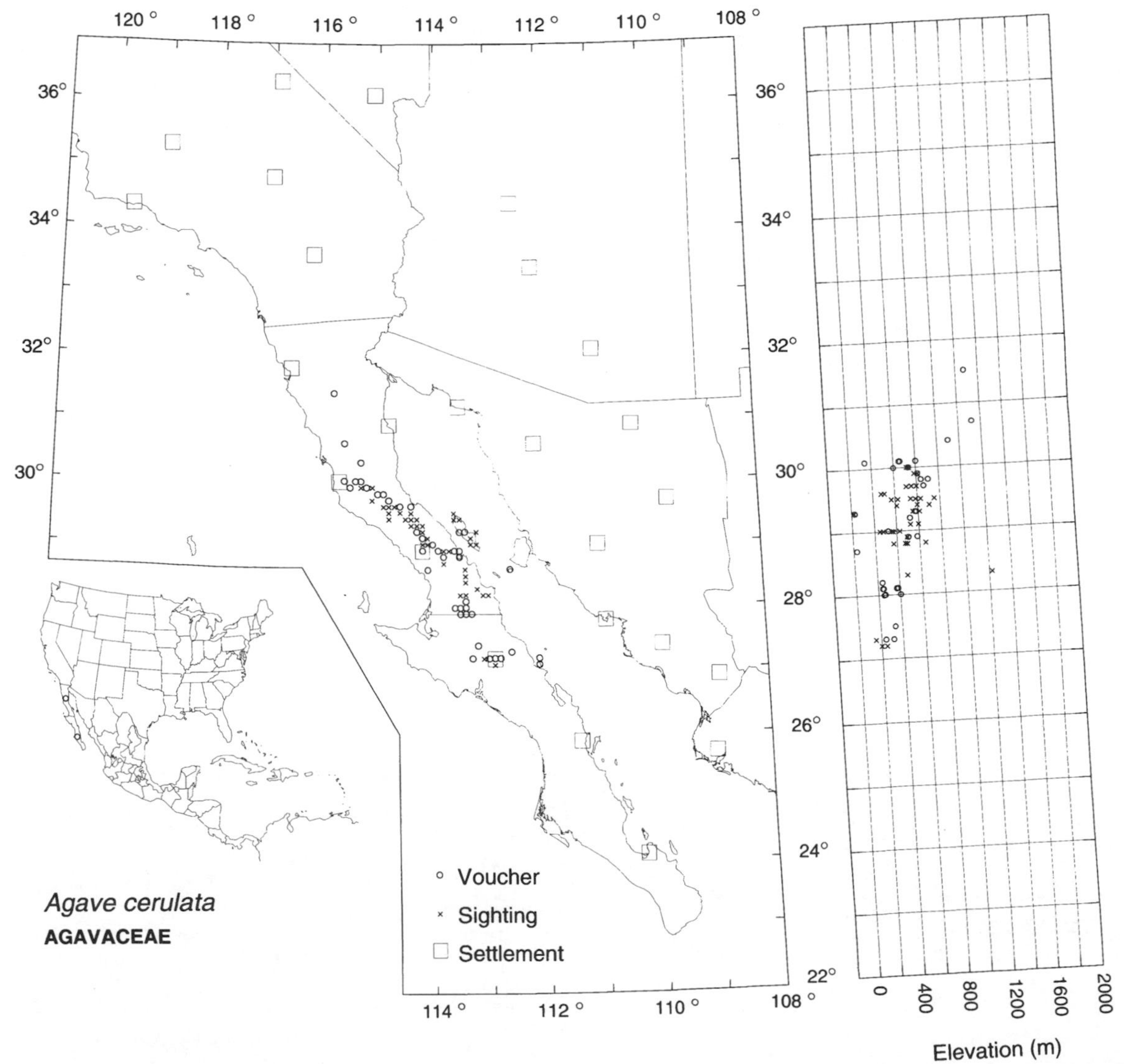

slightly below the middle. It inhabits the central portion of the peninsula. Subspecies *dentiens* (Trel.) Gentry of Isla San Esteban (and perhaps also Isla Angel de la Guarda and the adjacent peninsula), has long-acuminate, grayish leaves and broad panicles. It may be an isolated genotype on its way to complete speciation (Gentry 1982). Subspecies *subcerulata* Gentry has crenate leaves that are 15–30 cm long, 3–6 times as long as broad, and generally broadest above the middle. Subspecies *nelsonii* (Trel.) Gentry has somewhat undulate to nearly entire leaves that are 25–40 cm long, 3–6 times as long as broad, and broadest near or below the middle. Subspecies *subcerulata* occurs at the southeastern end of the overall range, subspecies *nelsonii* toward the northwestern end.

Figure 11. Agave cerulata *subsp.* dentiens *on Isla San Esteban. (Photograph by R. M. Turner.)*

Figure 12. Agave cerulata *subsp.* nelsonii *east of El Rosario, Baja California. (Photograph by R. M. Turner.)*

This is probably the most abundant agave in the Sonoran Desert (Gentry 1982). The plants are locally common on rocky slopes and plains, usually below 900 m. Although they prefer medium- to heavy-textured soils on volcanic rock, they grow on coarser granitic soils as well. *Agave cerulata* reaches 1,225 m on Cerro Lechuguilla and probably grows at high elevations in the nearby Sierra San Borja. The northernmost point (31.5°N) represents a collection by Reid Moran from near El Milagro.

Unlike its close relative *A. deserti*, most commonly found in arid continental habitats, *A. cerulata* grows both in the maritime zone, where fog and ocean breezes ameliorate the otherwise arid climate (Gentry 1982), and farther inland, where conditions are warmer and drier (Burgess 1988). Over most of its range, *A. cerulata* experiences biseasonal rainfall with the heaviest precipitation from November to April and occasional showers from July through September.

The leaf shape of *A. cerulata* changes geographically with climate (Burgess 1988). In general, the leaves are smallest where the climate is driest. Warmer summers, higher precipitation deficits, and more frequent warm-season storms are associated with narrower leaves. Higher surface-to-volume ratios are associated with warmer winters and cooler summers, or with extreme cold. Leaves with relatively thick bases may allow food storage at the expense of photosynthetic capacity in regions of higher maximum temperatures, lower summer rainfall, and large precipitation deficits (Burgess 1988).

Relationships with congeners indicate a history of continuing geographic boundary adjustments and occasional genetic exchanges. Examples include *A. cerulata* and *A. deserti* in the vicinity of San Borja; *A. deserti*, *A. moranii*, and *A. cerulata* subsp. *nelsonii* in the Valle Chico (30.6°N, 115.0°W); *A. shawii* and *A. cerulata* subsp. *nelsonii* east of El Rosario on the Pacific slope of the Sierra San Miguel; and *A. cerulata*, *A. shawii*, and *A. avellanidens* Trel. on the Pacific slope southwest of the Sierra Calmallí (north of 28°N, west of 113.5°W). In most of these populations, geography and morphology suggest that the anomalous individuals have been derived from introgression. In certain areas, *A. shawii* and *A. cerulata* meet without apparent hybridization, as north of Punta Prieta and near San Borja. At both locations, *A. shawii* is common on valley alluvium and *A. cerulata* dominates rocky, volcanic slopes. Apparently, substrate significantly influences the local distribution of these two taxa where climates are marginal.

Flowering is in March and April.

The Cochimí Indians of the central peninsula ate the stems, which contain a lower level of sapogenins than the leaves (Gentry 1982). The Seri made a beverage from the juice of macerated leaves (Gentry 1982; Felger and Moser 1985).

Agave chrysantha

Peebles

[=*A. palmeri* Engelm. var. *chrysantha* (Peebles) Little]

Rosettes of *Agave chrysantha* are 5–10 dm tall and 8–18 dm wide. The grayish to yellowish green leaves are 40–75 cm long, deeply guttered, and linear-lanceolate to lanceolate in outline. They are 5–8 times longer than wide. Leaf margins are straight to undulate and toothed. The larger teeth are 5–10 mm long and usually spaced 1–3 cm apart. The terminal spine is 25–45 mm long. The inflorescence is a panicle 4–7 m tall with 8–10 congested umbels of bright yellow flowers. The woody capsules are 35–50 mm long. The diploid number of chromosomes is 30 pairs (Pinkava and Baker 1985).

Where this species grows near *A. palmeri*, it can be difficult or impossible to identify vegetative specimens. In general, *A. palmeri* leaf margins are undulate, and the teeth are spaced at intervals of 0.5 cm. When the plants are in flower, the clear golden-yellow blossoms of *A. chrysantha* are readily distinguished from the reddish-tipped ones of *A. palmeri*. Other similar species include *A. parryi* Engelm., which has shorter leaves (20–40 cm) that are 1.5–6 times longer than wide, and *A. murpheyi* F. Gibson, which has narrower leaves (6–8 cm) with nearly parallel margins and shorter terminal spines (12–20 mm). A recently discovered species from east-central Arizona (Hodgson et al. 1994) can be separated from *A. chrysantha* by its terminally inflexed, more uniformly glaucous leaves and its larger panicle with fewer branches oriented more perpendicularly to the peduncle axis.

Agave chrysantha apparently hybridizes with *A. palmeri*, a close relative, in the Rincon Mountains and with the more distantly related *A. parryi* in the Sierra Ancha and Mazatzal Mountains (Gentry 1982). With these few exceptions, the species is well-marked taxonomically and geographically.

Locally common on rocky granitic or volcanic slopes, *A. chrysantha* is narrowly endemic to the mountains of central Arizona. The southernmost points are in the Santa Catalina and Rincon mountains. Based on its limited range and close relationship to *A. palmeri*, Gentry (1982) suggested that *A. chrysantha* may have arisen through introgression with *A. palmeri* and *A. parryi*, which sometimes occur together in scattered populations on isolated mountain ranges.

Most populations of *A. chrysantha* receive biseasonal rainfall with roughly equal amounts in late summer and winter. Winter frosts are common within the

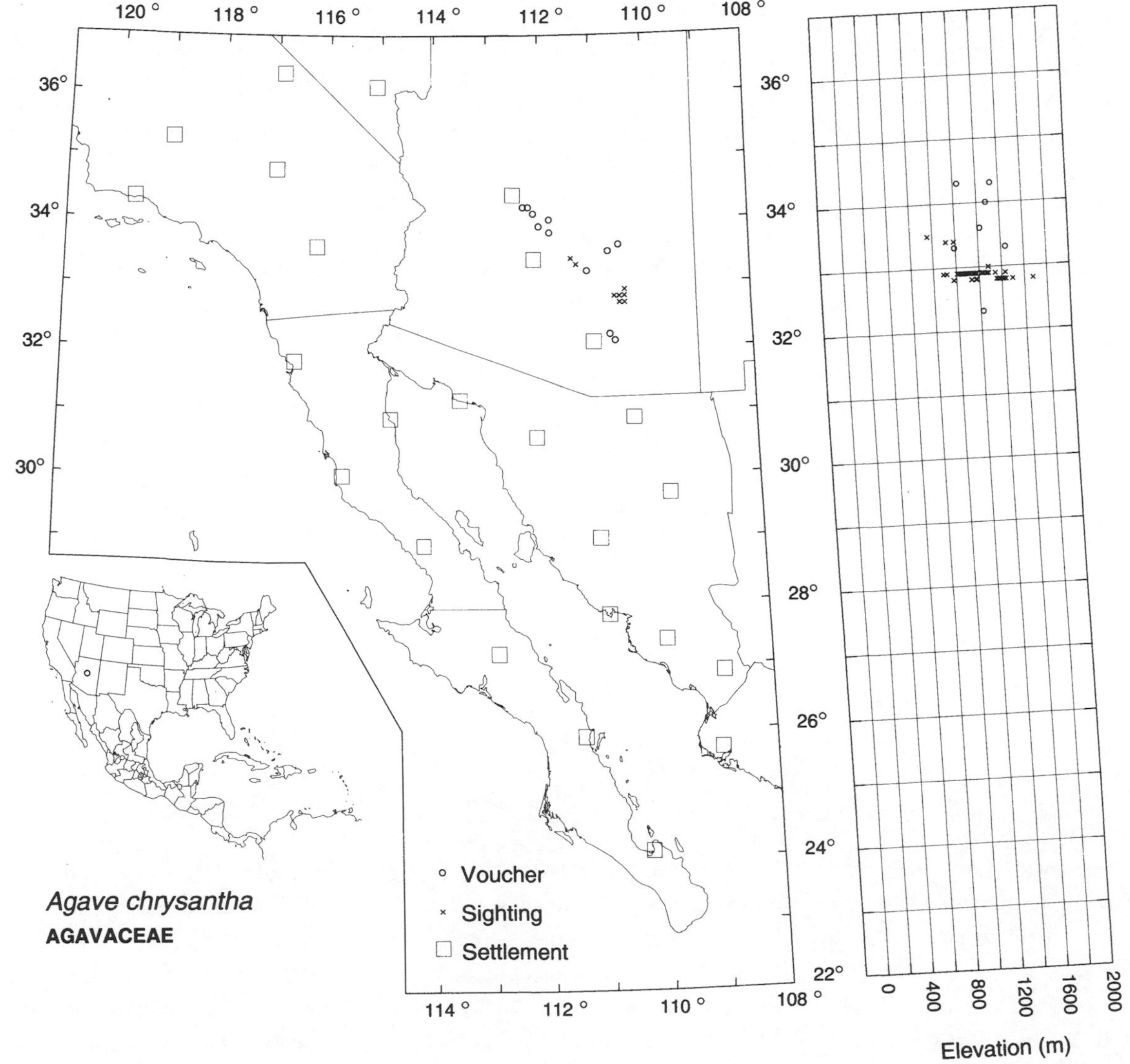

Agave chrysantha
AGAVACEAE

range of the species. The truncated northern limit seen on the elevational profile may be shaped by extreme low temperatures. Lower elevational limits are influenced by aridity, southern limits by interactions with *A. palmeri.*

The flowers appear in June and July. Their nectar is rich in fructose and glucose, poor in sucrose (Freeman et al. 1983). Most related species in group Ditepalae are pollinated by bats (Gentry 1982), but no nectar-feeding bats live within the range of *A. chrysantha* (Hoffmeister 1986). Reproduction in the wild is mainly by seed and offsets, rarely by bulbils (Gentry 1982). Plants may not reach reproductive age for forty years (Gentry 1982).

Agave chrysoglossa

I. M. Johnston

Agave chrysoglossa (figure 13) produces single, open rosettes of light green, linear-lanceolate leaves. Flat above and convex below, these leaves are 70–120 cm long and 4–7 cm wide. The margins are toothless. Borne in densely flowered spikes 2 to 4 m tall, the yellow flowers are made conspicuous by their long-exserted anthers.

The closely related *A. vilmoriniana* resembles this species but has a bulbil-bearing inflorescence and deeply guttered, arching leaves 7–10 cm wide. Other sympatric congeners include *A. felgeri,* which has smaller leaves less than 50 cm long, and *A. colorata,* which has toothed leaves and paniculate inflorescences.

Locally common on rocky slopes and bedrock outcrops, *A. chrysoglossa* is the coastal, lowland representative of group Amolae (Gentry 1982). The known populations are widely scattered, occurring in the vicinity of Guaymas and inland along the Río Yaqui near Bacanora and Sahuaripa. The Sahuaripa population apparently consists of morphological intermediates between *A. chrysoglossa* and *A. vilmoriniana* (Gentry 1982). Sierra Seri populations described by Felger and Moser (1985) appear to bridge a morphological gap between *A. chrysoglossa* and *A. pelona.* The Felger specimens cited by Gentry (1982) seem closer to *A. pelona,* which is

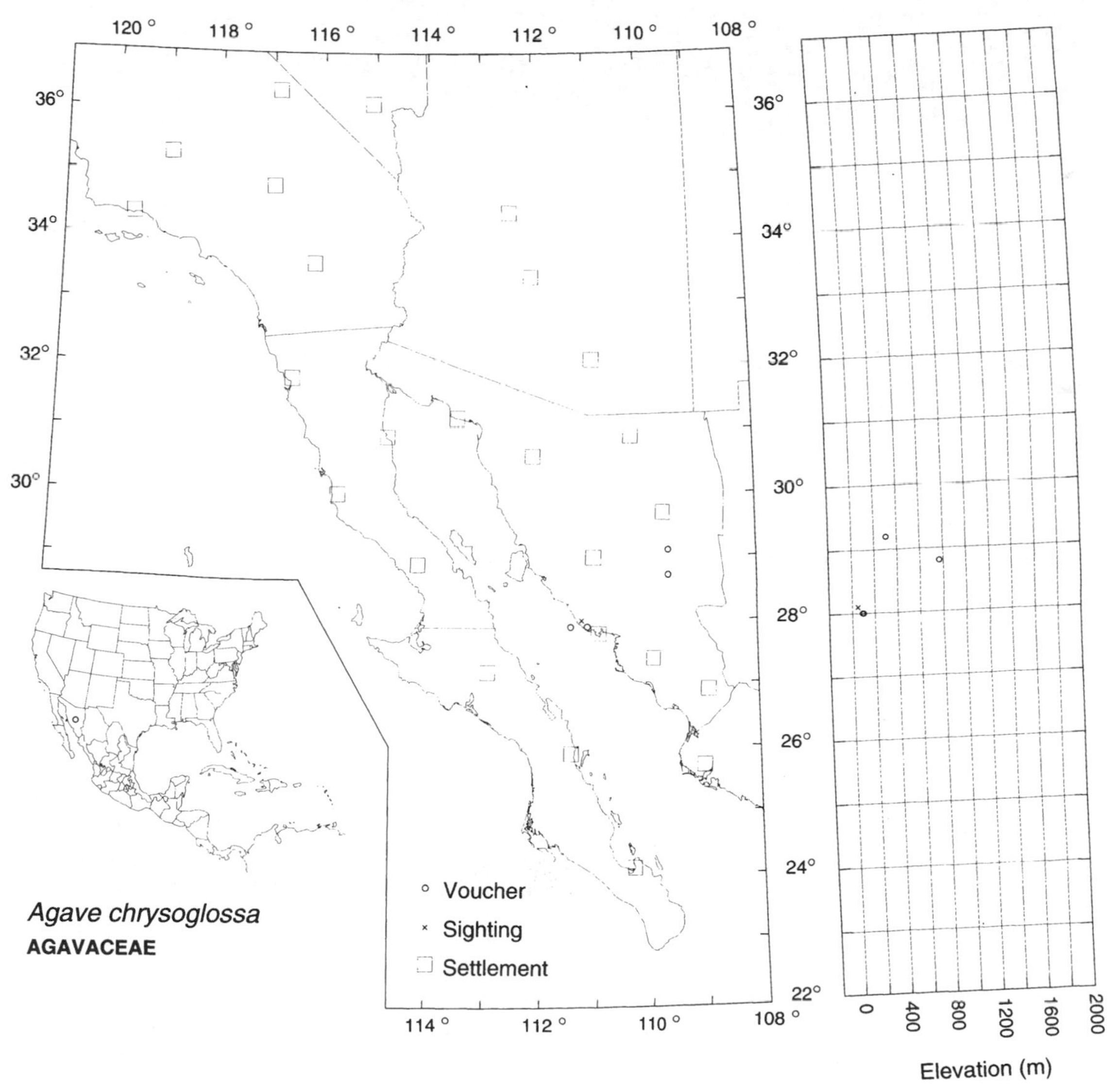

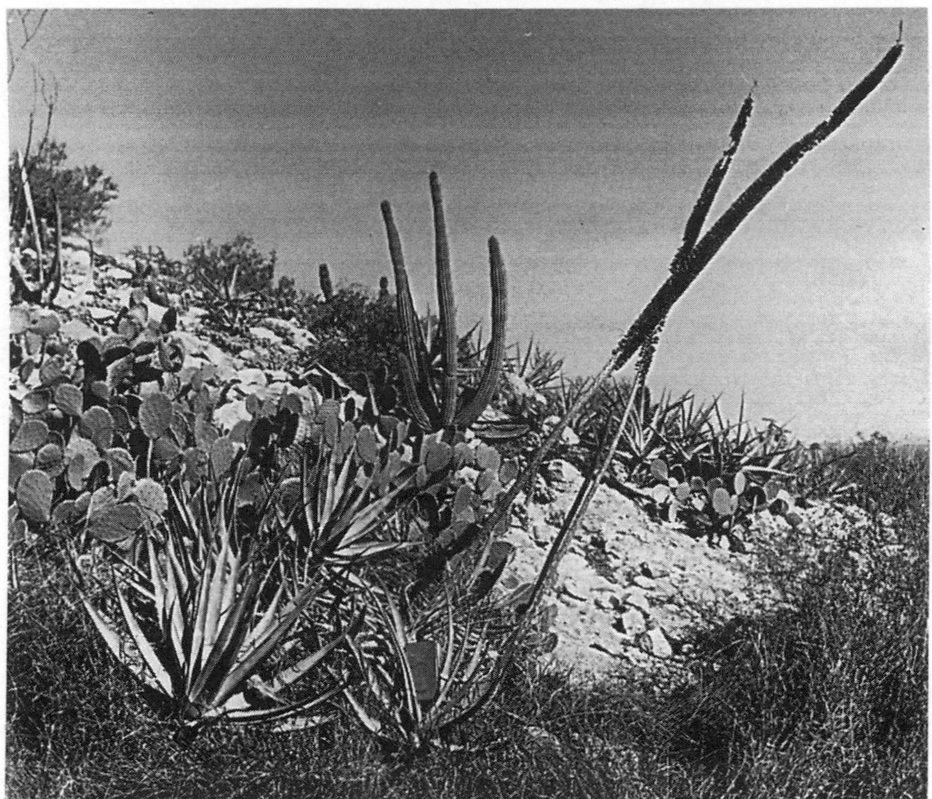

Figure 13. Agave chrysoglossa *on Isla San Pedro Nolasco. (Photograph by R. M. Turner.)*

where we have plotted them.

Agave chrysoglossa tolerates higher temperatures and is more xerophytic than other members of group Amolae. Coastal populations receive most of their annual rainfall between July and October, with occasional good rains in early winter.

Blooming from March to May, the flowers are visited by insects and birds, especially hummingbirds, with great pollination success. The plants are highly fecund; a single inflorescence may produce 500,000–750,000 seeds (Gentry 1982). Vegetative reproduction by offsets also occurs in a few populations.

The plants are sapogenous and are used locally for washing clothes (Gentry 1982). The Seri ate the baked stems and leaf bases despite the bitter taste (Felger and Moser 1985). This species is frost tolerant in Tucson and is occasionally cultivated there and elsewhere.

Agave colorata
Gentry

Mescal ceniza

The obovate to lanceolate leaves of *Agave colorata* form a compact, small to medium rosette. The leaves are 25–60 cm long and 12–18 cm wide, glaucous, and light gray with pinkish cross-bands. Each of the prominent crenations on the leaf margins has a brown or gray tooth 5 to 10 mm long. Panicles are 2–3 m tall with 15–20 densely flowered umbels. Reddish in bud, the flowers (50 to 70 mm long) are yellow upon opening.

Agave colorata is the Sinaloan thornscrub representative of group Ditepalae, which contains several species that enter the Sonoran Desert from the east. Of these, *A. shrevei* Gentry is most similar to *A. colorata* and may be only a variety of it. *Agave shrevei* subsp. *matapensis* Gentry enters the Sonoran desertscrub east of Hermosillo and can be distinguished from *A. colorata* by its generally narrower (8–14 cm) and unbanded gray or green leaves. Its panicle is more than 2.5 m tall and branches in the upper third to half. Very similar to *A. colorata* is *A. fortiflora* Gentry, a poorly understood species collected in the Sierra Jojoba (30.1°N, 112.0°W) and in the Sierrita de Lopez (29.4°N, 111.3°W). Its undulate-margined leaves are generally narrower (8–12 cm) and longer (50–100 cm) and its panicle taller (4–6 m). *Agave palmeri,* also in group Ditepalae, has narrower leaves (7–10 cm) with less undulate margins. In certain populations, as in the mountains northwest of Guaymas and at Bahía San Pedro, the leaves of mature *A. colorata* are long and lanceolate, and flowers may be essential for proper identification. The remaining agave species within the range of *A. colorata* all have much narrower leaves.

An uncommon plant of open, rocky sites in Sinaloan thornscrub, *A. colorata* extends as far north as 28.6°N, where the species was seen by Thomas R. Van Devender in Cañon la Pintada, Sierra Libre. The nearby voucher (28.5°N) is Laurence J. Toolin's from 61 km south of Hermosillo.

This species receives most of its yearly rainfall from midsummer into early autumn. Because cultivated plants in Arizona have proven surprisingly frost tolerant, it seems likely that summer aridity rather than winter cold shapes the northwestern limit of this species. Its mesic limits are probably set by genetic interactions with other members of group Ditepalae.

Flowers appear from March through June and are undoubtedly pollinated by bats and perhaps also hummingbirds (Gentry 1982). Plants flower at about 15 years of age (Gentry 1982). Reproduction is primarily by seed, occasionally by offsets (Gentry 1982).

The stems are occasionally baked and eaten by the Seri and other tribes. The Seri also used the plants in making wine (Felger and Moser 1985). Yaqui and Mayo Indians, in harvesting the plants for sugar, may have made serious inroads upon local populations (Gentry 1982). The plants make attractive ornamentals for desert gardens and are resistant to frost and agave weevils (*Scyphophorus* species).

Agave datylio

Simon ex Weber

Datilillo, mescalillo

The spreading green or yellow-green leaves of *Agave datylio* (figure 14) form small to medium rosettes 6–10 dm tall. Colonies may be formed through rhizomatous cloning. Channeled above, the linear-lanceolate leaves are 50–80 cm long and 3–4 cm wide. The small, dark brown teeth are widely spaced. The greenish yellow flowers are in 8–15 lateral umbels on a panicle 3–4 m tall.

Sympatric species include *A. aurea* and *A. promontorii,* both with arching leaves that are much longer and wider (7–17 cm wide). *Agave capensis* Gentry produces offsets by axillary budding rather than rhizomatous cloning, and its leaves tend to be wider (4–7 cm) and undulate. The glaucous gray, cross-banded leaves of *A. sobria* and *A. margaritae* Brandegee are quite distinct from the typical yellowish green leaves of *A. datylio* and are, moreover, somewhat wider (5–10 cm) and more lanceolate.

Gentry (1982) recognized two varieties: var. ***datylio*** of the Cape Region and var. ***vexans*** (Trel.) I. M. Johnston of the Sierra de la Giganta and the Magdalena Plain. Variety ***vexans*** differs from the species in its small rosettes and leaves (30–50 cm long) and is evidently a xerophytic ecotype (Gentry 1982). Our map combines the two.

Endemic to Baja California Sur, *A. datylio* is locally common on sandy plains and rocky granitic slopes. It is the only peninsular representative of group Rigidae (Gentry 1982) and might have evolved in situ following gulf-floor spreading.

Cultivated *A. datylio* plants suffer frost damage at –5°C (Gentry 1982); however,

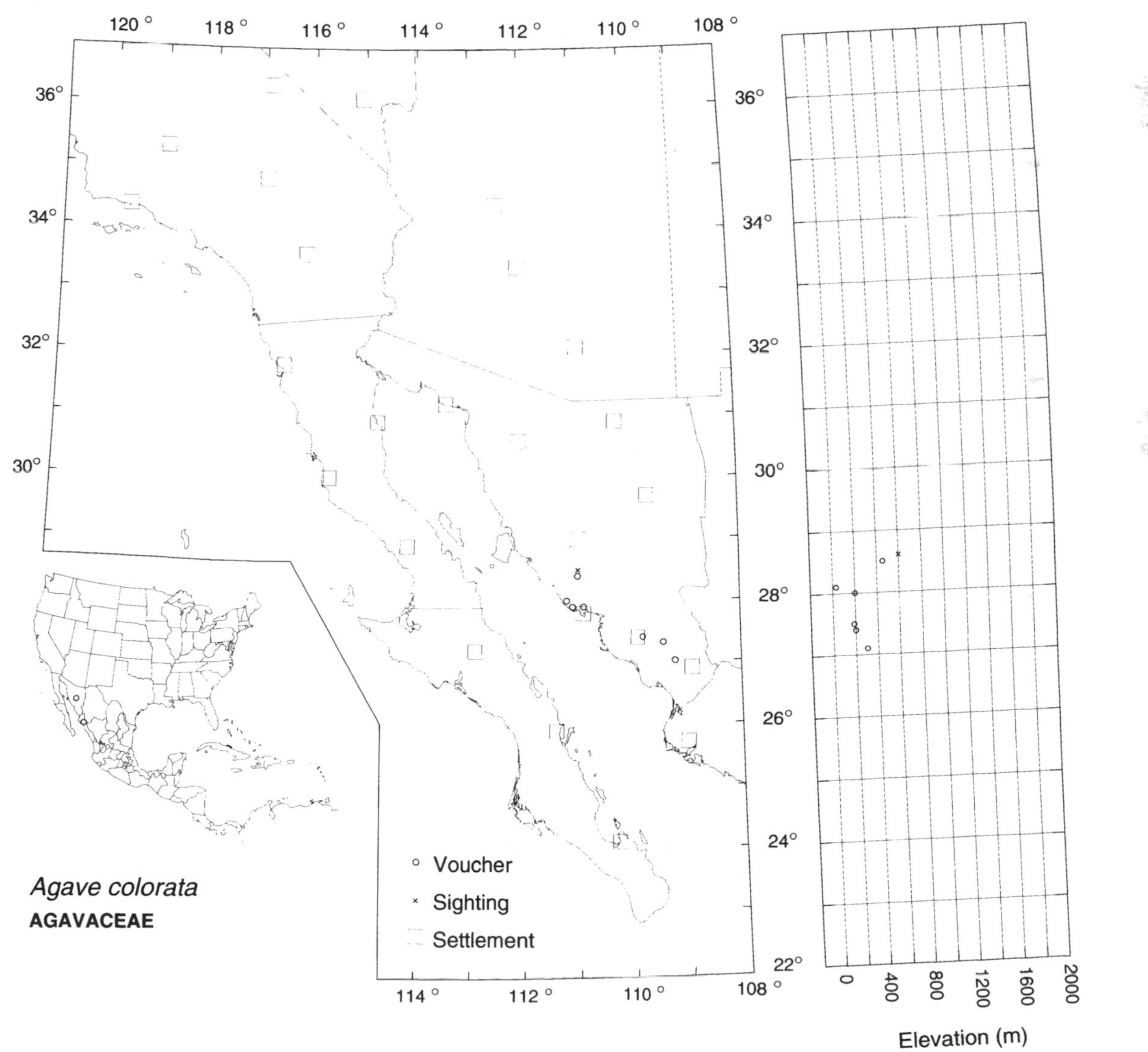

Agave colorata
AGAVACEAE

it seems unlikely that low temperatures limit the distribution of this species in the wild. Instead, the northern limit is probably shaped by lack of warm-season rain and restricted occurrence of sandy alluvial plains, the preferred habitat.

The plants flower from September to December.

Figure 14. Agave datylio *subsp.* vexans *at its northernmost locality near San Juanico, Baja California Sur. (Photograph by J. R. Hastings.)*

Agave deserti

Engelm.

Desert agave, amul

This species forms medium-sized, light gray rosettes 30–70 cm tall (figures 15, 16). They often produce copious offsets. The glaucous, narrowly lanceolate leaves are 25–70 cm long and 5–8 cm wide. Concave above and convex below, they are regularly armed with teeth 2 to 8 mm long. The panicle (2.5 to 6 m tall) has 6–15 lateral branches of yellow flowers 40 to 60 mm long. Cave (1964) reported a haploid

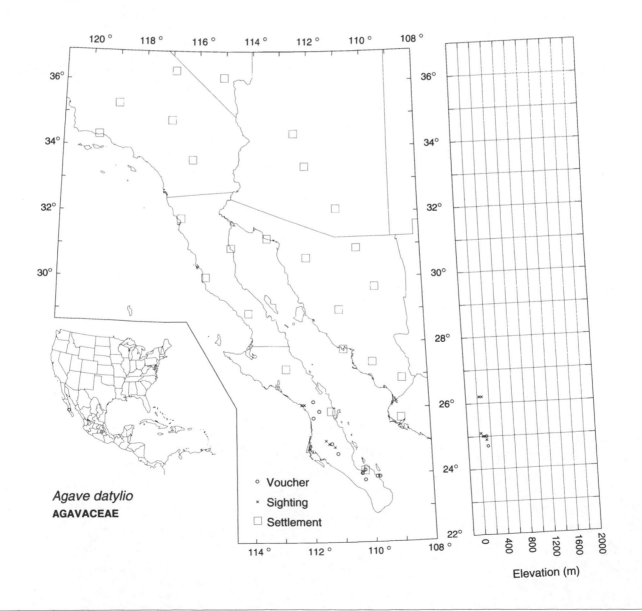

120° 118° 116° 114° 112° 110° 108°

36° 34° 32° 30° 28° 26° 24° 22°

114° 112° 110° 108°

Agave datylio
AGAVACEAE

○ Voucher
× Sighting
□ Settlement

0 400 800 1200 1600 2000

Elevation (m)

chromosome count of 59.

Agave mckelveyana, a smaller plant of somewhat higher elevations, has narrower leaves (3–5 cm). Where the ranges of *A. deserti* and *A. cerulata* meet, individuals with characters of both may be found. In general, *A. cerulata* leaves tend to be yellower, smaller, and narrower (Gentry 1982); also, their weakly attached marginal teeth are ringed with brown at the base. *Agave subsimplex,* a species of the Sonoran coast, has smaller leaves 3–6 cm wide (Gentry 1982). *Agave utahensis* Engelm. has narrower leaves (1.5–5 cm), and its shorter lateral branches produce a more clearly spicate inflorescence. The closely related *A. moranii* has large, single rosettes of long (70–120 cm) rigid leaves; also, *A. deserti* generally produces offsets, whereas *A. moranii* does not. *Agave deserti* and *A. moranii* appear to hybridize. The two species can be impossible to tell apart without fertile material (Gentry 1982).

Gentry (1982) recognized three taxa in this large and variable complex. Mapped together here are subsp. *deserti* of southern California and northwestern Baja California and the widespread subsp. *simplex* Gentry. Subspecies *deserti* suckers copiously to form large clones. Subspecies *simplex* Gentry generally produces single rosettes (rarely as many as three per clone). Mapped separately is subsp. *pringlei* (Engelm. ex Baker) Gentry, which has narrower, less glaucous leaves than its conspecifics. Also, its terminal spine is decurrent into the corneous margins along its distal parts. As presently understood, subsp. *pringlei* represents an extension of *A. deserti* into more mesic communities in the eastern Sierra Juárez and Sierra San Pedro Mártir (Gentry 1982). At San Matías Pass (31.3°N, 115.5°W), green, narrow-leaved rosettes typical of subsp. *pringlei* occur with rosettes of broader, glaucous leaves conforming to subsp. *deserti.* Although Gentry (1982) described this population as a hybrid swarm, the characters show a degree of coordinated segregation that suggests genetic exchange has been limited.

Figure 15. Agave deserti *subsp.* pringlei *in San Matías Pass, Baja California. (Photograph by J. R. Hastings.)*

Agave deserti is locally abundant on sandy flats and occasional on rocky slopes. The easternmost points (32.3°N, 111.5°W and 32.4°N, 111.5°W) represent vouchers from the Waterman and Silverbell mountains, respectively. The southernmost disjunct (30.1°N, 112.7°W) is based on a sighting by Tony L. Burgess north of Puerto Libertad, where *A. subsimplex* and what appears to be *A. deserti* coexist in a mixed population.

Agave deserti covers a broader geographic area and a wider elevational range (90–1,620 m) than most Sonoran Desert agaves yet skirts the hottest, most arid portions of the study area. This distributional pattern may result from physiological limitations: for example, internal food and water reserves are depleted faster at higher temperatures (Burgess 1985; Nobel and Hartsock 1986a). Upper elevational limits apparently reflect a combination of factors, among them intolerance of cold (Nobel and Hartsock 1986a), shade, and fire. Because acclimated seedlings can tolerate tissue temperatures as low as –8°C (Nobel and Smith 1983; Nobel 1984b) and mature plants will withstand temperatures as low as –15°C (Nobel and McDaniel 1988), it seems likely that extreme cold events shape the rather truncated northern limit seen in the elevational profile. The northern edge of *A. deserti* is roughly contiguous with the southern limit of *A. utahensis.* Where the two species meet in the Ivanpah Mountains, California, *A. deserti* grows on granite, *A. utahensis* on limestone (Robert Woodhouse, personal communication 1980). The southern limits are not precisely known. They seem to be heavily influenced by interactions with congeners.

The flowering season lasts from May to July (Gentry 1982). Birds, especially hummingbirds, and insects are probably the main pollinators.

Agave deserti has a number of physiological and morphological adaptations to its hot, arid environment (for reviews, see

Burgess 1985; Nobel 1985, 1988). The shallow root system (mean depth of 8 cm) responds quickly to rain—within twelve hours after a threshold event of 7 mm (Nobel 1976a). Most root growth occurs within a month of rain; during this time, total root length of a medium-sized plant may increase by 58% (Franco and Nobel 1990). The roots essentially function like rectifiers, readily removing water from wet soils but losing little to dry soils (Nobel and Sanderson 1984; Palta and Nobel 1989a).

As a CAM plant, *A. deserti* takes up carbon dioxide at night, permitting daytime stomatal closure and so fostering high water-use efficiency. Nobel (1976a) reported the extremely low transpiration ratio (mass of water transpired/mass of carbon dioxide fixed) of 18 for a representative winter day. When the soil is wet, 55% of nighttime transpiration is from water stored in the leaves. Recharge takes place during the day (Schulte and Nobel 1989). Because the amount of water stored per unit of leaf area is large, stomates can continue to open at night for some time after the soil dries out, as water from the leaves takes the place of water from the soil (Nobel 1976a; Nobel and Jordan 1983). Recharge of stored leaf water occurs during the week following rain (Schulte and Nobel 1989). When soil water is available for long periods, the plants can switch to the C_3 photosynthetic pathway, fixing carbon dioxide during the day. Although net photosynthesis is about the same either way (18.3 $\mu mol/cm^2/24$ hours for CAM versus 21.0 $\mu mol/cm^2/24$ hours for C_3), C_3 photosynthesis allows greater productivity when days are long and nights are short (Hartsock and Nobel 1976; Nobel 1989a).

For *A. deserti,* carbon dioxide uptake can be light limited, since the mean irradiance on the leaf surface is below the 90% saturating value even on clear summer days (Woodhouse et al. 1980). The radially symmetrical rosettes maximize the amount of photosynthetically active radiation that falls upon the leaves (Woodhouse et al. 1980).

The optimal temperature for carbon dioxide uptake is 15.2°C for plants acclimated to a day/night temperature regime

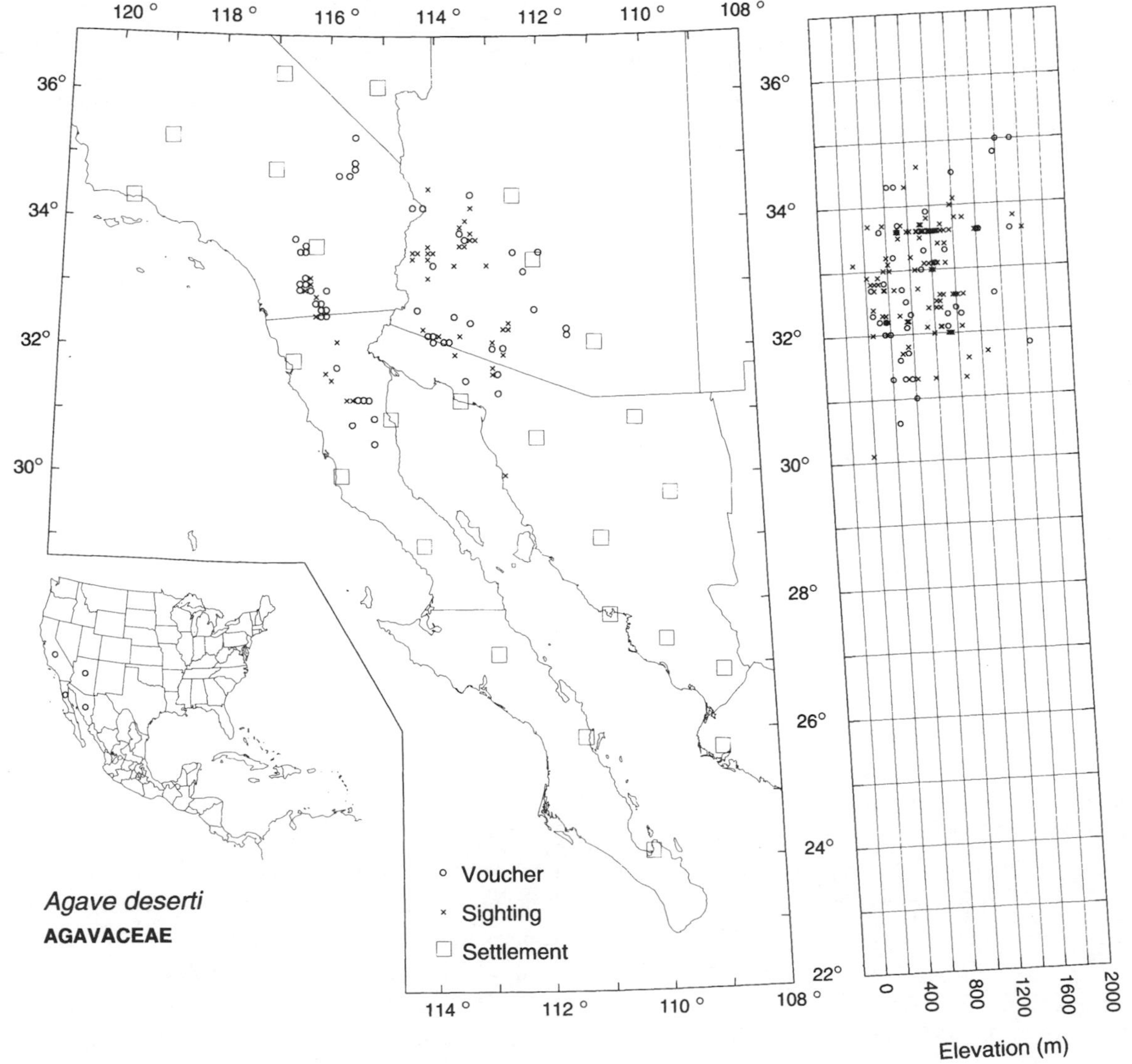

of 10/10°C and 17.8°C for those acclimated to a 30/30°C day/night regime (Nobel and Hartsock 1981). Net photosynthesis is higher at the cooler temperature regimes, as expected for a CAM plant (Nobel 1976a; Nobel and Hartsock 1981). Nocturnal acid accumulation is negligible in plants grown at a day/night regime of 10/0°C (Nobel and McDaniel 1988). At high temperatures (60°C), photosynthetic electron transport proceeds without disruption in acclimated plants, which no doubt facilitates survival during desert summers (Chetti and Nobel 1987). Despite relatively slow growth, *A. deserti* is surprisingly productive for a CAM plant; from early June to late October, the whole plant productivity (based on dry weight) in one study was 0.57 kg/m² of ground area (Nobel 1984a). Productivity in the wild is nitrogen limited (Nobel 1989b).

A rosette must attain a minimum size of about 1,000 g dry weight before it can flower (Tissue and Nobel 1990). Flowering depends on the number of wet days during the preceding two years (Nobel 1987). Within a given population, poor flowering years tend to alternate with good flowering years (Nobel 1987). The developing inflorescence grows an average of 7 cm/day, reaching full height in about 104 days (Nobel 1977a). Substantial translocation of carbon from leaves to inflorescence occurs at this time. Approximately 70% of the total nonstructural carbohydrate in the inflorescence comes from the leaves, the rest from the inflorescence (Tissue and Nobel 1990). Flowers and fruits lose significant amounts of water, most of which is apparently taken from the leaves (Nobel 1977a). The final dry weight of the inflorescence is about the same as the mature plant's total annual photosynthetic productivity, and this is why, even when water is plentiful, the plants cannot photosynthesize fast enough to support the developing inflorescence. Because the plant needs all its resources to make the inflorescence, the rosette must die shortly after flowering (Nobel 1977a).

Although an individual inflorescence produces 65,000 seeds, seedlings are extremely rare in the wild, and most repro-

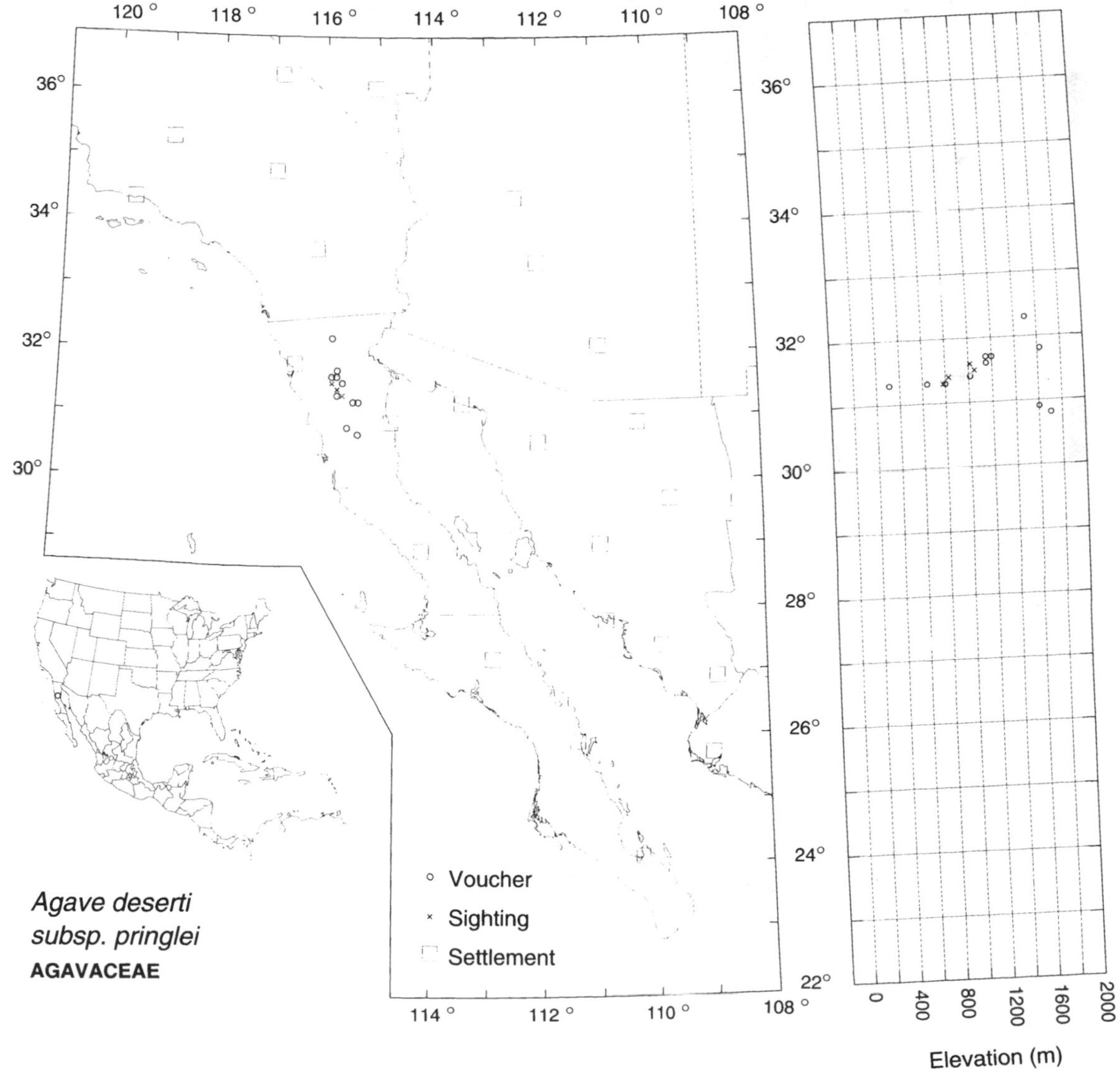

Agave deserti subsp. pringlei
AGAVACEAE

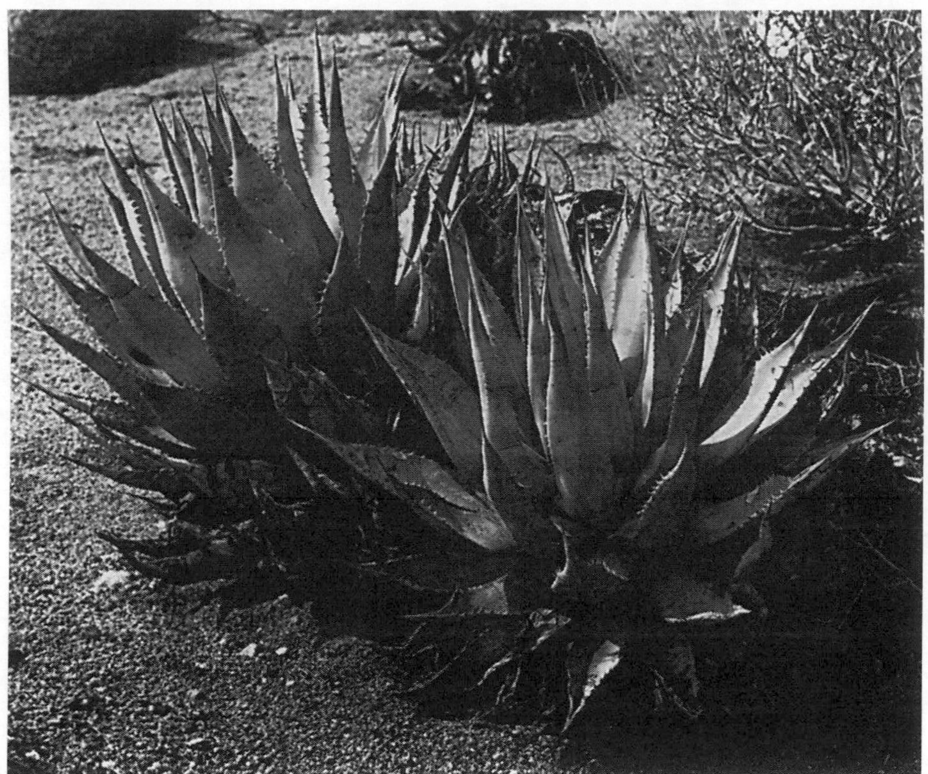

Figure 16. Agave deserti *subsp.* deserti *in San Matías Pass, Baja California. (Photograph by J. R. Hastings.)*

duction—at least of subsp. *deserti*—is vegetative (Jordan and Nobel 1979). The parent-ramet connection enhances survivorship by supplying water to the young plant for about the first 14 years (Raphael and Nobel 1986), and carbohydrates as well (Tissue and Nobel 1988). Unlike adults, which are quite resistant to high temperatures (Nobel and Smith 1983), seedlings require sheltered microhabitats if they are to escape damage caused by high temperatures (Nobel 1984b). In southeastern California, the perennial grass *Pleuraphis rigida* often provides the shade that is crucial for seedling establishment, but it may compete with seedlings for soil water (Franco and Nobel 1988; Nobel 1989c; Franco and Nobel 1990). Seedlings lose much water because of their high surface-to-volume ratio. Consequently, they must generate enough biomass during their first wet season to survive droughts when the soil water potential drops below −1.6 MPa (Jordan and Nobel 1979). In all but a few years, rainfall patterns do not favor the rapid growth necessary to reach that critical biomass, and most seedlings perish (Jordan and Nobel 1979).

In Deep Canyon, California, the sole episode of *A. deserti* seedling establishment in 17 years of observation (Jordan and Nobel 1979) demonstrates the critical role of warm-season rain. In 1967, unusual August rains totaling 39 mm were followed by smaller storms in November, December, February, March, and April for an additional 71 mm (Jordan and Nobel 1979). Another 13 mm fell in July. Thus, not only was the rainfall from the typical winter-spring frontal storms above normal, but the two summers following germination were exceptionally wet. (Such summer/fall rains are typically caused by an influx of moist tropical air or a tropical storm [Huning 1978].) This sequence of rainfall, although unusual, is probably required if *A. deserti* seedlings are to establish in the western part of the range.

The gradient in warm-season rainfall (decreasing from east to west) across the Sonoran Desert may determine the extensively clonal habit of subsp. *deserti* as opposed to the solitary or sparsely suckering habit of subsp. *simplex*. As opportunities for seedling establishment become rarer, genotypes with a longer period of fecundity are favored. In agaves, this habit is accomplished by the vegetative production of rosettes, which extends the life span of the genotype.

Agave deserti is a valued wildlife resource, providing food, moisture, and shelter for a variety of desert animals (Gentry 1982). Native Americans of southern California and Baja California ate all parts of the plant and used the leaves for fiber (Castetter et al. 1938; Gentry 1982).

Agave felgeri

Gentry

Mescalito

The small, green to yellow-green rosettes of *Agave felgeri* form dense, caespitose clones. Filiferous along the margins, the leaves are linear to narrowly lanceolate, 25–35 cm long and 0.7–1.5 cm wide. The inflorescence is a spike 1.5 to 2.5 m tall with flowers clustered toward the tip.

This species strongly resembles *A. schottii,* but their ranges do not overlap. All other agaves within its range have larger, wider leaves.

A plant of rocky hillsides, *A. felgeri* is generally rather rare and widely scattered, although it can be locally common. Its apparent range skirts the tropical humid margin of true desert vegetation. The northernmost voucher (29.5°N) is Howard Scott Gentry's from 24 km northwest of Hermosillo. The southernmost, also Gentry's, represents a collection made 24–29 km east of Navajoa. The disjunct population at 28.5°N represents a 1972 sighting by J. Rodney Hastings in the vicinity of Los Pocitos. All the remaining localities are in the vicinity of Bahía San Carlos and the rocky coast to the northwest.

Resort development has destroyed the type colony near San Carlos (Gentry 1982).

Flowers appear in the summer (May–August) and again in October. Like other species in group Filiferae, *A. felgeri* flowers sparsely and unreliably (Gentry 1982) and no doubt depends heavily on asexual reproduction by offsets. Further coastal development should take the status of this species into account.

Agave mckelveyana

Gentry

This species forms small, few-leaved rosettes 20–40 cm tall that may be single or clumped. Light green or yellowish green, the linear or lanceolate leaves are 20–35 cm long and 3–5 cm wide. The small to medium marginal teeth are mostly 1–3 cm apart and downflexed. Flowering panicles are 2–3 m tall with 10–19 lateral branches. The small flowers are yellow. The woody capsules are 30–45 mm high.

Agave mckelveyana is closely related to *A. deserti* subsp. *simplex* but does not intergrade with it (Gentry 1982). *Agave mckelveyana,* the smaller of the two, is green or yellowish green rather than glaucous gray and has firmly rather than feebly attached marginal teeth (Gentry 1982).

Narrowly endemic to west-central Arizona, *A. mckelveyana* grows on rocky slopes in interior chaparral and Great Basin conifer woodland, often partly hidden among shrubs. Its niche is quite distinct from that of the shade-intolerant *A. deserti,* which is confined to lower elevations within the desert (Gentry 1982; Burgess 1985). At climate stations within the range of *A. mckelveyana,* freezing occurs 40–115 days per year (Sellers and Hill 1974). Rainfall is biseasonal, with the wettest months in late summer and winter. About 45–65%

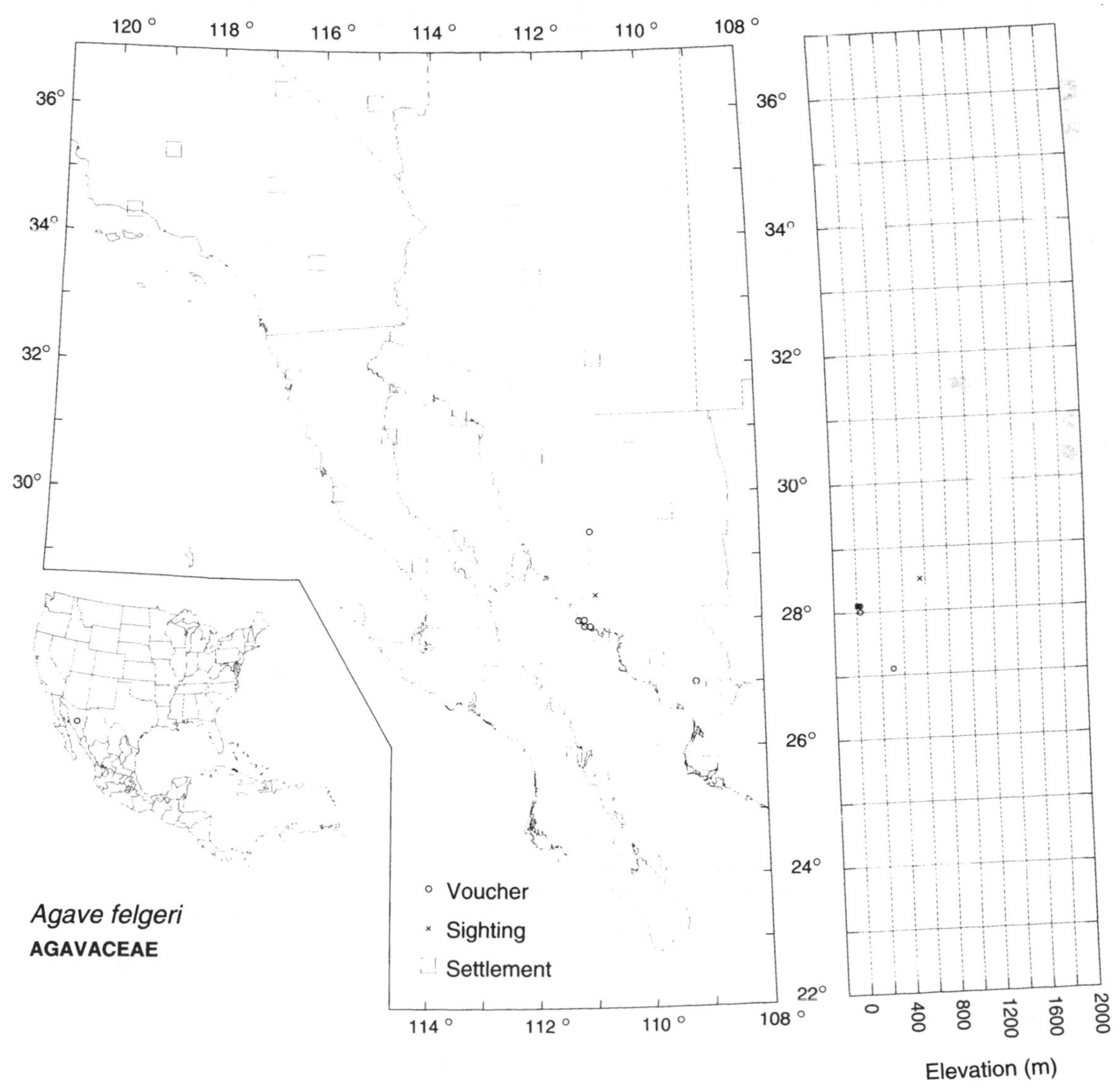

Agave felgeri
AGAVACEAE

of the annual precipitation falls from November through April.

The flowering season is May–July (Gentry 1982). Blooming starts at the bottom of the panicle and proceeds upward over a two- to three-week period. On a given lateral branch, all flowers bloom at once, with anthers dehiscing before stigmas are receptive. The nectar is rich in fructose and glucose but poor in sucrose (Reid et al. 1983), as is the case for *A. palmeri,* a bat-pollinated species. *Agave mckelveyana* grows outside the range of nectar-feeding bats and is pollinated mainly by hummingbirds, carpenter bees, and wasps (Sutherland 1982).

The plants reproduce by seed and by vegetative offsets (Gentry 1982). Fruit set is low, about 20%, and appears to be resource limited rather than pollinator or pollen limited (Sutherland 1982, 1987). If so, the "excess" flowers typical of this and other *Agave* species would not be the result of pollinator selectivity for large inflorescences, as suggested by Schaffer and Schaffer (1979). They would, rather, serve as pollen donors. Thus, the evolution of large inflorescences and consequent low ratios of fruits to flowers maximizes total plant fitness in that both fruit production and pollen donation are accommodated within the resources available (Sutherland 1982).

Agave moranii

Gentry

The large, single rosettes of *Agave moranii* are 1–1.5 m tall. The deeply guttered, lanceolate leaves are 70–120 cm long and 8–12 cm wide and are outlined with a narrow strip of thick, white tissue near the apex. Their large, gray marginal teeth are sinuously flexed or curved. The stout panicle, 4 to 5 m tall, has 20–30 sturdy lateral branches with dense umbels of yellow flowers. Large bracts are a conspicuous feature of the inflorescence.

Agave deserti var. *pringlei* hybridizes with *A. moranii* and, without fertile material, the two can be impossible to tell apart

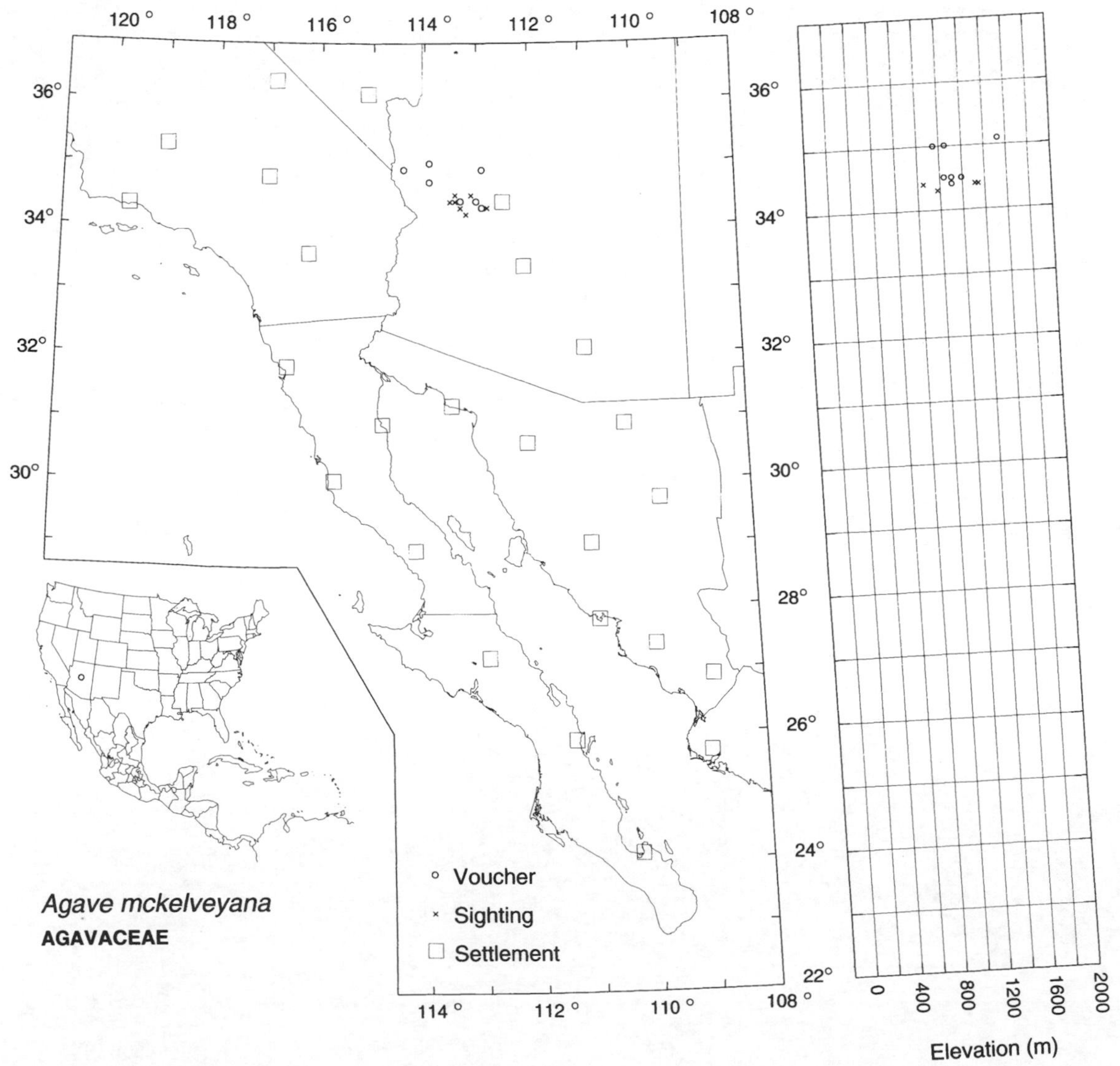

(Gentry 1982). Where they overlap, *A. moranii* is most frequent in the ecocline (van der Maarel 1990) between Californian chaparral and Sonoran desertscrub, usually between 500 and 1,900 m, while *A. deserti* var. *pringlei* tends to occur at somewhat higher elevations in chaparral (Gentry 1982). *Agave deserti* usually forms colonies via rhizomatous cloning, whereas rosettes of *A. moranii* are always solitary.

Scattered on rocky bajadas and plains, *A. moranii* is narrowly endemic to the southern and eastern slopes of the Sierra San Pedro Mártir. At the southwest end of Valle Chico (30.6°N, 115.1°W), the type locality, it grows with *A. cerulata* subsp. *nelsonii*. The two occupy different soil types and apparently do not hybridize here. Throughout the range of *A. moranii*, precipitation generally comes in winter; however, some populations experience occasional warm-season showers. In Arizona, cultivated *A. moranii* is quite sensitive to freezing and tends to rot during hot, wet summers. Its narrow range has probably been shaped by requirements for mild winters and relatively cool, dry summers. The flow of maritime Pacific air around the southern side of the Sierra San Pedro Mártir may provide these conditions.

Flowering is from late April through July. Gentry (1982) suggested that some panicles start to grow in the fall, pause over the winter, then elongate for flowering in the spring. In September 1980, Tony L. Burgess saw young inflorescences and recently matured capsules in the same population.

The flowers are reported to be edible. Cócopah Indians may have used the leaf fibers for baskets and cordage (Gentry 1982).

Agave palmeri

Engelm.

Lechuguilla

Growing 3–12 dm tall and 10–12 dm broad, the rosettes of *Agave palmeri* are pale green to glaucous green, sometimes reddish tinged. The lanceolate, acuminate

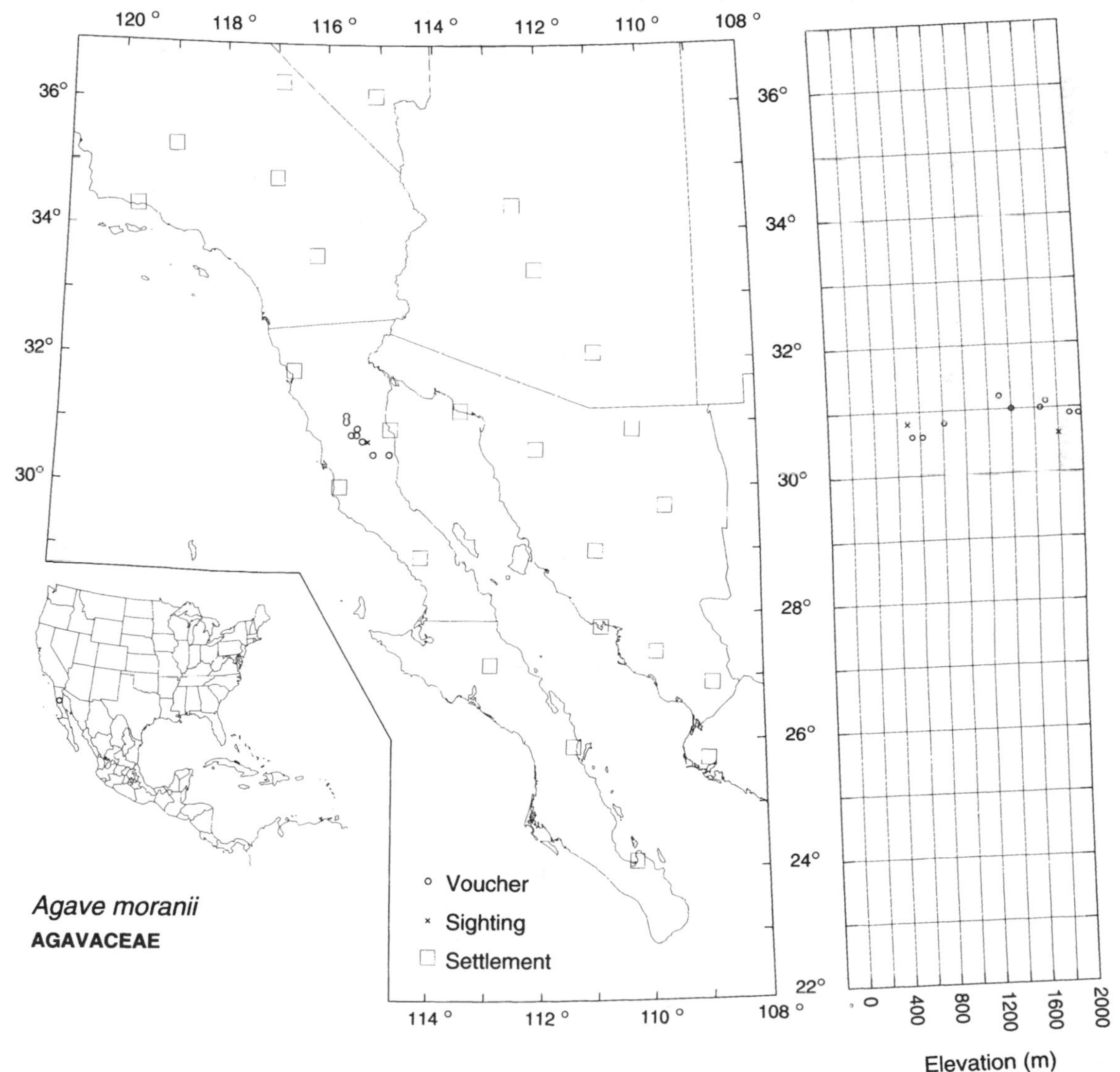

Agave moranii
AGAVACEAE

leaves are 35–75 cm long and 7–10 cm broad. Their undulate or nearly straight margins are beset with regular slender teeth. On some plants, smaller teeth alternate with larger ones. Flowering panicles are 3–5 m tall. The 8–12 branches have crowded pink to reddish brown flowers and woody oblong capsules. The diploid number of chromosomes is 30 pairs (Pinkava and Baker 1985).

Agave chrysantha and *A. palmeri* are vegetatively similar but geographically distinct. When flowers are present, *A. chrysantha* can be distinguished by its yellow-tipped buds. Where *A. palmeri* and *A. shrevei* overlap, individuals may combine characters of both. Most *A. shrevei* have relatively broad (2.5–4.5 times longer than wide), ovate, undulate leaves. The lanceolate leaves of *A. shrevei* subsp. *magna* Gentry are much larger (15–25 cm wide) than those of *A. palmeri*. Those of *A. murpheyi* F. Gibson are more linear and narrower (6–8 cm wide). In the Sierra Jojoba (30.1°N, 112.0°W), the poorly known *A. fortiflora* Gentry occurs near a southwestern outpost of *A. palmeri*. As biologists presently understand the species (Gentry 1982), *A. fortiflora* leaves are more strongly cross-banded, and the marginal teeth are more remote (1–3 cm apart).

Agave palmeri hybridizes with *A. chrysantha* to the north and with *A. shrevei* to the south (Gentry 1972, 1982). Its latitudinal limits are determined by genetic interactions with related taxa. The resulting cline of genotypes has probably not been very stable, given regional patterns of Pleistocene and Holocene change. The location of the transition from *A. palmeri* to *A. chrysantha* may be heavily influenced by pollinator availability. If a predicted decline in bat pollinators results in fewer *A. palmeri* (Howell and Roth 1981), then *A. chrysantha*, which is not dependent on bats for pollination, might spread southward into the former range of *A. palmeri*.

This is a plant of rocky slopes in semidesert grassland, Madrean evergreen woodland, and Madrean montane conifer forest, generally between 930 and 2,000 m. The overall range is rather extensive for an agave (Gentry 1982), with the species

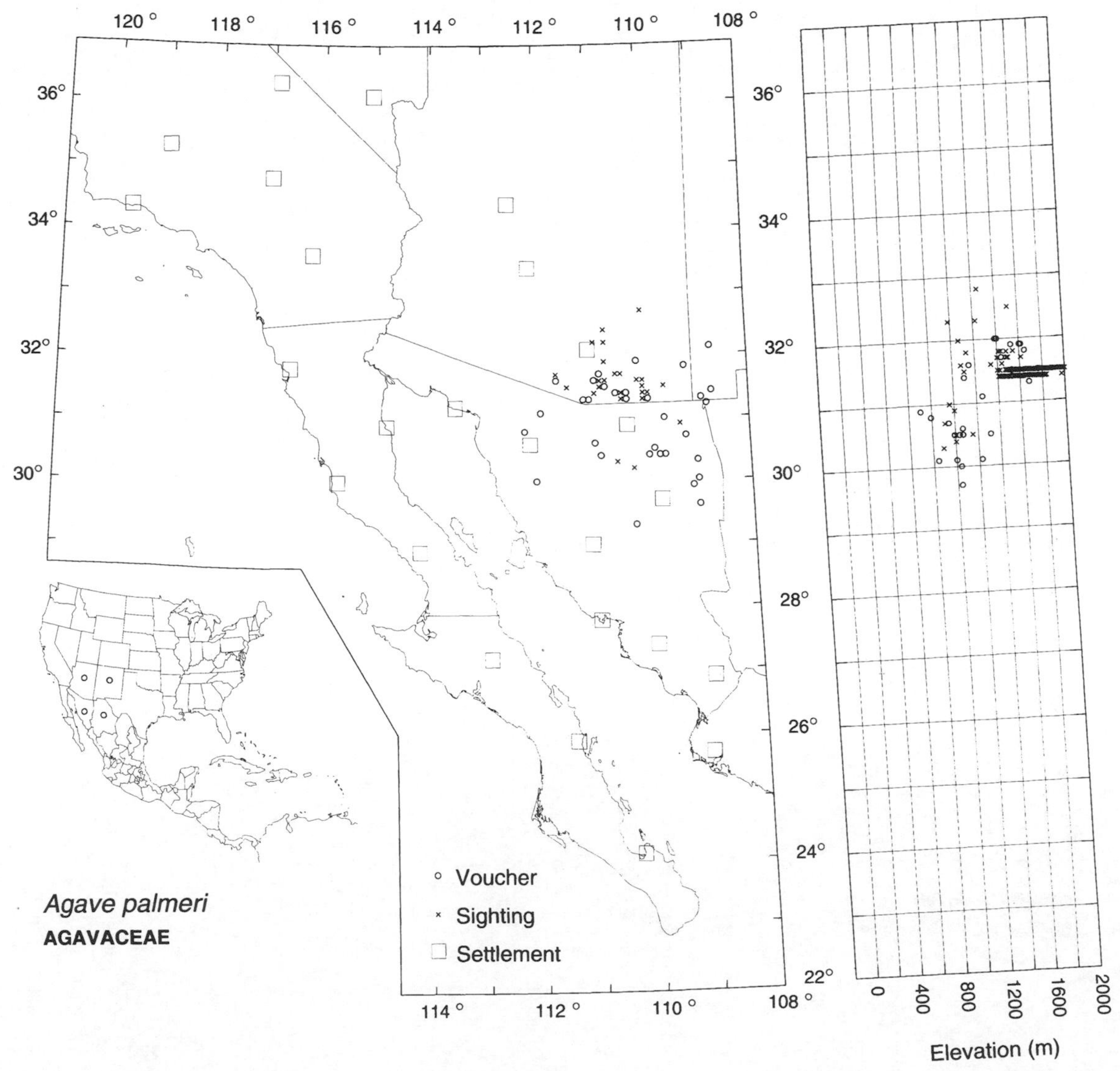

occupying portions of Arizona, New Mexico, Sonora, and Chihuahua. Within this region, the populations are typically widely scattered on isolated mountain ranges. Weather stations within the Arizona portion of the range show a temperate climate where freezing occurs an average of 37–152 nights per year (Sellers and Hill 1974). Rainfall is biseasonal: 53–75% falls between July and October, most of the remainder in winter (Hastings and Humphrey 1969; Sellers and Hill 1974). Freeze damage has been observed near the upper elevational limit, which is probably determined by cold tolerance (Nobel and Smith 1983). *Agave palmeri* enters the desert from above and at scattered locations in western Sonora, as in the Sierra Jojoba (30.1°N, 112.0°W, vouchered by Howard Scott Gentry) and in the Sierra La Gloria (30.9°N, 112.2°W, vouchered by Raymond M. Turner). These isolated populations are most likely remnants of what was, during the Pleistocene, a more continuous distribution.

The members of group Ditepalae exhibit a south-to-north, winter-to-spring flowering season that provides a steady flow of nectar for migrating *Leptonycteris* species (Gentry 1982). Adaptations for bat pollination include copious nocturnal production of nectar and pollen, a high concentration of dissolved solids in the nectar, a high percentage of protein in the pollen, the exposed position of the flowers (Schaffer and Schaffer 1977b), and the production of nectar rich in fructose and glucose but poor in sucrose (Freeman et al. 1983). Taken together, these traits suggest a high degree of coevolution between *Leptonycteris* and *Agave* species in group Ditepalae (Howell 1979; Howell and Roth 1981; Gentry 1982).

The main flowering season for *A. palmeri* is June and July (Gentry 1972). A few plants often bloom in September (Howell 1979). This lengthy period of flowering brackets the arrival and departure of the nectar-feeding bats (*Leptonycteris sanborni*) that are the primary pollinators (Howell 1979). Small flocks of these bats forage at a plant for several minutes, then move to others. While feeding on nectar, the bats become liberally dusted with pollen, which they readily transfer from plant to plant and from flower to flower (Howell 1979). Since *A. palmeri* plants are self-sterile, it is to their advantage to maximize pollen movement. This pollination pattern is accomplished with steeply declining rates of nectar production, forcing flocks to locate new plants instead of returning to those already worked that night (Howell 1979). The system is mutually advantageous: the clumped distribution and high nectar production of *A. palmeri* enable the bats to minimize their foraging effort (Howell 1979).

Reproduction is mostly by seed, occasionally by root suckers (Howell and Roth 1981; Gentry 1982). *Agave palmeri* is monocarpic: rosettes die shortly after flowering because much of their stored energy is required to produce the massive inflorescence (Howell and Roth 1981). Given adequate pollination, *A. palmeri* populations set enough seed to maintain themselves despite inflorescence consumption (Waring and Smith 1987), seed predation, and low recruitment (Howell and Roth 1981). When pollinators are scarce or absent, however, the monocarpic strategy represents wasted reproductive effort, since the individual sacrifices its life without producing seed (Howell and Roth 1981).

Agave palmeri is sometimes cultivated in desert gardens. Plants can be propagated from seed or offsets. The rosettes are cold tolerant to –10.4°C and heat tolerant to 61.9°C (Nobel and Smith 1983), but high summer temperatures, self-sterility, and absence of pollinators typically inhibit flower and capsule development at low elevations (Gentry 1982). Although *A. palmeri* is not an endangered species, seed set in the United States may have steadily declined since about 1950; this decline has been attributed to a decrease in the numbers of nectar-feeding bats (Howell and Roth 1981). The importance of bat pollination has been questioned (Cockrum and Petryszyn 1991). A marked decrease in *A. palmeri* numbers would precipitate further decline in populations of bats and other organisms dependent upon the bat-agave symbiosis (Howell and Roth 1981).

Agave pelona Gentry

Mescal pelón

The small to medium-sized rosettes of *Agave pelona* (figure 17) are 40–60 cm tall. Linear-lanceolate, dark green, and shiny, the leaves are 35–50 cm long and 3–5 cm wide. They may be tinged with purple or red. Each ends in a long, slender, white spine. The toothless leaf margins have a smooth, white border of firm tissue. The flowers are reddish and campanulate, crowded in a spikelike raceme 2–3 m tall. The oblong capsules are 25–30 mm long.

Leaves of the somewhat similar *A. felgeri* are narrower (0.7–1.5 cm wide), filamentous along the margins, and striped down the center. *Agave schottii* also has narrower leaves (0.7–1.2 cm wide) with filamentous margins. Leaves of *A. chrysoglossa* are larger (70–120 cm long, 4–7 cm wide). All other agaves within the range of *A. pelona* have marginal teeth.

Felger and Moser (1985) assigned specimens from the Sierra Seri (29.2–29.3°N, 112.1°W) to *A. chrysoglossa* based on their smaller, greener leaves and clonal habit. Even though *A. pelona* rosettes in the Sierra del Viejo are strictly solitary, these clonal specimens seem closer to *A. pelona* than *A. chrysoglossa,* and we have so mapped them here.

Locally common on limestone cliffs and rocky slopes, *A. pelona* is known from only a few locations in western Sonora: the Sierra del Viejo (30.3–30.4°N, 112.1–112.4°W), the Sierra Seri, and the Sierrita de Lopez (29.4°N, 111.3°W). This species, separated from its nearest relatives by several hundred kilometers, has evidently been isolated long enough to evolve morphological characters distinct from other species in its group (Marginatae) (Gentry 1982). Confusion about the Sierra Seri specimens suggests that *A. pelona* may be more closely related to taxa in Gentry's group Amolae.

The flowering season is March–May; according to Gentry (1982), flowering may be sparse or nonexistent in dry years. Individual rosettes die after flowering, and, since most populations do not proliferate by suckering (Gentry 1982), the species is heavily dependent upon sexual reproduction.

Agave pelona can be used in making mescal, although its small rosettes are inferior to those of the sympatric *A. zebra* for that purpose (Gentry 1982). Seri Indians roasted and ate the hearts (Felger and

Figure 17. Agave pelona *(left) and* Agave zebra *(right) at the summit of Sierra del Viejo, Sonora. (Photograph by R. M. Turner.)*

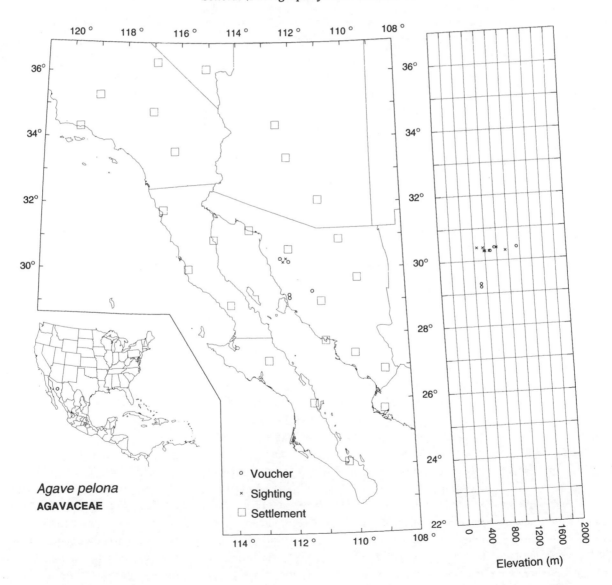

Agave pelona
AGAVACEAE

◦ Voucher
× Sighting
□ Settlement

Elevation (m)

Moser 1985). In Arizona, *A. pelona* is increasingly popular as an ornamental. It can withstand temperatures as low as −6°C as well as hot summers.

Agave schottii Engelm.

Shindagger, amole, amolillo

The densely caespitose rosettes of *Agave schottii* are green or yellowish green and may form spreading colonies a meter or more across. The linear leaves, measuring 25–40 cm long and 0.7–1.2 cm wide, are straight, incurved, or falcate. The cream-colored or brown margins lack teeth but are filiferous, with sparse, brittle threads. A spike 1.8–2.5 m tall, the inflorescence comprises tubular yellow flowers crowded on the upper third or half of a slender stalk.

Agave toumeyana Trel. and *A. parviflora* Torr. bear diagonal white markings on bright green leaves.

Two subspecific taxa, var. *schottii* and var. *treleasei* (Toumey) Kearney and Peebles, have been described. Our map combines them. Variety *treleasei*, distinguished by its wider (15–25 mm wide), deep green leaves, is known from the Santa Catalina and Ajo mountains, Pima County, Arizona. The restricted distribution suggests that it arose as an inviable hybrid or polyploid clone: for example, through hybridization between either *A. palmeri* or *A. chrysantha* and *A. schottii* var. *schottii;* through polyploidy in *A. schottii* var. *schottii;* or through hybridization and introgression between *A. deserti* and *A. schottii* var. *schottii* (Hodgson et al. 1994). More than one of these processes may have occurred within a given population (Reichenbacher 1985a).

Agave schottii clones are locally common on rocky slopes and often grow in shallow soil on bedrock outcrops. They are densest where the plants are protected from fire (Niering and Lowe 1984). Most populations lie between 1,000 and 1,500 m in semidesert grassland and Ma-

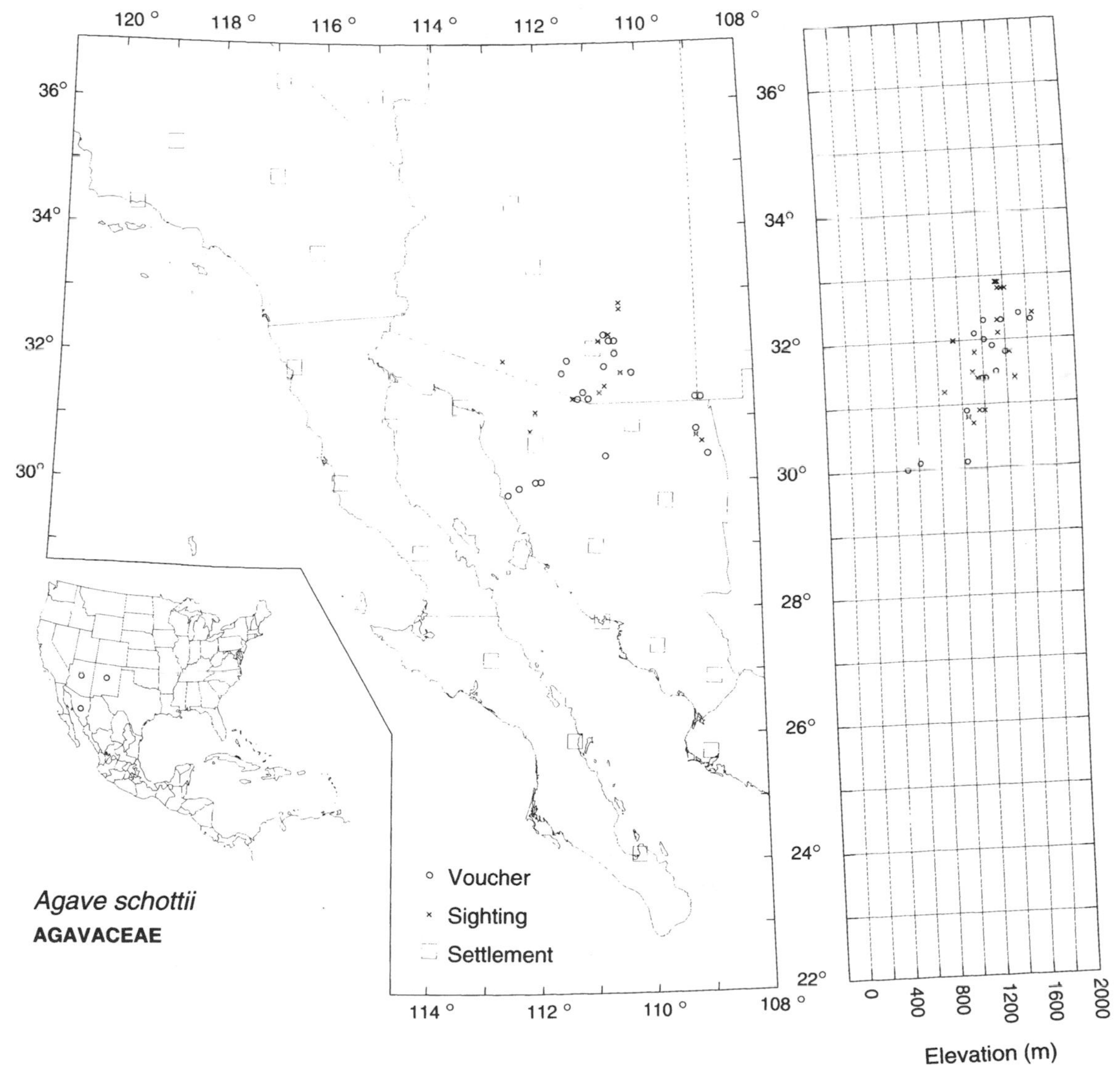

Agave schottii
AGAVACEAE

drean evergreen woodland, generally on isolated mountain ranges (for example, the Santa Catalina, Rincon, and Baboquivari mountains in southern Arizona). Several, however, are known within the desert proper, including the following outliers in western Sonora: Puerto Libertad (29.9°N, 112.6°W), Picú Pass (30.0°N, 112.4°W), and Rancho Primavera (30.1°N, 112.1°W), all vouchered by Howard Scott Gentry, and in the Sierra San Manuel (31.2°N, 112.1°W) and the Sierra La Gloria (30.9°N, 112.2°W), reported by Raymond M. Turner.

This species tolerates temperatures from –11 to 61°C and shows substantial ability to acclimate to heat (Nobel and Smith 1983). It typically grows under a biseasonal rainfall regime where late summer or early fall is the wettest period. A maritime influence may ameliorate the high summer temperatures at the arid outpost near Puerto Libertad. The upper elevational limits are probably determined more by competition in dense vegetation than by cold. Factors influencing the northern limit are unknown—neither cold nor aridity appears likely.

The flowers, present from April to August, are visited by hummingbirds, carpenter bees, bumblebees, honeybees, wasps, solitary bees (Schaffer and Schaffer 1979), ants (Schaffer et al. 1983), and bats (Howell 1972). Howell (1972) suggested that the spicate agave inflorescence evolved with bees as the main pollinators, whereas the paniculate inflorescence apparently coevolved with bat pollinators. In *A. schottii*, a spicate species, anthesis and nectar production are mainly nocturnal, suggesting adaptation for bat pollination, but the sweet fragrance and the low protein content in the pollen of the yellow flowers might favor pollination by bees, especially carpenter bees and bumblebees (Schaffer and Schaffer 1977b). Given its mixed signals and wide array of pollinators, *A. schottii* may be undergoing secondary modification from bat to bee pollination (Schaffer and Schaffer 1977b).

Both varieties proliferate abundantly by suckers. Variety *treleasei* is known to reproduce sexually (Hodgson et al. 1994).

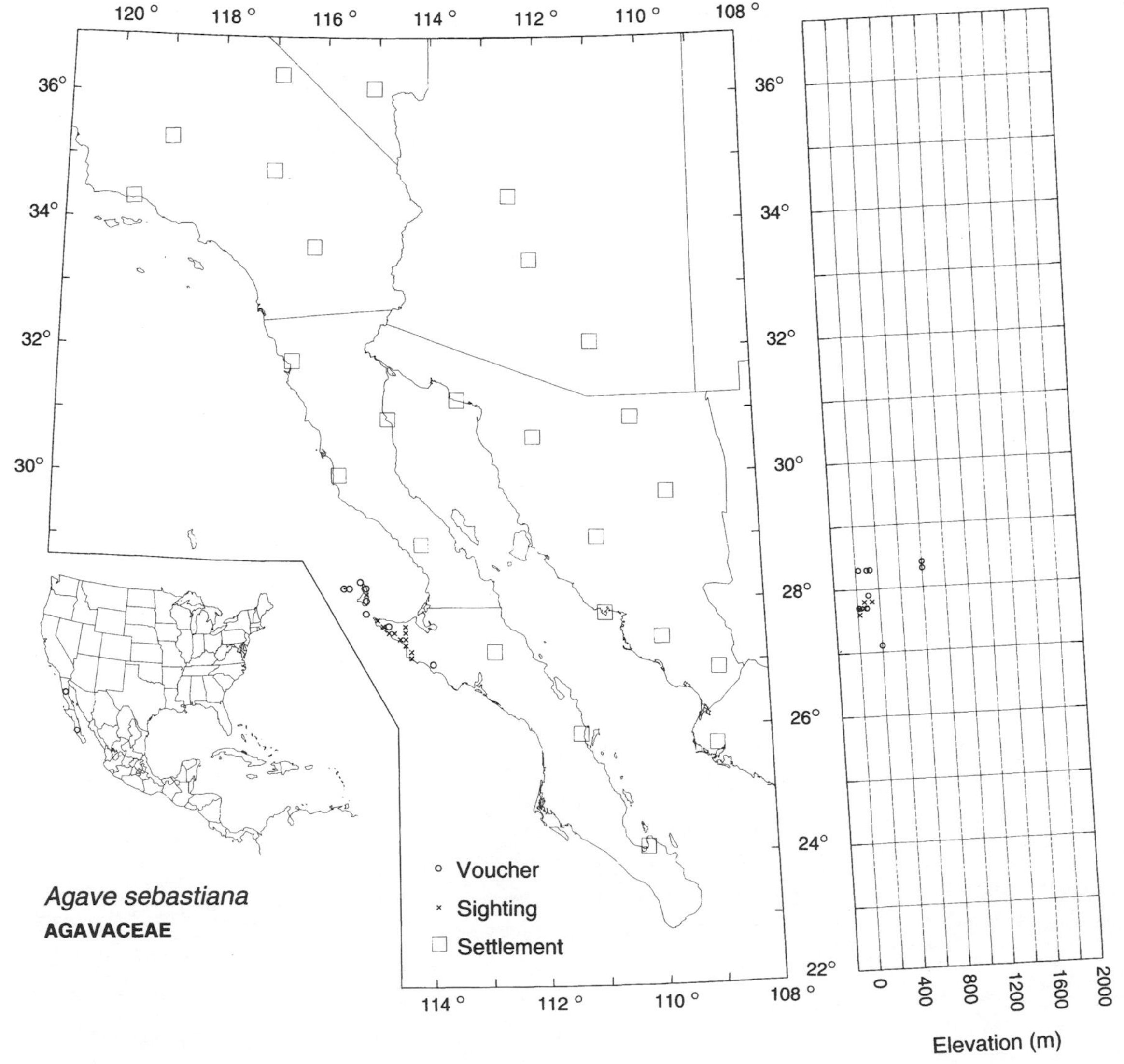

Agave sebastiana
AGAVACEAE

As in other agaves, the rosettes are monocarpic. Schaffer and Schaffer (1979) suggested that monocarpy in *A. schottii* var. *schottii* (and, by implication, in the genus as a whole) evolved as pollinators selected for larger and larger inflorescences. The flowering stalks eventually became so large that a rosette necessarily devoted all its food reserves to production of a single inflorescence (the so-called Big Bang) rather than using a proportion of its energy to produce smaller stalks at widely spaced intervals (Schaffer and Schaffer 1979).

This species has been used in Mexico for washing clothes (Standley 1920; Gentry 1972) and hair (Felger and Moser 1985). Because of sapogenins, the plants are not palatable to livestock (Gentry 1972). Wildlife, perhaps rodents, consume the leaves, caudices, and inflorescences of var. *treleasei* (Hodgson et al. 1994).

Agave sebastiana

E. Greene

The rosettes of *Agave sebastiana* are 6–12 dm tall and have closely imbricate leaves. Broadly linear to ovate, the pale green, glaucous leaves are 25–45 cm long and 8–24 cm broad. The slender reddish brown teeth are often downflexed. The congested panicle is 2–3 m tall with 8–12 yellow-flowered umbels in a rounded or nearly flat crown.

Agave sebastiana has a broader, flatter panicle than the closely related *A. shawii,* and its pale, glaucous leaves are also distinctive. *Agave vizcainoensis* Gentry has relatively narrower leaves (about four times longer than wide); a more elongate, less congested panicle; and less vegetative proliferation of rosettes.

Narrowly endemic to the coastal portion of the Vizcaíno Peninsula and adjacent islands (Islas Cedros, San Benito, and Natividad), *A. sebastiana* is occasional to locally common on slopes and plains from sea level to 600 m. This region, one of the driest in North America (Schmidt 1989), receives significantly more rain in winter than in summer. Along the coast, the usual long summer drought is ameliorated by cool, humid maritime air. *Agave vizcainoensis* replaces *A. sebastiana* in the Sierra Vizcaíno and the Picachos de Santa Clara. There is a cline in leaf and panicle characters between the two species, and some specimens that we have mapped as *A. sebastiana* may represent intergrades.

During much of the Miocene and Pliocene, the present-day Vizcaíno Peninsula was an island (Gentry 1972). This separation evidently provided enough isolation to promote divergence of *A. sebastiana* from the mainland stock that became *A. shawii.* Today, *A. sebastiana* is separated from coastal populations of the related *A. shawii* subsp. *goldmaniana* by the arid plains surrounding the Laguna Ojo de Liebre.

The plants form large colonies as new rosettes grow from the bases of old stems. As in *A. shawii,* the trunks of old plants recline upon the ground (Gentry 1982). Flowers appear from March to May (Gentry 1982).

In southern Arizona, cultivated *A. sebastiana* is especially prone to rot after hot, wet periods.

Agave shawii

Engelm.

Agave shawii subsp. *shawii* produces single or caespitose rosettes of small or medium size (figure 18). The tightly imbricated, ovate leaves are 20–50 cm long and 8–20 cm wide. Marginal teeth are variable in size and shape. The subcapitate inflorescence is a stout panicle 2–4 m high. Large purple, succulent bracts closely subtend the 8–14 lateral umbels. The flowers are yellow or reddish. Subspecies *goldmaniana* differs in several respects. The rosettes are larger and have longer leaves (40–70 cm long) that are lanceolate rather than ovate. The inflorescence (3 to 5 m tall) has 18–25 dense laterals, the whole forming a massive, pyramidal panicle. The diploid number of chromosomes for both subspecies is 30 pairs (Pinkava and Baker 1985).

The two subspecies can be difficult to distinguish where they intermingle but are otherwise clearly marked ecotypes that seem to be evolving independently (Gentry 1972). *Agave avellanidens* Trel., which replaces *A. shawii* southeast of Punta Prieta, can hardly be distinguished from subsp. *goldmaniana* in its vegetative state (Gentry 1972). When in flower, *A. avellanidens* has a longer panicle (4–6 m tall) and shorter, more numerous (25–35) flowering branches. *Agave sebastiana* has paler, glaucous leaves that are smaller (25–45 cm long) than those of *A. shawii*

Figure 18. Agave shawii, *with developing inflorescence, in Arroyo Aguajito east of El Rosario, Baja California. (Photograph by J. R. Hastings.)*

subsp. *goldmaniana* and also has a broader, flatter panicle. *Agave cerulata* leaves are smaller and narrower (2.5–8 cm wide).

Locally common on plains, open slopes, and coastal terraces, *A. shawii* grows on both igneous and sedimentary substrates (Gentry 1982), often on deep, somewhat sandy soils (Shreve 1964) or fine-textured loams. Subspecies *shawii*, a maritime plant in Californian coastalscrub (Pase and Brown 1982), reaches its southern limit at 30°N, while subsp. *goldmaniana*, a desert-adapted plant, occurs in the central peninsula, mostly south of 29.7°N. Outliers of subsp. *shawii* occur within the more southerly range of subsp. *goldmaniana* at about 29.4°N, 114.8°W (Gentry 1972). The population at Torrey Pines, California (32.9°N, 117.2°W), may have originated as cultivated plants (Reid Moran, personal communication 1992).

Both subspecies are confined to a winter rainfall regime in a region where yearly precipitation averages only 85–95 mm. Frequent coastal fogs doubtless ameliorate the summer droughts (Gentry 1972). In the higher, inland parts of its range, subsp. *goldmaniana* occasionally experiences warm-season rain. In the north, the range of subsp. *shawii* contracts into the most arid fringe of Californian coastalscrub. Tony L. Burgess observed sprouting rosettes following a fire near Colonet (31.3°N, 116.2°W). Sprouting is probably crucial for persistence in fire-prone communities. Shading and fire frequency doubtless prevent reproduction in more mesic Californian chaparral. Introgression with *A. cerulata* has probably occurred where the two species meet.

Inflorescences appear between February and June and, in response to unusual summer rains, occasionally in the fall (Gentry 1972). Hummingbirds, bats, and bees are among the visitors to the flowers (Humphrey 1974). Flowering occurs at an age of 20–40 years (Moran 1964; Gentry 1972).

In both subspecies, caulescent plants produce rosettes on their branched, creeping stems. Individual rosettes take root and become independent of the mother plant as the connecting stem dies.

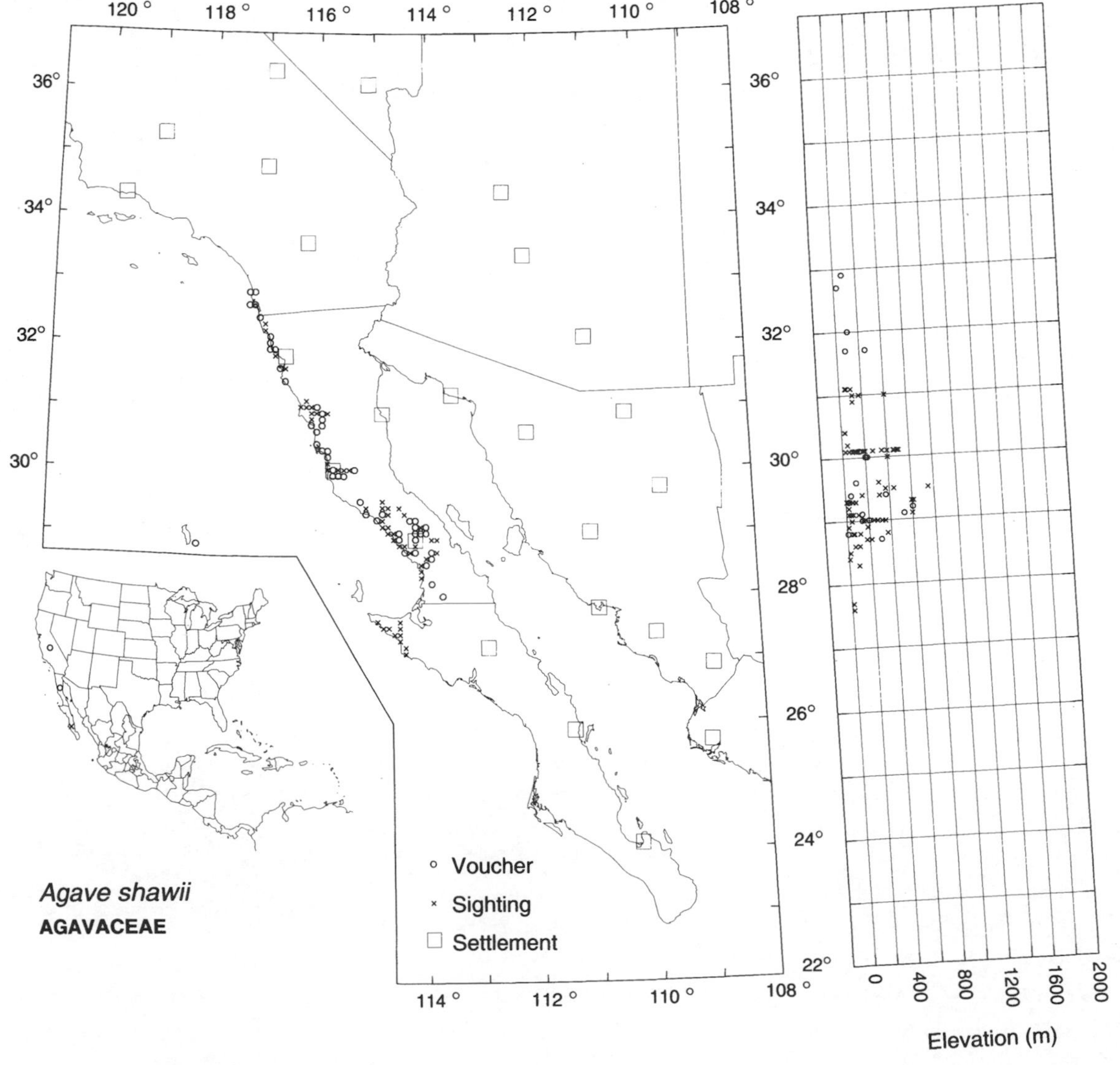

The resulting "fragmented supine clones" can be extensive (Gentry 1982). Subspecies *goldmaniana* shows highly variable cloning. On granitic bajadas near El Crucero (29.3°N, 114.2°W), virtually all individuals form tangled colonies, whereas populations farther south and closer to the coast (between Jesús María at 28.3°N, 114.0°W and Sierra San Andrés at 28.7°N, 114.3°W) are almost entirely composed of solitary rosettes.

Individual rosettes die after flowering. In populations that produce single rosettes, the entire plant dies. In populations where several or many rosettes arise from a common stem, the individual is a perennial that survives the death of single rosettes (Gentry 1972).

Native peoples ate the young flowering stalk and used the leaf fibers in cordage and basketry (Gentry 1972). Currently, the dried flower stalks are used in constructing fences and mud-plastered walls, and the developing stalks are cut and fed to livestock (Humphrey 1974).

Both subspecies are occasionally cultivated, especially in coastal California. The plants suffer frost damage at about –5°C (Gentry 1982), and internal tissues are heavily damaged at –8°C (Nobel and Smith 1983). In southern Arizona, subsp. *shawii* often dies during hot summers, but some clones of subsp. *goldmaniana* easily tolerate the climate. Plants are subject to destruction by the agave weevil, *Scyphophorus* species (Vaurie 1971). In Mexico, wild populations are being decimated by rapid development of coastal property (Gentry 1972; Haiman 1974).

Agave sobria

Brandegee

Mescal pardo, pardito

Small to medium in size, the open, caespitose rosettes of *Agave sobria* (figure 19) are light glaucous gray, bluish, or yellow-green. The linear to lanceolate leaves are 20–80 cm long and 5–10 cm wide and are often conspicuously cross-banded and

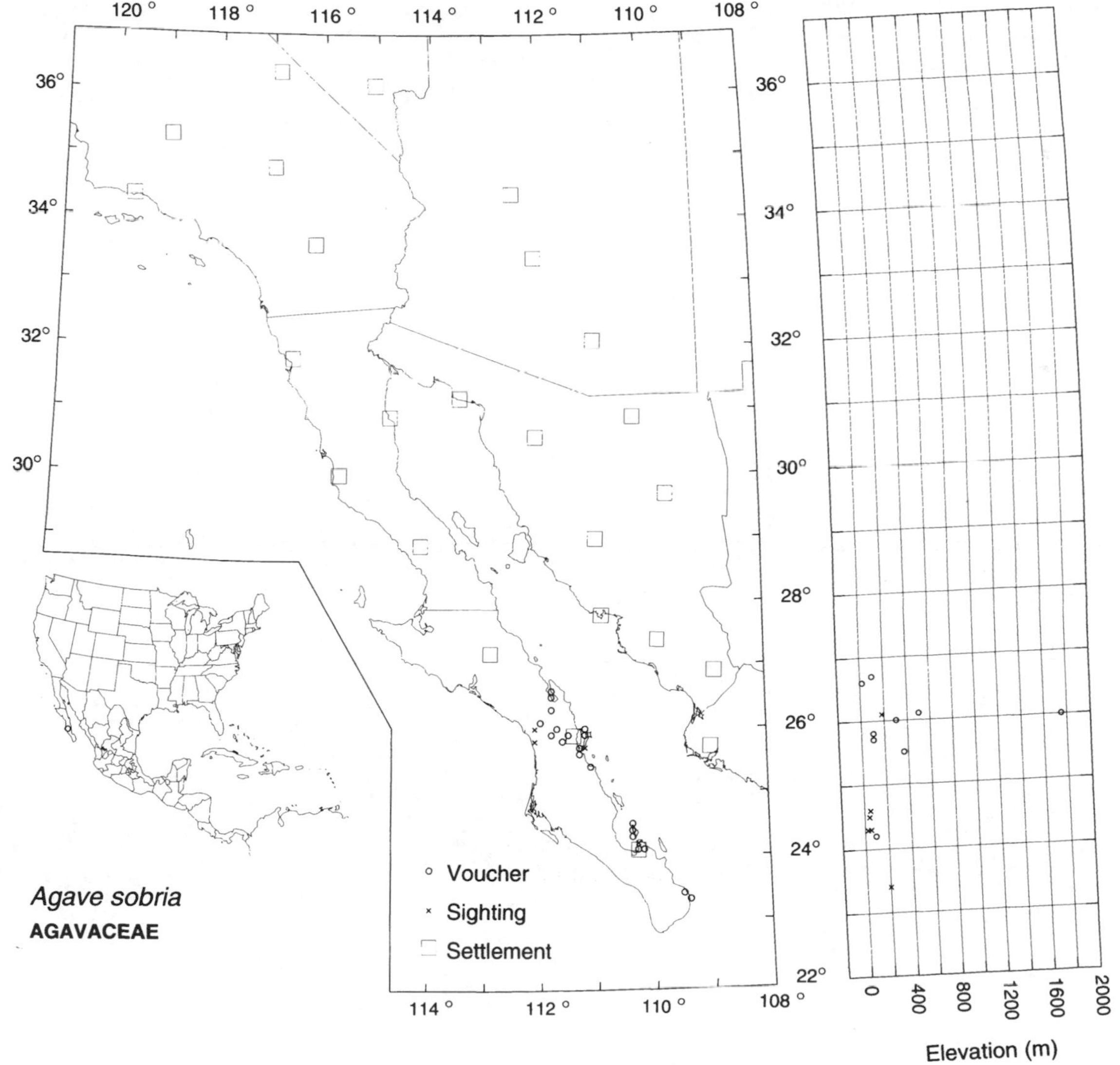

sometimes twisted. Marginal teeth are flexuous and spaced 3–4 cm apart except in subsp. *frailensis* Gentry. The 8–20 short lateral umbels make compact, nearly globose units on a slender stalk 2.5–4 m tall. Flowers are pale yellow.

The light glaucous color, remote teeth, and long, lanceolate leaves of this species are distinctive. The larger (63–110 cm long, 7–12 cm wide), usually arching leaves of *A. aurea* are seldom cross-banded. Also, rosettes of *A. aurea* do not form offsets, and their panicle branches are usually longer. Seldom cross-banded, the leaves of *A. capensis* Gentry are closely toothed (1–2 cm apart); the teeth have mammillate bases. Leaves from northern populations of *A. sobria* are similar in size (30–60 cm long) to *A. capensis* leaves; however, *A. capensis* has larger leaves than its nearer neighbor, *A. sobria* subsp. *frailensis*. The two taxa differ also in relative leaf width: 7–8 times longer than wide in *A. capensis*, 3–5 times longer than wide in *A. sobria* subsp. *frailensis*. Leaves of *A. cerulata* are somewhat smaller (2.5–7 cm wide) and have brown rings around the bases of the marginal teeth. *Agave datylio* leaves are narrower (3–4 cm wide). *Agave gigantensis* Gentry leaves are wider (11–16 cm), and the plants are not known to produce offsets.

Three subspecies have been recognized in this variable taxon (Gentry 1982). Subspecies *sobria* is distinguished by its longer (50–80 cm), relatively narrow leaves with entire to moderately undulate margins. Subspecies *roseana* leaves are 35–50 cm long and bear large, widely spaced teeth on mammillate projections. Subspecies *frailensis* leaves are usually smaller (20–30 cm long), with more closely spaced, smaller teeth borne on undulate margins.

In the Sierra de la Giganta, *A. sobria* is most frequent on rocky, north-facing slopes (Gentry 1982). Plants of more xeric exposures are typically small and depauperate. Subspecies *sobria* has a wide elevational range in the Sierra Giganta, but the others appear restricted to lower elevations. *Agave sobria* subsp. *roseana* is found in more arid communities northeast and east of La Paz, a distribution remarkably similar to that of *Acacia pacensis*. Subspe-

Figure 19. Agave sobria *subsp.* frailensis *north of Punta Frailes, Baja California Sur. (Photograph by R. M. Turner.)*

Figure 20. Agave subsimplex *near Libertad, Sonora. (Photograph by H. L. Shantz.)*

cies *frailensis* appears confined to volcanic hills northwest of Punta Frailes where it grows in low open scrub with *Alvordia brandegeei* Carter, *Acacia goldmanii*, *Cardiospermum spinosum* Radlk., and *Turnera diffusa* Willd. The composition and stature of this vegetation suggest that nutrient limitations as well as climate are important factors shaping the distribution of this narrowly endemic taxon. To the north, *A. sobria* is replaced by *A. cerulata* where cool-season rain becomes more dependable than warm-season rain. In higher, more mesic habitats in the Sierra Giganta, *A. sobria* is apparently replaced by *A. gigantensis*.

Across the range of the species, flowers appear from March through November. The plants reproduce freely by offsets and seed (Gentry 1982).

The specific epithet *sobria*, meaning "sober," was evidently applied ironically since the plants are used locally for mescal (Gentry 1982). This species makes an attractive ornamental for desert gardens. In Tucson, *A. sobria* subsp. *sobria* can tolerate temperatures as low as –5.6°C with little damage.

Agave subsimplex Trel.

Small and spreading, the glaucous rosettes of *Agave subsimplex* (figure 20) are gray or yellow-green, sometimes tinged with purple. The lanceolate to ovate leaves, 12–35 cm long and 3–6 cm wide, are shallowly channeled and variably toothed. The yellow to pink flowers are in 5–8 small umbels on a slender panicle 2–3.5 m tall.

Agave deserti has larger leaves 6.5–10 cm wide and 25–40 cm long. It may occur with *A. subsimplex* in a limited area north of Puerto Libertad, where rosettes of two

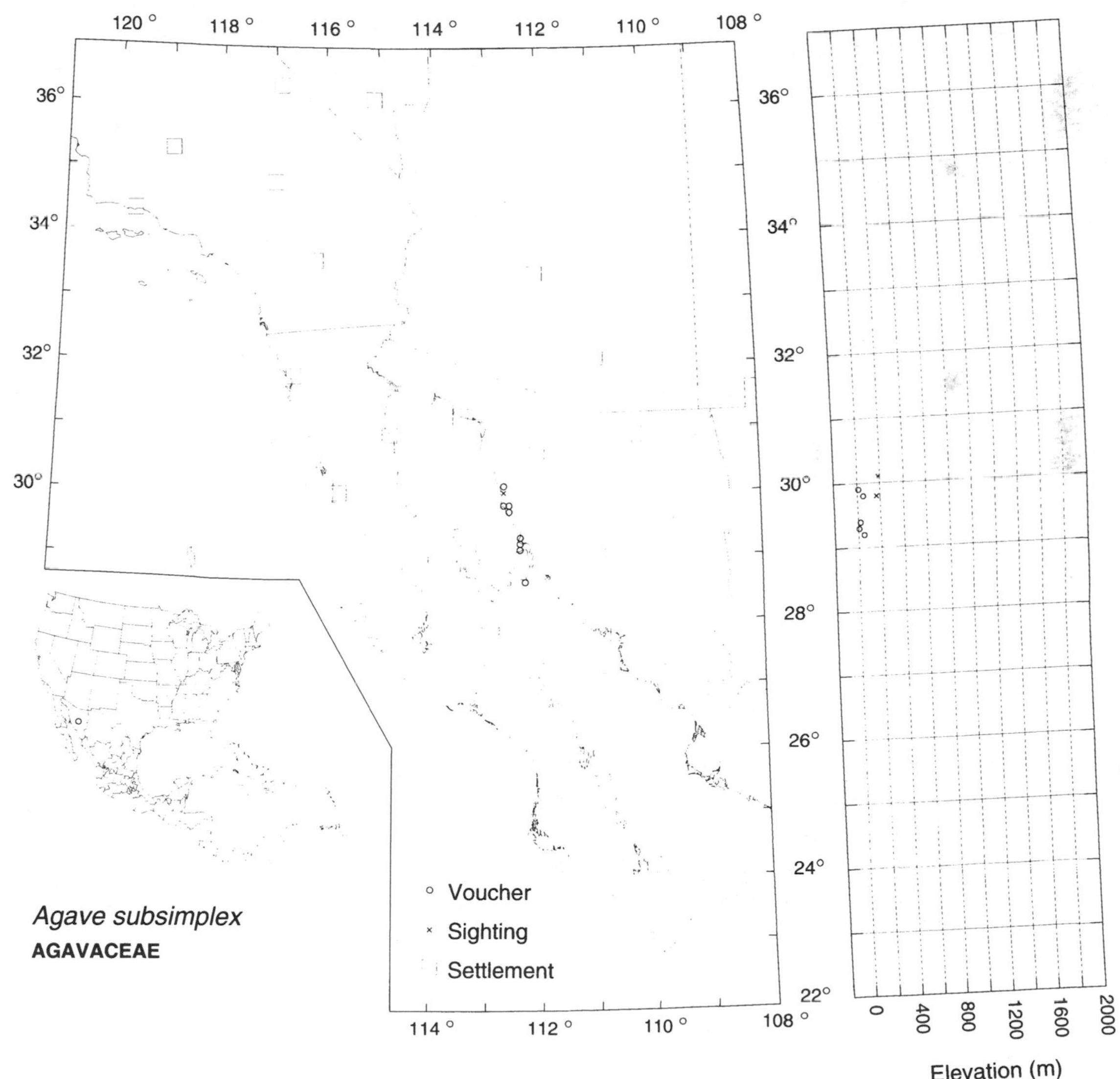

distinct sizes are mingled. The larger leaves of *A. colorata* are up to 12–18 cm wide. *Agave angustifolia* leaves are much longer (greater than 60 cm) and proportionally narrower. Both *A. felgeri* and *A. pelona* have narrower leaves that lack marginal teeth.

This species, locally common on rocky slopes and plains, is endemic to a short portion of the Sonoran coast and Isla Tiburón. Populations extend somewhat farther south along the Sonoran coast than our map shows (Richard Felger, personal communication 1991). Most lie below 250 m. Like *A. deserti* and *A. cerulata,* which it closely resembles, *A. subsimplex* is xeromorphic and appears well suited to its extremely arid environment. It seems restricted to an area influenced by maritime air from the Gulf of California. The susceptibility of cultivated plants to freeze damage shows that the upper, northern, and eastern limits are probably set by occasional incursions of unusually cold air. Scanty records from two weather stations within the range (Hastings and Humphrey 1969) indicate rainfall peaks in early winter and early autumn. The southern and perhaps eastern limits of *A. subsimplex* may correspond with a frontier beyond which the pervasive summer storms of Sonora seldom penetrate.

The blooming period is May–July (Gentry 1982).

The Seri baked and ate the stems and leaf bases and also employed them in making a sweet beverage (Gentry 1982; Felger and Moser 1985). Seeds and flower buds are used in necklaces (Felger and Moser 1985). This species is occasionally cultivated but has proven to be fairly frost sensitive.

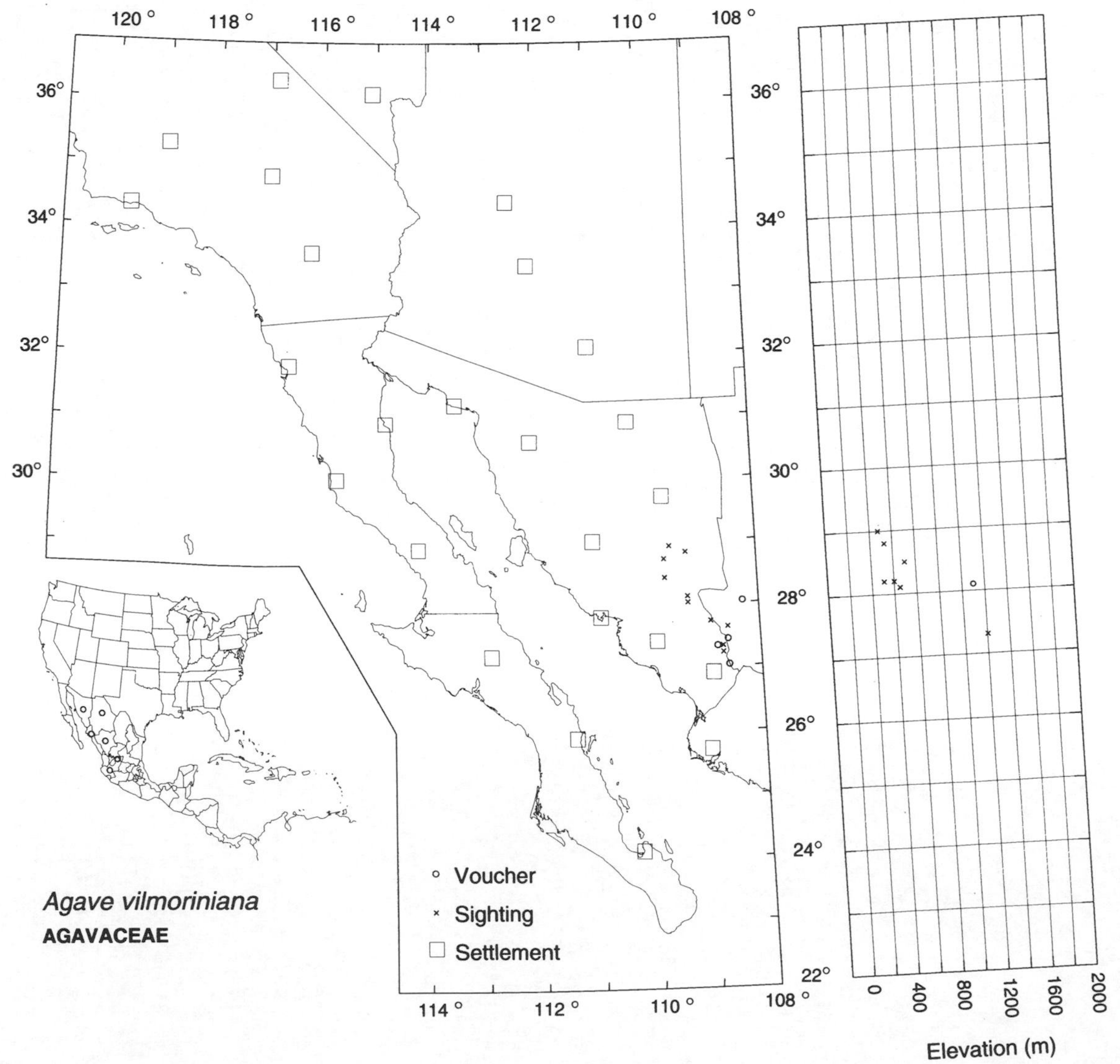

Agave vilmoriniana
AGAVACEAE

Agave vilmoriniana
Berger

[=*A. mayoensis* Gentry]

The light green to yellowish green rosettes of *Agave vilmoriniana* are 1 m tall and 2 m wide. The deeply guttered, arching leaves, 90–180 cm long and 7–10 cm wide, are linear-lanceolate and toothless along the margin. The inflorescence is a spike 3 to 5 m tall of crowded yellow flowers and, occasionally, bulbils.

The unarmed, gracefully arching, deeply guttered leaves are distinctive. The closely related *A. chrysoglossa* can be distinguished by its straighter, somewhat shorter (70–120 cm) and narrower (4–7 cm wide) leaves. Gentry (1982) reported apparent intergrades between *A. vilmoriniana* and *A. chrysoglossa* along the Río Yaqui west of Sahuaripa (29.2°N, 109.6°W). Some of our mapped sightings in this vicinity probably represent similar intergrades. All other agaves in the study area that lack marginal teeth have smaller, much narrower leaves.

Agave vilmoriniana forms extensive colonies on volcanic cliffs and on steep, rocky slopes (Gentry 1982). Growing as far south as Jalisco and Aguascalientes, it reaches the study area in the deep canyons of southeastern Sonora and adjacent Chihuahua in Sinaloan thornscrub, Sinaloan deciduous forest, and Madrean montane conifer forest. Freezing temperatures evidently do not limit this species to the north; cultivated plants tolerate temperatures as low as –9°C (Nobel and McDaniel 1988), and wild populations growing above 1,250 m experience occasional sharp frosts (Gentry 1982).

At its arid margins, *A. vilmoriniana* is replaced by *A. ocahui* Gentry and *A. chrysoglossa.* It is doubtful that our elevational profile reflects the true upper elevational limits.

Unlike other *Agave* species studied, *A. vilmoriniana* typically assimilates substantial amounts of carbon dioxide during the day. The proportion of diurnal carbon dioxide fixation varies with temperature and ranges from 36% at a day/night regime of 40/30°C to 92% under a 20/10°C regime (Nobel and McDaniel 1988). Well-watered small plants show growth rates of 4.6 mg dry weight/g dry weight/day; larger plants grow more slowly (Idso et al. 1986). Enrichment with carbon dioxide can offset the negative effect of drought on growth (Idso et al. 1986; Szarek et al. 1987).

Flowers appear from April to July and are visited by bees (Stephen Buchmann, personal communication 1989), hummingbirds, and perhaps bats (Gentry 1982). Plants flower at 7 to 15 years of age (Gentry 1982). Because reproduction in the wild is through seed and bulbil, a population may include both apomictic and sexual generations.

The cliff-face habitat and high sapogenin content may deter herbivores (Gentry 1982).

The leaf fibers are shaped into brushes and used in washing clothes (Gentry 1982). Crushed leaves supply the Tarahumara with detergent for baths, shampooing, and laundry. The Tarahumara also use the leaf juice to stupefy fish (Bye et al. 1975). An attractive ornamental, *A. vilmoriniana* withstands drought and frost and can be readily propagated from bulbils (Gentry 1982). In the summer, the plants should be given moderate water and should be protected from reflected sunlight (Duffield and Jones 1981). They are susceptible to soil grubs (Duffield and Jones 1981).

Agave zebra
Gentry

Strongly armed with large, curved, flattened teeth, the light gray, lanceolate, undulate leaves of *Agave zebra* (figure 17) are 50–80 cm long and 12–17 cm wide. They are cross-banded and deeply guttered. The narrow panicles, 6 to 8 m tall, have small yellow flowers on 7–14 lateral umbels. The rosettes are often solitary, but limited rhizomatous cloning occurs while the plants are small.

The species, a narrow endemic, can be confused with no other in its range. *Agave fortiflora* Gentry and *A. palmeri* have been collected in nearby mountains. Both differ from *A. zebra* in having narrower leaves (7–12 cm wide) that are less scabrous and less undulate.

Known only from the vicinity of Cerro Quituni and the Sierra del Viejo, *A. zebra* is locally common on rocky limestone slopes, where it often occurs near *A. pelona,* another narrow endemic found predominantly on limestone cliffs. Cultivated plants are cold hardy (Gentry 1982), so it seems unlikely that low temperatures limit its present distribution. Records from the nearest weather stations indicate biseasonal rainfall with the highest amounts in summer and a secondary peak in early winter. Since this climatic regime is typical of much of the Sonoran Desert, it seems likely that the species evolved in response to substrate rather than climate, especially since limestone is uncommon in the Sonoran Desert. From its foothold on limestone massifs, *A. zebra* evidently endured major shifts in climate and vegetation through the Pleistocene and Holocene in a manner similar to many endemic succulents of the Chihuahuan Desert (Van Devender and Burgess 1985).

Flowers appear from June through August (Gentry 1982).

The plants are worth greater attention as ornamentals.

Albizzia occidentalis

Brandegee

Arellano, bolillo, cico, palo fierro, trucha, palo escopeta

A tree up to 15 m tall, *Albizzia occidentalis* (figure 21) has smooth gray bark and bipinnate leaves. Each of the 2–4 pairs of pinnae has 2–6 pairs of leaflets 1.5–5 cm long. The terminal leaflets are usually the largest. The yellowish white flowers are in heads. Fruits are flat, 3–4 cm wide and 15–30 cm long.

Albizzia sinaloensis, known only from southern Sonora and Sinaloa, has smaller leaflets and more of them, and it has yellowish rather than gray bark.

This tree grows on mountain slopes and along washes in Baja California Sur, Sinaloa, and Nayarit. The disjunction between Sinaloa and the Cape Region is unusual and suggests that competitive interactions with *A. sinaloensis* may have kept *A. occidentalis* from expanding into Sonora. Aridity determines the northern peninsular limit.

Flowers appear in May and June. The fruits persist into the winter.

Figure 21. Albizzia occidentalis *with pendant fruits near Miraflores, Baja California Sur. (Photograph by J. R. Hastings.)*

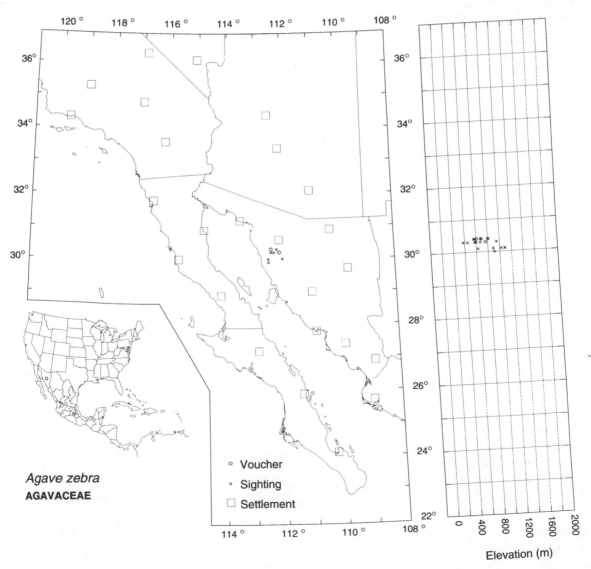

120° 118° 116° 114° 112° 110° 108°

36° 34° 32° 30° 28° 26° 24° 22°

114° 112° 110° 108°

Agave zebra

AGAVACEAE

◦ Voucher

× Sighting

□ Settlement

0 400 800 1200 1600 2000

Elevation (m)

Albizzia sinaloensis

Britton & Rose

Palo joso

Albizzia sinaloensis is a striking tree up to 20 m tall with smooth yellowish bark and twice-pinnate leaves with 4–9 pairs of pinnae. The leathery leaflets, 12–16 pairs per pinna, are 10–15 mm long and 2–5 mm wide. They are glabrous or nearly so above, finely puberulent below. The fruits are broadly linear pods up to 2.5 cm wide and 15 cm long.

This tree is known only from southern Sonora and Sinaloa. It grows along arroyo margins, especially where water flows at the surface or shallowly below (Gentry 1942). Our sparse distributional records suggest a requirement for substantial summer rainfall and warm winters.

The flowers appear in the spring as the leaves of the previous year are dropping. The fruits ripen and dehisce during the winter.

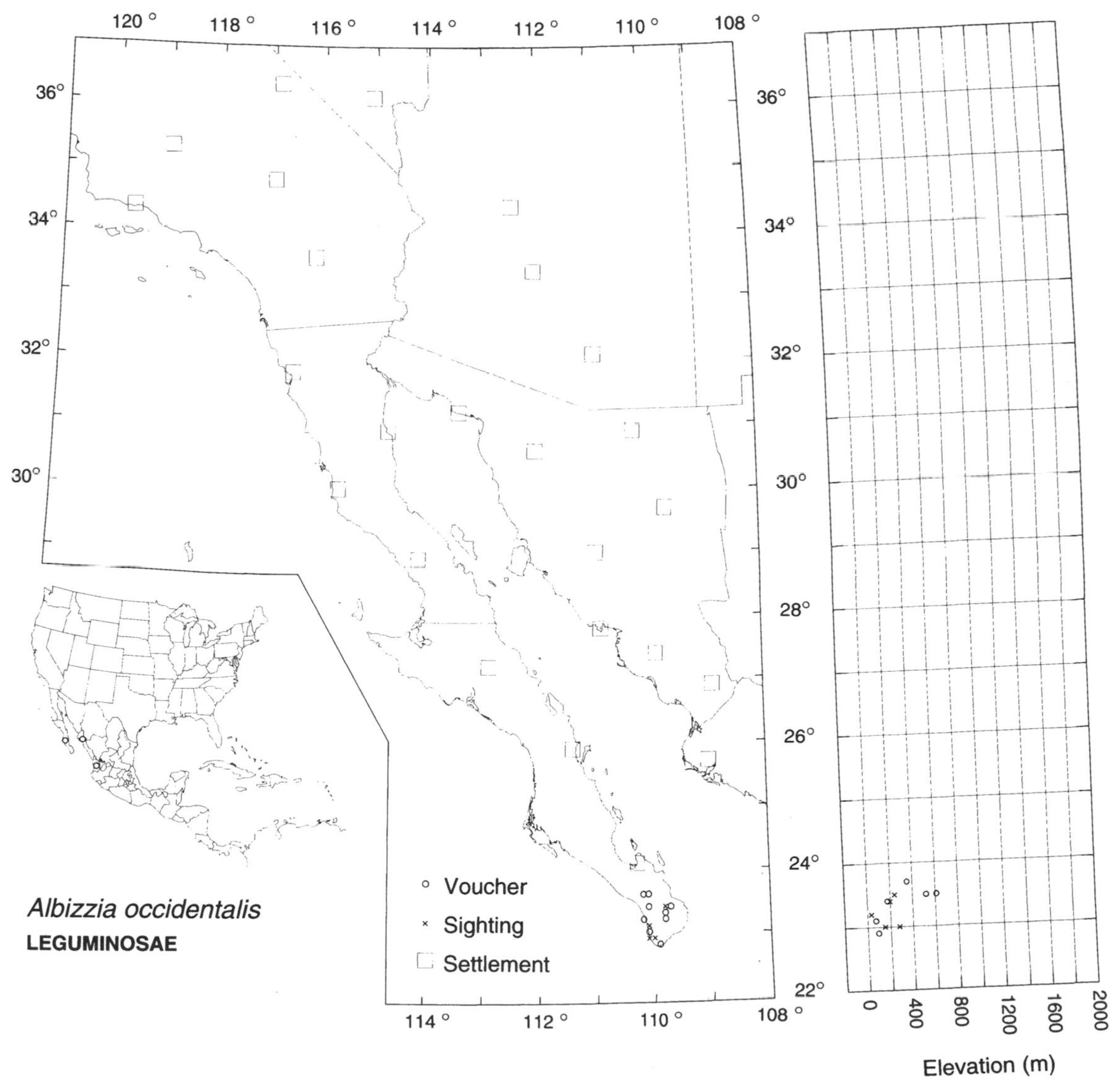

Albizzia occidentalis
LEGUMINOSAE

Alhagi maurorum
Medikus

[=*A. camelorum* Fisch.]

A low shrub up to 1 m tall, *Alhagi maurorum* has many rigid, spiny stems from extensive, deep-seated rhizomes. The simple, glabrous leaves are elliptic to oblanceolate and 8–20 mm long. The red to red-purple corollas, 7–9 mm long, are in 4- to 6-flowered axillary racemes. Fruits are flat, constricted pods 1–3 cm long. The diploid chromosome number is 16 (Lessani and Chariat-Panahi 1979).

The rhizomatous habit, simple leaves, green spines, and glabrous foliage set this species apart from other spiny shrubs in the Sonoran Desert region.

Arizona collections have usually been referred to as *A. camelorum* (Kearney and Peebles 1960). Davis (1970) reduced *A. camelorum* and *A. persarum* Boiss. & Buhse to synonymy under *A. pseudalhagi* (Bieb.) Desv. We follow Barneby (1989) in treating Arizona material as *A. maurorum.*

A native of Asia, *A. maurorum* has been sparingly introduced into California, Arizona, and Texas, where it is occasional in cultivated fields, on sandy plains, and on dry, rocky flats. In Arizona, *A. maurorum* is most abundant along the banks and bottomlands of the Little Colorado and Salt rivers (Parker 1972). Since construction of Glen Canyon Dam in 1963, the species has become established along the Colorado River in the Grand Canyon (Turner and Karpiscak 1980). Once established, *A. maurorum* is invasive and difficult to eradicate. Fewer than 20 years were required for it to occupy 24 km of canal near Gillespie Dam, Maricopa County, Arizona (Parker 1972).

Flowers appear from May to July; pods persist until October or November (Parker 1972).

Extending 0.75 to 1.5 m below the soil surface, the root system often taps a zone of moist soil, enabling plants to withstand drought (Ambasht 1963; Parker 1972). Seeds are dormant upon ripening. Germi-

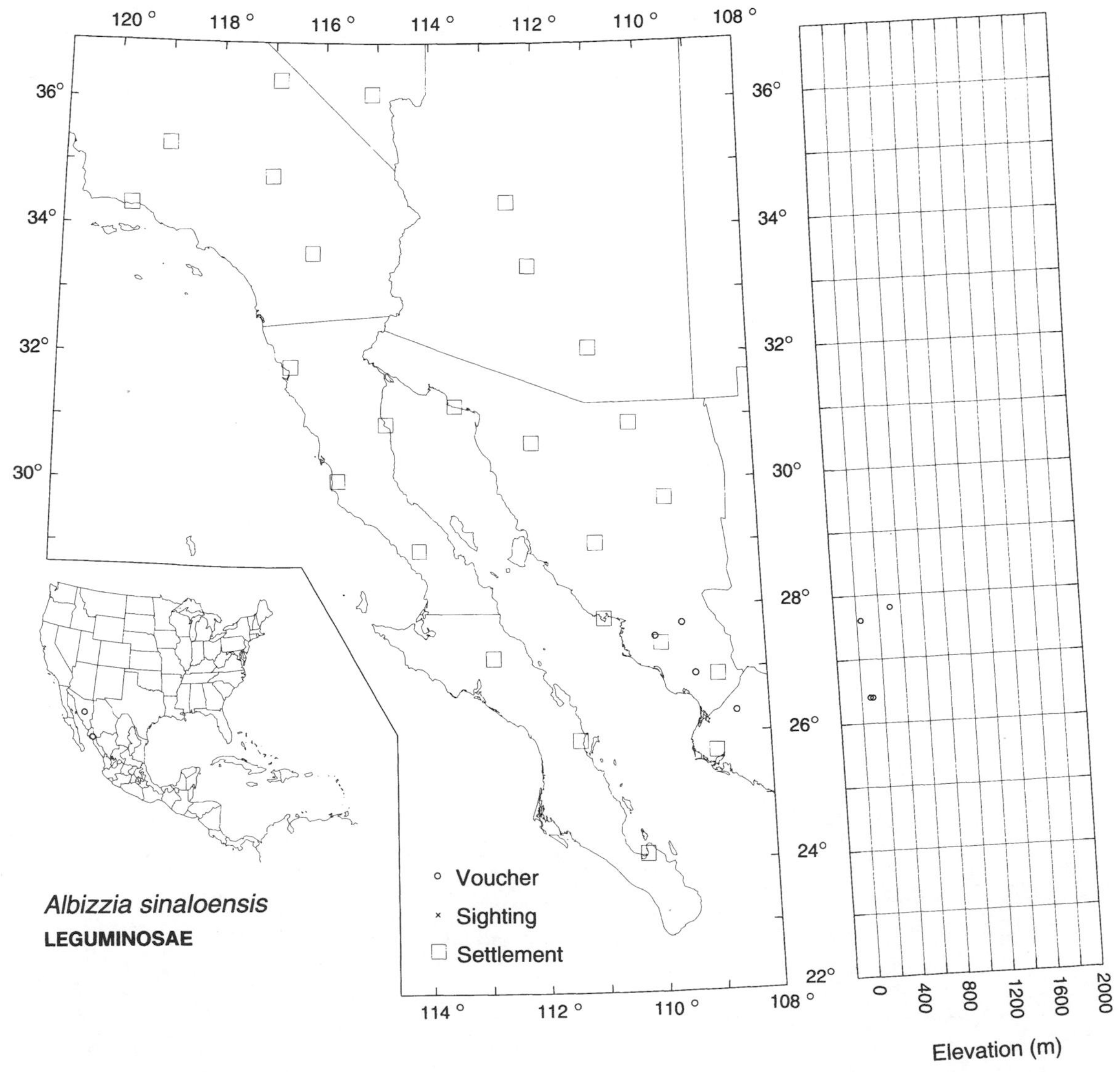

Albizzia sinaloensis
LEGUMINOSAE

nation is 4% five months after seeds are shed and rises to 25% subsequently (Ambasht 1963). In India, most reproduction is vegetative (Ambasht 1963). Clones enlarge as the deep-seated rhizomes branch and produce new shoots (Ambasht 1963; Parker 1972). Tilling the soil breaks up the rhizomes and fosters the spread of the plant, since each piece of root is capable of producing buds and new shoots (Ambasht 1963).

In India, camels and goats browse the plants (Ambasht 1963). Nomads in the deserts of the Near East use stem exudates in preparing sweetmeats (Barneby 1989).

Aloysia wrightii

(A. Gray) Heller

[=*Lippia wrightii* A. Gray]

Wright lippia, bee brush, oreganillo, altamisa, vara dulce

Aloysia wrightii is an aromatic, diffusely branched shrub up to 1.5 m tall with sharply 4-angled stems. The opposite, crenate leaves, 2–15 mm long and 2–13 mm wide, are ovate to orbicular. They are deeply veined on the upper surface and woolly with kinky hairs on the underside. The small, white flowers, about 4 mm long, are in dense, elongate spikes 2–7 cm long and 4–5 mm in diameter. The fruits are nutlets.

Aloysia lycioides Cham. has oblong or elliptic leaves 0.5–1.8 cm long and 3–7 mm broad. They are entire or few-toothed.

A plant of rocky or gravelly slopes and canyons, *A. wrightii* can be locally dominant and is often associated with calcareous substrates. Populations are disjunct on isolated mountain ranges. In California it is known from the Little San Bernardino, Clark, and Providence mountains (Munz 1974); in Utah from Washington County (Welsh et al. 1987).

Over most of its range, *A. wrightii* grows in warm-temperate regimes where winter temperatures regularly fall below freezing. In the northeastern part of the

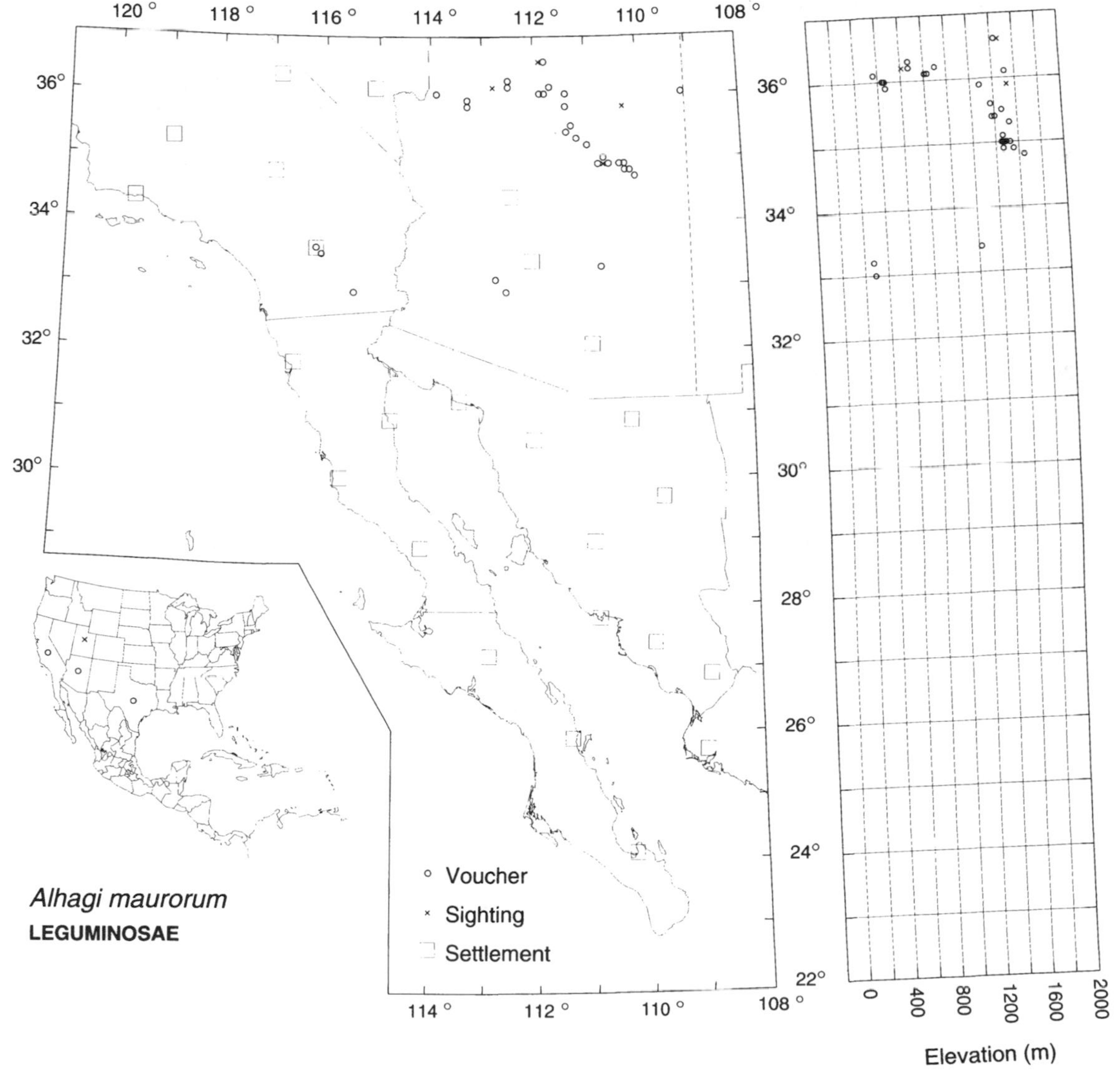

Alhagi maurorum
LEGUMINOSAE

Sonoran Desert, it experiences biseasonal rainfall with peaks in winter and summer. The westerly outposts probably occur where orographically induced summer storms are relatively dependable. At the drier end of its climatic range, *A. wrightii* is most abundant on north-facing slopes.

Flowers appear in spring (March–May), in summer (July–October), and at other times of year in response to rain. Leaves are drought deciduous.

The plants may live 72 years or more (Goldberg and Turner 1986).

Livestock browse the foliage. The flowers are a good source of nectar for honeybees (Kearney and Peebles 1960). *Aloysia wrightii* is occasionally grown as an ornamental in southern Arizona. Fresh seeds germinate without pretreatment (Nokes 1986). Plants can also be propagated from softwood tip cuttings taken in spring or early summer (Nokes 1986).

Alvaradoa amorphoides Liebm.

Belzinic-che, camarón, guachipil, palo de hormiga, pinkwing, Mexican alvaradoa

A shrub or small tree up to 5 m tall, *Alvaradoa amorphoides* has pinnately compound leaves 1–2 dm long. The 17–51 leaflets, 0.5–2 cm long, are alternate on the yellowish orange rachis. Their upper surface is dark green and glabrous to sparsely puberulent; the lower surface is pale green and finely appressed-puberulent. The dioecious flowers lack petals and are crowded in slender racemes.

When flowers and fruits are missing,

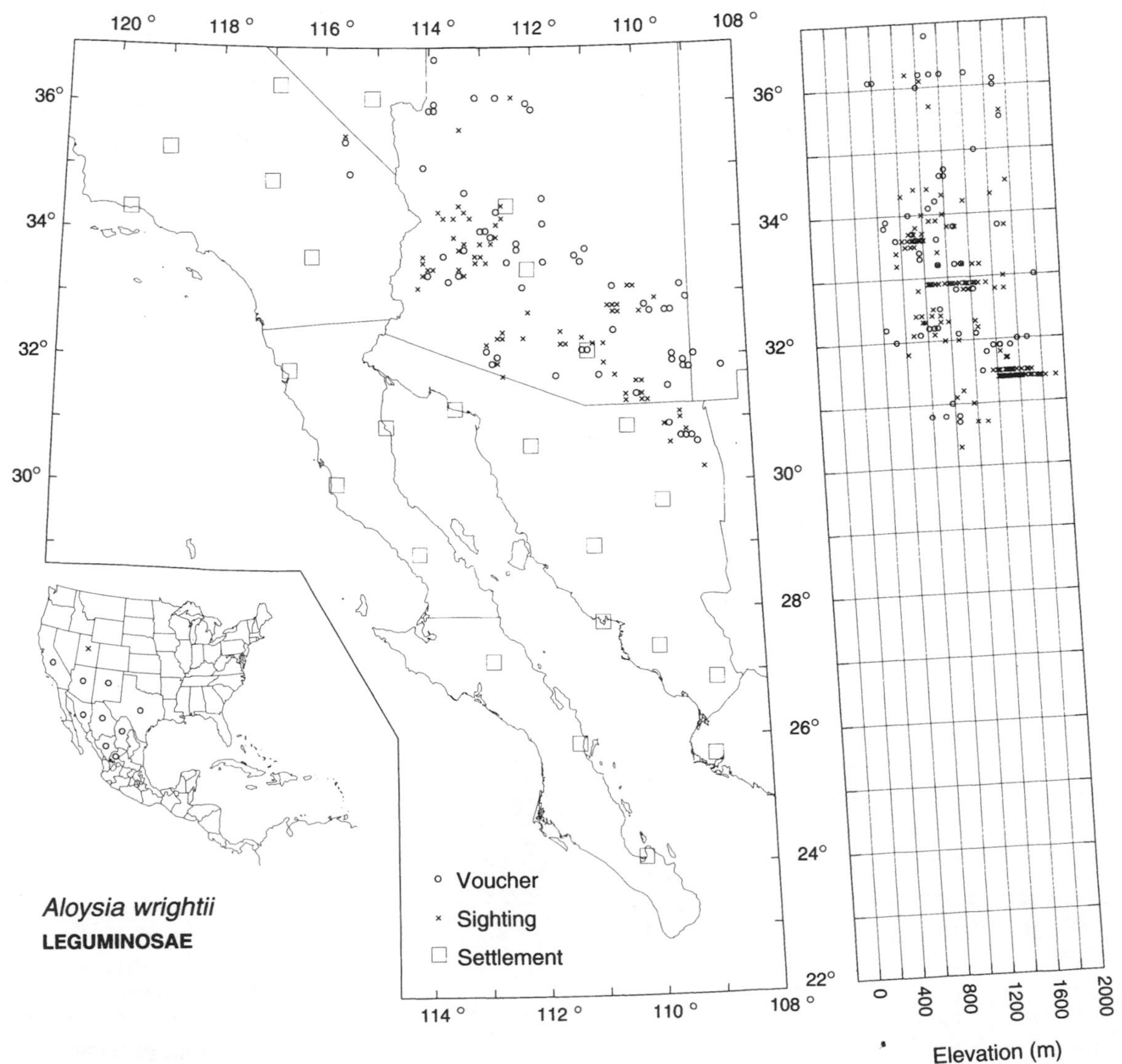

A. amorphoides may be easily mistaken for a legume.

This is a plant of rocky arroyo banks and hillslopes. The northernmost locality is a collection by Steven P. McLaughlin from northwest of Cumpas, Sonora.

Soto de Villatoro and coauthors (1974) and Pearl and coauthors (1973) reported on the chemistry of this species. Vales and Martinez (1983), Babos (1979), and Babos and Borhidi (1978) studied its anatomy.

Ambrosia ambrosioides (Cav.) Payne

[=*Franseria ambrosioides* Cav.]

Canyon ragweed, chicura

This shrub (1–2 m tall) has elongate leaves 4–18 cm long and 1.5–4 (occasionally 6) cm wide (figure 22). They are lanceolate-attenuate and coarsely toothed. Plants are monoecious, with staminate heads above the pistillate in terminal or axillary racemes. Fruits are burs 10–15 mm long armed with numerous hooked spines. The haploid chromosome number is 18 (Payne 1964).

Leaves of *A. ilicifolia* are sessile and conspicuously reticulate veined. Their marginal teeth are attenuated into short spines.

Payne (1964) allied *A. ambrosioides* with *A. ilicifolia*, *A. deltoidea*, and *A. chenopodiifolia*, regarding them as a small group of derivative species with unlobed leaves, glandular pubescence, and cocklebur-like fruits (Payne 1964). Seaman and Mabry (1979a), however, based upon sesquiterpene lactone chemistry, eliminated *A. ilicifolia* from this group and added *A. cordifolia*, *A. eriocentra*, and *A. dumosa*.

Ambrosia ambrosioides is the most widely distributed shrubby ambrosia in the area mapped in this atlas. It is locally common in sandy washes and disturbed areas such as roadsides and borrow pits

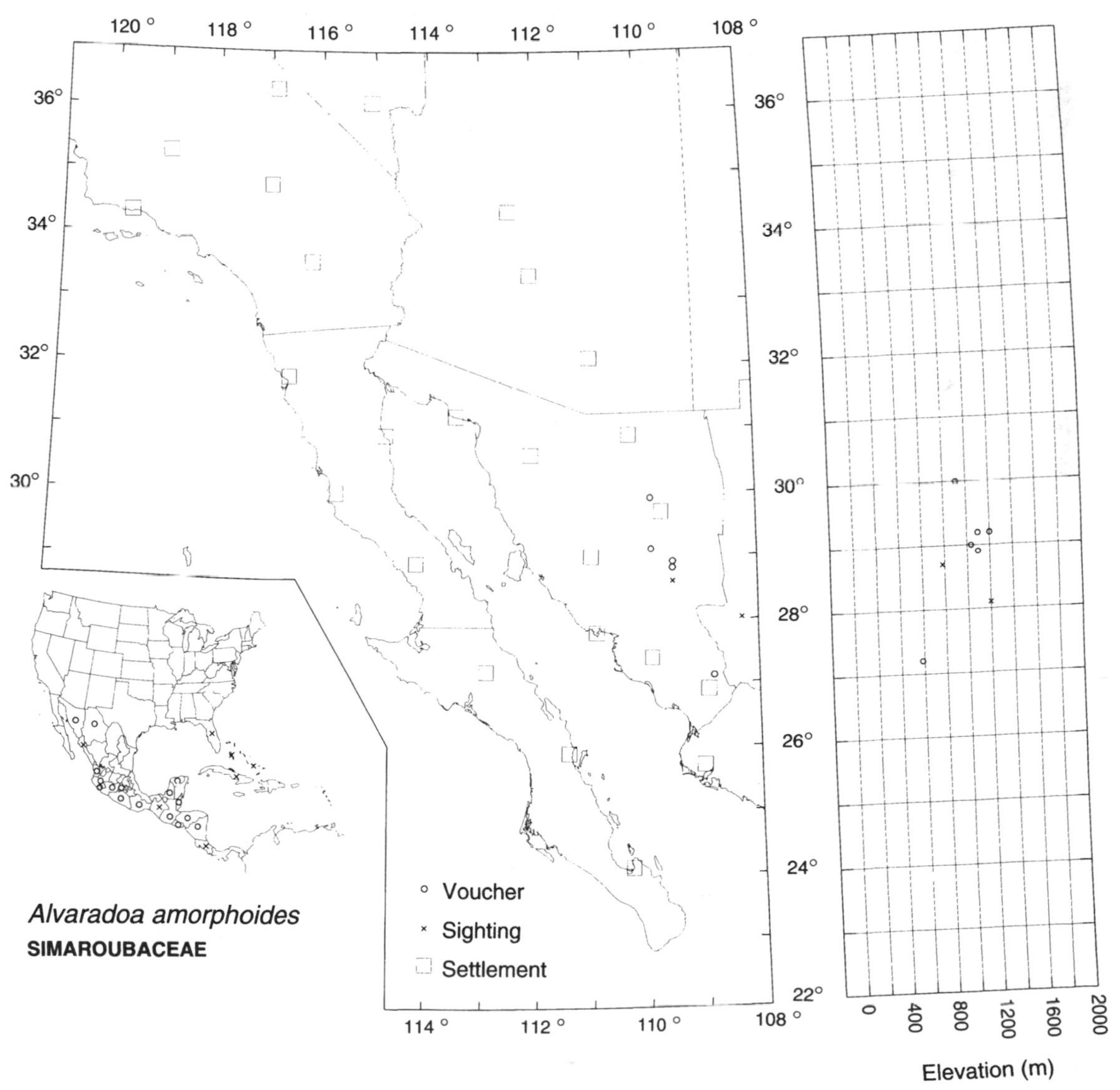

Alvaradoa amorphoides
SIMAROUBACEAE

and is occasional in rock crevices. We have seen no vouchers from California, although Munz (1974) noted *A. ambrosioides* from waste places near San Diego. Beauchamp (1986) did not include it in his flora of San Diego County.

Frost likely controls the northern and upper elevational limits: the stems freeze to the ground at –6.6°C (Anonymous 1979; Bowers 1980–81) but rapidly regrow from the root crown (Gentry 1942). The distribution indicates a need for periodic warm-season flooding. Over most of the range,

Figure 22. Ambrosia ambrosioides *along the Río Sonora near Sinoquipe. The large, evergreen leaves suggest the need for riparian habitats. (Photograph by J. R. Hastings.)*

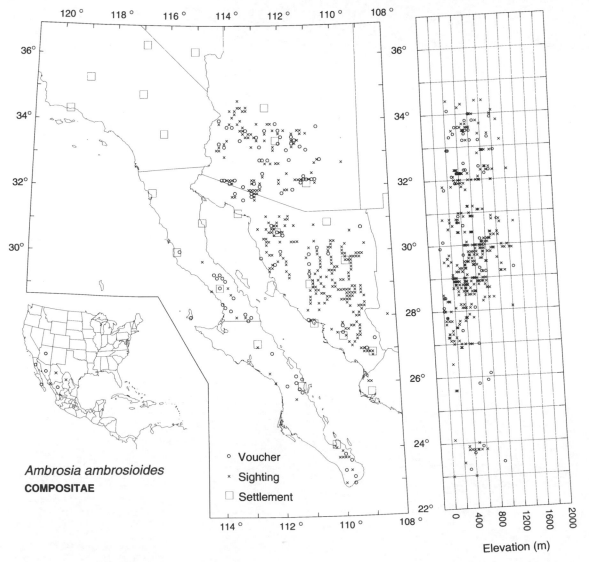

Ambrosia ambrosioides
COMPOSITAE

○ Voucher
× Sighting
□ Settlement

Elevation (m)

rainfall is biseasonal with peaks in winter and late summer to early fall.

Nominally evergreen, the leaves are shed after severe frost or during prolonged drought. Flowers appear mainly in spring (February–April), occasionally at other times of year. They are wind pollinated and can be a powerful allergen to susceptible persons. The pollen is often reported as "rabbitbrush" or "false ragweed" (Mary Kay O'Rourke, personal communication 1993).

The Seri smoked the dried leaves and made pigment and medicinal tea from the roots (Felger and Moser 1985).

Ambrosia bryantii

(Curran) Payne

[=*Franseria bryantii* Curran]

A low, spreading subshrub 2–5 dm tall, *Ambrosia bryantii* (figure 23) has white, nearly glabrous branchlets and dark green foliage. The ovate to oblong leaves, 2–3.5 cm long and 1–2.5 cm wide, are bi- or tripinnatifid. Staminate heads form a short raceme above the solitary pistillate head. Fruits are burs 5–6 mm long, armed with 4–7 terete, white spines 1.5–3 cm long. The haploid chromosome number is 18 (Payne et al. 1964).

No other *Ambrosia* species in the Sonoran Desert has long, white spines on the fruits.

Endemic to the Baja California peninsula, *A. bryantii* is locally common on hillsides, plains, mesas, and arroyos. The southern outlier (23.7°N) is based on Raymond M. Turner's 1964 sighting. The northern limit ends abruptly at 30°N. In the northwestern part of the range, winter is the season of reliable moisture, and the plants are restricted to coastal slopes where an influx of maritime air may ameliorate summer moisture stress. In the central and southern parts of its range, *A. bryantii* experiences biseasonal rainfall with a reliable peak in late summer or early autumn and a secondary, less predictable, pulse in winter.

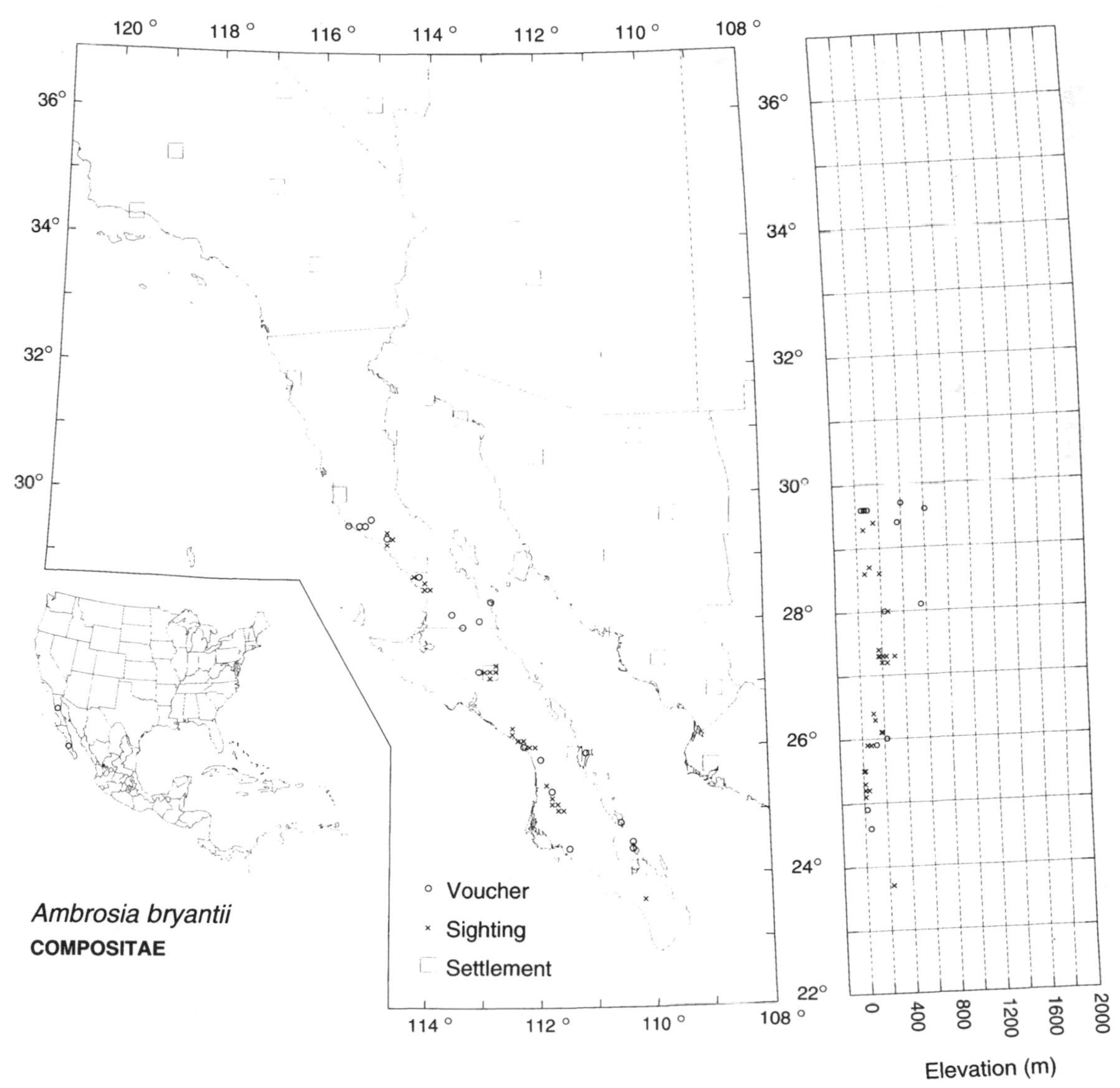

Ambrosia bryantii
COMPOSITAE

Figure 23. Ambrosia bryantii *is the only member of the genus with long spines on persistent fruits. (Photograph by J. R. Hastings taken south of Cataviñá, Baja California.)*

Payne (1964) assigned *A. bryantii* to a core group of species including *A. dumosa, A. camphorata,* and *A. magdalenae.* Based on sesquiterpene lactone chemistry, Seaman and Mabry (1979a) placed *A. bryantii* in the same group as *A. chamissonis* Less., *A. ilicifolia, A. camphorata, A. divaricata,* and *A. magdalenae.*

The wind-pollinated flowers appear from January to April. The fruits are unique in the genus: not only do they have unusually long, stiff spines, but they persist on the branches indefinitely, effectively arming the plants (Payne 1962). On old herbarium specimens some persistent burs contain apparently fertile achenes. If these persistent achenes remain viable, the plants would in essence have a dual dispersal mode in which some seeds fall immediately and others only much later (see Ellner and Shmida 1981 for analogous examples from Old World deserts).

Ambrosia bryantii has potential as an ornamental. Plants transplant easily, especially when dormant, if top growth is trimmed heavily.

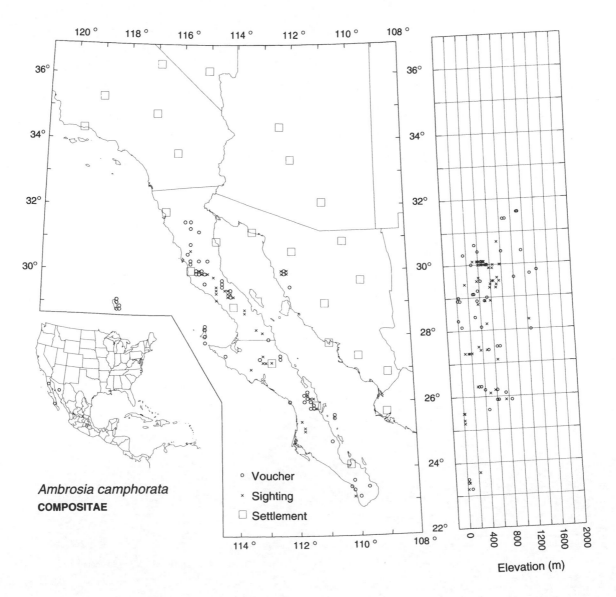

Ambrosia camphorata
COMPOSITAE

○ Voucher
× Sighting
□ Settlement

Elevation (m)

Ambrosia camphorata

(E. Greene) Payne

[=*Franseria camphorata* E. Greene]

Estafiate

This subshrub, woody only at the base, grows up to 6 dm tall. The bipinnatifid leaves, ovate to triangular in outline, are 1.5–3.5 cm wide and 3–8 cm long and have a distinct camphorous odor when crushed. They are white-tomentose beneath and floccose but green above. Staminate heads are in terminal racemes or panicles above the pistillate heads. In northern and Cape Region populations, the burlike, obovoid fruits are 6–7 mm long with 6–10 spines. Most southern populations have fruits 7–16 mm long with 25–60 spines (Seaman and Mabry 1979b). The haploid chromosome number is 18 for diploid populations, 36 for tetraploid forms (Payne et al. 1964; Seaman and Mabry 1979b).

Leaves of *A. dumosa* are white-tomentose above and below. *Ambrosia bryantii* is conspicuously armed with white spines. *Ambrosia magdalenae* leaves are green or only faintly cinereous above. The spines on its burlike fruits are hooked at the apex rather than straight as in *A. camphorata.*

Payne (1964) suggested that *A. camphorata* may have evolved from the less specialized *A. dumosa,* perhaps in reproductive isolation, but based on sesquiterpene lactone chemistry, Seaman and Mabry (1979a) aligned *A. camphorata* with *A. chamissonis, A. bryantii, A. ilicifolia, A. divaricata,* and *A. magdalenae.* Variations in ploidy level, fruit characters, and sesquiterpene lactones indicate considerable evolutionary divergence within this taxon, which may include at least two sibling species (Seaman and Mabry 1979b).

Locally abundant in heavy clay soil, *A. camphorata* is widely distributed on the peninsula and scattered on the mainland. Willard Payne found it as far south as San Luis Potosí (Payne et al. 1964). In keeping with its infraspecific variability, *A. camphorata* grows under a variety of rainfall

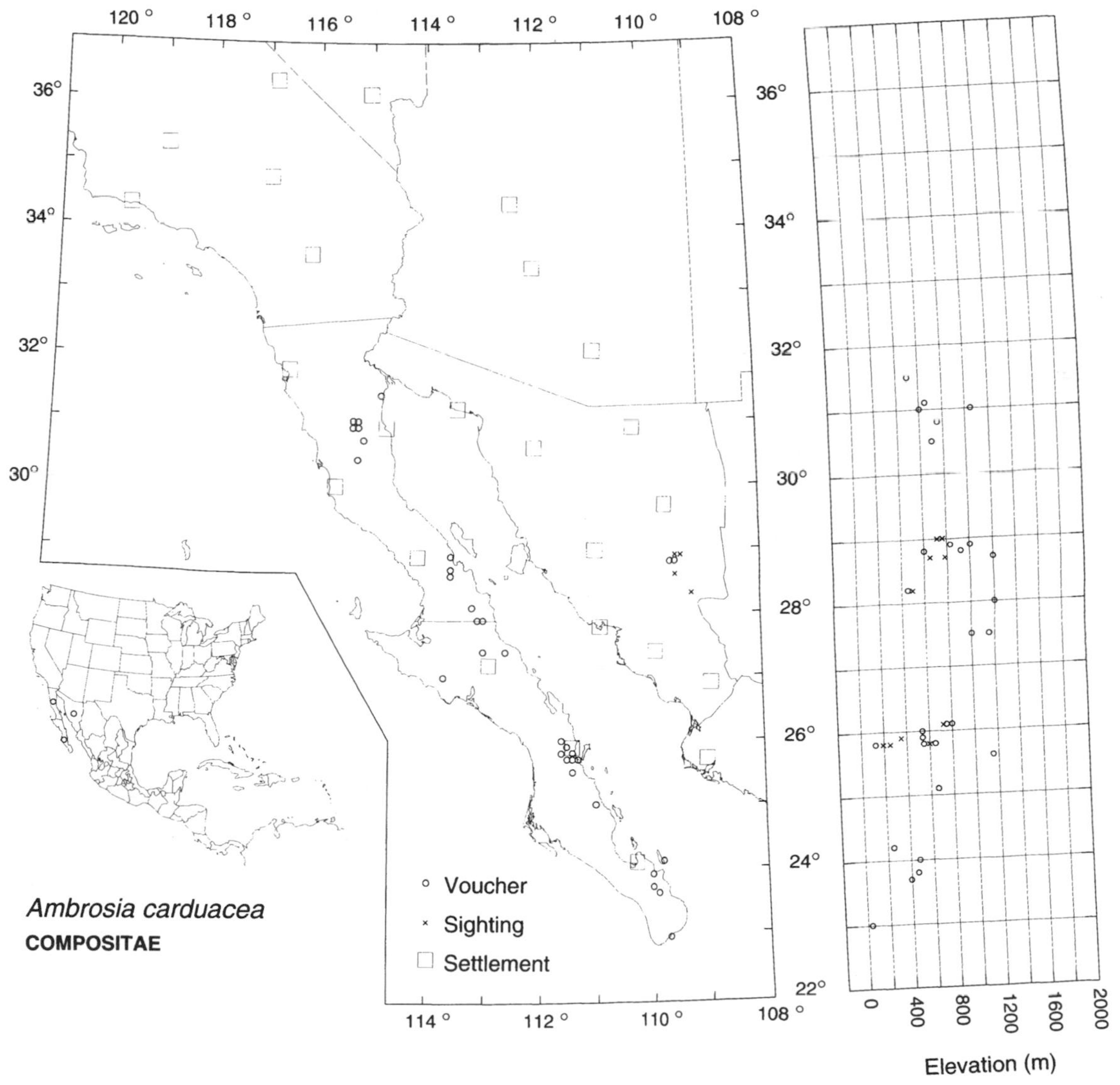

Ambrosia carduacea
COMPOSITAE

regimes, from winter rains in the north through biseasonal rainfall at midpeninsula to predominantly autumn rains in the Cape Region. The disjunct populations on the Sonoran mainland at 30.1°N and 30.2°N have been known to botanists since at least 1923. The point at 29.6°N, 112.2°W, represents a 1968 collection by Richard Felger. Four other shrubs in the genus also have disjunct Sonoran colonies: *A. chenopodiifolia, A. divaricata, A. magdalenae,* and *A. carduacea.*

Flowers are wind pollinated and appear from October to May. Leaves are drought deciduous.

Ambrosia carduacea

(E. Greene) Payne

[=*Franseria arborescens* Brandegee]

Ambrosia carduacea is a shrub or small tree up to 5 m tall. The ovate to lance-ovate leaves, deeply cleft into 3–7 coarsely serrate lobes, are 5–8 cm wide and 8–22 cm long. They are sparsely puberulent above, more densely so beneath. The inflorescence is a panicle with staminate heads above the pistillate. The fruits are elliptic burs 8–9 mm high armed with 7–12 subulate, hooked spines.

Its arborescent habit and large, pinnatifid leaves make *A. carduacea* readily recognizable on the Baja California peninsula. In Sonora, the similar shrub *Parthenium tomentosum* var. *stramonium* has entire to weakly repand, not deeply cleft, leaves.

Ambrosia carduacea is a species complex that shows considerable variation in leaf morphology (Payne 1964).

The disjunct points in Sonora represent six different collections and sightings made between 1965 and 1975. The two collections are Elinor Lehto's and Raymond M. Turner's. All the sightings are Turner's.

Occasional to common in canyons and along washes, this species demonstrates a wide ecological amplitude, growing from near sea level to 1,200 m and from Sonoran

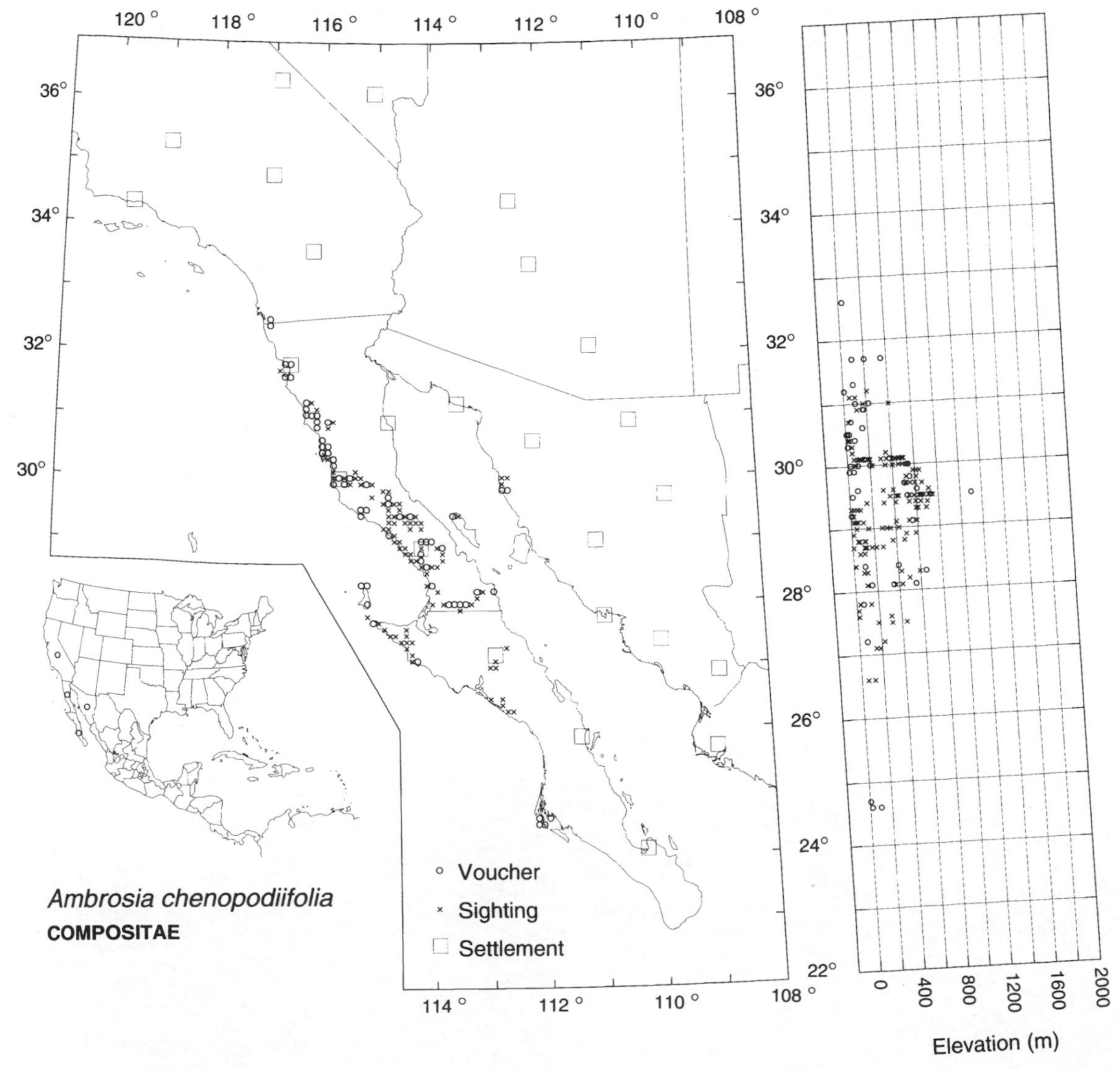

Ambrosia chenopodiifolia
COMPOSITAE

desertscrub to Sinaloan thornscrub. Several different ecotypes may be involved.

The wind-pollinated flowers appear from October to February.

Ambrosia chenopodiifolia

(Benth.) Payne

[=*Franseria chenopodiifolia* Benth.]

This *Ambrosia* species is a compact, rounded shrub 2–4 dm tall. Densely white-tomentose when young, the ovate, dentate leaves become glabrate in age and are 0.5–2 (occasionally 3) cm wide and 1–3 (occasionally 5) cm long. The spikelike or paniculate inflorescence has crowded staminate heads above the pistillate ones. Fruits are densely villous, eglandular burs 5–6 mm long, armed with 20–30 terete, uncinate spines. The haploid chromosome number is 36 (Payne et al. 1964).

Ambrosia deltoidea fruits are sparsely villous and distinctly glandular; otherwise, the plant looks very much like *A. chenopodiifolia*. (Rarely, *A. deltoidea* populations in Arizona contain aberrant individuals that key to *A. chenopodiifolia* [Thomas R. Van Devender, personal communication 1992].) The two may be chromosome races of one species, with *A. deltoidea* being diploid, *A. chenopodiifolia* tetraploid. Seaman and Mabry (1979a) suggested that *A. chenopodiifolia* may have arisen by polyploidy from *A. deltoidea*. According to Payne (1964), both belong to a small, derivative complex that also includes *A. eriocentra, A. ambrosioides,* and *A. ilicifolia*. Sesquiterpene lactone chemistry confirmed this relationship except for *A. ilicifolia* (Seaman and Mabry 1979a).

Often a dominant on hillsides, slopes, plains, and mesas, *A. chenopodiifolia* essentially replaces *A. deltoidea* on the Baja California peninsula. They occupy similar habitats and appear to fill the same niche. *Ambrosia deltoidea* is most abundant in areas of biseasonal precipitation where summers are wet and hot, whereas *A. chenopodiifolia* is limited to a predominantly winter rainfall climate in which summers are cooler and drier. Both produce flowers and new leaves in response to winter rains. In certain coastal localities south of Puerto Libertad, *A. chenopodiifolia* may be more abundant than *A. deltoidea*. Spread of the few Sonoran populations may be inhibited by competition with *A. deltoidea* (Cody et al. 1983). The scattered occurrences may also represent the remnants of a more widespread Pleistocene distribution.

Toward the northern end of its range, *A. chenopodiifolia* becomes restricted to the most xeric coastal areas, probably where fires are infrequent. *Bergerocactus emoryi* and *Agave shawii* have a similar pattern. *Ambrosia chenopodiifolia* is also restricted to the coast toward the south, perhaps showing a need for relatively dry summers with ameliorated temperatures.

The wind-pollinated flowers appear from January to May (occasionally as early as October and as late as June). Leaves are drought deciduous. *Ambrosia chenopodiifolia* sometimes acts as a nurse plant for cacti, especially *Mammillaria* and *Ferocactus* species, and for *Fouquieria columnaris* (Humphrey 1974). Goeden and Ricker (1976a) reviewed insect herbivores of the species.

Ambrosia cordifolia

(A. Gray) Payne

[=*Franseria cordifolia* A. Gray]

Chicurilla

This woody-based perennial has several stems up to 7–10 dm tall. The ovate-cordate leaves are 1–5 cm broad and 1.5–6.5 cm long with crenate margins. The underside is densely tomentulose, the upper side scaberulous or glabrate in age. Staminate heads are above the pistillate in terminal racemes up to 1.5 dm long. Fruits are hardened burs 6–8 mm long with 8–15 conical, hooked spines. The haploid number of chromosomes is 18 (Seaman and Mabry 1979a).

The ovate-cordate leaves distinguish this from other species in the genus. Various genera in the Malvaceae (for example, *Horsfordia, Abutilon, Sida*) with similar cordate leaves have stellate pubescence, showy flowers, and dehiscent capsules setting them apart from *A. cordifolia*.

Sesquiterpene lactone chemistry shows *A. cordifolia* belonging to the same group as *A. deltoidea, A. chenopodiifolia, A. ambrosioides, A. eriocentra,* and *A. dumosa* (Seaman and Mabry 1979a). Wollenweber and coauthors (1987) suggested that flavonoid patterns within the genus may be unique for each species.

Ambrosia cordifolia can be locally abundant in canyons and arroyos and on rocky slopes, often in the shade of larger shrubs and trees. It is most abundant in southern Sonora and northern Sinaloa, where it is an important understory shrub in open forests of *Ipomoea arborescens* and *Acacia cochliacantha* (Gentry 1942). The northernmost points on our map (33.5°N and 33.6°N) represent a sighting by Robert A. Darrow and a collection by Robert H. Peebles near Horse Mesa Dam, Maricopa County, Arizona. Although Standley (1926) reported this species from San Luis Potosí, and Payne (1964) recorded it from the Baja California peninsula, neither author cited vouchers, nor have we seen any.

At the northern end of its range, populations are disjunct on isolated mountain ranges. Farther south, the distribution is more continuous. Over most of its range, rainfall is concentrated in summer and early fall. In the absence of frost, *A. cordifolia* can grow in response to winter rains at a time when associated canopy species may be nearly leafless. The distribution may be influenced by the southward penetration of winter storms. Low temperatures probably limit the species to the north; near Tucson, Arizona, the plants freeze to the ground in the coldest winters. Aridity doubtless also checks the northward movement of the species.

Leaves are cold- and drought-deciduous. Flowers are wind pollinated and appear from December to April (rarely as early as November or as late as May).

In the Río Mayo area, *A. cordifolia* is good pasturage for livestock (Gentry 1942).

Ambrosia deltoidea

(Torr.) Payne

[=*Franseria deltoidea* Torr.]

Triangle-leaf bursage

This low shrub or suffrutescent perennial grows 3–6 dm tall. The narrowly triangular leaves, 1.5–3.2 cm long and 0.4–1.5 cm wide, are truncate to cuneate at the base and serrate to crenate along the margins. When young, they are densely woolly above and below, but the upper surface often becomes glabrate in age. Staminate heads are above the pistillate in terminal panicles or racemes. Fruits are ellipsoid or spherical burs with 15–30 flattened, straight spines. The fruit surface is sparsely villous and distinctly glandular. The haploid chromosome number is 18 (Payne et al. 1964).

The eglandular fruits of the very similar *A. chenopodiifolia* are densely lanate and possess terete, uncinate spines. It is difficult to distinguish *A. deltoidea* from *A. chenopodiifolia* without fruits, and even then, intermediate forms can be impossible to identify (Payne 1964). These closely similar taxa might be chromosome races of a single species. *Ambrosia eriocentra* is a taller shrub (up to 1 m) with pinnately lobed or incised leaves.

Ambrosia deltoidea belongs to a small, derivative complex of the Mojave and Sonoran deserts (Payne 1964). Other mem-

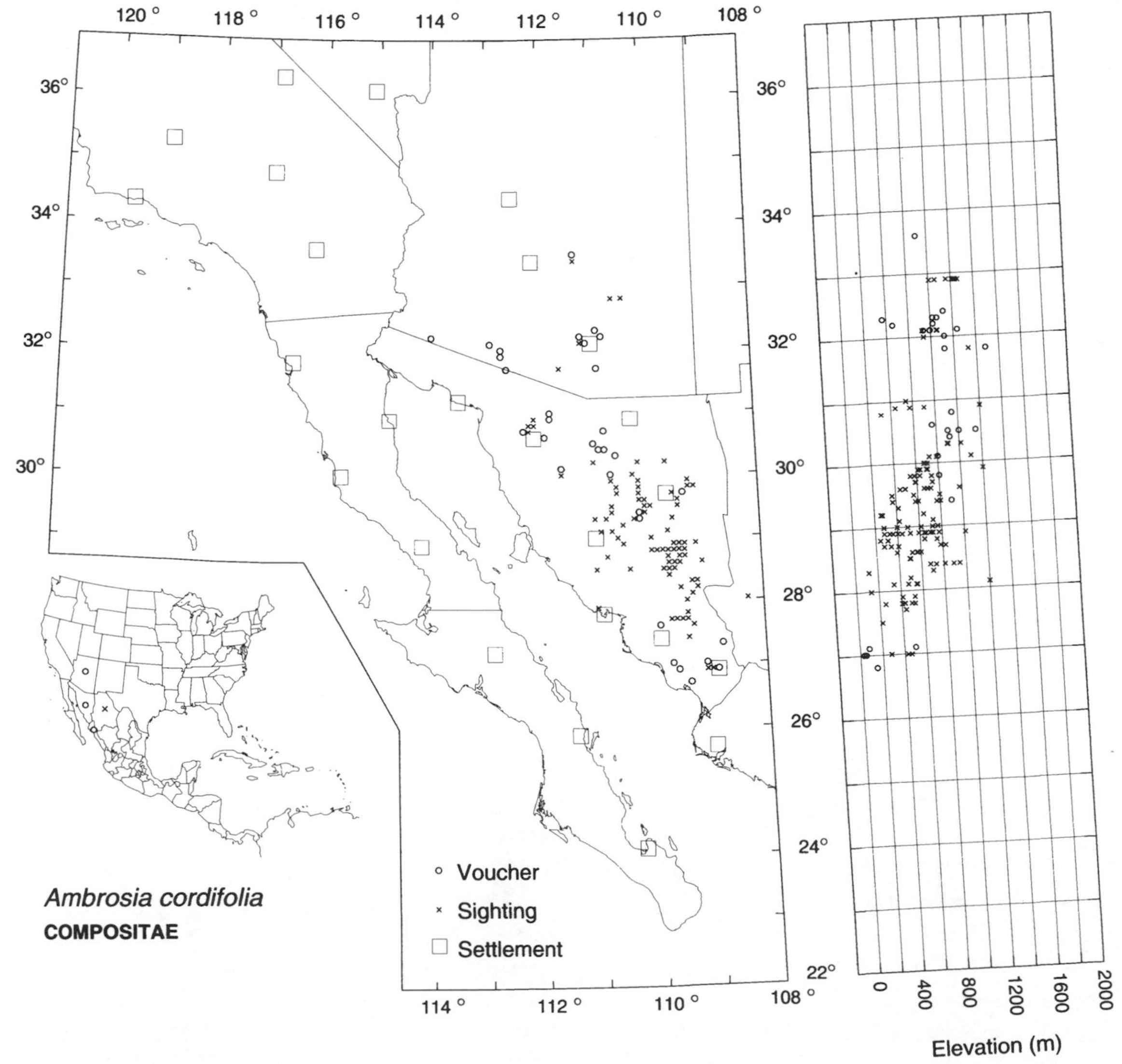

Ambrosia cordifolia
COMPOSITAE

bers include *A. chenopodiifolia, A. ambrosioides,* and *A. eriocentra* (Payne 1964; Seaman and Mabry 1979a). Wollenweber and coauthors (1987) discussed taxonomic relationships as revealed by flavonoid chemistry.

Ambrosia deltoidea is often regarded as the dominant bursage of the Arizona Upland, in contrast to *A. dumosa,* which is often characterized as a species of the Lower Colorado Valley and the Mojave Desert (for example, Benson and Darrow 1981). In fact, the zone of overlap between the two is rather large. Where they occur together, *A. deltoidea* typically occupies relatively moist runnels of the plains or rocky, upland sites, and *A. dumosa* is confined to the finer, drier soils of the lower bajadas.

The range of *A. deltoidea* is largely in Arizona and northwestern Sonora. We show one location in Baja California, based on a 1935 collection by Forrest Shreve near El Arco, and four in Baja California Sur near Santo Domingo. All appear to have been correctly identified.

Both distribution and phenology indicate that *A. deltoidea* depends on biseasonal rainfall. Its western limit does not extend beyond the area of dependable summer rain. Cold helps to shape the northern and upper elevational limits; competitive interactions with *A. eriocentra* may also have some influence. At its southern mesic limits, *A. deltoidea* is replaced by *A. cordifolia* where Sonoran desertscrub gives way to denser vegetation.

The flowering period is February–April; occasionally, some populations may bloom as early as December or as late as May. A cool-season rain trigger of at least 9 mm is required for flowering. Thereafter, degree-days above the base temperature (10°C) must reach about 472 for plants to bloom (Bowers and Dimmitt 1994). The flowers are wind pollinated. The shallowly rooted plants are sensitive to fluctuating soil moisture and commonly lose most of their leaves during the arid foresummer and in the dry period after summer rains. Renewed leafing after winter rains is relatively rapid; on Tumamoc Hill, Tucson, Arizona, cover of *A. deltoidea* doubled fol-

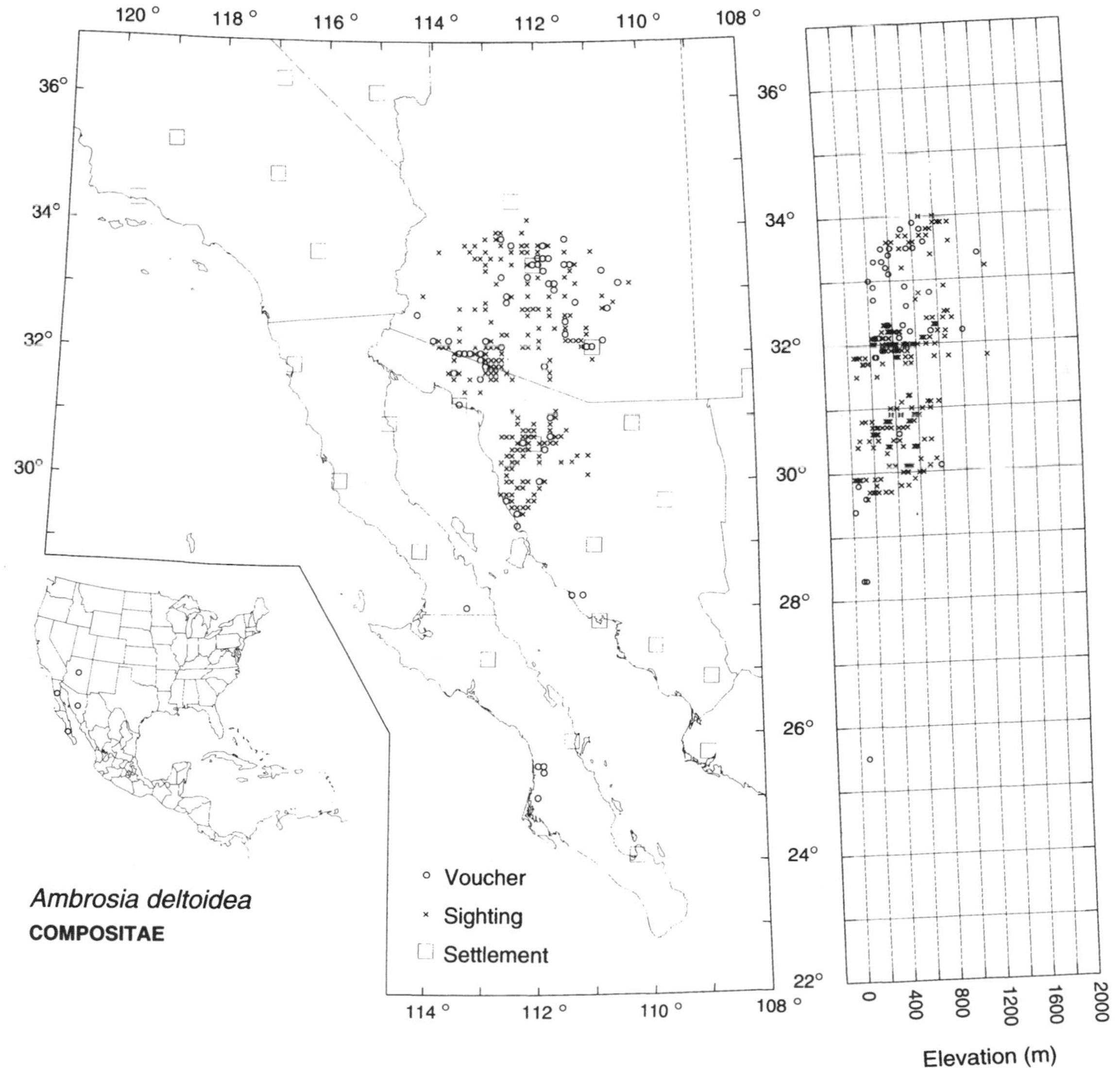

Ambrosia deltoidea
COMPOSITAE

lowing the winter rains of 1967–68 (Goldberg and Turner 1986). Recovery during the summer rainfall season is less pronounced.

Plant water potential of *A. deltoidea* varies widely on a daily and seasonal basis (Halvorson and Patten 1974; Szarek and Woodhouse 1976). Photosynthesis decreases in a nonlinear fashion as plant water potential drops (Szarek and Woodhouse 1977). Although *A. deltoidea* responds to rain at any season, its best growth and highest photosynthetic rates are in winter, when researchers in one study measured a maximum of 38 mg $CO_2/dm^2/hr$ (Szarek and Woodhouse 1976, 1977). Photosynthesis declined from March to May, when the plants flowered and fruited, then dropped to zero in June, when they were leafless. With refoliation following summer rains, the photosynthetic rate rose to 10 mg $CO_2/dm^2/hr$ (Szarek and Woodhouse 1977). Shading reduces carbon dioxide assimilation but prolongs the growing season (Smith et al. 1987). *Ambrosia deltoidea* maintains a relatively high water use efficiency throughout the year, probably because of seasonal changes in leaf conductance (Szarek and Woodhouse 1977). The rate of water extraction varies with temperature (Cable 1977a). Asymmetrical canopy growth, described by Rogers (1989), suggests that neighboring plants compete, probably for water.

At Tumamoc Hill, Tucson, Arizona, *A. deltoidea* grows on shallow bajada soils underlain by caliche and on deeper, sandy floodplain soils (Cannon 1911). The caliche layer restricts the roots to the top 30 cm of soil (but they may spread up to 1.6 m from the main stem); on the floodplain, the taproot extends down as much as 1.8 m (Cannon 1911). Most water is extracted within 50 cm of the plant crown (Sammis and Weeks 1977).

Seeds germinate following heavy autumn rains. Early survival is relatively high, but few individuals live to the maximum observed age of 50 years (Goldberg and Turner 1986). One cause of death (albeit a rare one) is fire (McLaughlin and Bowers 1982; Patten and Cave 1984). McLaughlin and Bowers (1982) reported

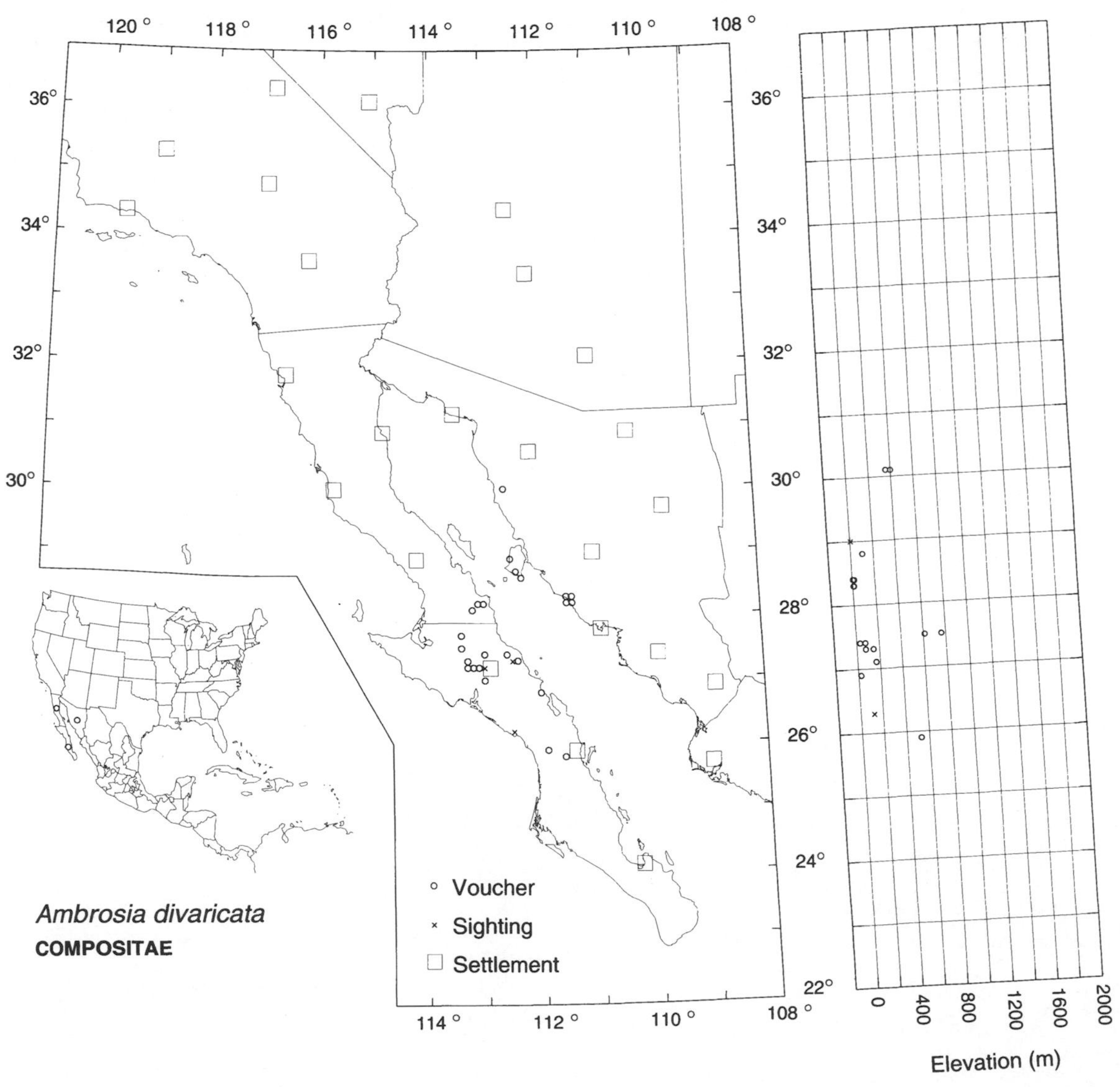

Ambrosia divaricata
COMPOSITAE

up to 92% mortality from an extensive desert burn. Resprouting from stems or roots was rare (McLaughlin and Bowers 1982).

Because the low, dense canopies provide protection from browsing animals and accumulate windblown litter, mature *A. deltoidea* act as nurse plants for other woody perennials, including *Cercidium microphyllum* and *Fouquieria splendens* (McAuliffe 1988; Franco and Nobel 1989). Conversely, production of annual plants is often reduced under *A. deltoidea* canopies (Patten 1978).

Ambrosia deltoidea deserves wider cultivation. Commercial growers have had difficulty in germinating seeds. Small plants can be dependably transplanted provided the shoots are substantially trimmed and adequate water is given for reestablishment. Gardeners can deter rabbits by tying freshly cut *A. deltoidea* stems around any recent transplant (Joseph McAuliffe, personal communication 1992). The pollen is often reported as "rabbit brush" or "false ragweed" (Mary K. O'Rourke, personal communication 1993). It is a powerful allergen to susceptible persons.

Ambrosia divaricata
(Brandegee) Payne

[=*Franseria divaricata* Brandegee]

Ambrosia divaricata is a divaricately branching shrub 3–6 dm tall. The 3- to 5-lobed, dentate leaves are ovate to suborbicular. Minutely canescent in youth, they become greener in age. Staminate and pistillate heads are on separate terminal racemes. The burlike fruits, 5–8 mm long, are subglobose to obovoid. Their 25–30 subulate spines are grooved on the upper side near the base but not appreciably flattened. The haploid number of chromosomes is 18 (Seaman and Mabry 1979a).

The very similar *A. magdalenae* has bipinnatifid rather than lobed leaves. *Ambrosia dumosa* leaves, also bipinnatifid, are densely white-tomentose. *Ambrosia bryantii* is conspicuously armed with slender white spines.

According to Payne (1963), the arrangement of heads is a primitive character. *Ambrosia divaricata,* like *A. magdalenae,* lacks sesquiterpene lactones, not surprising given the great morphological similarity between the two (Seaman and Mabry 1979b).

Locally common on sandy plains, in washes, and on gravelly slopes, *A. divaricata* is primarily peninsular in distribution with outlying populations on Isla Tiburón and the nearby Sonoran coast. The Sonoran vouchers are Raymond M. Turner's from the vicinity of Estero Tastiota, San Augustín Bay, and Cerro Libertad. On Isla Tiburón, the westernmost point (29.0°N) represents two collections by Richard Felger, and the southernmost point (28.8°N) is based on a collection by Reid Moran. Populations occur along much of the channel between the island and the mainland (Felger 1966). The point on Isla Turner (Isla Dátil), immediately south of Isla Tiburón, represents a collection by Reid Moran.

Many of the collection sites are near the periphery of *A. magdalenae,* which suggests that competitive interactions may shape the distribution. Climate in the range is frost free and decidedly arid. Rainfall is biseasonal, either with equal amounts of summer and winter rain or with a moderate predominance of summer moisture.

Leaves are drought deciduous. The wind-pollinated flowers appear from January to March (occasionally May).

Ambrosia dumosa
(A. Gray) Payne

[=*Franseria dumosa* A. Gray]

White bursage, hierba del burro

Ambrosia dumosa is a stiffly branched, rounded shrub or chamaephyte 0.2–0.6 m tall with white-tomentose leaves 0.8–2 cm long. Ovate in outline, the leaves are bipinnatifid with many small, fernlike divisions. Plants are monoecious; staminate and pistillate heads are mixed in racemes or spikes. The inflorescence is a primitive type in the genus (Payne 1963). Fruits are indurated burs with 30–40 flattened spines 1.5–2.2 mm long. The basic chromosome number is 18; diploids, tetraploids, hexaploids, and perhaps octoploids are known (Payne et al. 1964; Raven et al. 1968; Seaman and Mabry 1979c).

Several other woody *Ambrosia* species also have dissected leaves, but none is completely white-tomentose. Fruits of *A. dumosa* have subulate, flattened spines in contrast to the hooked, subterete spines of *A. magdalenae* and the conical ones of *A. camphorata.*

Apparently, *A. dumosa* spans all ploidy levels known for the genus, from diploid to octoploid (Payne 1964). Diploids and tetraploids occur throughout the range of the species, except in Baja California Sur and the Owens Valley, California, where tetraploids are absent. Hexaploids are known from the deserts of California (Raven et al. 1968), central Baja California, and northwestern Arizona (Seaman and Mabry 1979c). Where two chromosomal races overlap, they may be morphologically, ecologically, and phenologically distinct, or they may intergrade, making it impractical to distinguish them taxonomically (Raven et al. 1968). Sesquiterpene lactone chemistry suggests that the tetraploid form arose from hybridization of two morphologically identical but chemically distinct diploid elements, and that the hexaploid form most likely originated from the tetraploid group (Seaman and

Mabry 1979c).

Payne (1964) placed *A. dumosa* in a group of shrubby species with much divided, irregularly lobed leaves, among them *A. bryantii, A. camphorata,* and *A. magdalenae.* Based upon sesquiterpene lactone chemistry, Seaman and Mabry (1979a) aligned *A. dumosa* with *A. deltoidea, A. chenopodiifolia, A. ambrosioides, A. cordifolia,* and *A. eriocentra.*

With *Larrea tridentata, A. dumosa* dominates many square kilometers of gravelly plains and alluvial fans. It also grows on the broken surface of volcanic malpais; on rocky, igneous slopes; and on stable dunes and sand sheets. In the more arid portions of its range, where soil moisture varies little from one topographic situation to another, *A. dumosa* occupies most habitats (Shreve 1925). It colonizes dunes and may eventually stabilize them (Shreve 1937a; A. F. Johnson 1982). In some communities, *A. dumosa* pioneers on disturbed soils (Vasek 1980b; Prose et al. 1987). Some genotypes are relatively salt tolerant (Romney and Wallace 1980). The Gila River Valley in eastern Arizona has evidently served as a distributional corridor, as has the Colorado River in northern Arizona. The southernmost disjunct (24.2°N) represents a collection by Ira L. Wiggins.

Macrofossils from ancient packrat middens show that *A. dumosa* grew in southwestern Arizona between 15,700 years B.P. and 10,750 years B.P. (Van Devender 1990b) and in Death Valley, California, about 10,200 years B.P. (Wells and Woodcock 1985). It spread across the Mojave Desert between 9,200 and 5,000 years B.P. (Spaulding 1990) but may not have arrived along the lower Colorado River until 700 years ago (Cole 1986).

Currently, *A. dumosa* ranges over much of the Sonoran and Mojave deserts. It is scarce or absent where cool-season rainfall is low. *Ambrosia chenopodiifolia* replaces *A. dumosa* in the coastal climates of the Vizcaíno region, and *A. magdalenae* replaces it in Baja California Sur.

The flowers are wind pollinated. Given enough rain, plants may flower in spring, summer, or fall (Ackerman et al. 1980), but in most years the main flowering pe-

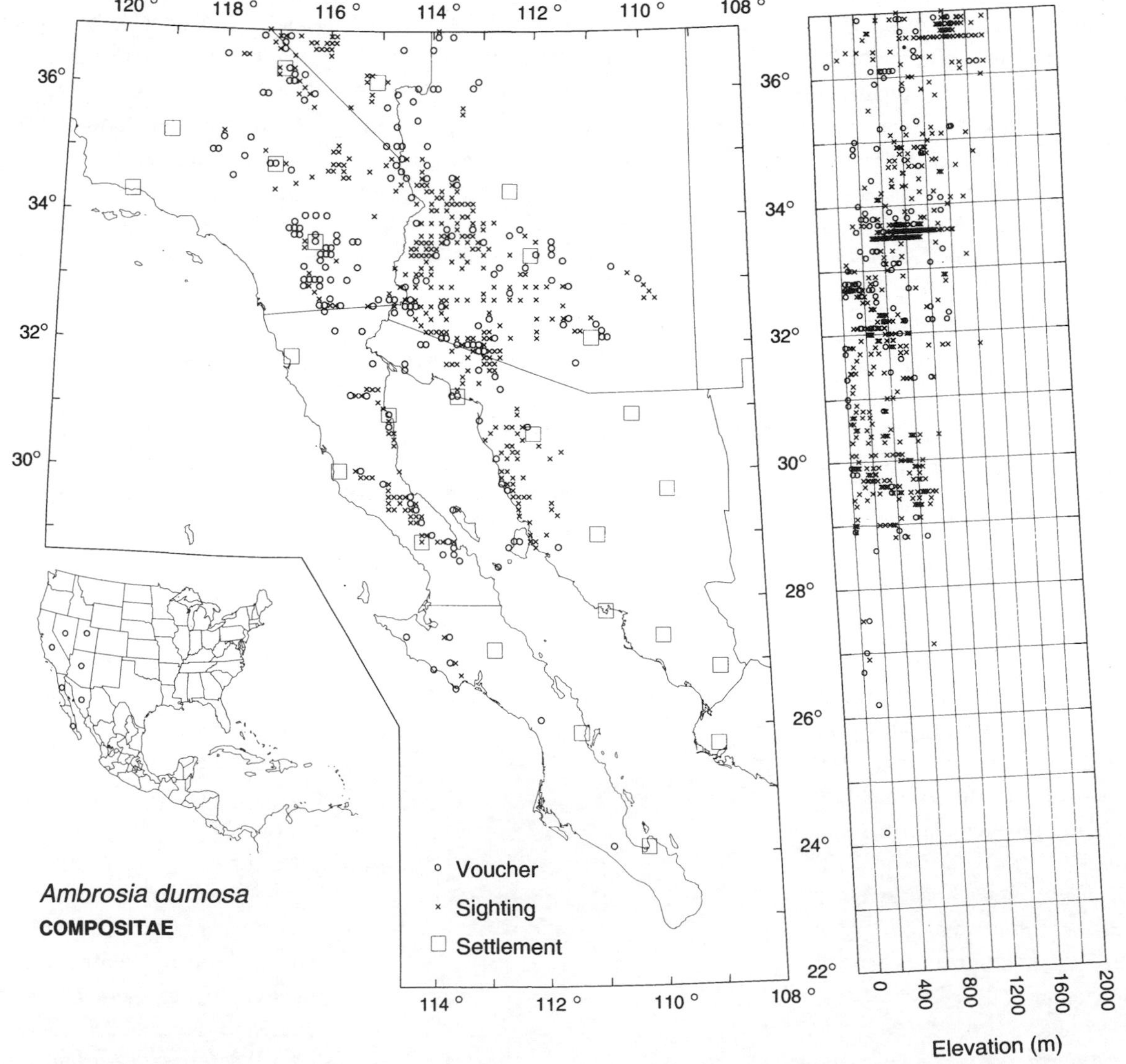

riod is early to mid spring (February–April). In one study conducted in southern Nevada, the earliest date of spring flowering was accelerated by higher than normal temperatures from January to March and was retarded by higher than normal rainfall in February and March (Turner and Randall 1987).

Leaves are drought- and cold-deciduous (Humphrey 1975; Ackerman et al. 1980; Comstock et al. 1988). In the Sonoran Desert, leaves and stems grow in response to both summer and winter rains (Humphrey 1975). In the Mojave Desert, where winter rain predominates, most growth is in the spring. Turner and Randall (1987) found that in southern Nevada, the earliest date of leafing over nine years varied from February 1 to March 27. Greater than normal rainfall in December and January retarded leafing, as did higher than normal temperatures in November and December (Turner and Randall 1987). Dead leaves may persist on the plant for many months (Humphrey 1975). Litter flux beneath the canopy is greater than that observed in other Mojave Desert perennials (Strojan et al. 1979).

Annual peaks in carbon assimilation (400–500 mg CO_2/g leaf dry weight/day) occur in spring, from March to May (Bamberg et al. 1976). Growth declines when soil temperatures fall below 28°C (Wallace et al. 1970). Aboveground net production in a southern Nevada study was strongly correlated with October–September rain and varied from 2.1 g/m^2 in a dry year to 12.7 g/m^2 in a wet one (Turner and Randall 1989). Below-ground biomass can be estimated from stem weights (Wallace et al. 1974). Average respiration rate has been estimated to be 0.21 mg carbon/g dry weight/hr (Wallace, Cha et al. 1980).

Competition between neighboring *A. dumosa,* whether adults or juveniles, appears minimal (Fonteyn and Mahall 1981; Wright and Howe 1987; Mahall and Callaway 1991); however, when *A. dumosa* and *Larrea tridentata* grow together, they may compete strongly for water (Fonteyn and Mahall 1981).

The seeds disperse via attachment to animals, dissemination by harvester ants (Clark and Comanor 1973), and tumbling in wind (Maddox and Carlquist 1985). In the northern Mojave Desert, germination has been observed from mid June to mid September when minimum air temperatures ranged from 12–20°C and maximum temperatures were between 26–34°C (Ackerman 1979; Hunter 1989). A rainfall of at least 25 mm is necessary to trigger germination at these rather high temperatures (Ackerman 1979). Since warm-season rain is infrequent over much of its range, *A. dumosa* germinates episodically. Sometimes the stem axis splits, producing independently rooted clones that can survive the death of the original crown (Muller 1953; Jones 1984).

Seedling survival apparently varies markedly from place to place. Ackerman (1979) reported low seedling survival in the northern Mojave Desert (4 of 49 seedlings survived one year, none survived two), but in the western Sonoran Desert, McAuliffe (1988) calculated much higher rates of survival and recruitment—30.8 individuals/ha/yr—on a site denuded of vegetation. In the northern Mojave Desert, Hunter (1989) found low seedling survival (3%) in undisturbed vegetation and much higher survival (58%) on a cleared site.

Muller (1953) stated that *A. dumosa* is "extremely long-lived." Matched photographs from the Grand Canyon show individuals at least a century old (Robert Webb, personal communication 1993). McAuliffe (1988), however, estimated average longevity in western Arizona to be 35.7 years. The plants do not tolerate burning, and fire is an occasional cause of mortality (Brown and Minnich 1986). Since *A. dumosa* serves as a nurse plant for *Larrea tridentata, Cercidium microphyllum,* and other woody dominants (McAuliffe 1986, 1988), its status directly affects the composition and density of many square kilometers of desert vegetation.

Ambrosia dumosa is grazed by jackrabbits (Wright and Howe 1987), cattle (Hughes 1982), and burros (Hanley and Brady 1977). Insect herbivores include fruit flies (Diptera: Tephritidae) (Silverman and Goeden 1980), lacebugs (Hemiptera: Tingidae) (Silverman and Goeden 1979), and plume moths (Lepidoptera: Pterophoridae) (Goeden and Ricker 1976b). Estimated recovery rates after anthropogenic disturbance range from 30 to 100 years (Artz 1989). In revegetation of disturbed soils, *A. dumosa* is more dependably established from transplants than from seed (Graves et al. 1978). Inoculation with endomycorrhizae improves growth (Melanson and Carter 1981). Stem cuttings root easily (Wieland et al. 1971).

Ambrosia eriocentra

(A. Gray) Payne

[=*Franseria eriocentra* A. Gray]

Woolly bursage

Ambrosia eriocentra is an upright shrub up to 1 m tall. The plants are monoecious, with staminate heads in short terminal spikes and pistillate heads in congested lateral spikes. The alternate leaves are lanceolate to oblanceolate with pinnatifid or incised margins. White-tomentulose beneath and greenish-glabrate above, they are 2–4 cm long and 5–18 mm wide. The burlike fruits are 7–8 mm long with 10–18 subulate spines. The haploid chromosome number is 18 (Payne 1964).

Ambrosia deltoidea leaves are serrate or crenate, not incised. Foliage of *Bernardia incana* Morton is closely felted with stellate hairs. *Hyptis emoryi* has opposite, crenate leaves.

Barely entering our area along the northern edge of the Sonoran Desert, this species is more characteristic of the eastern Mojave Desert. It grows most often in sandy washes. The Salt and Gila rivers have apparently served as distributional corridors. Benson and Darrow (1981) show the species extending farther up the

Colorado River than we do and also indicate two outlying populations in southwestern Arizona. We have seen no vouchers from these locations.

North of 35°N, *A. eriocentra* essentially replaces *A. ambrosioides.* It is more cold tolerant and ranges to higher elevations than other shrubs in the genus. Over most of its range, rainfall is biseasonal with at least half the annual total in winter.

Ambrosia eriocentra belongs to a small group of desert species that includes *A. ambrosioides, A. deltoidea,* and *A. chenopodiifolia* (Payne 1964; Seaman and Mabry 1979a). Payne (1964), based on inflorescence morphology, considered *A. eriocentra* to be the least specialized member of the group.

The flowering period is March to May, occasionally June. Flowers are wind pollinated. In southeastern California, the plants lose about 30% of their leaves during the dry summer months of July and August (Comstock et al. 1988). Minimum net photosynthesis drops to nearly zero at the same time, apparently a result of low plant water potentials (about –3.0 MPa). When fall rains bring about high plant water potentials, photosynthesis reaches a maximum of about 10 μmol/m²/sec well before the canopy is fully leafed out. Retaining some leaves during drought evidently lets the plants recover rapidly once autumn rains begin (Comstock et al. 1988).

Given its preference for washes, where it would be subject to occasional scour, *A. eriocentra* is probably a relatively short-lived colonizer, much like the *Hymenoclea* species with which it often occurs.

Larvae of the ragweed plume moth [*Adaina ambrosiae* (Murtfelt)] eat the leaf epidermis and mesophyll (Goeden and Ricker 1976b).

Ambrosia flexuosa

(A. Gray) Payne

[=*Franseria flexuosa* A. Gray]

This low, much-branched shrub 2–4 dm

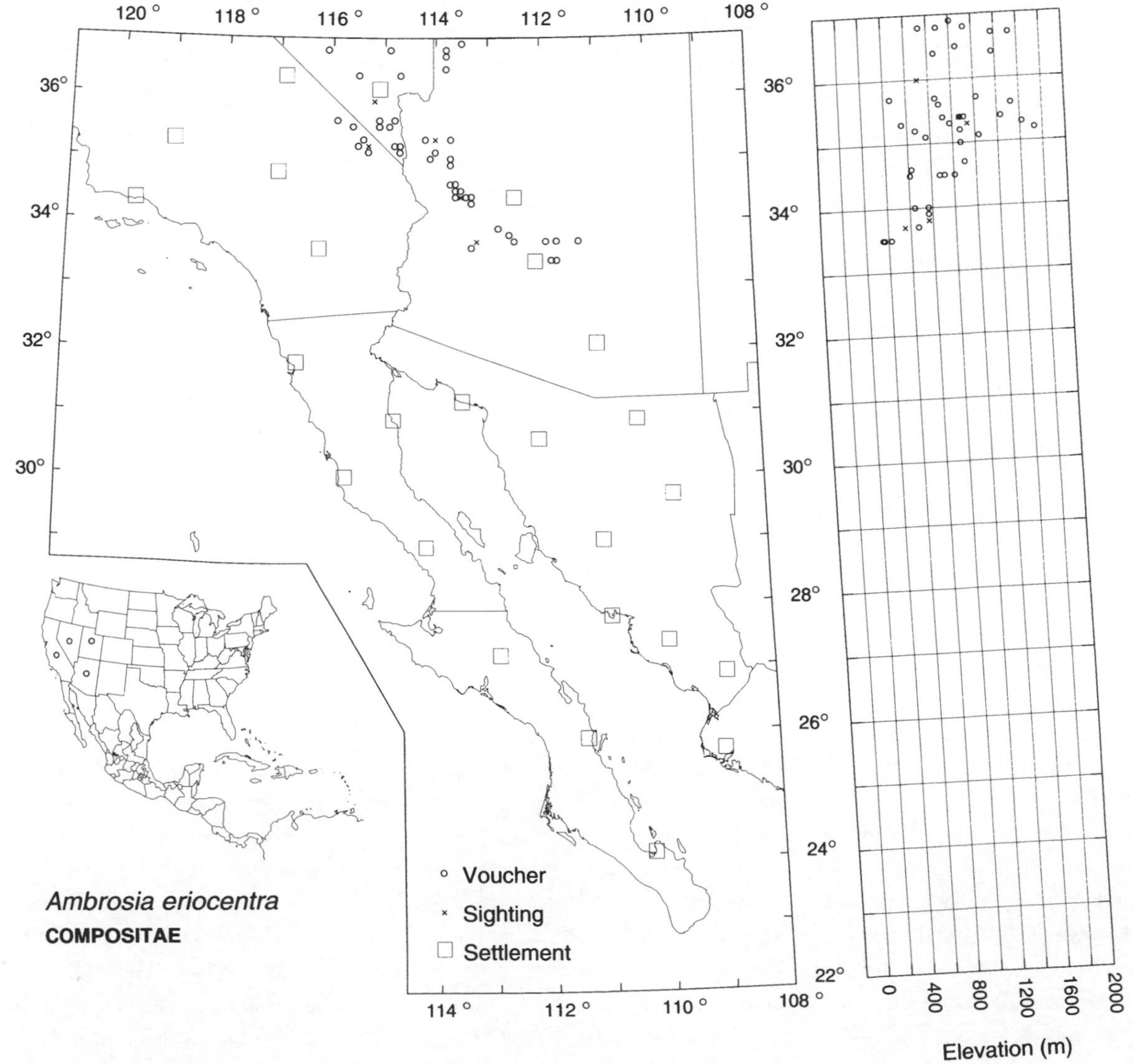

tall has flexuous, striate twigs and deltoid-lanceolate leaves. The leaves, 2–5 cm long and 3 cm wide, are somewhat hastate and sinuately lobed. Staminate heads are above the pistillate in narrow panicles or racemes. The burlike, subglobose fruits are 5–7 mm long and armed with subulate spines markedly flattened and channeled on the upper surface.

Ambrosia carduacea is a large shrub or small tree with larger leaves (5–8 cm wide and 8–22 cm long). *Ambrosia ambrosioides* has longer leaves (4–18 cm) and larger fruits (10–15 mm long).

According to Payne (1964), this species probably belongs with the subgeneric group comprising *A. ambrosioides, A. chenopodiifolia,* and other shrubs having unlobed leaves and glandular indument.

Locally common in canyons and on rocky hillsides, *A. flexuosa* is narrowly endemic to the granitic desert slopes of the Sierra Juárez and the Sierra San Pedro Mártir. Within its range, rainfall is biseasonal, with a predominance of winter moisture. The elevational range indicates some sensitivity to frost.

The leaves are drought deciduous. The wind-pollinated flowers appear March–May.

Ambrosia ilicifolia (A. Gray) Payne

[=*Franseria ilicifolia* A. Gray]

Hollyleaf bursage

A sprawling shrub 3–12 dm tall and wide, *Ambrosia ilicifolia* (figure 24) has a densely leafy canopy of sessile, spinose-dentate leaves. Ovate to lanceolate in outline, they are 3–8 cm long and 2–4.5 cm wide with cordate-clasping bases. Both surfaces are conspicuously reticulate veined. Staminate heads are above the pistillate in terminal panicles. Fruits are burs 15 mm long with 7–12 subulate, hooked spines. The haploid chromosome number

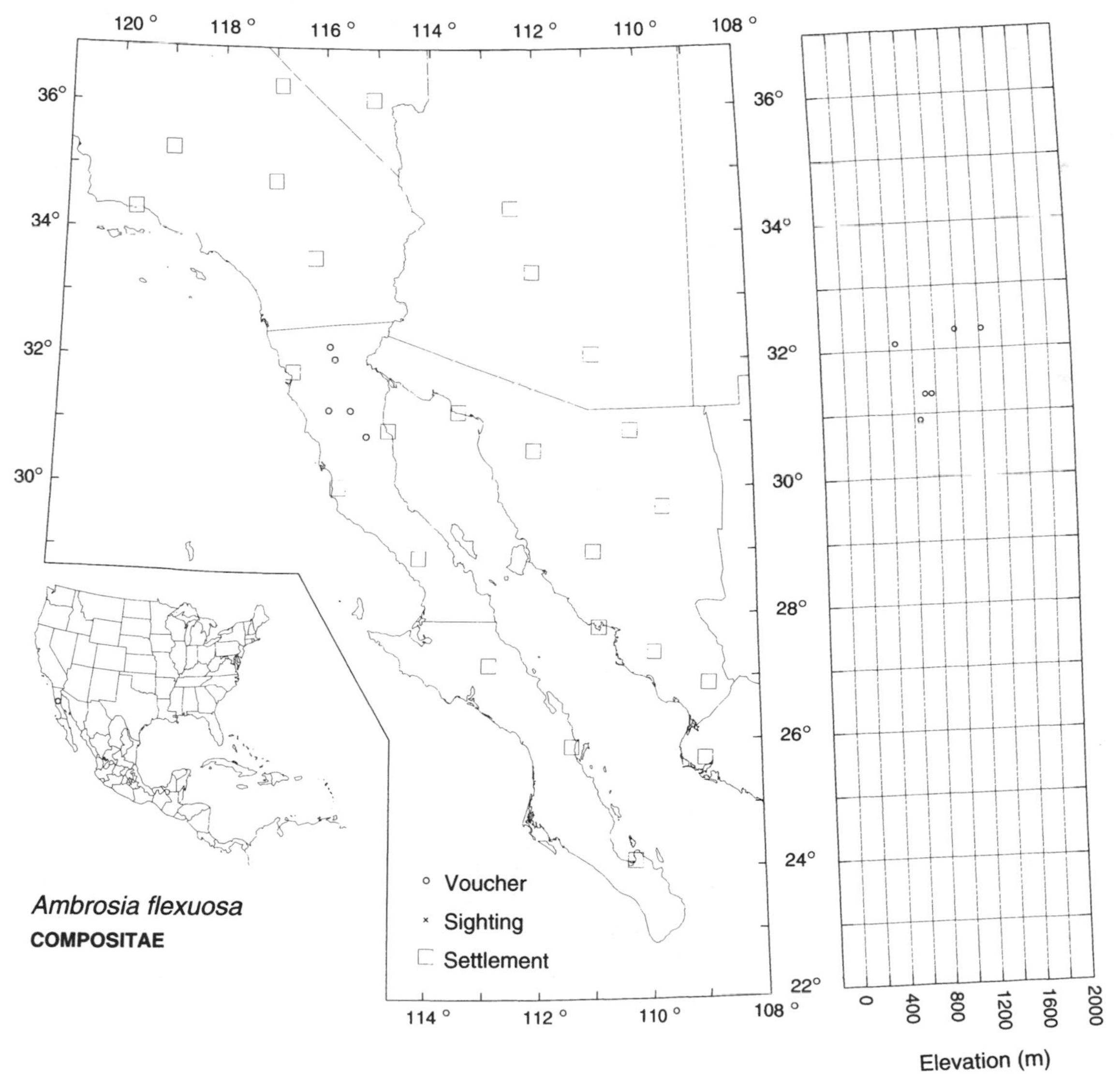

Ambrosia flexuosa
COMPOSITAE

Figure 24. Ambrosia ilicifolia *in flower on Isla San Esteban. (Photograph by R. M. Turner.)*

is 18 (Payne et al. 1964).

The old, whitened or grayish leaves that remain on the plants for many months are a distinctive feature of this species. Leaves of *A. ambrosioides* are coarsely dentate but not spinose. *Brickellia atractyloides* A. Gray has smaller leaves 1.5–2.5 cm long and 1–1.5 cm wide. *Xanthium* species are robust annuals with shallowly lobed, lanceolate or deltoid leaves.

Based on sesquiterpene lactone chemistry, Seaman and Mabry (1979a) allied *A. ilicifolia* with *A. chamissonis, A. bryantii, A. camphorata, A. divaricata,* and *A. magdalenae.*

Like *A. ambrosioides, A. ilicifolia* is restricted to desert washes, canyon bottoms, moist ledges, and boulder crevices where greater water availability supports its large leaf area. It is a narrow endemic that essentially replaces *A. ambrosioides* in the more arid portions of the Sonoran Desert. In the Tinajas Altas Mountains, Yuma County, Arizona, *A. ilicifolia* grew with *Pinus monophylla* Torr. & Frem. and *Juniperus californica* Carr. about 15,000 years B.P. (Van Devender 1990b). Apparently the northern part of the range has been rather stable since the late Pleistocene despite substantial climatic changes.

Temperatures are warm and rainfall biseasonal over the entire range. In most summers, temperatures are extremely high while moisture is only transiently available. The restriction to low elevations indicates a requirement for arid habitats

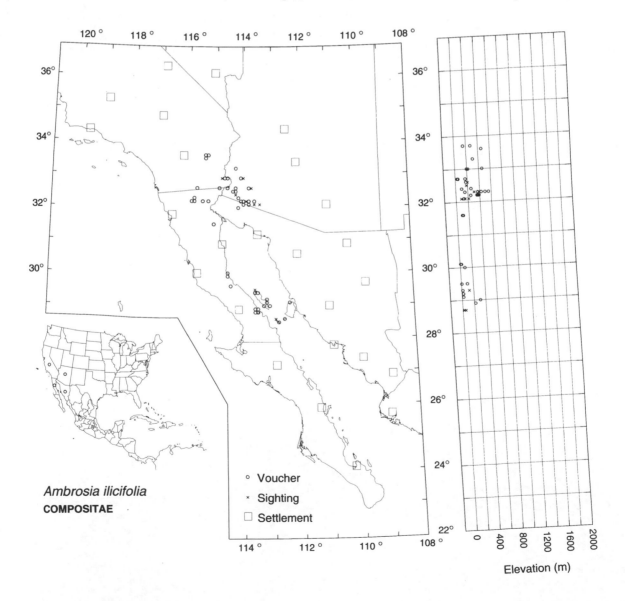

Ambrosia ilicifolia
COMPOSITAE

○ Voucher
× Sighting
□ Settlement

Elevation (m)

and perhaps a sensitivity to frost; however, the Pleistocene record suggests greater cold tolerance than is currently demonstrated.

Flowers appear from January through May and are wind pollinated. Leaves are evergreen with thick cuticles that no doubt prevent undue water loss.

Ambrosia magdalenae

(Brandegee) Payne

[=*Franseria magdalenae* Brandegee]

Ambrosia magdalenae is a widely branching shrub 2–6 dm tall with bright green leaves 3–7 cm long and 1–4.5 cm wide. Ovate to oblong-ovate and bipinnatifid into linear, dentate divisions, the leaves are tomentose beneath. The racemose inflorescence has staminate heads above the pistillate. The burlike fruits, 4–5 mm long, are armed with 20–40 subulate, hooked spines. The haploid number of chromosomes is 18 (diploids), 36 (tetraploids), or 51 (perhaps an aneuploid derivative of a hexaploid that has yet to be discovered) (Seaman and Mabry 1979a).

Ambrosia divaricata leaves are 3- to 5-lobed. *Ambrosia bryantii* is conspicuously armed with long, white spines on the persistent fruits. *Ambrosia dumosa* leaves are white-tomentose on both surfaces, and its fruit spines are straight. The fruit spines of *A. camphorata* are conical, and the odor of the crushed foliage is distinctively pungent.

According to Payne (1964), *A. magdalenae* belongs to a core group of shrubs with pinnately to tripinnately lobed leaves, including *A. dumosa,* the putative ancestor of *A. magdalenae.* Seaman and Mabry (1979a) aligned *A. magdalenae* with *A.*

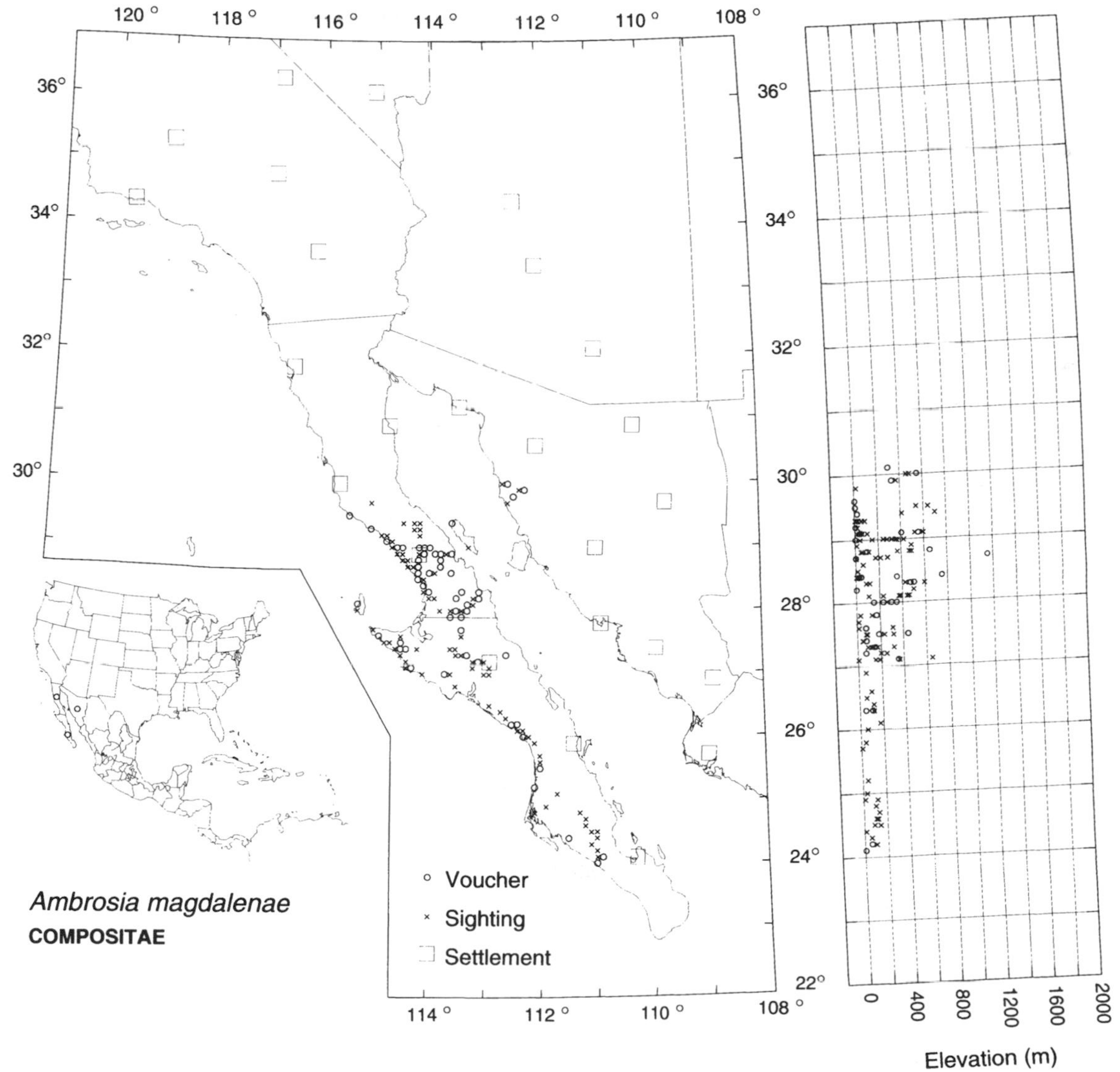

Ambrosia magdalenae
COMPOSITAE

chamissonis, A. bryantii, A. ilicifolia, A. camphorata, and *A. divaricata.* Like *A. divaricata,* which it closely resembles, *A. magdalenae* lacks sesquiterpene lactones (Seaman and Mabry 1979a). Diploids are more common in the southern part of the range, tetraploids in the northern. Collections from the La Paz area are morphologically distinct, having smaller heads and fewer fruit spines than more northern specimens (Seaman and Mabry 1979a).

Locally abundant on sandy plains on the Baja California peninsula, *A. magdalenae* generally grows below 800 m. Reid Moran collected it at 1,150 m on the Cerro de la Mina de San Juan (28.7°N, 113.6°W). The plant's range is restricted to arid maritime climates, with rainfall regimes varying from largely winter in the northwest to predominantly summer and early fall in the south. Several disjunct populations occur on the mainland. The Sonoran vouchers, all Raymond M. Turner's, were collected near Puerto Libertad, Sonora. Other peninsular species found in Sonora only in this area include *A. chenopodiifolia, A. divaricata, Fouquieria columnaris,* and *Viguiera laciniata.* Humphrey and Marx (1980) suggested that *F. columnaris* survives near Puerto Libertad because of unusually high humidity. Turner and Brown (1982) thought that anomalously low summer temperatures caused by offshore upwelling may enable these disjuncts to persist at Puerto Libertad.

Leaves are drought deciduous. The wind-pollinated flowers appear from October to April. The plants are unpalatable to native and domestic grazing animals (Humphrey 1974).

Anisacanthus thurberi

(Torr.) A. Gray

Desert honeysuckle, chuparosa

Anisacanthus thurberi is a shrub 2 (occasionally 2.5) m tall with exfoliating bark and two vertical lines of sparse pubescence on the stems. The opposite, lanceolate leaves, 4–6 cm long and 1–1.5 (occa-

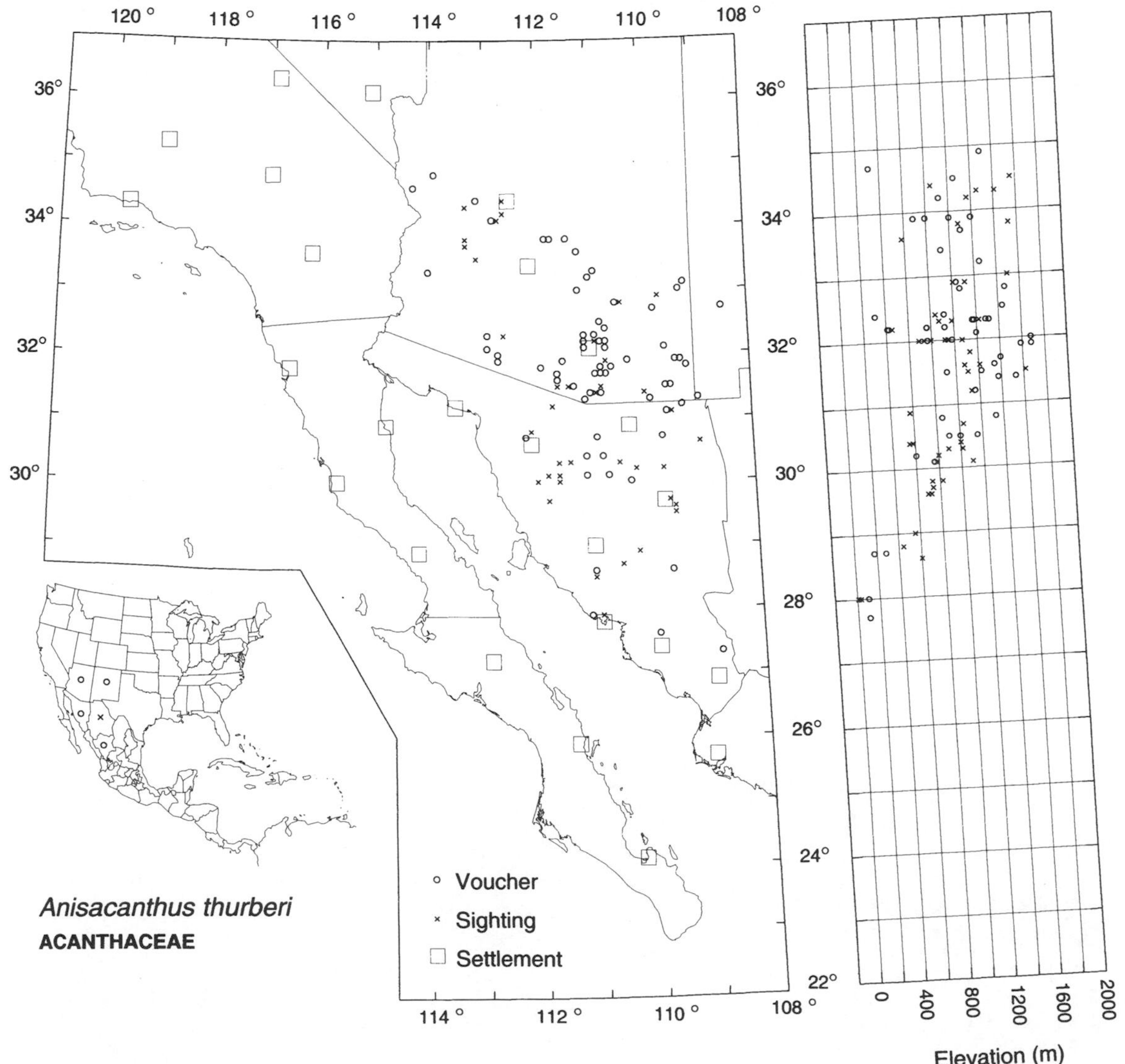

Anisacanthus thurberi
ACANTHACEAE

sionally 2) cm wide, are puberulent to glabrous. Usually brick red, occasionally yellow or orange, the tubular flowers are 2–3.5 cm long. Fruits are dehiscent, 2-valved capsules 12–14 mm long. The haploid chromosome number is 18 (Daniel et al. 1984).

The closely related *A. andersonii* T. F. Daniel, a shrub of Sonora and Chihuahua, can be distinguished by its larger flowers (45–60 mm long) and capsules (20–23 mm long) as well as by its evenly disposed (rather than fasciculate) leaves (Daniel 1982). *Justicia californica* and *J. candicans* (Nees) L. Benson, also shrubs of the Acanthaceae, have clear red or vermilion flowers, ovate leaves, and dense, short pubescence.

A plant of rocky canyon bottoms and sandy or gravelly washes, *A. thurberi* occurs widely above 600 m in the eastern half of the Sonoran Desert. Its distribution complements that of *J. californica,* which occupies the lower, more arid, western section of the desert. Both grow in watercourses and have reddish flowers that attract hummingbirds, so their distributions might reflect some competitive displacement. *Justicia californica,* a frost-tender plant (Anonymous 1979), is limited by cold temperatures, whereas *A. thurberi* is hardier and ranges to higher elevations and latitudes. The elevational profile indicates that *A. thurberi* is restricted to lower elevations in the south, suggesting a need for relatively open vegetation. The species is limited to areas with dependable summer rain. Its absence from summer-wet regions of Baja California Sur may reflect incompatibility with tropical summer temperatures. Failure of dispersal is also a possibility.

Anisacanthus thurberi blooms mainly in spring (March–June) and occasionally in fall (October–November). Leaves are cold deciduous and appear early in spring with the first flowers. Summer rains stimulate stem growth. The tubular, brick-red flowers seem well adapted for hummingbird pollination. Pollen and nectar are also accessible to solitary bees and perhaps other insects.

Whisenant and Ueckert (1981) reported on germination characteristics of

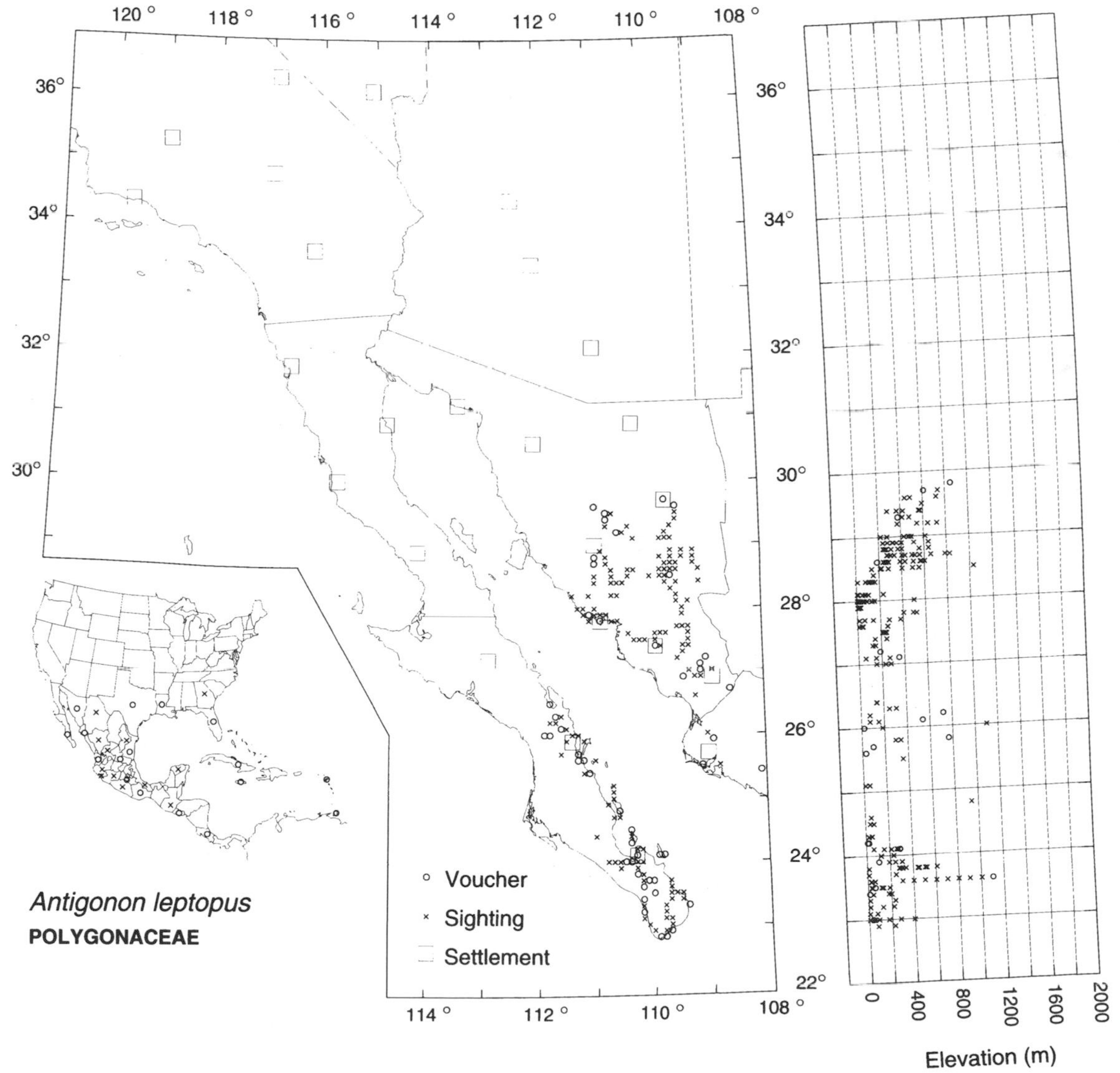

A. wrightii (Torr.) A. Gray, a shrub of Texas and adjacent Mexico.

Cattle and sheep browse the plants, especially when other forage is scarce (Kearney and Peebles 1960).

Antigonon leptopus

Hook. & Arn.

Queen's wreath, coral vine, Confederate vine, privy-vine, cadeña de amor, bellisima, flor de San Diego, rosa de mayo, corona de la reina, San Miguelito, corona, flor de San Miguel, coronilla, cuamecate, coamécatl

This scrambling or climbing perennial vine grows from enlarged, somewhat tuberlike roots. The deeply cordate leaves, 4–20 cm long and 2–12 cm wide, are ovate and reticulate veined. The deep pink (rarely white) flowers are in drooping axillary racemes 10–50 cm long. The three cordate, accrescent outer sepals, each about 1.5 cm long, enclose three inner sepals that clasp an ovoid, trigonous achene. The sepals tend to be red in Baja California Sur, light pink in Sonora.

No other vine in the area covered by this atlas has flowers of three outer and three inner sepals. In *Exogonium bracteatum* (Cav.) Choisy, the bright pink or reddish floral bracts are 2–3.5 cm long, and the flowers are pentamerous, not trimerous.

Antigonon leptopus is a tropical species that reaches its northern limit in the Sonoran Desert. It is grown as an ornamental at many localities not plotted on our map. Where native, it scrambles on fences, trees, and shrubs along roadsides and

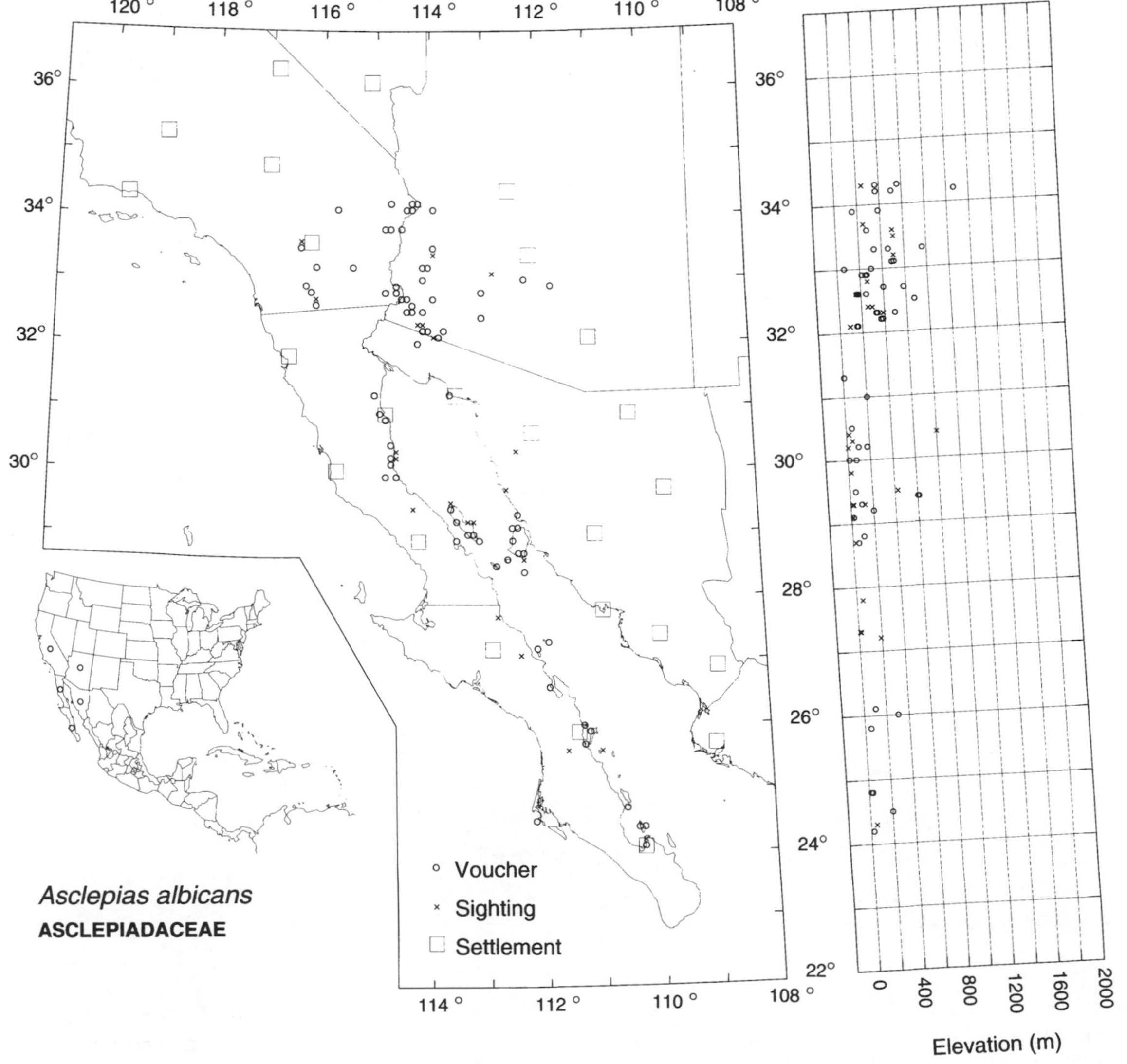

Asclepias albicans
ASCLEPIADACEAE

washes and in canyons. In southern Texas it occasionally volunteers around old homesteads (Correll and Johnston 1970). In the vicinity of Tucson, where the species is a common ornamental, the stems are annual, dying back to the ground with the first hard frost, then resuming growth in late spring. After the start of summer rains, growth becomes luxuriant. Low winter temperature extremes likely determine the northern limit of this species, and aridity may shape its western limit.

Flowers generally appear from November to April (Wiggins 1964), but flowering individuals can be found in any month. Honeybees and hummingbirds visit the blossoms (Correll and Johnston 1970).

The tendril is a modified inflorescence axis; its development has been investigated by Shah and Dave (1971). Carlquist (1985) described the stem histology.

Antigonon leptopus is widely available in commercial plant nurseries. Its relatively low water use makes it a good vine for desert regions. Winter dieback should be expected where temperatures drop below freezing, and the roots should be mulched where temperatures go below -7°C (Duffield and Jones 1981). The tubers are said to be edible and to have a nutlike flavor (Standley 1922). Indians of the peninsula ground and ate the parched seeds (Aschmann 1959). *Antigonon leptopus* is an important honey plant in India (Nair and Singh 1974) and the West Indies (Rindfleisch 1979).

Asclepias albicans

S. Watson

Wax milkweed, white-stem milkweed

Asclepias albicans (figure 25) is a woody perennial with clustered, wandlike stems 1–4 m tall. Generally, the stems are leafless. When present, the leaves are linear and 1–3 cm long. The stems exude copious milky latex when cut and are coated with a flaky, waxlike material. The flowers are white, tinged with purple or greenish brown. They are 5–6 mm long and are in axillary or terminal umbels. Fruits are waxy, attenuate follicles 8–12 cm long and 8–10 mm in diameter. Seeds are densely tufted with silky hairs.

Pedilanthus macrocarpus, a clump-forming perennial, has succulent rather than woody stems. *Asclepias subulata* has shorter, more numerous stems in dense clumps 1–1.5 (rarely 2) m tall. *Asclepias subaphylla* Woodson has longer (5–8 cm), more persistent leaves than *A. albicans* (Woodson 1945). *Asclepias masonii* Woodson, perhaps only a subspecies of *A. albicans* (Woodson 1945), is restricted to Isla Santa Margarita and the vicinity of Bahía Magdalena. The two differ in floral characters.

Where *A. albicans* and *A. subulata* grow together on the peninsular gulf coast, apparent hybrids occur (Steven P. McLaughlin, personal communication 1993). Over most of their range, the two are well marked.

Asclepias albicans is locally common on steep, rocky slopes and occasional on cliffs, bajadas, washes, and sandy flats. Unlike *A. subulata,* which is often found on dunes, at roadsides, and in borrow pits, *A. albicans* is not a plant of disturbed sites. Wiggins's (1964) report of *A. albicans* from Sinaloa was in error; the specimens in question are *A. subaphylla.*

Asclepias albicans is spottily distributed, having failed to occupy much apparently suitable habitat. It evidently requires open sites where competition from other

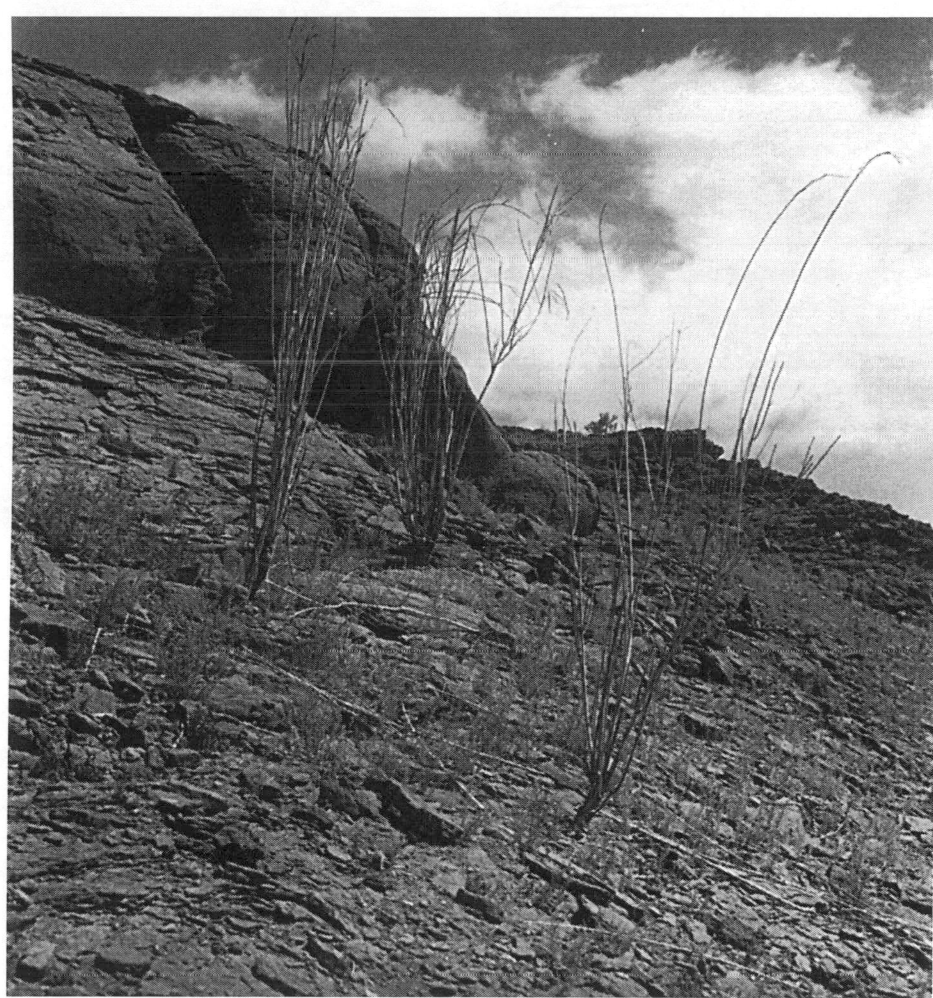

Figure 25. Asclepias albicans *on rocky slope overlooking Bahía Refugio, Isla Angel de la Guarda. (Photograph by R. M. Turner.)*

plants is minimal. Throughout the range, rainfall is biseasonal and summers are hot. Because plants grown at Tucson, Arizona, suffered only minor frost damage during the severe freeze of December 1978 (Steven P. McLaughlin, personal communication 1993), it seems unlikely that low temperatures determine the northern limit of this species. The plants are susceptible to fungi that thrive in warm, moist soils (Rotkis and Alcorn 1984), which may exert a strong influence on the distribution of the species.

Flowers appear from March to May, occasionally June.

The latex contains 0.88 to 5.40% natural rubber on a dry weight basis (Hall and Long 1921; Beckett et al. 1938; Buehrer and Benson 1945). No commercial use of this rubber has been made.

Asclepias subulata Decne.

Desert milkweed, rush milkweed, yamate, cadenilla bronca, jumete

A clump-forming perennial 1–1.5 (rarely 2) m tall, *Asclepias subulata* has numerous slender, reedlike stems that grow from the base and exude milky latex when cut. The threadlike leaves, 2–3 cm long and less than 1 mm wide, are ephemeral. The whitish yellow flowers are in umbels arranged in panicles. Fruits are follicles 8–13 cm long and 9–12 mm in diameter. Seeds are densely tufted with silky hairs.

Asclepias albicans, A. masonii Woodson, and *A. subaphylla* Woodson all have taller, thicker, waxy stems often woody well above the base. *Pedilanthus macrocarpus* has stouter, more succulent stems and vestigial leaves greater than 1 mm wide.

Where *A. subulata* grows with *A. albicans,* apparent hybrids may occur (Steven P. McLaughlin, personal communication 1993). In general, however, the two are well marked morphologically and ecologically.

An opportunistic species, *A. subulata* is

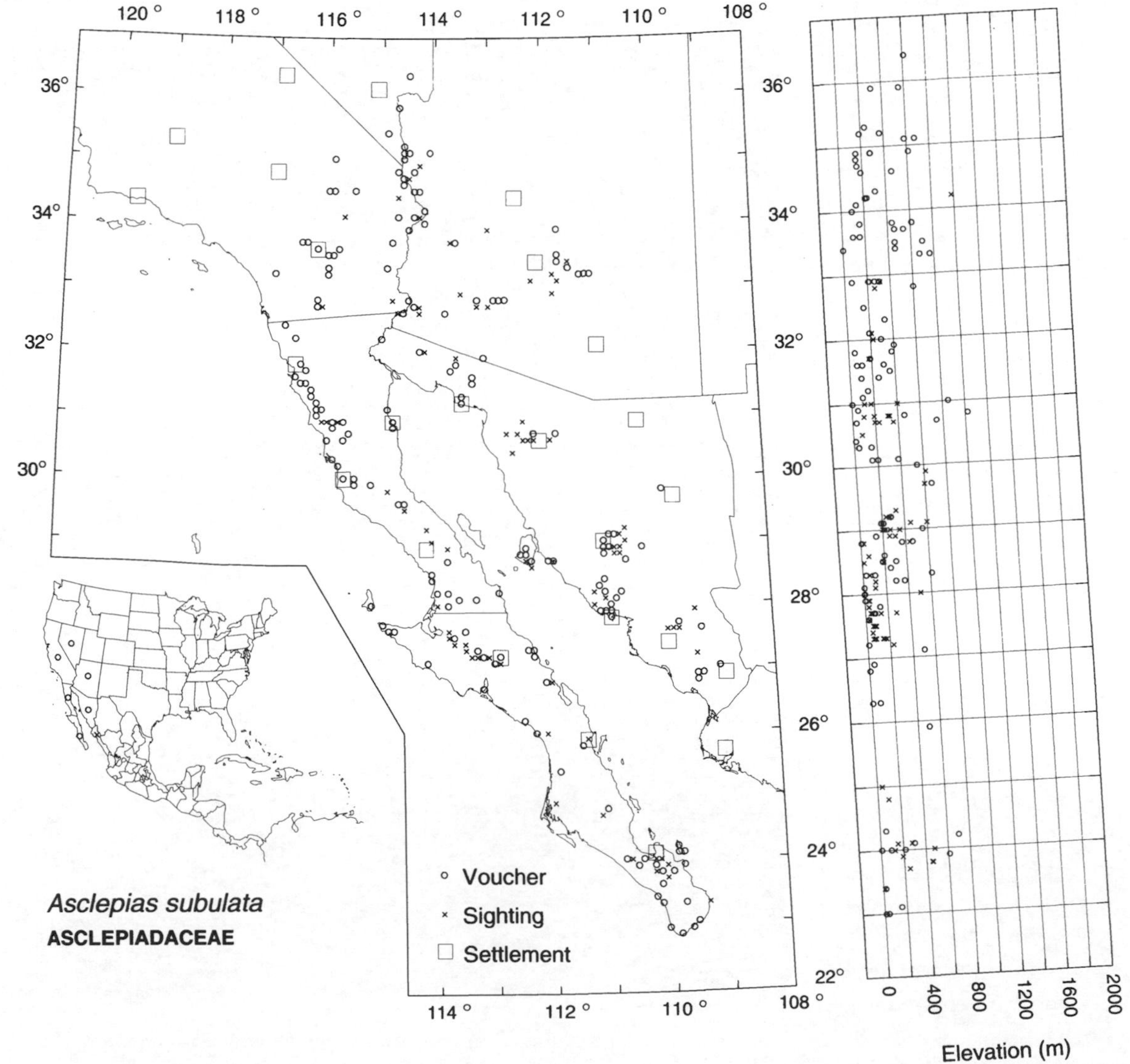

locally common along roads, in washes, in borrow pits, and on sandy flats. It is an early successional species, in contrast to *A. albicans,* which is more typical of mature, stable communities. *Asclepias subulata* occurs in a variety of rainfall regimes within the Sonoran Desert and adjacent regions. Its distribution appears rather erratic to the east, where summer rain predominates. Low temperatures may restrict the upward and northward movement of this species: at Tucson, Arizona, about two-thirds of the *A. subulata* seedlings grown in field plots were killed by the severe freeze of December 1978 (Steven P. McLaughlin, personal communication 1993).

Flowering specimens have been collected in every month of the year across the range of the species. In southeastern California, the main bloom is May and June, followed by a lesser bloom in September (Beckett and Stitt 1935).

The roots apparently vary according to habitat. In the Vizcaíno region of Baja California, plants on relatively fine textured soil had single taproots well over 1 m long, sometimes with minor lateral roots within 12 cm of the surface. A different population on gravelly alluvium had more contorted roots that occasionally gave rise to surface sprouts (Tony L. Burgess, personal observation 1987).

Although it is difficult to propagate *A. subulata* by cuttings, root-crown division, or budding, seed germinates readily, and seedlings can be easily transplanted (Beckett and Stitt 1935). Growth of young plants is rapid; three-year-old plants in field plots averaged 1.2 m high with 149 stems (Beckett and Stitt 1935).

The latex contains natural rubber, from 0.5 to 6.5% of the plant dry weight (Hall and Long 1921; Beckett and Stitt 1935). Although the U. S. Department of Agriculture investigated this species as a potential rubber plant (Beckett and Stitt 1935), no commercial use of the rubber was ever made. The latex also contains cardenolides and lignan (Jolad et al. 1986). The species has been studied as a source of "biocrude" (petroleum substitute) (Proksch et al. 1981; McLaughlin and Hoffmann 1982).

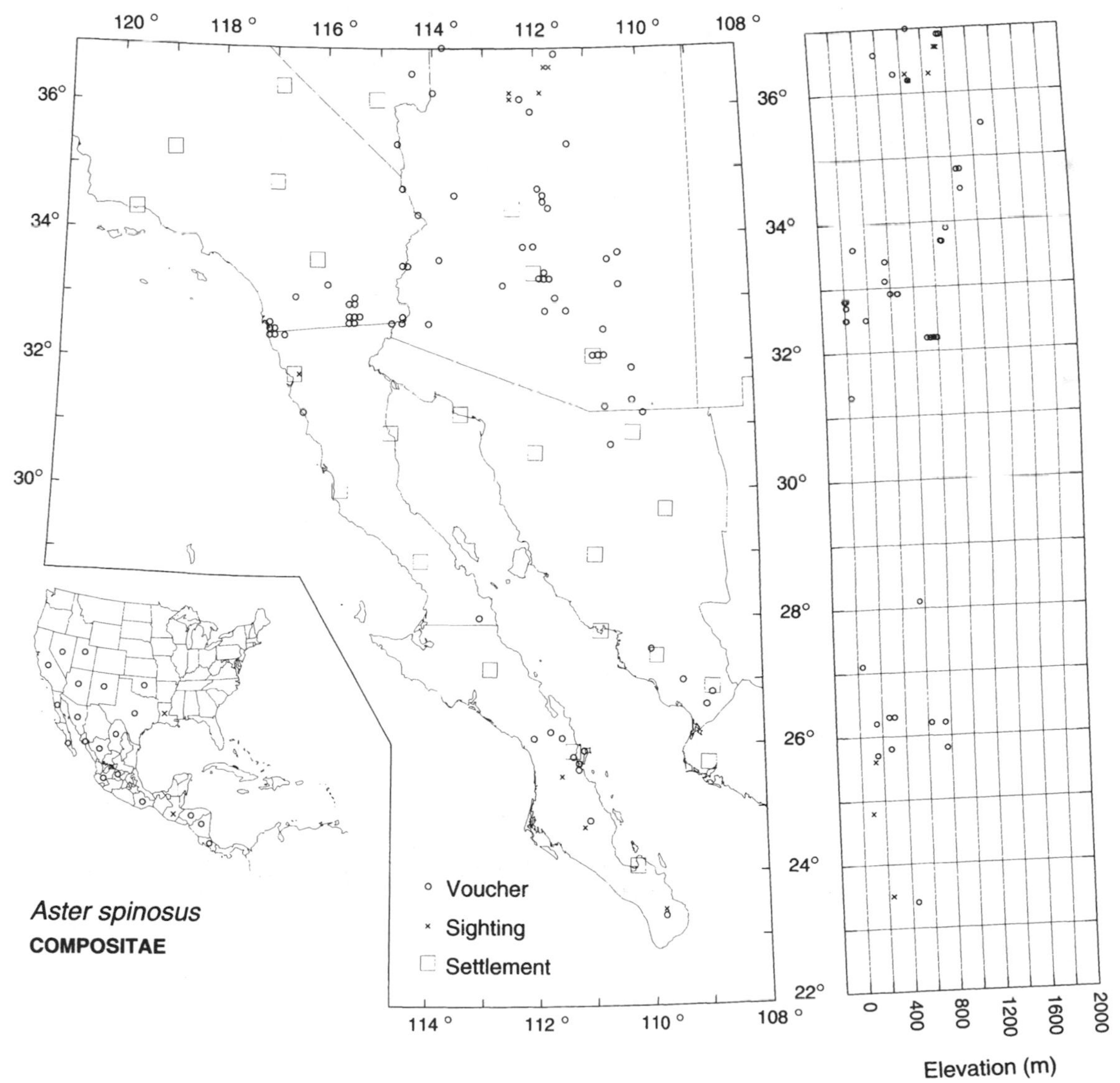

Aster spinosus
COMPOSITAE

Seri Indians used the roots to treat headache, toothache, and heart pain (Felger and Moser 1985). The milky latex has been used elsewhere in Mexico as an emetic and purgative (Standley 1924).

Aster spinosus

Benth.

Spiny aster, Mexican devil-weed

Aster spinosus, a perennial herb or suffrutescent shrub up to 3 m tall, has pale green, striate, broomlike stems that grow from wide-spreading rhizomes. The stiff, axillary branchlets are rigidly spiny. Young shoot leaves are linear to spatulate and 1–5 cm long; the scalelike older leaves are 1–5 mm long. Flower heads, up to 2 cm wide and high, are in panicles. Ray flowers are white, disk flowers yellow. The pappus is composed of capillary bristles. The plant has a haploid chromosome count of 9 (Turner et al. 1961).

Bebbia juncea (Benth.) E. Greene lacks spines, as does *A. intricatus* (A. Gray) Blake, a lower plant up to 1 m tall. Also, *A. intricatus* is discoid.

Sundberg (1987) placed *A. spinosus* in *Erigeron* section *Leucosyris* but did not propose a new combination. The *Erigeron*-like characteristics include golden-brown nerves on the phyllaries and short style branches (Sundberg 1987).

Aster spinosus grows along ditch banks, in moist river bottoms, and in pastures and cultivated fields, often where the soil is alkaline or saline. It can form hedgelike thickets and in some places is a troublesome weed (Parker 1972; Mutz et al. 1979). The broad geographic distribution indicates tolerance for a variety of tropical to warm temperate climates, provided edaphic and soil-moisture needs are met.

Flowers appear from April to October, leaves from January to April. Much of the year the plants are leafless and the green stems are the photosynthetic organ (Mayeux et al. 1979). The characteristically large stands result from extensive vegetative propagation after initial colonization by wind-borne seeds (Mayeux et al. 1979).

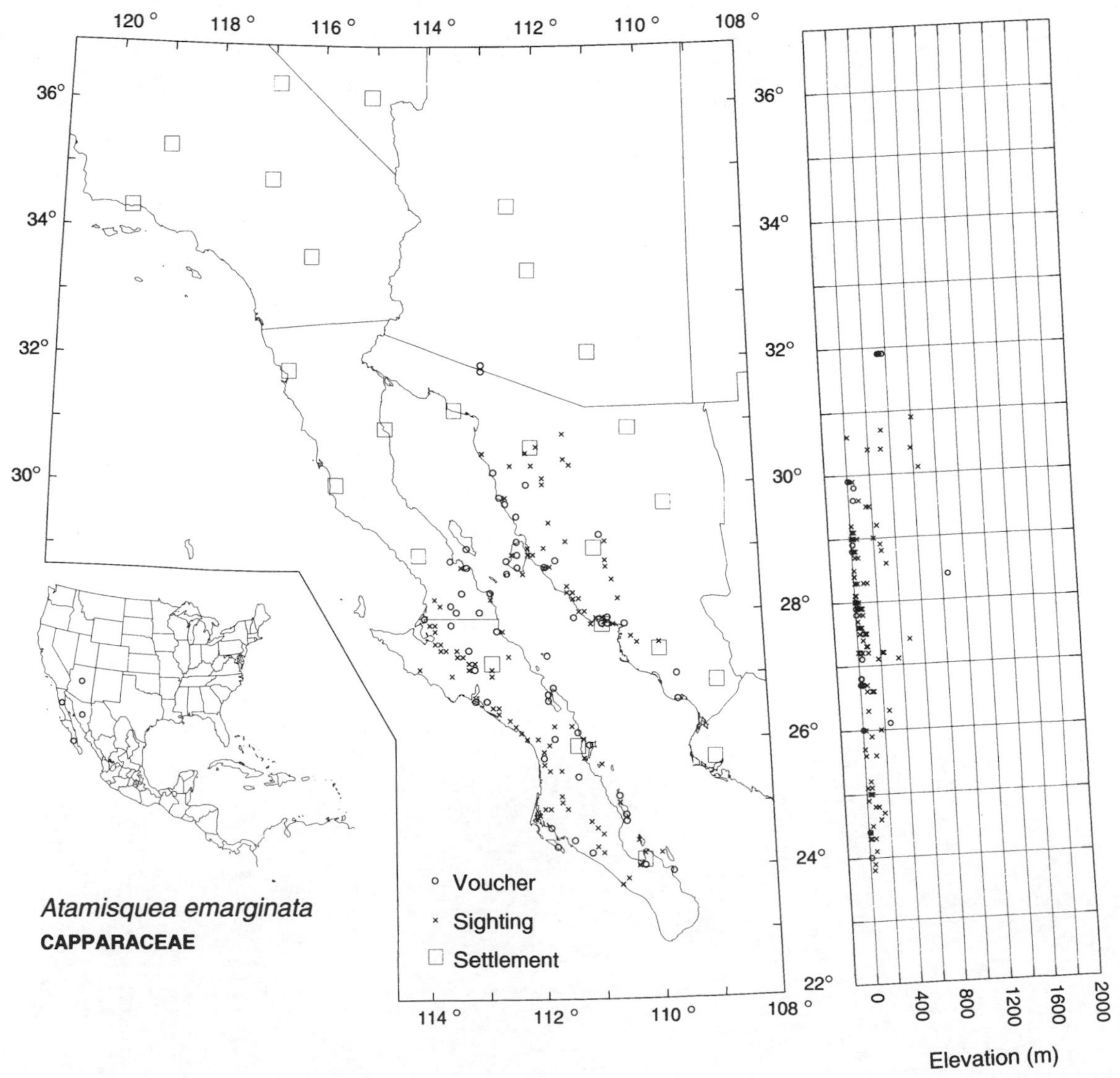

Atamisquea emarginata
CAPPARACEAE

Atamisquea emarginata Miers

[=*Capparis atamisquea* Kuntze]

Palo zorrillo

Atamisquea emarginata is a rigidly branched shrub 1–8 m high. The lateral stems form near-right angles with the main branches. Twigs, sepals, fruits, and undersides of leaves are densely lepidote (pubescent with round scales). The alternate, leathery leaves are linear to oblong and 8–15 mm long. Often their margins are rolled upward. The creamy or yellowish 4-petaled flowers, 7–8 mm in diameter, are in few-flowered fascicles. Fruits are ovoid drupes containing one or two seeds, each partially embedded in a pulpy aril.

No other shrub in the area mapped in this atlas combines lepidote pubescence with oblong, involute leaves.

A plant of wide, shallow washes and gravelly or sandy plains and bajadas, *A. emarginata* is sometimes locally common, more often rare to occasional. The collections at 32.0°N (Charles T. Mason, Jr., in 1959) and 31.9°N (C. F. Harbison in 1939) are from Organ Pipe Cactus National Monument. *Atamisquea emarginata* occurs in Argentina and may have dispersed from South America to North America (Raven 1963). The arillate seeds seem likely candidates for long-distance dispersal via birds.

This is essentially a plant of arid maritime climates. Over most of the range, rainfall is biseasonal with the largest peak from July through September. On parts of the Vizcaíno Plain, near the northwestern limit, winter rains predominate. Cold temperatures and declining summer rainfall seem to be major factors determining the northern limit. Upper elevational and southern limits are most likely shaped by a requirement for open vegetation.

Leaves are evergreen (Shreve 1964). The main flowering period is March to May; flowers occasionally appear in other months, probably in response to rain.

Atriplex barclayana (Benth.) Dietr.

Coast saltbush, chamiso

A prostrate or sprawling plant 2–15 dm tall, *Atriplex barclayana* may be woody or herbaceous. At first densely white-scurfy, the stems become reddish or striped in age. The petiolate leaves, 0.5–5 cm long, are densely white-scurfy above and below and may be obovate or elliptic, toothed or entire. The plants are generally dioecious, with staminate or pistillate flowers in glomerules on terminal panicles. The fruits, 2.5–4 mm long, are flat to subglobose and have entire or dentate margins. The haploid chromosome number is 18 (Nobs 1978).

The much smaller, cordate-clasping leaves of *A. julacea* are markedly crowded and involute. The opposite, fasciculate leaves of *Frankenia palmeri* are strongly revolute and nearly terete.

Wiggins (1964), following Hall and Clements (1923), segregated the Sonoran Desert material into five varieties on the basis of fruit characters. Several of these taxa are largely sympatric, and our map combines all five.

Atriplex barclayana is frequent on sandy soils of the coastal strand, most often on dunes and beaches, where it is occasionally inundated. Yensen and coworkers (1983), noting its occurrence in coastal salt marshes, classified it as a "high-zoned halophyte." It also grows along washes, on alkaline flats, and on rocky slopes, especially where basalt has weathered into clay. Although most populations are near the coast and below 300 m, some also occur inland and as high as 970 m (Reid Moran voucher from Cerro Mechudo, 24.8°N, 110.7°W).

Present on both sides of the gulf, *A. barclayana* has a wider range than *A. julacea,* which occupies the same niche but is nearly confined to the Pacific side of the peninsula. *Atriplex barclayana* replaces *A. leucophylla* (Moq.) D. Dietr., a perennial herb of the California coast, at about 28°N (Johnson 1977). The seeds may be adapted for ocean dispersal, as is the case for *Abronia maritima* Nutt. (Wilson 1976), another halophyte of coastal dunes.

The distribution suggests intolerance of frost. The species grows in relatively warm and arid maritime climates. Contrasting rainfall and temperature regimes have doubtless contributed to its ecotypic diversification.

Because plants of the coastal strand are exposed not only to occasional inundation by sea water but also to saline soils and salt spray, salt tolerance is crucial (Barbour and DeJong 1977). In *A. barclayana* the bladderlike hairs of the foliage no doubt accumulate and exude salt, preventing a buildup of toxic salts in the parenchyma and vascular tissues (Mozafar and Goodin 1970).

As a C_4 plant, *A. barclayana* is well suited to the physiologically and climatically arid environment of coastal dunes (Johnson 1977; DeJong and Barbour 1979). C_4 plants make more efficient use of water, particularly in saline conditions, and more efficient use of nitrogen, especially at high temperatures, than do C_3 plants (Osmond et al. 1980).

Flowering specimens have been collected in every month. Occasional plants flower the first year (Ezcurra et al. 1988). Leaves are evergreen. This species is probably short lived (Ezcurra et al. 1988).

Atriplex barclayana shows promise as a fodder plant that can be irrigated with seawater (Aronson et al. 1988).

Atriplex canescens

(Pursh) Nutt.

[=*A. linearis* S. Watson]

Four-wing saltbush, cenizo, chamiso, chamiza, costilla de vaca, saladillo

Atriplex canescens is an evergreen shrub 1–2 m high with oblanceolate leaves 2–4 cm long and 2–7 mm wide. Plants are usually dioecious, with inconspicuous flowers in terminal panicles. The 4-winged fruits, derived from two pistillate bracts, are 6–14 mm long and may be entire to toothed or, rarely, laciniate. The diploid chromosome number is 18 (McArthur 1977).

The distinctive scurfy pubescence of the foliage readily separates *Atriplex* species from other desert shrubs. Within the genus, its 4-winged utricles set *A. canescens* apart from other species in the Sonoran Desert region. Also, *A. hymenelytra* leaves are silvery and irregularly toothed; *A. lentiformis* leaves are broadly ovate or hastate; and *A. polycarpa* leaves are small and fasciculate.

Atriplex canescens is a species complex with several ecologically diverse and morphologically similar subspecies. These include var. *laciniata* Parish, var. *macilenta* Jepson, subsp. *linearis* (S. Watson) Hall & Clements, and subsp. *macropoda* (Rose & Standley) Hall & Clements (Wiggins 1964). These various taxa are loosely connected by hybridization and polyploidy. Subspecies *linearis*, for example, which resembles *A. polycarpa* and lies within its range, may be of hybrid origin. Biologists have produced viable seed by crossing *A. canescens* with *A. confertifolia* (Torr. & Frem.) S. Watson, *A. corrugata* S. Watson, *Ceratoides lanata* (Pursh) J. T. Howell, *Sarcobatus vermiculatus* (Hook.) Torr., *Grayia spinosa* (Hook.) Moq., and *G. brandegei* A. Gray (Blauer et al. 1976).

Atriplex canescens often grows in nearly pure stands on fine-grained soils and also on dunes, plains, and rocky slopes on a variety of soil types and in several plant communities. Measurements of sodium adsorption ratio, electrical conductivity, and pH suggest that the plants are not as well

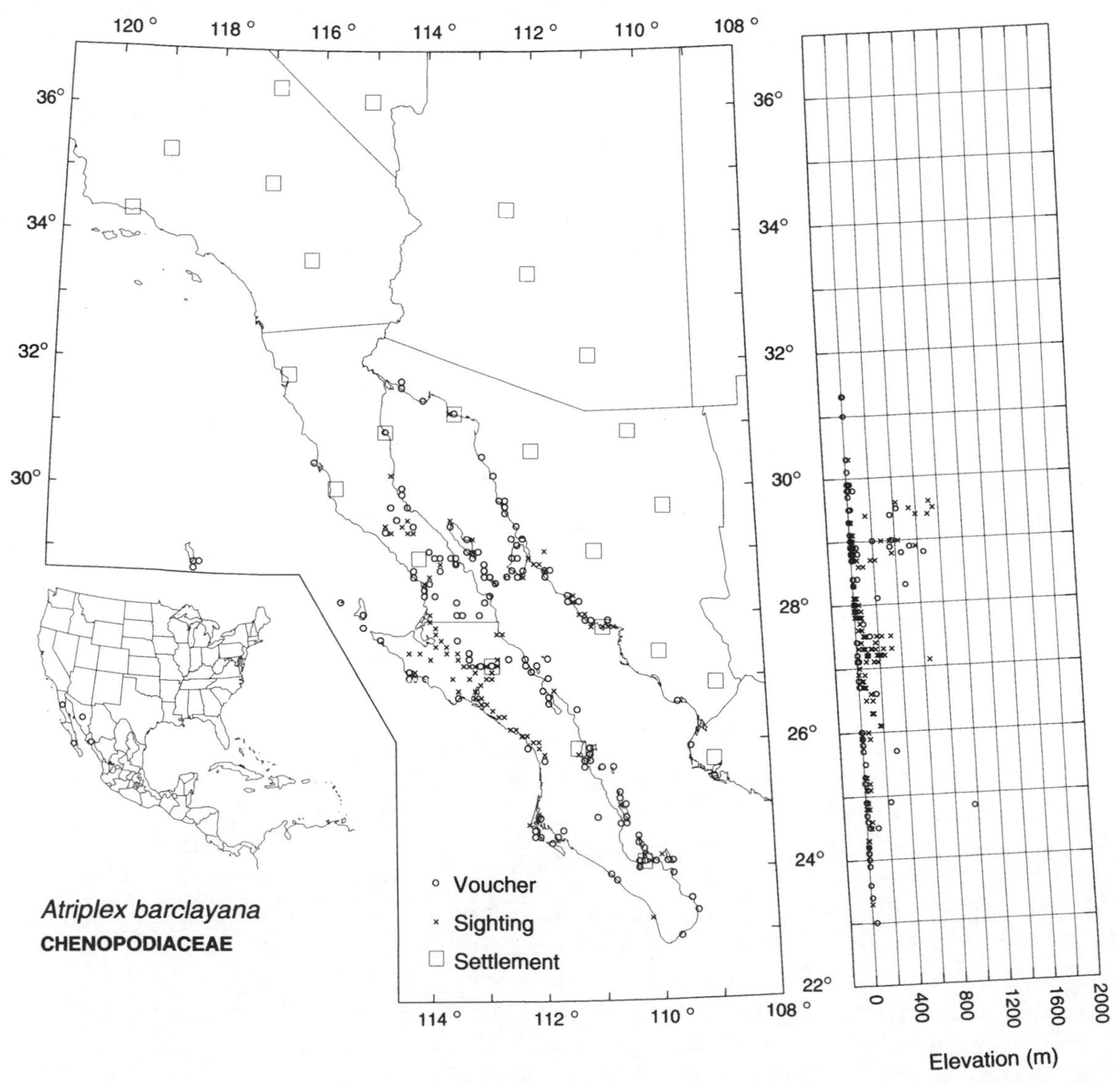

adapted to saline or alkaline conditions as are other species in the genus (Hodgkinson 1987).

One map combines the distributions of all subspecies except for *A. canescens* subsp. *linearis*, which is mapped separately. The collection from Isla Espíritu Santo (24.4°N) was by Ira L. Wiggins in 1979. The sighting of subsp. *linearis* at 1,225 m is from Zabriskie (1979).

This species is the most widespread perennial *Atriplex* in North America. Its different ecotypes and races have adapted to almost every temperature and precipitation regime except humid or tropical ones. Within its range, *A. canescens* subsp. *linearis,* which has linear leaves 2–5 cm long and 1–3 mm wide, occupies the warmer, more arid areas where rainfall is biseasonal. Cold tolerance varies among ecotypes (Van Epps 1975).

In the Sonoran Desert, the main flowering period is March–September (Wiggins 1964); however, individuals may flower sporadically throughout the year. *Atriplex canescens* is wind pollinated (Blackwell and Powell 1981). Utricles mature 14–16 weeks after flowering and often persist until the following spring (Blauer et al. 1976). The leaves are evergreen; given enough rain, new leaves appear from spring through autumn (Ackerman et al. 1980). The carbohydrate reserves in basal stems and taproots are lowest as twigs elongate in the spring, highest in early fall when fruits are ripe and seeds are ready to shatter (Menke and Trlica 1981).

The success of *A. canescens* is due in part to its physiology. As a C_4 plant with Kranz-type anatomy, it has high water use efficiency, particularly in saline conditions, and high nitrogen use efficiency, particularly in high-temperature environments (Osmond et al. 1980). However, when soil water potential declines to –4 MPa, photosynthesis falls to zero and the competitive advantage of the C_4 pathway is lost (Kleinkopf et al. 1980).

Another factor in its success is karyotypic response to environment. Four ploidy levels—2n, 4n, 6n, and 12n—are known (Stutz and Sanderson 1979). They differ in leaf length and width (Dunford 1985), ratio of male to female plants (Bar-

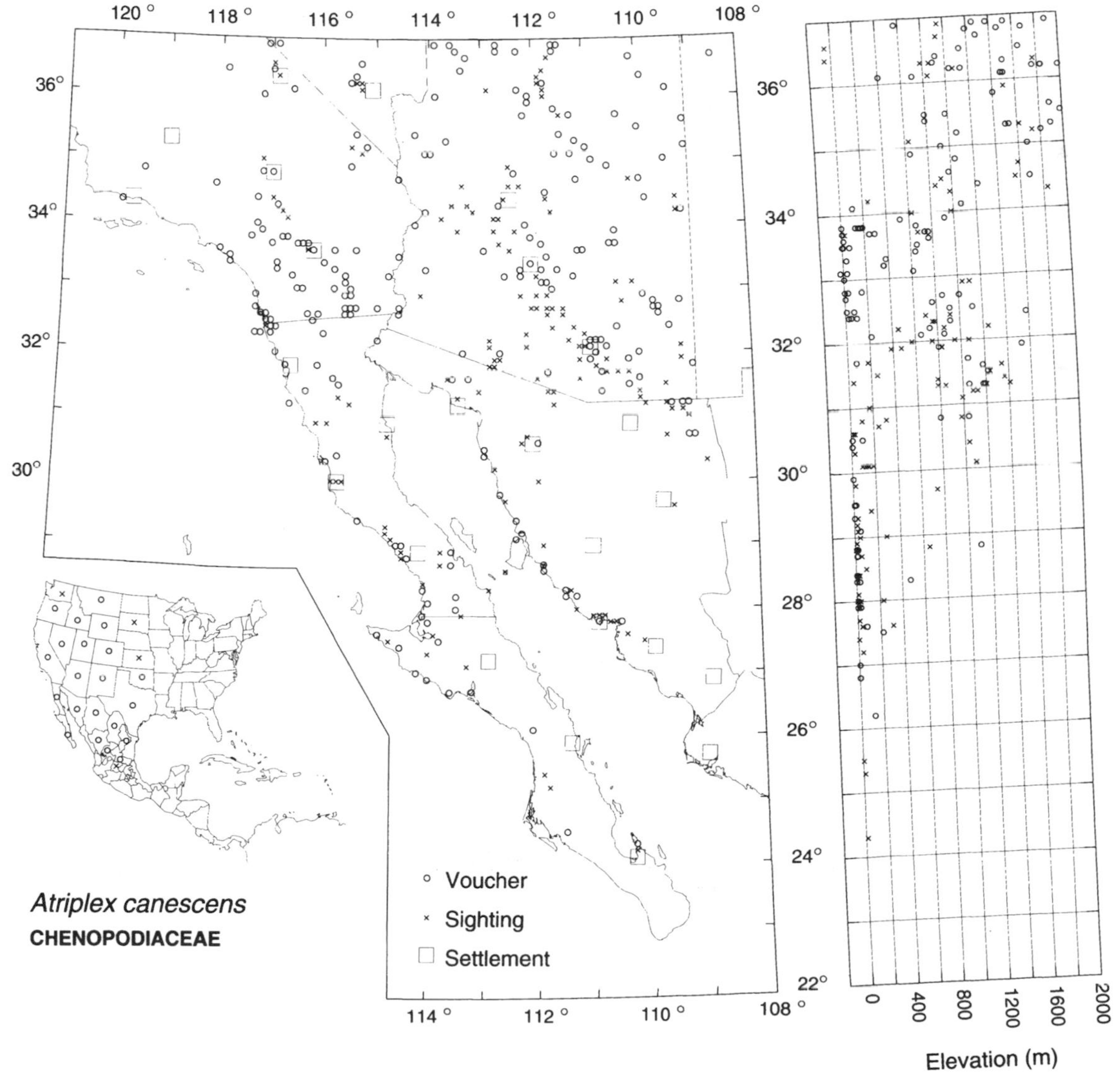

Atriplex canescens
CHENOPODIACEAE

row 1987), and habitat (Stutz and Sanderson 1979; Dunford 1984). The tetraploid form, the most widespread of the four genotypes, apparently arose through autopolyploidy (McArthur et al. 1986). Much of its variation is apparently introduced by means of hybridization with associated species of *Atriplex* (Stutz and Sanderson 1979). Diploid *A. canescens,* a drought-resistant form, grows on coarse sandy soils and dunes from southern Arizona into south-central New Mexico (Stutz and Sanderson 1979; Dunford 1984). Scattered hexaploid populations also occur in the area mapped in this atlas. They apparently arise in response to repeated environmental stresses, to which they are better adapted than the parental tetraploid form (Stutz and Sanderson 1979). In New Mexico and Texas, hexaploids grow on silty floodplain soils (Dunford 1984). The 12n plants are apparently autoallopolyploids resulting from hybridization of *A. canescens* and *A. polycarpa.* Found in the western Sonoran Desert and the Mojave Desert, these 12n plants tend to be very drought resistant and tolerant of both sandy and clay soils (Stutz and Sanderson 1979).

Atriplex canescens can change its sexual expression, thus altering the proportion of male and female plants in a population (McArthur 1977; Tiedemann et al. 1987). Winter cold and competition for water, for example, can turn pistillate plants into staminate plants (McArthur and Freeman 1982). Because internal water stress is seasonally higher in female than in male plants, sex switching could minimize stress at times of critical water deficit (McArthur and Freeman 1982). Also, because pollen production requires less energy than seed production, female plants tend to switch sex after a heavy seed-bearing year, probably to accumulate carbohydrate reserves (McArthur and Freeman 1982).

In one study in Rock Valley, Nevada, *A. canescens* root depth corresponded to penetration of annual rainfall (about 10 cm) (Wallace, Romney et al. 1980). Mojave Desert plants allot about three times as much biomass to stems as to roots (Wallace et al. 1974). Association with

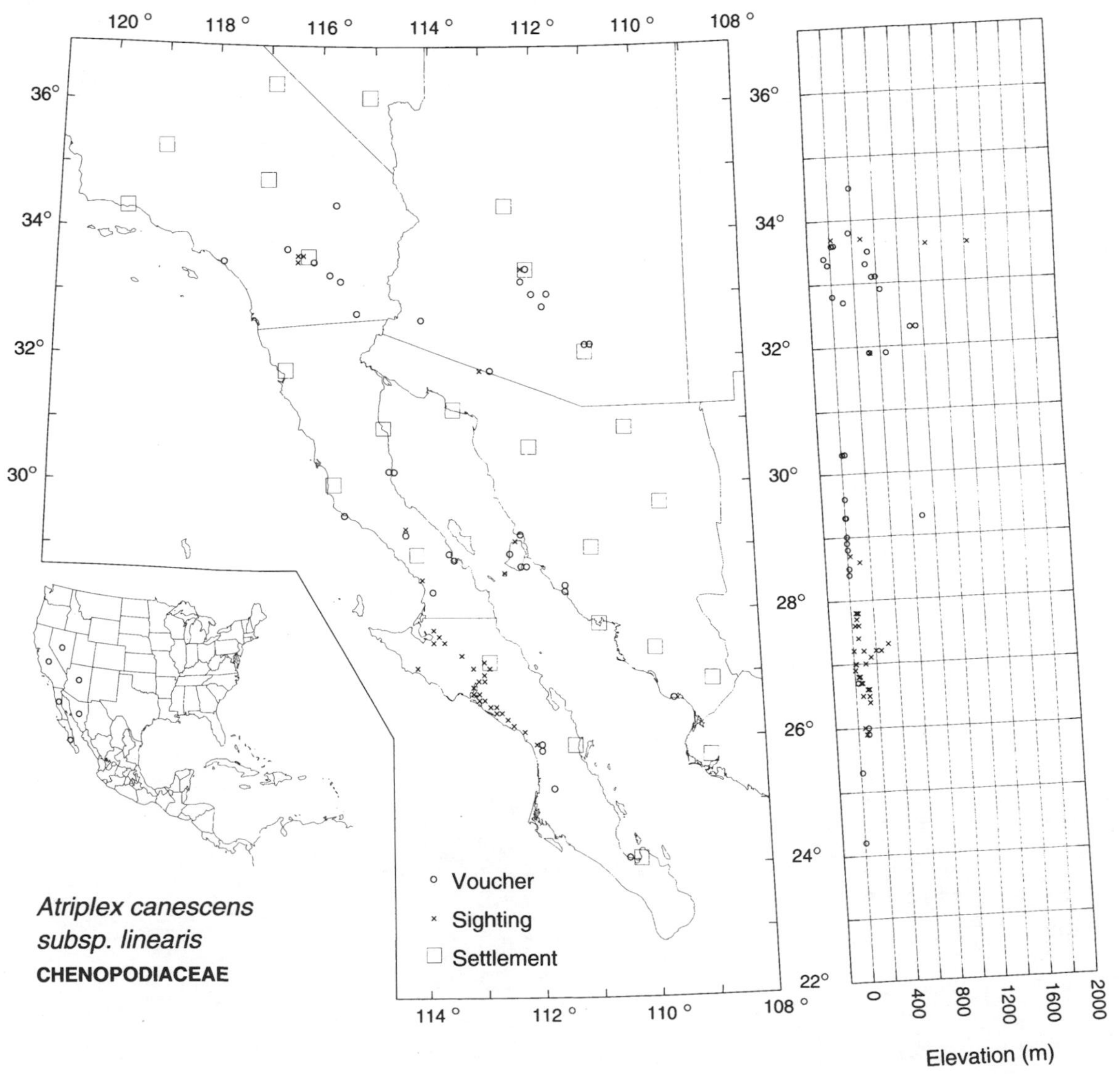

Atriplex canescens
subsp. linearis
CHENOPODIACEAE

vesicular-arbuscular mycorrhizae stimulates growth (Williams et al. 1974) but apparently is not obligatory (Lindsey et al. 1984).

Seeds collected in the fall undergo an after-ripening process that lasts for about 10 months (Springfield 1970). Removing the fruiting bracts, which contain a germination inhibitor, produces germination rates as high as 100% (Al Charchafchi and Clor 1989). Germination in laboratory trials is not improved by treatment with thiourea, hydrogen peroxide, citric acid, or sulfuric acid (Springfield 1970). Cold stratification improves germination of some but not all seed collections (Springfield 1970). Optimum germination temperature apparently varies according to ecotype: seeds collected near Isleta, New Mexico, germinated well at temperatures of 12.8–23.9°C (Springfield 1970), those from West Texas at 15–18°C (Potter et al. 1986). Not all ecotypes can germinate and become established under relatively dry conditions (Springfield 1970; Potter et al. 1986). In the wild, *A. canescens* germinates in response to small amounts of spring or summer rain, but seedlings become established only in years of favorable rainfall (Hennessy et al. 1984).

Matched photographs from the Grand Canyon show individuals aged at least 100 years (Robert Webb, personal communication 1993).

Atriplex canescens is highly prized as browse for all classes of livestock. The winter foliage contains up to 24% protein (Garza and Fulbright 1988). As a secondary or facultative absorber of selenium, *A. canescens* can be mildly poisonous to livestock where selenium occurs in the soil (Blauer et al. 1976). The species is valued for reclamation of mine spoils, where it may enhance secondary succession (Booth 1985), and for revegetation of rangelands (Briggs 1984). The Soil Conservation Service has developed cultivars for both purposes (Briggs 1984). Other strains have been bred for improved forage and fuel production (Sankary and Goodin 1988). *Atriplex canescens* can be propagated from stem cuttings as well as by seed (McArthur et al. 1984).

Seri Indians made an emetic tea from the leaves and used the leafy branches as thatch (Felger and Moser 1985). The pollen is a significant allergen (Mary Kay O'Rourke, personal communication 1992).

Atriplex hymenelytra

(Torr.) S. Watson

Desert holly

A compact shrub 2–12 dm tall, *Atriplex hymenelytra* has deltoid, irregularly toothed leaves 1.5–4.5 cm wide and long. Silvery lepidote when young, the leaves turn red to purplish in age. The plants are dioecious, with inconspicuous flowers in dense, leafy panicles. The fruits are utricles with orbicular bracts 6–10 mm long enclosing the seed, which is 1.5–2 mm long. The diploid chromosome number is 18 (Bassett 1969).

The silvery, irregularly toothed leaves are distinctive. Leaves of *A. canescens* and *A. polycarpa* are elliptic and entire, those of *A. lentiformis* subhastate.

This species grows on rocky slopes and gravel fans, often within runnels. It tolerates moderately saline soils (0.5% soluble salt) (Osmond et al. 1980). In Death Valley, *A. hymenelytra* sometimes grows in sparse but nearly pure stands (Hunt 1966; Gulmon and Mooney 1977).

The eastern disjunct in Arizona (111.7°W) represents a collection by M. French Gilman. The southern disjunct in Baja California is based on Richard S. Cowan's 1963 collection from the vicinity of Bahía de los Ángeles. Wiggins (1964) listed this species for Sonora; we have seen no Sonoran specimens.

Atriplex hymenelytra thrives in the hottest, most arid sections of the Sonoran and Mojave deserts. From its distribution, it appears to be the most drought-tolerant member of the genus in North America. Summers throughout its range are so hot that even where they are also wet, winter is the season when soil moisture is most available.

New shoots and leaves appear from December to April in response to winter rains. The wind-pollinated flowers appear between November and March (Pearcy et al. 1974). Seeds ripen and drop by late spring (usually in May). In summer, leaf color and texture change from pale green and semisucculent to white or gray and brittle. Substantial leaf drop occurs throughout the summer; however, the plants seldom lose all their leaves. Sexual expression and flowering phenology are plastic: when clones of *A. hymenelytra* from Death Valley were grown in the cool, humid climate of coastal California, all were monoecious rather than dioecious and flowered in summer rather than in midwinter (Osmond et al. 1980).

The leaf surfaces are covered with bladderlike hairs that contain salts of inorganic sodium and chlorine ions and organic oxalate ions (Bennert and Schmidt 1983). When the bladders collapse in old age, the salts crystallize on the leaf surface. By removing salt from the leaf, salt-filled bladders evidently prevent accumulation of toxic salts in the parenchyma and vascular tissues (Mozafar and Goodin 1970).

As the water content of the leaves decreases throughout the summer, their salt concentration increases, as does their reflectance. New leaves produced in winter reflect less than 35% of incident radiation at 550 nm; by autumn, reflectance gradually increases to 60% (Mooney, Ehleringer et al. 1977). Leaves show high absorptance during the cool season, when photosynthetic activity is greatest, and high reflectivity during the hot, dry season, when photosynthesis is low. At all seasons of the year, the leaves are steeply angled, thus reducing incident radiation except at low sun angles. Despite their steep angle, the leaves are close to light saturation most of the day, so potential carbon gain is reduced little if at all. At the same time, steep leaf angles reduce heat load, thus improving water use efficiency (Mooney,

Ehleringer et al. 1977). High reflectance and steep leaf angle enable *A. hymenelytra* to remain evergreen in its extremely hot and arid climate (Mooney, Ehleringer et al. 1977).

Despite its wide-reaching root system, *A. hymenelytra* does not maintain uniform water relations throughout the year but exhibits distinct seasonal and diurnal variations in leaf water content, xylem water potential, osmotic potential, and pressure potential (Bennert and Mooney 1979). The plants combine strong regulation of water loss by stomata with low efficiency of the water transport system. They cannot prevent depression of plant water potential as transpiration increases (Sanchez-Diaz and Mooney 1979). In one study, under natural conditions the dawn water potential of *A. hymenelytra* was –2.75 MPa in July (Sanchez-Diaz and Mooney 1979); even lower water potentials (up to –4.35 Mpa) have been measured in August (Bennert and Mooney 1979). The main solutes involved in regulating plant water status are sodium, chlorine, potassium, and oxalate (Bennert and Schmidt 1984).

Phenological observations and gas-exchange measurements show that *A. hymenelytra* grows best under the moderate temperatures of early spring (Pearcy et al. 1974). Photosynthetic rates were substantially reduced in July (4 μmol/dm^2/min) as compared to March (8.1 μmol/dm^2/min). Because leaf conductances were also substantially lower in summer than in spring, whereas xylem sap tensions were higher, Pearcy and coresearchers (1974) concluded that increased water stress is probably involved in the decreased rates of carbon dioxide uptake in July.

In *A. hymenelytra,* the combination of C_4 photosynthesis with a relatively low rate of leaf conductance results in a high ratio of photosynthesis to transpiration at only moderate rates of carbon dioxide uptake. Particularly during dry springs, this high water use efficiency may be more advantageous than a high rate of carbon dioxide uptake. Because *A. hymenelytra* is evergreen and relatively long lived, it can photosynthesize over a long period of

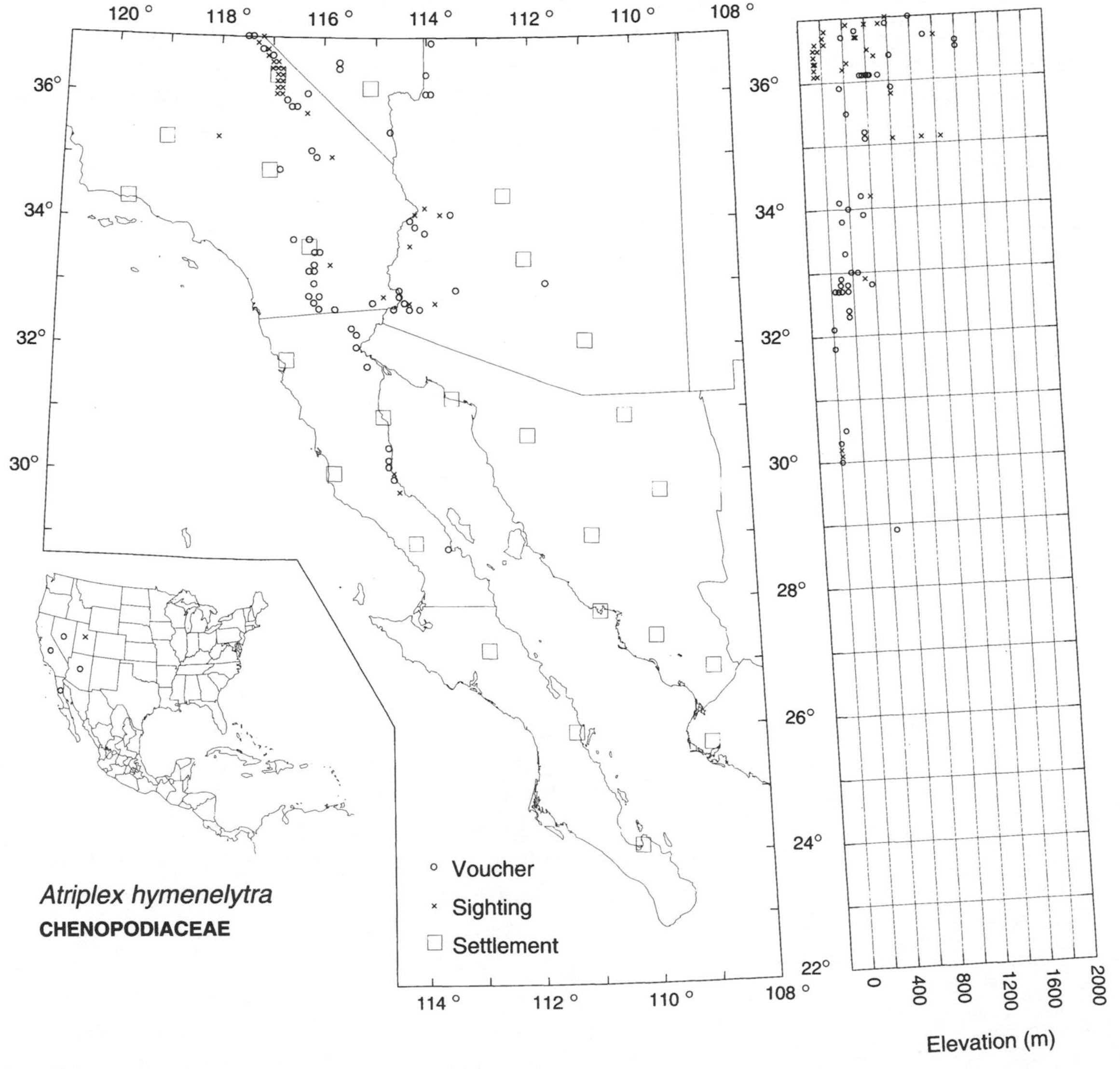

time, thus compensating for low rates of carbon dioxide uptake (Pearcy et al. 1974). Maintenance of internal carbon dioxide homeostasis may also be critical (Osmond et al. 1980).

Atriplex julacea

S. Watson

Growing 1–3 dm tall, this compact subshrub has crowded or overlapping leaves, 2–4 mm long, on brittle stems (figure 26). The fleshy, scurfy leaves are broadly triangular and cordate clasping. Their margins

Figure 26. Atriplex julacea *along the Pacific shore at the mouth of Arroyo San José, Baja California. (Photograph by R. M. Turner.)*

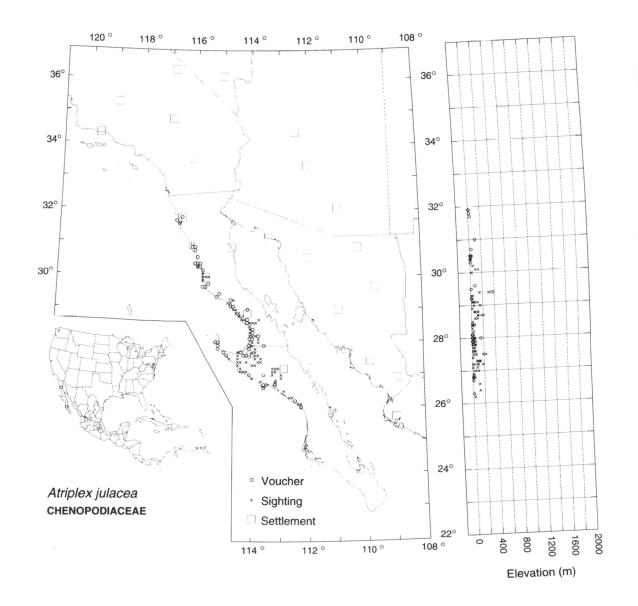

Atriplex julacea
CHENOPODIACEAE

○ Voucher
× Sighting
□ Settlement

Elevation (m)

are rolled inward. The inconspicuous flowers are solitary or clustered in axillary glomerules. The ovate fruits, 3.5–6 mm long and broad, are ornamented with corky knobs and ridges. The haploid chromosome number is 18 (Nobs 1978).

Atriplex barclayana has petiolate rather than clasping leaves. *Frankenia palmeri* has revolute, opposite, fasciculate leaves.

Atriplex julacea is intermediate between the relatively primitive *A. polycarpa* and the more advanced *A. barclayana;* in fact, small, very woody plants of *A. julacea* resemble miniature *A. polycarpa* (Hall and Clements 1923).

A peninsular endemic, *A. julacea* overlaps slightly with its near relative *A. barclayana,* which occupies a similar niche but a wider geographic and elevational range. Basically a halophyte of the coastal strand, where it is occasional to common on dunes and sandy flats, *A. julacea* also grows inland along broad washes and on slopes. It dominates large areas of the western Vizcaíno Plain. Most populations grow below 200 m. Climatically, the species is restricted to arid maritime climates with spring rain and dry, relatively cool summers.

Atriplex julacea, like many other coastal dune plants, has small, densely pubescent, succulent leaves with inrolled margins (Johnson 1977). Dense pubescence decreases absorptance, thereby lowering leaf temperature and reducing water loss (Ehleringer et al. 1976). Inrolled margins and small leaf size also lessen water loss (Johnson 1977), and succulence appears to be a response to salinity (Boyce 1954). Like *A. barclayana,* this species doubtless functions in its highly saline environment by accumulating and exuding salt in the bladderlike hairs on the leaf surface.

The wind-pollinated flowers appear on dioecious plants in spring (February–April) and fall (October–November). Leaves are apparently evergreen.

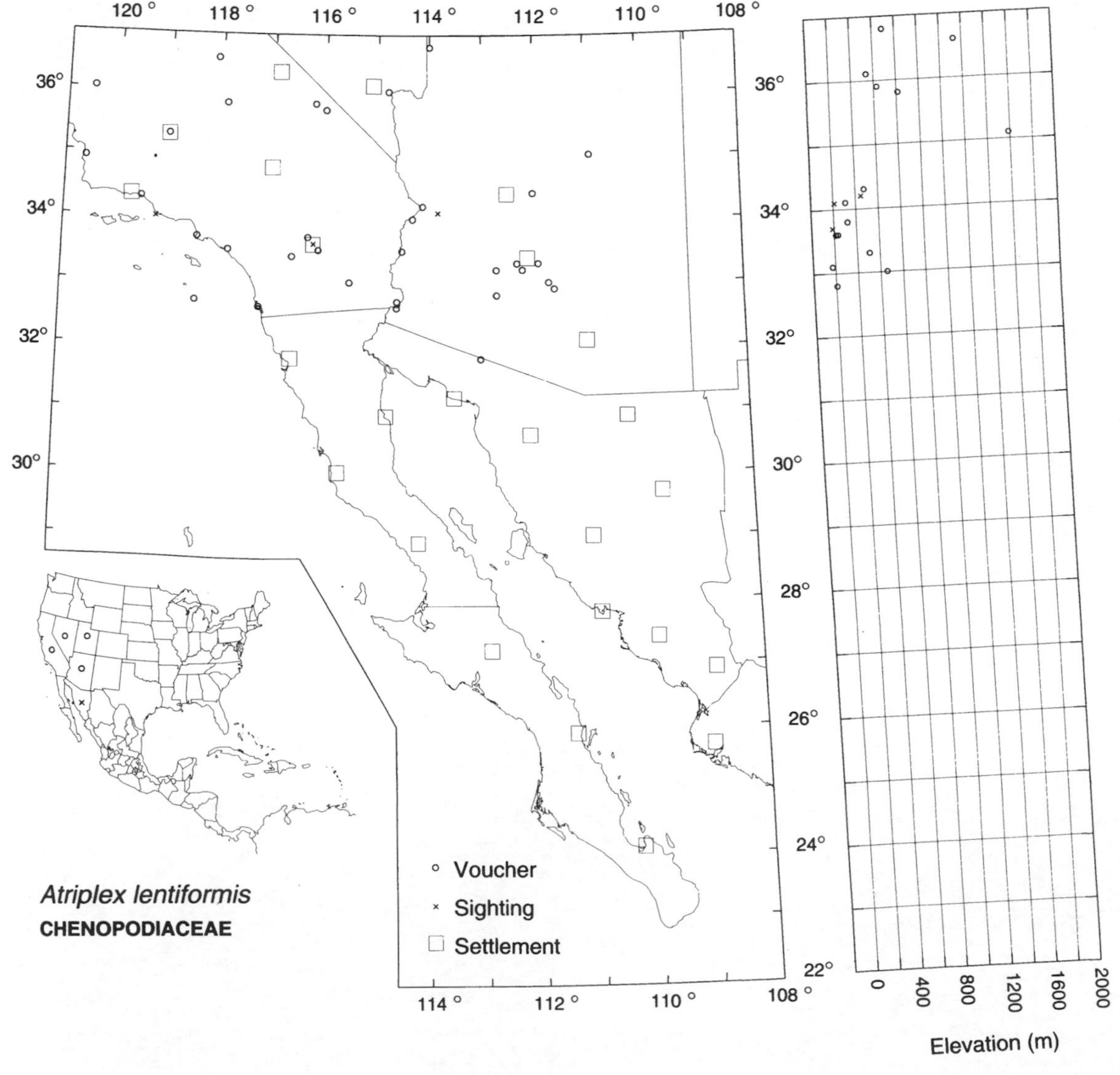

Atriplex lentiformis
CHENOPODIACEAE

Atriplex lentiformis

(Torr.) S. Watson

Quail brush, lens scale

Atriplex lentiformis is a dome-shaped shrub 1–3 m high with whitish or pale gray branches. Plants are generally dioecious, occasionally monoecious, with inconspicuous flowers in staminate and pistillate panicles. The entire, ovate leaves, 1.5–4 cm long and 0.4–2 cm broad, are truncate to cuneate or hastate at the base. They are scurfy pubescent and appear bluish or grayish green. The lens-shaped fruits develop within a pair of bracts that tightly enclose the seed. They are orbicular to somewhat oblong and 2–5 mm in diameter. The haploid chromosome number is 9 (Nobs 1978).

Leaves of *A. hymenelytra* are similar in size but irregularly toothed; leaves of *A. canescens* are somewhat smaller and elliptic to oblanceolate rather than ovate.

The taxon in the Sonoran Desert region is *A. lentiformis* subsp. *lentiformis*. Subspecies *breweri* (S. Watson) Hall & Clem., a plant of coastal marshes and California valley grasslands, has larger leaves (3–5 cm long and 1.5–4.5 cm wide) and longer fruiting bracts (4–7 mm) (Munz 1974).

A halophytic phreatophyte, *A. lentiformis* is most abundant where water tables are close to the surface and soil salinity is relatively high, as on floodplains, playas, and valley bottoms (Benson and Darrow 1981) and in sewage effluent (Correll and Correll 1972). Its erratic distribution in the area mapped in this atlas is partly a measure of the spotty occurrence of its preferred habitat and partly, perhaps, a result of infrequent collection.

Wiggins (1980) cited *A. lentiformis* without locality from "central Baja California." The only purported voucher of *A. lentiformis* from Baja California that we have seen (Peta Mudie, at 28.4°N, 113.9°W) is a sterile specimen. This collection was recently annotated to *A. barclayana* based on growth form and leaf shape (Geoffrey A. Levin, personal communication 1993). It seems likely that *A. lentiformis* might occur on the Colorado River delta in northwestern Baja California.

Most populations apparently flower in summer and fall (July–November); occasional plants may bloom in spring (April–June). Leaves and stems grow fastest in summer (Pearcy and Harrison 1974). The flowers are wind pollinated. The ratio of male to female plants within a population varies from year to year (Freeman and McArthur 1984). *Atriplex lentiformis* fixes nitrogen (Malik et al. 1991).

Atriplex lentiformis has the Kranz-type leaf anatomy of a C_4 plant (Pearcy and Harrison 1974). Across the range of the species, there are ecotypic differences in acclimation (Pearcy 1976, 1977). In comparing photosynthesis between coastal and desert populations, Pearcy and Harrison (1974) found that the maximum rates of carbon dioxide uptake in coastal plants (15.8 μmol/dm^2/min) was about the same as for desert plants (16.5 μmol/dm^2/min). However, the two populations differed greatly in their photosynthetic response to temperature: the thermal optimum for photosynthesis at the coast was 32°C, in the desert 44°C. Moreover, below 36°C, rates of carbon dioxide uptake were higher in coastal plants than in desert plants, whereas above 36°C, the situation was reversed. Plants grown at high temperatures show increased thermal stability of the photosynthetic apparatus, which accounts in part for the greater acclimation of desert plants (Pearcy 1977; Pearcy et al. 1977).

Because of its lush, gray foliage, this is the best *Atriplex* species for ornamental use, especially when given adequate water and pruning (Duffield and Jones 1981). Seeds must be buried to germinate. Best germination is obtained with alternating rather than constant temperatures (Young et al. 1980). Semihardwood stem cuttings root easily (Everett et al. 1978). Horticultural varieties are commercially available (Duffield and Jones 1981).

Cahuilla Indians of southern California ground the seeds, then boiled the meal in salted water. Pima Indians prepared the seeds for winter storage by roasting, drying, and parching; they also boiled young shoots as a pot herb (Standley 1923). The species shows good commercial potential as a livestock fodder that can be irrigated with seawater (Pasternak et al. 1985; Watson et al. 1987).

Atriplex polycarpa

(Torr.) S. Watson

Desert saltbush, desert sage, cattle spinach, all-scale, chamizo, cenizo

An intricately branched shrub 0.5–2 m high, *Atriplex polycarpa* has small, oblong, fasciculate leaves mostly less than 1 cm long and up to 3 mm wide. The plants are dioecious with inconspicuous flowers in dense terminal panicles. The fruits are utricles 2–4 mm long with dentate margins and, usually, tuberculate faces. Diploid chromosome counts of 18 (Nobs 1978) and 72 (Bassett and Crompton 1971) have been reported.

Leaves of *A. canescens* are linear or oblanceolate rather than oblong and are usually longer. *Atriplex lentiformis* has ovate or hastate leaves up to 4 cm long, and *A. hymenelytra* has irregularly toothed leaves up to 4.5 cm in length.

Often dominant on valley bottoms and playa margins, *A. polycarpa* is also occasional on rocky slopes and dunes and along washes. The plants tolerate moderately saline soils (Chatterton and McKell 1969) but are not restricted to them: in Death Valley, California, *A. polycarpa* grows in sparse, essentially pure stands on nonsaline gravel fans (Hunt 1966).

The disjunct voucher from Isla Espíritu Santo (24.4°N) is Ivan M. Johnston's, made in 1921. Although the easternmost occurrence at 109.5°W is close to the New Mexico border, botanists have not identified *A. polycarpa* in that state (Wagner and Aldon 1978). The species ranges from regions of winter rain to regions where biseasonal rain combines with hot summer temperatures. The effects of cold—frost or perhaps soil too cool to allow moisture extraction—truncate the upper elevational limit. Extensive geographic overlap with *A. canescens* suggests a complex history of interactions between the two.

In the San Joaquin Valley, California, new growth appears from April through June (Sankary and Barbour 1972). Individuals flower sporadically throughout the year; in the area mapped in this atlas, most flower from September through November. The flowers are wind pollinated.

Atriplex polycarpa is a C_4 plant with typical Kranz-type anatomy (Osmond et al. 1980). Advantages conferred by C_4 photosynthesis include high water use efficiency, particularly in saline conditions, and high nitrogen use efficiency, particularly in high-temperature environments (Osmond et al. 1980). These attributes help compensate for the low water and nitrogen status typical of *A. polycarpa* habitats. Even when present in low concentrations, nitrogen does not limit growth in *A. polycarpa*. The plants may have developed a unique nitrogen absorption-metabolism system (Chatterton et al. 1971). Like *A. lentiformis, A. polycarpa* acclimates to changes in environmental temperature (Chatterton et al. 1970).

The plants can be quite salt tolerant: in one experiment, adults tolerated soil salinities of up to 5% (Sankary and Barbour 1972); in another, greenhouse-grown plants survived salt concentrations equal to seawater (Wallace et al. 1982). The topsoil in *A. polycarpa* communities frequently has a low electrical conductivity, usually below that considered saline. At a depth of 30–60 cm, however, these same soils often contain enough soluble salts to be classified as saline (Chatterton and McKell 1969). *Atriplex polycarpa* populations differ in salt tolerance, and even within a population salt tolerance may vary considerably (Chatterton and McKell 1969).

The mechanism of salt tolerance in *A. polycarpa* is both anatomical and physiological (Chatterton et al. 1971). Bladderlike trichomes on the leaves contain salts that crystallize on the leaf surface

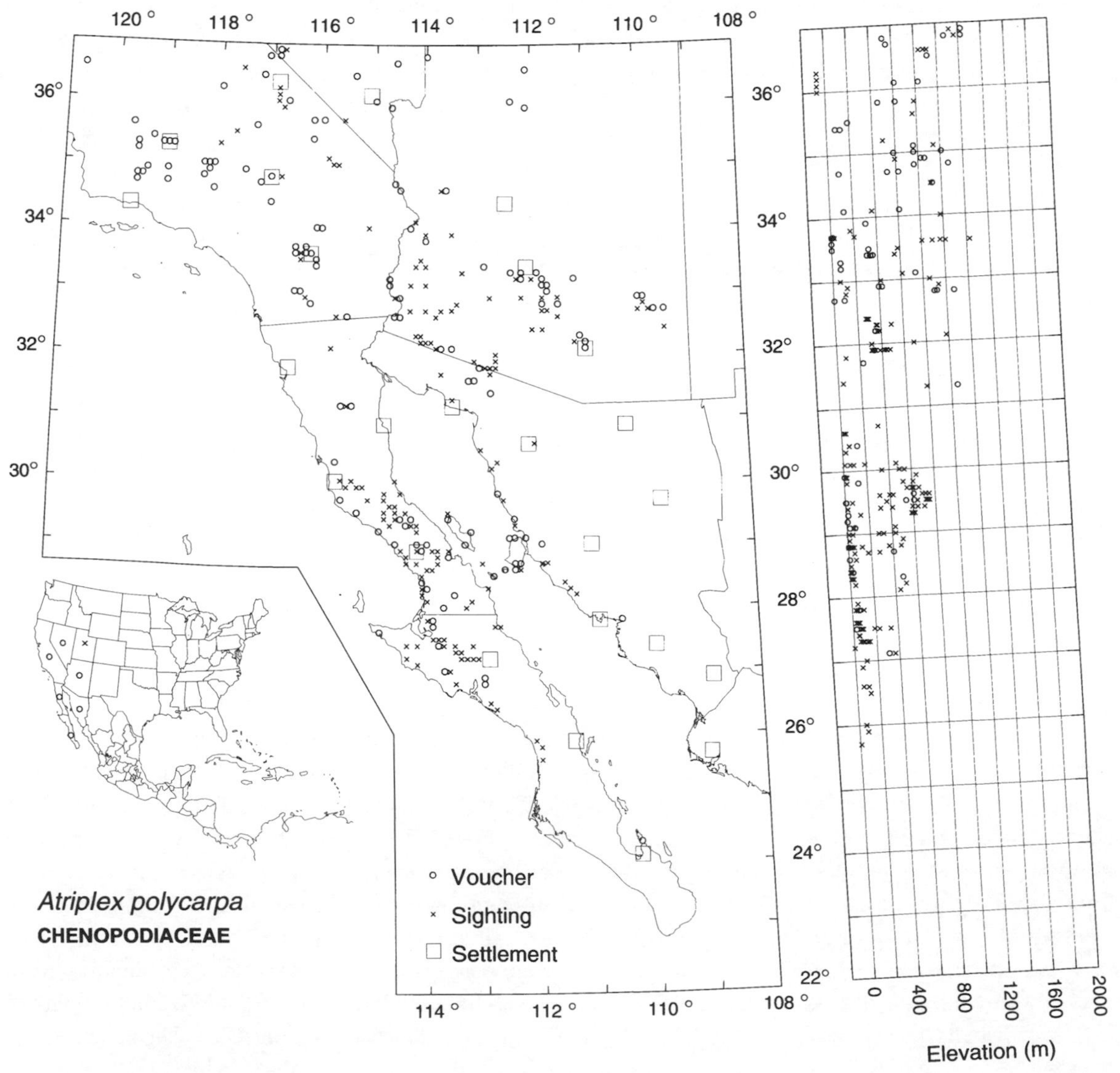

when the bladders collapse. In *A. halimus* L. (and presumably in other species of *Atriplex*), these trichomes play a significant role in removing salt from the leaf and in preventing salt accumulation in the parenchyma and vascular tissues (Mozafar and Goodin 1970). Sodium is the primary cation in *A. polycarpa* leaves (Wallace et al. 1973).

Salt poisoning in some plants may be due less to the direct osmotic effects of salt itself than to salt-induced accumulation of amino acids and other intermediate products of protein metabolism, which have deleterious effects at high concentrations. Chatterton and coworkers (1971) hypothesized that in *A. polycarpa,* the protein might form labile bonds with anions and cations in the plant cells, thus reducing their ability to disturb normal cell metabolism and thereby conferring increased salt tolerance.

Atriplex polycarpa is relatively drought resistant. Mature plants can endure soil dry enough to cause permanent wilting in sunflowers and can recover from shoot water potentials as low as –6.9 MPa (Sankary and Barbour 1972). Diurnal fluctuation of shoot water potential is relatively minor; even under summer drought conditions in the southern San Joaquin Valley, California, it varied only 1 MPa over the course of a day. The difference between shoot water potential in April and August was also minor (Sankary and Barbour 1972). *Atriplex polycarpa* can maintain turgor at low plant water potentials, allowing access to a greater volume of soil water (Monson and Smith 1982).

Saponin (a germination inhibitor) in the foliage and seed coats may account for the low germination in greenhouse and field plot experiments (Askham and Cornelius 1971, 1972). Germination can be speeded up and increased by stratification or treatment with activated carbon (Graves et al. 1975). The seeds germinate over a wide range of temperatures (3–33°C); the optima (9–15°C) are close to the prevailing temperatures at the time of dispersal (Sankary and Barbour 1972). Some ecotypic variation in the optimum may occur across the range of the species. Germination is reduced when the sodium content of the soil reaches 5,000 ppm and is completely inhibited at 20,000 ppm (Chatterton and McKell 1969). Seeds germinate readily when watered with solutions containing 0.25, 25, and 50 ppm boron (Chatterton et al. 1969).

Establishment of *A. polycarpa* in the wild can be inhibited by competition with introduced annuals, particularly *Bromus rubens* L. and *Erodium cicutarium* (L.) L'Her. (Sankary and Barbour 1972). *Atriplex polycarpa* seedlings are more likely to reach maturity on soils that are too saline for the exotics (Sankary and Barbour 1972). Seedling survival is retarded by deep burial (2.5 cm or greater), perhaps because the seedlings exhaust their food reserves before they can begin photosynthesis (Nord et al. 1971).

Atriplex polycarpa is by far the most important browse plant in its range, particularly on moderately saline soil (Kearney and Peebles 1960). Due to heavy grazing, the range of the species in the San Joaquin Valley has greatly contracted over the past 100 years (Sankary and Barbour 1972). Much of the area formerly covered by dense stands of the plants in the Coachella and Gila River valleys was under cultivation by 1924 (Shantz and Piemeisel 1924).

Seri Indians used the wood for fuel. Mashed leaves and twigs added to water made a solution for shampooing or for washing clothes (Felger and Moser 1985).

Avicennia germinans

(L.) L.

Black mangrove, honey mangrove, mangle blanco, mangle prieto, mangle negro, culumate, chifle de vaca, mangle bobo, palo de sal, manglecito, puyeque, mangle salado, saladillo, arbol de sal, istaten

A shrub or tree up to 25 m tall (more often up to 16 m in the area mapped in this atlas), *Avicennia germinans* has leathery, opposite leaves 4.5–15 cm long and 2–4 cm wide. They are green and glabrate above, densely white-puberulent beneath. The whitish, campanulate-rotate flowers, 12–20 mm long, are in dense spikes at branch tips and in leaf axils.

Three other mangroves (*Laguncularia racemosa, Rhizophora mangle,* and *Conocarpus erecta* L.) grow in the area mapped in this atlas. (See Walsh 1979 for a comparison of their physiological and ecological attributes.) Leaves of *Laguncularia* and *Rhizophora* are glabrous above and below. *Conocarpus* leaves have two small glands at the juncture of the petiole and blade.

Across its wide range, *A. germinans* varies greatly in foliage pubescence and other characters. Although various forms have been described as species or varieties, they intergrade considerably and are not segregated geographically (Moldenke 1960). The anatomical variation often can be correlated with environmental features such as soil salinity (Schnetter 1978, 1985), as can morphological characters such as height, branching pattern, biomass allocation, leaf life span, leaf shape, and leaf length-to-width ratio (Soto and Corrales 1987; Soto 1988). Populations on opposite sides of the Gulf of Mexico–Caribbean Sea vary in isozyme patterns but have not been segregated taxonomically (McMillan 1986). The similar mangrove of Africa, sometimes reported as *A. germinans,* is *A. africana* P. Beauv. (Moldenke 1960).

Avicennia germinans can dominate plant communities of brackish or salt water in inlets, bays, and coastal mud flats. In Sonora and elsewhere, the various mangroves (*Rhizophora, Avicennia, Laguncularia, Conocarpus*) often grow in more or less distinct zones (Felger 1966; Rabinowitz 1978; Odum and McIvor 1990). Along coastal Sonora, cover in mangrove thickets can reach 100% (Felger 1966).

On the mainland, *A. germinans* grows as far north as Puerto Lobos (30.3°N). In 1935 Phillip Lichty found it at Puerto Libertad (29.9°N). We have not seen it there in recent years. The northernmost peninsular sighting is Wiggins's from Bahía de los Ángeles (personal communication 1970). According to Richard Felger (personal communication 1989), *A. germinans* grows along the coast between Guaymas and the Sonora-Sinaloa border.

On the Texas coast, *A. germinans* appears to be limited by cold: in 1983 and 1989, extended subfreezing temperatures caused 80–85% mortality (Sherrod and McMillan 1985; Lonard and Judd 1991). Earlier, in response to generally mild and moist conditions in the 1970s, the species expanded its Texas range. Clearly, its northern limits are far from static (Sherrod and McMillan 1981, 1985). McMillan and Sherrod (1986) reported that in the Gulf of Mexico, northern populations of *A. germinans* have greater chill tolerance than do southern populations. The same is true of Caribbean mangroves (Markley et al. 1982). In Florida, recovery from frost damage appears to be faster where salinity is lower (Lugo and Zucca 1977). Florida plants often resprout from the roots after frost (Odum and McIvor 1990). In the area under consideration, *A. germinans* also seems limited by cold; however, the species does not extend as far north along the Pacific coast as we would expect, which suggests that the Sonoran Desert populations are more frost sensitive than those in the northern Gulf of Mexico. If salinity does indeed affect cold recovery, distributional limits in the Gulf of California may reflect the combined interaction of cold and salinity stresses.

On the Pacific coast of Baja California, *A. germinans* is the southernmost of the three mangroves; on the gulf side of the peninsula, it is the northernmost. Perhaps this species cannot disperse and establish as well as the other mangroves along the Pacific coast, where wave energy is higher and tidal amplitude is lower than in the Sea of Cortez. Alternatively, the cooler summer temperatures of the Pacific coast may affect *A. germinans* more adversely than they do *R. mangle* or *L. racemosa*.

Avicennia germinans has been collected in flower in every month from November through July, and it seems likely that plants flower and fruit sporadically throughout the year. Bees are probably

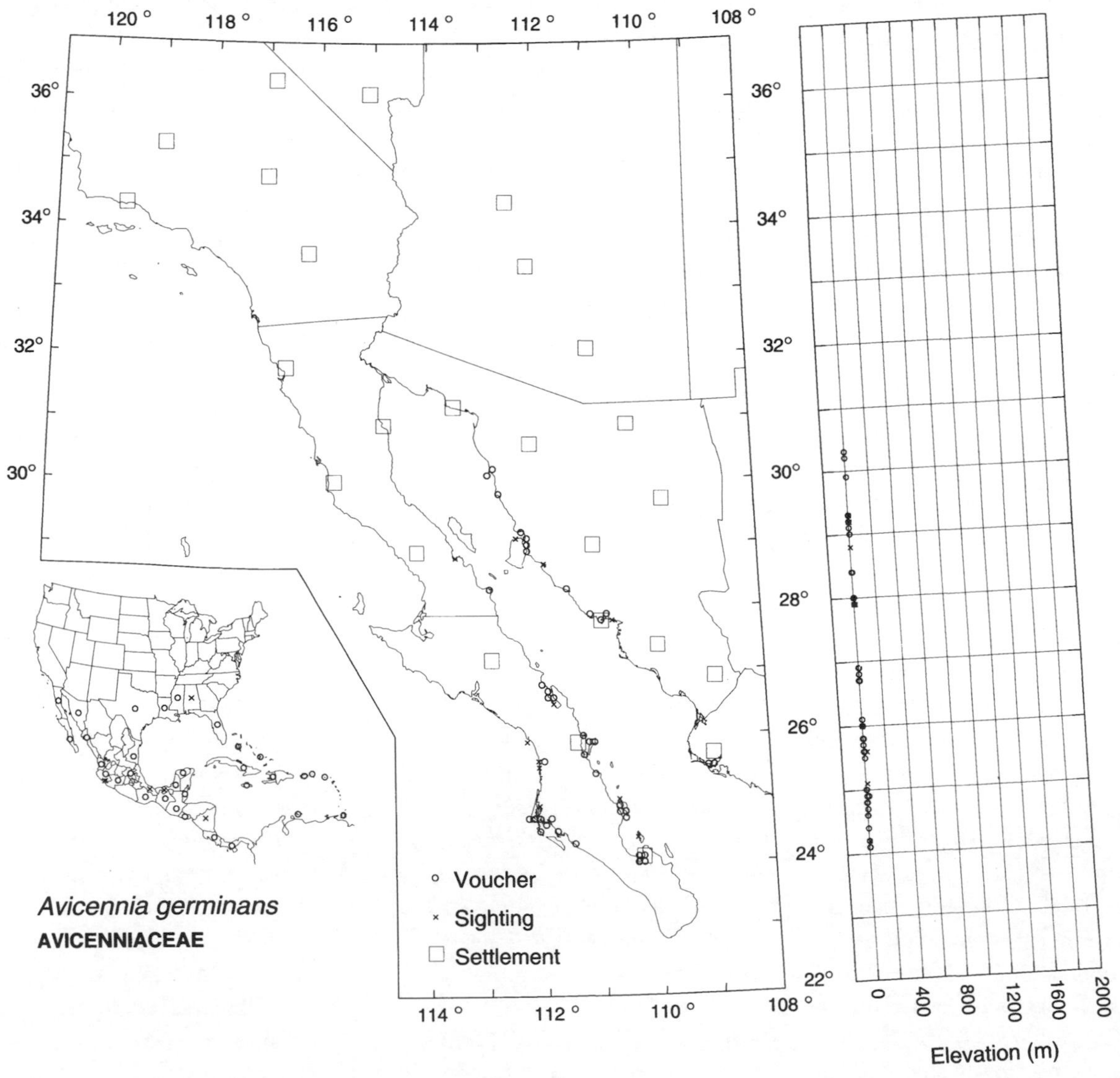

Avicennia germinans
AVICENNIACEAE

the most important pollinators (Tomlinson 1986). Ants use the nectar but do not pollinate the flowers (Rico-Gray 1989).

Subterranean roots of *Avicennia* species typically grow in an anaerobic environment. Pneumatophores, pencil-like structures that rise up to 30 cm from the subterranean roots, provide the necessary gas exchange (Tomlinson 1986; McKee et al. 1988). When pneumatophores are eliminated, the maximum oxygen content of the subterranean roots drops from 10–18% to 1% or less (Scholander et al. 1955). For seedlings, metabolic adaptations may be as important as internal oxygen diffusion in tolerance of flooding (McKee and Mendelssohn 1987).

Like all mangroves, *A. germinans* grows in saline environments. All the available evidence suggests that *A. germinans* excretes rather than excludes salt (Scholander et al. 1966; Mallery and Teas 1984). The xylem sap contains 4–8 mg NaCl/ml (Scholander et al. 1966), a relatively high concentration, showing that not all the salt in the environment is excluded at the roots (Tomlinson 1986). Most of the xylem-sap sodium is excreted metabolically via salt glands, multicellular hairs on both the upper- and the underside of the leaves (Tomlinson 1986). The precise mechanism of salt secretion through these hairs is not understood; however, the process requires energy and can be stopped by metabolic inhibitors (Tomlinson 1986).

In northern Venezuela, photosynthetic rates vary seasonally: total net diurnal carbon dioxide uptake in the dry season is 61% of the rainy season value (J. A. C. Smith et al. 1989). Seasonal variation in photosynthesis corresponds with fluctuations in the moisture and salt content of the soil, which in turn vary with inundation and evaporation (J. A. C. Smith et al. 1989). Changes in soil moisture status are reflected in large seasonal shifts in dawn xylem tension, from 1.34 MPa in the rainy season to 5.50 MPa in the dry season (J. A. C. Smith et al. 1989). In contrast to mature plants, flooded seedlings show substantial reduction in net photosynthesis, presumably a consequence of decreased leaf area (Pezeshki et al. 1990).

Fruits of *A. germinans* are "incipiently viviparous" (Tomlinson 1986); as in other mangroves, the embryo is also the unit of dispersal. *Avicennia* propagules can withstand immersion in salt water for up to 110 days (Rabinowitz 1978). Because they will not sink of their own accord during this period, they must be stranded above the high-tide line for about seven days for rooting to occur. This characteristic restricts *Avicennia* species to higher ground within the tidal zone where inundation is less frequent and may account in part for the typical zonation of the three species where they grow together (Rabinowitz 1978). Other factors may also influence zonation in mixed mangrove communities: *A. germinans* tolerates more anaerobic substrates than does *R. mangle* (Thibodeau and Nickerson 1986); *R. mangle* and *L. racemosa* interact with *A. germinans* to lower its percent cover (Lopez-Portillo and Ezcurra 1989); and seed predation in some locations is high enough to exert an influence on the local distribution and cover of *A. germinans* (T. J. Smith et al. 1989).

McMillan (1971) found that rooting of propagules was not limited by salinity, as earlier authors had suggested, but that turbulence, water depths greater than 5 cm, and high air temperatures (39–40°C for 48 hours) all inhibited seedling establishment. He concluded that "vivipary as a niche property allows floating of seedlings until proper conditions for establishment are contacted and promotes rapid root development in soil at water depths less than 5 cm. . . . The total of niche properties insures seedling establishment in the Texas habitat under winter and early spring conditions" (McMillan 1971:929).

Wood of *A. germinans* has been widely used for many purposes, and the bark is employed in tanning (Standley 1924).

Bauhinia divaricata

L.

[=*B. peninsularis* Brandegee]

Orchid tree, pata de vaca, pie de vaca, pato de chivo, pata de cabra, pie de cabra, barba de mantel, calzoncillo

A tree up to 9 m tall over much of its range, *Bauhinia divaricata* is an open shrub 2–3 m tall in the Sonoran Desert region. The leaves, 3–6 cm wide and long, are rounded, deeply lobed, and parallel veined. The showy white flowers (2 cm long) are in racemes. Both perfect and dioecious flowers can occur on the same plant (Tucker 1988). Oblong and flat, the elastically dehiscent pods are 1–1.8 cm wide and 5–8 cm long. The diploid chromosome number is 28 (Wunderlin 1973).

Originally described as *B. peninsularis*, a peninsular endemic, this plant has since been subsumed under *B. divaricata*, a widespread species that is highly variable in leaf size, shape, and pubescence (Wunderlin 1983). Although several geographic races have been segregated as species, considerable intergradation occurs, and botanists have found no completely satisfactory way to separate them taxonomically (Wunderlin 1983). The large, bilobed leaves are distinctive.

Bauhinia divaricata reaches the region included in this atlas only in the Cape Region of Baja California Sur, where it is common in coarse soils of the coastal plain. Throughout most of its range, *B. divaricata* is an early successional plant of disturbed sites in tropical deciduous forest (Wunderlin 1983). In the Cape Region, however, it grows in stable Sinaloan thornscrub. The disjunct populations here may be relicts that predate gulf-floor spreading. It is not found on the gulf islands (Moran 1983b).

Flowers appear throughout the year, most abundantly in the rainy season (Wunderlin 1983). Bats, butterflies, and moths are common pollinators in the genus *Bauhinia* (Wunderlin 1983).

Indians used the wood for making bows (Standley 1922).

Bergerocactus emoryi

(Engelm.) Britton & Rose

[=*Cereus emoryi* Engelm.]

Bergerocactus emoryi (figure 27) is a thicket-forming cactus that branches prolifically from the base. The densely spiny stems, 0.5–2 m tall and 2–5 cm thick, are mostly erect or ascending but may be procumbent. Ribs usually number 16–18 but may be as few as 14 or as many as 21 (Moran 1966). Each areole has 35–50 needlelike spines, the central ones 3–7 cm long. New spines at the stem tip are yellow; they gradually turn black or brown as they age. Flowers are lemon-yellow and about 2–6 cm broad. The spiny, globular fruits, 2–4.5 cm in diameter, are flushed with purplish red. The haploid chromosome number is 22 (Pinkava et al. 1973).

Bergerocactus emoryi looks like no other native cactus in its range and can be recognized by its distinctive spines and thicket-forming growth.

Benson (1982) treated this species as *Cereus emoryi* Engelm. We follow Gibson and Horak (1978). The anatomy of flowers, fruits, seeds, and pollen suggest that it is closely related to tribe Pachycereae and may represent a connection between Pachycereae and Echinocereae (Gibson and Horak 1978). Moran (1962a) described a natural hybrid of *B. emoryi* with *Myrtillocactus cochal.* A natural hybrid between *B. emoryi* and *Pachycereus pringlei* was described in 1900 as *Cereus orcuttii* K. Brandegee (Moran 1962b).

Bergerocactus emoryi grows on hillsides and bluffs near the coast, mostly from sea level to 500 m. Throughout its range, rain falls mainly in winter, and temperatures are moderated by maritime air. Toward its northern limit, it is more closely restricted to sea level. These northern coastal sites are among the most xeric available, which suggests that *B. emoryi* does not compete well in more mesic habitats at higher elevations (see also Cody 1984). Cold temperatures and increasing rainfall might also deter upward penetration. An occasional to common component of Californian coast-

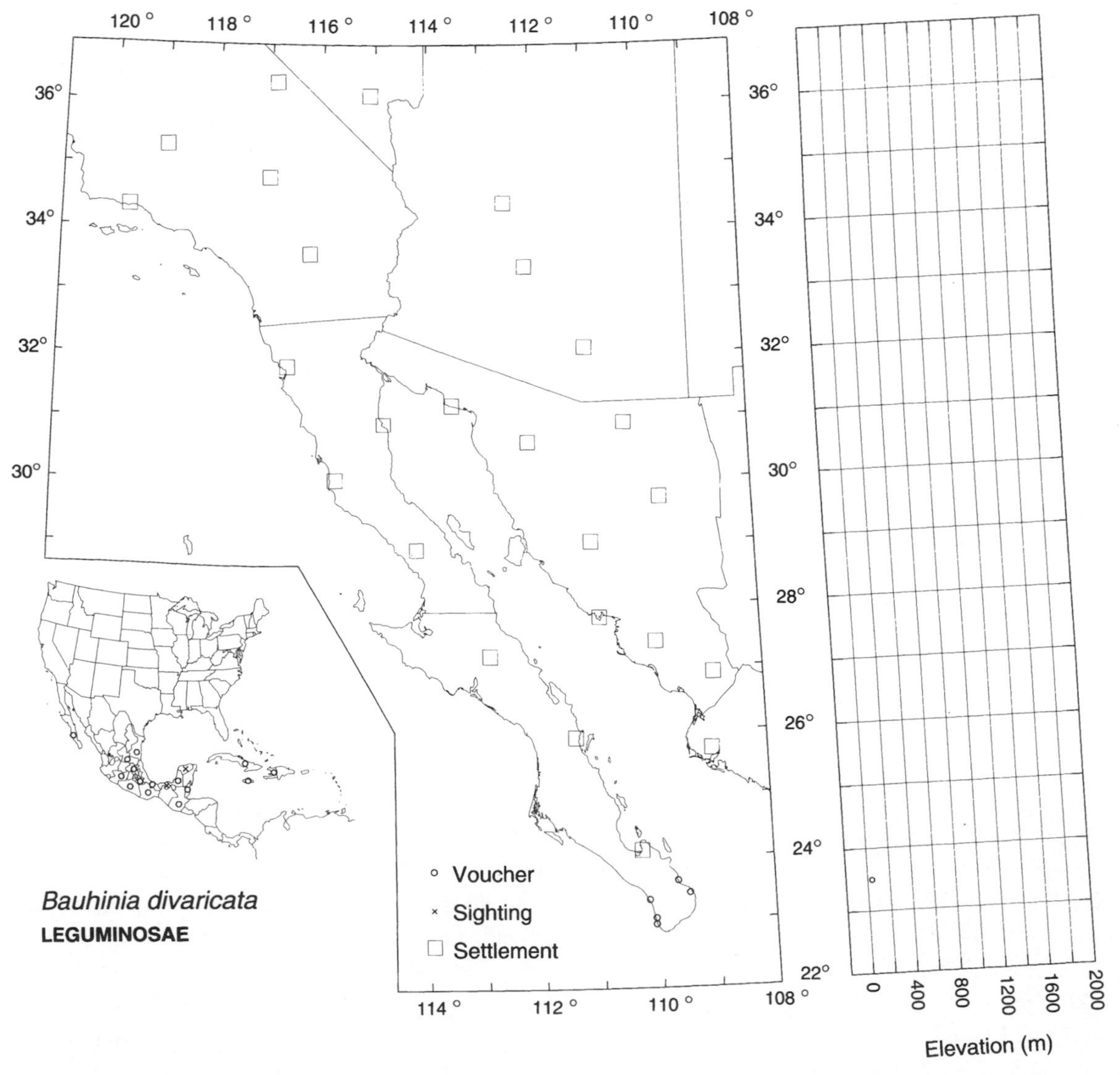

alscrub and Californian chaparral, *B. emoryi* barely enters the Sonoran Desert at the southern end of its range. Shreve (1936) considered it endemic to the transition zone between Sonoran desertscrub and Californian chaparral along with *Rosa minutifolia* Engelm. in Parry, *Aesculus parryi, Opuntia prolifera,* and several others. The small population at Torrey Pines, California (32.0°N, 117.3°W), may have been planted (Reid Moran, personal com-

Figure 27. Illuminated from behind, the yellowish spines of Bergerocactus emoryi *are prominent. (Photograph by J. R. Hastings taken east of El Rosario, Baja California.)*

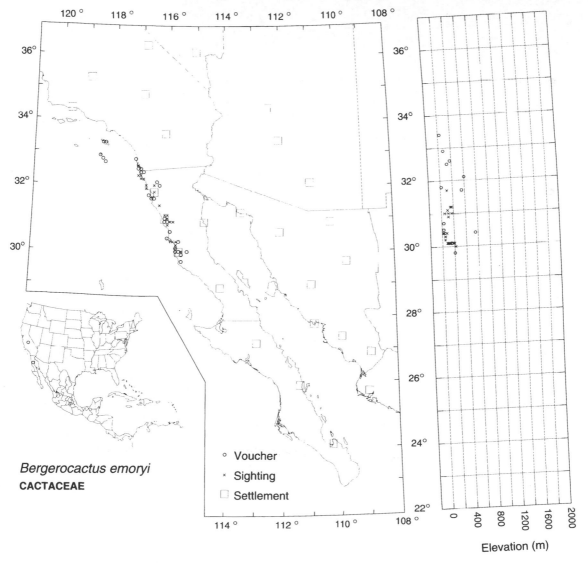

Bergerocactus emoryi
CACTACEAE

○ Voucher
× Sighting
□ Settlement

Elevation (m)

munication 1987). The reported sighting on Cedros Island (Hastings et al. 1972) was Edward L. Greene's. He most likely confused *O. prolifera* with *B. emoryi* (Moran 1965).

Cody (1984) noted that in Californian coastalscrub and chaparral (where *B. emoryi* is not limited by water, light, or temperature), severe crowding from shrubs produces competition for physical space. At the northern end of their range, the plants grow in coastalscrub but fail to enter chaparral, where the only common cactus is *Opuntia parryi* Engelm., a thicket-forming cholla with taller and thinner stems (Cody 1984). Because its thinner stems confer a higher ratio of photosynthetic surface area to volume, *O. parryi* should grow more rapidly than *B. emoryi* and be better able to compete with chaparral shrubs (Cody 1984).

Flowers appear throughout the spring on stems 1–5 years old. When ripe, the fruits extrude a ribbon of pulp and seeds over several days (Moran 1966). According to Munz (1974), prostrate stems may root, eventually producing new plants.

At the end of each year's growth, the stems become slightly constricted, forming annual segments (Moran 1966). Individual stems live about ten years (Moran 1966); botanists have not yet determined the life span of the entire plant. In Californian coastalscrub, *B. emoryi* frequently burns, but afterward new stems sprout from the tuberous roots.

Collectors of cacti and succulents value *B. emoryi;* as a result, commercial acquisitions have severely depleted populations on the California mainland (Moran 1966). Clearing of coastal lands in Baja California and California has eliminated many stands, as well. Cultivated plants do best where summers are relatively cool. They should be kept dry during hot weather.

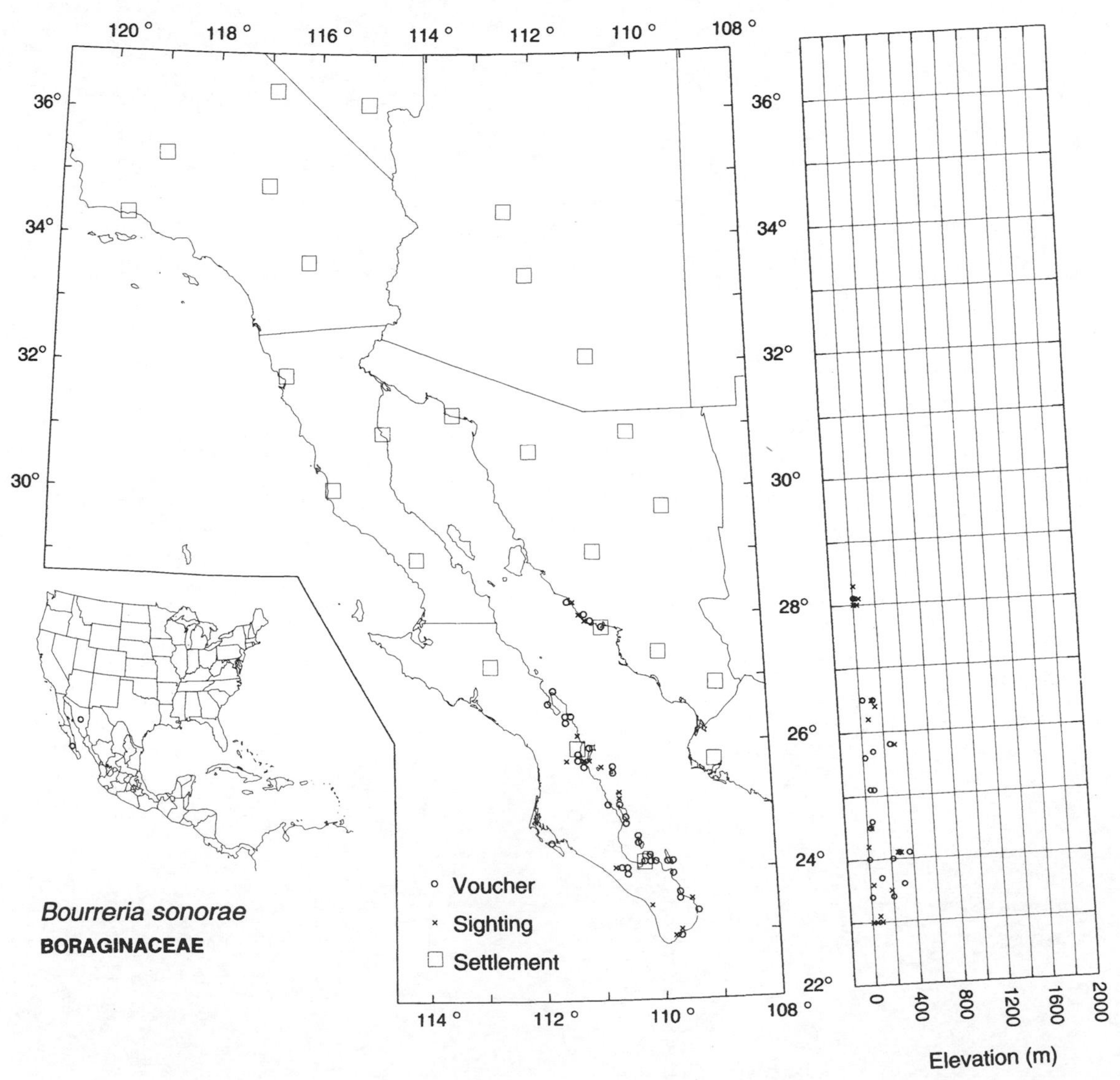

Bourreria sonorae

S. Watson

Bourreria sonorae is an openly branched shrub or small tree 3–6 m tall. The crowded, oblong-oblanceolate to obovate leaves are on short spur branches. Harshly scabrous above, the flabelliform leaves are velvety pubescent and conspicuously veined beneath and are 8–25 mm wide and 1.5–4.5 cm long. The salverform white flowers, 10–12 mm long, are in sparse cymes. The black drupes are globose and 6–10 mm in diameter.

Its distinctive sandpapery pubescence distinguishes *B. sonorae* from the other drought-deciduous, small-leaved shrubs in its range.

Occasional in rocky canyons and arroyos, *B. sonorae* generally grows below 400 m. It occurs on all the gulf islands from Isla Carmen southward (Moran 1983b). The range occupies arid, tropical, maritime climates with warm-season rainfall. Except for populations in the vicinity of Guaymas, Sonora, the species is restricted to Baja California Sur and adjacent gulf islands. The species may have originated on the peninsula after Miocene gulf-floor spreading. If so, the Guaymas-area populations would be the result of long-distance dispersal. The drupaceous fruits are probably eaten and transported by birds.

Most populations bloom between October and February.

According to Standley (1924), the fruits are edible.

Brahea armata

S. Watson

[=*Erythea armata* (S. Watson) S. Watson]

Big blue hesper palm, blue palm, palma blanca

A large palm up to 15 m tall, *Brahea armata* has a stout trunk and heavily blue-glaucous leaves 1–2 m wide on meter-long

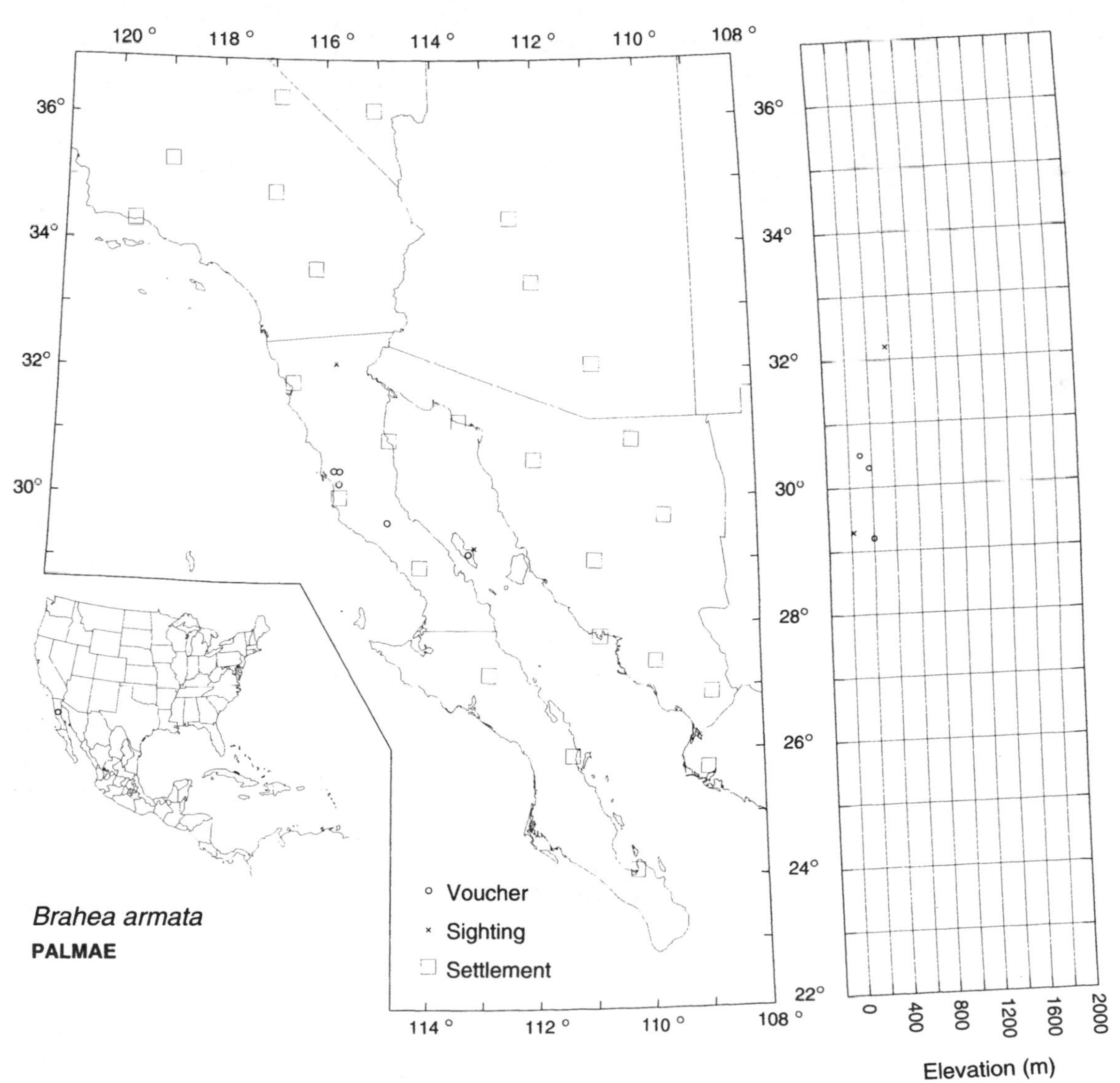

Brahea armata
PALMAE

petioles. The leaves form a persistent shag but are often burned or cut (Bailey 1937). The branches of the inflorescence, up to 5 m long, extend well beyond the crown and nearly reach the ground in young specimens. The flowers are small and inconspicuous. The brown, ovoid to globose fruits are 18–24 mm long and slightly fleshy at maturity.

Washingtonia filifera has green, not blue, leaves. *Washingtonia robusta* H. Wendl. also has green leaves and is a much taller tree with a slender trunk that flares abruptly near the base.

Brahea armata is locally common in canyon bottoms and arroyos, sometimes with *W. filifera* or *W. robusta.* At higher elevations, it also grows in bedrock crevices (Moran 1977a). This is the most widespread endemic palm of the northern peninsula and has its counterpart in *B. brandegeei,* endemic to the southern peninsula.

Flowers appear in February and March. Young plants only a meter tall may flower (Bailey 1937).

Especially when young, *B. armata* is an attractive ornamental (Bailey 1937; Moran 1977a; Duffield and Jones 1981). The plants are drought tolerant and thrive in partial shade to full sun. They are hardy to −10°C. Growers obtain best growth by applying occasional deep irrigation (Duffield and Jones 1981).

The Cócopah Indians roasted and ate the seeds (Felger and Moser 1985).

Brahea brandegeei (Purpus) H. E. Moore

[=*Erythea brandegeei* Purpus]

Brandegee palm, San José hesper palm, palma de Tlaco, palma negra, palmilla

A slender palm 10–12 m tall with an open, loose crown, *Brahea brandegeei* has a shag of dead leaves that persists until cut or burned. Dull green above and conspicuously glaucous beneath, the fan-shaped leaves are 1–1.3 m broad and deeply cut into slender, bifid segments. The long, slender petioles are 1 m or more long, 3–4 cm broad at the lower end, and armed with

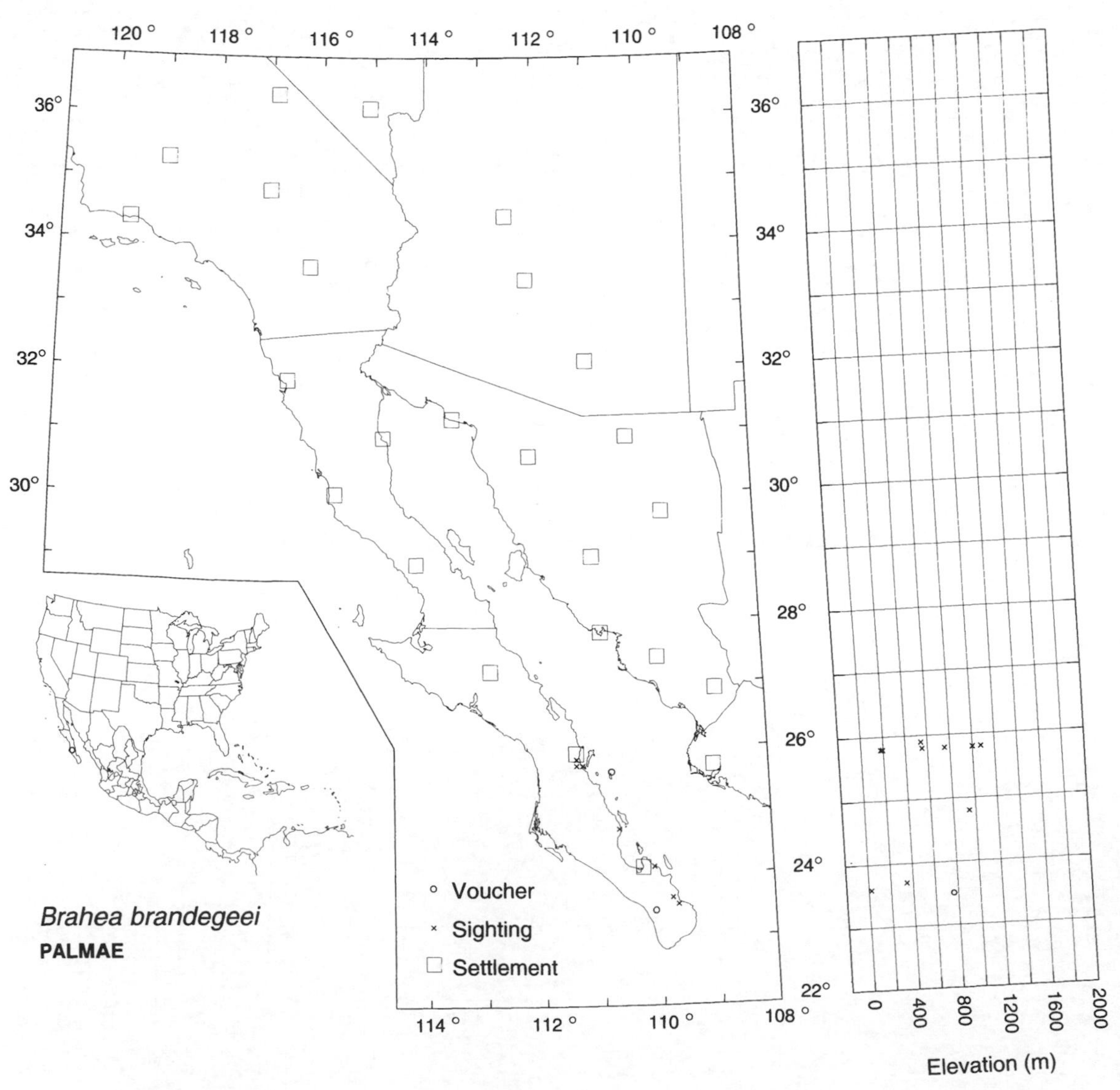

Brahea brandegeei
PALMAE

marginal teeth 5–7 mm long. Inflorescences are shorter than the leaves and are often obscured by them. The oblong to globose fruit, 15–22 mm long, has a papery, puffy skin.

Washingtonia robusta H. Wendl. is much taller (up to 25 m) with inflorescences that extend beyond the crown. Where it sheaths the trunk, the petiole is split in *Washingtonia*, entire in *Brahea*.

Locally common in canyons and on north-facing slopes, *B. brandegeei* grows up to an elevation of 1,100 m on the east coast of Baja California Sur. It is the endemic palm of the southern peninsula, as *B. armata* is of the northern peninsula. Bailey (1937) believed that wild populations were declining as a result of floods and timber cutting.

The leaf buds are edible (Standley 1920), but harvesting them kills the tree. The fruits are also edible (Moran 1977a).

Brahea roezlii

Linden

[=*Erythea roezlii* (Linden) Beccari]

Short-arm blue hesper palm

Brahea roezlii is a slender palm 10–18 m tall with a trunk 30 cm in diameter and, on young trees, a shag of old leaves. The bluish, glaucous (occasionally green) fan-shaped leaves are 1–1.75 m long and form a dense crown. The petioles, 1–1.5 m long and 3–4 cm broad at the middle, have wide-spaced, slender teeth 8–20 mm long along the margins. Extending well beyond the leaves, the inflorescence is 2–3 m long and often persists after the globose, wrinkled fruits have fallen. The diploid chromosome number is 36 (Sharma 1970).

Sabal uresana has no marginal teeth on the petiole. *Washingtonia robusta* H. Wendl. is taller (up to 25 m), with a trunk that flares abruptly at the base. The petiole base where it sheaths the trunk is split in *Washingtonia*, entire in *Brahea*.

Brahea roezlii grows in canyons and on shady cliffs near sea level. It is apparently an extremely narrow endemic, restricted

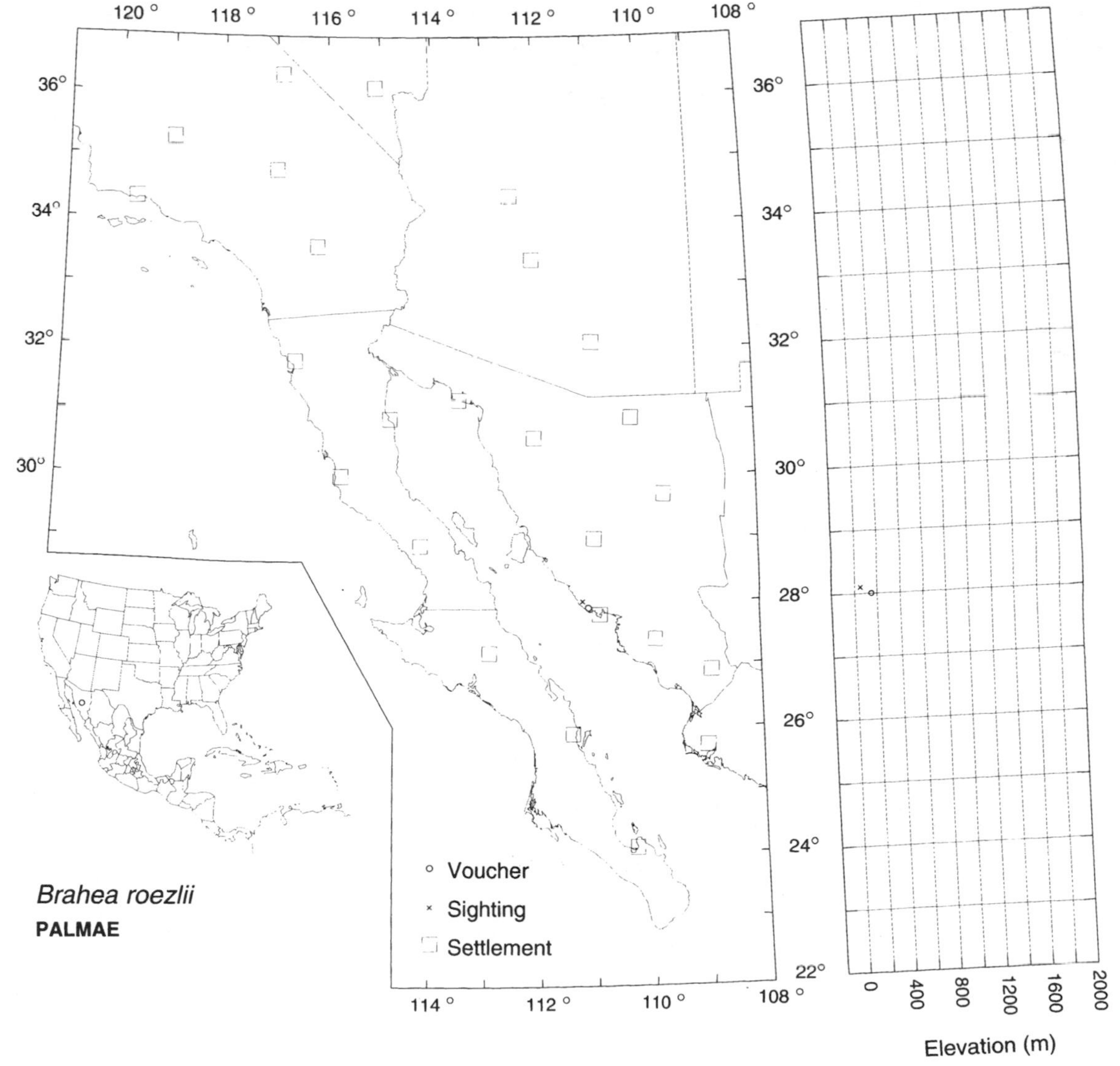

to the Guaymas area, where it is locally common but infrequently collected. Bailey (1937) reported this species from Cataviñá, Baja California, but cited no vouchers. Wiggins (1964) cited the range of the species as "from near Magdalena to Guaymas"; however, the palms in the vicinity of Magdalena appear to be *B. prominens* L. H. Bailey, not *B. roezlii* (Richard S. Felger, personal communication 1993).

Flowering specimens have been collected in May.

Brongniartia alamosana Rydb.

Palo piojo

This small tree grows to 7 m or more in height. The leaves are odd-pinnate with 9–13 elliptic, lanceolate, or ovate leaflets 2–5 cm long. Conspicuous oval stipules, 12–28 mm long, are present as the leaves develop but fall soon after the leaves mature. The flowers, solitary or paired at leaf-bearing nodes, are at first brick-red, then deep purple.

The grayish to grayish brown bark, streaked with irregular near-white vertical fissures and light-colored lenticels, is so striking that the plant can be readily identified even when leafless.

Efforts to delineate this species and *B. palmeri* Rose have been confused. Both were collected in 1890 by Edward Palmer at Alamos, Sonora, and were originally described as shrubs (Rose 1891; Rydberg 1923). Later, Gentry (1942) correctly noted that near Alamos, the taller *B. alamosana* typically reaches 6–7 m in height. But to add to the confusion, Gentry's specimen number 4778, cited as *B. palmeri* (Gentry 1942), is actually *B. alamosana*.

Once thought to be closely limited to southern Sonora (Gentry 1942), the species is now considered one of the most widespread species in the genus and is found southward in Mexico through the

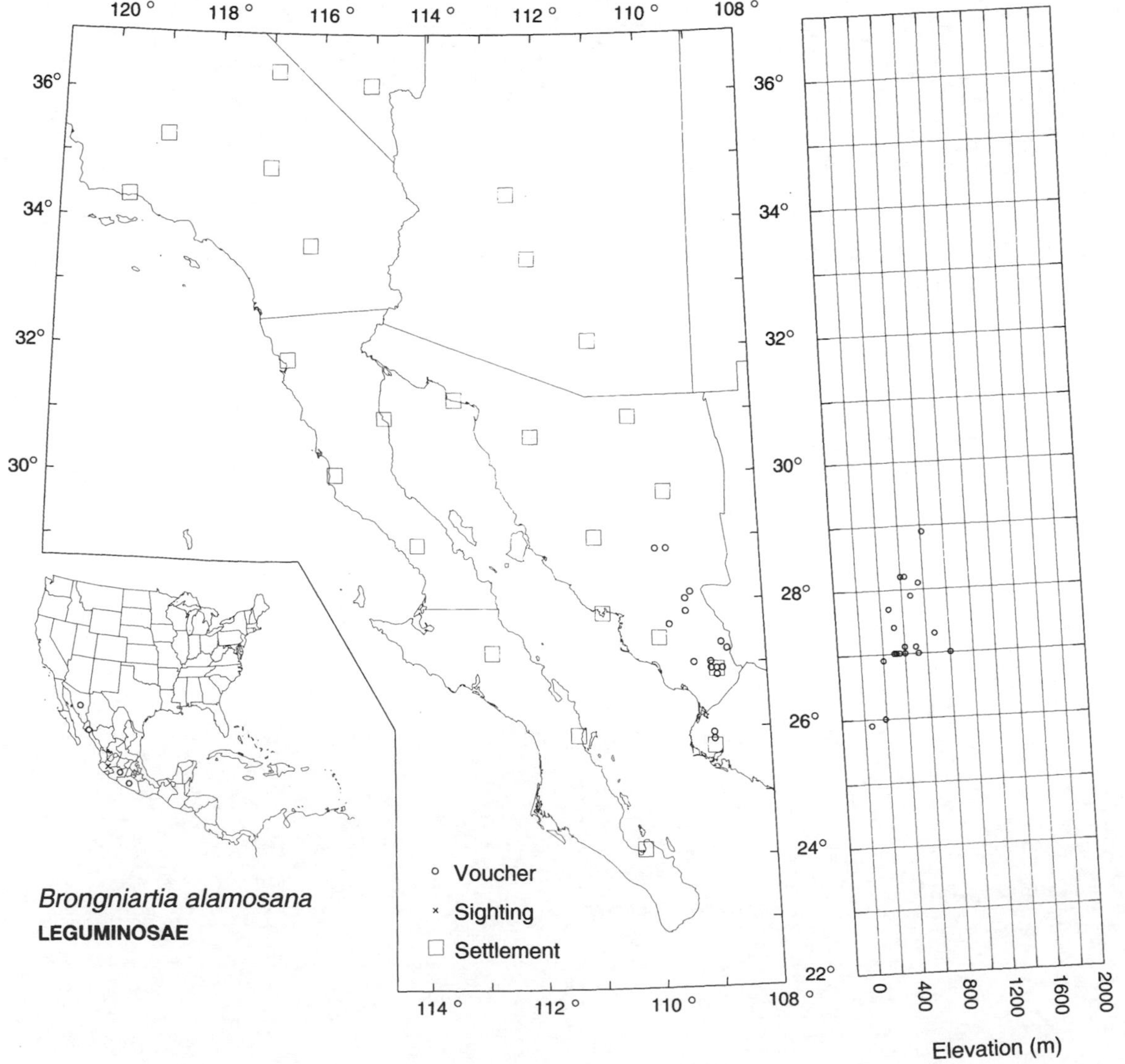

Brongniartia alamosana
LEGUMINOSAE

states of Sinaloa, Jalisco, Michoacán, and Guerrero (Dorado 1992). Its lower elevational limit is shaped by aridity, its upper elevational and northern limits by cold.

The flowers appear first in May when the plants are virtually leafless and continue to appear throughout the rainy season (Dorado 1992). The leaves fall soon after the summer rains stop. The conspicuous pods persist after leaf fall and, when ripe, burst with a loud report, hurling the seeds through the air. Hummingbirds visit the flowers.

Brongniartia minutifolia S. Watson

[=*B. shrevei* Wiggins]

Brongniartia minutifolia (figure 28) is a shrub up to 1 m tall with several slender branches from the base. The once-compound leaves, 6–10 cm long, consist of 35–55 narrow leaflets less than 1 mm wide and 3–10 mm long. The yellow flowers are single in the leaf axils. The 2- to 5-seeded pods are broadly elliptic, 12–15 mm wide and 2.5–3 cm long.

Dorado (1992) considered *B. shrevei* conspecific with *B. minutifolia* S. Watson, found in the Chihuahuan Desert.

In Sonora the species is known only from rocky limestone outcrops east of Hermosillo, Sonora. Plants grown near Tucson, Arizona, were damaged when temperatures fell below –4°C (Mark A. Dimmitt, personal communication 1993).

Flowers appear from August through December.

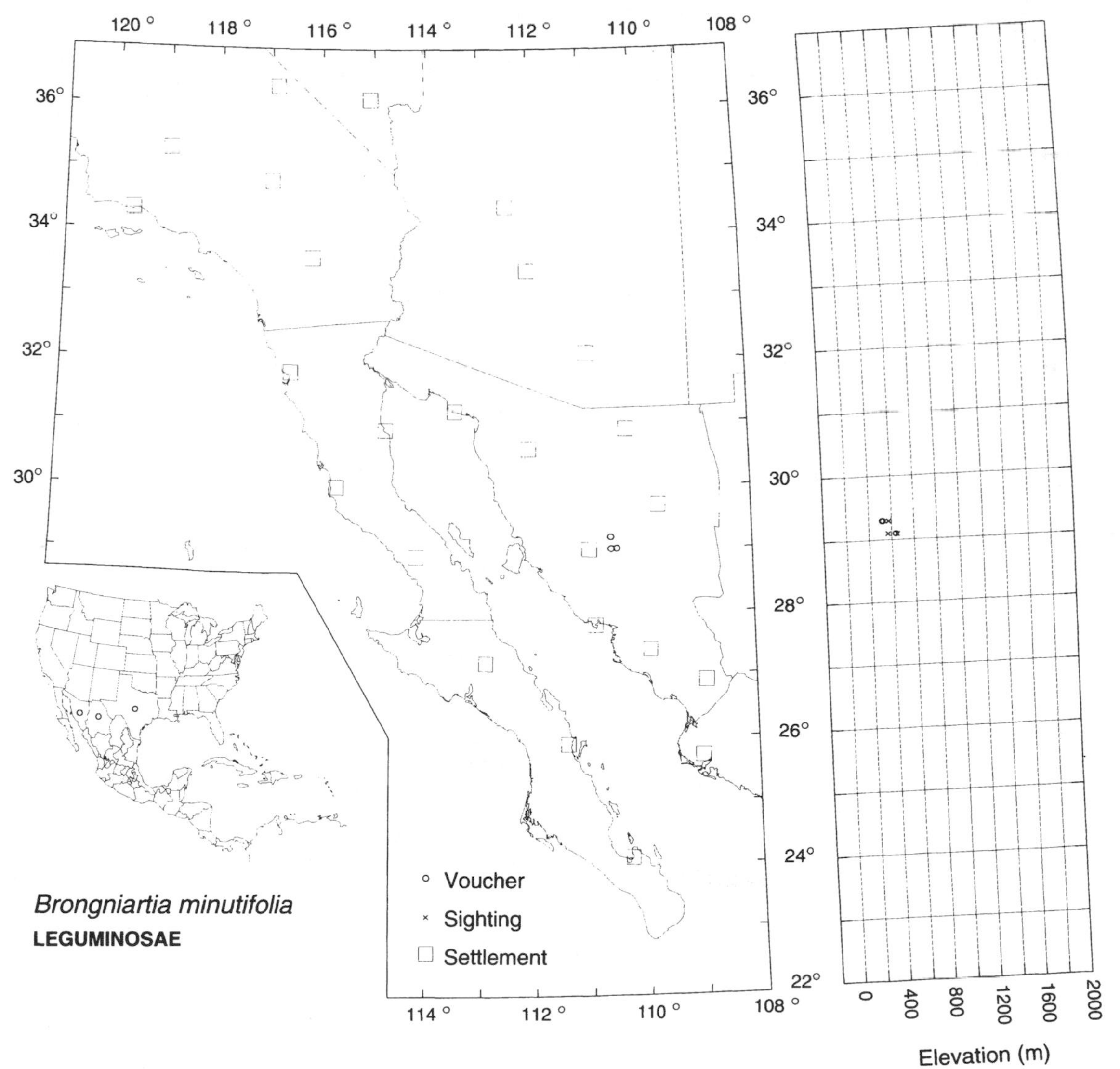

Brongniartia minutifolia
LEGUMINOSAE

Figure 28. Brongniartia minutifolia *on a limestone hill east of Hermosillo, Sonora. (Photograph by J. R. Hastings.)*

Figure 29. Bursera fagaroides, *here leafless in response to winter drought, has scattered fruits. (Photograph by J. R. Hastings taken north of Hermosillo, Sonora.)*

Brongniartia nudiflora

S. Watson

[=*B. palmeri* Rose]

Brongniartia peninsularis

Rose

Brongniartia tenuifolia

Standley

These openly branched shrubs 1–2 m (sometimes 3 m) tall have once-pinnate leaves with 7–17 narrow leaflets 1–4 cm long. The leaflets have a sharp but weak spine 2–3 mm long at the apex. The flowers, 1.5–1.8 cm long, change from brick-red to yellow with age, except that the brick-red flowers turn to purple as they dry in a plant press. Dorado (1992) described *Brongniartia tenuifolia* as having yellow or reddish yellow flowers. The stipitate, broadly oblong pods are 3–6 cm long.

The specimens of *B. tenuifolia* and *B. peninsularis* that we have seen are closely similar to the type specimen of *B. palmeri* (=*B. nudiflora*) collected at Alamos, Sonora, in 1890 (Palmer #300). We have mapped these three taxa together, recognizing that their status is uncertain. Phillip Jenkins (personal communication 1994) notes that Standley (1923) considered *B. palmeri* equivalent to *B. nudiflora* S. Watson (first collected in Jalisco). McVaugh (1987) reached the same conclusion but with reservations. If the two taxa are indeed synonymous, the first published name, *B. nudiflora,* is the correct one.

The rigid, spine-tipped leaves of these plants look somewhat like those of *Jacquinia pungens;* however, *J. pungens* leaves are simple. Distinguishing among these three species is often made possible by differences in leaflet size and shape as well as the appearance of the small appendages (stipels) at the base of leaflets (Turner and Busman 1984). Leaflets of *B. nudiflora* are 1–2 cm long and 2 times longer than broad; those of *B. tenuifolia* and *B. peninsularis* are 1–4 cm long and 3–4 times longer than broad. Stipels are needlelike and 1–3 mm long in *B. tenuifolia* but are absent and often replaced by clusters of minute, dark brown hairs in *B. peninsularis.*

These shrubs are common in Madrean evergreen woodland and occasional in Sinaloan thornscrub. Some of the northernmost plants are on open grassy ridges. *Brongniartia peninsularis* has traditionally been considered a peninsular endemic, while *B. nudiflora* and *B. tenuifolia* have been regarded as mainland species. Considered as a group, they occur from the vicinity of Rancho La Brisca, Sonora, southward to Durango, and in Baja California Sur. The plants are especially common between Tezopaco and Movas, Sonora, and in the area west of Bahía Concepción, Baja California Sur. The group is not known from any Gulf of California islands (Moran 1983b). Throughout the range, climates are semiarid or arid and virtually frost free. Rain falls predominantly in summer. The elevational profile shows compression between arid and mesic factors.

The flowers and the large, paired, leaflike stipules appear in spring before the leaves. Later, when leaves develop between the stipules, what appeared to be a many-flowered, leafless inflorescence becomes a leafy stem with flowers single or paired in the leaf axils. Such seasonal dimorphism is common in the genus (Arroyo 1981).

Bursera fagaroides

(H.B.K.) Engler

[=*B. odorata* Brandegee, *B. confusa* (Rose) Engler in Engler & Prantl]

Torote papelillo, cuajiote, cuajiote amarillo, cuajiote blanco, cuajiote colorado, cuajiote verde, chutama, palo mulato

This shrub or tree grows up to 6 m tall and has papery bark that exfoliates in large sheets (figure 29). The leaves, 3–15 cm long, are once-pinnate with 3–13 (occasionally 17) elliptic lanceolate or ovate leaflets. The leaflets are 12–20 mm long and 5–10 mm wide. In Baja California Sur

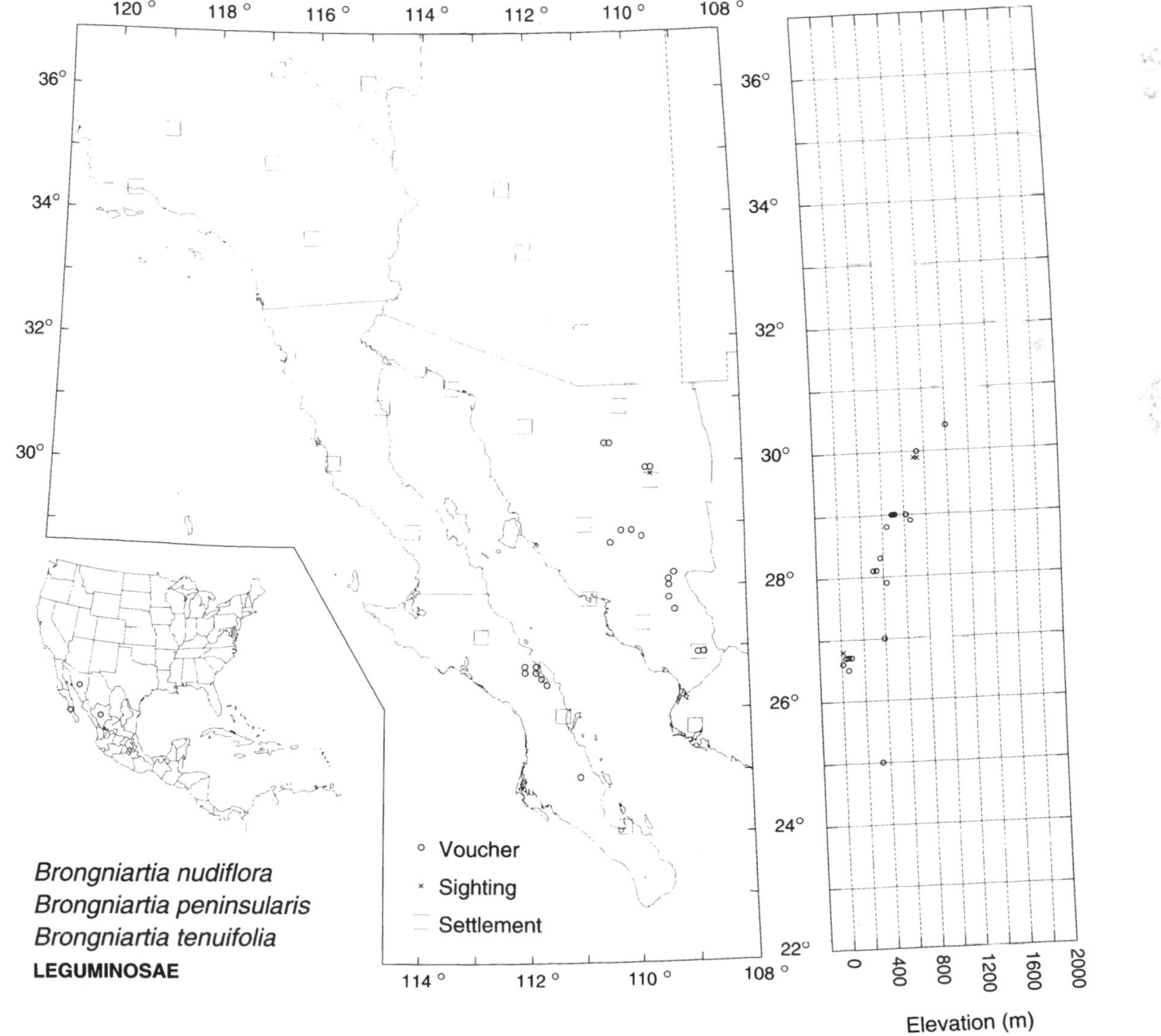

Brongniartia nudiflora
Brongniartia peninsularis
Brongniartia tenuifolia
LEGUMINOSAE

they are entire and glabrous; in Sonora they are often irregularly dentate. The flowers (4 mm long) are in few-flowered clusters. The 3-valved drupes are ovoid or globose and 5–8 mm long.

Leaves of *B. laxiflora* are pubescent. Leaflets of *B. fragilis* are evenly serrate. Those of *B. microphylla* are narrower (1–2.5 mm wide). *Bursera grandifolia* has much larger leaves (up to 4 dm long) and leaflets (up to 16 cm long).

Bursera fagaroides, B. confusa, and *B. odorata* form a taxonomically difficult group. Bullock (1936) concluded that *B. confusa* was hardly more than a form of *B. fagaroides.* Standley (1923) considered *B. odorata* and *B. confusa* to be conspecific, but Wiggins (1964) and Johnson (1992) segregated them. McVaugh and Rzedowski (1965) suggested that *B. confusa* may have originated from hybrids involving the *B. fagaroides* complex and *B. multijuga* Engler. They described *B. fagaroides* as "a complex of weakly differentiated plants"; according to their treatment, the taxon occurring within the region discussed in this monograph is *B. fagaroides* var. *elongata* McVaugh and Rzedowski. In view of this taxonomic confusion, we have mapped *B. fagaroides* var. *elongata, B. odorata,* and *B. confusa* together, an alignment also promoted by Cody and coworkers (1983).

These trees are locally abundant in valleys and on slopes throughout Sinaloan thornscrub and Sinaloan deciduous forest (Gentry 1942; Johnson 1992), reflecting their evolutionary affinities with the tropical deciduous forest of western Mexico (McVaugh and Rzedowski 1965). Their current distribution suggests dependence on summer rain and sensitivity to low temperatures. The wide elevational range of this complex in Sonora (sea level to 1,350 m) corresponds well with its broad distribution on the Pacific slope of Mexico. It occurs in the Gulf of California on Isla Tiburón (Moran 1983b). The Arizona outlier, documented by several collections made between 1927 and 1935, has not been seen for many years, despite several attempts to relocate the population (Charles T. Mason, Jr., personal communication 1992). Apparently this relict stand

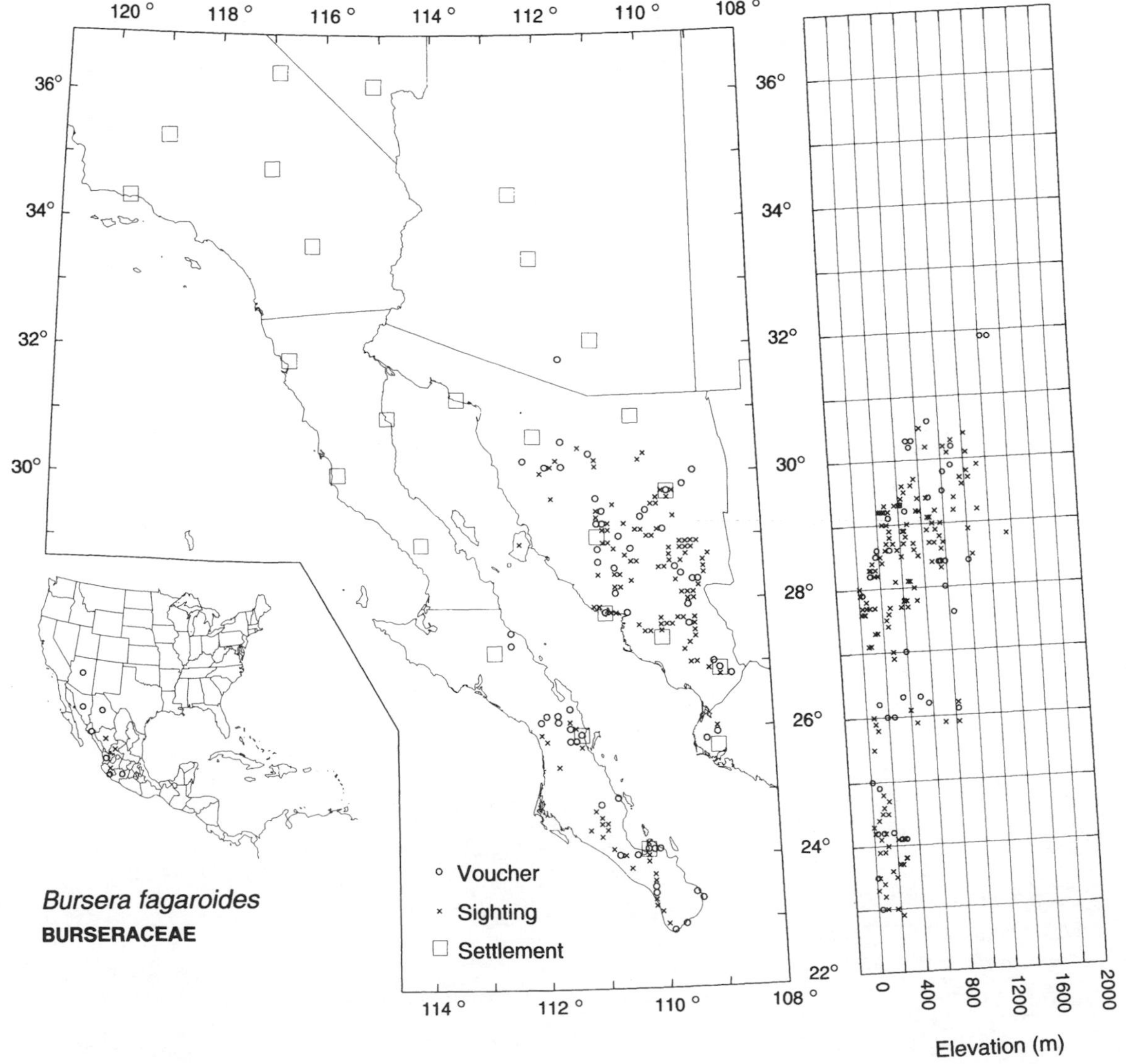

Bursera fagaroides
BURSERACEAE

has died out. Its relatively high elevation (1,220–1,280 m) may have ensured adequate summer rain yet also resulted in occasional severe freezes.

The exfoliating bark may serve at least two functions. When young, the thin bark admits light to photosynthetic tissue underneath. As the bark ages, it becomes more opaque. Exfoliation of the opaque layer evidently gives the photosynthetic layer continued access to light (Rzedowski and Kruse 1979). In addition, the shedding of bark discourages the establishment of crustose lichens, which would block light (Rzedowski and Kruse 1979).

Flowers appear with the new leaves in June and may be produced through August. The drought-deciduous leaves fall by October. The trees mature their fruits over a period of months instead of simultaneously like certain other *Bursera* species (Rzedowski and Kruse 1979). The arillate seeds suggest bird dispersal.

The gum is used for treating scorpion and insect stings and for mending broken dishes. The bark is employed in tanning (Standley 1923).

Bursera fragilis

S. Watson

[=*B. lancifolia* (Schlecht.) Engler]

Torote prieto, torote jolopete, incienso, tacamaca

A tree up to 8 m tall, *Bursera fragilis* has straw-colored to pale gray bark that peels off in papery sheets. The leaves, up to 15 cm long, are once-pinnate with 3–9 glabrous, evenly serrate leaflets 6–20 mm broad and 2.5–7 cm long. The flowers are rather large for the genus (up to 1 cm long) and are in panicles 6 cm long. The 3-valved drupes are ovoid and 6–10 mm long.

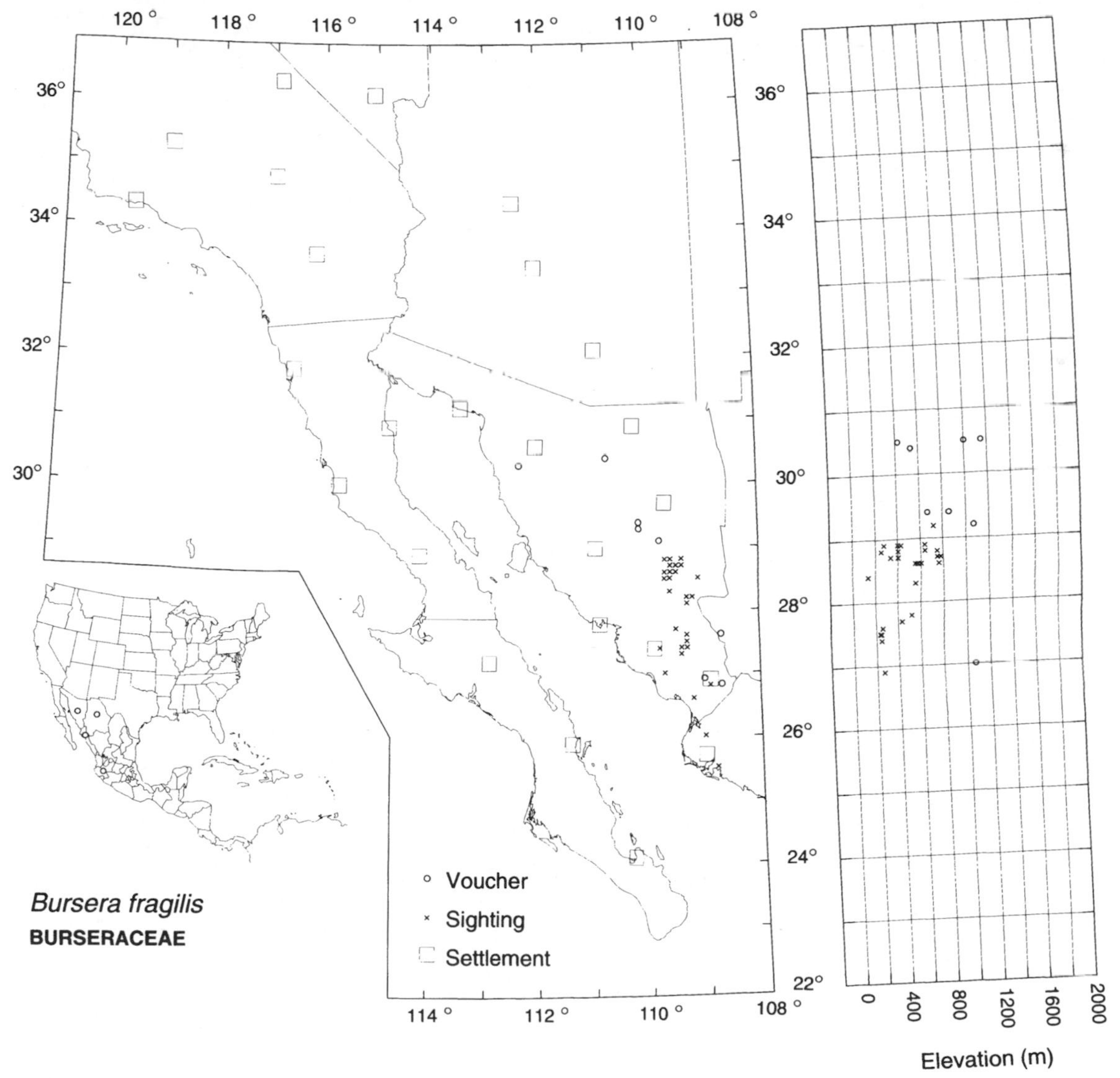

Bursera fragilis
BURSERACEAE

Leaflets of the *B. fagaroides* complex are smaller and either entire or irregularly dentate. Those of the *B. laxiflora* complex are pubescent. *Bursera grandifolia* has much larger leaves (up to 4 dm long) and leaflets (up to 9 cm wide and 16 cm long).

Johnson (1992) regarded *B. fragilis* and *B. lancifolia* as synonymous. The two exhibit much overlap in leaf, leaflet, and fruit size and few if any consistently distinguishing characters (Johnson 1992).

Bursera fragilis is common in thornforest and short-tree forest, often on dry ridges and rocky hills. Two remarkable disjunctions place this species well within the Sonoran Desert: one from Palm Canyon 27.5 km southeast of Magdalena (30.5°N, 110.8°W), vouchered by Thomas R. Van Devender in 1979; the other in the Sierra del Viejo (30.4°N, 112.4°W), collected by Howard Scott Gentry. Both areas have a notable array of disjuncts, many of which apparently require significant summer rain and mild winter temperatures. An unusually severe freeze in December 1978 caused significant damage to *B. fragilis* populations in western Sonora (Jones 1979).

The thin sheets of bark admit light to photosynthetic tissue underneath, then exfoliate as they become more opaque. Bark exfoliation also prevents establishment of crustose lichens, which would block light (Rzedowski and Kruse 1979).

June and July are the months of bloom. Leaves appear with the onset of summer rains in July and last until October (Gentry 1942; Johnson 1992).

The gum is used as a poultice for broken bones and bruises (Gentry 1942).

Bursera grandifolia

(Schlecht.) Engler

Palo mulato, chutama, chicopun, jiote blanco

The yellow- or orange-brown outer bark of this tree (up to 12 m tall) contrasts markedly with its green inner bark (McVaugh and Rzedowski 1965). The finely pubescent leaves, up to 4 dm long, are once-

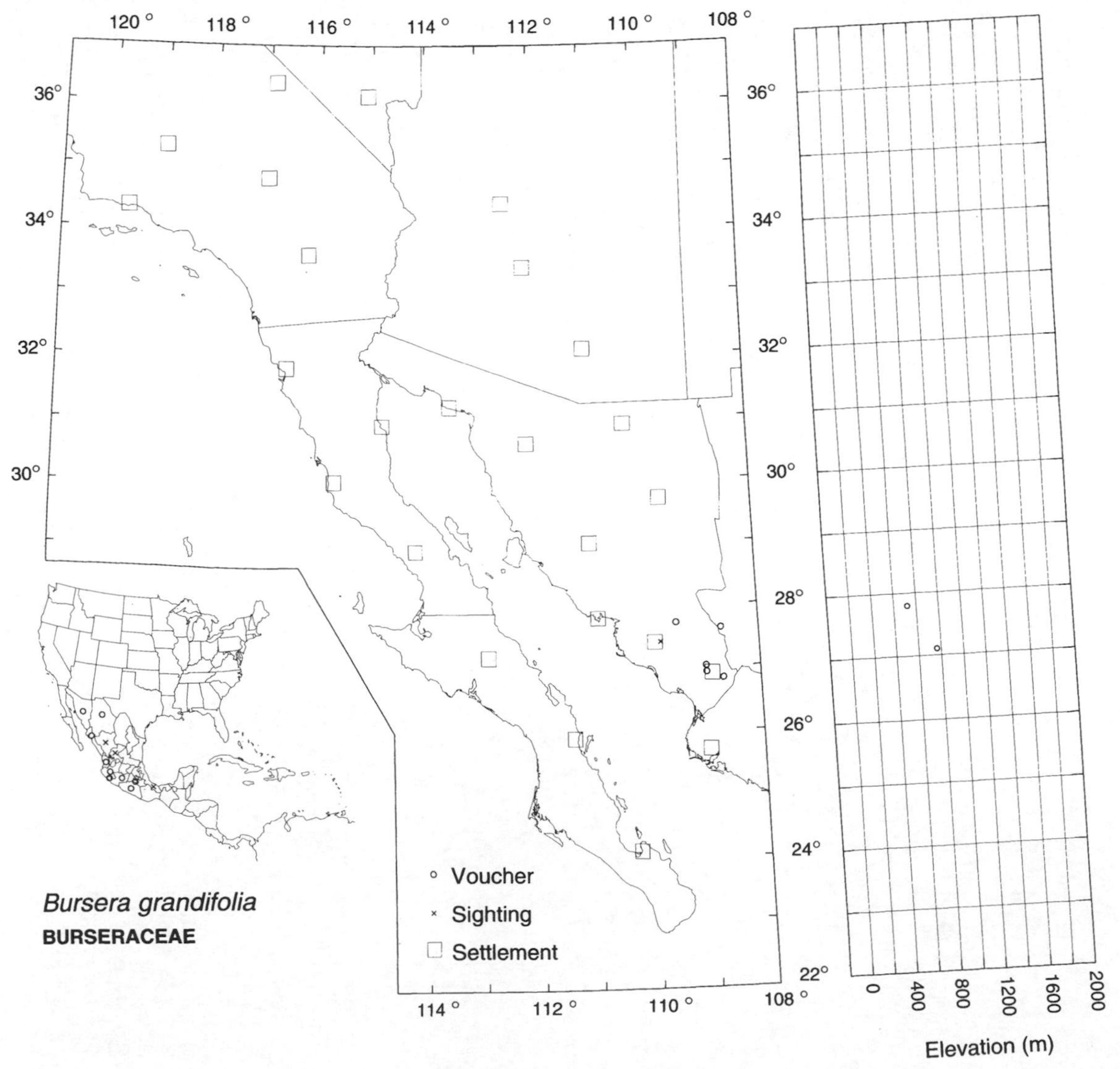

pinnate with 3–7 entire leaflets, each as much as 9 cm wide and 16 cm long. The many small flowers are in panicles 10 cm long. The minutely puberulent, 3-valved drupes are 8–9 mm long.

No other *Bursera* species in the area encompassed by this atlas has such large leaves. *Bursera arborea* (Rose) Riley and *B. simaruba* (L.) Sargent leaflets may be almost as large, but they are glabrous or sparsely pubescent (Johnson 1992).

Bursera grandifolia is frequent in Sinaloan deciduous forest in canyons and on slopes, typically as widely scattered individuals (Gentry 1942; Johnson 1992). The northernmost point represents a 1969 collection by Raymond M. Turner 12 km southwest of Tezopaco. Reaching well into central Mexico, the species has its northern limit in Sonora and does not penetrate into the desert. The large leaves mark it as a mesophyte and emphasize its tropical affinities.

Leafing and flowering is from July to September (Gentry 1942; Wiggins 1964). Seed germinates in 4–14 days (Johnson 1992).

This species is suitable for living fences. Stem cuttings root easily. Because the root system is shallow, fairly large trees can be transplanted as long as they are moved when leafless. Their stems must be protected from sunburn.

The gum is used for glue and caulk (Standley 1923), and the bark is used for medicinal tea (Gentry 1942).

Bursera hindsiana

(Benth.) Engler

[=*B. epinnata* (Rose) Engler, *B. cerasifolia* Brandegee]

Red elephant tree, copal, torote prieto

This shrub or small tree up to 8 m tall has reddish twigs, reddish gray bark, and leaves on short shoots. On any plant, all leaves may be unifoliate, or simple and pinnate leaves may be mixed. The ovate simple leaves are 1–2 cm wide and 4–6 cm long. The pinnate leaves have 3 (rarely 5–7) leaflets 1–2.5 cm wide and 1.5–4.5 cm long. Leaves or leaflets are crenate and

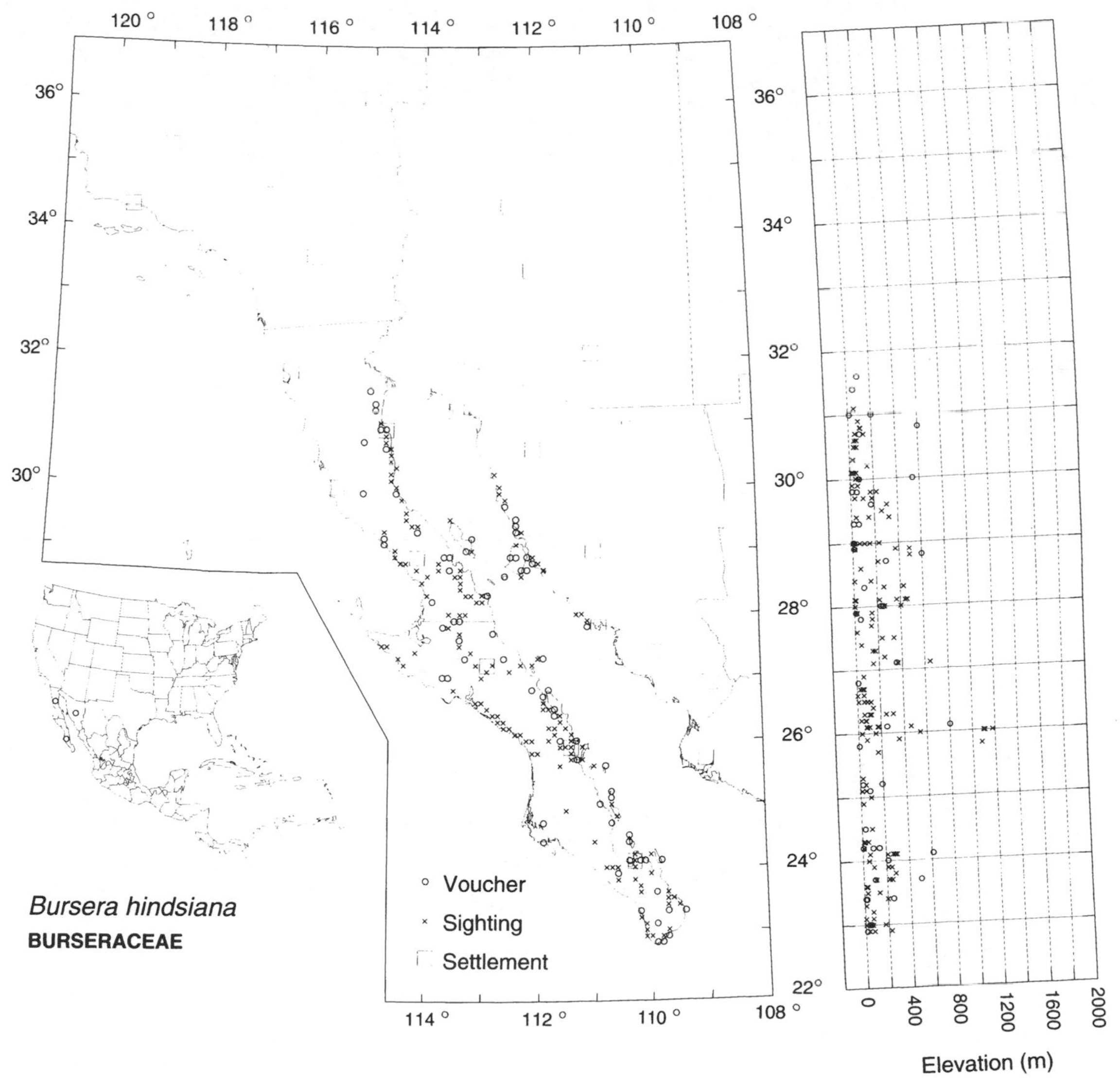

Bursera hindsiana
BURSERACEAE

may be pubescent or glabrate. The small, creamy flowers are 4-petaled and inconspicuous. The 2-valved, leathery drupes are 5–12 mm long. The ovoid seed is half-buried in an orange aril.

Following Cody and coworkers (1983), we have mapped *B. epinnata* and *B. cerasifolia* with *B. hindsiana.* The differences among them are not great: vegetative *B. epinnata* is often indistinguishable from *B. hindsiana,* and *B. cerasifolia* appears to be little more than a consistently simple-leaved form of *B. hindsiana. Bursera cerasifolia* is endemic to the Cape Region of Baja California Sur, and *B. epinnata* is largely restricted to the Vizcaíno Region and Baja California Sur.

No other *Bursera* species within the range of this species complex has simple leaves or combines simple and pinnate leaves on the same plant.

A tree of arroyos, washes, gentle slopes, and rocky hillsides, *B. hindsiana* often shares dominance with other sarcocaulescents, including *Fouquieria diguetii, B. microphylla, Jatropha cuneata* and *J. cinerea* (Shreve 1964). The species is known from the Baja California peninsula and Sonora, including most of the gulf islands (Moran 1983b) and Isla Socorro (Reid Moran, personal communication 1992). It is widespread on the peninsula and restricted to coastal sites in Sonora, a pattern also seen in *Stenocereus gummosus* and *Fouquieria diguetii,* among others. These species may have crossed from the peninsula to the mainland via the midriff islands (Cody et al. 1983).

Bursera hindsiana grows mostly in warm and arid maritime climates. Across its range, rainfall varies from equally biseasonal to mainly summer, and freezing temperatures are extremely rare or absent.

Like *B. microphylla,* this is an arid-adapted member of a basically subtropical and tropical genus. Rzedowski and Kruse (1979) concluded that simple leaves are derivative in the genus *Bursera* and may be an adaptation to drier habitats. Shreve (1964) categorized *B. hindsiana* as a sarcocaulescent tree. The water content of the trunk is modest (0.9 kg H_2O/m^3) (Nilsen et al. 1990).

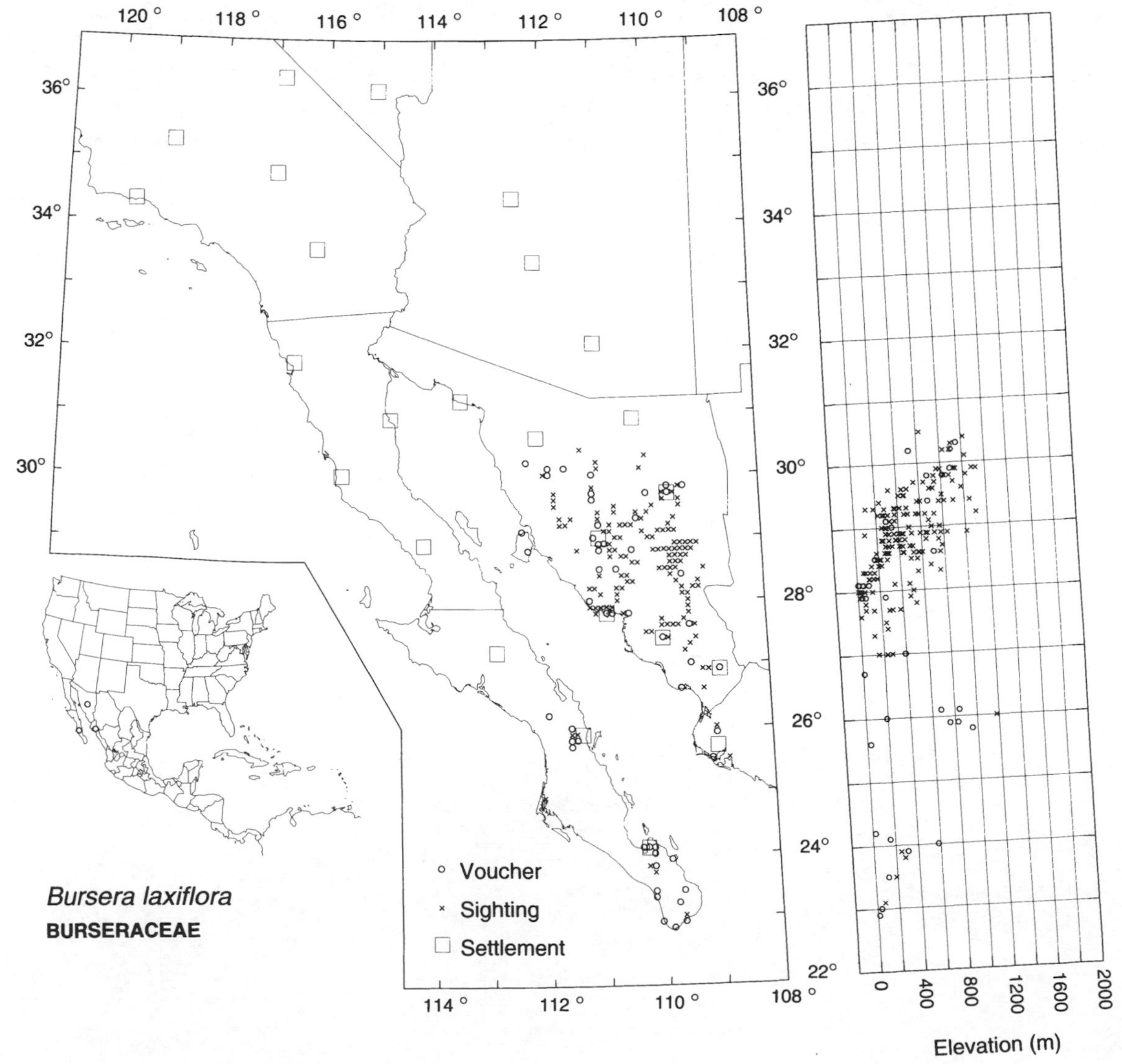

Bursera laxiflora
BURSERACEAE

Leaves are drought deciduous. Summer rains stimulate leaf and stem growth. If rains are ample, leaves may remain throughout the fall or even into the following spring (Humphrey 1975). Wiggins (1964) reported the flowering season as August to December; Humphrey (1975) noted flowers in July, as well. Flowering fails in some years (Humphrey 1975). Fruits ripen throughout fall and winter (Bates 1987). Plants cultivated near Tucson, Arizona, suffered frost damage at –6°C (Mark A. Dimmitt, personal communication 1992).

The Seri use this tree in a variety of ways, especially for carving small storage boxes and the wooden figurines known in Spanish as *santos* (Felger and Moser 1985). The fruits are an important food for birds (Bates 1987).

Bursera laxiflora

S. Watson

[=*B. filicifolia* Brandegee]

Copal, torote prieto, torote papelio, palo mulato

This shrub or small tree grows up to 10 m tall. The bark of the trunk is light or dark gray, that of the twigs, dark red or pale yellow. The leaves, 1–12 cm long, are pinnately or bipinnately compound with 5–15 obovate, ovate, or oblong leaflets 3–20 mm long. Few-toothed or entire, they are glabrous (subsp. *laxiflora*) or minutely pubescent on both surfaces (subsp. *filicifolia*). The leaflets are 6–12 mm long and 1–2.5 mm wide. The inconspicuous flowers are on slender pedicels. Fruits are 2-valved, ovoid drupes up to 1 cm long. An orange or red aril partly covers the seed.

Leaves of *B. fagaroides* are glabrous. Those of *B. grandifolia* are much larger, up to 4 dm long. *Bursera hindsiana* leaves are often simple or, if compound, have wider, strongly crenate leaflets.

Bullock (1936) suggested that *B. laxiflora* and *B. filicifolia* are conspecific. Felger and Lowe (1970) segregated them as subsp. *laxiflora* of Sonora and Sinaloa and subsp. *filicifolia* (Brandegee) Felger & Lowe of Baja California Sur. We have mapped the two subspecies together.

Rare to common on rocky hills and outwash slopes, *B. laxiflora* is a characteristic thornscrub tree that enters the desert where summer rainfall is sufficient and winter temperatures are not limiting. The only gulf population is on Isla Tiburón (Moran 1983b). Subspecies *filicifolia* grows in the Cape Region and on the Magdalena Plain. Scattered populations of subsp. *laxiflora* occur on the peninsula in the vicinity of 26°N. The geographic ranges of the *B. laxiflora* and *B. fagaroides* complexes are very similar.

Flowers appear in June, July, and August. Leaves last throughout the summer rainy season; a second flush of leaves may appear during winter rains (Gentry 1942). Seeds germinate in 1–2 weeks (Johnson 1992).

The gum was used to treat toothache and other ailments (Gentry 1942). The Seri used tea made from the bark to treat sore throats, coughs, scorpion stings, and black widow spider bites (Felger and Moser 1985).

Bursera microphylla

A. Gray

Elephant tree, torote, torote colorado, copal

A shrub or small tree up to 8 m tall, *Bursera microphylla* (figure 30) has aromatic leaves on cherry-red branches. The bark of the main trunk is whitish and exfoliates in thin sheets. The trunk and lower branches are thickened out of proportion to the height of the plant. The odd-pinnate leaves, 3–8 cm long, have 7–35 small, linear, glabrous leaflets and are alternate or clustered on short shoots. The leaflets are 6–12 mm long and 1–2.5 mm wide. The inconspicuous flowers are single or in few-flowered clusters. Fruits are 3-angled, dark blue or purple drupes containing a single seed.

Leaflets in the *B. laxiflora/filicifolia* complex are pubescent on both surfaces and toothed. Fruits in this group are 2-angled. Leaflets in the *B. odorata/confusa/fagaroides* complex are larger (4–10 mm wide and 1.2–4 cm long) and less numerous (3–9, usually 7). Fruits in this group are 3-angled.

The genus *Bursera* is affiliated with the tropics and subtropics; *B. microphylla* is one of several species that have radiated into more arid environments. Its distribution is nearly coincident with the extent of the Sonoran Desert (Shreve 1964). Unlike *B. fagaroides,* its counterpart in Sinaloan thornscrub, *B. microphylla* grows on virtually all the gulf islands (Cody et al. 1983; Moran 1983b), which are presumably too arid to support the relatively mesophytic *B. fagaroides.* The peninsular and mainland populations have distinct terpene compositions, which suggests a long period of separation (Mooney and Emboden 1968). Benson and Darrow (1981) reported *B. microphylla* from as far south as Zacatecas, Morelos, and Puebla; we follow McVaugh and Rzedowski (1965), who assigned these southern entities to *B. morelos* Ramirez and *B. multiflora* (Rose) Engler.

Bursera microphylla is rare to locally common in washes, on gravelly plains, and on rocky limestone or igneous slopes. At the northern end of its range, it typically grows in the warmest microhabitats, often south-facing slopes of low desert mountain ranges. The northernmost outpost in Arizona (33.7°N, 113.4°W), a sighting by Paul R. Krausman and John J. Hervert, is in the Harquahala Mountains on steep canyon slopes and in the canyon bottom. In central Sonora and Baja California Sur, where frost is infrequent, *B. microphylla* is more evenly distributed.

At Organ Pipe Cactus National Monument and at Signal Peak, Arizona, mature plants froze to the ground in December 1978 with minimum temperatures of

−6.1°C and −4.4°C (Anonymous 1979; Bowers 1980–81), and it seems likely that cold limits the northward and upward penetration of the species. The eastern limit may be determined by competition with *B. fagaroides* and *B. laxiflora,* close relatives growing to the east in Sinaloan deciduous forest and Sinaloan thornscrub. Fossil midden assemblages suggest that *B. microphylla* may have arrived relatively recently in Arizona, perhaps 5,800 years B.P. (Van Devender 1987, 1990b).

The thin sheets of bark admit light to photosynthetic tissue underneath, then can be shed as they age and become more opaque. Bark exfoliation also prevents establishment of crustose lichens, which would block light (Rzedowski and Kruse 1979).

Flowers appear before the leaves, typically in July. (Occasional plants flower in other months.) Summer rains stimulate leaf production and stem growth (Humphrey 1975). Given enough rain, new leaves may appear in other seasons, as well (Felger and Moser 1985). In warm winters, leaves may remain until spring drought (Humphrey 1975) or, in some cases, for two to three years (Shreve 1964). Frost kills leaves and stem tips. On the Sonoran coast, fruits ripen in the fall (Bates 1987).

Bursera microphylla stores water in the conductive and parenchymal tissues of the trunk, lower limbs, and wood (Webber 1936). Shreve (1964) classified it as a sarcocaulescent tree. Recent measurements placed the trunk water content at 1.4 kg/m³ (Nilsen et al. 1990). Water-storage cells begin to enlarge in the seedling stage (Shreve 1933, 1964), first in roots, later in shoots (Newland et al. 1980; Johnson 1992). The sarcocaulescent habit acts as a buffer against variation in environmental water balance (Nilsen et al. 1990) and is especially common in the Central Gulf Coast subdivision and parts of the Vizcaíno region (Shreve 1964), where a year or more may pass without measurable rainfall (Hastings 1964).

The seeds germinate within two weeks (Newland et al. 1980; Johnson 1992). Seedlings are abundant at first, followed

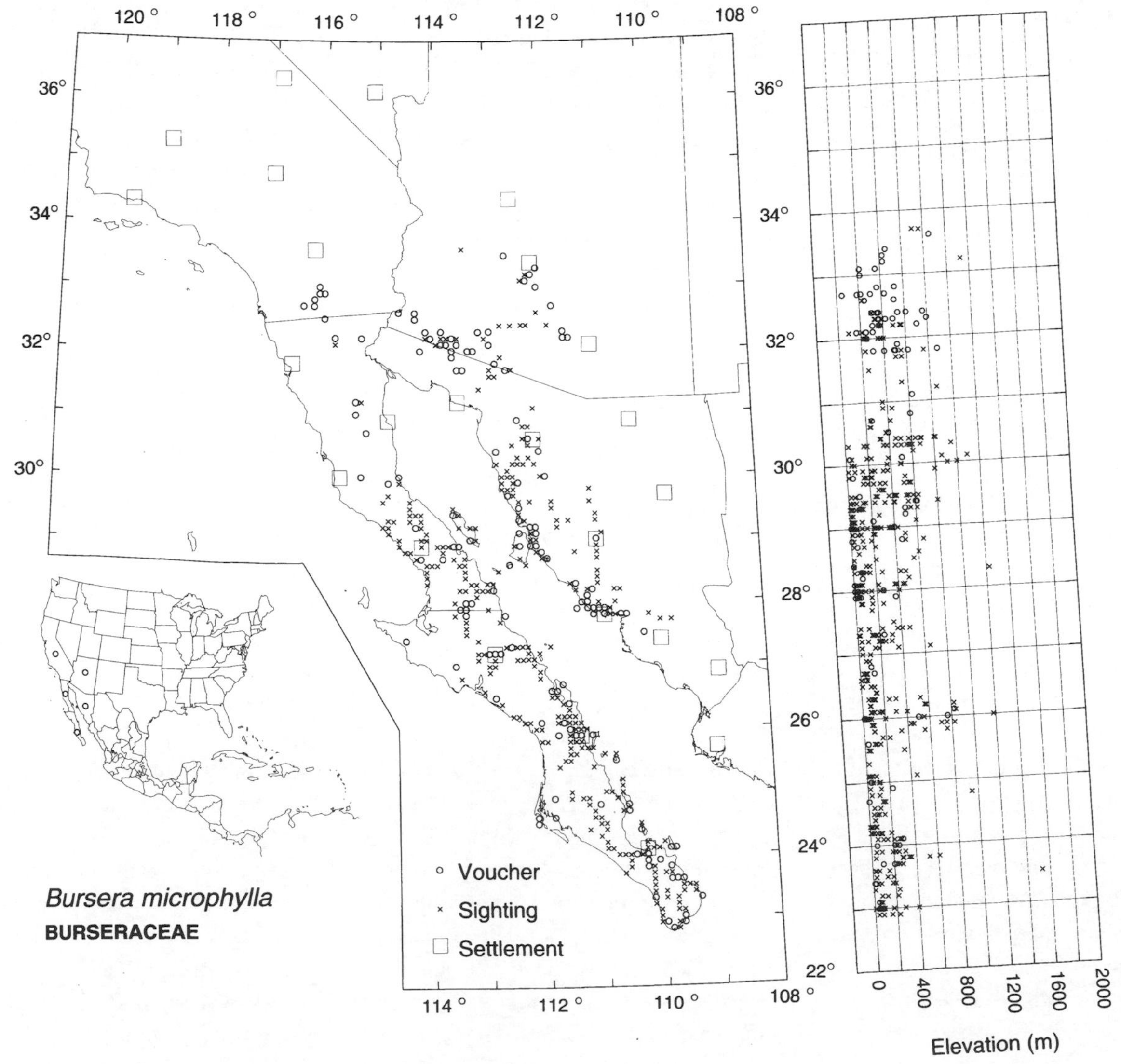

Bursera microphylla
BURSERACEAE

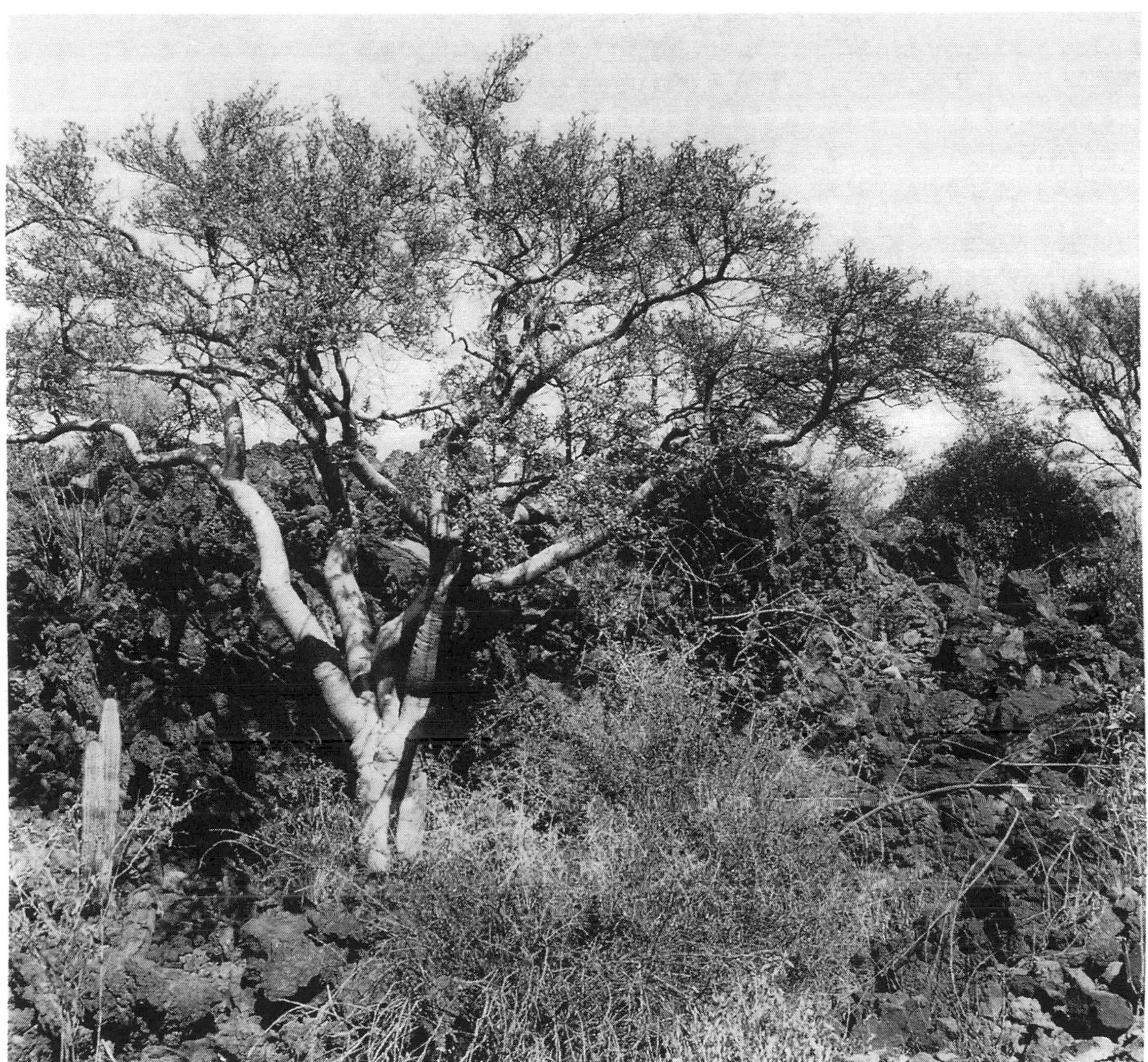

Figure 30. Bursera microphylla *in the Tres Vírgenes lava flow, Baja California. (Photograph by J. R. Hastings.)*

by high mortality in the 10–40 cm height class (Shreve 1964). Cultivated seedlings are subject to fungus infections (Newland et al. 1980).

The bark contains tannins and was gathered in Sonora for export (Kearney and Peebles 1960). Parts of the plant can be used medicinally (Moore 1989). Felger and Moser (1985) reported that the Seri used *B. microphylla* for fuel, shampoo, medicine, and paint. The fruits are an important food for gray vireos and other birds throughout the winter and early spring (Bates 1987).

Caesalpinia arenosa
Wiggins

Caesalpinia arenosa is an openly branched shrub up to 2 m tall. The twice-compound leaves have 1–3 pairs of pinnae with 2–3 pairs of slightly thickened leaflets 5–12 mm long and 3–7 mm wide. The leaflets are sparsely punctate-glandular and finely pubescent on both surfaces. The yellow petals, 6–8 mm long, are dotted with reddish glands. The obovate-lunate fruits are stipitate-glandular and up to 3.5 cm long.

One needs to see flowers or fruits to distinguish this species from *C. californica* and *C. pannosa*. Pods of *C. californica* lack glands, and calyces and pedicels of *C. pannosa* are stipitate-glandular.

This plant of dunes and sandy washes, endemic to Baja California Sur, is found from Bahía Concepción to Los Frailes. No gulf island locations are known (Moran 1983b). Throughout the range, climates are essentially frost free. Rain is concentrated in late summer and early fall. Our maps suggest competitive displacement involving *C. pannosa* and perhaps *C. californica*.

Flowers appear throughout the year.

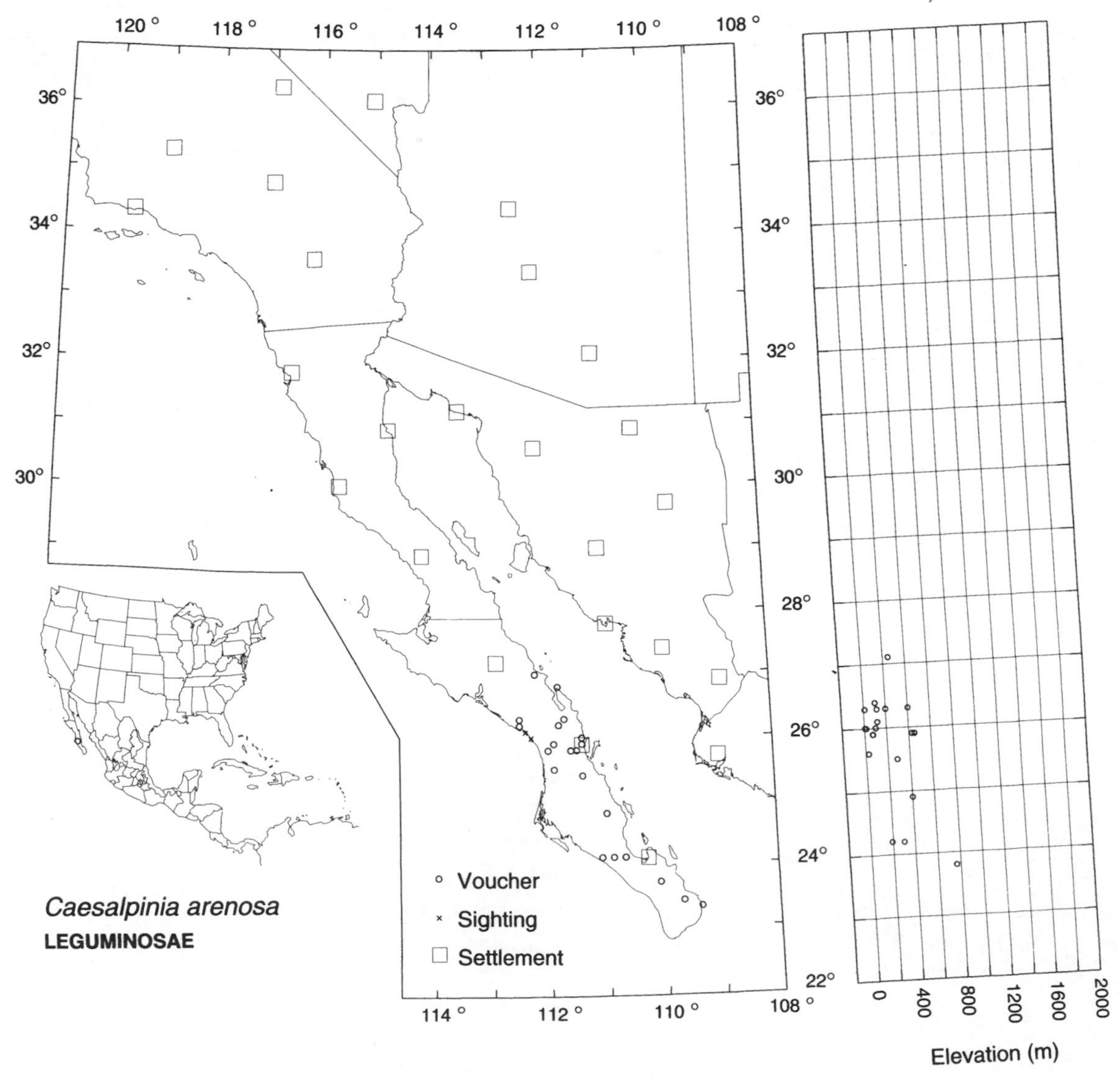

Caesalpinia arenosa
LEGUMINOSAE

Caesalpinia caladenia
Standley

A spineless, gray-barked shrub or small tree 5–6 m tall, *Caesalpinia caladenia* is odd-pinnate with 3–9 pairs of pinnae and 2–4 pairs of leaflets 10–30 mm long. The yellow petals are 10–12 mm long. The velvety, stipitate-glandular pods are 4.5–6.5 cm long and 12–16 mm wide.

Caesalpinia caladenia has generally larger leaflets with more prominent veins than *C. palmeri.*

Found from Sonora to Colima on rocky canyon slopes and hillsides, *C. caladenia* grows east of the desert proper.

The flowering season is February–April. When the flowers first appear, the plants are usually leafless.

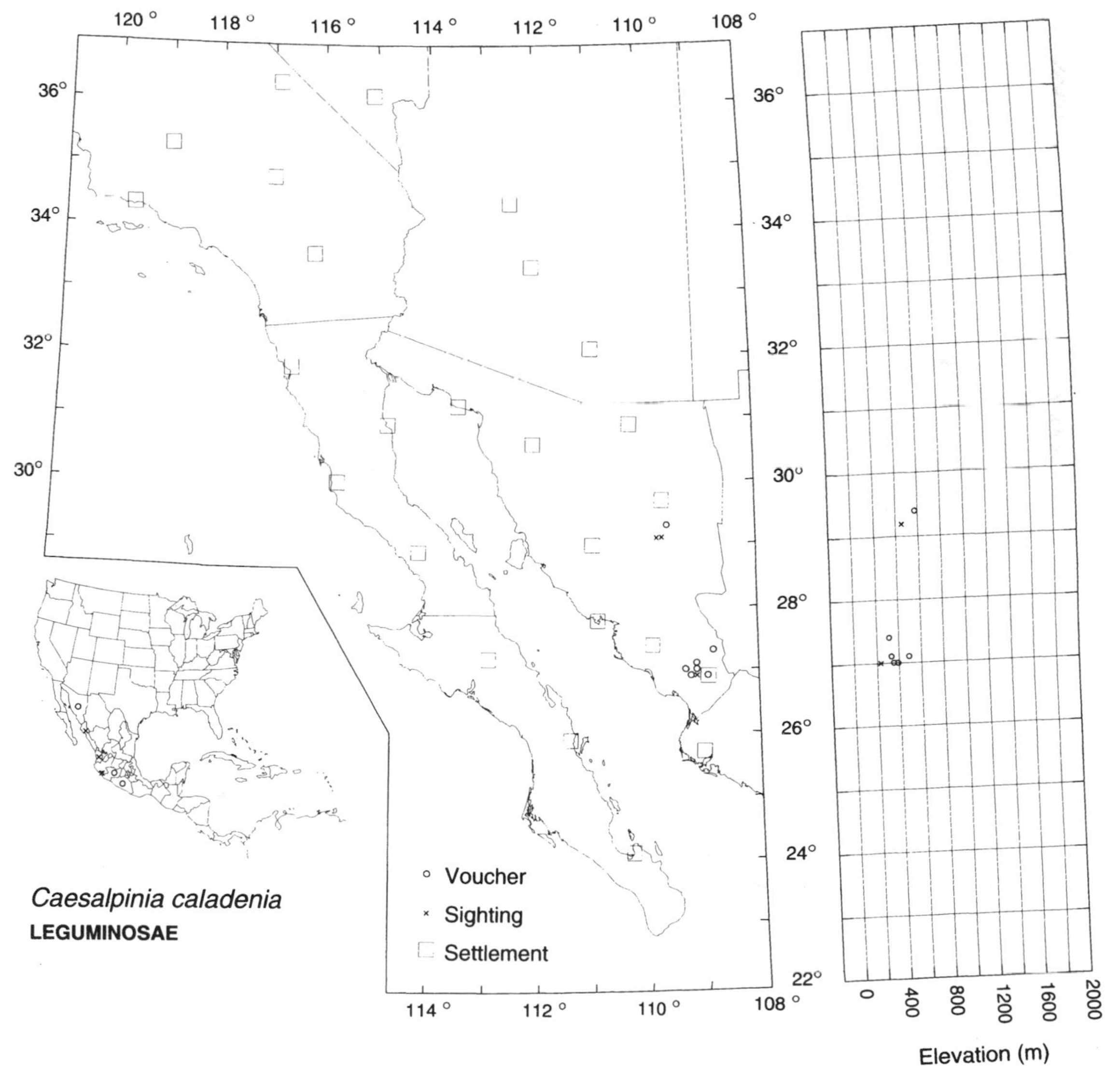

Caesalpinia caladenia
LEGUMINOSAE

Caesalpinia californica
(A. Gray) Standley

Caesalpinia californica is a glabrous, unarmed shrub up to 2 m tall. The odd-pinnate leaves have 3, 5, or 7 pinnae with 3–4 pairs of oblong, glabrous leaflets 6–18 mm long. The red-orange to yellow petals are 7–9 mm long and dotted with dark red, sessile glands. The puberulent or glabrate pods, 4–5 cm long, are oblong-lunate in outline.

This species can be distinguished from *C. arenosa* and *C. pannosa* by its glabrous leaflets (densely pubescent in *C. arenosa*) and eglandular pods (gland dotted in *C. pannosa* and *C. arenosa*).

Endemic to Baja California Sur, *C. californica* grows on dry hillsides and along washes at widely scattered sites from 26°N southward. It is relatively abundant between San Lucas and Todos Santos. The northernmost collections are Reid Moran's. Most of the collection sites, greatly influenced by Pacific maritime air, have relatively moderate summer and winter temperatures.

Flowers appear in spring and occasionally in fall after rains (Wiggins 1964).

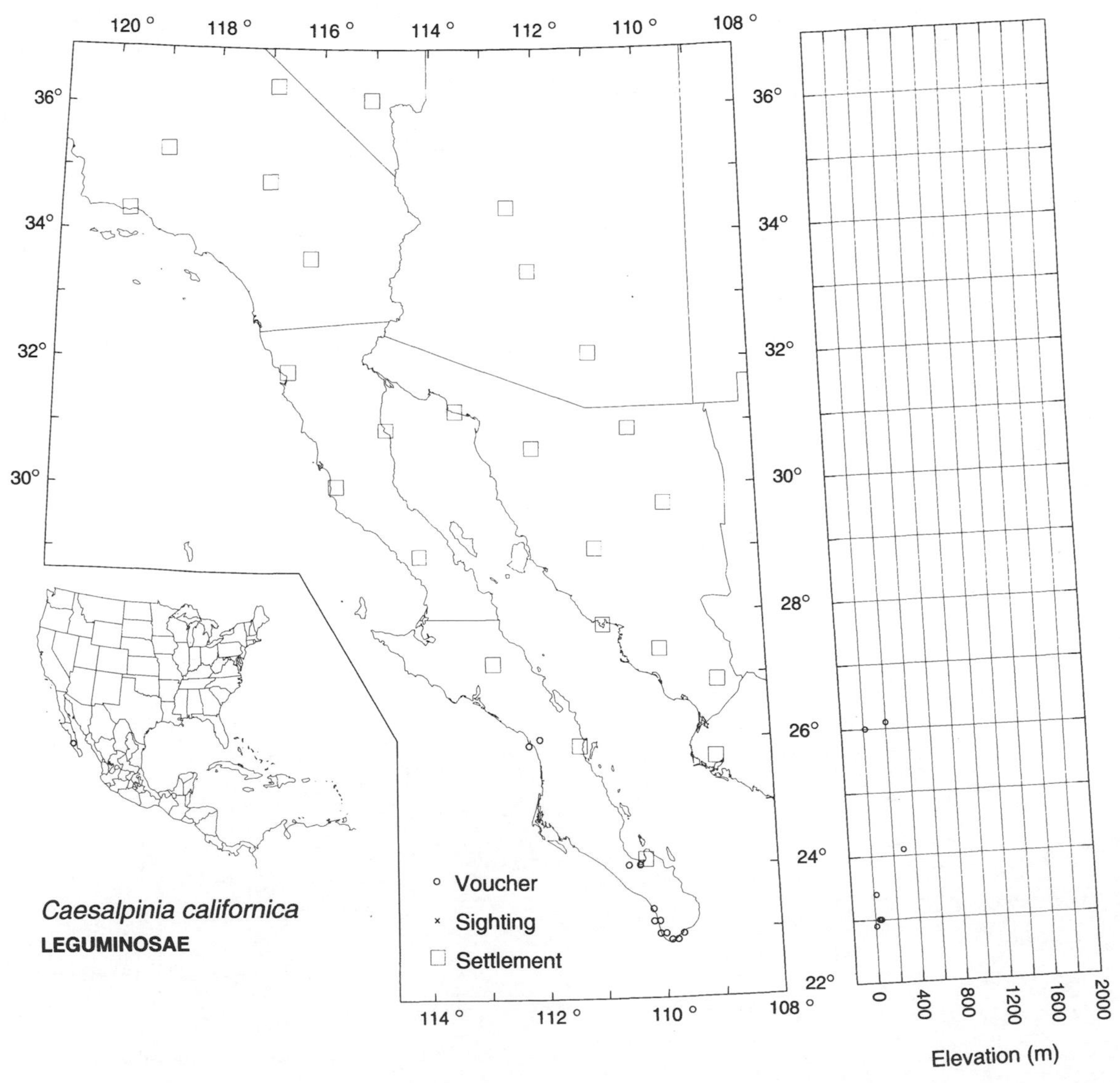

Caesalpinia californica
LEGUMINOSAE

Caesalpinia palmeri

S. Watson

Piojo, piojito, palo piojo, polilla

Caesalpinia palmeri is a shrub or small tree up to 3 m tall with an open crown. Branchlets have dark gray bark dotted with light gray lenticels. The elliptic to rounded leaflets, 6–10 mm long, are in 2–5 pairs on the 3, 5, or 7 pinnae. The bright yellow petals are 8–9 mm long. The oblong-lunate pods are 2.5–4 cm long. When ripe, they burst open with a resounding report, flinging the seeds some distance.

This shrub, known only from Sonora and Sinaloa, closely resembles *C. arenosa*, a peninsular endemic, and *C. pumila*, which has an even number of pinnae and larger leaflets (5–20 mm long).

Caesalpinia palmeri grows on rocky hillsides and sandy slopes in the transition from Sonoran desertscrub to Sinaloan thornscrub. The northernmost sighting is Raymond M. Turner's, from the south base of Sierra San Manuel. The species experiences biseasonal rainfall with the greatest concentration in summer.

Flowers appear from February through May and again in October, evidently in response to winter and summer rains.

The plants can be grown as ornamentals where minimum winter temperatures exceed –6.5°C (Johnson 1993). The Seri have used the red sap for painting and the wood for crafts (Felger and Moser 1985).

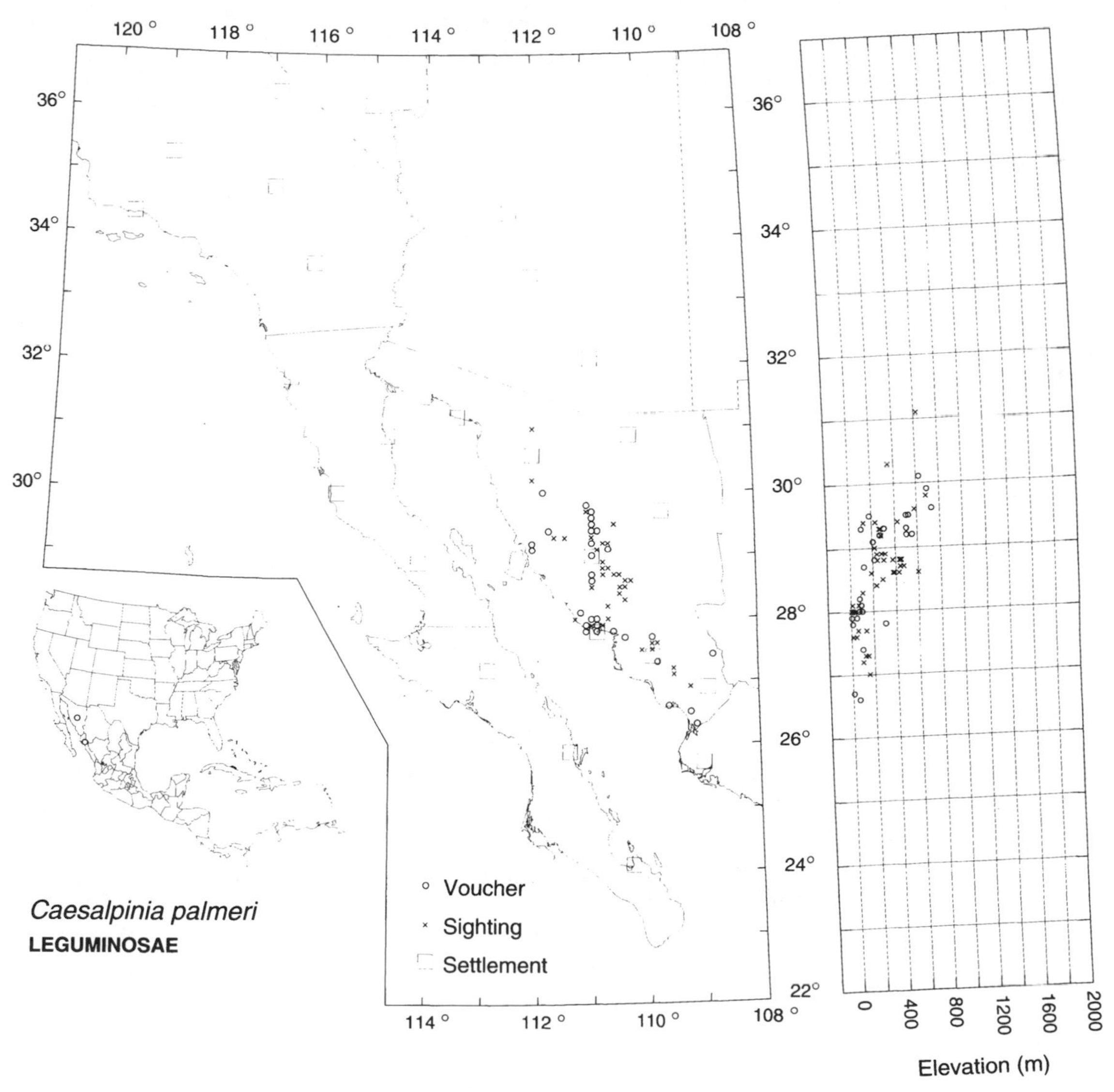

Caesalpinia pannosa
Brandegee

Caesalpinia pannosa is an unarmed shrub up to 1.5 m tall. Its odd-pinnate leaves have 3, 5, or 7 pinnae, each with 2–5 pairs of leaflets 3–15 mm long. The yellow petals, 6–7 mm long, are dotted with sessile red glands. The pedicels are jointed below the middle. The pods, 2.5–3.5 cm long, are marked with reddish brown to black sessile glands.

Similar species include *C. californica,* which has eglandular pods; *C. arenosa,* which has stipitate-glandular pods; and *C. placida,* which has crenulate leaflets and pedicels jointed above the middle.

A shrub of arroyos and dry hillsides, this Baja California Sur endemic is found south of 26°N and also occurs on three Gulf of California islands (Moran 1983b). Throughout its range, rainfall is concentrated in summer and early fall. Our maps suggest that interactions with *C. placida, C. californica,* and *C. arenosa* have resulted in geographic displacement.

Viscin threads connect several pollen grains into a single pollination unit in this and several other *Caesalpinia* species (Cruden and Jensen 1979).

The flowers appear from October through May.

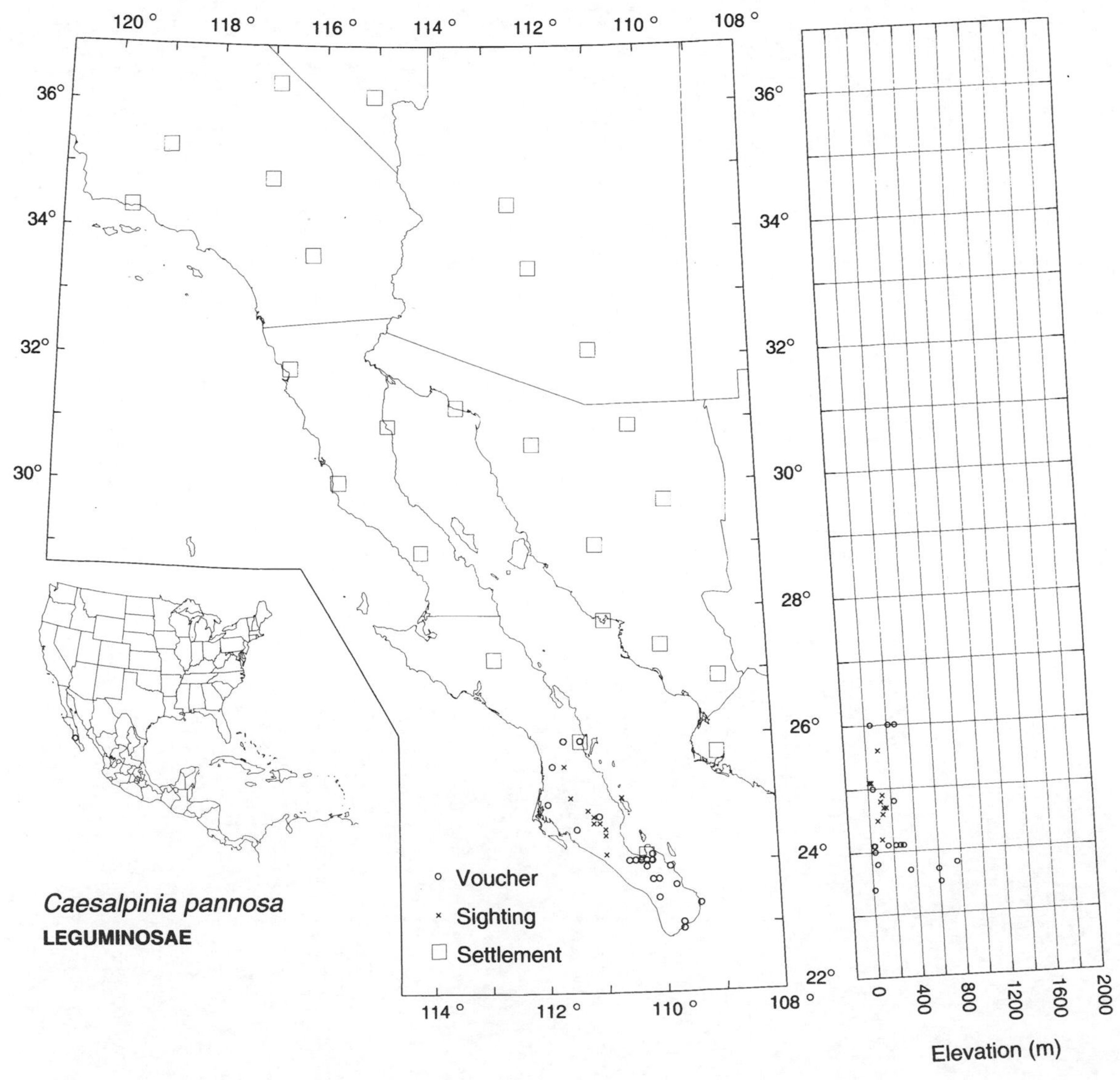

Caesalpinia pannosa
LEGUMINOSAE

Caesalpinia placida
Brandegee

Caesalpinia placida is a shrub 2 m tall with dark brown bark and an open crown. The compound leaves have 2 or 3 pinnae with 3–7 pairs of papillate leaflets 4–7 mm long. The leaflet margins are crenulate toward the apex. The yellow petals are 7–9 mm long and densely gland dotted below. The oblong-falcate pods, up to 4 cm long, are stipitate-glandular.

The crenulate, papillate leaflets set *C. placida* apart from similar species in the genus.

A shrub of rocky canyons and hillsides, *C. placida* grows only in Baja California Sur from Comondú and Loreto south to La Paz and Espíritu Santo Island (Moran 1983b).

The sighting at 25.1°N was made by Annetta Carter at 550 m, the highest locality for this species. Our limited data suggest that *C. placida* is associated with warm summers and a predominantly summer rainfall regime.

Flowers appear from October through April.

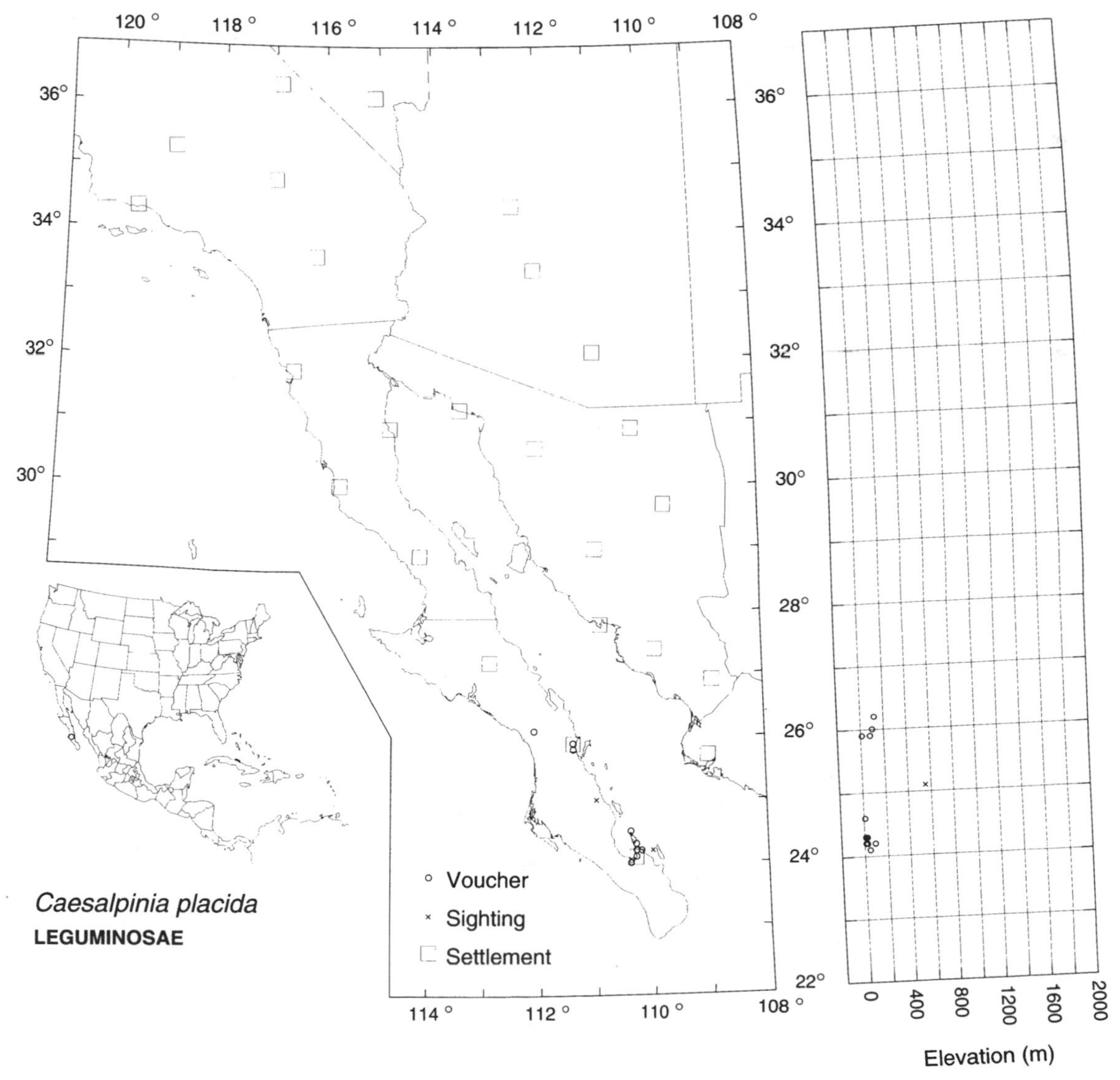

Caesalpinia placida
LEGUMINOSAE

Caesalpinia platyloba
S. Watson

Palo colorado, arellano

Caesalpinia platyloba is a small tree 4–8 m tall. The large (15–20 cm long), twice-compound leaves each bear 2–4 pairs of pinnae. On poor sites, the leaves may be smaller. The leaflets, which occur in 4–7 pairs per pinna, are 2–6 cm long and obliquely unequal at the base. The yellow petals are 8–9 mm long. Measuring 5–10 cm long, the dark reddish brown fruits are mainly indehiscent.

Small-leaved *C. platyloba* may look like *C. pumila.* One may use the fruits, glabrous in *C. pumila* and usually finely puberulent in *C. platyloba,* to distinguish them.

This small tree is widespread in Mexico but in the region of this atlas is restricted to plains and hillsides east of the desert or along arroyos within it.

Flowers appear from July to September (Wiggins 1964). The leaves turn bright red in autumn, then drop, leaving the branches bare through winter and spring. New leaves appear with summer rains. Fruits persist on the leafless trees well into the winter.

The wood is widely used for construction in the Sonoran Desert region.

Caesalpinia pulcherrima
(L.) Swartz

Bird-of-paradise flower, Barbados flower, dwarf poinciana, peacock flower, barba del sol, flor de San Francisco, tabachín, chacmol, flor de guacamaya, flor del camarón, poinciana enana, tetezo

Caesalpinia pulcherrima is a suffrutescent perennial, shrub, or small tree. Young stems may have weak, recurved prickles. The leaves, 20–30 cm long, are twice-pinnate with 3–9 pairs of pinnae, each with 5–12 pairs of oblong leaflets. Bright green above and pale beneath, the leaflets are 5–10 mm wide and 7–20 mm long. The

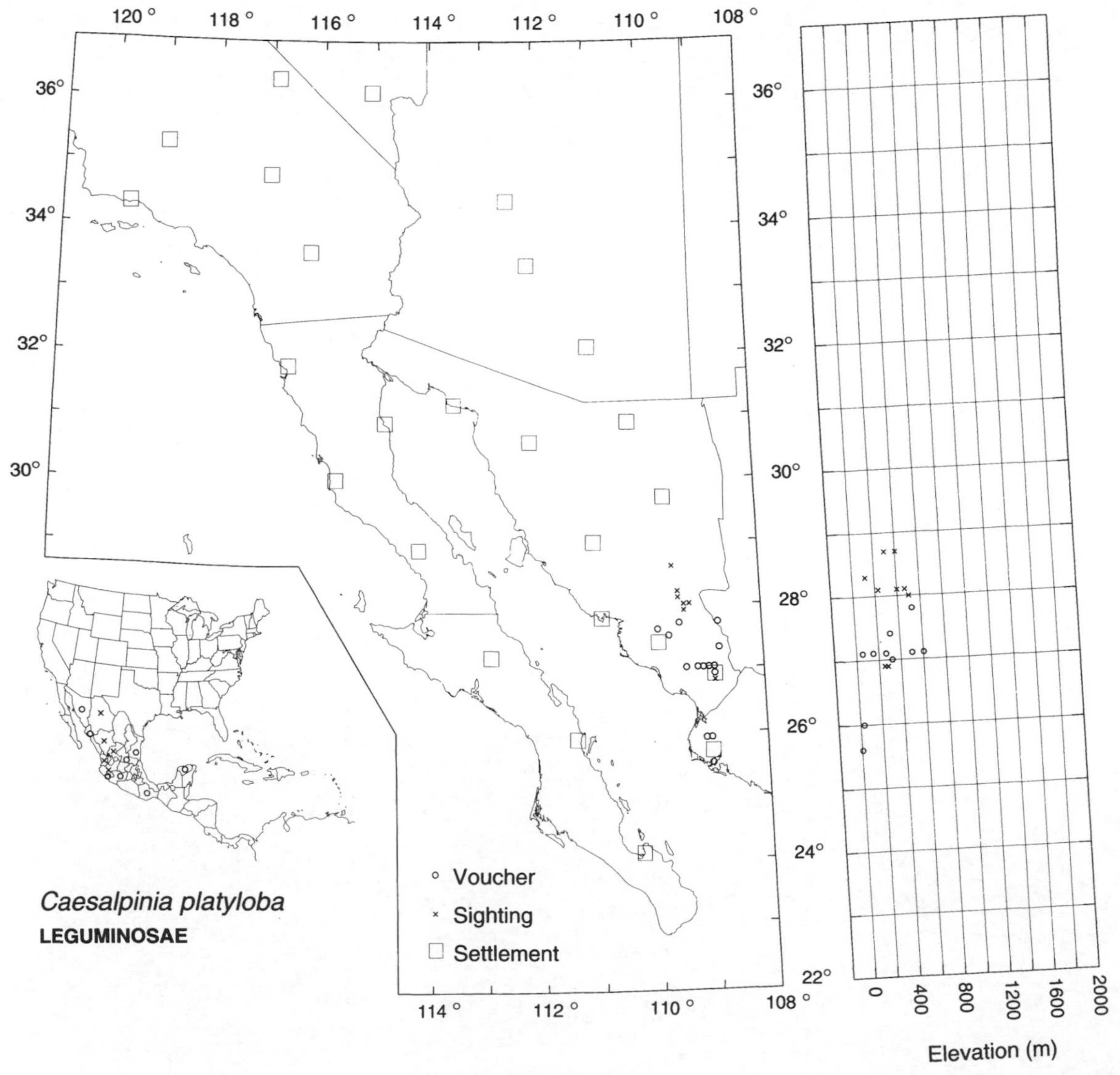

Caesalpinia platyloba
LEGUMINOSAE

showy vermilion and yellow flowers, about 3 cm long, are in terminal or axillary racemes. Four of the clawed petals are fan shaped; the fifth forms a narrow tube. The bright red stamens are two to three times as long as the flowers. Most populations have male and hermaphroditic flowers (Cruden and Jensen 1979). Fruits are oblong pods 7–13 cm long. The haploid chromosome number is 12 (Tixier 1965).

Tucker and coauthors (1985) described the floral ontogeny of *C. pulcherrima*. According to Lim and Ng (1977), the roots do not form rhizobial nodules.

Most other *Caesalpinia* species in the Sonoran Desert region have smaller leaves (10 cm long or less in length) and shorter stamens (barely or not exceeding the petals). None is prickly. *Caesalpinia gilliesii* Wall. in Hook. has a markedly glandular inflorescence.

Caesalpinia pulcherrima is locally common on slopes, in washes, and along roads. Widely cultivated in the tropics and subtropics of the Old and New World, it has naturalized throughout the warmer parts of Mexico (Standley 1922; Wiggins 1964). Its place of origin is unknown.

The points in Baja California Sur represent an 1890 collection by T. S. Brandegee from San José del Cabo and a sighting by Ira L. Wiggins near La Paz (personal communication 1971). Although *C. pulcherrima* is cultivated at least as far north as Phoenix, Arizona, it has not naturalized north of 30.4°N or west of 111.0°W. Lack of sufficient summer rain may determine the western limit in Sonora. Low winter temperatures and aridity may combine to prevent naturalization in Arizona.

The main season of bloom is summer (July–October), but flowers may appear as early as April and as late as November. In western Mexico, hummingbirds and butterflies pollinate the blossoms (Gentry 1942; Cruden and Hermann-Parker 1979). Several floral traits suggest coevolution with butterflies: nectar is secreted when butterflies are active; the modified upper petal is just wide enough to accommodate the butterfly proboscis; and the concentration of amino acids in the nectar resembles that of other butterfly-pollinated flowers (Cruden and Hermann-Parker

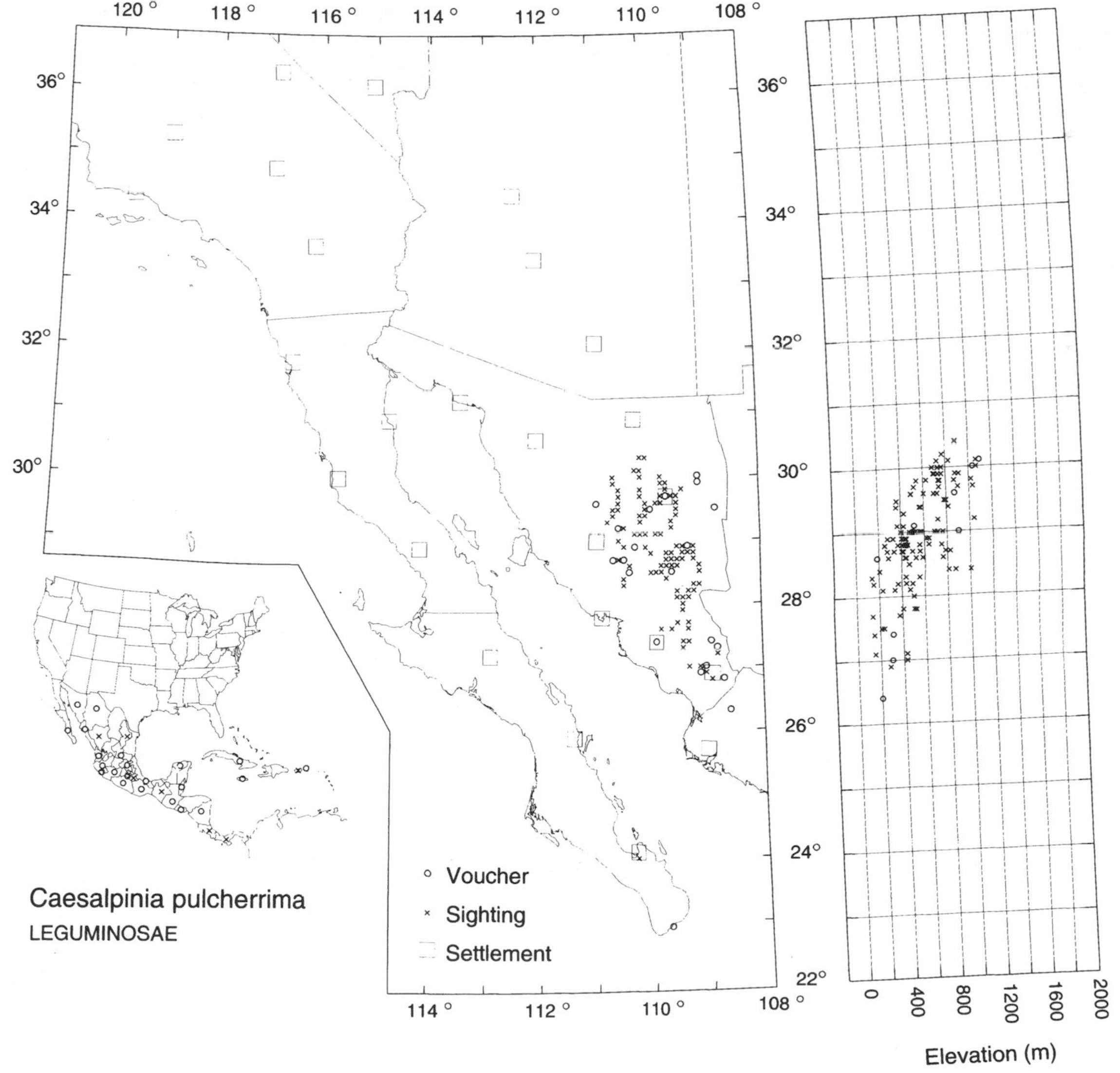

1979). Viscin threads connect several pollen grains into a single pollination unit (Cruden and Jensen 1979) and also attach pollen units to potential pollinators. Butterflies carry the pollen mainly on their wings and to some extent on their bodies; large butterflies carry more pollen and are therefore more effective pollinators than small ones (Cruden and Hermann-Parker 1979; Cruden and Jensen 1979).

Most pods contain many seeds, but some have only a few. Shaanker and Ganeshaiah (1988) suggested that bimodal seed distribution may result from a conflict between the interests of the maternal plant (many seeds per pod) and the interests of the offspring (few seeds per pod).

An attractive ornamental during the warm season, *C. pulcherrima* can be grown in full to reflected sunlight with moderate to ample water (Duffield and Jones 1981). The stems freeze to the ground at –4°C (Johnson 1993), and the roots should be mulched where winter temperatures go much lower (Duffield and Jones 1981).

The fragrant flowers are a good source of nectar for honeybees. Tarahumara Indians eat the young seeds raw (Gentry 1942). The fruits, which contain tannins, are used for tanning and dyeing. Leaves and flowers have a variety of medicinal uses, including treatment for fever, mouth ulcers, colds, insect stings, and rattlesnake bites (Standley 1922; Gentry 1942).

Caesalpinia pumila
(Britton & Rose) Hermann

An unarmed, open-crowned shrub up to 3 m tall with red-brown or dark gray bark conspicuously dotted with white lenticels, *Caesalpinia pumila* has 2–5 pairs of leaflets on each of 2–3 pairs of pinnae. The obovate to nearly orbicular leaflets, 0.5–2 cm long, are yellow-green and distinctly pinnate veined. The yellow petals are 8–9 mm long. The pods (2–3.5 cm long) are veiny and glabrous.

Caesalpinia palmeri has smaller leaflets and an odd number of pinnae.

This shrub, endemic to the tropical, semiarid margin of the southeastern So-

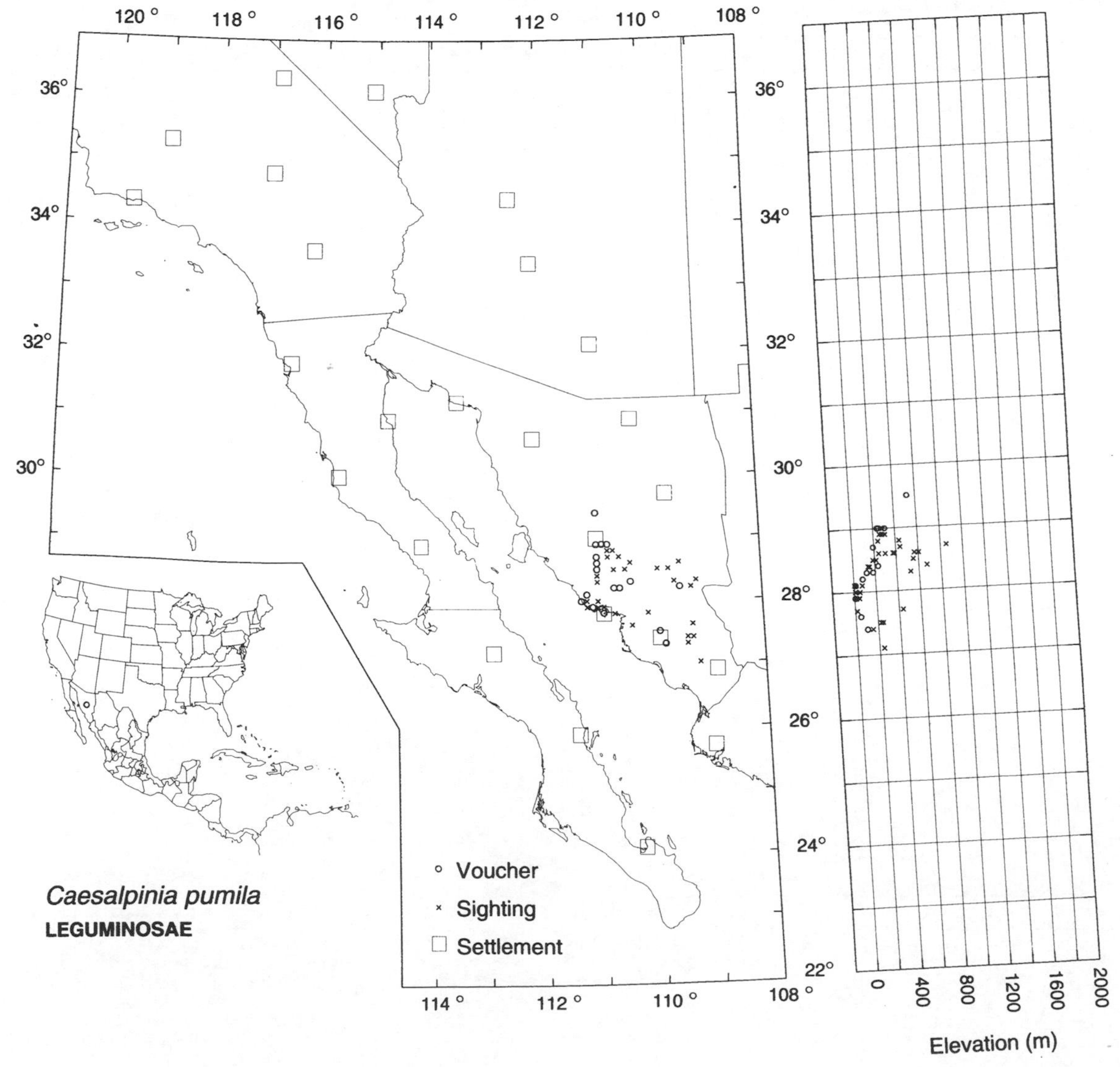

noran Desert, grows on hillsides, gravelly plains, and arroyo margins, especially on soils derived from granite (Wiggins 1964). It is common in Sonoran desertscrub from Carbó to Guaymas, Sonora, less so from Guaymas south to Navojoa. Throughout its range, frost is rare and rainfall is concentrated in the summer.

The leaves are drought deciduous. The flowers appear after the summer rains begin.

Caesalpinia virgata

E. M. Fisher

[=*Hoffmanseggia microphylla* Torr.]

The clumped, rushlike stems of *Caesalpinia virgata* are 1–2 m tall. They may be suffrutescent or woody, and they remain green into old age. The bipinnate leaves have three widely spaced pinnae. The terminal pinna is 15–40 mm long with 7–15 pairs of leaflets; the two lateral pinnae are 5–10 mm long with 4–9 pairs. The yellow flowers are 5–6 mm wide. The fruits are flat, lunate, dehiscent pods measuring 1.5–2.5 cm long and 7–8 mm broad. They are dotted with stipitate, reddish glands. A haploid chromosome number of 12 has been reported (Bell 1965). The roots are not nodulated (Eskew and Ting 1978).

In *Hoffmanseggia intricata,* the three pinnae are more or less equal in length and the stems are more highly branched. Leaves of *Senna purpusii* are once-pinnate. No other *Caesalpinia* has persistently green stems.

Isely (1975) transferred this species from *Hoffmanseggia* to *Caesalpinia* largely because of fruit characters. According to his treatment, *Hoffmanseggia* in the narrow sense is a genus of perennial, rhizomatous herbs with (usually) indehiscent pods.

Locally common in gullies, slopes, canyons, and washes, *C. virgata* grows in

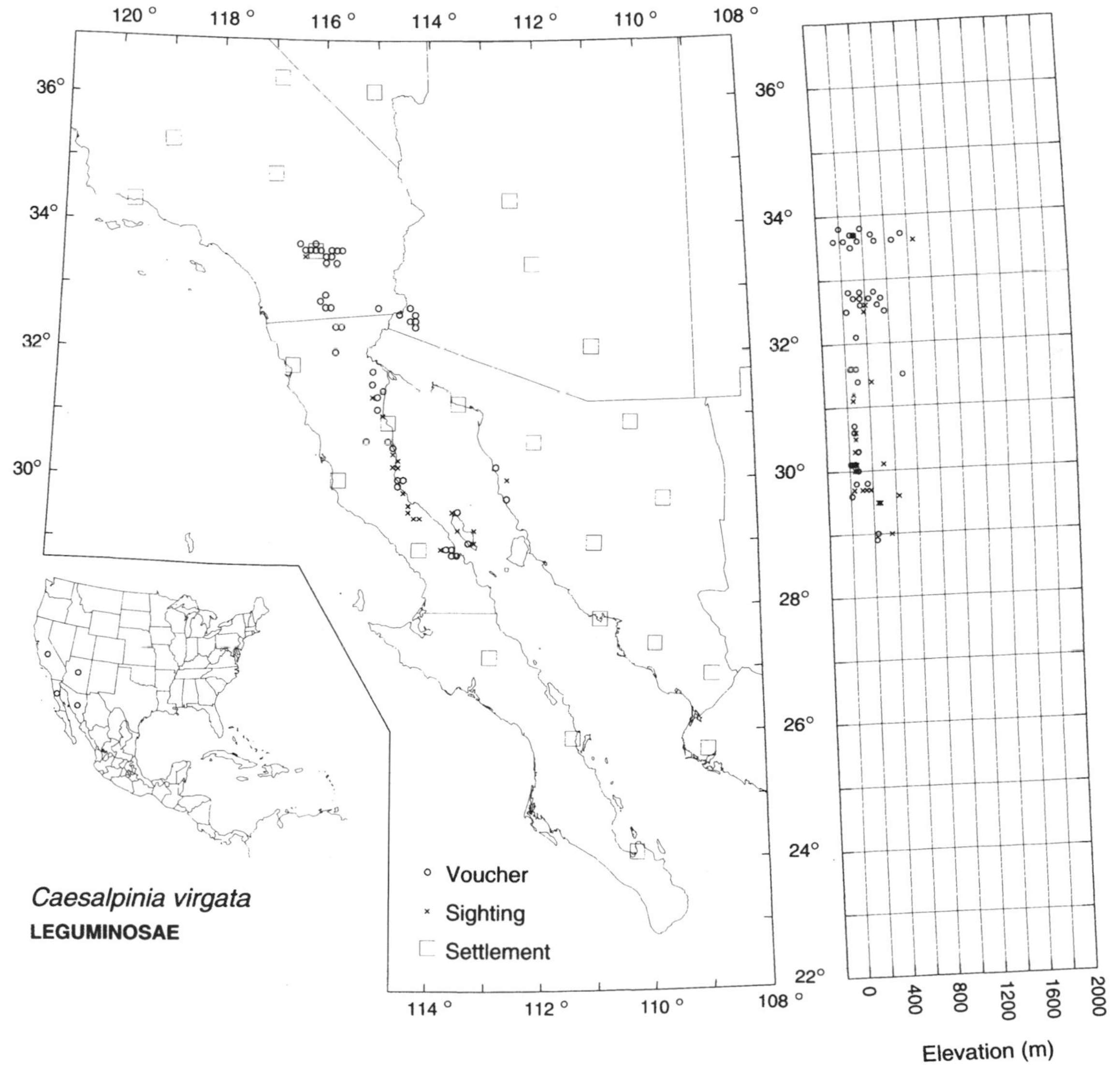

Caesalpinia virgata
LEGUMINOSAE

some of the most arid parts of the area mapped in this atlas. It occurs on only one gulf island, Isla Angel de la Guarda (Moran 1983b). The three disjunct populations on the Sonoran coast represent seven collections and several sightings. Across the range of *C. virgata,* rainfall is erratic and biseasonal. The species boundaries indicate sensitivity to frost and a requirement for occasional warm-season rain.

Most populations bloom in spring (March–May) and again in fall (September–October). Occasional plants flower in other months, particularly after rain.

The drought-deciduous habit, chlorophyllous bark, and rushlike stems enable *C. virgata* to survive in very arid regions.

Calliandra brandegeei

(Britton & Rose) Gentry

The twigs, petioles, rachises, and peduncles of this small shrub are glandular-pubescent with brown hairs and glands. Leaves are twice-pinnate with 4–6 pairs of pinnae, each with 10–22 pairs of oblong leaflets, 5–9 mm long. The purplish red flowers, clustered in globose heads, are made conspicuous by the filaments, which are 2 cm long. The flattened pods are 1–1.8 cm wide and 7–10 cm long.

No other *Calliandra* species in its range has glandular, brown pubescence.

Narrowly endemic to the Cape Region of Baja California Sur, *C. brandegeei* grows on canyon slopes above 1,000 m. Wiggins (1964) reported it from the Sierra de la Giganta, but we have seen no vouchers from that far north. *Calliandra brandegeei* may have speciated when changes in sea level during the Pliocene isolated the Cape Region from the remainder of the peninsula, forming an island (Gastil et al. 1983). It is absent from the gulf islands (Moran 1983b).

The flowering season is August–October (Wiggins 1964).

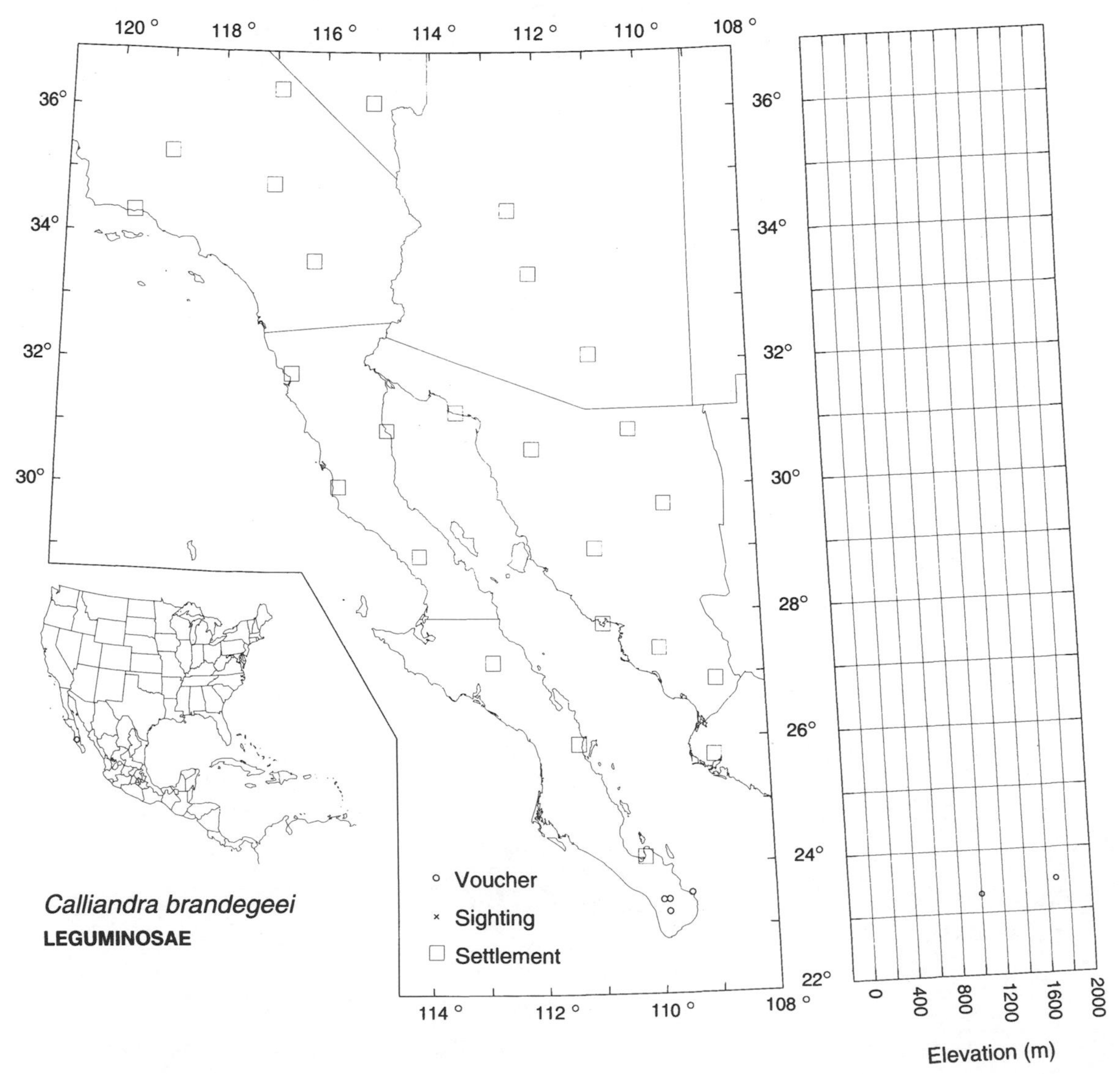

Calliandra brandegeei
LEGUMINOSAE

Calliandra californica
Benth.

Tabardillo, zapotillo

This straggling, divaricately branched shrub grows up to 1 m tall. The leaves are twice-pinnate with 1–4 pairs of pinnae, each with 5–15 pairs of oblong leaflets (3–6 mm long) that are glabrate and shining above, dull and pale beneath. The capitate inflorescence, deep rose to purplish red, is made conspicuous by the showy filaments. Fruits are flattened pods 5–7 mm wide and 4–6 cm long.

In Baja California, *C. eriophylla* occurs north of 31°N, well beyond the range of *C. californica*. *Calliandra peninsularis* has 18–25 pairs of leaflets per pinna. *Calliandra brandegeei* has glandular brown pubescence on young twigs and rachises.

Often common on washes, hillsides, and plains, this is the calliandra most often seen on the peninsula. Disjuncts on Islas Angel de la Guarda and Tiburón represent sightings by Reid Moran and Richard Felger. The single mainland population, first vouchered by Tony L. Burgess in 1983, is at Algodones Bay. The species is found in a variety of climates. In the northern part of the range, rain falls mainly in winter, and frosts are occasional; in the southern part, rains are concentrated in summer and early fall, and frost is absent.

In more mesic times certain peninsular species (for example, *Euphorbia magdalenae* and *Fouquieria diguetii*) may have spread to the mainland via the midriff islands (Cody et al. 1983). Some of them evidently migrated southward along the Sonoran coast and later, with the return of arid conditions, retreated to more favorable sites, such as low mountains in the Guaymas region. The present distribution of *C. californica* fits this hypothesis. Nevertheless, although the single mainland population grows in an area often visited by botanists, it was not found until 1983.

Flowers appear year-round in response to rain. They are visited by hummingbirds, which may be the major pollinators. Leaves are drought- and cold-deciduous.

The stems freeze at temperatures

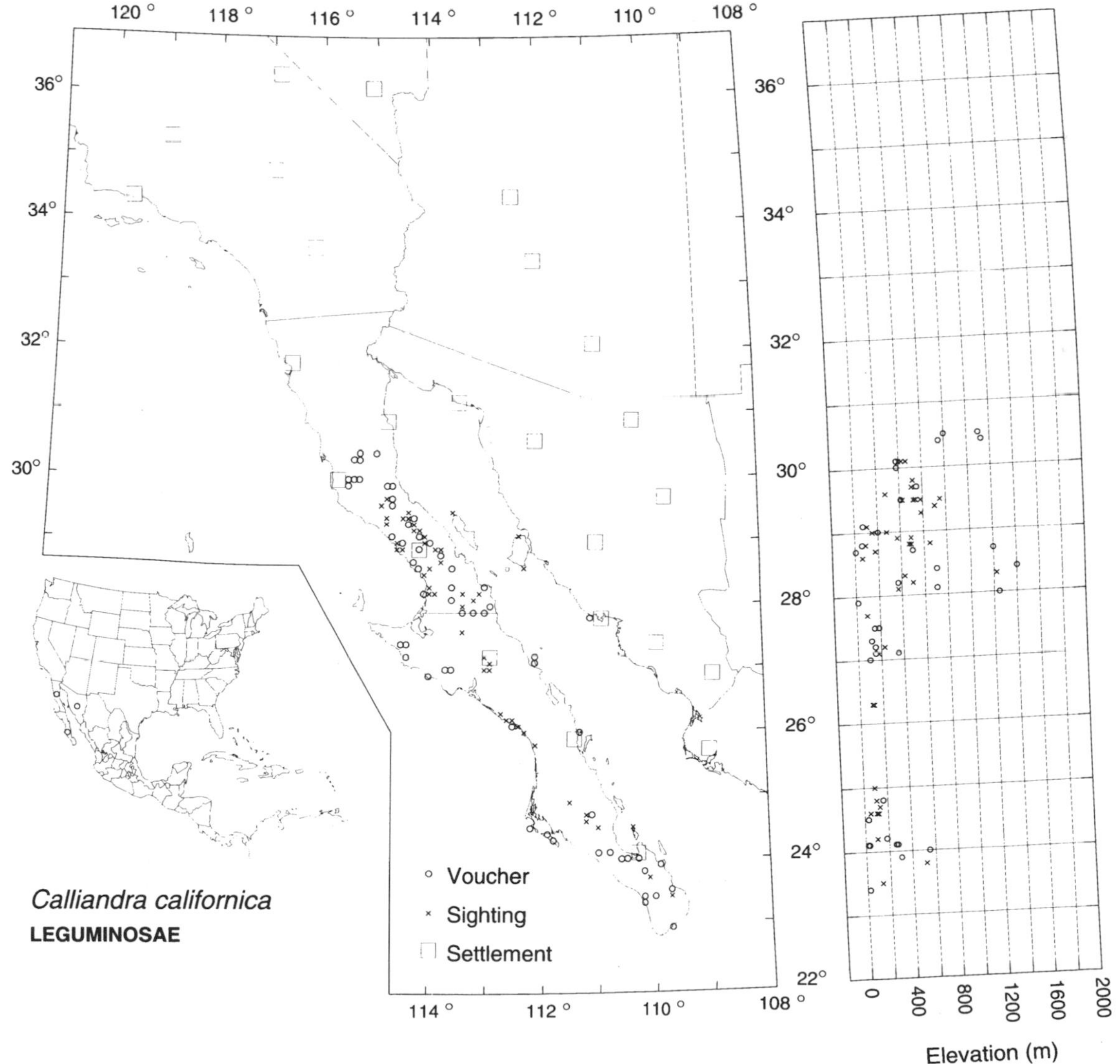

Calliandra californica
LEGUMINOSAE

below –4°C (Johnson 1993), limiting the ornamental use of this species to warm-desert gardens. In the horticultural trade, it is not always distinguished from *C. peninsularis* (Gregory D. Starr, personal communication 1992).

Calliandra eriophylla

Benth.

Fairy duster, huajillo, cabeza angel, cabelleto de angel

This spreading shrub grows up to 1 m high and has unarmed light gray or whitish stems. The widely spaced leaves are twice-pinnate with 2–4 pairs of pinnae, each with 7–9 (occasionally 10) pairs of leaflets 2–3 mm long. The showy flowers are in dense, spherical heads 4–5 cm in diameter. Corollas are only 5–6 mm long and inconspicuous: it is the pink, rose, or reddish purple stamens, up to 1.5 cm long, that make the showy display. Fruits are linear, velvety pods 5–7 mm wide and 3–7 cm long with thickened margins.

In Arizona and California, no other shrub has long, pink stamens. In the absence of flowers, the low stature and unarmed whitish or light gray stems help separate it from acacias and mimosas. *Zapoteca formosa* (including *C. schottii* and *C. rosei*) also has long stamens, but these are white, and its leaves are much larger, with leaflets 5–9 mm long. *Calliandra californica, C. brandegei* and *C. peninsularis,* all peninsular endemics, occur well south of *C. eriophylla.*

According to Isely (1972), the closest relative of *C. eriophylla* is *C. conferta* A. Gray of Texas, Coahuila, Nuevo Leon, and Tamaulipas. Isely (1972) described intermediate forms from western Texas and adjacent New Mexico as *C. eriophylla* var. *chamaedrys.* The plants of the area included in this atlas are var. *eriophylla.* They show a cline in pinnae number, from 1–2 pairs in the west to 2–4 pairs in the east (Isely 1972). Although our map shows *C. eriophylla* in Texas, Coahuila, San Luis Potosí, and Tamaulipas, these populations may actually be *C. conferta* A. Gray

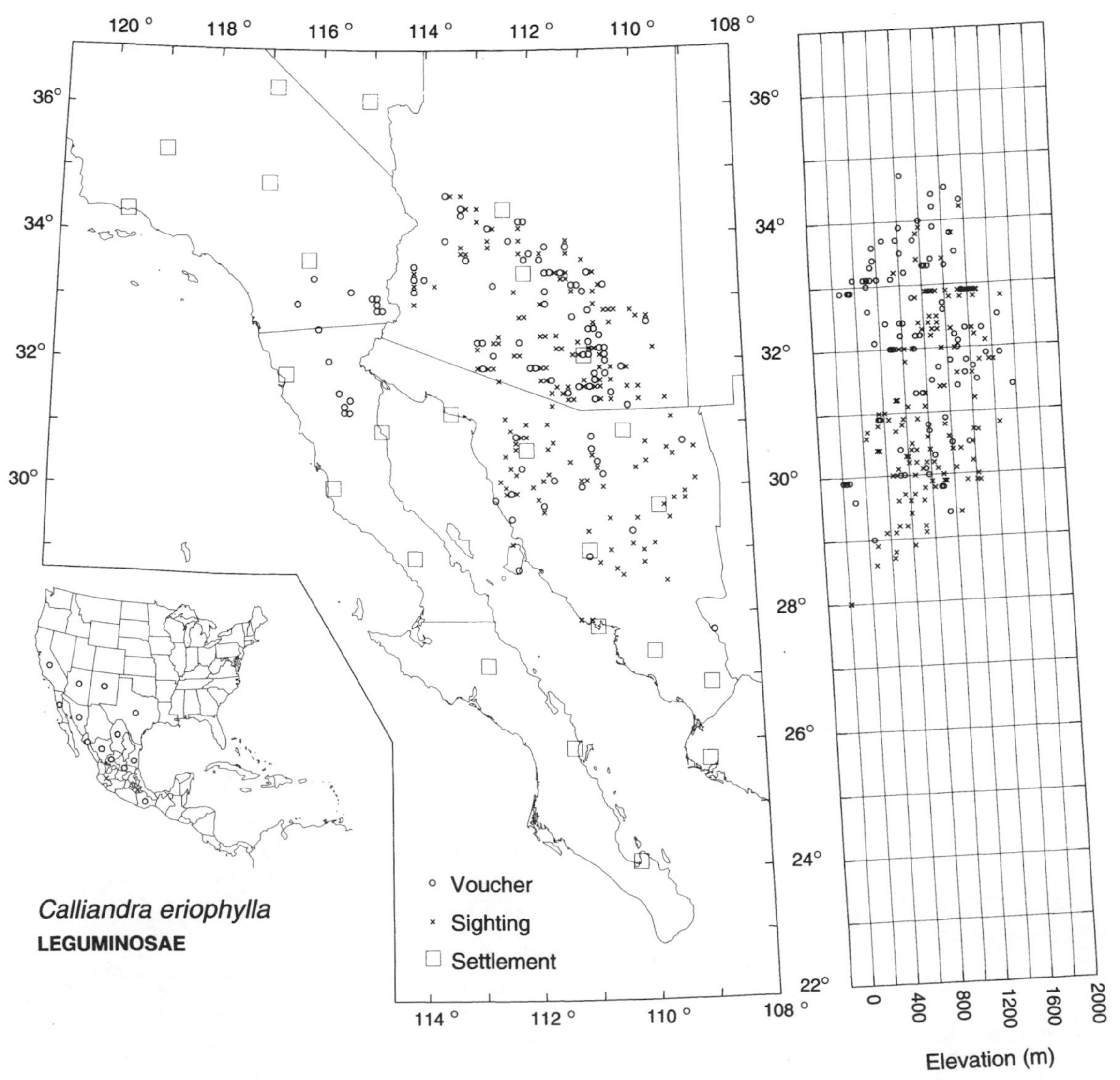

(Correll and Johnston 1970).

The widely ranging *C. eriophylla* grows along washes in the drier parts of its range and on slopes and mesas elsewhere. In semidesert grassland it is often a dominant woody plant on open slopes. Typically, it is low, almost creeping, toward its upper limits, erect and bushy at lower elevations. The disjunct populations between 31 and 32°N in Baja California are near similarly disjunct localities for *Acacia constricta* and *Mimosa aculeaticarpa.*

Over most of its range, rainfall is biseasonal with a predominance of summer moisture. The western and northern populations receive more dependable winter rain, but the species boundaries show a need for summer storms, probably for seedling establishment. Winter frost is occasional in the higher and northerly parts of the range.

Spring flowering starts in February at low elevations, in March or April higher up. Plants occasionally bloom again in September and October. Leaves are generally cold deciduous, falling in winter and reappearing with or shortly before the flowers.

The infrequent flower visitors include *Apis mellifera, Xylocopa,* dipterans, and lepidopterans (Simpson 1977; Simpson and Neff 1987). Plentiful nectar (2.17 mg total sugars/flower/24 hours) is produced (Simpson 1977), so it seems odd that the flowers are not more heavily used than they appear to be. According to Stephen Buchmann (personal communication 1989), the pollen is produced in packets that adhere to the wings of butterflies, which presumably transfer them from plant to plant.

Calliandra eriophylla is grown as an ornamental and will survive temperatures as low as −12°C (Johnson 1993). The plants are valuable browse for livestock and deer and, because of their rhizomatous habit, are important in erosion control (Benson and Darrow 1981).

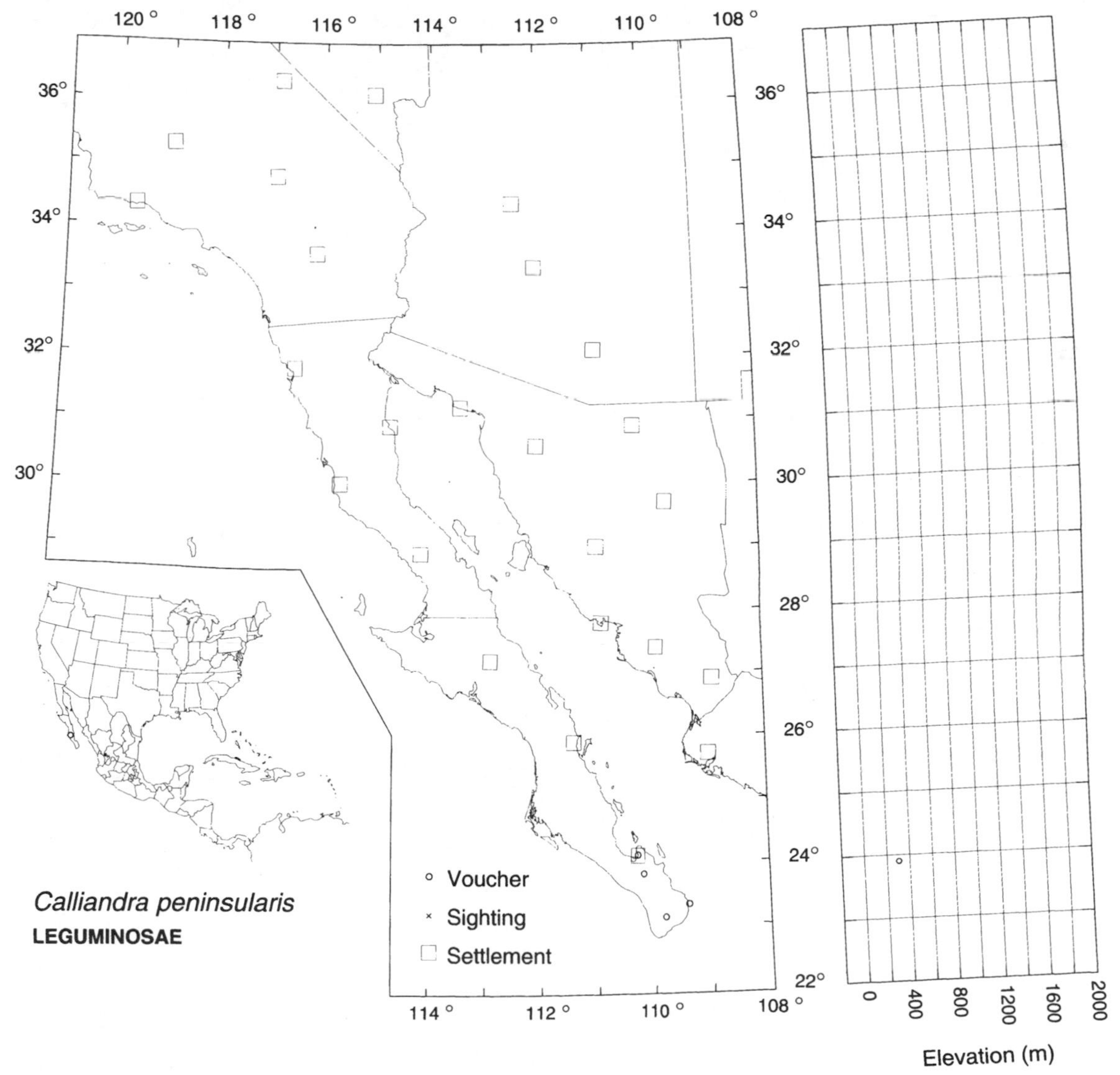

Calliandra peninsularis
LEGUMINOSAE

Calliandra peninsularis
Rose

Tabardillo, zapotillo

Calliandra peninsularis is a spindly shrub with stiff, slender twigs and puberulent foliage. The twice-pinnate leaves have 5–6 pairs of pinnae, each with 18–25 pairs of linear-oblong leaflets 4–6 mm long. The individual flowers are small, but their bright red or purplish filaments (up to 2 cm long) make a showy display. Fruits are tapered pods 6–9 cm long and 8 mm wide.

Calliandra californica has 5–15 pairs of leaflets per pinna. *Calliandra brandegeei* has brown, glandular pubescence.

A plant of dry watercourses and rocky hillsides, *C. peninsularis* is narrowly endemic to La Paz and the Cape Region of Baja California Sur. In contrast to *C. brandegeei,* which is limited to the mountains of the Cape Region, *C. peninsularis* appears to be a lowland plant.

The flowering season is November–March (Wiggins 1964).

The plants are commonly grown as ornamentals in warm-desert gardens and are often sold in nurseries as *C. californica* (Gregory D. Starr, personal communication 1992). The roots have been used as a remedy for fever (Standley 1922).

Calliandra rupestris
Brandegee

This unarmed shrub or small tree grows up to 4 m tall and has a single pair of pinnae, each with two pairs of conspicuously veined leaflets. Three of the leaflets are 1–3 cm long; the fourth is only 2–7 mm long or, rarely, absent. The dark purple, axillary flower heads are made conspicuous by the stamens, which are 3–5 cm long.

The large, unequal leaflets and unarmed stems are distinctive.

Standley (1922) considered *C. rupestris* synonymous with *C. emarginata* (Humbl. & Bonpl.) Benth., a variable

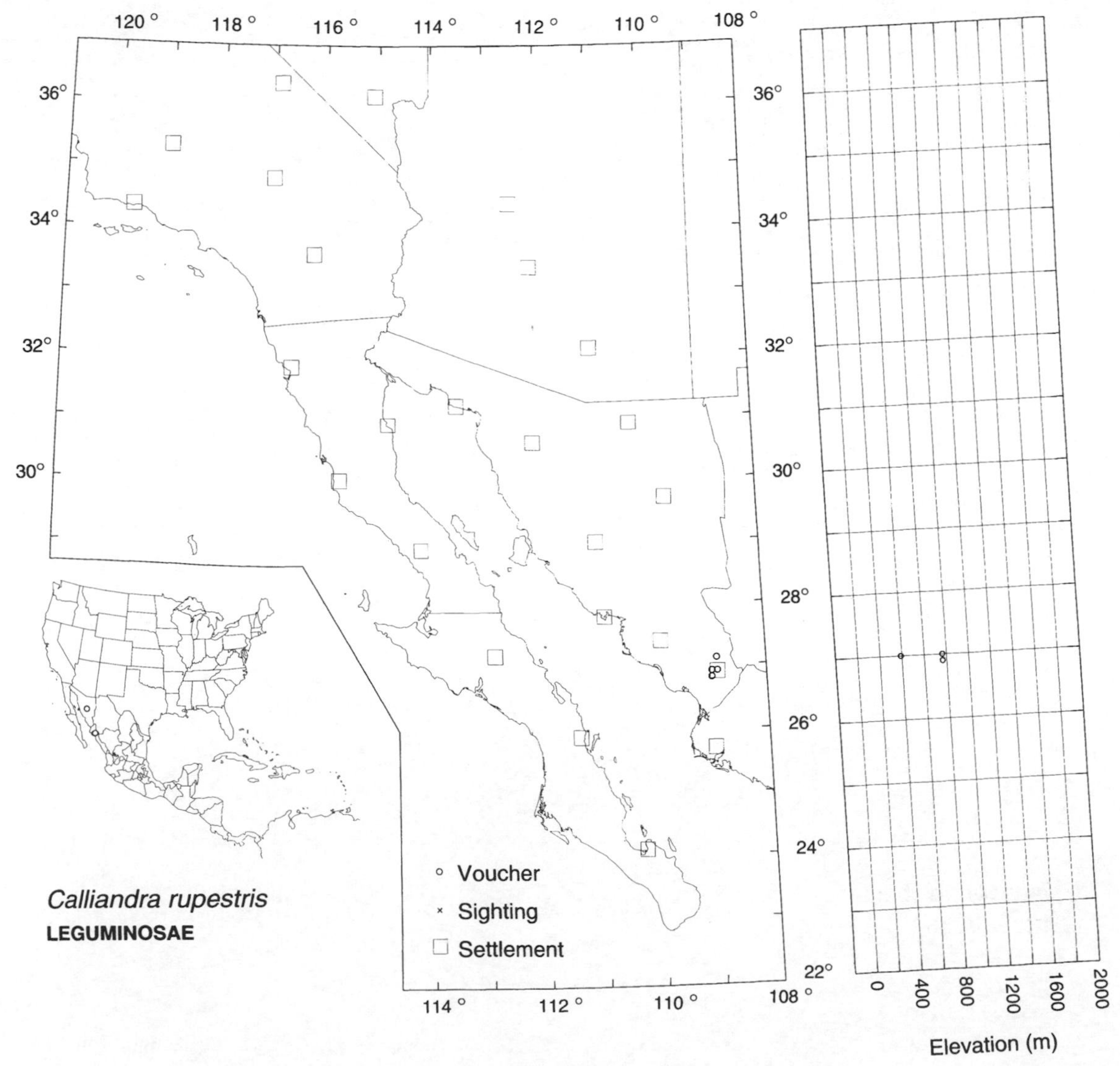

Calliandra rupestris
LEGUMINOSAE

taxon widespread in western Mexico.

An occasional element of Sinaloan deciduous forest (Gentry 1942), *C. rupestris* is narrowly endemic to southern Sonora and adjacent Sinaloa.

The plants flower abundantly in June and July (Gentry 1942).

Canotia holacantha

Torr.

Crucifixion thorn, corona de cristo, junco

A shrub or small tree 2–6 (occasionally 10) m tall, *Canotia holacantha* has green bark and slender, striate, rushlike, yellow-green branches. When young, the branches are flexible; with age they become rigid, spinescent, and brown. The small, scalelike leaves are soon deciduous. The greenish yellow flowers, 5–6 mm long, are in small axillary clusters. Fruits are woody, 5-valved capsules 12–14 mm long. Each valve splits at the apex into two long, slender horns.

Without fruits or leaves, *C. holacantha* superficially resembles several other green-barked shrubs and trees. *Cercidium microphyllum* branchlets are divergent and rigid, those of *C. holacantha* upright, parallel, and flexible. *Cercidium floridum* has small spines at the nodes. *Parkinsonia aculeata* has characteristic "streamers" (the rachillae of the primary leaves) and paired spines at the nodes. Other species commonly called crucifixion thorn are *Koeberlinia spinosa,* with rigid, divergent stems, and *Castela emoryi,* with stiff, stout ones.

In central Arizona, where *C. holacantha* is occasionally a dominant, the plants show a marked preference for limestone and other deeply permeable substrates. They replace *Cercidium microphyllum* on rocky hillslopes and mesas in the northern portion of the Gila River drainage (Benson and Darrow 1981). The widely disjunct Sonoran stations may be Pleistocene relicts. The southernmost collections, made on Isla Tiburón (28.9°N and 29.0°N), are Richard S. Felger's. Throughout its range, the species receives biseaso-

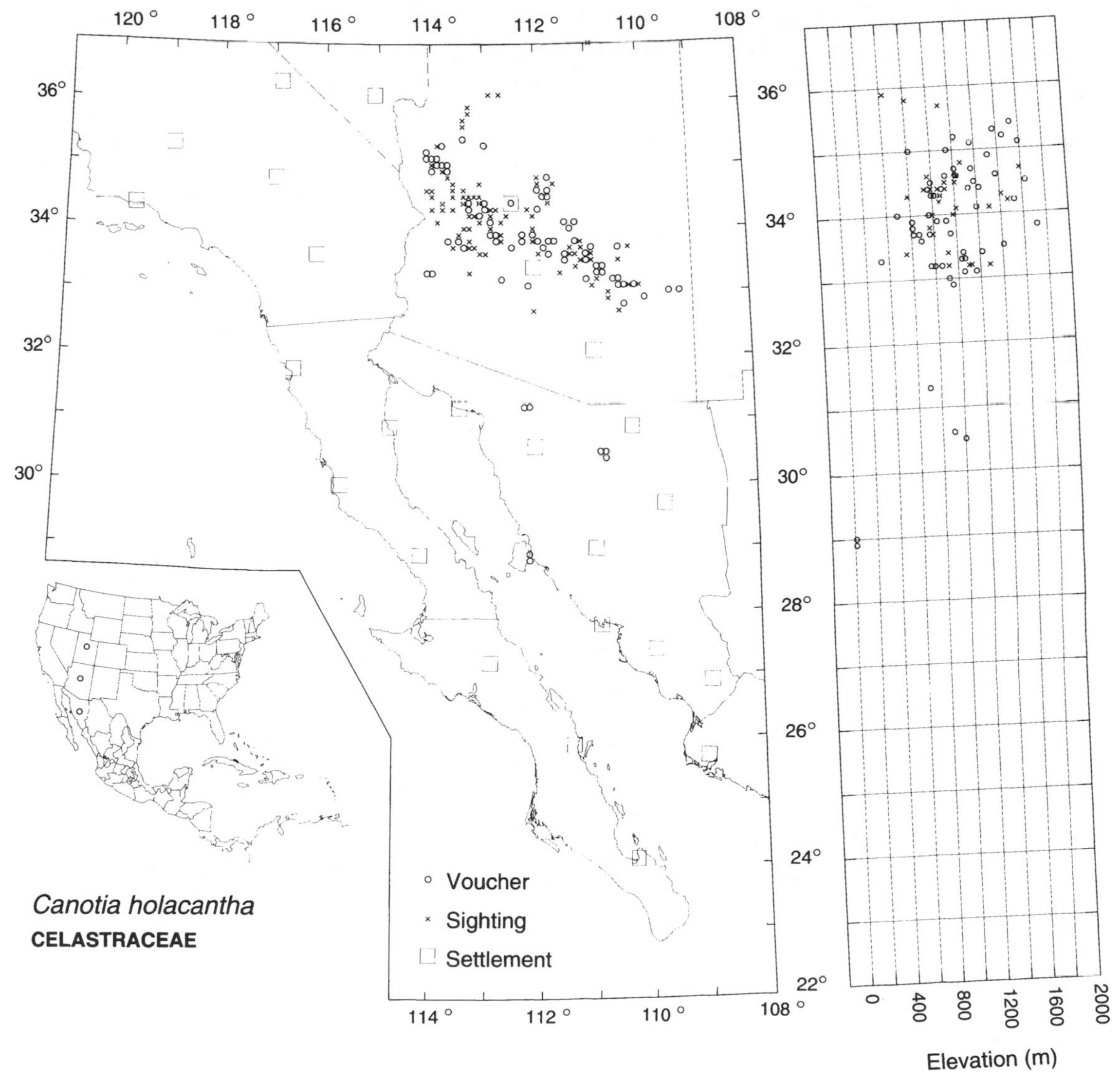

Canotia holacantha
CELASTRACEAE

nal rainfall with approximately equal amounts in summer and winter. Winter frosts are probably common in most populations.

The main period of bloom is June and July, with some flowering as early as May and as late as October. Fruits mature during summer and typically persist until the following spring.

The normally leafless condition helps conserve water. Other water-conserving adaptations in *C. holacantha* include a thick cuticle (visible as flakes of wax in stem grooves) and sunken stomata (Gibson 1979). The green bark is photosynthetic.

Flowers are much visited by bees.

Carnegiea gigantea

(Engelm.) Britton & Rose

[=*Cereus giganteus* Engelm.]

Saguaro, giant cactus

This large columnar cactus grows to a height of 12 m or more, with (on older plants) one to several branches originating 2–3 m above the base (figure 31). The main stem, up to 40 cm in diameter, has 12 to 25 vertical ridges, or ribs. Clusters of 15–30 spines up to 3.8 cm long are spaced about 2–3 cm apart on the ribs. The white flowers are 5–6 cm across. The fleshy fruits, 6–10 cm long, turn red or purple at maturity and split open, revealing many tiny black seeds. The haploid chromosome number is 11 (Pinkava and McLeod 1971).

This is the largest cactus in Arizona and California. *Pachycereus pringlei* is a stouter, more massive plant that branches closer to the ground and has fewer (10–15) vertical ribs.

Benson (1982) referred to this plant as *Cereus*. We follow Gibson and Horak (1978) in retaining it in *Carnegiea*. The species shows strong relationships with *Neobuxbaumia* of southern Mexico (Gibson and Horak 1978).

Carnegiea gigantea is common on rocky hillsides and outwash slopes throughout its range. Dense stands some-

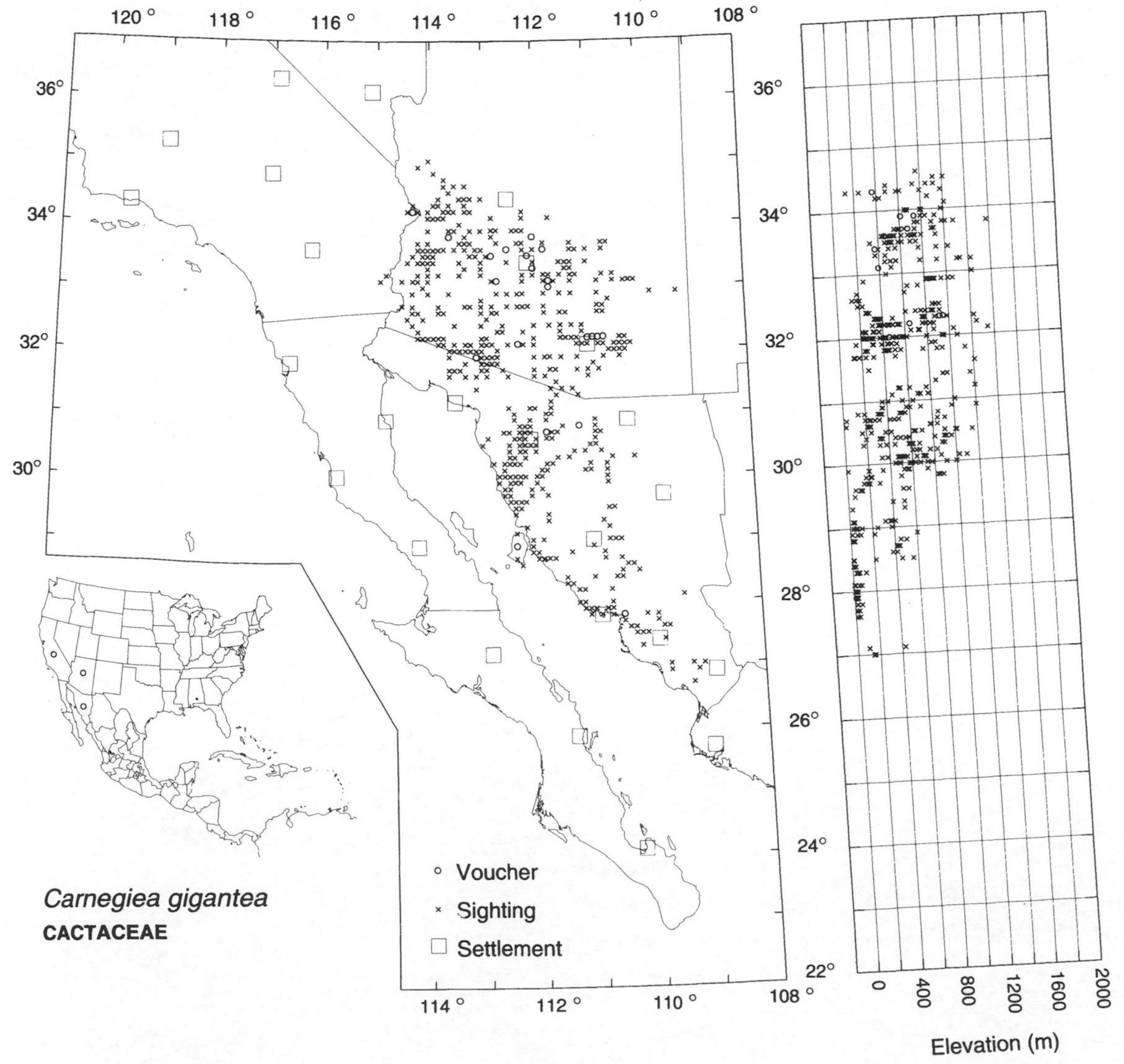

Carnegiea gigantea
CACTACEAE

times grow on sandy flats (as near Los Vidrios, Sonora) or on fine alluvium (as, in years past, near Chandler, Arizona) (Shreve 1964). Especially toward their northern limit, the plants are generally more abundant on south-facing than north-facing slopes. The densest stands are in Arizona roughly east of 112.8°W.

The northernmost of the large columnar cacti, *C. gigantea* grows in most parts of the Sonoran Desert where summer rainfall is substantial, with the exception of Baja California Sur and eastern Sonora. Its northwestern limit is roughly the Colorado River (a few small populations are in the Chocolate, Whipple, and Picacho mountains, California). Like many Sonoran Desert cacti, *Carnegiea* depends largely on warm-season rain (Cannon 1916a; MacDougal 1924; Hastings 1961), and west of the Colorado River, the amount of summer rain drops too low for survival. Because the mature plants use little if any soil moisture when temperatures are low, the increased winter rainfall to the west does not compensate for the dry summers.

Low temperatures limit *Carnegiea* along the northern, northeastern, and upper elevational margins of its range (Shreve 1911a; Steenbergh and Lowe 1977; Nobel 1978, 1980b, 1980c). The limiting factors in eastern Sonora have not been identified; perhaps the increase in density and cover of associated species discourage seedling establishment. Toward the southern end of its range, competition with more mesic tropical vegetation restricts *Carnegiea* to progressively lower elevations.

This species was present in the Puerto Blanco Mountains of southwestern Arizona 10,500 years ago (Van Devender 1987) and in the Hornaday Mountains, Sonora, 9,400 years ago (Van Devender, Burgess et al. 1990), shortly after the last glacial retreat. During the early to middle Holocene, it spread to the northern part of its present range, where it grew with *Acacia greggii, Cercidium floridum, Prosopis velutina,* and *Juniperus californica* Carr. long before current associates such as *Cercidium microphyllum, Olneya tesota,* and *Larrea tridentata* arrived (Van Devender 1990b).

The vertical ribs enable *Carnegiea* stems to expand as moisture is absorbed and to contract as it is used (Spalding 1905). Rib spacing and depth vary from one side of the plant to another and affect surface/volume ratios as well as plant circumference (Geller and Nobel 1984). During drought, stem water is preferentially concentrated in chlorenchyma, allowing the plants to survive water losses of up to 81% (Barcikowski and Nobel 1984).

Carnegiea stems grow mostly during rainy periods in July, August, and September. Flowers appear from late April to early June. They open in late afternoon and remain open until the afternoon of the following day, then close permanently. A stem might produce more than 100 flowers (Steenbergh and Lowe 1977); the average per plant is 295 (Schmidt and Buchmann 1986). Flowers develop earlier and in greater numbers on the east side of each stem (Johnson 1924), presumably a result of greater accumulation of heat on that side. Fruits ripen from late May until mid July (Steenbergh and Lowe 1977).

Bats visit the flowers at night, and bees, other insects, and birds visit them during the day (Alcorn et al. 1961). Pollination occurs mainly at night and in the early morning (Alcorn et al. 1959; Schmidt and Buchmann 1986). To set fruit, flowers must receive pollen from another plant or from flowers on another arm of the same plant (Alcorn et al. 1959). In one study (Schmidt and Buchmann 1986), honeybees were the main visitors to the flowers, and, so efficient were they at harvesting pollen, virtually all of it was gone by 9:00 or 10:00 A.M. Before honeybees were introduced (about 100 years ago), intermittent lack of native pollinators may have sporadically limited fruit set (Schmidt and Buchmann 1986).

Seeds are shed near the start of the summer rainy period. Dispersal agents include coyotes, peccaries, and doves (Steenbergh and Lowe 1977; Olin et al. 1989). If germination occurs at all, it will

Figure 31. Carnegiea gigantea *at the Desert Laboratory, Tucson, Arizona. (Photograph by D. T. MacDougal, courtesy of the Arizona Historical Society/Tucson.)*

probably happen within a few weeks of fruit ripening and will follow the first adequate summer rain (Steenbergh and Lowe 1969, 1977). The few seeds that overwinter lose their viability before warm, moist conditions return (Steenbergh and Lowe 1977). The seeds require light for germination and will not germinate if buried deeper than a few millimeters (Alcorn and Kurtz 1959; McDonough 1964; Steenbergh and Lowe 1977).

Nurse plants or rocks improve seedling establishment by protecting the young plants against heat (Turner et al. 1966; Despain 1974; Franco and Nobel 1989), cold (Steenbergh and Lowe 1976; Nobel 1980b), and herbivores (Niering et al. 1963; Steenbergh and Lowe 1977). Not all associates of *Carnegiea* serve equally well as nurse plants. Hutto and coworkers (1986) found that at Organ Pipe Cactus National Monument, *Prosopis velutina* and *Cercidium microphyllum* were more common nurse plants than *Larrea tridentata.*

Once a *Carnegiea* plant is big enough to survive on its own, it may compete with its nurse plant for soil moisture. *Carnegiea* roots are shallowly placed and able to intercept rainwater as it percolates through the soil column (Vandermeer 1980; McAuliffe 1984a). Several *Carnegiea* individuals associated with a single *Cercidium microphyllum* may use up to half the summer rain that falls within reach of their roots (McAuliffe and Janzen 1986). Eventually, the *Cercidium* nurse plant declines and dies (Vandermeer 1980; McAuliffe 1984a). Young *Carnegiea* subjected to drought produce absorbing roots within a few hours of rain (Helbsing and Kreeb 1985).

Carnegiea reaches sexual maturity at 30–35 years (Steenbergh and Lowe 1977). At first, a plant produces only a few fruits. The number increases rapidly over the next few decades, and by the time the cactus is about 4–5 m tall (50–70 years old), it can produce as many as 100 fruits per year. The first branches may appear at 50–70 years, and, within 3 years, they may produce flowers and fruits (Steenbergh and Lowe 1977). With the development of each new branch, the reproductive capacity of the plant rises dramatically. Reproductive capacity in more arid habitats may be reduced by complete failure of flowering during dry years (Brum 1973).

Carnegiea photosynthesizes via Crassulacean acid metabolism (CAM) (Helbsing and Kreeb 1985; Nobel and Hartsock 1986b). The optimal temperature for nocturnal carbon dioxide uptake is 14°C at a day/night temperature regime of 10/10°C. It shifts to 21°C when the plants are grown at 30/30°C (Nobel and Hartsock 1981). *Carnegiea* tolerates day/night temperature regimes up to 50/40°C; carbon dioxide uptake continues after an hour of exposure to 57°C (Smith et al. 1984). Complete acclimation to high temperatures requires about ten days (Kee and Nobel 1986). Low nitrogen can limit rates of carbon fixation and growth (Nobel 1983a).

Branch production varies from place to place and seems to be related to ambient moisture. Toward the western, drier limit and on dry sites in more mesic parts of the range, branches are few or none (Yeaton et al. 1980; Steenbergh and Lowe 1983). Branching, by increasing the amount of surface area, substantially augments carbon dioxide uptake (by as much as 52% for a hypothetical five-stemmed plant at Tucson, Arizona) (Geller and Nobel 1986). (Because volume increases at the same time, the surface-to-volume ratio critical to long-term growth is not altered much.) On flat ground, more than twice as many branches grow on the south-facing side of the main stem as on the north, east, or west sides (Yeaton et al. 1980; Geller and Nobel 1986). Apparently, the greater photosynthetic capacity brought about by additional branches largely offsets the energy costs of increased reproduction.

Annual height growth varies with plant size. During the first few years, a seedling may grow as little as 1–2 mm per year. The growth rate then increases rapidly until a height of 2–4 m, when it may be 100 times greater than at the seedling stage (Steenbergh and Lowe 1983). As plants reach reproductive age, they divert energy into flowers and fruits, and vegetative growth is proportionately reduced (Steenbergh and Lowe 1977). Thereafter, growth lessens each year until branches start to form, then remains more or less constant until a slight decline toward the end.

Because a columnar cactus adds and retains a growth increment each year, researchers can determine its age based on its height. Biologists have made this calculation for several *Carnegiea* populations (Shreve 1910; Hastings and Alcorn 1961; Steenbergh and Lowe 1983; Turner 1990), and growth rates (and, hence, ages at given heights) clearly vary widely with habitat and geographic location. Some individuals live 200 years or more (Steenbergh and Lowe 1983), although most probably die several decades earlier.

Where injured, the plants produce a highly ligniferous callus tissue containing numerous phenols. Dopamine, the principal phenolic constituent of healthy cortical tissue, increases markedly at the site of wounding. The blackened tissue found at wound sites contains an abundance of melanin, which has been converted from dopamine (Steelink et al. 1967). Various yeasts and *Drosophila* species are associated with necrotic *Carnegiea* tissue (Ganter et al. 1986; Starmer et al. 1988).

Carnegiea receives more attention in the lay press than any other Sonoran Desert plant, including periodic predictions of decline (Anonymous 1902; Lansford 1967). Although new plants failed to establish in some populations for extended periods (Shreve 1910; Niering et al. 1963), these same populations later included many new individuals (Steenbergh and Lowe 1983; Turner and Bowers 1988). Reduced establishment rates have been blamed on a number of circumstances: climate (Steenbergh and Lowe 1969, 1976, 1977; Jordan and Nobel 1982; Turner 1990), rodents (Niering et al. 1963; Turner et al. 1969), bacteria (Alcorn and May 1962), livestock grazing (Niering and Whittaker 1965), nurse-plant scarcity (Vandermeer 1980), and insects (Steenbergh and Lowe 1977). Following successful establishment, other forces cause mortality, among them wind (Shreve 1964), lightning (Steenbergh 1972), fire (McLaughlin and Bowers 1982; Rogers 1985), catastrophic freezes (Steenbergh and Lowe 1977; Niering and Lowe 1984), and flicker (*Colaptes auratus chrysoides*) nests (McAuliffe and Hendricks 1988). The fungus *Poria carnegiea* penetrates the roots of standing plants, causing root decay (Gilbertson and Canfield 1972).

Native peoples of the Sonoran Desert, especially the Tohono O'odham (Papago), used *Carnegiea* for food, as a building material, and in many other ways (Crosswhite 1980; Fontana 1980).

Greenhouse-grown seedlings and field-dug mature plants are sold commercially. Considerable care is necessary if plants over 2 m tall are to successfully reestablish, and many larger transplants decline and die within 5–10 years.

Castela emoryi

(A. Gray) Moran & Felger

[=*Holacantha emoryi* A. Gray]

Crucifixion thorn, corona de Cristo

Castela emoryi is a leafless shrub or tree 2–4 m tall with stout, rigid, thorn-tipped branches. (The juvenile plants have entire to repand leaves 1–1.5 cm long.) The inconspicuous, greenish yellow flowers are 6–8 mm wide. Male and female flowers are on separate plants. Each mature fruit comprises several divergent carpels about 6 mm long. The dense fruit clusters persist on the branches for several years. The diploid chromosome number is 26 (Raven 1967).

Koeberlinia spinosa branches are dark rather than light green, and its fruits are black berries. *Canotia holacantha* branches are striate and flexible, and the fruits are woody capsules tipped with ten slender horns.

Castela emoryi has long been treated in regional manuals as *Holacantha emoryi* A.

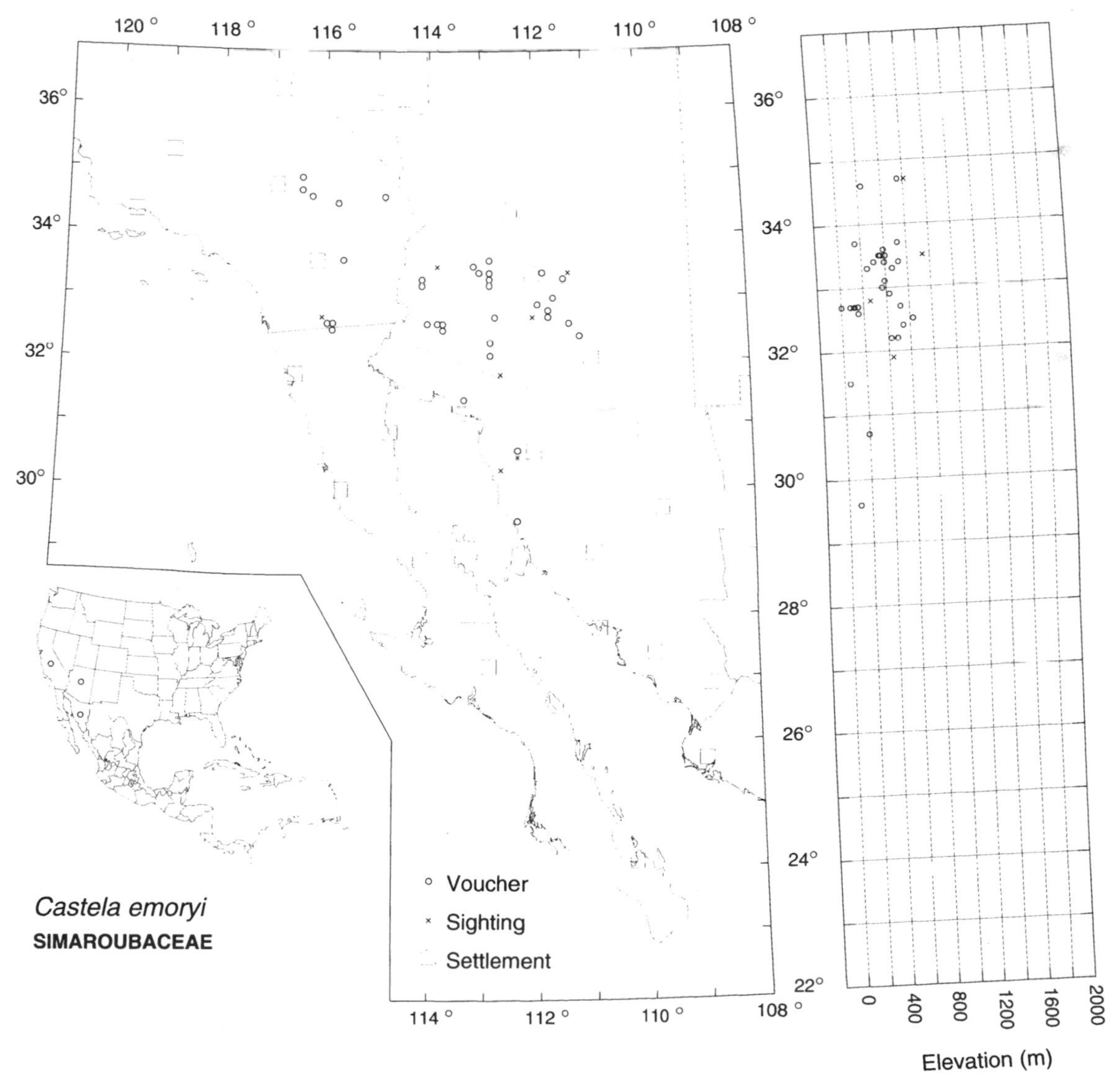

Gray (for example, Kearney and Peebles 1960). However, the discovery of *Castela polyandra,* a closely related species, made the genera *Castela* and *Holacantha* seem less distinct than formerly (Moran and Felger 1968), and recent treatments generally reduce *Holacantha* to a section of *Castela* (Benson and Darrow 1981, for example).

Castela emoryi is locally common on silty soil of plains and alluvial bottomlands, occasional on dunes. Generally, the colonies are small and widely scattered (Shreve 1964), but a notably dense population of about 1,000 plants occurs in southern California near 32.7°N, 115.9°W. The Bureau of Land Management (U.S. Department of the Interior) has created the Crucifixion-thorn Natural Area to preserve them. The distribution covers climates in which the biseasonal rainfall varies from winter dominant in the northwest to summer dominant in the east. Temperature regimes are generally subtropical, with hot summers and rare frosts.

Flowers appear from April to July. Seeds remain in their capsules on the tree for five to seven years. After dispersal, germination is further delayed by the thick capsule wall (Shreve 1964). Seedlings rapidly develop a deep taproot and, if moisture is adequate, grow quickly (Shreve 1964). Growth of adults is slow and sporadic, and most individuals do not flower until well advanced in years (Shreve 1943). Based upon the low level of seedling establishment and the apparently poor adjustment of the plants to current conditions, Shreve (1943) suggested that *C. emoryi* is a "waning genus." Fossil remains dated to 9,750 years B.P. have been reported from packrat middens from the Kofa Mountains, Arizona (Van Devender 1990b). The steep, rocky midden site is quite different from the habitats in which *C. emoryi* grows now.

Extracts from the branches can be used as an insecticide or to inhibit intestinal protozoa (Moore 1989).

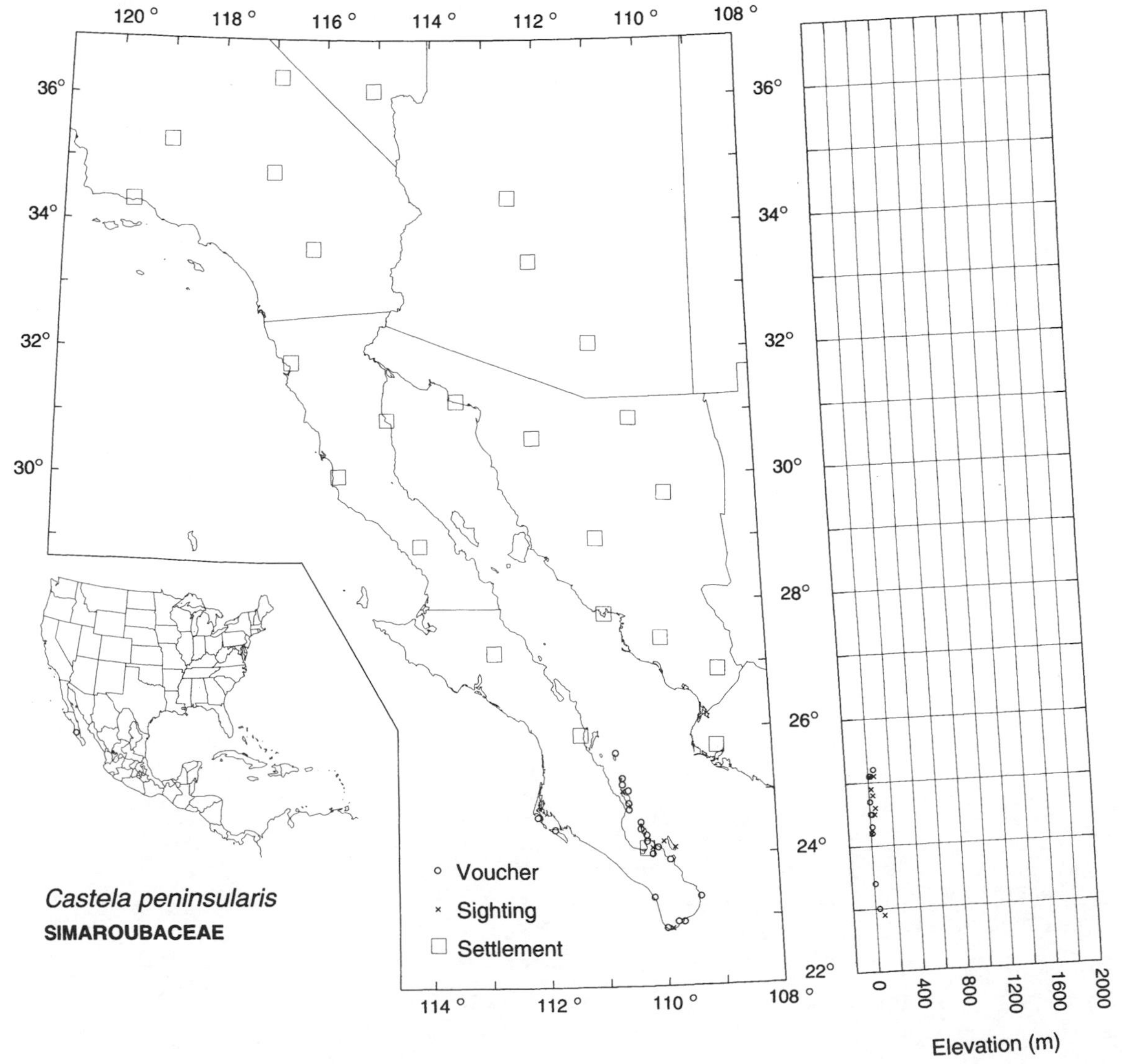

Castela peninsularis
SIMAROUBACEAE

Castela peninsularis
Rose

Amargoso

Castela peninsularis is an intricately branched shrub with numerous spinescent branchlets and simple, alternate leaves up to 2 cm long. The elliptic to obovate leaves are thick-leathery and have entire to few-toothed margins. Male and female flowers are on separate plants. The orange to red fruits are about 8 mm in diameter and bitter.

The plants are leafless much of the year. New leaves appear after rains, then drop when the soil dries out. Flowering is in spring and late summer (Wiggins 1964).

Castela peninsularis grows only in Baja California Sur, from the vicinity of Magdalena Bay around the cape to Santa Catalina Island (Cronquist 1944; Moran 1983b). It is confined to low elevations between sea level and about 100 m. The range appears restricted to arid, frost-free maritime climates where rainfall is concentrated in late summer and early fall.

Castela polyandra
Moran & Felger

A stiffly branched shrub up to 2 m tall and 5 m wide, *Castela polyandra* (figure 32) is leafless much of the year. The spine-tipped, zigzag branches become glabrous with age. The elliptic to cuneate leaves are 5–24 mm long and 4–15 mm wide. Male and female flowers are on separate plants in compact axillary panicles up to 1 cm long. The male flowers are 8–11 mm wide, the female, 11–14 mm wide. Both have red petals and sepals. The pale orange to red drupes, 11–14 mm long, are single or in clusters of 2–4. The haploid chromosome

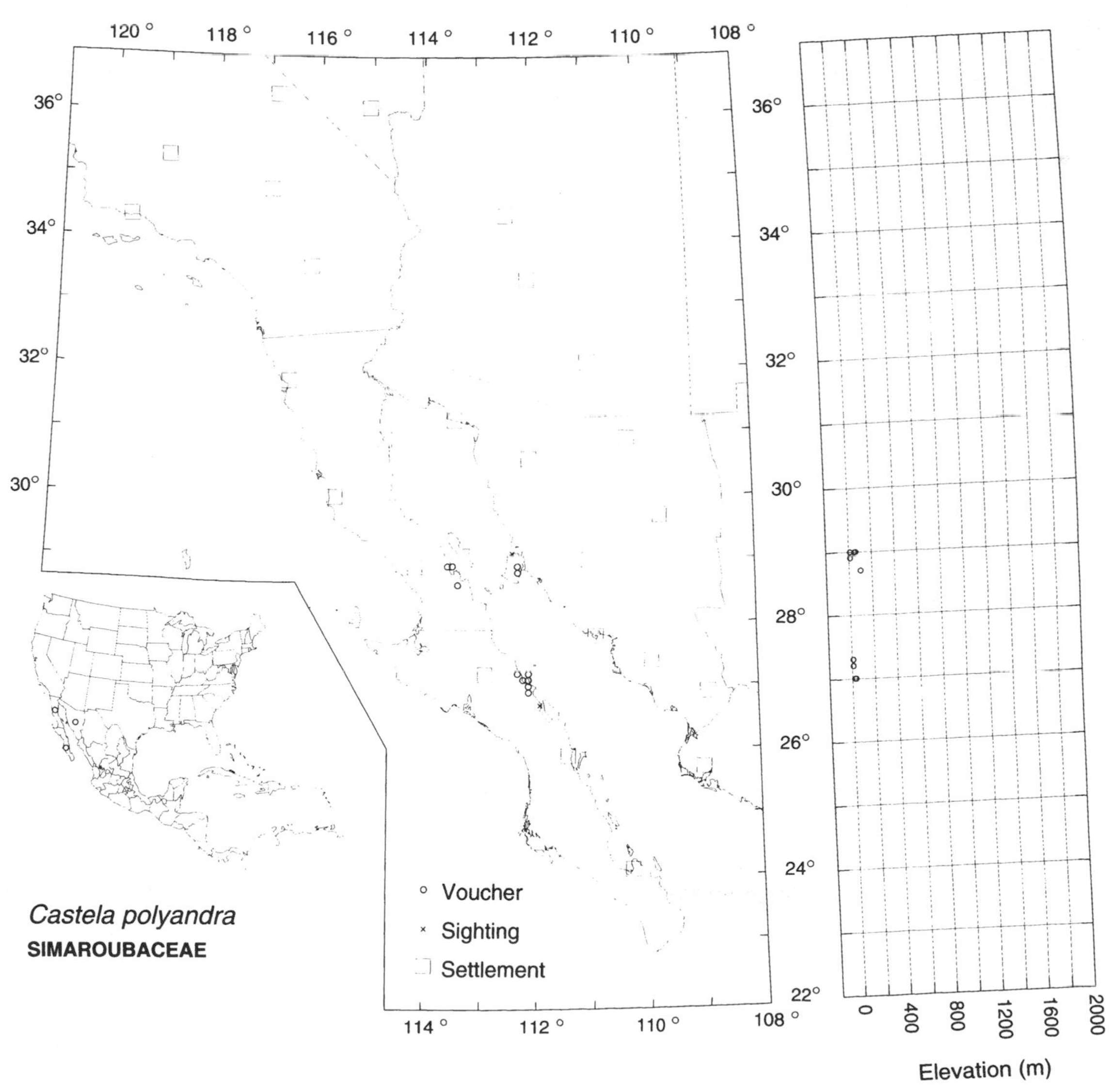

Castela polyandra
SIMAROUBACEAE

Figure 32. Close-up of Castela polyandra, *showing the stiff, compact, spine-tipped branches. (Photograph by J. R. Hastings taken near Bahía de los Angeles, Baja California.)*

number is 13 (Moran and Felger 1968).

First collected in the early 1930s, *C. polyandra* was not described for nearly 40 years for lack of flowering specimens (Moran and Felger 1968). The species is intermediate between the genera *Holacantha* and *Castela,* having the branching pattern and petal dimorphism of *Holacantha* and the 4-parted flowers and deciduous fruits of *Castela.*

Castela peninsularis has spinose spur shoots, *C. polyandra* does not.

In the few places where it is known, *C. polyandra* grows almost exclusively on east-sloping, gravelly bajadas. The plants can be locally abundant (Moran and Felger 1968). Populations occur on Tiburón and San Marcos islands (Moran 1983b). Climates within the range have hot summers, mild winters, and biseasonal rainfall about evenly split between summer and winter.

Ceiba acuminata

(S. Watson) Rose

Ceiba, kapok, pochote

This large or medium-sized tree up to 20 m tall has light gray bark and many conical spines on the trunk. The trees may be slightly buttressed at the base. Leaves are palmately compound with, usually, 7 lance-elliptic or oblanceolate leaflets 3–15 cm long and 1–6 cm wide. Flowers are large (10–15 cm long), yellow, and silky-pubescent. The woody capsules, 12–18 cm long, split into five valves. The dense silky hairs inside are the kapok of the common name. The tuberous roots can become quite large (Nabhan and Felger 1985).

Bark of *Bombax palmeri* S. Watson is green rather than gray and lacks conical spines. *Ceiba pentandra* (L.) Gaertn., a much larger tree up to 40 m tall, is extensively buttressed at the base.

Barely entering the Sonoran Desert, *C. acuminata* is infrequent in Sinaloan thornscrub on plains, hillslopes, and canyon bottoms (Wiggins 1964). To the south, the trees are abundant and dominant in Sinaloan deciduous forest (Gentry 1942). Although Wiggins (1980) lists *C. acuminata* for the Baja California peninsula, we have seen no specimens from there. Turnage and Hinckley (1938) reported stem-kill of cultivated plants at Tucson in the severe freeze of January 1937, when the minimum dropped to –9.4°C. Frost sensitivity may limit the northward distribution of this species.

The trees are leafless much of the year. Appearing once summer rains are under way, the leaves last through summer and drop in autumn. Leaf fall is usually complete by October. Flowers appear while the trees are still leafless, generally in spring (March–May) but occasionally as late as July (Krizman 1972). During winter and spring, the silky fibers in the capsules blow away, carrying the seeds with them (Baker 1983; Augspurger 1986). These phenological patterns are characteristic of many thornforest species (Krizman 1972). Blooming before the canopy leafs out exposes flowers to potential pollinators, and seed dispersal is also enhanced by the absence of leaves (Janzen 1967; Baker 1983).

The tuberous roots and the brief duration of the leaves suggest a strategy in which photosynthetic rates are high, allocation of carbon to roots is substantial, and flower development occurs at the expense of root and stem reserves.

The large size, abundant nectar, exposed anthers, and nocturnal opening of the flowers all suggest adaptation for bat pollination; bats, including *Leptonycteris sanborni,* are indeed the primary pollinators (Baker et al. 1971). Minor pollinators include hummingbirds and large moths. Many nonpollinating birds, bees, wasps, moths, and skippers also visit the flowers (Baker et al. 1971).

The oil-rich seeds are edible and have a

nutlike flavor. Yaqui and other Indians harvest the tuberous roots at the end of the rainy season and roast them in hot coals (Nabhan and Felger 1985). The capsule fibers have been used as a kapok substitute. (The primary source of commercial kapok is *C. pentandra.*)

Celtis iguanea

(Jacq.) Sarg.

Granjeno, garabato, palo de arco, chaparro blanco, vainoro

Celtis iguanea is a shrub or small tree 2–13 m tall with stout, recurved spines 7–14 mm long on arched, scandent branches. The oval to ovate leaves, 3–5 cm wide and 3.5–7 cm long, are somewhat leathery, slightly revolute, and entire to finely serrate. The inconspicuous flowers are yellow-green. The globose, yellowish or orangish drupes are 7–12 mm long.

Celtis pallida has slender, straight spines and smaller, more coarsely toothed leaves. *Pisonia capitata* (S. Watson) Standley has clavate, stipitate-glandular fruits and opposite rather than alternate leaves. *Ziziphus amole* has green branches and shorter (2–8 mm) spines.

Celtis iguanea, a dominant along arroyos and in valley bottoms, often grows with *Pisonia capitata* and *Vallesia glabra,* also large, evergreen shrubs (Gentry 1942). Widespread in tropical America, *C. iguanea* reaches its northern limit in the Sinaloan deciduous forest and Madrean evergreen woodland of southern Sonora.

The northernmost point (28.5°N) represents a 1973 sighting by Rod Hastings in Arroyo de los Alisos. Although Standley (1922) included Texas in the distribution of the species, Correll and Johnston

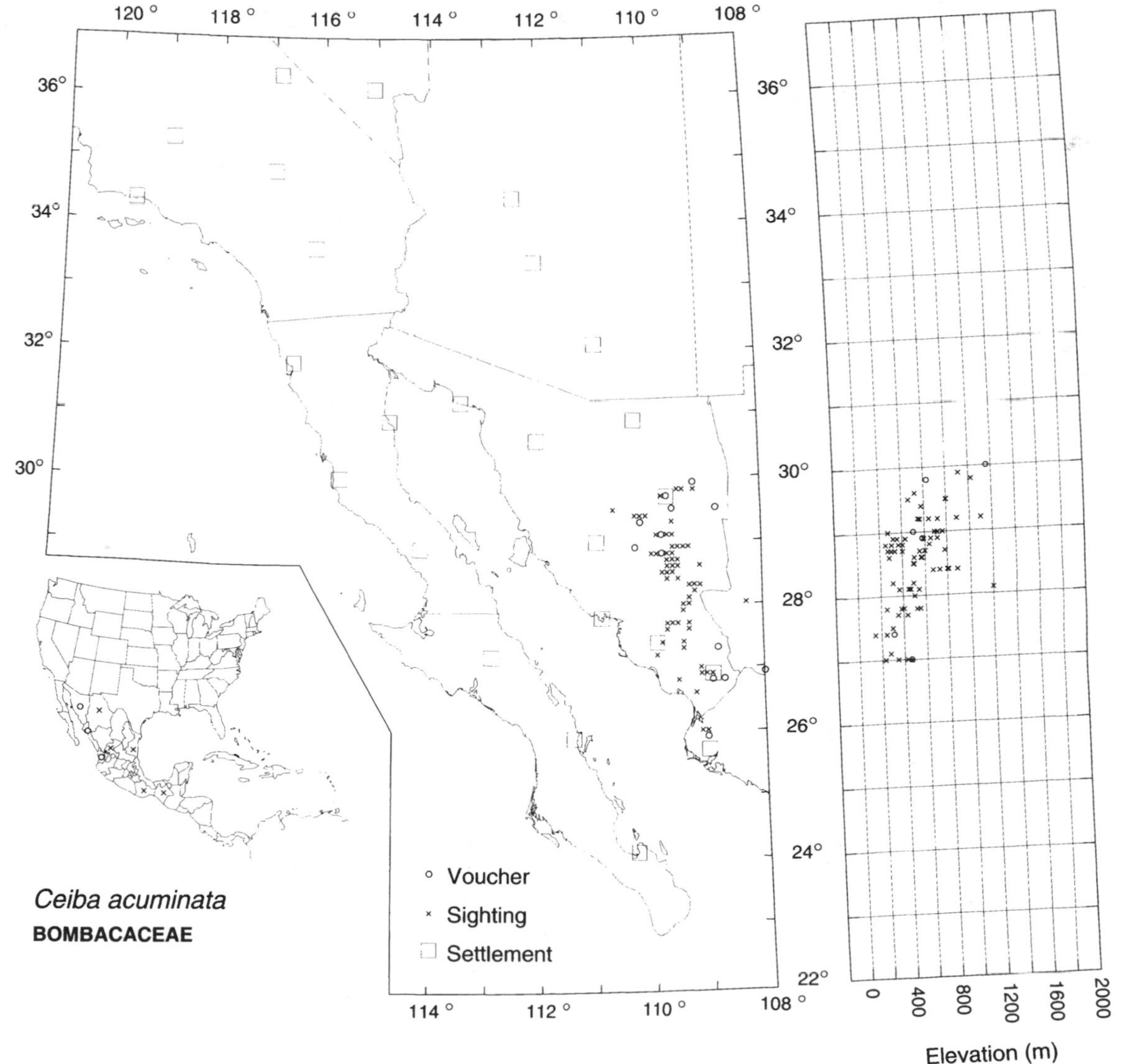

Ceiba acuminata
BOMBACACEAE

(1970), citing the absence of vouchers, excluded it from their flora of the state.

Flowers appear March–April and again in summer. The evergreen habit of the leaves may restrict the plants to riparian zones where moisture is readily available year-round.

People, coyotes, foxes, coatimundis, birds, and iguanas eat the fruits (Standley 1922; Gentry 1942).

Celtis pallida

Torr.

[=*C. spinosa* Sprengel var. *pallida* (Torr.) M. C. Johnston, *C. tala* Gillies var. *pallida* (Torr.) Planch.]

Desert hackberry, spiny hackberry, granjeno, huasteco, capul, garabato, garambullo, rompecapa, acebuche, bainoro, palo de águila

A densely branched shrub 1–6 m high, *Celtis pallida* has paired, straight spines and short, lateral thorn-tipped branches. The leaves, 1–3 cm long and 0.6–2 cm wide, vary from subentire to serrate and ovate to elliptic. The inconspicuous, greenish yellow flowers are in small cymes. Perfect, staminate, and pistillate flowers may be found together. Fruits are yellow or orange drupes 5–8 mm in diameter.

In the northern Sonoran Desert, the paired spines at the nodes distinguish *C. pallida* from most other thorny, simple-leaved shrubs, but in the southern part of the Sonoran Desert region, several species have paired spines. *Ziziphus amole* leaves are larger (3–6 cm long and 2–3.5 cm wide). *Celtis iguanea* has rather large leaves and stout, recurved spines. *Randia thurberi* has leaves in clusters on spur shoots.

Although various authors have used

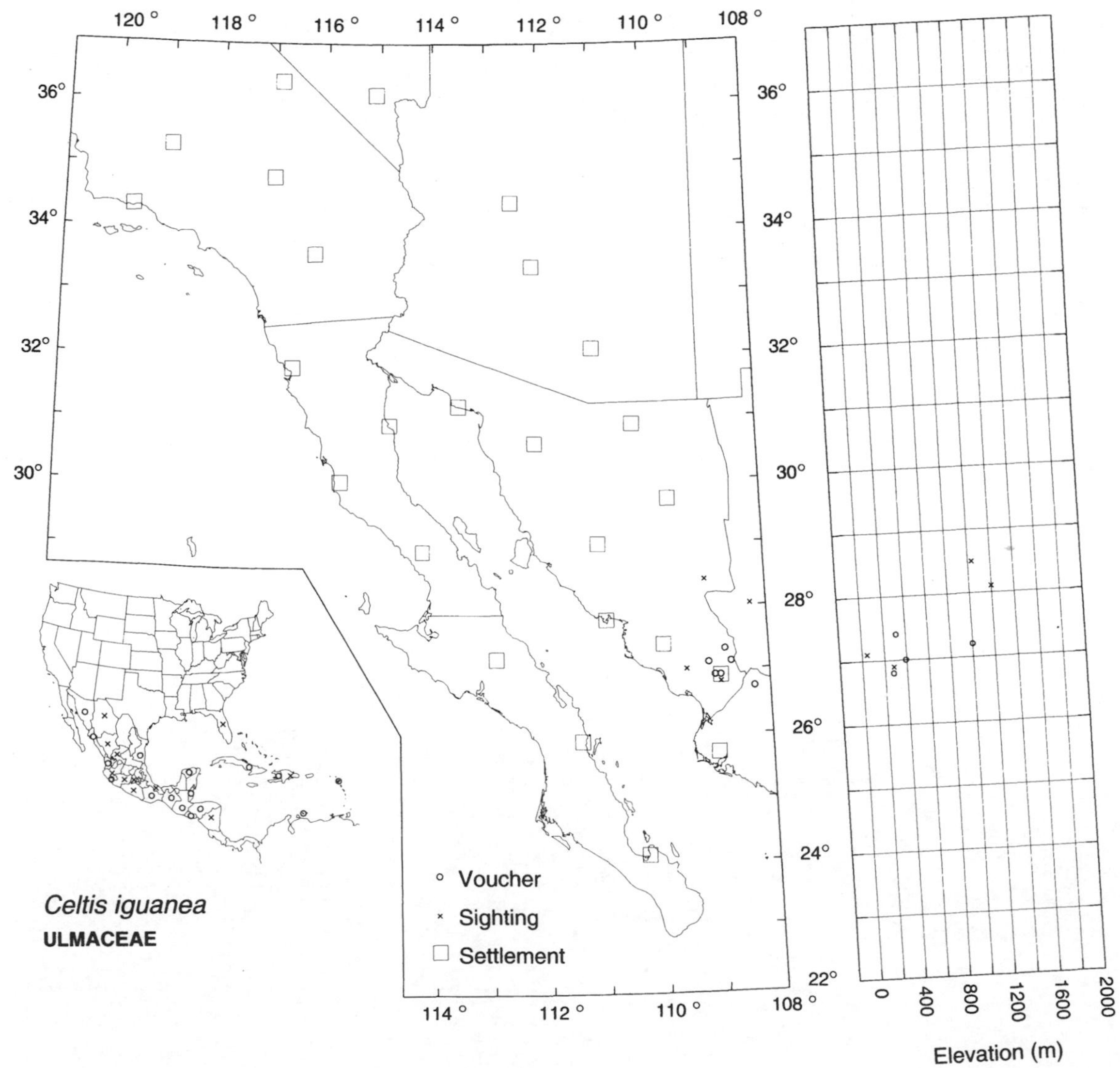

different names for this plant (see, for example, Benson 1943; Johnston 1957; Kearney and Peebles 1960; Wiggins 1964), all have regarded it as the northern representative of a widespread tropical and subtropical species found from the southwestern United States to Argentina. According to recent flavonoid analysis and pollen fertility studies (Romanczuk and De Martinez 1978), the entity in the Sonoran Desert is *C. pallida* subsp. *pallida*. As defined by Romanczuk and De Martinez (1978), *C. spinosa* and *C. tala* do not occur outside South America.

Celtis pallida is common along washes and on rocky and gravelly slopes and occasionally dominates bajadas at the upper margin of the Sonoran Desert. In southern Texas, where the species is common in brush-invaded grassland, the plants are often considered range pests (Scifres et al. 1979; Van Auken and Bush 1985). On floodplain terraces of the San Antonio River in southern Texas, *C. pallida* is a plant of early successional stages (Van Auken and Bush 1985). In the Sonoran Desert region, however, it grows in mature, stable associations in Sonoran desertscrub and semidesert grassland.

Although *C. pallida* tolerates colder temperatures than many Sonoran Desert shrubs, cold does affect the plants, as shown by the truncated upper limit of the elevational profile. Low temperatures may determine the northern limit primarily through their effect on seedlings. The geographic range suggests a need for warm-season rain. Toward their western limit, the plants compensate for climatic aridity by growing in riparian habitats.

Celtis pallida flowers in spring (March and April) and again in summer and fall (July–October). Ripening throughout the summer and fall, the fruits are eaten by a variety of birds (Fulbright et al. 1986) and by coyotes, foxes, and javelinas. All no doubt disperse the seeds. The leaves are generally evergreen but may drop during prolonged drought (Shreve 1964) or when temperatures fall below −10°C. Even when frost kills the leaves, the limbs apparently suffer little or no damage.

Matched photographs show that *C. pallida* lives for at least 88 years (Ray-

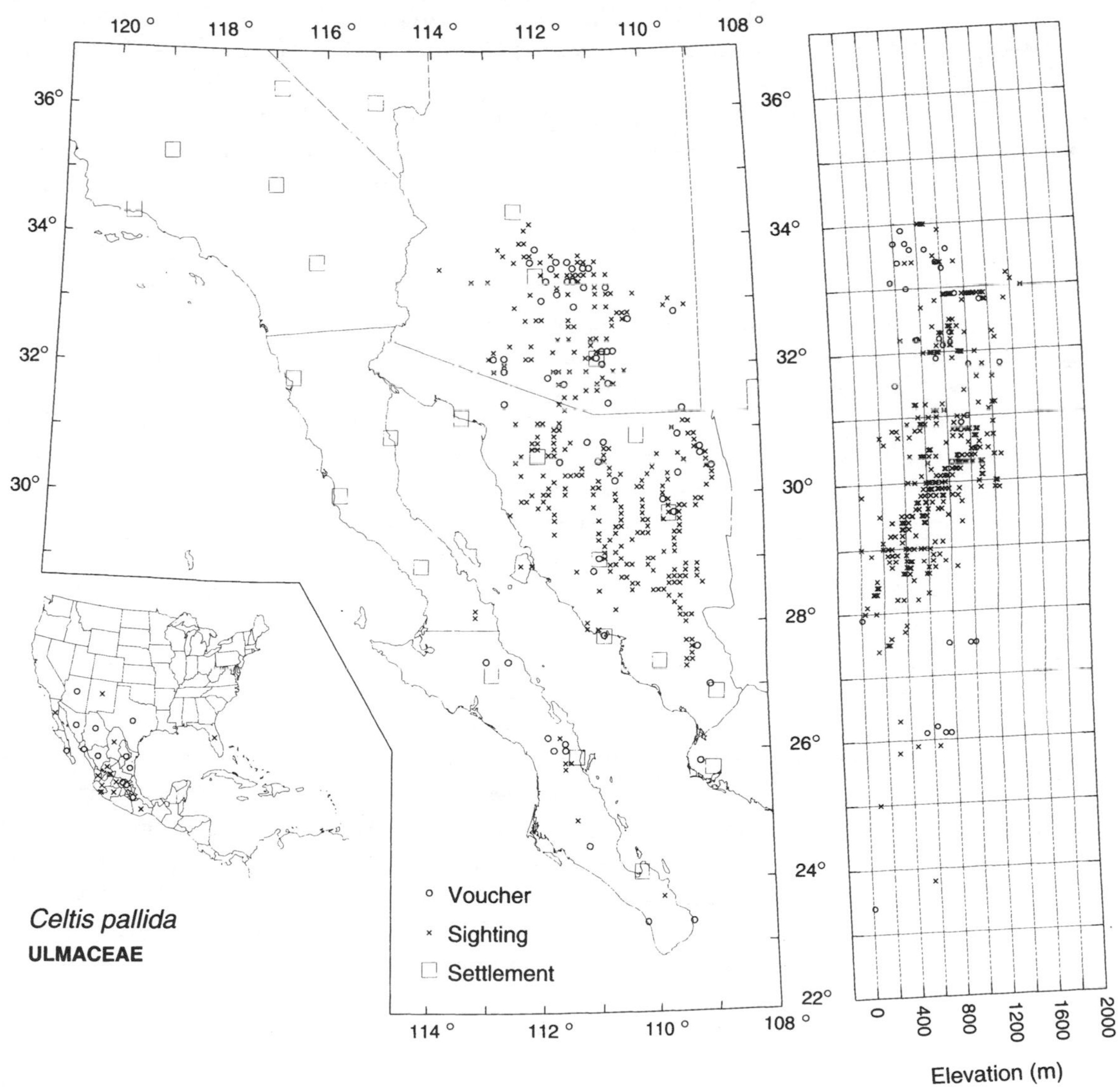

mond M. Turner, unpublished photographs).

Evans and coworkers (1981) found that, in the vicinity of Tucson, Arizona, the roots extended no more than 1 m below the soil surface, and most lay within the top 0.6 m. Given the shallow root system, it is not surprising that evapotranspiration throughout the year corresponds closely to soil moisture. Sammis (1974) found that from October to April, when soil water potential was high (greater than –0.1 MPa), evapotranspiration was also high, and the plants initiated flowers and new growth. As soil water potential declined to a minimum of –5 MPa in September, plant diffusion resistance increased and some leaves fell (Sammis 1974).

Wheeler and coauthors (1989) described the wood anatomy.

This species requires no special care in arid-land gardens and is tolerant of drought, heat, wind, and poor soil. Supplemental irrigation is necessary where annual rainfall is less than 25 cm (Duffield and Jones 1981). The seeds exhibit dormancy and often have an after-ripening period. Germination can be increased from 1% to 49% by a combination of treatments involving mechanical scarification, gibberellic acid, moist heat, and prechilling (Fulbright et al. 1986). Nokes (1986) recommended stratifying seeds at 5°C for 60–90 days. Vora (1989), by contrast, found that no pretreatment was necessary.

Celtis pallida provides browse for deer, nesting sites for white-winged doves, cover for quail, and food for many species of birds (Fulbright et al. 1986) and several mammals. According to Standley (1922), the wood is used for fuel and fence posts.

Cercidium floridum
Benth.

[=*C. peninsulare* Rose, *Parkinsonia torreyana* S. Watson]

Blue paloverde, paloverde

Cercidium floridum is a tree up to 12 m tall with small straight spines borne singly at the nodes. The bark of twigs and young branches is bluish green, that of older trunks often gray. Leaves are pinnate with a single pair of pinnae, each with 2–4 pairs of obovate leaflets 4–8 mm long. The yellow, somewhat irregular flowers are in terminal racemes. Petals are 9–12 mm long. The straw-colored, oblong pods are 4–10 cm long. The haploid chromosome number for the species is 14 (Turner and Fearing 1960).

We follow Carter (1974b) in recognizing two subspecies: the widespread subsp. *floridum* and the endemic subsp. *peninsulare* (Rose) Carter, restricted to Baja California Sur south of about 27.3°N. *Cercidium floridum* hybridizes with *C. microphyllum* in the northwestern region of range overlap (Carter and Rem 1974).

Cercidium floridum is larger than most of the other trees with which it grows. *Cercidium microphyllum* has thorn-tipped stems rather than nodal spines and yellowish rather than blue-green bark and foliage. It generally grows on bajadas and hillslopes rather than along washes.

A tree of desert watercourses, *C. floridum* is absent from the northeast coast of Baja California, where *C. microphyllum* replaces it in the wash habitat. *Cercidium floridum* grows higher on desert mountains than *C. microphyllum* and extends farther south on the plains of Sinaloa. From early through middle Holocene times, *C. floridum* grew on rocky slopes of the Puerto Blanco and Tinajas Altas mountains, Arizona, and the Hornaday Mountains, Sonora (Van Devender 1990b). Packrat midden fossils from the Whipple Mountains, California, near the present northern limit of the species, show that it arrived there in the middle Holocene (4,240 years B.P.) (Van Devender 1990b). The truncated northern and upper limits of the elevational profile are shaped by cold. Throughout the range, rainfall is biseasonal, varying from mainly winter in the northwest to mainly summer in the south.

The stem parenchyma contains an abundance of calcium oxalate crystals. Starch is abundant in the xylem. Oil is also present as a reserve food (Scott 1935b). Based on various anatomical features, Carlquist (1989a) concluded that *C. floridum* should require more water than *C. microphyllum*, a finding in agreement with field observations of their habitat preferences.

The entire crown is a mass of yellow flowers when *C. floridum* blooms in late March or April. At least in the northern part of the range, the trees are leafless much of the year. Leaves appear after summer rains start and drop with the onset of cool temperatures in autumn (R. M. Turner 1963).

Although *C. floridum* does not fix nitrogen (Eskew and Ting 1978; Felker and Clark 1981), soil nitrogen under the canopy (243 g/m^2) was about the same as under *Prosopis velutina*, which does (Barth and Klemmedson 1982). Foliage nitrogen is highest in spring, lowest during winter dormancy (Barth and Klemmedson 1986). Plant litter peaks in late winter and early spring as fruits and small branches are shed, then declines throughout the warm season from accelerated decomposition (Barth and Klemmedson 1986). Soil carbon in one study was 2 kg/m^2 (Barth and Klemmedson 1982).

Several different species of native bees and the introduced honeybee, *Apis mellifera*, are the main pollinators of *C. floridum* and *C. microphyllum*. Flower color and ultraviolet patterns enhance pollinator constancy and help forestall hybridization (Jones 1978), as do staggered blooming times (*C. floridum* is one to two weeks earlier than *C. microphyllum*).

The seeds are heavily used by bruchid beetles and rodents (Mitchell 1977), even though the seed coats are a formidable barrier (C. D. Johnson 1982). Individual plants differ in susceptibility to bruchid beetle attack (Siemens and Johnson 1990). Seedlings establish readily in depositional environments (Turner and Bowers 1988).

This tree is often planted for shade but seems especially susceptible to mite infestations in urban settings (Duffield and Jones 1981). Transplantation from open ground is rarely successful, but plants will readily start from scarified seed sown in deep containers (Duffield and Jones 1981). Gardeners can achieve fastest growth and best appearance by applying supplemental irrigation (Duffield and Jones 1981). The plants resist freezing at temperatures above −12°C (Johnson 1993). Indians ate the seeds, flowers, and immature pods (Felger and Moser 1985).

Cercidium microphyllum

(Torr.) Rose & I. M. Johnston

[=*Parkinsonia microphyllum* Torr.]

Foothill paloverde, yellow paloverde, littleleaf paloverde, paloverde, dipúa

This small tree grows up to 6 m tall with smooth, green bark on all twigs and branches except the base of the trunk, which is gray. The stem tips are spinose. The leaf is a pair of diverging rachises, each with 3 to 5 pairs of leaflets 1 mm wide. The flowers are bicolored, with four yellow petals and one white one. The tan or straw-colored pods are 4–8 cm long. The haploid chromosome number is 14 (Turner and Fearing 1960).

Cercidium floridum and *Parkinsonia aculeata* have spines at the nodes rather than thorn-tipped stems.

The plants may hybridize with *C. floridum* in the northern part of their range and with *C. praecox* in the south (Carter 1974a, 1974b). For the most part, flower color and ultraviolet patterns enhance pollinator constancy and help forestall hybridization, as do staggered blooming

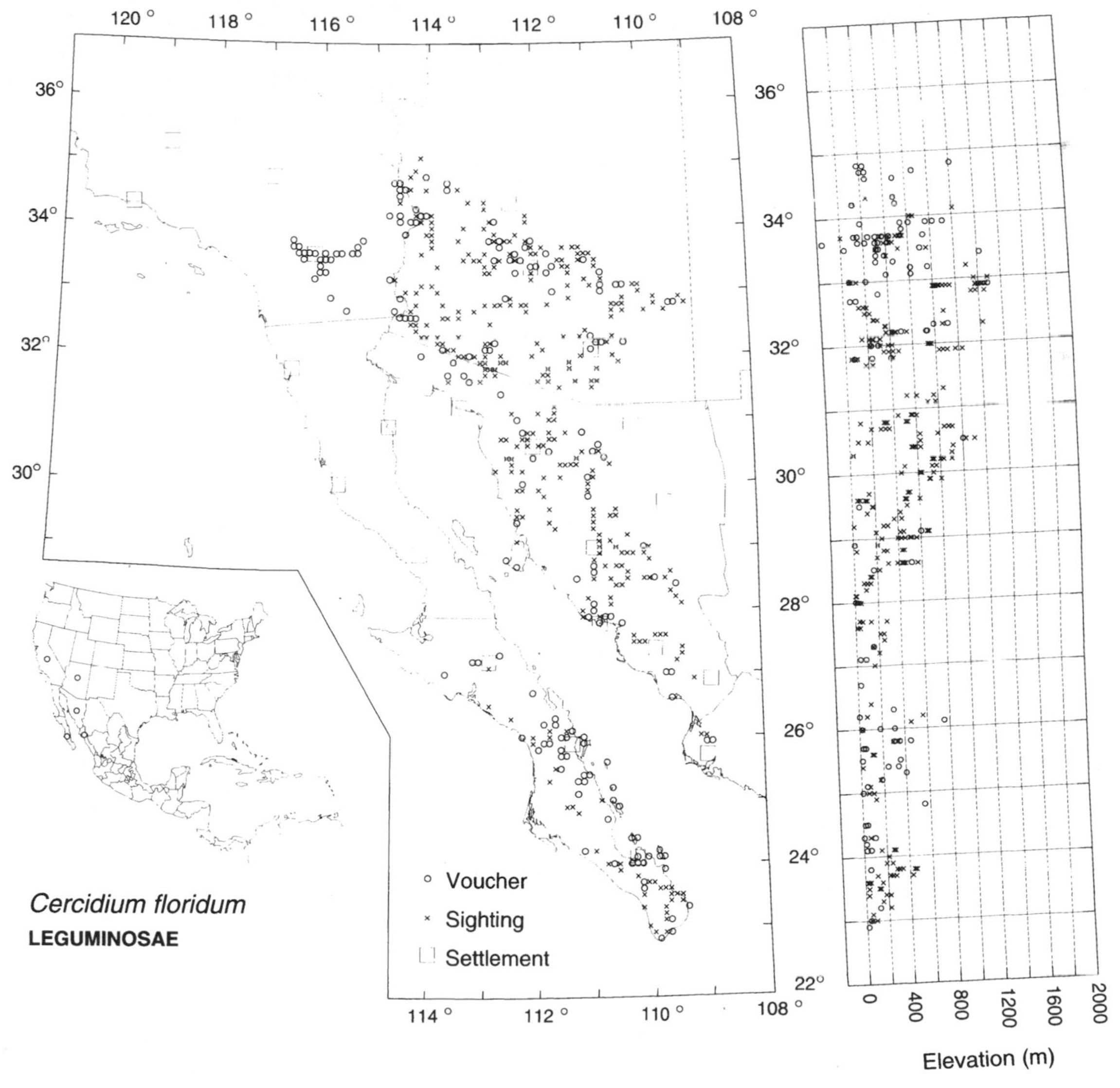

Cercidium floridum
LEGUMINOSAE

times (Carter and Rem 1974; Jones 1978).

Cercidium microphyllum, a Sonoran Desert endemic, is abundant on bajadas, plains, and hillslopes throughout much of Arizona, Sonora, and the Baja California peninsula. It is all but absent from the Pacific side of the peninsula and from the valleys of the Río San Miguel and Río Sonora on the eastern edge of the desert. The trees provide sheltered microhabitats in which *Carnegiea gigantea* and many other plants become established, and they are critical to the structure and pattern of Sonoran desertscrub wherever they grow (Patten 1978; McAuliffe 1984a; Franco and Nobel 1989).

The elevational profile suggests that the northern and upper limits are truncated by cold. Over most of the range, rainfall is biseasonal. The species boundaries show that regular summer moisture is a requirement. The southern limits appear to be strongly influenced by interactions with *C. praecox.*

Fossilized packrat middens from the Puerto Blanco Mountains, Arizona, show that *C. microphyllum* arrived there about 5,240 years B.P., several thousand years later than *C. floridum* or *Carnegiea gigantea,* two common associates (Van Devender 1987, 1990b). To the south, in the Hornaday Mountains, Sonora, *C. microphyllum* apparently arrived about 9,400 years B.P., some 1,000 years before *C. floridum* (Van Devender, Burgess et al. 1990).

The plants are drought deciduous. New leaves are produced in response to summer and winter rains (Shreve 1914; R. M. Turner 1963; Humphrey 1975). They drop before the plants start to flower in late April. At Tucson, Arizona, the flowering trigger is a day length of 11 hours, and the base temperature for flowering is 10°C (Bowers and Dimmitt 1994). After day length reaches 11 hours, the temperature sum required for flowering is about 719 degree-days above 10°C (Bowers and Dimmitt 1994). Flowering may continue into late May (McGinnies 1983). After dry winters, flowering may fail altogether (Shreve 1964; R. M. Turner 1963). Seeds ripen and fall roughly 6 weeks after the end of the flowering period (Shreve 1964), providing a new crop of seeds just

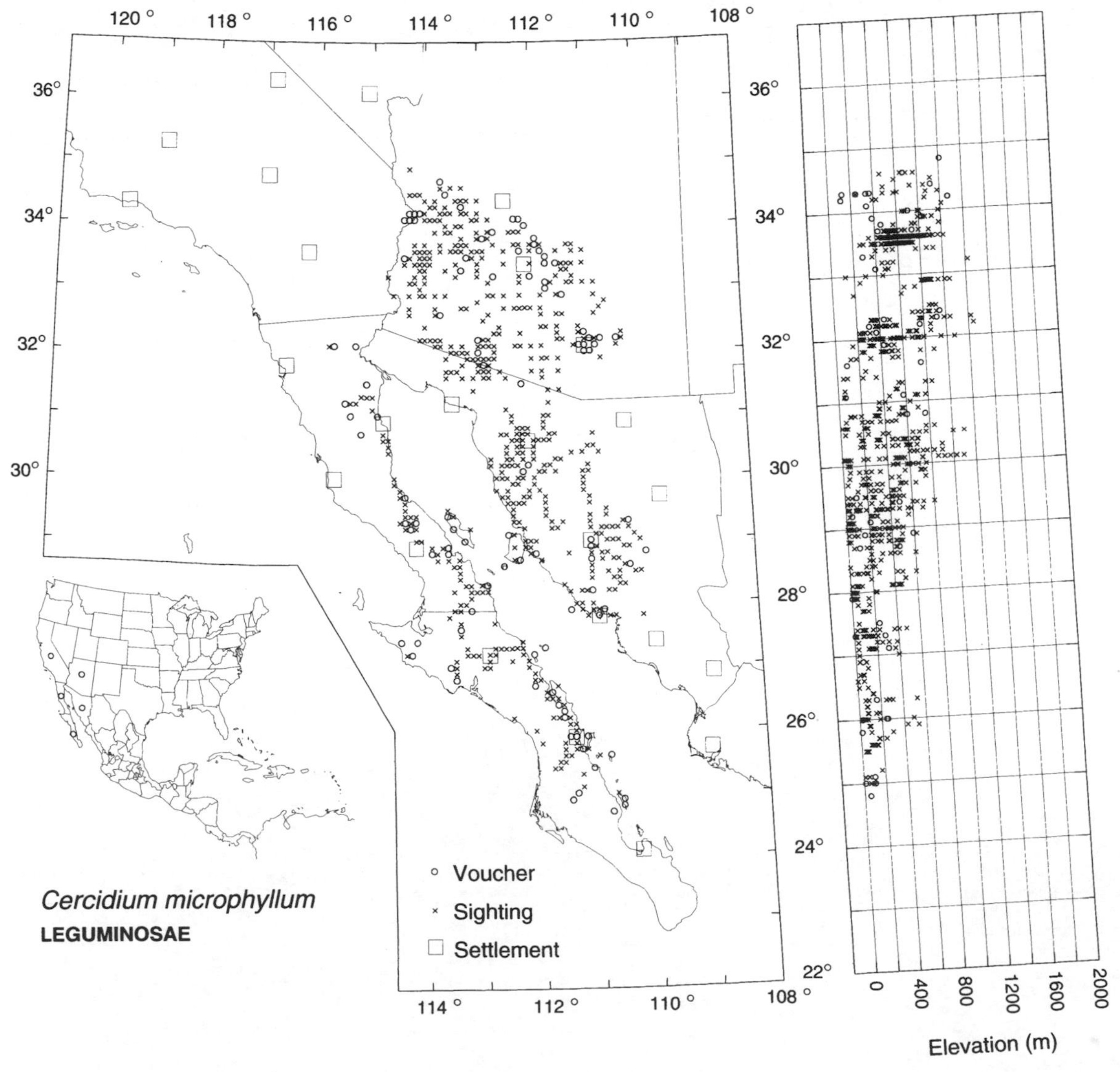

Cercidium microphyllum
LEGUMINOSAE

before the onset of summer rains.

Cercidium microphyllum seeds last several years or more in the soil (Bowers 1994). They germinate only in the summer when the soil is warm and moist (Shreve 1911b, 1917). The minimum rain trigger for seedling emergence is 17 mm (Bowers 1994). Air temperatures during the germination period must reach or exceed 20°C (Bowers 1994). The hard seed coat must be cracked or scarified for germination to occur. Poole (1958) reported high germination rates in the laboratory (up to 88%) for untreated seeds; however, bruchid beetles might have provided the requisite scarification. Seeds cached by rodents often germinate in clusters (McAuliffe 1990).

Seedling mortality is heaviest during the arid late summer and fall. Winter freezes and seasonal drought of late spring and early summer also result in heavy losses (Shreve 1911b, 1917). Except in rocky habitats, mammalian herbivory limits seedling establishment, and, as a result, young plants often grow within the canopies of *Ambrosia* species (McAuliffe 1986).

The green-barked stems account for about 74% of the total photosynthetic productivity, the leaves about 24% (Szarek and Woodhouse 1978b). (Flowers and fruits account for the remainder.) Uptake of carbon dioxide is low because the stem photosynthetic apparatus (chlorenchyma) is inefficient compared to the leaf photosynthetic apparatus (parenchyma) (Gibson 1983). Photosynthesis does not occur when plant water potentials drop below about –3 MPa (Szarek and Woodhouse 1978a). Although leaves are usually present throughout the summer, photosynthetic rates then are minimal compared to the fall, when they are especially rapid (Szarek and Woodhouse 1978b). Both starch and oil are present as reserve foods (Scott 1935b).

During periods of extended drought, entire branches may die. Drought pruning reduces water loss and may limit plant size in a given habitat (Shreve 1914). Establishment may be greatly curtailed during prolonged drought (Shreve 1911b; Turner 1990), but the long life span (400 years or more) (Shreve 1911b) buffers population shifts. Fire, a rare event in the Sonoran Desert, killed 63% of *C. microphyllum* trees in a Sonoran desertscrub community in south-central Arizona (McLaughlin and Bowers 1982).

Cercidium microphyllum can be an attractive ornamental in desert gardens. Once transplants have become established, little or no supplemental irrigation is needed where annual rainfall exceeds 25 cm; however, growth is faster with occasional watering (Duffield and Jones 1981). The plants survive freezing temperatures as low as –9.5°C (Johnson 1993).

The branches, even when leafless, are an important livestock feed during drought. The seeds, immature pods, and flowers are edible (Carter 1974b; Felger and Moser 1985).

Cercidium praecox

(Ruiz, Lopez & Pavon) Harms

Palo brea, mantecosa

The bark of *Cercidium praecox* (figure 33) is green, and the trunk is unbranched to a height of 1 or even 2 m. The main branches spread nearly horizontally in a flat, wide-spreading crown. Twigs have small, straight spines at the nodes. The leaves have 1–2 pairs of pinnae, each with 4–8 pairs of oblong leaflets 2–8 mm long. They are puberulent on both surfaces when young, sometimes glabrate in age. The golden-yellow petals are 8–15 mm long. The oblong pods (5–8 cm long) are narrowed at the ends.

The bright green bark and characteristic branching pattern are distinctive.

Natural hybridization with *C. microphyllum* (Carter 1974a, 1974b; Carter and Rem 1974) has caused much taxonomic confusion. The hybrid has been called *C. sonorae* Rose & I. M. Johnston and *C.*

Figure 33. The lower branches of Cercidium praecox *usually extend horizontally from the main stem. (Photograph by J. R. Hastings taken north of Loreto, Baja California Sur.)*

molle I. M. Johnston. Shreve (1964) and Wiggins (1964) used the name *C. sonorae* for *C. praecox*.

This is the only Sonoran Desert *Cercidium* species with a disjunct distribution between North and South America. Of the several subspecies (Burkart and Carter 1976) known from South America, only subsp. *praecox* occurs in the Sonoran Desert region.

Common throughout the valleys of eastern Sonora, *C. praecox* reaches its northernmost limit on terraces of the Río Batepito. It occurs in Baja California Sur from the vicinity of San Ignacio southward to Cabo San Lucas. Within the region of our interest, aridity, frosts, and interactions with *C. microphyllum* help to shape the northern and western limits. Except in the vicinity of San Ignacio, where rainfall is evenly biseasonal, *C. praecox* generally grows where rainfall is concentrated in summer. Over most of its range, frosts are uncommon to absent.

The flowering period is spring (March–May). Fruits develop before the leaves appear.

The common name *palo brea,* meaning "tar tree," refers to a waxy substance that coats the bark. The wax is scraped from branches, melted, and used as glue. *Cercidium praecox* is grown as a drought-tolerant ornamental where winter temperatures do not drop below –6.5°C (Johnson 1993).

Chilopsis linearis

(Cav.) Sweet.

Desert willow, mimbre

A large shrub or small tree up to 9 m tall, *Chilopsis linearis* has fissured gray bark and lanceolate leaves 8–14 cm long and 2–4 mm wide. The funnelform, two-lipped flowers, up to 3 cm long, are white to pink. Their tubes are streaked and spotted with lavender. The linear capsules, up to 2 dm long and 5 mm broad, contain many flat seeds tufted with hairs at each end.

The true willows (*Salix* species) in the Sonoran Desert region have inconspicuous flowers in catkins and serrulate leaves.

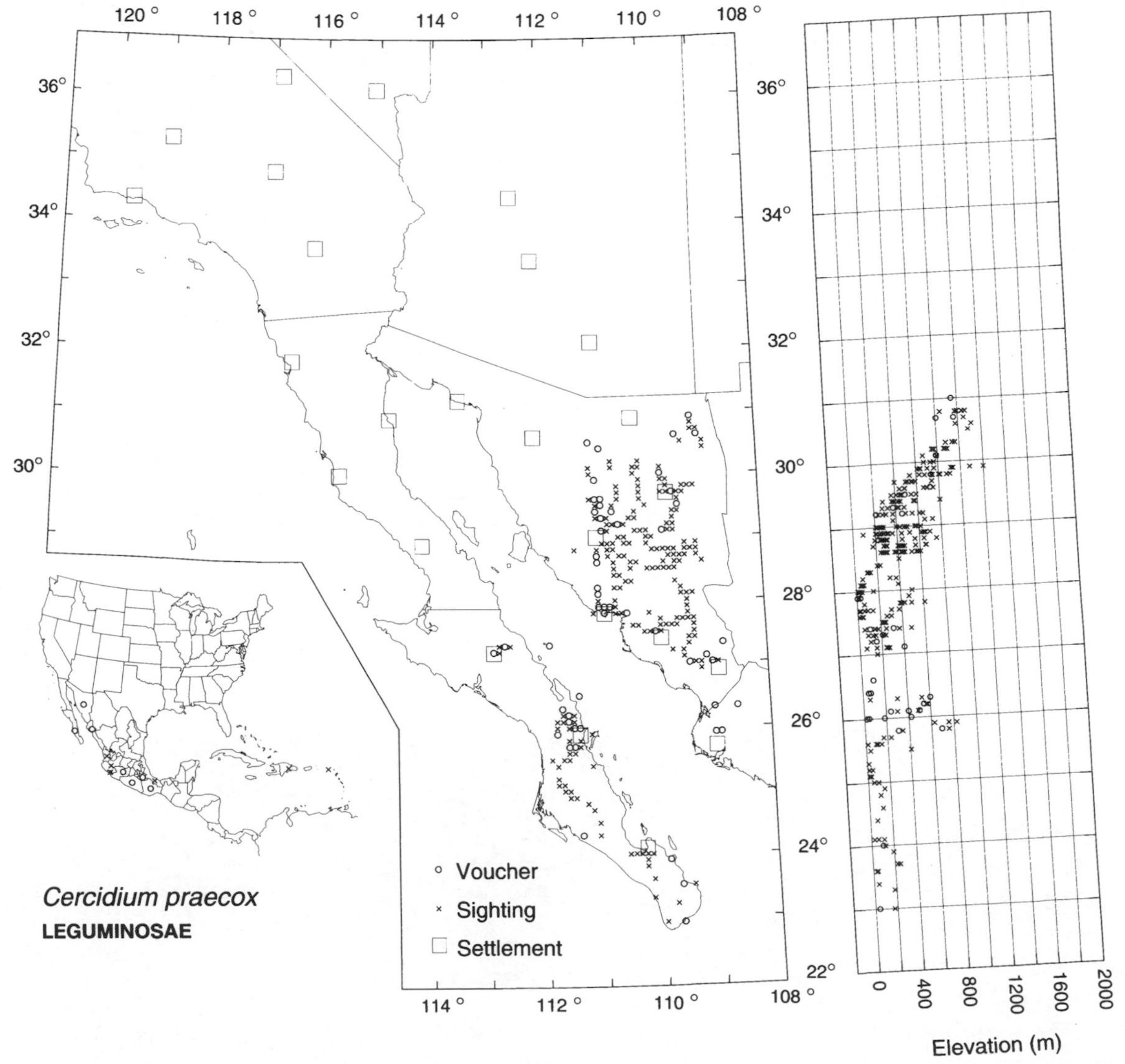

Cercidium praecox
LEGUMINOSAE

The prevailing taxon in the area covered by this atlas is subsp. *arcuata* (Fosberg) Henrickson. Subspecies *linearis* var. *linearis* and subsp. *linearis* var. *tomenticaulis* Henrickson are largely confined to the Chihuahuan Desert (Henrickson 1985). Brooks and McGregor (1979) reported *C. linearis* as a new addition to the flora of Kansas.

Within the desert, *C. linearis* grows mostly in large watercourses where water is continuously available at depth. At higher elevations (up to 1,500 m), *C. linearis* can be found on roadcuts, as well. Rarely, it grows on unstable dunes as at the Kelso Dunes (San Bernardino County, California) and in the Cuatro Ciénegas Basin (Coahuila, Mexico). Its soft wood, flexible trunks, and sprouting ability enhance survival on dunes, where plants are vulnerable to excavation and burial (Bowers 1986).

In the region covered in this atlas, *C. linearis* receives biseasonal rainfall, ranging from mainly winter in the northwest to mostly summer in the east. Temperature regimes are generally warm-temperate, and frosts are occasional. The species seems unable to occupy truly subtropical climates to any great extent.

Chilopsis linearis is cold deciduous and begins to lose leaves once minimum temperatures drop to 5°C, during early autumn over much of its range. After the first hard frost, all leaves drop (DuPree and Ludwig 1978). Leaf renewal begins between March and May. Flowers appear from April to August (Kearney and Peebles 1960), apparently in response to winter and summer rains. In the more arid, western parts of its range, *C. linearis* is drought deciduous. Leaves drop in June and July with the onset of summer heat and drought and are renewed with summer rains in August (Nilsen et al. 1984).

Chilopsis linearis maintains positive net photosynthesis at all but the lowest levels of leaf water potential (Odening et al. 1974), but photosynthesis decreases rapidly with relatively small decreases in tissue water potential. In laboratory experiments, net photosynthesis was zero at –3.5 MPa, a relatively high water potential for a desert plant (Odening et al.

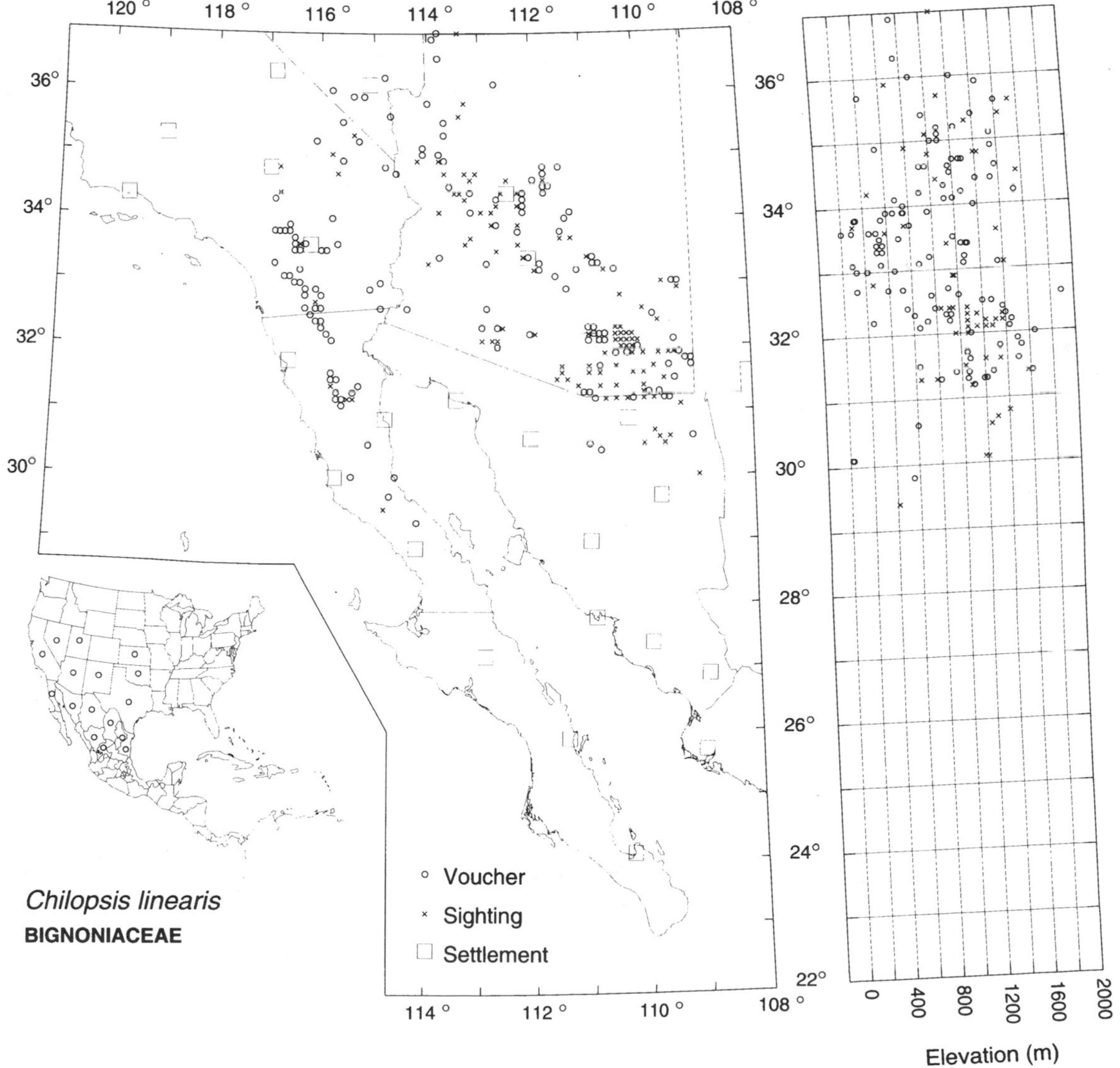

Chilopsis linearis
BIGNONIACEAE

1974). Maximum net photosynthesis is achieved at moderate temperatures (daytime 15–30°C, nighttime 7–15°C), showing that the photosynthetic mechanism is poorly adapted to extended high temperatures (Strain and Chase 1966).

The restriction of *C. linearis* to washes (and to dunes, where water is also available at depth) is crucial for its survival in the desert. As long as moisture is available, *C. linearis* can maintain relatively high water potentials even during extreme climatic drought (Odening et al. 1974). The relatively high rates of photosynthesis at moderate temperatures allow a reasonable level of annual productivity. In one study near Las Cruces, New Mexico, for instance, leaf biomass was 9 kg/ha in a wet year, 3 kg/ha in a dry one (DuPree and Ludwig 1978).

Several anatomical features of the leaves, such as a thick cuticle projecting over the stomates, lower the rate of water loss (Scott 1935a; Odening et al. 1974). The trees are deeply rooted (up to 16 m in one case) (Dupree and Ludwig 1978). In the most arid parts of their range, the rate of evapotranspiration in summer is so great that the trees drop their leaves, thus lowering water loss (Odening et al. 1974; Nilsen et al. 1984).

The bee *Bombus sonorus* is the main pollinator in southeastern Arizona (Whitham 1977). Hummingbirds visit the flowers occasionally (Brown et al. 1981). The styles fold shut when touched, preventing self-pollination (Stephen L. Buchmann, personal communication 1992). Fruit and seed production are pollen limited, not resource limited (Petersen et al. 1982). Lady-beetles and small flies visit nectaries on the flower buds and leaves (Pemberton 1988).

Chilopsis linearis is widely grown as an ornamental in the southwestern United States. Different color forms are available commercially (Duffield and Jones 1981). Seeds germinate readily (Scott 1935a), and semihardwood cuttings taken in late summer, or dormant cuttings made in winter, root easily (Nokes 1986). Given ample water, seedlings grow as much as 1 m per year for the first few years (Duffield and Jones 1981).

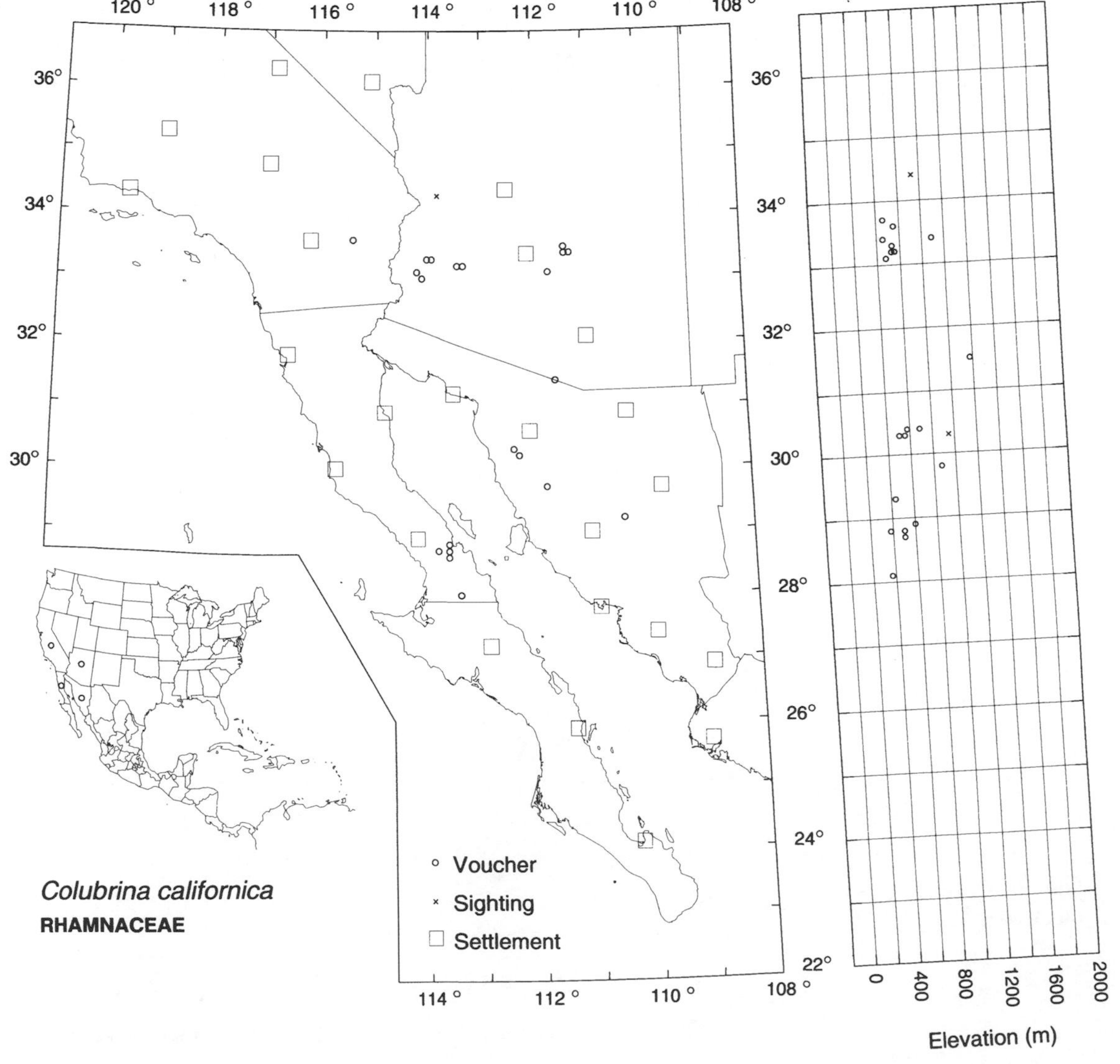

The wood is sometimes used for fence posts, the slender branches for baskets. Dried flowers are sold in Mexico to make a decoction for coughs (Standley 1924). The powdered leaves and bark are antiseptic and antifungal (Moore 1989).

Colubrina californica

I. M. Johnston

[= *C. texensis* (Torr. & A. Gray) A. Gray var. *californica* (I. M. Johnston) L. Benson]

California colubrina, snakewood

An intricately branched shrub 1–2.5 m tall, *Colubrina californica* has grayish-tomentose, spinescent branches and thickish, pubescent leaves. Alternate or fasciculate, entire or denticulate, the oblong-ovate leaves (12–30 mm long and 5–15 mm wide) are tipped with spines at the apex. The small, inconspicuous flowers are in axillary clusters 5–10 mm long. Fruits are globose, 3-celled capsules 8–9 mm long with one seed per cell.

Benson and Darrow (1981) regarded *C. californica* as a variety of *C. texensis.* We follow Johnston (1971) in treating it separately.

The fruits of *Lycium, Condalia,* and *Ziziphus* species are drupes, not woody capsules. *Adelia obovata* leaves are not tipped with spines, and those of *A. virgata* are usually shallowly emarginate. Leaves of *Phaulothamnus spinescens* are spatulate or oblanceolate and have a distinct glaucous bloom. *Canotia holacantha, Castela emoryi,* and *Koeberlinia spinosa* all have green-barked branches and twigs. Leaves of *Colubrina viridis* are mostly glabrous.

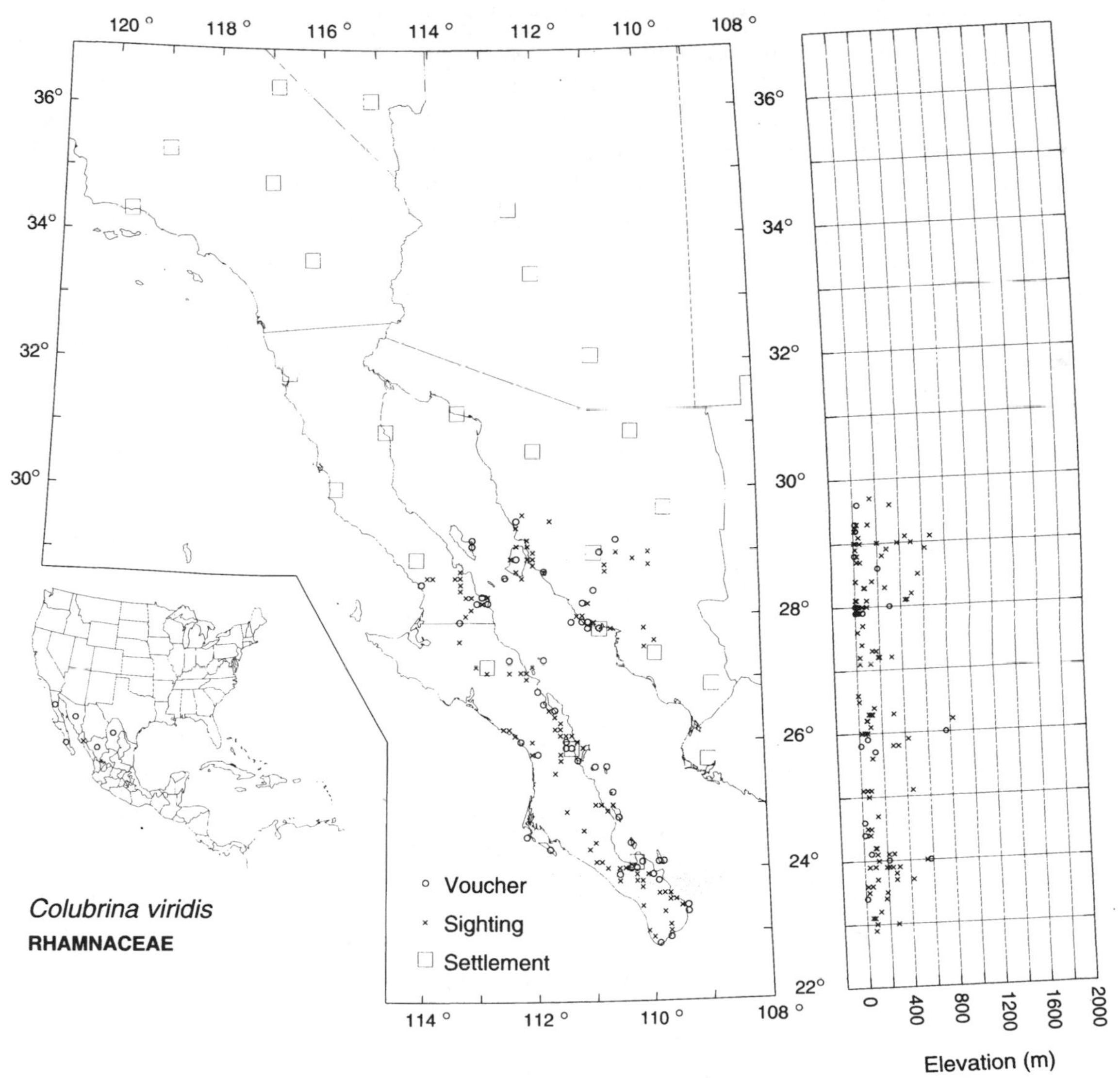

Colubrina viridis
RHAMNACEAE

Uncommon on rocky or gravelly slopes, *C. californica* is widely scattered in the Sonoran and Mojave deserts. Our single California locality (33.7°N) represents a collection made in 1941 by Annie M. Alexander in the Eagle Mountains, Riverside County. Munz (1974) also reported the species from Joshua Tree National Monument and the Chuckwalla Mountains. The northernmost locality (34.4°N) is based on a sighting made by Mary Butterwick in 1979 in southern Mojave County. The widely disjunct grouping in Baja California is remarkable, particularly as it overlaps the much more widely distributed *C. viridis*. As Benson and Darrow (1981) noted, *C. californica* has the appearance of a relict. The peninsular disjuncts may be the result of Pleistocene dispersal down the east coast of Baja California or across the midriff islands.

At the Baja California and western Arizona localities, the biseasonal rainfall is evenly divided between winter and summer. Because summers in this region are hot and winters are warm, soil moisture is more usable in winter than in summer. At the southeastern sites, most of the rain falls during the summer. Throughout the range, frost is uncommon or rare.

Flowers appear in spring or summer after rains. The capsules may persist for up to six months after ripening (Johnston 1971).

Colubrina viridis
(M. E. Jones) M. C. Johnston

[=*C. glabra* S. Watson]

Granadita, palo colorado

Colubrina viridis is a gray-barked, stiffly branched shrub 2–3 (occasionally 5) m tall. The younger branches may be tapered and spinescent. The bright green, glabrous leaves, 5–16 mm wide and 5–30 mm long, are obovate to suborbicular. The flowers (5 mm wide) are single or in axillary, few-flowered glomerules. The fruit is a globose capsule 4–6 mm in diameter.

Lycium, Condalia, and *Ziziphus* species have drupes, not woody capsules. Leaves of *Adelia obovata* are densely pilose, and those of *A. virgata* are usually shallowly emarginate. The thickish leaves of *Phaulothamnus spinescens* are spatulate or oblanceolate and have a distinctive glaucous bloom. *Colubrina californica* is densely and persistently pubescent on leaves, twigs, and calyces.

Wiggins (1964) treated this species as *C. glabra*. We follow Johnston (1963a) in considering *C. glabra* a synonym of *C. viridis*.

A plant of bajadas, plains, hillsides, and arroyos, *C. viridis* is rather widespread below 1,000 m in the southern half of the Sonoran Desert. It can be locally common, as on Isla Tiburón (Felger and Moser 1985). Except in Baja California, where their ranges barely overlap, *C. viridis* and *C. californica* are geographically separated.

In most parts of the range, frosts are rare or absent, and rainfall is concentrated in summer and early fall. Toward the north, rainfall varies from half in winter to most in winter.

Flowers may be found at any time of year. Peak flowering is in early spring (February–March) and late summer (August–September). The leaves appear quickly after rains and fall when the soil dries out.

The hard wood is an excellent fuel and was also employed by the Seri Indians in making harpoon shafts and pry bars (Felger and Moser 1985).

Condalia brandegei
I. M. Johnston

This intricately branched shrub grows 1–3 m tall and has stout main branches and sturdy, spinose branchlets. The dark gray bark is dotted with red-brown lenticels. The alternate or fasciculate leaves, 12–20 mm long and 8–14 mm wide, are entire and obovate to elliptic. Dark green and shining above, they are pale and prominently veined on the underside. The flowers, 2–3 mm wide, are single or in axillary clusters. The black to reddish black drupes are 6–8 mm long and about 5 mm in diameter.

Lycium leaves are not prominently veined. *Condalia globosa* leaves are smaller and spatulate rather than elliptic or obovate. *Ziziphus obtusifolia* has conspicuously light-gray stems.

This species is closely related to *Condalia hookeri* M. C. Johnston of southern Texas and northeastern Mexico (Johnston 1962a). Johnston (1962a) suggested that *C. brandegei* diverged from the *C. hookeri* complex after gulf-floor spreading separated the peninsula from the mainland.

A narrow endemic of the Baja California peninsula, *Condalia brandegei* grows on bajada slopes and in canyons. Its spotty distribution suggests a population broken up by increasing aridity, and during Pleistocene pluvials, the species may have extended to lower elevations and covered a wider area. Present-day populations receive biseasonal rainfall with somewhat more in winter than in summer.

Flowers appear in spring (February–April) (Wiggins 1964).

Condalia globosa

I. M. Johnston

Bitter condalia, crucerilla

Condalia globosa is a large, densely branched shrub 2–5 (occasionally 6) m tall with thorn-tipped branchlets that diverge from the main branches at right angles. The spatulate, entire leaves, 3–12 mm long and 1.6–5 mm wide, may be alternate or fasciculate. The lower surface has 3–4 pairs of prominent veins. The flowers (2 mm wide) are in axillary clusters. Fruits are globose drupes 3–5 mm long. They are black and often very bitter at maturity.

Without flowers and fruits, it can be difficult to separate *C. globosa* var. *pubescens* from *C. warnockii* var. *kearneyana*. Further study may show them to be conspecific (Richard S. Felger, personal communication 1992); however, they do occupy climatically distinct regions. *Condalia brandegei* has larger leaves (12–20 mm long and 8–14 mm broad) that are elliptic or obovate rather than spatulate. *Ziziphus obtusifolia* has white stems and larger leaves not prominently veined beneath. *Phaulothamnus spinescens* has thick, smooth leaves with a distinct glaucous bloom. *Lycium* leaves are not prominently veined.

The two varieties, var. *globosa* and var. *pubescens* I. M. Johnston, differ mainly in pubescence of leaves, twigs, pedicels, and calyces. Benson and Darrow (1954) erroneously assigned the California populations of *C. globosa* var. *pubescens* to *C. spathulata* A. Gray, a shrub of southern Texas and adjacent Mexico (Johnston 1962a).

Occasional to common in sandy washes, rare on rocky slopes, *C. globosa* grows below 1,200 m southward from about 33°N. In the northeastern Sonoran Desert, *C. warnockii* var. *kearneyana* largely replaces it. The range includes a variety of subtropical and tropical arid climates with infrequent frost. Rainfall varies from almost entirely summer in the south to almost completely winter in north-central Baja California. According to John-

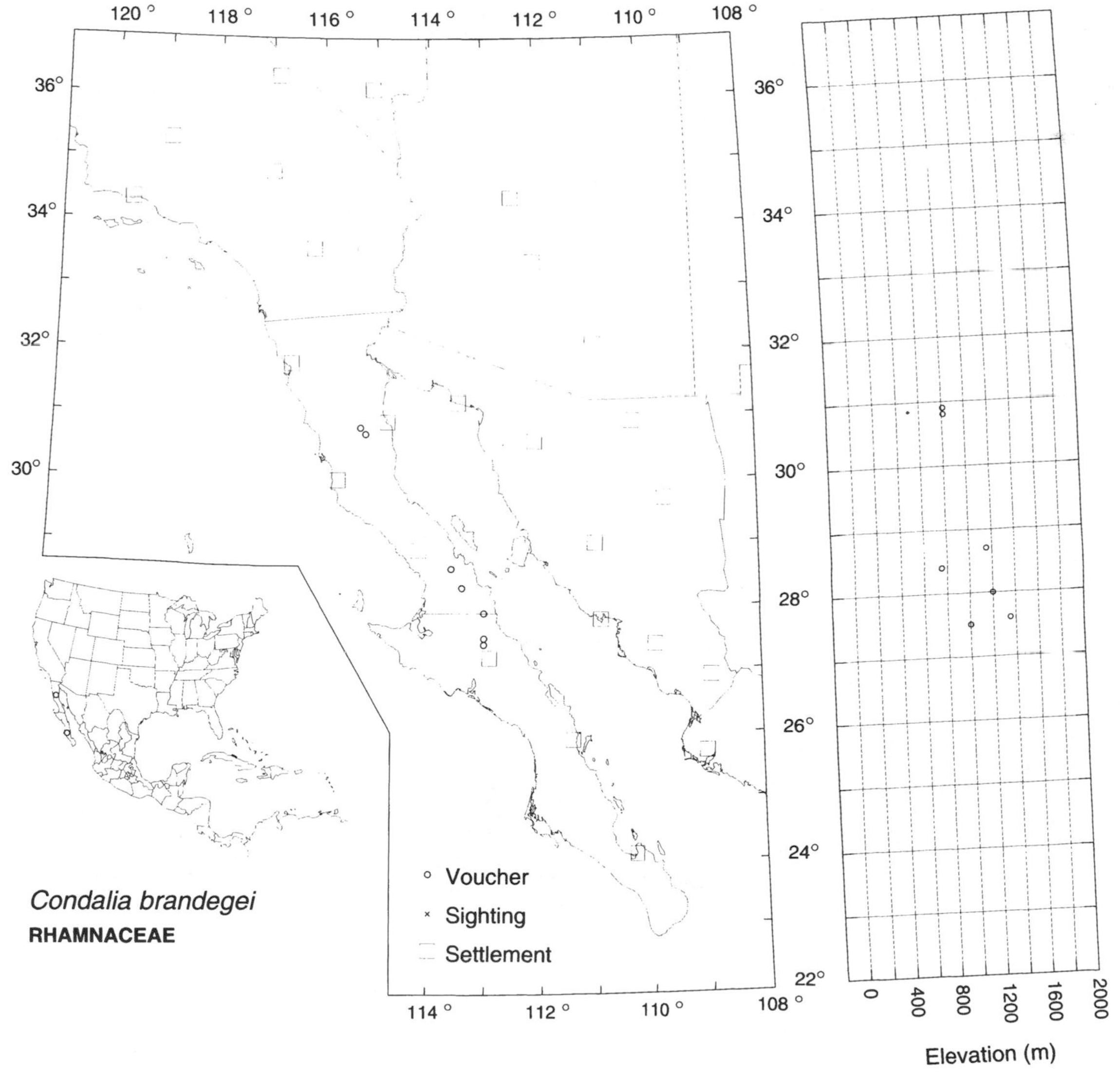

ston (1962a), the present-day species of *Condalia* probably lived during the Miocene in essentially their current form. If so, *C. globosa* was probably on the Baja California peninsula before its separation from the mainland.

Flowers appear throughout the year in response to rain. Peak flowering is March–May and October–December. A variety of bees and lepidopterans visit the sweet-scented flowers.

The wood makes an excellent fuel. Seri Indians used the mashed leaves and thorn-tipped twigs in tattooing (Felger and Moser 1985).

Condalia warnockii M. C. Johnston

var. *kearneyana* M. C. Johnston

[=*C. spathulata* auth., not A. Gray]

Mexican crucillo, squawbush, guichutilla

Condalia warnockii var. *kearneyana* is a rounded shrub up to 1.5 (occasionally 3) m tall with spinose stems and branchlets. From a distance, the plants look dark green or even blackish due to the crowded leaves and dark bark. Alternate or fasciculate, the spatulate, entire leaves are 1.3–5 mm long and 0.5–2 mm broad. They have 2–4 pairs of prominent veins on the underside. The flowers, 2–3 mm wide, are single or in axillary clusters. Fruits are black or reddish black drupes 4–6 mm long. A haploid chromosome count of 24 has been reported for *C. warnockii* (Ward 1984).

Condalia correllii M. C. Johnston has larger leaves (4–6 mm wide) with less prominent, narrower veins. *Ziziphus obtusifolia* has whitish stems and larger leaves. *Lycium* leaves are not prominently veined.

This species has often gone under the name *C. spathulata,* as in Kearney and Peebles (1960). Johnston (1962a) reserved *C. spathulata* for the closely related plants of southern Texas and adja-

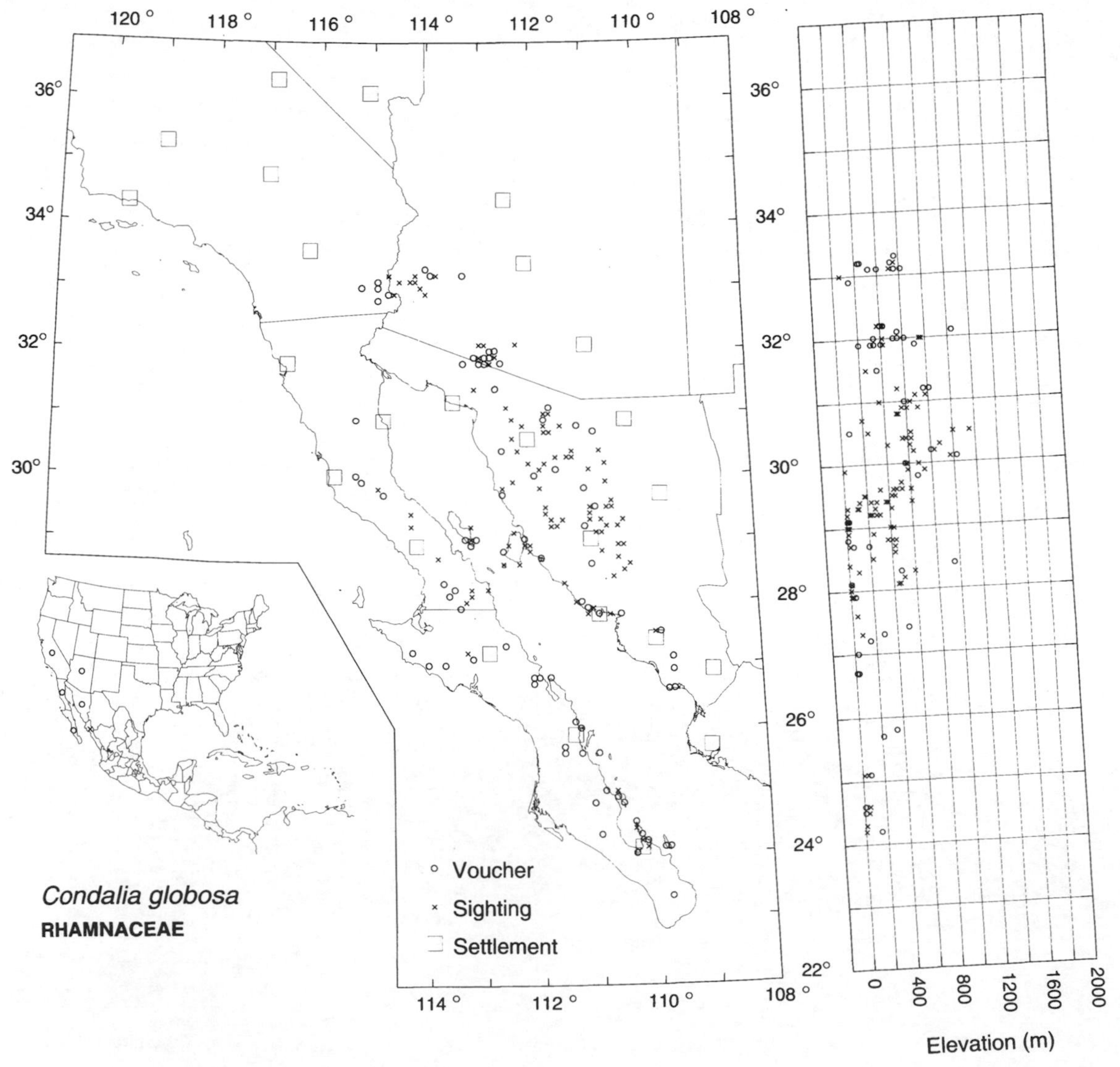

Condalia globosa
RHAMNACEAE

cent Mexico and described a new species, *C. warnockii,* to accommodate the remaining populations. The variety in the region mapped in this atlas, *C. warnockii* var. *kearneyana,* apparently does not overlap in distribution with var. *warnockii* of New Mexico, Texas, and northern Mexico (Benson and Darrow 1981).

Within its rather narrow range, *C. warnockii* var. *kearneyana* can be common on calcareous, sandy or gravelly slopes and flats between 500 and 1,700 m. The range seems restricted to climates with relatively mild winters and biseasonal rainfall dominated by summer storms.

Flowers appear in response to rain between March and November. They are scented and attract insects (Gentry 1942). Leaves are generally evergreen but may be shed during unusual drought (Shreve 1964).

Cordia curassavica

(Jacq.) Roem. & Schult

[=*C. brevispicata* Mart. & Gal.]

Bolita prieta

A straggling shrub 1–3 m tall, *Cordia curassavica* has musty-scented, glandular foliage. The linear-oblanceolate to oblong leaves are 3–18 mm wide and 2–6.5 cm long and are conspicuously veined beneath. The white, funnelform flowers, 4–5 mm long and 3.5–5 mm wide, are in crowded spikes. Fruits are bright red drupes 4–5 mm in diameter.

Cordia parvifolia flowers are in cymes, not spikes, and its leaves are ovate to obovate. The leaves of *Bourreria sonorae* have harshly scabrous pubescence. *Melochia tomentosa* has stellate-tomentose foliage.

This extremely variable species ranges from Baja California Sur and Sonora to South America and the West Indies (Miller 1988). Much of the variation is phenotypic (Miller 1988). Wiggins (1964) treated the plants in the Sonoran Desert region as *C. brevispicata.* We follow Johnston (1949) and Miller (1988).

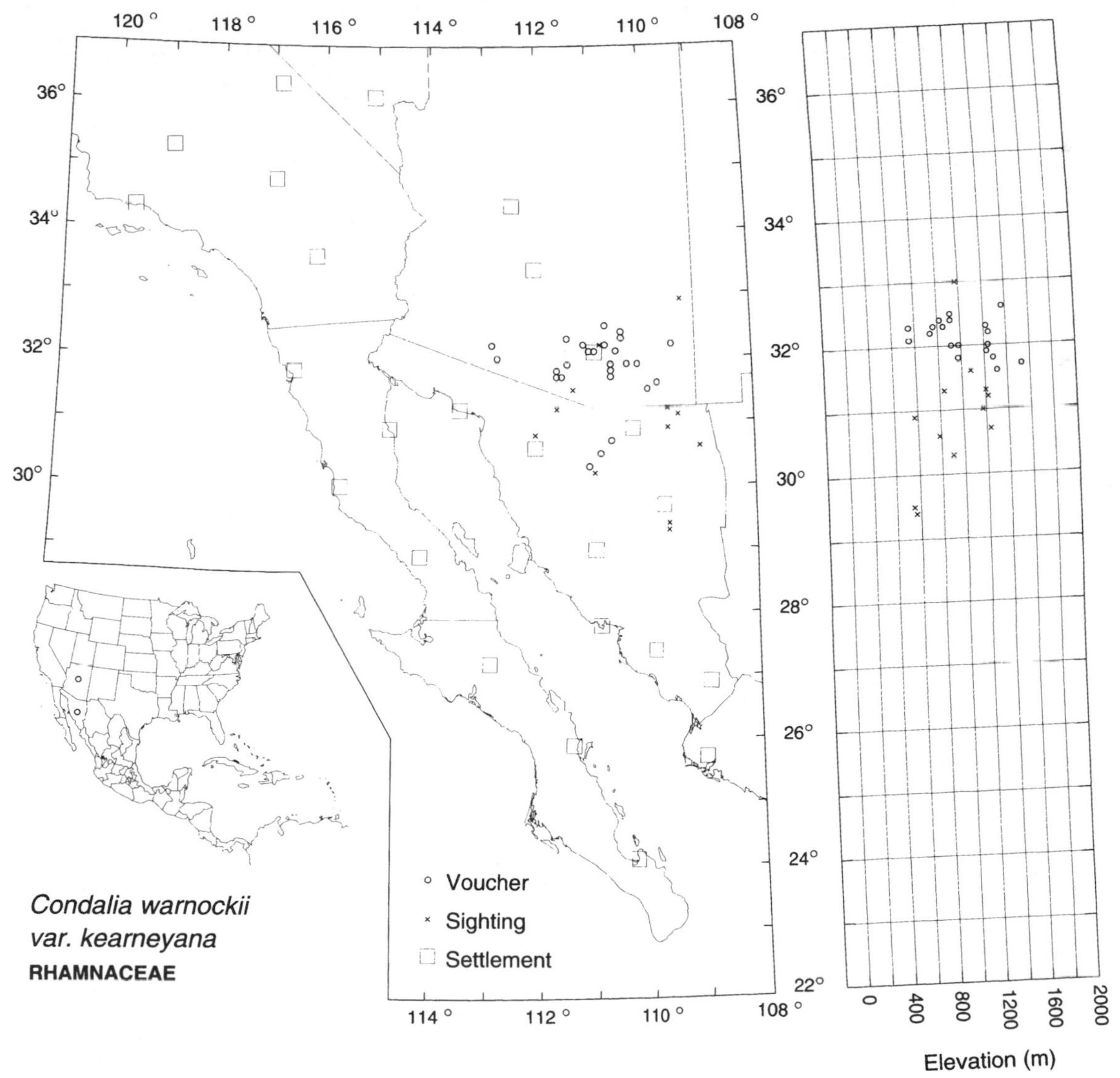

Condalia warnockii var. kearneyana
RHAMNACEAE

Growing in gravelly or rocky canyons and arroyos, *C. curassavica* is limited in the area of this atlas to the peninsula and three widely disjunct populations on the mainland coast. In Costa Rica, it is a predominant shrub in secondary succession, especially where human activities cause continual disturbance (Opler et al. 1975). The two Sinaloan disjuncts represent collections by Howard Scott Gentry: one from Bahía Topolobampo (25.5°N) in 1954, the other from Isla Tachechille (24.9°N) in 1945. Gulf of California populations occur only on Cerralvo and Espíritu Santo islands (Moran 1983b). The range is restricted to virtually frost-free climates. Toward the south, rain is concentrated in late summer and early fall; in the central peninsula, it is more or less equally distributed between winter and summer.

Plants bloom throughout the year. The flowers are heterostylous and strongly outcrossing (Opler et al. 1975). They attract large numbers of small butterflies as well as wasps, bees, and flies. The seeds are dispersed by birds (Opler et al. 1975).

Cordia curassavica is a serious weed in Malaysia (Simmonds 1980).

Cordia parvifolia
A. DC.

Vara prieta, San Juanito, palo prieto

Cordia parvifolia is an open shrub 1–3 m tall with smooth, dark gray bark. Except on young twigs, the obovate or ovate leaves, 3–10 (occasionally 15) mm wide and 1–3 cm long, are fasciculate on short spur branches. They are harshly scabrous, with veins impressed above and conspicuous beneath. The funnelform, white flowers turn purplish or brownish in age. They are 1.5–2 cm long and up to 3 cm wide and are in few-flowered, headlike cymes. The drupes, about 6–9 mm in di-

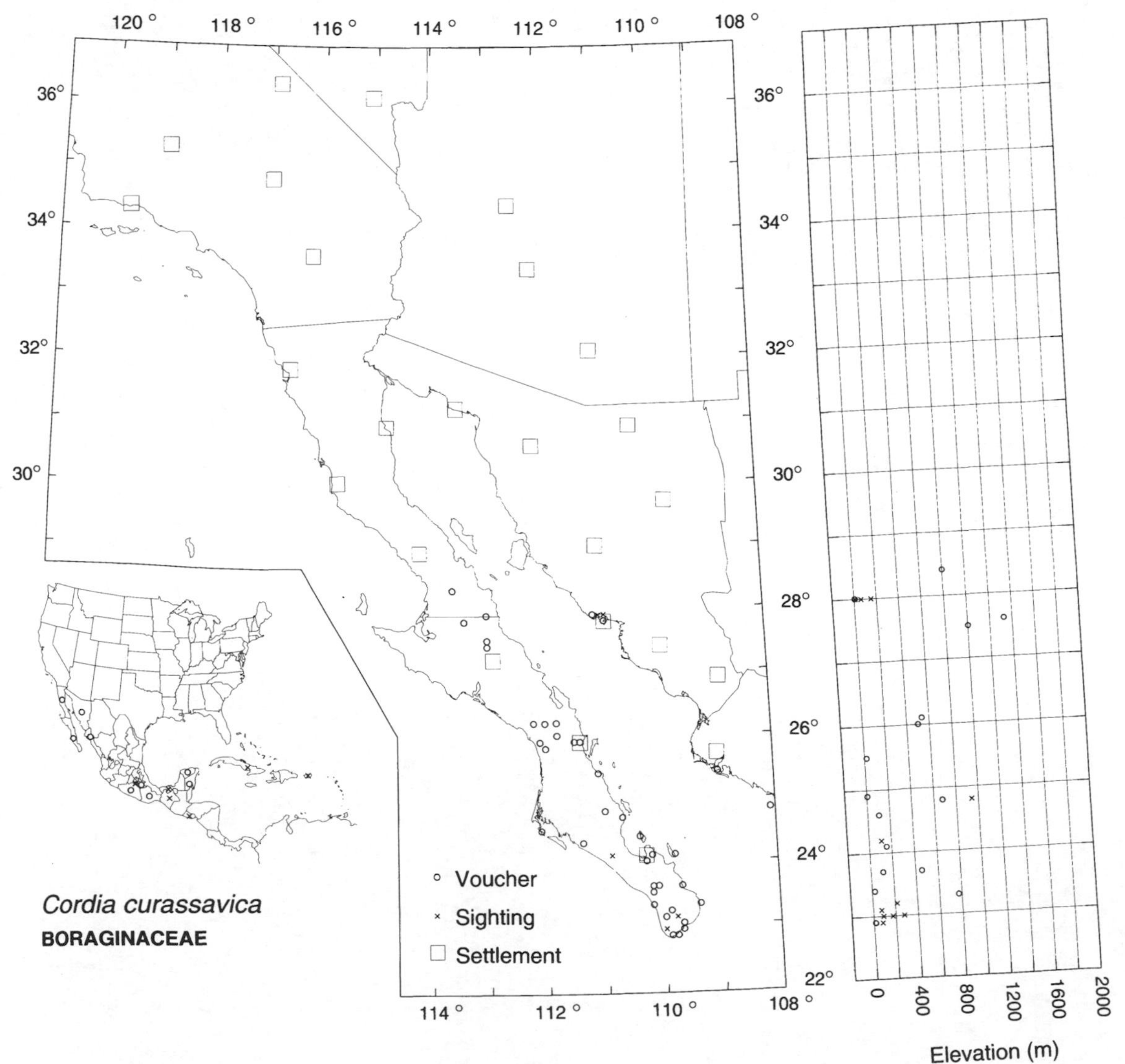

Cordia curassavica
BORAGINACEAE

ameter, are enclosed in the enlarged calyx.

Cithyarexylum flabelliforme has opposite, fan-shaped leaves.

Cordia parvifolia grows on hillslopes, along arroyos, in silty bottomlands, and on rocky plains. The northernmost point on our map (32.0°N) represents several specimens collected about 24 km south of Tucson (K. F. Parker in 1951 and 1952 and C. F. Allfillisch in 1951). Howell and McClintock (1960) noted that a single, old, many-stemmed plant still grew there. The species has not been collected in Arizona since and apparently no longer occurs in the state.

Widely distributed on the Sonoran mainland, *Cordia parvifolia* grows only in a narrow band on the peninsula. The only gulf occurrence is on Isla Tiburón (Moran 1983b). If the Baja California Sur population is relictual, it most likely dates back to the early Miocene before the peninsula separated from the mainland. Alternatively, the species may have jumped the gulf; certainly the drupaceous fruits could be dispersed by birds. Climates over much of the range are marginally tropical and arid. The peninsular range experiences biseasonal rainfall with about equal amounts in summer and winter. Sonoran sites receive more summer rain. Frost is rare over most of the range, except at the former Arizona locality.

Flowers appear nearly year-round. Peak flowering is in late summer (August–October) and spring (March–April).

This shrub is coming to be widely used as a drought-tolerant ornamental. Frost-tolerant strains from the Chihuahuan Desert are probably more common in the horticultural trade than the more tender Sonoran Desert strains (Gregory D. Starr, personal communication 1992). It is easily propagated from stem cuttings (Jones et al. 1980).

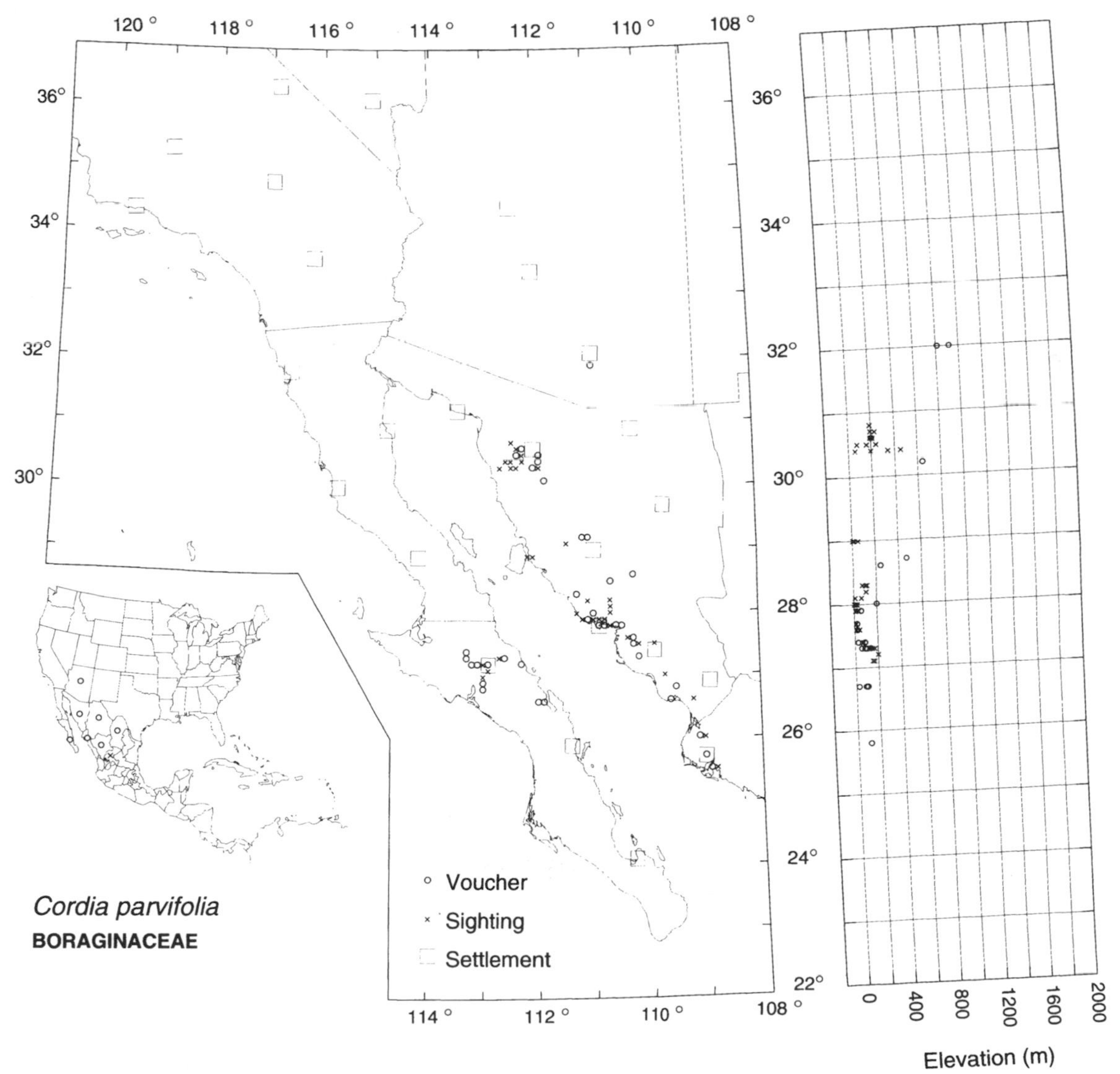

Cordia parvifolia
BORAGINACEAE

Cordia sonorae

Rose

Asta, palo de asta, amapa bola, amapa blanca

A small tree up to 7 m high, *Cordia sonorae* has dark gray, smooth or slightly fissured bark and an open crown. The simple leaves, 1.5–4.5 cm wide and 5–11 cm long, are glabrous and elliptic to oblong. The white, funnelform flowers, in dense cymes, are 2–3 cm long and 2.5–3.5 cm wide.

Leaves of *C. parvifolia* are much smaller, only 3–15 mm wide and 1–3 cm long, and are fasciculate on spur branches. *Bourreria sonorae* has markedly scabrous leaves on short spur branches. *Hintonia latiflora* has opposite leaves.

Its large, simple leaves suggest that *C. sonorae* is a mesophyte, and in the Sonoran Desert it is indeed restricted to watercourses (Shreve 1964). Farther south, in Sinaloan thornscrub, it grows on hillslopes as well (Gentry 1942). The northernmost sighting (29.9°N) is Raymond M. Turner's. This species has not been found on any gulf islands (Moran 1983b). The range and elevational profile are typical of a tropical species at its frigid and arid limits.

Cordia sonorae blooms from March to May, making a showy display when the plant itself (indeed, the entire forest canopy) is leafless (Gentry 1942). The leaves appear with the onset of summer rains and often persist through winter (Shreve 1964).

The flowers may be pollinated by moths, as noted by Opler and Janzen (1983) for *Cordia alliodora* (Ruiz & Pav.) Cham. in Costa Rica. Barajas-Morales (1983) described the wood anatomy.

The plants are sometimes grown as ornamentals. Seeds germinate readily (Gregory D. Starr, personal communication 1992).

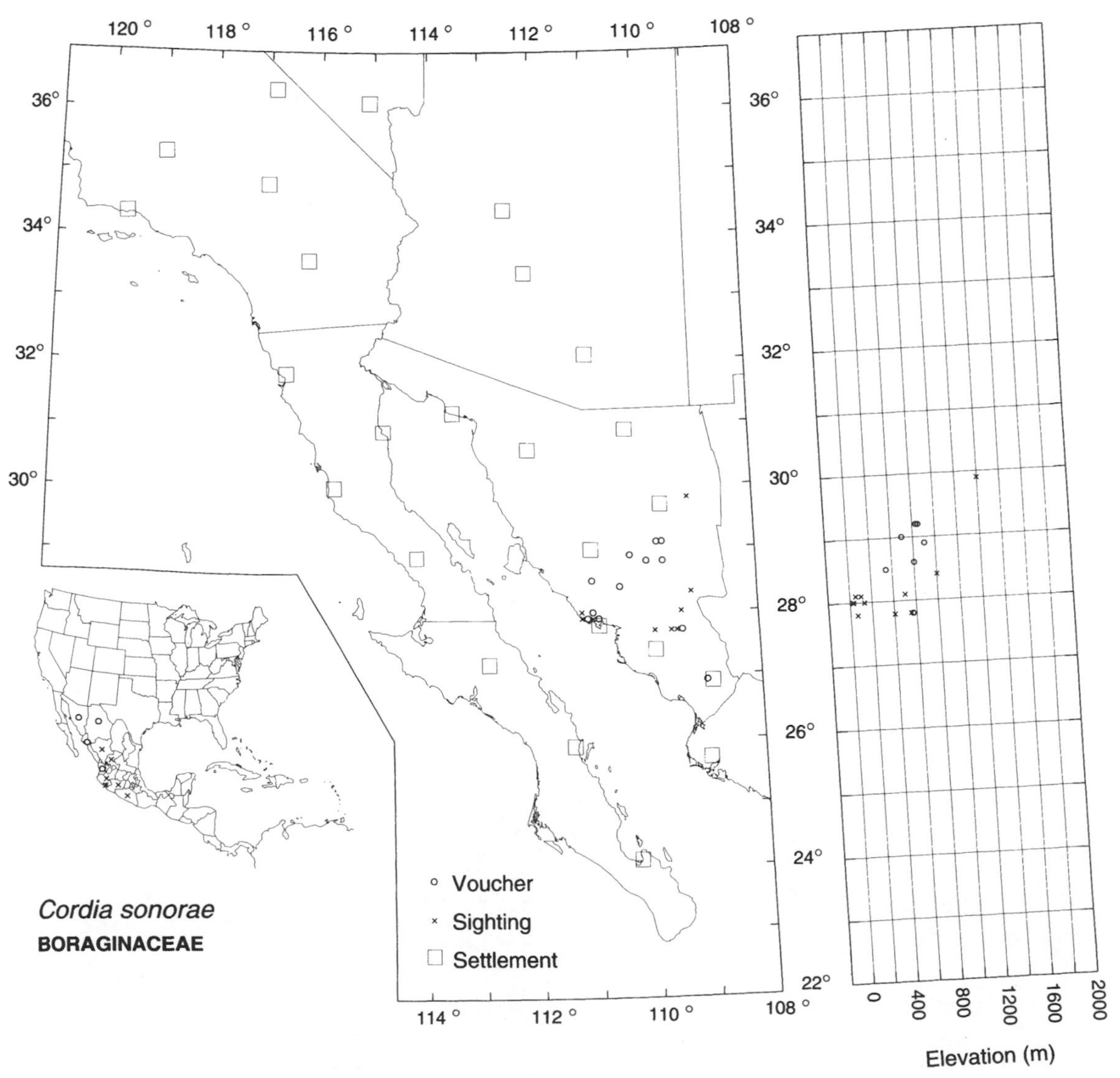

Coursetia glandulosa

A. Gray

[=*C. microphylla* A. Gray]

Samo prieto, samota, cousamo, chino, chipile, chipilillo, tepechipile

Coursetia glandulosa is an open shrub 1.5–5 m tall with flexible branches. The leaves, 2–5 cm long, are once-pinnate with 4–9 pairs of leaflets. The flowers, 11–13 mm long, are white or pale yellow (usually tinged or veined with lavender or red) and are borne in short, axillary racemes. Fruits are constricted pods 2–5 cm long and 5 mm broad. The haploid chromosome number is 8 (Lavin 1988).

Leaves of *Eysenhardtia orthocarpa* are punctate-glandular and odd- rather than even-pinnate. *Diphysa occidentalis* is also odd-pinnate, but its leaves, although minutely punctate-glandular above, are larger.

We follow Wiggins (1964) in considering *C. microphylla* to be conspecific with *C. glandulosa*. Leaflet length varies clinally from 50 mm at the southern end of the range to 15 mm at the northern end (Lavin 1988).

At the southern end of its range, *C. glandulosa* is a common understory shrub in tropical dry forests (Lavin 1988). According to Gentry (1942), it is one of the most abundant plants in foothills of the Río Mayo watershed. In the more arid parts of the Sonoran Desert, the plants grow along washes, in moister areas on rocky slopes. The northernmost points on our map represent collections and sightings made in or near the Superstition Mountains, Maricopa County. The westernmost sighting (Robert Darrow, Kofa Mountains, Yuma County) places the species in a very arid region indeed. Isla Tiburón is the only gulf locality for *C. glandulosa* (Moran 1983b). The species has failed to occupy apparently suitable habitat at the northern end of its range.

The plants are hardy to temperatures as low as –6.5°C (Johnson 1988) and require warm-season rain. *Coursetia glan-*

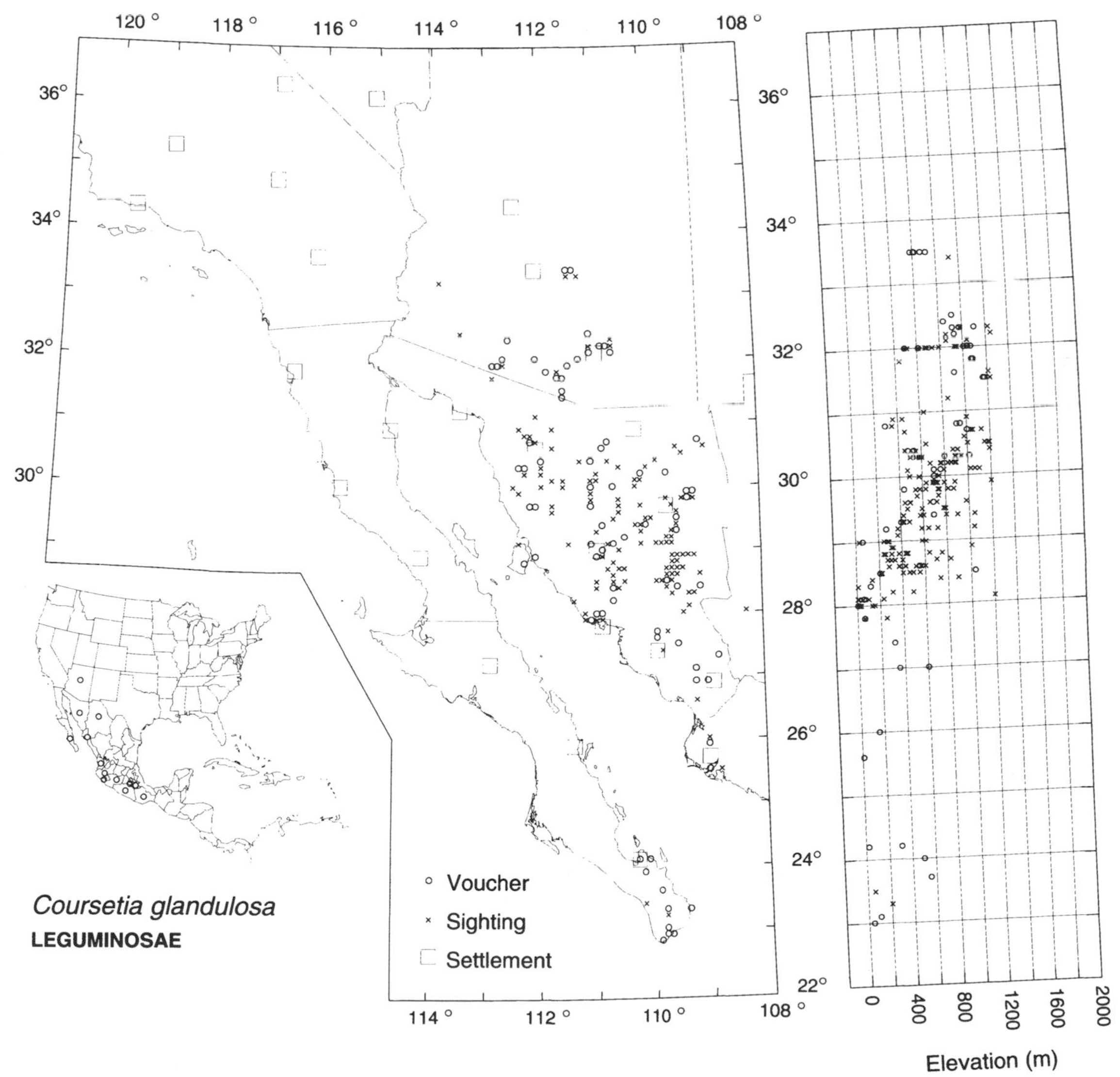

Coursetia glandulosa
LEGUMINOSAE

dulosa is evidently best regarded as a thornscrub species that can grow in the desert where conditions are neither too cold nor too arid. The widely disjunct population in the Cape Region, Baja California Sur, may date back to the early Miocene before the peninsula separated from the mainland.

In Arizona the plants are winter deciduous. Flowers appear shortly before or with the new leaves, as early as February in the Sonoran Desert region. Peak bloom is in March and April.

Pollen shed inside the keel accumulates on the brushlike stigma. Self-pollination is therefore theoretically possible, but in view of the low fertility of self-pollinated plants, it seems likely that the species is largely outcrossing (Lavin 1988). Pollinators include solitary bees in the genera *Xylocopa, Centris,* and *Trigona* (Lavin 1988). In the wild, propagation by root suckers sometimes creates dense colonies (Lavin 1988).

Seri Indians used the wood for bows, digging sticks, cradleboards, and other utilitarian objects. The scale insect *Tachardiella* deposits orange lac on the stems. This gummy substance is used by the Seri for hafting harpoon heads (Felger and Moser 1985). Sonoran pharmacies once sold lac as "goma Sonora" for use in treating colds, fevers, and tuberculosis (Standley 1922).

In cultivation, growth rates are moderate with supplemental irrigation (Johnson 1988). In greenhouse trials, seed germinated readily (Lavin 1988).

Cyrtocarpa edulis

(Brandegee) Standley

Plum-tree, ciruelo

Cyrtocarpa edulis (figure 34) is a thick-stemmed tree up to 10 m tall with smooth, light gray bark and a rounded, open crown. The alternate, short-pubescent leaves, 5–10 cm long, are pinnately compound with 7–11 elliptic to oval leaflets 1–3 cm long. The inconspicuous, whitish or

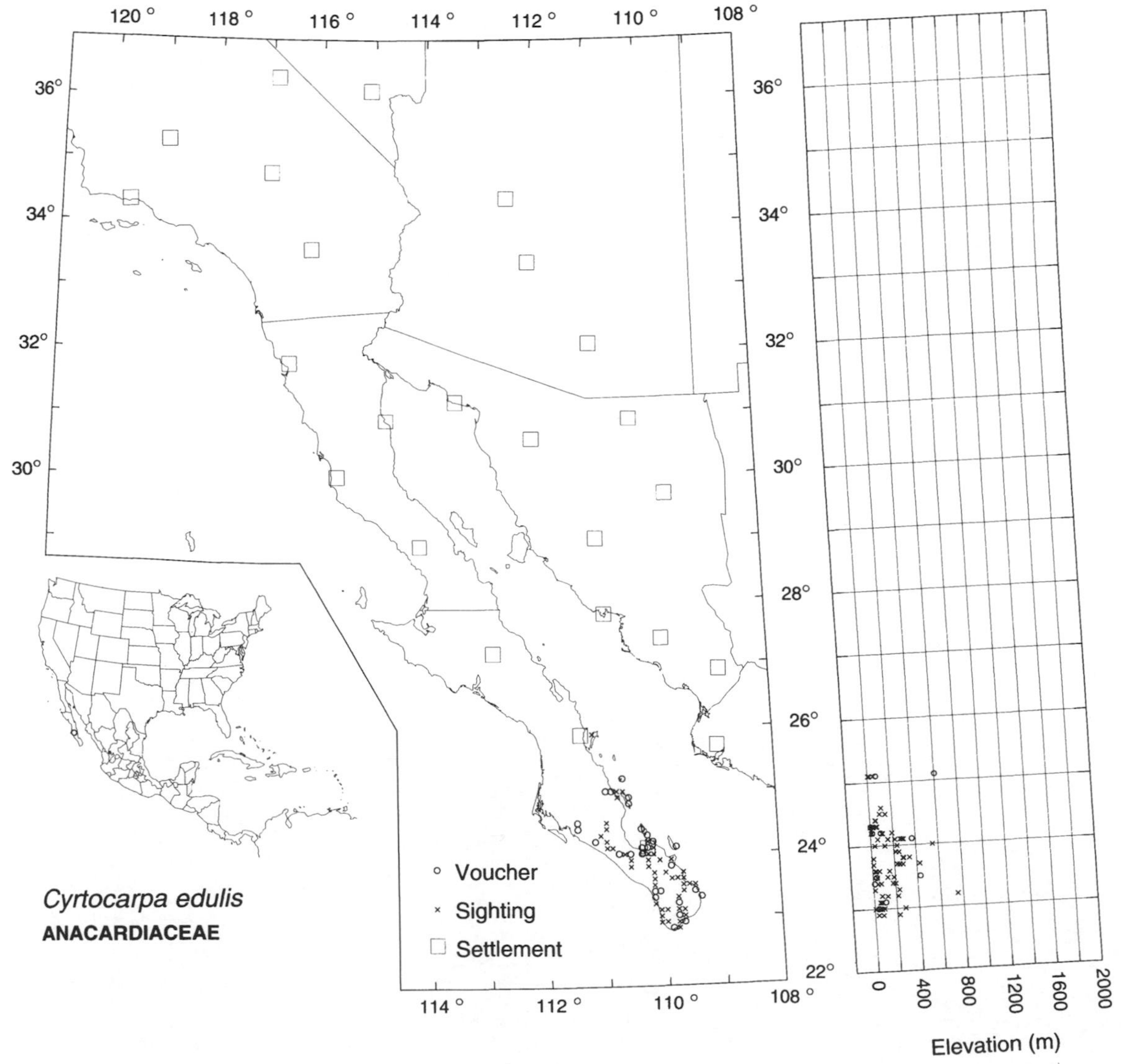

greenish flowers are in axillary panicles. Fruits are yellow, velvety, ovoid drupes 1.5–2 cm long with bony seeds.

Pachycormus discolor has a stout trunk and thick, markedly crooked branches, and its white or yellowish bark exfoliates in sheets. *Bursera microphylla* and *B. odorata* Brandegee, both small trees with exfoliating bark, have glabrous leaflets. Leaves of *B. hindsiana* are simple or trifoliate.

This narrow endemic is restricted to Baja California Sur and adjacent islands. It can be locally dominant on sandy soil of plains and gentle slopes. Such short-tree woodlands sometimes resemble "an unkempt apple orchard" (Coyle and Roberts 1975:116). Moran reported this species from Isla Carmen (26.0°N), the northernmost outpost. The distribution suggests an intolerance of freezing and a need for relatively substantial summer and early autumn rains.

The swollen trunk is a water-storage organ, an adaptation seen in several other characteristic Sonoran Desert genera, including *Bursera, Pachycormus,* and *Fouquieria.* Wood water content in these species ranges from 60.9% to 80.9% (Shreve 1933). Shreve (1964) referred to this as the sarcocaulescent life-form and noted that such plants could better be regarded as drought escaping than drought resistant (Shreve 1933). Although the sarcocaulescent life-form is uncommon in the Sonoran Desert flora as a whole, sarcocaulescents are locally dominant in the Vizcaíno region (*Fouquieria columnaris* and *Pachycormus discolor*) and in the central Gulf Coast region (*Bursera microphylla, B. hindsiana, Jatropha cinerea*) (Shreve 1964).

Flowers appear in April and May. Despite the abundant water presumably available in the wood, *C. edulis* leaves are drought deciduous (Shreve 1937b). Evidently the moisture in the cortex and periderm is used to sustain the plant over long periods of drought rather than to maintain evergreen leaves. The fruits can be found throughout the summer and into autumn. They have a thin exocarp, a fleshy mesocarp, and a bony endocarp with 2–5 small openings (opercula). The extremely hard tissue of the endocarp protects the embryo and would also deter germination if not for the opercula, which serve as doors through which the embryo can emerge (Mitchell and Daly 1991).

The edible fruits may be sweet or bitter (Standley 1923).

Figure 34. A small Cyrtocarpa edulis *near La Paz, Baja California Sur. (Photograph by J. R. Hastings.)*

Dalea bicolor Humb. & Bonpl. ex Willd. var. *orcuttiana* Barneby

[=*D. megalostachys* (Rose) Wiggins, *D. orcuttii* S. Watson]

Damiana, hierba del borrego

This shrub grows up to 1 m tall; it has green to gray foliage and conspicuous glands on the branches. The leaves, 1–3.5 cm long, are once-pinnate with 4–10 pairs of leaflets. The small rose-purple flowers are in ovoid or oblong heads 1.5–6 cm long. The haploid number of chromosomes is 7 (Spellenberg 1973).

Dalea purpusii has fewer leaflets and plumose calyx teeth.

Dalea bicolor var. *orcuttiana* is a common plant of steep, rocky slopes and desert washes on the central Baja California peninsula. It grows from San Matías Pass to Cerro Mechudo. The sole Sonoran location is near Desemboque. The range includes a variety of maritime arid climates. In the south, winters are frost free, and most of the rain falls in summer; in the north, frosts are infrequent except at the highest sites, and rain falls mainly in winter. The plants suffer frost damage at temperatures below –6.5 to –4.0°C (Johnson 1993).

Leaves are drought deciduous; indeed, the entire aspect of the plant is xerophytic (Barneby 1977). The fragrant flowers appear after winter rains in the north (February–May), after summer rains in the south (August–October) (Barneby 1977).

Mexicans sometimes make a tea from the herbage to treat stomachaches (Reid Moran, personal communication 1992). *Dalea bicolor* var. *orcuttiana* is occasionally grown as a drought-tolerant ornamental. It can be propagated from stem cuttings (Starr 1988).

Dalea formosa

Torr.

Feather dalea, feather plume, feather peabush, yerba de Alonso García

An intricately branched shrub up to 0.5 m tall, *Dalea formosa* has light gray bark and odd-pinnate leaves 1 cm long. The 7–15 narrowly obovate, glabrous leaflets are 2–3 mm long. The flowers are usually bicolored with a yellowish or creamy banner and rose-purple or magenta wings and keel. They are about 15 mm long and are in few-flowered spikes. The pods are 3 mm long.

Dalea formosa differs from *D. pulchra*

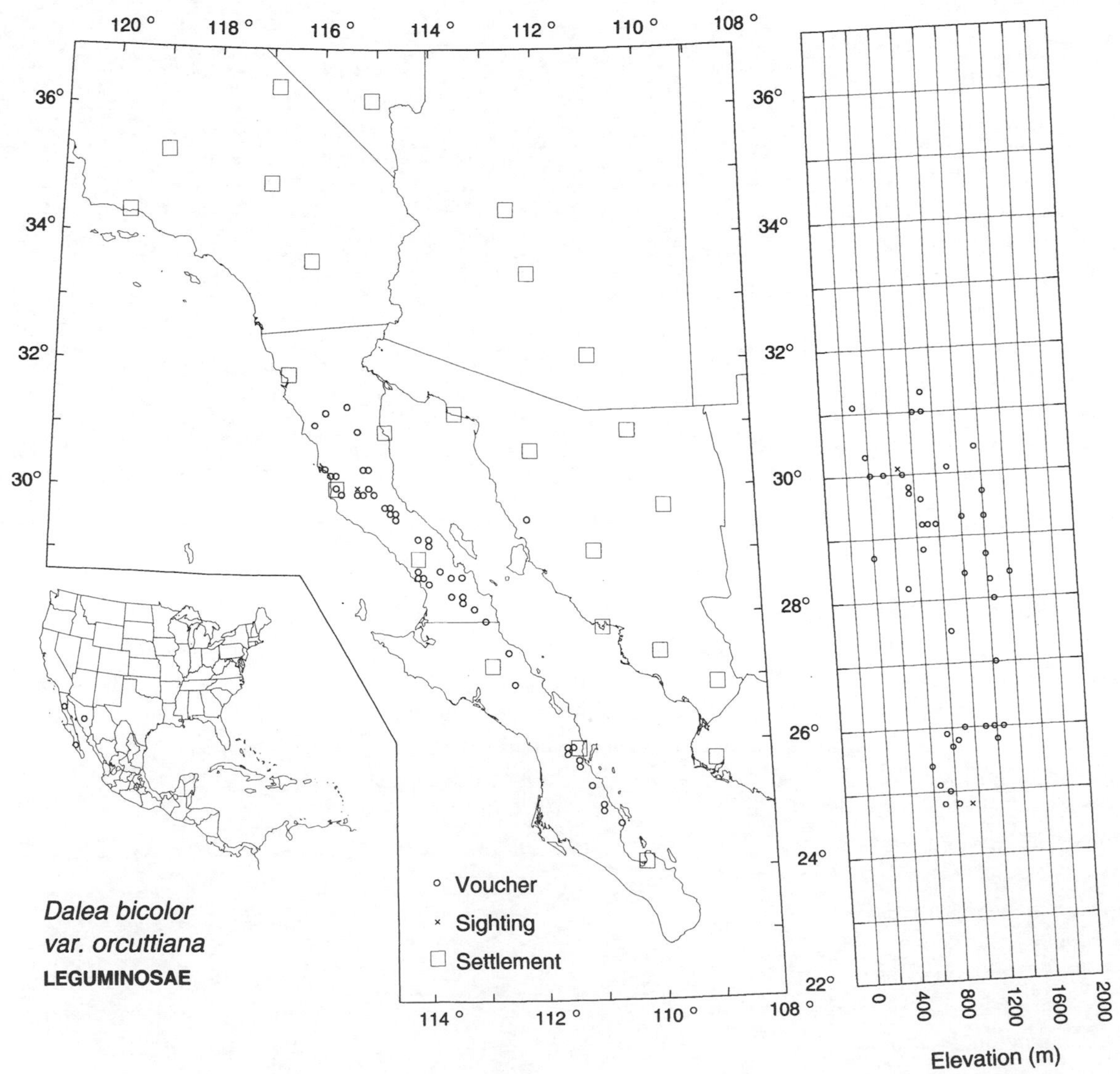

Dalea bicolor var. orcuttiana
LEGUMINOSAE

and *D. versicolor* in its smaller stature, sparsely flowered spikes and glabrous leaves.

The species has diploids (n=7), tetraploids (n=14), and hexaploids (n=21) (Spellenberg 1981). Sonoran Desert plants are mainly diploid. The polyploids, restricted to the Chihuahuan Desert and its immediate borders, seem adapted to more xeric conditions than the diploids (Spellenberg 1981). *Dalea formosa* might have arisen from a prototype much like *D. versicolor* (Barneby 1977).

The plants are found on rocky hills and canyon slopes, especially on limestone substrate. At its upper elevational limit in southern Colorado and New Mexico, *D. formosa* endures freezing temperatures and frozen ground (Barneby 1977). In the Sonoran Desert region, it grows in warm-temperate climates where winter frost is common. Johnson (1993) reports survival to temperatures as low as –12°C. In most years, the plants receive more summer than winter rain. Most characteristic of oak woodland and desert grassland, *D. formosa* enters the desert at its upper margin and at scattered points within it. The northernmost point on our map (35.8°N), a collection by Virginia T. Cotter, is at Tuzigoot National Monument.

Flowers appear from March to September.

Fresh, untreated seeds germinate readily. Plants can be grown from semi-hardwood cuttings taken in summer or early fall (Nokes 1986). Although not yet extensively tested as an ornamental, *D. formosa* should grow well in cultivation in southern Arizona (Starr 1988). Fragrant tea can be made from the flowering branches (Moore 1989).

Dalea pulchra

Gentry

[=*D. greggii* auth., not A. Gray]

This *Dalea* species is an erect, bushy, suffrutescent herb or subshrub 5–8 dm high. The leaves (1.5 cm long) are once-pinnate

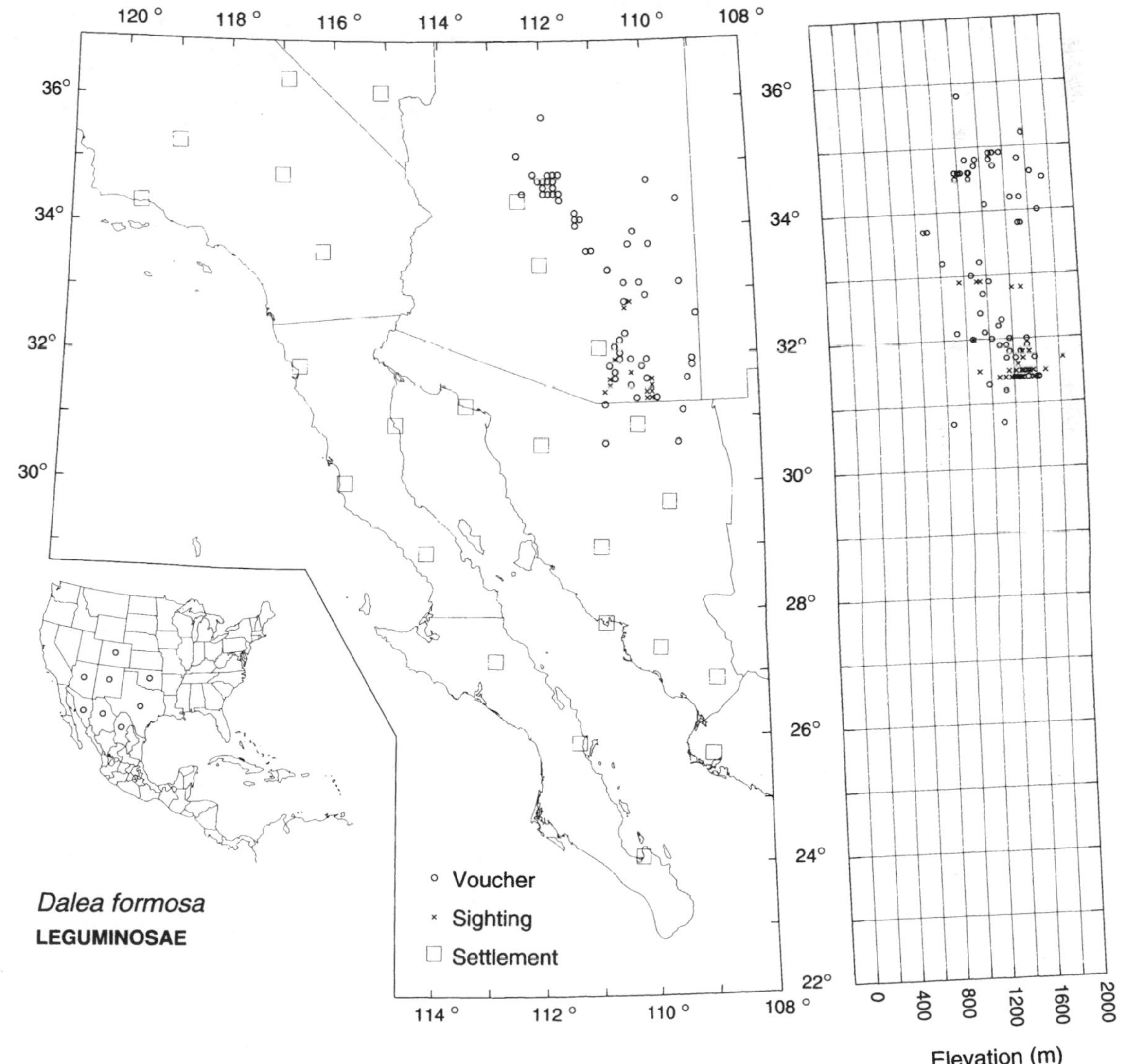

Dalea formosa
LEGUMINOSAE

with 5–7 (occasionally as few as 3 or as many as 9) obovate leaflets 2–5 mm long. The leaflet margins are often folded under. Leaves and stems are silvery from dense, silky pubescence. The magenta to rose-purple flowers, 7–9 mm long, are in dense, terminal spikes about 1 cm long and up to 1.5 cm broad. The obovate pods are 2.5–3 mm long. The haploid chromosome number is 7 (Spellenberg 1973).

This species has been called *Dalea greggii* A. Gray, a name first misapplied to Arizona plants by Asa Gray and Per Axel Rydberg (Gentry 1950) and later by various authors (for example, Kearney and Peebles [1960] and Wiggins [1964]). True *D. greggii* is a sprawling or prostrate shrub with small flower heads that grows in foothill grasslands in northeastern Mexico and adjacent Texas (Gentry 1950).

Dalea formosa is a miniature, intricately branched shrub up to 5 dm high with glabrous leaves and sparsely flowered spikes. *Dalea versicolor,* a small, often sprawling shrub of higher elevations, has sparsely pubescent leaves and noticeably bicolored flowers.

The plants are locally common on rocky hills and canyon slopes in Madrean evergreen woodland, semidesert grassland, and the upper margin of Sonoran desertscrub. Populations occur as local colonies on isolated mountain ranges. The northern limit is the Santa Catalina Mountains, Pima County, Arizona. The southernmost point on our map, vouchered by Howard Scott Gentry, is from 19 km north of Tesopaco, Sonora. Throughout the range, rain mostly falls in summer, but occasional winter storms can bring substantial moisture, especially in the more northerly sites. In unusually cold winters, the stems freeze back half their length or more, so it seems likely that intolerance to cold determines the upper elevational limit (about 1,500 m). The plants survive winter temperatures as low as –9.5°C (Johnson 1993).

Flowers appear as early as February and as late as May, but the main bloom in most years is in March and April. The leaves are drought deciduous (Barneby 1977) and also drop from severe cold. In most years, the flower spikes are on short,

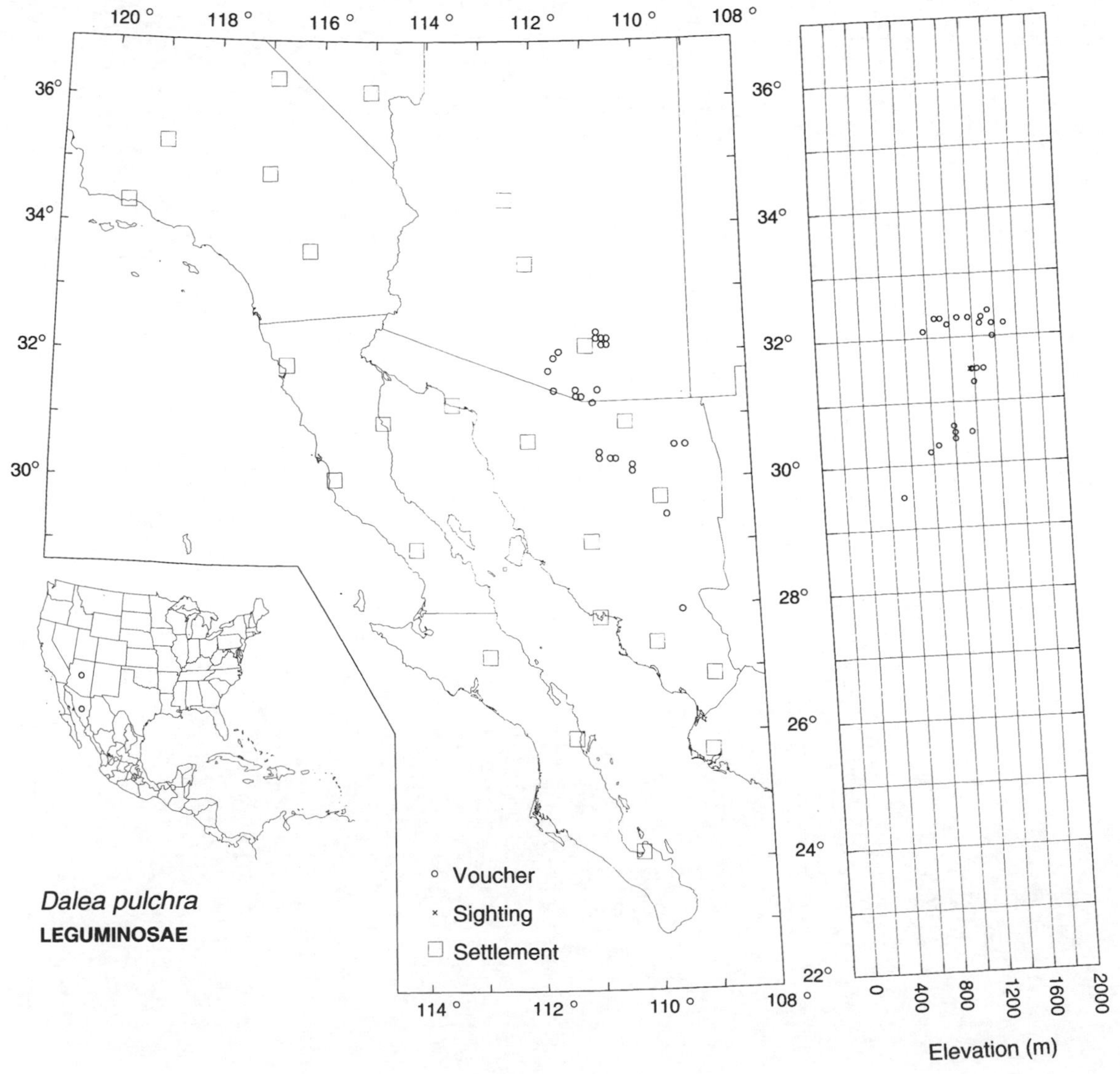

axillary stems from which the primary leaves have fallen. After unusually wet winters, the plants are more luxuriant, with persistent leaves and flowering spikes on leafy, elongated, lateral branchlets (Barneby 1977).

Bees in the genus *Xylocopa* are frequent visitors and are probably the major pollinators.

Dalea pulchra makes an attractive ornamental and has been cultivated in southern Arizona since 1970 (Starr 1988). Best growth is achieved in full sun and a slightly heavy soil. Young plants should be protected from rabbits (Starr 1988).

Dalea purpusii

Brandegee

A rounded shrub 4–8 dm tall, *Dalea purpusii* is rigidly and intricately branched and sparsely armed with stout thorns. The leaves, 8–25 mm long, have 1–7 obovate leaflets thinly pubescent with silky hairs. The bicolored flowers have creamy banners and purplish wings and keels and are in loose spikes 1–3 cm long.

The only other shrubby dalea in its range is *D. bicolor* var. *orcuttiana,* which differs in its more numerous leaflets, denser spikes, and smaller flowers (Barneby 1977).

Dalea purpusii is rare and local on arid hillsides and in bouldery washes on the Baja California peninsula (Barneby 1977). The Sierra de la Giganta collections (25.9°N–26.2°N) are Annetta Carter's. Forrest Shreve, Ira Wiggins, and Howard Scott Gentry documented the Tres Vírgenes population at 27.4°N. The northernmost point represents the type collection by C. A. Purpus from Calmallí. Frost is rare throughout the range. Southern populations experience mostly summer and early autumn rainfall, the northernmost sites about equal amounts of winter and summer rainfall.

Flowers appear between December and April, apparently in response to rains. In the Sierra de la Giganta, most of the pri-

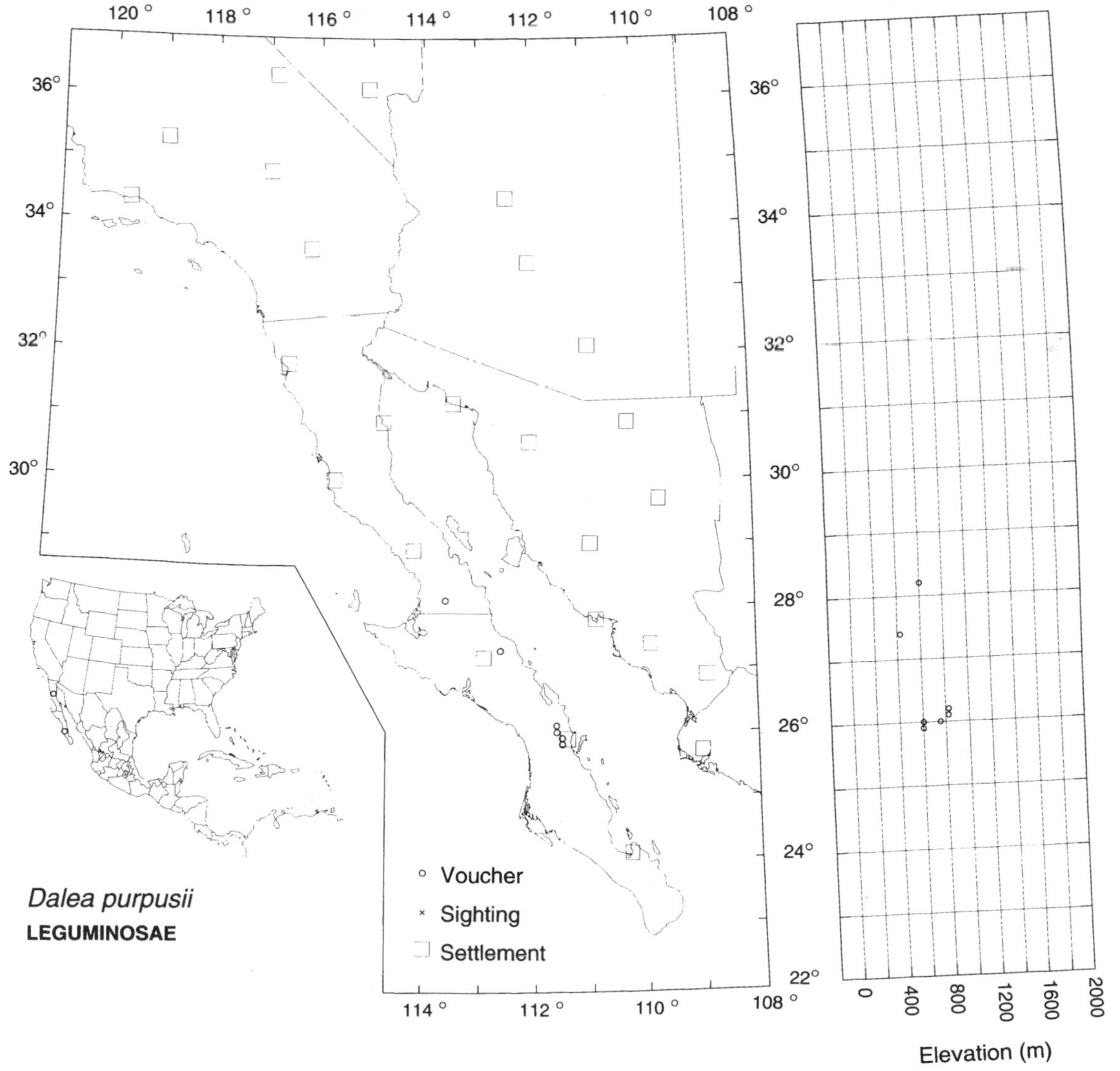

Dalea purpusii
LEGUMINOSAE

mary leaves fall before the inflorescence develops (Barneby 1977).

Dalea versicolor
Zucc.

var. *sessilis*
(A. Gray) Barneby

[=*D. wislizenii* A. Gray var. *sessilis* A. Gray, *D. wislizenii* subsp. *sessilis* (A. Gray) Gentry]

A shrub 5–12 dm high, *Dalea versicolor* var. *sessilis* is often somewhat sprawling in habit. The once-pinnate, sparsely puberulent leaves, 1–3 cm long, have 11–23 oblong leaflets 2–4 mm long. The flowers (1 cm long) are in dense, short spikes and have yellowish or creamy banners and rose-purple or magenta wings and keels. The haploid chromosome number is 7 (Spellenberg 1973).

Barneby (1977) recognized seven varieties in *D. versicolor.* He suggested that the species may have originated in the highlands of the central Sierra Madre, then diverged as it migrated northward and southeastward, giving rise to genetically separate but morphologically similar forms.

Dalea versicolor var. *sessilis* grows on rocky hills and canyon slopes in semidesert grassland and Madrean evergreen woodland. It descends to the upper margin of the Sonoran Desert in canyons and on shaded slopes. The populations are isolated from one another on widely scattered mountain ranges. Frosts are common over much of the known distribution. The biseasonal rainfall comes mostly in summer.

The northernmost point on our map (35.1°N), widely disjunct from the main distribution, is from Dean Peak in the Hualapai Mountains, Mohave County, Arizona (Mary Butterwick's collection). The southernmost populations (27°N–28°N) were vouchered by Howard Scott Gentry in 1935 and 1948 and Paul S. Martin in 1985. It seems likely that the species is

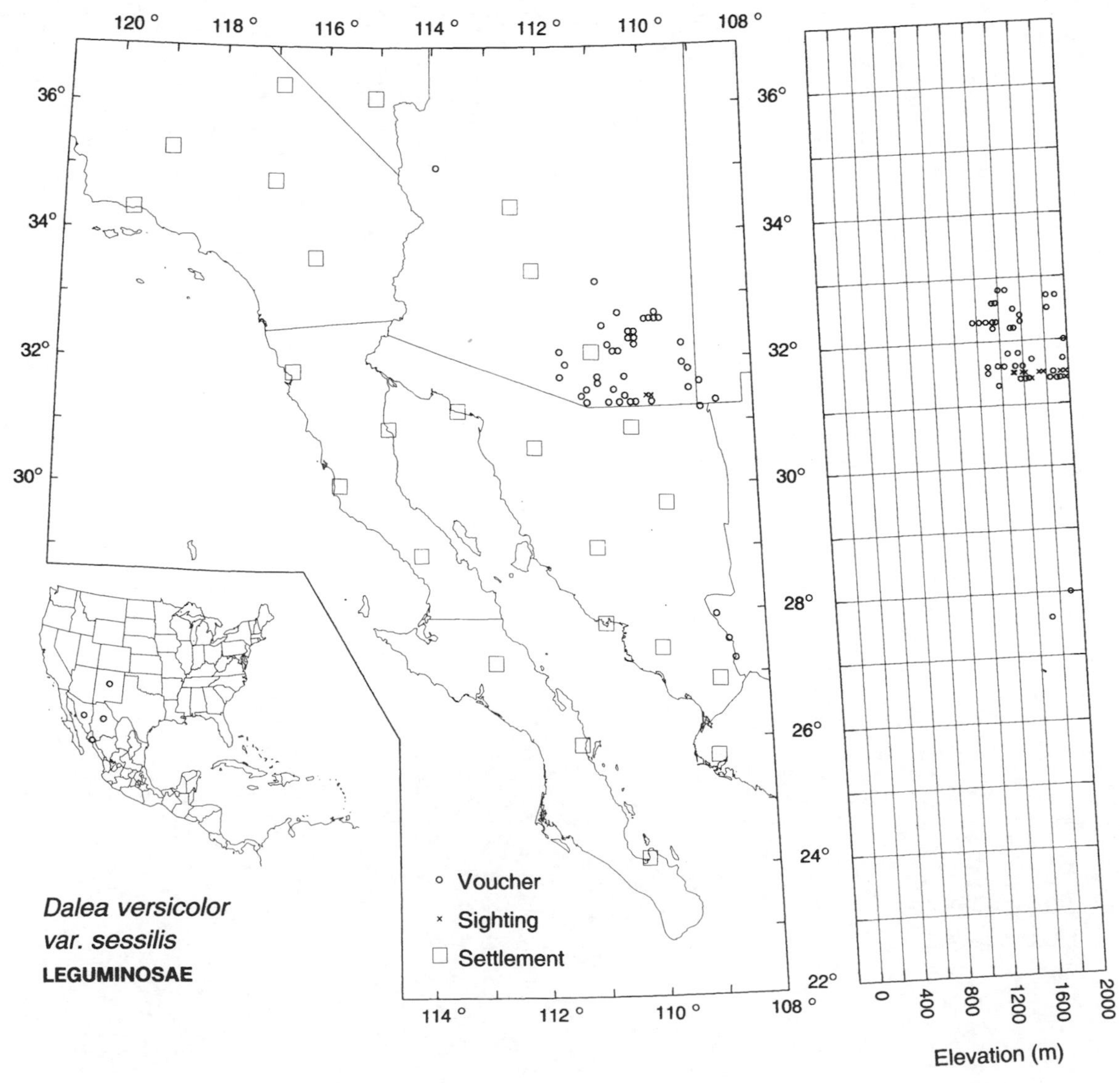

Dalea versicolor var. sessilis
LEGUMINOSAE

more widespread in the Madrean evergreen woodland of Sonora than our map indicates.

The plants flower from late August to November and again in April and May (Barneby 1977). The autumnal form has leafy, virgate stems that branch into panicles of flower spikes. The showier spring form has a succession of subsessile flower heads representing axillary short shoots (Gentry 1950; Barneby 1977).

Dalea versicolor var. *sessilis* shows promise as a groundcover in southern Arizona. Plants grow rapidly from cuttings and do well in full sun or filtered shade (Starr 1988).

Desmanthus covillei
(Britton & Rose) Wiggins ex Turner

[=*D. covillei* (Britton & Rose) Wiggins ex Turner var. *arizonicus* Turner, *D. palmeri* (Britton & Rose) Wiggins ex Turner]

Desmanthus covillei is an unarmed shrub up to 2.5 m tall with many ascending, undivided branches. The twice-compound leaves have 8–17 pairs of leaflets 4–8 mm long on 1–3 (rarely 4) pairs of pinnae. The white or cream-colored flower heads, 0.9 to 2.3 cm long, may have hermaphroditic, male, and sterile flowers. The pods are 5.5–13 cm long.

Desmanthus fruticosus has more pinnae and longer leaflets.

Desmanthus covillei grows throughout much of Sonora south of 30°N and in Sinaloa, where it is a conspicuous roadside weed (Luckow 1989). It is found in the Gulf of California only on Isla Tiburón. In Baja California Sur it is closely limited to the Sierra de la Giganta; in Arizona, to the Tucson and Baboquivari mountains. The range covers parts of the eastern Sonoran Desert where summer rain is more frequent than winter rain, although occasional winter storms also bring moisture. Interactions with *D. fruticosus* may limit the peninsular distribution.

The leaves are drought deciduous. Flowers are present from August to

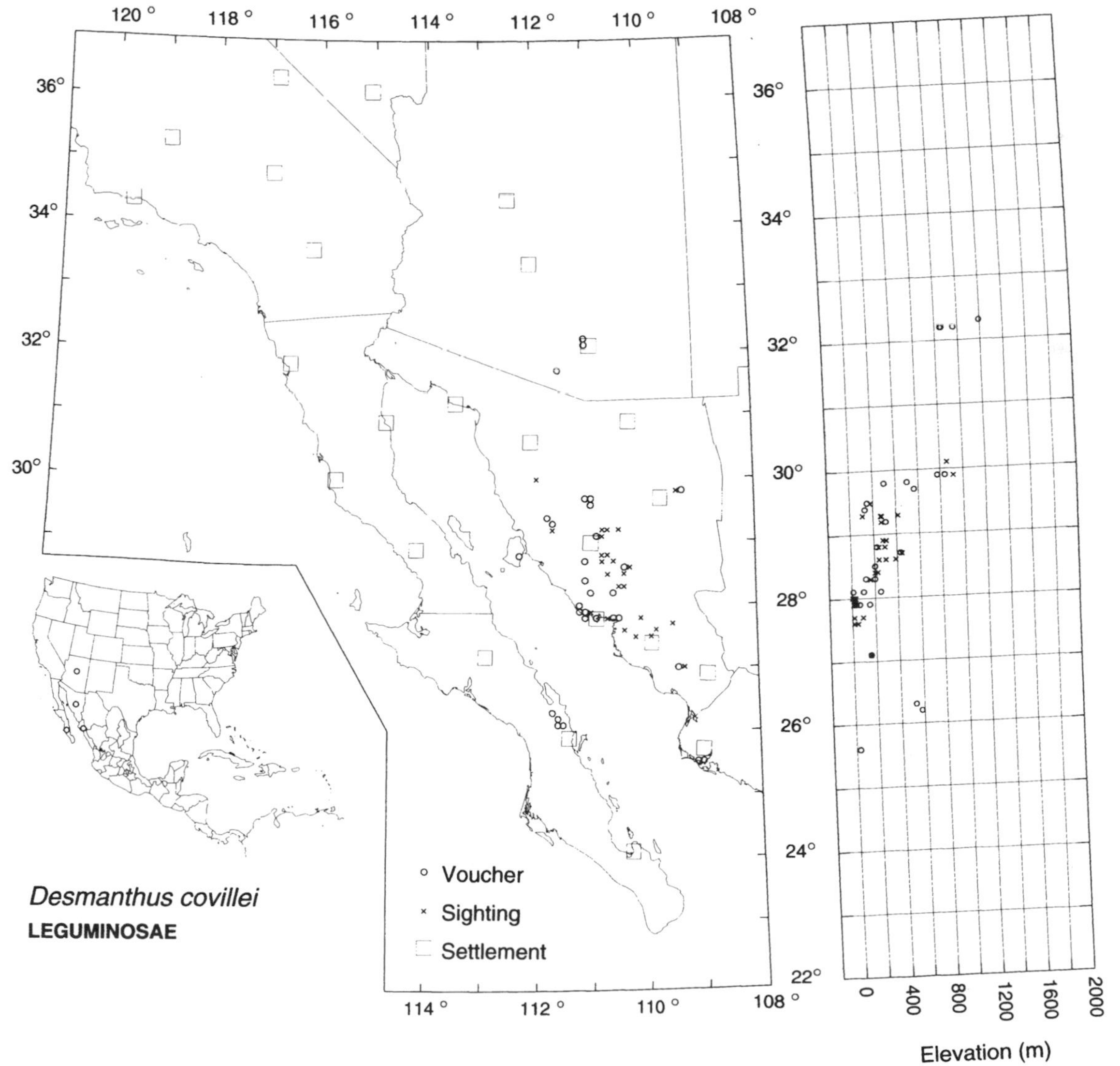

Desmanthus covillei
LEGUMINOSAE

October, fruits from September to December. Bruchid beetles attack the seeds (Luckow and Johnson 1987). The species has been reported to fix nitrogen (Date 1991).

Desmanthus fruticosus

Rose

Desmanthus fruticosus is a small, weak tree up to 4 m tall with a single feebly branched trunk and remarkably flexible stems. The twice-pinnate leaves have 2–9 pairs of pinnae and 10–20 pairs of leaflets 6–12 mm long. The numerous small flowers are in axillary heads or short spikes. A head may contain hermaphroditic, male, and sterile flowers. The fruits are 5–6 mm wide and 5–9 cm long. Seeds are diagonal in the pods.

Desmanthus covillei has smaller leaves with fewer pinnae and leaflets. For additional differences, see Luckow (1989).

Primarily a plant of lowland habitats, *D. fruticosus* grows on the Baja California peninsula from 28.5°N southward to Cabo San Lucas and in Sonora from the vicinity of Kino Bay to Guaymas. It is found on most of the gulf islands (Moran 1983b). The species grows in arid maritime climates with warm summers and peak rainfall during summer and early autumn. Winter storms can be a critical source of moisture near Bahía Magdalena and on the midriff islands.

Flowering is in late summer and fall (August–November) (Wiggins 1964). The drought-deciduous leaves typically last from early summer to early fall, persisting even after the pods dehisce. Bruchid beetles attack the seeds (Luckow and Johnson 1987). The species has been reported to fix nitrogen (Date 1991).

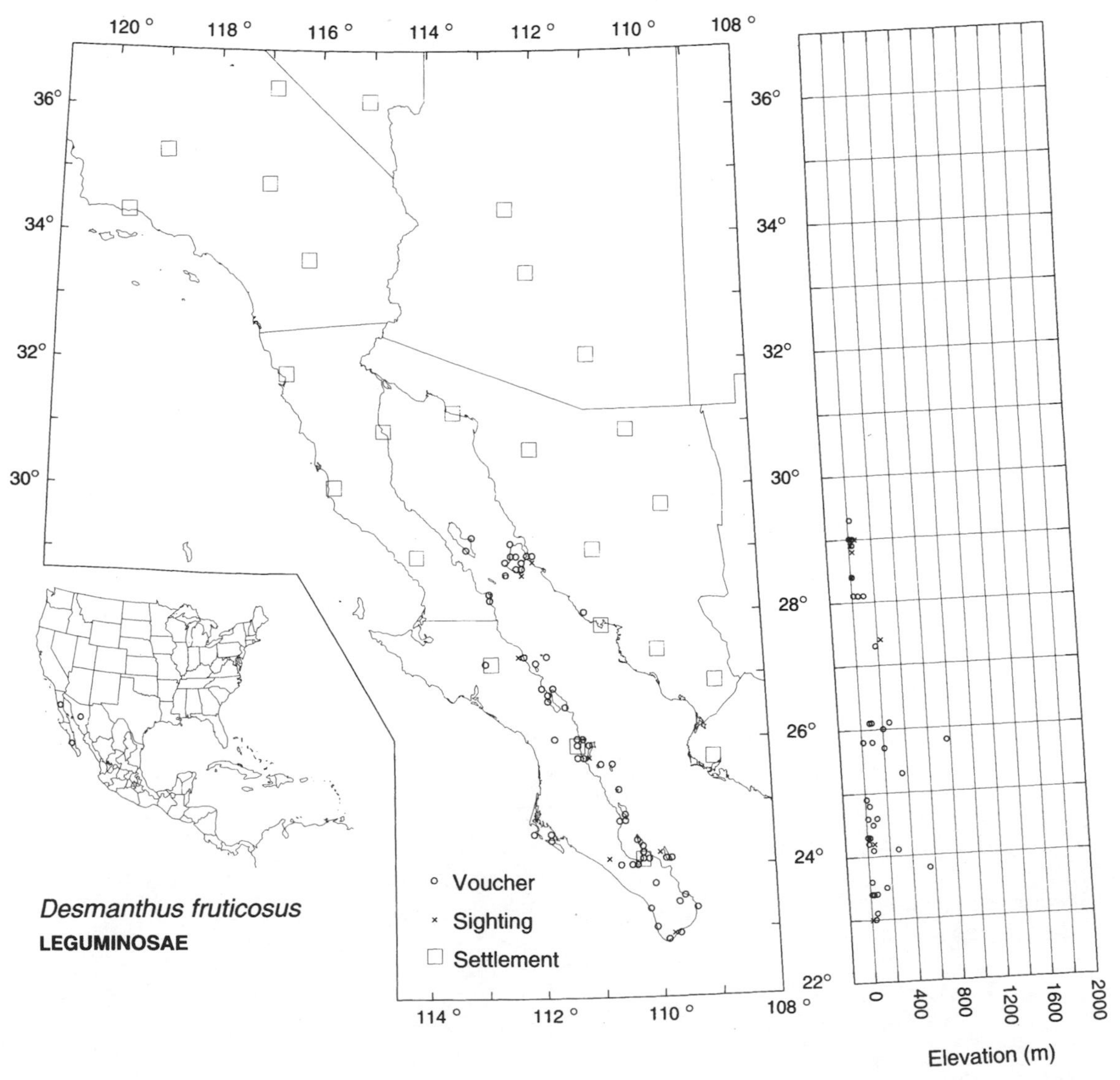

Desmanthus fruticosus
LEGUMINOSAE

Diphysa occidentalis

Rose

Guiloche

This unarmed shrub or small tree grows up to 8 m high and has thin, scaly bark and an open canopy. The odd-pinnate leaves have 4–8 pairs of elliptic leaflets 6–20 mm long plus the terminal leaflet. On the upper leaflet surface, numerous small, clear glands can be seen with 10× magnification. The flowers are 12–15 mm long. The fruits are 4–7 cm long, with an inflated, wrinkled, parchmentlike pericarp.

Diphysa suberosa has deeply furrowed, corky bark.

This species grows from Ures, Sonora, as far south as Guerrero. It often occurs in open stands on slopes but may grow in valleys and along arroyo margins as well. Our northernmost locality at 29.4°N is based upon a collection by Ira L. Wiggins. In Sonora, this species experiences relatively dependable summer rains that follow a hot and dry foresummer.

The plants are leafless in winter and spring. Flowers are seen mainly in June and July. The indehiscent fruits fall soon after ripening.

The pole-size stems are used for house, corral, and tool construction (Gentry 1942).

Diphysa racemosa

Rose

Flor de iguana, guiloche

This shrub or small tree grows up to 5 m high and has densely glandular-viscid foliage. The once-pinnate leaves have 9–17 broadly ovate to oblong leaflets 3–7 mm long. The flowers (1 cm long) are in racemes 10–20 cm long. The papery pods, 6–8 cm long and 7–8 mm wide, are turgid and somewhat constricted between the seeds.

A plant of Sinaloan thornscrub and Sinaloan deciduous forest, *D. racemosa* is common in abandoned fields (Gentry

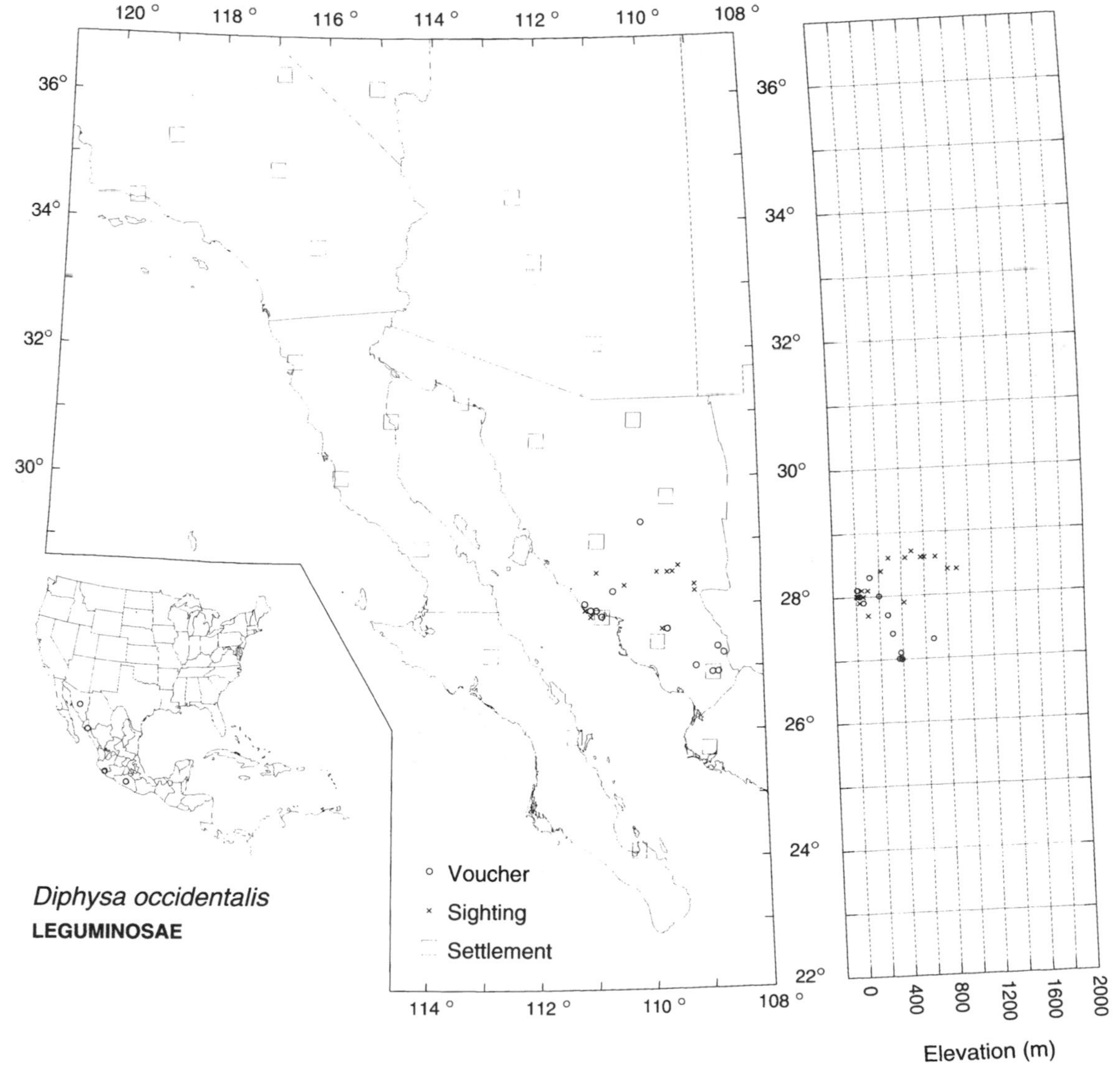

Diphysa occidentalis
LEGUMINOSAE

1942). It ranges from Sonora to Chiapas along the Pacific slope of Mexico and apparently requires more rain than *D. occidentalis.*

The flowers appear from April to July (Wiggins 1964).

Diphysa suberosa

S. Watson

Corcho, colchol, palo santo

This small, slender tree or shrub has soft, pithy, corklike bark that splits deeply into irregular vertical ridges. The once-pinnate leaves comprise 4–8 pairs of leaflets (plus 1 terminal leaflet) 6–20 mm long. The flowers are 12–15 mm long. The fruit (7 cm long) has an inflated parchmentlike pericarp.

Diphysa suberosa has more conspicuous leaf venation than *D. occidentalis.* The characteristic bark is also distinctive.

Diphysa suberosa grows mainly in more mesic habitats east of the Sonoran Desert, from the area east of Ures, Sonora, south to Jalisco. Scattered plants enter the desert along relatively moist drainageways.

Flowers appear from April to July (Wiggins 1964).

Ditaxis brandegei

(Millsp.) Rose & Standley

[=*Argythamnia brandegei* Millsp.]

Ditaxis brandegei is an openly branched shrub up to 2.5 m tall with slightly glaucous, purplish stems and leaves. The ser-

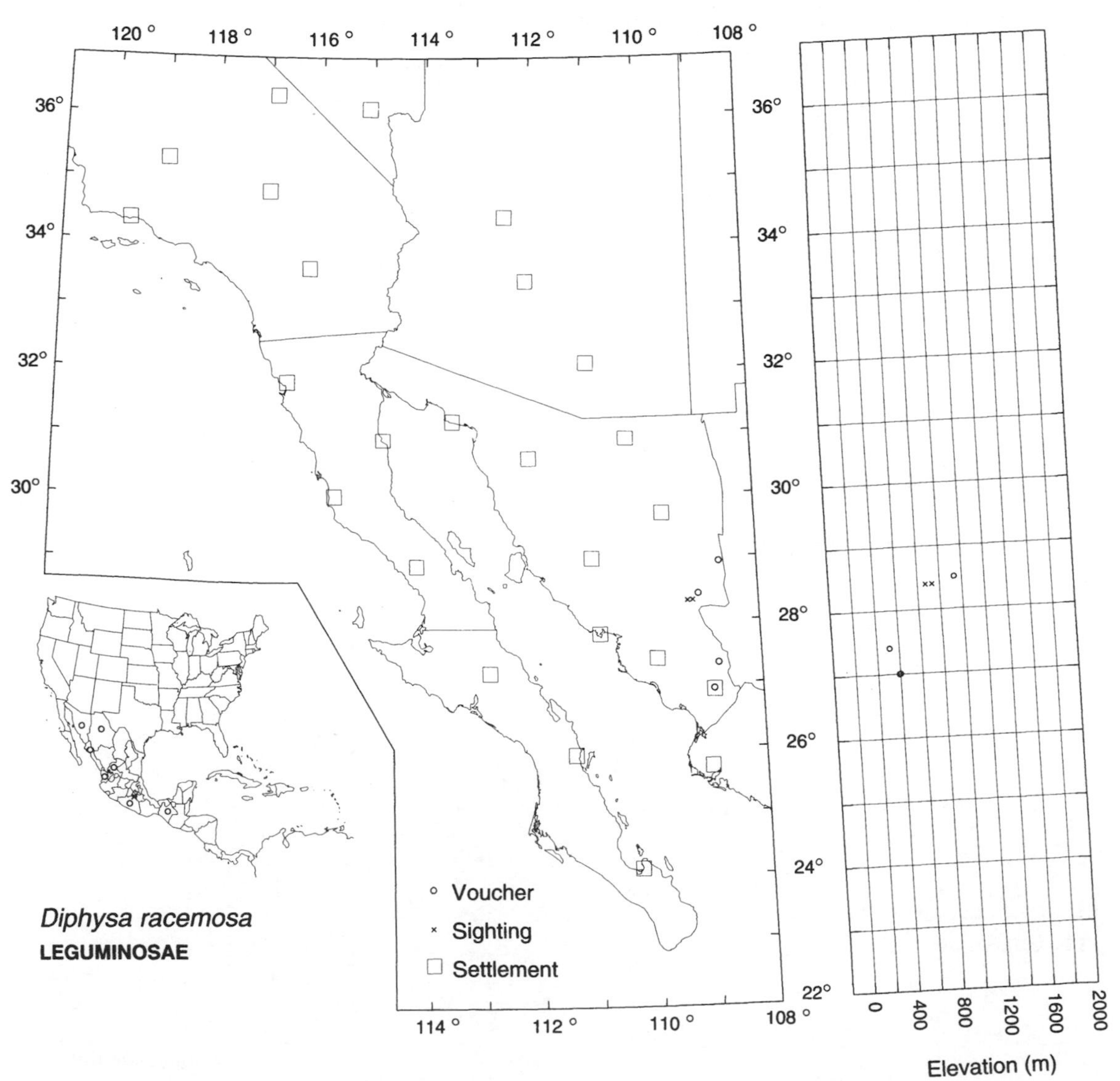

Diphysa racemosa
LEGUMINOSAE

rulate, lanceolate leaves, crowded at the tips of the bare branches, are 0.5–2 cm wide and 2–8 cm long and may be glabrous (in var. *brandegei*) or sparsely pubescent with coarse, forked hairs (in var. *intonsa* I. M. Johnston). The plants are monoecious, with small flowers in axillary racemes, staminate flowers above the pistillate. The 3-angled capsules are 8–10 mm broad and are glabrous in var. *brandegei*, coarsely pubescent in var. *intonsa*.

Ditaxis lanceolata (Benth.) Pax & K. Hoffm., a smaller plant with capsules 3–5 mm wide, has densely pubescent foliage that is neither glaucous nor purplish.

Croizat (1945) included *Ditaxis* in *Argythamnia*. We follow Wiggins (1964) and Munz (1974) in keeping it as a separate genus.

The plants grow on rocky slopes, along washes, and on alkaline flats. The northern populations are all var. *intonsa*, the southern ones predominantly var. *brandegei*. Throughout the range, summers are hot and winters are mild. The southern populations receive much of their rain during summer. Occasional winter storms supply additional moisture. Climates of the northern sites, decidedly more arid, exhibit erratic rainfall about equally divided between late summer and winter. Restriction to elevations below 510 m suggests sensitivity to frost.

The plants bloom from October to April.

Dodonaea angustifolia L. f.

[=*D. viscosa* Jacq. var. *angustifolia* (L.f.) Benth.]

Hopbush, switch-sorrel, airía, pirimu, granadina, guayabillo, jarilla, hierba de la cucaracha, varal, munditos, tarachico, chapuliztle, cuerno de cabra

Dodonaea angustifolia is a shrub 1–3 m high with sticky, linear-oblanceolate leaves 5–9 cm long and 4–8 mm broad. The resin on the foliage is secreted by multicellular, hemispherical, peltate glands (West 1984). The inconspicuous

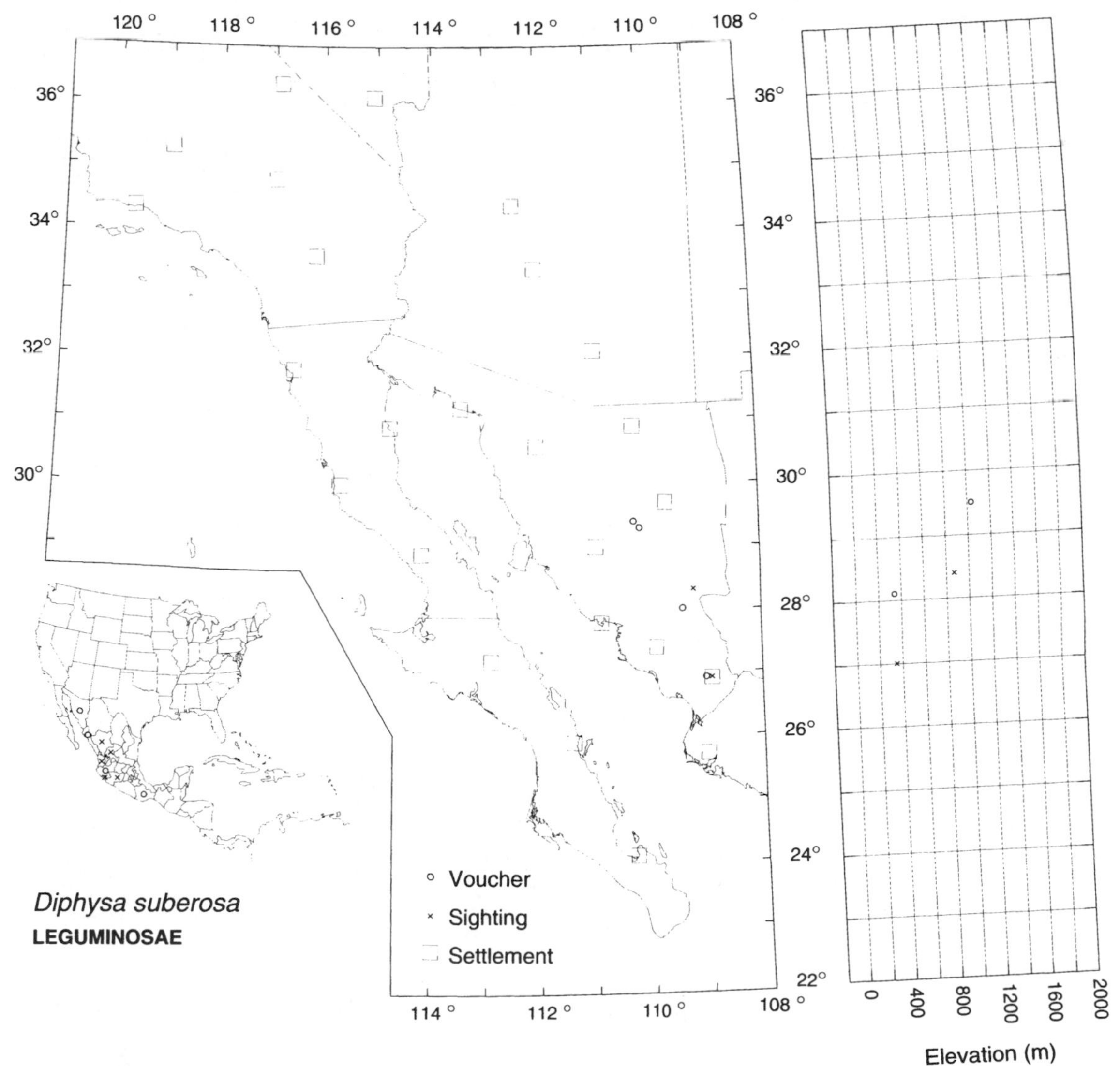

Diphysa suberosa
LEGUMINOSAE

flowers lack petals. Most flowers are unisexual, although female flowers may have staminodes (Leenhouts 1983). Most plants are monoecious, some dioecious. Fruits are 3- to 4-winged samaras 11–13 mm long and 15–20 mm in diameter. The diploid number of chromosomes is 28 (Ferrucci 1981).

Sapium biloculare has serrate leaves and milky sap. The narrow-leaved *Salix* species in the Sonoran Desert region have neither viscid foliage nor winged fruits.

This taxon has been treated in regional manuals as *D. viscosa* var. *angustifolia* (for example, Kearney and Peebles 1960; Wiggins 1964; Benson and Darrow 1981). Leenhouts (1983) considered this species to be *D. angustifolia* on the basis of its breeding system. It is an inland species found widely throughout the world in the tropics and subtropics (Leenhouts 1983).

In the area mapped in this atlas, *D. angustifolia* is locally abundant on rocky or gravelly slopes, often on acidic soils derived from hydrothermally altered rock. It tends to increase on overgrazed ranges (Kearney and Peebles 1960; Harrington 1991) and colonizes old burns at the upper margin of the Sonoran Desert. Since germinability of *D. angustifolia* in Australia is not enhanced by fire (Hodgkinson and Oxley 1990), it seems possible that invasion of old burns represents a response to disturbance.

Absence from western Arizona and southeastern California suggests intolerance of summer aridity. (In Baja California, Pacific maritime air ameliorates the arid summers.) Plants suffer frost damage at –4.4°C (Kinnison 1979), and cold temperatures no doubt keep them from spreading to higher elevations as well as farther north and east.

Leaves are evergreen. Flowering lasts from February through October (Wiggins 1964). Copious pollen production (Reddi et al. 1980) and inconspicuous, apetalous flowers suggest wind pollination. The winged fruits are also dispersed by wind.

Under natural conditions, *Dodonaea* species show some features of Crassulacean acid metabolism (Madhusudana-Rao et al. 1979). The same individuals can also

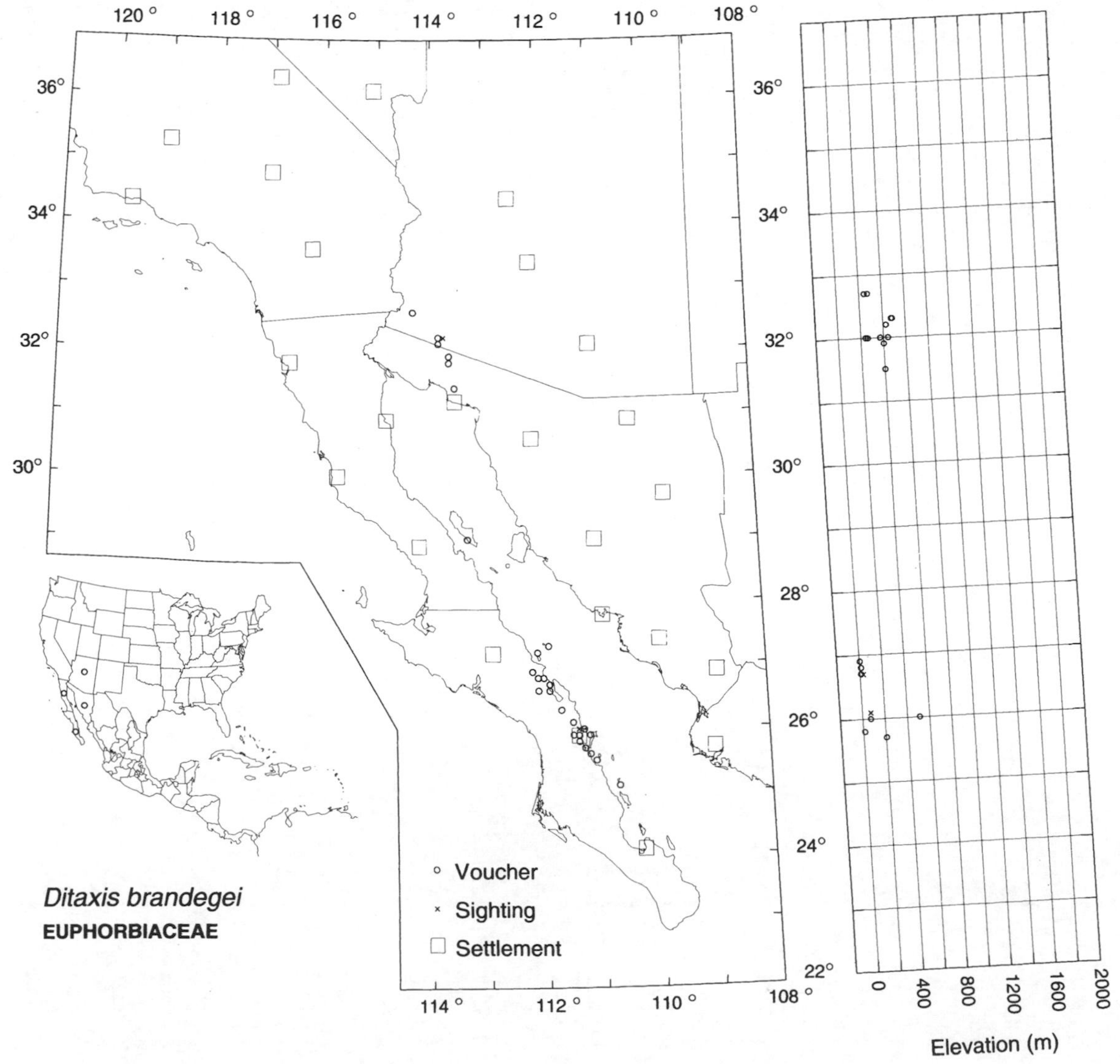

Ditaxis brandegei
EUPHORBIACEAE

fix carbon dioxide in the light, which suggests that they switch from one photosynthetic mode to another according to environmental conditions. Although laboratory-grown plants tolerate both water and salinity stress (Salunkhe and Karadge 1989), wild plants in the Sonoran Desert show no particular affinity for saline habitats. Seasonal variation in transpiration can be correlated with changes in the chemical composition of epicuticular wax, accounting in part for the adaptability of *D. angustifolia* to arid and semiarid climates (Rao and Reddy 1980).

The plants are often grown as ornamentals. Several cultivars are commercially available (Duffield and Jones 1981; Mielke 1986). Although tolerant of alkaline, rocky, or heavy soils, the plants do best in improved garden soil and profit by regular, widely spaced irrigation (Duffield and Jones 1981).

Dodonaea angustifolia is used in revegetation of copper mine wastes (Norem et al. 1982). The foliage contains biologically active compounds such as saponins and, in tropical and subtropical regions, has traditionally been used to treat inflammation, swellings, rheumatism, and pain (Wagner et al. 1987; Mata et al. 1991). Seri Indians used an infusion of leafy branches as a hot compress on sore or aching legs (Felger and Moser 1985).

Echinocactus horizonthalonius Lemaire

Turk's head, bisnaga meloncillo, manca caballo, manca mula

This single-stemmed cactus grows 40–50 cm tall and 12.5–15 cm wide and has 7–13 (commonly 8) ribs and rather dense, interlaced spine clusters. Each cluster comprises 3 central spines, the shortest strongly recurved, and 5 radial spines, all more or less flattened at the base and 2–3 cm long. In var. *nicholii* L. Benson, the Sonoran Desert taxon, they are nearly black or dark gray with an underlayer of red.

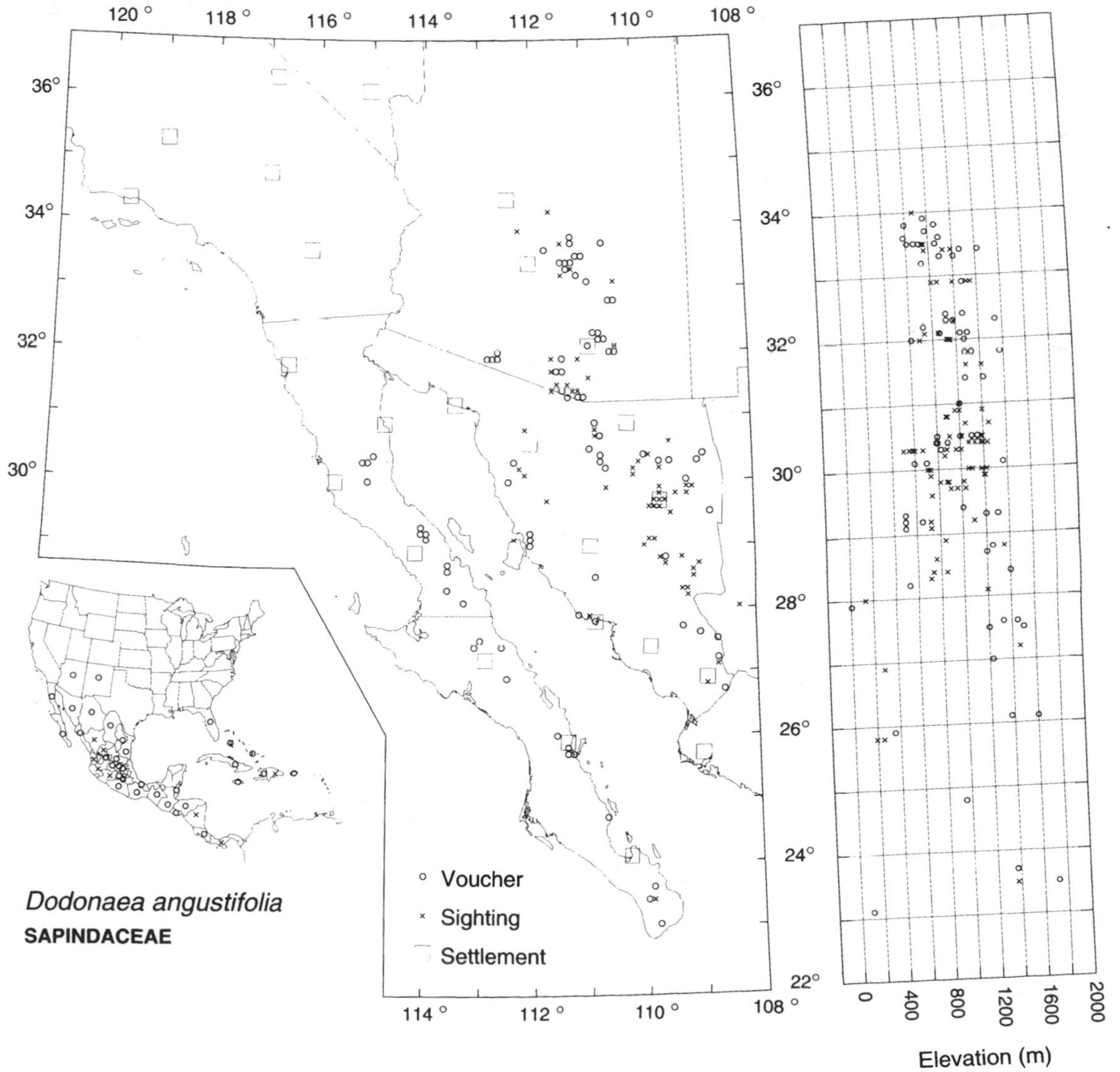

Dodonaea angustifolia
SAPINDACEAE

The flowers, 5–6.5 cm in diameter, have pink, oblanceolate petals. At maturity, the fruits (25 mm long) are dry and densely white-woolly. The haploid chromosome number is 11 (Weedin and Powell 1978a).

Echinocactus polycephalus has multiple rather than single stems. *Ferocactus* species are taller, more massive cacti with glabrous fruits and red, yellow, or orange flowers.

Echinocactus horizonthalonius is apparently restricted to limestone or calcareous soils (Correll and Johnston 1970; Benson 1982). Variety *horizonthalonius* is rather widely distributed in the Chihuahuan Desert. The prevalent taxon in the Sonoran Desert, var. *nicholii,* occurs spottily. Sonoran Desert populations experience biseasonal rainfall with more in summer than in winter.

During the Pleistocene, *E. horizonthalonius* may have migrated from the Chihuahuan Desert region into the Sonoran Desert region during a period of relatively high temperatures and low rainfall (Yatskievych and Fischer 1984). If so, return of pluvial conditions would have fragmented its distribution into disjunct populations, enabling the Sonoran Desert plants to diverge from the typical form (Yatskievych and Fischer 1984).

Packrat-midden fossils from the Waterman Mountains, Arizona, show that 22,400 and 11,500 years B.P., *E. horizonthalonius* grew with *Pinus monophylla, Vauquelinia californica,* and *Yucca brevifolia* (Van Devender 1990b). The cactus is still there, but now it grows with *Cercidium microphyllum* and *Carnegiea gigantea.* In view of this adaptability to substantial climatic and vegetational change, it is surprising that the species is not more widespread. The scarcity of limestone mountains in the eastern Sonoran Desert seems the most likely limiting factor.

The flowers, which appear in May and June, are probably pollinated largely by bees.

The small populations have probably suffered much attrition from cactus collectors. In parts of the range, climatic change or grazing has produced a profound decline in reproductive success (Reid et al. 1983). One small population near Marana,

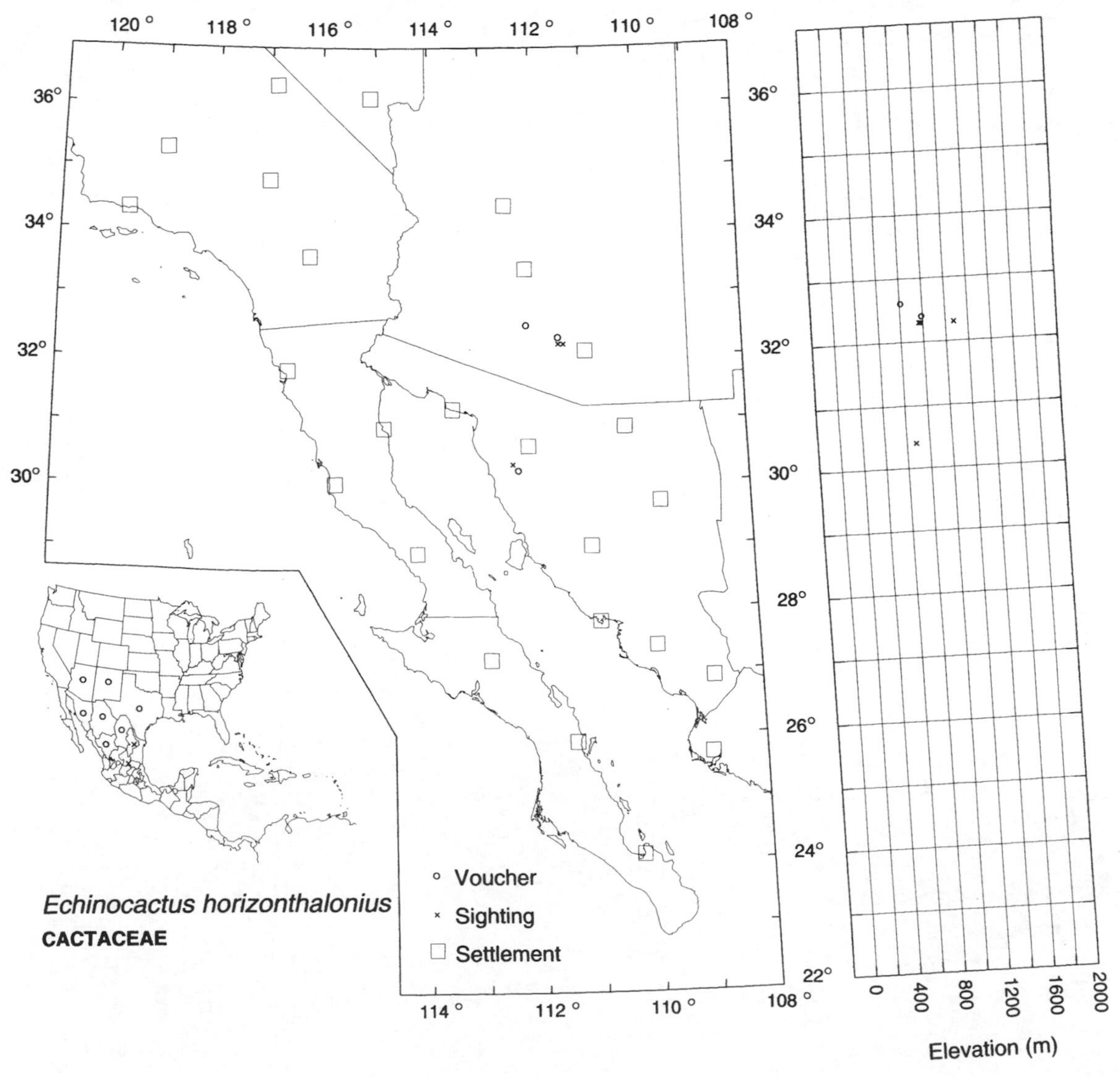

Arizona, has been wiped out by limestone quarrying. Benson (1982) considered var. *nicholii* to be in immediate peril of extinction.

Echinocactus polycephalus

Engelm. & Bigel.

Cottontop cactus, many-headed barrel cactus

Echinocactus polycephalus (figure 35) is a clump-forming cactus with solitary or multiple (up to 30) spheroidal or cylindrical stems from the base. The clumps are usually about 0.6 m tall and up to 1.2 m across. Individual stems are 10–20 cm in diameter with 13–21 densely spiny ribs. Each spine cluster comprises four spreading central spines 6–7.5 cm long and six to eight radial spines about half that long. Flattened and felty canescent when young, the spines are red or pink after the felt peels off. The flowers, produced at the stem apex, are yellow and about 5 cm across. Fruits are dry and densely matted with white, woolly hairs. The haploid chromosome number is 11 (Pinkava et al. 1977).

This species can be distinguished from *E. horizonthalonius* by its denser, stouter spines, multiple stems, and clump-forming habit. These same characteristics also separate it from the *Ferocactus* species in the area mapped in this atlas, as do the woolly fruits.

The prevalent taxon in the Sonoran Desert region is var. *polycephalus.* It is replaced to the northeast by var. *xeranthemoides* Coult., which is almost completely restricted to the Colorado Plateau from Lake Mead and the Vermillion Cliffs eastward. The Grand Canyon appears to have been a distributional corridor for var. *xeranthemoides.* The two varieties intergrade in the western end of the Grand Canyon but are otherwise distinct geographically (Chamberland 1991). The most closely related species, *E. parryi* Engelm., is restricted to a small area in northern Chihuahua (Chamberland 1991).

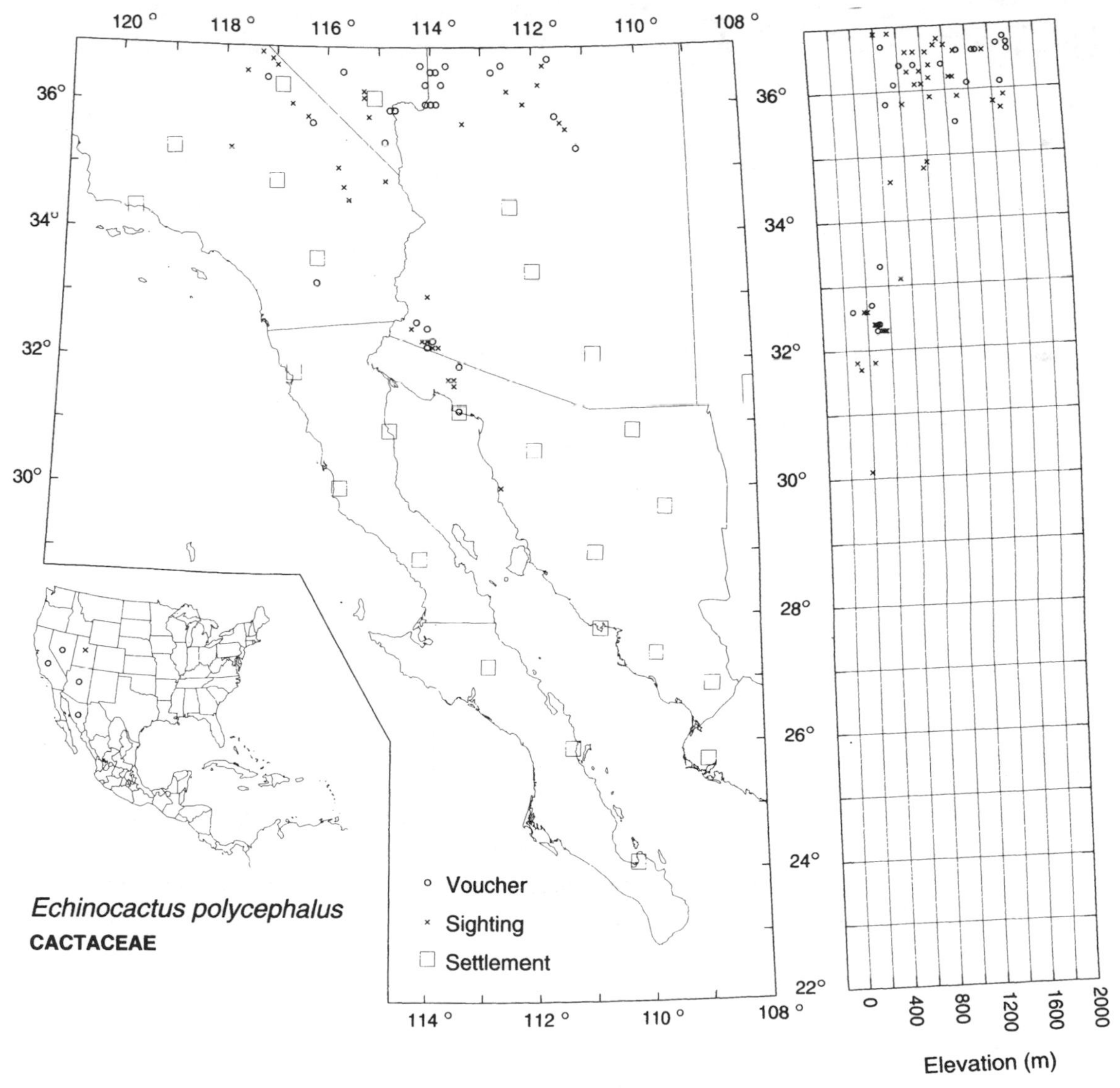

Echinocactus polycephalus
CACTACEAE

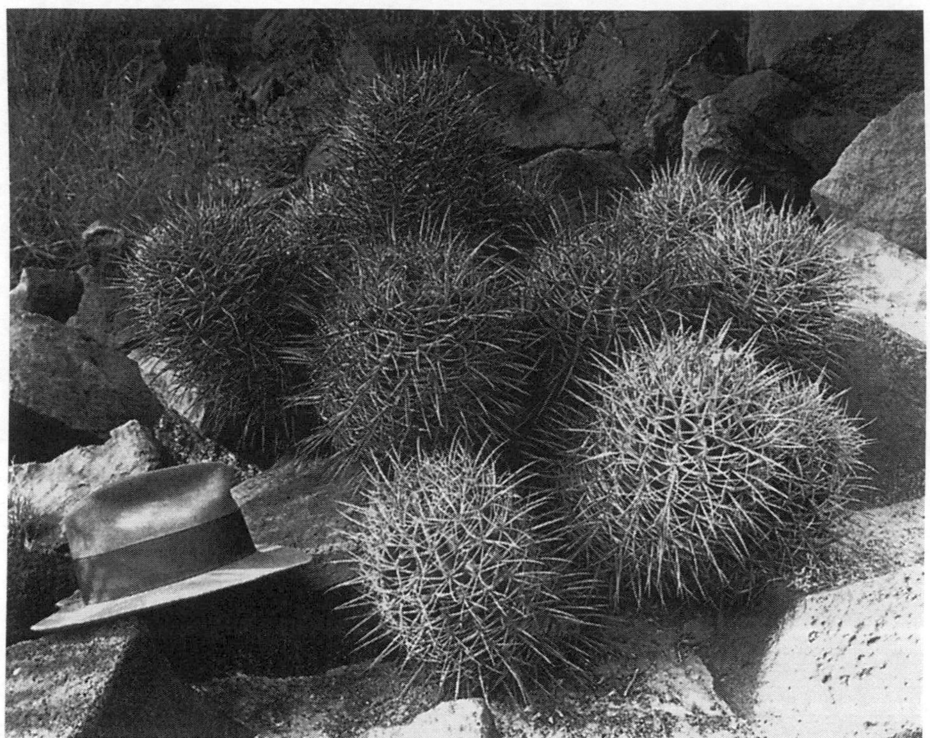

Figure 35. Echinocactus polycephalus *among lava rocks at Puerto Peñasco, Sonora. (Photograph by L. M. Huey, courtesy of University of Arizona Library Special Collections Department.)*

The plants are locally common on rocky slopes and gravelly flats and occasionally grow in clay soil of valley bottoms. Variety *xeranthemoides* often grows on south-facing canyon ledges (Benson 1982), where it may find protection from low temperatures. The outlier of var. *polycephalus* along the Sonoran coast at 30.1°N is a sighting by Raymond M. Turner.

In the Sonoran Desert region, *E. polycephalus* receives biseasonal rainfall over most of its range. Because of the prevailing high summer temperatures, soil moisture should be more available in winter than summer, but in the northern part of the range, where frost occurs regularly, cold undoubtedly limits the usefulness of winter soil moisture. The expanded elevational distribution in the northern, colder part of the range is unusual. Our maps suggest that interactions between *E. polycephalus* and *Ferocactus cylindraceus,* and perhaps also between *E. polycephalus* and *F. emoryi,* may have influenced the southern boundary.

Unpublished midden records cited by Chamberland (1991) show that both subspecies have occurred within their present ranges for at least 30,000 years. In the eastern Mojave Desert, *E. polycephalus* has been found in packrat middens dated 10,450 and 14,800 years B.P. (Spaulding 1983), and it is common in late Pleistocene middens from Rampart Cave in the lower Grand Canyon (Phillips and Van Devender 1974).

Both varieties bloom in July, and seeds mature about a month later (Chamberland 1991). The flowers are most likely bee pollinated. The fruit dries slowly and may persist on the plant for several months. Fruit detachment leaves a hole through which the small seeds of var. *xeranthemoides* pass easily. The larger seeds of var. *polycephalus* tend to remain wedged together until the fruit is torn open. Birds and packrats disperse the seeds (Chamberland 1991). Restriction to arid climates suggests that the species has few opportunities for growth and seedling establishment. Although individuals can live for 100 years, most are probably not so long lived (Robert Webb, personal communication 1993).

Encelia farinosa
A. Gray

Brittlebush, incienso, hierba del vaso, hierba de las animas, hierba cenisa

A rounded, much-branched shrub up to 1.5 m tall, *Encelia farinosa* (figure 36) has whitish-pubescent stems that become glabrate in age. When broken, they exude golden resin. The alternate, lanceolate to ovate leaves, 2–5 cm wide and 3–10 cm long, are generally entire (occasionally repand or undulate) and subglabrate to densely white-farinose. They are crowded toward the branch tips on older stems. The flower heads (2 cm wide) are in loose panicles on long, naked branchlets. In var. *farinosa,* both disk and ray flowers are yellow. Varieties *radians* Brandegee ex S. F. Blake and *phenicodonta* (S. F. Blake) I. M. Johnston have purplish disks. Leaves of var. *radians* are somewhat larger and less pubescent than in the other varieties. The flattened achenes are silky-villous on the margins. Haploid chromosome counts of 17 (Turner and Flyr 1966) and 18 (Solbrig et al. 1972) have been reported, leading Solbrig and coworkers (1972) to suggest that there may be more than one chromosome number in the genus.

Encelia virginensis A. Nels., *E. palmeri* Vasey & Rose, and *E. halimifolia* Cav. all have one to several heads on pubescent rather than glabrous peduncles. Disk flowers of *E. virginensis* are yellow, those of *E. palmeri* and *E. halimifolia* purplish. *Encelia californica* Nutt. has green leaves. Leaf bases of *E. palmeri* are truncate to cordate, those of *E. farinosa* typically obtuse or acute. On the Baja California peninsula, *E. halimifolia* and *E. palmeri* typically grow on dunes and sand sheets, *E. farinosa* on stable substrates. In *Viguiera* species, the lower leaves, at least, are opposite.

Wisdom and Rodriguez (1982), based on sesquiterpene lactone and chromene chemistry, suggested that there may be two chemical races in *E. farinosa* in California and Arizona.

In the northern part of its range, *E. farinosa* is often a dominant on rocky slopes; to the south it frequently grows in pure stands on the coarse loam or sandy soils of level plains (Shreve 1964). In the western Sonoran Desert, the plants colonize old burns (Brown and Minnich 1986). The Isla Patos population at 29.4°N, 112.5°W, vouchered by I. M. Johnston in 1921, is gone (Felger 1966), apparently eliminated by guano harvesters (Gentry 1949a).

Fossil midden assemblages from the Puerto Blanco Mountains of southwestern Arizona show that *E. farinosa* was locally dominant on rocky slopes by 10,540 years B.P. (Van Devender 1987). Middens from the Tinajas Altas Mountains, Arizona, record *E. farinosa* with *Juniperus californica* and *Pinus monophylla* at 43,200 years B.P., 15,700 years B.P., and 11,000 years B.P. (Van Devender 1990b). The species was present in the southern Mojave Desert by 9,500 years B.P. (Spaulding 1990) and in the Grand Canyon since at least 1,350 years B.P. (Cole 1990). The Grand Canyon apparently served as an eastward migrational corridor.

Across its range, *E. farinosa* grows under summer, winter, or bimodal rainfall regimes, always with a long dry season, usually in late spring and early summer (Cunningham and Strain 1969). Most populations occur below 1,300 m. The species is limited upslope and to the north by low temperatures. On Tumamoc Hill, Tucson, Arizona, the leaves and stem tips suffer frost damage at –2.5°C.

Variety *farinosa* occurs throughout the range of the species. Variety ***radians*** is restricted to the Cape Region, Baja California Sur, and var. ***phenicodonta*** (originally described by Blake [1913] as a form) is most abundant in the central peninsula, the lower Colorado River Valley, and coastal Sonora. We have mapped var. ***phenicodonta*** separately. Kyhos (1971) suggested that effective moisture is a crucial

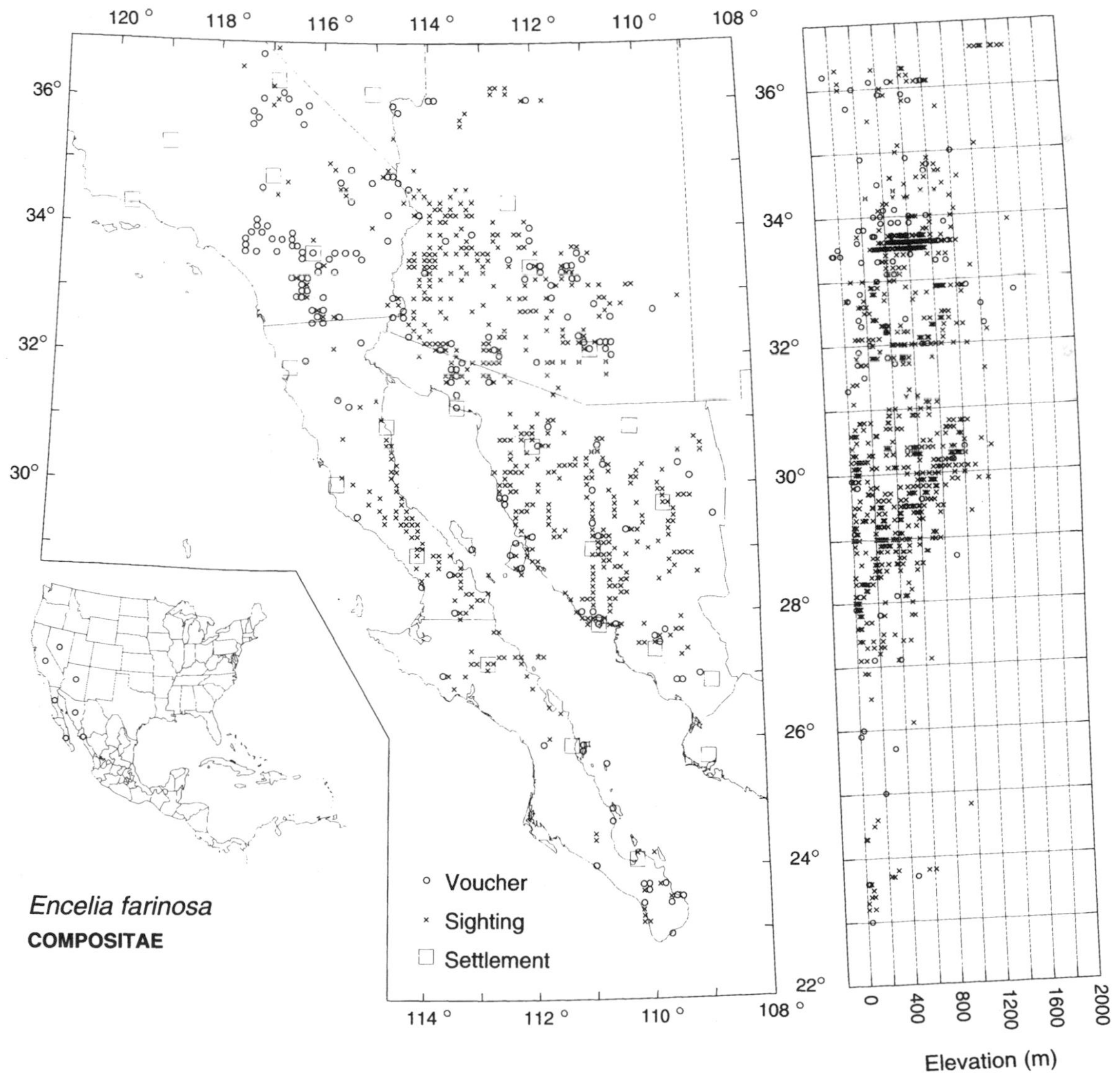

factor in the distribution of var. *farinosa* and var. *phenicodonta.* Because most populations of var. *phenicodonta* grow at lower elevations than those of var. *farinosa,* it seems likely that differing sensitivity to cold also controls their topographic distribution, with the more frost-tolerant var. *farinosa* occurring farther inland as well as at higher elevations and latitudes. The two varieties sometimes grow in mixed populations (Kyhos 1971).

Spring (February–May) is the main blooming season, but given enough rain, flowers may appear from October through January (rarely as early as August). Flowering requires a cool-season rain trigger of at least 20 mm. Thereafter, degree-days above the base temperature (10°C) must accumulate to about 415 (Bowers and Dimmitt 1994). Flowers are sparse or nonexistent in dry years. The plants are obligate outcrossers (Simpson 1977). Varieties *farinosa* and *phenicodonta* interbreed where sympatric, and seeds from either variety will produce both yellow- and purple-disked plants (Kyhos 1971).

In the vicinity of Tucson, Arizona, the major flower visitors during one study were butterflies and moths (Simpson 1977). In another study in the lower Colorado River Valley, a malachid beetle, *Tanaops abdominalis* Le Conte, was the most important pollinator (Kyhos 1971). *Tanaops* displayed no preference for either var. *phenicodonta* or var. *farinosa.* Variety *phenicodonta* seems to mimic other *Encelia* species with purplish disks (for example, *E. palmeri, E. halimifolia,* and *E. ventorum* Brandegee). Convergence on a bold, readily recognized floral pattern may facilitate pollination for mimics. If this is the case in *Encelia,* the models should be rich in pollen or nectar rewards, the mimic poor.

New leaves appear after winter rains when soil water potential is high (Cunningham and Strain 1969). These nearly glabrous, blue-green leaves are up to 50 cm^2 in area (Shreve 1923, 1924; Cunningham and Strain 1969). As the soil dries out, most leaves are lost, generally at a leaf water potential of –3.6 MPa (Walter and Stadelmann 1974). The few that remain (typically retained throughout the dry season)

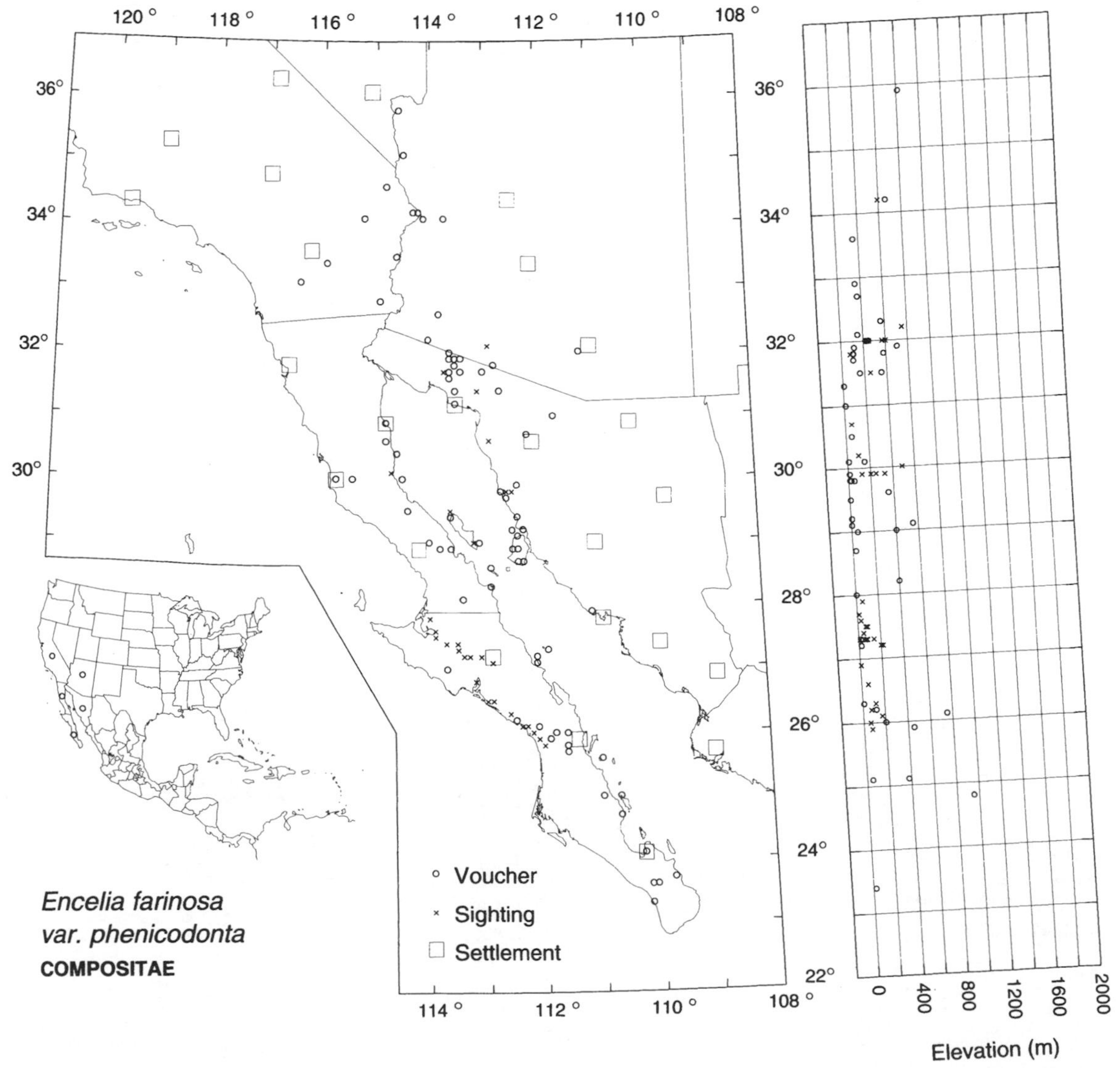

are smaller (a minimum of 0.5 cm²) and so densely pubescent they appear white or gray (Shreve 1923, 1924; Cunningham and Strain 1969). The gray leaves are thicker than the blue-green ones, have a more compact mesophyll and less intercellular space (Cunningham and Strain 1969), and have dry rather than turgid hairs (Ehleringer and Mooney 1978). New leaves are not produced again until the apical meristem undergoes drought-induced dormancy at a leaf water potential of about –4.0 MPa (Ehleringer 1982). Thus, pubescence development is irreversible during any one growing season (Ehleringer 1982). Under extreme drought, when leaf water potentials fall to –4.05 MPa (Walter and Stadelmann 1974), all leaves are lost. The terminal buds and leaf initials can withstand osmotic potentials as low as –5.0 MPa (Walter and Stadelmann 1974). Plants remain dormant until adequate rainfall renews leaf production (Shreve 1924; Odening et al. 1974). Plants of ridge and wash habitats manage water uptake differently. Those on ridges use water in a more conservative manner, thus retaining their leaves longer (Monson et al. 1992).

The adaptation of *E. farinosa* to arid environments is primarily morphological rather than physiological (Ehleringer and Bjorkman 1982). Unlike many desert plants, for example, it cannot adjust the optimum temperature for photosynthesis as environmental conditions change (Ehleringer and Bjorkman 1982; cf. Nobel et al. 1978), nor can it maintain turgor at low leaf water potentials (Nilsen et al. 1984). Instead, photosynthesis and transpiration fluctuate with changes in leaf morphology. When the air-filled hairs of dry-season leaves are densest, they may reflect up to 70% of incoming solar radiation between 400 and 700 nm (the photosynthetically active range). As a result, leaf temperatures are lower than expected for their relatively large size, and transpiration rates are also lower (Ehleringer and Bjorkman 1978; Ehleringer and Mooney 1978; Smith 1978). The green, cool-season leaves, which absorb as much as 81% of incoming solar radiation, are much less resistant to heat and drought stress (Ehleringer et al. 1976). Ehleringer (1983) found that a mutant glabrous form compensated for lack of pubescence with steeper leaf angles and higher leaf conductance to water vapor, but, even so, the plant lost its leaves earlier than normal pubescent individuals.

The optimum leaf temperature for photosynthesis is 25°C. At leaf temperatures greater than 35°C, photosynthesis drops precipitously (Ehleringer and Mooney 1978). Relative carbon gain is lowest from August to September, highest from February to April (Comstock et al. 1988). During the cool season, reduced leaf pubescence permits an unusually high rate of photosynthesis for a C_3 plant: >48 mg $CO_2/dm^2/hr$ (Ehleringer and Bjorkman 1982), or 15 mg CO_2/gm dry weight/hr (Strain 1969). The low ratio of internal to external leaf area, by lowering leaf resistance to water vapor diffusion, also encourages higher rates of photosynthesis when soils are moist (Smith and Nobel 1977a). During the dry season, dense pubescence allows only a low level of photosynthesis, about 3 mg CO_2/gm dry weight/hr (Strain 1969; Ehleringer and Mooney 1978), but even then, the densely pubescent leaves have a higher rate of photosynthesis than they would if they were glabrous (Cunningham and Strain 1969; Ehleringer and Mooney 1978). Leaf osmotic adjustments allow the stomates to remain partly open during brief growing-season droughts, thus fostering continued low levels of photosynthesis (Ehleringer and Cook 1984). The photosynthetic activity of the flower heads (about 1.5% of net leaf photosynthesis) might be enough to offset the respirational costs associated with flowering (Werk and Ehleringer 1983).

When the soil is moist, the plants produce abundant annual rootlets (Cannon 1912). These are presumably shed when the soil dries out. Annual roots seem to suppress the growth of herbaceous spe-

Figure 36. Encelia farinosa *in the Pinacate region, Sonora. The compact, hemispheric crown is typical. (Photograph by R. M. Turner.)*

cies in the vicinity (Cannon 1911; Muller and Muller 1956). On Tumamoc Hill, Tucson, Arizona, the perennial roots penetrate as much as 55 cm below the soil surface and branch sparingly at a depth of 15–30 cm (Cannon 1911). The combination of shallow, ephemeral roots and deeper, perennial roots lets the plants use both intermittent soil moisture near the surface and deeper soil moisture resulting from percolation of heavier rains. Ehleringer (1984) has shown that adjacent *E. farinosa* plants compete for water. When pure stands were thinned, the remaining plants showed higher leaf water potentials, higher leaf conductance, greater leaf area, faster growth, and higher flower and seed output than controls (Ehleringer 1984).

This species has a persistent seed bank, that is, seeds can remain viable in the soil for a year or more without germinating (Bowers 1994). In the northern Sonoran Desert, *E. farinosa* germinates from October to April following a rain trigger of 19 mm or more. Average daily temperatures during the emergence period must not drop below 6.0°C nor rise above 18.5°C (Bowers 1994). Shreve (1937c) reported greenhouse germination at soil temperatures of 15.5–24.0°C, which correspond to the natural soil temperatures during the winter rainy season. A report of germination in summer (Went 1948) is probably in error.

Seedling survival over the dry season is generally low. On a plot 100 m² on Tumamoc Hill, for example, 510 seedlings appeared following the heavy winter rains of 1977–1978 (Goldberg and Turner 1986), but only 13 survived until 1984 (Turner, unpublished data). Most probably perished within the first six months.

At an ungrazed site in the Sonoran Desert, adults underwent wide fluctuations in numbers and size over a 23-year period (Turner 1990). Maximum observed longevity on Tumamoc Hill is 32 years; few individuals survive more than 7 years (Goldberg and Turner 1986). Climatic events strongly influence adult mortality. Between 1985 and 1989, some plants on Tumamoc Hill that lost all or most of their leaves from unusually low winter temperatures did not, because of drought, regain full leafiness, and many finally died. Apparently their commitment to predominantly cool-season photosynthesis leaves them vulnerable when normal winter growth is impossible. The architecture of the plants also contributes to short life span. Unlike the long-lived *Larrea tridentata,* for instance, *E. farinosa* does not form clones by stem division, and the death of the root crown eliminates the entire individual (Muller 1953). Larvae of a specialist insect herbivore, *Trirhabda geminata,* feed exclusively on *E. farinosa* foliage (Wisdom 1985); biologists have not determined whether these larvae contribute to plant mortality.

Encelia farinosa is an attractive, fast-growing ornamental in desert cities, where it is commonly cultivated (Duffield and Jones 1981). The plants prefer a light, gravelly soil with good drainage (Duffield and Jones 1981). Some pruning may be required if top growth is frozen (Duffield and Jones 1981). Under frequent irrigation, growth tends to be lush, even rank; less heavily irrigated plants have a thriftier appearance and may be less susceptible to frost damage.

The golden sap has been burned for incense in Mexican churches and used by Indians for glue (Kearney and Peebles 1960). A medicinal tea can be made from leaves and stems (Moore 1989).

Encelia frutescens

(A. Gray) A. Gray

Green brittlebush

A rounded, much-branched shrub 1–1.6 m tall, *Encelia frutescens* has white branches and dark green foliage. The alternate, ovate to oblong leaves, 6–16 mm wide and 10–30 mm long, are sparsely scabrous with stout hairs from pustulate bases. The heads, 1–2.5 cm broad, are single on long peduncles. Ray flowers are usually lacking. The black, flattened achenes are silky-villous along the margins. Haploid chromosome counts of 17 (Jackson 1960) and 18 (Solbrig et al. 1972) have been reported, leading Solbrig and coworkers (1972) to suggest that there may be more than one chromosome number in the genus.

Flowers of *Viguiera parishii* are radiate, and its lower leaves, at least, are opposite. *Encelia farinosa* and *E. virginensis* A. Nels. also have radiate flowers and gray or blue-green foliage.

Benson and Darrow (1981), following Blake (1913), included *E. virginensis* under *E. frutescens.* They recognized var. *frutescens,* var. *virginensis* (A. Nels.) Blake, and var. *actonii* (Elmer) Blake. Our map depicts *E. frutescens* in the narrow sense.

Encelia frutescens is common to abundant in washes and along roads from sea level to nearly 1,500 m. It grows in several plant communities, including Sonoran desertscrub and Madrean evergreen woodland. The plants can rapidly colonize disturbed areas (Vasek 1980b). The southernmost point (28.3°N, 111.4°W) represents a 1979 sighting by Raymond M. Turner at San Augustín Bay, Sonora. Wiggins (1980) reported this species from the vicinity of Punta Prieta in Baja California, but we have seen no vouchers from the peninsula.

Throughout the range, rainfall is biseasonal and varies from mostly winter in the west to mostly summer in the east. The restriction to lower elevations toward the southern limit in central Sonora, an unusual pattern for a Sonoran Desert perennial, may show a need for aridity combined with some critical threshold of winter moisture. The overall geographic distribution does not lend itself to a simple scenario of climatic limitation.

Flowers appear from September to May in response to rain. The drought-deciduous leaves, also produced in response to rain, last 1 to 4 months (Comstock and Ehleringer 1986).

Photosynthesis of *E. frutescens* is quite

responsive to changes in leaf water potential (Comstock and Ehleringer 1984). In one series of studies, the photosynthetic rate declined linearly from 41.6 to 1.7 μmol/m²/sec as leaf water potential decreased from a high of –1.5 to a low of –4.0 MPa (Comstock and Ehleringer 1984, 1986). By maintaining photosynthesis at moderately low leaf water potentials, *E. frutescens* can exploit an intermittent rainfall regime (Comstock and Ehleringer 1984). When soil moisture is high, the plants grow rapidly and flower abundantly. When soil water resources have been exhausted, leaf water potential drops, leading to leaf abscission and dormancy (Comstock and Ehleringer 1984, 1986).

Like *E. farinosa,* this species shows no adjustment in the temperature optimum for photosynthesis (30°C) (Comstock and Ehleringer 1984). Photosynthetic rates decline steeply below 18°C and above 41°C. When the air temperature exceeds the optimum, *E. frutescens* can prevent excessive leaf temperatures only by high transpiration (as demonstrated experimentally by Ehleringer and Cook [1990]). This characteristic restricts it to habitats where soil water is relatively abundant, such as desert washes (Comstock and Ehleringer 1984; Ehleringer 1988) and roadsides.

Ephedra aspera

Engelm. ex S. Watson

[=*E. nevadensis* S. Watson var. *aspera* (Engelm.) L. Benson]

Boundary ephedra, popotillo, cañatillo, pitamo real, sanguinaria, tepopote

Ephedra aspera is a twiggy, green-stemmed shrub up to 1.5 m tall. The rigid branches are opposite or whorled and may be minutely roughened. They are pale to dark green when young and often turn yellow in age. The scalelike and (usually) opposite leaves eventually turn fibrous and drop, leaving the brown or

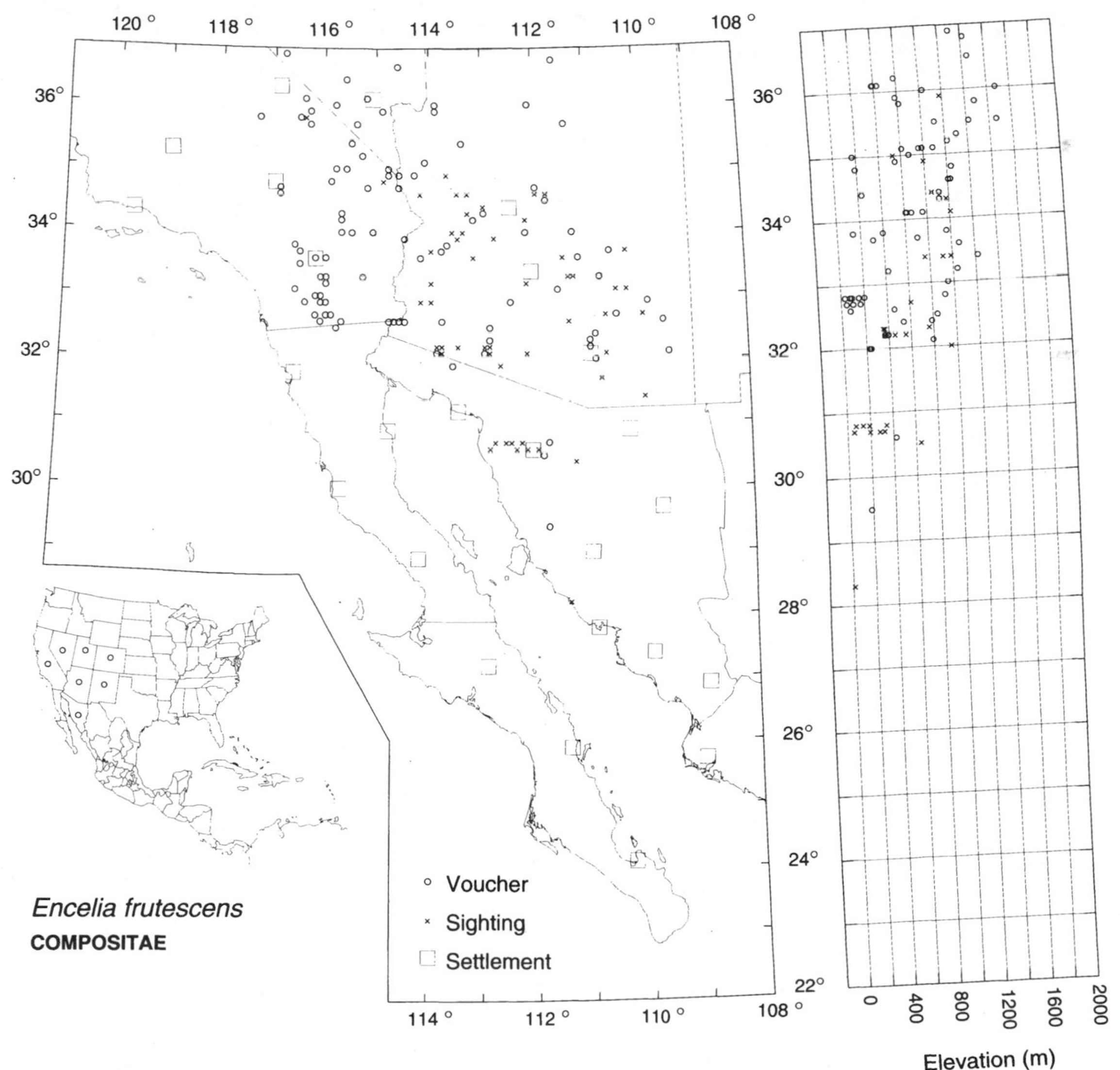

Encelia frutescens
COMPOSITAE

black leaf bases behind. Plants are generally dioecious, occasionally monoecious. The female cones, 6–10 mm long, are sessile or short stalked. Carlquist (1989b) described the wood and bark anatomy.

Ephedra nevadensis S. Watson has smooth, bluish green stems and stalked cones. *Ephedra californica* has leaves in threes, as does *E. trifurca*, which has spinose branch tips, in addition.

Benson and Darrow (1981) treated this as a variety of *E. nevadensis*. We follow Wiggins (1964) and Munz (1974) in separating it.

Ephedra aspera is occasional to common on hills, flats, and slopes, often on rocky or gravelly substrates. The westernmost disjunct in California (35.1°N, 118.3°W) represents a 1908 collection by LeRoy Abrams. Benson and Darrow (1981) included Nevada and Utah within the range of this taxon. However, neither Welsh and coauthors (1987) nor Albee and coauthors (1988) included *E. aspera* in the Utah flora. The most widespread of Sonoran Desert ephedras, this species grows throughout arid North America from sea level to 1,500 m.

The species is absent from areas that combine warm, dry winters with dependably wet summers. Otherwise, it thrives in a variety of climates across its range. The Pacific coastal populations experience cool, dry summers and infrequent winter rains. Farther north and east, *E. aspera* receives biseasonal rainfall that shifts from predominantly winter in the west to mostly summer in the east.

Packrat middens from near Picacho Peak, California, record *E. aspera* from 12,700 to about 8,000 years B.P., but it is not in the area at present (Cole 1986). Middens from the south side of the Hornaday Mountains, Sonora, indicate that *E. aspera* was locally dominant 10,000 years B.P. but absent after 4,400 years B.P. (Van Devender, Burgess et al. 1990). The species is now widely scattered on north-facing exposures in this small, granitic range. Apparently, *E. aspera* was much more common at lower elevations during the latest Wisconsin and early Holocene periods than is the case today. The range probably became fragmented over the last 6,000 years.

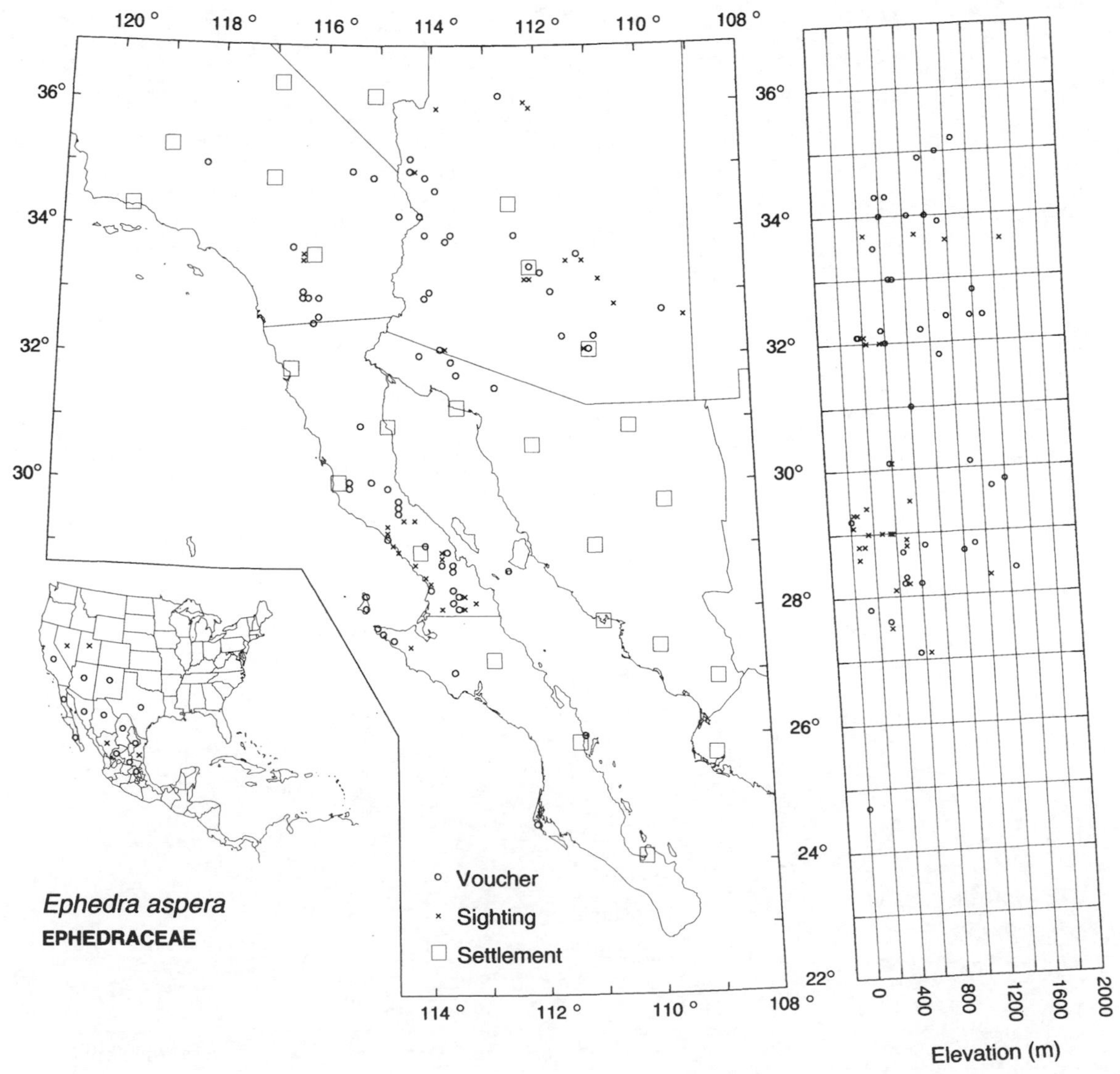

This plant is well adapted to arid environments. The scalelike leaves conserve water, as does the placement of the stomates in stem furrows (Cutler 1939). The chlorophyllous stems are the main photosynthetic organ (Cannon 1908).

New shoots appear in response to rain between November and May. Male and female cones are produced in spring, generally February to April, rarely as early as January or as late as June. The plants are wind pollinated. Fertilization is double in this species (Friedman 1990).

Germination requirements are probably similar to those of *E. nevadensis*, which reaches peak germination (68%) at the rather wide temperature range of 5–20°C (Young et al. 1977). Seedlings are rarely observed in the wild (Cutler 1939). Plants most often reproduce by stems that become buried and form adventitious roots and offshoots (Cutler 1939). Matched photographs taken many decades apart show that plants may reach full size in 20 years, then change little in size over the next 80 years. The life span is at least 100 years (Robert H. Webb, personal communication 1993).

Ephedra aspera is a valuable winter browse (Kearney and Peebles 1960). The stems and dried flowers contain pseudoephedrine and tannins. Tea made from them has been used to treat syphilis, pneumonia, kidney disease, and other ailments (Standley 1920; Kearney and Peebles 1960).

Ephedra californica

S. Watson

[=*E. funerea* Cov. & Morton]

Mormon tea, joint fir, California ephedra, canutillo

This twiggy, spreading shrub grows 5–12 dm tall (figure 37). The pale green, smooth or slightly scabrous stems are waxy in the grooves. The scalelike leaves are in threes at the nodes. The ovulate cones, 7–10 mm long, may be single or whorled. Plants are generally dioecious. Carlquist (1989b) described the wood and bark anatomy.

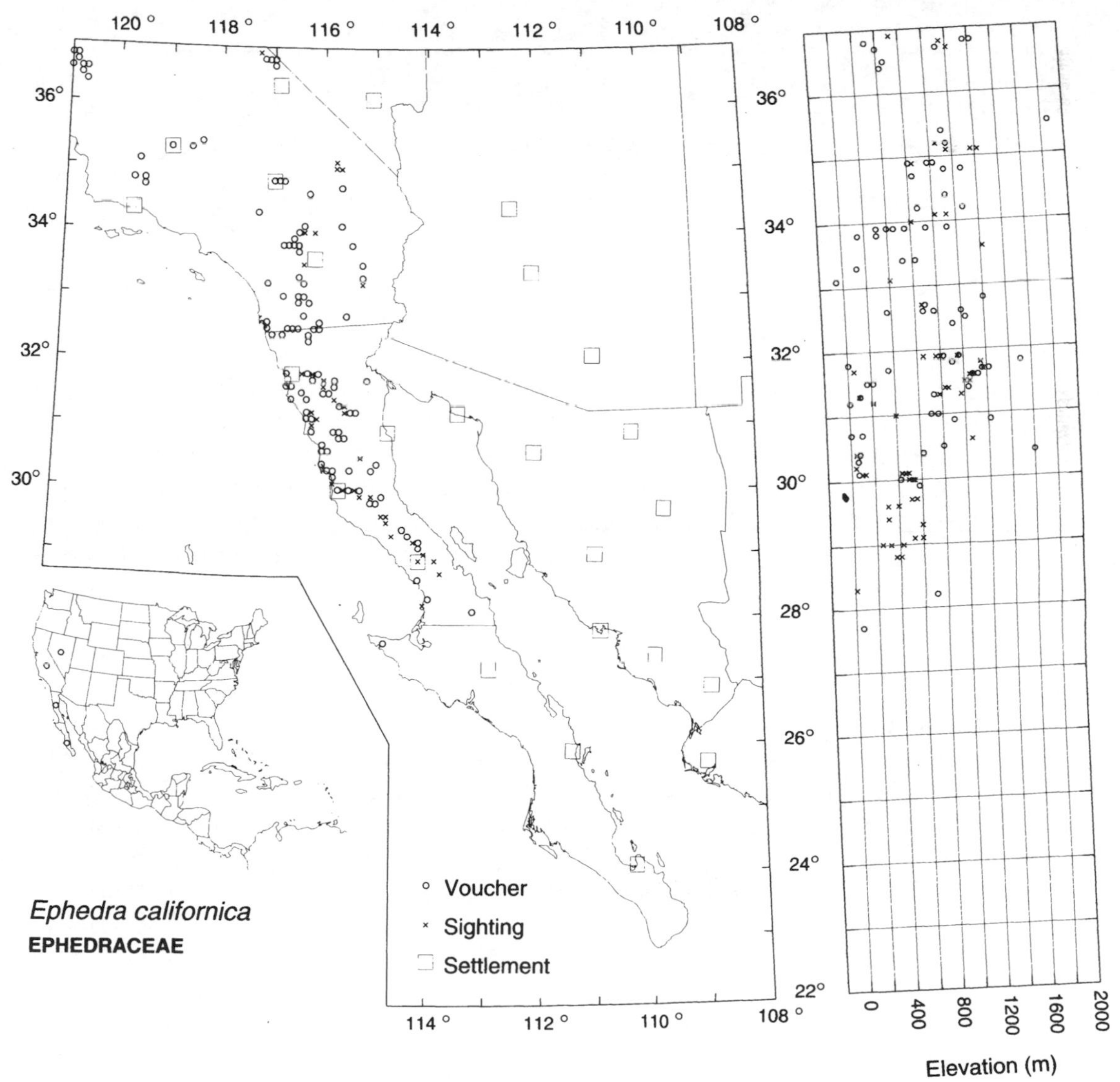

Figure 37. Ephedra californica *near El Paseo San Matías, Baja California. (Photograph by J. R. Hastings.)*

Ephedra nevadensis S. Watson and *E. aspera* have opposite leaves. *Ephedra trifurca* has spine-tipped stems and persistent, dark brown leaf bases.

We follow Benson and Darrow (1981), who recognize two taxa: the widespread var. *californica,* found throughout Southern California and the northern Baja California peninsula, and the more narrowly distributed var. *funerea* (Cov. & Morton) L. Benson, limited to Death Valley, California, and adjacent Nye and Clark counties, Nevada.

Occasional to abundant on rocky or gravelly hills and plains, *E. californica* also grows on stable dunes, both inland and coastal. The species is characteristic of the California Floristic Province and enters the Sonoran Desert in southeastern California and in the Vizcaíno region of the central Baja California peninsula. Across its range, *E. californica* grows in a variety of plant communities, including Mojave and Sonoran desertscrub, Californian chaparral, and Californian valley grassland. Its broad elevational range, from sea level to 1,860 m, suggests tolerance of high temperatures in summer and subfreezing temperatures in winter. The single factor common to this diverse array of habitats and temperature regimes appears to be a need for winter rain. *Ephedra californica* reaches its southern limits in the more arid parts of the Vizcaíno region, where summers are dry but relatively cool compared to the rest of the Sonoran Desert. Summers in the northeastern part of the range can be quite hot.

The wind-pollinated cones appear mainly from February through May, occasionally in other months. The drought-deciduous leaves presumably appear in response to rains. Stems grow mainly in spring (February–May), although the cambium may be reactivated following unseasonal rain (Alfieri and Mottola 1983).

Ephedra trifurca

Torr. in Emory

Mexican tea, longleaf ephedra, cañatilla, tepopote, popotillo, itama real

Ephedra trifurca, a twiggy, green-stemmed shrub 0.5–2 m tall, has persistent scalelike leaves in threes on smooth, spine-tipped stems. In age the leaves become white and shredded. The female cones, 10–14 mm long, are whorled or single and may be sessile or short stalked. The plants are generally dioecious. Carlquist (1989b) described the wood and bark anatomy.

Ephedra aspera and *E. nevadensis* S. Watson have leaves in twos. *Ephedra californica* has blunt-tipped rather than spinose branches.

This shrub is occasional to common on gravelly slopes and plains, sandy washes, and stable dunes. The northernmost point (35.0°N) represents a collection by Selma Broem from the highway between Kingman and Oatman, Arizona.

The species grows where rainfall is biseasonal and frosts are infrequent to common. The western limits, falling where summer and early autumn storms are sporadic, suggest that *E. trifurca* needs occasional warm-season moisture, perhaps for seedling establishment. The northern limit may have been shaped by interactions with *E. torreyana* S. Watson or *E. viridis* Cov., which have nearly contiguous distributions.

Cones and green leaves can be found from December through August. Aerodynamic characteristics of the stems, ovules, and pollen grains facilitate wind pollination (Niklas and Buchmann 1986; Niklas et al. 1986; Buchmann et al. 1989). Where they grow together, *E. trifurca* and *E. nevadensis* avoid hybridization by virtue of their different pollen-capture morphologies (Niklas and Buchmann 1986).

Maximum observed longevity is 50 years (Goldberg and Turner 1986). Although Cutler (1939) said that seedlings are rarely observed in the wild, juvenile plants are not uncommon where moisture conditions are relatively favorable, as on dunes. On the Jornada Experimental Range, New Mexico, the average root-to-shoot ratio for mature plants was 0.4 (Ludwig et al. 1975).

Domestic livestock rarely graze on *E. trifurca* (Cutler 1939). The dried flowers and stems have been used in treating syphilis and other diseases (Kearney and Peebles 1960). Honeybees (*Apis mellifera*) collect the pollen (Schmidt and Johnson 1984).

Errazurizia benthami

(Brandegee) I. M. Johnston

[=*Dalea benthami* Brandegee]

This shrub grows up to 0.5 m tall; it has white-tomentose foliage conspicuously dotted with oily glands. The once-compound leaves, 2–5 cm long, have 7–11 leaflets 4–10 mm long. The light yellow flowers are 7–8 mm long. The pod (10 mm long) is pubescent and prominently glandular. The haploid chromosome number is 14 (Reveal and Moran 1977).

Errazurizia megacarpa has longer leaves (3–8 cm) and lacks sharp, persistent stipules. *Errazurizia* species can be distinguished from *Dalea* species by the minute, mammillate protuberances located on the glands of the former.

A shrub of sandy or clay soils, typically in open vegetation, this peninsular endemic is most frequent on the Vizcaíno Peninsula. It is common between Bahía Tortugas and Punta Eugenio. The range includes Isla Cedros and Isla Margarita (Barneby 1977). *Errazurizia benthami* is nearly confined to the most arid stretch of the Pacific coast. The sporadic rains are most likely during winter but may occur from late summer through spring when storms stray from their normal paths. Summers are dry and rather cool compared to other parts of the Sonoran Desert.

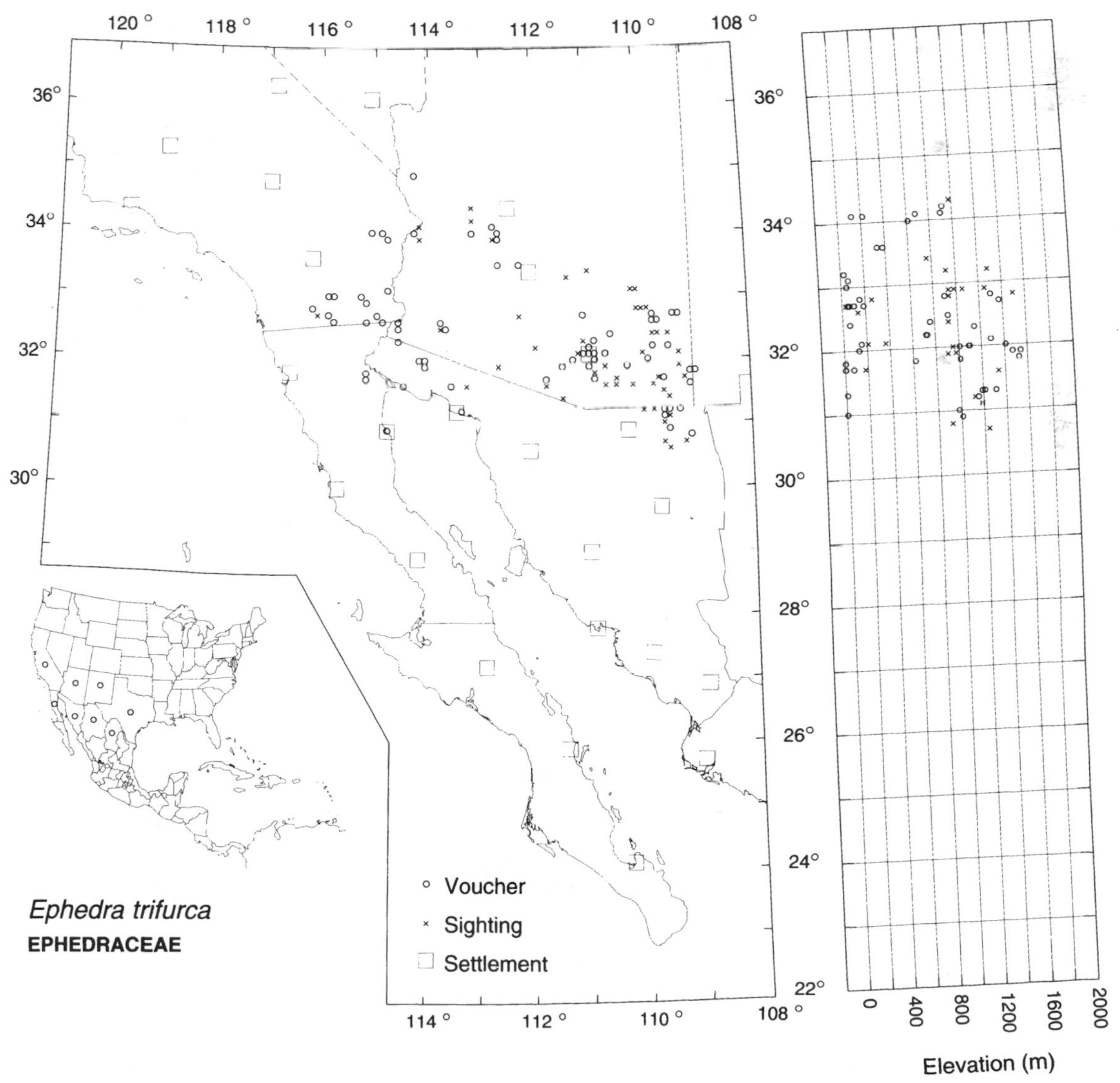

Ephedra trifurca
EPHEDRACEAE

Flowers appear from February through April (Barneby 1977). The plants are leafless during summer drought. Plants dug from bouldery, sandy alluvium near San Ignacio Lagoon had extensive generalized root systems with brittle tap and lateral roots.

Errazurizia megacarpa

(S. Watson) I. M. Johnston

[=*Dalea megacarpa* S. Watson]

This shrub grows up to 1 m tall; it has white-tomentose foliage and conspicuous oily glands that make persistent, reddish brown stains. The once-compound leaves, 3–8 cm long, have 9–13 round and notched leaflets 4–10 mm long. The yellow flowers are 6–7 mm long. The pods (11 mm long) are ovoid, canescent, and prominently glandular.

Errazurizia species differ from *Dalea* species in having minute, nipplelike protuberances on the glands. *Errazurizia benthami* is a smaller plant with shorter leaves (2–5 cm) and spinose stipules.

Errazurizia megacarpa grows in washes and on dunes. It is endemic to the Sonoran Desert and occurs mainly near the coast on both sides of the Gulf of California. It is found on Islas Angel de la Guarda, Tiburón, and San Marcos (Moran 1983b). Throughout the range, the scanty rain is biseasonal. Late summer and early fall storms are followed by late fall drought. Winter brings occasional rains, followed by late spring drought through June or July. Summers are hot, and winters are warm to mild.

Flowers appear mainly in the spring but may be found throughout the year.

The Seri used the seeds for necklaces (Felger and Moser 1985).

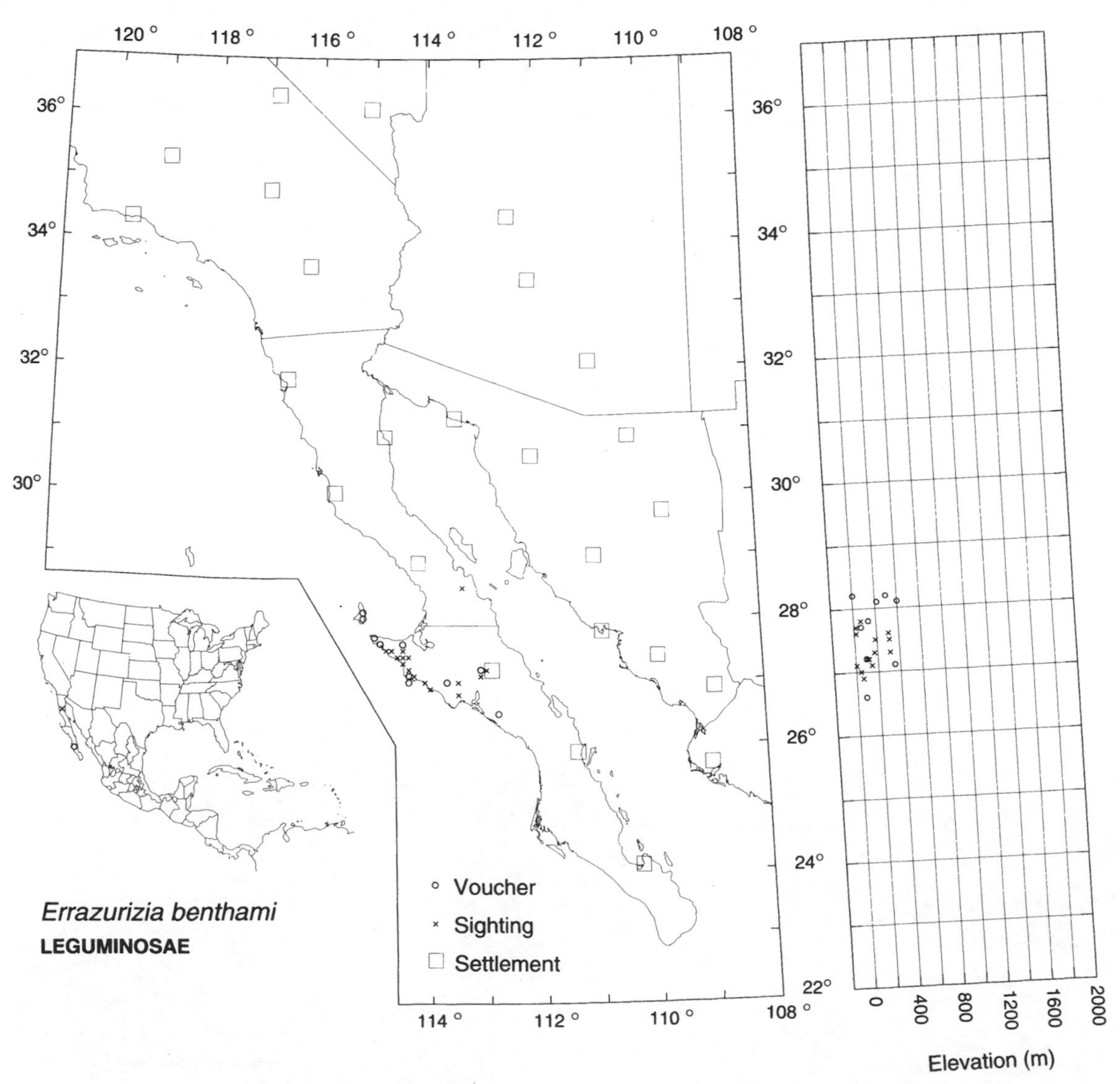

Errazurizia benthami
LEGUMINOSAE

Erythrina flabelliformis
Kearney

Coral bean, coral tree, chilicote, coralina, zumpantla, colorín, peonia

At the northern end of its range, *Erythrina flabelliformis* is a shrub up to 1.5 m tall; farther south it is a tree up to 9 m tall. The light tan stems, leafless much of the year, have scattered spines that, on large plants, may develop into conical tubercles. The leaves are compound with three triangular leaflets, each 4–7 cm long and 5–10 cm broad. The bright red flowers, in terminal racemes, are 2–5 cm long. The elongate, folded banner gives the flower a tubular appearance. The brown, torulose pods, 1.2–2.5 dm long, contain 3–8 red or orange, ellipsoid seeds 12–14 mm long. The diploid chromosome number is 42 (Krukoff 1969).

No other shrub in the northern Sonoran Desert has trifoliate leaves, showy red flowers, and spiny stems. The torulose pods, often persistent on the leafless stems, are distinctive.

Throughout its range, *E. flabelliformis* grows on rocky slopes and cliffs, and in the north, it is often found among boulders in streambeds. It tolerates a wide variety of soil types, including those derived from limestone and quartz monzonite (Conn and Snyder-Conn 1981). The outlier in Baja California Sur at 26.9°N, 112.5°W represents a 1964 collection by Reid Moran from 5 km west of Misión Guadalupe. Except for the population near Guaymas, the species is absent from the Sonoran coast, probably because of insufficient summer rainfall. Its sole Gulf of California occurrence is on Isla Cerralvo (Moran 1983b).

The stems freeze at −2.2 to −4.4°C (Johnson 1988, 1993). The northernmost populations (Santa Catalina Mountains, Pima County, Arizona) experience frequent frost in winter. Although their boulder habitat provides some protection (Conn and Snyder-Conn 1981), they do freeze to the ground in the coldest winters. Because the species has failed to occupy apparently suitable habitat to the

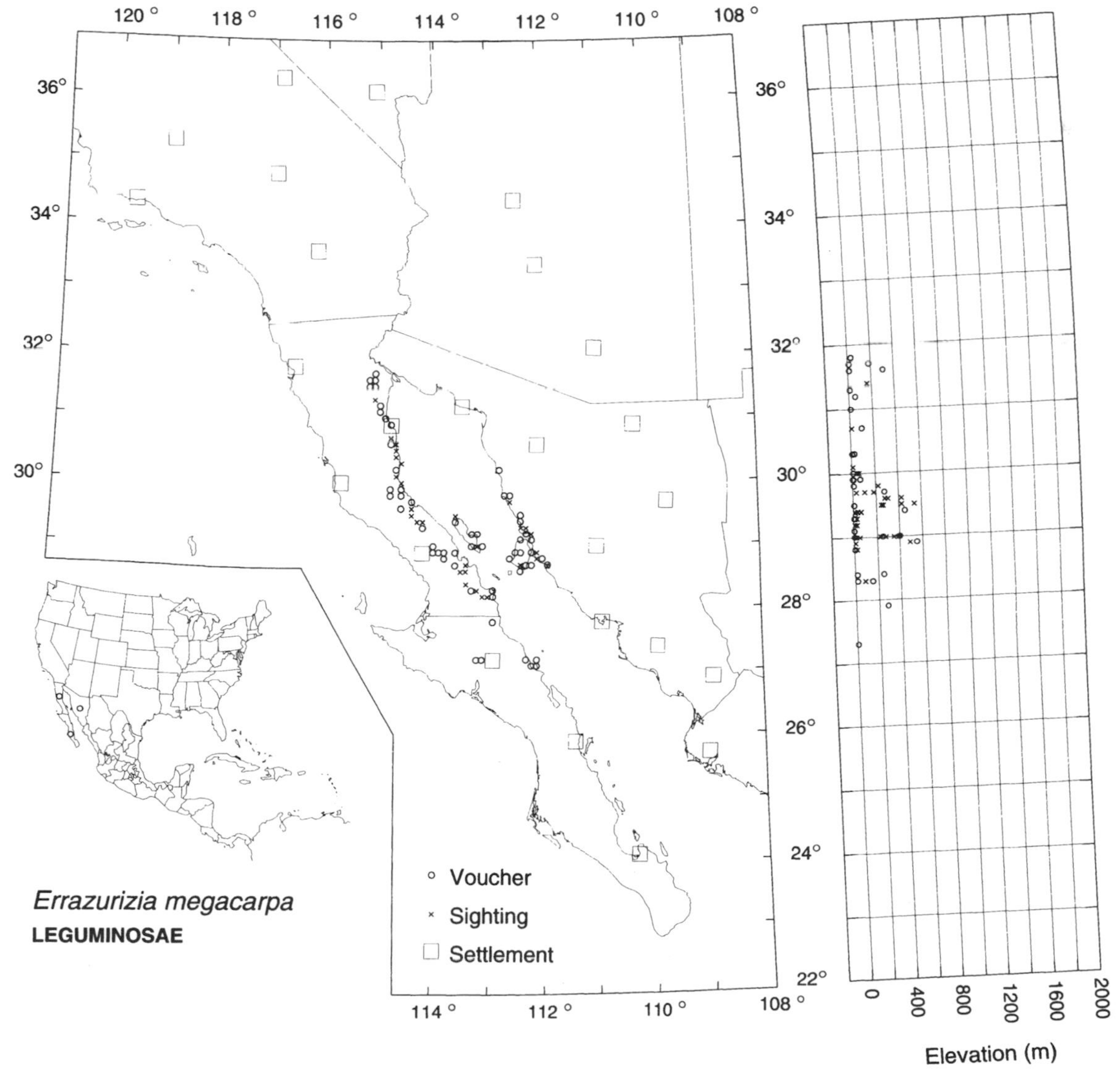

north, it seems unlikely that freezing temperatures alone determine its present northern limit. Although *E. flabelliformis* grows in Sinaloan thornscrub over most of its range, in southern Arizona it is most frequent in Madrean evergreen woodland, where summer moisture is sufficient and winter temperatures are not so low as to be limiting. The thickened, somewhat tuberous roots provide reserves for regrowth after frost. The tuberous-rooted, shrubby life-form has allowed *E. flabelliformis* to extend much farther north than most thornscrub species.

In the northern Sonoran Desert, flowering is from mid June to late July (Conn and Snyder-Conn 1981). Farther south, blooming begins as early as February and continues through July. Leaf buds appear about the same time as flower buds. The leaves grow slowly until the onset of summer rains, then increase rapidly in size. Leaves drop from late September to mid October, most likely as a result of water stress. The pods persist on the stems for many months after ripening.

The genus as a whole is adapted to bird pollination (Toledo 1974), and *E. flabelliformis* is doubtless pollinated mainly by hummingbirds. Ants that visit extrafloral nectaries on the calyx and petioles, presumably for sugar, might protect the plants from insect predators (Sherbrooke and Scheerens 1979).

When in leaf, *E. flabelliformis* is a mesophyte. The boulder habitat, by providing greater runoff, shade, and soil temperature stability, favors soil water absorption over the entire growing period (Conn and Snyder-Conn 1981).

Seeds typically remain on the ground for months after the pods dehisce. Most likely they wash downslope and lodge in boulder cracks, where they are eventually buried by soil and other debris (Conn and Snyder-Conn 1981). They must be scarified to germinate. In greenhouse trials, highest germination rates were found for buried seeds at temperatures typical of midsummer (37/20°C day/night) or spring (23/7°C day/night) (Conn and Snyder-Conn 1981). In nature, seeds probably germinate in response to summer rains.

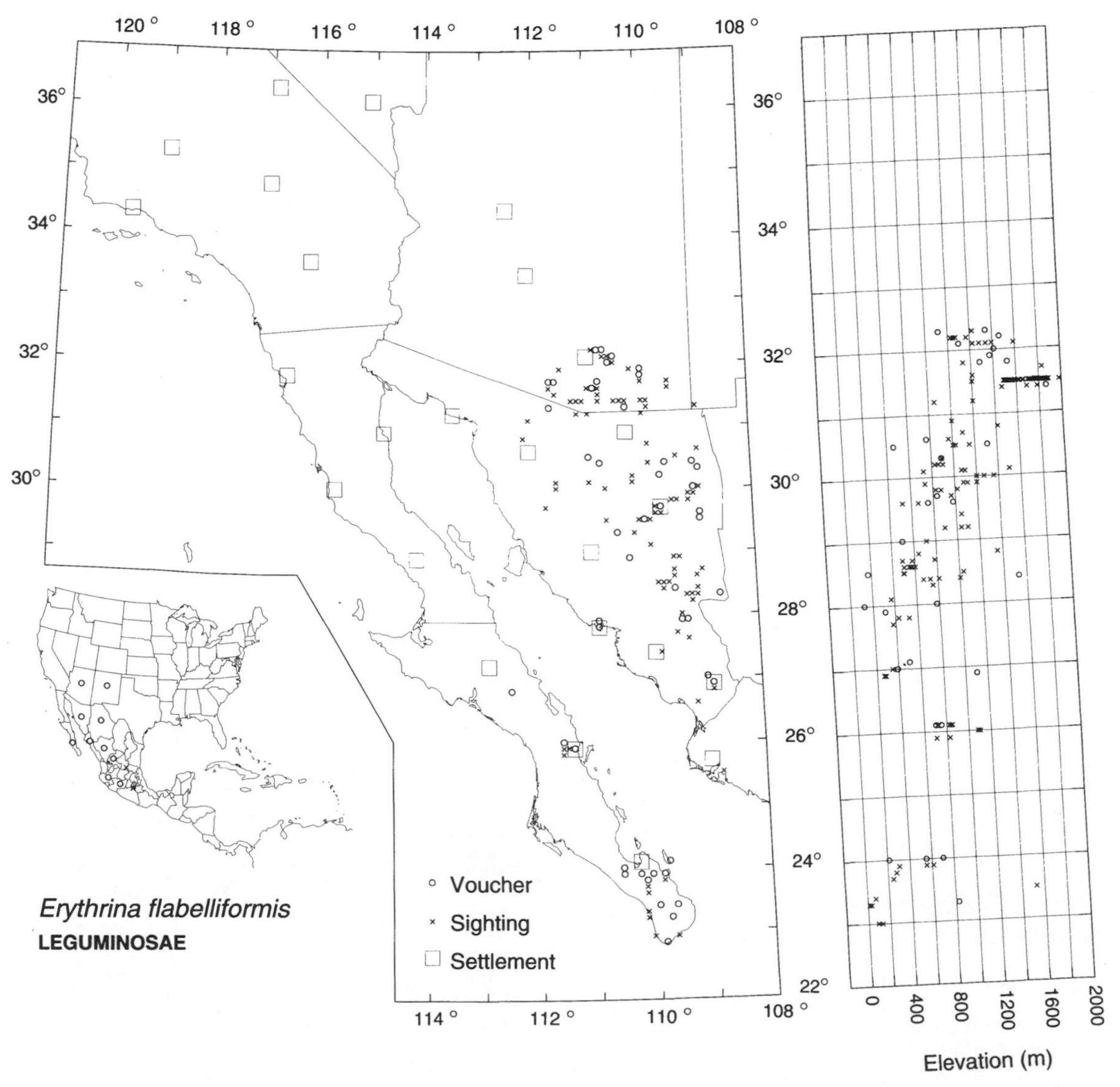

Erythrina flabelliformis
LEGUMINOSAE

The Yaqui Indians carved the soft wood and made a rust-colored dye from the crushed bark (Pennington 1963). The high alkaloid content of the seeds makes them potentially toxic. The Seri Indians made a tea from the cooked seeds to treat diarrhea (Felger and Moser 1985), and the Tarahumara made an emetic drink from the toasted and crushed seeds and also used the crushed seeds for toothache (Pennington 1963). The wood is used to make corks for bottles and gourd flasks (Gentry 1942).

Although leafless much of the year, this species has some potential as an ornamental. Plants can be grown from scarified seed or cuttings (Johnson 1988) and are easily transplanted when leafless. In warm areas and protected sites, they may develop into small trees (Johnson 1988).

Esenbeckia flava

Brandegee

Palo amarillo

Esenbeckia flava is a shrub or small tree up to 8 m tall with dark gray bark and tomentulose foliage. The alternate, simple leaves, 5–10 cm long and 2.5–5 cm wide, are elliptic to obovate with entire or somewhat sinuate margins. The white, 4- or 5-parted flowers, 13–15 mm across, are in terminal and axillary panicles up to 20 cm long. The echinate capsules are 3–5 cm broad.

The punctate glands on the large, simple leaves are distinctive, as are the spiky fruits.

Growing along arroyos and on rocky hillsides and canyon slopes, *E. flava* is endemic to Baja California Sur. The plants are most frequent in Sinaloan thornscrub of the Cape Region. North of 26°N, the known populations occur at or above 500 m. Farther south, where aridity is less likely to be limiting, the species occurs down to sea level.

Flowers appear from August to October, fruits from September to May. The leaves are apparently drought deciduous.

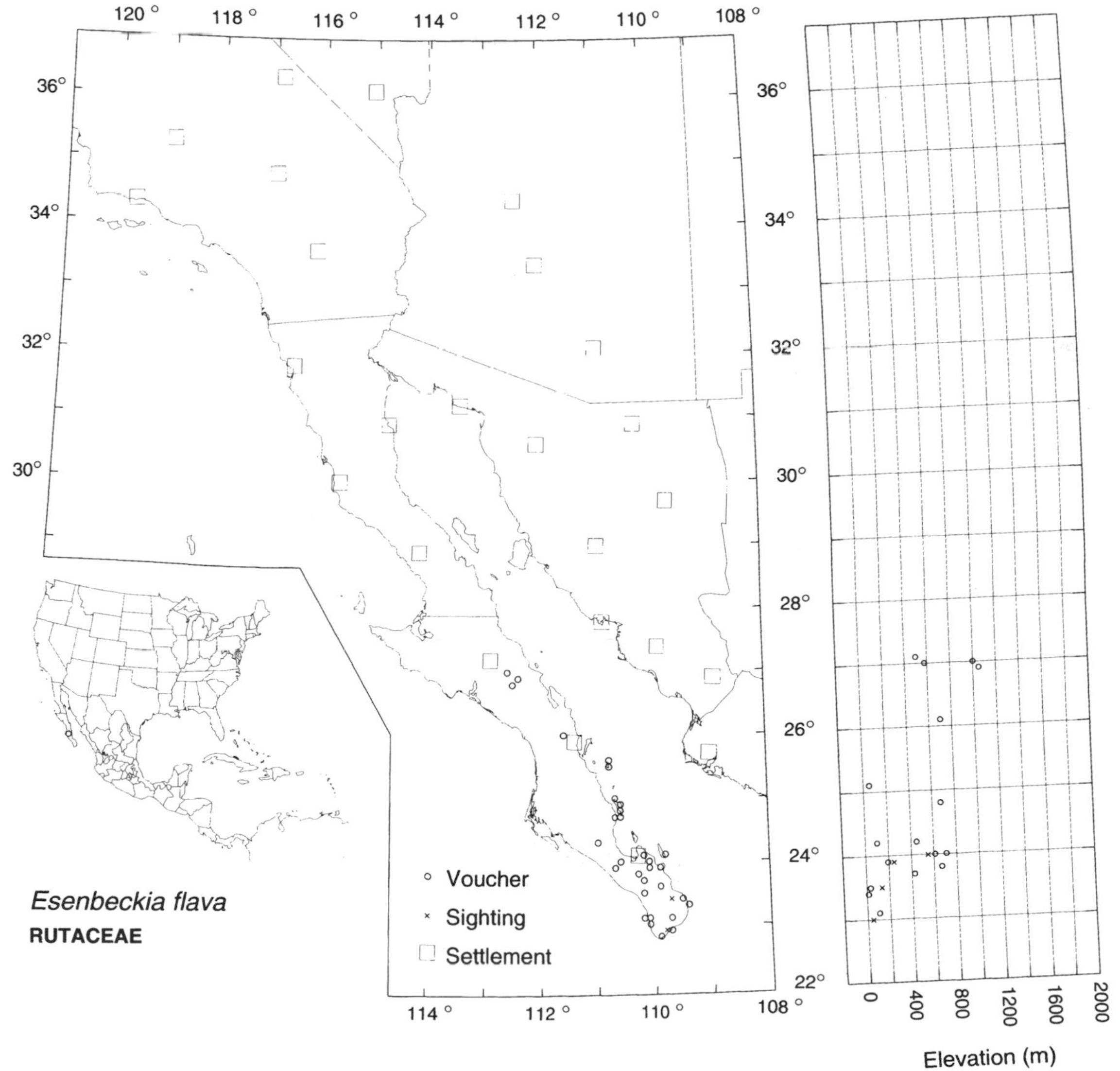

Esenbeckia flava
RUTACEAE

Esenbeckia hartmannii
Robins. & Fern.

Samota

This shrub grows 2–5 m tall and has leaves crowded toward the branch tips. The entire, obovate to elliptic leaves, 1.5–6 (occasionally 8) cm long and 1–3 cm broad, are rather thick and are punctate with pellucid glands. The white, 4- to 5-parted flowers are in short, terminal panicles. The warty capsules are 1.5–2.5 cm broad.

The simple, pellucid-punctate leaves and warty capsules are characteristic.

Rare to common on rocky slopes and plains and along washes, *E. hartmannii* is a Sinaloan thornscrub species most frequent in the Río Bavispe, Río Moctezuma, and Río Yaqui drainages. It appears to be narrowly limited by summer aridity at lower elevations and perhaps by cold higher up.

Flowers appear from June through August, rarely as early as May or as late as September. Leaves are presumably drought deciduous.

Euphorbia californica
Benth.

Sipehui

This tortuously branched shrub grows 0.5–2 m tall and has glabrous foliage. The leaves, 7–19 mm long, are orbicular to oblong and entire. The lower margin of the blade usually continues across the upper side of the petiole. Clustered on spur shoots, the involucres are 3–5 mm wide, including the narrow white or greenish appendages. Fruits are 3-lobed capsules 3–4 mm high and 4–5 mm across.

Euphorbia misera has finely puberulent leaves and twigs, and its leaf blades

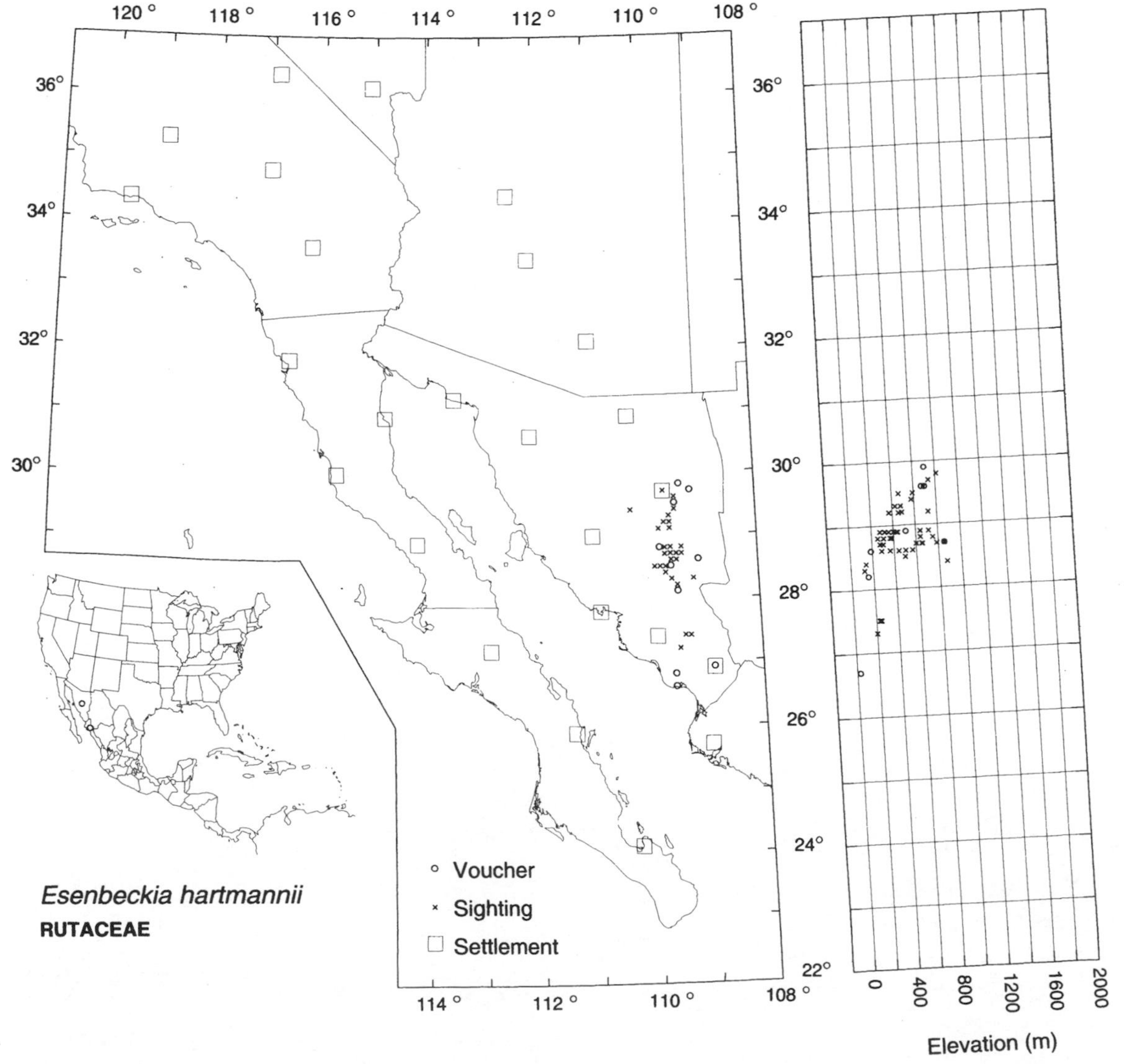

Esenbeckia hartmannii
RUTACEAE

are not continuous across the upper petiole. *Jatropha* species do not have milky sap.

Wiggins (1964) recognized two varieties. Variety *californica,* a plant of sandy or gravelly washes, hillsides, and plains, occurs near Hermosillo, Sonora, in Baja California Sur, and in Sinaloa. Variety *hindsiana* (Benth.) Wiggins, which has thicker leaves and stouter petioles, is known from rocky, granitic ridges near Cabo San Lucas, Baja California Sur (Wiggins 1964). The peninsular disjuncts at 27.1 and 27.2°N represent sightings by Raymond M. Turner and Tony L. Burgess. The only gulf island occurrence is on Isla Espíritu Santo (Moran 1983b).

Over most of the range, winters are warm and rainfall is biseasonal, mainly in the warm season. The northern limits are roughly contiguous with the southern limits of *E. misera,* suggesting competitive interactions between the two species.

Plants flower sporadically throughout the year, presumably in response to rain. Leaves are drought deciduous.

Euphorbia ceroderma

I. M. Johnston

Euphorbia ceroderma is a shrub 5–10 dm tall with many wandlike, waxy stems in dense clumps. The plants are leafless. In age the stem tips become spinescent. The yellowish flowers are in small clusters along the stem.

Stems of *Pedilanthus macrocarpus* are fleshy and turgid, those of *E. ceroderma* rigid and woody. *Euphorbia xanti, E. colletioides* Benth., *E. californica,* and *E. misera* are leafy at least part of the year and do not have spine-tipped stems.

Euphorbia ceroderma belongs to section *Trichosterigma* within subgenus *Agaloma.* Its closest relatives are *E. antisyphilitica* Zucc. of the Chihuahuan Desert and *E. rossiana* Pax of the Tehuacan Valley in Puebla and Oaxaca (Huft 1984). All three share a clumped habit and leafless stems; however, neither *E. antisyphilitica* nor *E. rossiana* is woody and spinescent. Oddly,

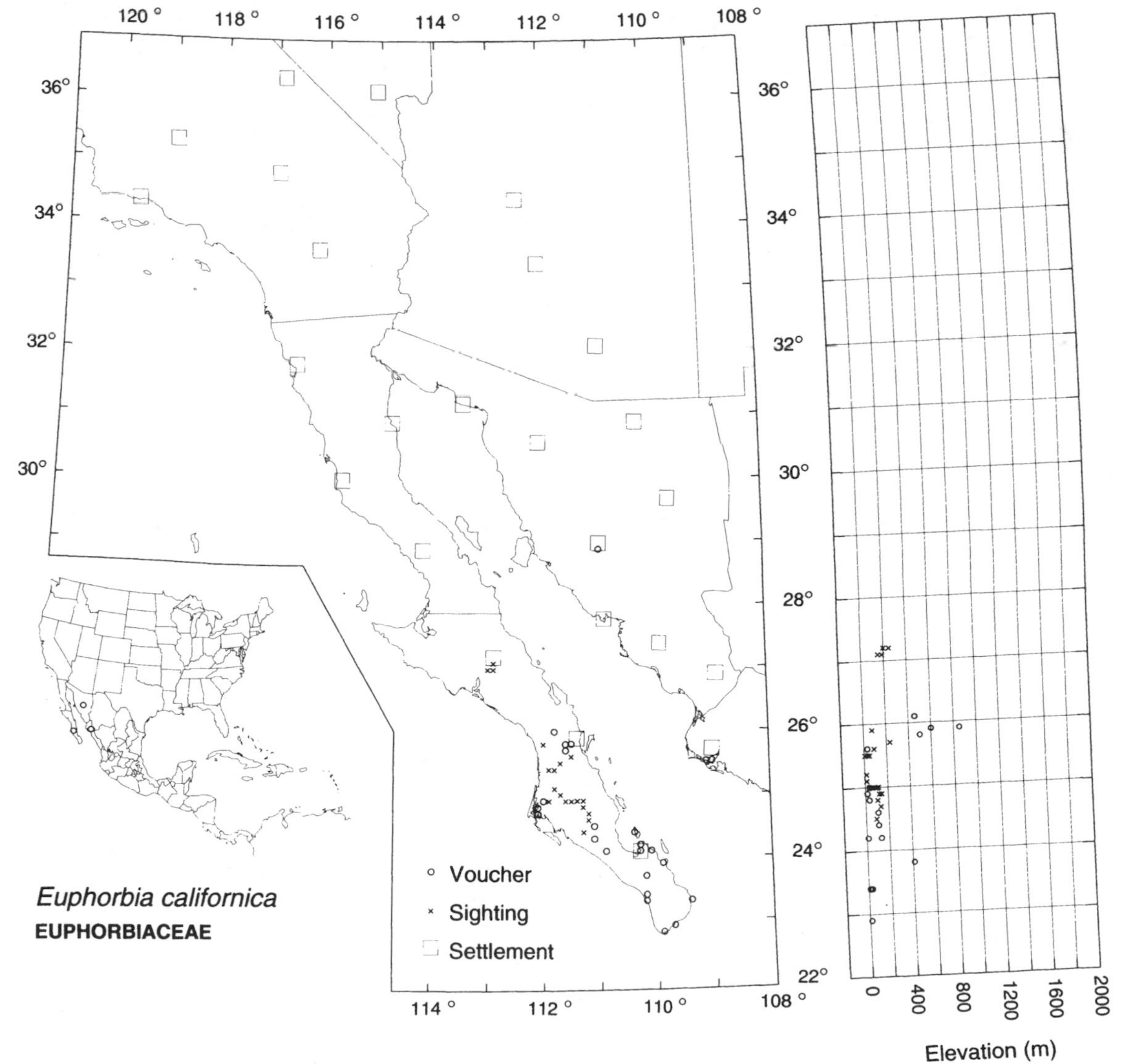

Euphorbia californica
EUPHORBIACEAE

E. ceroderma much more closely resembles certain *Euphorbia* species native to the Namibian coast, doubtless a result of convergent evolution.

The omission of *E. ceroderma* from regional floristic manuals (e.g., Wiggins 1964, 1980) makes it difficult to identify. Our map may underestimate its frequency. Sonoran and peninsular populations experience biseasonal rainfall, most of it in late summer or early fall. Winters are warm and virtually frost free across the range.

If autochthonous, *E. ceroderma* is probably a very old species, given its taxonomic distance from its closest relatives. The species conceivably predates the gulf-floor spreading that began in the Miocene. If so, the disjunct populations on either side of the gulf could represent relics of an earlier, broader distribution.

The plants flower rarely; our meager records indicate that September and October are the usual blooming period.

Under cultivation in Tucson, some plants freeze only at the tips and recover readily, while others freeze to the ground and regrow hesitantly, if at all.

Euphorbia magdalenae Benth.

[=*Chamaesyce magdalenae* Millsp.]

Euphorbia magdalenae is a subshrub 2–10 dm tall with a spreading or almost prostrate habit and intricately branched, slender, reddish stems. Stipules are united into a single deciduous scale, 0.3–0.5 mm high. The opposite, oblong-obovate leaves are 2.5–7 mm wide and 5–13 mm long. The small (less than 1 mm long) involucres are on short lateral branches or in the axils of the upper leaves. The brownish or black glands of the flowers have white petal-like appendages. The capsules are 2 mm long.

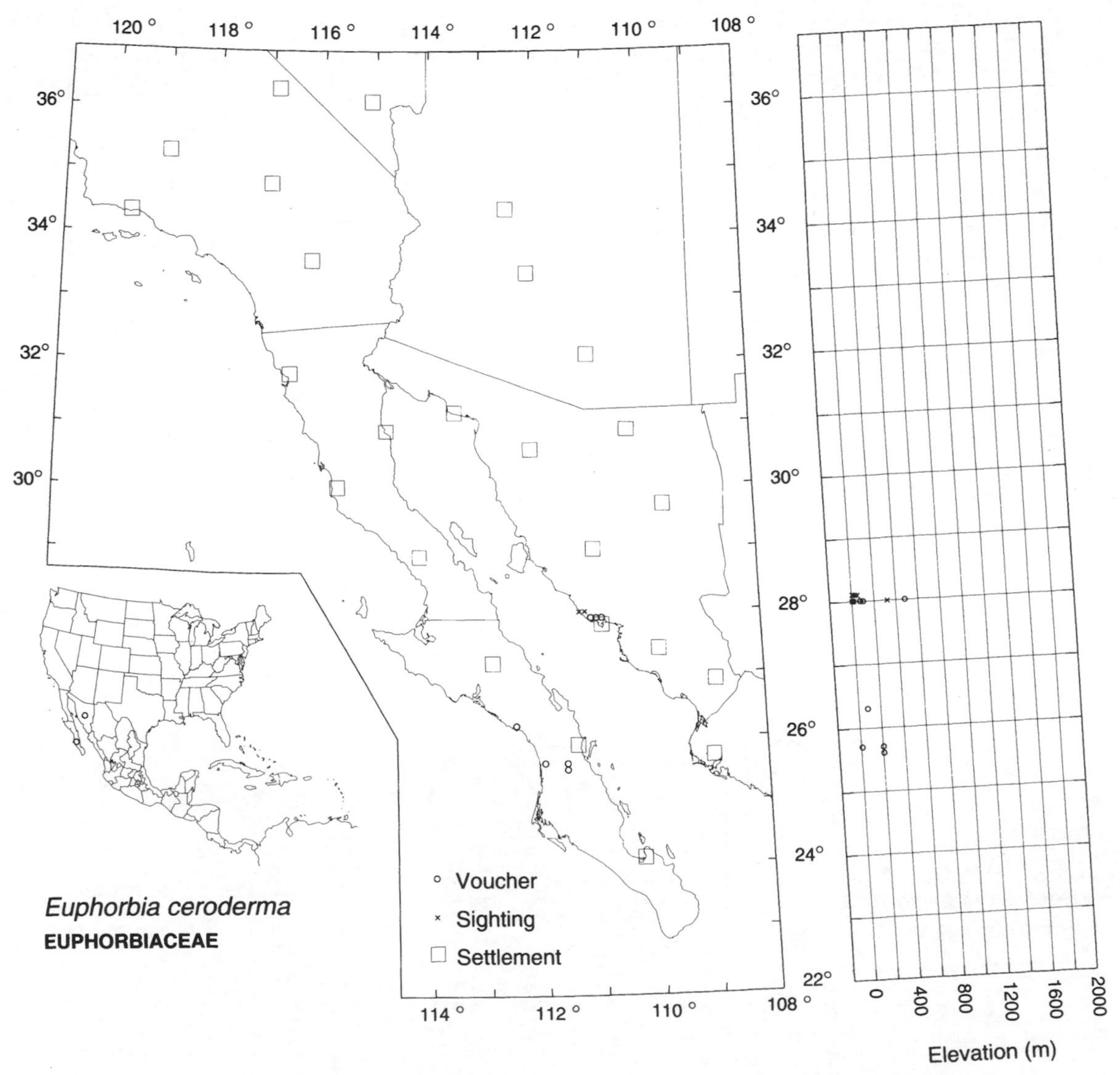

Euphorbia ceroderma
EUPHORBIACEAE

Euphorbia tomentulosa has distinct rather than united stipules and a repeatedly dichotomous branching pattern. *Euphorbia misera* is a much larger shrub with thick, knobby branches. *Euphorbia xanti* is a green-stemmed, scandent shrub up to 2 m tall.

The plants grow in open Sonoran desertscrub along gravelly washes and on rocky hillsides, usually below 300 m. In Baja California, the northernmost collection (29.0°N) is Edward Palmer's from Bahía de Los Ángeles. Richard S. Felger collected it on Isla Tiburón in 1968 and reported it from Isla San Pedro Nolasco in 1965. Cody and coworkers (1983) suggested that during a relatively moist period this species crossed from the peninsula to the Sonoran mainland via the midriff islands, then, with the return of drier conditions, retreated southward, leaving disjuncts near Guaymas, Sonora, and on Isla Tiburón.

Climates are arid with biseasonal rainfall. In the southern peninsula, rain is most frequent during summer and early fall. Near the northern limits, summers are often dry and rain is most likely in winter. Our maps suggest displacement interactions between *E. magdalenae* and *E. tomentulosa.*

The main flowering period is November–April. Leaves may appear with or after the flowers. Plants lose their leaves during long droughts.

Euphorbia misera
Benth.

Cliff spurge

This straggling shrub grows 0.5–1.5 m tall with sparse foliage and tortuous, thick branches (figure 38). In Sonoran populations, the branches are mahogany colored; in peninsular populations, the branches are gray. It has milky sap. The orbicular, oblong, or cordate leaves, 5–15 mm long, are mostly on short spur shoots. Leaves and young twigs are finely puberulent. Flowers are in solitary involucres 3–7 (occasionally up to 12) mm wide. The purplish glands have white, petal-like append-

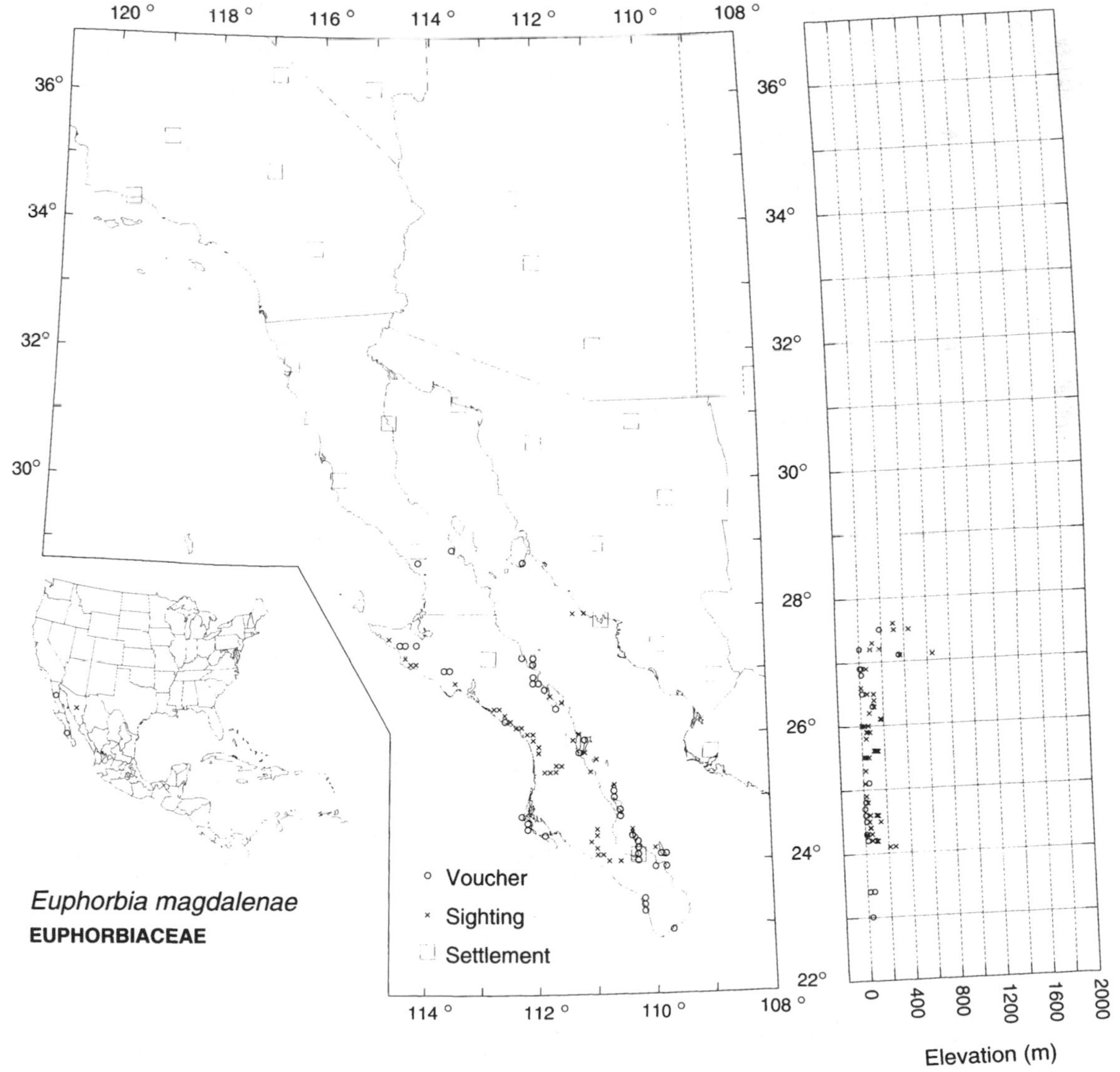

Euphorbia magdalenae
EUPHORBIACEAE

Figure 38. Close-up of Euphorbia misera *near Puerto Libertad, Sonora. Note inflorescence at center of photograph. (Photograph by R. M. Turner.)*

ages. The globose, 3-lobed capsules are 3.5–5 mm long.

Euphorbia californica has glabrous foliage. Also, the leaf blade margin is continuous across the upper petiole.

Euphorbia misera is locally common on gravelly flats, hillsides, and coastal bluffs and along washes. The northernmost outlier (33.9°N) is at Whitewater, California (Munz 1974). Sally Walker collected it in the Cape Region, Baja California Sur. If the scattered populations in northwestern Sonora are advance disjuncts, we would expect the species to continue moving north until limited by low temperatures, as *Jatropha cuneata* and *Solanum hindsianum* may have done.

The species grows in a variety of arid climates. The northwestern range along the Pacific coast experiences dry, relatively cool summers and winter rains. For southern and eastern populations, rain is equally likely in the cool or the warm season, and rainfall varies considerably from year to year. Frosts are uncommon to absent throughout the range. Interactions with *E. californica* have probably influenced the southern limits.

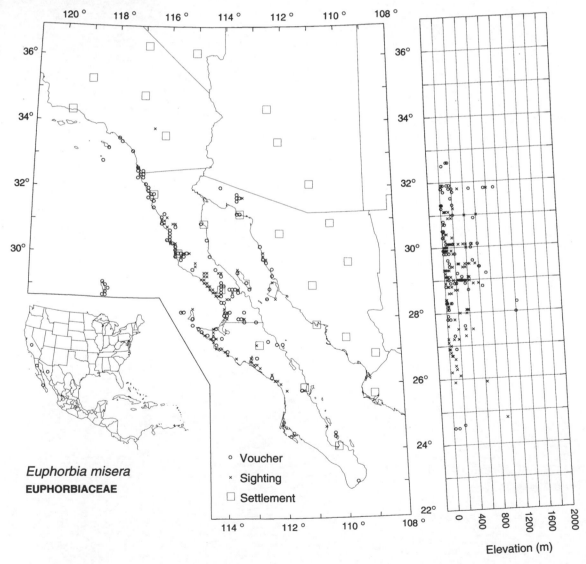

Euphorbia misera
EUPHORBIACEAE

○ Voucher
× Sighting
□ Settlement

Elevation (m)

Plants flower sporadically throughout the year. They are obligate outcrossers (Charles Hansen, personal communication 1993). Leaves appear soon after winter or summer rains and remain on the plant until the soil dries out (Humphrey 1975).

Seri Indians made a tea from the roots for dysentery, stomachache, and venereal disease (Felger and Moser 1985). The plants are occasionally cultivated and can be propagated from seed or cuttings (Charles Hansen, personal communication 1993).

Euphorbia tomentulosa

S. Watson

[=*Chamaesyce tomentulosa* Millsp.]

This low, rounded shrub 3–7 dm tall has repeatedly forked stems and a pair of lance-subulate stipules at each node. Young twigs are reddish, older ones gray or white. The opposite, serrate leaves, 5–17 mm wide and long, are oval to suborbicular. The inflorescences, bunched toward the branch tips, are of crowded involucres. The petal-like appendages are white, pinkish, or purplish. The 3-angled, globose capsules are 2–2.8 mm long.

Euphorbia magdalenae has narrower, oblong-ovate leaves 2.5–7 mm wide and stipules united into a single narrow scale. *Euphorbia xanti* is a clambering, green-branched shrub up to 2 m tall. *Euphorbia misera* has thick, knobby branches with alternate, ovate or orbicular leaves on short shoots.

Rare to common on cliffs, rocky slopes, stable dunes, and plains and along washes, *E. tomentulosa* is widespread in the arid parts of the Baja California peninsula and spotty in Sonora. It grows from sea level to 1,225 m. The species may have migrated from the peninsula to the mainland via the midriff islands (Cody et al. 1983). According to Wiggins (1964), its Sonoran range extends northward to Altar (30.7°N, 111.8°W).

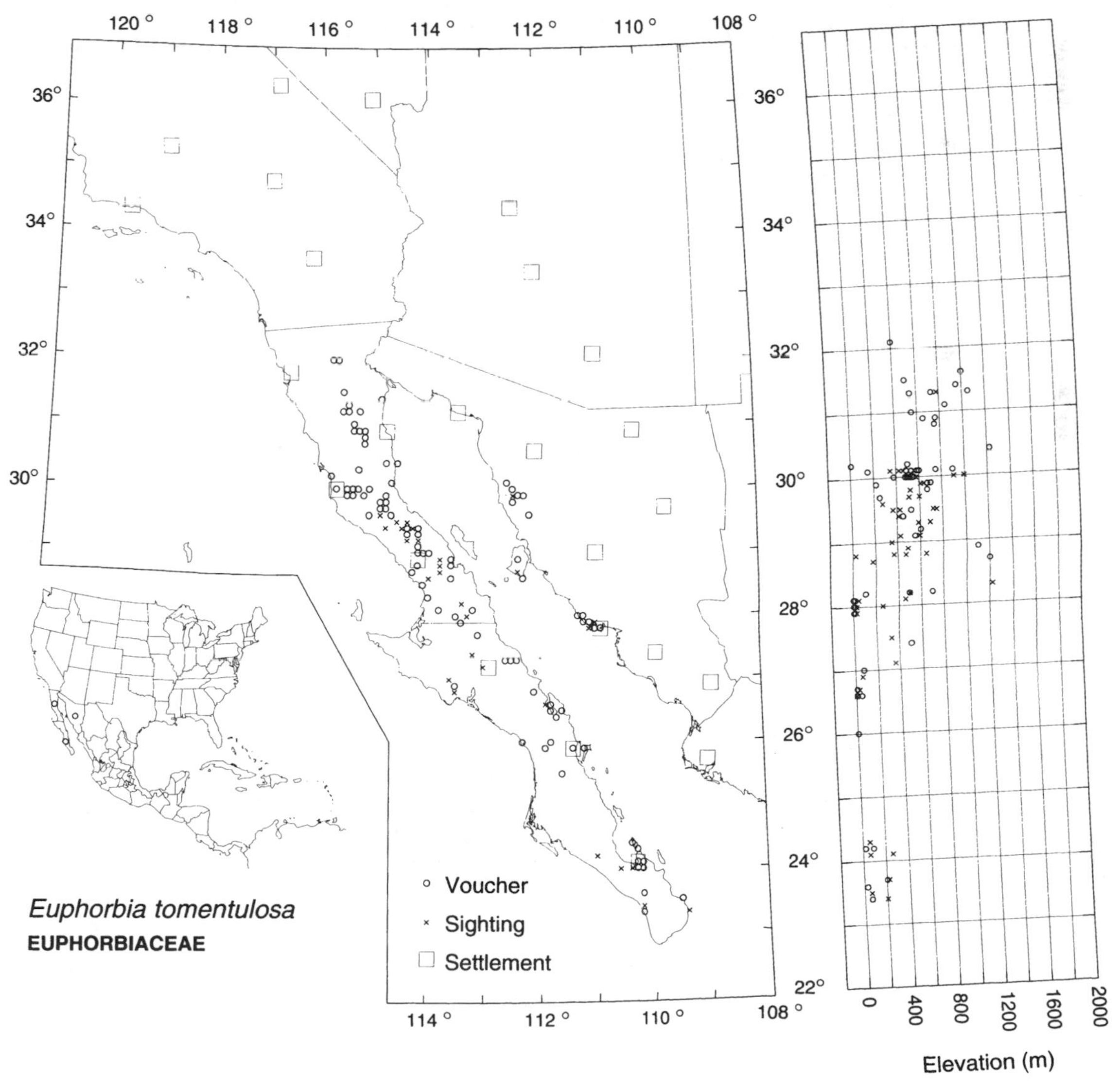

Euphorbia tomentulosa
EUPHORBIACEAE

Rain is biseasonal over most of the range. In the northwest, winter rain predominates; near the eastern and southern limits, most of the rain is in summer. Frosts are uncommon to absent.

Flowering occurs year-round in response to rain. During protracted drought, the plant keeps only a few immature leaves.

Euphorbia xanti

Engelm. ex Boiss. in A. DC.

Jumetón, liga

Euphorbia xanti is a scandent shrub 0.5–2 m tall with slender, straight branches that often clamber among other shrubs and trees. The linear or elliptic leaves, 1.5–3.5 cm long and 1–24 mm wide, are opposite or whorled. Involucres are in terminal cymes. The black or purplish floral glands have white or pink petal-like appendages 3–4 mm long. The globose, 3-lobed capsules are 3–4 mm long and 4–5.5 mm across.

Euphorbia plicata S. Watson and *E. colletioides* Benth. have larger leaves (2–8 cm long) and smaller floral appendages (1.0–1.5 mm long). *Euphorbia misera* has thick, knobby branches and is not a clambering plant. *Euphorbia tomentulosa* and *E. magdalenae* are irregularly branched subshrubs under 1 m tall with small oval, suborbicular, or oblong leaves that are never whorled.

Euphorbia xanti is locally common in washes, on sandy flats, and on rocky slopes, most often in the shelter of other shrubs and trees. Cody and coworkers (1983) suggested that the species migrated from the peninsula to the mainland via the midriff islands. The northernmost point on the mainland is about 22.5 km east of Puerto Libertad, Sonora. Ac-

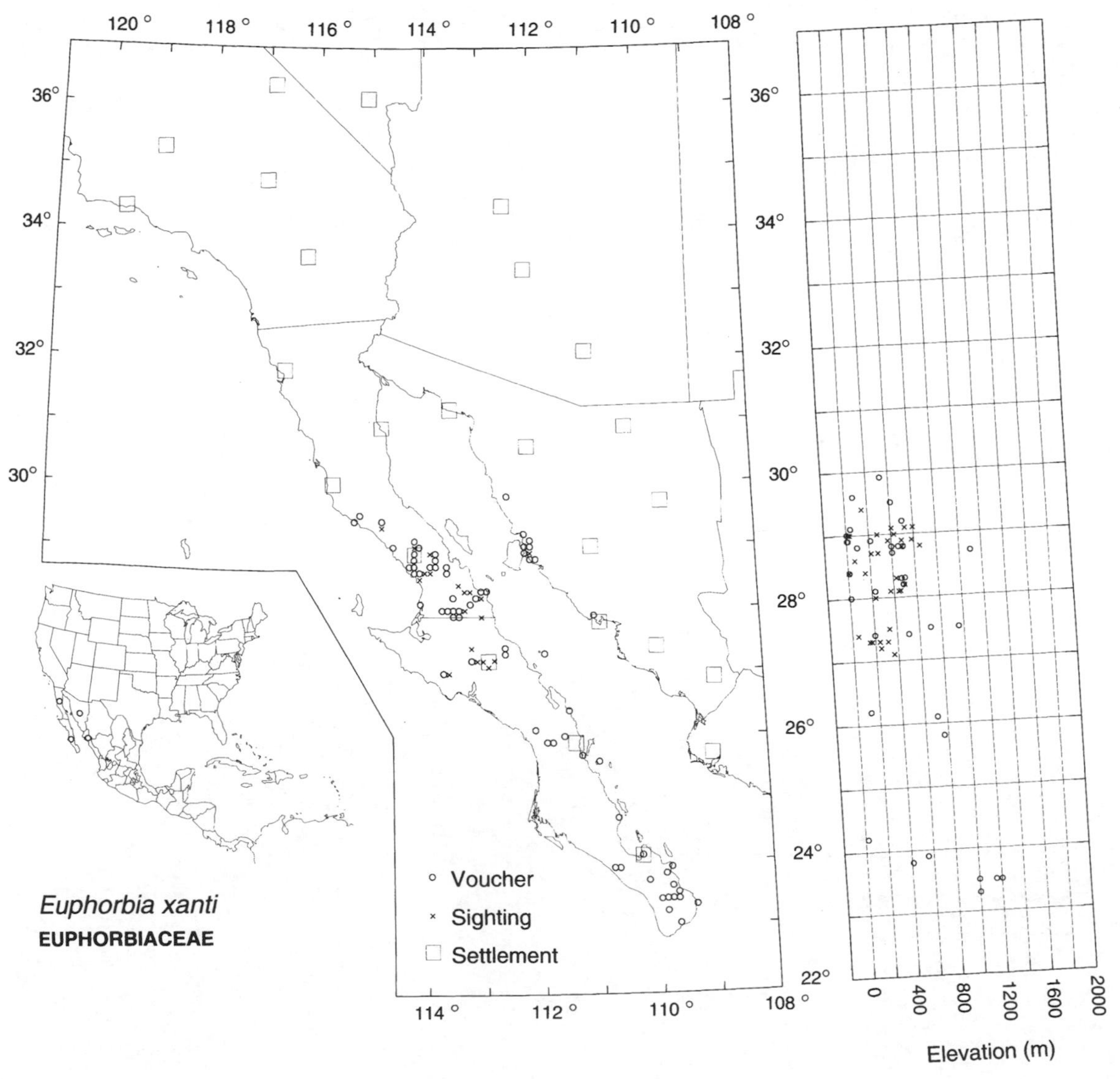

Euphorbia xanti
EUPHORBIACEAE

cording to Richard Felger (personal communication 1992), the plants grow from the vicinity of Bahía Kino northward to El Desemboque.

The species occurs in rainfall regimes ranging from predominantly cool-season in its northern peninsular range to almost entirely warm-season in the Cape Region.

Flowers appear from October through May, usually after rains. Stems and leaves grow in response to summer and winter rains. Some or all of the leaves may drop in the arid foresummer. Rather large clumps can be formed from root sprouting.

The plants are not self-compatible (Charles Hansen, personal communication 1992). Hummingbirds have been seen at the flowers, but it is doubtful if they achieve much pollination.

Euphorbia xanti is occasionally grown as an ornamental in southern California and Arizona. Although frost sensitive, the stems regrow quickly if water is supplied during warm weather. The plants are easily propagated from stem cuttings or seed (Charles Hansen, personal communication 1992).

Eysenhardtia orthocarpa

(A. Gray) S. Watson

Kidneywood, palo dulce

Eysenhardtia orthocarpa (figure 39) is a shrub or small tree 1.5–6 m tall. The odd-pinnate leaves, 5–15 cm long, have 21–41 oblong leaflets 8–20 mm long and may be finely puberulent to glabrate. The underside is punctate with minute, dark glands. The white flowers, 5–7 mm long, are in dense racemes 5–12 cm long. The pods are 1-seeded and indehiscent.

The closely related *E. polystachya* (Ortega) Sarg. does not occur in the Sonoran Desert region. *Eysenhardtia reticulata*

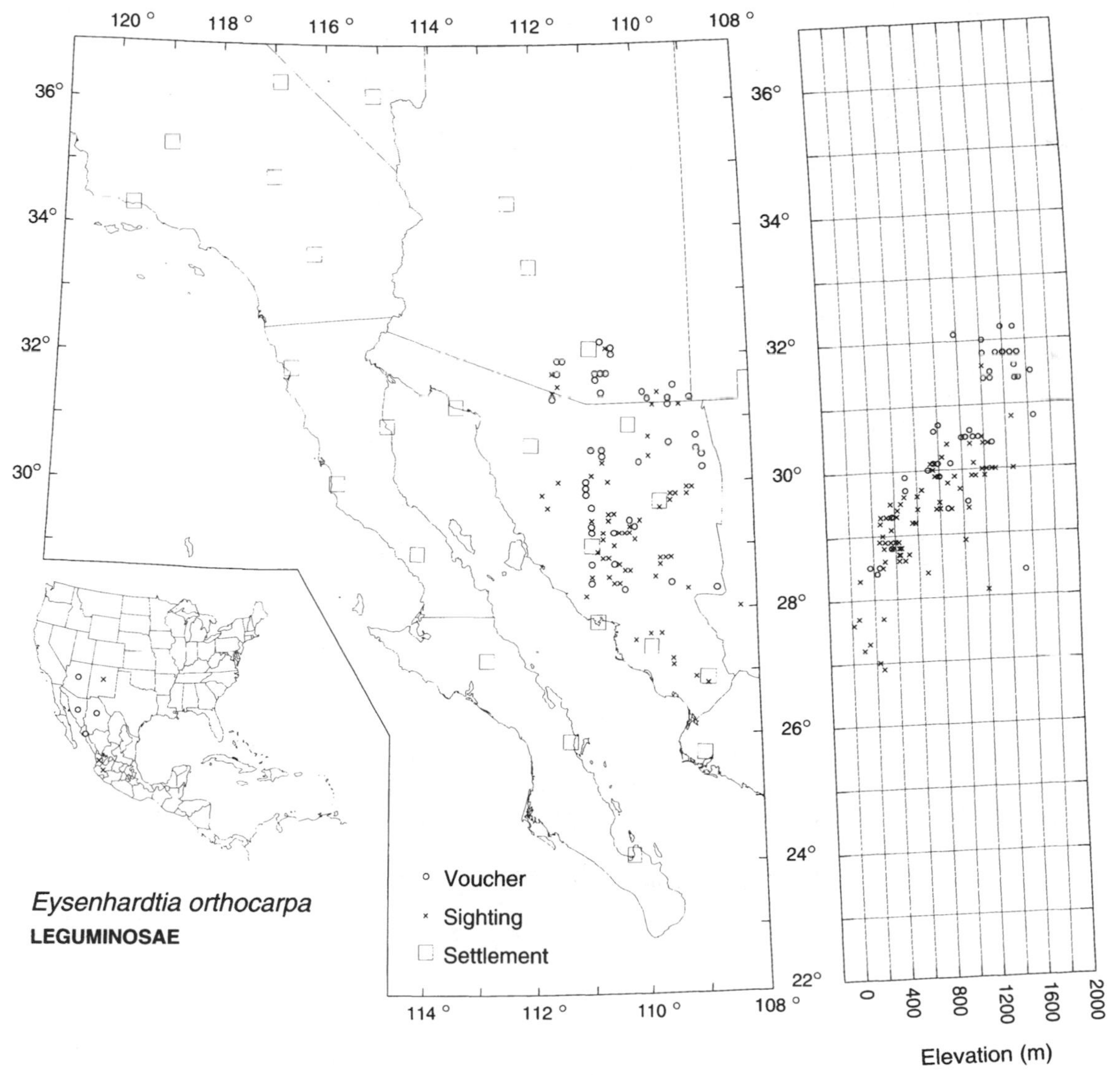

Eysenhardtia orthocarpa
LEGUMINOSAE

Pennell, found in southern Sonora and farther south, has glabrous, strongly reticulate, conspicuously glandular leaflets. The various species of *Brongniartia* and *Diphysa* do not have conspicuously glandular-punctate leaves, nor does *Coursetia glandulosa. Amorpha fruticosa* L. differs in its larger leaflets (1.5–5 cm long and 0.9–2 cm wide) and the skunk-like odor of its foliage.

The minor amount of morphological differentiation within *Eysenhardtia* suggests that species are in the early stages of evolutionary divergence (Lang and Isely 1982). *Eysenhardtia orthocarpa* and *E. polystachya* have been combined by some authors, for example, Standley (1922), Kearney and Peebles (1960), and Benson and Darrow (1981). We follow Lang and Isely (1982) in separating them. The two varieties are var. *orthocarpa*, with minutely pubescent to canescent leaflets, and var. *tenuifolia* Lang, which is glabrous (Lang and Isely 1982). Variety *orthocarpa* is a plant of Madrean evergreen woodland, var. *tenuifolia* of Sonoran desertscrub and Sinaloan thornscrub.

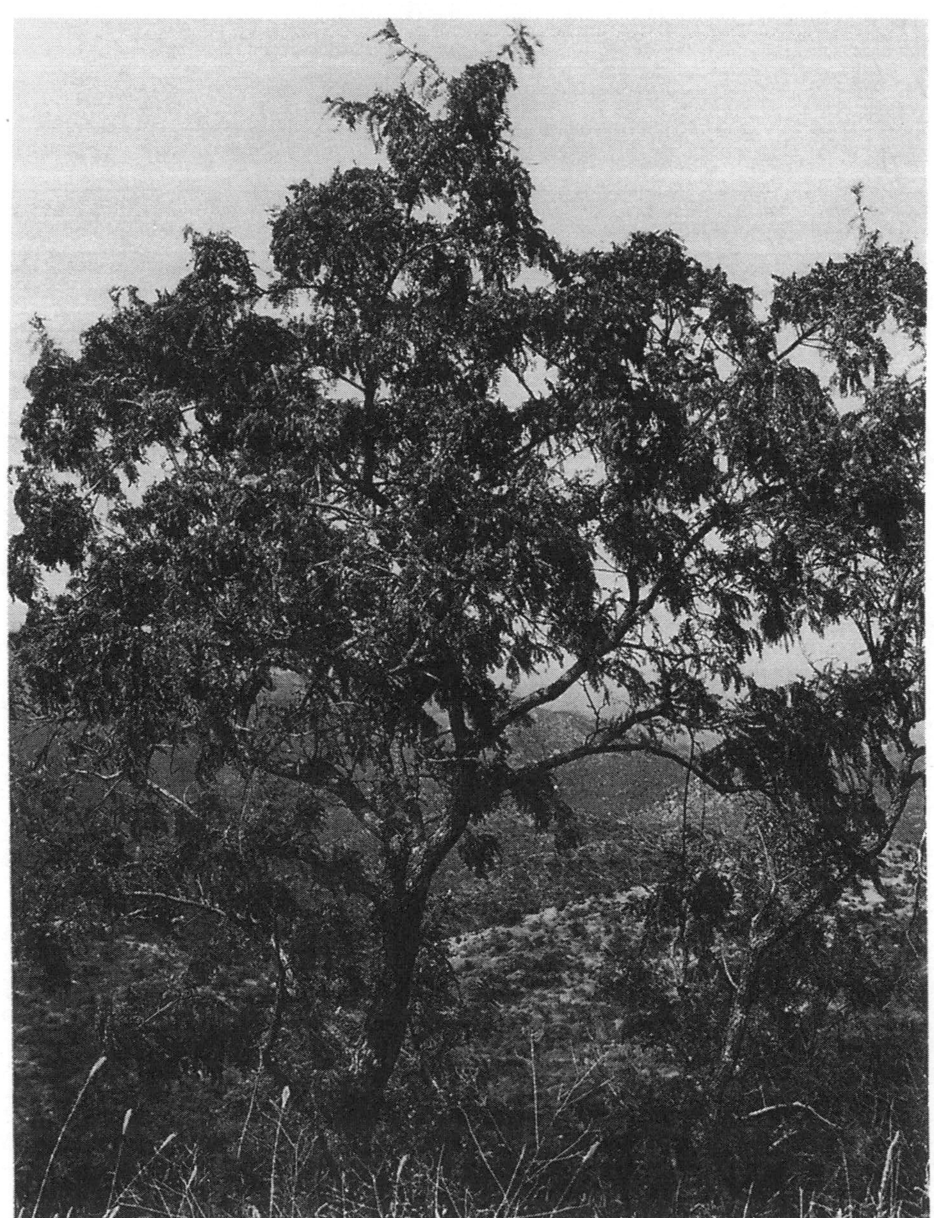

Figure 39. Eysenhardtia orthocarpa *east of Ures, Sonora, at 1,200 m. This site, typical for the species, is transitional between thornscrub and oak woodland. (Photograph by J. R. Hastings.)*

At the northern end of its range, *E. orthocarpa* grows along watercourses and on canyon slopes and hillsides at the upper margin of Sonoran desertscrub and lower edge of Madrean evergreen woodland. These northern populations are disjunct on isolated mountain ranges. Farther south, where *E. orthocarpa* drops below the woodland, its distribution is more continuous. In the Santa Catalina Mountains, Pima County, Arizona, the plants suffer frost damage in the coldest winters. Johnson (1988, 1993) reported that they are hardy to –6.5°C. The range appears to be confined within an area of dependable summer rainfall and relatively mild winters.

Flowers appear sporadically from April to September, apparently in response to winter and summer rains. Leaves are cold deciduous at higher latitudes and elevations and are drought deciduous throughout the range.

Infusions of *E. polystachya* wood have been widely used in Mexico to treat urinary disease (Standley 1922). The strong wood is used for tool handles and dwellings (Gentry 1942). The plants are suitable for ornamental use and can be propagated from cuttings or by seed. Seeds need not be scarified first (Johnson 1988). Growth is rapid with supplemental irrigation (Johnson 1988).

Ferocactus cylindraceus

(Engelm.) Orcutt

[=*F. acanthodes* (Lem.) Britton & Rose]

Barrel cactus, bisnaga, compass plant

This single-stemmed, cylindrical cactus is 1–3 m tall and 30–40 cm in diameter and has 18–30 ribs nearly obscured by dense straw-colored, yellowish, or reddish spines. Rib spacing and depth vary with compass direction (Geller and Nobel

1984). The spines in each cluster usually grade from fine bristlelike radials to stout centrals. Two of the 4–7 central spines may be distinctly flatter than the others. The lowermost, largest central spine, 7–17 cm long, is curved or (sometimes) hooked. The yellow flowers may be tinged with red, especially on the outermost sepals. The haploid chromosome number is 11 (Pinkava et al. 1977).

The principal central spine of *F. wislizeni* is much more flattened and markedly hooked. *Ferocactus gracilis* H. Gates, common in the Vizcaíno region, has red flowers, and its spines are more clearly differentiated into stout, colored centrals and fine, white radials. *Ferocactus emoryi* lacks the bristlelike radial spines characteristic of *F. cylindraceus*. *Ferocactus viridescens* (Torr. & A. Gray) Britton & Rose is usually a small plant (less than 30 cm tall) with spines less than 5 cm long.

Taylor (1979) proposed *F. cylindraceus* as a replacement for the name *F. acanthodes*. He recognized var. *cylindraceus*, var. *lecontei* (Engelm.) H. Bravo-H., var. *eastwoodiae* (L. Benson) N. P. Taylor, and var. *tortulispinus* (H. Gates) H. Bravo-H. Our sightings seldom distinguish among the four, so we have combined them here.

Variety *cylindraceus* and var. *lecontei* grow on gravelly or rocky hillsides, canyon walls, alluvial fans, and wash margins. Variety *eastwoodiae* is found most often on inaccessible ledges (Benson 1982). The plants are sensitive to salinity and show little tolerance to boron (Nobel 1983a; Berry and Nobel 1985).

This species grows on both sides of the Gulf of California. Many of our points in the Vizcaíno region may actually be *F. gracilis*, which surrounds it on the peninsula (Taylor 1984). The southernmost disjunct in Sonora (28.5°N) is from a sighting by Raymond M. Turner near Estero Tastiota. Variety *eastwoodiae* is restricted to a small area on the northeastern margin of the Sonoran Desert (Benson 1982). Variety *tortulispinus* is endemic to Baja California, where it grows from El Crucero to Calamajué and Laguna Chapala (Taylor 1984). Fossil material of *F. cylindraceus* from Picacho Peak in central Arizona has been dated to 13,200 years B.P. (Van De-

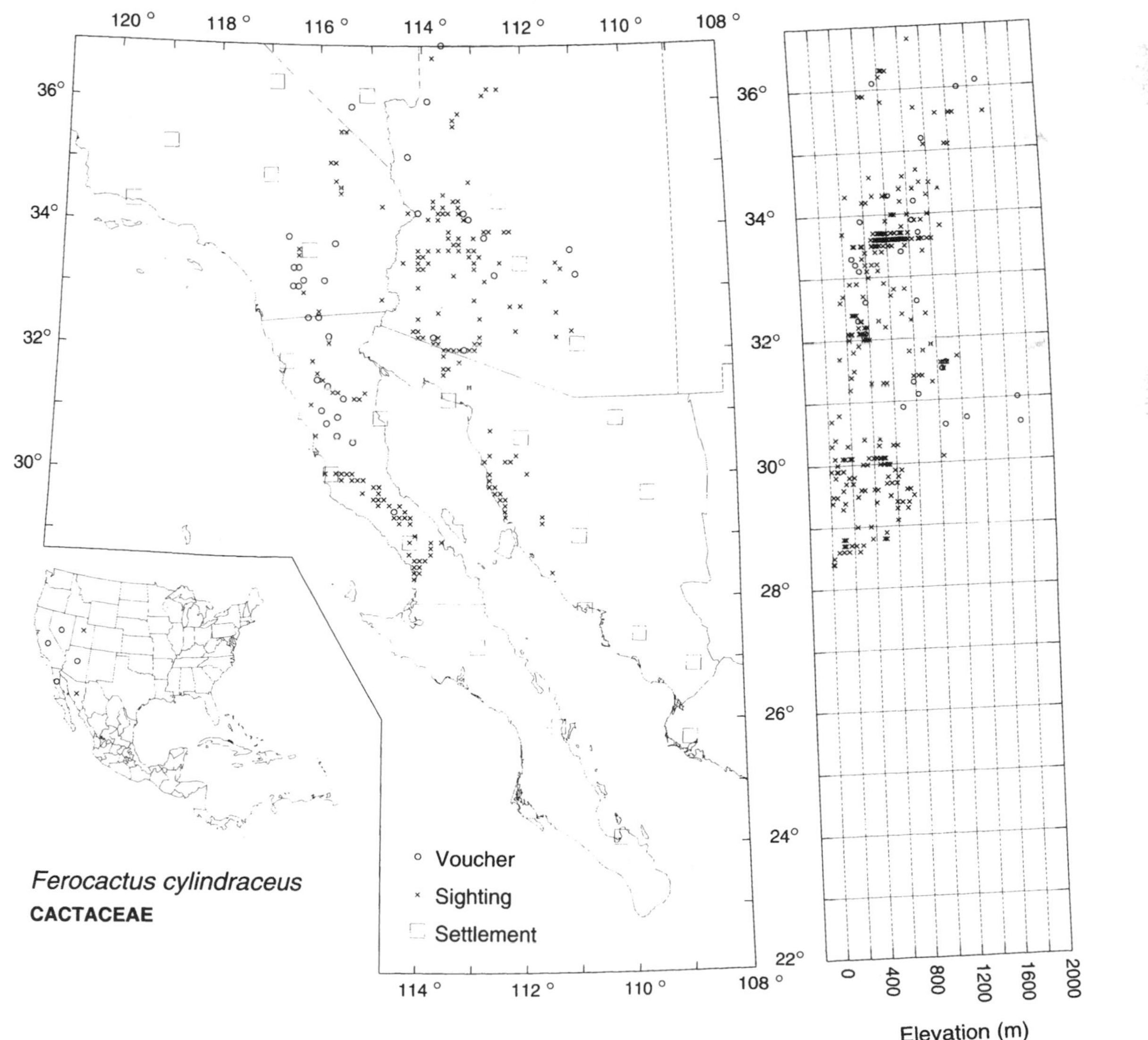

Ferocactus cylindraceus
CACTACEAE

vender 1990b); fossil material from Picacho Peak, California, to 12,400 years B.P. (Cole 1986).

Ferocactus cylindraceus is limited in its northward and upward distribution by winter temperatures (Nobel 1980e). At the northern boundary in the Beaver Dam Mountains, Utah, the plants minimize winter frost damage by growing on south-facing slopes and tilting toward the south (Ehleringer and House 1984). Seedlings can withstand temperatures no lower than −8°C and often grow in protected microsites (Nobel 1984b). Competition for light may exclude *F. cylindraceus* from Californian coastalscrub and Californian chaparral. In the most arid portions of our region, not enough rain falls to allow seedling establishment (Nobel and Hartsock 1986a).

Self-shading at the stem apex apparently plays a major role in determining the relative distribution of *Ferocactus* species in the Sonoran Desert region (Nobel 1980e). In *F. cylindraceus,* spines and pubescence shade 65–95% of the stem apex, with the most shading at the highest elevations and latitudes (Nobel 1980e). In winter, minimum apical temperatures are significantly higher in *F. cylindraceus* than in *F. emoryi* and *F. wislizeni,* enabling it to grow farther north (Nobel 1980e) and at higher elevations.

The plants usually bloom during late spring or early summer, occasionally in late summer. Solitary bees are probably the most important pollinators. Ants visit extrafloral nectaries on the plant apex and may protect the cactus from insect herbivores (Ruffner and Clark 1986; Pemberton 1988). The same phenomenon has been reported for *F. gracilis* (Blom and Clark 1980).

As a CAM plant, *F. cylindraceus* minimizes transpirational water loss by nighttime stomatal opening. Shallow root depth (about 8–15 cm) allows rapid response (within 24 hours) to infrequent rains, and the succulent stems permit considerable water storage (Nobel 1977b). Water stored in the stem permits nighttime stomatal opening for up to 40 days after the soil water potential drops below that of the stem, even though plants are unable to extract water from the soil (Nobel 1977b). About one-third of the water transpired at night comes from storage tissues (Schulte et al. 1989).

The roots essentially function like rectifiers, readily removing water from wet soils but losing little to dry soils (Palta and Nobel 1989b). After rains, the perennial roots produce ephemeral rain roots (Jordan and Nobel 1984). As the soil dries out, both types of roots undergo reduced respiration. For rain roots, respiration becomes zero at a soil water potential of −1.0 MPa. At this point, respiration is irreversibly inhibited, and the roots are unable to rehydrate (Palta and Nobel 1989b). For established roots, the respiration rate becomes zero at −1.6 MPa, and even after the soil dries out, full recovery occurs within a day of soil wetting (Palta and Nobel 1989b). Larger plants explore less ground area per unit of shoot surface area than do smaller plants (Hunt and Nobel 1987). The optimum soil temperature for water uptake increases with root age (Lopez and Nobel 1991). High soil carbon dioxide inhibits root growth, perhaps restricting the plants to well-drained soils (Nobel and Palta 1989; Nobel 1990).

The ribbed structure of the stems enables them to swell upon taking up water and to shrink during dry periods. About 6.1% of the water that enters the stem is stored (Nobel 1977b). At the beginning of a dry period, stomates are active and the plants draw on stored water. Transpirational water loss at such times is about 0.71 g/cm^2, or about 15.4 kg/yr/plant. As seasonal drought intensifies, water lost in transpiration is not replenished by water uptake from the soil. Stomatal activity ceases, and stem osmotic pressure may drop to −0.6 MPa, more than twice that measured during wet periods (Nobel 1977b). Even so, mature plants are not particularly susceptible to drought and can survive stem water loss of up to 81% (Barcikowski and Nobel 1984). During drought, a greater fraction of stem water is lost from water storage tissue than from chlorenchyma (Barcikowski and Nobel 1984).

Laboratory experiments showed that at a day/night temperature regime of 23/14°C, net carbon dioxide uptake over 24 hours was 118 nmol/m^2, the highest value. Net carbon dioxide uptake was approximately halved at day/night temperatures of 11/5°C and 32/23°C (Nobel 1986). Optimum temperature for carbon dioxide uptake was quite low, near 12.4°C (Nobel 1977b). When plants were grown at 10°C, the thermal optimum for photosynthesis was 12°C. The optimum shifted to 22°C when plants were grown at 30°C (Nobel and Hartsock 1981). Stomatal opening on cool nights conserves water but supposedly restricts the species to regions of winter rain, where cool nights and moist soils are most likely to coincide (Nobel 1977b). Net carbon dioxide uptake is severely limited by droughts of 10–30 days and temperatures below 0°C (Nobel and Hartsock 1986a).

Ferocactus cylindraceus is quite tolerant of heat. In the laboratory, plants can be acclimated to withstand day/night temperatures of 50/40°C (Kee and Nobel 1986). Under natural conditions, the dense spines reduce the daytime rise in stem temperature. Without spines, the stem surface would undergo diurnal temperature changes of 23°C in winter, 41°C in summer. With spines, the daily variation in one study was 17°C in winter, 25°C in summer (Lewis and Nobel 1977). Ribbing also reduces stem surface temperatures. Normally ribbed plants have 54% more area for convective heat loss than would hypothetical unribbed plants (Lewis and Nobel 1977).

Jordan and Nobel (1981) found that seed germination in the laboratory was greatest at 29°C. In nature, this would correspond to summer or early fall temperatures. For 70 days after germination, the bulbous hypocotyl is the major photosynthetic organ. Seedlings grow rapidly for the first fifty days, more slowly (0.28 cm/100 days) thereafter. Under natural conditions, a one-year-old seedling may be 1.5 cm tall with a volume/surface ratio of 0.27

cm^3/cm^2 (Jordan and Nobel 1981). Seedlings grow about 9 mm/yr after the first year (Jordan and Nobel 1982).

Their relatively small volume/surface ratio makes the seedlings vulnerable to drought. The larger the ratio, the longer they can survive a drought. Lethal water stress in one study occurred when 84% of the tissue water volume was lost. Jordan and Nobel (1981) calculated that a 210-day-old seedling would have a volume/surface ratio of 0.19 cm^3/cm^2 and would lose 84% of its volume after a drought of 110 days. In Deep Canyon, Santa Rosa Mountains, California, drought is frequent enough that only eight years between 1962 and 1980 were suitable for establishment (Jordan and Nobel 1981). Other populations also have shown intermittent seedling establishment (Jordan and Nobel 1982).

Although cultivated seedlings can be acclimated to withstand tissue temperatures over 60°C (Nobel 1984b), soil surface temperatures may reach lethal levels under natural conditions, making establishment more likely in sheltered microsites (Nobel 1984b, 1989c). In the northwestern Sonoran Desert, the seedlings often use the perennial grass *Pleuraphis rigida* as a nurse plant. Maximum soil surface temperatures may be reduced by as much as 10°C under its canopy (Nobel 1989c). Higher soil nitrogen under *P. rigida* also facilitates seedling establishment (Franco and Nobel 1989).

During seasonal drought, bighorn sheep survive without surface water by eating *Ferocactus* tissue (Warrick and Krausman 1989). Seri Indians considered the juice of *F. cylindraceus* unpalatable (Felger and Moser 1985), but pounding the soft pulp provides a considerable quantity of moisture in an emergency (Benson 1982). This procedure destroys the plant and cannot be recommended except in cases of dire need. The Seri ground the seeds for gruel (Felger and Moser 1985). Candy makers boil the soft tissue of the pith and cortex, then add sugar and flavorings (Benson 1982).

The plants are occasionally grown as ornamentals. Propagation by tissue culture is possible (Ault and Blackmon 1987). Variety *eastwoodiae* is threatened and should not be collected from the wild (Benson 1982).

Ferocactus emoryi

Britton & Rose

[=*F. covillei* Britton & Rose]

Emory barrel, bisnaga

This ovoid or barrel-shaped cactus is 0.6–1.5 (occasionally 2.5) m tall and 30–100 cm in diameter and has 15–30 ribs (figure 40). In each spine cluster, the single, strongly curved (but not hooked) central spine, red or ashy gray, is surrounded by 7–9 thick radial spines of the same color and texture. Bristlelike spines are lacking. Flowers are red or yellow, about 7.5 cm long and broad. The spineless fruits, 2.5–4.5 cm long and 2.5–3 cm in diameter, are yellow at maturity.

Ferocactus peninsulae (F. A. C. Weber) Britton & Rose, *F. wislizeni,* and *F. cylindraceus* usually have bristlelike radial spines or, if not, have more than 10 spines per cluster. In *F. diguetii* (Weber) Britton & Rose, the 4–8 spines are not clearly differentiated into radials and centrals.

According to Taylor (1984), *F. emoryi* is the correct name for this species. He recognized var. *emoryi* from southern Arizona, Sonora, and northern Sinaloa and var. *rectispinus* (Engelm.) N. P. Taylor from Baja California Sur. Variety *rectispinus* has longer central spines (up to 26 cm long). *Ferocactus peninsulae* and *F. em-*

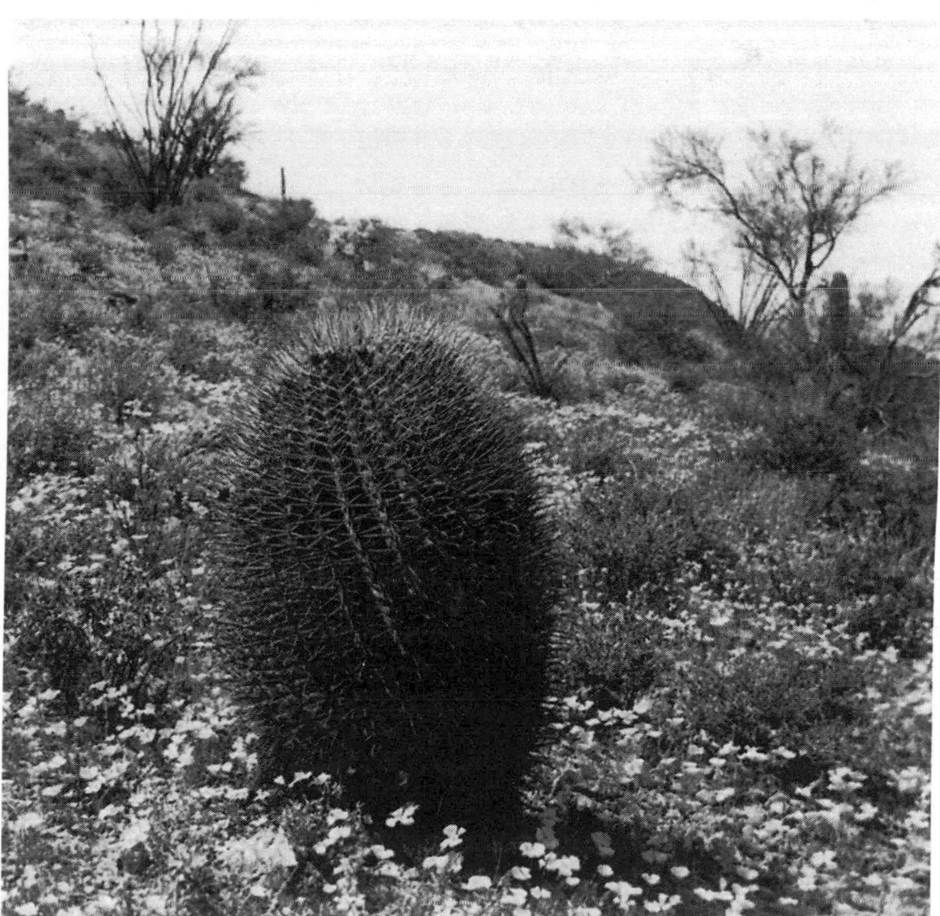

Figure 40. Ferocactus emoryi *with* Eschscholzia californica *south of Ajo, Arizona. (Photograph by L. M. Huey, courtesy of University of Arizona Library Special Collections Department.)*

oryi var. *rectispinus* are sympatric, and much introgression between them has occurred (Taylor 1984).

Ferocactus emoryi grows in gravelly, rocky, or sandy soil on hillsides, wash margins, alluvial fans, mesas, or flats (Benson 1982). The eastern disjunct at 32.3°N, 110.7°W represents a collection made by Lyman Benson in 1939 in Redington Canyon, Santa Catalina Mountains. Mary Butterwick made the northernmost sighting (33.2°N) in the Sierra Estrella. The overall distribution suggests intolerance of prolonged low temperatures and tolerance of aridity as long as summer rainfall is adequate.

Our maps show *F. emoryi* between *F. wislizeni* in the eastern part of the Sonoran Desert region and *F. cylindraceus* in the northwestern section. Across its range, *F. emoryi* has lower amounts of summer rainfall than most populations of *F. wislizeni* and more dependable summer rains than most populations of *F. cylindraceus.*

Ferocactus emoryi shows less tissue tolerance to cold and less ability to acclimate to cold than *F. cylindraceus* or *F. wislizeni* (Nobel 1982a). Moreover, because the stem apex of *F. emoryi* is scarcely protected against cold temperatures (only 16–18% of the stem tip is self-shaded by spines or pubescence), it should be more vulnerable to low temperatures than either *F. wislizeni* (46–52% self-shaded) or *F. cylindraceus* (65–95% self-shaded) (Nobel 1980e). Our elevational profiles show that as Nobel (1980e) predicted, the upper elevational limits of these *Ferocactus* species do differ substantially—1,220 m for *F. emoryi,* 1,494 m for *F. wislizeni,* and 1,750 m for *F. cylindraceus.* Fossil midden assemblages from the Puerto Blanco Mountains of southwestern Arizona show *F. emoryi* replacing *F. wislizeni* and *F. cylindraceus* in the late Holocene (about 3,400 years B.P.), apparently in response to warmer, drier conditions, then declining in abundance to its present status (Van Devender 1987).

Blooming from June to September, *F. emoryi* is probably pollinated by medium-sized solitary bees as described for *F. wislizeni* by Grant and Grant (1979a).

Seri Indians believed that *F. emoryi*

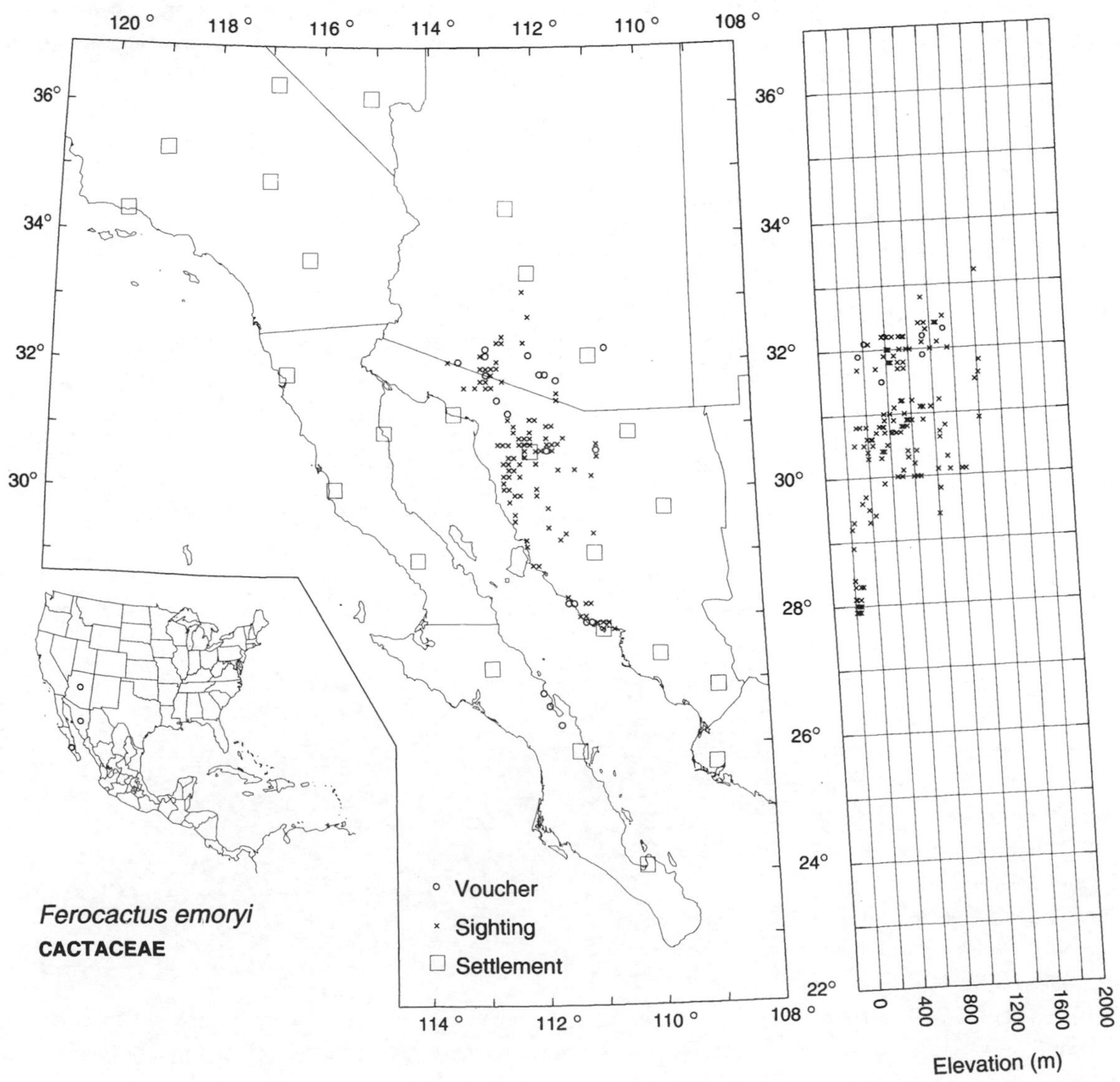

juice was poisonous and therefore was not suitable as an emergency water supply. They did eat flower buds, flowers, fruits, and seeds (Felger and Moser 1985).

Ferocactus wislizeni

(Engelm.) Britton & Rose

Barrel cactus, bisnaga, compass plant

Ferocactus wislizeni (figure 41) is a barrel-shaped or columnar cactus 0.6–1.6 (rarely up to 3) m tall and 45–83 cm in diameter with 20–30 ribs. The stems are unbranched except in case of injury and may weigh 80–100 kg (MacDougal 1912). The spine clusters are of 4 sturdy central spines and 12–20 bristlelike or needlelike radial spines. The cup-shaped flowers are red, orange, or yellow and are about 5 cm in diameter (Grant and Grant 1979a). The yellow fruits may persist for up to a year (Humphrey 1936). The haploid chromosome number is 11 (Pinkava et al. 1973).

The larger, flattened, strongly hooked central spine distinguishes *F. wislizeni* var. *wislizeni* from *F. cylindraceus,* which usually has nearly straight or somewhat curved central spines. *Ferocactus emoryi* has 10 or fewer spines per cluster and always lacks bristlelike spines.

Taylor (1984) recognized three varieties. Variety *wislizeni,* the northern taxon, has thicker stems; whitish, bristlelike radial spines; and a strongly flattened and hooked central spine. Variety ***herrerae*** (J. G. Ortega) N. P. Taylor, the southern taxon, has thinner stems; about 13 ribs, often spiraled; fewer or no radial spines, at least on larger stems; and a central spine that is neither flattened nor hooked. The two intergrade in southern Sonora (Lindsay 1955; Taylor 1984). The central spine of var. ***tiburonensis*** G. Lindsay, endemic to Isla Tiburón, is not hooked, and bristlelike radial spines are lacking.

Ferocactus wislizeni grows in a variety of habitats, including rocky, gravelly, or sandy soil of hills, flats, canyons, wash margins, and alluvial fans (Benson 1982). The plants are limited to the north by cold (Nobel 1980e, 1982a). In Texas, the

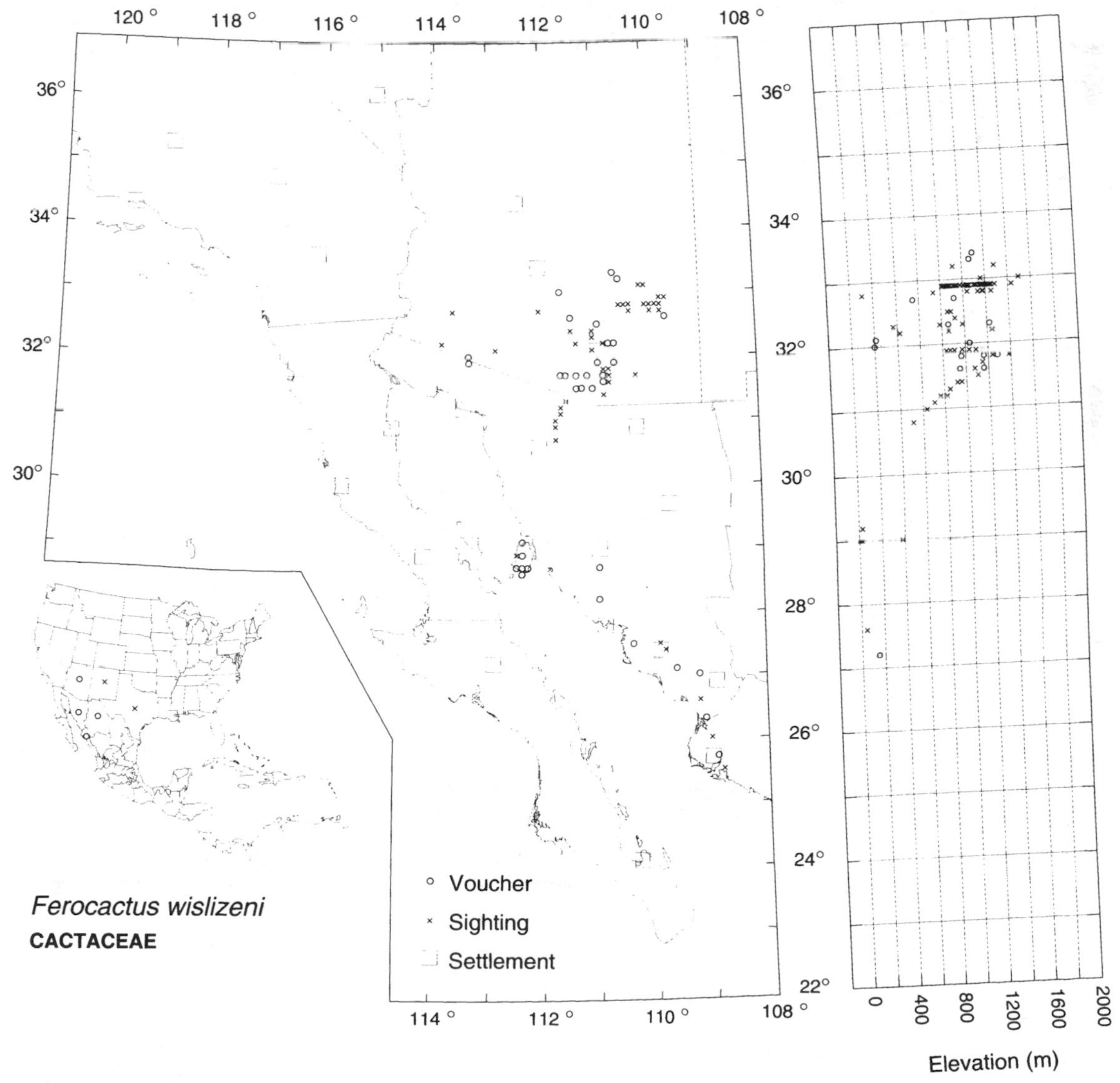

Ferocactus wislizeni
CACTACEAE

abrupt eastern limit of the species occurs on igneous substrates, perhaps because the prevalent calcareous substrates to the east are too xeric. The western limits appear to be shaped by summer aridity and interactions with *F. emoryi* and perhaps *F. cylindraceus.* The distribution of var. *wislizeni* coincides with the region of biseasonal precipitation where summer rainfall normally exceeds winter rainfall. Variety *herrerae* experiences much more summer rainfall than the other two varieties.

Self-shading of the stem apex by spines and pubescence increases with elevation and latitude, from 46% at lower elevations and latitudes to 52% at higher elevations and latitudes (Nobel 1980e). The frigid limits of *F. wislizeni, F. emoryi,* and *F. cylindraceus* correspond with their degree of apical shading. *Ferocactus cylindraceus,* the most heavily pubescent, is found farthest north and in the coldest sites; *F. emoryi,* the least pubescent, in lower, more southerly regions; and *F. wislizeni* in intermediate areas (Nobel 1980e).

Ferocactus wislizeni var. *tiburonensis* flowers in spring (Wiggins 1964), var. *wislizeni* in late summer. The flowers are pollinated mainly by medium-sized solitary bees, including *Megachile sidalceae, Lithurge echinocacti, Diadasia australis,* and *Idiomelissodes duplocincta* (Zavortink 1975; Grant and Grant 1979a). At maturity the stigma is exserted above the level of the anthers. Bees land on the stigma, crawl into the stamens, and emerge covered with pollen (Grant and Grant 1979a). Flowers last for about four days and are at least partly self-sterile (McGregor and Alcorn 1959). Ants visit extrafloral nectaries on the apex of the plant.

Botanists have advanced several explanations for the noticeable southward tilt of mature *F. wislizeni* plants. One is that the higher temperatures at a south-tilting stem apex could stimulate flower development (Nobel 1981a). Another is that due to unfavorable conditions for growth, the south side forms fewer cells than the north side, causing the plant to tilt (Humphrey 1936). A third possibility is that the southwest side is under greater water stress and grows more slowly, so that the plant tilts toward the southwest as it grows (Walter 1979).

Mature plants are heat tolerant (Smith et al. 1984) and extremely drought resistant (MacDougal 1912, 1915). An uprooted specimen taken into the laboratory lost 54% of its original weight before it died. Another lost 48% of its weight over 13 months and still produced flower buds at the usual time of year (MacDougal 1912). Juveniles and seedlings are no doubt much more susceptible to drought than mature plants, as Jordan and Nobel (1981) have shown for *F. cylindraceus.*

Ferocactus wislizeni seedlings grow extremely slowly. A seedling approximately 4 years old was 4 cm tall and 6 cm in diameter; 4 years later it was 11 cm tall and 12 cm in diameter (Shreve 1917). For the first 10–15 years, growth in diameter nearly equals growth in height (Shreve 1935).

Seedlings are more likely to become established in protected sites, as under vegetation or rocks (Humphrey 1936). Establishment is evidently rare and episodic. Over seven years, researchers discovered only one *F. wislizeni* seedling on a closely observed plot 557 m² in size (Shreve 1917). Maximum observed life span on Tumamoc Hill, Tucson, Arizona, was 65 years (Goldberg and Turner 1986). Shreve (1935) estimated that 85 years are required for plants to reach 1.2 m in height and an additional 35 to 45 years for them to reach 1.8 m. In certain portions of its range, *F. wislizeni* is not reproducing well, perhaps as a result of grazing or climatic change (Reid et al. 1983).

Cactus candy is occasionally made by boiling the succulent tissue, then adding sugar and flavorings. Seri Indians commonly obtained drinking liquid from *F. wislizeni* (Felger and Moser 1985). They also cooked the flowers for food, made gruel from the ground seeds, and used the hollowed-out cactus as a honey container (Felger and Moser 1985).

The plants are occasionally grown as ornamentals.

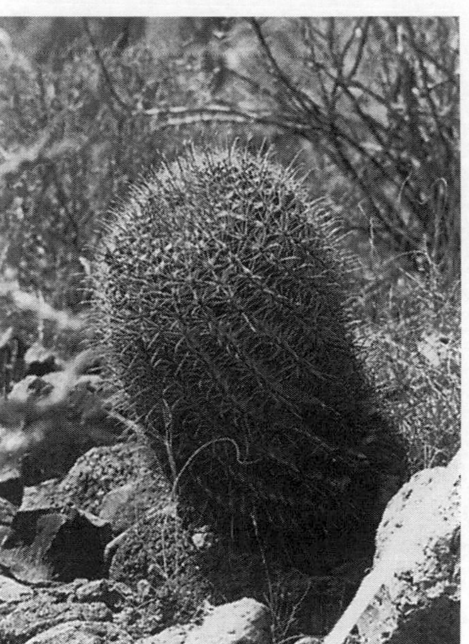

Figure 41. Ferocactus wislizeni *at the Desert Laboratory, Tucson, Arizona. (Photograph by J. R. Hastings.)*

Forchammeria watsoni

Rose

Palo San Juan, jito, palo jito

Forchammeria watsoni, a tree 3–8 m high, has smooth, light gray bark and a dense, rounded canopy. The simple (sometimes trifoliate), leathery leaves, 5–12 cm long and 4–12 mm wide, are linear or oblong with revolute margins. Plants are dioecious with small, unisexual flowers in racemes 1.5–6 cm long. The dryish, orange or purple drupes are 8–10 mm long.

Its dark green, linear, leathery leaves and dense canopy make *F. watsoni* easy to tell apart from other Sonoran Desert trees.

This species reaches its best development on the coastal plains of southern Sonora and Baja California Sur, where it is occasional to common. The Sinaloa sightings at 25.6°N are Howard Scott Gentry's. The Isla Tiburón sighting is David E. Brown's. In the 1970s, *F. watsoni* could be seen along the highway between Hermosillo and Ures, but the area has since been cleared for agriculture (David E. Brown, personal communication 1992). The dense cluster of points on either side of the gulf suggests long-distance dispersal from the peninsula to the mainland via the midriff islands (Cody et al. 1983).

The range is confined to essentially frost-free climates with warm, dry springs. At Sonoran sites and in the more southerly portion of the peninsular range, rain is concentrated in summer and early fall. The northern end of the peninsular range occasionally receives appreciable moisture from winter storms.

Forchammeria watsoni flowers in March and April. It seems likely that the blossoms are wind pollinated. Although some leaves drop at flowering, they are largely evergreen (Shreve 1964).

The leathery leaves are drought resistant. Shreve (1964) noted that little deadwood can be found on the plants, in contrast to such trees as *Cercidium* species and *Olneya tesota,* which generally lose twigs, branches, and even limbs in response to drought (Shreve 1964). This species could more properly be called "drought escaping" than "drought resistant" since its wood contains 47.5–60.6% water, a relatively high proportion for a tree (Shreve 1933).

The plants are often heavily browsed by livestock.

Fouquieria burragei
Rose

Palo adán, ocotillo blanco, pichilinge

Fouquieria burragei is a shrub 1–3 m tall with crooked, sparsely branched stems from several short, twisted trunks up to 8 dm tall and 1.5 dm in diameter. The dark, greenish bronze bark is waxy and exfoliates in sheets. The spiny stems have long-shoot leaves 17–35 mm long and 3–5.5 mm wide and short-shoot leaves 12–17 mm long and 6–10 mm wide. The rose-red to white, salverform corollas, 6–10 mm long and 3.5–4.5 mm wide, are in elongate, narrow, upright inflorescences that are panic-

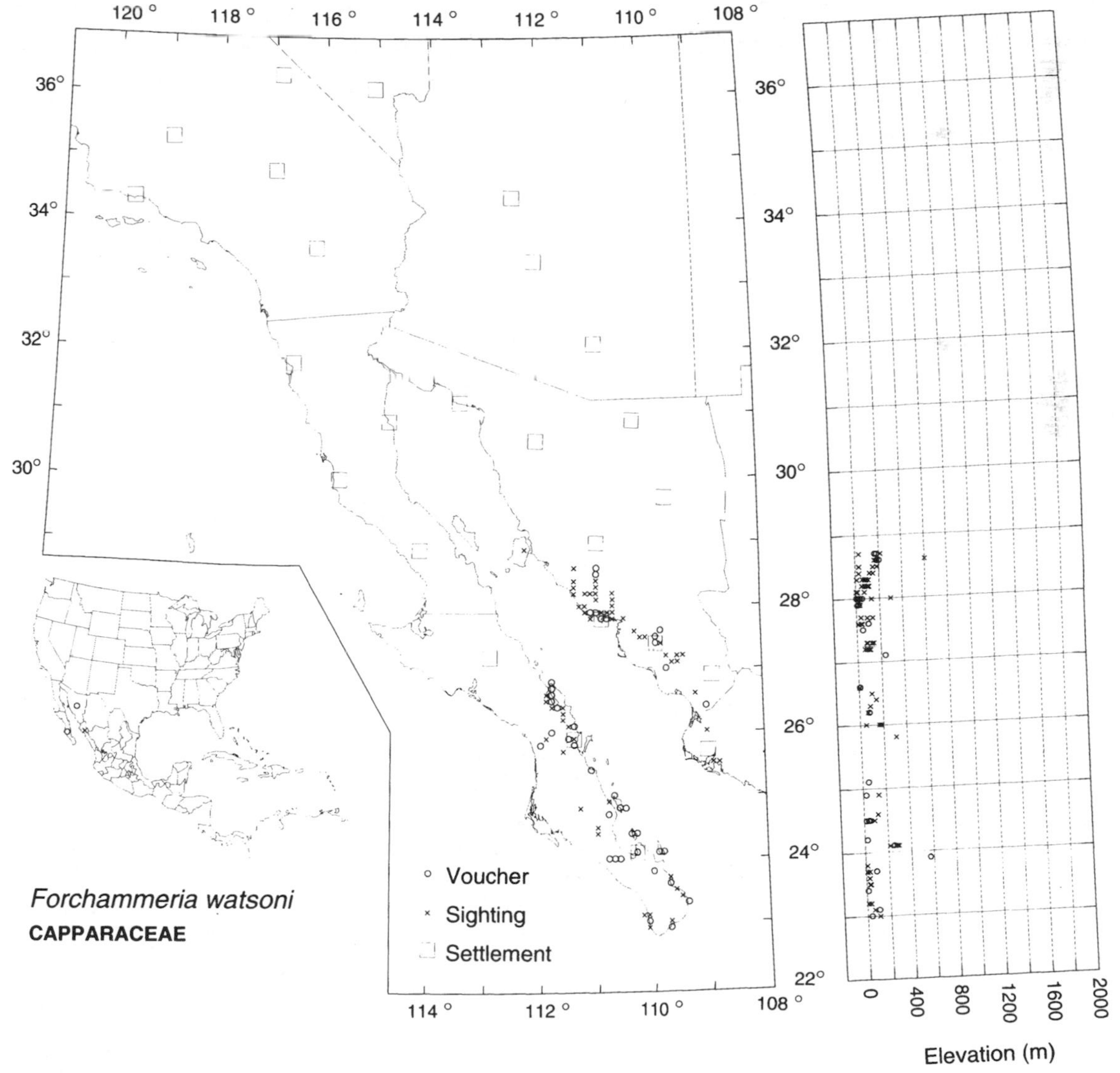

Forchammeria watsoni
CAPPARACEAE

ulate toward the base and racemose above. The 14–23 stamens are long-exserted.

Fouquieria diguetii has bright red flowers and 10 stamens barely or not exserted from the corolla.

Fouquieria macdougalii (2n=24), *F. diguetii* (2n=48), and *F. burragei* (2n=72) form a polyploid series (Henrickson 1972). *Fouquieria burragei* is clearly of amphiploid origin. The closely similar *F. diguetii* may be one parent, *F. splendens* or an extinct diploid species the other (Henrickson 1972).

Growing on rocky slopes in sandy to clay soils, *F. burragei* grows in scattered populations in Baja California Sur, mostly along the gulf coast below 200 m. Throughout the range, climates are tropical and arid. Occasional winter storms supplement the predominantly late summer and early fall rain. Most of the scattered populations are probably the remnants of a once-continuous distribution. Some may represent "founder populations" (Henrickson 1972). Variation in corolla size, corolla color, and degree of stamen exsertion from site to site is a result of limited gene flow between populations (Henrickson 1972).

Fouquieria burragei flowers from July through January in response to rain. June–September is the most reliable flowering season (Wiggins 1964). Leaves also appear after rains, then drop once the soil dries out. Long-shoot leaves appear on new stem growth in the summer, short-shoot leaves in the axils of the spines any time after rain.

Henrickson (1972) suggested that the short corolla tubes allow small insects to exploit floral resources, whereas the long-exserted stamens facilitate pollination by hummingbirds.

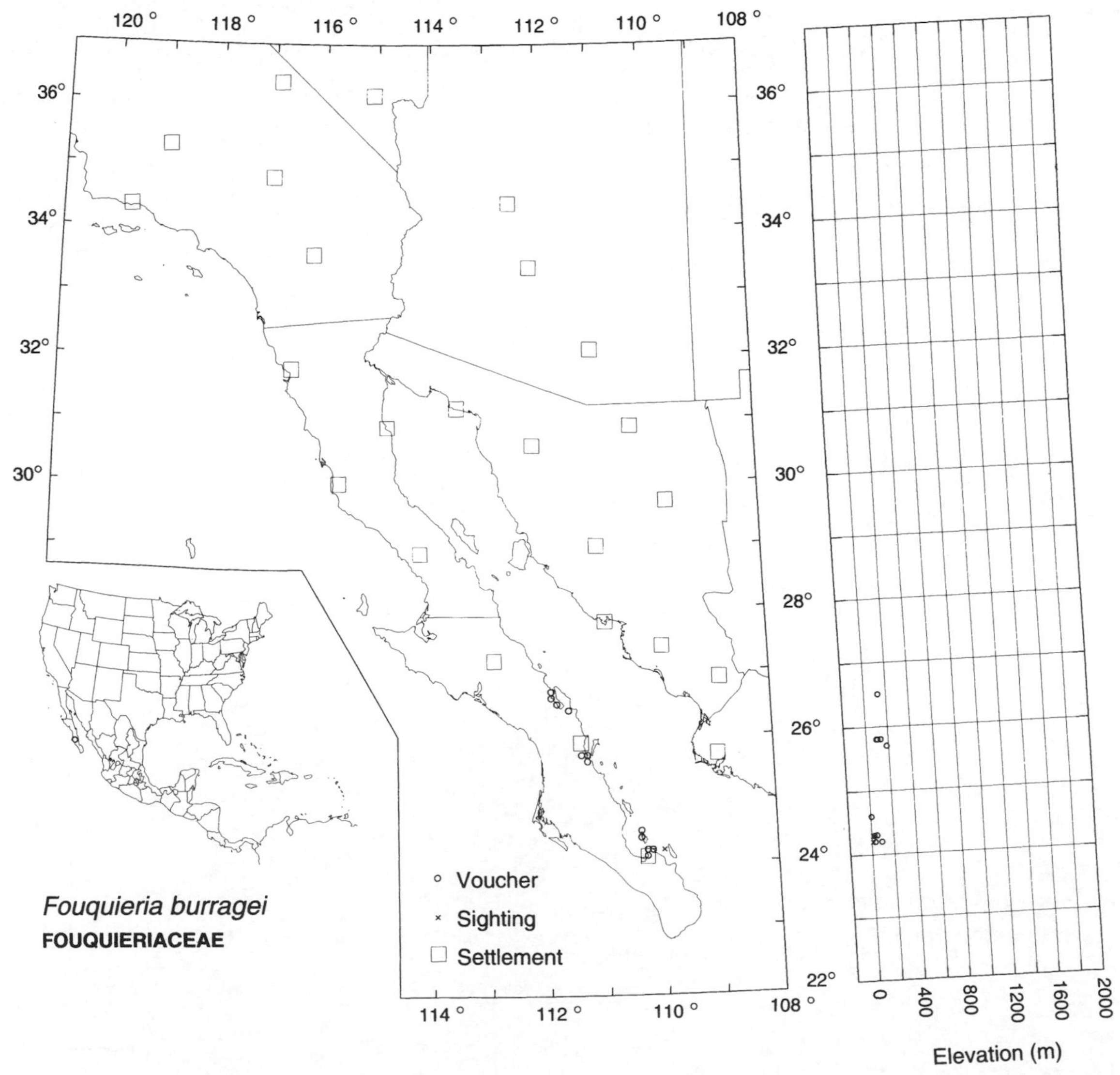

Fouquieria burragei
FOUQUIERIACEAE

Fouquieria columnaris

(Kell.) Kell. ex Curran

[=*Idria columnaris* Kell.]

Cirio, boojum tree

This tree grows up to 20 m tall with numerous short side branches and a massive, tapering, white-barked trunk as much as 0.5 m in diameter near the base (Humphrey 1970; Henrickson 1972). The tallest known plant was estimated to be 26.37 m (Humphrey 1991). The plants look like giant, inverted parsnips or carrots, and the many side branches heighten the illusion with their resemblance to rootlets (figure 42). The alternate, obovate or oblanceolate leaves are 1–2 cm long. The pale yellow flowers, 6–8 mm long, are in large panicles at the top of the trunk. The 3-valved capsules are 12–16 mm long. The haploid chromosome number is 36 (Henrickson 1972).

The original "boojum" was a make-believe creature invented by children's author Charles Dodgson (better known as Lewis Carroll) in the nineteenth century (Dodgson 1966). Godfrey Sykes, a well-read and well-traveled Englishman, immediately thought of the storybook boojum when he saw *F. columnaris* for the first time in 1923, and the name stuck (Sykes 1982).

Fouquieria columnaris is locally common on granitic or volcanic soils, usually on well-drained hillsides or alluvial plains. The plants occupy a relatively restricted range between 27.5°N and 30.5°N on the Baja California peninsula, with isolated populations on Isla Angel de la Guarda and near Puerto Libertad, Sonora. Most populations grow where there is a strong Pacific maritime influence, and in Baja California, they are seldom if ever found east of the main peninsular mountain crest. Throughout the range, winter rain is dependable, summer or early autumn rain occasional (Warren 1979). The tightly confined range in Sonora has been attributed to anomalously cool summers, perhaps a result of local upwelling in offshore waters (Robinson 1973; Humphrey

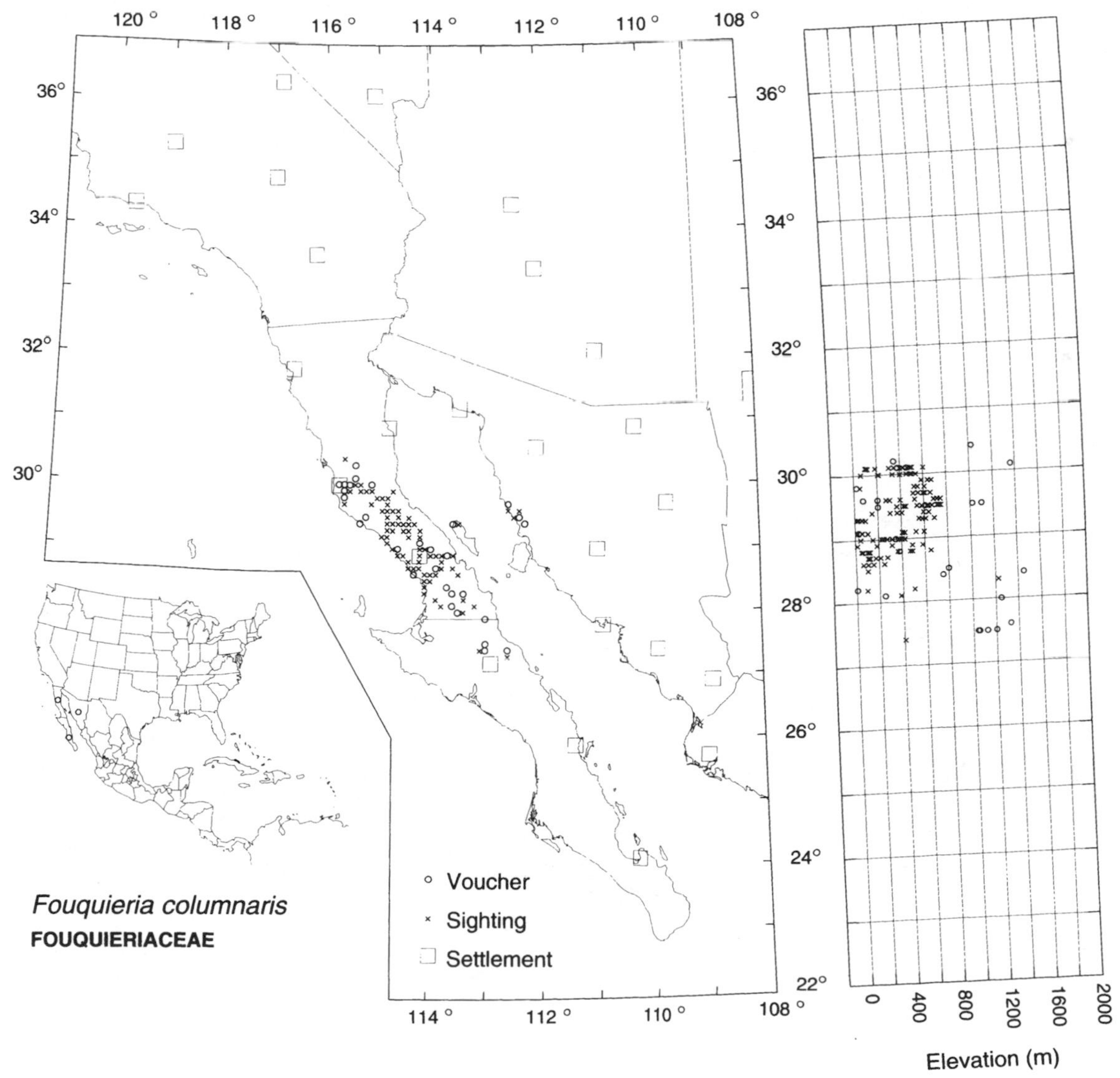

Fouquieria columnaris
FOUQUIERIACEAE

1981; Turner and Brown 1982; Badan-Dangon et al. 1985), and to local high humidity (Humphrey and Marx 1980). Judging from transplants at Tucson, Arizona, larger plants can survive temperatures as low as –9°C (Robert R. Humphrey, personal communication 1992). Seedlings may die at –5.5°C (Humphrey 1974). The periodic fires typical of Californian coastalscrub may limit the species along the Pacific coast to the north.

New leaves are produced in the winter rainy season and also after heavy summer and fall rains. They drop in spring. The main stem elongates only in winter and only after sufficient rain. Little or no apical growth is typical of most years. Side branches elongate mainly in winter but may grow in any season given enough rain (Humphrey 1974). Botanists have made several morphological and anatomical studies of this species (Humphrey 1931, 1935; Henrickson 1969). The roots harbor endomycorrhizal fungi (Rose 1981).

The leaves are the only site of exogenous carbon dioxide assimilation. Evidently a layer of cork prevents carbon dioxide from reaching the chlorophyllous tissue in the bark (Franco-Vizcaíno et al. 1990). Endogenous carbon dioxide may be recycled, however, thus maintaining energy reserves and permitting rapid production of leaves after rain (Franco-Vizcaíno et al. 1990). Nilsen and coworkers (1990) measured a trunk water content of 3.6 kg H_2O/m^3, comparable to that of cacti and higher than that of associated succulents such as *Bursera microphylla* and *Pachycormus discolor.* The succulent stems, acting as buffers, help maintain leaf turgor (Nilsen et al. 1990).

The sweetly scented flowers attract a variety of insects, including 15 species of bees as well as various beetles, ants, and butterflies (Humphrey and Werner 1969). The petals are wrapped around the filaments and the stigma, leaving the anthers exposed. Large insect visitors pry open the petals to obtain nectar and coincidentally contact the stigmatic surface (Henrickson 1972).

Flowers appear in August (occasionally July) and September with little regard to rainfall (Henrickson 1972; Humphrey 1974). Toward the end of the flowering period, some flowers may be newly opened as seeds are being shed (Humphrey 1974). This trait ensures a protracted contribution to the seed bank during autumn (October and early November). Most seeds probably germinate during the winter rains. In the laboratory, they germinate readily without pretreatment (Hum-

Figure 42. Fouquieria columnaris *near Puerto Libertad, Sonora. (Photograph by R. M. Turner.)*

phrey 1974). Seedlings are mostly found beneath shrubs, in rock crevices, or beside rocks (Humphrey 1974).

Biologists have completed no thorough studies of height growth in the wild. Three seedlings grown outdoors at Tucson, Arizona, attained an average height of 48 mm in 8 months. The main stems were swollen and spindle shaped, with about 10 lateral branches covering the apical meristems (Humphrey 1974). Under natural conditions, the plants grow fastest (4–5 cm/yr) at a height of 2.5–3.5 m. Exactly matched photographs show that plants near Puerto Libertad, Sonora, grew 1.3–3.0 m between 1932 and 1994 (Raymond M. Turner, unpublished data). Yearly measurements of height growth suggest that plants 15 m tall may be 500–600 years old (Humphrey and Humphrey 1990; Raymond M. Turner, unpublished data). A demographic study in Baja California showed *F. columnaris* to be declining in abundance on older geomorphic surfaces and increasing on younger ones (McAuliffe 1991).

The plants are grown as ornamentals in desert landscapes. Propagation from seed is relatively easy (Newland et al. 1980).

Fouquieria diguetii

(Van Tieghem) I. M. Johnston

[=*F. peninsularis* Nash]

Palo adán, ocotillo

Fouquieria diguetii (figure 43) is a shrub or small tree 1–8 m tall with a short, thick trunk and spiny, tortuous branches. On older plants, the trunks are often diagonally striped in gray and greenish bronze. The glabrous, leathery leaves are elliptic to ovate. Short-shoot leaves 12–26 mm long are in axillary fascicles. Long-shoot leaves 16–44 mm long are on elongating stem tips. The red, tubular flowers, 1.5–2

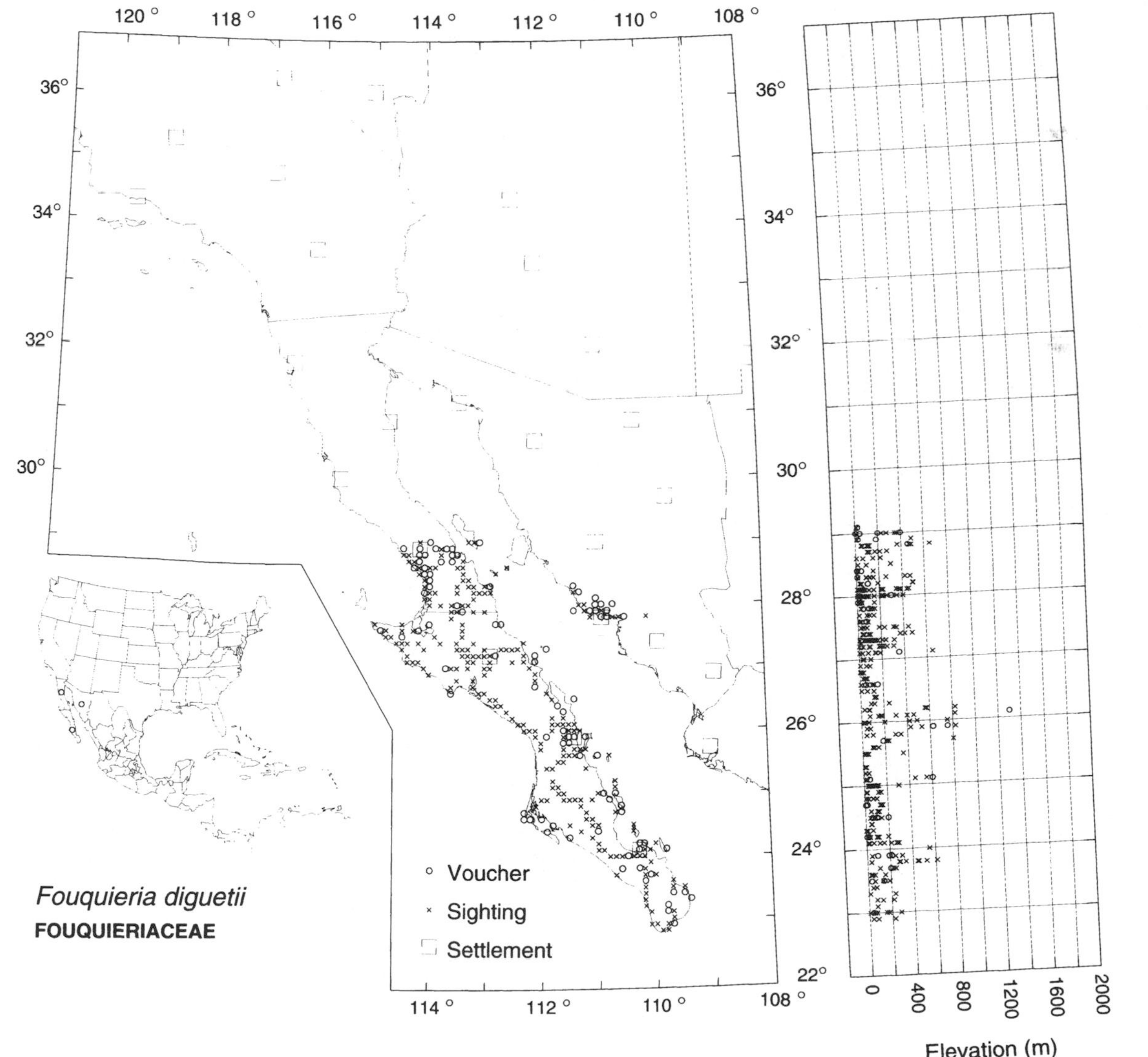

Figure 43. Lichen-draped Fouquieria diguetii *with* Pedilanthus macrocarpus *and* Stenocereus gummosus *northwest of El Arco, Baja California. (Photograph by J. R. Hastings.)*

Figure 44. Fouquieria macdougalii, *leafless and in flower near Moctezuma, Sonora. (Photograph by J. R. Hastings.)*

cm long, are in conical panicles 5–15 cm long and 2–5 cm broad. The 10 stamens are barely or not exserted from the corolla. Fruits are small capsules containing winged, flattened seeds. The diploid chromosome number is 48 (Henrickson 1972).

The corymbose panicles of *F. macdougalii* are broader than long, and the trunk has exfoliating, bronze bark. The smaller flowers of *F. burragei* are pinkish or whitish rather than red. *Fouquieria splendens* has whiplike branches and reaches peak bloom after winter rains.

This species is the common *Fouquieria* of the Baja California peninsula from 29°N to the Cape Region. It grows on a wide variety of soil types on alluvial plains and hills, mostly below 900 m but sometimes up to 1,300 m. The range covers a variety of climates, ranging from very arid with mostly winter rain on the Vizcaíno Plain to semiarid with late summer and early fall rain in the Cape Region. The abrupt northern limit of this species falls near the southern limit of *F. splendens;* the zone of overlap occupies some 160 km (Aschmann 1959).

The disjuncts on the Sonoran coast may have originally migrated across the gulf via the midriff islands, then followed the coastline southward (Cody et al. 1983). Competition with the closely related and ecologically similar *F. macdougalii* may have kept *F. diguetii* from expanding its range into mainland climates similar to those occupied in the Cape Region (Cody et al. 1983).

Most plants flower and fruit without regard to rainfall of the previous winter and spring (Humphrey 1974). Unlike other *Fouquieria* species, which tend to have a concentrated period of bloom, *F. diguetii* produces sporadic solitary inflorescences over a long period (generally February to May), thereby forcing outcrossing and greatly extending the time when flowers are available. Hummingbirds visit the flowers and are presumably the main pollinators (Henrickson 1972). The drought-deciduous leaves appear after substantial rains and drop once the soil dries out.

In its leaf anatomy and taxonomic position, *F. diguetii* exemplifies a trend toward greater xeromorphy in the Fouquieriaceae (Henrickson 1972). The growth form is plastic and varies considerably, from a little-branched shrub in the most arid parts of its range to a much-branched small tree in moister regions (Henrickson 1972). Inflorescence morphology is also variable: in wet seasons, the panicle is highly branched, but in dry seasons it may be reduced to a raceme (Henrickson 1972).

Fouquieria macdougalii Nash

Tree ocotillo, paloverde, jaboncillo, chunari, torotillo, torote verde, torote espinosa, chimulí

Fouquieria macdougalii (figure 44) is a small tree up to 6 (occasionally 11) m high with divergent, much-branched, spiny stems from one or several basal trunks 1–2 m high and 15–40 cm in diameter. The light bronze bark is thin and exfoliates in sheets. The bright red to scarlet flowers, 18–26 mm long and about 4 mm wide at the throat, are in corymbose panicles. Long-shoot leaves are 30–55 mm long and 5–11 mm wide, short-shoot leaves 12–30 mm long and 7–14 mm wide. The diploid chromosome number is 24 (Henrickson 1972).

Fouquieria macdougalii differs from *F. splendens* in its treelike habit, pendulous terminal branches, and lax, corymbose panicles. *Fouquieria diguetii* has more erect conical panicles and (on older plants) diagonally striped bark.

The plants are locally common on a wide range of soil types on rocky flats and slopes, lava beds, and sandy playas (Henrickson 1972). The abrupt western limit is probably determined by decreasing summer rainfall, the northern limit by low temperatures. Shreve (1964) noted that in the valleys of the Sonora, Moctezuma, and

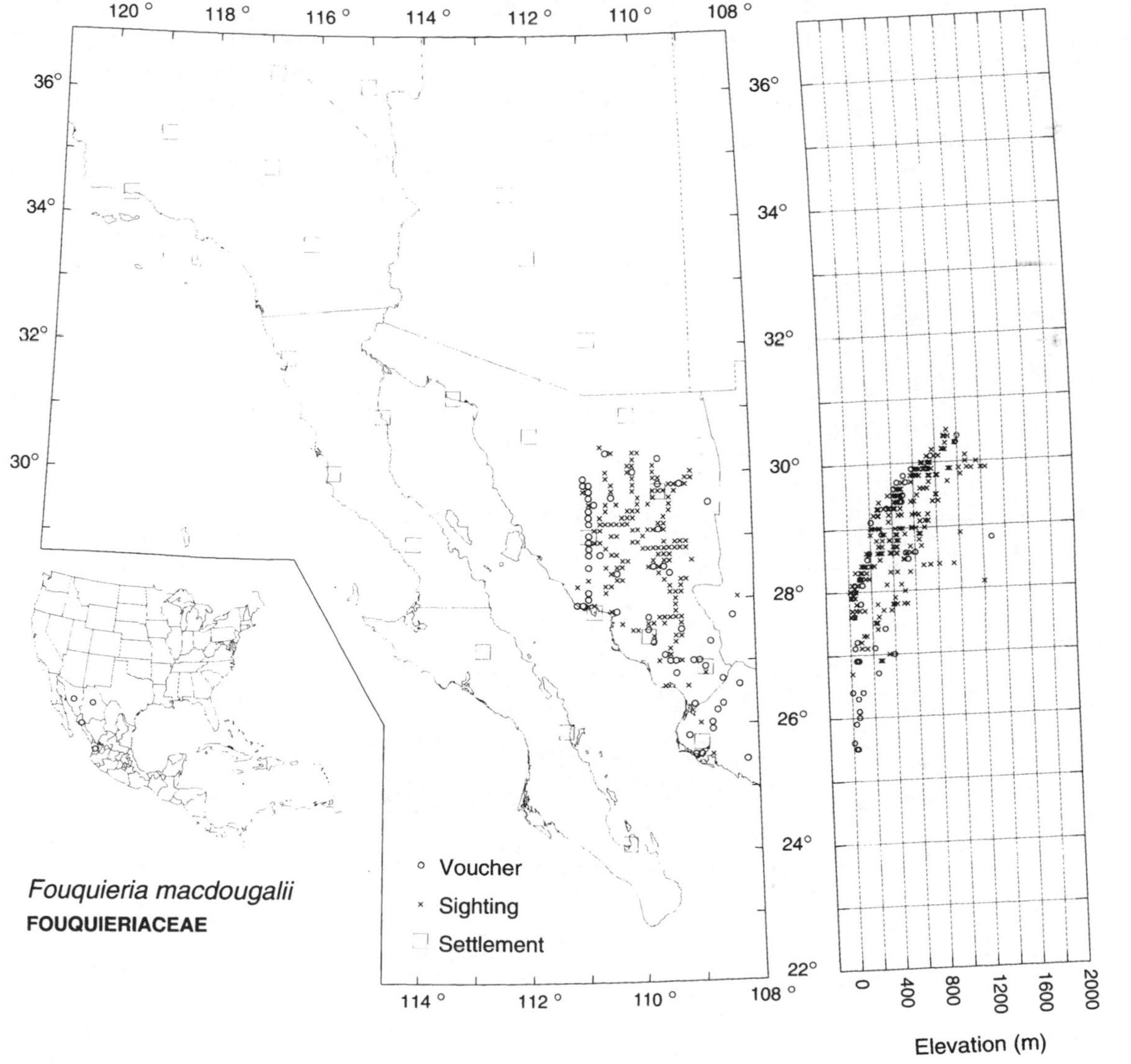

Fouquieria macdougalii
FOUQUIERIACEAE

Bavispe rivers, *F. macdougalii* and other tender species grew only on south-facing slopes above the influence of temperature inversion. The elevational profile shows a sloping lower limit shaped by aridity and a similarly sloping upper limit that may result from an inability to persist in more mesophytic vegetation.

Fouquieria macdougalii blooms after rains, usually between July and October, occasionally in other months. With heavy rain, the plant produces large, much-branched paniculate inflorescences. In dry seasons, the inflorescence may be reduced to a raceme (Henrickson 1972). Elongating stems produce long-shoot leaves in the summer rainy season. Short-shoot leaves appear after heavy rains at any time of year. Leaves are drought deciduous, and plants are leafless much of the year. The wood contains 62.5–67.7% water, a rather high value for a shrub. *Fouquieria macdougalii* could more properly be termed "drought escaping" than "drought resistant" (Shreve 1933).

Hummingbirds appear to be the main pollinators (Henrickson 1972). The position of the stigmas above the anthers probably favors outbreeding, but the flowers are self-compatible to some extent (Henrickson 1972).

According to Standley (1923), the bark is used as a substitute for soap, especially in washing woolens.

Fouquieria splendens

Engelm.

Ocotillo, coachwhip, slimwood, candle bush, albarda, barda, ocotillo del corral

Fouquieria splendens (figure 45) is a candelabra-form shrub 2–10 m tall with a stocky trunk and 6–30 (rarely up to 100) spinose, wandlike, ascending stems. The bark of the basal stems is rough and fissured. The waxy leaves, 1.4–4 cm long, are obovate to elliptic. The red flowers, 1.5–2 cm long, are in dense, conical panicles 5–25 cm long and 4–8 cm wide at the stem tips. The 3-valved capsules are 10–15 mm long.

Fouquieria diguetii is more treelike in habit, and its stems are thicker and more tortuous. It blooms in summer and fall. *Fouquieria macdougalii* has an obvious trunk and peeling bark.

Henrickson (1972) recognized three subspecies: subsp. *campanulata* (Nash) Henrickson and subsp. *breviflora* Henrickson, both endemic to the Chihuahuan Desert, and subsp. *splendens*, widespread throughout the Sonoran and Chihuahuan deserts and their margins. For all three, the diploid chromosome number is 24 (Henrickson 1972). We map and discuss only subsp. *splendens*. Apparent hybrids between *F. splendens* and *F. diguetii* occur north of Calmallí. The two species overlap in Baja California (Aschmann 1959).

Fouquieria splendens subsp. *splendens* is common on rocky outwash slopes and plains and can be abundant on limestone, especially toward its upper elevational limits. In the drier parts of the Sonoran Desert, it is common on fine and coarse soils, occasional on loose sand (Shreve 1964). The plants form a nearly continuous cover in parts of the Chihuahuan Desert (Gentry 1949b). In southern Nevada, they are reduced in stature and vigor, perhaps showing that they are near their geographic and ecologic limit (Bradley 1966). That *F. splendens* appears relatively late in the Holocene fossil record is curious, considering the present-day dominance and ubiquity of this species. Midden assemblages from the Puerto Blanco Mountains show it in southwestern Arizona no sooner than 3,400 years B.P., and middens from Picacho Peak, California, show *F. splendens* having arrived fewer than 700 years ago in a series of deposits that spanned the last 13,000 years (Cole 1986).

Our southernmost record is Andy Méling's sighting (reported through Reid Moran) on the trail south of El Potrero. Reid Moran made the collection at 26.9°N between Misión Guadalupe and San Pe-

Figure 45. Fouquieria splendens *during a transient leafy phase. (Photograph by J. R. Hastings taken north of San Felipe, Baja California.)*

dro and the one at 27.1°N at Rancho Manglito and Rancho Santa Cruz in the Sierra Santa Lucía. According to Ira Wiggins (personal communication 1972), the species occurs throughout the eastern foothills and bajadas of the Sierra Juárez and San Pedro Mártir, from about 30°N to 32.5°N.

The entire range has a biseasonal rainfall regime, varying from dependable winter moisture with occasional summer storms in the west to about twice as much summer as winter rain in the east. Compared to many Sonoran Desert shrubs, the plants are frost tolerant and grow to relatively high altitudes. The elevational profile shows a truncated southern limit rather than the more gradual slope seen in many other arid-adapted species, which suggests that the southern limits are probably heavily influenced by interactions with *F. diguetii* and *F. macdougalii.*

Fouquieria splendens alternates between brief periods (three to four weeks) of rain-induced leafiness and longer periods of drought-induced leaflessness. Except perhaps in warm microclimates, plants generally remain leafless during the coldest part of winter (Humphrey 1975). As many as five or six leafy periods can occur during the year (Shreve 1964), although two or three are probably more usual (Darrow 1943). Embryonic leaves appear within 24 hours of significant rain and are usually fully developed within 5 days (Humphrey 1975). Throughout its range, *F. splendens* blooms primarily in spring (March–May). Occasional plants flower in late summer or autumn. At Tucson, Arizona, spring flowering seems to be triggered by the first cool-season rainfall of 10 mm or more (Bowers and Dimmitt 1994). Thereafter, degree-days above the base temperature (10°C) must accumulate to 515 for the plants to bloom (Bowers and Dimmitt 1994). In more arid regions, smaller rains may suffice to trigger spring bloom. Spring flowering lasts 50–60 days (Waser 1979). Unseasonably cool weather delays the onset of flowering (Waser 1979). The roots harbor endomycorrhizal fungi (Rose 1981).

An east-west cline in flower and inflorescence size may reflect regional climate

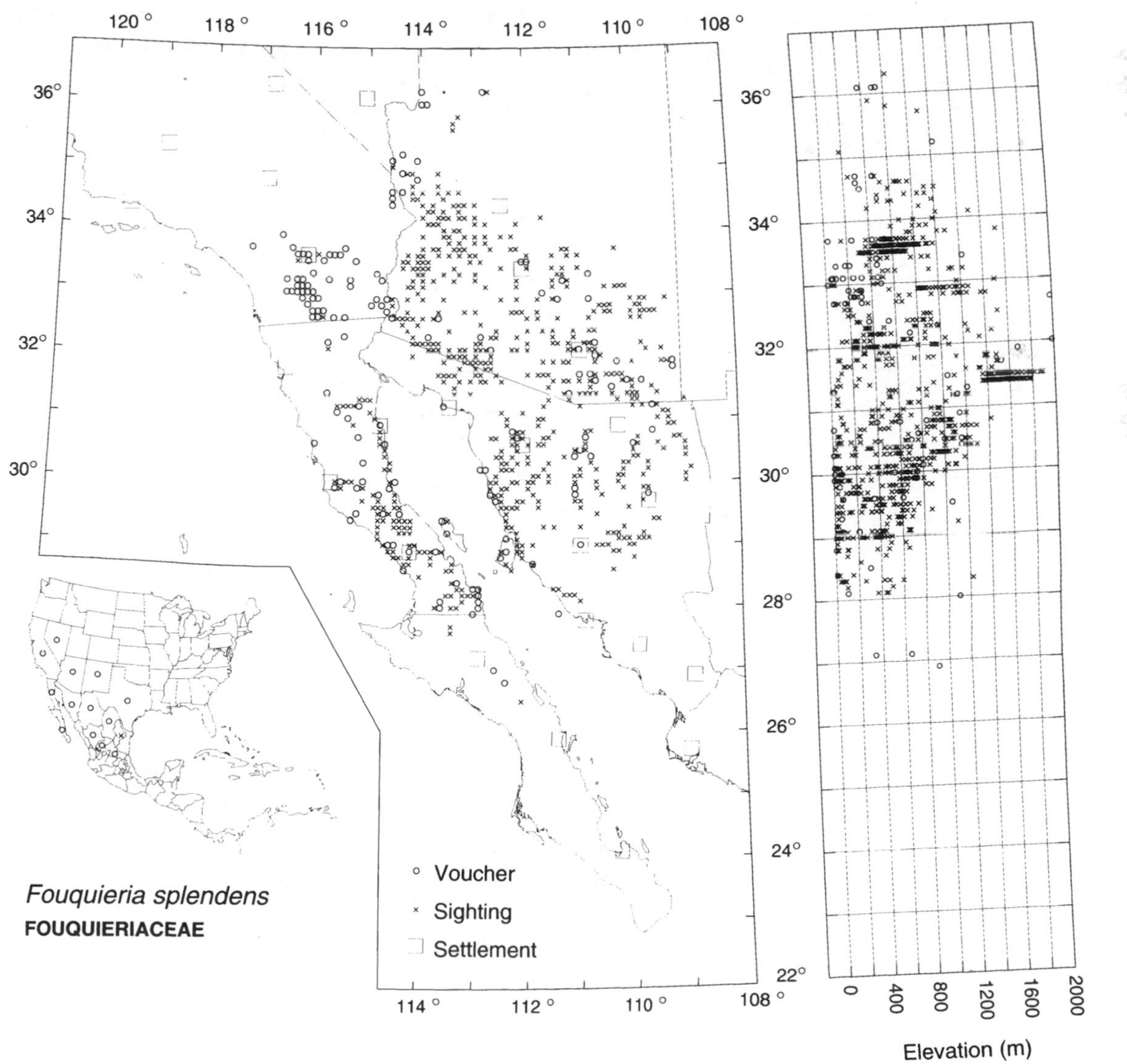

Fouquieria splendens
FOUQUIERIACEAE

(Henrickson 1972). In the western portion of the range, where winter rain is common, leaves usually appear during or just before the flowers, providing energy for larger flowers and inflorescences. To the east, the plants bloom during the dry season, so flower production relies on stored water and energy, leading to smaller corollas and panicles (Henrickson 1972).

Short-shoot leaves, clustered in the axil of spine and stem, appear any time after heavy rain, often within 24–48 hours (Cannon 1905; Darrow 1943; Killingbeck 1990). The plant mobilizes stem rather than root reserves to produce new leaves (Killingbeck 1990). Long-shoot leaves appear in summer only on actively growing stem tips. The petioles of long-shoot leaves persist after the blades drop, becoming rigid, conical spines (Robinson 1904; Humphrey 1931). Axillary leaf buds are protected by a cap derived from leaf primordia (Henrickson 1977). Henrickson (1969) and Lersten and Carvey (1974) further discussed the leaf and stem anatomy of this and other *Fouquieria* species.

The stems are moderately succulent, making *F. splendens* intermediate in structure between typical desert shrubs and true succulents (Humphrey 1935). The water stored in the stems may permit rapid development of leaves after rains (Scott 1932). Apical stem (long-shoot) growth occurs only in summer and, except in seedlings, does not happen every year. The best predictor of apical stem growth is the combined spring and summer rainfall during a given growing season, with spring rainfall being the more important (Darrow 1943). As plants mature, long-shoot production becomes increasingly sporadic and ceases when branches reach 3–5 m in length (Darrow 1943; see also Plates 89a and 89b in Hastings and Turner 1965b).

Cannon (1905) early noted that the bark contains chlorophyll. Because a layer of impermeable cork prevents atmospheric gas exchange (Nedoff et al. 1985), stem bark can fix only endogenous carbon dioxide (Mooney and Strain 1964). The chlorophyllous tissue of the bark fixes 0.25–0.45 mg/CO_2/dm^2 stem surface/hr, a small fraction of the photosynthetic contribution of the leaves, which is 5.23–6.45 mg CO_2/dm^2 leaf surface/hr (Mooney and Strain 1964). Though bark photosynthesis does not contribute significantly to the overall seasonal economy of *F. splendens,* it may produce the photosynthate needed for rapid development of ephemeral roots and leaves after rains (Mooney and Strain 1964).

Hummingbirds and carpenter bees (*Xylocopa californica*) are the main pollinators in the Sonoran Desert region (Waser 1979). Other flower visitors include small solitary bees, orioles, finches, verdins, warblers, syrphid flies, and butterflies. Waser (1979) suggested that the flowering time of *F. splendens* has been determined by hummingbirds, which traverse large portions of its range when they migrate northward in the spring. According to Simpson (1977), the plants are self-compatible when hand pollinated. Extrafloral nectaries occur on the flower buds (Pemberton 1988).

In one study, the plants averaged 200 fruits/plant/yr (Waser 1979). Seeds germinate soon after the onset of summer rains (Shreve 1917) and require a rain trigger of at least 25 mm (Bowers 1994). The seeds do not persist in the soil for a year and are apparently consumed by predators if summer rains are inadequate or delayed (Bowers 1994). At 20–25°C, germination approaches 90%, but the seeds can germinate within a much wider range of temperatures (10–40°C) (Freeman 1973; Freeman et al. 1977). Seedling mortality is extremely high; only one seedling in thousands survives to the following summer (Shreve 1917). Mortality among seedlings that survive to the second year is relatively low (Shreve 1917). The plants reach maturity at 60–100 years and may live as long as 150–200 years (Darrow 1943).

Fouquieria splendens transplants readily and is widely grown as an ornamental. The plants prefer rocky soil with good drainage (Duffield and Jones 1981). Newly planted specimens require weekly irrigation. Established plants need supplemental water only in dry years or where annual rainfall is less than 18 cm (Duffield and Jones 1981). Stem cuttings root easily (Nokes 1986) and have traditionally been used in the southwestern United States and northern Mexico as "living fences" around gardens and corrals.

The stems were used for support of thatched roofs (Benson and Darrow 1981) and, by the Seri Indians, as framework for brush houses (Felger and Moser 1985). The Cahuilla Indians of southern California reportedly ate the flowers and capsules. Apache Indians bathed in a decoction of the roots to relieve fatigue and also applied the powdered root to painful swellings (Kearney and Peebles 1960). A tincture of the inner bark can be used medicinally as well (Moore 1989).

Frankenia palmeri

S. Watson

Hierba reuma, saladito

This low, spreading shrub is 1–3 dm tall with brittle stems and grayish leaves in dense, opposite fascicles. The sessile, linear leaves, 2–4 mm long, are strongly revolute and nearly round in cross section. The inconspicuous white flowers are solitary or fasciculate in the axils. Fruits are small, ovoid capsules.

The oblanceolate leaves of *F. salina* (Molina) I. M. Johnston (=*F. grandiflora* Cham. & Schlecht.) are twice as wide as those of *F. palmeri* and may be plane or slightly revolute. *Atriplex julacea* has grayish, triangular, cordate-clasping leaves. *Atriplex barclayana* has larger, ovate to elliptic leaves.

Except for a few outlying populations in northern Baja California and southern California, *F. palmeri* is virtually endemic to the Sonoran Desert. It dominates many hundreds of square kilometers on the coastal plains of the Baja California peninsula and Sonora, often in fine-textured saline or subsaline soils, and also grows on sandy soils of stable dunes and washes. Farther inland, it occurs in alkaline soil, as on margins of salt lakes and playas (Wiggins 1964; Yensen et al. 1983). The plants are highly salt tolerant and can survive irrigation with water containing 30 parts per thousand salt (Schaefer 1988). Salts taken up by the roots are excreted by salt glands on the foliage, thus maintaining homeostatic balance in saline habitats (Whalen 1987). High humidity ameliorates the dry summers experienced by coastal populations (Humphrey 1975). The plants are cold tolerant to –10°C (Schaefer 1988).

Flowering specimens have been collected in every month but September. According to Wiggins (1964), the main period of bloom is November to May. Summer and winter rains stimulate leaf and stem growth (Humphrey 1975). The leaves are evergreen even during protracted drought (Humphrey 1975). Plants reproduce by seed and by forming adventitious roots at the nodes (Whalen 1987).

Exactly matched photographs taken decades apart show that the plants live at least 57 years (Raymond M. Turner, unpublished photographs).

The Seri used the resin as a glue and the dry wood for roasting sea turtle meat. They also made tea from the roots for use as a cold remedy (Felger and Moser 1985).

Guaiacum coulteri
A. Gray

Guayacán, arbol santo, palo santo

Guaiacum coulteri (figure 46) is a tightly and intricately branched shrub or small tree up to 8 m tall with bright green foliage. The crowded, opposite leaves, 3–6 cm long, are evenly pinnate with 6–10 leaflets 3–6 mm wide and 1–2.5 cm long. The deep violet-blue flowers are 16–30 mm wide. The narrowly winged capsules are 5-celled and 1–1.5 cm long and broad. The ellipsoid seeds have a pale orange or yellow aril.

Zanthoxylum sonorense is a spiny shrub

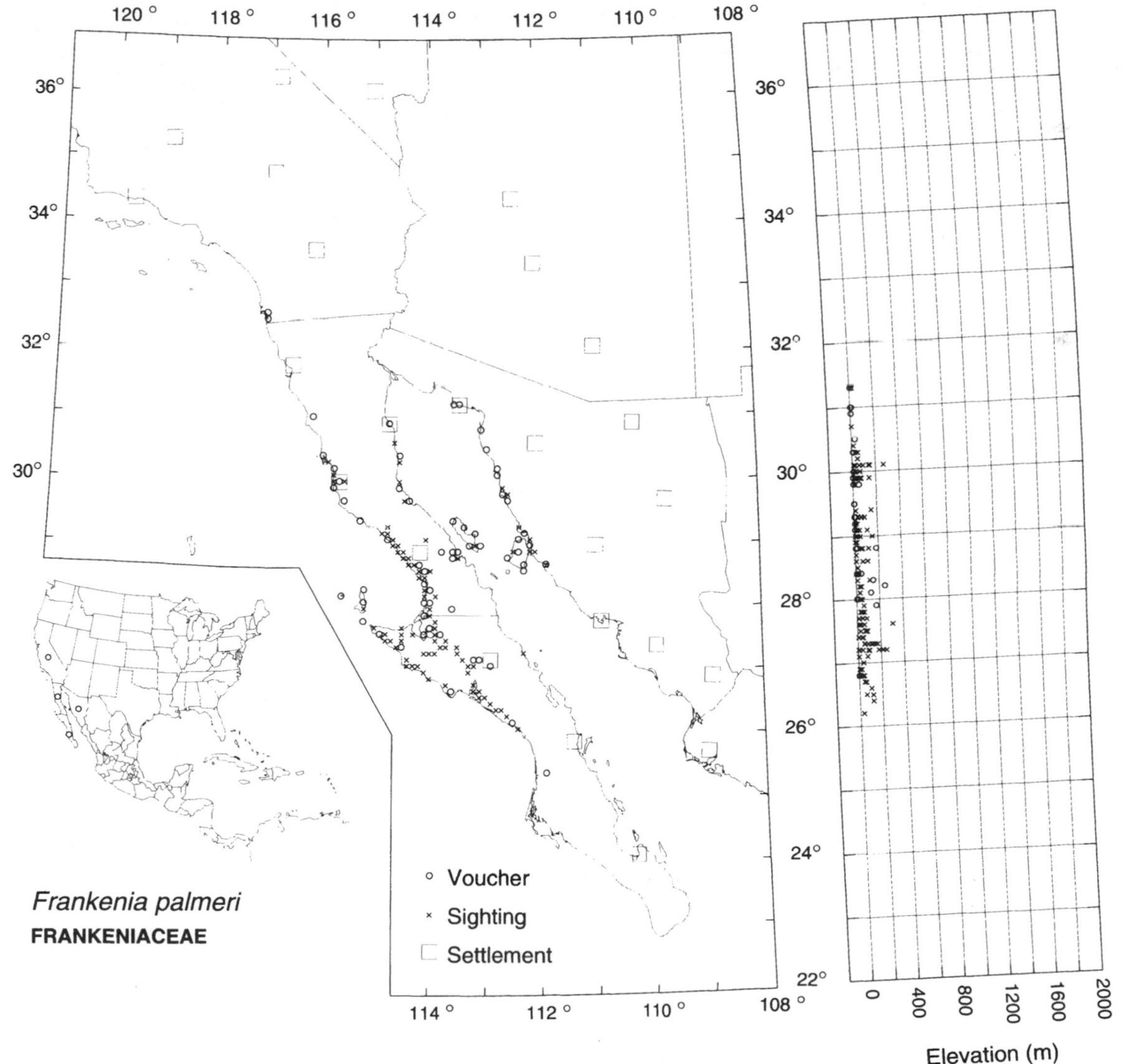

Frankenia palmeri
FRANKENIACEAE

Figure 46. Guaiacum coulteri *north of Hermosillo, Sonora. (Photograph by J. R. Hastings.)*

with odd-pinnate, pellucid-punctate leaves. *Guaiacum unijugum* Brandegee, known only from the east side of the Cape Region, is a smaller shrub (up to 2 m tall) with only 2 leaflets per leaf.

Two varieties have been recognized: var. *coulteri,* from Sonora to Guerrero, Nayarit, and Oaxaca, and var. *palmeri* (Vail) I. M. Johnston, in Sonora and Sinaloa (Wiggins 1964). Variety *palmeri* differs in its pubescent ovaries and fruits. Our map combines them.

Guaiacum coulteri grows on gravelly bajadas, plains, and gentle slopes. It is seldom abundant. The species reaches its northern limit in Sonora. The distribution indicates sensitivity to frost and a need for adequate summer rain. The disjunct local-

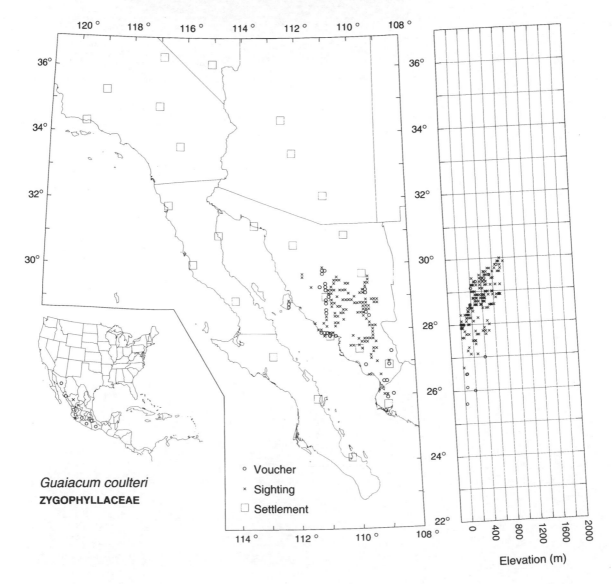

Guaiacum coulteri
ZYGOPHYLLACEAE

° Voucher
× Sighting
□ Settlement

Elevation (m)

ities on Isla Tiburón (based on Richard S. Felger's collections and sightings) are curious. Given its absence on the adjacent mainland and its importance to Seri Indians, we wonder if the Seri have not introduced it, either purposely or accidentally, to the island.

Peak flowering is in May, June, and July, but flowers can be found from April to November. Leaves are drought deciduous, often dropping in October. New leaves appear in early spring. Sometimes the first leaves sprout from the trunk rather than the branch tips (Gentry 1942). The seeds are probably dispersed by birds, as is the case for *G. sanctum* L. (Wendelken and Martin 1987).

The stem resin is a key ingredient of Seri blue, a pigment used in face-painting and in decorating beads, pottery, arrows, and many other objects (Felger and Moser 1985). Seri Indians used the mashed and moistened roots as a shampoo. They drank a tea made from the roots for dysentery (Felger and Moser 1985). The hard, durable wood is used for fuel, railroad ties, and other purposes (Standley 1923).

Guazuma ulmifolia
Lamarck

Pricklenut, guácima, guácimo, tablote, caulote, majagua de toro, palote negro, tapaculo, vacima

An unarmed tree or shrub up to 20 m tall, *Guazuma ulmifolia* has a dense crown and checked, fissured bark. The alternate, broadly ovate, serrulate leaves (2–6 cm wide and 4–16 cm long) are on stout reddish-tomentose petioles. Frequently both leaf surfaces are stellate-tomentose, although the upper side may become dark green and glabrate. The small flowers, in axillary cymes, are creamy or yellowish

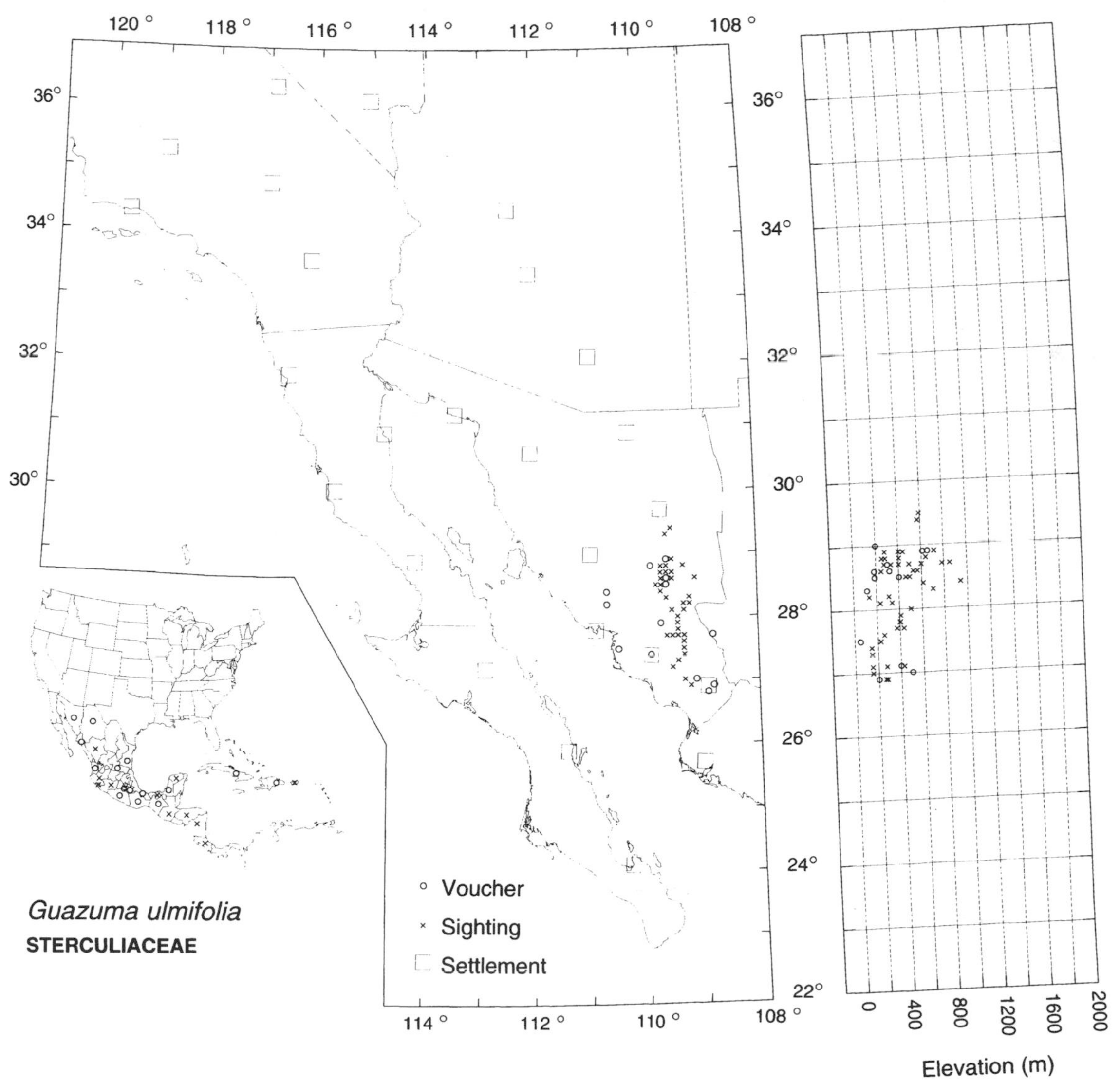

Guazuma ulmifolia
STERCULIACEAE

green and sweet-scented. The woody cylindrical or spherical capsules are 2–4 cm long.

Hintonia latiflora has opposite leaves and branches conspicuously dotted with lenticels. *Cordia sonorae* lacks conspicuous reddish tomentum.

This widespread tree of the tropics and subtropics quickly colonizes newly cleared areas and roadsides. It also grows on streambanks and in canyon bottoms, where it can be a dominant (Gentry 1942), and along roads. In the area mapped in this atlas, *G. ulmifolia* is found largely in Sinaloan thornscrub and Sinaloan deciduous forest. Considering its wide distribution in the western hemisphere, its absence from Baja California Sur is difficult to explain. The plants are frost sensitive (Turnage and Hinckley 1938) and are probably limited to the north by cold temperatures.

Near Alamos, Sonora, *G. ulmifolia* blooms in the summer (June–July), and fruits mature by the following spring (March–May) (Krizman 1972). The small flowers might be wind pollinated (Janzen 1982). The leaves are drought deciduous and in lowland Costa Rica drop about 6–7 weeks after the last substantial rain (Reich and Borchert 1984). The plants remain leafless until summer rains end the dry season (Reich and Borchert 1984). Dried leaves sometimes remain on the plants throughout the winter, particularly along drainages (Krizman 1972). Young plants may grow 2.5–3 m in four years. Older ones resprout from the stumps (Gentry 1942; Janzen 1982).

In laboratory trials, the seeds germinated best after scarification with sulfuric acid or hot water (Acuna and Garwood 1987; Maruyama et al. 1989). In the wild, they germinate readily after passing through the digestive tracts of horses (Janzen 1982). Deer, squirrels, agouti, and peccaries eat the fruits and disperse the seeds (Janzen 1975), as does the lizard *Ctenosaura similis* (Traveset 1990). During the Pleistocene, horses or other large mammals may have dispersed the seeds (Janzen 1982; Janzen and Martin 1982). The wide distribution of the species can be ascribed at least in part to humans and

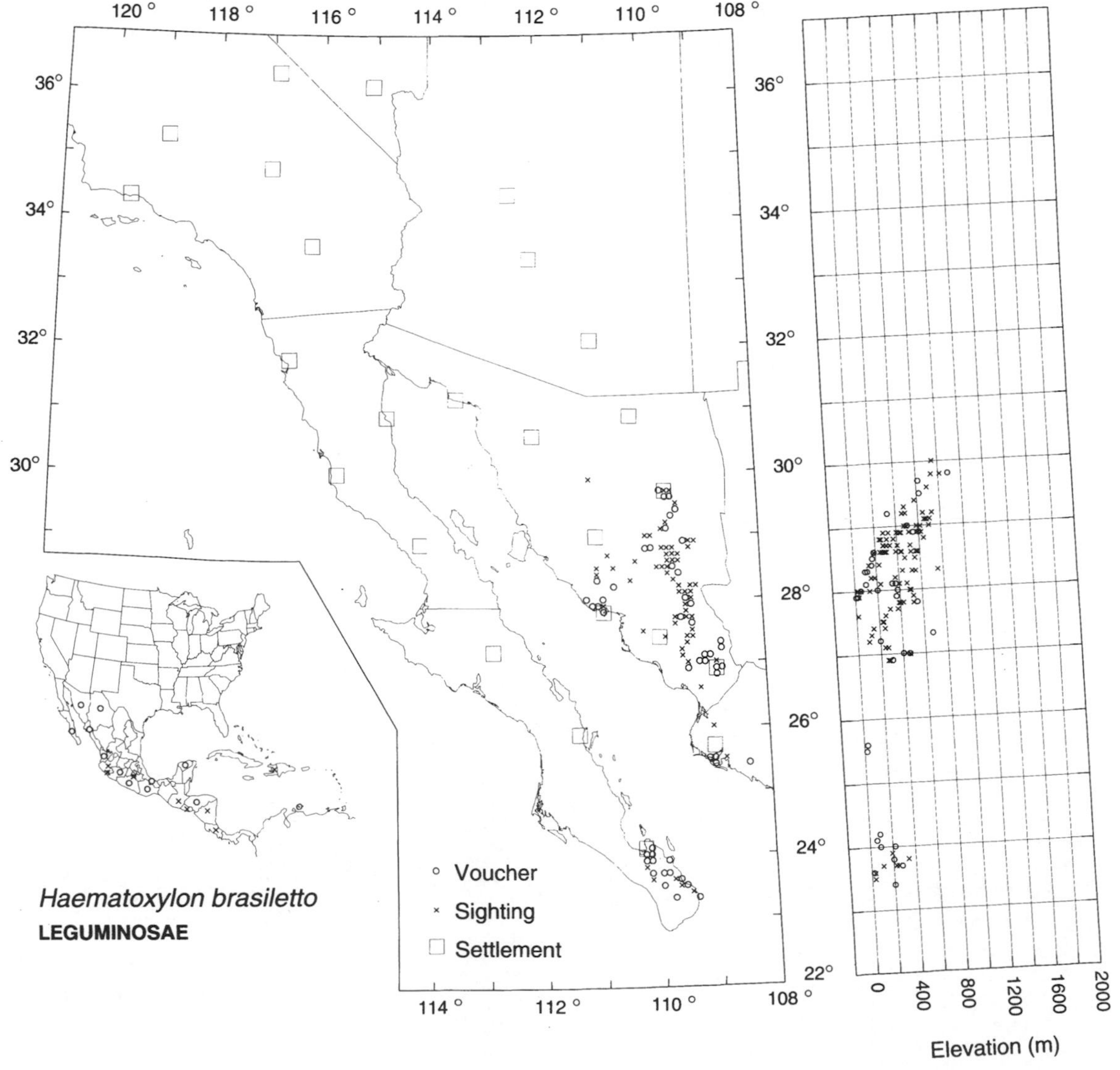

Haematoxylon brasiletto
LEGUMINOSAE

their livestock. Continual clearing of thornscrub for agriculture has provided a steady supply of disturbed areas for colonization, and the usefulness of the fruits, both raw and cooked, and of other parts of the tree, as well, has probably led to its accidental or even deliberate introduction in suitable areas.

Bruchid beetles attack the seeds with varying severity (Janzen 1975). When mammalian seed dispersers are removed, bruchid beetle predation increases (Herrera 1989).

The rather strong wood has been used for a variety of purposes, including boat ribs and barrel staves. The stem fibers have been made into rope. Fruits are eaten raw when young and, when mature, are ground into flour or used as a substitute for coffee. The bark has been used in treating leprosy and many other diseases (Standley 1923). In laboratory tests, *G. ulmifolia* showed significant antibacterial activity (Caceres et al. 1990). Foliage and fruits are valuable fodder for cattle and horses (Janzen 1982).

Haematoxylon brasiletto

Karsten

Brasil, palo de brasil, palo de tinta, palo de Campeche, azulillo, brasiletto, huchachago

A shrub or tree 1.5–7 m tall, *Haematoxylon brasiletto* has reddish brown, zigzag branches with rigid spines 0.5–2 cm long. The leaves, fascicled on spur shoots, are evenly pinnate with 1–3 pairs of obovate-cuneate leaflets 5–25 mm long. The leaflets are leathery and strikingly veined. The yellow flowers, 12–20 mm wide, are in few-flowered axillary racemes 3–6 cm long. The elliptic to oblong pods are 2–6 cm long and 8–15 mm wide.

No other woody legume in its range has leathery, cuneate leaflets with conspicuous pinnate venation. The reddish brown, zigzag stems are also distinctive.

Haematoxylon brasiletto is common to abundant on gravelly or rocky ridges and slopes. The plants are shrubs in the Sonoran Desert region, moderate-sized trees farther south. Raymond M. Turner made the northernmost sighting (30.0°N), 22.4 km south of Benjamin Hill. This observation, disjunct from the main distribution by some 150 km, may be in error. A widespread dominant of dry tropical forests, *H. brasiletto* reaches its northern, arid limit in central Sonora in Sinaloan thornscrub and its western, arid limit near the coast in Sonoran desertscrub. The northern limit is not determined entirely by frost, since the plants do not sustain damage until the temperature drops to between –4.4 and –5.5°C (Johnson 1988, 1993). The disjunct distribution in the Cape Region of Baja California Sur may predate gulf-floor spreading. Many other thornscrub elements display a similar pattern, among them *Cercidium praecox, Erythrina flabelliformis, Karwinskia humboldtiana, Lysiloma microphyllum, Senna atomaria,* and *Tecoma stans.*

Flowers appear at any time of year in response to rain (Gentry 1942). In the area covered by this study the cold- and drought-deciduous leaves drop between late September and mid October and appear again in late spring (April–May) (Shreve 1964; Johnson 1988).

The wood is used for fuel, construction, and dye (Gentry 1942) and in the treatment of jaundice and erysipelas (Standley 1922).

Haematoxylon brasiletto is cultivated as an ornamental in warm areas. Scarified seed germinates readily; growth with supplemental irrigation is moderate (Johnson 1988).

Harfordia macroptera

(Benth.) E. Greene & Parry

Harfordia macroptera is a sprawling shrub 2–6 dm tall with brittle, straggling, vinelike branches 1–6 dm long. The somewhat fleshy leaves are oblong-spatulate and 6–15 mm long. The inconspicuous flowers, 1–1.2 mm long at anthesis, are solitary or few in the axils. The minute bract subtending the flower enlarges in fruit, becoming an inflated, bilobed sac 1–1.5 cm long. It is conspicuously reticulate with red veins.

Cardiospermum species, pinnate-leaved vines with well-developed or spinelike tendrils, also have inflated fruits, but these are neither bilobed nor conspicuously red veined. None of the woody *Eriogonum* species on the Baja California peninsula has straggling stems or inflated fruits.

Reveal (1989) recognized three morphologically weak but geographically distinct varieties. Variety *fruticosa* (E. Greene) Reveal is limited to Isla Cedros; var. *macroptera* is known only from Isla Santa Margarita and Isla Magdalena but is expected from adjacent peninsular stations; and var. *galioides* (E. Greene) Reveal is found on the Baja California peninsula.

A peninsular endemic, *H. macroptera* is rare to common on rocky hillsides, bajadas, and bluffs. The two southern disjuncts at 24.7°N and 24.8°N represent collections made by T. S. Brandegee and Reid Moran. The plants grow in Californian coastalscrub at the northern end of their range and in Sonoran desertscrub at the southern end. Shreve (1936) considered the species endemic to the transition zone between Californian chaparral and Sonoran desertscrub along with *Rosa minutifolia* Engelm. in Parry, *Aesculus parryi, Bergerocactus emoryi,* and several others. *Harfordia macroptera* extends farther south than these. It is not known from the gulf islands (Moran 1983b).

Most populations experience arid maritime climates with winter rain and dry, relatively cool summers. Pacific maritime air greatly influences the entire range. Inland populations get occasional summer rain. Variety *macroptera* is more likely to receive late summer than winter moisture.

Flowers appear from October–May (rarely as late as July). The inflated fruits seem well adapted for wind dispersal. The balloonlike bract may persist around the achene for as long as three years (Reveal 1989).

Hintonia latiflora

(Sess. & Moc.) Bullock

[=*Coutarea latiflora* Sess. & Moc.]

Copalquín, quina, copalchi, campanilla, palo amargo

An unarmed shrub or small tree 1–6 m high, *Hintonia latiflora* has gray or brown bark conspicuously dotted with numerous white lenticels. The ovate, opposite leaves, 1.5–6 cm wide and 3–10 cm long, are dark green with prominent lateral veins. The white, tubular flowers, 6–8 cm long, are solitary in the leaf axils. The oval capsules, 1.5–2 cm wide, are finely tuberculate and prominently dotted with white lenticels.

Guazuma ulmifolia leaves are alternate rather than opposite, and their petioles are reddish-tomentose.

Growing on canyon slopes and hillsides, *H. latiflora* is scattered in thornscrub and thornforest communities from Sonora to Central America. It reaches its arid and frigid limits in central Sonora, where it is characteristic of Sinaloan thornscrub.

Flowers cover the canopy in July and August (Gentry 1942). The long tubes and white color indicate that moths are the most probable pollinators.

A decoction of the bark is used to treat fevers (Gentry 1942). The species shows

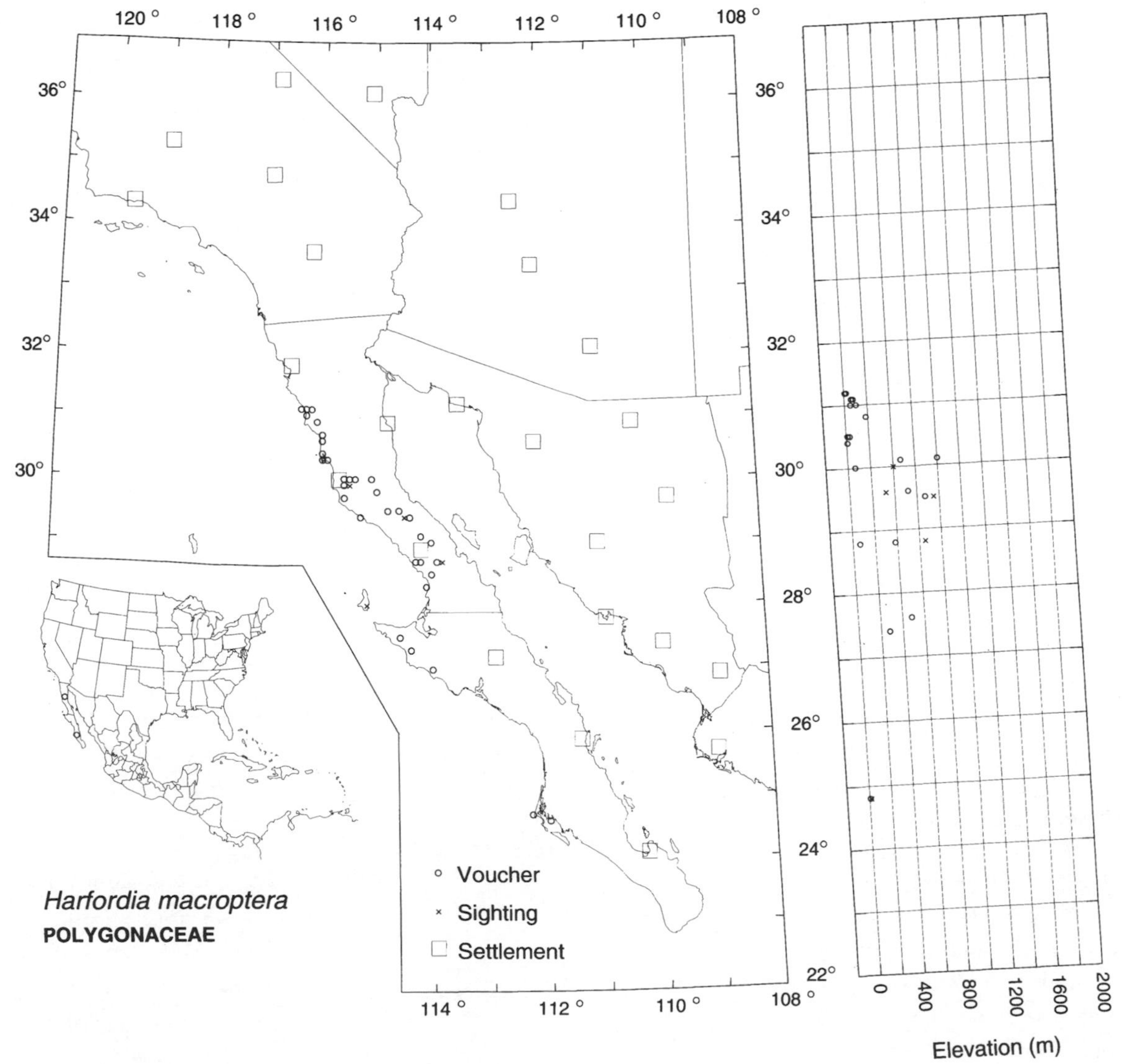

Harfordia macroptera
POLYGONACEAE

significant antimalarial and antidiabetic activity (Perez et al. 1984; Noster and Kraus 1990).

Hoffmanseggia intricata
Brandegee

This intricately branched shrub grows up to 0.6 m tall and has spinescent stem tips. The bipinnate leaves have three more or less equal pinnae each with 4–10 pairs of oblong, glandular-ciliate leaflets 2–3.5 mm long. The yellow flowers are veined with red and gland dotted on the back. The lunate pods, 3.5–5 mm wide and 10–20 mm long, are reticulate veined and gland dotted.

In *Caesalpinia virgata,* the terminal pinna is 2–3 times as long as the lateral pinnae. *Senna purpusii* has thicker branches, and its leaves are once-pinnate and fleshy.

Growing on rocky or sandy soils, *H. intricata* is largely restricted to the central gulf coasts of Sonora and the Baja California peninsula. Most populations experience warm, arid, maritime climates with biseasonal rainfall. In southern sites, rain is most dependable in late summer. Rainfall levels decrease to the north, and rain is rare at the Pacific coast site.

Leaves are drought deciduous. Flowers appear in spring (March–May) and again in the fall (October).

The Seri made a reddish dye from the boiled, mashed roots (Felger and Moser 1985).

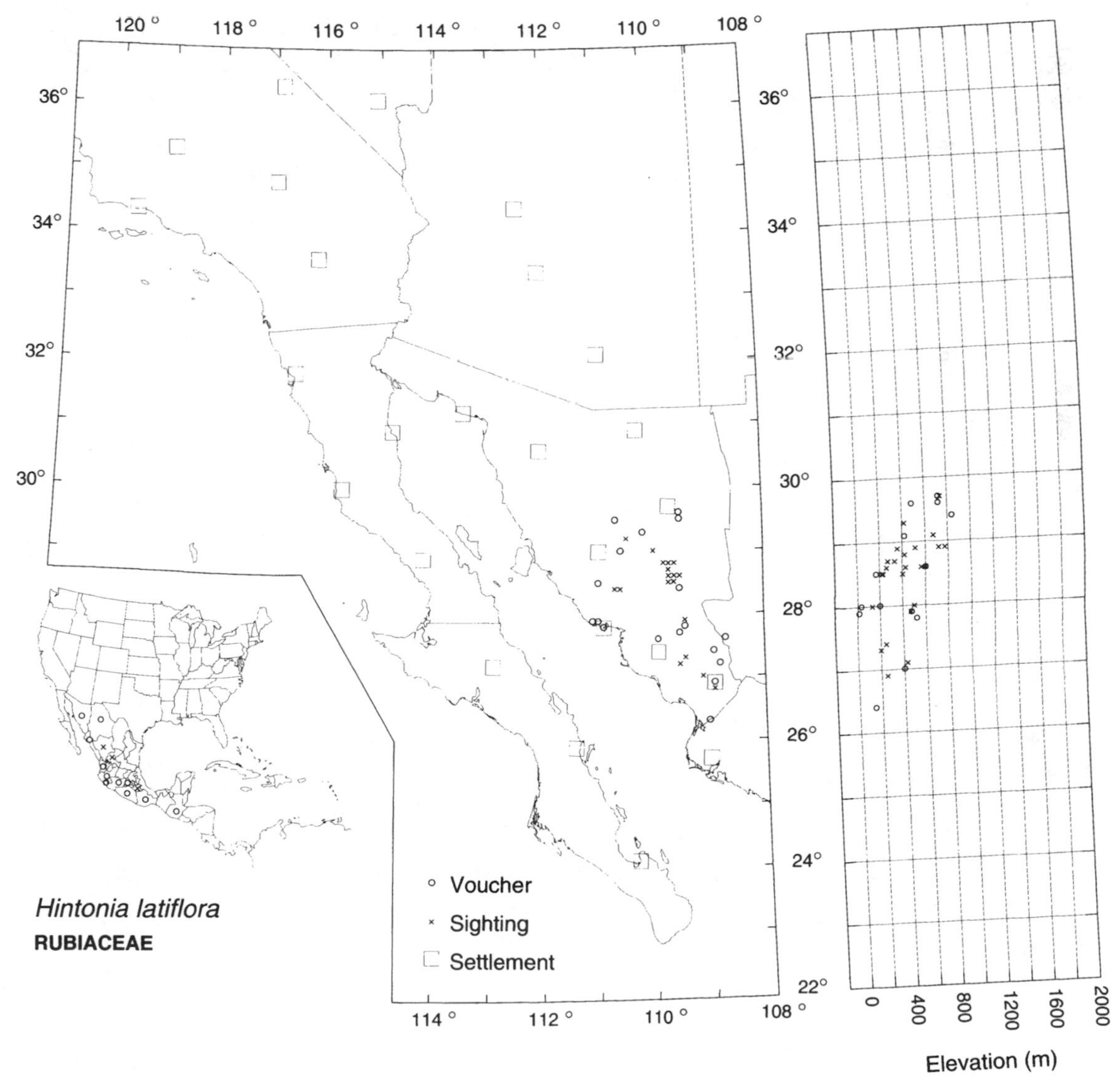

Hintonia latiflora
RUBIACEAE

Horsfordia alata

(S. Watson) A. Gray

Malva blanca, mariola

Horsfordia alata (figure 47) is a sparingly branched, erect shrub up to 3 m tall with densely and harshly stellate-tomentose foliage. The ovate or ovate-lanceolate leaves are 2–9 cm long. The pink flowers, 20–30 mm across, may be single or in leafy panicles. At maturity, the 10- to 12-celled capsules separate into individual carpels about 8 mm long.

Horsfordia newberryi and *H. rotundifolia* S. Watson both have yellowish tomentum and yellow flowers. *Abutilon incanum* has velvety pubescence, orange-yellow or deep orange petals, and 4- to 6-celled capsules.

This slender shrub is found on washes and bedrock outcrops. The northernmost point is a collection by Marjorie D. Clary from the south side of the Coachella Valley (Munz 1974). The eastern disjunct at 111.4°W is based on a sighting by Thomas R. Van Devender at Picacho Peak, Arizona. *Horsfordia alata* is restricted to arid climates where rainfall is biseasonal. Al-

Figure 47. Horsfordia alata *in the Pinacate region, Sonora. The open crown and single erect stem are typical. (Photograph by R. M. Turner.)*

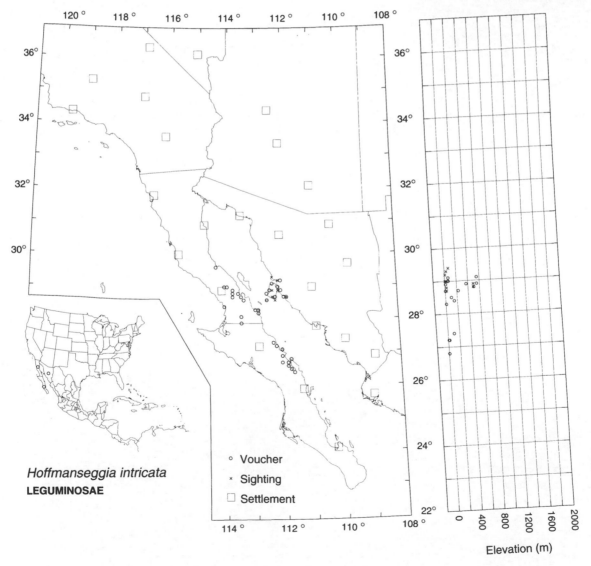

Hoffmanseggia intricata
LEGUMINOSAE

○ Voucher
× Sighting
□ Settlement

Elevation (m)

though their mapped distributions overlap, *H. alata* and *H. newberryi* seldom occupy the same sites. *Horsfordia alata* is probably the more frost sensitive of the two.

Flowers appear from March to October (Kearney and Peebles 1960).

Seri Indians soak the outer bark of the root to make a medicine for eye and mouth sores (Felger and Moser 1985).

Horsfordia newberryi

(S. Watson) A. Gray

This erect, sparingly branched shrub grows up to 3 m tall and has densely yellowish-tomentose foliage. The thick, lance-ovate to narrowly lanceolate leaves are 1–4 cm wide and 3–10 (occasionally up to 15) cm long. The yellow flowers are 8–10 mm long. Fruits are 10- to 12-celled capsules that separate into individual carpels 9–10 mm long.

Horsfordia alata has grayish tomentum, ovate leaves, and pink flowers. *Horsfordia rotundifolia* S. Watson, found in Baja California Sur and near Guaymas, is under 1.5 m tall and has ovate to suborbicular leaves less than 3.5 cm long. No *Abutilon* species in the Sonoran Desert region has dense yellowish pubescence.

This plant of washes and rocky hillsides is seldom abundant. The northernmost disjunct (35.8°N) is based on Thomas R. Van Devender's 1973 sighting from Diamond Creek to Peach Springs Wash. Climates are arid, and rainfall is biseasonal throughout the range. *Horsfordia newberryi* grows at higher latitudes and altitudes than *H. alata* and *H. rotundifolia* and is probably more frost tolerant.

Flowers appear from March to October, probably in response to rains. Plants may bloom the first year.

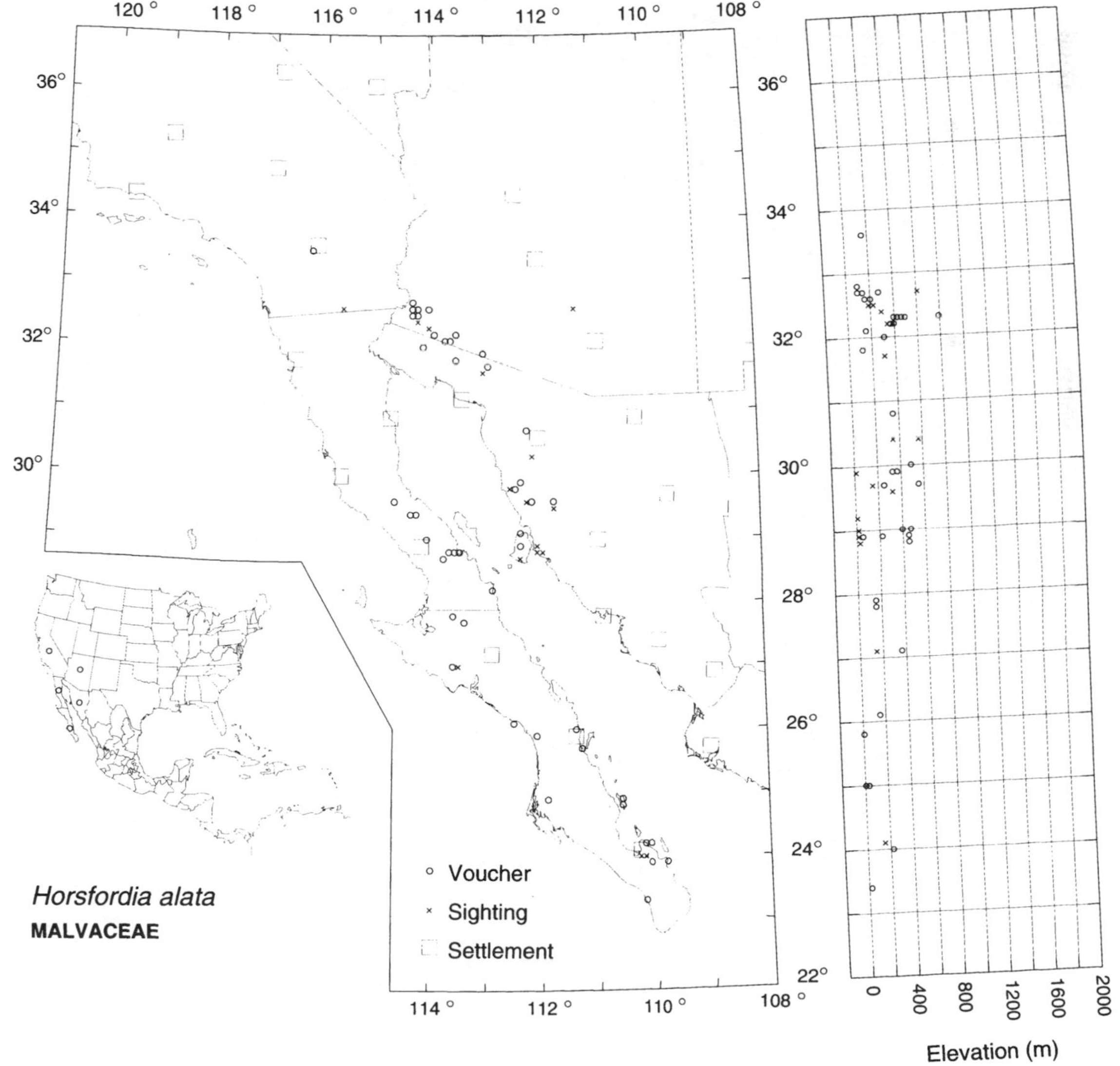

Horsfordia alata
MALVACEAE

Hymenoclea monogyra

Torr. & A. Gray ex A. Gray

Burrobush, romerillo, jecota

A bushy shrub 1–4 m tall, *Hymenoclea monogyra* has alternate, narrow, threadlike leaves 2–5 cm long, the lowermost generally entire. The inconspicuous flowers are unisexual. Pistillate and staminate heads cluster in the upper axils of the branches. Rarely, plants are dioecious. The spindle-shaped fruit, 4–5 mm long, has a single whorl of narrow, elliptic wings.

Hymenoclea salsola blooms in the spring. The fruits have fan-shaped or circular wings in one to several series. The lowermost leaves are generally pinnatifid (Benson and Darrow 1981).

Locally abundant on floodplains and along arroyos and washes, *H. monogyra* thrives in the disturbance created by occasional floods. A seedling requirement for adequate warmth in late fall and winter, when the seeds are dispersed, may determine much of the current range. Seedling establishment may depend on occasional floods between late fall and spring. The northern limit in Arizona follows the 1,300 m contour fairly closely. Low temperatures may limit upslope and northward movement. *Hymenoclea salsola,* which grows farther north in Nevada and Utah, is apparently more cold tolerant.

Hymenoclea monogyra blooms from September to November. The flowers are wind pollinated. The fruits are evidently modified for wind dispersal and can also be transported by floods (Peterson and Payne 1973).

As an early-successional floodplain species, *H. monogyra* is no doubt short lived. In its prolific production of lightweight seeds and its short life span, the species could be considered to be r-selected in the sense of MacArthur and Wilson (1967). Propagating readily from buried stems and rootstocks, the plants regrow quickly after floods (Campbell and Green 1968).

In Mexico, *H. monogyra* is used locally as a remedy for abdominal pains (Standley 1926).

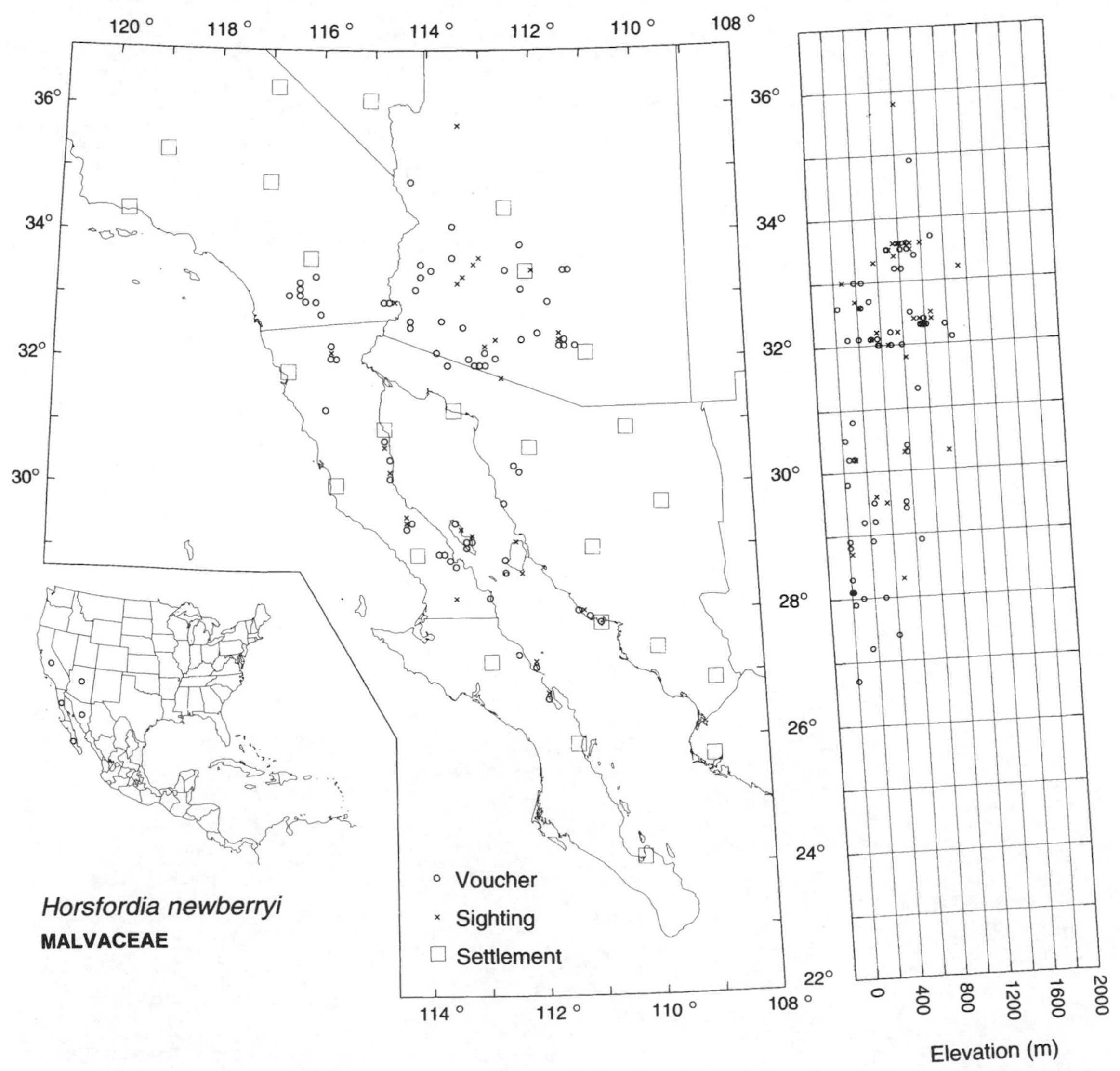

Hymenoclea salsola

Torr. & A. Gray ex A. Gray

[=*H. pentalepis* Rydb.]

Burrobush, white burrobush, jecota

A shrub 0.5–2 m tall, *Hymenoclea salsola* has alternate, linear leaves 1–4 cm long, the lowermost often pinnatifid with threadlike divisions. The inconspicuous flowers are unisexual. Pistillate and staminate heads are in terminal spikes or in small clusters along the ends of the branches. The haploid chromosome number of var. *pentalepis* is 18 (Pinkava and Keil 1977). The spindle-shaped fruits, about 6 mm long, have one to several whorls of fan-shaped or circular wings.

Hymenoclea monogyra flowers in fall. The fruits have narrow wings in a single whorl. The lowermost leaves are entire (Benson and Darrow 1981). *Hymenoclea platyspina* Seaman, known from San Felipe in northeastern Baja California, has pinnatifid leaves throughout. The fruits have spinose wings (Seaman 1975).

Until recently, botanists have regarded *Hymenoclea salsola* as three different species (e.g., Kearney and Peebles 1960). We follow Peterson and Payne (1973, 1974), who treat it as a single species with, in the area of this study, three subspecific taxa: var. *salsola*, var. *pentalepis* (Rydb.) L. Benson, and var. *patula* (A. Nels.) Peterson & Payne. The most primitive taxon, var. *pentalepis*, occurs primarily in the southern part of the range; var. *patula*, the most advanced taxon, in the northern part; and var. *salsola*, the intermediate, in the center (Peterson and Payne 1973). This complex presents "a near classic picture of one species with two derivative varieties, occurring in different climatic areas and with different adaptive peaks, overlapping in a broad area of ancestral intermediacy and of morphological intergradation" (Peterson and Payne 1973:253).

Locally common in washes and on alluvial fans, *H. salsola* also has colonized disturbed habitats such as abandoned townsites (Wells 1961; Vasek et al. 1975; Webb and Wilshire 1980; Webb et al. 1987), old

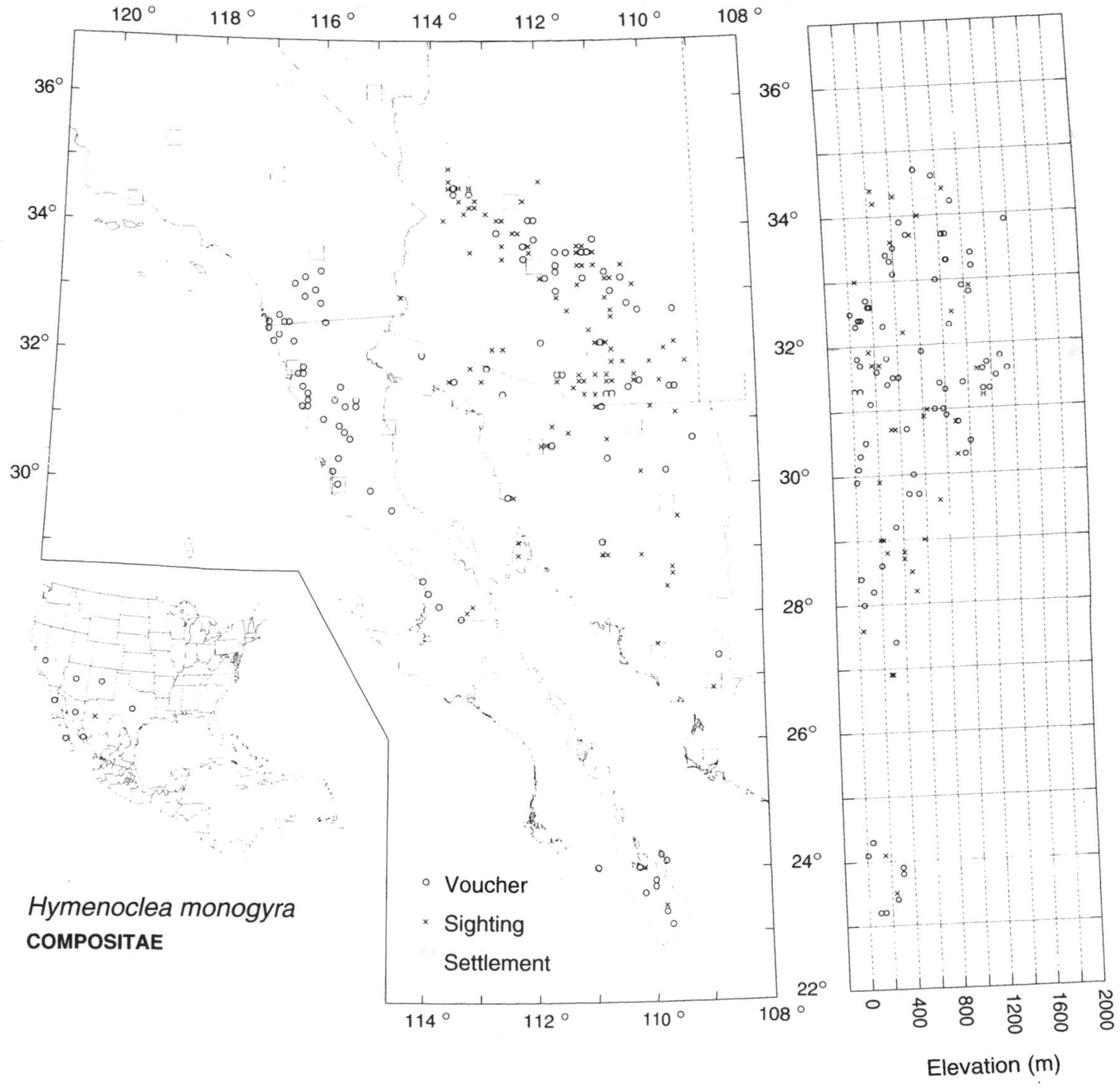

Hymenoclea monogyra
COMPOSITAE

fields (Carpenter et al. 1986), and abandoned military camps (Prose et al. 1987). The distribution suggests tolerance of aridity and frost. All sites receive some winter rain. Warm-season rainfall varies across the range from substantial in the east to slight in the west.

The species apparently occupied the southern part of its present range throughout much of the Holocene. Fossilized remains dating to 11,000 years B.P. have been found in packrat middens from Picacho Peak, California (Cole 1986). The species appeared in mid Holocene middens from the Butler Mountains, Arizona (Van Devender 1990b), and at 2,300 years B.P. in the Hornaday Mountains, Sonora (Van Devender, Burgess et al. 1990).

Flowers appear from March to May and are wind pollinated. The winged fruits are modified for wind dispersal and can also be transported by floods (Peterson and Payne 1973; Maddox and Carlquist 1985). In the Mojave Desert, the plants lose about 50% of their twigs and leaves during the dry months of July, August, and September. Winter rains bring about full leafiness by April of the following year (Comstock et al. 1988).

Leaves and green twigs both contribute significantly to the carbon budget of the plants. Light-saturated photosynthesis of twigs is 36.9 $\mu mol/m^2/sec$, that of leaves about 0.62 times greater (Comstock and Ehleringer 1988). Maximum photosynthesis is in April as canopies reach their fullest extent. Photosynthesis drops to nearly zero between July and September as a result of drought-induced leaf and twig drop (Comstock et al. 1988).

The seeds mature in late spring or summer. In the eastern part of the range, summer or late fall storms provide the first opportunity for germination. Germination in the western part of the range may depend on unusual events, such as late summer tropical storms.

Webb and coworkers (1987), following Grime's (1979) life-history classification, described *H. salsola* as a "stress-tolerant ruderal" based on its short life span, small stature, high seed production, and ability to colonize disturbed sites rapidly. The plants can resprout after burning

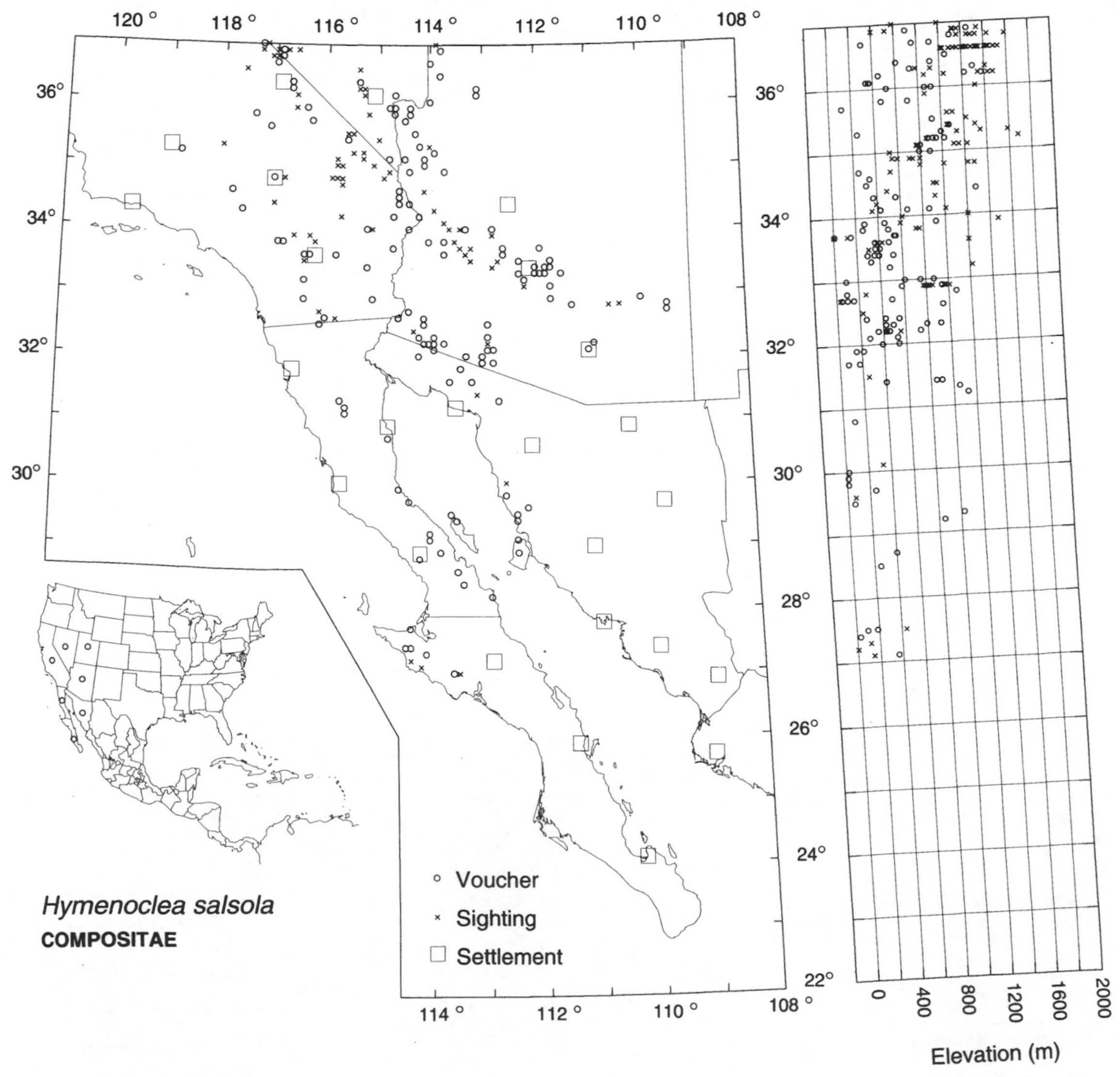

Hymenoclea salsola
COMPOSITAE

(O'Leary and Minnich 1981).

Seri Indians used the wood for firing pottery. They also made a tea from twigs and herbage for relieving lung pain, reducing swelling, and curing skin rashes (Felger and Moser 1985). A variety of insects eat the foliage (Goeden and Ricker 1986). The plants show some promise for revegetation of disturbed sites (Kay et al. 1977) and can be grown from seeds or cuttings. In one study, optimal germination (75%) was at 20°C, while below 10°C germination was poor (0–4%). Optimal planting depth was 1 cm (Kay et al. 1977). Cuttings root best when treated with 200 ppm indoleacetic acid and grown in vermiculite under continuous light with moderate bottom heat (Chase and Strain 1966).

Hyptis emoryi
Torr.

Desert lavender, bee sage, salvia

Hyptis emoryi, a straggling shrub 0.5–3 m tall, has pale gray bark on older branches and dense pubescence on younger ones. The opposite, ovate, crenate leaves, 1.5–4.5 cm long and 8–25 mm wide, are woolly with dendritic hairs. (Variety *amplifolia* of the southeastern peninsula has leaves 6–8.5 cm long and 30–35 mm wide.) The tubular flowers, clustered in the axils of the upper leaves, are lavender or bluish purple and 3–4.5 mm long. The 4-lobed nutlets, 1.5–2 mm long, are contained within the tubular calyx. The haploid chromosome number is 16 (Baker and Parfitt 1986).

Botanists have segregated this species into the densely pubescent var. *emoryi,* distinct from var. *palmeri* (S. Watson) I. M. Johnston and var. *amplifolia* I. M. Johnston, both with green upper leaf surfaces (Wiggins 1964). Our map combines them.

The opposite, crenate leaves and dense, dendritic pubescence distinguish *H. emoryi* from most other shrubs in its range with the exception of *H. laniflora, H. tephrodes* A. Gray, and *H. albida* H.B.K. Telling these species apart can be difficult even with fertile material, and identifying sterile specimens is nearly hopeless.

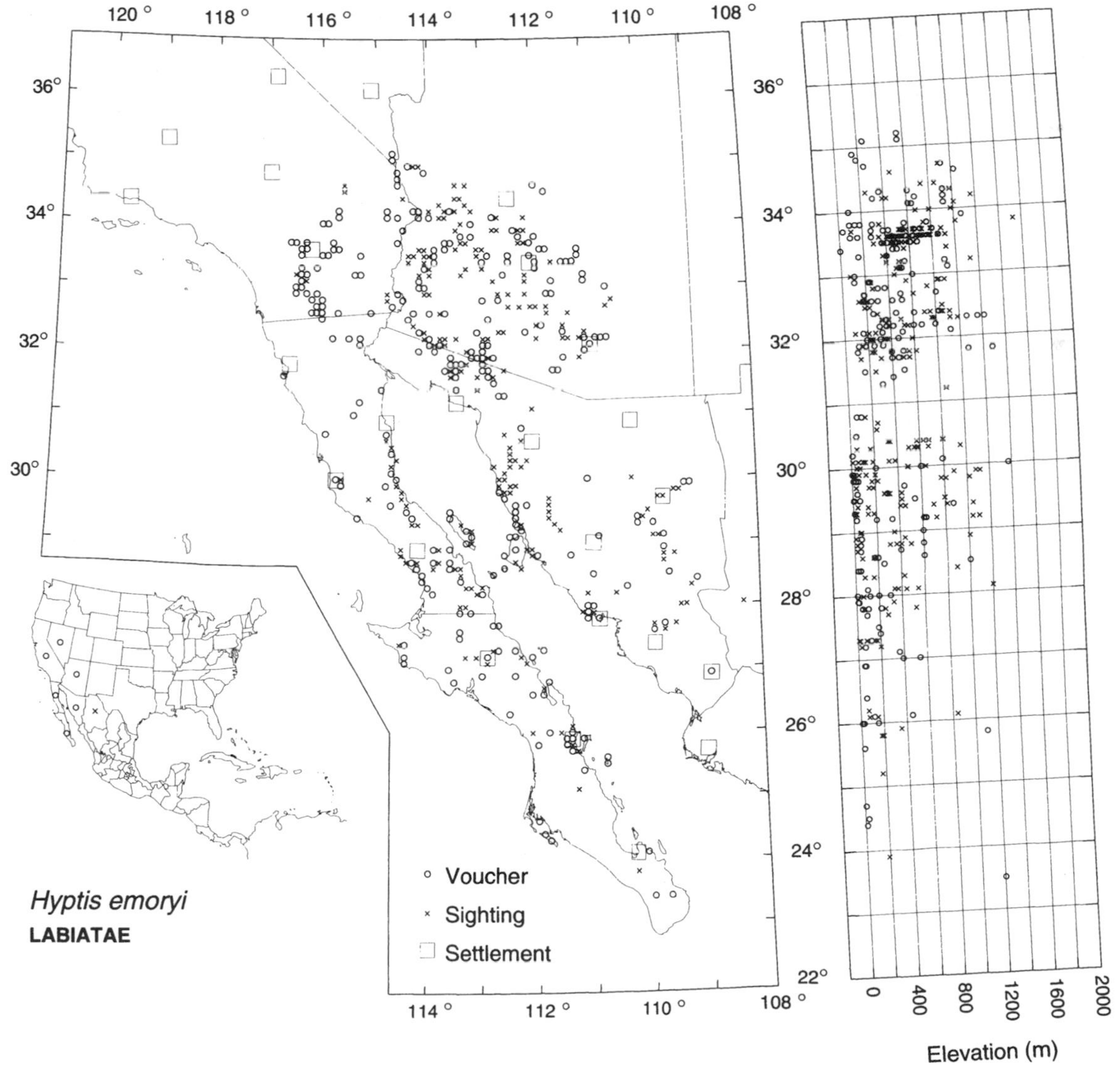

Hyptis emoryi
LABIATAE

Hyptis emoryi grows on rocky slopes, often among boulders, and, in the drier parts of its range, along sandy washes. The sighting at 28.1°N, 108.3°W was made in 1986 by Paul S. Martin in the Candamena Barranca, Chihuahua. Both the Chihuahuan population and the southernmost Sonoran population (collected by Howard S. Gentry in 1934 at Alamos) occur outside the Sonoran Desert. The broad range includes a variety of rainfall regimes. In the coldest winters, mature plants near Tucson suffer frost damage to leaves and upper stems, and some plants may be killed to the ground (Anonymous 1979; Bowers 1980–81). It seems likely that low temperatures control the northern and upper elevational limits.

Fossil midden assemblages from the Puerto Blanco Mountains of southwestern Arizona (Van Devender 1987) and the Hornaday Mountains of northwestern Sonora (Van Devender, Burgess et al. 1990) suggest that *H. emoryi* has been a component of local desertscrub associations for about 8,000 to 9,000 years. The earliest record in the Whipple Mountains, California, dates to 4,200 years B.P. (Van Devender 1990b), indicating a middle Holocene expansion into the northern part of the range.

The plants are evergreen, although in the more arid, western parts of the range, a large proportion of the leaves drop during April and May. Flowers appear from October through May and occasionally in the summer. In the northern Sonoran Desert, leaves and stems grow from February to May (Smith and Nobel 1977b). Honeybees, various native bees (including *Xylocopa* species), and hummingbirds visit the flowers. Bees are probably the main pollinators.

Nilsen and coauthors (1984) characterized *H. emoryi* as a drought-avoiding phreatophyte. Low stomatal conductance and seasonal leaf-drop slow water loss during dry seasons. Plants produce both small, thick "sun" leaves with a high ratio of internal to external leaf area, and large, thin "shade" leaves with a low ratio of internal to external leaf area (Nobel 1976b). Leaf pubescence is densest and leaves are smallest during the summer and fall,

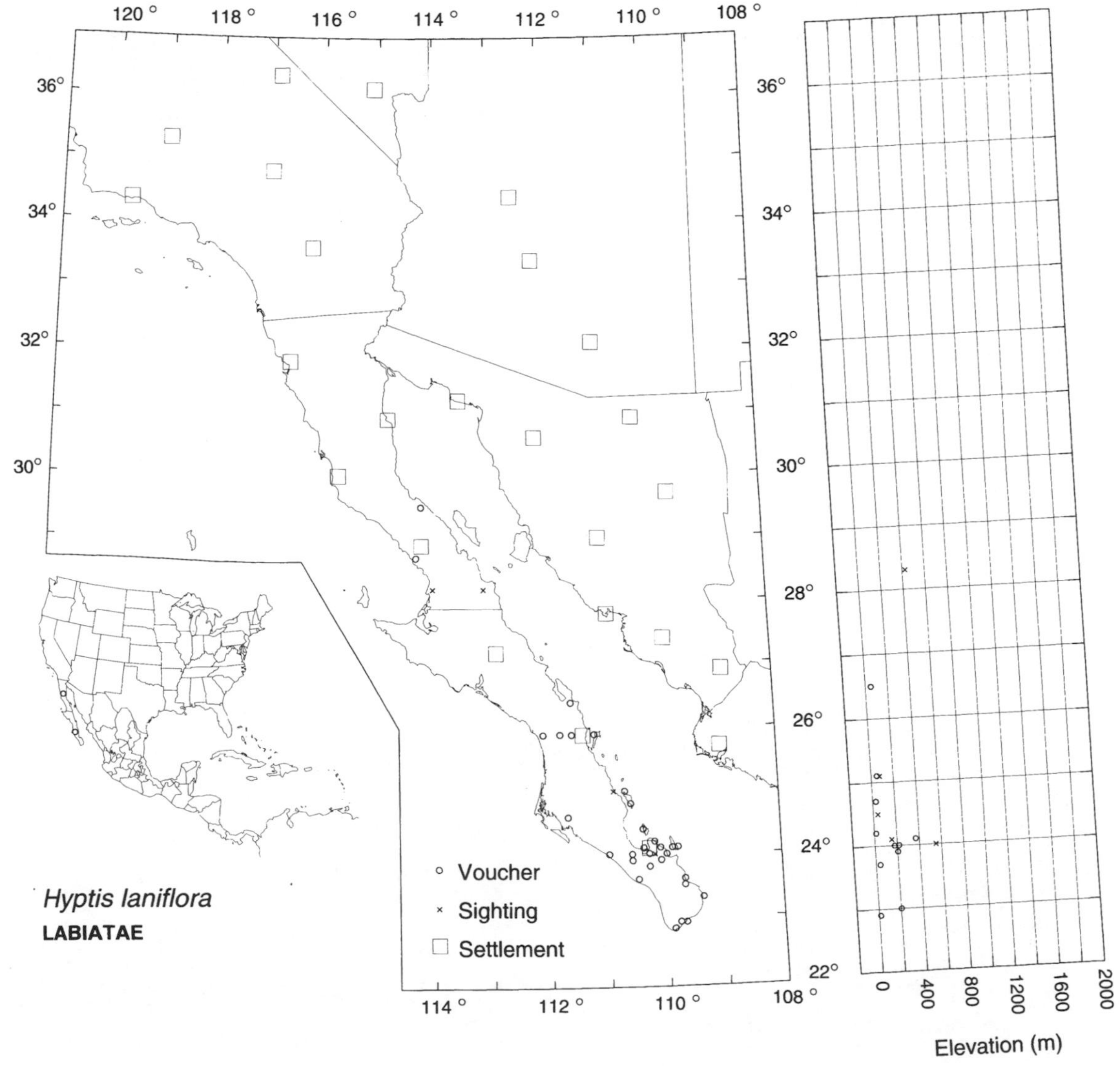

when soil moisture is generally low. During the spring, when soil moisture is greater, leaf size increases and pubescence becomes more sparse (Smith and Nobel 1977a). Gradual change in leaf size and pubescence over the growing season maximizes photosynthesis when water is most available because sparsely pubescent leaves absorb more solar radiation than densely pubescent ones and therefore have a higher photosynthetic rate (Smith and Nobel 1977a, 1977b).

Hyptis emoryi is occasionally grown as an ornamental. Its flowers are valued as a nectar source for honeybees, and domestic livestock browse the foliage (Benson and Darrow 1981). A medicinal tea for stomach problems can be made from the flowers and leaves (Moore 1989). A tumor-inhibitory agent has been isolated from this species (Sheth et al. 1972).

Figure 48. Leafless Ipomoea arborescens *with numerous flower buds. (Photograph by J. R. Hastings taken east of Ures, Sonora.)*

Hyptis laniflora

Benth.

Salvia

Hyptis laniflora, an openly branched shrub 2–5 m tall, has densely white-pubescent twigs and leaves. The opposite, ovate or suborbicular leaves are crenate-serrate and 1–3 cm wide and 2–6 cm long. The blue-purple, tubular flowers, in axillary clusters, are 6–8 mm long. Their peduncles are 1–4 cm long. The four nutlets are enclosed within the calyx, which is 3.5–4 mm long.

Hyptis collina Brandegee and *H. decipiens* M. E. Jones have sparsely pubescent, green leaves. Flowers of *H. emoryi* and *H. tephrodes* A. Gray are sessile or on peduncles less than 1 cm long.

This shrub of washes, rocky slopes, and sandy plains is endemic to the Baja California peninsula. Ira L. Wiggins made the northernmost collection (29.6°N, 114.2°W) about 24 km north of Misión de Calamajué. Most of the range lies in arid tropical maritime climates where rain falls mostly in late summer and early fall. Occasional winter storms may also bring moisture. Although the main distribution suggests reliance on summer rain and winter warmth, the northern outposts occur where summers are dry and winters are cool. This suggests that interactions with *H. emoryi* may have influenced the present distribution to a greater extent than have the effects of climate.

Hyptis laniflora flowers from autumn though spring, generally September–May.

According to Standley (1924), a decoction of the plant has been used to treat fevers.

Ipomoea arborescens

(Humb. & Bonpl.) G. Don.

Tree morning-glory, palo blanco, palo del muerto, casahuate, casahuate blanco, palo santo, palo bobo, ozote, palo cabra

Ipomoea arborescens (figure 48) is a tree up to 12 m tall with smooth, white bark. The large, white, funnelform flowers, 4–5 cm wide and long, are in few- to many-flowered panicles. The leathery, ovate capsules are 4–5 cm long. Leaves are ovate and variable in size, 3–8 cm broad and 8–20 cm long. The stems produce white latex (McPherson 1982). The haploid chromosome number is 15 (Ting et al. 1957; Shibata 1962).

Variability in *I. arborescens* is great enough that the species could be separated into several taxonomic entities (Wiggins 1964). Variety *pachylutea* Gentry has yellowish bark and longer, more pubescent leaves.

The smooth, white bark and large, ovate leaves are distinctive. *Ceiba acuminata* has conical spines on the trunk and digitately compound leaves. *Tabebuia palmeri* and *Bombax palmeri* S. Watson have digitately compound leaves, and the latter has green bark.

Ipomoea arborescens grows on rocky hillsides and gravelly plains. Basically a plant of Sinaloan thornscrub and Sinaloan deciduous forest (McPherson 1982), it enters the Sonoran Desert where moisture and temperature are not limiting. Variety *pachylutea* may be more common at higher elevations than the typical form (Gentry 1942). In one study, plants grown in central Arizona showed frost damage to twigs at –4.4°C (Kinnison 1979), and it seems likely that occasional, severe freezes determine the northern and upper limits. To the west, the plant is probably limited by low summer rainfall. The sharp western boundary may be an artifact of our data set.

At the northern end of the range, flowering occurs from November through April. Drought seems necessary for the onset of flowering in *I. wolcottiana* Rose, a tree of tropical deciduous forest in central Mexico (Bullock et al. 1987), and may be a requirement for *I. arborescens,* as well. Leaves appear after the flowers and persist throughout the summer, usually dropping by October. In southern Sonora, flowers appear in winter (November–January), while leaves are not produced until the start of summer rains in late June or early July (Krizman 1972). Wood anatomy was described by Carlquist and Hanson (1991).

The nectar-feeding bats *Choeronycteris mexicana, Glossophaga soricina,* and *Leptonycteris sanborni* are presumably the most important pollinators in the Sonoran Desert region (Butanda-Cervera et al. 1978; Hevly 1979). The flowers provide nectar at a time of year when few other bat-pollinated flowers are in bloom (Hevly 1979).

Ipomoea arborescens is reputed to be poisonous to horses and cattle. In Sinaloa the bark is used as a remedy for bites of rattlesnakes and other poisonous animals and for diseases of the spleen (Standley 1924).

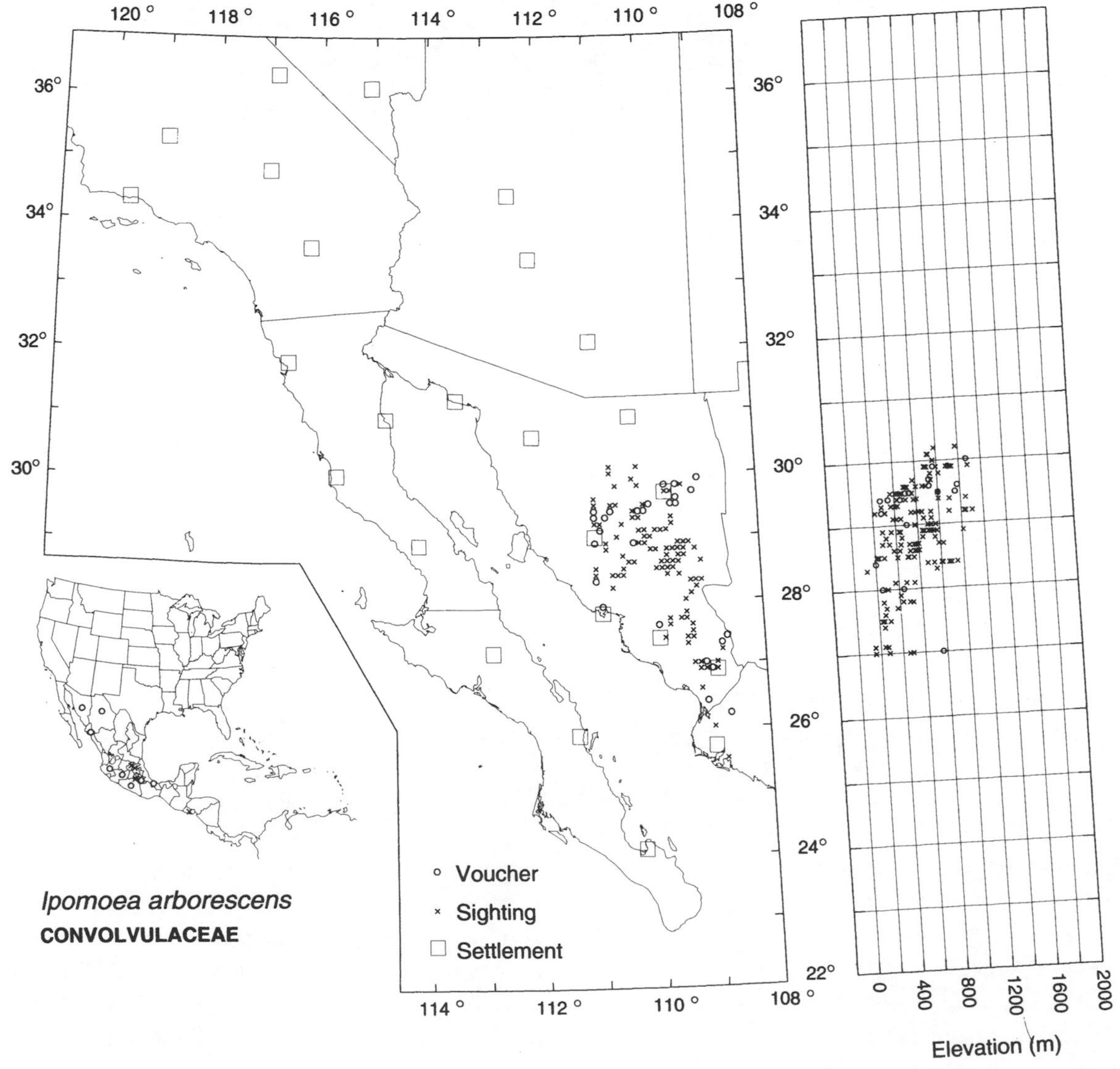

Ipomoea arborescens
CONVOLVULACEAE

Jacquinia macrocarpa
Cav.

subsp. *pungens*
(A. Gray) Stahl

[=*J. pungens* A. Gray]

San Juan, San Juanico, San Juanito, flor de niño, palo de las animas, pinicua, rosadilla, mata-peje, sacaté

Jacquinia macrocarpa subsp. *pungens* is a shrub 1–4 (occasionally 6) m high with sharply pointed, elliptical to lanceolate leaves 3–6 cm long. The stiff, waxy flowers, 7 mm long, are yellow to orange, mildly fragrant, and grow in short racemes. The globose, fleshy fruits, 1.5–2 cm in diameter, are reddish brown when ripe. A diploid chromosome number of 36 has been reported for *J. macrocarpa* subsp. *macrocarpa* under the name *J. aurantiaca* (Faure 1968).

The rigid, sharply pointed leaves combined with the shrubby habit and globose fruits are distinctive.

Jacquinia macrocarpa is a highly variable species found from Panama to Sonora (Stahl 1989). Subspecies *pungens* grows in the Sonoran Desert region (Stahl 1989).

In the Sonoran Desert, this shrub is found along arroyos and on gravelly plains and arid hillsides (Wiggins 1964; Felger and Moser 1985). The northernmost sightings (29.7°N, 29.8°N) are Raymond M. Turner's. The plants are restricted to lower elevations than many other tropical species that enter the Sonoran Desert region. The distribution suggests a need for drier, more open phases of Sinaloan thornscrub. Aridity probably shapes the northern lower elevational limits. Frost probably truncates the northern upper elevational limits.

Jacquinia macrocarpa is evergreen in the desert (Shreve 1964; Felger and Moser 1985). Farther south, in tropical deciduous forest, the plants lose their leaves about four weeks after the rainy season begins (spring in lowland Costa Rica). Starch reserves fall steadily during this period of rainy-season dormancy, and a plant can lose as much as 50% of the

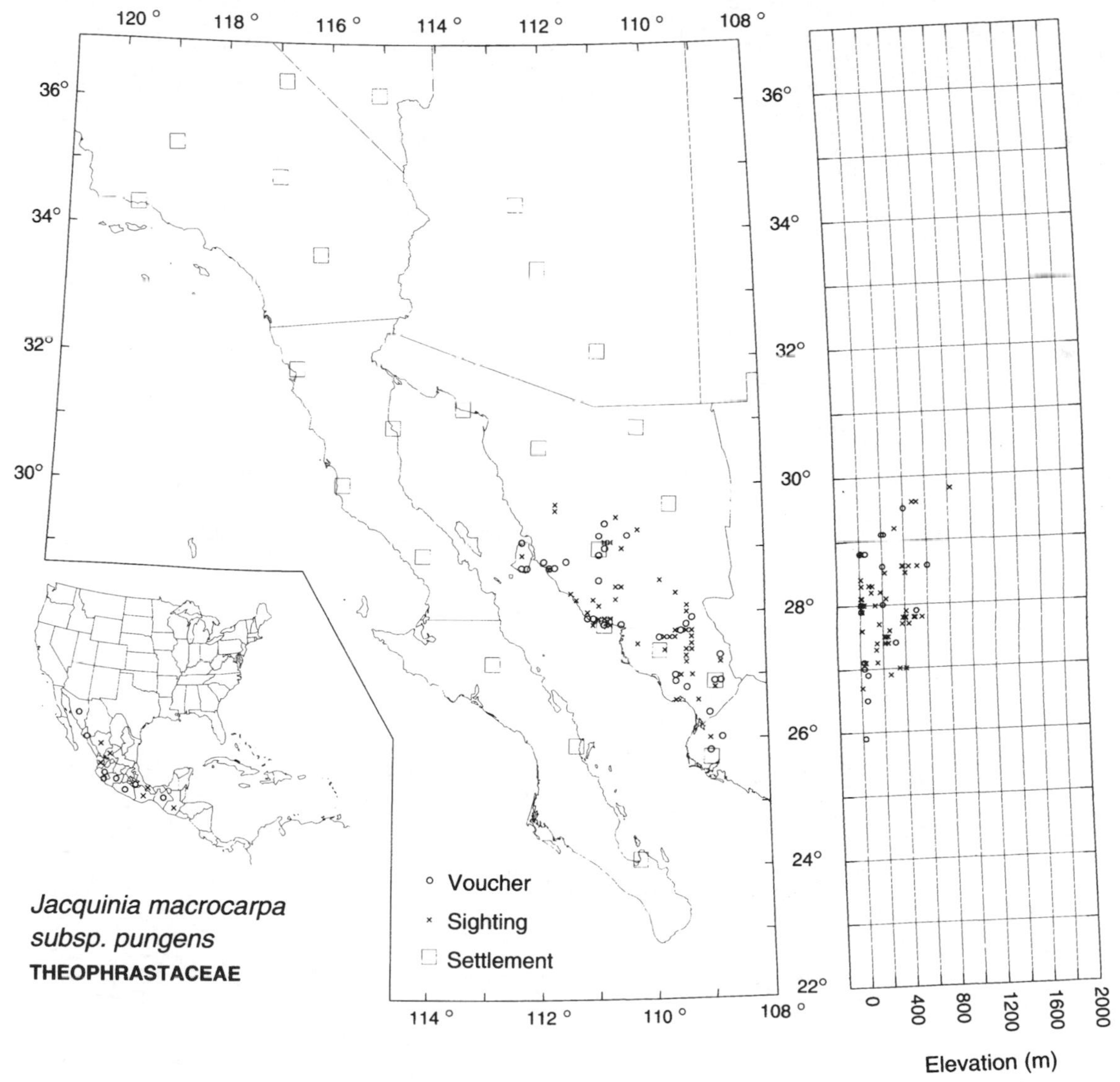

Jacquinia macrocarpa subsp. pungens
THEOPHRASTACEAE

carbohydrate in its stems (Janzen and Wilson 1974). As leaf-drop begins in the forest canopy (late fall in Costa Rica), *J. macrocarpa* produces a new crop of leaves (Janzen 1970, 1983b). In the Sonoran Desert, peak flowering occurs in May (Wiggins 1964), but flowering may occur sporadically at other times (Felger and Moser 1985).

Hummingbirds are likely pollinators (Janzen 1970). Arreguin-Sanchez and coauthors (1986) described the pollen morphology.

The fruits turn sweet as they ripen in late summer (Janzen 1970). The minimal seed predation compensates for the low number of seeds per fruit (about 9–10) (Janzen 1970). Birds and rodents are probably the main consumers of ripe fruits. The seeds probably pass unharmed through their digestive tracts (Janzen 1970).

If they are to survive the first dry season, seedlings must develop a deep taproot that can absorb stored soil moisture (Janzen 1970). The roots penetrate down to 50 cm before branching laterally, enabling the plants to sustain dry-season growth by drawing on stored soil moisture (Janzen 1970).

Leaf toxins discourage herbivores (Janzen 1970; Okunade and Wiemer 1985). The leaves and fruits have been used extensively as a fish poison along the west coast of Mexico (Standley 1924) and in lowland Veracruz (Janzen 1970). Seri Indians of Sonora eat small amounts of the fruit pulp (Felger and Moser 1985) and use the dry fruits as rattles (Standley 1924). Flowers are strung as necklaces and used to dye palm leaves and baskets yellow (Standley 1924; Felger and Moser 1985). An infusion of the flowers was used externally for headache and earache.

Jatropha cardiophylla

(Torr.) Muell. Arg.

Limberbush, sangre-de-cristo, sangre-de-drago, sangregrado, torote

Jatropha cardiophylla (figure 49) is a red-barked shrub with many wandlike, flex-

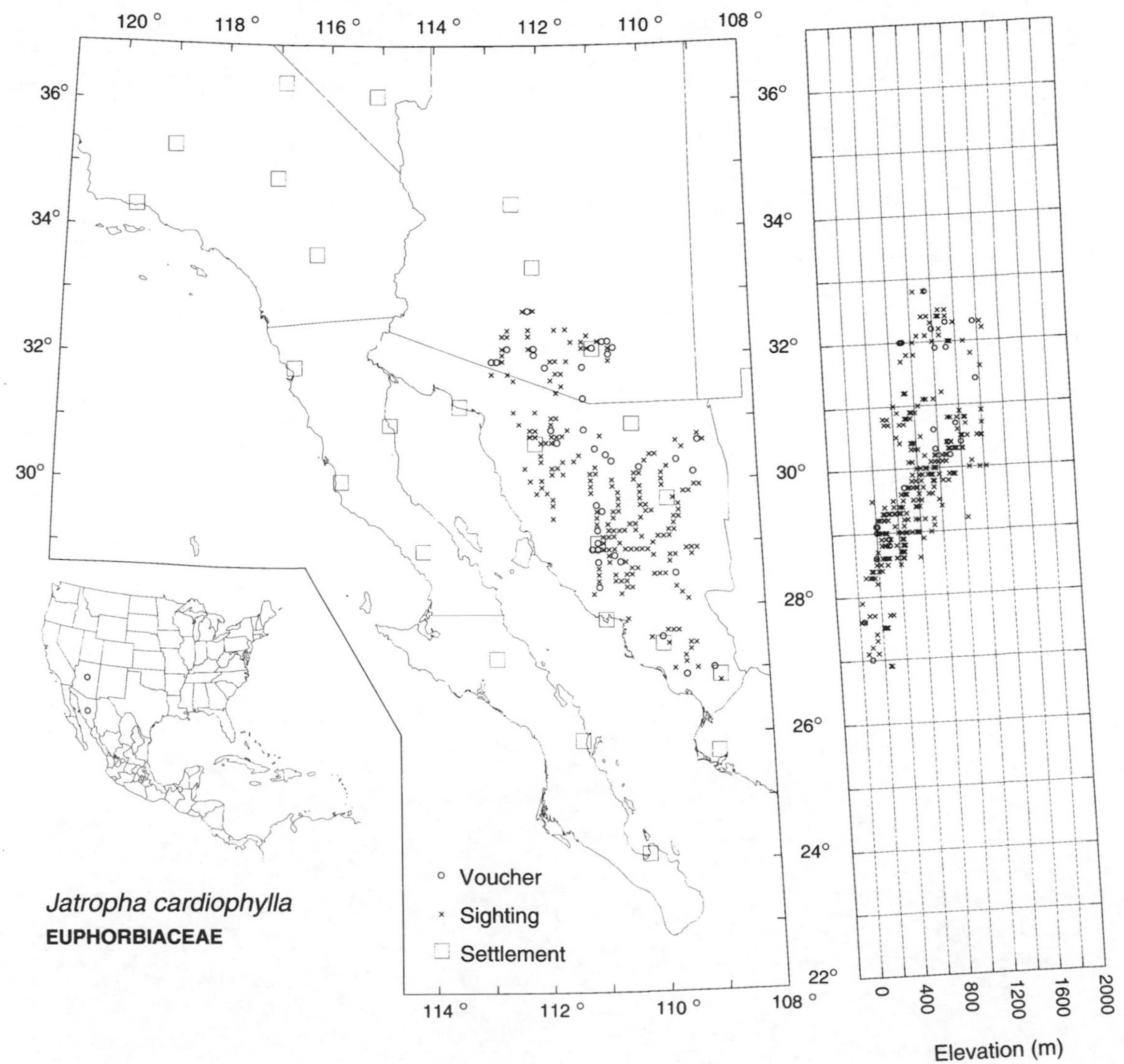

Jatropha cardiophylla
EUPHORBIACEAE

ible stems from the base. Cut branches and roots may exude bloodlike sap. The glabrous, cordate-deltoid, crenate leaves, fascicled on short shoots or single on long shoots, are up to 2.5 cm long and wide. Plants are reported to be monoecious; however, on Tumamoc Hill, Tucson, Arizona, most are dioecious. The white, urceolate flowers, in dichotomously branched cymes, are 4–7 mm long. The diploid chromosome number is 22 (Dehgan and Webster 1979).

Jatropha cinerea is a taller shrub with thick branch tips and brownish or yellowish bark. Its leaves are felty-pubescent beneath when young. *Jatropha cordata* is a small tree with a somewhat swollen trunk, slender branch tips, and pale red to yellow peeling bark.

Jatropha cardiophylla grows on rocky slopes, frequently under other shrubs. It is endemic to the eastern subtropical margin of the Sonoran Desert. In the southern part of its range, where frost is rare, the plants can become small trees (Shreve 1964). To the north, where frost is more frequent and severe freezing is occasional, plants seldom grow taller than 1.5 m, and in the coldest winters unprotected plants may be killed to the ground. Undoubtedly, cold determines the northern and upper elevational limits. To the west, the species may be limited by decreasing summer rainfall. In the more arid parts of the Sonoran Desert, *J. cardiophylla* is replaced by *J. cuneata* and *J. cinerea*. In central Sonora it is sympatric with *J. cordata*, which ranges much farther south and has little frost tolerance (Shreve 1964).

The leaves are drought deciduous. At Tucson they appear during the arid foresummer (early to late June). They increase rapidly in size with the onset of summer rains. In a dry summer, they may remain small. Leaf drop begins in early September and is complete by October. Stems elongate only during the summer rainy season. The first flower buds appear about the same time as the new leaves. Flowering lasts into August.

Flower visitors include wasps, small lepidoptera, and flies (Simpson 1977; Simpson and Neff 1987). In the vicinity of Tucson, bombyliid flies in the genus *Phthiria* are probable pollinators. Reproduction by means of rhizomes may be more common than reproduction by seed.

In nature, long shoots are produced only under the most favorable conditions. Shoot dimorphism in *J. cardiophylla* is evidently controlled by auxin, which in turn responds to environmental cues such as soil moisture and light (Dehgan and Webster 1979).

The stems contain 3% rubber on a dry weight basis (Hall and Long 1921). This rubber has not been used commercially. The roots contain about 5% tannic acid and have been used in tanning (Standley 1923). The sap has been used to staunch bleeding from slight wounds (Kearney and Peebles 1960) and to make an astringent tincture (Moore 1989).

Figure 49. Jatropha cardiophylla *in flower at the Desert Laboratory, Tucson, Arizona. (Photograph by J. R. Hastings.)*

Jatropha cinerea

(Ort.) Muell. Arg. in A. DC.

Jatropha canescens

Muell. Arg.

Jatropha giffordiana

Dehgan & Webster

Ashy limberbush, sangregrado, lomboy, lomboy blanco, torotillo, torotito

This widely branching shrub or small tree grows up to 7 m tall and has smooth, yellowish, grayish, or brownish bark that adheres tightly to the flexible stems. When cut, the branches and roots ooze bloodlike sap. The ovate, suborbicular, or reniform leaves, 3–6.5 cm across, are cordate at the base and entire to shallowly 3-lobed. Above, they are bright green to cinereous; beneath, somewhat paler. Young leaves are felty-pubescent beneath. Stipules are inconspicuous, sessile glands. The tubular flowers are 6–8 mm long. The plants are either monoecious (Wiggins 1964) or dioecious (Dehgan and Webster 1978). The globose or 3-lobed capsules are about

10 mm long and up to 2.8 cm wide. The diploid chromosome number is 22 (Dehgan and Webster 1979).

Jatropha vernicosa has pale, yellow sap and glandular-serrulate leaves. *Jatropha cordata* has an erect, treelike habit; slender twigs; and exfoliating bark. *Jatropha purpurea* Rose & Pax, of Baja California Sur and Sinaloa, has glabrous leaves and dissected stipules. *Jatropha moranii* Dehgan & Webster, endemic to the Cape Region of Baja California Sur, has 5-lobed leaves ciliate with stipitate glands.

Jatropha cinerea is a species complex with great variation in habit and morphology (Dehgan and Webster 1979). McVaugh (1945) distinguished four races on the basis of leaf shape and pubescence but did not name them. Dehgan and Webster (1978) segregated the material in the study area into three species: *J. giffordiana*, endemic to the Cape Region; *J. canescens*, known from Isla Magdalena, Baja California Sur, and from Guaymas, Sonora, south into Sinaloa; and *J. cinerea*, widespread in Sonora and on the Baja California peninsula and barely entering southern Arizona. Few of our records distinguish among the three, and we have mapped them together.

The plants are locally common on plains, hillsides, washes, and roadsides. The northernmost cluster of points is at Senita Basin, Arizona, and in adjacent Sonora. At this latitude, severe freezes occasionally kill the plants to the ground. Low temperatures likely keep them from moving northward. Most populations grow below 700 m (rarely as high as 1,300 m), which also suggests sensitivity to frost. The rather sharp eastern limit in Sonora nearly coincides with the 500 m contour. This pattern may show frost sensitivity, or it may show where *J. cordata* replaces *J. cinerea*. Across the range, rainfall seasonality varies from almost exclusively winter in the northwest to predominantly summer and early fall toward the south.

Jatropha cinerea may flower in any month. The main flowering period is August through November (Wiggins 1964). Plants are generally leafless in May and June. The drought-deciduous leaves typically appear after the onset of summer

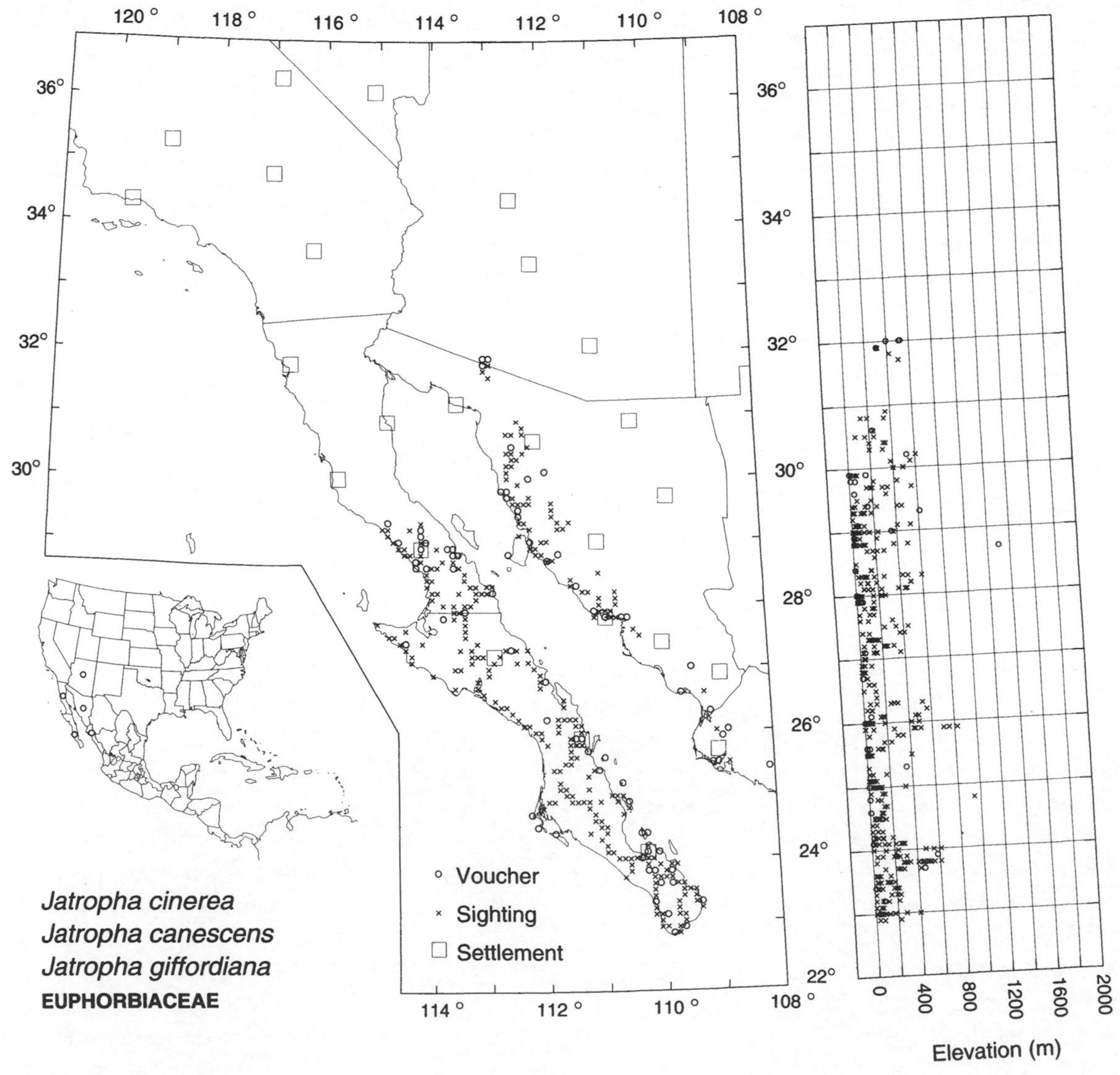

Jatropha cinerea
Jatropha canescens
Jatropha giffordiana
EUPHORBIACEAE

rains and may remain on the plants through the winter and early spring. No leaves are produced when the summer rains fail (Shreve 1964). Early in the growing season short shoots produce fascicles of leaves, then, if rainfall is sufficient, long shoots with alternate leaves appear. Appearance of long shoots seems tied to production of auxin, which is in turn linked to the length of the growing season and the amount of available moisture (Dehgan and Webster 1979).

Seri Indians made utensils from the wood and mashed the roots to make a tea for dysentery. The sap was used to poison arrow points (Felger and Moser 1985). The roots can be inoculated with mycorrhizae (Rose 1981).

Jatropha cordata

(Ort.) Muell. Arg.

Mata muchachos, jiotillo, torota blanca, copalillo, mata mala, miguelito, sapo

Jatropha cordata (figure 50), a small tree 2–8 m high, has shining, red twigs and pale red to yellowish bark that peels off in papery sheets to reveal greenish inner bark beneath. The glabrous, serrate leaves, 2.5–6 cm long and 1.5–4.5 cm wide, are ovate to cordate. The stipules are glands 1–2 mm long. Plants are monoecious, with inconspicuous white flowers in lax cymes. The diploid chromosome number is 22 (Dehgan and Webster 1979).

Jatropha cinerea is a shrub with tightly adherent bark. *Jatropha purpurea* Rose & Pax has shallowly 3-lobed leaves and dissected stipules. Material of *J. cordata* from Sonora and Chihuahua is glabrous or essentially so; specimens from farther south are more densely pubescent (McVaugh 1945).

Often a dominant on volcanic slopes and mesas, *J. cordata* is also common on plains (Shreve 1964). In Tucson unprotected plants are killed to the ground by frost. Cold temperatures evidently limit the northward movement of the species (Shreve 1964).

At the northernmost stations for this species, frost is not (or only rarely) limiting, and, at higher elevations, summer

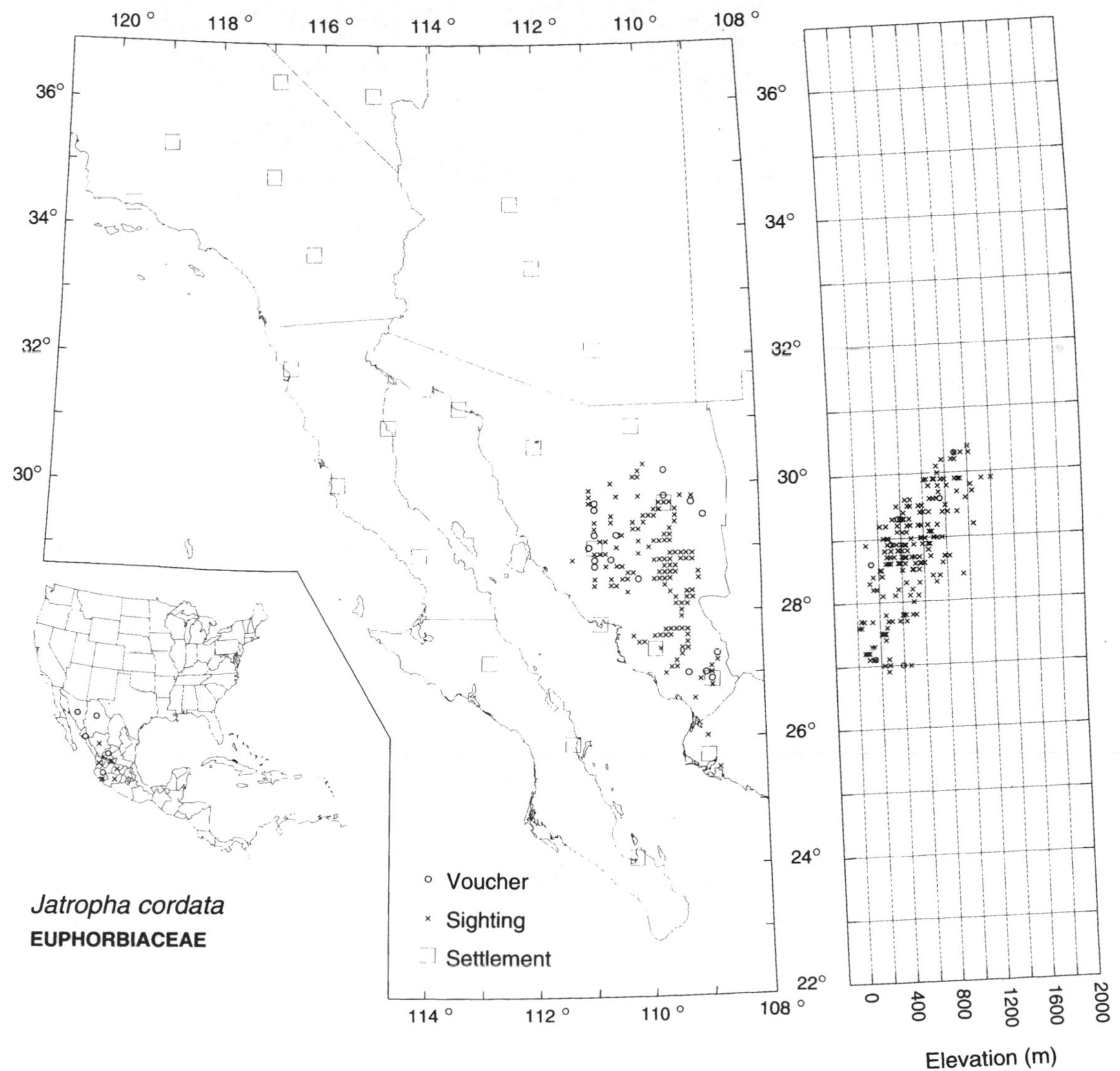

Jatropha cordata
EUPHORBIACEAE

rain is adequate for growth. This trade-off between the need for summer rain and the need for protection from frost may account for the tightness and steepness of the points on our elevational profile.

Jatropha cordata flowers from August to September (Wiggins 1964). Leaves appear at the start of the summer rains and remain until the arid foresummer of the following year (Shreve 1964). Popham (1947) discussed the developmental anatomy of the seedlings.

The bruised leaves are applied to sores (Standley 1923).

Figure 50. Leafless Jatropha cordata *north of Hermosillo, Sonora. The loose, parchmentlike bark on the stems is characteristic. (Photograph by J. R. Hastings.)*

Jatropha cuneata
Wiggins & Rollins

[=*J. spathulata* (Ort.) Muell. Arg.]

Limberbush, leatherplant, sangregrado, sangre-de-drago, tecote prieto, torote amarillo, matacora

An intricately branched shrub up to 2 m tall, *Jatropha cuneata* (figure 51) has stout, flexible stems that exude clear sap when broken. The cuneate-obovate to spatulate leaves, 5–18 mm long, are fascicled on spur shoots. Plants are monoecious, with male or female flowers in separate terminal cymes. The white, tubular flowers are about 7 mm long. The diploid chromosome number is 22 (Dehgan and Webster 1979).

The other *Jatropha* species in the Sonoran Desert region have deltoid to ovate leaves. *Euphorbia misera* has milky rather than clear sap. Branchlets of *Adelia virgata* are usually spinescent, and leaves of *A. obovata* are densely pubescent.

Jatropha cuneata is locally abundant on gravelly plains and rocky slopes, often with other sarcocaulescents such as *Bursera microphylla*, *B. hindsiana*, and *E. misera*. It occasionally colonizes road shoulders (Dehgan and Webster 1979). The southernmost collection in Sinaloa is Howard Scott Gentry's from Cerro de Navachiste near Bahía Topolobampo. A 10,000-year sequence of packrat middens from the Hornaday Mountains, Sonora, shows *J. cuneata* arriving only 1,700 years B.P. (Van Devender, Burgess et al. 1990).

The range is restricted to subtropical arid climates that receive warm-season rain on a moderately regular basis. Substantial cool-season rain may also occur. Cultivated plants in Tucson are killed by severe freezes, and sensitivity to cold no doubt limits the northward distribution. The eastern limit in Sonora is at uniformly low elevations where frost is not likely to be limiting. Competition with *J. cardiophylla* and *J. cordata*, which need more summer rain than *J. cuneata*, perhaps determines this line.

Flowering occurs from July into September (Wiggins 1964; Humphrey 1975). Leaves generally appear at the start of summer rains and drop as the soil dries out in late spring of the following year (Shreve 1964). Frost also causes leafdrop. Stem elongation is seasonal, interrupted by periods of drought- or cold-induced dormancy. A distinct morphological articulation occurs between each growth increment (Dehgan and Webster 1979).

Matched photographs taken several decades apart show that *J. cuneata* can live for at least 55 years (Raymond M. Turner, unpublished photographs).

The stems have been used for making baskets, the bark for tanning and dye. The astringent juice has been used medicinally for skin eruptions, dysentery, shampoo, and sore throats (Standley 1923). Seri Indians used the stems for splints in all their basketry and for making head-rings (Felger and Moser 1985).

Figure 51. Leafless Jatropha cuneata *near Bahía de los Angeles, Baja California. (Photograph by J. R. Hastings.)*

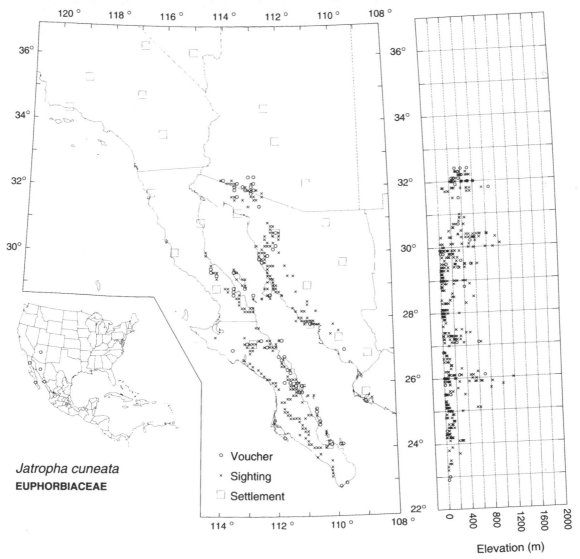

120° 118° 116° 114° 112° 110° 108°

36° 34° 32° 30° 28° 26° 24° 22°

Jatropha cuneata

EUPHORBIACEAE

○ Voucher

× Sighting

□ Settlement

114° 112° 110° 108°

0 400 800 1200 1600 2000

Elevation (m)

Jatropha vernicosa
Brandegee

Sangregrado, sangre-de-drago, lomboy colorado

Jatropha vernicosa is a shrub 2–4 m tall with limber stems, pale yellow sap, smooth bark, and cordate, glandular-serrulate leaves. Acute (rarely acuminate) at the apex, the leaves (3–5 cm wide and 5–7 cm long) are dark green and lustrous above, paler beneath. Plants are dioecious with inconspicuous staminate flowers in loose cymes and pistillate flowers solitary at the branch tips. The glabrous, spherical capsules are about 2 cm in diameter.

Jatropha cinerea has entire leaves and clear sap. *Jatropha moranii* Dehgan & Webster is less than 1 m tall and has shallowly 5-lobed leaves.

Standley (1923) regarded *J. vernicosa,* endemic to Baja California Sur, as doubtfully distinct from *J. cordata,* a mainland species. The differences between them are slight, if any.

Jatropha vernicosa grows on mountain ridges and steep hillsides. Annetta Carter (personal communication 1988) noted that it replaces *J. cinerea* above 600 m. On a broader scale, the mapped distribution of *J. vernicosa* is entirely contained within that of *J. cinerea.* The points clustered around 26.0°N are Annetta Carter's. The northernmost vouchers are Reid Moran's from near San Sebastián.

The plants flower mainly in July and August, occasionally through December. Leaves are produced in response to summer rains and may last through winter or may be lost as soil moisture declines in fall.

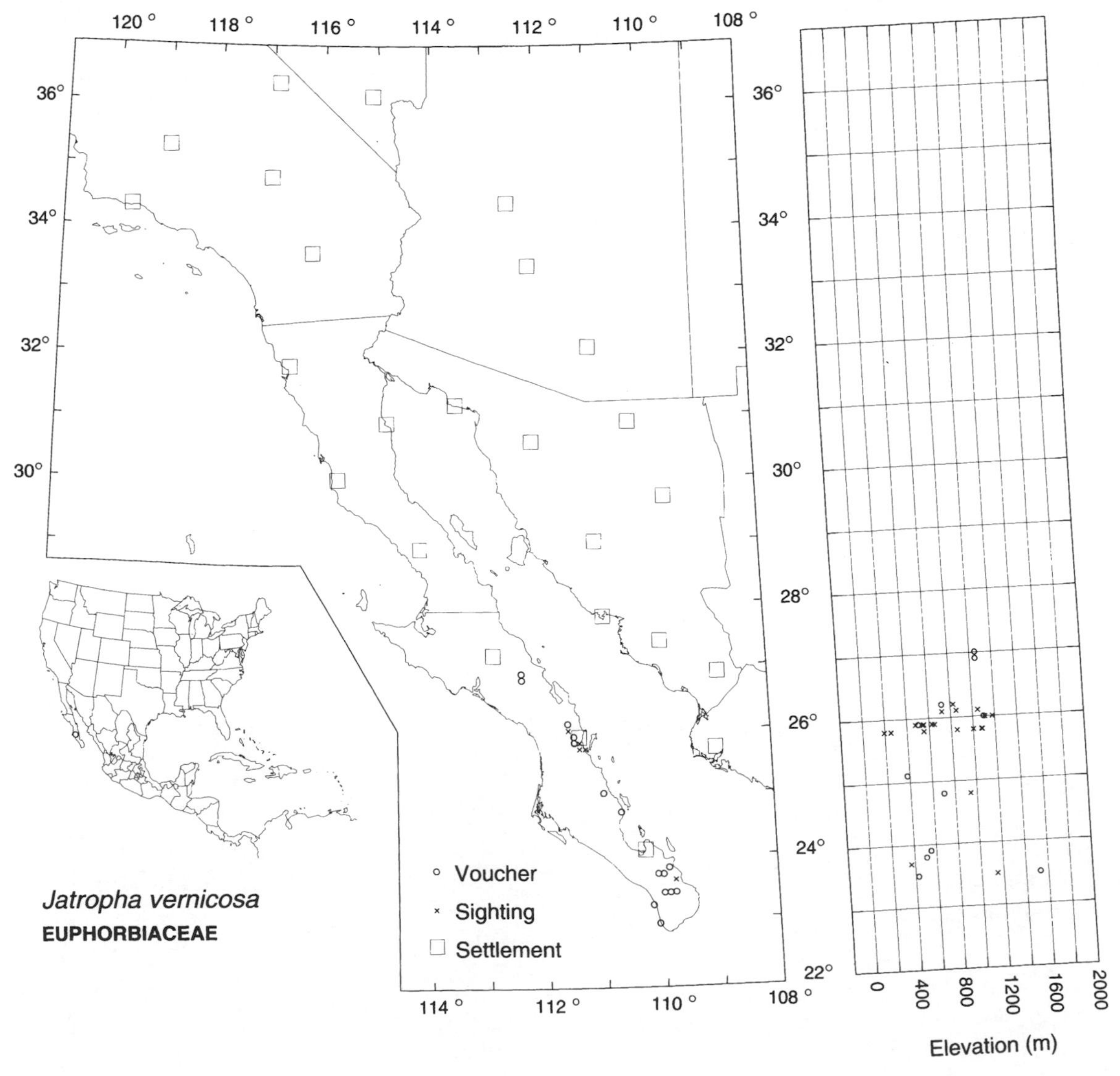

Justicia californica

(Benth.) D. Gibson

[=*Beloperone californica* Benth.]

Chuparosa

A rounded shrub up to 2 m high and nearly as wide, *Justicia californica* has pallid, softly pubescent branches. The ovate, opposite leaves, 1–6.5 cm long and 0.5–4 cm wide, are rounded or cordate at the base. The bright red (rarely yellow), tubular flowers, 21–40 mm long, are in short, loose racemes. The capsules (2 mm long) contain 4 round seeds. The diploid chromosome number is 28 (Grant 1955).

Justicia candicans (Nees) L. Benson (=*Jacobinia ovata* A. Gray) has straw-colored or brown stems, and its red flowers are often streaked with white. *Anisacanthus thurberi* has brick-red or orange flowers, exfoliating bark, and lanceolate to oblong leaves.

Leonard (1958) subsumed the genus *Beloperone* under *Justicia*. Gibson (1972) made the new combination for *J. californica* and also transferred *Jacobinia* and several other genera into *Justicia*.

Frequent in gravelly or sandy washes and occasional on gentle slopes, *J. californica* is very nearly restricted to the Sonoran Desert. The southernmost point on the mainland (24.9°N) is a collection by Howard S. Gentry from Isla Tachechilte. No localities are known between Isla Tachechilte and the nearest population in southern Sonora (Tom Daniel, personal communication 1987). The easternmost stations in Arizona are in the Tortolita Mountains and the Tucson Mountains, Pima County. The westernmost localities, vouchered by Reid Moran, lie outside the desert proper in southern California at Lakeside, San Diego County, and in northern Baja California at Arroyo de las Palmas.

The northernmost plants suffer frost damage during the coldest winters (Anonymous 1979, Bowers 1980–81), and it seems likely that low temperatures limit the species to the north and northeast (Daniel 1984). Farther south, where winters are warmer, the species extends into the Sierra Madre Occidental, as at San Bernardo. Apparently it is uncommon in

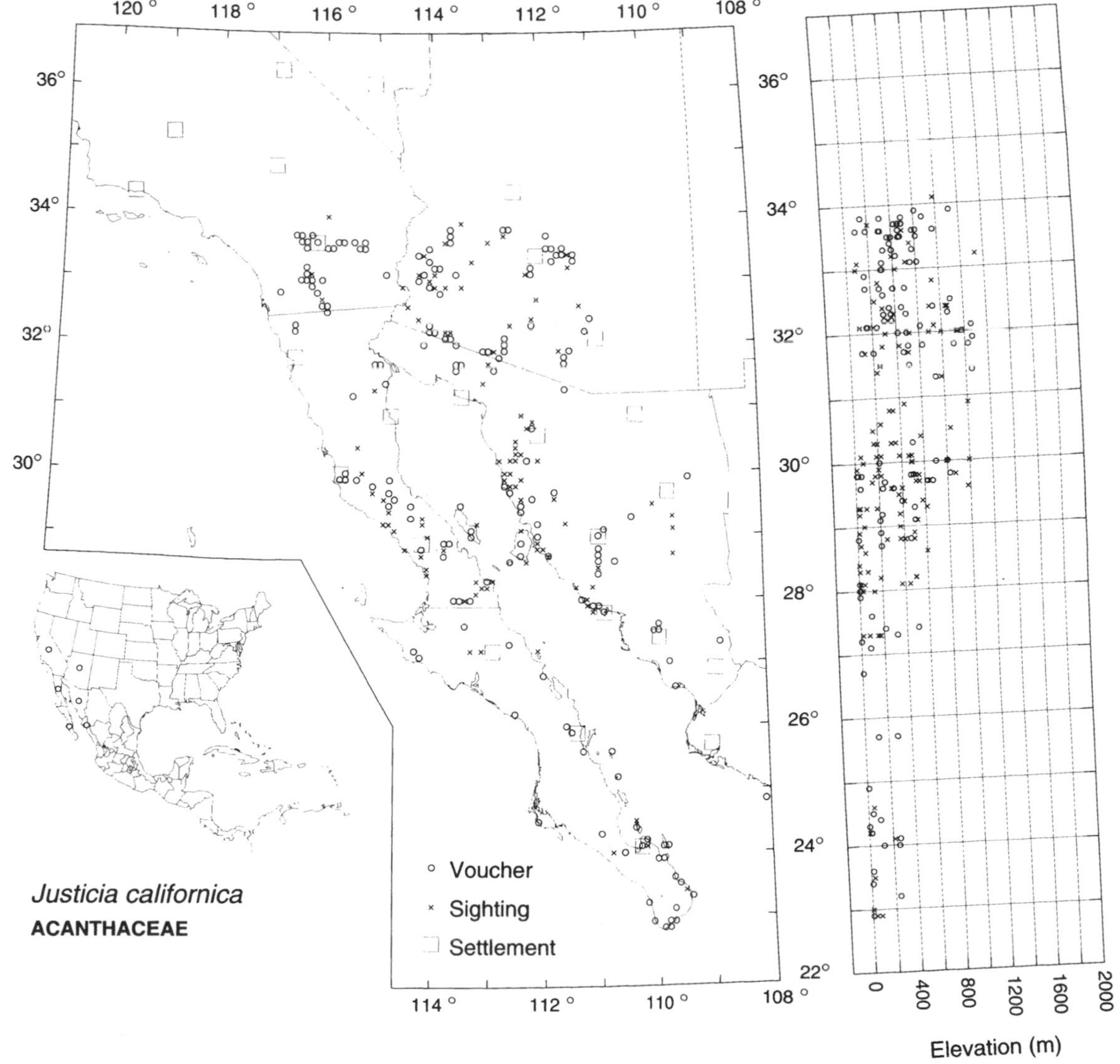

Justicia californica
ACANTHACEAE

this region (Gentry 1942).

Flowers appear from October to June (Daniel 1984). In southeastern California, the plants tend to bloom in late winter and early spring. To the south, summer and fall flowering, apparently in response to rain, is also common. Leaves are drought- and cold-deciduous. Especially in the northern part of their range, the plants are typically leafless from October through February. The first flowers often appear before the leaves. New stem and leaf growth occurs in response to both summer and winter rains (Humphrey 1975).

Hummingbirds are probably the most important pollinators. Along the lower Colorado River, *J. californica* is a major resource for hummingbirds during their late winter and early spring migration.

Justicia californica is used in landscaping to a limited extent. It responds well to irrigation and, although somewhat frost sensitive, recovers rapidly from winter setbacks (Duffield and Jones 1981).

Karwinskia humboldtiana

(Roem. & Schult.) Zucc.

Karwinskia parvifolia

Rose

Buckthorn, coyotillo, tullidora, capulincillo, capulincillo cimarrón, capulín, palo negrito, margarita, cacachila, china, cacohila silvestra, frutillo negrito, cochila, margarita del cerro

A leafy shrub or small tree 1–8 m tall, *Karwinskia humboldtiana* has smooth, brown bark dotted with numerous white lenticels. The glabrous, elliptic to ovate leaves are 4–8 cm long. On the underside, the veins are conspicuously marked with short, black lines. The small flowers are in sessile or short-pedunculate umbels. The shining, black drupes are 7–10 mm long.

Karwinskia parvifolia, a shrub up to 5 m tall, has somewhat smaller leaves (1–4.5 cm long) with black dashes almost entirely confined to the margins (rather than

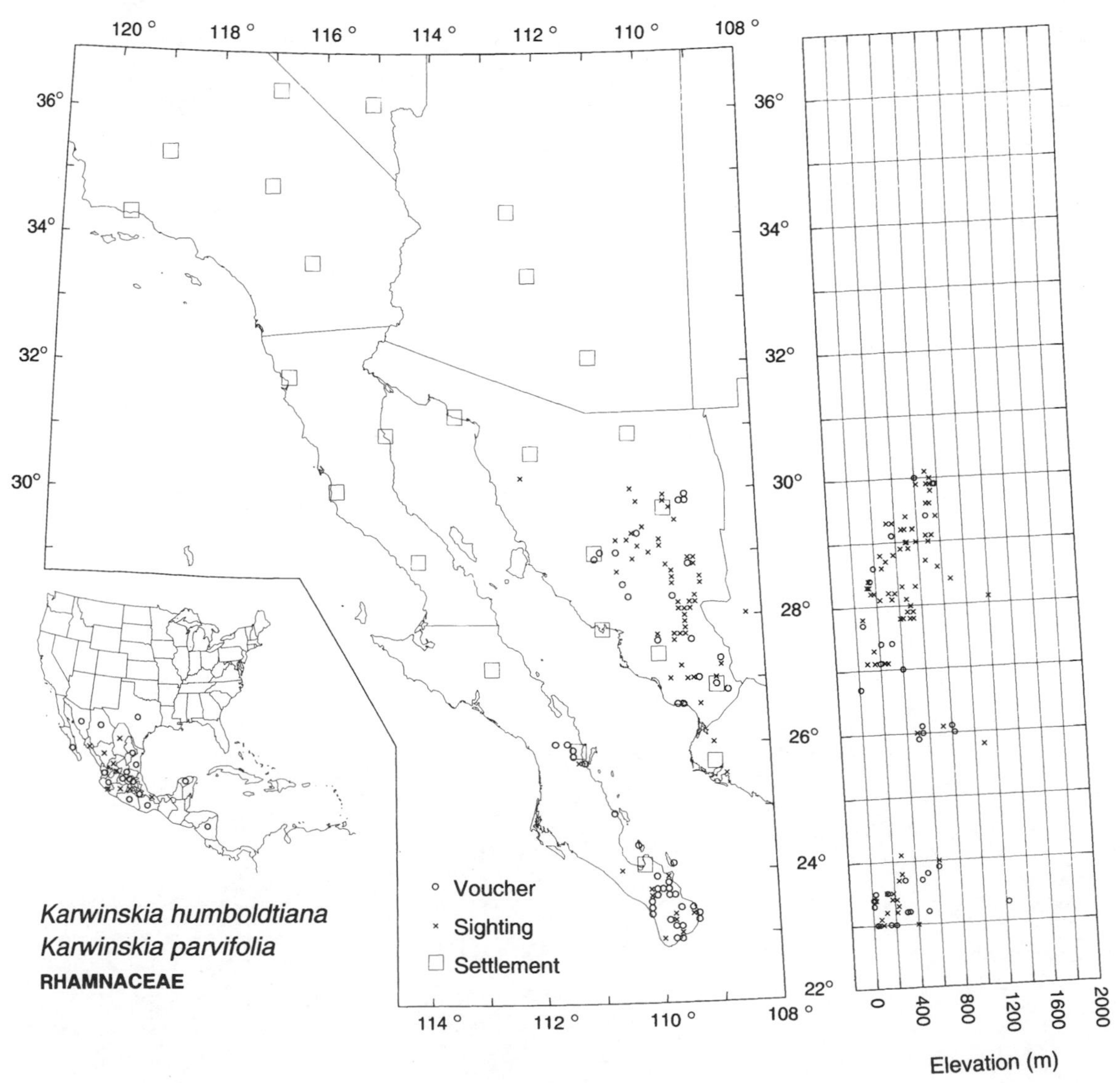

Karwinskia humboldtiana
Karwinskia parvifolia
RHAMNACEAE

the veins). Botanists have recognized two taxa: var. *pubescens* (Standley) Wiggins, which has dense, minute hairs on leaves, twigs, and flowers; and var. *parvifolia,* which has glabrous foliage and flowers.

The dashed black lines on the underside of the leaves are distinctive.

The taxonomy of this group is confused. *Karwinskia humboldtiana,* named in 1832, is a widespread, variable species that has proven difficult to delimit (Johnston 1966). *Karwinskia parvifolia* was segregated from it in 1895 on the basis of a collection from Agiabampo, Sonora, but in 1923, Standley relegated *K. parvifolia* to synonymy under *K. humboldtiana.* At the same time, he proposed *K. pubescens* Standley, possibly "only a form of *K. humboldtiana.*" Wiggins (1964) resurrected *K. parvifolia* and relegated *K. pubescens* to varietal status under it. He treated *K. humboldtiana* more narrowly than previous authors had done.

We have mapped *K. humboldtiana* and *K. parvifolia* together because we cannot consistently distinguish them in the field. Particularly in central Sonora, leaf size, leaf markings, and pubescence are all unreliable.

Most characteristic of Sinaloan thornscrub and Sinaloan deciduous forest, these species can be locally common along arroyos and on rocky or gravelly slopes (Gentry 1942). The Sonoran disjunct from the Sierra del Viejo (30.3°N) represents a sighting by Howard Scott Gentry. The mapped distribution suggests reliance on summer rainfall and intolerance of freezing temperatures. Lonard and Judd (1991), however, reported that *K. humboldtiana* in the lower Rio Grande Valley, Texas, recovered from a severe freeze. Our populations may be cold-sensitive ecotypes. Competition may exclude them from woodlands at higher elevations.

Lux and Earl (1989) investigated the leaf anatomy.

Eaten in sufficient quantity, the leaves and fruits are poisonous to livestock and humans, resulting in paralysis and, often, death (Marsh and Clawson 1928; Corbett 1991). The roots show antimicrobial activity (Mitscher et al. 1985); in Mexico, an in-

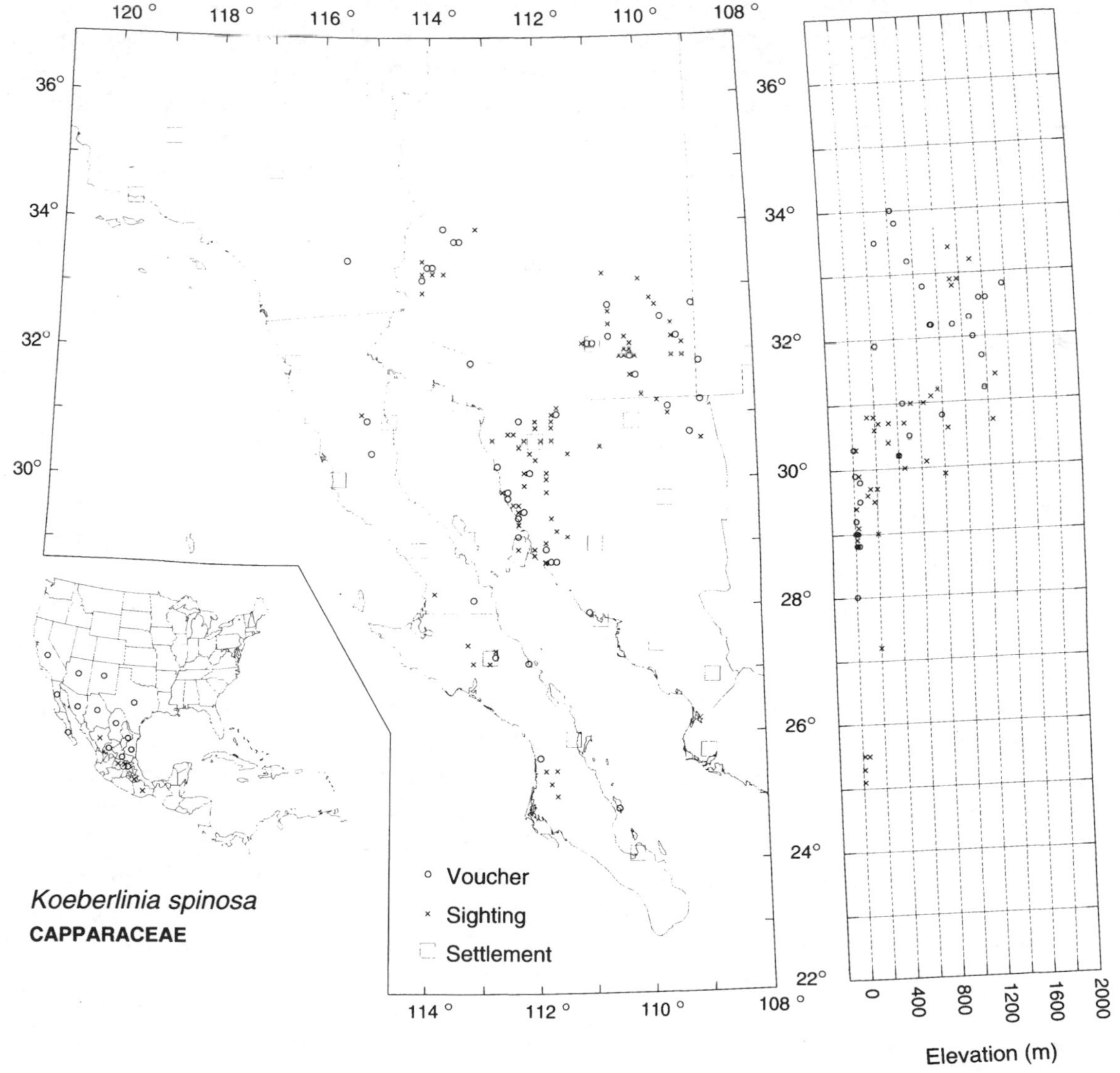

Koeberlinia spinosa
CAPPARACEAE

fusion of the leaves was used to treat fever, toothache, and neuralgia, and the powdered bark was used as a laxative (Marsh and Clawson 1928).

Koeberlinia spinosa

Zucc.

All-thorn, crucifixion-thorn, crown of thorns, abrojo, corona de Cristo, junco

An intricately branched shrub up to 4.5 m tall, *Koeberlinia spinosa* has rigid, spinose, divergent branches with dark green bark. The leaves are minute and scalelike. The inconspicuous flowers are in small, umbelliform clusters. The black and shining berries are 3–4 mm in diameter. The haploid chromosome number is 22 (Weedin and Powell 1978b). A diploid chromosome count of approximately 88 has also been reported (Goldblatt 1976).

Stems of *Ziziphus obtusifolia* var. *canescens* are conspicuously gray-hairy. The striate branches of *Canotia holacantha* are erect or nearly so and not particularly rigid. *Castela emoryi* has stout branches and persistent dry capsules.

Gibson (1979) suggested that the species be reassigned from Koeberliniaceae to or near the Capparaceae (=Capparidaceae) based on stem anatomy. The two varieties are var. *spinosa,* which extends from southeastern Arizona to west Texas and south into Mexico, and var. *tenuispina* Kearney & Peebles, known from disjunct locations in western Arizona and southeastern California. Variety *spinosa* flowers in late summer and seldom grows taller than 1.8 m; var. *tenuispina* flowers in March and April and grows up to 4.5 m tall (Kearney and Peebles 1960).

The plants grow on sandy or gravelly plains, along arroyos, and on rocky slopes. Where abundant, they may form impenetrable thickets. The disjunction in southeastern California is based upon E. C. Jaeger's 1939 collection from the north base of the Chocolate Mountains, California. The isolated collection at 31.9°N, 113.3°W is Richard Felger's from Playa Diaz in the Pinacate Region, Sonora. Variety *tenuispina* grows in arid subtropical climates where rainfall is almost evenly biseasonal. Variety *spinosa* grows in warm-temperate, semiarid areas where rain falls mainly in summer.

Koeberlinia spinosa populations may once have been more closely connected. Pleistocene vegetation shifts would have severed these connections, leaving the pattern of disjunct populations presently observed and permitting divergence at the varietal level. The species also occurs in Bolivia. Movement may have been from the southern to the northern hemisphere in post-Miocene times (Raven 1963).

The plants are considered range pests in some locations (Kearney and Peebles 1960; Benson and Darrow 1981). Seri Indians burned the wood to disinfect their houses of diseases and made tea from the flowers to treat dizziness and intestinal disorders (Felger and Moser 1985).

Laguncularia racemosa

(L.) Gaertn.

White mangrove, white buttonwood, mangle blanco, mangle bobo, mangle amarillo, patabán, mangle prieto, mangle chino

A shrub or tree up to 10 m tall in the study area, *Laguncularia racemosa* has leathery, opposite leaves 1–4 cm wide and 2–12 cm long. The petiole has two small glands where it joins the blade. The small, inconspicuous flowers (in lax, clustered spikes) are perfect or polygamous. The fruit, a leathery drupe about 15 mm long, retains its pericarp until the radicle has emerged.

Rhizophora mangle and *Avicennia germinans* lack conspicuous petiolar glands. *Conocarpus erecta* L. leaves are alternate rather than opposite.

Like other mangroves, *L. racemosa* is characteristic of brackish or salt water in inlets, bays, and coastal mudflats. In Sonora and elsewhere, the various species of mangrove (genera *Rhizophora, Avicennia, Laguncularia, Conocarpus*) often grow in more or less distinct zones (Felger 1966; Rabinowitz 1978; Odum and McIvor 1990). Along coastal Sonora, cover in mangrove thickets can reach 100% (Felger 1966).

The northernmost points on our map represent collections made along the Infiernillo channel by Richard Felger in 1968 and 1972. Populations grow along the Sonoran coast from Guaymas into northwestern Sinaloa (Richard Felger, personal communication 1992). Our maps indicate almost identical northern limits for *L. racemosa* and *R. mangle.* Occasional severe freezes determine the northern limit for both species (Felger and Moser 1985). Data on the chilling response of *L. racemosa* seedlings in the Caribbean and Gulf of Mexico (Markley et al. 1982) suggest that the northern populations in the Sonoran Desert region may be more cold tolerant than those farther south.

Laguncularia racemosa blooms July–October in the study area (Wiggins 1964). The leaves are evergreen.

The plants do not require saline conditions to survive; in one study, seedlings irrigated with pure water did not significantly differ in leaf conductance or net carbon assimilation from those irrigated with saline water (Pezeshki et al. 1990). Researchers have not yet clearly established the mechanism of salt tolerance. According to Scholander and coauthors (1966), *L. racemosa* excludes salts almost entirely (the sodium chloride content of the sap in that study was only 1.2–1.5 mg/ml), but Tomlinson (1986) suggested that excretion of salt crystals by leaves helps maintain the salt balance of the plant.

Subterranean roots of mangroves often grow in anaerobic environments. *Avicennia germinans* and *R. mangle* adapt by producing aerial roots, which serve as the site of gas exchange for the root system (Gill and Tomlinson 1977). Such specialized roots are not always produced in *L. racemosa,* which has numerous lenticels on the trunk. When present, these "peg" roots are blunt-tipped, erect cylinders up to 20 cm tall (Tomlinson 1986).

The embryo (rather than the ungerminated seed) is the dispersal unit (Rabinowitz 1978). After dispersal, the propagules may float for some time before rooting in the substrate. Germination usually occurs within 11–16 days after the propagule falls from the parent plant (Rabinowitz 1978). The propagules must be stranded for about 5 days for rooting to occur. This characteristic restricts *L. racemosa* to areas within the tidal zone where inundation is not a daily occurrence. The various stranding requirements help account for the typical zonation of *Rhizophora, Avicennia,* and *Laguncularia* species where they grow together (Rabinowitz 1978). Competition may also influence zonation in mixed mangrove communities; *R. mangle* and *L. racemosa* apparently interact to lower the percent cover of *A. germinans* (Lopez-Portillo and Ezcurra 1989).

Walsh (1979) further compared the physiological and ecological attributes of the various mangroves.

The bark is rich in tannins and has been used for tanning skins. It is also used as an astringent and a tonic (Standley 1924). Seri Indians used the wood for oars and harpoon shafts (Felger and Moser 1985). In Puerto Rico, *L. racemosa* is harvested for lumber (Wadsworth 1959).

Larrea tridentata

(Moc. & Ses.) Cav.

[=*L. divaricata* Cav. subsp. *tridentata* (Ses. & Moc. ex DC.) Felger & Lowe, *Covillea tridentata* (DC.) Vail]

Creosote bush, greasewood, gobernadora, hediondilla, guamis

This aromatic shrub, 0.5–3.5 m tall, has many slender stems arising vertically or somewhat obliquely from the root crown (figure 52). The obliquely lanceolate or falcate leaves are formed of two glabrous leaflets, each 4–10 mm long, fused at the base. Young leaves are often resinous and

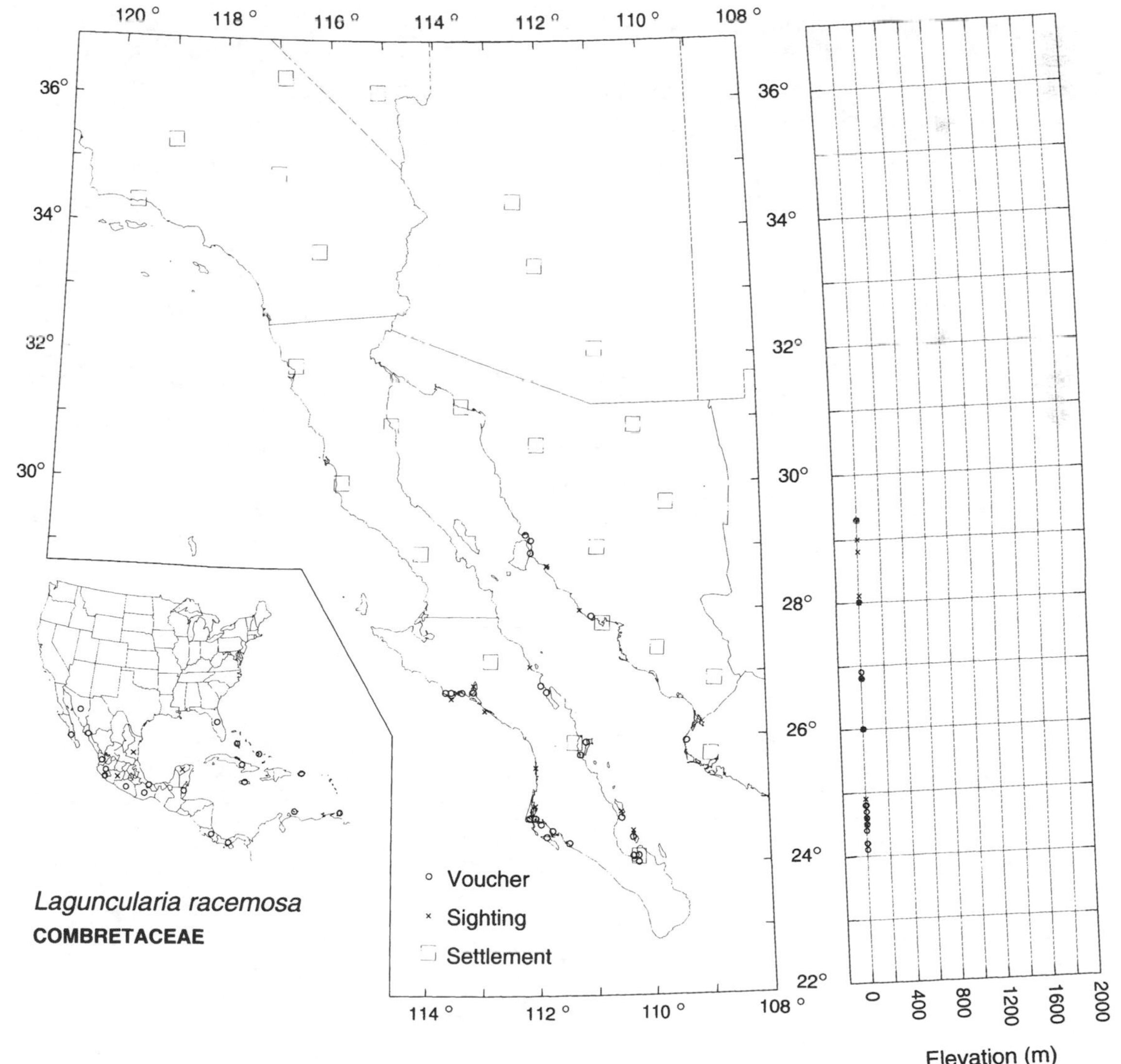

Laguncularia racemosa
COMBRETACEAE

shining. The yellow petals are about 7–11 mm long. The silky-pilose fruit, 6–8 mm long, separates at maturity into five indehiscent, 1-seeded sections (mericarps).

Botanists have recognized two varieties: var. *tridentata,* which occurs throughout the range of the species, and var. *arenaria* L. Benson, known from stable dunes in Imperial Valley, California. *Larrea tridentata* may be derived from *L. divaricata* Cav., a South American diploid (Hunziker et al. 1977). The plants of the two continents differ in minor leaf and stipule characters (Porter 1963) as well as branching rate (Barbour 1969).

Larrea tridentata is represented in North America by three chromosome races: diploid (n=13), tetraploid (n=26), and hexaploid (n=52). Studies of the external phenolic resin of the leaves of *Larrea* species revealed a single chemical type for all three ploidy levels, suggesting an autoploid origin for the tetraploid and hexaploid races. The North American diploid race shows closer ties to Peruvian than Argentinean populations (Sakakibara et al. 1976).

The diploid, tetraploid, and hexaploid races grow in the Chihuahuan, Sonoran, and Mojave deserts, respectively (Yang 1967a, 1967b, 1970; Yang and Lowe 1968; Barbour 1969). Hunziker and coworkers (1977) interpreted this distribution pattern to mean that ancestral *L. tridentata* migrated successively through the Chihuahuan, Sonoran, and Mojave deserts, differentiating into the three known races in order of increasing chromosome number. *Larrea tridentata* was present in the Tinajas Altas Mountains, Arizona, as early as 18,700 years B.P. (Van Devender 1987) and apparently entered the northern Chihuahuan Desert only 4,340 years B.P. (Van Devender and Toolin 1983). It first appeared near its present northern limit in the Mojave Desert about 5,400 years B.P. (Spaulding 1990). If the species did migrate from east to west, it must have done so well before the Holocene.

The most common and widespread shrub in the warm deserts of North America, *L. tridentata* is a characteristic dominant, with or without a few subordinate species, on valley bottoms and plains,

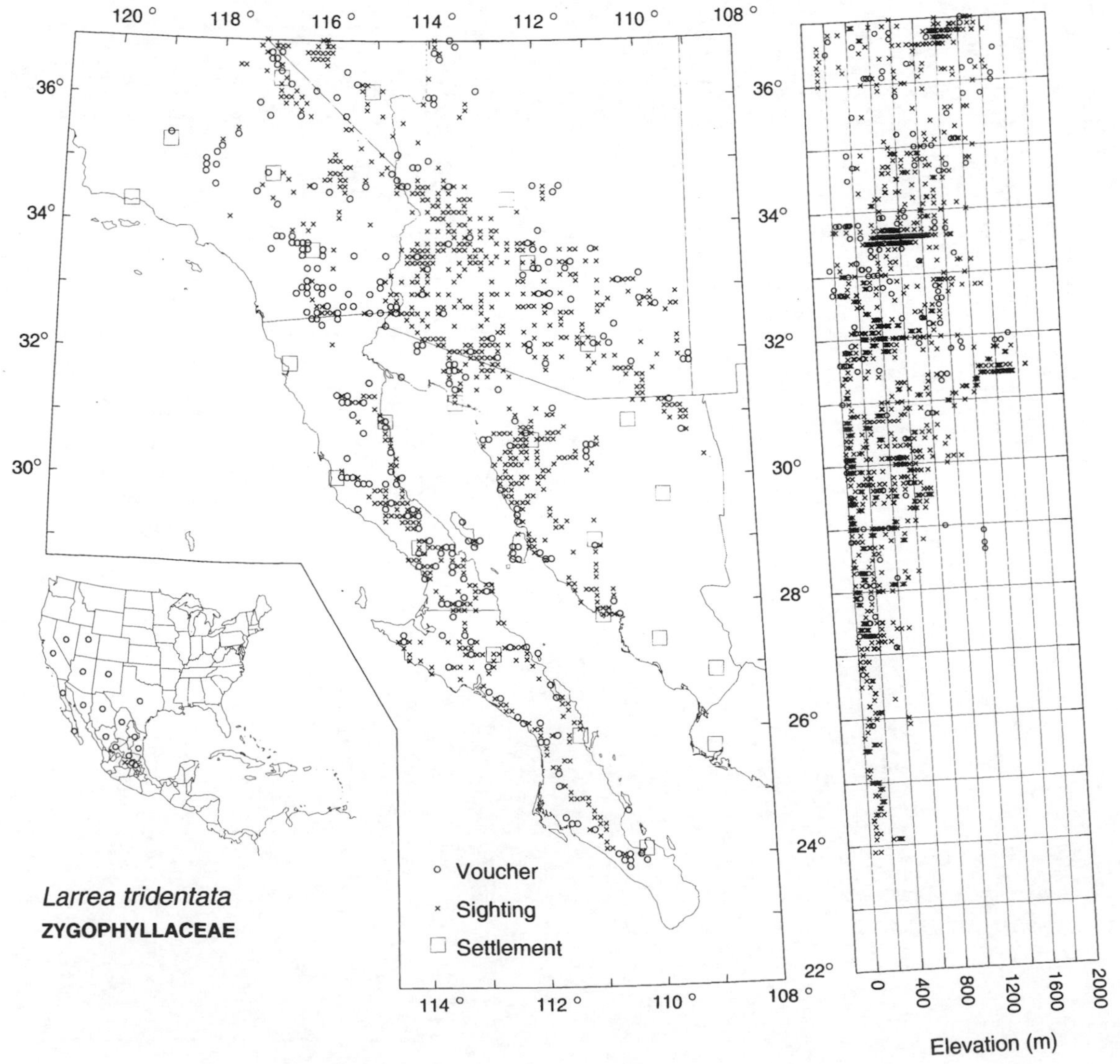

often in sandy or gravelly soils. In the vicinity of Tucson, Arizona, it tends to be most abundant on Holocene terraces and fans with little soil development and rare or absent on sites with well-developed argillic horizons near the surface (McAuliffe 1991). The plants often grow on soils with a prominent calcic horizon (caliche) (Hallmark and Allen 1975; Musick 1977; Wierenga et al. 1987). Where rainfall is exceedingly low, caliche blocks the downward percolation of water, a phenomenon that may favor growth by enhancing plant water status (Cunningham and Burk 1973) or inhibit it by obstructing root penetration (Shreve and Mallery 1933). Under more favorable rainfall conditions, moisture is more available in soils lacking caliche (Cunningham and Burk 1973).

The wide range of this species brings it in contact with several different communities (Shreve 1940, 1942): semidesert grassland and interior chaparral in Arizona; Californian chaparral in northern Baja California; Sinaloan thornscrub in Baja California Sur; and Great Basin desertscrub and Great Basin conifer woodland in California. The isolated station on the Río Bavispe (30.4°N, 108.9°W) is from near Bácerac (White 1949). Except for this outpost of the Chihuahuan Desert race, the species is absent from most of eastern Sonora (Shreve 1940).

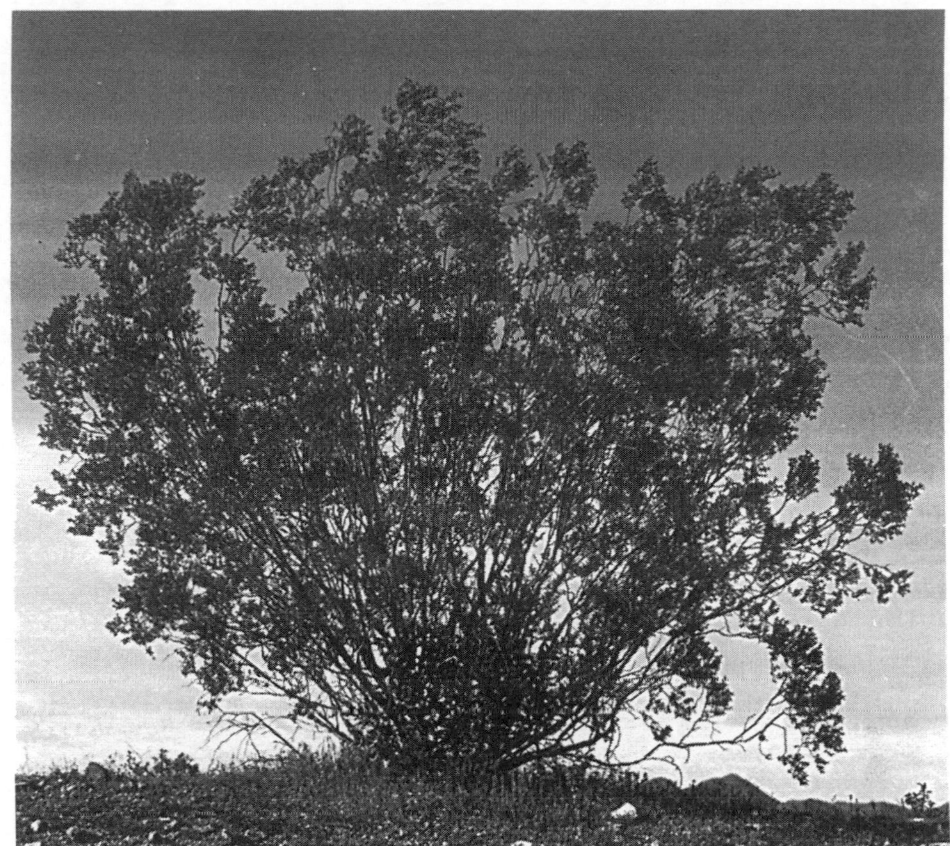

Figure 52. Larrea tridentata *near Phoenix, Arizona. (Photograph by R. M. Turner.)*

Although *L. tridentata* is more cold tolerant than many Sonoran Desert plants, it is susceptible to extreme freezing conditions, and winter cold determines the northern limit. Entire stands were killed to the ground in southern Utah during one unusually cold week in January 1937 when nightly temperatures descended below −18°C and the minimum for the week reached −24°C (Cottam 1937). These plants quickly revived, however (Fosberg 1938). In New Mexico, *L. tridentata* lost leaves after a freeze of −28°C but later resprouted (Valentine and Gerard 1968). In the Mojave Desert of California, the ponding of cold air during winter nights apparently excludes *L. tridentata* from closed basins, with temperatures of −17.8°C representing the lethal threshold (Beatley 1974a). Excessive rainfall rather than cold apparently determines the upper elevational limits in the Mojave Desert, where seed germinability is severely reduced when rainfall levels between mid November and late March exceed 150 mm (Beatley 1974a). High levels of rainfall lessen flower production in both the Mojave and Chihuahuan deserts (Valentine and Gerard 1968; Beatley 1974a; Cunningham et al. 1979), and this factor may also limit upslope movement of the plants. Researchers do not yet fully understand the factors limiting *L. tridentata* to the south (Shreve 1940; Rzedowski and Leal 1958). The southern boundary seems to fall where the annual range of monthly average temperatures is less than 7°C (Garcia et al. 1961). Competitive exclusion from Sinaloan thornscrub may restrict *L. tridentata* to maritime habitats near the southern end of its range.

Stems elongate between February and November, with the most rapid growth in July and August in response to rains. Before July, growth depends more on previous winter rains than spring rains (Burk and Dick-Peddie 1973). Low temperatures inhibit growth during cold periods in the Chihuahuan Desert (Chew and Chew 1965; Burk and Dick-Peddie 1973), but in warm-winter parts of the Sonoran Desert, the plants are never dormant and respond rapidly to minor rainfall events (Oechel et al. 1972). Stem and leaf growth invariably precedes flowering (Ackerman et al. 1980; Abe 1982; Turner and Randall 1987). The plants are drought deciduous, but it is rare to see one completely lacking leaves.

Flowers appear at any season following adequate rain (Chew and Chew 1965). A rain trigger of at least 12 mm is required for flowering. Thereafter, degree-days above the base temperature (10°C) must reach about 443 for plants to bloom (Bowers and Dimmitt 1994). The flowers last a few hours to a day (Simpson et al. 1977a). In the Mojave Desert, where winter rain predominates, peak flowering is in spring (Stark and Love 1969; Boyd and Brum 1983a) followed by seed dispersal between May and early August (Ackerman et al. 1980; Boyd and Brum 1983b). In the eastern Sonoran Desert, flowering is in response to winter or summer rains (Barbour et al. 1977), although spring flowering is more dependable (Abe 1982). In the western Sonoran Desert, *L. tridentata* flowers in spring after winter rains, and fruits mature in early summer (Oechel et al. 1972). In the Chihuahuan Desert, fruit production is much heavier in late summer or fall than in spring (Chew and Chew 1965; Valentine and Gerard 1968; Burk and Dick-Peddie 1973). The mericarps are adapted for wind dispersal by tumbling (Maddox and Carlquist 1985).

Larrea tridentata is extremely resistant to high temperatures and low tissue water potential. Temperature and moisture responses vary among geographical races (Yang 1967b), but all three can maintain photosynthetic activity under a wide range of conditions. Photosynthetic and dark respiration rates reflect a plant's acclimation to a wide temperature range (Strain and Chase 1966; Strain 1969; Armond et al. 1978). Most photosynthetic production is directed into growth; little, if any, is allocated to long-term storage (Oechel et al. 1972). The stem and foliage architecture minimizes self-shading when conditions for photosynthesis are most favorable and may also improve water use efficiency (Neufeld et al. 1988). The leaflets may fold together during periods of moisture stress and may also alter their angle to minimize direct solar radiation (Ezcurra et al. 1992).

Photosynthesis is closely related to internal water potential (Oechel et al. 1972). The extensive root system buffers rapid decreases in plant water potential by sampling a variety of soil microsites. This buffering activity integrates the water status of the plant and, incidentally, can give rise to unusual diurnal patterns of stem water potential (Syvertson et al. 1975). As internal water potential decreases, leaves and sometimes branches are shed, thus maintaining an appropriate balance between root absorption, active meristems, and transpiring area (Chew and Chew 1965; Nilsen et al. 1984). Other mechanisms for maintaining leaf turgor include continual adjustments in leaf anatomy and leaf osmotic potential (Runyon 1934; Bennert and Mooney 1979; Nilsen et al. 1984; Meinzer et al. 1986). Leaf water potentials as low as –8.0 MPa have been reported (Odening et al. 1974). The relatively small size of the leaflets keeps them closely coupled to air temperatures, thus reducing transpiration (Smith 1978). Monson and Smith (1982) found that during very dry periods, the dawn water potential of *L. tridentata* was lower than the turgor-loss threshold, which led them to postulate that metabolic adaptations other than adjustment of water relations were critical for drought survival. The plants are most tolerant of drought and high temperatures when the onset is gradual. The leaf resin is an ideal antitranspirant because it diminishes transpiration more than assimilation (Meinzer et al. 1990).

About 50% of the root system lies within the top 20 cm of the soil (Wallace, Romney et al. 1980). A relatively high oxygen requirement may be a limiting factor in poorly aerated soils (Lunt et al. 1973). In the laboratory, root growth was ten times higher at 30–35°C than at 15–20°C (Cannon 1916b), which suggests that in nature roots grow mainly in summer. Cool soil temperatures probably limit root penetration and water absorption. Root growth is inhibited if the pH is greater than 8 and the concentration of sodium chloride is higher than 1,500 ppm (Barbour 1968). The roots are probably associated with nitrogen-fixing organisms and endomycorrhizae (Mattson 1980). Inoculation with mycorrhizae enhanced seedling survival in Death Valley, California (Sheps 1973).

In root chambers, extracts from living *L. tridentata* roots inhibited root growth in both *Ambrosia dumosa* and *L. tridentata*, suggesting that competitive interactions may occur in nature (Mahall and Callaway 1991). In the field, removal experiments suggested that *L. tridentata* and *A. dumosa* compete for water (Fonteyn and Mahall 1981), as do *L. tridentata* and *Muhlenbergia porteri* Scribn. (Welsh and Beck 1976) and, perhaps, *L. tridentata* and *Opuntia leptocaulis* (Yeaton 1978).

Many animals visit the flowers, but as a group, bees are by far the most abundant pollinators (Hurd and Linsley 1975; Simpson et al. 1977a). Scales at the base of the stamens function as nectaries. During years of copious flowering, flower numbers far exceed pollinator numbers, and the majority of flowers are never visited (Simpson et al. 1977a; Boyd and Brum 1983a). Self-pollination produces a small percentage of viable seed. Individuals become reproductively mature as early as 13 years, although significant numbers of fruits are not produced until plants reach 18 or 20 years (Chew and Chew 1965).

Across the range of the species, germination is in late summer or fall (Went and Westergaard 1949; Sheps 1973; Beatley 1974a, 1974b; Rivera and Freeman 1979; Zedler 1981; Boyd and Brum 1983b). There is no significant seed bank (Boyd and Brum 1983b). Germination of seeds enclosed within the mericarp is reduced and delayed compared to that of naked seeds (Knipe and Herbel 1966; Tipton 1983). In the laboratory, germination is best at 23°C under conditions of darkness, low sodium, and low osmotic pressure. Germination can be improved by first leaching the seeds with running water, exposing them to low temperatures, and subjecting them to wetting and drying cycles (Barbour 1968; Tipton 1985). High temperatures (above 40°C) reduce germination (Rivera and Freeman 1979). The seeds germinate poorly at a pH of 8 and above, which may account for the absence of *L. tridentata* on saline or sodic soils (Lajtha et al. 1987).

The seedlings rarely become established in openings between other shrubs. In one study, most juvenile plants were associated with *Ambrosia dumosa*, which evidently protected them from rabbits (McAuliffe 1988). *Larrea tridentata* seedlings planted beneath mature *L. tridentata* plants suffer higher mortality than those farther away (Sheps 1973; Boyd and Brum 1983b). In one study, *L. tridentata* became established on bare soil mounds abandoned by kangaroo rats (*Dipodomys spectabilis*) as the rodents' preferred grassland site was altered by the shrub's increase (Chew and Whitford 1992). Walters and Freeman (1983) reported seedling growth rates. Seedling stem diameter enlarges by 0.5–1.1 mm/yr (Vasek 1980a). The average radial crown growth in mature plants is 0.3–1.6 mm/yr (Sternberg 1976; Vasek 1980a; McAuliffe 1988).

Researchers can determine the age of plant crowns by counting annual growth rings in stem xylem (Chew and Chew 1965; McAuliffe 1988). New branches are produced at the edge of the crown but not at the center (Sternberg 1976; Vasek 1980a); after 40 to 90 years, death of the oldest central branches results in circular groups of isolated shrublets. As age increases, the circle of genetically identical individuals increases in size and may reach a breadth of several meters. Vasek (1980a), by extrapolation from young clones with known growth rates, estimated that the largest clones approach 11,700 years in age.

Larrea tridentata leaves contain more nitrogen than most evergreen leaves. Perhaps the leaves serve as nitrogen storage sites, making this important element readily available during periods of stress (Freeman 1982). Reabsorption efficiency of nitrogen and phosphorus from senescent leaves is relatively high, especially during drought stress, which helps the plants cope with limited soil nutrients (Lajtha 1987).

Cunningham and Reynolds (1978) modeled primary production and carbon allocation in *L. tridentata.* Dry matter production was about 1,000 kg/ha/yr in a relatively mesic part of the range (Chew and Chew 1965), about 20–54 kg/ha/yr in more arid sites (Bamberg et al. 1976; Ludwig and Flavill 1981). The ratio of root-to-shoot biomass is 0.2–0.5 in seedlings (Walters and Freeman 1983), about 0.5 in mature plants (Ludwig et al. 1975). In field plots, growth increases with added nitrogen (Ettershank et al. 1978) but not with added phosphorus (Lajtha 1987). In laboratory experiments, Lajtha and Klein (1988) observed positive growth responses to both nitrogen and phosphorus. Artificial shade reduces growth and fruit production (Smith et al. 1987).

Fire is a rare cause of mortality. The plants sometimes resprout after burning (O'Leary and Minnich 1981; McLaughlin and Bowers 1982; Brown and Minnich 1986). Pruning by pocket gophers (*Thomomys bottae*) (Hunter et al. 1980) and attack by buprestid and cerambycid borers (Valentine and Gerard 1968) may also kill the plants. In the arid, western part of its range, *L. tridentata* is often slow to recolonize sites where it has been killed (Vasek 1980b; Webb and Wilshire 1980; Turner 1990). In the northern Chihuahuan Desert, populations expanded dramatically during the last century (Stein and Ludwig 1979).

Compounds in the foliage may repel leaf-chewing insects by acting upon the protein-digesting system within the gut (Rhoades 1977). Alternatively, herbivores may be deterred by toxicity rather than digestibility (Meyer and Karasov 1989). In spite of its chemical defenses, over 40 species of insects are wholly or partly dependent on *L. tridentata* for food. Among the most abundant are 17 species of gall-forming insects (mostly of the genus *Asphondylia*) (Waring 1986; Waring and Price 1990). Sap-sucking insects, which account for the majority of arthropods on *L. tridentata,* are more responsive to differences in foliar nitrogen than are leaf-chewing insects (Lightfoot and Whitford 1987).

The plants are occasionally grown as ornamentals. Under occasional irrigation, they may grow rapidly or may fail to grow at all (Duffield and Jones 1981). Before transplanting young individuals from the ground, the soil should be soaked (Duffield and Jones 1981), and the shoots should be heavily pruned (Tipton and McWilliams 1979).

The Seri Indians made adhesive and sealant from lac formed on stems attacked by the scale insect *Tachardiella* species (Felger and Moser 1985). Dried leaves and flowers, often sold as "chaparral," are used for skin abrasions, liver ailments, and other maladies (Moore 1989). Timmermann (1977) and Segura and Calzado (1981) have reviewed other medicinal uses. *Larrea tridentata* resins also have potential as agricultural fungicides, cellulose stabilizers, adhesives, antioxidants, and polymer stabilizers, among other proposed uses (Fernandez et al. 1979; Belmares et al. 1981; Garza and del Rosario 1981).

Lophocereus schottii

(Engelm.) Britton & Rose

[=*Cereus schottii* Engelm.]

Senita, sina, pitayita, hombre viejo, cabeza de viejo, pitahaya barbona, garambullo, musaro, old man cactus

A columnar cactus 3–8 m high, *Lophocereus schottii* (figure 53) has several to many stems (5–12 cm wide) from the base (or, in one variety, from a treelike trunk). On young stems and the lower portions of older stems, the 5–10 ribs are sparsely armed with stout, gray spines as much as 1.2 cm long. The upper parts of mature stems are obscured by bristlelike, twisted spines up to 7.5 cm long. The pink, lavender, or white flowers are funnelform and 2.5–3.8 cm long. The globose, spineless fruits, 2–3 cm in diameter, are red and fleshy at maturity.

No other columnar cactus in the Sonoran Desert has dense, twisted spines on the upper stems. *Lophocereus gatesii* M. E. Jones, endemic to Baja California Sur, differs in its more numerous ribs (10–15), smaller stature (less than 2 m), and longer (2–6 cm) spines.

Benson (1982) treated this species as *Cereus schottii.* We follow Gibson and Horak (1978), who retained it in the genus

Lophocereus on the basis of stem anatomy and biochemistry.

Size, habit, rib number, spine coverage, flower color, flower size, and fruit size are variable (Lindsay 1963). Most of the variation appears ecotypic, not broadly geographical, but three named taxa (combined on our map) do display geographically correlated variation (Lindsay 1963). Variety *schottii,* which occurs from southwestern Arizona into northwestern Sonora and Baja California Sur, has 4–7 ribs per stem and lower branches more than 10 cm thick. Variety *australis* (K. Brandegee) Borg, which replaces var. *schottii* in the Cape Region, and var. *tenuis* Lindsay, which replaces var. *schottii* south of Hermosillo, Sonora, have more ribs (6–13) on thinner branches (less than 10 cm thick). Variety *australis* branches well above the base, var. *tenuis* at the base. Botanists have described two bizarre local forms of var. *schottii:* forma *mieckleyanus* Lindsay near Rancho Unión (28.2°N, 113.3°W), and forma *monstrosus* Gates near El Arco (28.0°N, 113.4°W) and northwest of Rancho Santa Inés (28.9°N, 114.8°W). (The Rancho Santa Inés population may have been deliberately planted [Clark and Blom 1982].) Notable for their knobby, discontinuous ribs, these two forms (sometimes referred to as "totem pole cacti") are probably more frequent in cultivation than in the wild.

Occasional to common on stable dunes, wash borders, silty flats, and rocky hillsides, *L. schottii* often grows with other columnar cacti, including *Stenocereus thurberi, Pachycereus pringlei,* and *P. pecten-aboriginum.* The northernmost Arizona voucher (32.3°N) is Norman M. Simmons's from northeast of Sinita Tank (Simmons 1966). The northernmost points on the peninsula (32.3°N, 32.4°N) are from Cañon de los Torrentos (Moran 1977b). The disjunct sightings in eastern Sonora are Raymond M. Turner's from 3 km north of Onavas (28.5°N) and near San Pedro de la Cueva (29.3°N). The highest point on the elevational profile, 1,200 m, represents a collection by Reid Moran from Cerro de la Mina de San Juan (28.7°N, 113.6°W).

The range of *L. schottii* encompasses

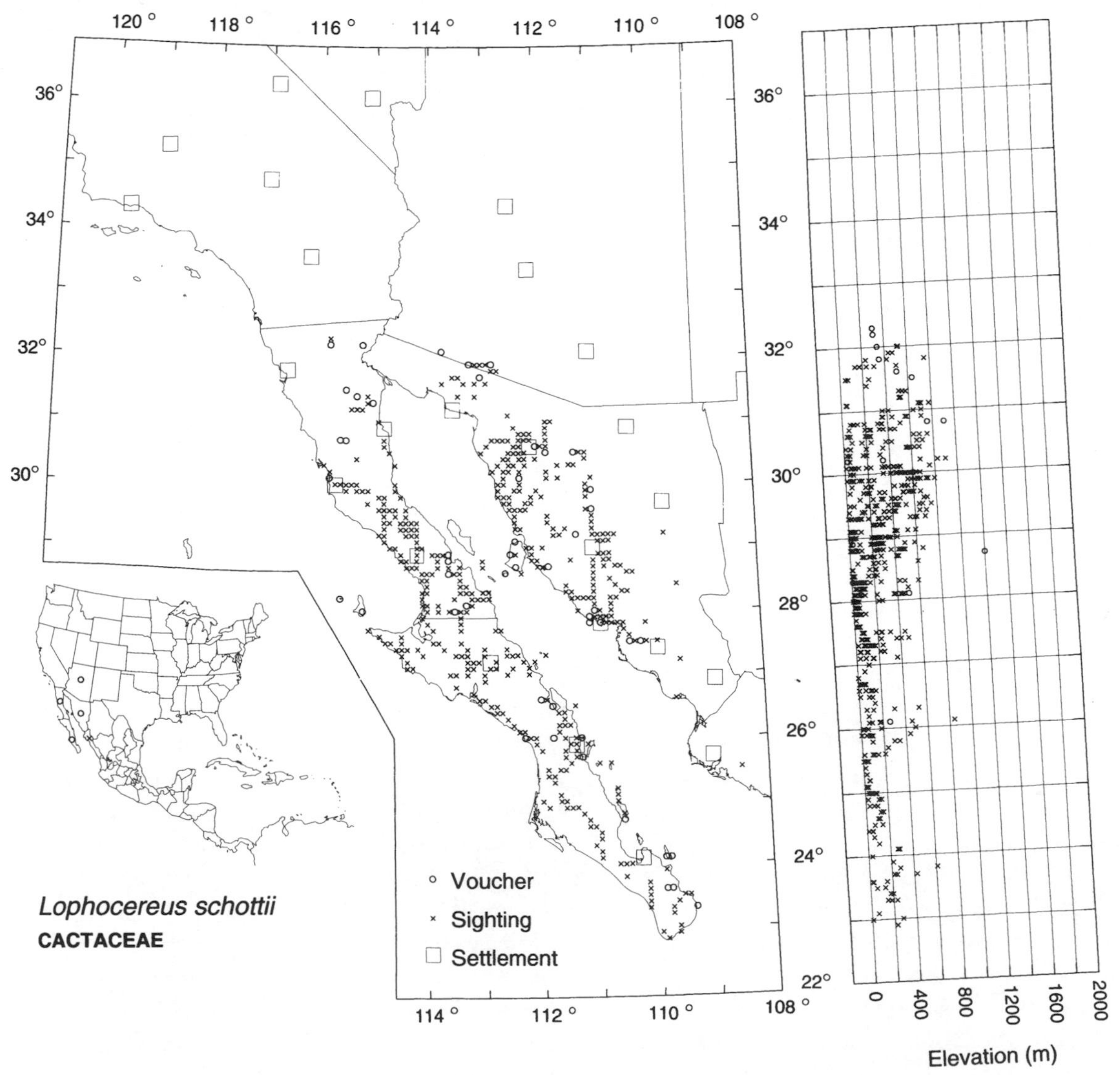

Lophocereus schottii
CACTACEAE

the more tropical arid parts of the Sonoran Desert. Rainfall seasonality ranges from mostly winter in the northern Vizcaíno region of Baja California through biseasonal over most of the range to mostly late summer and fall near the southern limits. The northern limit is doubtless determined by low temperature, as shown by frost damage to plants in the northernmost populations (Nobel 1980c; Parker 1988b). The chlorenchyma loses membrane integrity, hence cell viability, at –7.2°C (Nobel 1982a), a somewhat higher temperature than for the more frost-tolerant *Carnegiea gigantea* and *Stenocereus thurberi.* The lack of apical pubescence, combined with its greater cold sensitivity, restricts *L. schottii* to more southerly latitudes than *C. gigantea* or *S. thurberi* (Nobel 1982a).

Flowers appear from April to August (Wiggins 1964). Fruits ripen from July to October (Humphrey 1975). At Punta Cirio, Sonora, stems grow in response to both summer and winter rain (Humphrey 1975). At Organ Pipe Cactus National Monument, where winters are colder, stem growth occurs primarily in the warm months (Parker 1988b).

Lophocereus schottii shows a north-south cline in stem diameter, with thicker stems occurring at the northern, colder end of its range (Felger and Lowe 1967; Nobel 1980c). Although Felger and Lowe (1967) suggested that the larger stem mass should provide a decided selective advantage at low temperatures, Nobel (1980c) concluded that increased diameter in fact has little effect on minimum temperatures at the stem surface. Computer simulation showed that thinner stems might have lower surface temperatures during the summer (Nobel 1980c; Smith et al. 1984). Thin-stemmed populations experience generally warmer summers than do thick-stemmed populations; however, maximum temperatures are not as high in thin-stemmed as in thick-stemmed populations. It is difficult to conclude whether thinner stems do indeed promote survival at the highest air temperatures.

The photosynthetic surface area of many cacti is so reduced that carbon dioxide uptake may be limited by the amount of photosynthetically active radiation that falls on the stems (Nobel 1981a, 1983b). Self-shading, whether by spines or branches, can further reduce absorption of photosynthetically active radiation. The profuse branching of *L. schottii* may be an adaptation to maximize absorption of photosynthetically active radiation (Geller and Nobel 1986, 1987). Its particular combination of rib number and rib depth allows for maximal carbon dioxide uptake under 20–30% shading (Geller and Nobel 1984). Varieties *tenuis* and *schottii,* which branch at the base, generally grow in open desertscrub where there is no competition for sunlight. Variety *australis,* which branches above the base, grows in denser, taller thornscrub communities where tree canopies may well limit the light available to the understory. (Cody [1984] has discussed this phenomenon in *S. thurberi.*)

Large *L. schottii* plants have a greater photosynthetic surface area and water capacity than do small ones and therefore grow faster. The mean growth rate is 0.06 m/yr for plants under 1 m in height, 0.88 m/yr for plants 5 m or taller (Parker 1988b). Growth rates tend to be highest after unusually wet winters and lowest after relatively dry ones. A winter of frequent or severe freezes retards growth for two years (Parker 1988b).

The flowers may be pollinated mainly by hawkmoths (Gibson and Horak 1978). Ants that visit extrafloral nectaries on the areoles may provide protection from herbivorous insects (William H. Clark, personal communication 1993).

In southwestern Arizona, the average plant produces 100 fruits per year, each containing about 200 seeds (Parker 1989). The fruit is edible (Standley 1924), and the seeds are doubtless spread by birds, mammals, and insects. Because the stems reproduce vegetatively, several clumps may consist of a single clone (Benson 1982). At Organ Pipe Cactus National Monument, *L. schottii* reproduces more frequently by stem layering and stem dispersal than by seed, perhaps because frost prevents seedling establishment in most years (Parker and Hamrick 1992). Genetic diversity nevertheless remains rea-

Figure 53. Lophocereus schottii, *showing the bristlelike, twisted spines on older stems, near Cataviñá, Baja California. (Photograph by J. R. Hastings.)*

sonably high due mainly to the long flowering period and the long distances that pollen travels (Parker and Hamrick 1992). Shreve (1935) estimated that few individuals live longer than 75 years. Long-term (13 years) measurements of annual growth indicate that the plants would reach heights of 0.1 m in 14 years, 1 m in 27 years, and 10 m in 42 years (Parker 1988b).

Lophocereus schottii is grown as an ornamental in warm-winter areas. In Tucson, it requires some frost protection. Fishermen in Baja California sometimes used the chopped stems as a fish poison (Aschmann 1959).

Lotus rigidus

(Benth.) E. Greene

Lotus rigidus is a sparsely branched chamaephyte 3–9 dm tall. The stems are woody near the base. Leaves are odd-pinnate with 3–5 oblong to oblanceolate leaflets 5–15 mm long and 1.2–5 mm wide. Stem internodes greatly exceed the leaves in length. The yellow flowers, 15–20 mm long and often tinged with red, are in few-flowered umbels. The elastically dehiscent pods are 2.5–4.5 cm long and 3–5 mm broad. The haploid chromosome number is 7 (Raven et al. 1965).

Most species in the genus are annual or, if perennial, prostrate or decumbent (Ottley 1944). Two exceptions are *L. haydoni* (Orcutt) E. Greene of southeastern California and northern Baja California and *L. scoparius* Nutt. in Torr. & A. Gray var. *brevialatus* Ottley in southern California from San Diego County to Kern County. *Lotus scoparius* has a broomlike habit, and *L. haydoni* has minute leaflets (2–3 mm long).

The plants are occasional to common on rocky hillsides and along washes. Most characteristic of Sonoran desertscrub, *L. rigidus* also grows above the desert up to 1,940 m. Its distribution in areas of significant winter and spring rainfall implies a need for cool-season moisture. The broad elevational range suggests frost tolerance.

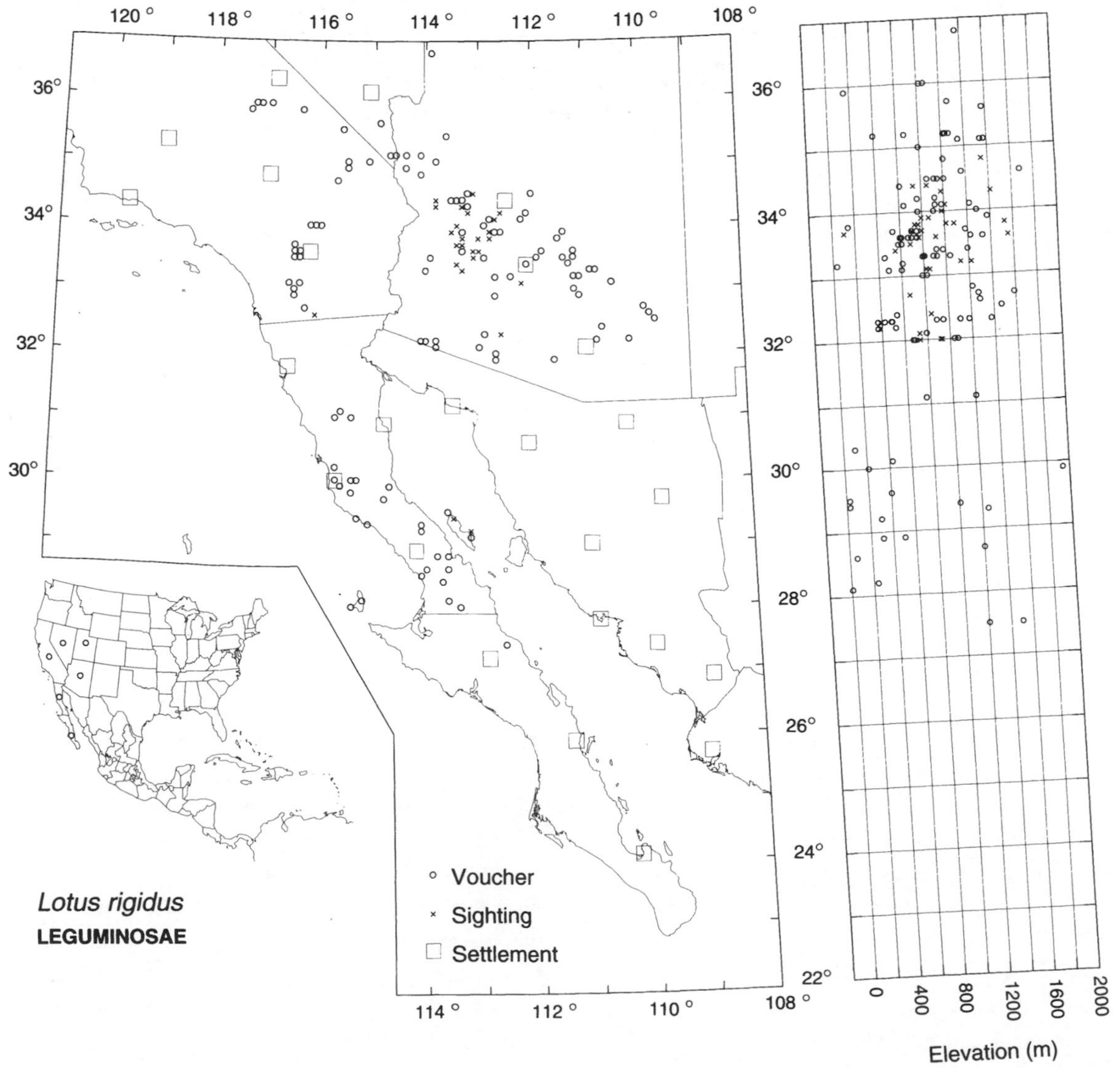

Flowers may be found in almost any month except perhaps July and August. The roots can form nitrogen-fixing nodules (Eskew and Ting 1978).

Lysiloma candidum

T. S. Brandegee

[=*L. candida T. S. Brandegee;* see Thompson 1980 for an explanation of the change in spelling]

Palo blanco

A tree up to 10 m tall, *Lysiloma candidum* (figure 54) has smooth white bark on all except the smallest twigs and branches. The twice-compound leaves, on short petioles 0.5–1.5 cm long, have 1 pair (occasionally 2 pairs) of pinnae, each with 2–8 pairs of leaflets 10–22 mm long. New leaves have large rounded stipules. The white flowers are in globose heads. When young, the pods are copper-red and spirally coiled. At maturity, they are papery and straight, up to 20 cm long and 3 cm wide.

No other leguminous tree in the Sonoran Desert has smooth, white bark except *Acacia willardiana,* which has petioles 6–30 cm long.

Common along washes and on rocky slopes, *L. candidum* is endemic to the Baja California peninsula except for an isolated outpost at Ensenada Grande north of Guaymas, Sonora, which Wiggins (1964, 1980) apparently overlooked. The species occurs on many gulf islands off the east coast of Baja California Sur (Moran 1983b). Plants often grow near the sea but may occur as high as 825 m in the Sierra Giganta (Annetta Carter, personal communication 1979). The plants are frost hardy to −4°C (Johnson 1993). The species is restricted to frost-free arid to semiarid climates where winters are warm and rain falls mainly in the warm season.

The leaves persist throughout the year. Flowers appear in spring just before the new leaves.

Cattle relish the immature pods (Wiggins 1964). The bark has been widely used

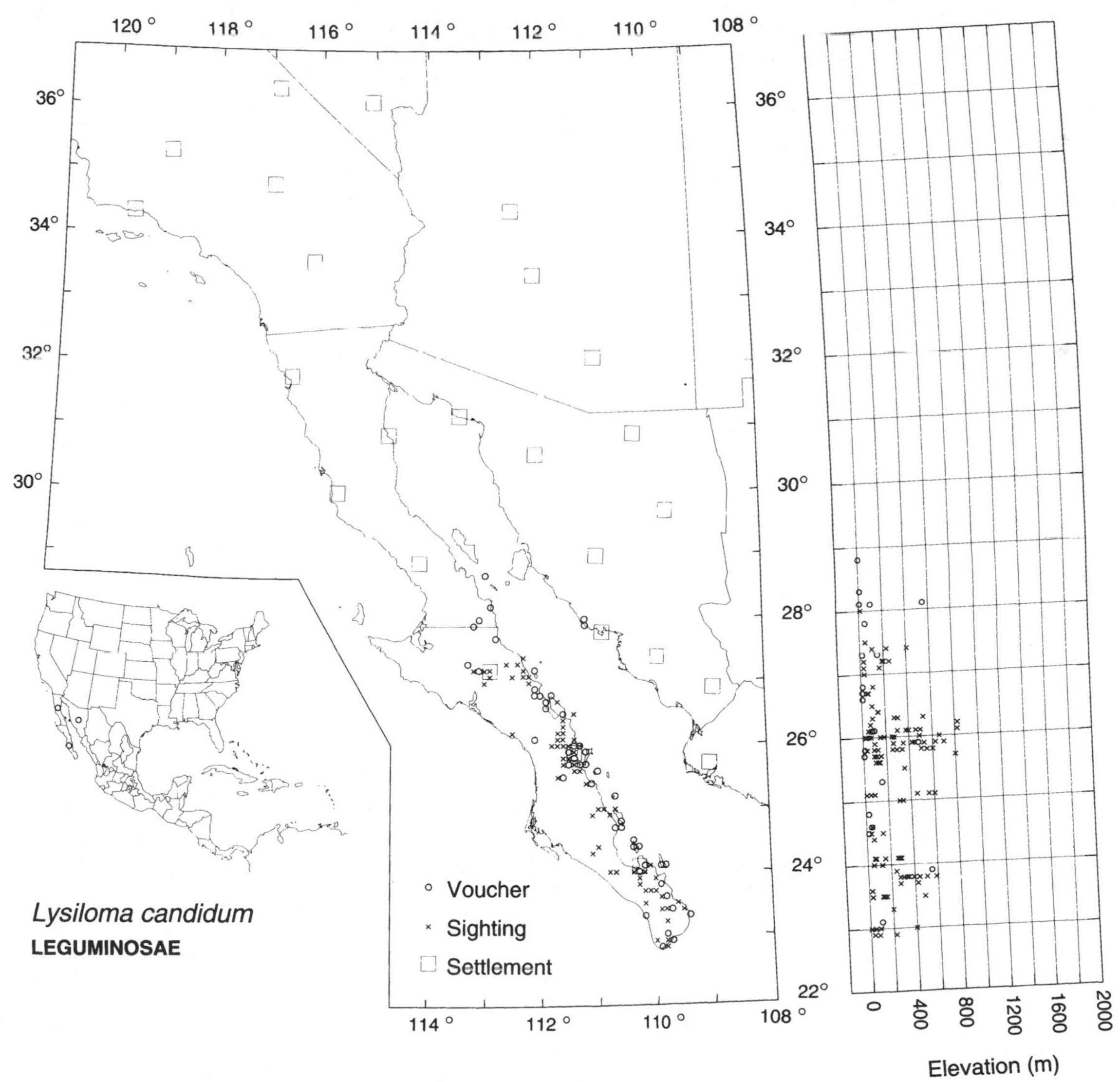

for tanning and was at one time exported. *Lysiloma candidum* is occasionally grown as an ornamental in desert areas where frosts are rare.

Figure 54. Lysiloma candidum *in an arroyo near Santa Rosalía, Baja California Sur. (Photograph by J. R. Hastings.)*

Lysiloma microphyllum Benth.

[=*L. divaricata* (Jacq.) Macbr., *L. microphylla* Benth.; see Thompson 1980 for an explanation of the change in spelling]

Mauto, manta

Lysiloma microphyllum (figure 55) is a large shrub or small tree up to 15 m tall with conspicuous ovate stipules. The twice-compound leaves, on petioles less than 2.5 cm long, have 4–9 pairs of pinnae, each with 20–30 pairs of leaflets (3–8 mm long). The white flowers are in globose

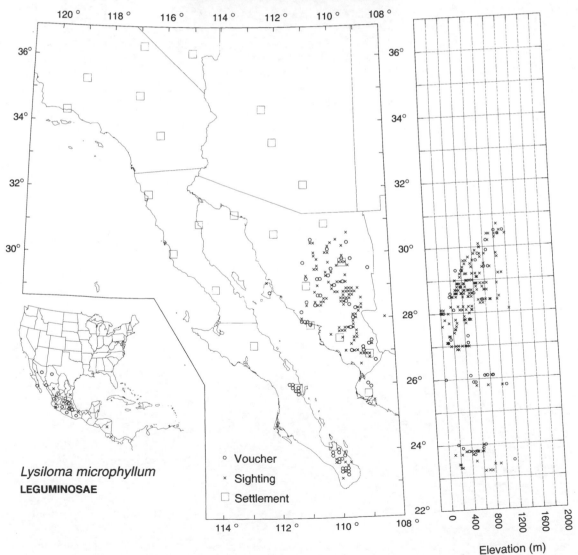

Lysiloma microphyllum
LEGUMINOSAE

○ Voucher
× Sighting
□ Settlement

Elevation (m)

heads. The linear-oblong pods, 8–15 cm long, dehisce readily at maturity.

Leaves of *Acacia coulteri* are on petioles 2.5–5.0 cm long.

Lysiloma microphyllum is widespread throughout much of tropical Mexico and has been reported from Nicaragua and Costa Rica (Standley 1922). In Baja California Sur, it grows in the Cape Region and in the Sierra de la Giganta. In Sonora, it enters the desert mainly along moist drainages. On the lower slopes of the Sierra Madre Occidental in Sonora, it is a dominant on less acidic soils (Goldberg 1985). The lower elevational limit shows a well-defined slope upward to the north. This pattern is typical for species that reach their arid limits at the southern edge of the Sonoran Desert.

The plants are usually leafless through the winter, although a few leaves may remain until the spring drought, when all trees become bare. The flowers appear in late summer. The fruits mature in late fall or winter.

The wood is used in construction, the bark in tanning.

Lysiloma watsoni

Rose

[=*L. microphylla* Benth. var. *thornberi* (Britton & Rose) Isely, *L. thornberi* Britton & Rose (in part)]

Tepeguaje

This shrub or small tree grows as tall as 15 m and has dark gray bark roughened by vertical and horizontal fissures (figure 56). The twice-compound leaves, subtended by conspicuous stipules, are 10–20 cm long with 4–8 pairs of pinnae. The leaflets (4–10 mm long) are in 15–45 pairs on pinnae 5–9 cm long. The whitish or creamy flowers are in dense cylindrical racemes. The reddish pods, up to 22 cm long, are oblong-linear.

Lysiloma watsoni can be distinguished from other leguminous trees by the irregular horizontal branches, which extend 7–10 m from the trunk on large plants.

A common component of thornscrub communities in Sonora, Sinaloa, Chihuahua, and Durango, this species enters the Sonoran Desert along mesic arroyo corridors. In Arizona it grows only in south-facing canyons of the Rincon Mountains, Pima County, between 1,025 and 1,465 m. At these sites, damaging freezes restrict plant height to 2–3 m. The lower elevational limit shows a well-defined slope upward to the north, which is typical for species reaching their arid limits at the southern edge of the Sonoran Desert.

Flowers appear in April or May. The

Figure 55. (left) Lysiloma microphyllum *in the mountains above El Novillo, Sonora. In this autumn view, the fruits are not fully ripened, and many leaves have fallen. (Photograph by J. R. Hastings.)*

Figure 56. (above) *An early spring view of* Lysiloma watsoni *near Alamos, Sonora. Few other woody species in this area retain their leaves throughout the winter and arid foresummer. (Photograph by R. M. Turner.)*

fruits dehisce readily upon ripening in late summer or fall.

Lysiloma watsoni makes an attractive ornamental for desert gardens. The plants are hardy to temperatures as low as −6.5°C (Johnson 1993). They may require frost protection in colder areas. Recovery after freezing is rapid (Duffield and Jones 1981). Best growth is obtained with moderate irrigation (Duffield and Jones 1981).

The dense, reddish brown wood is used for furniture in rural parts of Mexico. The seeds are nutritious (Thorn et al. 1983).

Malosma laurina

Nutt. ex Abrams

[=*Rhus laurina* Nutt.]

Laurel sumac

A large, rounded shrub or small tree 2–5 m tall, *Malosma laurina* (figure 57) has lance-oblong, bicolored leaves 4–10 cm long and 2–4 cm wide. The leaves are folded upward along the midrib. The small, white, inconspicuous flowers are on numerous slender branchlets in terminal panicles. The whitish drupes are 2–3 mm in diameter.

Leaves of *Rhus integrifolia* are flat, not folded up along the midrib, and those of *R. ovata* are ovate rather than lance-oblong. Neither has glabrous, white fruits.

Young (1974b), based on significant differences in wood anatomy between *Rhus* and *Malosma*, concluded that they are not closely related and should be treated as distinct genera. Wood of most *Rhus* species is ring-porous, but *M. laurina* has diffuse-porous wood, the general tendency among chaparral shrubs (Young 1974b).

A characteristic plant of Californian chaparral and Californian coastal scrub, *M. laurina* grows on hillsides, in canyons, and along rocky washes on a variety of substrates including sandstone, conglomerate, shale, and volcanics. It is favored on

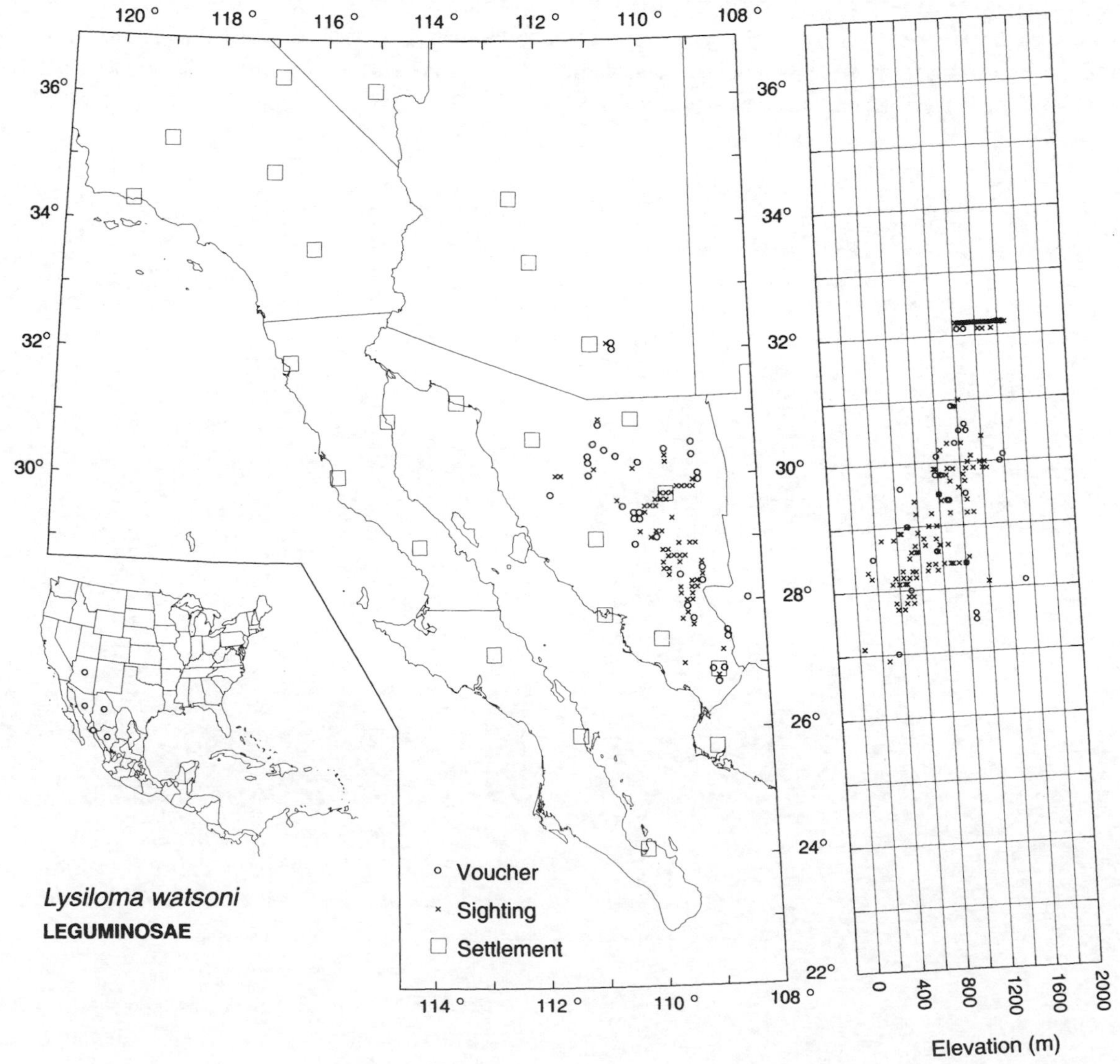

soils with high exchangeable potassium levels (Westman 1981a). The species barely enters the desert except along the Pacific coast of the Baja California peninsula from about 27.5°N to 29.5°N. The southernmost outliers represent collections made by Annetta Carter and Howard Scott Gentry in the Sierra Victoria. Cody and coauthors (1983) noted that these and other disjuncts in Baja California Sur may be relicts of a wider distribution in more mesic times. The presence of the population on Guadalupe Island suggests that the seeds can disperse long distances, however, so the disjunct populations may not all be relictual.

Figure 57. Malosma laurina *in a large arroyo east of El Rosario, Baja California. (Photograph by J. R. Hastings.)*

Throughout most of the Sonoran Desert region, *M. laurina* receives rain

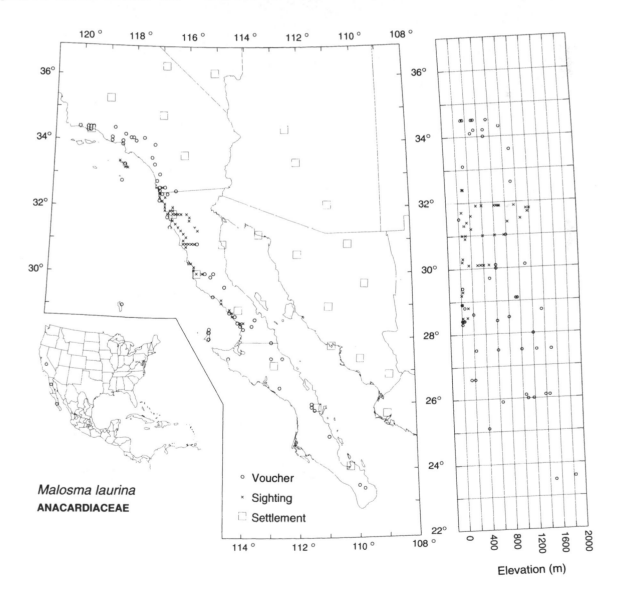

Malosma laurina
ANACARDIACEAE

○ Voucher
× Sighting
□ Settlement

Elevation (m)

mainly in winter or spring. South of 28°N, warm-season rainfall increases, and populations in the Cape Region typically receive most of their rain from July to October. In the most arid parts of the central peninsula, warm-season drought stress is ameliorated by relatively cool, moist air created by offshore upwelling. The plants are susceptible to frost (Munz 1974) and are less abundant in the colder parts of their range (Westman 1981a). Frost, combined with exclusion from more mesic vegetation, probably determines the northern upper limits of this species. Toward the southern end of its range, *M. laurina* grows to higher elevations. It is not clear why the species is absent from arid habitats along the southern coast.

The leaves are evergreen. Flowers appear March–August (Wiggins 1964). Like many chaparral species, *M. laurina* sprouts from the stem base or woody crown after fire. Seedlings also appear after fire, and it seems likely that the seeds require heat for germination (Westman 1981b). After fire, seedlings show much lower survival rates than resprouts exhibit, evidently because the resprouts can tap deep soil moisture reserves during summer drought, whereas the shallowly rooted seedlings cannot (Saruwatari and Davis 1989; Thomas and Davis 1989).

Mascagnia macroptera

(Sess. & Moc.) Niedenzu

Gallinita, bataneni, matanene, bejuco prieto, doncella amarilla

A trailing vine or scandent shrub up to 2 m tall, *Mascagnia macroptera* has opposite, ovate to oblong leaves 5–22 mm wide and 2.5–7 cm long. The yellow petals, 6–12 mm long, are orbicular and short clawed and vary notably in size within a single flower. The sepals have large, oval glands. The fruits are samaras (4.5–5 cm wide) with lateral, flabelliform, papery wings. The haploid chromosome number is 10 (Baker and Parfitt 1986).

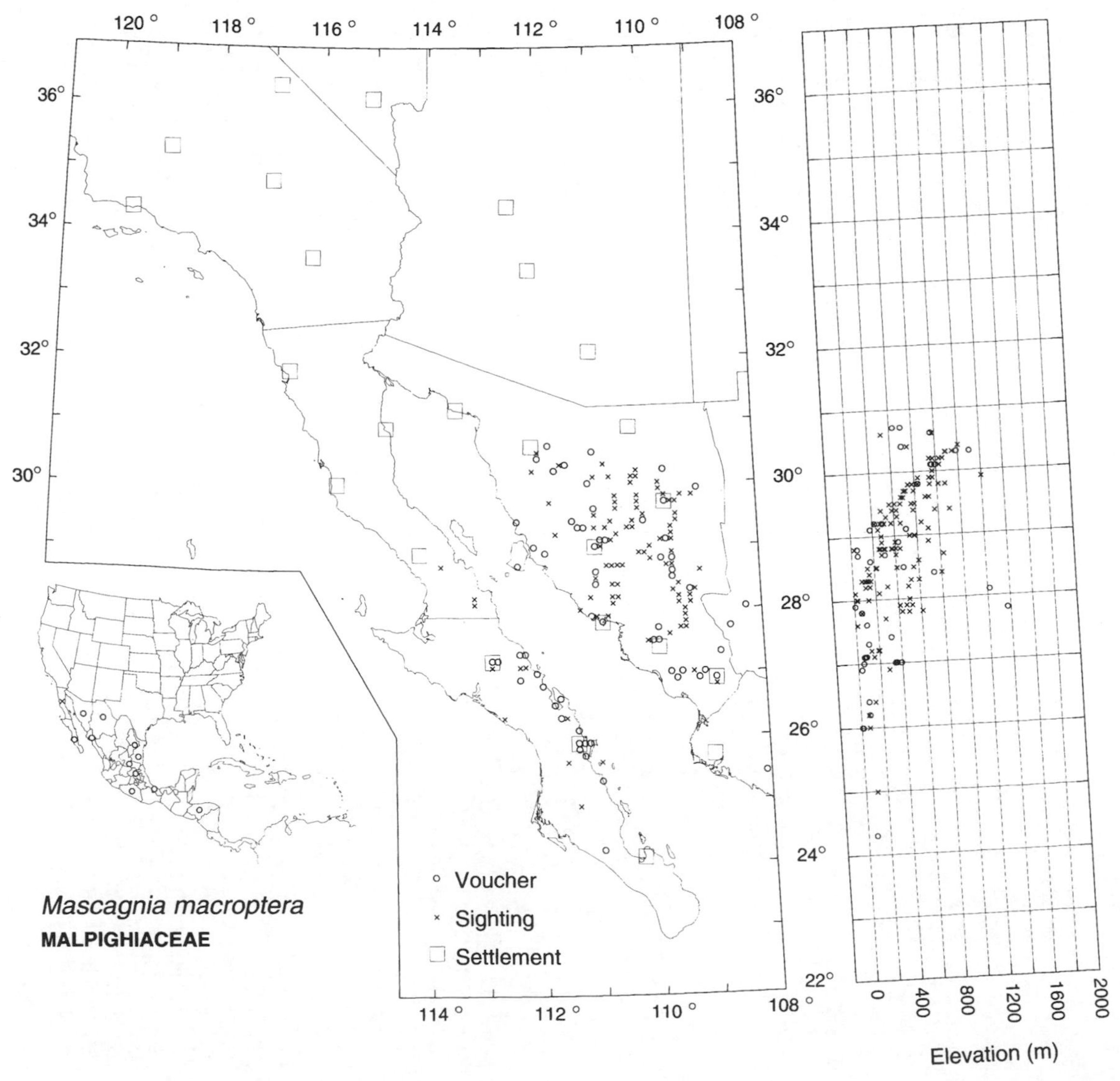

Mascagnia macroptera
MALPIGHIACEAE

In fruit, *M. macroptera* can hardly be confused with any other vine in the Sonoran Desert.

The plants are occasional in disturbed habitats such as roadsides and washes. The distribution reflects a need for summer rainfall and few frosts. The apparent absence of the species from the Cape Region of Baja California Sur is puzzling.

Mascagnia macroptera blooms after rain at any time of year (Wiggins 1964).

In *Janusia gracilis* A. Gray, another Sonoran Desert Malpighiaceae, the floral glands produce an oil (a free fatty acid) that is gathered for larval food by solitary bees of the genus *Centris*, the only pollinators (Buchmann 1987a). It seems likely that the floral glands in *M. macroptera* serve a similar purpose.

The seeds presumably disperse as the samaras tumble across the ground in the wind.

Seri Indians wove the stems into a summertime head covering and made a tea from the roots to treat colds and diarrhea (Felger and Moser 1985). Mexicans sometimes used the leaves as a poultice for bruises or sores (Standley 1923).

Melochia tomentosa L.

[=*M. speciosa* S. Watson]

Broom-wood, malvavisco, bretonica, malva, malva de los cerros, malva rosa, varita de San José

Melochia tomentosa is a shrub or subshrub up to 2.5 m tall with stellate-tomentulose foliage and twigs. The alternate, ovate to suborbicular leaves, 1.5–4 cm wide and 2.5–6 cm long, are dentate and deeply veined above. The 5-petaled, pink to rose-purple flowers, 16–20 mm across, are in dense glomerules. On a sin-

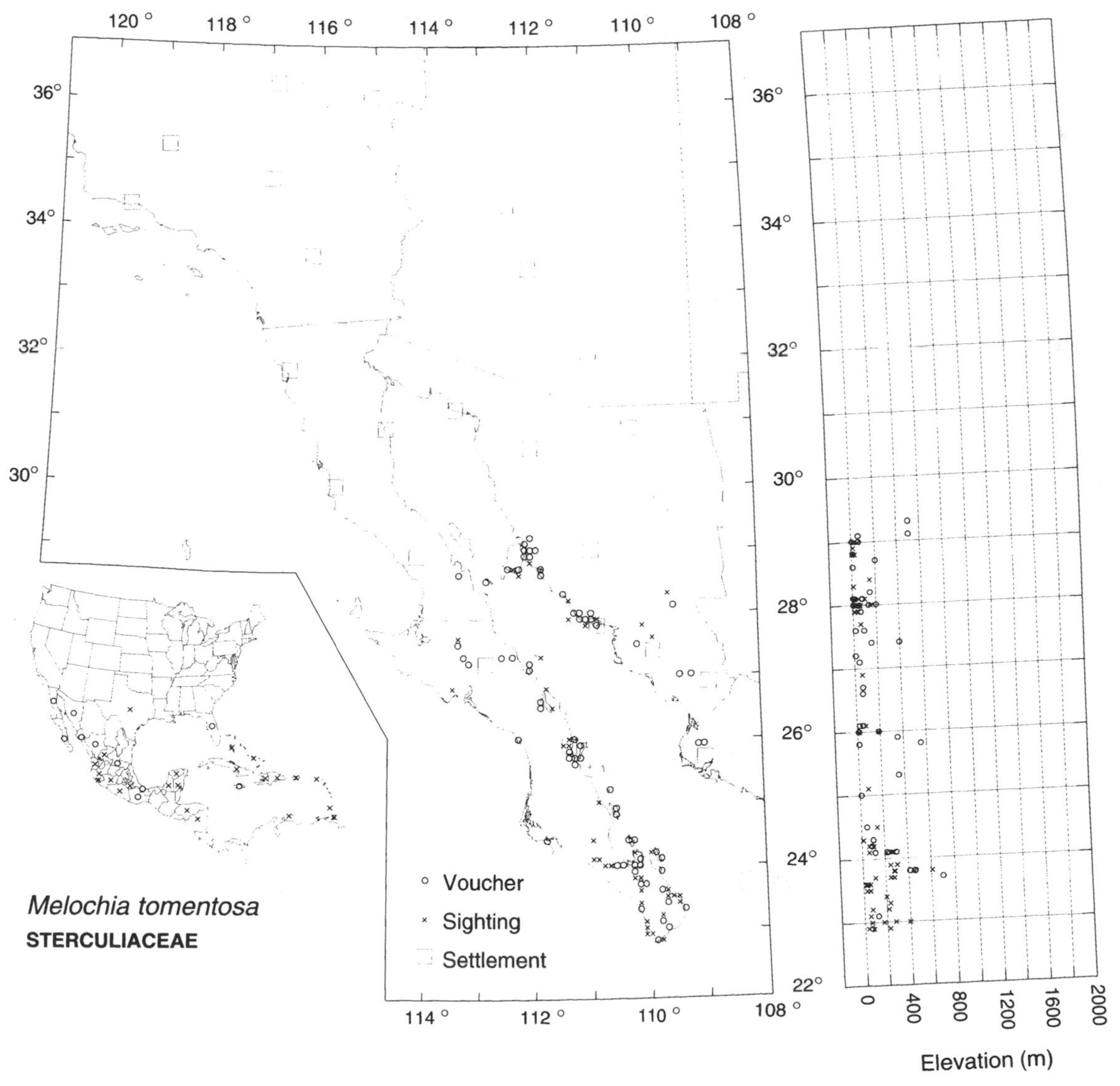

gle plant the flowers may be longistylous, with pistils exceeding the stamens, or brevistylous, with stamens surpassing the pistils (Goldberg 1967). The pyramidal capsules are 7–9 mm long with broadly winged angles. The haploid chromosome number is 9 (Bates 1976).

Cordia curassavica and *C. parvifolia* have white rather than pinkish flowers and lack stellate tomentum.

Considerable variation occurs in leaf and flower size, pubescence, and fruit shape (Standley 1923). Goldberg (1967) recognized four varieties, three of which (*frutescens* [Jacq.] DC., *tomentosa,* and *speciosa* [S. Watson] Goldberg) occur in the study area. Wiggins (1964) treated var. *speciosa* as *M. speciosa* S. Watson. The three varieties are combined on our map.

In the area mapped in this atlas, *M. tomentosa* grows in Sonoran desertscrub and Sinaloan thornscrub, often along roadsides and in old fields, occasionally on rocky hillsides and mesas. In southern Texas it grows in open woodlands and brushlands (Correll and Johnston 1970). Basically a weedy, early successional species of the tropics and subtropics, *M. tomentosa* barely reaches the warm-winter areas of the temperate zone. The distribution in the Sonoran Desert region (centered along both coasts of the southern Gulf of California) suggests a need for warm-season rain. The preference for maritime habitats at the northern end of the range suggests sensitivity to frost and intolerance of dense vegetation.

Plants bloom throughout the year (Goldberg 1967). The leaves are incompletely drought deciduous; only the younger ones remain under extreme moisture stress. Pollination of *M. tomentosa* has not been specifically studied; heterostyly is generally believed to promote outcrossing (Proctor and Yeo 1972; Faegri and van der Pijl 1979).

The Seri made a reddish brown dye from the roots (Felger and Moser 1985). In Sonora cattle often browse the plants (Gentry 1942).

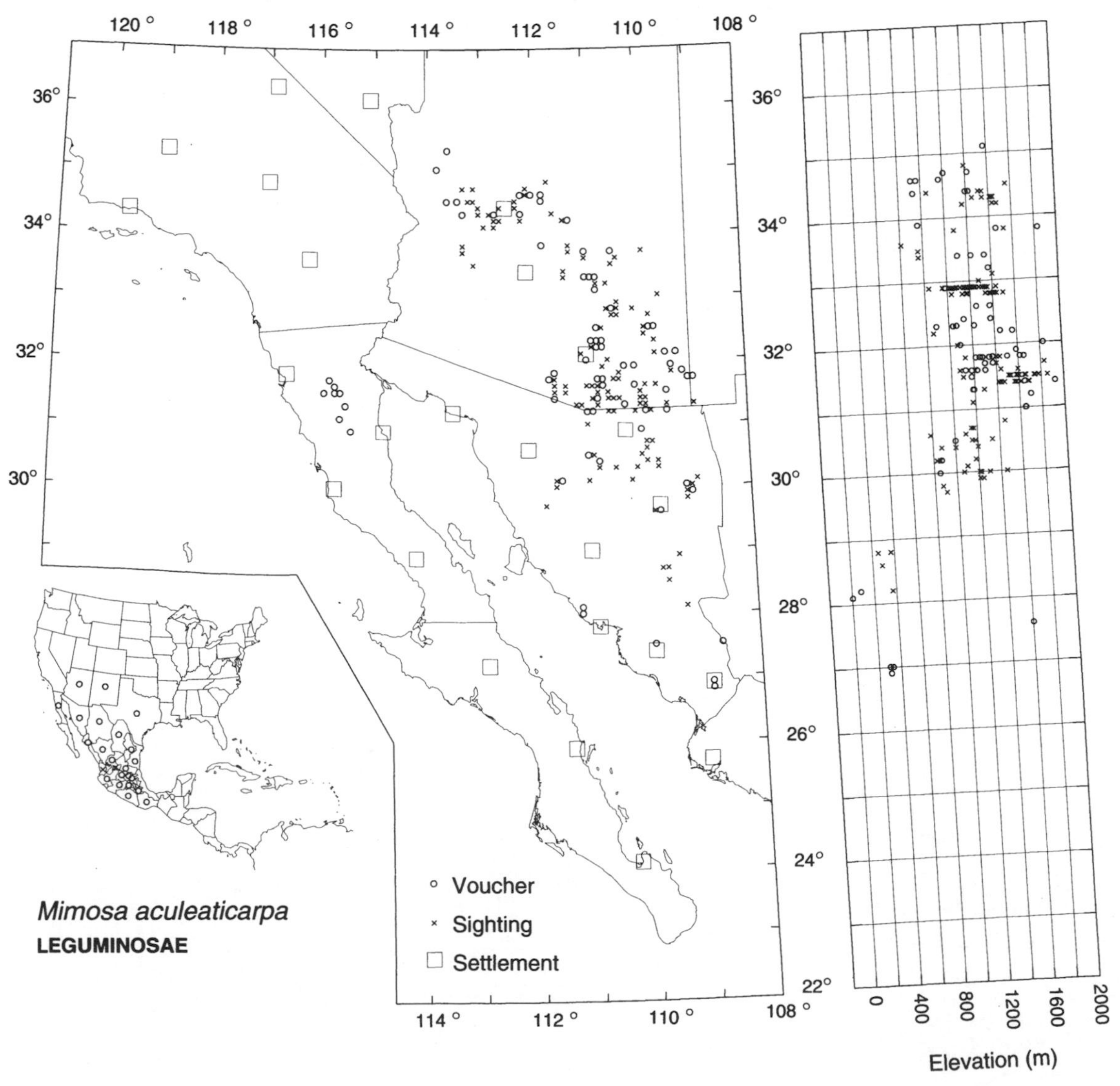

Mimosa aculeaticarpa
LEGUMINOSAE

Mimosa aculeaticarpa

(Gomez) Ort.

[=*M. biuncifera* Benth.]

Wait-a-minute, cat claw, cat-claw mimosa, uña de gato, garruño, gatuño

Mimosa aculeaticarpa is a shrub 0.6–2 m high with paired recurved thorns at the nodes. The twice-pinnate leaves have 4–7 pairs of pinnae, each with 6–13 pairs of leaflets 2 mm long. The white or pinkish flowers are in capitate clusters about 15 mm broad. The falcate pods are 3–5 mm wide and 3–4 cm long. At maturity, the pod walls separate from the prickly margins. The diploid chromosome number is 52 (Turner and Fearing 1960).

Acacia greggii also has recurved spines, but these are scattered rather than paired, and its creamy flowers are in spikes, not heads. *Mimosa dysocarpa* has scattered prickles, and its spicate flowers are white and pink. *Acacia constricta* has straight or slightly curved spines and yellow flowers. *Calliandra eriophylla* lacks spines and is rarely more than 1 m tall.

Until recently, botanists have treated the plants in the study area as the highly variable *M. biuncifera* (Isely 1971). Turner (1959) recognized two varieties in Texas: var. *biuncifera* and var. *lindheimeri* (A. Gray) Robins. We follow Barneby (1991) in referring our plants to *M. aculeaticarpa,* a widespread species from Chiapas northward to Arizona, New Mexico, and Texas. It is represented in the Sonoran Desert region by var. *biuncifera* (Benth.) Barneby (Barneby 1991).

Mimosa aculeaticarpa var. *biuncifera* grows in thickets on hill and canyon slopes and along washes. Mainly a shrub of semidesert grassland and Madrean evergreen woodland, it enters the desert along washes. A few scattered populations occur well within the desert. The remarkable disjuncts in the Sierra Juárez of Baja California grow near disjunct *Calliandra eriophylla* and *Acacia constricta.* Plants survive freezing temperatures down to −18°C (Johnson 1993). The distribution shows a need for warm-season rain and a

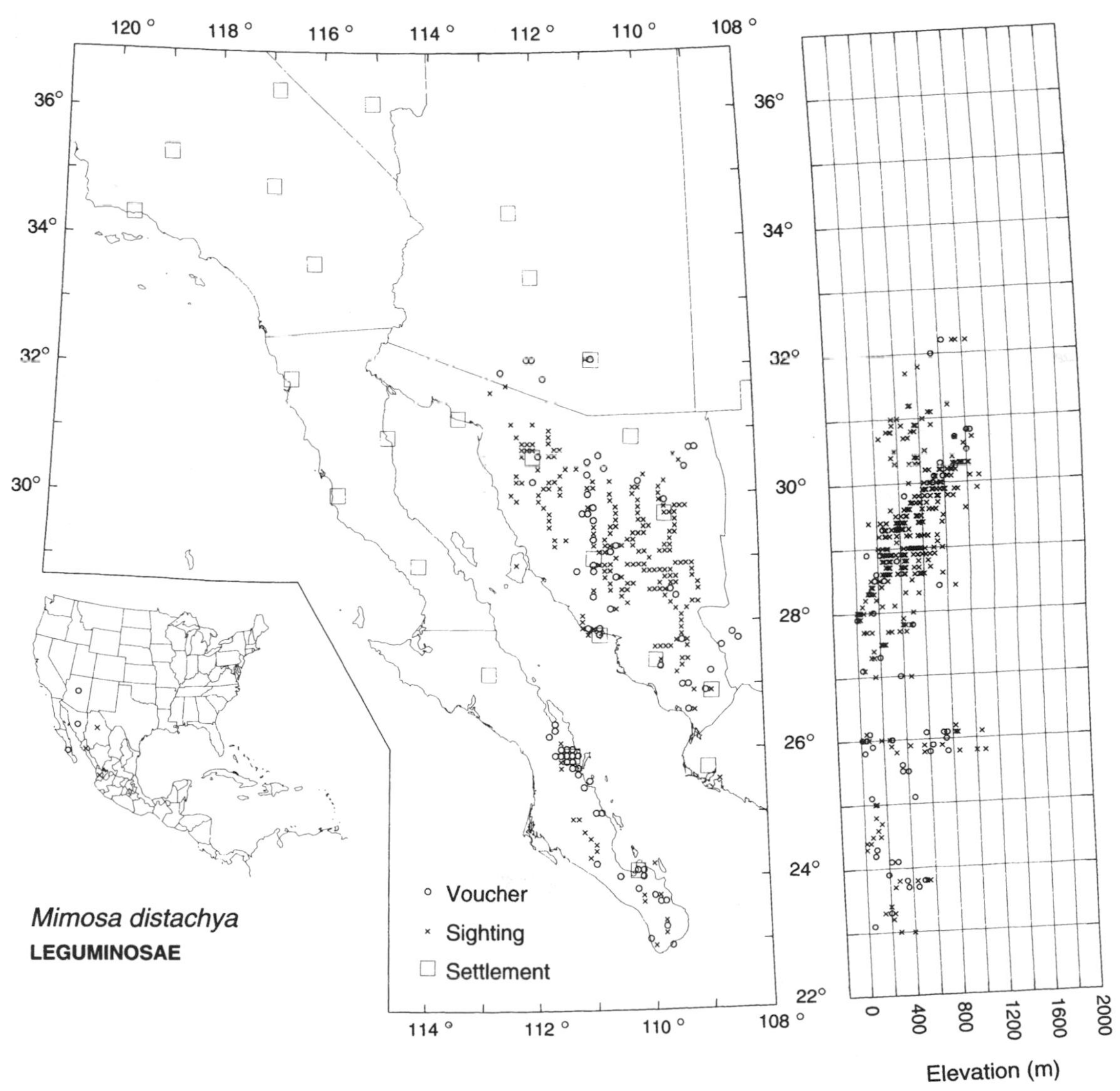

tolerance of frosts.

The plants flower between May and August (occasionally as early as April or as late as September) and attract a variety of insects including bees, wasps, beeflies, and small butterflies.

During the past century, *M. aculeaticarpa* has proliferated in semidesert grassland (Hastings and Turner 1965b). In West Texas, where the plants are considered invasive on rangeland, they often establish under the canopies of *Juniperus* trees (McPherson et al. 1988).

Mimosa distachya

Cav.

[=*M. brandegeei* Robins., *M. laxiflora* Benth., *M. purpurascens* Robins.]

Garabatillo, curca, cuilón, iguano, gastuña

Mimosa distachya is a straggly shrub or small tree 1.5–7 m tall with flat, curved prickles sparingly scattered along the branches. The 2–5 pairs of pinnae each have 2–6 pairs of widely spaced leaflets 6–12 mm long. The spicate inflorescences are lavender to white. The linear-oblong pods, 3–5 cm long, break into 1-seeded segments at maturity.

The leaflets are large for mimosas in the Sonoran Desert region. *Zapoteca formosa* looks something like this species but is unarmed.

Two varieties occur in the Sonoran Desert region: var. *distachya* and var. *laxiflora* (Benth.) Barneby. They differ in characters of flower and fruit (Barneby 1991).

Mimosa distachya is a shrub of canyons, washes, alluvial flats, and rocky slopes. Widespread and common in Sonora and Baja California Sur, it tends to be scattered and occasional at its northern limit in Arizona. Phillip Jenkins made the collections near Tucson. The sighting on Isla Tiburón is Richard Felger's. The species is notable for its absence from other

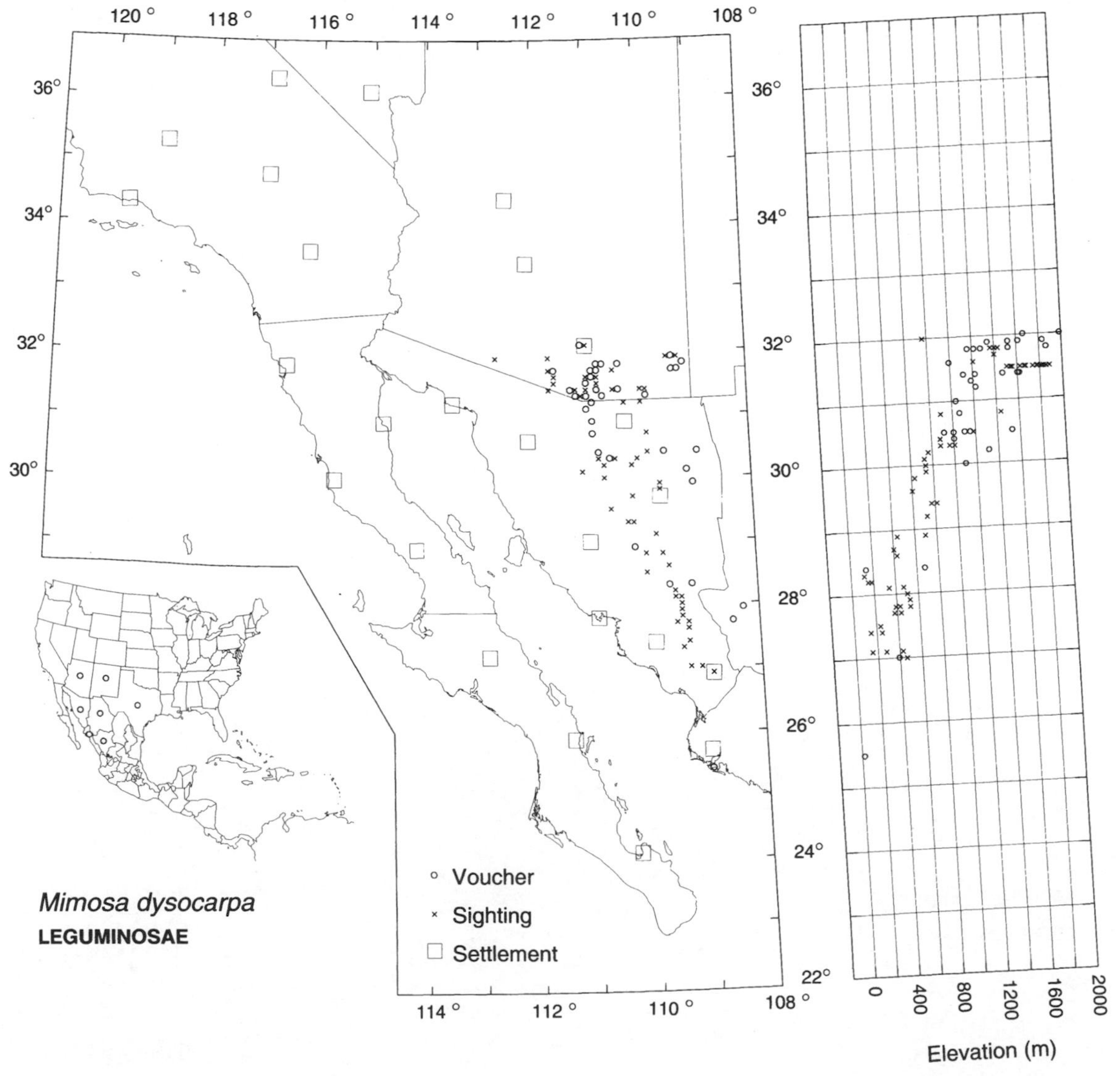

Mimosa dysocarpa
LEGUMINOSAE

gulf islands (Moran 1983b). The range is restricted to areas that seldom if ever experience frost. Summer rain is relatively dependable across the range. The northern and upper elevational limits are more or less contiguous with the arid and lower limits of *M. aculeaticarpa*.

The leaves are drought- and cold-deciduous. The main flowering season is August–October, but flowers may appear as early as July and as late as November.

Mimosa dysocarpa

Benth.

Gatuño, velvet-pod mimosa

A shrub 1–2 m high, *Mimosa dysocarpa* has abundant scattered prickles on the stems and rachises. The twice-pinnate leaves have 5–10 pairs of pinnae, each with 6–12 pairs of oblong leaflets 3–5 mm long. The leaflets are divided by the midveins into strongly unequal parts. The rose-pink flowers are in cylindrical spikes 1–2 cm long. As the fresh flowers fade to white, the spikes become strikingly bicolored. The flattened pods are 3–5 mm wide.

Isely (1971) recognized two varieties in the United States: var. ***wrightii*** (A. Gray) Kearney & Peebles and var. ***dysocarpa***. We follow Barneby (1991) in combining them.

Mimosa aculeaticarpa has paired spines and creamy, globose heads. ***Mimosa grahamii*** has creamy flowers in globose heads and a combination of scattered prickles and paired or single spines. Both ***Acacia angustissima*** and ***Lysiloma microphyllum*** have creamy or white flowers in globose heads, and neither is armed.

J. J. Thornber made the northernmost collection (32.2°N) in 1906 in the Tucson Mountains, Pima County, Arizona. The nearby sighting was Robert Darrow's (Benson and Darrow 1981). The southernmost point (a collection by Howard S. Gentry) is from the Sierra Navachiste

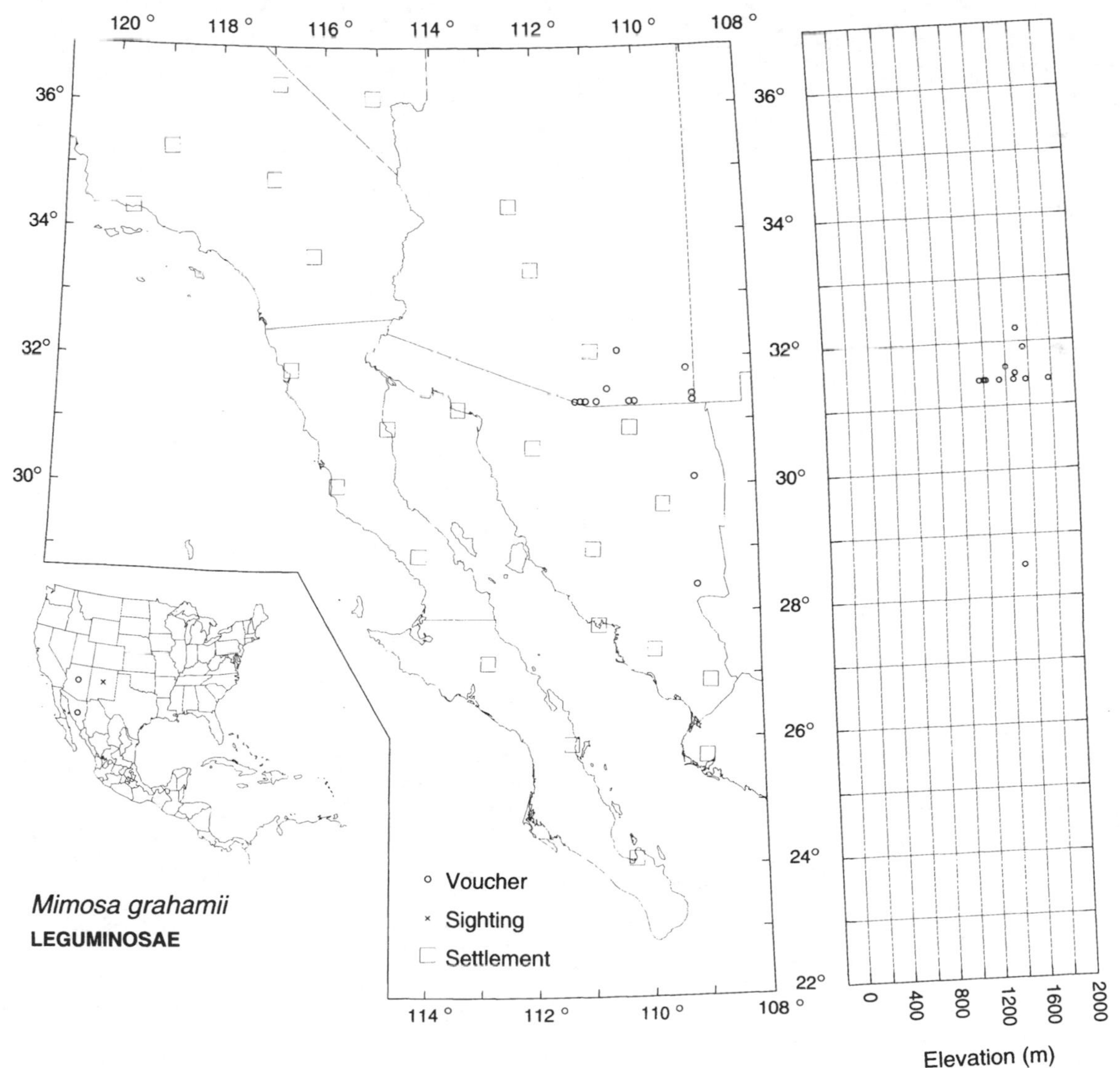

Mimosa grahamii
LEGUMINOSAE

near Bahía Topolobampo, Sinaloa. The species reaches its northwestern limit in the Ajo Mountains, Pima County, Arizona (32.0°N, 112.7°W), the westernmost range to receive significant summer rain (Turnage and Mallery 1941), and favors areas of heavy summer rain throughout the region mapped in this atlas. The progressive southward restriction to lower elevations may reflect a need for relatively open vegetation.

The plants flower from July to September, occasionally October. Leaves fall with the first frost and are renewed late in the spring.

Mimosa dysocarpa makes an attractive ornamental for arid regions. The plants tolerate freezing temperatures down to −9.5°C (Johnson 1993). In Tucson, Arizona, cultivated plants tend to freeze back slightly every winter, but if pruned, they flower abundantly the following summer. Germination is improved by scarifying the seed coat mechanically or chemically (Nokes 1986). Semihardwood cuttings taken in summer and early fall root easily when treated with a rooting hormone and intermittent mist (Nokes 1986).

Mimosa grahamii

A. Gray

Graham mimosa

A low shrub 3–6 dm high, *Mimosa grahamii* has internodal prickles and paired stipular spines. The leaf rachis is also prickly, as are the margins of the pods. The twice-pinnate leaves have 6–8 pairs of pinnae, each with 8–15 pairs of oblong leaflets 4–6 mm long. The creamy flowers are in heads 1.5 cm in diameter. The pods are 7–9 mm wide and 3–4.5 cm long.

Isely (1971) recognized two varieties in the southwestern United States: var. *lemmonii* (A. Gray) Kearney & Peebles, with

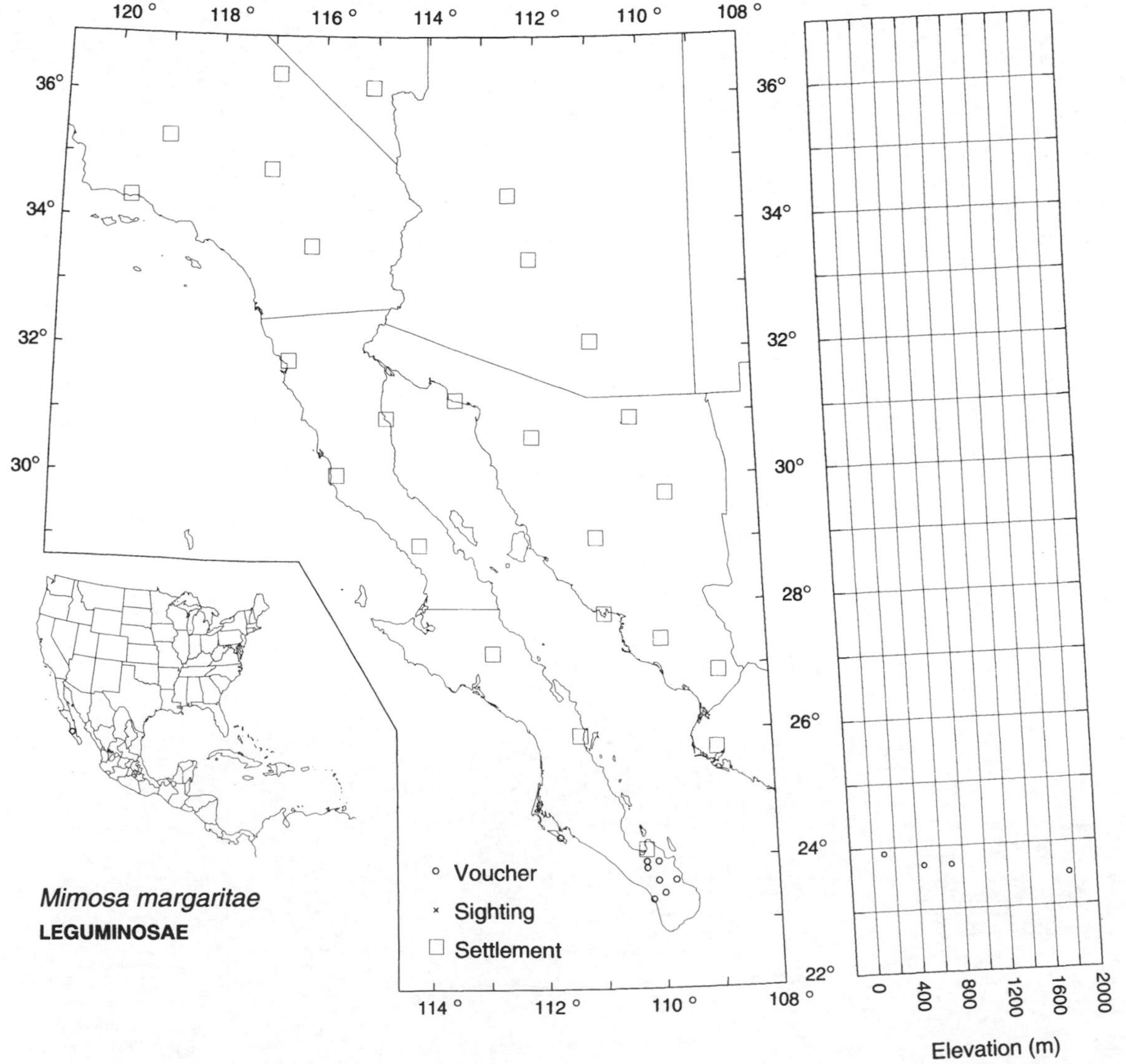

pubescent foliage and corollas, and the largely glabrous var. *grahamii*. Apparently, their ranges are similar, and we follow Barneby (1991) in combining them.

No other mimosa in its range has both internodal prickles and paired stipular spines. *Acacia angustissima* lacks spines and prickles, as does *Lysiloma microphyllum*. *Acacia millefolia* has 20–30 pairs of leaflets per pinna.

Occasional in shaded canyons, on rocky slopes, and on road cuts, *M. grahamii* is most typical of Madrean evergreen woodland and semidesert grassland. It enters the upper margin of the Sonoran Desert along washes. The northernmost point (32.2°N) is in the Rincon Mountains, Pima County, Arizona (a collection by Janice E. Bowers). The southernmost locality (28.5°N) is near Yécora, Sonora (a collection by Laurence J. Toolin).

Flowers appear in April and May, occasionally August.

Mimosa margaritae

Rose in Britton & Rose

Mimosa margaritae is a slenderly branched shrub with reddish brown bark. The leaves have 1 pair of pinnae, each with 3–12 pairs of leaflets, 9 mm long, which are ciliate with stout yellowish prickles. The whitish or pale lavender flowers are in globose heads about 1 cm across. The linear-oblong pods, 2–3 cm long and about 5 mm wide, are sparsely prickly on the margins and sides.

The closely related *M. xanti* has larger leaflets (8–15 mm long).

This shrub is found only in Baja California Sur from the vicinity of La Paz southward and on Isla Margarita. It ranges from sea level to 1,750 m. At lower elevations, the plants are common along sandy arroyos.

Flower appear in April and May (Wiggins 1980) and again from September through December.

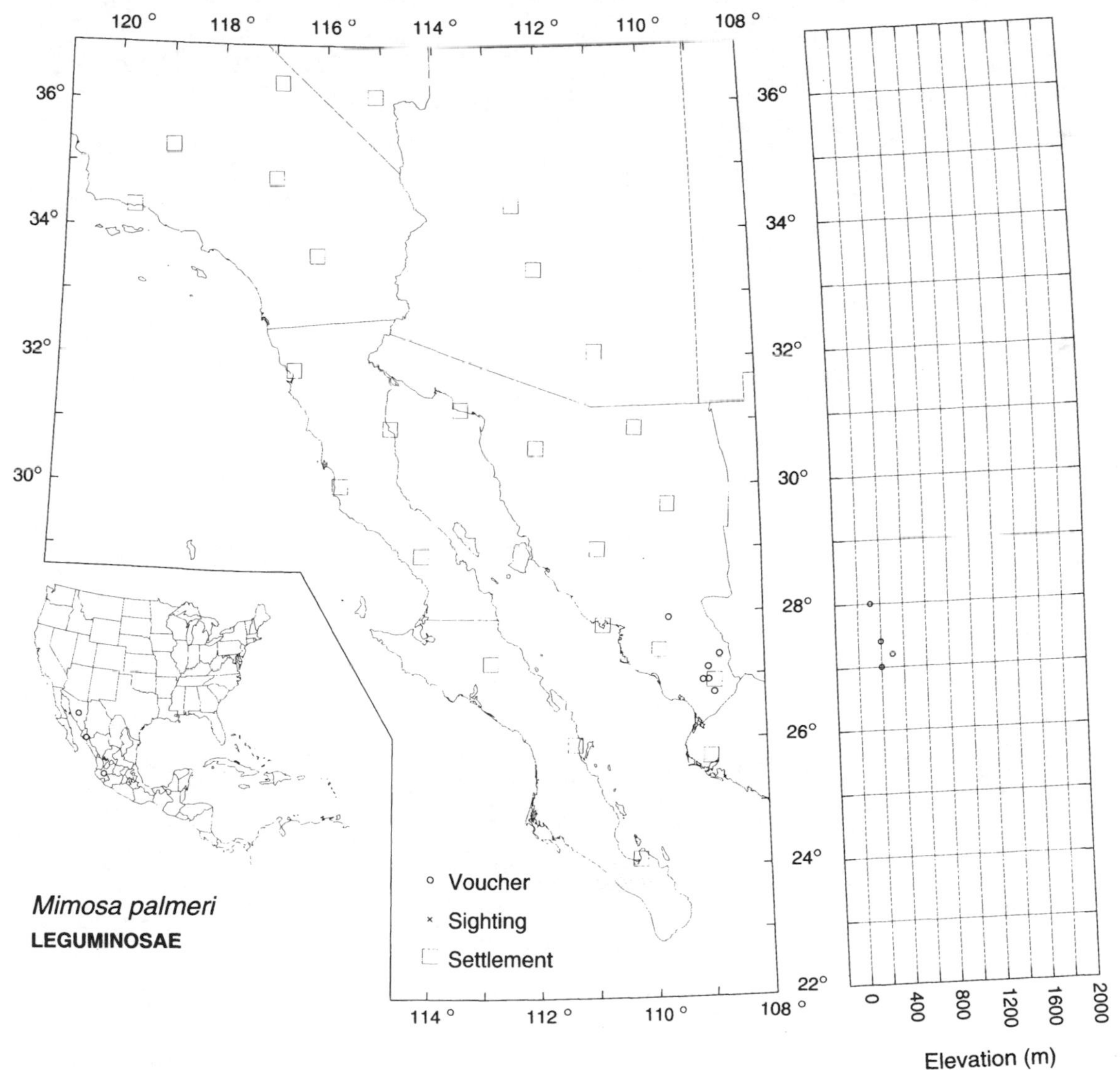

Mimosa palmeri
LEGUMINOSAE

Mimosa palmeri

Rose

Chopo

Mimosa palmeri is a slender tree up to 7 m tall with straight to recurved internodal prickles, most unpaired. The twice-pinnate leaves have 9–20 pairs of pinnae, each with 5–15 pairs of leaflets 2–4 mm long. The leaflet blades are divided into nearly equal halves by the midveins. The pink flowers are in spikes 4–6 cm long.

Mimosa dysocarpa is a shrub with leaflets divided by the midveins into strongly unequal parts.

This plant grows in Sinaloan thorn-scrub and Sinaloan deciduous forest from southern Sonora to Sinaloa and Jalisco. The northernmost point represents a Frank Reichenbacher collection.

The plants bloom copiously from August to early October.

The bark is chewed for treatment of tooth decay and gum disease (Gentry 1942).

Mimosa xanti

A. Gray

Celosa

Mimosa xanti is a shrub up to 2 m tall with scattered straight prickles on the stems and stout, appressed marginal prickles on the leaflets. The leaves have a single pair of pinnae, each with 3–12 pairs of leaflets 8–15 mm long. The pink flowers are in heads 1.5–2 cm in diameter. The oblong pods, 2–2.5 cm long, break into 2–4 segments at maturity.

Mimosa margaritae has smaller leaflets (about 9 mm long).

In the study area, *M. xanti* grows in

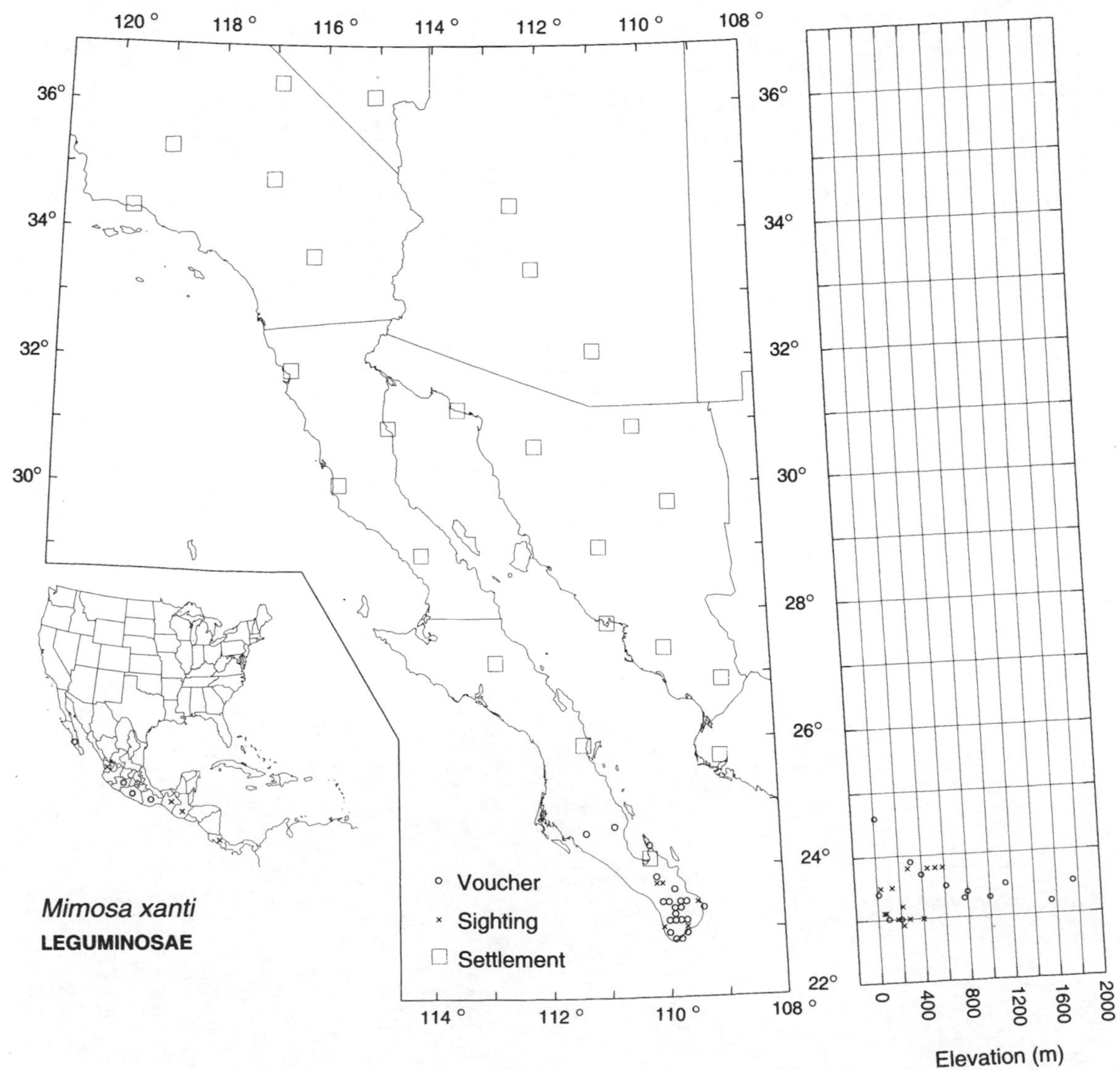

Baja California Sur and on Isla Espíritu Santo (Moran 1983b). It is also found from southern Mexico to Costa Rica (Barneby 1991). It is a common component of Madrean evergreen woodland in the Sierra de la Laguna (León de la Luz and Domínguez-Cadena 1989).

The flowers appear from March to May (Wiggins 1964).

Mortonia sempervirens A. Gray

[=*M. scabrella* A. Gray, *M. utahensis* (Cov.) A. Nels.]

Sandpaper bush, tickweed, mortonia

This densely branched shrub, up to 2 m tall, has crowded, overlapping, elliptic leaves 5–15 mm long and 3–12 mm broad. The foliage is harshly scaberulous. The whitish flowers, 4–7 mm wide, are in short, terminal panicles. The fruits are cylindrical nuts 4–6 mm long.

The sandpaperlike texture of the crowded leaves is distinctive. *Petalonyx* species are under 1 m tall and often are suffrutescent rather than woody.

Our map combines subsp. *utahensis* Cov. ex A. Gray and subsp. *scabrella* (A. Gray) Prigge. The former, restricted to southern Nevada and adjacent areas of California, Utah, and Arizona, has been treated as *M. utahensis* (Cov.) A. Nels. (Munz 1974). *Mortonia sempervirens* subsp. *scabrella* grows from southeastern Arizona to Texas and adjacent areas of northern Mexico.

Often growing in nearly pure stands on gravelly or rocky slopes and mesas, *M. scabrella* shows a marked preference for limestone and limestone-derived soils in the study area. It enters the upper margin of the Sonoran Desert in the Santa Cata-

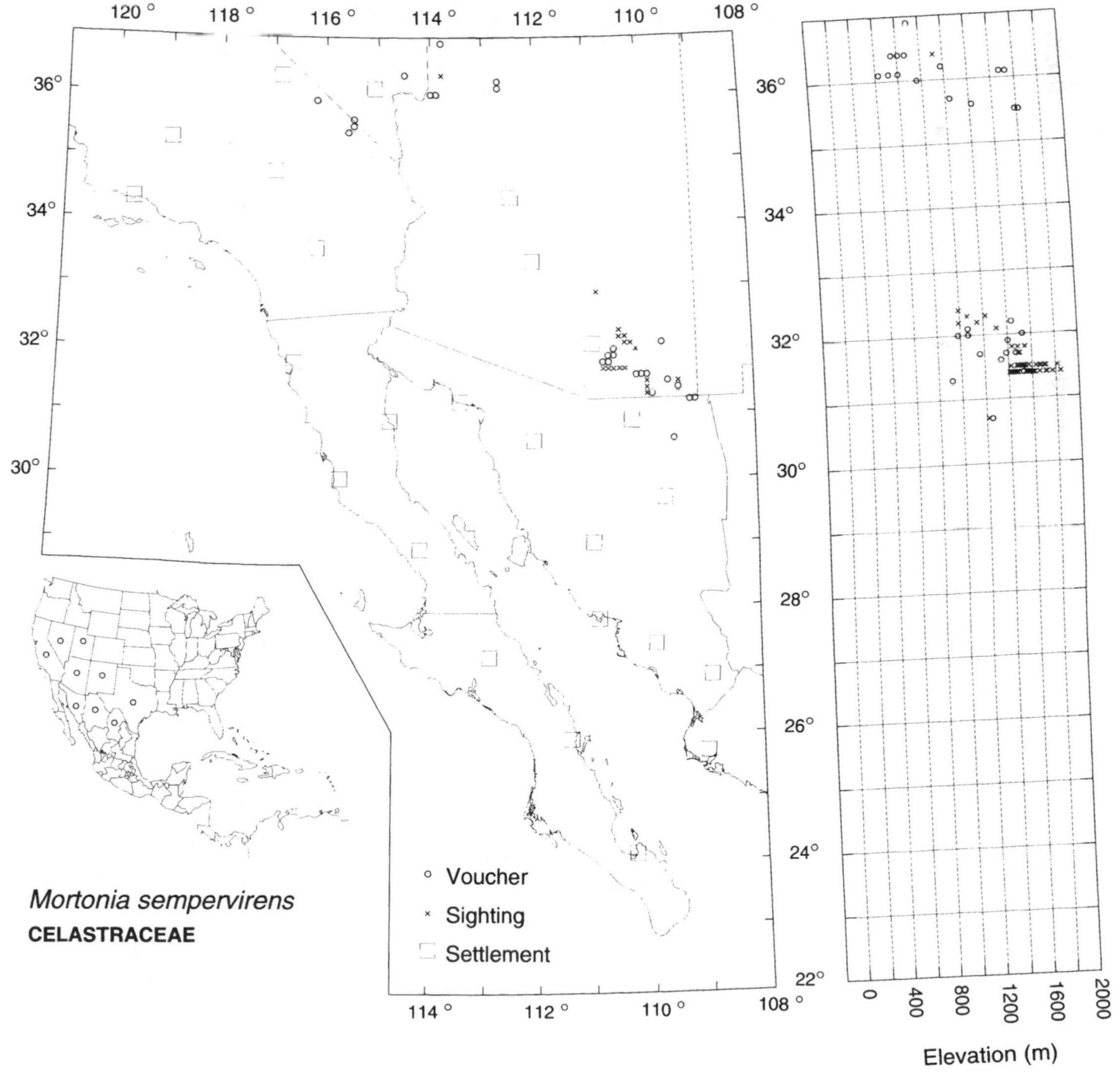

Mortonia sempervirens
CELASTRACEAE

lina and Rincon mountains, Pima County, Arizona, and in the Tortilla Mountains, Pinal County, Arizona. The actual elevational range (425–2,090 m) is somewhat greater than we show. The species occupies semiarid warm-temperate climates where some summer rain falls.

The current wide separation of the two varieties may date back to the Holocene. Packrat middens from southern Nevada record subsp. *utahensis* as a dominant between 10,000 and 8,200 years B.P.; it is rare in the area now (Spaulding et al. 1990). Subspecies *utahensis* is also common in late Pleistocene packrat middens from the western Grand Canyon (Phillips and Van Devender 1974).

Flowers appear from March to September and are visited by a variety of insects, including chrysomelid beetles, wasps, bees, and butterflies.

Myrtillocactus cochal (Orcutt) Britton & Rose

Candelabra cactus, cochal

Myrtillocactus cochal (figure 58) is a columnar cactus up to 4 m tall with many short, 6- to 8-ribbed branches from a stocky trunk. The sparse spine clusters are of 5 stout radial spines 8–12 mm long and a single spreading central spine 12–20 mm long. Flowers are green, tinged with purple, and about 2.5 cm long and broad. The globose fruits, 12–20 mm wide, are scarlet (often tinged yellow or brown). The haploid chromosome number is 11 (Pinkava et al. 1985).

Bergerocactus emoryi grows in sprawling thickets and has dense yellow spines. *Lophocereus schottii* has dense gray spines near the tips of the older stems. *Lophocereus gatesii* M. E. Jones has 10–15 ribs per branch and lacks an obvious trunk. *Stenocereus thurberi* is taller (to 8 m) and has more ribs (12–19 per branch). *Stenocereus gummosus* also forms sprawling thickets and has spine clusters with 8–12 radials and 3–6 centrals.

Locally common on hillsides and

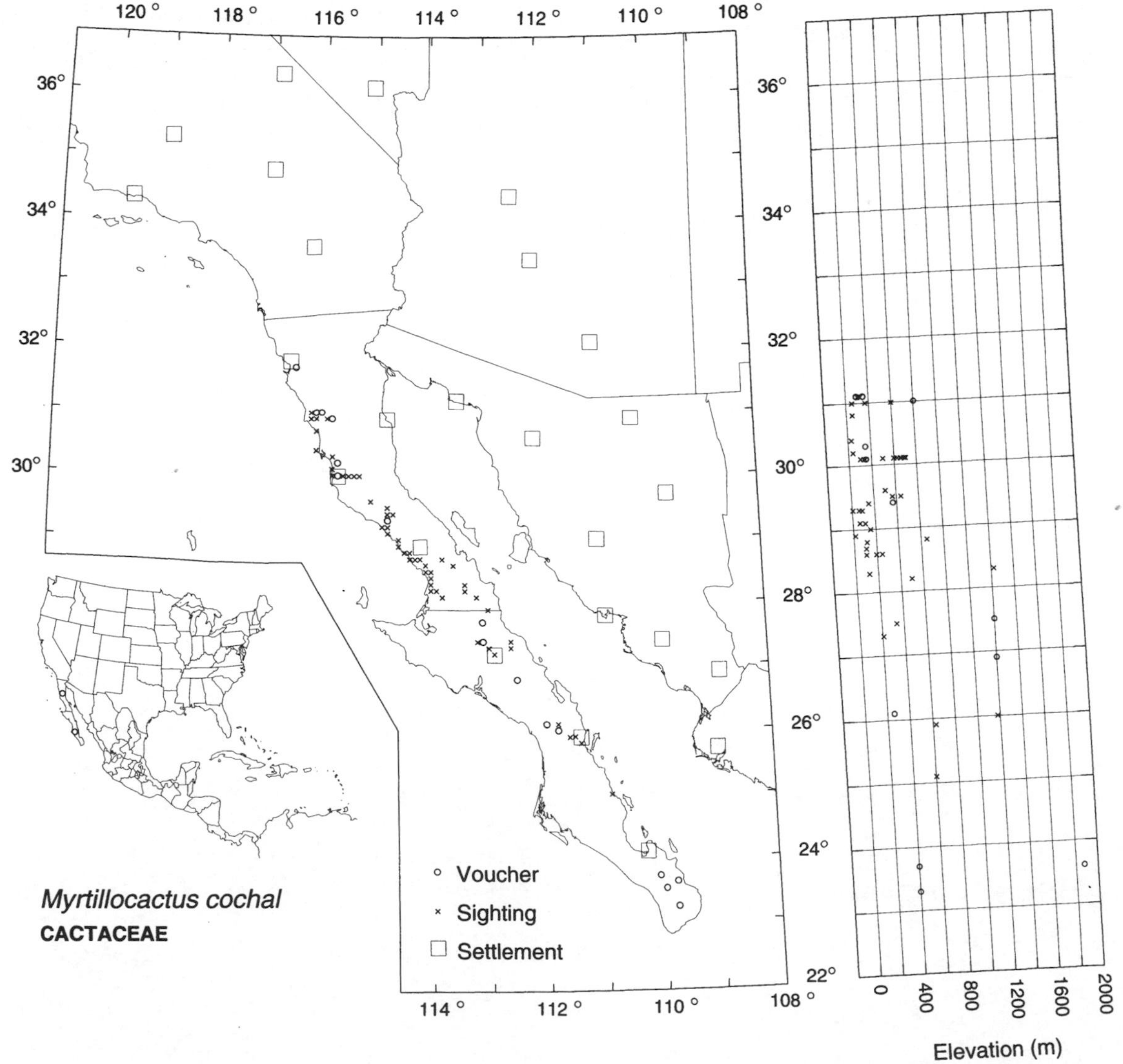

mesas, *M. cochal* grows nearly throughout the length of the Baja California peninsula. In the northern half, it is confined to the Pacific slope, where rain falls mainly in winter and summers are relatively cool. In the southern half it grows on the warmer gulf side, where summer and autumn rains predominate. The highest point on our graph (1,875 m) represents a 1959 collection made by Reid Moran on a peak west of La Laguna (23.6°N, 110.0°W). At higher latitudes, the plants tend to grow at lower elevations, probably because the stems are easily damaged by frost.

In Californian coastalscrub and Californian chaparral communities, *M. cochal* is not limited by water, light, or temperature; however, competition for physical space may be severe (Cody 1984). The plants respond by increasing their number of branches as they increase in height. In this way, the bulk of the plant rises above the surrounding vegetation (Cody 1984). Plants in dense Californian coastalscrub tend to be killed by wildfires.

Plants bloom throughout the year (Wiggins 1964). The flowers open in the evening and close a few hours after sunrise the following day (Gibson 1988a). Geometrid moths are likely pollinators at night; by day hummingbirds and bees also visit the blossoms (Gibson 1988a). Nectar-feeding bats (particularly *Choeronycteris mexicana*) may also pollinate the flowers (Hevly 1979).

The slightly acid fruits are reported to be edible (Wiggins 1964), and the wood of the stems is used for fuel (Standley 1924). The fruit fly *Drosophila mojavensis* feeds on the rotting tissue of dying plants and on a variety of yeasts found in this moist microenvironment (Starmer et al. 1976). *Myrtillocactus cochal* is occasionally grown in pots but is less common in cultivation than the more frost-tolerant *M. geometrizans* (Mart.) Console of the southern Chihuahuan Desert.

Figure 58. Myrtillocactus cochal *near El Rosario, Baja California. (Photograph by J. R. Hastings.)*

Figure 59. Olneya tesota *in the Pinacate region, Sonora. (Photograph by V. D. Roth.)*

Olneya tesota
A. Gray

Ironwood, palo de hierro, tésota, palo fierro, comitín

Olneya tesota (figure 59) is a tree 5–10 m tall with spiny, green branchlets and light gray, fissured bark. The leaves, once-pinnate with 4–10 (occasionally 12) pairs of obovate to oblong, bluish green leaflets, are 3–10 cm long. The pink to violet flowers, 9–10 mm long, are in axillary racemes 2–6 cm long. The torulose pods, 8–9 mm wide and 3–6 cm long, contain a few shining brown seeds. A diploid chromosome count of 18 has been reported (Turner and Fearing 1960). The roots form nitrogen-fixing nodules (Felker and Clark 1981).

The dark green foliage, light gray bark, and once-pinnate leaves are distinctive.

Olneya tesota is often a dominant on rocky or gravelly slopes and plains in the eastern part of its range. To the west, it is frequently restricted to washes. The geographic limits correspond closely to the outline of the Sonoran Desert (Shreve 1964). The southernmost point on the mainland (27.0°N) is from Chinobampo (Gentry 1942). In Baja California, the species is sparsely distributed on the Pacific side of the peninsula where summers are dry and relatively cool, but it is common near the gulf.

Packrat midden assemblages from the Wellton Hills, Arizona, suggest that *O. tesota* did not arrive in the northern Sonoran Desert until 8,700 years ago (Van Devender 1990b). Until the late Holocene, it dominated xeric, south-facing slopes; since that time it has become restricted to flats and washes (Van Devender 1987; Van Devender, Burgess et al. 1990).

In the northern Sonoran Desert, *O. tesota* grows on benches and bajadas above the valley bottoms, where cold air ponds on winter nights (Shreve 1964). When temperatures drop below freezing, adults lose some or many leaves, but the main branches and trunks appear unharmed. Insofar as low temperatures determine the northern and eastern limits of this spe-

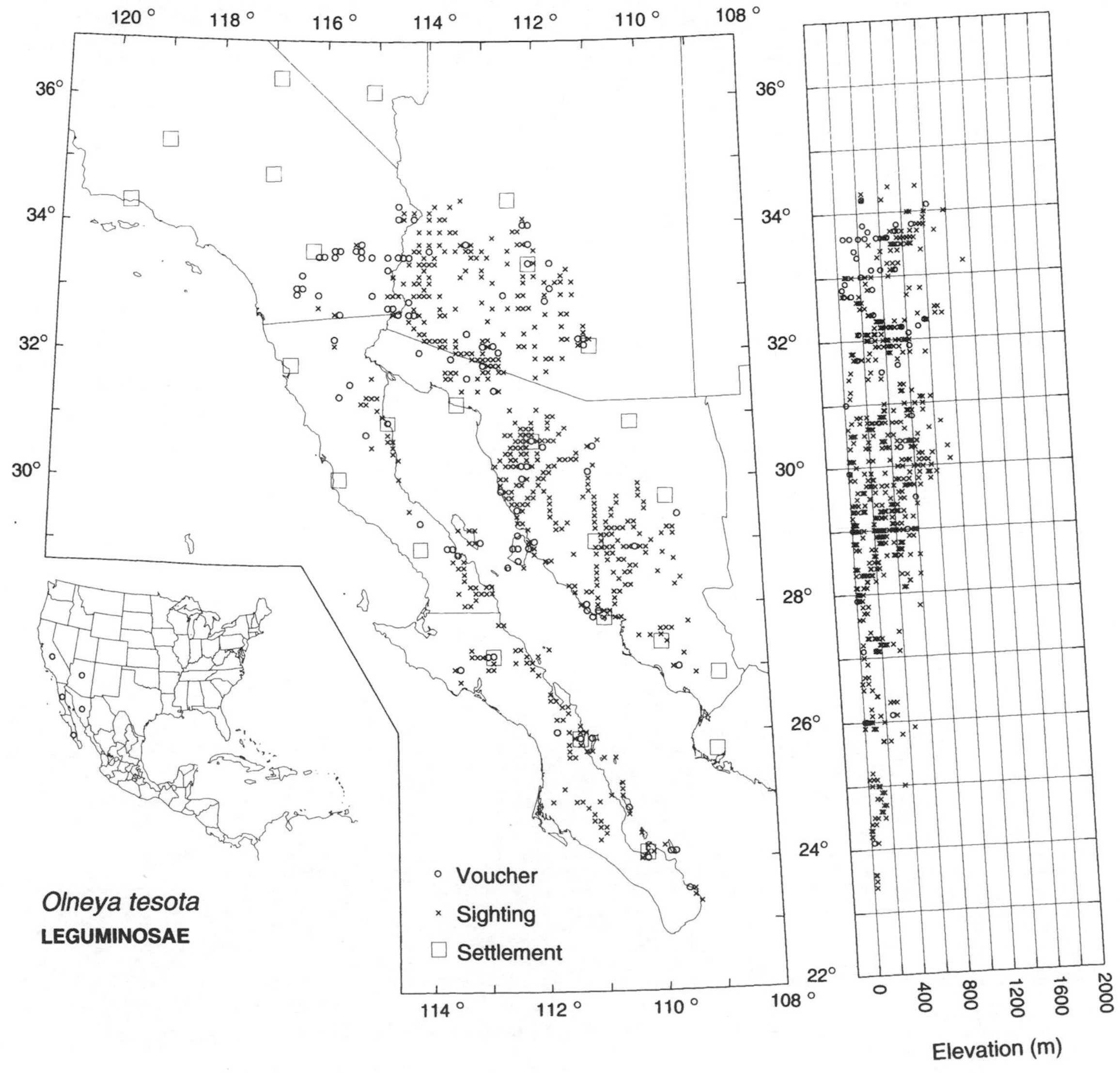

cies, they likely operate through their effect on seedlings. The distribution suggests a need for occasional warm-season rain, probably for seedling establishment.

Across its range, *O. tesota* blooms from March into May or June. At a given locality, flowering lasts only a few weeks. Fruits begin to ripen in May or June and fall four to eight weeks later (Shreve 1964; R. M. Turner 1963). The trees are essentially evergreen but may lose some leaves as soil moisture declines in the arid foresummer (Shreve 1964; Nilsen et al. 1984); after heavy frost in winter; and just before flowering. In late spring to early summer, new leaves appear as old leaves drop (R. M. Turner 1963; Humphrey 1975). Stem tips elongate during the summer rainy season (Humphrey 1975). Stem diameters increase following both summer and winter rains (R. M. Turner 1963).

Because the trees are deeply rooted, they generally do not lack for soil moisture. Moisture stress varies only slightly from February to May (Klikoff 1967), and the plants maintain relatively high and constant values of dawn plant water potential (Monson and Smith 1982). Nevertheless, *O. tesota* has low conductance values compared to other desert perennials in the same habitat (Nilsen et al. 1984), which suggests that drought avoidance is at times necessary despite the deep roots. Indeed, mature trees lose small limbs or accessory trunks as a result of drought (Shreve 1964). *Olneya tesota* responds to favorable soil moisture in summer or winter (Szarek and Woodhouse 1976). In one study, maximum gross photosynthesis was 8 mg $CO_2/dm^2/hr$, a relatively low rate (Szarek and Woodhouse 1976). Winter rates were lower still. Gross production is about 7.42 kg dry weight/plant/yr (Szarek and Woodhouse 1977).

The flowers attract a variety of potential pollinators, including carpenter bees (*Xylocopa* species), honeybees (*Apis mellifera*), and hummingbirds. In the vicinity of Tucson, Arizona, the solitary bee *Centris pallida* is by far the most abundant visitor (Simpson 1977).

The seeds germinate within several weeks of dispersal once summer rains start (Shreve 1964). Went (1957) stated that the seeds have an impermeable seed coat, but, as Poole (1958) demonstrated, 77% of unscarified seeds germinated. Several factors deter seedling recruitment: the trees bloom abundantly in only two years out of five; rodents prey heavily upon the fallen seeds (Shreve 1964); and young seedlings are susceptible to drought.

Olneya tesota is occasionally grown as an ornamental. The plants suffer frost damage at temperatures below –6.5°C (Johnson 1993) and are slow growing (Duffield and Jones 1981). They prefer loose, sandy or gravelly soil and perform best with periodic irrigation (Duffield and Jones 1981).

The seeds contain 17–21% protein (Becker 1983). The Seri Indians cooked the seeds in several changes of water, then ground them or ate them whole. *Olneya* leaves are a major food for packrats (*Neotoma lepida*) on arid islands in the Gulf of California (Vaughan and Schwartz 1980). The Seri have long used the extremely hard wood for boat paddles, clubs, harpoon shafts, pestles, and so forth (Felger and Moser 1985). More recently, Seris and others have used the wood for figurines. Ironwood burns hot and makes an excellent fuel.

Opuntia arbuscula

Engelm.

Pencil cholla

A much-branched cactus up to 1.2 (rarely 3) m tall, *Opuntia arbuscula* (figure 60) has a trunk up to 30 cm high. The cylindrical joints, 6–9 (occasionally 12) mm in diameter, are sparsely armed with reddish or tan spines 1–4 cm long, usually one (occasionally 2–4) per areole. The green, yellow, or terra-cotta flowers are 2–3.5 cm wide. At maturity the obovoid fruits are green, tinged with purple or red, and 2–4 cm long. Although fleshy, they are not juicy. The haploid chromosome number is 33 (Pinkava and McLeod 1971).

Figure 60. Close-up of Opuntia arbuscula *near Altar, Sonora. (Photograph by J. R. Hastings.)*

The stems of *O. leptocaulis* are about half as thick. The tubercles of *O. ramosissima* are flattened and diamond shaped. *Opuntia kleiniae* DC. is less densely branched, and its tubercles are more prominent. At maturity, its fruits are vermilion.

Opuntia arbuscula grows in sandy and gravelly soils of washes, valleys, and plains. The range covers the eastern part of the Sonoran Desert where the biseasonal rainfall regime has a predominance of summer rain. The northern and upper elevational limits are probably determined by the duration and severity of freezing temperatures, the westward limit by increasingly unreliable summer rainfall.

Peak flowering is in May and June (Kearney and Peebles 1960), but some plants may flower at other months. Fruits persist throughout the winter. They are occasionally sterile. They rarely form short chains of two or three fruits (Benson 1982).

The Tohono O'odham boil the young joints and buds as a vegetable (Kearney and Peebles 1960). Seri Indians occasionally ate the fresh fruits (Felger and Moser 1985).

Opuntia basilaris

Engelm. & Bigel.

Beavertail cactus

A low, spreading cactus up to 3 dm tall and 1 m across, *Opuntia basilaris* usually lacks spines but has abundant glochids. (Variety *treleasei* [Coult.] Toumey, with yellow or brown spines up to 3 cm long, occurs west of the Sonoran Desert.) The blue-green, ashy gray, or purplish joints, 5–20 (occasionally 32.5) cm long and 4–10 (occasionally 15) cm wide, are obovate to spatulate. The magenta flowers are 5–7.5 cm long and wide. The fruits lack spines and are dry at maturity. The base chromosome number is 11 (Pinkava and McLeod 1971); triploids and tetraploids are known (Pinkava et al. 1977; Pinkava and Parfitt 1982). Freeman (1970, 1973) described the developmental anatomy.

Opuntia santa-rita, a much taller plant, also has spineless, bluish or purplish joints, but these are orbicular.

Our map probably includes sightings or specimens for all five varieties treated by Benson (1982). The predominant taxon in the Sonoran Desert is var. *basilaris* (Benson 1982).

Opuntia basilaris grows in sandy, gravelly, or rocky soil on plains, valleys, washes, and slopes. In general, *O. basilaris* grows in hot deserts where winter rains predominate. Since the plants can take up carbon dioxide even at high temperatures (Hanscom and Ting 1978), it seems unlikely that summer rain is detrimental to them. Warm winter temperatures and erratic winter rain may determine the southern limits. The eastern limits are roughly contiguous with portions of the western limits of *O. engelmannii/O. phaeacantha, O. fulgida,* and *O. leptocaulis,* perhaps showing competitive displacement where summer moisture is more dependable. The distributional gap in southeastern California (between 33.4–34.5°N and 114.6–116.0°W) may result from a paucity of collections. The United States military has, until recently, denied access to much of this area. Wiggins (1980) reported this

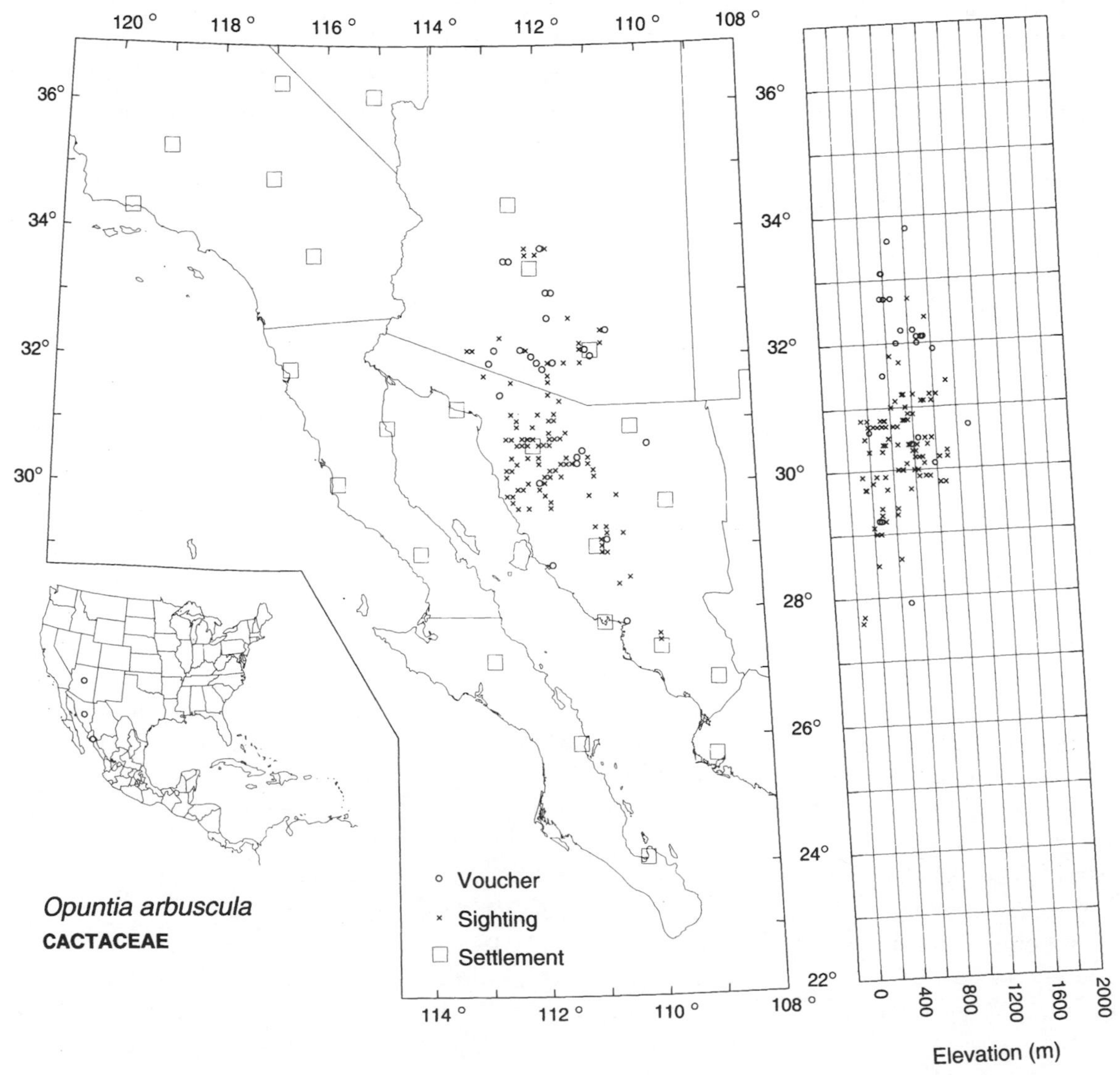

species for northern Baja California.

Opuntia basilaris has occupied much of its present range since the late Pleistocene (with some shifts in elevation). Packrat middens record *O. basilaris* at Death Valley about 19,500 years B.P. (Wells and Woodcock 1985) and at Picacho Peak, California, from 10,400 years ago to the present (Cole 1986). Apparently, the species was common in the lower Grand Canyon during the late Pleistocene (Phillips 1977). It reached its present northern limit in the Mojave Desert by 6,800 years B.P. (Spaulding 1990).

Flowers appear from March to June. They are pollinated largely by bees (Grant and Grant 1979c). Extrafloral nectaries on newly formed pads and flower buds are tended by ants, which may deter insect herbivores (Pemberton 1988).

Seasonal patterns of acid metabolism, gas exchange, and water use are well adapted to the very arid habitats where *O. basilaris* grows. The plants have net carbon gain only when water stress is low (Szarek and Ting 1974). During much of the year in Death Valley, California, for instance, *O. basilaris* experiences drought-induced photoinhibition (Adams et al. 1987). Typically, the greatest rates of carbon dioxide uptake are from late January to late March. The terminal pads often face north-south. This maximizes absorption of solar radiation during winter, when soils are more likely to be wet (Nobel 1982b). As long as plant water potential is greater than –0.5 MPa to –0.6 MPa, stomates open during the day as well as at night. Once it drops as low as –0.8 to –1.0 MPa, they open at night only (Szarek and Ting 1974, 1975). (Note, however, that Hanscom and Ting [1977] found no daytime stomatal opening in irrigated plants.) During drought, the stomates close, which, along with the impervious cuticle, greatly reduces gas exchange and transpiration (Szarek et al. 1973). While the stomates are closed, most of the respiratory carbon dioxide is retained within the joints. Recycling of endogenously produced carbon dioxide keeps the plants in a moderately active metabolic state, enabling them to respond rapidly to slight rainfall (Szarek et al. 1973; Szarek and

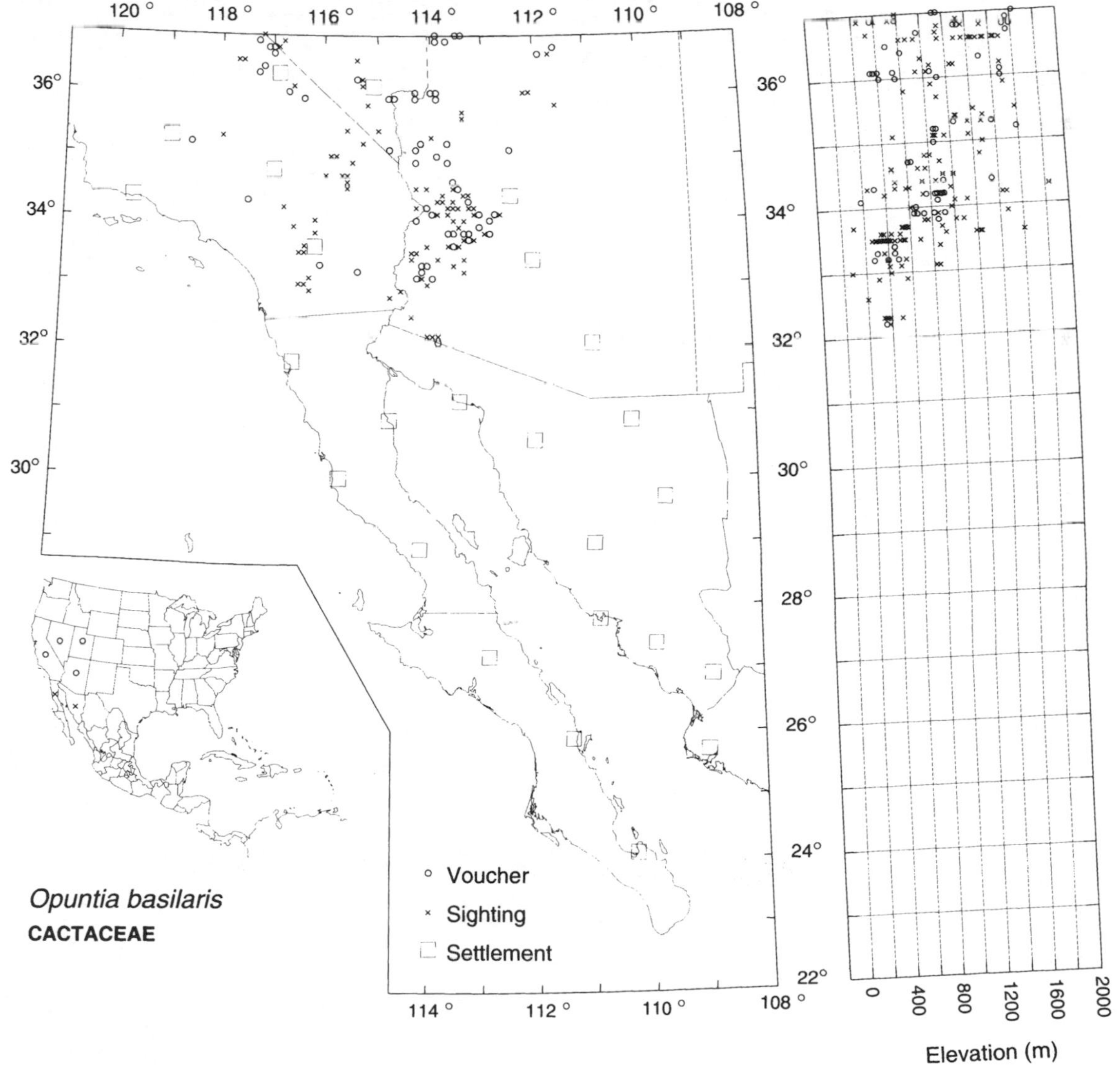

Ting 1974).

The chlorenchyma (photosynthetic tissue) has a higher osmotic pressure than the parenchyma (water-storage tissue) and therefore is more resistant to water loss (Barcikowski and Nobel 1984). During drought, water is lost first from the parenchyma. The high water content of the chlorenchyma allows stomates to open at night even after water can no longer be extracted from the soil (Barcikowski and Nobel 1984). It also helps keep the photosynthetic tissue alive during severe drought (Barcikowski and Nobel 1984). Mucilage and associated solutes in the parenchyma moderate the release of water to cells during the initial phase of a drought (Nobel et al. 1992). As long as night temperatures are lower than day temperatures, carbon fixation by crassulacean acid metabolism will use less water than daytime fixation by the C_3 pathway (Gulmon and Bloom 1979).

As might be expected, *O. basilaris* tolerates extremely high temperatures (Gulmon and Bloom 1979). Plants acclimated to a day/night temperature regime of 50/40°C can tolerate 62.6°C (Smith et al. 1984). This marked potential for acclimation ensures survival during the summer and allows the plants to fix carbon when temperatures are high and the soil is moist (Smith et al. 1984).

This cactus is commonly cultivated as an ornamental in arid landscapes.

Opuntia bigelovii

Engelm.

Teddy-bear cholla, cholla guera

A treelike cactus 1–1.5 (rarely up to 2.5) m tall, *Opuntia bigelovii* (figure 61) has a single trunk with many short, lateral branches near the top. The readily detachable joints are densely armed with straw-colored spines 1.5–2.5 cm long. The pale green or yellowish flowers are 2.5–4 cm across. Often sterile, the fruits are fleshy and yellow at maturity. The haploid chromosome number is 11 (Pinkava and McLeod 1971). Variety *bigelovii* is mostly triploid; diploids are known from a few

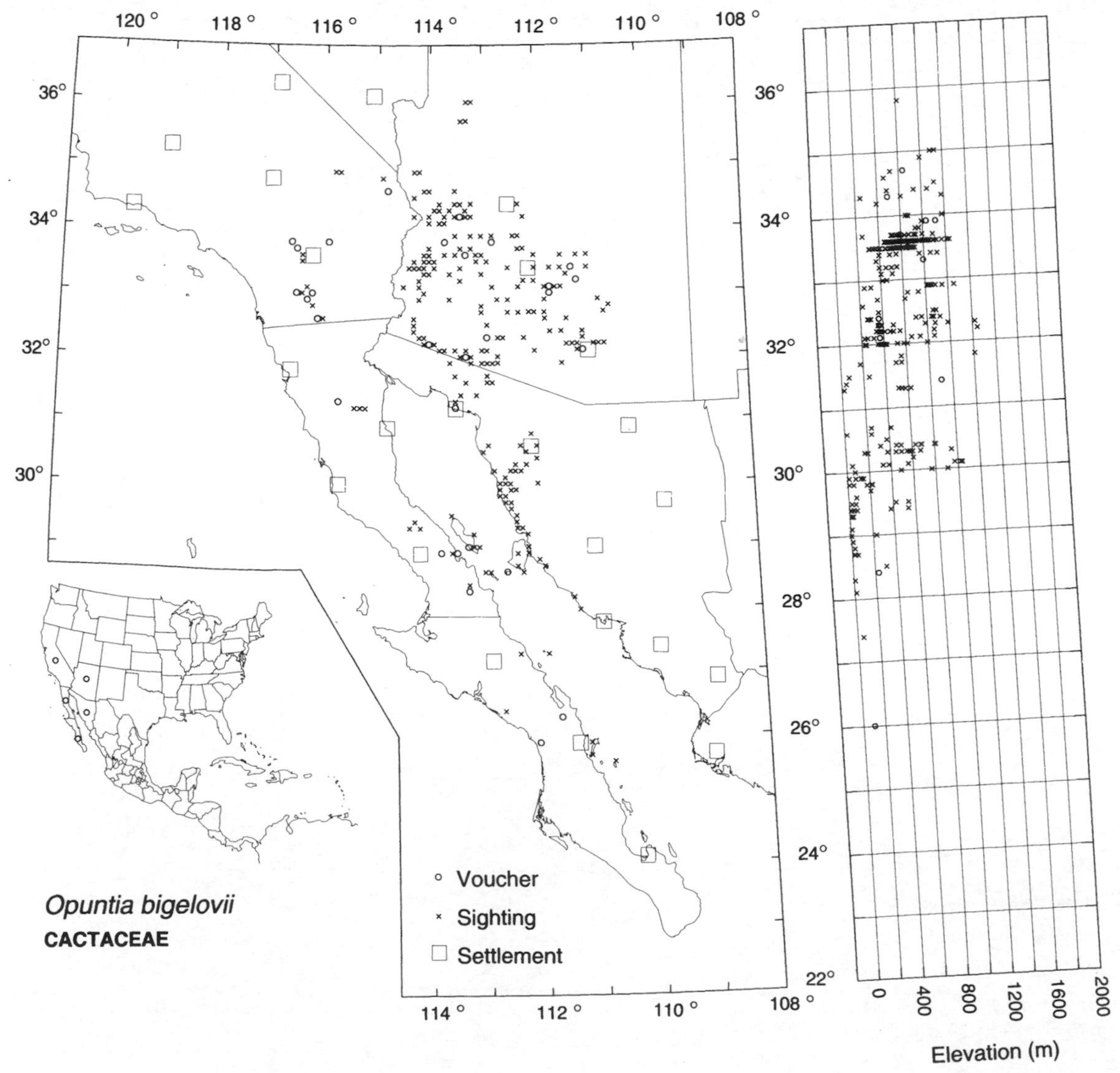

populations (Pinkava et al. 1985).

Benson (1982) recognized two varieties: var. *bigelovii* and var. *hoffmannii* Fosberg. The latter differs in its more diffuse habit, thinner stems, and less dense armature and is narrowly distributed in eastern San Diego County, California (Benson 1982). Our map combines them.

Opuntia fulgida trunks are branched and rebranched, in contrast to the postlike trunks of *O. bigelovii*. Unlike *O. fulgida*, *O. bigelovii* does not form chains of 5–22 fruits.

Opuntia bigelovii often forms dense stands on rocky, south- or west-facing slopes and also grows on gravelly flats and rolling hills. The range encompasses the more arid parts of the Sonoran Desert with biseasonal rainfall. The distributional gap between the Colorado River delta and the Salton Sink occurs in the hottest and driest part of the range.

Packrat middens record the arrival of *O. bigelovii* at the Puerto Blanco Mountains, Arizona, by 10,500 years B.P. (Van Devender 1987) and in the Hornaday Mountains, Sonora, by 9,400 years B.P. (Van Devender, Burgess et al. 1990). These dates suggest that the species has occupied much of its present range only since the early to middle Holocene.

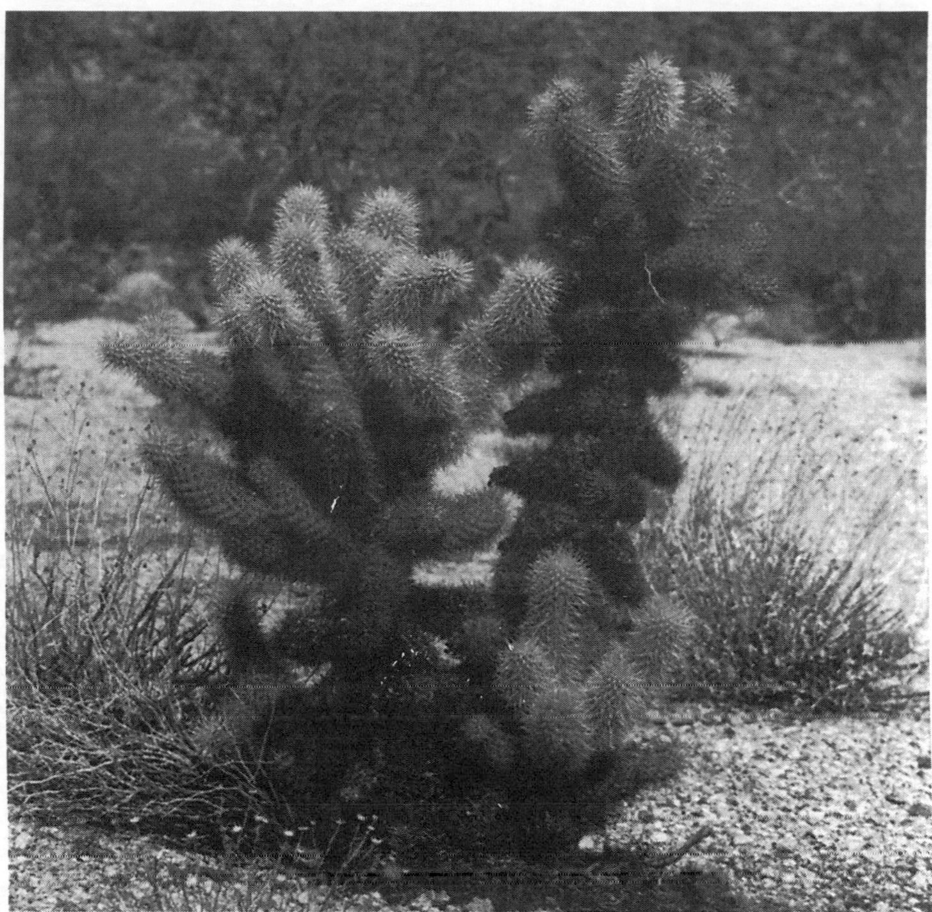

Figure 61. Opuntia bigelovii, *showing how the light-colored spines darken with age. (Photograph by J. R. Hastings taken north of San Felipe, Baja California.)*

Flowers appear from February to May (Kearney and Peebles 1960). Most of the flowers eventually produce sterile fruits. Reproduction by rooting of fallen joints is more common than reproduction by seed (Benson 1982), at least in the typical triploid plant, but biologists have identified diploids with higher rates of seed set (Pinkava et al. 1985). The joints grow vigorously in response to summer or winter rains, and a few new stems may be formed even during protracted droughts (Humphrey 1975). Ants that visit extrafloral nectaries on the newly formed joints and flower buds may reduce or deter insect herbivory (Pemberton 1988).

The plants are tolerant of high temperatures and sensitive to low ones. With acclimation, the heat tolerance of rooted joints can be increased from 52°C to 60°C (Didden-Zopfy and Nobel 1982; Nobel et al. 1986). Similar acclimation occurs seasonally in nature (Didden-Zopfy and Nobel 1982). Tolerance of high temperatures is crucial to the survival and establishment of fallen joints, which experience high soil temperatures (Didden-Zopfy and Nobel 1982). In one study, substantial freezing damage occurred to the chlorenchyma cells at –7.3°C, a relatively high value (Nobel 1982a). In one study, when plants were grown at 10°C, the optimum temperature for photosynthetic carbon dioxide uptake was 11°C. The thermal optimum shifted to 22°C when plants were grown at 30°C (Nobel and Hartsock 1981).

The photosynthetic surface area of many cacti is so reduced that carbon dioxide uptake may be limited by the amount of photosynthetically active radiation that falls on the stems (Nobel 1981a, 1983b). Self-shading, whether by spines or branches, can further reduce absorption of photosynthetically active radiation. Nevertheless, a mature cactus may devote as much as 16% of its biomass to spines. Nobel argued that "the overriding importance of protection from herbivory apparently has warranted this substantial allocation of biomass to spines" (Nobel 1983b:159). Dense spines do confer some degree of heat tolerance (Gibbs and Patten 1970) and also provide a measure of frost protection, enabling the species to extend farther north than more sparsely armed species.

A large proportion of *O. bigelovii* joints fall to the ground while still alive. Some are carried away in the fur or skin of animals. Many root in place near the parent plant. Such stands are genetically identical; indeed, the entire population of *O. bigelovii* in the United States may consist of a single vegetative strain (Benson 1982). In the Mojave Desert, *O. bigelovii* rapidly colonizes cleared sites through rooting of dispersed joints (Vasek 1980b) and, to a limited extent, from seed (Vasek 1983).

Matched photographs (Hastings and Turner 1965b) indicate that *O. bigelovii* populations may remain stable for 60 years, and it seems likely that individuals live 60 years or somewhat longer. Vasek (1980b) described the species as a "long-lived opportunist" with a life span of one or a few centuries. On undisturbed sites, the various height classes reach a dynamic equilibrium (Vasek 1980b). Survivors in senescent stands appear to be randomly distributed (McDonough 1965).

Packrats (*Neotoma* species) use the joints and even small plants in constructing nests (Brown et al. 1972; Vasek 1980b). After burning off the spines, Seri Indians cooked and ate the stems. They also made a diuretic tea from the roots (Felger and Moser 1985).

Opuntia chlorotica

Engelm. & Bigel.

Pancake pear, silver-dollar cactus

Opuntia chlorotica is an erect, arborescent or shrubby cactus 1–2 m tall with a densely spiny trunk 3–5 dm tall. The orbiculate to broadly oval pads, 12.5–20 cm across, have yellowish, downward-pointing spines over most of their surface. The yellow flowers are 4–6 cm across. Fruits lack spines (but bear glochids) and turn grayish purple at maturity. The haploid chromosome number is 11 (Pinkava and McLeod 1971).

Intermediates between *O. chlorotica* and other taxa, including *O. santa-rita* and *O. engelmannii,* do occur (Parfitt and Clark 1978; Benson 1982). In general, however, the species is well marked. *Opuntia engelmannii* seldom has a distinct trunk, and the pad spines are less numerous and, usually, white. *Opuntia santa-rita* shares with *O. chlorotica* an erect habit and orbicular pads. The pads of

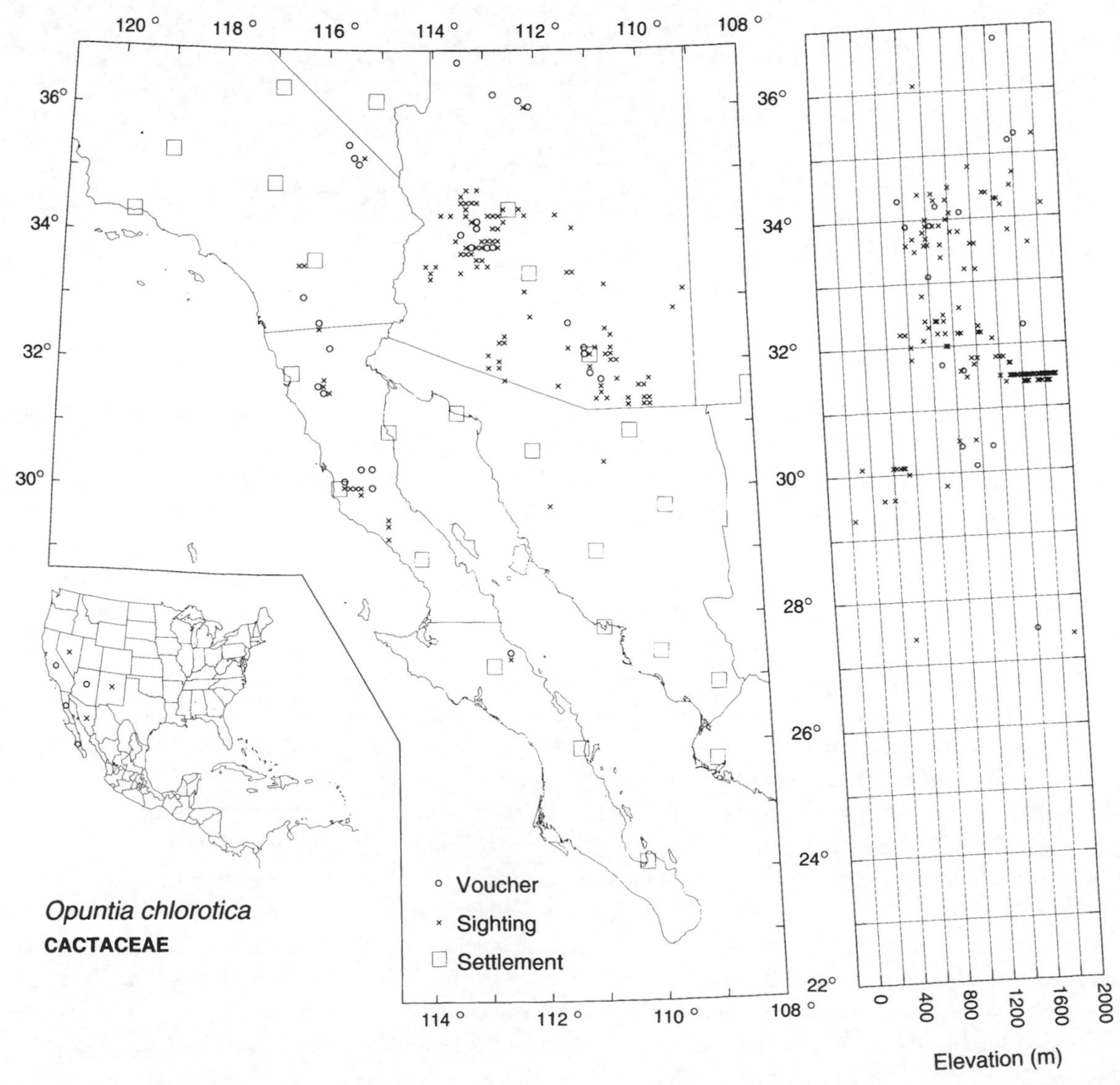

O. chlorotica are bluish, those of *O. santa-rita* tinged with purple. Also, *O. santa-rita* does not have the yellow, deflexed spines characteristic of *O. chlorotica.*

Opuntia chlorotica grows in rocky habitats, especially on ledges and steep slopes. Occasionally, as near Aguila, Arizona, the plants thrive on sandy flats. In western Arizona, they are common in semidesert grassland. The southernmost collection on our map (27.5°N) is Reid Moran's from the Cerro Azufre. The nearby sighting is also his. Robert H. Peebles made the northernmost collection (36.8°N). The two disjunct populations in Sonora (at 29.8°N and 30.5°N) represent sightings by Raymond M. Turner and Thomas R. Van Devender, respectively. The plant's distribution appears to follow the mesic borders of the northern Sonoran Desert. It is absent from stations below 500 m except in Baja California between latitudes of 29.3°N and 30.1°N. The elevational profile shows well-developed frost tolerance.

A sequence of packrat midden deposits in the Puerto Blanco Mountains, Arizona, records *O. chlorotica* from 14,120 years B.P. to 8,000 years B.P. (Van Devender 1987). Today, the species is still in the area but at somewhat higher elevations. Evidently, *O. chlorotica* was more widespread at low elevations during full glacial periods than it is today. Its range may have contracted during the Holocene.

Flowers appear in the spring (Kearney and Peebles 1960), and fruits mature by late summer or fall (Benson 1982). New pads develop at the same time as the flowers. Solitary bees in genera such as *Diadasia, Halictus, Lasioglossum, Augochlorella,* and *Ashmeadiella,* as well as sap beetles (Nitidulidae), were among the flower visitors in the Cerbat Mountains, Mohave County, Arizona. *Diadasia* species were the most effective pollinators. Smaller bees contacted the stigma less often and were less important as pollinators (Parfitt and Pickett 1980).

The photosynthetic surface area of many cacti is so reduced that carbon dioxide uptake may be limited by the amount of photosynthetically active radiation that falls on the stems (Nobel 1981a, 1983b). Terminal pads of *O. chlorotica* are oriented to maximize the absorption of photosynthetically active radiation during the season of active growth (Nobel 1981a). In the Mojave Desert, where soil is wet in winter and early spring, a significant proportion of *O. chlorotica* pads faced north-south. They reaped 104% more photosynthetically active radiation in winter and early spring than did pads oriented east-west (Nobel 1981a). At the Sonoran Desert site, where 66% of the rain fell in late summer, more pads faced east-west than north-south. East-west-facing pads absorbed 34% more photosynthetically active radiation annually than did north-south-facing pads (Nobel 1981a).

Opuntia chlorotica is occasionally grown as an ornamental.

Opuntia cholla

Weber

Opuntia cholla (figure 62) is an arborescent cactus 1–5 m tall with a sturdy trunk as much as 1.5 m tall. The cylindrical joints, 0.5–3 dm long and 2–3 cm in diameter, are whorled on ascending or spreading branches. The spines, yellowish when young, dark gray in age, grow in clusters of 5–13 (mostly 7–10) per areole. They do not obscure the stem. The peach-pink to pale purplish flowers are 2.5–3 cm across. The fruits, 2–3 cm in diameter and 3–5 cm long, are fleshy but not juicy and are spineless when mature. The haploid chromosome number is 11 (Pinkava and Parfitt 1982).

Opuntia cholla is the peninsular counterpart of *O. fulgida,* a common mainland species. Where they grow together (Isla

Figure 62. Opuntia cholla, *showing the ovoid fruits that readily sprout after falling to the ground. (Photograph by J. R. Hastings taken east of El Rosario, Baja California.)*

San Lorenzo) they can be separated, with difficulty, by subtle differences in flower color, fruit shape, and size.

This peninsular endemic is most abundant on mesas, plains, and valley bottoms. The northernmost sighting is Reid Moran's (31.9°N), and the outlier at Punta Eugenia (27.8°N, 115.0°W) is Raymond Turner's. Moran (1983b) also reported it from Isla San Lorenzo. Its wide occurrence on the peninsula suggests that *O. cholla* has a broad range of climatic tolerances. The elevational profile is truncated at 800 m, which suggests that maritime air moderates temperature extremes across much of the range. The spines may be too sparse to prevent tissue temperatures from reaching lethal levels. If so, this vulnerability to temperature extremes might account for its absence from northeastern Baja California.

Flowers appear in April and May. The fruits form chains that persist for several years (Wiggins 1964). Seeds are frequently lacking (Wiggins 1980), but the fruits themselves readily take root after they fall (Wiggins 1964).

Opuntia cholla likely shares many physiological, demographic, and morphological features with the closely related *O. fulgida*. One feature might be shallow roots capable of absorbing the moisture from relatively light rains. Like *O. fulgida*, *O. cholla* reproduces vegetatively for the most part (Wiggins 1964). Reliance on asexual reproduction tends to result in even-age stands in *O. fulgida* (Tschirley and Wagle 1964) and probably in *O. cholla*, as well.

Opuntia ciribe

Engelm. ex Coult.

A shrubby, densely branched cactus 0.5–2 m tall, *Opuntia ciribe* has cylindrical, light green or blue-green joints that are obscured by the straw colored or glistening white spines, which are 1–3 cm long. The flowers (2.5–3 cm wide) are greenish to yellowish, tinged with purple. The fleshy, yellow fruits, 2–4 cm long, are excavated

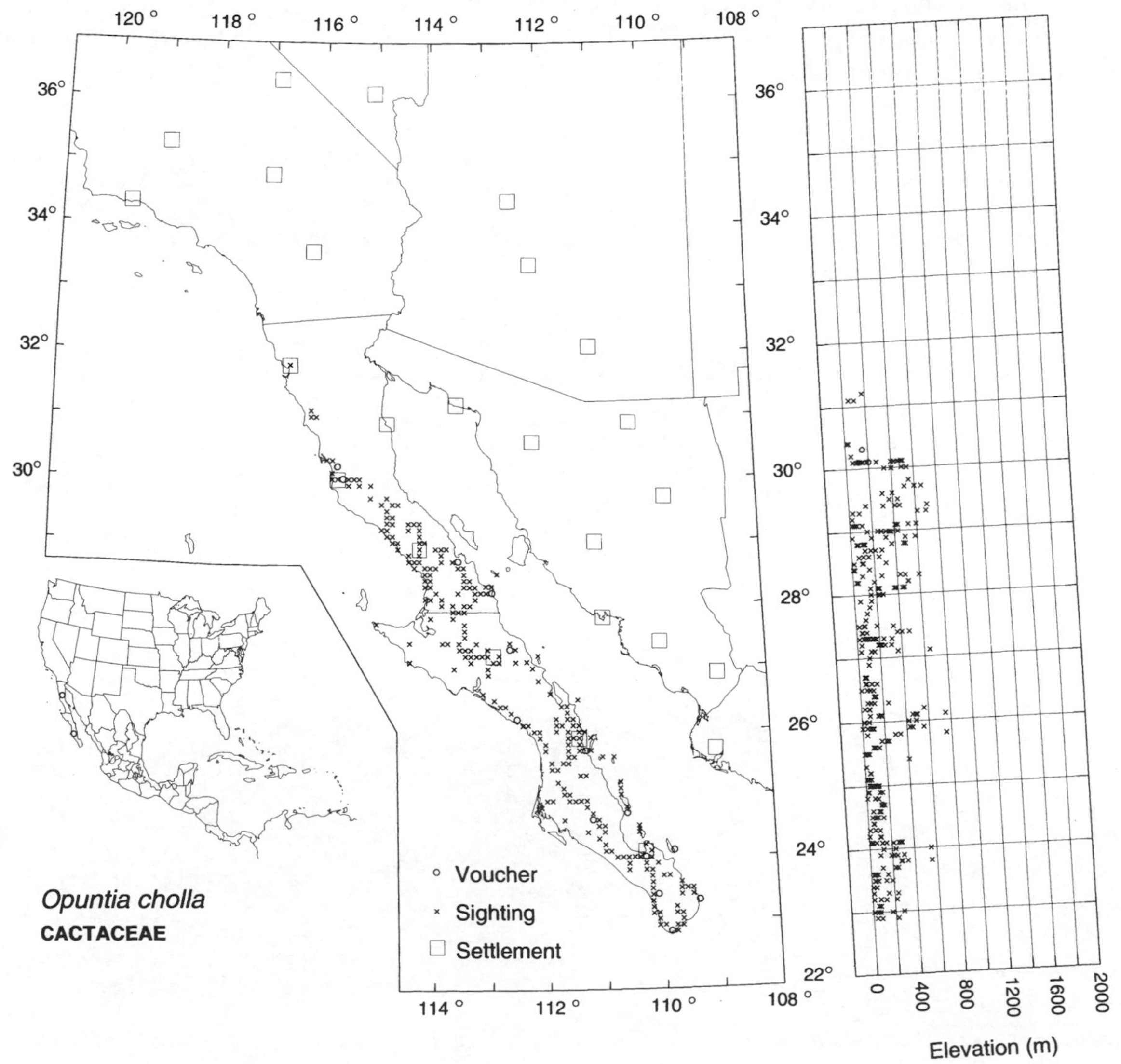

at the apex and sometimes form persistent chains. Most contain no more than a few seeds. The diploid number of chromosomes is 22 (Yuasa et al. 1974).

Opuntia ciribe can be difficult to distinguish from *O. cholla, O. prolifera, O. echinocarpa* Engelm. & Bigel., and *O. acanthocarpa* Engelm. & Bigel. In general none of these is as densely armed and plentifully branched as *O. ciribe.* Of this group, only *O. ciribe* and *O. cholla* produce chains of fruit. *Opuntia prolifera,* also chain-fruited, has dark green stems and yellow or red-brown spine sheaths.

Ira L. Wiggins vouchered the disjunct population in the Cape Region. George Lindsay made the collections on Isla San Esteban (28.7°N) and Isla Santa Cruz (25.3°N) in 1947. The distribution covers the more arid parts of the Baja California peninsula. Rainfall shifts from mainly winter to mainly late summer from north to south across the range.

Endemic to the central peninsula and a few gulf islands, *O. ciribe* may be locally abundant, probably due to its colonial habit. The fruits and readily detachable young joints take root after they fall to the ground, giving rise to new plants (Wiggins 1964). Reproduction by seed is evidently rare, and it seems likely that the existing populations represent a limited number of genotypes.

The plants flower in April and May (Wiggins 1964).

Opuntia emoryi
Engelm.

Opuntia kunzei
Rose

Opuntia parishii
Orcutt

[=*O. stanlyi* Engelm. var. *stanlyi,* var. *kunzei* (Rose) L. Benson, var. *parishii* (Orcutt) L. Benson, var. *peeblesiana* L. Benson]

Devil cholla

Opuntia emoryi, Opuntia kunzei (figure

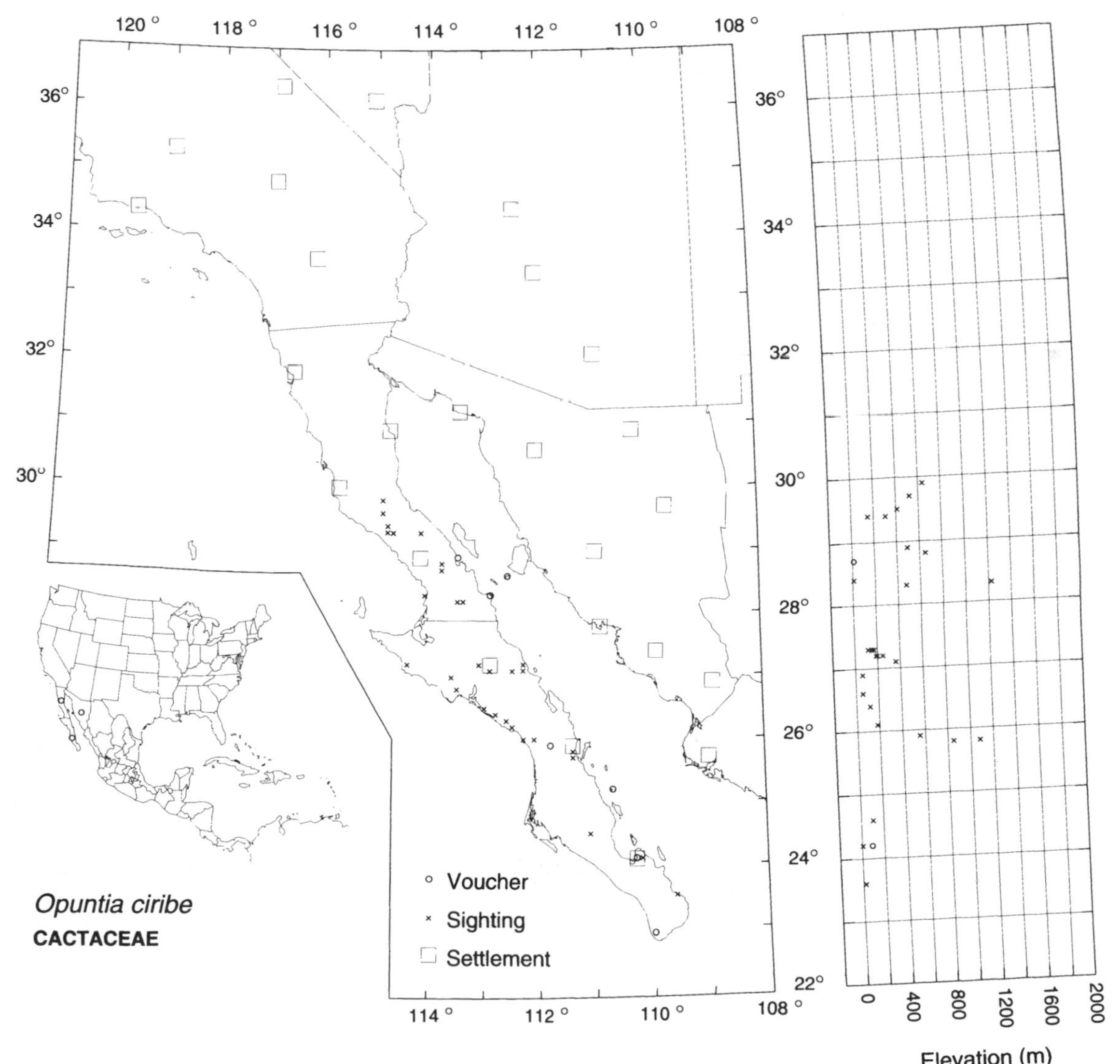

Opuntia ciribe
CACTACEAE

63), and *Opuntia parishii* grow as clumps or mats seldom more than 15–30 cm tall but several meters in diameter. The joints (5–15 cm long) are narrowly club shaped with large, conspicuous tubercles. The spines, up to 5 cm long, are 16–33 per areole, mostly on the upper stem. Flowers are yellow (sometimes reddish) and are 2.5–5 cm wide and 5–7.5 cm long. The fruits are densely spiny (or, occasionally, spineless or weakly spiny), fleshy, and yellow. The haploid chromosome number is 22 (Pinkava et al. 1985).

In Baja California, *O. emoryi* occurs well north of the superficially similar *O. santamaria* (Baxter) Wiggins and *O. invicta*. *Echinocereus* stems are cylindroid rather than club shaped; moreover, no *Echinocereus* species in the Sonoran Desert region forms mats several meters or more in diameter.

The name *O. stanlyi* was not validly published. *Opuntia emoryi* is the oldest taxonomically valid name (Pinkava and Parfitt 1988). Benson (1982) recognized four varieties in *O. stanlyi:* var. *stanlyi,* var. *peeblesiana,* var. *kunzei,* and var. *parishii.* Pinkava and Parfitt (1988) considered these distinct at the species level; however, because many of our records were identified only as *O. stanlyi,* we have mapped them together here. A new combination has not yet been made for *O. stanlyi* var. *peeblesiana.* The several taxa are, for the most part, well-defined geographically (Benson 1982).

These cacti grow on silty, sandy, or gravelly flats, mesas, and washes, typically in clumps or dense, low thickets. The southernmost point on our map (30.9°N) is near Tubutama. The voucher from northwestern Baja California is Grady Webster's (1961). The three species span a variety of climates, from warm-temperate to subtropical, and from semiarid with predominantly warm-season rain to arid with mostly winter rain.

A 10,000-year sequence of packrat middens from the Hornaday Mountains, Sonora, records the earliest *O. emoryi* about 6,100 years B.P. (Van Devender, Burgess et al. 1990).

Plants bloom from April to September. Peak flowering is in May and June. These

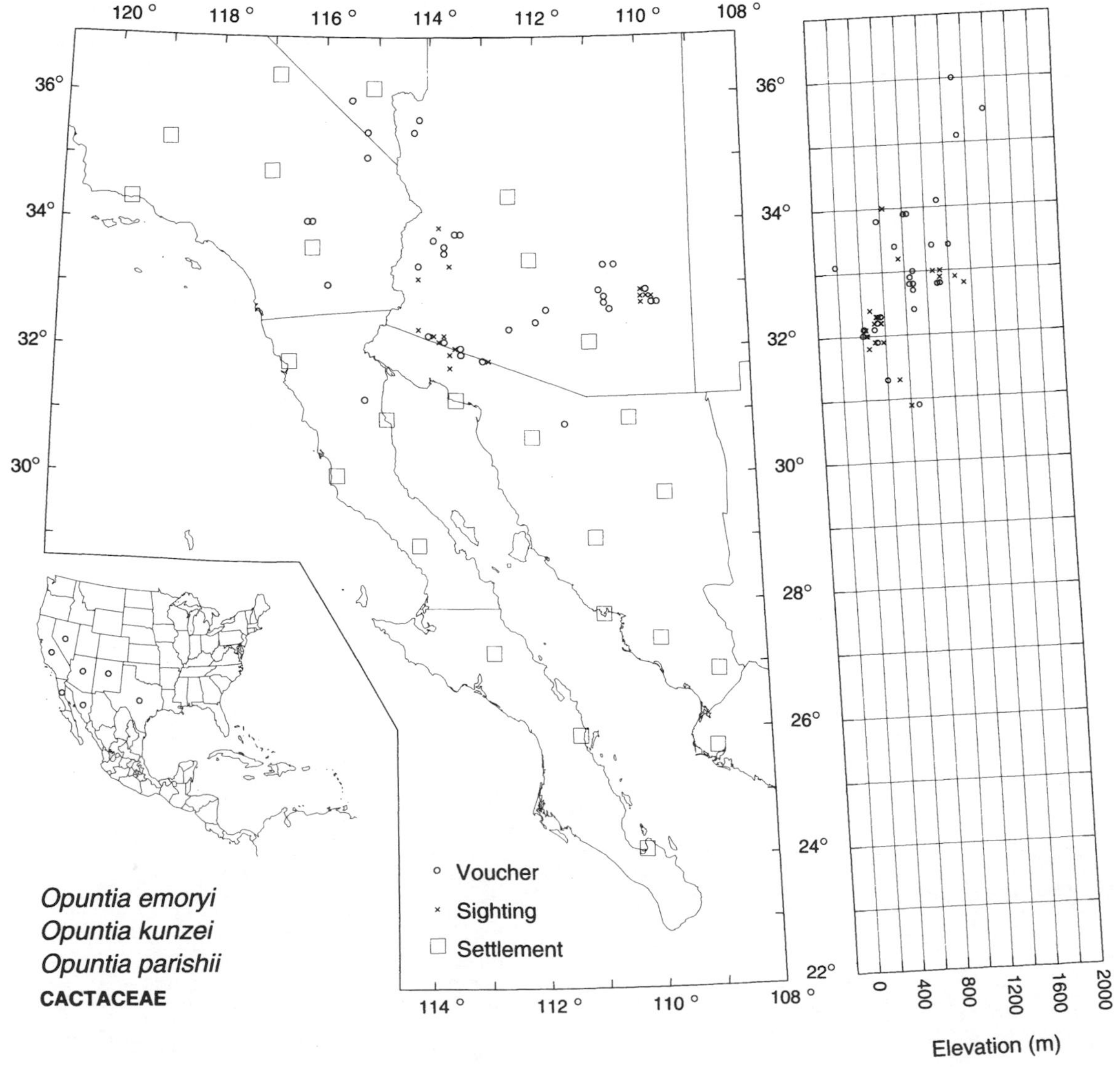

Opuntia emoryi
Opuntia kunzei
Opuntia parishii
CACTACEAE

species contain beta-phenethylamines, a class of alkaloids (Meyer et al. 1980). The clones can be long lived: an *O. emoryi* clump shown in a 1907 photograph taken by D. T. MacDougal was still present in 1987 (Raymond M. Turner, unpublished photograph).

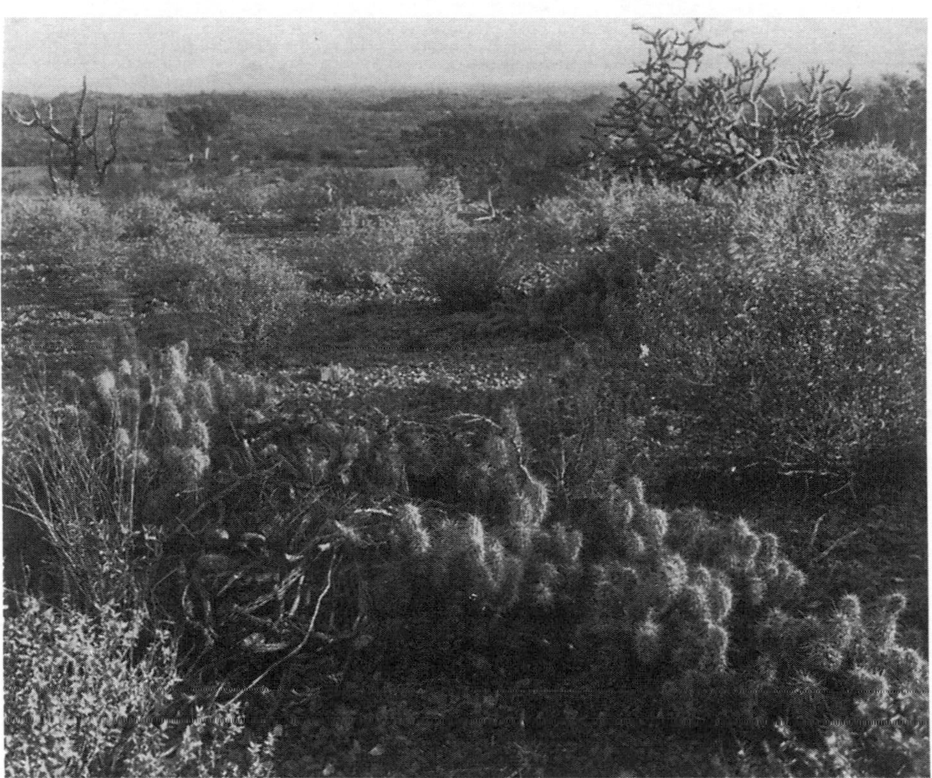

Figure 63. Opuntia kunzei *west of Sonoyta, Sonora. (Photograph by D. T. MacDougal, courtesy of the Arizona Historical Society/Tucson.)*

Opuntia engelmannii

Salm-Dyck.

[=*O. phaeacantha* var. *discata* (Griffiths) Benson & Walkington]

Opuntia phaeacantha var. *major*

Engelm.

Engelmann prickly pear, nopal, abrojo, joconostle, vela de coyote

These large, sprawling cacti grow 30–60 cm high; they have yellow flowers 6–8 cm across and fleshy, reddish purple fruits 4.5–7 cm long. The pads of *O. engelmannii* are markedly larger, up to 30 cm long and 22.5 cm broad, and are elliptic to orbicular-elliptic in shape with white to ashy gray spines distributed over most of the pad. The broadly obovate or nearly orbiculate pads of *O. phaeacantha* var. *major* are 12.5–25 cm long and 10–20 cm broad with dark brown spines usually restricted to the upper one-third to one-half of the pad. The haploid chromosome number is 33 (Pinkava and McLeod 1971).

In southern California, *O. littoralis* (Engelm.) Cockll. has more narrowly elliptic or obovate joints less than 16 cm wide. The spines are usually elliptic to circular rather than flattened in cross section. In northern Arizona and adjacent Nevada and California, *O. littoralis* var. *martiniana* (L. Benson) L. Benson is more glaucous than either *O. engelmannii* or *O. phaeacantha*. The glaucous, bluish joints of *O. macrorhiza* Engelm. are less than 10 cm long. Joints of *O. santa-rita* are purplish to lavender and circular. *Opuntia wilcoxii* Britton & Rose of southern Sonora has finely puberulent joints and a more upright habit.

Opuntia engelmannii was known for many years as *O. phaeacantha* var. *discata* (Benson 1982). Grant and Grant (1979b), arguing that var. *discata* "differs markedly" from var. *major*, elevated it to specific rank as *O. discata* Griffiths. More recently, Parfitt and Pinkava (1988) resurrected *O. engelmannii* as the correct name for this taxon. Benson (1982) recognized ten varieties of *O. phaeacantha:* of these, var. *discata* and var. *major* are the most common in our region. Because many of our records do not distinguish between these two, our map combines vouchers and sightings for both.

Where *O. engelmannii* and *O. phaeacantha* occur together, hybrid swarms are common. Although individual plants can be difficult to identify as one species or the other, "pure" strains can be readily distinguished. Introgression occurs among the contiguous varieties of *O. phaeacantha* and other species, as well (Parfitt and Clark 1978; Grant and Grant 1979b; Benson 1982). McLeod (1974) suggested that hybridization among members of the *O. compressa* Macbr. complex (diploid, n=11) and the *O. violacea* complex (tetraploid, n=22) gave rise to the *O. phaeacantha* complex (hexaploid, n=33). Benson (1982) described *O. littoralis* var. *martiniana* as a "connecting link" in a five-species complex that included *O. santa-rita.* He also noted that, in the Sonoran Desert, plants of *O. engelmannii*/*O. phaeacantha* may contain features of *O. macrorhiza.*

Both species are sympatric throughout much of their range and often can be found together. They grow in a variety of habitats, including plains, hillsides, washes, arroyos, and mesas. From central Texas westward, these two species most commonly grow in semidesert grassland and Madrean evergreen woodland. They reach their arid limits in the Sonoran Desert region. Most populations receive biseasonal rainfall, mainly in summer in the east and in winter in the west. Where summer moisture is scarce, *O. basilaris* replaces *O. engelmannii*/*O. phaeacantha.* To the south in Sonora, they give way to other platyopuntias, including *O. wilcoxii.*

Appearing in April and May, the flowers open around 8:00 A.M. and close about 8 hours later (Parfitt and Pickett 1980). Each lasts one day. New pads bearing fugacious leaves are produced at the same time as the flowers.

The flowers are self-compatible in part (Osborn et al. 1988). Flower visitors include a variety of solitary bees (among them Anthophoridae, Halictidae, and Megachilidae) and sap beetles (Nitidulidae). Of these, the large anthophorids appear to be the most effective pollinators. As they walk on the anther platform collecting pollen, one pollen-loaded leg is in continual contact with the stigma. The smaller bees and beetles touch the stigma much less frequently (Grant et al. 1979; Parfitt and Pickett 1980; Osborn et al. 1988).

As in all cacti, photosynthesis is by crassulacean acid metabolism (Szarek and Troughton 1976). The photosynthetic surface area of many cacti is so reduced that carbon dioxide uptake may be limited by the amount of photosynthetically active radiation that falls on the stems (Nobel 1981a, 1983b). In the Sonoran Desert, terminal pads of *O. engelmannii* face predominantly east-west, thus maximizing absorption of solar radiation during the summer rainy season (Nobel 1982b).

Carbon dioxide exchange rates vary according to temperature. In one study undertaken near the northeastern edge of the Sonoran Desert, carbon dioxide influx rates were greatest at low temperatures (November–January), while efflux rates were greatest at high temperatures (May and June) (Nisbet and Patten 1974). The authors concluded that carbohydrate reserves would probably be depleted by the high efflux rates if the plant could not absorb carbon dioxide at higher temperatures during summer than winter (Nisbet and Patten 1974). Seasonal temperature acclimation allows *O. engelmannii* to maintain a positive carbon balance over a relatively wide range of temperatures (Nisbet and Patten 1974).

Conde (1975), Ross (1982), and Rivera and Smith (1979) described the floral and stem anatomy.

The ripe fruits are eaten and the seeds

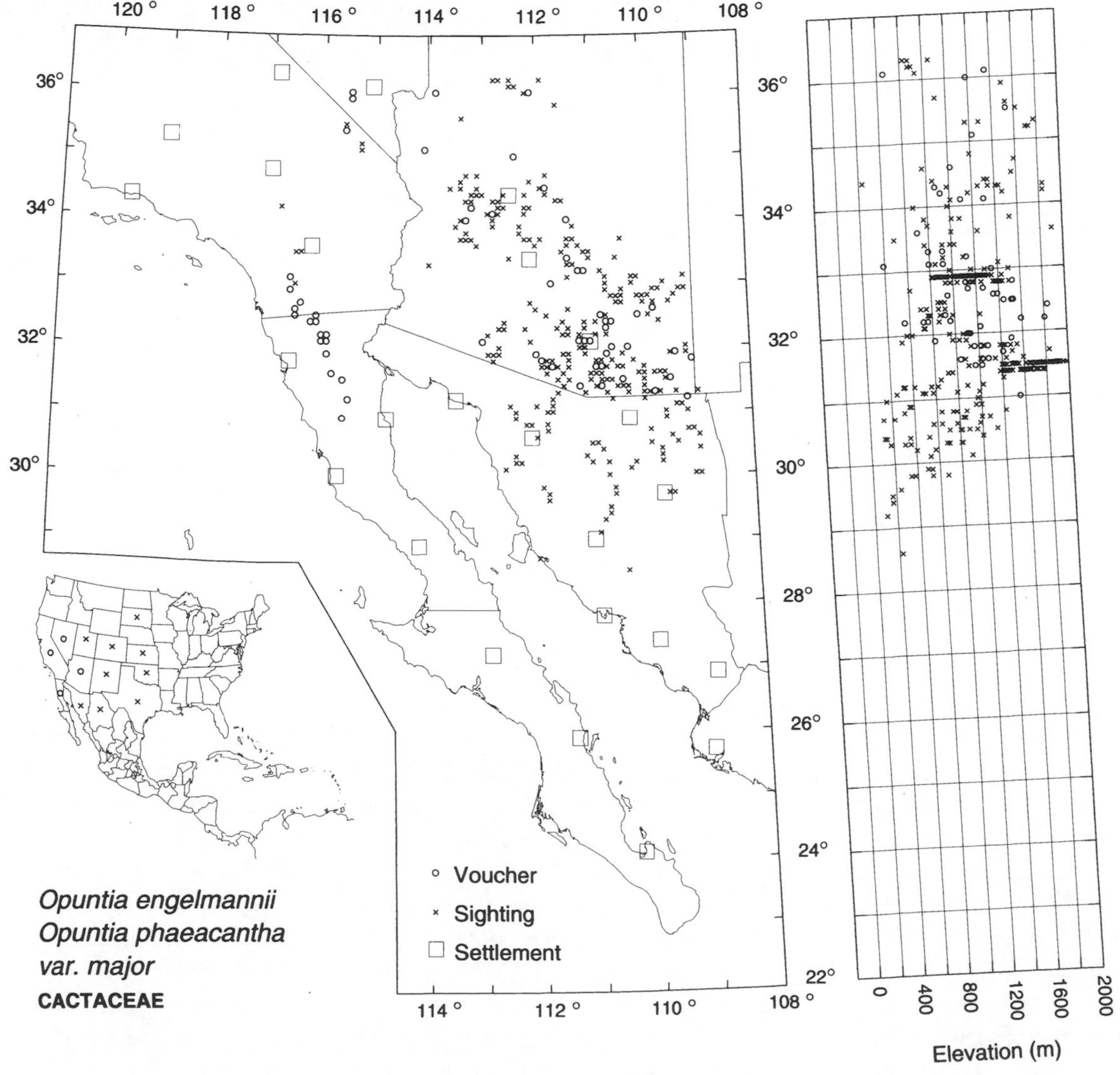

Opuntia engelmannii
Opuntia phaeacantha var. *major*
CACTACEAE

presumably dispersed by a wide variety of animals, including coyote (Short 1979), black bear, javelina, and desert tortoise. The seeds evidently pass unharmed through the guts of these and other fruit-eating animals.

Platyopuntias are among the most rapidly growing cacti in the Sonoran Desert (Shreve 1935). *Opuntia engelmannii* lives at least 30 years, perhaps longer (Shreve 1935). The homopteran *Dactylopius confusus* can cause severe damage to the pads (Gilreath and Smith 1988). Fire is an occasional cause of mortality. Burned plants often resprout but are more susceptible to insect and rodent damage (Bunting et al. 1980; McLaughlin and Bowers 1982).

The fruits were a dependable summer food for many native tribes. The Tohono O'odham devised an elaborate taxonomy based on fruit color, time of ripening, and storage qualities. The fruits are often eaten by bears. Javelinas selectively feed on pads with few spines and low concentrations of sodium oxalate crystals (Theimer and Bateman 1992). Heteromyid rodents rely on the seeds in summer (McLoskey 1983). Rotting stems support a variety of yeasts and fruit flies (*Drosophila* species) (Ganter et al. 1986).

Opuntia engelmannii and *O. phaeacantha* are common ornamentals in warm-desert regions. When grown as potential fodder plants, they responded to applications of nitrogen and phosphorus (Nobel et al. 1987).

Opuntia fulgida
Engelm.

Jumping cholla, chain-fruit cholla, cholla brincadora, velas de coyote

Opuntia fulgida is an arborescent cactus 1–3 (occasionally 4.5) m tall with a branching trunk 30–60 cm tall. The terminal joints, up to 15 cm long and 5 cm in diameter, detach readily. Spines of var. *fulgida* obscure the joint. Those of var. *mammillata* (Schott) Coult. are sparse. The white, pink, or magenta flowers are about 2 cm wide. Fruits are 2.5 cm in diameter and form chains of up to 20 fruits.

Opuntia bigelovii has a single trunk

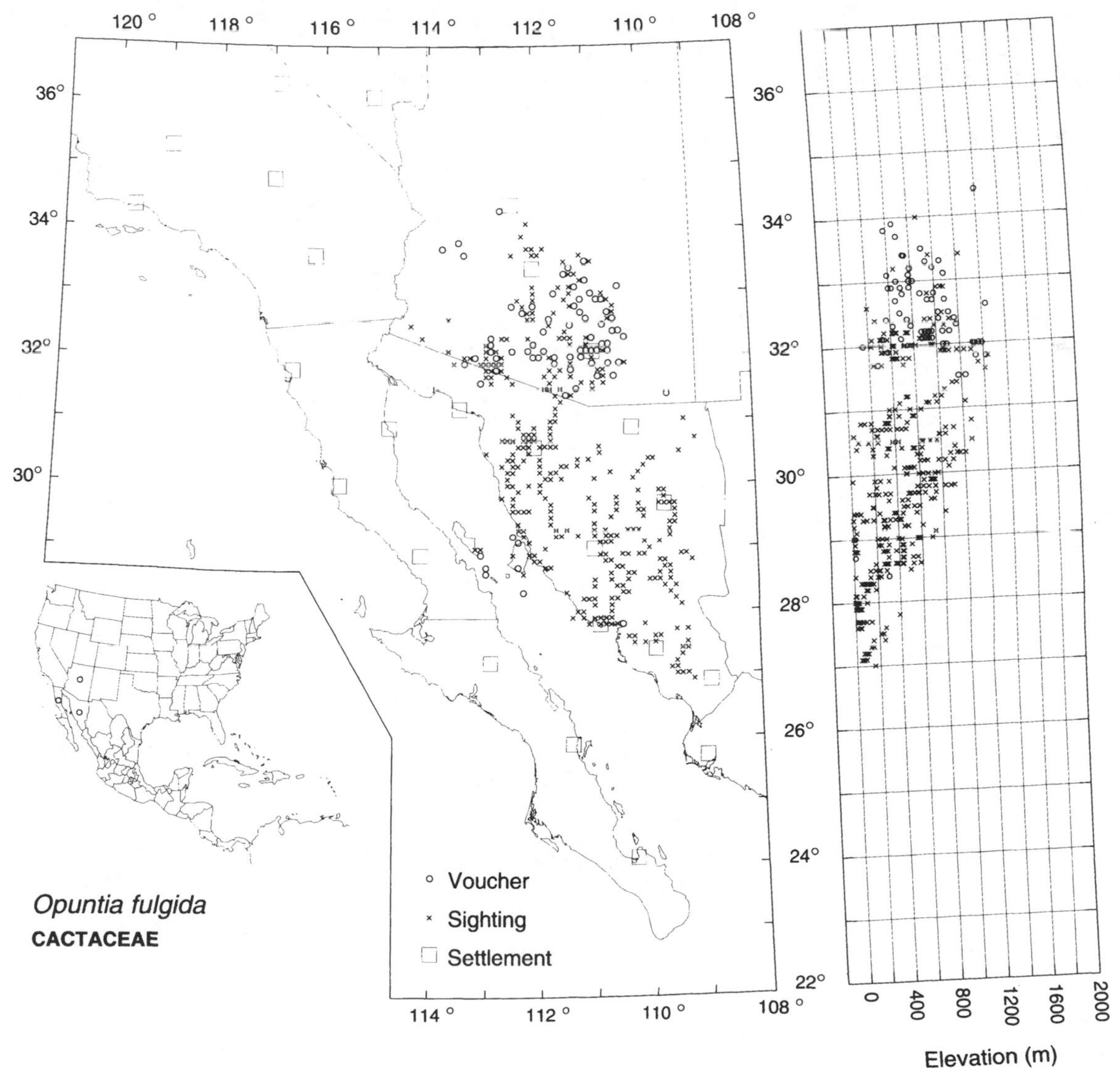

Opuntia fulgida
CACTACEAE

with short lateral branches. Its fruits do not form chains. *Opuntia cholla,* a close relative, is confined to the Baja California peninsula and Isla San Lorenzo.

Generally hexaploid (6n=66), *O. fulgida* hybridizes with the diploid *O. spinosior* (Engelm. & Bigel.) Toumey (2n=22) where they grow together (Gibson and Nobel 1986). The hybrid, a clonal microspecies, has been described as *O. kelvinensis* V. & K. Grant (Grant and Grant 1971) but is more accurately called *O.* × *kelvinensis* Baker & Pinkava (Baker and Pinkava 1987). Biologists have also identified diploid and triploid forms of *O. fulgida* (Pinkava et al. 1985).

This species is the mainland counterpart of *O. cholla.* Their ranges overlap on Isla San Lorenzo (Moran 1983b). *Opuntia fulgida* typically grows in even-age stands on level terrain, often in sandy or gravelly soil. Occasional plants can be found on steep, rocky hillsides, a habitat usually occupied by *O. bigelovii.* Varieties *mammillata* and *fulgida* often grow together. The range of the former is discontinuous and contained within that of the latter (Benson 1982). The two varieties are well marked and apparently maintain themselves largely through asexual reproduction (Benson 1982).

The range of *O. fulgida* covers the eastern Sonoran Desert where summer storms are relatively dependable. Progressive restriction to lower elevations toward the south seems to reflect a need for relatively open vegetation. Packrat middens first record *O. fulgida* in the Puerto Blanco Mountains, Arizona, about 3,200 years B.P. (Van Devender 1987). It may have reached much of its northern range only during the late Holocene.

Like many cacti, *O. fulgida* has a heteromorphic root system (Cannon 1911). The anchoring roots extend down about 35 cm; the absorbing roots, which take advantage of light rains, radiate up to 3 m at depths of 4–10 cm (Cannon 1911). Since *Opuntia* roots take up water only from wet soil (water potential above –1.0 MPa to –1.5 MPa), rapid absorption is a necessity (Szarek et al. 1973). The roots can be infected by symbiotic microflora (Bloss 1988).

Flowers appear from June to September and are vespertine and short lived (Grant and Grant 1971). Pollinators are most likely medium-sized native bees and the large carpenter bees (*Xylocopa* species) (Grant and Hurd 1979). Fertile fruits containing up to 200 seeds are produced occasionally (Johnson 1918), but sterile ones are more common (Benson 1982). Fruits tend to remain attached to the plant, and since new flowers develop from the areoles of old fruits, long chains can be formed after many years. Reproduction by seed is seldom if ever observed in nature (Tschirley 1963), but the fruits themselves sometimes take root after they fall (Tschirley 1963; Benson 1982). Fallen joints also take root and form new plants. Clearly, *O. fulgida* populations are maintained almost entirely by asexual reproduction (Tschirley 1963; Benson 1982). Accumulated joints beneath the plants provide microsites for establishment of *Mammillaria microcarpa* Engelm., *Echinocereus engelmannii* (Parry) Rümpler, and other small cacti (McAuliffe 1984b).

Joints that fall during the period from late winter to midsummer initiate shoots quickly, but those that fall during midsummer to late winter show shoot dormancy. This shoot dormancy may be a mechanism to evade extremely high soil temperatures during the summer (Tschirley 1963). Fallen joints remain physiologically active for only a few months (Holthe and Szarek 1985). They stop fixing carbon in early summer, a metabolic shift that may be coupled with root production (Holthe and Szarek 1985).

Tschirley (1963) obtained the best germination (58–61%) of cleaned, scarified seed at 25–30°C. In nature the seeds may need to pass through the digestive tracts of mammals or birds before they can germinate (Johnson 1918; Tschirley 1963), or the original dispersal agents may be extinct.

The natural life span has been estimated at 40 years (Tschirley and Wagle 1964) or 60–80 years (Shreve 1935). The reliance on asexual reproduction results in stands of relatively even age. As the oldest individuals in a senescent stand die, they are not replaced. A population may thus eventually eliminate itself from a given locality (Tschirley and Wagle 1964; Martin and Turner 1977). Explanations for local die-off have included prevalence of the bacterial pathogen *Erwinea carnegieana* (Tschirley and Wagle 1964) and the occurrence of infrequent severe freezes (Steenbergh and Lowe 1977).

Certain stands have relatively large, sweet fruits that compare favorably with those of domesticated platyopuntias such as *O. ficus-indica.* Seri Indians harvest these stands for food (Nabhan and Felger 1985). They also collect black gum nodules from the stem for food and medicine (Felger and Moser 1985).

Opuntia gosseliniana
Weber

Opuntia macrocentra
Engelm.

Opuntia santa-rita
(Griffiths & Hare) Rose

[=*O. violacea* Engelm. var. *violacea,* var. *gosseliniana* (Weber) L. Benson, var. *macrocentra* L. Benson, var. *santa-rita* (Griffiths & Hare) L. Benson]

Purple prickly pear, duraznilla

Opuntia gosseliniana and *O. macrocentra* are sprawling shrubs up to 1 m tall. *Opuntia santa-rita* is arborescent (up to 2 m tall) with a short trunk. The orbicular or obovate joints, 10–20 cm across, may be tinged with reddish purple. The few spines (usually 1–3 per areole) are clustered near the upper edge of the joint (*O. santa-rita*) or scattered over the entire surface (*O. gosseliniana* and *O. macrocentra*). They may be black, reddish brown, or pink. The yellow flowers are 6–7.5 cm across. At maturity, the fleshy fruits are red, lilac, or purple. The base chromosome number is 11 (Pinkava et al. 1977). Tetraploids are known (Pinkava et al. 1992).

Opuntia chlorotica has bluish joints densely armed with yellow, downward-pointing spines. *Opuntia basilaris* has obovate to spatulate joints and occurs farther west.

Of the five varieties of *O. violacea* that Benson (1982) recognized, four grow in the area mapped in this atlas: var. *violacea*, var. *gosseliniana*, var. *macrocentra*, and var. *santa-rita*. Pinkava and Parfitt (1988) considered var. *gosseliniana*, var. *macrocentra*, and var. *santa-rita* to be distinct at the species level and var. *violacea* to be unworthy of recognition. Because many of our records were identified only as *O. violacea*, our map combines vouchers and sightings of all four taxa.

In Arizona, *O. macrocentra* and *O. santa-rita* are more typical of semidesert grassland than Sonoran desertscrub (Benson 1982). The populations on the gulf coast of Sonora (*O. gosseliniana*) are evidently better adapted to arid climates (Benson 1950; Felger and Moser 1985). Lack of summer rain may keep *O. gosseliniana* from penetrating into northwesternmost Sonora or into the southwestern corner of Arizona.

Most populations bloom between March and June. Peak flowering is in April. Occasional plants flower in the summer (July–August), probably in response to rains. The reddish purple color of the pads, imparted by betalain pigments, intensifies during drought or cold.

Seri Indians ate the fruits and made face paint from them (Felger and Moser 1985). The plants are commonly cultivated as ornamentals.

Opuntia invicta
Brandegee

This caespitose cactus grows 2–4.5 dm tall and forms clumps up to 2 m across (figure 64). The low, branching stems are obovoid

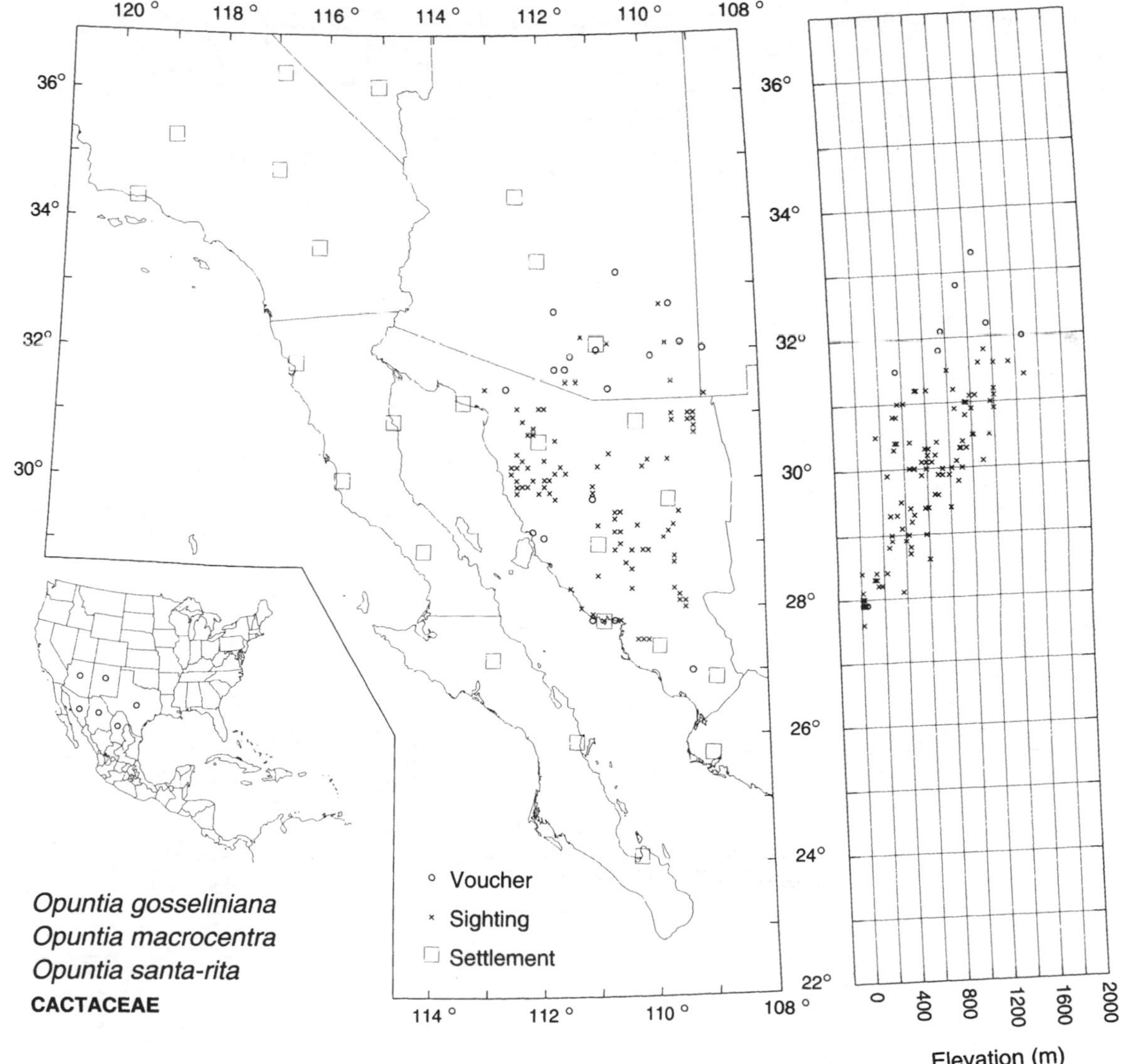

Opuntia gosseliniana
Opuntia macrocentra
Opuntia santa-rita
CACTACEAE

to clavate and are strongly tuberculate. The grayish spines (1–5 cm long, 10–25 per areole) are rigid and may be spreading or deflexed. Flowers are yellow and 4–6 cm wide. The spiny, obovoid fruits are somewhat fleshy at maturity. The diploid number of chromosomes is 22 (Yuasa et al. 1974).

The resemblance of this plant to *Echinocereus brandegeei* (Coult.) K. Schum. is striking. *Echinocereus brandegeei* has a

Figure 64. Opuntia invicta *near Las Cruces, Baja California Sur. (Photograph by J. R. Hastings.)*

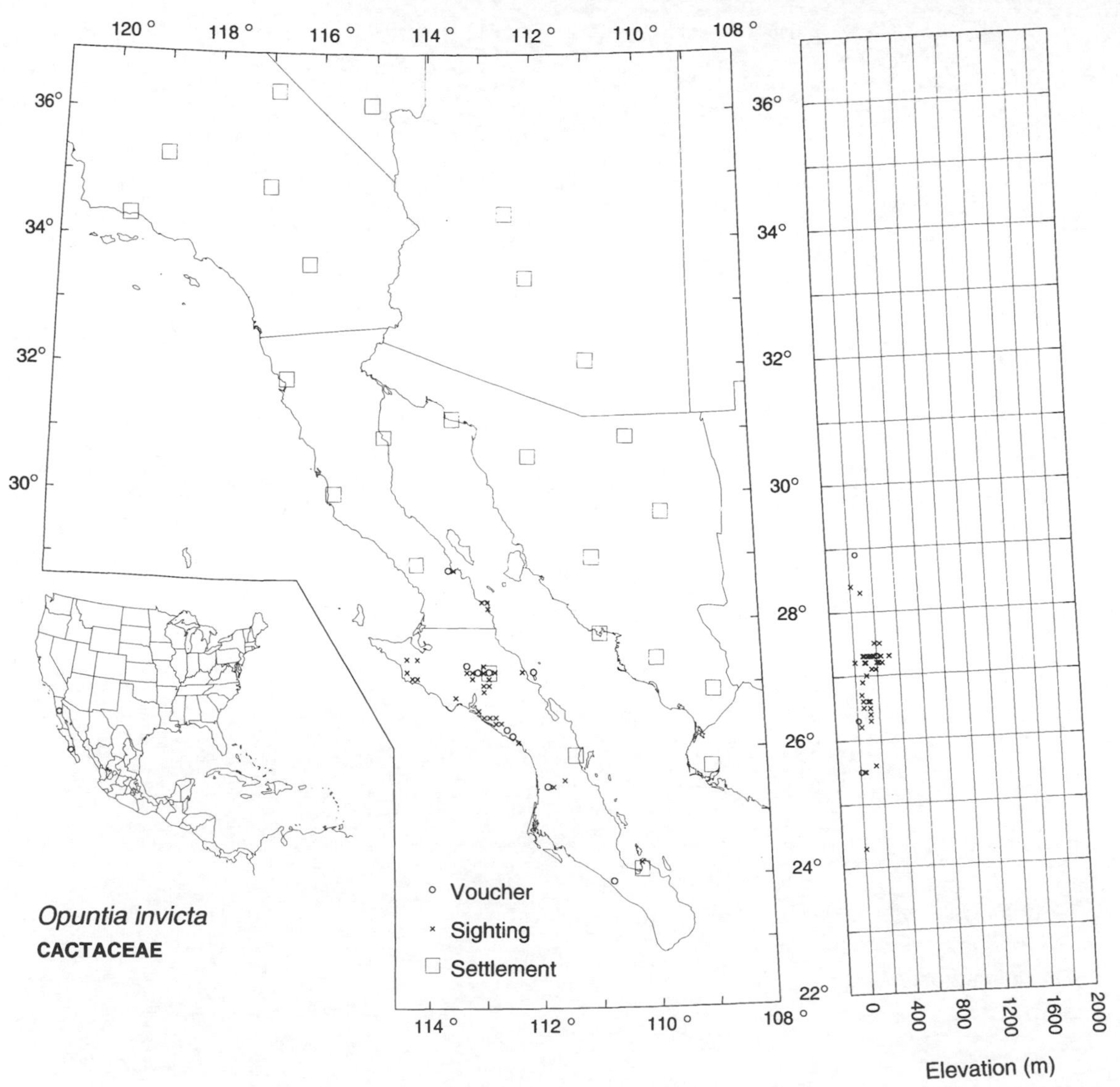

paler epidermis, more ribs, fewer tubercles, and narrower, less flattened spines. Of the other peninsular cylindropuntias, only *O. santamaria* (Baxter) Wigg. of Isla Magdalena simulates *E. brandegeei.*

Found in sandy or silty soil on coastal plains or among rocks on bajadas and mesas, *O. invicta* is most frequent on the southern Vizcaíno Plain, with outlying populations to the north and south. Reid Moran vouchered the northernmost population at Bahía de los Ángeles in 1959. The southernmost voucher represents a 1931 collection by Ira L. Wiggins. The species occurs in arid, relatively humid coastal areas where rainfall is biseasonal and erratic.

Flowers appear in April and May (Wiggins 1964) and, rarely, in October. This species contains beta-phenethylamines, a class of alkaloids (Meyer et al. 1980).

Opuntia leptocaulis DC.

Desert Christmas cactus, pipestem cactus, tesajo, tasajillo, agujilla

A bushy cactus 0.5–1 m tall, *Opuntia leptocaulis* often spreads to a meter or more across. The long, cylindrical joints, 3–6 mm in diameter, have short (less than 1 cm) or long (2.5–5 cm) spines. These glabrous stems are yellow-green in color. The greenish yellow, cream, or bronze flowers are 1–1.5 cm wide. The spineless fruits, about 12 mm long, are bright red and fleshy when ripe. Sonoran Desert plants are diploid or triploid, with a base number of 11 (Pinkava et al. 1977, 1992). Chihuahuan Desert plants include diploids and tetraploids (Pinkava et al. 1977, 1992).

Opuntia tesajo, a peninsular endemic, has puberulent or minutely scurfy joints and, usually, a more erect habit. The joints of *O. leptocaulis* detach readily; those of *O. tesajo* do not. *Opuntia ramosissima* differs in its prominent, diamond-shaped tubercles, its more numerous spines, and its dry, spiny, burlike fruits. Stems of *O. ar-*

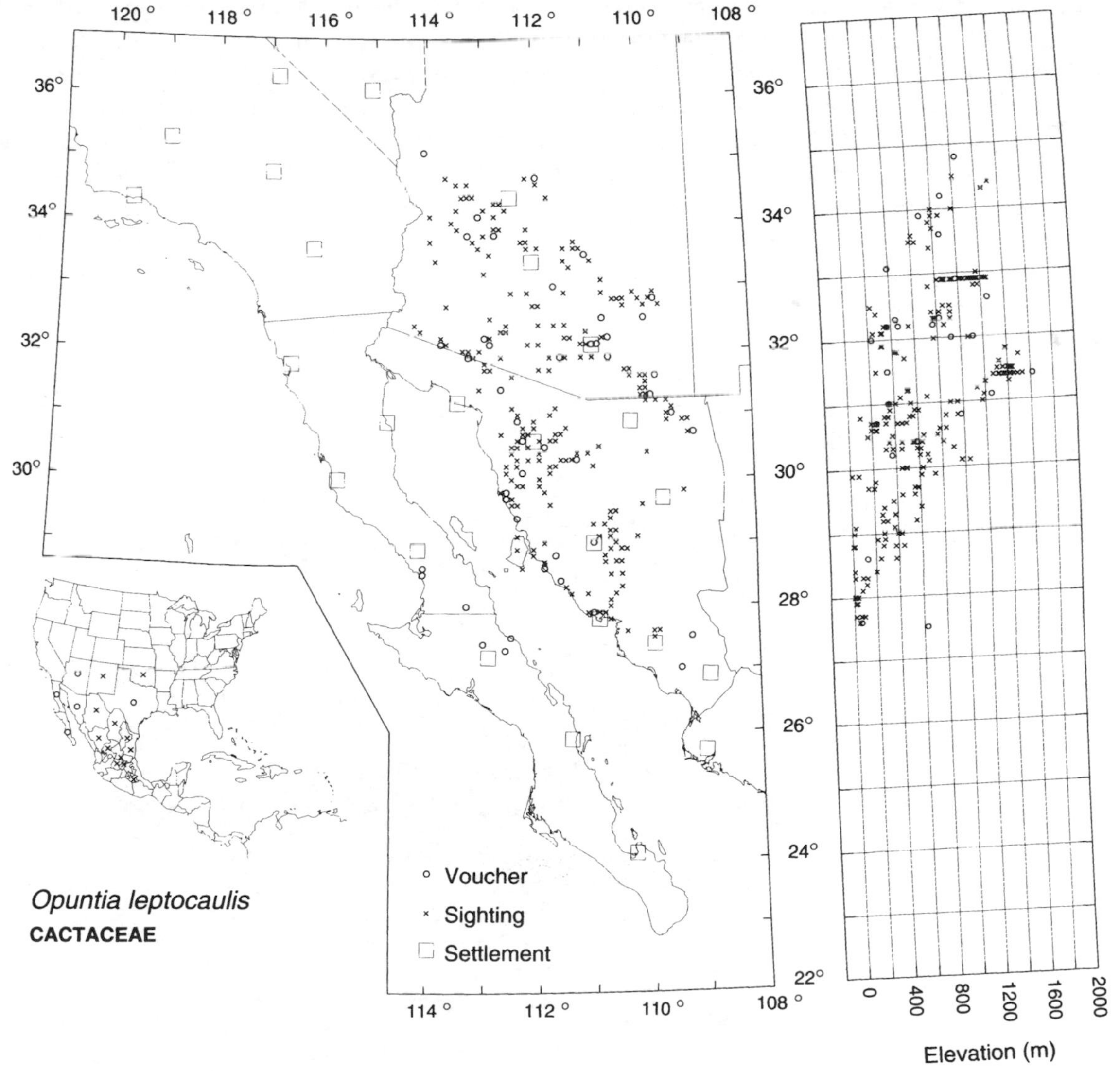

Opuntia leptocaulis
CACTACEAE

buscula are about twice as thick (6–9 mm). Moreover, *O. arbuscula* has a distinct trunk and resembles a miniature tree. Stems of *O. kleiniae* var. *tetracantha* (Toumey) W. T. Marshall are thicker still, 0.5–1 cm in diameter.

The short- and long-spined forms have been described as different varieties. According to Benson (1982), the present meager information does not justify varietal segregation. *Opuntia leptocaulis* hybridizes with *O. kleiniae,* producing plants with intermediate characters (Pinkava and Parfitt 1982).

Conde (1975) described the anatomy and compared it with that of other cylindropuntias.

Sometimes forming impenetrable thickets, *O. leptocaulis* grows on gentle hillslopes, mesas, plains, valleys, and bottomland washes. Four very narrow-stemmed cylindropuntias occur in the Sonoran Desert (*O. leptocaulis, O. ramosissima, O. tesajo,* and *O. cineracea* Wiggins), and although their ranges overlap to some extent, they appear to form a geographic replacement series. *Opuntia tesajo* is limited to the central Baja California peninsula. *Opuntia cineracea* is restricted to the San Felipe Desert in northeastern Baja California. *Opuntia ramosissima* is primarily a plant of the Mojave and Colorado deserts. *Opuntia leptocaulis,* the most widely distributed of the group, is found across the northern and eastern Sonoran Desert and eastward across the Chihuahuan Desert into central Texas. The peninsular populations may belong to an undescribed species (Mark Baker, personal communication 1989).

The scarcity of *O. leptocaulis* west of the Colorado River shows a need for fairly substantial summer rain. In the eastern parts of their range, the plants experience frequent winter freezes, so it is unlikely that frost significantly influences their upper elevational or northern limits. The southward restriction to low elevations probably represents exclusion from denser vegetation.

Packrat middens from the Puerto Blanco Mountains, Arizona, do not record *O. leptocaulis* until about 1,000 years B.P. (Van Devender 1987).

In the Sonoran Desert, *O. leptocaulis* flowers in May and June and occasionally again in summer, perhaps in response to summer rain (Shillington and Yang 1975). Fruits mature slowly throughout summer and fall and persist for months after ripening.

Because the flowers open late in the afternoon, hawk moths (Sphingidae) would seem to be likely pollinators. Simpson and Neff (1987), however, reported that moths are rarely observed at the flowers. Frequent visitors include hummingbirds, *Apis mellifera,* and *Diadasia rinconis,* all capable of transferring pollen (Simpson and Neff 1987).

According to Yeaton (1978), *O. leptocaulis* and *Larrea tridentata* in the northern Chihuahuan Desert succeed one another cyclically. The underlying mechanism of this phenomenon, which may be relatively local, needs further study.

Opuntia leptocaulis propagates from fallen branches and fruits. Early survival is relatively high (Goldberg and Turner 1986), perhaps an effect of the nurse-plant phenomenon. Seedlings also benefit from nurse plants. Individuals are short lived for shrubs, probably having a life span of about 50 years (Goldberg and Turner 1986). The plants resprout only slightly after burning, and fire results in high mortality (Bunting et al. 1980).

Seri Indians harvested the fruits and ate them fresh after removing the glochids (Felger and Moser 1985). Everitt and Alaniz (1981) analyzed the nutritional content of the fruits.

Opuntia molesta

Brandegee

This openly branched shrub grows 1–2 m tall; it has cylindrical green joints that do not detach readily and flattened spines 2–5 cm long (figure 65). The purplish bronze flowers are 3–5 cm wide. Yellow at maturity, the fleshy fruits are 2.5–3.5 cm long and deeply excavated at the apex.

The readily detachable joints of *O. cholla* are sparsely armed. *Opuntia ciribe* joints are nearly hidden by short, dense spines.

Wiggins (1964) considered *O. clavellina* Engelm. ex Coult. and *O. calmalliana* Coult. to be synonymous with this species. We do not, and our map shows *O. molesta* in its narrow sense.

The plants are locally common among

Figure 65. Close-up of Opuntia molesta *near Punta Prieta, Baja California Sur. (Photograph by J. R. Hastings.)*

boulders and on gravelly plains and hillsides. Their range is nearly confined to Shreve's (1964) Vizcaíno Subdivision of the Sonoran Desert. The climate of this area is strongly influenced by Pacific maritime air, which moderates temperature extremes, especially in summer. Rain falls mostly during winter, but occasional late-summer storms occur in the western part of the range. Like many cacti, this one is poorly represented in herbaria. Most of the sightings on our map are Raymond M. Turner's.

The plants bloom in April and May (Wiggins 1964).

Along with four other cylindropuntias (*O. bigelovii, O. cholla, O. echinocarpa,* and *O. prolifera*), this species is host for a cerambycid boring beetle, *Moneilema michelbacheri* (Blom 1987).

Opuntia prolifera

Engelm.

Coastal cholla

Opuntia prolifera is a shrubby or arborescent cactus 1–2.5 m tall with dark green, cylindrical joints 3–5 cm thick and 3–15 cm long. Older stems are woody and up to 10 cm thick. Younger stems are armed with interlocking, reddish brown to yellowish or gray spines 8–25 mm long. The flowers are rose to red-purple. The fleshy green fruits, spineless at maturity, form short, persistent chains. Usually they lack seeds. The base chromosome number is 11. Diploid, triploid (Yuasa et al. 1974), and hexaploid (Pinkava et al. 1992) individuals are known.

Joints of *O. ciribe* are thinner (1.5–3.5 cm across) and are light green or blue-green and glaucous. *Opuntia cholla* has fewer spines on its somewhat thicker stems.

A thicket-forming cactus of coastal bluffs, valley bottoms, and low hills, *O. prolifera* grows from southern California to the northern Magdalena Plain as well as on several islands off the Pacific coast of California and Baja California. Like *Agave*

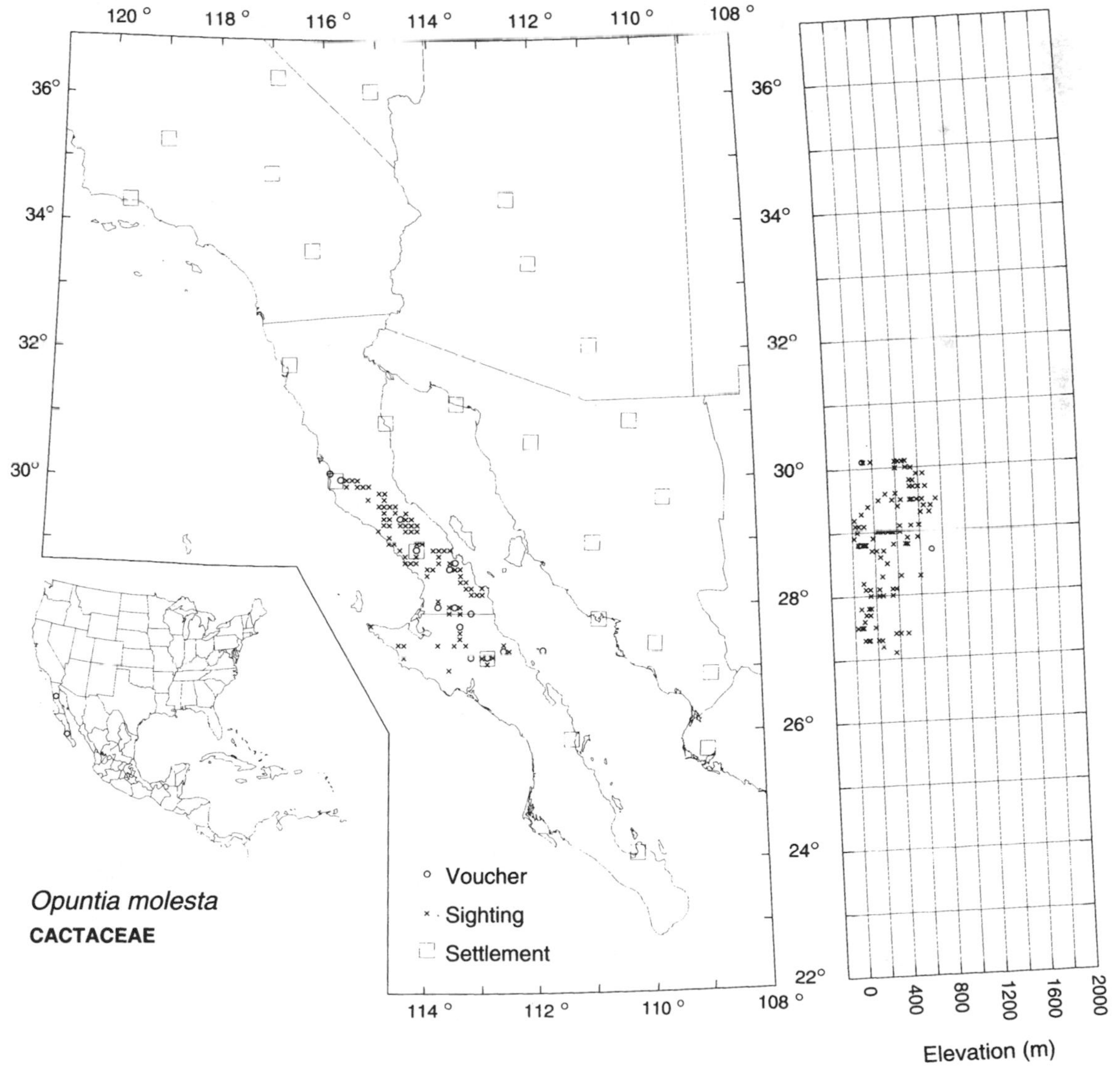

Opuntia molesta
CACTACEAE

shawii and *Bergerocactus emoryi,* this species spans the transition from Californian coastalscrub to Sonoran desertscrub in Baja California. The plants thrive under a winter rainfall regime, largely avoiding the parts of the peninsula where rain is limited to the warm season. Throughout the range, Pacific maritime air provides relatively cool summers.

The joints detach readily. Fallen joints and fallen fruits both take root and produce new plants. Stands of *O. prolifera* are typically very dense, with the largest, oldest plants surrounded by a thicket of offshoots (Benson 1982). Sexual reproduction is evidently unknown in this species. The few seeds that are produced result from apomixis and involve no genetic recombination (Benson 1982).

The flowering period is April–July (Wiggins 1964), occasionally again in September–October.

Opuntia ramosissima

Engelm.

Diamond cholla

Opuntia ramosissima is a bushy cactus up to 0.6 m tall (rarely 1.5 m) of variable habit, either sprawling or erect. The narrow, pencil-like terminal joints are about 6–10 mm in diameter and 5–10 cm long, with flattened, diamond-shaped tubercles. The spines, 4.5–5 cm long, have a loose, papery, yellow sheath. The greenish yellow to apricot to reddish brown flowers are about 1–2 cm across. Fruits are usually dry, spiny, and burlike, occasionally spineless. The base chromosome number is 11 (Pinkava et al. 1973). Both diploids and tetraploids are known (Pinkava et al. 1985).

Opuntia leptocaulis, O. arbuscula, and *O. tesajo* all have fleshy fruits and lack diamond-shaped tubercles.

Apparently, the closest relative of *O. ramosissima* is *O. anteojoensis* Pinkava, a local endemic of the central Chihuahuan Desert (Pinkava 1976).

The plants grow in fine or sandy soils

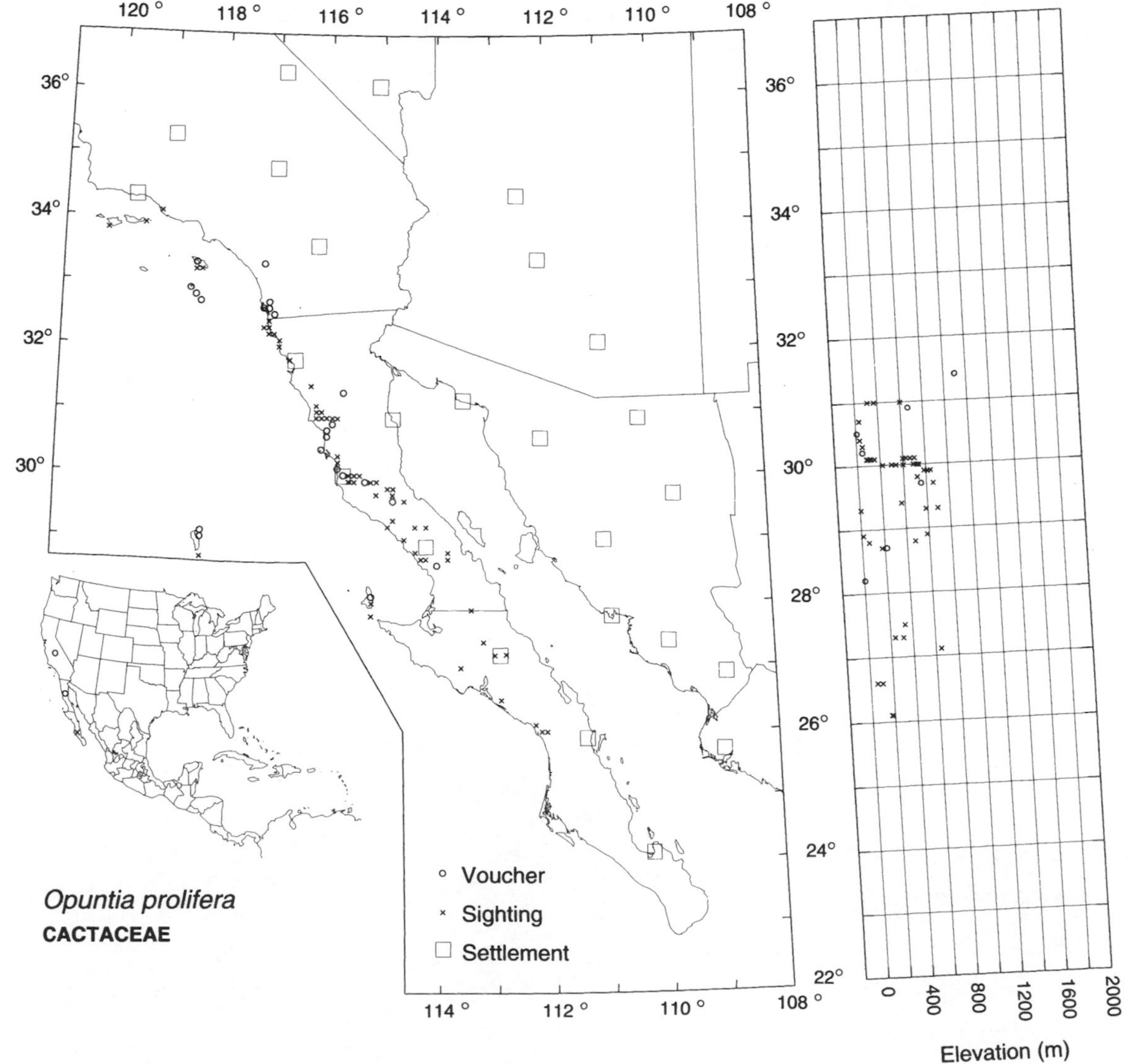

Opuntia prolifera
CACTACEAE

along washes and on flats. Although the distribution of *O. ramosissima* overlaps somewhat with that of *O. leptocaulis* and *O. tesajo,* the three are largely allopatric. *Opuntia ramosissima* is somewhat frost tolerant. Extensive tissue damage does not develop until temperatures drop to –4°C or below (Nobel 1982a). The species thrives in the absence of summer rain. Its rather abrupt eastern limit is not determined by any geographical feature. The elevational profiles strongly suggest that interactions at the arid limit of *O. leptocaulis* have determined the eastern boundary. Interactions with *O. tesajo* may shape the peninsular limit.

Packrat middens record *O. ramosissima* in the Specter Range, Nevada, near its present northern limit, about 11,700 years B.P. The species reached its present southern limit in Sonora by about 6,100 years B.P. (Van Devender, Burgess et al. 1990).

The plants flower in April and May and may retain their fruits until the following spring.

Opuntia rosarica

Lindsay

Opuntia rosarica is a low, bushy cactus up to 1 m tall and wide with cylindrical joints and brown spines 1–3.5 cm long. The tubercles often merge to form undulate, vertical ribs. The flowers, 3–3.5 cm wide, are yellow but may be tinged with red or pink and are plentiful at the branch tips. The dry, spiny fruits are 1–1.5 cm long. The diploid number of chromosomes is 22 (Yuasa et al. 1974).

This is the only cholla in its range with confluent tubercles. The yellow flowers and brown spines are also distinctive.

Rare to occasional on hills, mesas, and valley floors, *O. rosarica* is narrowly endemic to Baja California, where it grows in the transition zone from Sonoran desertscrub to Californian coastalscrub (Wiggins 1964). Other transition-zone endemics include *Rosa minutifolia* Engelm. in Parry (Shreve 1936) and *Agave moranii.*

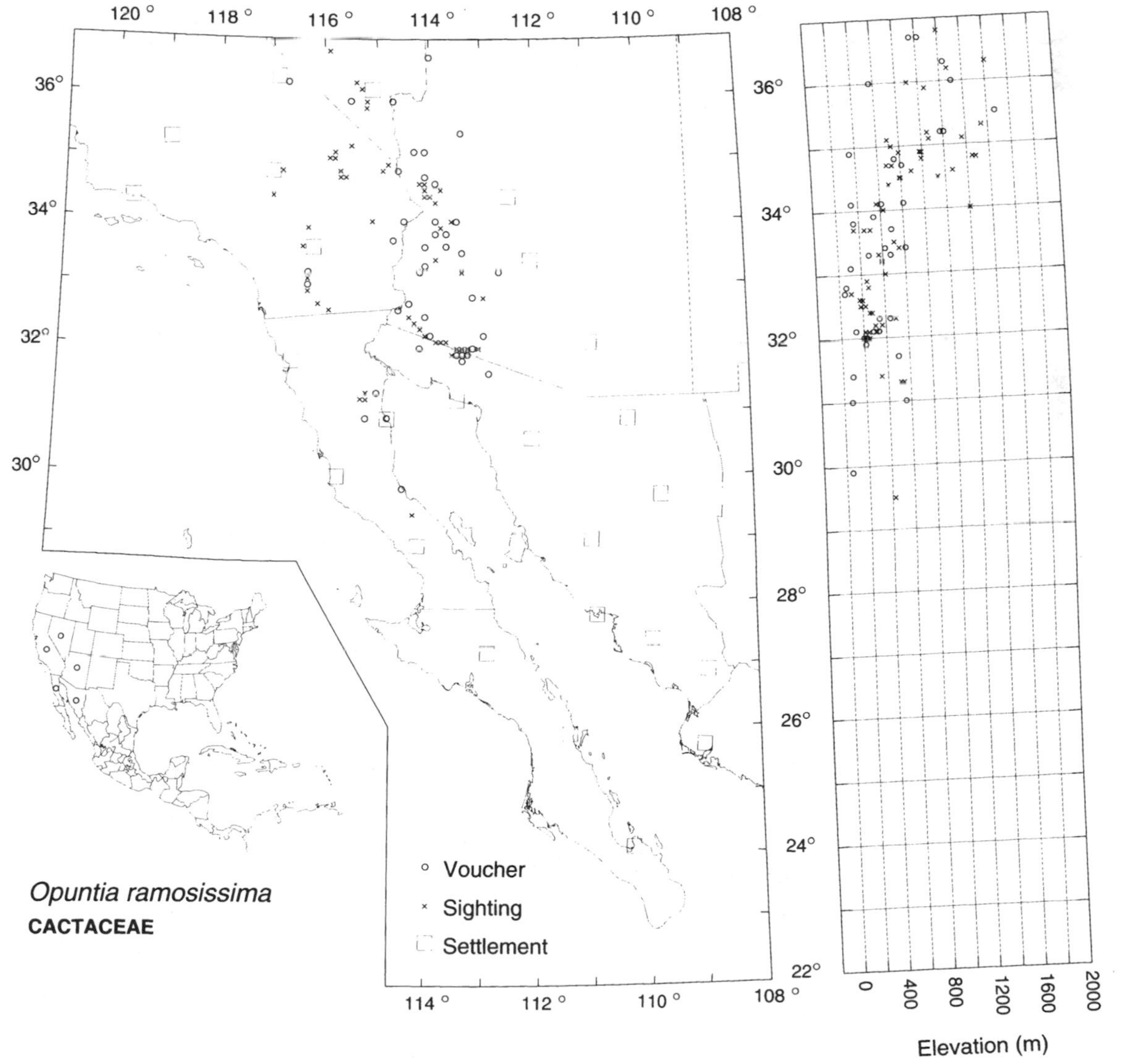

Opuntia ramosissima
CACTACEAE

This species is probably under the strong influence of Pacific maritime climates throughout its range. Near El Rosario, temperatures are moderated by maritime air, and rain falls almost entirely in winter. Toward the gulf and at higher elevations, late summer storms are more common, but winter remains the season of the most dependable soil moisture.

Flowers appear from June through August (Wiggins 1964).

Opuntia tesajo
Engelm. ex Coult.

Pencil cholla, tesajo

Opuntia tesajo (figure 66) is a much-branched cactus up to 1.5 m tall with slender cylindrical joints. The terminal joints, 2–5 (occasionally 10) cm long and 7–10 mm in diameter, are puberulent or minutely scurfy and gray-green. The sparse, needlelike spines, up to 6 cm long, are gray or purple with yellow tips and straw-colored spine sheaths. The yellow flowers are 2 cm wide. The fruits, fleshy and bright red at maturity, are 2–2.5 cm long.

Opuntia leptocaulis has longer, spineless obovoid fruits and yellow-green, glabrous stems. *Opuntia ramosissima* has diamond-shaped tubercles. *Opuntia cineracea* Wiggins has larger flowers (3–3.5 cm) and white spine sheaths and is restricted to the San Felipe region in northeastern Baja California.

Occasional to common along washes and on hillsides, often scrambling among other plants, *O. tesajo* is endemic to the Baja California peninsula. It is apparently much less common in the southern half of the peninsula than in the northern half.

The flowers appear in May and June.

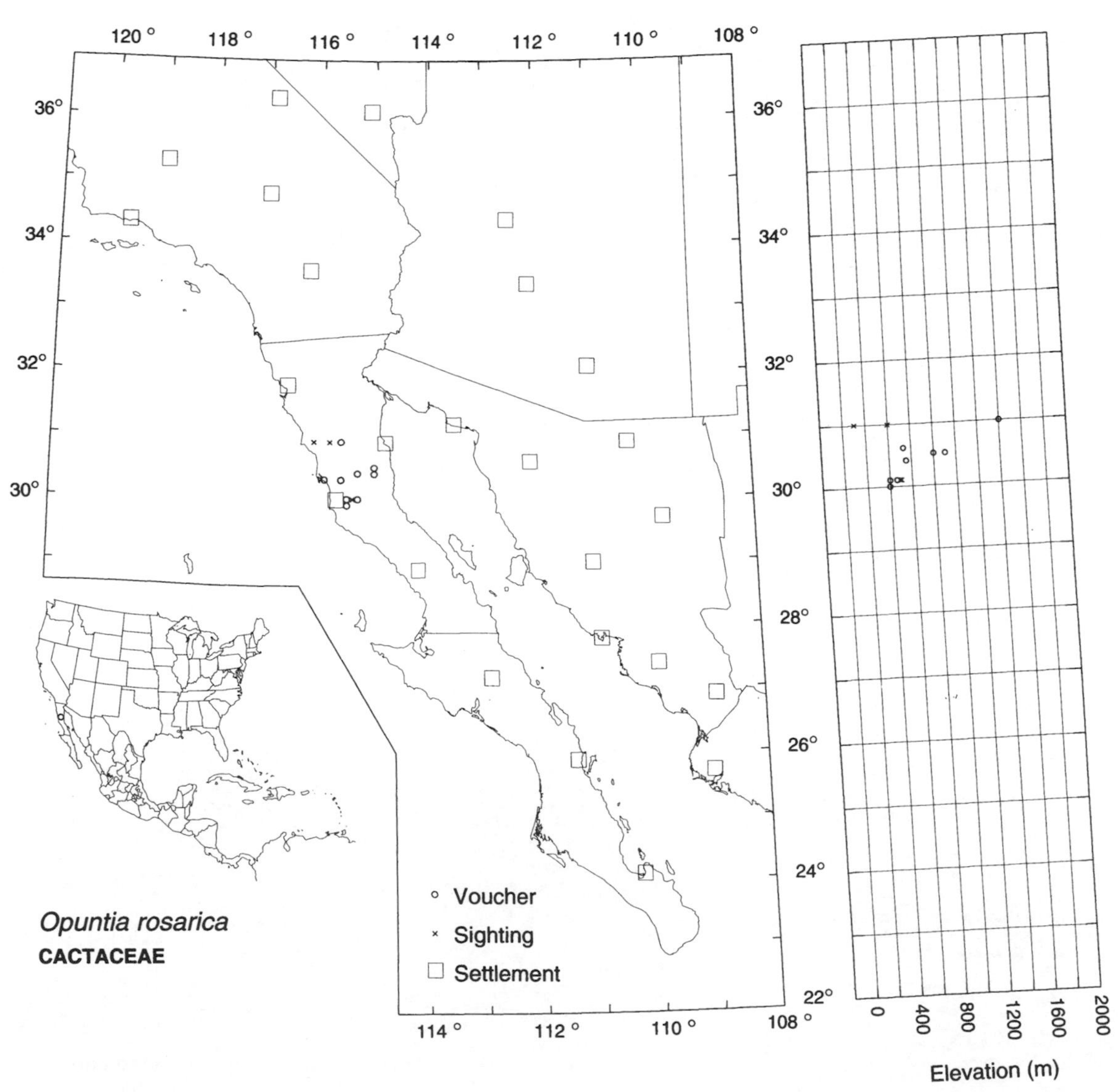

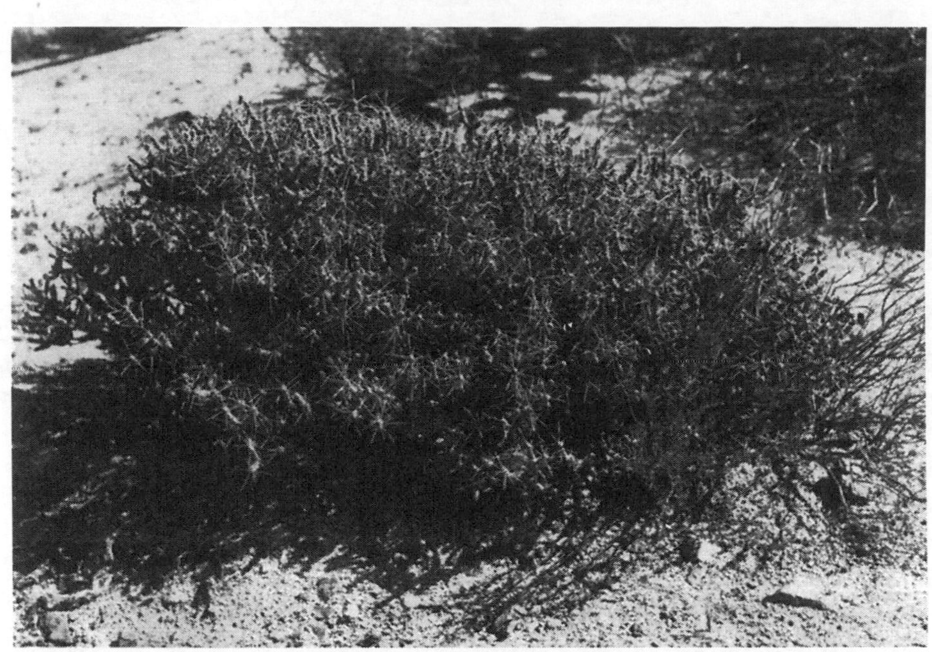

Figure 66. Opuntia tesajo *south of Cataviñá, Baja California. (Photograph by J. R. Hastings.)*

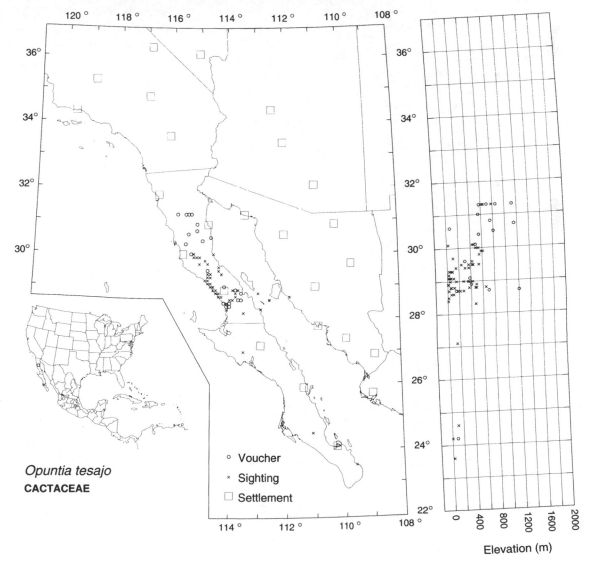

Opuntia tesajo
CACTACEAE

○ Voucher
× Sighting
□ Settlement

Elevation (m)

Pachycereus pecten-aboriginum

(Engelm.) Britton & Rose

Hairbrush cactus, hecho, cardón barbón, cardón espino, cardón hecho

Pachycereus pecten-aboriginum (figure 67) is an arborescent, columnar cactus up to 12 m tall with a slender trunk 1.2–5 m tall. Each of the many erect or ascending branches, 1.5–2.5 dm in diameter, has 10–12 ribs. The spines, up to 1 cm long on mature plants, sometimes much longer on young ones, are in clusters of 8–12. The whitish flowers, 5–7.5 cm long, are produced within 1 m of the branch tips. Dense tan or brown bristles up to 6 cm long make the fruits burlike. Exclusive of bristles, the fruits are 6–7.5 cm across. The haploid chromosome number is 11 (Pinkava et al. 1977).

Pachycereus pringlei is a more massive plant with a stout trunk and thicker branches. Its fruit bristles are generally less than 1 cm long. *Carnegiea gigantea* also tends to be more massive and has thicker branches (3–6.5 dm in diameter). *Stenocereus thurberi* has numerous stems arising from the base or from a short trunk. Neither *Carnegiea* nor *Stenocereus* has burlike fruits.

Figure 67. Pachycereus pecten-aboriginum *near Mazatán, Sonora. (Photograph by J. R. Hastings.)*

Pachycereus pecten-aboriginum is a plant of Sinaloan thornscrub that reaches the southernmost part of the Sonoran Desert. At the northern end of its range, the plants are most abundant on silty flats and gentle slopes (Wiggins 1964). Farther south, they often dominate Sinaloan thornscrub on hills and canyon slopes as well as on flats (Gentry 1942).

The northernmost point on our map (29.9°N) represents a sighting by Raymond M. Turner from the Río San Miguel drainage. Although low temperatures at the stem apex apparently limit the northward movement of *C. gigantea* and *S. thurberi, P. pecten-aboriginum* may not be similarly constrained (Nobel 1980c). Decreasing summer precipitation likely plays a crucial role in setting the northern boundary on the peninsula and on the mainland.

The plants bloom from January into late spring. Fruits ripen in June and July (Krizman 1972). Sometimes old fruits persist on the plants until the following spring. Seeds germinate during the summer rains (Krizman 1972). Bats pollinate the flowers (Ted Fleming, personal communication 1993). Hummingbirds and woodpeckers are also common flower visitors (Gentry 1942).

As the surrounding plant community becomes taller, *P. pecten-aboriginum* is increasingly shaded and receives less photosynthetically active radiation. Taller stems would circumvent this problem, and, in fact, the average height of *P. pecten-aboriginum* more than doubles between 28°N and 25°N, as does the surrounding vegetation (Nobel 1980a).

Gentry (1942) reported that Indians of the Río Mayo watershed made jam from the fruit pulp and ground the seeds into an oil-rich paste. Use of the bristly fruits as combs is the source of the specific epithet *pecten-aboriginum* (Standley 1924).

Pachycereus pringlei

(S. Watson) Britton & Rose

Cardón, sahuaso, cardón pelón

This cactus is the most massive one in the Sonoran Desert, reaching heights of 15–20 m and producing stems up to 1.5 m in diameter (figure 68). In older plants the branches often surpass the main axis in height. The stems have 10–15 vertical ribs. The spines, 20–30 per cluster, are stout and short (1–3 cm) on older plants, up to 12 cm long on younger ones. The large, white flowers are 6–8 cm long. The fruits are globose, about 5 cm in diameter and covered at first with tawny, felty areoles. At maturity, they have closely spaced, slender, yellow spines. The haploid chromosome number is 22 (Pinkava et al. 1973).

Pachycereus pringlei is readily told from *Carnegiea gigantea* even at a distance because it branches nearer the ground and its branches rise at a more acute angle. Also, *Carnegiea* has more (12–25) ribs.

This cactus hybridizes with *Bergerocactus emoryi* where their ranges meet (Moran 1962b).

Pachycereus pringlei is widespread throughout most of the Baja California peninsula south of 31.2°N. It is present on all the gulf islands, where it has undergone considerable differentiation (for example, dwarfed plants on Isla San Pedro Mártir [Moran 1968] and ground-level branching on Isla Santa Cruz [Cody et al. 1983]). We have found it in Sonora south almost to Ciudad Obregón (27.6°N). Gibson and Horak (1978) reported it from as far south as Nayarit. In Sonora, most plants grow within 50–75 km of the coast in a narrow, nearly uninterrupted band. The colony near Pitiquito, formerly thought to be isolated (Shreve 1964), is actually connected to the coastal plants through populations on Sierra Alamo, Sierra Aquituni, Cerro San Clemente, Sierra del Viejo, and Sierra Picú.

The cactus ranges from sea level up to 950 m. It grows mainly in areas where warm-season rainfall predominates, although in central Baja California it is

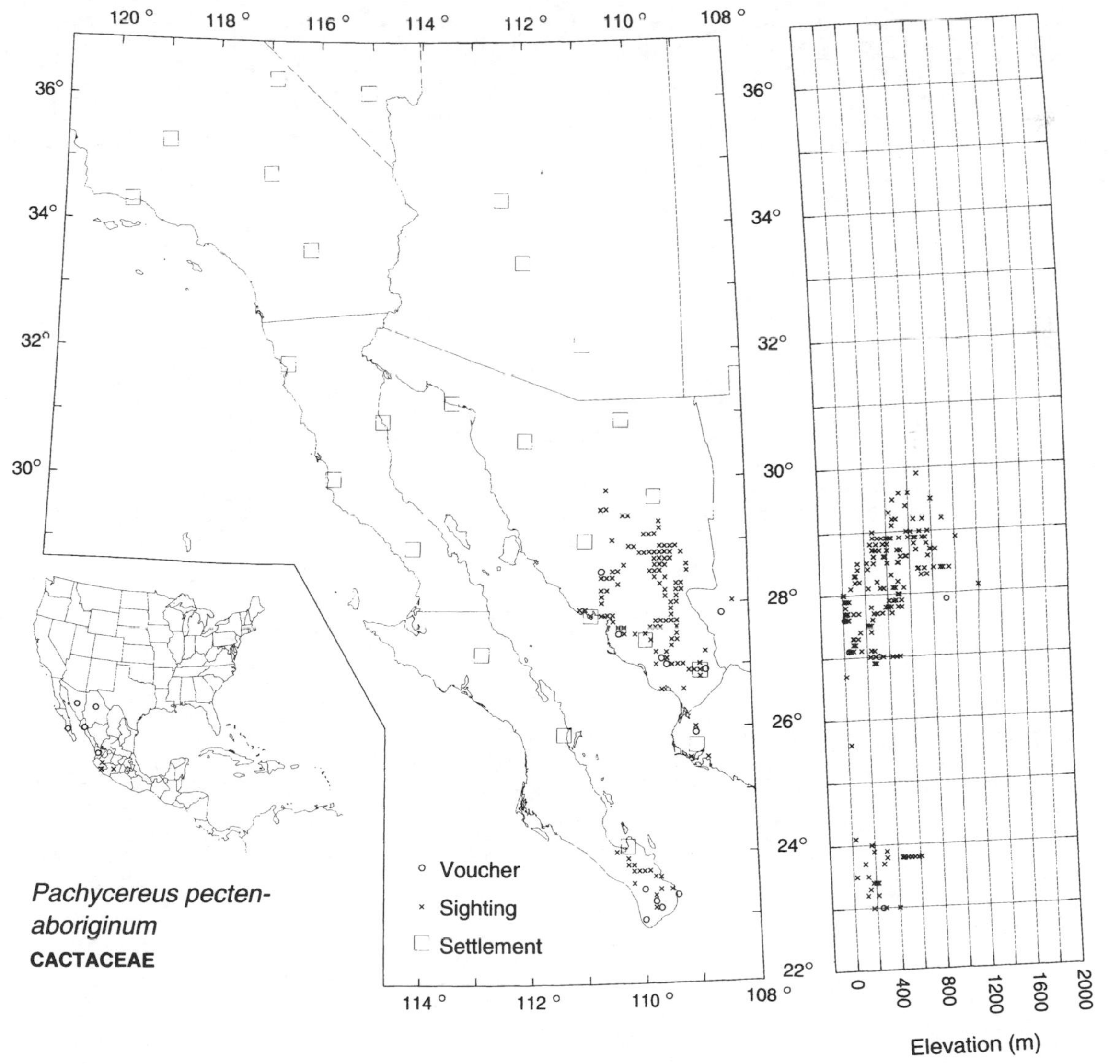

Pachycereus pecten-aboriginum
CACTACEAE

found in areas of mainly winter rainfall. In Tucson, Arizona, cultivated plants are often damaged by frost, and freezing almost certainly sets the northern limits.

Flowers appear from March through July (Shreve 1964; Moran 1968; Felger and Moser 1985; Cancino et al. 1993), mostly on the east, south, and west sides of the plants (Moran 1968) and generally within 30–40 cm of the apex (Shreve 1964). They open in late afternoon or evening and close about the middle of the next day. Bats, birds, and insects pollinate them (Moran 1962c). Near La Paz, Baja California Sur, ripe fruits remain on the plants from May to September (Cancino et al. 1993). The ripe fruits, which contain about 800 seeds apiece (Cancino et al. 1993), split open in summer (Moran 1968). Seeds germinate best after air temperature has exceeded 40°C for extended periods (Cancino et al. 1993). Germination in the wild is probably rather low when soil temperatures reach or exceed 70°C (Cancino et al. 1993).

Pachycereus pringlei photosynthesizes via crassulacean acid metabolism (Franco-Vizcaíno et al. 1990). The photosynthetic surface area of many cacti is so reduced that carbon dioxide uptake may be limited by the amount of photosynthetically active radiation that falls on the stems (Nobel 1981a, 1983b). Self-shading, whether by spines or branches, can further reduce absorption of solar radiation. The need to maximize absorption of photosynthetically active radiation may have determined the branching pattern in this and other columnar cacti (Geller and Nobel 1986). The ribs allow the plant body to expand and contract as moisture is accumulated and used (Moran 1968).

When wounded, the tissues turn red, then black because of the initial presence of the alkaloid dopamine and its later transformation to melanin (Gibson and Horak 1978). Gibson and Horak (1978) have conducted anatomical studies.

The occurrence of this species in the predominantly winter-rain areas of Baja California raises puzzling questions about seed viability and persistence in the seed bank. Recruitment in this region probably depends upon rare warm-season storms.

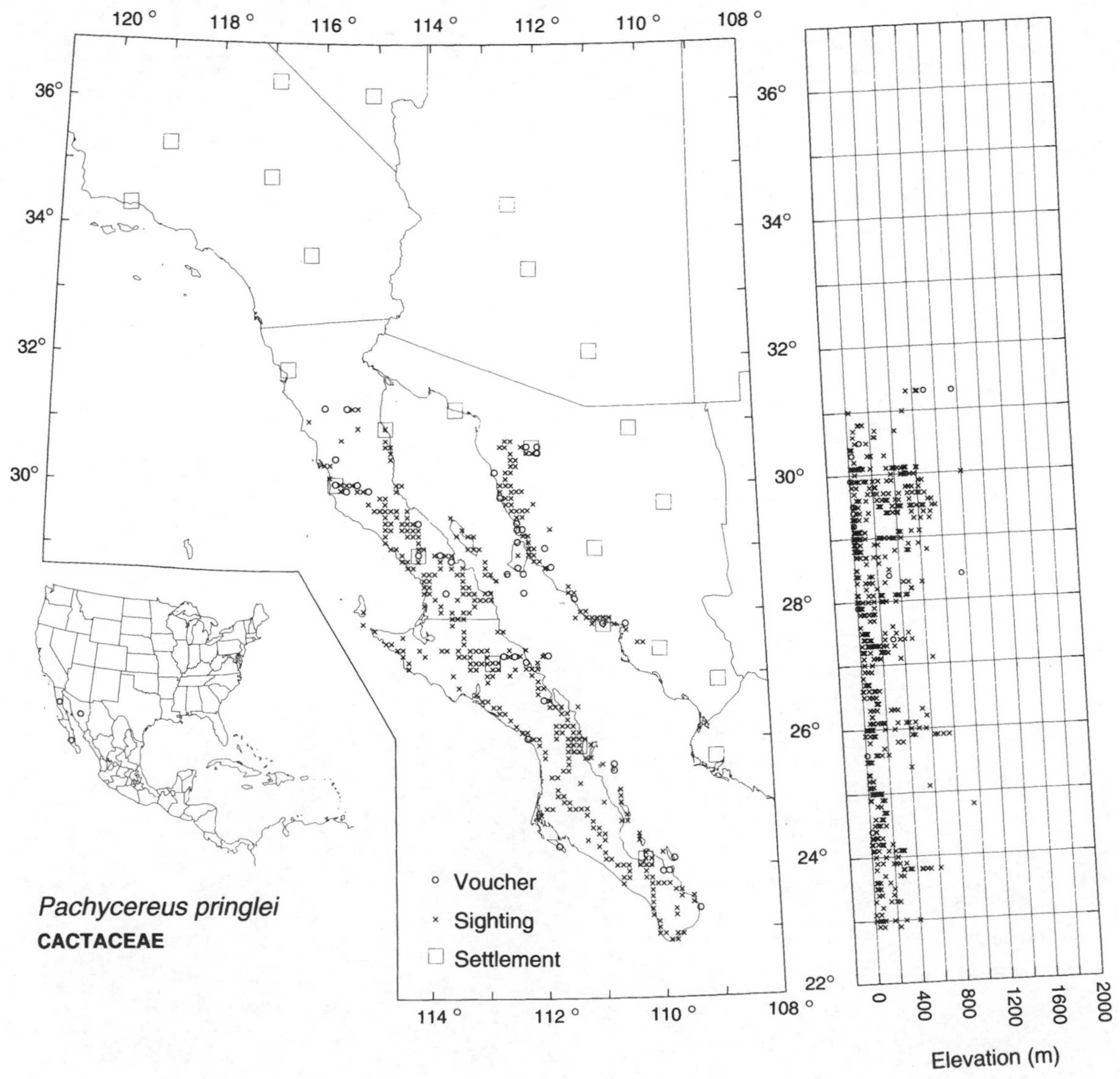

Pachycereus pringlei
CACTACEAE

Figure 68. Large Pachycereus pringlei *near Puerto Libertad, Sonora. The measuring rod is 12 feet tall. (Photograph by J. R. Hastings.)*

Estimates of age, based on known growth rates, indicate that plants may live longer than 300 years (Raymond M. Turner, unpublished data). Populations may fluctuate widely in numbers and age composition. One noteworthy example on a small island near Guaymas, Sonora, involves a population composed mainly of a few old plants in 1903. By 1961, the population instead included many young plants (Hastings and Turner 1965b). In 1964 the population consisted of 5,836 plants, or 8,000 cacti per hectare. Many of these had become established between 1930 and 1960 (Raymond M. Turner, unpublished data). The species is rare on the nearby mainland. In Baja California, *P. pringlei* is declining in abundance on older geomorphic surfaces and increasing on younger ones (McAuliffe 1991).

The Seri Indians ate the seeds and pulp, either fresh or preserved. They used the dry ribs in house construction (Felger and Moser 1985).

Pachycormus discolor

(Benth.) Cov.

Elephant tree, torote, torote blanco, copalquín, arbol elefante

This tree, 3–8 m tall, has a short, stout trunk and thick, crooked branches (figure 69). The habit varies from grotesquely twisted and sprawling shrubs, especially on exposed or impoverished sites, to slender, erect trees in more favorable locations (Shreve 1964). The thick, fleshy bark is white or yellowish white and exfoliates in thin sheets. Wounded bark exudes a milky sap that hardens into a gum or resin (Standley 1923). The leaves, up to 8 cm long and 2–3 cm wide, are pinnate with 3–11 elliptic or oblong leaflets bright green above and ashy-puberulent beneath. The small, rosy or whitish flowers are in terminal panicles. Fruits are utricles 3 mm long.

Bark of *Cyrtocarpa edulis* does not exfoliate. Leaves of *Bursera hindsiana* are simple or trifoliate, those of *B. microphylla* and *B. fagaroides* glabrous. All *Bursera* leaves have a strong resinous odor when crushed.

Biologists have recognized three varieties. According to Wiggins (1964), var. *discolor* occurs in the Sierra de la Giganta and on Magdalena and Santa Margarita islands, var. *pubescens* (S. Watson) Gentry from the Sierra de la Giganta north to 30°N, and var. *veatchiana* (Kell.) Gentry on Cedros Island and the western edge of the Vizcaíno Plain. Our map combines them.

Figure 69. Pachycormus discolor *near Punta Prieta, Baja California. (Photograph by J. R. Hastings.)*

Endemic to the Baja California peninsula, *P. discolor* is locally common on rocky slopes, especially on volcanically derived clay (Shreve 1964). The plants also grow on granitic, metamorphic, and tuffaceous substrates. *Pachycormus discolor* is most frequent in the Vizcaíno Subdivision (Shreve 1964), where it experiences sporadic winter rain and heavy summer fog. In this area, the species occurs well inland, but farther south, where winter rainfall decreases, most populations are found near the coast. Plants suffered damage from mild frost in January 1937 (Turnage and Hinckley 1938), and low temperatures may restrict their northward and upward movement. Unlike many other species centered in the Vizcaíno Subdivision, *P. discolor* also grows on the east coast of the peninsula, where summers are hot. In view of such climatic tolerances, its absence from the west coast of Sonora is surprising.

The swollen trunks and limbs are water-storage organs (Nilsen et al. 1990). Like other sarcocaulescents (for example, *Bursera microphylla, B. hindsiana, Jatropha cuneata*), the plants could more properly be regarded as drought escaping than drought resistant (Shreve 1933). Crosswhite and Crosswhite (1984) noted that a leafy sarcocaulescent functions like any leafy tree, whereas a leafless sarcocaulescent, in using stored water to endure drought, functions like a cactus. Nilsen and coworkers (1990) demonstrated that the stored water acts as a buffer that permits maintenance of leaf turgor as soil moisture drops. The turgor pressure of stored water also provides mechanical support, since the wood is weak (Gibson 1981).

Pachycormus discolor photosynthesizes via the C_3 metabolic pathway (Franco-Vizcaíno et al. 1990). Despite the extensive photosynthetic tissue on the trunks and stems, no measurable gas exchange occurs across the bark. Enough light apparently reaches the bark chloroplasts to allow endogenous recycling of carbon dioxide, thus maintaining energy reserves and permitting rapid production of leaves in response to rain (Franco-Vizcaíno et al. 1990).

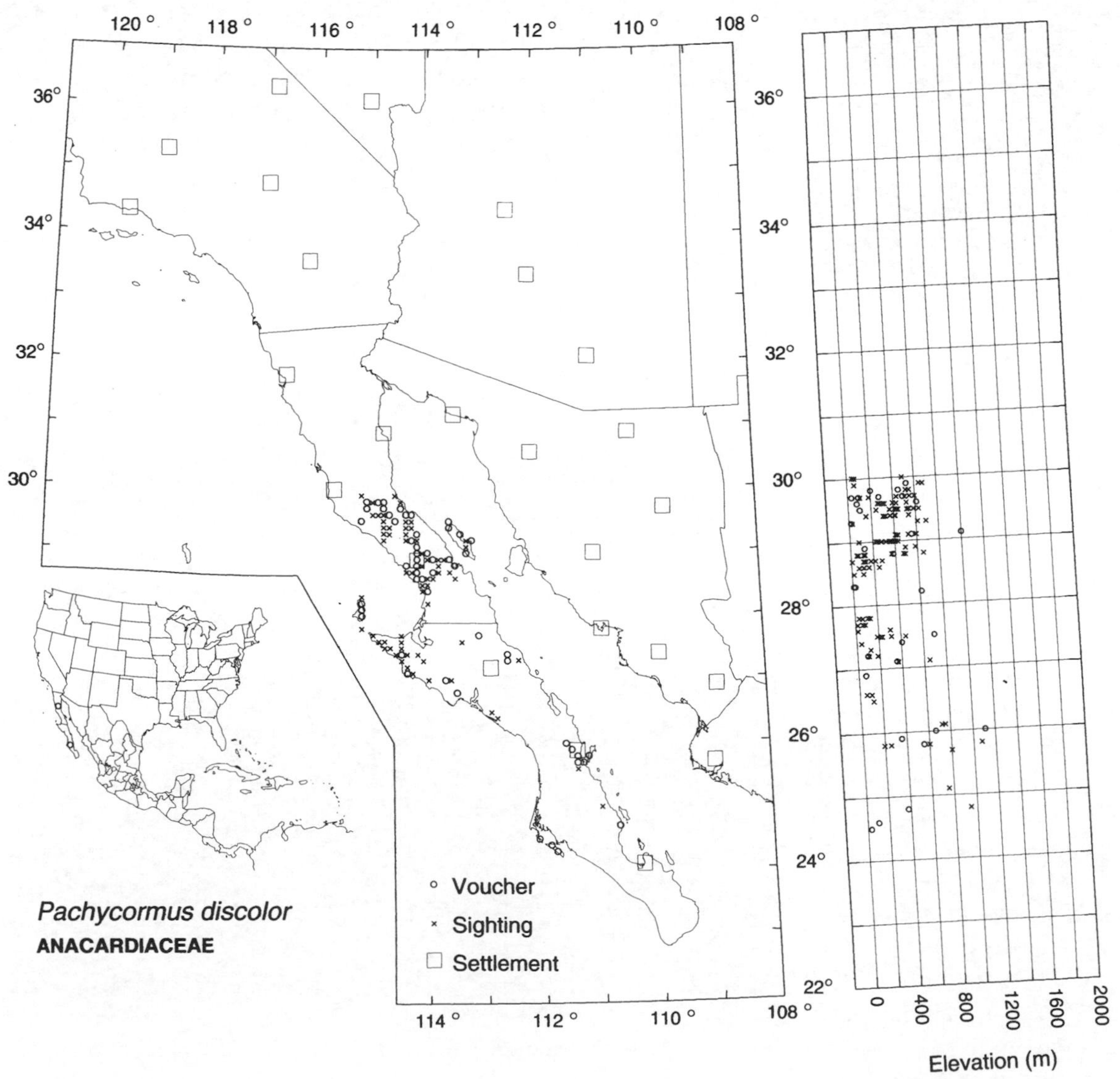

Pachycormus discolor
ANACARDIACEAE

Cuscuta veatchiana Brandegee, a twining parasite, attacks the trees in rainy years and may kill heavily infested individuals (Shreve 1964).

The plants are leafless much of the year (Standley 1923). Leaves appear after winter rains begin and are shed in April and May soon after the rains cease (Shreve 1964; Humphrey 1974). Their arrangement on long and short shoots may facilitate rapid leaf production in response to rain (Gibson 1981). Flowering lasts from May (occasionally as early as March) to September (occasionally as late as October or November). Peak bloom is in midsummer when the plants are leafless (Humphrey 1974). The flowers are visited by hummingbirds.

The bark has been used for tanning (Standley 1923). *Pachycormus discolor* can be grown from seed or small transplants.

Parkinsonia aculeata L.

Mexican palo verde, horse bean, Jerusalem thorn, bacapore, retama, bagota, guacóporo, junco, espinillo, mezquite verde

A tree up to 12 m tall, *Parkinsonia aculeata* has smooth, yellowish green bark on the upper branches and brown, rough bark on the trunk and main limbs. Young twigs have paired, nodal spines. The leaves are obscurely twice-pinnate, with an extremely reduced primary rachis. The 1–3 flattened pinnae, up to 30 cm long, have 10–40 pairs of fugacious leaflets 2–8 mm long. The yellow flowers are about 2 cm wide. The topmost petal turns orange or red in age. The tan pods, constricted between the seeds, are 5–10 cm long and 9–12 mm wide. The diploid number of chromosomes is 28 (Miege 1962). Jeune (1983) reported on leaf development. Carlquist (1989a) described the wood anatomy.

The flattened pinnae and the brown bark on the trunk and main branches set *P. aculeata* apart from the *Cercidium* spe-

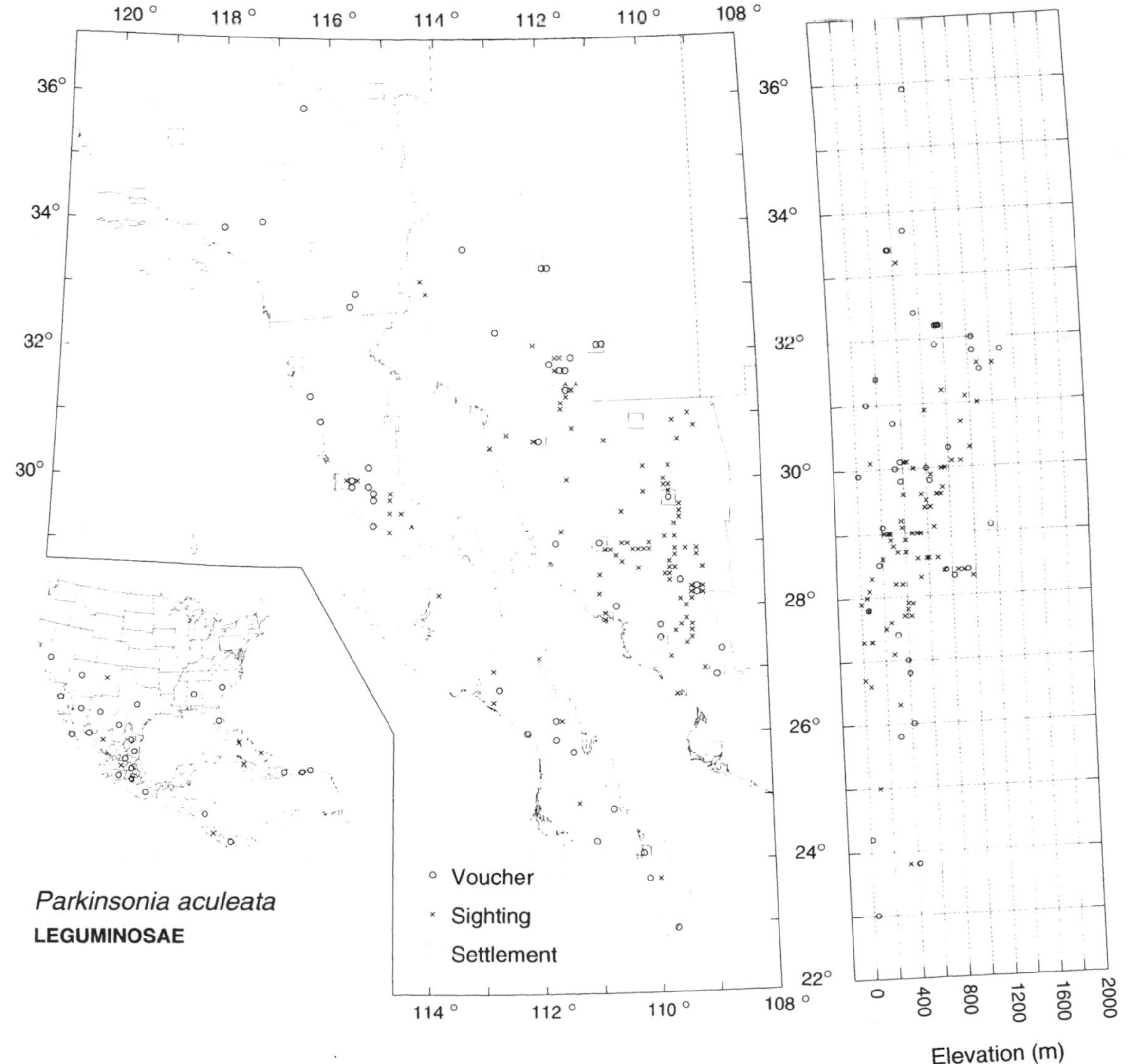

Parkinsonia aculeata
LEGUMINOSAE

cies in the Sonoran Desert region. Biologists have identified hybrids between *P. aculeata* and *Cercidium* species (Carter and Rem 1974). Brenan (1963), based on a reevaluation of African *Parkinsonia*, thought the two genera should not be segregated, but Carter (1974b) felt that more data were needed to justify their combination.

In the more arid parts of its range, *P. aculeata* is restricted to arroyos. Elsewhere it grows on sandy plains and at roadsides and other waste places, including low-lying areas where water accumulates after storms. The northernmost point on our map (35.9°N) represents a collection by Dennis Schramm from the Black Mountains, California, where it is undoubtedly adventive (Schramm 1982).

Native to the New World, *P. aculeata* is grown as an ornamental in tropical and subtropical regions around the world. It frequently naturalizes; on our map, most of the mainland locations north of 32°N represent adventives. *Parkinsonia aculeata* may be indigenous to the Coyote and Baboquivari mountains of southern Arizona (Kearney and Peebles 1960).

In the Sonoran Desert, the plants flower from March to May and occasionally again in summer or fall. Blooming lasts 3–4 weeks at a given locality (Dimmitt 1987). Near Tucson, Arizona, *P. aculeata* flowers at the same times as *C. microphyllum*, and occasional natural hybrids occur (Dimmitt 1987). The leaflets are drought deciduous and perhaps cold deciduous, as well. Although the leaf rachises are also deciduous, a tree seldom if ever loses all of them at once.

Parkinsonia aculeata is widely available commercially. Where annual rainfall is 25 cm or more, no supplemental irrigation is needed (Duffield and Jones 1981). Seeds germinate without pretreatment (Nokes 1986), but germination can be improved with scarification using sulfuric acid (Vora 1989). Plants can also be grown from semihardwood cuttings taken in summer (Nokes 1986). A complex hybrid between *P. aculeata*, *Cercidium microphyllum*, and *C. floridum* shows great promise as a landscape plant for arid regions (Dimmitt 1987).

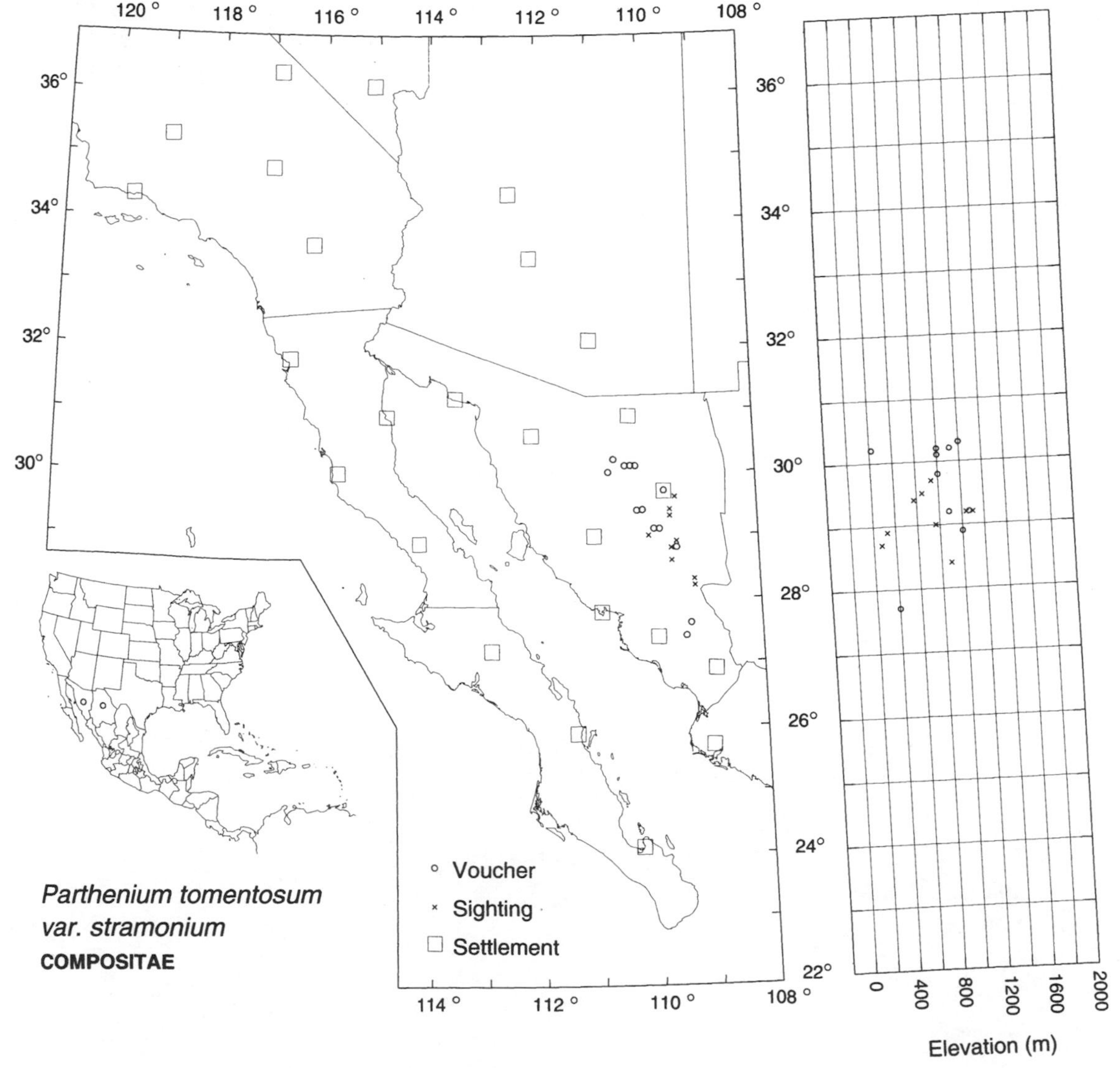

Parthenium tomentosum var. stramonium
COMPOSITAE

According to Standley (1922), the wood has been used in making paper, and an infusion of the leaves has been used to treat fever and epilepsy and to induce abortions.

Parthenium tomentosum
DC.

var. *stramonium*
(E. Greene) Rollins

[=*P. stramonium* E. Greene]

Otatillo, huasaraco

Parthenium tomentosum var. *stramonium* is a shrub 1–4 m tall with gray bark. The triangular leaves, 2–10 cm wide and 8–30 cm long, may be entire to crenulate. They are white-tomentose beneath, green and glabrescent above. The numerous small heads are in broad panicles. Each fruit consists of an achene with its attached ray flower and the two adjacent disk flowers. The diploid chromosome number is 36 (Rollins 1950).

Standley (1926) and Wiggins (1964) segregated *P. stramonium* from *P. tomentosum*. We follow Rollins (1950), who recognized *P. tomentosum* var. *tomentosum* of Puebla and Oaxaca and *P. tomentosum* var. *stramonium* of Sonora and Chihuahua. The two are similar morphologically and produce highly fertile seed when artificially crossed (Rollins 1950). The hybrids between *P. tomentosum* and *P. argentatum* A. Gray (guayule, a potential rubber crop) and between *P. tomentosum* and *P. incanum* H. B. K. are also fertile (Rollins 1946).

Parthenium tomentosum var. *stramonium* has a unique sesquiterpene lactone, an ambrosanolide not known from any other genus in the family, as well as some of the same parthenolides and xanthanolides found in other members of section Parthenichaeta (Rodriguez 1977). The distribution of these and other sesquiterpene lactones in the genus suggests speciation partly as a result of herbivore pressures (Rodriguez 1977).

Endemic to the mountainous interior of Sonora and adjacent Chihuahua, *P. tomentosum* var. *stramonium* forms local but extensive colonies in Sinaloan thornscrub on canyon slopes. It enters the Sonoran Desert at the northern end of its range, often on south-facing slopes above the level of nighttime temperature inversion (Shreve 1964).

Because the lowermost leaves are shed continually, branches are usually leafless except at the tip (Rollins 1946). Plants may bloom in their first year (Rollins 1950). Flowers appear sporadically throughout the year, presumably in response to rain. *Parthenium tomentosum* is largely self-incompatible, unlike the closely related *P. argentatum* and *P. incanum*, both facultatively apomictic (Rollins 1950). The plants also reproduce by root sprouts (Rollins 1950), making dense clones. Under greenhouse conditions, plants grown from seed reached 1–3 m in height by 18 months of age (Rollins 1946).

A decoction of the heartwood is used to treat a variety of illnesses (Gentry 1942). This species does not contain rubber but may provide useful genes when hybridized with rubber-bearing species of *Parthenium* (*P. argentatum* in particular) (Rollins 1946, 1950).

Pedilanthus macrocarpus
Benth.

Candelilla, gallito, zapate del diablo

Pedilanthus macrocarpus (figure 70), a clump-forming shrub up to 1 m high, has several to many turgid, succulent, waxy stems growing from the base. The stems are 6–15 mm in diameter and exude copious milky latex when cut. The oblong to ovate leaves are 1–2.5 cm long. The inflorescence is of bright red, boot-shaped involucres 2–2.5 cm wide and about 1 cm long. The red, 3-lobed capsule is 1.5–2 cm broad. Dressler (1957) obtained a haploid chromosome count of approximately 17.

The genus *Pedilanthus* was derived from a member of subgenus *Agaloma* in

Figure 70. Pedilanthus macrocarpus *near Punta Prieta, Baja California. (Photograph by J. R. Hastings.)*

Euphorbia (Dressler 1957). Its specialized involucre developed by fusion of the petaloid appendages and glands characteristic of *Euphorbia* species. The involucral tube contains the flowers, and the spur contains the glands (Dressler 1957).

Euphorbia ceroderma has thorn-tipped, woody stems and lacks boot-shaped involucres. Stems of *Asclepias subulata* are rarely more than several millimeters thick. Stems of *A. albicans* are as much as 2.5 m tall and are distinctly woody for much of their length.

On the Baja California peninsula, *P. macrocarpus* is locally common on sandy or loamy soils and in rocky areas. In Sonora and Sinaloa, it is restricted to the silty flats of the coastal plain (Dressler 1957). On our map, the southernmost point on the mainland (25.8°N) is at Los Mochis, Sinaloa (a collection by D. Taylor). The distribution suggests a need for summer rain, at least periodically. The northern limits seem to be shaped more by aridity than by frost.

The genus *Pedilanthus* probably arose on the Pacific slope of tropical Mexico and radiated from there (Dressler 1957). At some point, perhaps in the middle Miocene, the ancestor of *P. macrocarpus* became isolated on the peninsula, where, under an arid or semiarid climate, selection for increased stem succulence and reduced leaves took place, resulting in *P. macrocarpus*, an extreme xerophyte in the genus (Dressler 1957). Dressler (1957) suggested that the species spread to the Sonoran mainland in geologically recent times, perhaps in the late Pliocene or Pleistocene. Sternburg and Rodriguez (1982), however, based on an analysis of alkanes in the stem cuticular wax, concluded that the species was present in Sonora before gulf-floor spreading.

The plants flower from September to May (Dressler 1957). The small leaves are ephemeral, and most plants are leafless much of the year. Although fertile seeds are produced when pollen is artificially transferred from anther to stigma in the same flower, self-pollination is rare in nature because the style and stigma are well exserted by the time the staminate flowers mature (Dressler 1957). Humming-

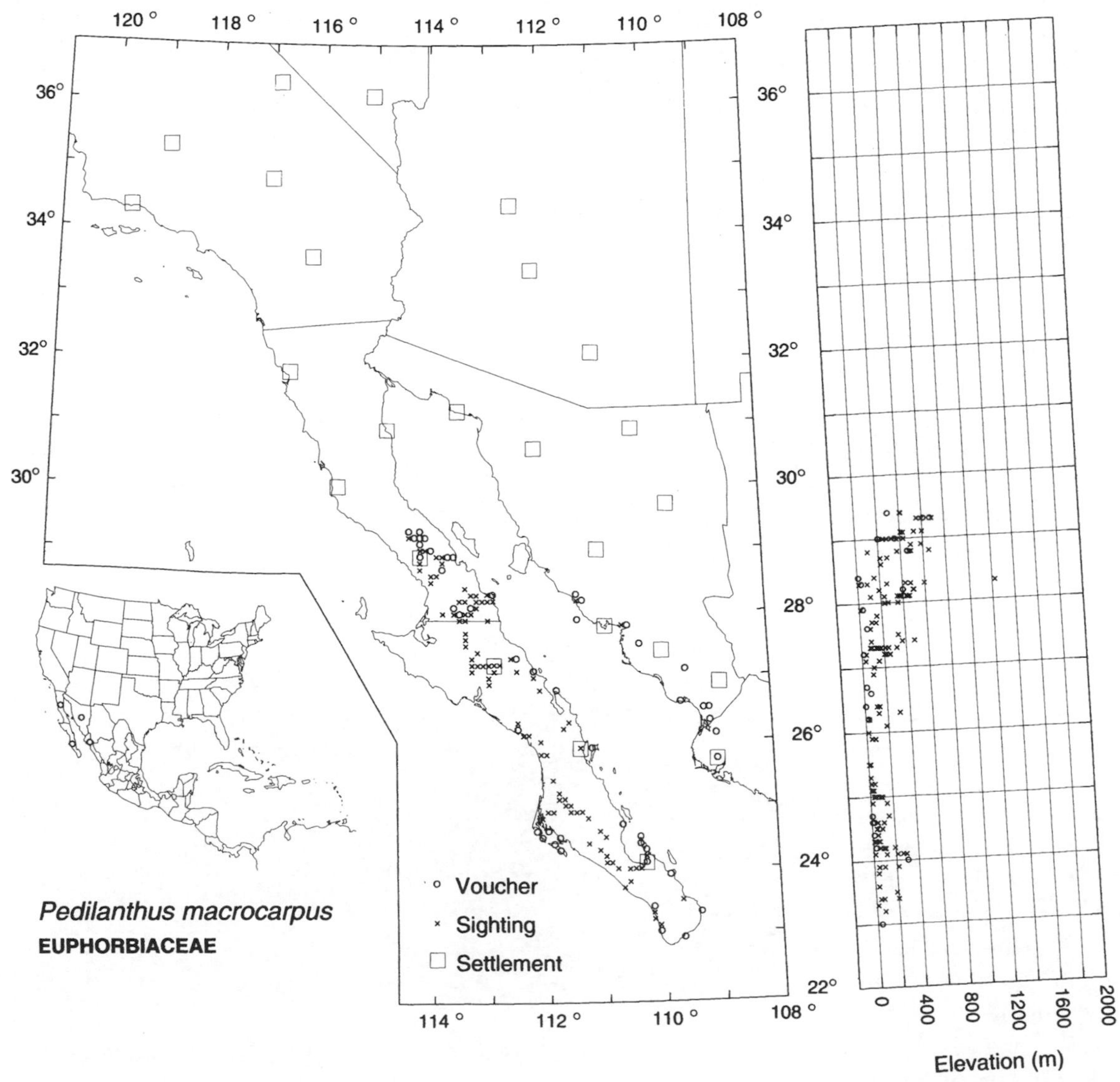

Pedilanthus macrocarpus
EUPHORBIACEAE

birds, the major pollinator, were probably a strong selective force in the evolution of the tubular involucre and red color of ancestral *Pedilanthus* (Dressler 1957).

Root sprouting sometimes forms extensive colonies (Dressler 1957).

The fresh latex contains 6–10% good-quality natural rubber by weight (Sternburg and Rodriguez 1982). Wild stands were harvested for natural rubber in the early 1940s (Wiggins 1980). The plants are an attractive ornamental in arid regions. They are susceptible to stem rot and need protection from freezing.

Petalonyx linearis

E. Greene

Sandpaper plant

Petalonyx linearis, a rounded shrub up to 1 m tall, has numerous erect, grayish green stems and sessile, linear to oblanceolate leaves 1–3 cm long and 2.5–7 mm wide. The foliage is markedly scabrous with short, spreading hairs. The small, white flowers are in capitate spikes at the branch tips. The haploid chromosome number is 23 (Davis and Thompson 1967). Hufford (1989) studied the floral anatomy.

Mortonia sempervirens and *Bourreria sonorae,* also scabrous shrubs, are much larger, woodier plants up to 2 m and 3–6 m tall, respectively. *Bourreria* leaves are flabelliform, those of *Mortonia* have margins that are rolled under. Petals form a tube in *P. thurberi* but are distinct in *P. linearis.* In addition, *P. thurberi* leaves are broadest at the base and (except in the narrowly endemic *P. thurberi* subsp. *gilmanii* [Munz] Davis & Thompson) pubescent with downturned hairs, whereas those of *P. linearis* are broadest about the middle with spreading hairs.

Seldom abundant, *P. linearis* usually grows in sandy soils (Davis and Thompson 1967). The points at 33.6°N and 33.7°N are in Deep Canyon, Santa Rosa Mountains, California (Zabriskie 1979). The disjunct population at 34.3°N is from

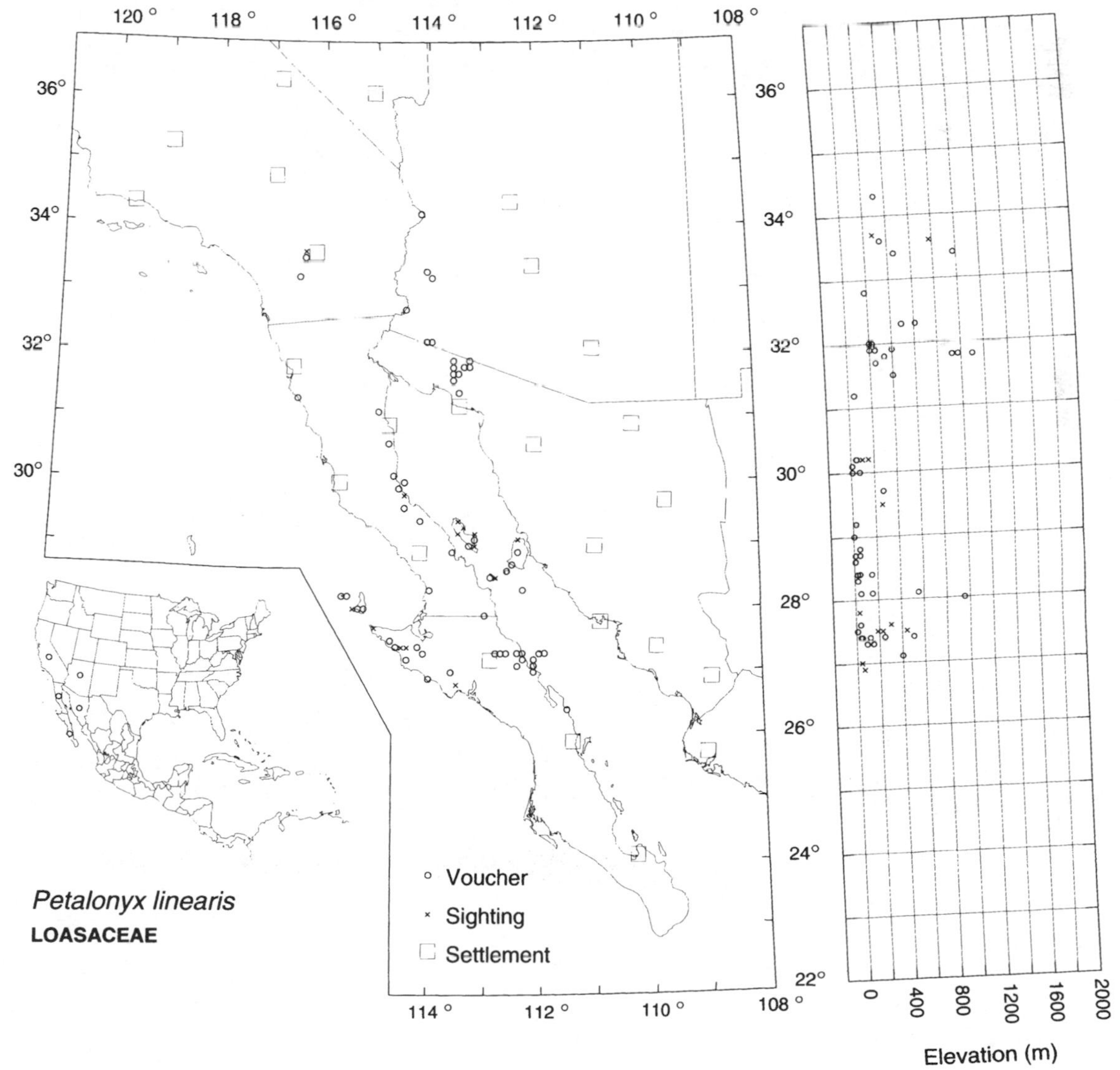

the Whipple Mountains, San Bernardino County, California (vouchered by Carl Wolf in 1932). The species may have evolved on coastal dunes on the peninsula, then hopped northward as new, inland dune sites became available with the drying trend of the past 8,000 years. In a 13,000-year midden sequence from Picacho Peak, California, *P. linearis* first appeared about 630 years B.P. (Cole 1986). Its presence on Isla Tiburón but not on the adjacent mainland is noteworthy.

Most populations, except for those on the Pacific coast, experience biseasonal rainfall. Throughout the range, winter is generally the season of most dependable soil moisture. The northern and upper elevational limits both suggest a preference for infrequent frost. *Petalonyx thurberi* is nearly contiguous with *P. linearis,* which suggests a history of competitive interactions.

Plants flower year-round, but peak bloom is January–April (Wiggins 1964). The leaves are essentially evergreen. They may drop during extreme drought. Davis and Thompson (1967) concluded that the flowers are largely self-pollinating. They argued that "the autogamous breeding habit and restriction to the geologically new desert habitats suggest a derivative status for *P. linearis*" (Davis and Thompson 1967:6).

Petalonyx thurberi

A. Gray

Sandpaper plant

This small, spreading chamaephyte, woody only at the base, is about 1 m tall. Its sessile, lanceolate to deltoid leaves, 1.5–3.5 cm long and 3–8 mm wide, become smaller toward the top of the stem. Except in *Petalonyx thurberi* subsp. *gilmanii,* which occurs outside the area mapped in this atlas, the foliage is rough with many stiff, downward-turned hairs. The slightly fragrant, white flowers, about 5 mm long, are in dense racemes 1–4 cm long. The haploid chromosome number is 23 (Davis

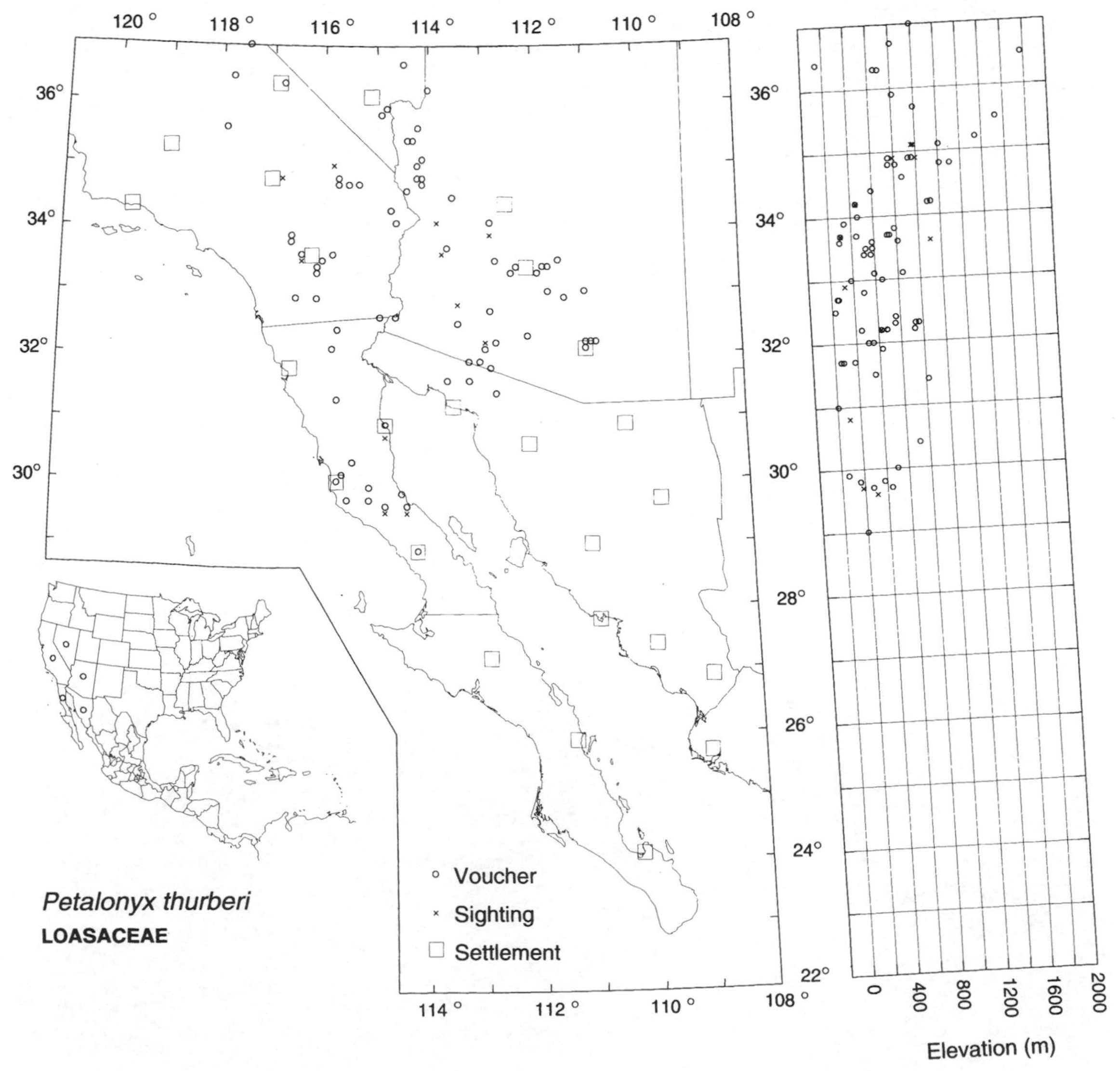

Petalonyx thurberi
LOASACEAE

and Thompson 1967).

Common in sandy washes and on dunes, *P. thurberi* occasionally grows with *P. nitidus* S. Watson and *P. parryi* A. Gray (Davis and Thompson 1967). *Petalonyx linearis* has distinct rather than connivent petals, and its leaves are not strongly graduated in size. *Petalonyx parryi* is taller—up to 1.5 m high—and has woody stems throughout. *Petalonyx nitidus* occurs largely outside the Sonoran Desert region; its leaves are short-petiolate, not sessile. Other shrubs with markedly scabrous foliage (*Mortonia sempervirens, Bourreria sonorae*) are much larger and strongly woody throughout. (When it occurs on dunes, however, *P. thurberi* can be notably robust, as much as 2 m high [Felger 1980], not unusual for plants on deep sand [Bowers 1982, 1984].)

Davis and Thompson (1967) recognized two subspecies: the relatively widespread subsp. *thurberi,* with harsh pubescence, and the narrowly endemic subsp. *gilmanii,* with soft pubescence. The latter is apparently restricted to Inyo County, California (Munz 1974). Benson and Darrow (1981) treated these two taxa as varieties.

The distribution encompasses arid areas where winter is the season of most dependable moisture. Summer rainfall declines from east to west across the range. The western part receives occasional summer storms. The elevational profile suggests greater cold tolerance than for *P. linearis.*

Plants bloom intermittently throughout the year with peak flowering in May and June. Davis and Thompson (1967) concluded that the plants are outcrossing. Major pollinators are probably solitary bees in the genus *Perdita* (Davis and Thompson 1967).

Peucephyllum schottii A. Gray

Pygmy cedar, romero

Peucephyllum schottii (figure 71) is a shrub 1–3 m high with linear leaves 1.5–2

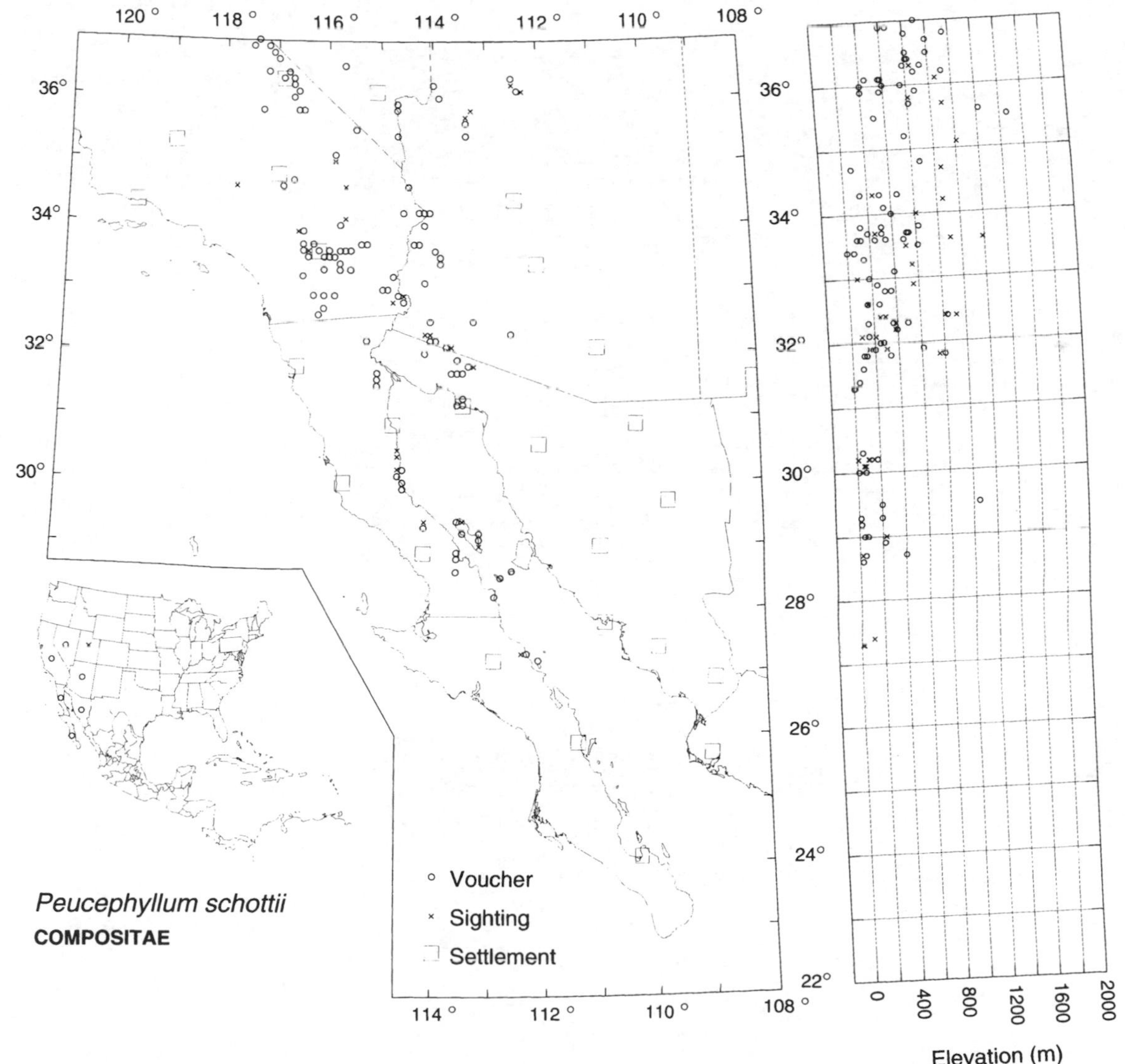

Peucephyllum schottii
COMPOSITAE

Figure 71. A large Peucephyllum schottii *near Bahía de los Angeles, Baja California. (Photograph by J. R. Hastings.)*

cm long crowded at the branch tips. The bell-shaped flower heads, solitary at the branch tips, are 10 mm tall and have 10–15 yellow disk flowers each. The haploid number of chromosomes is 10 (Turner et al. 1973).

The needlelike, evergreen leaves clustered at the branch tips set this species apart from other shrubs in its range.

Peucephyllum schottii is locally common on cliffs and rocky slopes and in washes. The first version of this atlas (Hastings et al. 1972) placed an outlying population in south-central Arizona about 29 km northwest of Tucson. The collection was actually made northwest of Tyson, Arizona (33.8°N, 114.4°W), in Yuma County, well within the range of the species. Packrat midden deposits record *P. schottii* in the Whipple Mountains, California, from 11,500 years B.P. to the present (Van Devender 1990b) and at Picacho Peak, California, from 11,000 years B.P. to the present (Cole 1986). Spaulding (1990) found the species in middens from the Skeleton Hills, Nevada (36.7°N, 116.2°W), near its present northern limit, at 9,200 years B.P. From these and other midden records, *P. schottii* appears to have spread rapidly northward during the Holocene and to have reached its present limits by the middle Holocene.

Because the uppermost branches suffer frost damage when temperatures fall below −2°C (Stark and Love 1969), cold temperatures likely limit the northward and eastward distribution. Where the plants grow on cliffs, as in the Grand Canyon, nocturnal radiation may buffer low temperatures. Along the gulf coast, *P. schottii* occurs as disjunct populations. Cody and coauthors (1983:93) suggested

Figure 72. Phaulothamnus spinescens (center) *has an intricate tangle of basal branches and an outer crown of erect, spinescent stems. The evergreen, spatulate leaves are fascicled along short branchlets (left). (Photograph by R. M. Turner taken at Rancho Americano, Sonora.)*

that "this distribution might date from the time when present Mojave Desert taxa extended much farther south down the eastern peninsula during wetter and especially cooler climates." Across its range, *P. schottii* experiences temperate to nearly tropical temperature regimes. Most populations receive biseasonal rainfall. Winter is generally the season of dependable soil moisture.

Stark and Love (1969) found that in Death Valley, California, *P. schottii* was not able to withstand drought as well as *Larrea tridentata* and *Atriplex hymenelytra*, two xerophytes with which it often occurs, and was also less able to control water loss. For these reasons, it typically grows where runoff keeps soil moisture reserves fairly high (Stark and Love 1969).

Matched photographs taken in the Grand Canyon show that *P. schottii* can live to be at least 100 years old (Robert Webb, personal communication 1993).

Phaulothamnus spinescens

A. Gray

Snake-eyes, mal de ojo, putia, tutuqui

This diffusely branched shrub grows up to 4 m tall and has spinescent gray branches (figure 72). The oblong to spatulate leaves, 1–5 cm long, are glaucous and yellow-green. On young stems, leaves are alternate; on older stems, fasciculate. The inflorescence is a few-flowered raceme 0.5–5 cm long. The fleshy fruits, 4–5 mm in diameter, are white and translucent. The single black seed makes them resemble eyes. Skvarla and Nowicke (1982) described the pollen anatomy.

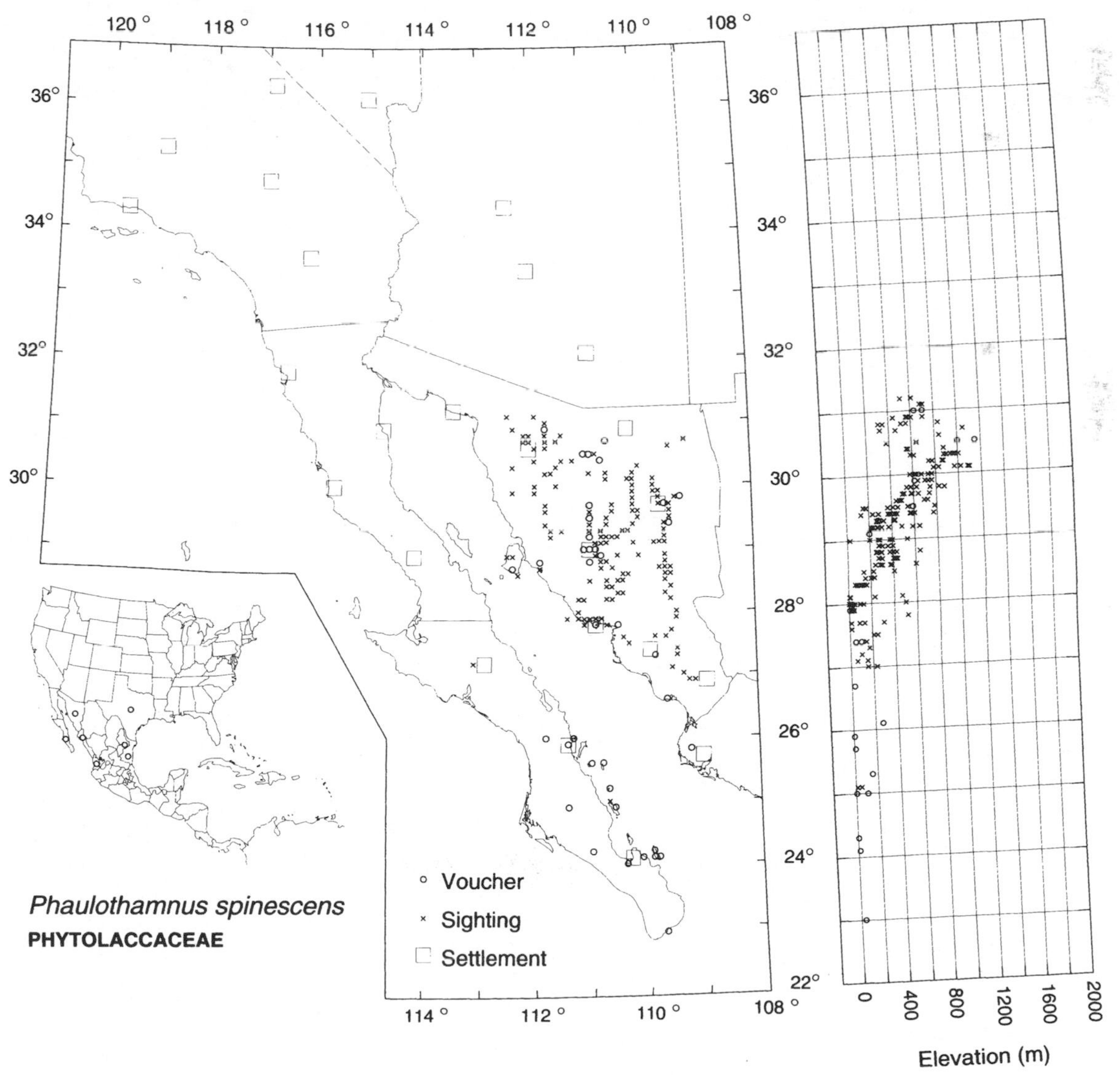

Phaulothamnus spinescens
PHYTOLACCACEAE

Without fruits, *Phaulothamnus spinescens* may not be readily distinguished from *Lycium* species.

Phaulothamnus spinescens grows widely scattered on bajadas and along washes. The isolated observation in Baja California Sur at 27.3°N was made 18.5 km southwest of San Ignacio by Raymond M. Turner in 1967. The distribution indicates a dependence on summer moisture and sensitivity to frost.

Piscidia mollis

Rose

Palo blanco

Piscidia mollis, a tree 3–9 m tall, has once-pinnate, densely cinereous-tomentulose leaves up to 25 cm long with 7–13 oval to ovate leaflets. The distinctive pods, 2–10 cm long and 3.5–5 cm wide, are indehiscent and prominently winged. This sturdy, white-barked tree, with its almost evergreen, gray foliage, has from a distance the appearance of a live oak.

No other tree in the range of this plant has winged pods, white bark, and leathery, gray foliage.

This plant grows in northern Sinaloa and in Sonora south of Moctezuma and east of Hermosillo. Further exploration might yield additional localities in eastern Sinaloa (Rudd 1969). In the Sonoran Desert, *P. mollis* is restricted to stream courses. In mountains southeast of the desert, the trees grow on hillsides and valley floors.

The leaves persist through the winter, then fall in mid May to June or July when the flowers appear.

The wood of this tree is highly valued as fuel, and the leaves are reportedly used to poison fish (Gentry 1942). Another species in this genus, *Piscidia erythrina* L.,

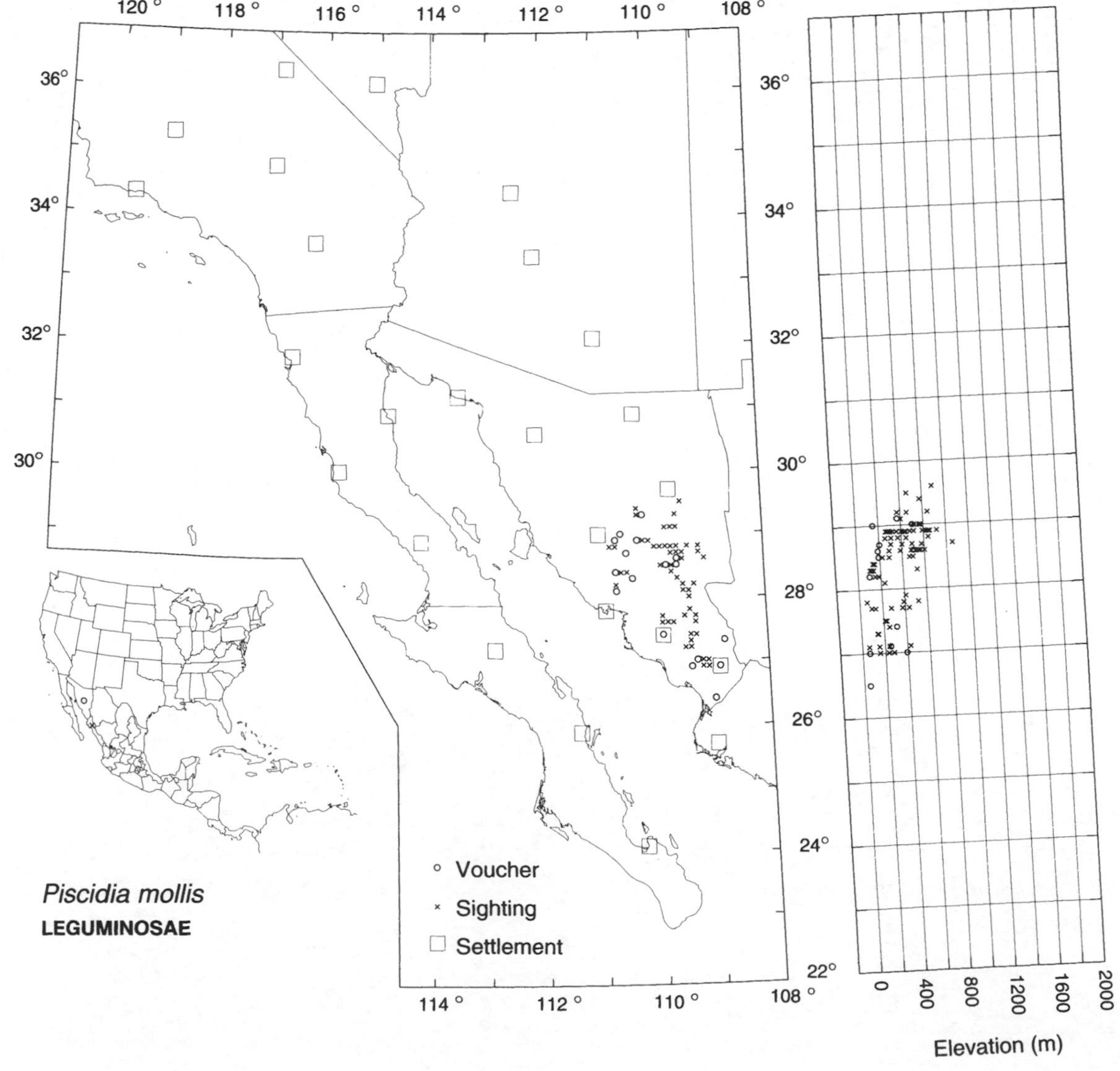

contains rotenone and other toxic isoflavonoids in its roots (Ingham et al. 1989).

Pithecellobium confine Standley

[=*Pithecollobium confine* Standley]

Palo fierro, ejotón

Pithecellobium confine, a shrub up to 4 m tall, has rigid, divaricate, zigzag branches with slender, straight, stipular spines 6–12 mm long. The twice-pinnate leaves have 2–5 pairs of leaflets 2.5–10 mm long on a single (rarely 2) pair of pinnae. The flowers, in small, dense heads about 1.5 cm across, are pale purple to rose. The indehiscent, turgid, woody fruits, 5–12 cm long and up to 3.5 cm in diameter, become black in age and often persist on the plants.

The fruits of *Acacia brandegeana* and *Prosopis palmeri* are not as broad, are not woody, and do not turn black.

The shrubs are most common along washes and on rocky slopes. Although Wiggins (1980) described *P. confine* as a peninsular endemic, it does grow along the coast of Sonora from El Desemboque del Arroyo San Ignacio to Guaymas. It occurs on Tiburón and other Gulf of California islands (Moran 1983b; Felger and Moser 1985) and on Natividad Island (27.9°N, 115.2°W), where it was collected by Reid Moran. The range lies mostly within maritime arid climates where summer rainfall is fairly dependable.

Flowers appear in April and May. Leaves are present at all seasons, as are at least a few fruits.

The roasted, ground seeds have been used as an adulterant for coffee and chocolate. The ground pods have been used to make a tea to relieve colds, coughs, and sore throats (Felger and Moser 1985) and to treat rattlesnake bites (Hicks 1966).

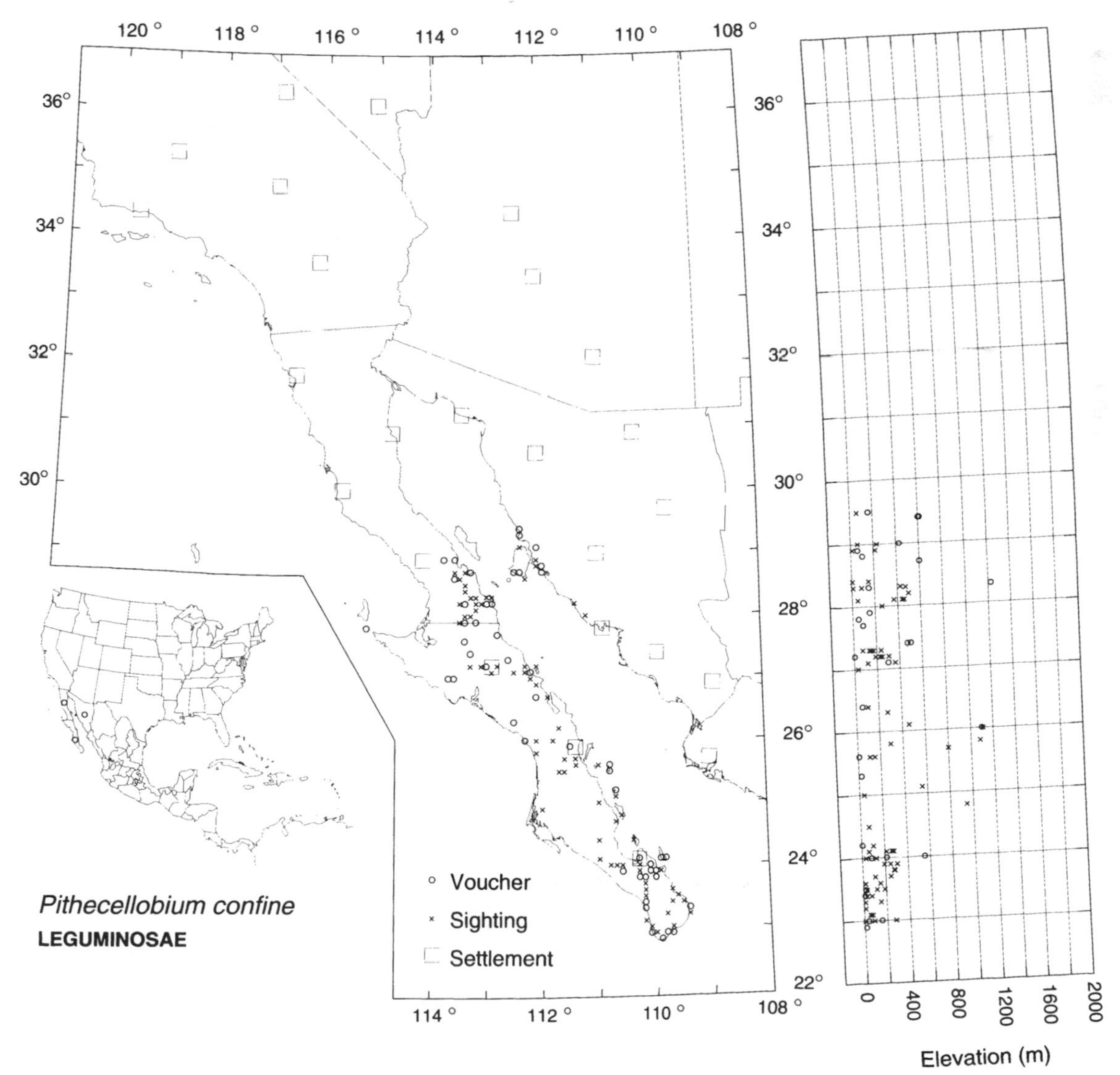

Pithecellobium confine
LEGUMINOSAE

Pithecellobium leucospermum
Brandegee

[=*Pithecollobium undulatum* (Britton & Rose) Gentry]

Palo fierro, guayacán, chucaro, ebano, palo pinto

This shrub or small tree grows up to 6 m tall; it has smooth, brown bark mottled with lighter patches. Straight axillary thorns may be present but are often lacking. The bipinnate leaves, 5–10 cm long, have 2–4 pairs of pinnae, each with 5–12 pairs of oblong-obovate leaflets 10–20 mm long. Leaflets have 3–4 prominent veins arising from their base. A gland is present near the middle of the petiole. The greenish white flowers are in heads 2 cm wide. The linear pods, often constricted between the seeds, are 10–18 cm long.

Pithecellobium leucospermum can be distinguished from various species of *Acacia, Lysiloma, Desmanthus,* and *Pithecellobium* by a combination of characters involving petiolar gland position, spine shape, and leaflet size, number, and venation (Turner and Busman 1984).

We follow McVaugh (1987) in treating *P. undulatum* as synonymous with *P. leucospermum.* Plants in our region that have been referred to *P. tortum* Mart., a South American species, are best regarded as *P. leucospermum* (McVaugh 1987).

This plant of Sinaloan deciduous forests enters the Sonoran Desert region along the Río Yaqui valley in the vicinity of Tónichi, Sonora. Nowhere abundant, it is found southward through Mexico to Central America.

Flowers appear in July after the summer rains begin. The fruits ripen in the fall.

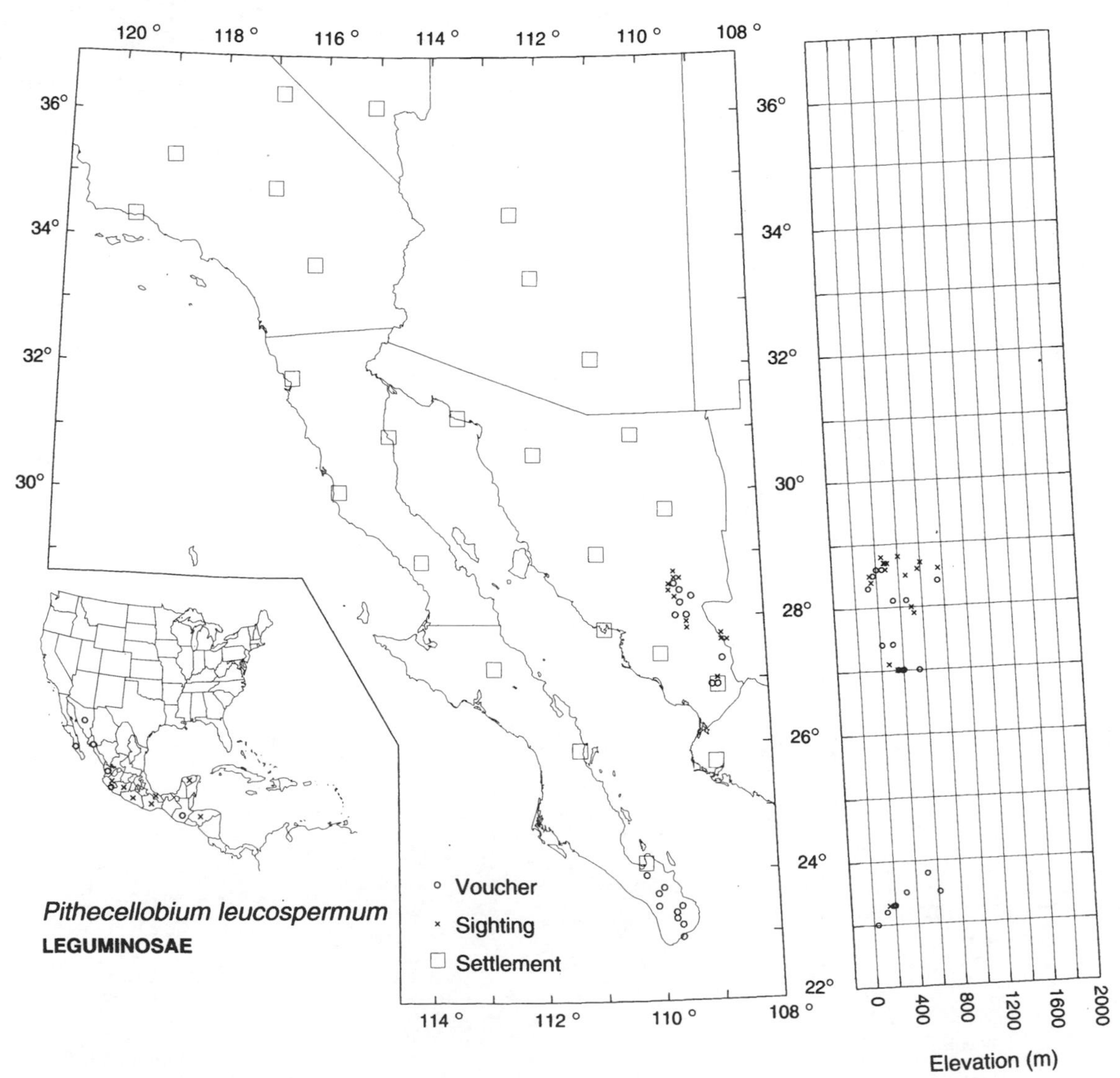

Pithecellobium leucospermum
LEGUMINOSAE

Pithecellobium mexicanum

Rose

[=*Pithecollobium mexicanum* Rose]

Chino, palo chino, palo chinu, joso

Pithecellobium mexicanum is a tree up to 10 m tall with paired stipular spines 1–4 mm long and twice-pinnate leaves. The leaflets, 3–8 mm long, are in 3–10 pairs on each of 2–4 pairs of pinnae. The yellow flower heads are in racemes or spikes. Fruits are oblong pods 6–10 cm long and 2–2.5 cm wide.

Neither *Acacia greggii* nor *A. occidentalis* has paired stipular spines.

This plant grows along arroyo margins throughout its Sonoran Desert range. The species is centered in the transition from Sinaloan deciduous forest to Sonoran desertscrub.

The trees stand out boldly in the spring when the abundant yellow blossoms cover their crowns.

The hard, reddish wood has been used for furniture, and the bark is excellent for tanning. *Pithecellobium mexicanum* survives freezing temperatures down to -9°C (Johnson 1993) and is cultivated as an ornamental in the warmer parts of Arizona.

Pithecellobium sonorae

S. Watson

[=*Pithecollobium sonorae* S. Watson]

Palo jocono, uña de gato, palo gato

Pithecellobium sonorae (figure 73) is a tree up to 8 m tall with a narrow, compact crown and smooth gray bark interrupted at irregular intervals by conical spines and horizontal ridges. The twice-pinnate leaves have 1–4 pairs of pinnae, each with 10–20 pairs of leaflets 3–8 mm long. The flowers are in small, solitary heads on paired, stoutish peduncles. The broadly

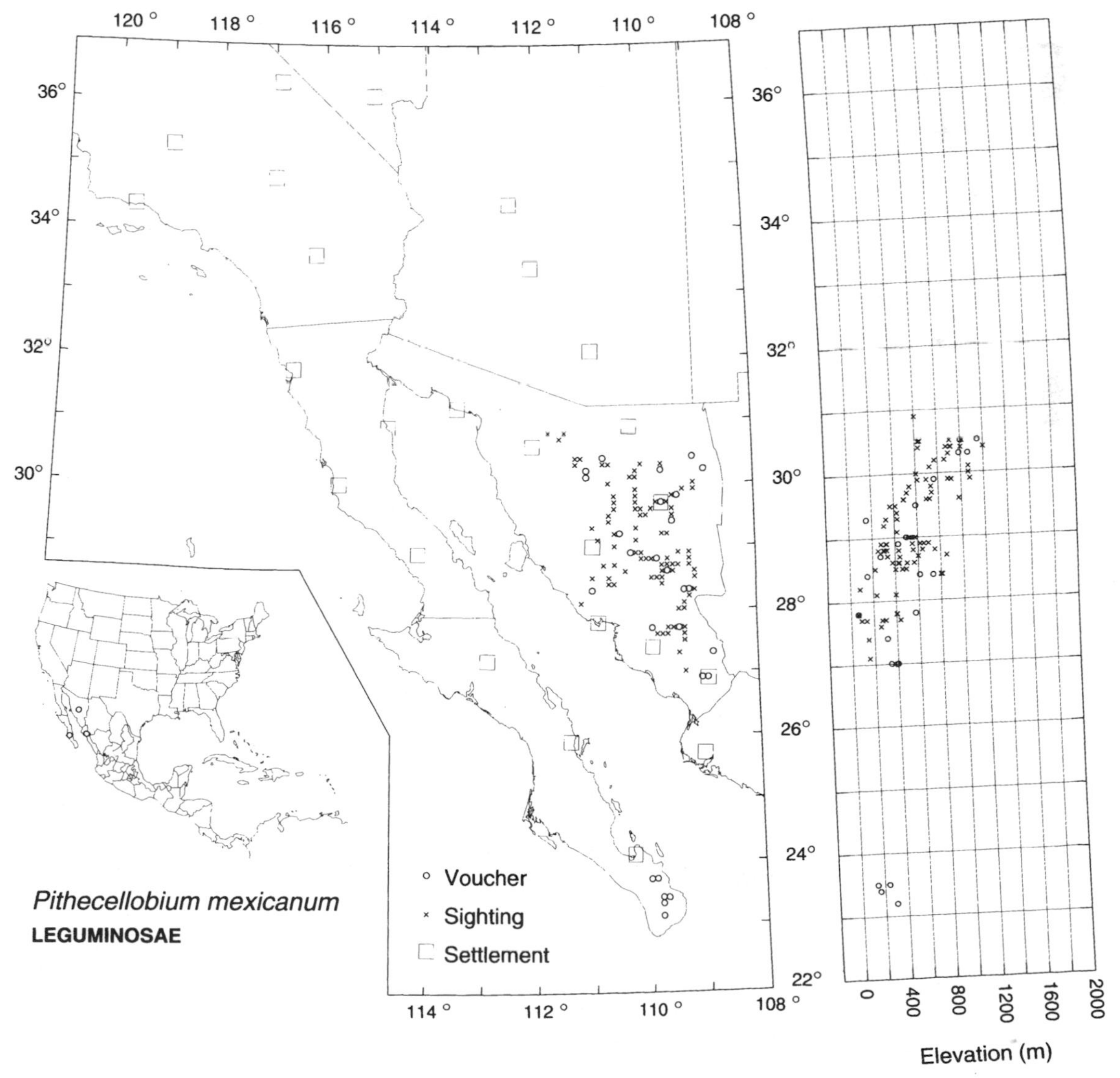

oblong pods are 6–18 cm long.

The combination of paired stipular spines, zigzag stems, and ridged, spiny trunks is distinctive.

In Sonoran desertscrub, *P. sonorae* is frequent along arroyos, usually on calcareous soils. In Sinaloan thornscrub and Sinaloan deciduous forest, the plants grow on drier, more open sites. Saplings often grow in thickets several meters across. Our records show that the species is confined to the state of Sonora south of Hermosillo. We have not seen specimens from Sinaloa although the species is listed from that state (Standley 1922; Wiggins 1964).

Flowers appear in summer. The mature fruits persist for several months before the seeds are shed.

Pleuraphis rigida

Thurb.

[=*Hilaria rigida* (Thurb.) Benth. ex Scribn.]

Big galleta

Pleuraphis rigida is a clumped, perennial, rhizomatous grass up to 60 (occasionally 100) cm tall. The rigid, freely branching culms are felty-pubescent near the base. The zigzag rachis has spikelets in tight groups of 3, forming a terminal spike 5–10 cm long. The haploid chromosome number is 54 (Reeder 1977).

Pleuraphis mutica Buckl. (=*H. mutica* [Buckl.] Benth.) is usually shorter and has completely glabrous stems.

Pleuraphis is sufficiently distinct from *Hilaria* to warrant recognition at the generic level (Reeder and Reeder 1988).

The plants are most common on sandy flats, washes, and dunes and also grow on rocky slopes and gravelly plains. They colonize active sand dunes (Shreve 1937a) and form up to 50% of the ground cover on stable dunes (Zabriskie and Zabriskie 1976). Tony L. Burgess vouchered the southernmost population on our map. The species occupies arid climates where rainfall is biseasonal. Winter rains are occasional throughout the range. Summer rains, which generally follow a foresum-

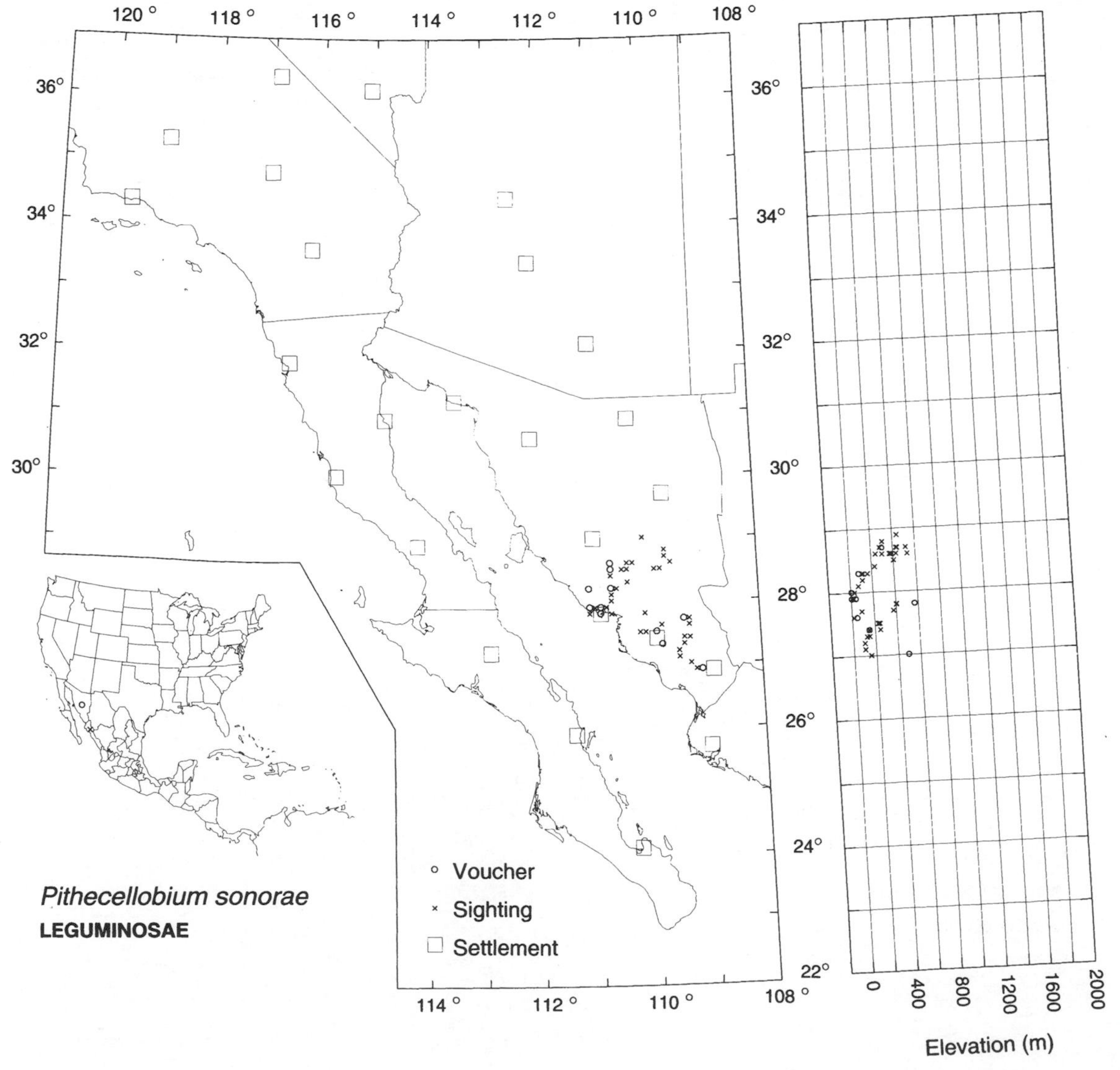

Pithecellobium sonorae
LEGUMINOSAE

mer drought, decrease in number and duration from southeast to northwest across the range.

Packrat middens record *P. rigida* in the Butler Mountains, Arizona, at 10,400 years B.P.; in the Tinajas Altas Mountains, Arizona, at 10,000 years B.P.; and in the Whipple Mountains, California, at 10,000 years B.P. (Van Devender, Toolin et al. 1990). The species is not known from Pleistocene midden deposits, and one may presume that it expanded northward from a more restricted distribution during the early Holocene.

In Arizona, *P. rigida* flowers sporadically from February to September (Kearney and Peebles 1960), apparently in response to rain. Leaves and stems also grow after rains in spring and summer (Nobel 1980d). Most perennial grasses renew growth from ground-level buds, but *P. rigida* functions as a chamaephyte, renewing growth from buds on aboveground stems.

Pleuraphis rigida is a C_4 species. The temperature optimum for carbon dioxide uptake, 29–43°C, shifts markedly in response to changes in ambient air temperature (Nobel 1980d). The rate of carbon dioxide uptake is very high (67 μmol/m^2/sec) (Nobel 1980d). Substantial carbon dioxide uptake over a wide range of temperatures is critical for survival in the arid climates where *P. rigida* grows.

On stable dunes, the shallow roots tap a ground area of 0.095 to 2.11 m^2 (Nobel 1981b). When *P. rigida* stands are thinned, soil water becomes available to the remaining plants for a longer period of time, increasing their longevity, new stem growth, and root biomass (Robberecht et al. 1983; Nobel and Franco 1986).

Nobel (1981b) estimated that individual clumps may live hundreds of years. Matched photographs taken in the Grand Canyon show plants at least 100 years old (Robert Webb, personal communication 1993). During prolonged drought, mortality may be high (Martin and Turner 1977). In serving as a nurse plant for *Agave deserti* and *Ferocactus cylindraceus,* this robust perennial grass is an important struc-

Figure 73. Pithecellobium sonorae *south of Torres, Sonora. Inset (right) shows highly modified stipular spines on the main stem of the plant. (Photograph by J. R. Hastings.)*

turing element in certain Sonoran Desert plant communities (Franco and Nobel 1988, 1989).

Especially when actively growing, *P. rigida* is an important forage plant on desert ranges (Humphrey 1953; Hughes 1982). The Tucson-based Plant Materials Center of the Soil Conservation Service is developing strains for landscaping in arid climates. Germination rates are 10–20% in most seed lots (Scott Lambert, personal communication 1992). Clumps divided into tillers and planted in June grow vigorously after the onset of summer rains and require little or no irrigation once they are established (Scott Lambert, personal communication 1992).

Prosopidastrum mexicanum (Dressler) Burk.

[=*Prosopis globosa* Gill. var. *mexicana* Dressler]

This sprawling shrub, 2–3 m tall, has prominent yellow stripes on the stems and weakly recurved stipular spines. Leaves are twice-pinnate with 2–3 pairs of leaflets 5 mm long on a single pair of pinnae. Fruits dehisce when the carpel walls break away from the thickened pod margins.

Dressler (1956) treated this species as *Prosopis*. We follow Burkart (1964).

Prosopidastrum mexicanum is narrowly endemic to Baja California. Its distribution is typical of a group of endemics, including *Acanthogilia gloriosa* and *Xylonagra arborea,* that occupy arid maritime climates with relatively cool, dry summers. The only other member of this small genus is a very closely related species restricted to southern Argentina (Burkart 1976).

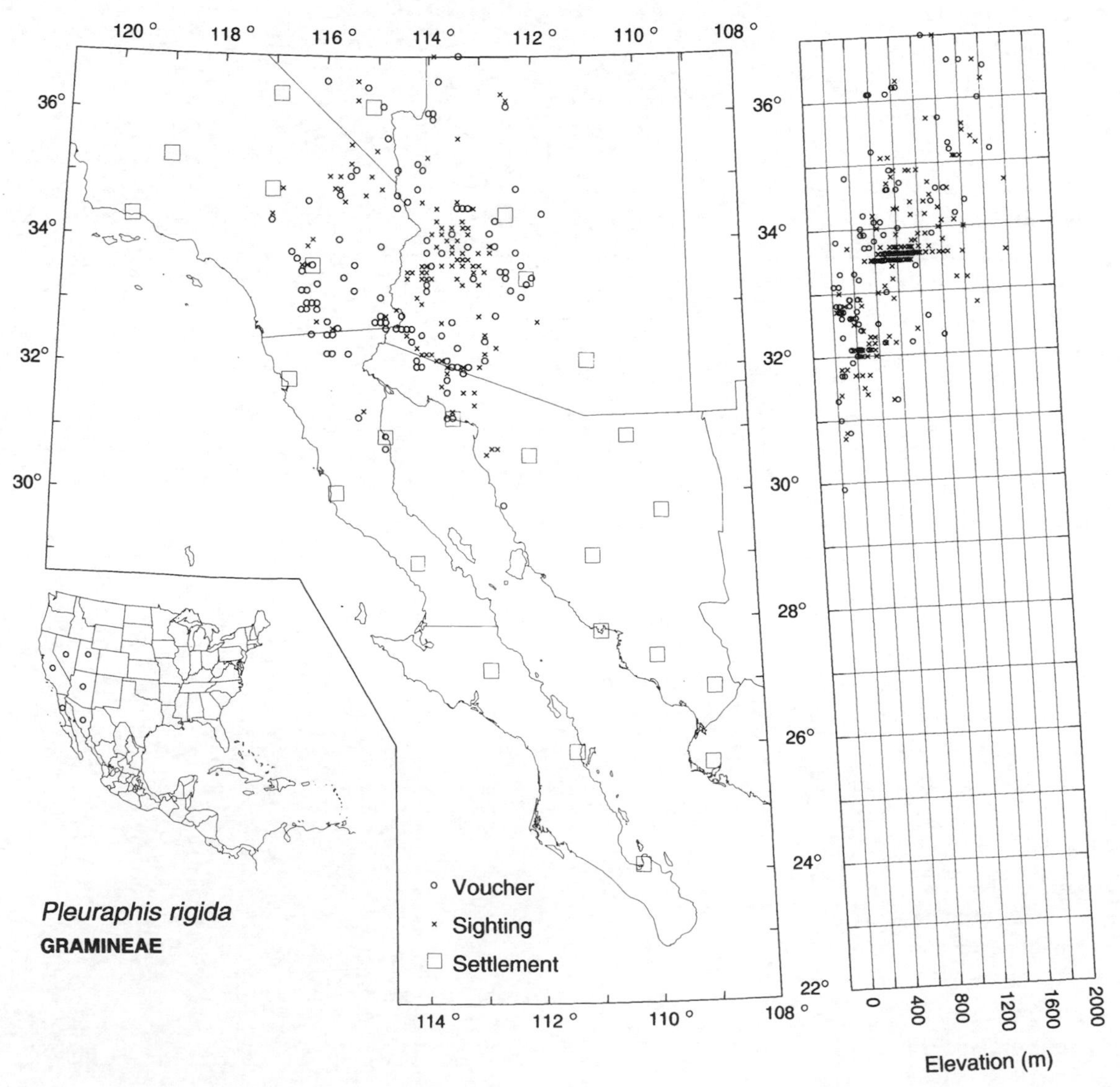

Pleuraphis rigida
GRAMINEAE

Prosopis articulata

S. Watson

[=*P. juliflora* (Swartz) DC. var. *articulata* (S. Watson) Wiggins]

Mesquite amargo, mesquit, mesquite

Prosopis articulata is a spiny shrub or small tree up to 5 m tall. The single pair of pinnae (rarely 2 pairs) has 6–20 pairs of finely pubescent to glabrous leaflets 2.5–12 mm long. The distance between leaflets is less than their own width. The pinkish flowers are in racemes equal to or longer than the leaves. The flattened, bitter, dry fruits are 10–24 cm long and have undulate margins.

No other *Prosopis* species in the Sonoran Desert region has pinkish flowers. Where they overlap, *P. articulata, P. velutina,* and *P. glandulosa* var. *torreyana* often intergrade, making identification difficult at times. Fruits of *P. articulata* are more deeply constricted between the seeds than those of *P. velutina.* In typical *P. glandulosa* var. *torreyana,* the glabrous or ciliate leaflets are larger and more widely spaced than those of *P. articulata.*

Prosopis articulata is often scattered but can form dense stands, as on heavy alluvial soil near Vicam, Sonora. It is common near Guaymas. Our map omits many sightings of questionable status. Seed dispersal by domestic livestock has probably influenced interactions with other *Prosopis* species and altered its pre-European distribution. The Hermosillo record is from Burkart (1976). Temperatures within the range are moderated by maritime air. Winters are essentially frost free. Most rain falls between July and October, but occasional winter storms may also deliver moisture.

Flowers appear from April to July, fruits through October or November. Palacios and Bravo (1974/75) described the seed anatomy.

In cultivation, *P. articulata* tolerates irrigation with seawater (Rhodes and Felker 1988). This trait—in combination with its high growth rate (Felker, Cannell et al. 1981), tolerance of light frost (Felker

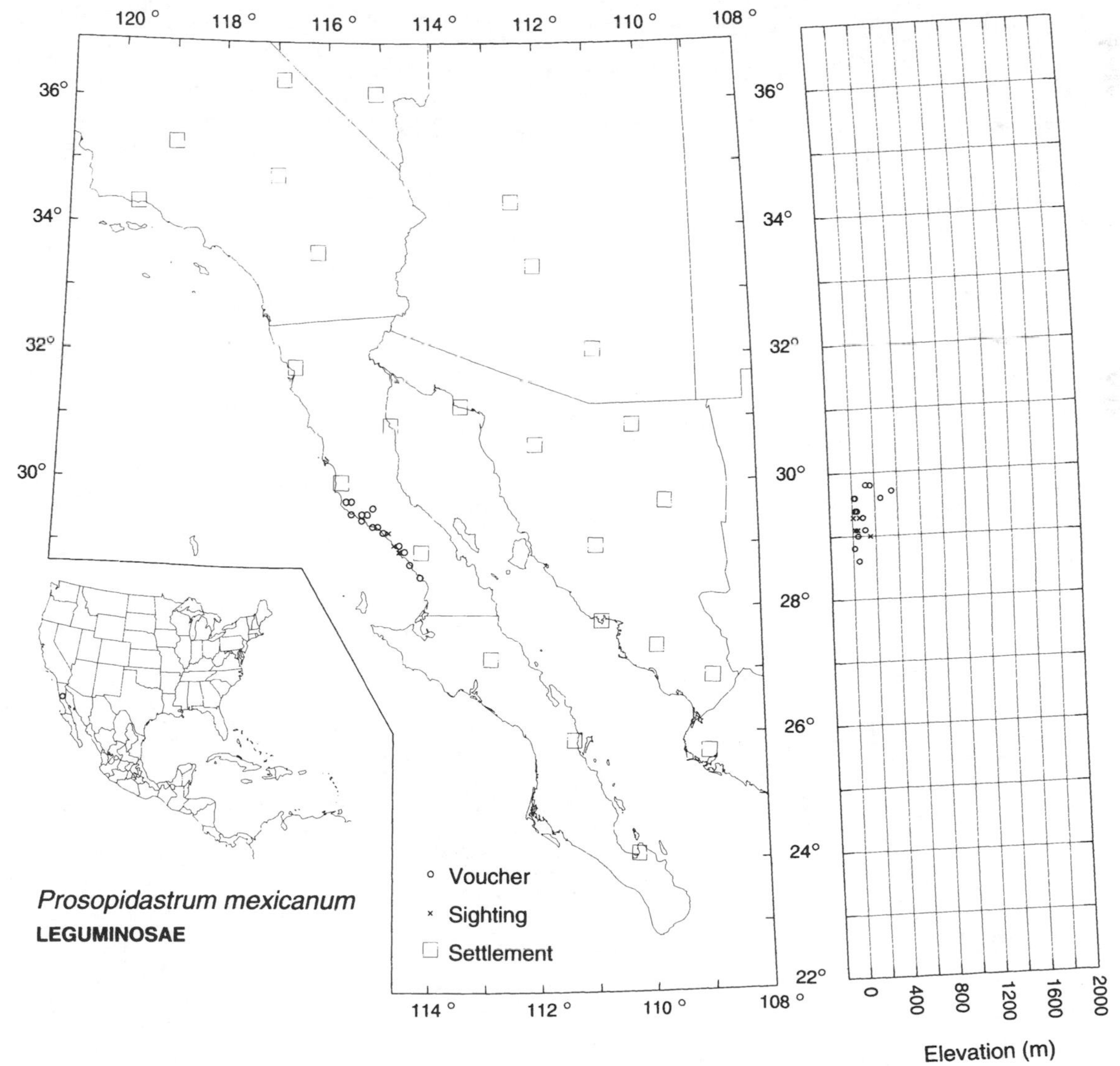

et al. 1982), and ability to fix nitrogen (Felker and Clark 1980)—gives it good potential for biomass production in arid lands (Rhodes and Felker 1988).

Prosopis glandulosa
Torrey

var. *torreyana*
(L. Benson) M. C. Johnston

[=*P. juliflora* (Sw.) DC. var. *torreyana* L. Benson]

Honey mesquite, mesquite, algarroba, chachaca

This large spiny shrub or small tree grows up to 9 m tall and has 8–20 pairs of leaflets on a single pair of pinnae (rarely 2 pairs) (figure 74). The linear-oblong leaflets, 15–22 mm long, are 7–9 times as long as broad and are usually spaced 5–6 mm apart. They may be glabrous or ciliate with short, stiff hairs along the margins. The small, greenish yellow flowers are in spikelike racemes 5–12 cm long. The compressed pods, 1–2.5 dm long and 1–1.5 cm wide, are straw colored when mature. The haploid chromosome number is 14 (Hunziker et al. 1975; Solbrig et al. 1977).

Where they overlap, *P. glandulosa*, *P. velutina*, and *P. articulata* may intergrade, making it difficult to tell them apart. In typical *P. glandulosa* var. *torreyana*, the pinnae are mostly 1 pair per leaf, compared to 1 or 2 pairs in *P. velutina*. The leaflets are larger and more widely spaced in *P. glandulosa* var. *torreyana* than in *P. velutina;* in addition, they are glabrous or ciliate in *P. glandulosa* var. *torreyana*, densely velutinous in *P. velutina*. The fruits are

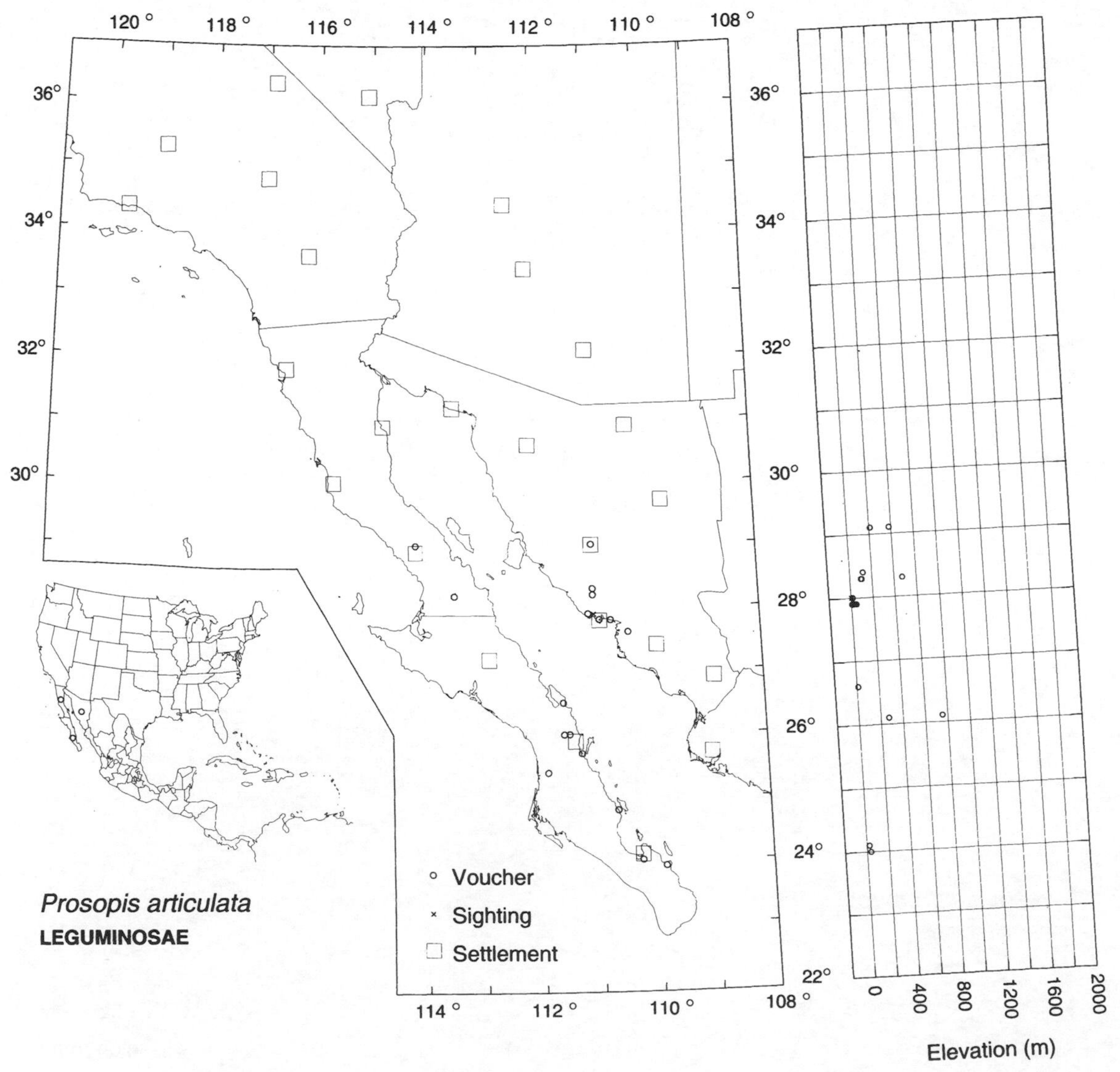

Prosopis articulata
LEGUMINOSAE

more deeply constricted between the seeds in *P. articulata* than in *P. glandulosa* var. *torreyana*.

Burkart (1976) reported *P. glandulosa* var. *torreyana* as restricted to the lower Colorado River Valley and southward. Our treatment follows that of Johnston (1962b). The broad variability where *Prosopis* species overlap buffers populations from environmental shocks and helps avoid decimation during periods of environmental extremes (Peacock and McMillan 1965; Hilu et al. 1982).

Prosopis glandulosa var. *torreyana* is common in washes and bottomlands, on heavy soils of uplands, and in coarse soils of sandy flats. Rapid growth and adventitious roots enable it to colonize active

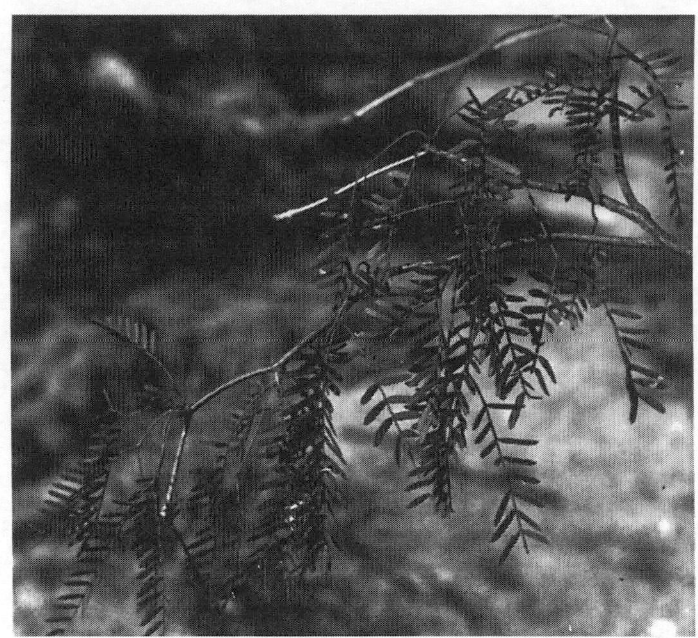

Figure 74. Prosopis glandulosa *var.* torreyana *near San Felipe, Baja California. (Photograph by J. R. Hastings.)*

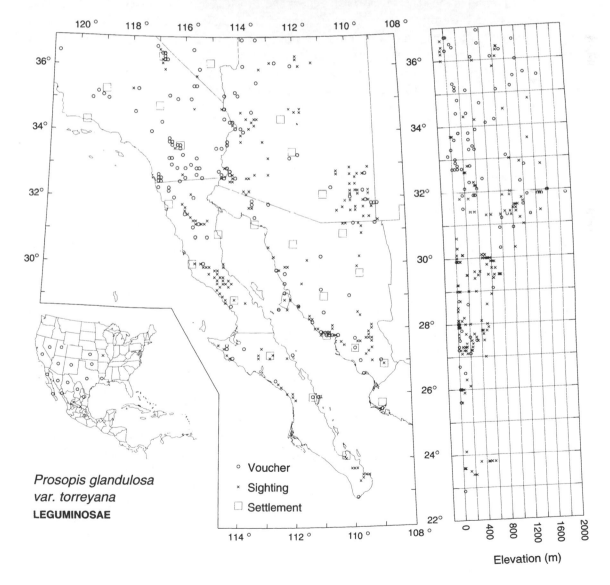

Prosopis glandulosa var. torreyana

LEGUMINOSAE

○ Voucher
× Sighting
□ Settlement

Elevation (m)

dunes (Bowers 1982). In some areas, the plants form coppice dunes (Graham-Gadzia and Ludwig 1983).

The species is native from Trans-Pecos Texas to California and southward to Sonora, Sinaloa, Chihuahua, and Baja California Sur. It grows on several Gulf of California islands (Moran 1983b). Reports of this species from Kansas, Oklahoma, Louisiana, Coahuila, Nuevo Leon, San Luis Potosí, and Zacatecas (Johnston 1962b; Isely 1972, 1973) may represent intentional or inadvertent introductions, as seems to be the case in central California (Holland 1987; Holland and Anderson 1988) and Missouri (Benson 1941). In south-central Arizona, in north-central Sonora, and along the Colorado River in the Grand Canyon, *P. velutina* interrupts the range of *P. glandulosa* (Benson 1941; Isely 1973; Turner and Karpiscak 1980; Benson and Darrow 1981). Winter temperatures probably determine the northern limit (Peacock and McMillan 1965).

Prosopis glandulosa has a long history in packrat middens from the Chihuahuan Desert. It was present in the Big Bend area of Texas (29.2°N) from 44,900 to 20,500 years B.P. but apparently did not become common until about 10,000 years B.P. (Van Devender 1990a). By 10,700 years B.P., *P. glandulosa* had reached the Hueco Mountains, Texas (31.7°N), and by 7,300 years B.P., it had arrived in the Sacramento Mountains, New Mexico (32.8°N) (Van Devender 1990a). Preliminary analysis of middens from Cataviñá, Baja California, indicates that *P. glandulosa* became common after chaparral species disappeared (Thomas R. Van Devender, personal communication 1993).

The plants may suffer considerable water stress in summer, even where they tap underground water supplies. Mechanisms for drought tolerance include diurnal osmotic adjustment and maintenance of high conductance in the morning (Nilsen et al. 1983, 1984). The ability to maintain high conductance values at low water potentials leads to notably high productivity for a desert plant: 3,650 kg/ha in the Salton Sink, California (Sharifi et al. 1982). In the Salton Sink, Sharifi and coworkers (1982) found that new woody tissues accounted for 51.5% of the net primary production, while leaves accounted for only 33.6%. Unripe fruits do produce some photosynthate; in years of heavy fruit load, however, complete pod development also requires reserve carbohydrates (Wilson et al. 1974; Fick and Sosebee 1981).

The roots produce nitrogen-fixing nodules (Palacios and Bravo 1974/1975; Felker and Clark 1980, 1982). Different strains of nodulating rhizobia dominate at different soil depths (Jenkins et al. 1989).

Judging from related species, seed germination is probably enhanced by relatively high temperatures (Peacock and McMillan 1965; Scifres and Brock 1971, 1972; Mooney, Ehleringer et al. 1977). Once established, young plants may become sexually mature in as few as 3 years (Haas et al. 1973). On coppice dunes in southern New Mexico, plant age was significantly related to canopy volume and to dune age (Graham-Gadzia and Ludwig 1983). Matched photographs from the Grand Canyon show individuals at least 100 years old (Robert Webb, personal communication 1993).

In *P. glandulosa* var. *torreyana*, phenological patterns seem relatively independent of precipitation (Nilsen et al. 1983, 1987). In West Texas, spring leaf initiation is a function of chilling requirements followed by warming minimum temperatures (Goen and Dahl 1982). In southern California, foliage grows during early spring (February–March), when temperatures are low, and again in summer (July–September), when temperatures are high (Sharifi et al. 1983). The rate of daily maximum photosynthesis is significantly higher in early leaves than in late leaves (Wan and Sosebee 1990). Flower production occurs soon after the flush of spring leaves. In some years, a second flowering occurs following the summer leaf flush (Sharifi et al. 1983; Nilsen et al. 1987). On average, each inflorescence contains 100 flowers but produces only one mature pod. Of the 12 seeds produced per pod in one study, 5 were killed by bruchid beetles (Nilsen et al. 1987).

Ants visit extrafloral nectaries on the leaf rachises and petioles (Pemberton 1988). The seeds are dispersed by javelina (Everitt and Gonzalez 1981) and livestock. Since the mid nineteenth century, this species has increased tenfold in parts of southwestern New Mexico (Buffington and Herbel 1965). A similar phenomenon has occurred with *P. velutina* in Arizona (Parker and Martin 1952) and *P. glandulosa* var. *glandulosa* in Texas (Archer et al. 1988).

This species has been widely used by native peoples (Felger and Moser 1971; Felger 1977). Ripe pods contain about 80% total carbohydrate. The seeds contain about 82% of the total protein in the pod (Harden and Zolfaghari 1988). The wood is heavily used today for fuel and fence posts.

Prosopis palmeri

S. Watson

Palo fierro, palo de hierro

Prosopis palmeri (figure 75) is a spiny shrub or tree up to 8 m tall with 3–10 pairs of leaflets on 1 pair of pinnae. The leaflets, 4 to 8 mm long, are oblong and finely puberulent to glabrate. The leaves are fascicculate on short shoots, which on older stems are black and up to 2 cm long (Burkart 1976). Flowers are in dense spikes 2–5 cm long. The fruits, 5–12 cm long and about 1 cm wide, are somewhat compressed but may occasionally be nearly round in cross section.

Both *P. palmeri* and *P. pubescens* belong to a subgeneric group (Strombocarpa) having paired stipular spines. (Other *Prosopis* species in the area mapped in this atlas have axillary spines.) The natural products chemistry of *P. palmeri* also indicates strong links with section Strombocarpa (Simpson et al. 1975).

Prosopis palmeri grows only in Baja

California Sur between 24.3°N and 26.8°N, usually along washes, in canyons, and on playas. Its geographic distribution suggests that it is a paleoendemic (Rzedowski 1988). The plants occur on no Gulf of California islands (Moran 1983b). Climates within the range are essentially frost free, with temperatures moderated by maritime air. Most rain falls between July and October. Winter storms also deliver moisture on occasion.

Figure 75. Prosopis palmeri *near Comondú, Baja California Sur. (Photograph by J. R. Hastings.)*

Prosopis pubescens
Bentham

Screwbean, screwpod mesquite, Fremont screwbean, tornillo

Prosopis pubescens is a tall shrub or tree 2–10 m high; it bears 5–9 pairs of leaflets 4–12 mm long on a single pair of pinnae. The stipular spines, 0.2–2.0 cm long, have bases that extend down the stem. The creamy or yellowish flowers are in dense spikes 2–5 cm long. The ripe pods are tightly coiled in spirals 5–7 mm wide and 3–6 cm long. The haploid chromosome number is 14 (Hunziker et al. 1975).

Fruits of *P. palmeri* are straight to

120° 118° 116° 114° 112° 110° 108°
36° 34° 32° 30° 28° 26° 24° 22°
114° 112° 110° 108°

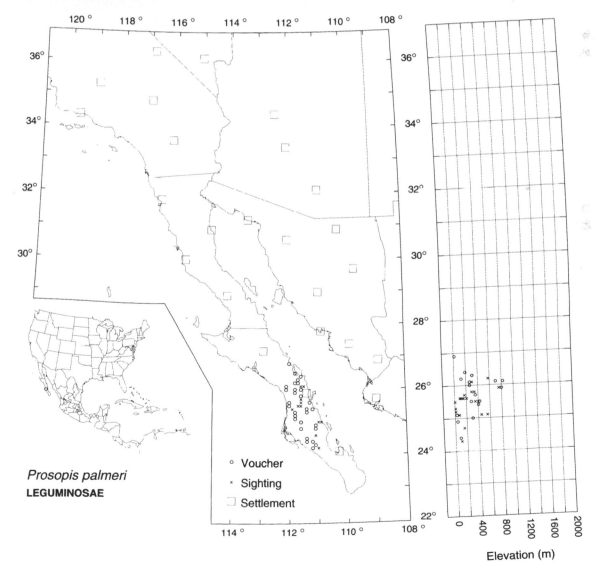

Prosopis palmeri
LEGUMINOSAE

° Voucher
× Sighting
□ Settlement

0 400 800 1200 1600 2000
Elevation (m)

slightly curved rather than tightly coiled. The ranges of *P. palmeri* and *P. pubescens* do not overlap (Rzedowski 1988).

Prosopis pubescens typically grows in well-watered habitats along watercourses and near ponds and seeps. Its range has shrunk as Sonoran Desert rivers have become dry or altered by changes in land use (Engel-Wilson and Ohmart 1979).

The leaves fall in winter. New leaves appear in spring before the flowers. Blooming starts in May and continues weakly through summer and fall. Fruits are produced throughout the summer (Ezcurra et al. 1988).

In one study, seedlings died when grown in sand cultures with a concentration of sodium chloride as low as 1.2% (Rhodes and Felker 1988).

The wood is used for fence posts, house construction, tool handles, and fuel. The plants are sometimes grown as ornamentals in southern Arizona (Brookbank 1978). Seeds are eaten by the American coot (*Fulica americana*) (Eley and Harris 1976).

Prosopis velutina Wooton

[=*P. juliflora* (Sw.) DC. var. *velutina* (Wooton) Sarg.]

Mesquite, velvet mesquite, algarroba, chachaca

This shrub or tree grows up to 17 m tall and has 9–30 pairs of leaflets 4–13 mm long on 1 or 2 pairs of pinnae. The foliage and fruits are densely pubescent. The paired spines, 1–2 cm long and borne on branchlets in the leaf axils, are usually somewhat unequal in the pair and are generally much reduced on older stems. The greenish yellow flowers are in spikelike

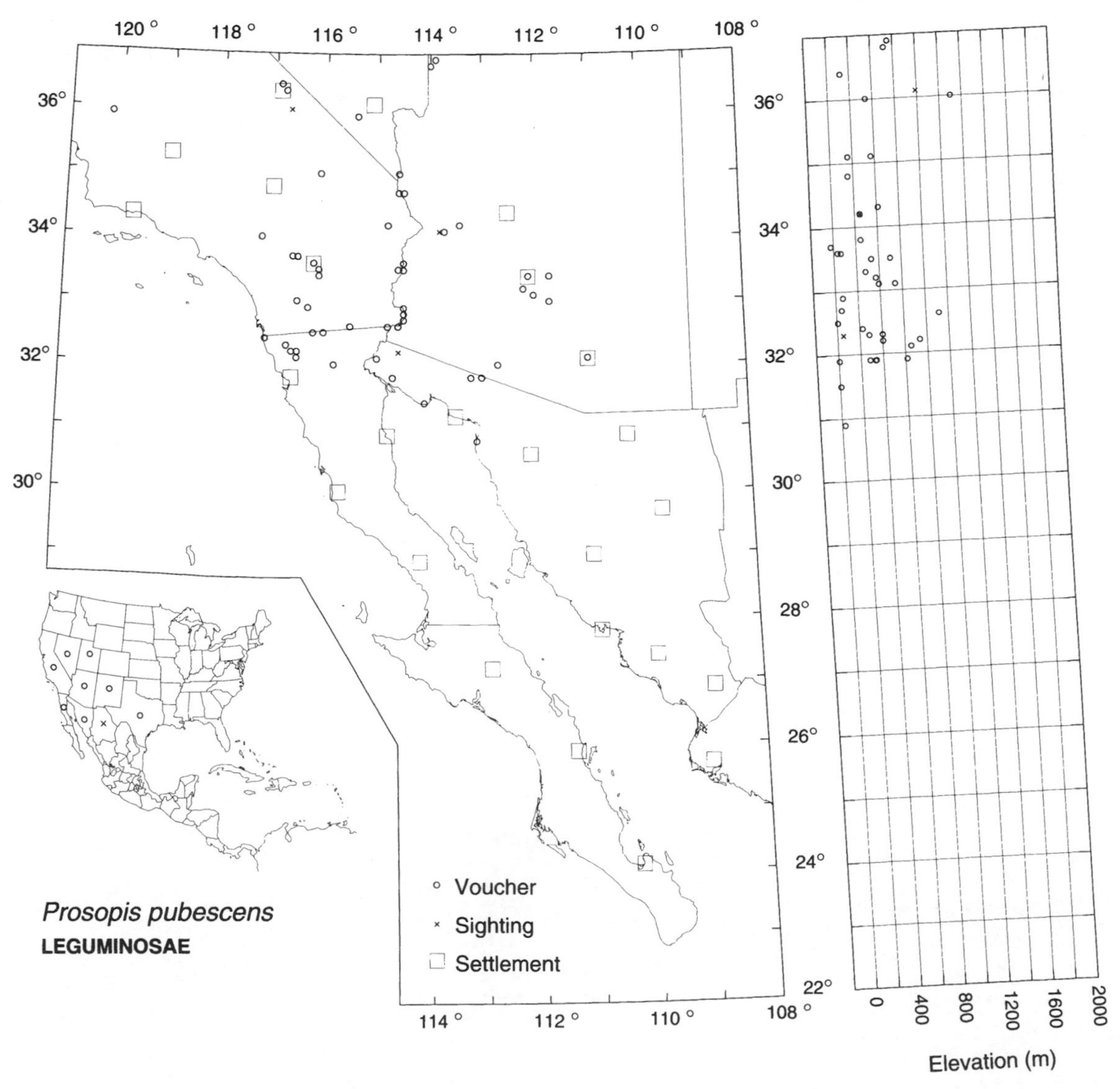

Prosopis pubescens
LEGUMINOSAE

racemes 5–12 cm long. The compressed pods, 1–2 dm long and 1–1.5 cm wide, are straw colored when mature. The haploid chromosome number in the typical diploid race is 14 (Hunziker et al. 1975; Solbrig et al. 1977). There is also a tetraploid race with n=28 (Cherubini 1954).

Prosopis pubescens has stipular spines and tightly coiled fruits. Where they overlap, *P. glandulosa, P. velutina,* and *P. articulata* may intergrade, making it difficult to tell them apart. In typical *P. glandulosa* var. *torreyana,* the pinnae are mostly 1 pair per leaf and the leaflets are larger, more widely spaced, and glabrous or ciliate. Fruits are more deeply constricted between the seeds in *P. articulata* than in *P. velutina.*

Some members of this species complex are taxonomically confusing. Plants near Guaymas, Sonora, and La Paz, Baja California Sur, for example, combine characters of *P. glandulosa, P. articulata, P. juliflora,* and *P. velutina* (Johnston 1962b).

In the eastern Sonoran Desert, where summer rainfall is relatively dependable, *P. velutina* is common along watercourses and frequent on upland sites. Along the northern and eastern margins of its range, it is densest above the desert. In December 1978 a severe freeze in the San Pedro Valley, Arizona, killed many plants to the ground (Glinski and Brown 1982). At one time, dense *P. velutina* forests were common along the major watercourses of the Sonoran Desert region, but fuelwood cutting and agricultural clearing have greatly reduced their extent. The northernmost point on our map is from Tanner's Crossing near Cameron, Arizona (Benson and Darrow 1981).

During the early Holocene, *P. velutina* became widespread in southern Arizona, often in sites now too dry for it. It grew on slopes in the Waterman Mountains, Arizona, around 11,700 years B.P., but around 3,800 years B.P. it retreated to nearby washes (Van Devender 1990b). In the Puerto Blanco Mountains, Arizona, *P. velutina* grew on slopes from 10,500 years B.P. to 1,000 years B.P. but is now restricted to washes (Van Devender 1987). The species apparently continued to expand its northern and upper elevational limits through the middle and late Holocene, arriving in the Hornaday Mountains, Sonora, by 10,000 years B.P. and at Eagle Eye Peak in west-central Arizona about 4,500 years B.P. (Thomas R. Van Devender, personal communication 1993).

Flowers appear dependably in spring after the leaves. In most years, a second bloom occurs in summer, but if the summer flush of new leaves is weak or fails, flowering also fails (R. M. Turner 1963; Solbrig and Cantino 1975). An environmental stimulus other than water availability, such as photoperiod, triggers the spring bloom (Solbrig and Cantino 1975). The intensity of flower production during the summer bloom is directly related to the amount of summer rainfall (Glendening and Paulsen 1955; R. M. Turner 1963; Solbrig and Cantino 1975; Simpson et al. 1977b). Although 220 to 240 flowers are produced per inflorescence, few develop into mature fruits. Only 26 flowers in 10,000 set fruit and only about 7 of these actually reach maturity (Solbrig and Cantino 1975), a yield of 0.07%. Glendening and Paulsen (1955) estimated that one individual produced 142,000 seeds each year. Fruit drop is a protracted process lasting 4 or more months. When two good crops are produced in a year, seeds may be added to the seed bank from June through January (Simpson et al. 1977b).

Germinability is a function of complicated interactions between the seed, the surrounding fruit tissue, and the environment. The mature fruit of *P. velutina* and other species in section Algarobia (e.g., *P. glandulosa, P. articulata*) comprises three layers. The outer layer is thin and skinlike; the middle layer, which contains much starch and sugar, is thick and spongy (Becker and Grosjean 1980); and the inner layer (the endocarp), which is impermeable to water, is thin, hard, and leathery. Seeds do not germinate unless the endocarp is broken (Solbrig and Cantino 1975). Once it has fractured or split, the decomposing endocarp apparently releases substances that increase the permeability of the seed coat and thereby enhance germinability (Glendening and Paulsen 1955). Since seed coats take varying lengths of time to become permeable, germination from a given cohort is staggered (Glendening and Paulsen 1955).

The seed coat becomes progressively less permeable to water as the seeds mature and dry. When removed from the pods, only 6–7% of dry seeds will germinate without scarification (Glendening and Paulsen 1955). Seeds may remain viable for up to 50 years when stored under artificial conditions (Martin 1948; Glendening and Paulsen 1955). Viability is greatly reduced by soil burial. Seeds in one study remained viable for 10 years (Tschirley and Martin 1960); in another they deteriorated so rapidly that none germinated after 3 years (Glendening and Paulsen 1955).

High temperatures enhance germination. Glendening and Paulsen (1955) found that 30°C is near the optimum and that germination percentages fall at higher and lower temperatures. Siegel and Brock (1990), however, achieved 100% germination from 21°C to 38°C. Most seeds probably germinate during the summer rainy period, after which viability diminishes rapidly (Tschirley and Martin 1960). Mooney, Simpson, and Solbrig (1977) suggested that the high temperature requirement for germination probably resulted from evolution under a summer rainfall regime and can be viewed as a means of preventing germination during winter rains.

For optimum germination, seeds must be buried by a thin layer of soil (1–2 cm) (Glendening and Paulsen 1955). Many seeds are probably buried by rodents such as *Dipodomys* (Glendening and Paulsen 1955) and *Perognathus* species. These and other rodents eat stored seeds and recently emerged seedlings. The lagomorphs *Lepus* and *Sylvilagus* also consume seedlings, especially during the arid foresummer (Paulsen 1950). The large numbers of seedlings destroyed by rodents and lagomorphs may strongly influence *Prosopis* demography.

Four species of bruchid beetles, two of

them obligately restricted to *P. velutina,* eat the fruits and seeds (Center and Johnson 1976; Kingsolver et al. 1977). Other bruchid predators are apparently deterred by the smoothness and indehiscence of the pods (Center and Johnson 1974; Kingsolver et al. 1977).

Although each flower produces only a small amount of nectar, there are approximately 80 flowers per inflorescence, which makes the nectar reward for a single visit greater than for almost any other plant in the desertscrub ecosystem (Simpson et al. 1977b). Because of the large number of stamens (approximately 800 per inflorescence), pollen production per floral unit is greater than for most other insect-pollinated desertscrub species in North America (Simpson et al. 1977b). The plants are self-incompatible (Simpson et al. 1977b).

Prosopis velutina roots extend deeply (more than 50 m) and spread widely (10–15 m) (Fisher et al. 1959; Phillips 1963; Cable 1977b). They produce nitrogen-fixing nodules (Palacios and Bravo 1974/1975; Felker and Clark 1980). The plants can contribute substantially to nitrogen and carbon levels in the soil (Barth and Klemmedson 1982; Tiedemann and Klemmedson 1986). A study of seasonal soil water use near Tucson, Arizona (Cable 1977b), showed that *P. velutina* experienced recharge during the winter rainy season. This recharge was followed by high moisture extraction during the arid foresummer as leaves, stems, flowers, and fruits developed. A second period of high moisture extraction occurred during the summer rainy period. Times of low water extraction coincided with autumn, when little or no rain fell, and winter, when the trees were leafless.

Seedling root and shoot growth is also attuned to high temperatures. *Prosopis velutina* is unusual in growing quickly at both high daytime temperatures (30°C) and high nighttime temperatures (26°C) (Hull 1956). In fact, growth of roots and shoots is strongly reduced when nighttime temperatures are only 9–10°C lower than the maximum daytime temperatures. The seedling taproots grow rapidly during the summer rainy season (Glendening

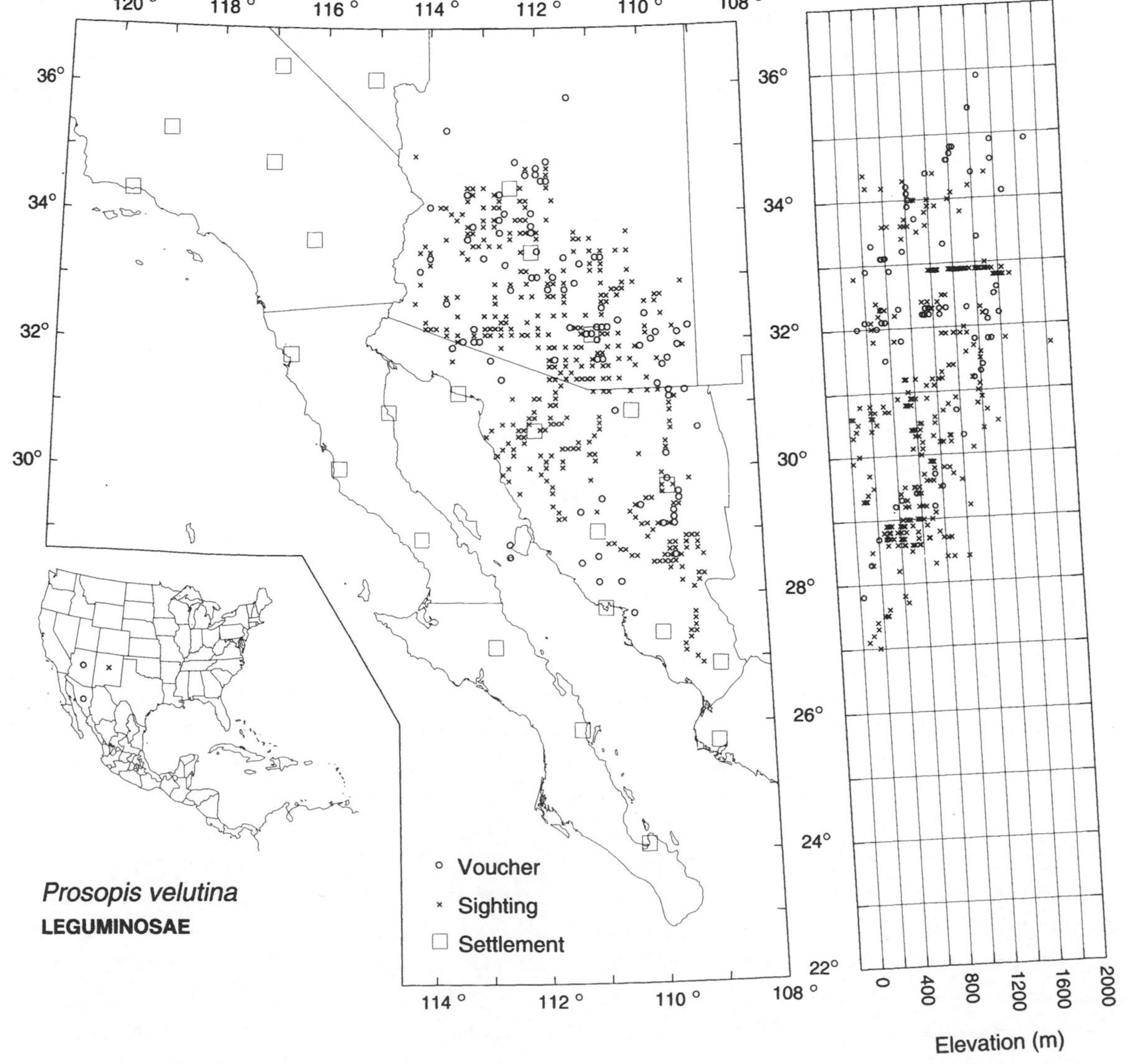

Prosopis velutina
LEGUMINOSAE

and Paulsen 1955). Solbrig and Cantino (1975) suggested that the high temperature optimum for root growth helps maximize the depth to which the taproot grows during summer rains, which in turn maximizes the probability of survival until the next rain. Seedlings grown in a nitrogen-free sand culture demonstrated good tolerance for a salinity of 6,000 mg/l of sodium chloride but did poorly at the next higher level of 12,000 mg/l (Felker, Clark et al. 1981).

Since the early 1900s, *P. velutina* has expanded into former grassland sites above Sonoran desertscrub (Glendening and Paulsen 1955; Humphrey 1958; Hastings and Turner 1965b). This expansion has been attributed to the introduction of domestic livestock in the late 1800s (Glendening and Paulsen 1955; Humphrey 1958). The plants are well adapted to invade and occupy new habitats. The highly viable seeds, much sought by domestic livestock and native animals, pass unharmed through mammalian digestive systems, then are deposited in dung, often far from their source. Under certain conditions *P. velutina* may become established in great numbers even in the absence of domestic herbivores (Turner 1990).

Exact age determination is difficult because the annual rings are obscure (Ferguson and Wright 1962), but if one may judge from the great size of some trees and the slow growth typical of the species (R. M. Turner 1963), *P. velutina* probably becomes several hundred years old.

The plants are frost hardy down to −15°C. They provide filtered shade in yards and gardens and make attractive ornamentals (Johnson 1993).

Prosopis velutina was widely used by native peoples for food, shelter, weapons, fibers, medicine, and fuel (Bell and Castetter 1937; Felger 1977). It is still an important fuel throughout its range, although fuelwood harvesting of this species is now strictly controlled in some areas. The species shows some promise for revegetating asbestos mill waste in the southwestern United States (Perry et al. 1987). *Prosopis velutina* is the only known host for *Incisitermes banksi,* a rare drywood termite (Nutting 1979).

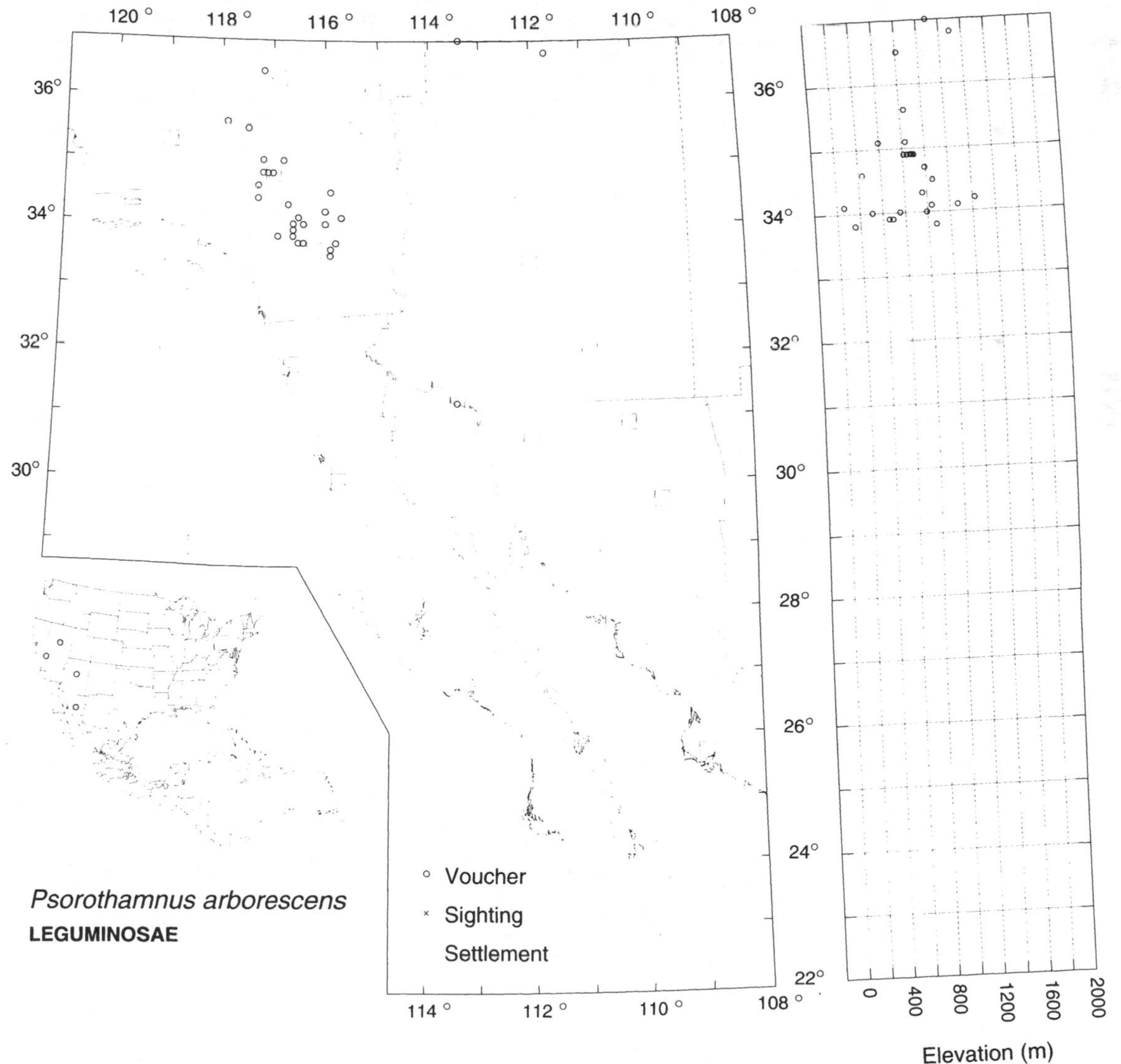

Psorothamnus arborescens
LEGUMINOSAE

Psorothamnus arborescens

(Torrey) Barneby

[=*Dalea arborescens* Torrey, *D. fremontii* var. *saundersii* (Parish) Munz, *D. fremontii* var. *minutifolia* (Parish) L. Benson, *D. fremontii* var. *simplifolia* (Parish) L. Benson, *D. fremontii* var. *pubescens* (Parish) L. Benson]

Mojave dalea, indigo bush

This stiff and irregularly branched shrub 3–10 dm tall may be thorny (var. *arborescens*) or unarmed. The foliage may be green and glabrate or gray with silky pubescence. The leaves, 1.3–3.5 (occasionally 5.5) cm long, are once-pinnate with 5–7 (occasionally 17) ovate to linear leaflets. In one variety, the leaflets are occasionally reduced to simple phyllodia. The indigo-blue to violet-purple flowers, 6–10 mm long, are in loosely flowered racemes 1.5–12 cm long. The ellipsoid pods (8–10 mm long) are dotted with large, blisterlike glands. Chromosome counts are given by Raven and coworkers (1965) and by Barneby (1977).

Psorothamnus schottii is taller with (usually) simple leaves that are linear rather than elliptic as in *P. arborescens* var. *simplifolius*. Pods of *P. fremontii* are covered with small orange glands that fuse into ridges, and its stems are seldom (or not markedly) thorny.

Barneby (1977) delimited four subspecific taxa: var. ***arborescens***, var. ***pubescens*** (Parish) Barneby, var. ***minutifolius*** (Parish) Barneby, and var. ***simplifolius*** (Parish) Barneby. Our map combines them. Variety *simplifolius* has been treated as a distinct species, but transitional forms between it and var. *minutifolius* suggest that it is best retained in *P. arborescens* (Barneby 1977). The unifoliate form of var. *simplifolius* grows where *P. arborescens* and the closely related *P. schottii* overlap in distribution. Variety *simplifolius* might be a result of introgression between *P. arborescens* and *P. schottii* (Barneby 1977).

The four varieties are fairly well separated geographically. Variety *arborescens* grows widely in the Mojave Desert. There

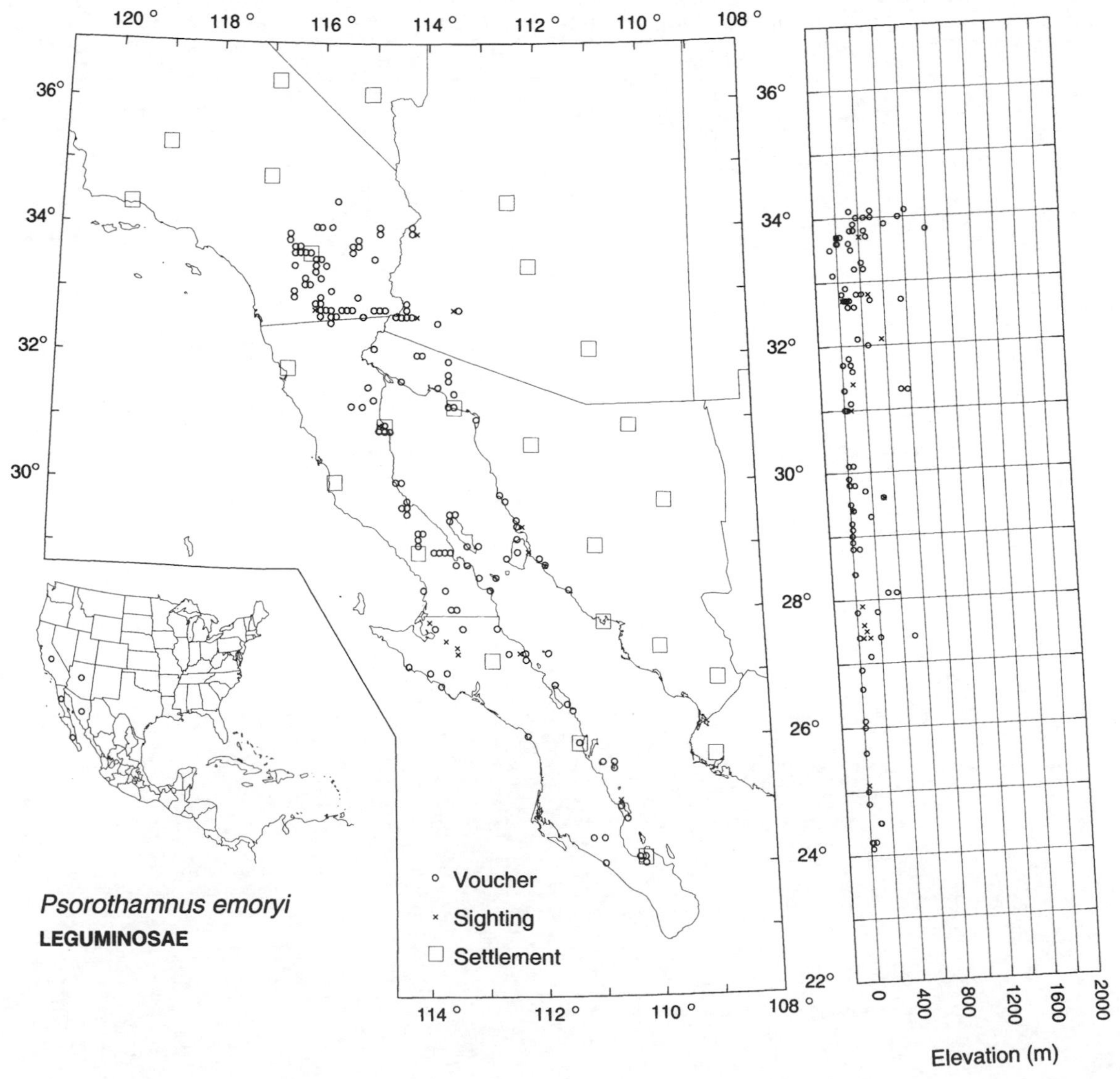

is an outlying population at Puerto Peñasco, Sonora (31.3°N, vouchered by Lawrence Cook in 1934). Variety *minutifolius* is found in the eastern Mojave Desert and adjacent Nevada, var. *simplifolius* in the Coachella Valley and San Bernardino Mountains, southern California. Variety *pubescens* is known only from a limited area along the Colorado River and tributaries in northern Arizona.

The plants grow on gravelly plains and hillslopes in areas of predominantly winter rainfall. The distribution seems to be squeezed between that of *P. schottii* and that of *P. fremontii*.

Flowers appear February–June, rarely in July and August.

Psorothamnus emoryi

(A. Gray) Rydb.

[=*Dalea emoryi* A. Gray]

Indigo bush, white dalea

Psorothamnus emoryi varies from a densely branched chamaephyte to a dwarf shrub up to 1 m high with densely felty-pubescent foliage and abundant orange glands on stems and petioles. Leaves are odd- pinnate with 3–9 leaflets. The lateral leaflets are obovate and 4–8 mm long; the terminal leaflet is oblanceolate and up to 14 mm long. The magenta or violet flowers are in dense heads about 1.5 cm across. The haploid number of chromosomes is 10 (Raven et al. 1965).

Errazurizia megacarpa has yellow flowers and orbicular leaflets.

Barneby (1977) considered *Dalea tinctoria* Brandegee var. *tinctoria, D. tinctoria* var. *arenaria* Brandegee, *P. dentatus* Rydb., and *D. juncea* (Rydb.) Wiggins to be synonyms of *P. emoryi,* a variable species with two varieties. The most widespread of these, var. *emoryi,* grows in Sonora and as far south as Mulegé in Baja California Sur. Magenta flowers, a definitely shrubby habit, and a relatively long calyx (4.3–7.2 mm) distinguish it from var. *arenarius* (Brandegee) Barneby, which has violet flowers, an herbaceous or semi-woody habit, and a relatively short calyx (3–4 mm). Variety *arenarius* is virtually restricted to Baja California Sur. The two varieties overlap on the peninsula between 26.5°N and 28.5°N. Genetic drift is segregating them but has not yet established discrete species (Barneby 1977).

Psorothamnus emoryi grows on stable dunes, along washes, and on sandy flats. The range encompasses arid climates around the Gulf of California and the lower Colorado River Valley. Frost is uncommon in the northern part of this range and absent in the south. Rainfall is biseasonal and shifts from winter dominance in the north and west to warm-season dominance in the southern peninsula.

Flowers appear from April to June and sometimes again in the fall. In summer the leaves drop, leaving a tangle of branches (Barneby 1977).

Seri Indians boiled the foliage and stems in saltwater to make a yellowish dye and used the branches in thatching brush houses (Felger and Moser 1985).

Psorothamnus fremontii

(Torrey) Barneby

[=*Dalea fremontii* var. *johnsoni* (S. Watson) Munz, *D. johnsoni* S. Watson in King var. *minutifolia*]

Fremont dalea, indigo bush

Psorothamnus fremontii is a low, straggling shrub 3–10 dm tall with stiff branches that become gnarled and twisted in age. The foliage is grayish- or silvery-pubescent. The leaves, 1–4 cm long, are once-pinnate with 3–7 (occasionally 9) obovate to linear leaflets 3–25 mm long. The vivid magenta-purple flowers 6–9 mm long are in loosely flowered racemes 2–9 (occasionally 11) cm long. Pods, 7–10 mm long, are marked with many small orange glands that coalesce into lengthwise ridges, making the surface appear resinous and wrinkled.

Pods of *P. arborescens* are dotted with large, blisterlike glands, and its stems are usually thorny.

This species may not be distinct from *P. arborescens* (Barneby 1977). Barneby (1977) defined var. *fremontii* and var. *attenuatus* Barneby based on leaflet size and shape. The latter has been treated as *Dalea fremontii* var. *johnsoni* (Kearney and Peebles 1960) and as *D. johnsoni* var. *minutifolia* (Munz and Keck 1959). Where the varieties overlap, as in southern Nevada, it can be difficult to identify individual specimens.

Psorothamnus fremontii enters the Sonoran Desert in western Arizona along the Bill Williams River, which may have served as a distributional corridor from the Colorado River. In general, var. *fremontii* grows on sedimentary rocks, var. *attenuatus* on granitic and volcanic rocks (Barneby 1977). Climates over much of the range are warm-temperate, receiving occasional to frequent frost. Rainfall is biseasonal, with winter-spring moisture more dependable than that of summer-fall.

Packrat middens record *P. fremontii* about 10,000 years B.P. near Ash Meadows, Nevada (Spaulding 1983). By 8,300 years B.P., the species had reached the present northern limit of Mojave Desert vegetation in the northern Eureka Valley of Death Valley National Monument (Spaulding 1990). The species has evidently occupied much of its present range throughout most of the Holocene.

Flowers appear from March to June.

Psorothamnus schottii

(Torrey) Barneby

[=*Dalea schottii* Torrey]

Mesa dalea, indigo bush

This intricately branched shrub, 1–2.5 m high, has flexible, green branchlets. The linear, revolute leaves, 1–2 mm wide and 1.5–3 cm long, are simple and entire. The vivid blue flowers, 7 mm long, are in loose racemes 5–7 cm long. The elliptic pods are 8–12 mm long and are dotted with red glands. The haploid chromosome number is 10 (B. L. Turner 1963). This species produces nitrogen-fixing root nodules (Eskew and Ting 1978).

Psorothamnus spinosus is a bushy tree with silvery or gray foliage. *Psorothamnus arborescens* var. *simplifolius* is a smaller shrub with pinnate or simple leaves. When simple, the leaf blades of *P. arborescens* are shorter than in *P. schottii* and are elliptic, not linear.

Psorothamnus schottii grows in a variety of habitats, including gravelly benches, alluvial fans, sandy or rocky washes, and coppice dunes (Barneby 1977). The single Arizona location is based on a collection by J. G. Lemmon near Yuma in 1881 (Barneby 1977). The distribution is restricted to the western half of Shreve's (1964) Lower Colorado Valley Subdivision of the Sonoran Desert. In this area, frost is uncommon and rainfall is erratically biseasonal, with winter as the season of most dependable rainfall.

Flowering is from November to May. Where *P. schottii* occurs with *P. spinosus*, the last *P. schottii* flowers appear about a month before the first blossoms of *P. spinosus* (Barneby 1977).

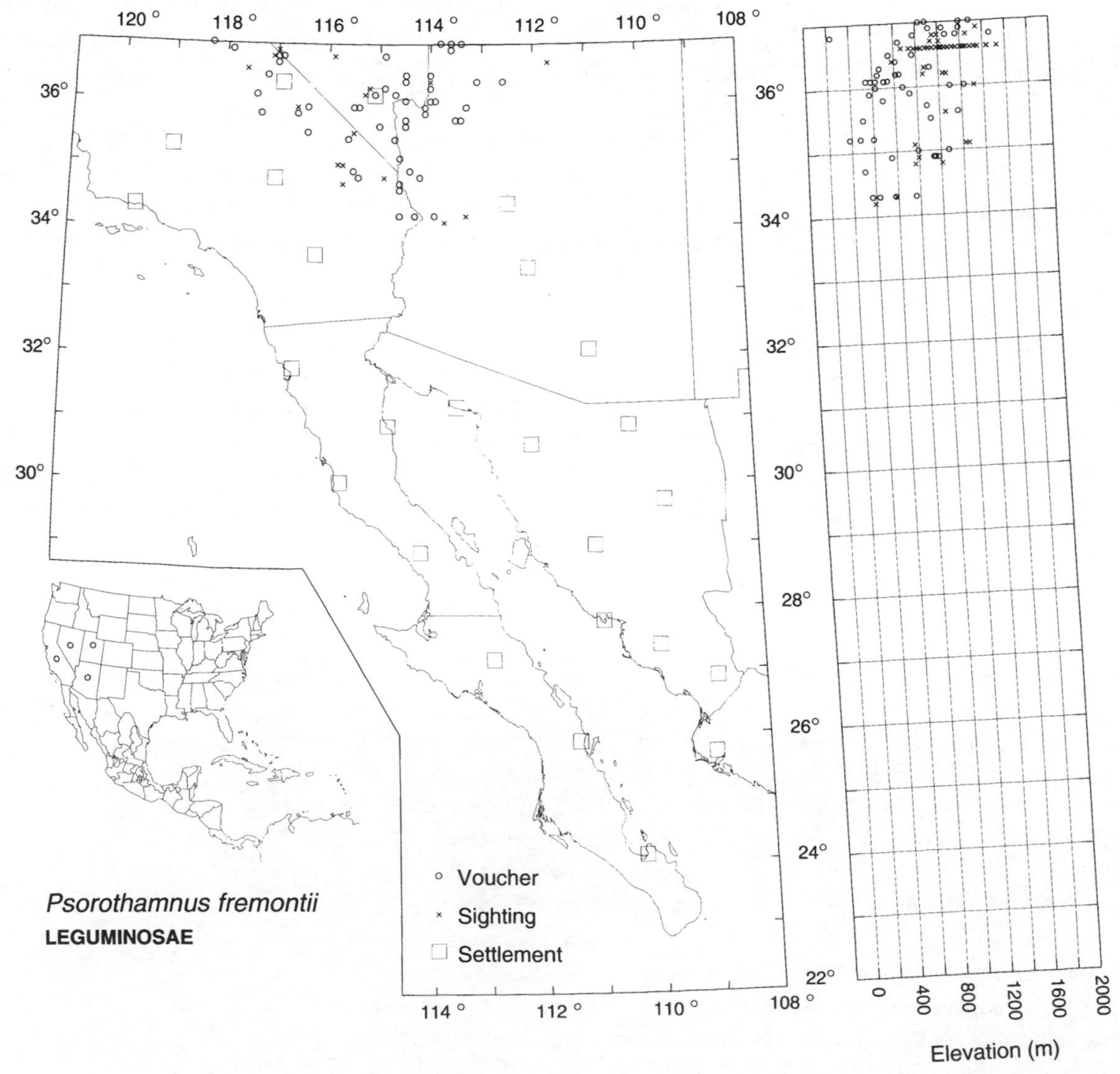

Psorothamnus fremontii
LEGUMINOSAE

Psorothamnus spinosus

(A. Gray) Barneby

[=*Dalea spinosa* A. Gray]

Smoke tree, smoke thorn, indigo bush, indigo thorn, mangle, corona de Cristo

Psorothamnus spinosus (figure 76), a small tree 2–8 m tall with simple, oblanceolate leaves 6–9 mm long, is almost entirely gray or silvery with dense, appressed pubescence. The sharp-tipped branches are stiff and repeatedly forked. The dark purple to deep blue flowers, 9–10 mm long, are in rather dense racemes up to 1.5 cm long. The diploid chromosome number is 10 (Munz 1974).

Other desert trees (*Cercidium floridum, C. microphyllum, Olneya tesota, Prosopis* species) are either green stemmed or pinnate leaved, or both. *Psorothamnus schottii,* a large shrub, has green stems and longer (5–7 cm) racemes. *Castela emoryi* also has green stems and, in addition, persistent, woody capsules.

Psorothamnus spinosus is locally dominant in and along large washes. The easternmost point represents a collection by Robert H. Peebles 10 km west of Gila Bend. The distributional gap in southeasternmost California is probably more apparent than real; much of the area is under military control and therefore is closed to the public. Although infrequent above 500 m (Benson and Darrow 1981), *P. spinosa* does range as high as 780 m. At Organ Pipe Cactus National Monument in southwestern Arizona, many plants suffered frost damage in December 1978 when minimum temperatures dropped to −6°C (see also Johnson 1993). The species is probably limited to the north and east by low temperatures (Bradley 1966). Throughout the range, early summer is often dry and very hot. Winter moisture is more dependable than the occasional late-summer and fall storms, but warm-season rains may be necessary for seedling establishment.

Psorothamnus spinosus blooms reliably except in years of extreme drought (Barneby 1977). Buds form in April or May, and the plants flower through June and

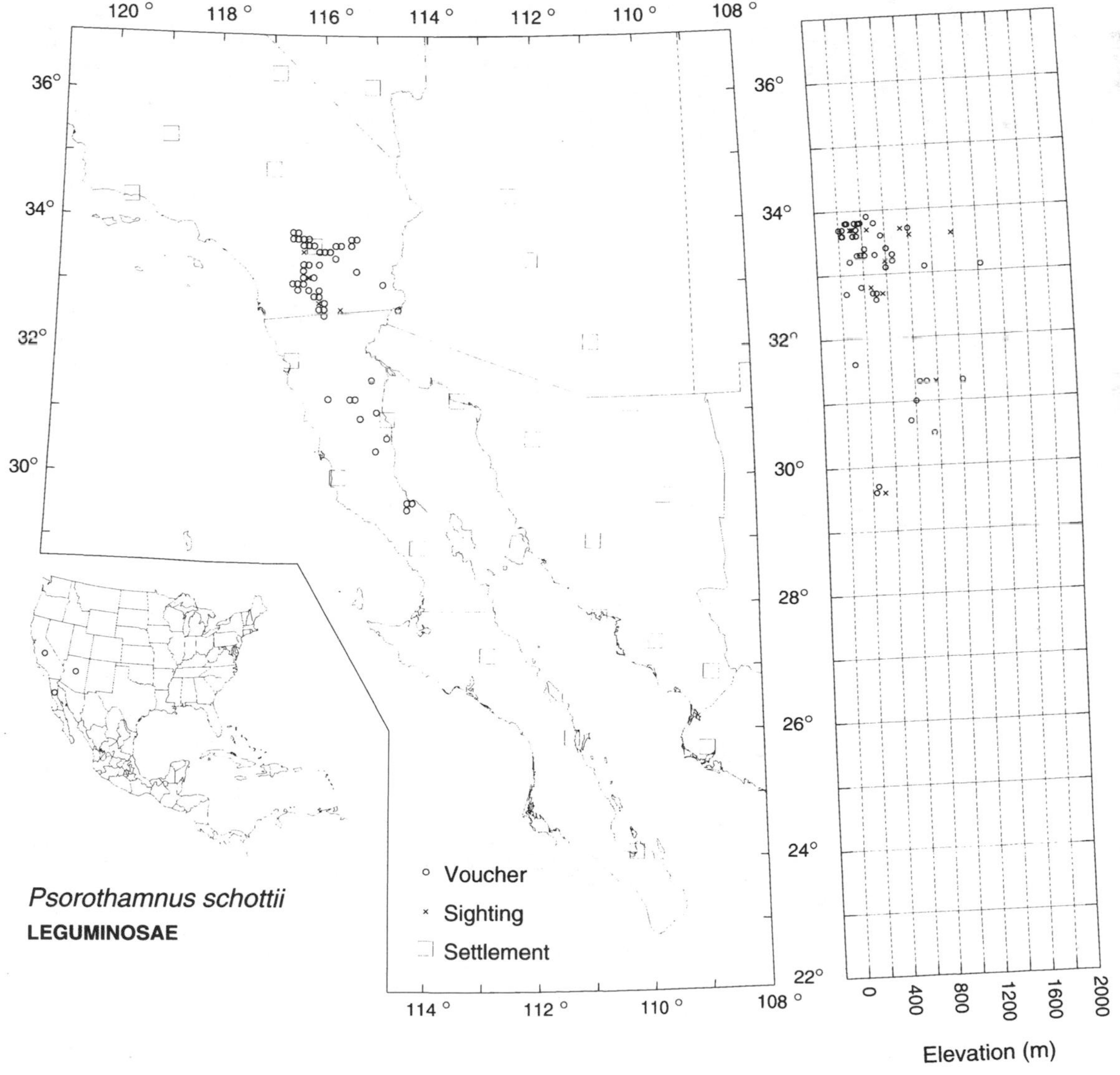

into early July. Sporadic flowering may occur in other months. The seeds germinate after fall or winter rains. They probably require scarification—for example, abrasion in the sand and gravel of washes (Went 1957; Barneby 1977). The large, oblanceolate seedling leaves permit rapid growth while the soil is moist (Barneby 1977). Apparently they are shed with the onset of seasonal drought. The much smaller adult leaves also appear in response to autumn and winter rains and

Figure 76. Psorothamnus spinosus *in an arroyo west of Sonoyta, Sonora. (Photograph by D. T. MacDougal, courtesy of the Arizona Historical Society/Tucson.)*

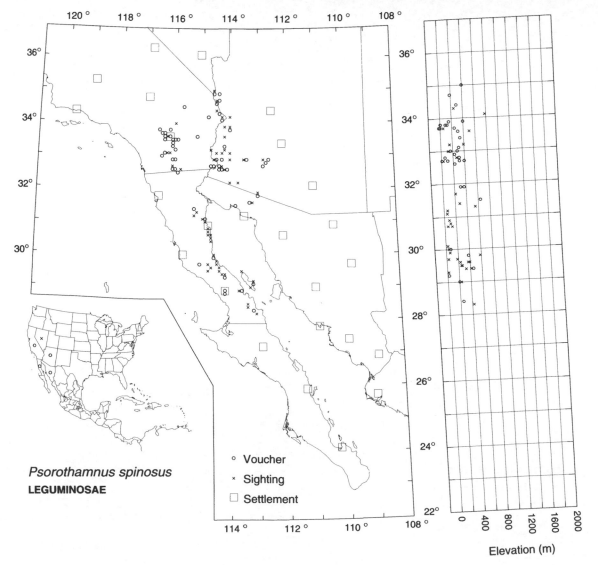

Psorothamnus spinosus
LEGUMINOSAE

○ Voucher
× Sighting
□ Settlement

Elevation (m)

drop once the soil dries out. Plants grown in deep sand under greenhouse conditions required only four waterings in two and one-half years (Went 1957). Seedling survival is facilitated by food reserves (sugar, starch, and oil) in root nodules and cortex (Scott 1938). The root nodules fix nitrogen (Eskew and Ting 1978).

Most of the photosynthesis is carried out by the branchlets, which are chlorophyllous beneath their dense pubescence (Barneby 1977). Maximum photosynthesis occurs at 39°C, and positive rates are maintained up to 51°C. At temperatures of 27–35°C, photosynthesis rates drop substantially (Nilsen et al. 1989). The degree of leaf pubescence is controlled by temperature (Went 1957).

Biologists have made no studies of seedling establishment or life span. Seedlings are no doubt vulnerable to burial and scour in their wash habitat. The adults are said to mature rapidly and perish early (Barneby 1977).

Randia echinocarpa

Sess. & Moc. in DC.

Papache, papache picudo, cirián chino, chacua, grangel

Randia echinocarpa is a rigid shrub 2–3 m tall with diverging branchlets and 4 whorled spines 1–3 cm long at each stem tip. The oval, rhombic, or obovate leaves (3–8.5 cm long and 1.5–5 cm wide) are also clustered at the stem tips. The tubular, white or pale yellow flowers, 25–32 mm long, may be single or clustered. The fruits, 4.5–10 cm wide, are subglobose and conspicuously tuberculate with flattened horns up to 3 cm long. Fruits and foliage are puberulent.

No other species combines terminal spines and leaf clusters with large, tuberculate fruits.

Rare to common along watercourses and on hillsides, *R. echinocarpa* is a plant of thornscrub and thornforest. Its large elevational range (40–1,200 m) is not surprising, given its wide distribution along

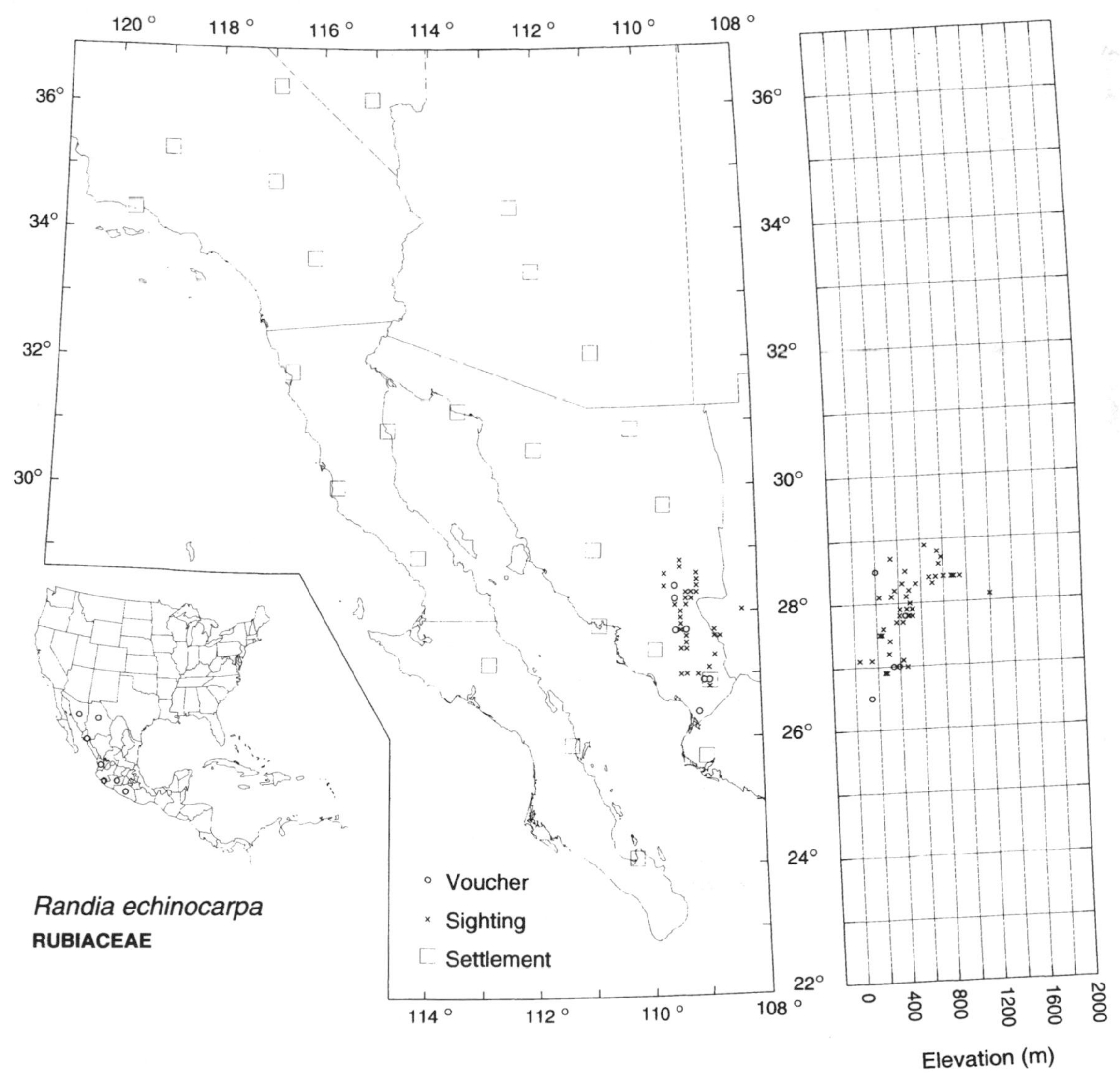

Randia echinocarpa
RUBIACEAE

the Pacific slope of Mexico.

Flowers appear from June to August. Fruits ripen in the winter (Gentry 1942). In exceptionally wet years, the leaves may remain year-round, but generally they drop in spring (Gentry 1942). Birds eat the leathery fruits (Gentry 1942) and may disperse the seeds, which are embedded in black, slimy pulp. The large size, copious pulp, and sturdy endocarp of the fruits suggest that they evolved with now-extinct Pleistocene megafauna as their main dispersal agents (ground sloths, for example) (Janzen and Martin 1982).

The fruits are edible, although bitter-sweet, and are often gathered while green to forestall competition with birds (Gentry 1942). In Sinaloa, they are considered a remedy for malaria (Standley 1926). The Tarahumara Indians used the bark in making an alcoholic beverage from corn and other plants (Pennington 1963).

Randia laevigata

Standley

Crucecilla de la sierra, sapuche

This sparsely branched, unarmed shrub or small tree grows up to 3 m tall and has elliptic, rhombic, ovate, or obovate leaves 10–25 cm long and 5–12 cm wide. The leaves are opposite and glabrate and are marked beneath with 10–13 conspicuous lateral veins. Flowers are yellow and tubular. The obovate fruits are 6–10 cm long.

No other *Randia* species in the area mapped by this atlas is unarmed. The large, opposite, multiple-veined leaves are distinctive.

Randia laevigata grows on rocky hillsides and canyon walls, generally in Sinaloan thornscrub (Shreve 1964). In the Río Mayo region, the scattered colonies occur just below the oaks on open slopes (Gentry 1942).

Flowers appear from June through August, and fruits ripen in November (Gentry 1942; Wiggins 1964). The leaves are presumed to be drought deciduous.

The fruit has been used to treat bron-

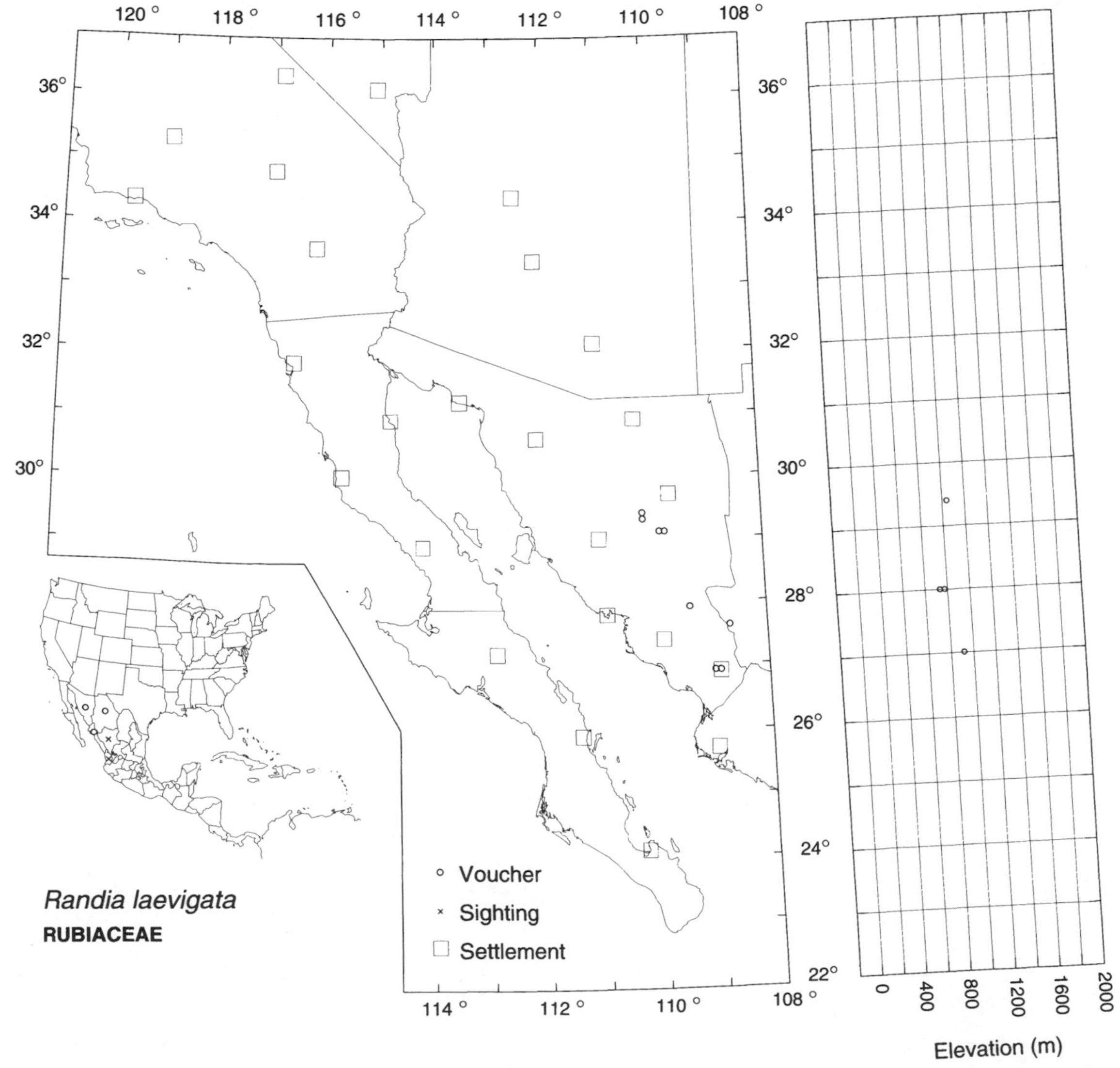

chitis (Standley 1926). The Tarahumara Indians used the moistened, pounded bark in fermenting the alcoholic beverage known as tesguino (Pennington 1963).

Randia megacarpa
Brandegee

This rigidly branched shrub 3–6 m tall has 4 stout spines, 5–15 mm long, at the branch tips. The ovate to obovate leaves, also tufted at the branch tips, are 2.5–8.5 cm long and 1.5–7.5 cm wide. They are short-pilose above and densely tomentose beneath. The tubular, white flowers are 18–20 mm long. The pulpy, globose fruits, 3–5 cm in diameter, are on a short stipe.

Randia armata (Swartz) DC. has leaves bright green above and puberulent along the veins beneath. *Randia obcordata* has paired rather than whorled spines.

This Baja California Sur endemic grows along arroyos and on canyon slopes and rocky hills. Although *R. megacarpa* does enter the desert around 26°N, its closest affinities are with Sinaloan thornscrub. The distribution is confined to areas with warm winters and fairly dependable rains arriving mostly in late summer or fall.

Flowers appear in July and August (Wiggins 1964), occasionally later.

Randia obcordata
S. Watson

Papache borracho, papachillo

The wandlike branches of this shrub, which grows 2–4 m tall, are conspicuously warty with leaf-bearing spurs. The stout, paired spines are 4–10 mm long. The glabrous, tufted leaves (4–20 mm long and wide) may be flabelliform, cuneate-obovate, or obcordate. The small, white, tubular flowers are solitary and sessile. The fruits, 5–9 mm in diameter, are black and lustrous at maturity with a thick, hard pericarp.

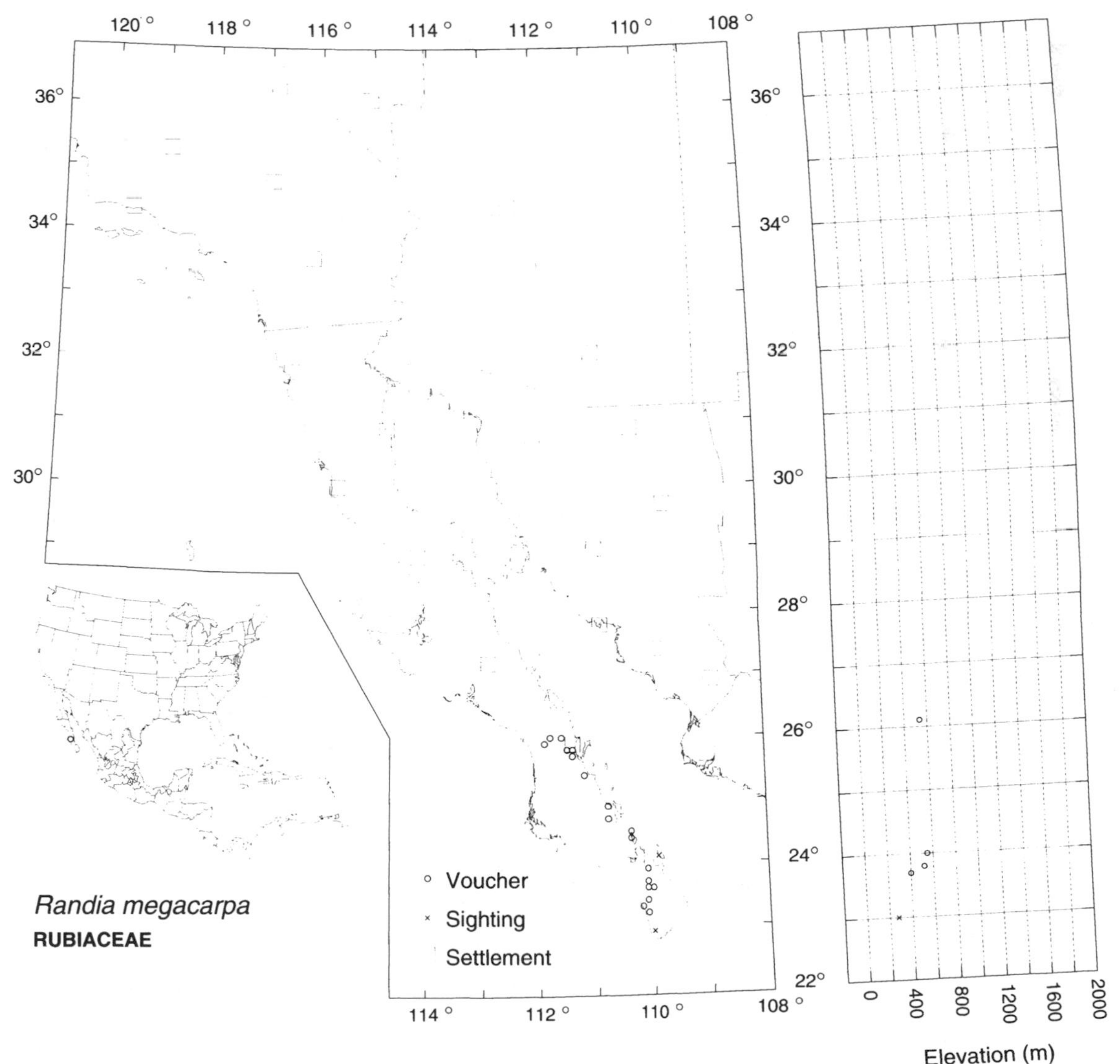

Randia megacarpa
RUBIACEAE

Randia armata, R. megacarpa, and *R. sonorensis* have whorled spines. *Randia thurberi* and *R. obcordata* can be hard to tell apart without fruits; generally, *R. thurberi* has divaricately spreading branches and small oval, orbicular, spatulate, or obovate leaves.

Randia obcordata may be locally common on rocky hillsides, as in the Cedros Valley, Sonora (Gentry 1942). Because sterile plants are difficult to identify, the species may be more frequent than we show. The Guaymas voucher (27.9°N) represents an 1887 collection by Edward N. Palmer.

This plant of Sinaloan thornscrub and Sinaloan deciduous forest reaches its northern limit near Cumpas, Sonora, where freezing occurs nearly every winter (Shreve 1964). Northern populations probably rely on microclimates (for example, rocky slopes and lava flows) for protection from low temperatures (Shreve 1964). The scattered populations on our map may represent relicts of a once widespread thornscrub formation that, under increasing aridity, retreated to relatively mesic sites.

Fruiting specimens have been collected in January, March, May, July, September, November, and December, which suggests that the plants flower intermittently throughout the year, probably in response to rain. The leaves are drought deciduous (Gentry 1942).

The fruits are sickening to humans, thus the common name *borracho,* "drunk" (Gentry 1942).

Randia sonorensis
Wiggins

This shrub grows 2–3 m tall; it has slender, divaricate twigs tipped with 4 subulate spines 6–15 mm long. The twigs are pubescent with rusty-brown hairs. The densely puberulent, elliptic to suborbicular leaves, 6–22 mm wide and 8–33 mm long, may be alternate on long shoots or fasciculate at the stem tips. Flowers are

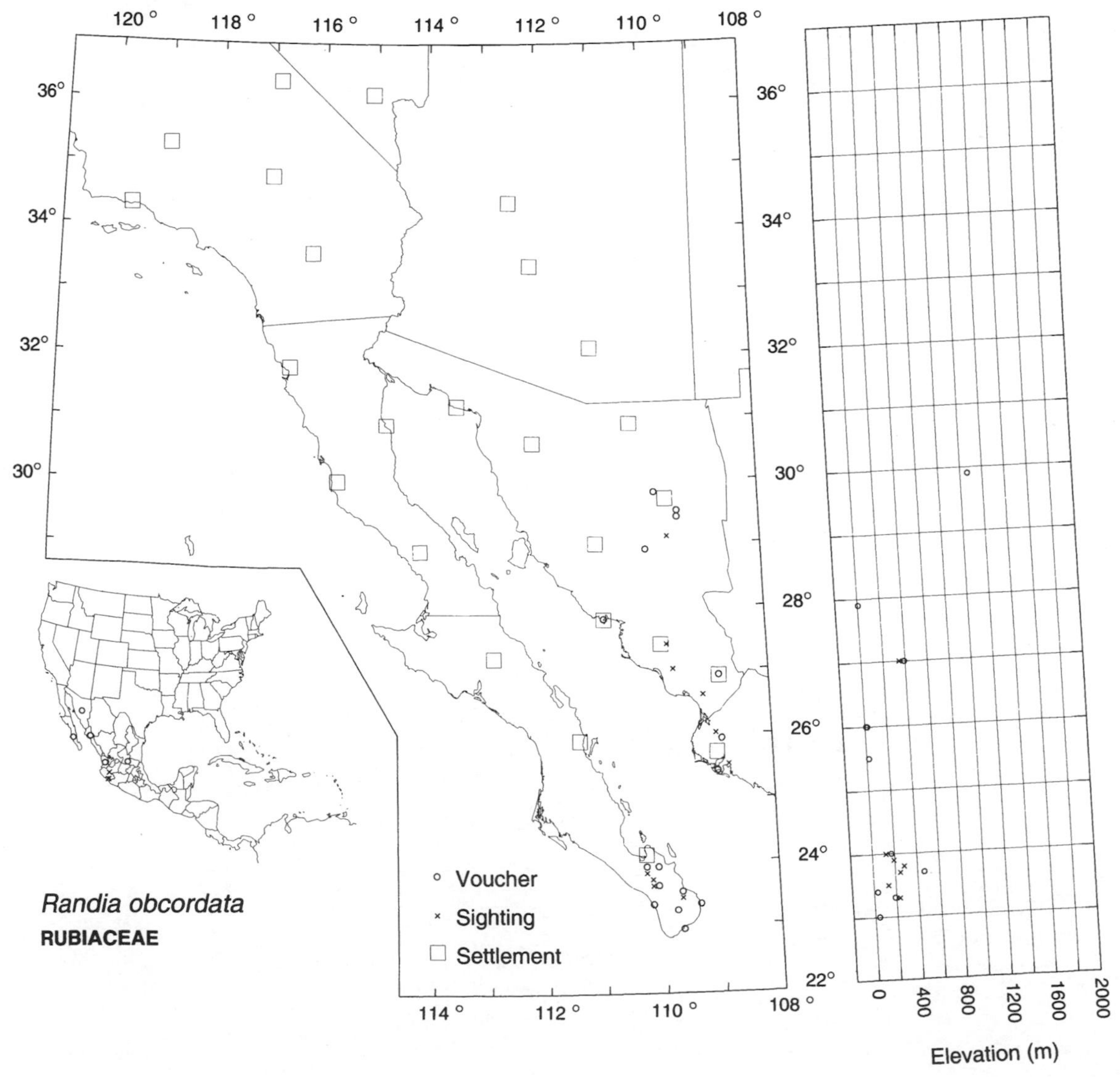

yellowish white or white. The globose, pulpy fruits are 2–3.5 cm in diameter.

No other *Randia* species in its range combines whorled spines with relatively small, fasciculate leaves.

Rare to common along washes and on rocky slopes, *R. sonorensis* is nearly confined to Sinaloan thornscrub in eastern Sonora and adjacent Chihuahua. The northernmost point (30.6°N) represents a collection by Raymond M. Turner at the west base of the Sierra el Tigre.

The reduced leaves suggest modification for increasing aridity in this and several other *Randia* species (for example, *R. obcordata* and *R. thurberi*), yet these plants barely if at all enter the desert proper. Their present distribution is shaped by both winter temperature and rainfall requirements.

The leaves are drought deciduous. The flowering season is May through August; blooming begins at the end of the dry season before the leaves are fully developed. Ripe fruits can be found from September to November.

Randia thurberi

S. Watson

Papache

Randia thurberi is a compact shrub (1–2.5 m tall) with stout, divaricate branches armed with paired spines 8–15 mm long. The leaves (6–45 mm long and clustered on short lateral spurs) are oval, orbicular, spatulate, or obovate. The white flowers open at dusk. The globose fruits are 12–25 mm in diameter.

At maturity, the fruit is smooth and greenish yellow in *R. thurberi,* black and rugulose in *R. obcordata.* Branching and leaf characters are not always useful

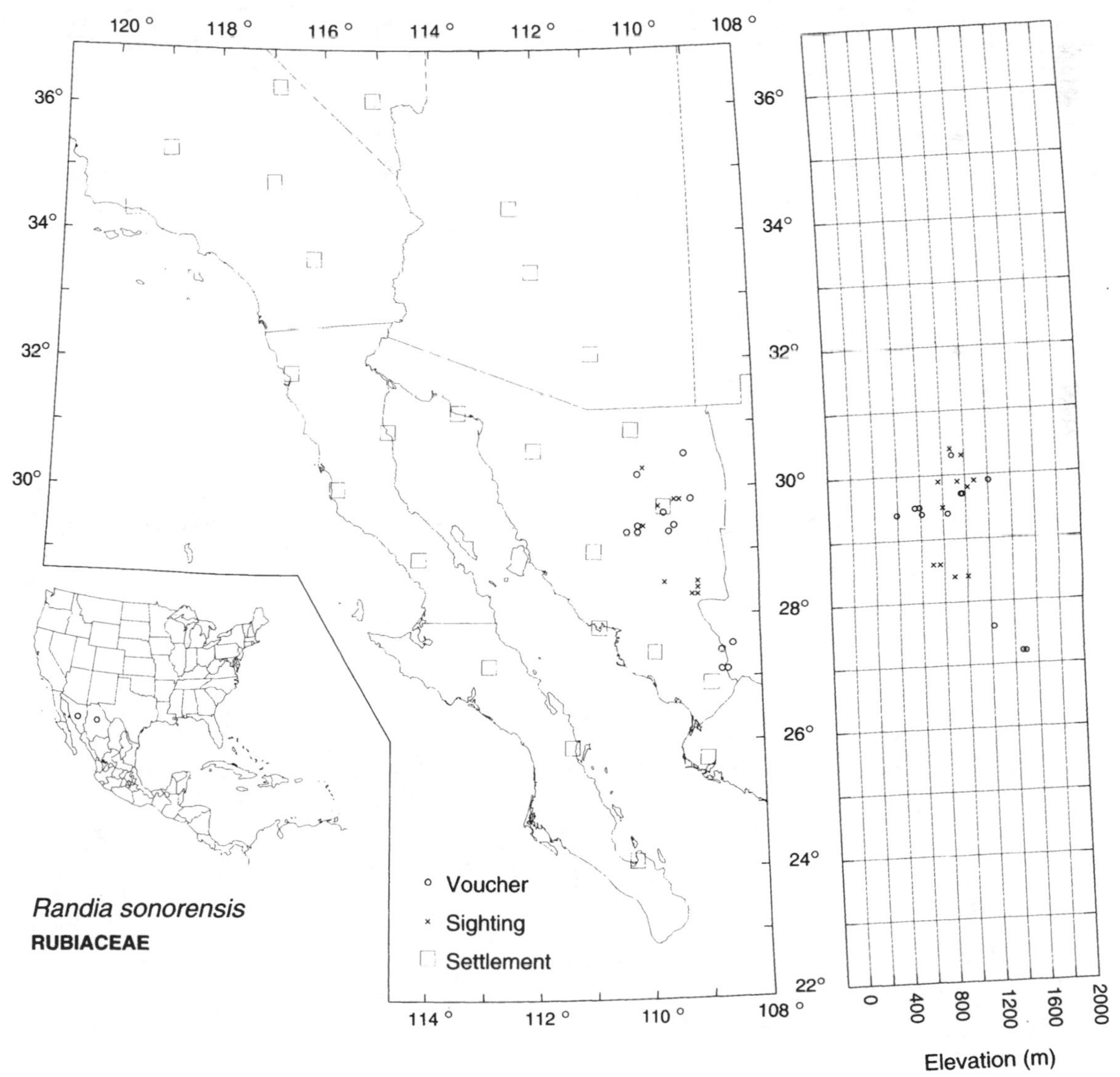

Randia sonorensis
RUBIACEAE

in distinguishing the two.

This plant of washes and rocky slopes is the common *Randia* species of central Sonora and the only one at all frequent in the desert proper. The Sinaloan record is Raymond M. Turner's from Topolobampo in 1964. Some of the sightings on our map may actually be *R. obcordata.*

The northern limit appears to be truncated by cold. The distribution encompasses the transition from Sinaloan thornscrub to Sonoran desertscrub.

Flowers appear from June to August and occasionally in winter. Fruits can be found from August to April. Birds reportedly eat the fruits (Standley 1926) and are probably important dispersal agents.

The Seri, especially children, eat the sweet pulp of the fruits (Felger and Moser 1985).

Rhizophora mangle

L.

Red mangrove, mangle, mangle dulce, mangle colorado, mangle tinto, candelón, mangle salado, mangle zapatero, mangle gateador, mangle rojo

In the Sonoran Desert region *Rhizophora mangle* is a shrub or small tree up to 5 m tall (Felger 1966), but it can reach 25 m in the tropics (Gill and Tomlinson 1969). A characteristic network of arching prop roots provides anchorage and aeration. The opposite, glabrous, elliptic, leathery leaves are 5–15 cm long. The yellowish white, inconspicuous flowers are clustered on axillary peduncles (Wiggins 1964). The conical, baccate fruits are about 2.5 cm long. The diploid chromosome number is 36 (Yoshioka et al. 1984).

Leaves of *Avicennia germinans* are minutely white-hairy underneath; leaves of *Laguncularia racemosa* and *Conocarpus erecta* have a pair of glands on the petiole. See Walsh 1979 for an extensive comparison of the physiological and ecological attributes of the various mangroves.

Rhizophora mangle grows in the brackish or saline intertidal zone of inlets, bays,

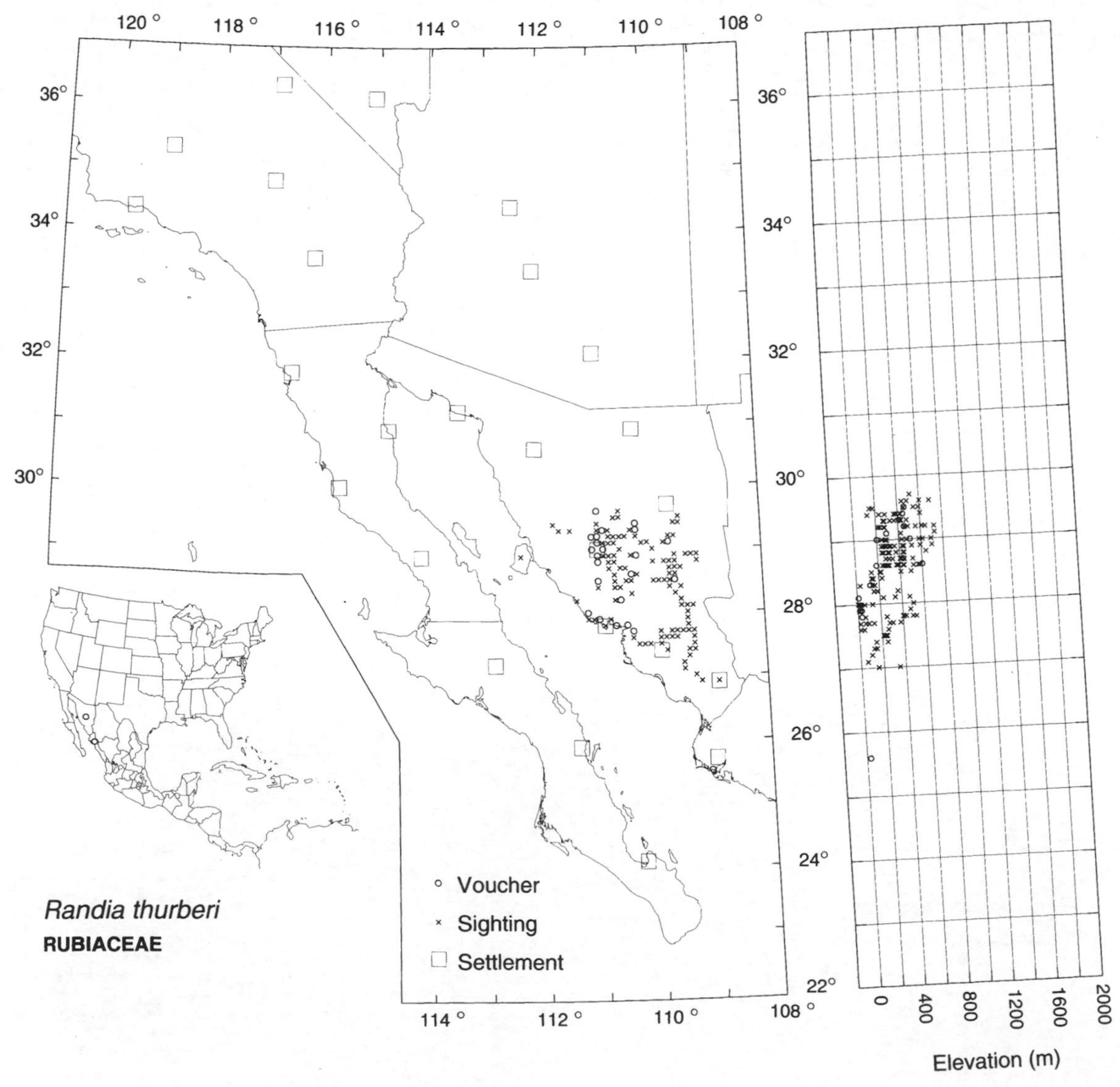

Randia thurberi
RUBIACEAE

and coastal mudflats. It is found in a variety of substrates, including quartz sand, marl, peat, and mud, and tolerates a wide range of salinities from fresh- to seawater (Stern and Voit 1959; Gill and Tomlinson 1977). In Sonora and elsewhere, the various species of mangrove (*Rhizophora, Avicennia, Laguncularia, Conocarpus*) often grow in more or less distinct zones (Felger 1966; Rabinowitz 1978; Odum and McIvor 1990). Along coastal Sonora, cover in mangrove thickets can reach 100% (Felger 1966).

The northernmost point in Sonora (29.3°N) represents collections made by Richard Felger. The species grows along the Sonoran coast to the Sinaloan border (Richard Felger, personal communication 1993). We have seen several vouchers from the disjunct population on the gulf coast of Baja California (28.9°N), including those by Ira L. Wiggins in 1959 and Peta Mudie in 1973. It seems likely that the peninsular population was established by propagules that drifted across the gulf from Isla Tiburón or the Sonoran mainland.

In January 1971, an unusually severe freeze killed many individuals at Estero Sargento (Felger and Moser 1985). Occasional freezes probably determine the northern limit (Felger and Moser 1985). As in the Caribbean and Gulf of Mexico (Markley et al. 1982), northern populations may be more cold tolerant than southern populations. The worldwide distribution of mangroves correlates well with sea-surface temperatures, which apparently become limiting at the 24°C isotherm (Tomlinson 1986).

In the Sonoran Desert, flowers appear from March to November (Wiggins 1964). Farther south, all reproductive stages can be observed throughout the year. Because the apical buds do not become dormant, leaf production and leaf drop are continuous (Gill and Tomlinson 1971).

The flowers are probably wind pollinated (Tomlinson et al. 1979; Juncosa and Tomlinson 1987). The propagules mature in four to six months (Gill and Tomlinson 1971).

In *R. mangle* the zygote develops into a

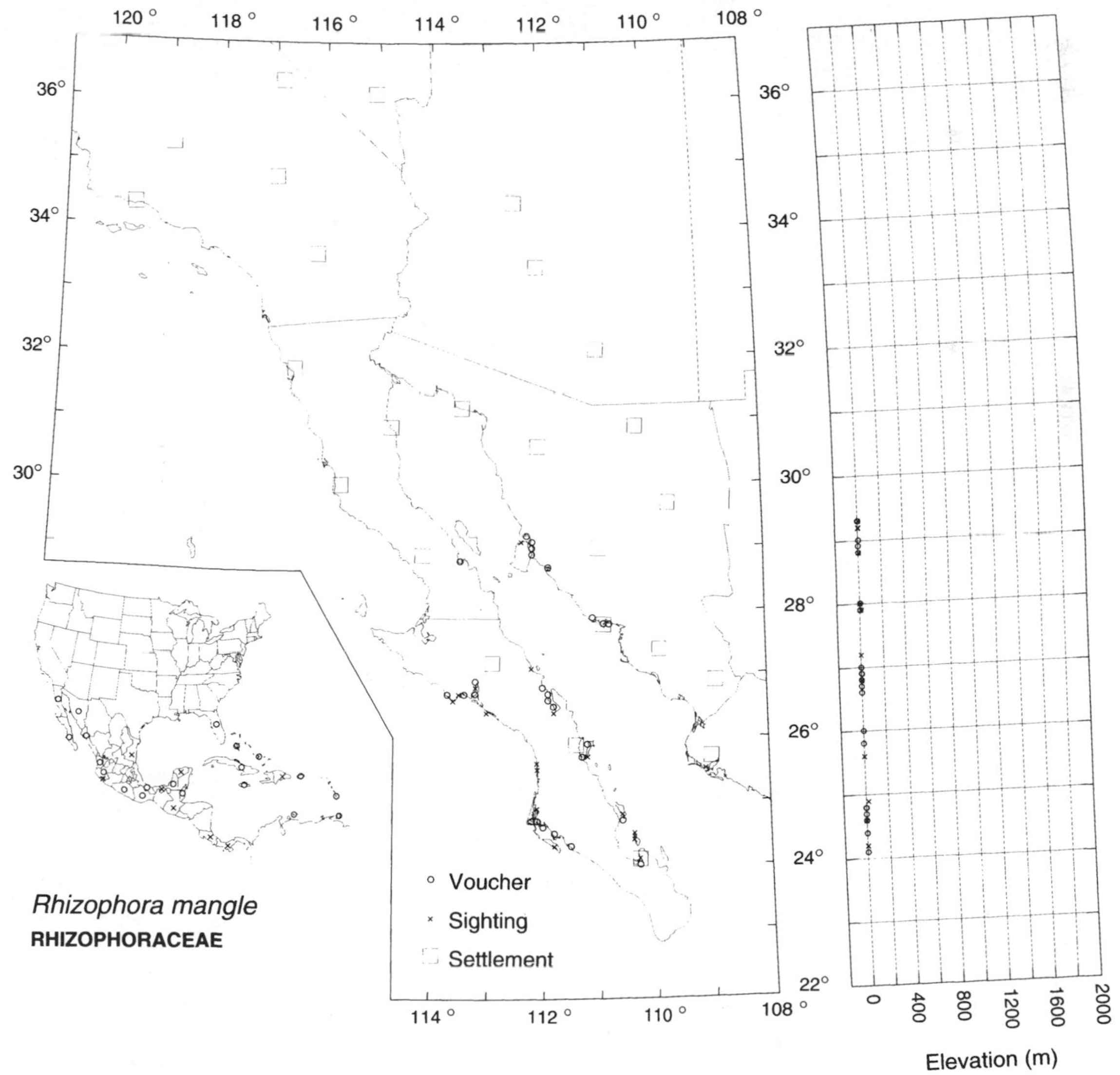

Rhizophora mangle
RHIZOPHORACEAE

seedling without an intervening seed stage. The dispersal unit, or propagule, is a hypocotyl up to 20 cm long and a short plumule 0.5 cm long (Gill and Tomlinson 1969). The hypocotyl emerges while the fruit is attached to the parent plant. After the propagule falls, it may stick immediately in the substrate, or it may float for some time before rooting or decaying. The propagules are remarkably hardy and long lived and may last a year or more while floating in fresh- or saltwater.

Propagule characteristics may account at least in part for the species zonation of mixed mangrove communities. The propagules of *A. germinans* and *L. racemosa* do not sink and must be stranded beyond the reach of tides if they are to root. These two species are therefore restricted to higher ground within the tidal zone. Because *R. mangle* propagules need not be stranded, they can root in deeper water (Rabinowitz 1978). Competition may also influence zonation (Lopez-Portillo and Ezcurra 1989), as may hydrogen sulfide concentrations in the soil (Nickerson and Thibodeau 1984).

The subterranean roots grow in an anaerobic environment (McKee et al. 1988) and must obtain oxygen via the aerial, or prop, roots. Gas exchange within the prop roots enables subterranean roots to maintain a high oxygen concentration of 15–18% (Scholander et al. 1955). The aerial roots are anatomically highly modified (Tomlinson 1986).

Unlike *Avicennia germinans,* which excretes excess salts from specialized structures, *R. mangle* excludes salts at the soil-root boundary (Mallery and Teas 1984; Tomlinson 1986). Salt concentrations of its xylem sap are less than one one-hundredth that of seawater (Tomlinson 1986). Nevertheless, some salts do accumulate, and the plant must dispose of these salts. A small proportion is probably lost through the leaf surface by cuticular excretion. Loss of plant parts, especially leaves, also reduces salt content. Water storage cells in the leaves play a role in osmotic pressure regulation (Camilleri and Ribi 1983; Mallery and Teas 1984; Rada et al. 1989).

The leaves accumulate malic acid,

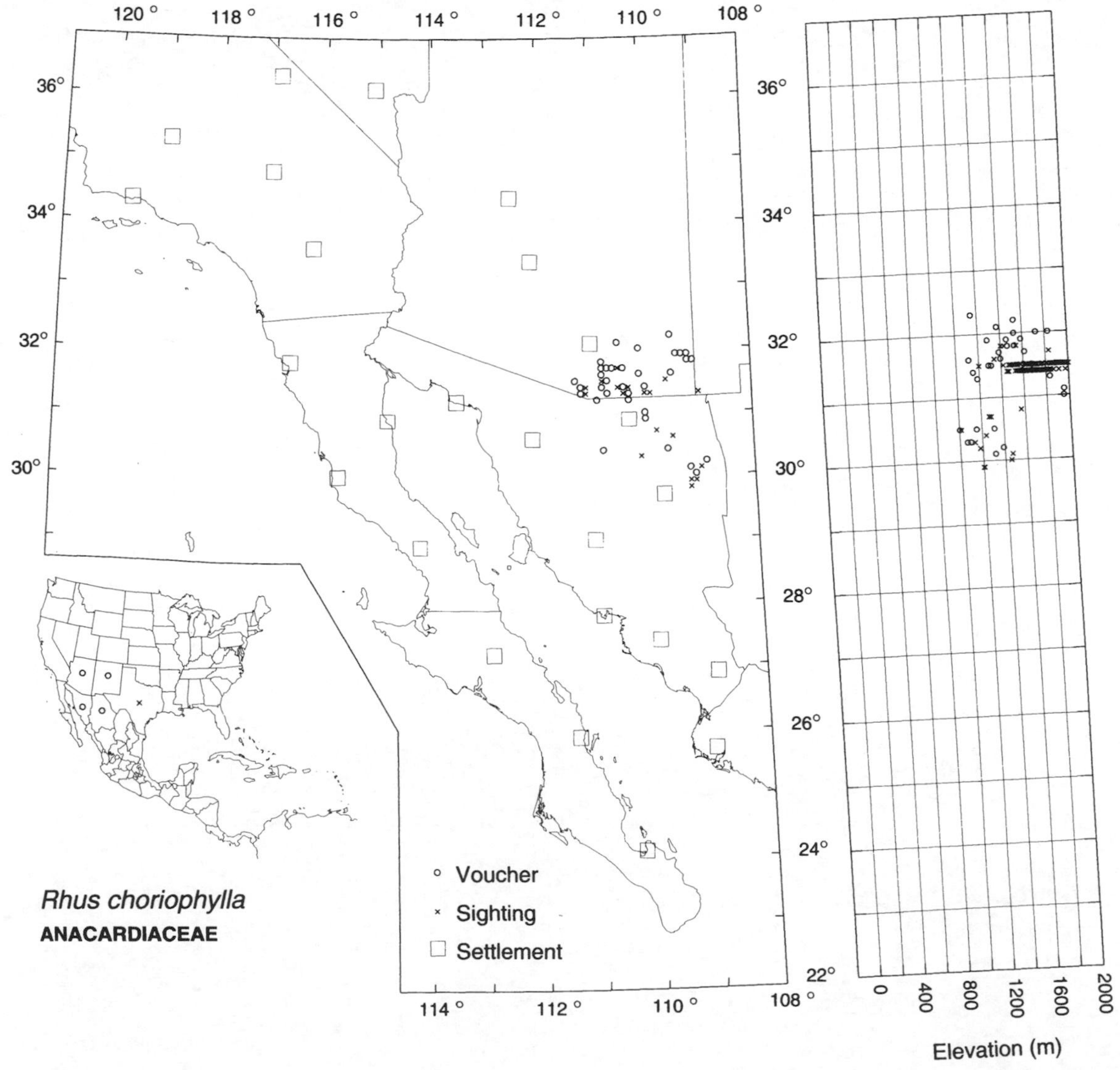

Rhus choriophylla
ANACARDIACEAE

which is typical of plants that photosynthesize by crassulacean acid metabolism. But because *R. mangle* does not show net fixation of carbon dioxide at night, the plants probably cycle between CAM, which fixes internally respirated carbon dioxide at night, and C_3 photosynthesis (Werner and Stelzer 1990).

The wood has been used for wharves, docks, and oars. The tannin-rich bark is used for tanning and, with various mineral salts, for making olive, brown, and slate dyes. The bark has been used to decrease fever, ease sore throats, and stop hemorrhages. The somewhat bitter fruits were eaten by Indians in time of hardship (Standley 1924). The groves provide habitat for oysters (*Ostrea columbiensis*), clams (primarily *Chione* species), and crabs (*Callenictes bellicosus*), all of which were harvested by Seri Indians (Felger and Moser 1985).

Rhus choriophylla

Woot. & Standl.

[=*R. virens* A. Gray var. *choriophylla* (Woot. & Standl.) L. Benson]

Evergreen sumac, tough-leaved sumac

Rhus choriophylla is an evergreen shrub up to 2 m tall. The leathery leaves are once-pinnate with 3–5 ovate, entire leaflets 2.5–4.5 cm long. The white flowers, in terminal or axillary clusters, are about 3 mm long. The fruits are red, pubescent drupes 6–7 mm long.

Juglans major (Torr.) Heller is a tree with deciduous, serrate leaflets. *Sapindus saponaria,* also a deciduous tree, has winged rachises.

Some authors (for example, Benson and Darrow [1981]) treat this plant as a variety of *R. virens* Lindh. The two species are very close, the primary differences being the glabrate leaflets and axillary inflorescences of *R. choriophylla* (Correll and Johnston 1970). With *R. andrieuxii* Engelm. of southern Mexico, they form a closely allied series (Barkley 1937).

Most frequent in semidesert grassland and Madrean evergreen woodland, especially on limestone, *R. choriophylla* enters the Sonoran Desert in Palm Canyon near Cucurpe, Sonora (30.6°N). The western limit is at Arivaca, Pima County, Arizona. The species seems restricted to temperate climates where average rainfall is higher than in most Sonoran Desert locales.

The scattered populations are no doubt remnants of a more continuous Pleistocene distribution. The elevational range represented in herbarium collections (1,000–2,175 m) is somewhat greater than we show.

Flowers appear mainly from August to October, occasionally in spring.

This handsome, drought-resistant, frost-hardy plant is well worth cultivation (Duffield and Jones 1981). The seeds require scarification (Nokes 1986). Cuttings can be rooted if treated with rooting hormone (Tipton 1990). Texas Indians smoked the leaves of the closely related *R. virens* (Standley 1923).

Rhus integrifolia

(Nutt.) Benth. & Hook.

Lemonadeberry

Rhus integrifolia is a shrub 1–3 m tall with small, pink flowers borne on short, compound spikes. The oblong-elliptic, leathery leaves, 5–8 cm long, may be entire or spinose-dentate; rarely, they are trifoliately lobed at the base. The leaves are silvery when young, glabrate in age. Fruits are white, viscid drupes 1–1.5 cm long.

Leaves of *R. ovata* are folded upward along the midrib. *Malosma laurina* leaves are longer (up to 10 cm), less rigid, and somewhat folded along the midrib. *Rhus lentii* Kell., endemic to Cedros Island and the Vizcaíno Peninsula, has broadly ovate to orbicular leaves that are glaucous beneath.

Biologists have identified natural hybrids between *R. ovata* and *R. integrifolia* (Young 1974a). A small percentage of artificial hybrids produce viable seed (Young 1974a).

Rhus integrifolia grows on coastal bluffs, on hillsides, and along washes. In coastal California, it prefers sandstone and conglomerate and appears to avoid granite, diorite, limestone, serpentine, and loose sand (Westman 1981a). In Baja California, it is rare from San Quintín to Isla Cedros (Wiggins 1964). The disjunct point at 28.8°N is based on Reid Moran's 1972 collection. The population on Isla Cedros, var. *cedrosensis* Barkley, is endemic (Wiggins 1964, 1980).

During the Miocene, *R. integrifolia* covered a much broader area than at present, ranging from central and southern California to central Nevada and central Arizona (Young 1974a). The warming, drying trend of the Pliocene pushed the species coastward. Hybrids between *R. ovata* and *R. integrifolia* near Warner Springs and Oak Grove (both in San Diego County, California) are likely the result of ancient hybridization when the two species grew together (Young 1974a). The current range covers relatively frost-free areas strongly influenced by maritime air. Most populations receive cool-season rain exclusively.

The leaves are evergreen. Flowers appear from October through May in the Sonoran Desert region (Wiggins 1964).

Like many other chaparral species, *R. integrifolia* is adapted to fire. Seeds require exposure to heat for germination (Young 1974a). (Soaking in sulfuric acid also promotes germination.) After fire, establishment from seeds is more or less continuous (Lloret and Zedler 1991). The roots sprout readily after fire (Westman et al. 1981).

The plants do not conserve winter moisture for use during the hot, dry summer but survive by physiological control of water loss (Poole and Miller 1975).

Populations include both hermaphroditic and male sterile plants. Seed production varies greatly from plant to plant and is especially high on male-sterile plants (Lloret and Zedler 1991).

Animals, especially birds, eat the fruits and are important dispersal agents (Lloret and Zedler 1991). California Indians used the fresh, dried, or roasted drupes to make a beverage (Barkley 1937). The wood is used for fuel (Standley 1923).

Rhus kearneyi

Barkley

Kearney sumac

Rhus kearneyi is a shrub or small tree up to 5.5 m high with compact, terminal panicles of small, white or pink flowers. The flat, broadly oblong or oval leaves, 2.5–6 cm long and 1.5–3 cm wide, are cordate or subcordate. Fruits are red, pubescent drupes 8–11 mm long and 4–5 mm thick.

Rhus kearneyi closely resembles *R. integrifolia* and was confused with it at first (Moran 1969). The puberulence of twigs and bracts is sparser in *R. kearneyi,* however, and its leaves are cordate or subcordate at the base rather than rounded or obtuse (Moran 1969). Barkley (1937) considered *R. kearneyi* to be most closely related to *R. standleyi* Barkley of southeastern Mexico and to *R. ovata*.

Until the 1960s, *R. kearneyi* was known only from the Tinajas Altas Mountains, Yuma County, Arizona. Since then, it has been found in the adjacent Gila Mountains, in the Sierra del Viejo of northwestern Sonora, and at several locations on the Baja California peninsula. In 1969, Moran described subsp. *borjaensis* and subsp. *virginum* from the peninsular populations. The southernmost record is Annetta Carter's.

The disjunct peninsular populations may have been connected during Pleisto-

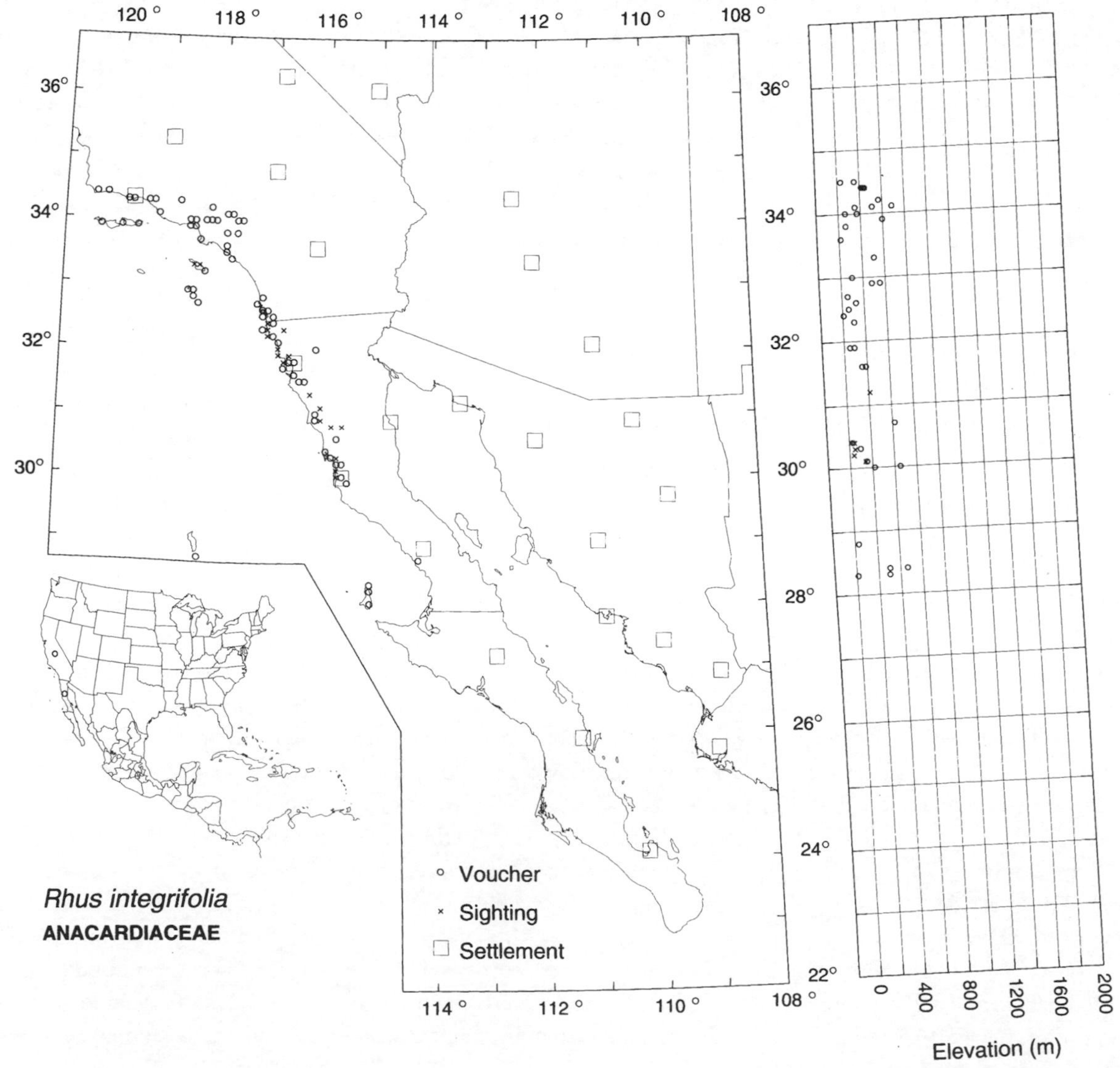

cene pluvials. Incipient speciation has probably occurred since their separation. The Sonoran population may have been established by migration around the head of the Gulf of California during cooler, wetter periods (Yatskievych and Fischer 1984). In more recent times, all populations of *R. kearneyi* have received biseasonal rain. On the peninsula, the rainfall regime shifts from predominantly winter in the north to predominantly late summer or fall in the south.

Rhus kearneyi subsp. *kearneyi* flowers in the Tinajas Altas Mountains from November to March; subsp. *borjaensis* in the Sierra San Borja, Baja California, in March and April; and subsp. *virginum* in the Sierra de las Tres Vírgenes, Sierra Santa Lucía, and Sierra de la Giganta, Baja California Sur, in the autumn. Leaves of all three subspecies are evergreen.

Rhus microphylla

Engelm.

Littleleaf sumac, desert sumac, scrub sumac, correosa, agritos, agrillo, lima de la sierra

This shrub grows 1–2 m high and has weakly spinescent branchlets. The small, leathery leaves are once-pinnate with 5–9 leaflets 6–18 mm long and 3–7 mm wide. The inconspicuous, white flowers are in terminal and axillary clusters. Fruits are red, ovoid drupes about 6 mm long.

As currently defined, *R. microphylla* is a large, polymorphic species. The variation may represent early stages of speciation (Barkley 1937).

Pachycormus discolor has similar foliage, but its range is quite distinct. *Rhus choriophylla* is also pinnate leaved but has much larger leaflets.

Hastings and coworkers (1972) erroneously reported *R. microphylla* from the Baja California peninsula based on unnumbered specimens collected by Pond. Young (1975) later annotated these to *P. discolor.*

Rhus microphylla is frequent on lime-

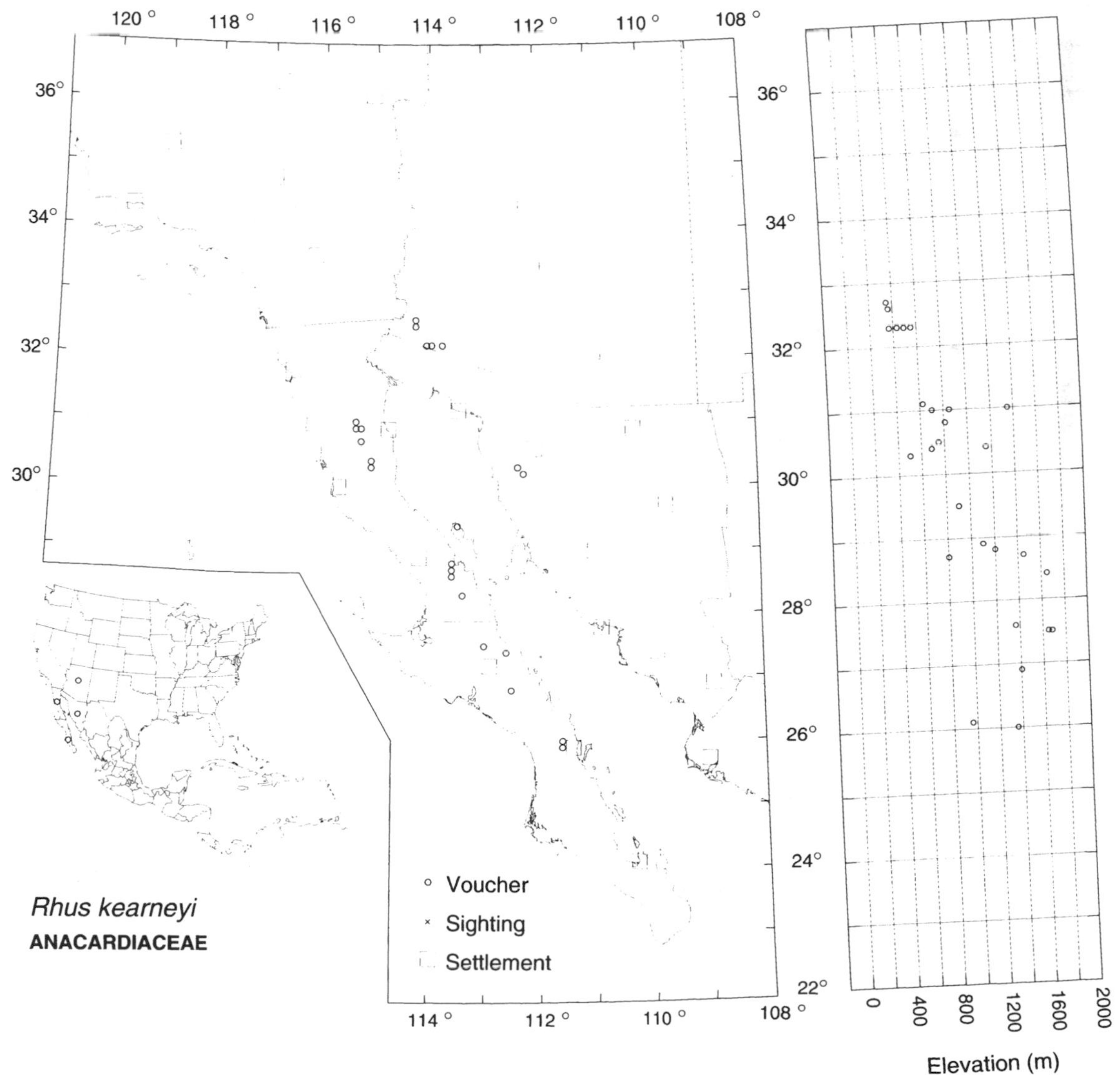

Rhus kearneyi
ANACARDIACEAE

stone, gypsum, and sand in the temperate parts of the Chihuahuan Desert. In the region mapped by this atlas it typically grows on limestone and calcareous soils with other Chihuahuan Desert shrubs, including *Flourensia cernua* DC., *Acacia neovernicosa*, and *Mortonia sempervirens*.

The leaves are winter deciduous. New leaves appear in May and are kept until November or December. Plants bloom in April and May before or as the leaves appear. Occasionally they may flower in response to summer rains.

In its wood anatomy, *R. microphylla* is similar to *R. trilobata* Nutt. and *R. aromatica* Ait.: it shows resin canals in the rays and flamelike clustering of small vessels in the latewood (Young 1974b).

The species has invaded some Texas rangelands (McPherson et al. 1988). It is occasionally cultivated as a drought-tolerant ornamental.

Rhus ovata

S. Watson

Sugar sumac, sugar bush, mountain laurel

Rhus ovata is a shrub up to 4.5 m high with leathery, simple, entire, ovate leaves. The leaves, 3–5.5 cm wide and 5–9 cm long, are on petioles 1–2 cm long and are folded upward along the midrib. The small, cream-colored flowers are in dense panicles. Fruits are sticky drupes 6–7 mm in diameter.

Rhus kearneyi differs mainly in its flat leaves and shorter petioles (less than 1 cm).

Where the ranges of *R. ovata* and *R. integrifolia* overlap, hybrids between them occur (Young 1974a). Viable seed can be produced by crossing *R. ovata* with *R. integrifolia* and from artificial backcrosses (Young 1974a).

Frequent in Californian chaparral on mountains and hillsides, *R. ovata* descends into the desert along washes. It is absent from the desert mountain ranges along the Colorado River; none has enough height or mass to support chapar-

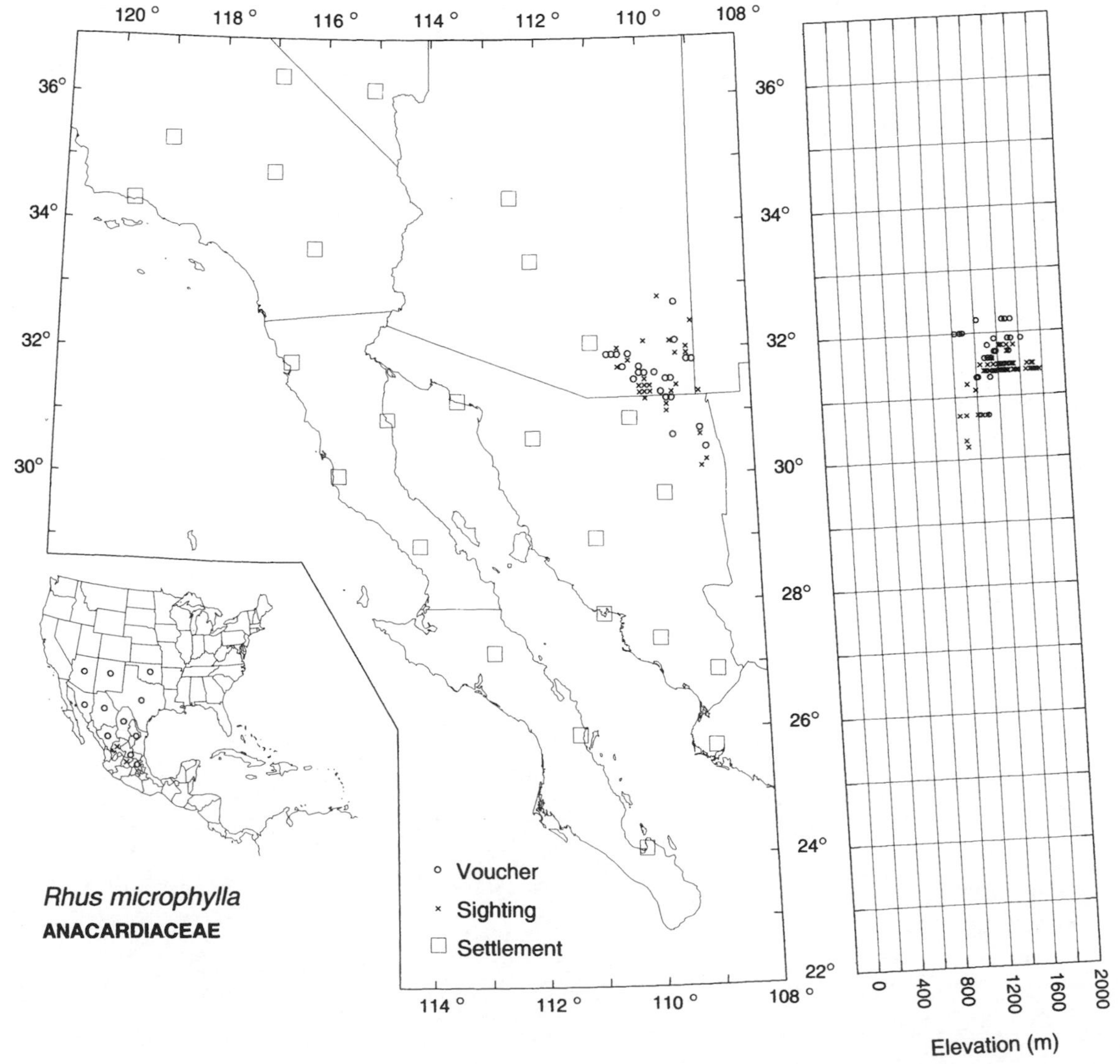

Rhus microphylla
ANACARDIACEAE

ral communities. The southernmost peninsular point on our map, a collection by Reid Moran, is at Volcán Tres Vírgenes. Most populations are of scattered individuals, although the plants may be locally abundant (Cable 1975). In southern California and northern Baja California, *R. ovata* tolerates summer drought. In central Arizona, where summers are hotter and annual rainfall lower, biseasonal rains are probably crucial for the survival of the species (Young 1974a).

Fossils of *R. ovata* from southern Nevada show that it was more widely distributed in the Miocene and Pliocene than at present. Its Miocene range probably included central and southern California, central Nevada, and central and western Arizona (Young 1974a). With the warming, drying trend of the Pliocene and the uplift of the Sierra Nevada and other mountain ranges in California, *R. ovata* was pushed toward the coast, leaving a disjunct distribution in central Arizona (Young 1974a).

Leaves are evergreen. Flowers appear from February to May (Wiggins 1964).

As an evergreen, sclerophyllous shrub, *R. ovata* adapts to seasonal drought by physiological control of water loss rather than by shedding its leaves (Poole and Miller 1975). On the basis of daily and seasonal patterns of water potential and leaf resistance to water loss, Poole and Miller (1975) concluded that, in Californian chaparral, *R. ovata* may be restricted to sites where soil moisture is available year-round, such as rock outcrops. Survival in the desert no doubt depends greatly on the exploitation of favorable microsites. The plants have anatomical features associated with adaptation to aridity, such as numerous narrow vessel elements (Young 1974b).

Like many other chaparral species, *R. ovata* is adapted to fire. Root crowns sprout vigorously after burning (Cable 1975), and seeds germinate best after exposure to temperatures of 60–120°C (Stone and Juhren 1951). (Heat, by cracking the inner of two seed coats, allows water absorption [Stone and Juhren 1951].) In one study, germination of burned seeds ranged from 30 to 52% but was quite low,

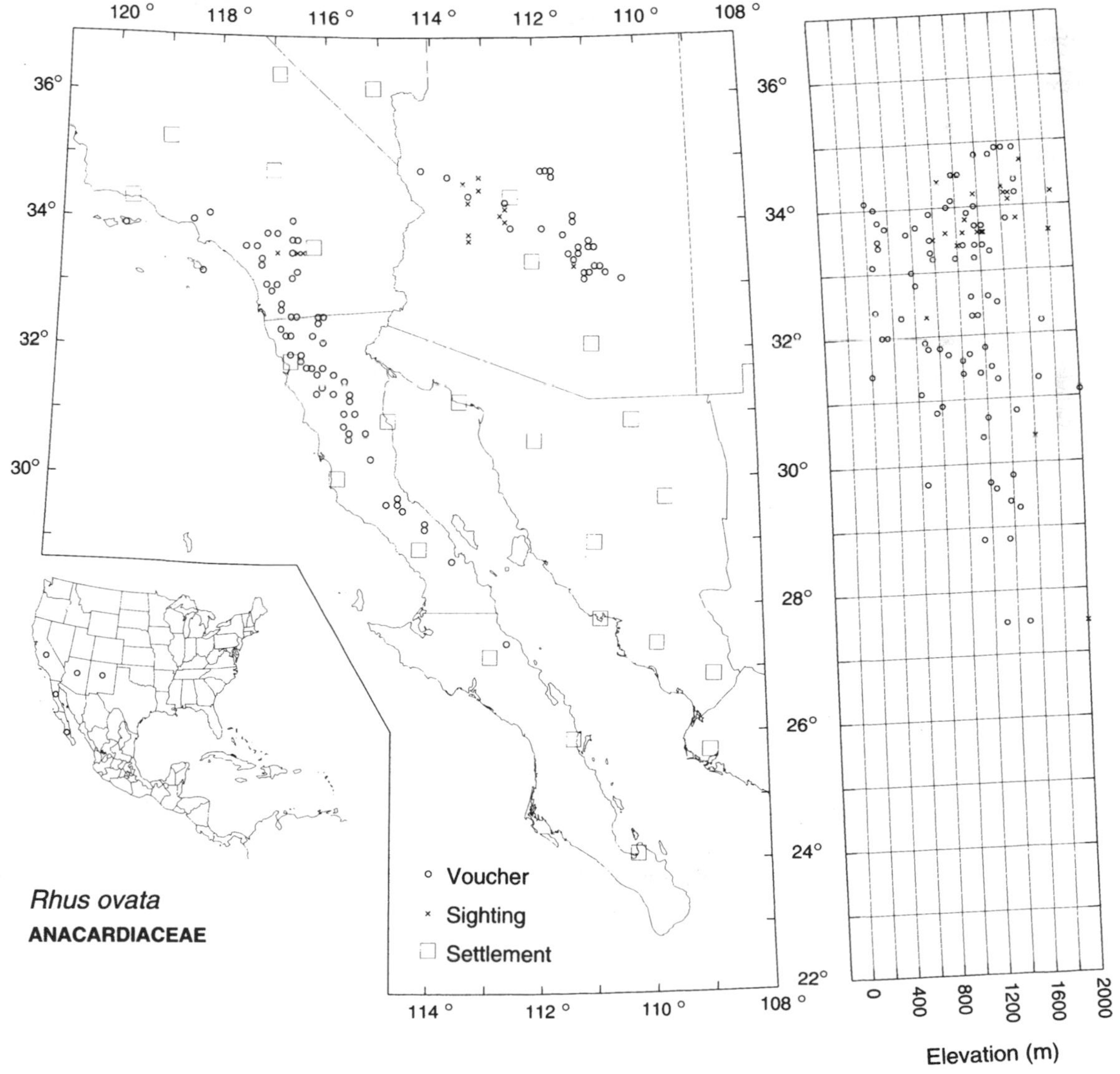

Rhus ovata
ANACARDIACEAE

about 1%, for untreated seeds (Stone and Juhren 1951).

Seedlings are rare in the wild except on burned areas and on edges of trails, roads, and firebreaks, where exposed soils can easily reach germination temperatures in summer (Stone and Juhren 1951).

This species makes an attractive ornamental for desert gardens, but it is susceptible to damping-off by fungus in the summer (Duffield and Jones 1981). The plants may die if overwatered or grown in poorly drained soils. Heaviest irrigation should be given in the winter (Duffield and Jones 1981).

The sweet, waxy exudate on the fruits has been used as a substitute for sugar. The Cahuilla Indians of California boiled and ate the flowers and used a decoction of the leaves to relieve chest pains and coughs (Standley 1923).

Ruellia californica

(Rose) I. M. Johnston

Rama parda

Ruellia californica, a shrub up to 3 m high, has whitish older branches and hirtellous branch tips. The ovate to lanceolate leaves, 1–3 cm long and 0.5–2 cm wide, are usually glandular-puberulent on both surfaces. The lavender or purple funnelform flowers, 4.5–5.5 cm long and 1–1.5 cm across, are single in the leaf axils. The 4-seeded, stipitate capsules are 12–15 mm long. The haploid number of chromosomes is 17 (Daniel et al. 1984).

Ruellia leucantha Brandegee has white flowers and longer leaves (3–6 cm). *Ruellia cordata* Brandegee, known only from Comondú, Baja California Sur, has orbicular leaves with cordate bases. The closely related *R. peninsularis* differs from *R. californica* mainly in pubescence: *R. peninsularis* is glabrous and glutinous, *R. californica* puberulent. (However, *R. californica* is sticky, and its leaves may become glabrate in age.) Their relationships deserve further study.

Ruellia californica is common to abun-

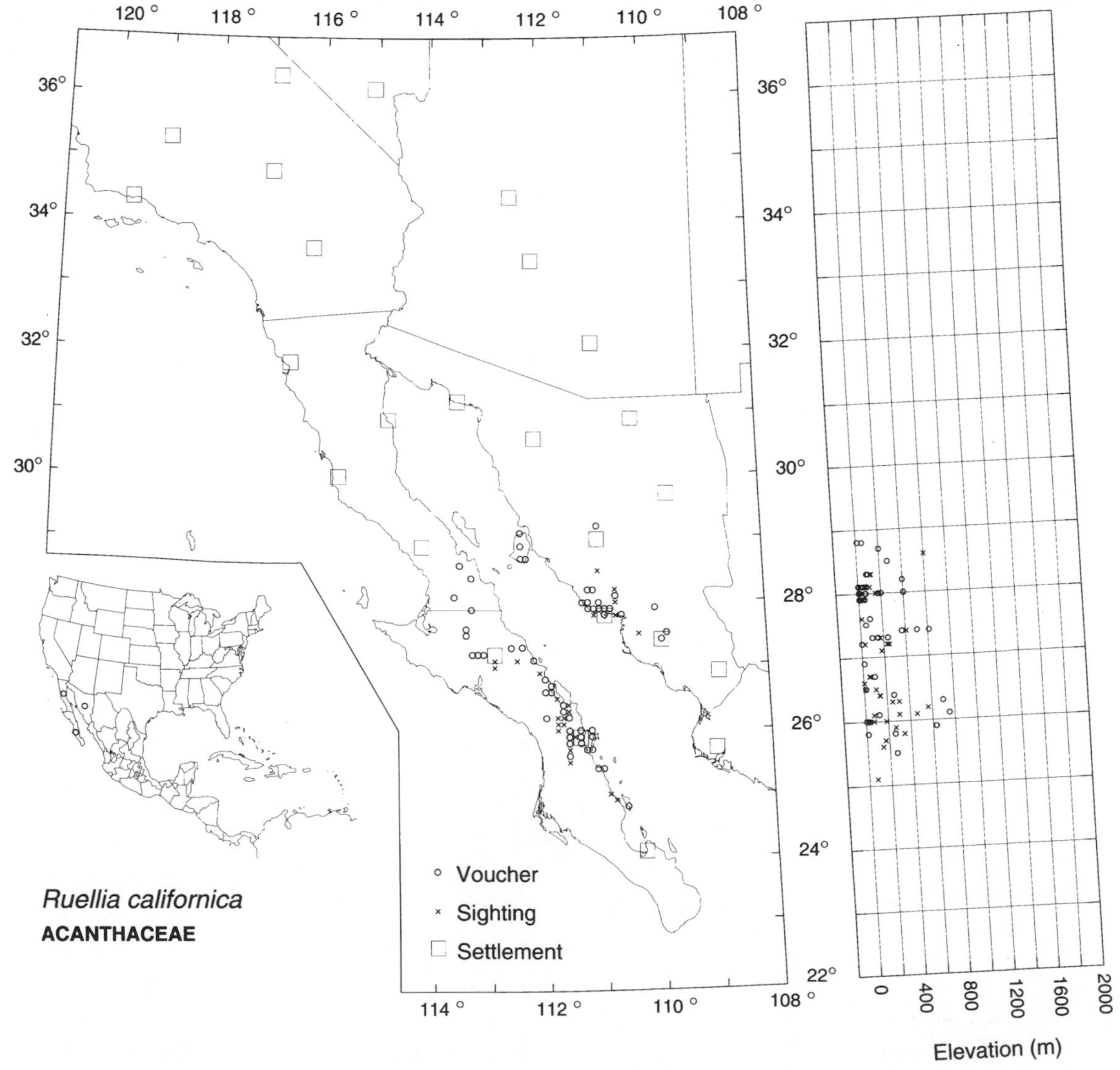

Ruellia californica
ACANTHACEAE

dant on dry, rocky slopes and along arroyos, often under larger shrubs and trees. The two species are on both sides of the Gulf of California. *Ruellia peninsularis* is widespread south of 25°N, whereas *R. californica* extends much farther north on the peninsula. *Ruellia californica* grows in regions of biseasonal rainfall. The dependability of August and September storms increases from north to south across the range.

The main bloom is from March to May and again from October to December. Flowers appear in other months as well, probably in response to rain. The ripe capsules dehisce explosively when moistened.

This species is an attractive, low-water-use ornamental in warm-winter areas. The Seri make a medicinal tea and shampoo from the leaves (Felger and Moser 1985).

Ruellia peninsularis

(Rose) I. M. Johnston

Rama parda

A much-branched shrub up to 1.3 m high, *Ruellia peninsularis* has gray or white stems and glabrous, glutinous foliage. The ovate to ovate-oblong leaves are 1–3 cm long and 0.5–2 cm wide. The purple funnelform flowers, single in the leaf axils, are 3–5 cm long and about 1 cm wide. The capsules, 15–20 mm long, contain 4 flat, suborbicular seeds. The haploid number of chromosomes is 17 (Daniel et al. 1984).

Ruellia californica is hirtellous and glandular-puberulent (sometimes becoming glabrate in age). *Ruellia leucantha* has white flowers and larger leaves (3–6 cm).

Occasional to common on dry, gravelly slopes, *R. peninsularis* is known from Baja California Sur and, in Sonora, on Isla Tiburón and in the Guaymas area. On the peninsula, *R. peninsularis* occupies somewhat wetter climates than *R. californica.* Although they receive some winter moisture, most *R. peninsularis* populations depend heavily on August and September rains.

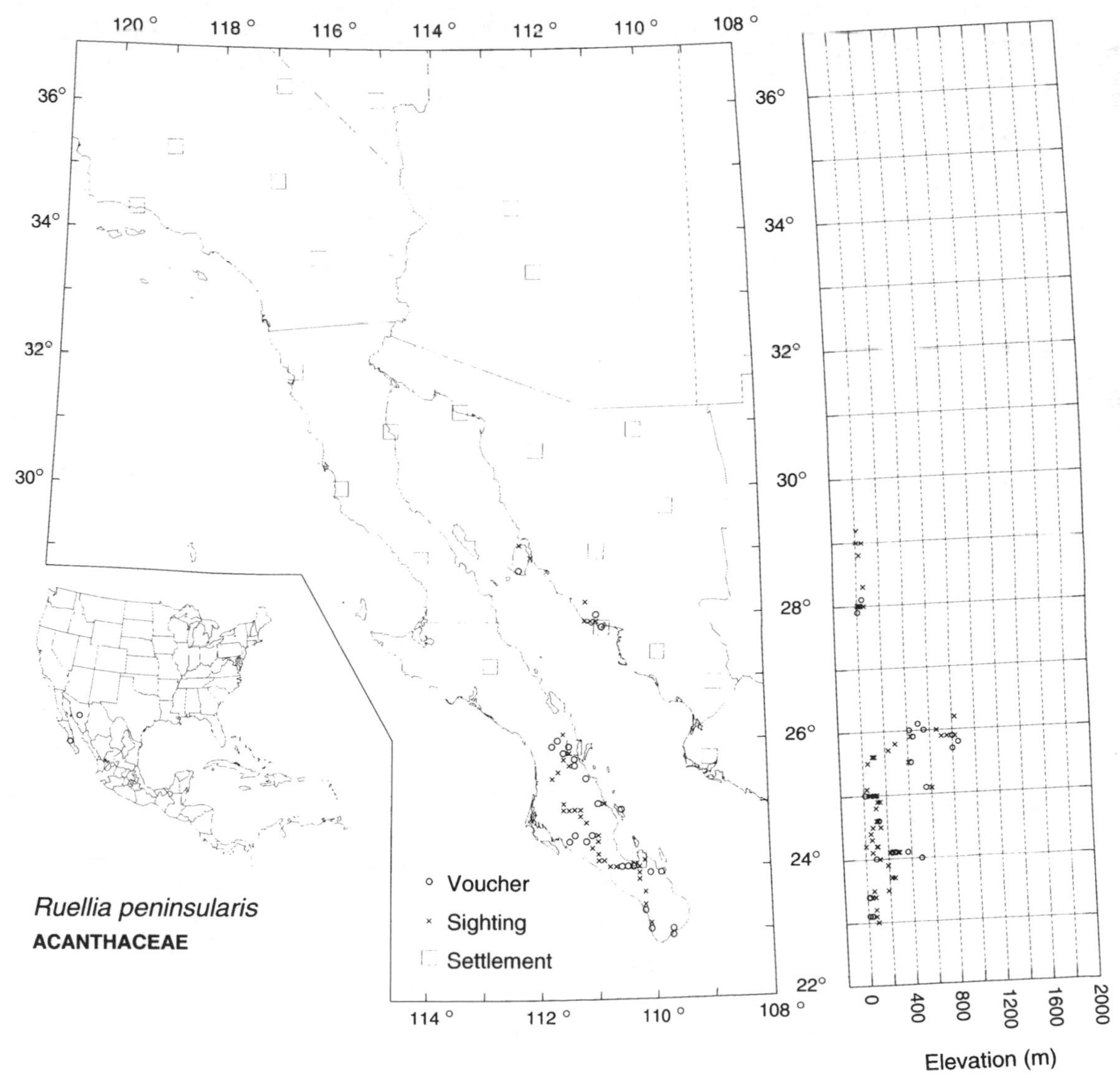

Ruellia peninsularis
ACANTHACEAE

Flowers appear from October to April. Leaves are presumed to be drought deciduous.

This species is occasionally grown as a low-water-use ornamental in warm-winter areas.

Sabal uresana

Trel.

Palma, tahcu

A slender palm 5–10 m tall, *Sabal uresana* has a naked, faintly ringed trunk and a small crown. The blue-glaucous, fan-shaped leaves, 1–2 m across, are on spineless petioles 2–3 m long and 5–6 cm broad. The inflorescence, 2–3 m long, is horizontal in youth, drooping in age. The diploid number of chromosomes is 36 (Eichhorn 1957).

Washingtonia robusta Wendl. is taller (up to 25 m) with an abruptly flaring base and prominent spines along the petioles. *Brahea roezlii* has marginal teeth on the petiole.

Locally common in canyons and on cliffs, this rather widely distributed palm occurs from sea level to 1,200 m. Young stands are often dense, especially along streams, while older plants are generally scattered (Gentry 1942).

Flowers appear in early spring. Bee pollination is likely for the entire genus (Zona 1990).

The trunks were often used in building houses, the leaves for thatch. The terminal buds are eaten and employed in basketry (Gentry 1942). Gentry (1942) believed that *Sabal* populations in Sonora were diminishing as a result of drought and overutilization.

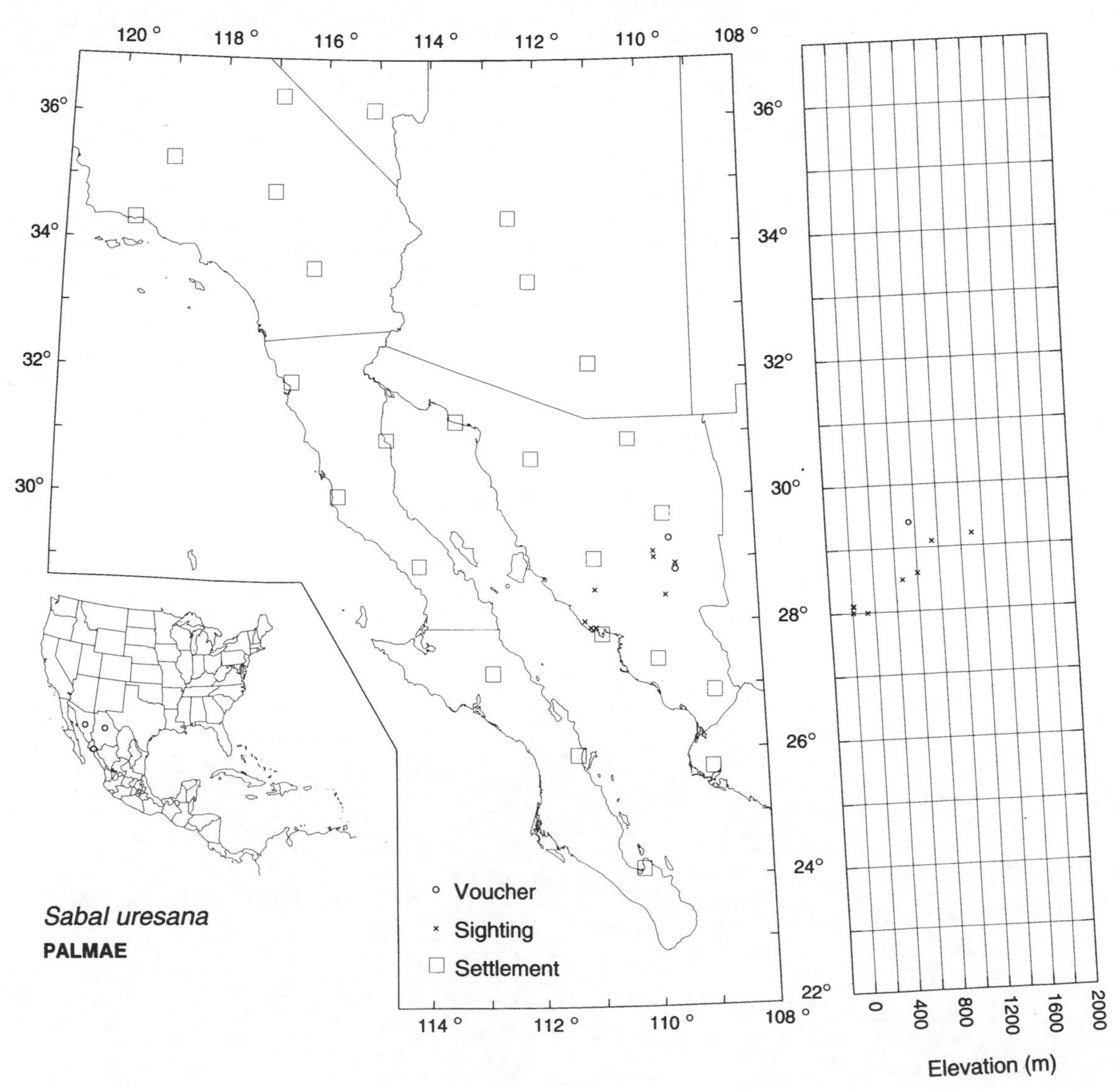

Sapindus saponaria

L.

Soapberry, western soapberry, cherioni, jaboncillo, palo blanco, amole, matamuchacho, tehuistle, tzatzupa, guayul, ojo de loro, arbolillo, amolio

A shrub or tree up to 10 m tall, *Sapindus saponaria* has once-pinnate leaves with 13–19 lanceolate, entire leaflets 4–11 cm long and 1–2 cm wide. The rachis is winged. The inconspicuous, whitish flowers are in large, dense panicles and may be perfect or unisexual. The amber or yellowish berries are 1–1.5 cm in diameter.

Leaflets of *Juglans major* (Torr.) Heller are serrate rather than entire, and the leaf rachis is not winged.

Our map combines var. *drummondii* (Hook. & Arn.) L. Benson and var. *saponaria*. Leaflets of var. *saponaria* are larger and more obtuse than those of var. *drummondii;* also, var. *saponaria* is evergreen, whereas var. *drummondii* is winter deciduous. The range of the species extends to the Old World, where it has probably been introduced.

Sapindus saponaria grows along watercourses and on canyon slopes. Especially in sandy soils, var. *drummondii* sprouts from the roots to form extensive thickets. In rural Texas, *S. saponaria* grows along fencerows and in old fields (Correll and Johnston 1970). Both varieties appear to be confined to the mesic fringes of the desert in areas where warm-season rainfall is fairly dependable.

Flowers appear from May to August in the north (Wiggins 1964) and in January and February in southern Sonora (Gentry 1942). Fruits mature in the summer and fall and may persist on the tree until the following spring. Leaves are winter deciduous or evergreen.

The fruits contain saponins and produce lather when macerated. They have been used for washing and to treat arthritis, fevers, rheumatism, and kidney disease. In parts of Mexico, fish were stupefied by throwing the fruits into the water (Standley 1923; Moore 1989).

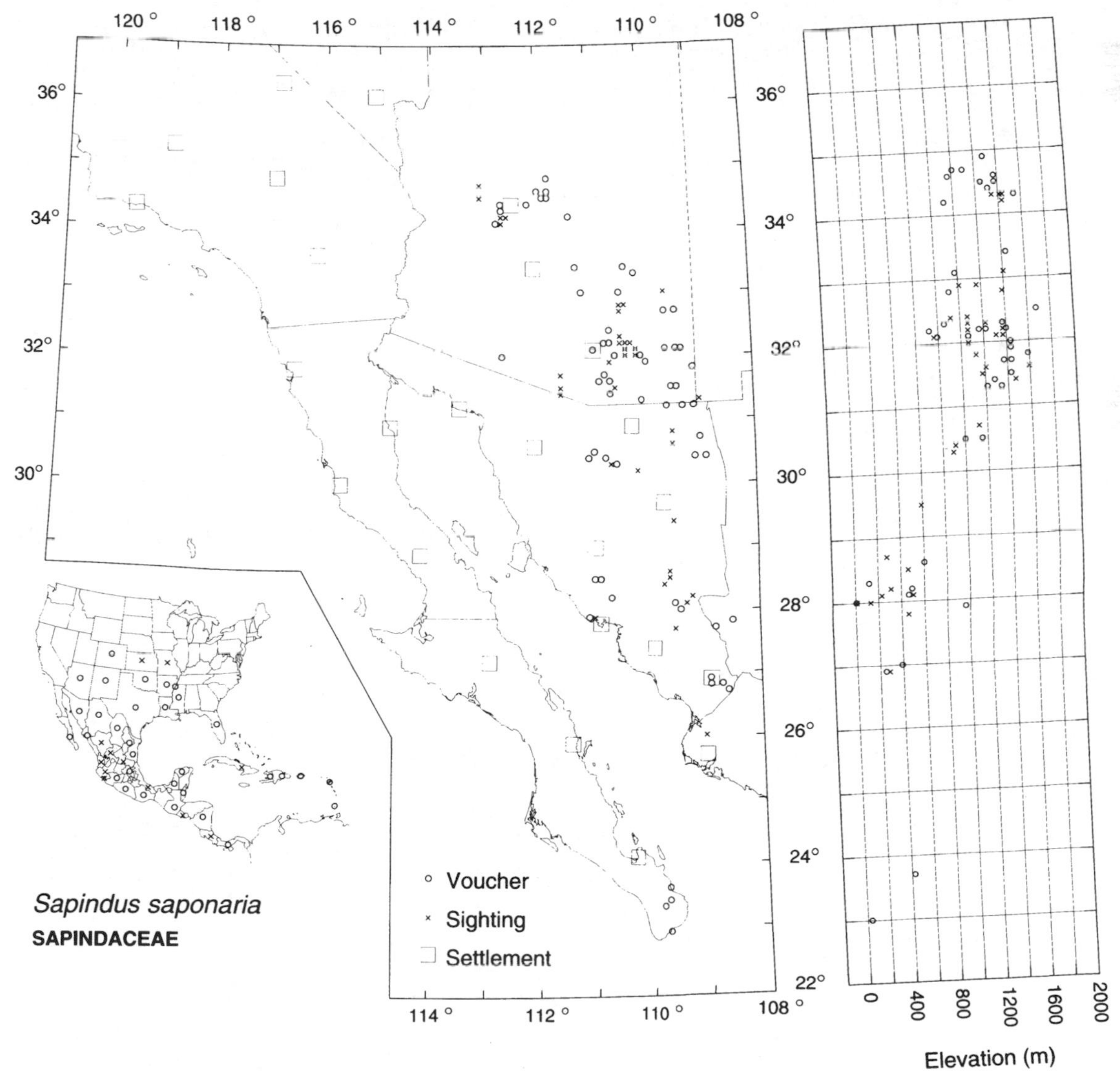

Sapindus saponaria
SAPINDACEAE

The plants show promise as ornamentals (Munson 1984; Khatamian and Abuelgasim 1986). Germination is ordinarily poor because of embryo dormancy and hard seed coats. In one study researchers obtained best results (89% germination) by scarifying and stratifying seeds collected in late autumn or early winter (Munson 1984). Stem cuttings taken in June can be rooted after treatment with rooting hormones (Khatamian and Abuelgasim 1986).

Sapium biloculare

(S. Watson) Pax

Mexican jumping bean, yerba de fleche, yerba mala, mago

A shrub or small tree up to 6 m tall, *Sapium biloculare* has smooth, whitish or grayish bark and milky sap. The leathery, serrulate leaves, 3–7 cm long and 5–15 mm wide, are lance-linear to lanceolate and have conspicuous glands along the margins. The inconspicuous flowers are on a zigzag rachis 2.5–6 cm long with staminate flowers above and pistillate flowers below. The bilobed capsules are about 1 cm wide. The haploid number of chromosomes is 11 (Bell 1965).

Vauquelinia californica lacks milky juice and gland-bearing leaf margins, and it grows in more mesic habitats over a range that hardly overlaps that of *S. biloculare*. *Dodonaea angustifolia* also lacks milky juice and has entire, not serrulate, leaves.

Sapium biloculare is locally dominant in sandy washes and on rocky and gravelly slopes. The plants suffer frost damage at –6.1°C (Bowers 1980–81), and low temperature likely limits them on the north. The northernmost points on our map (33.0°N) occur at relatively low elevations (235–325 m), where severe frost is infrequent. A need for dependable summer rains may determine the rather truncated

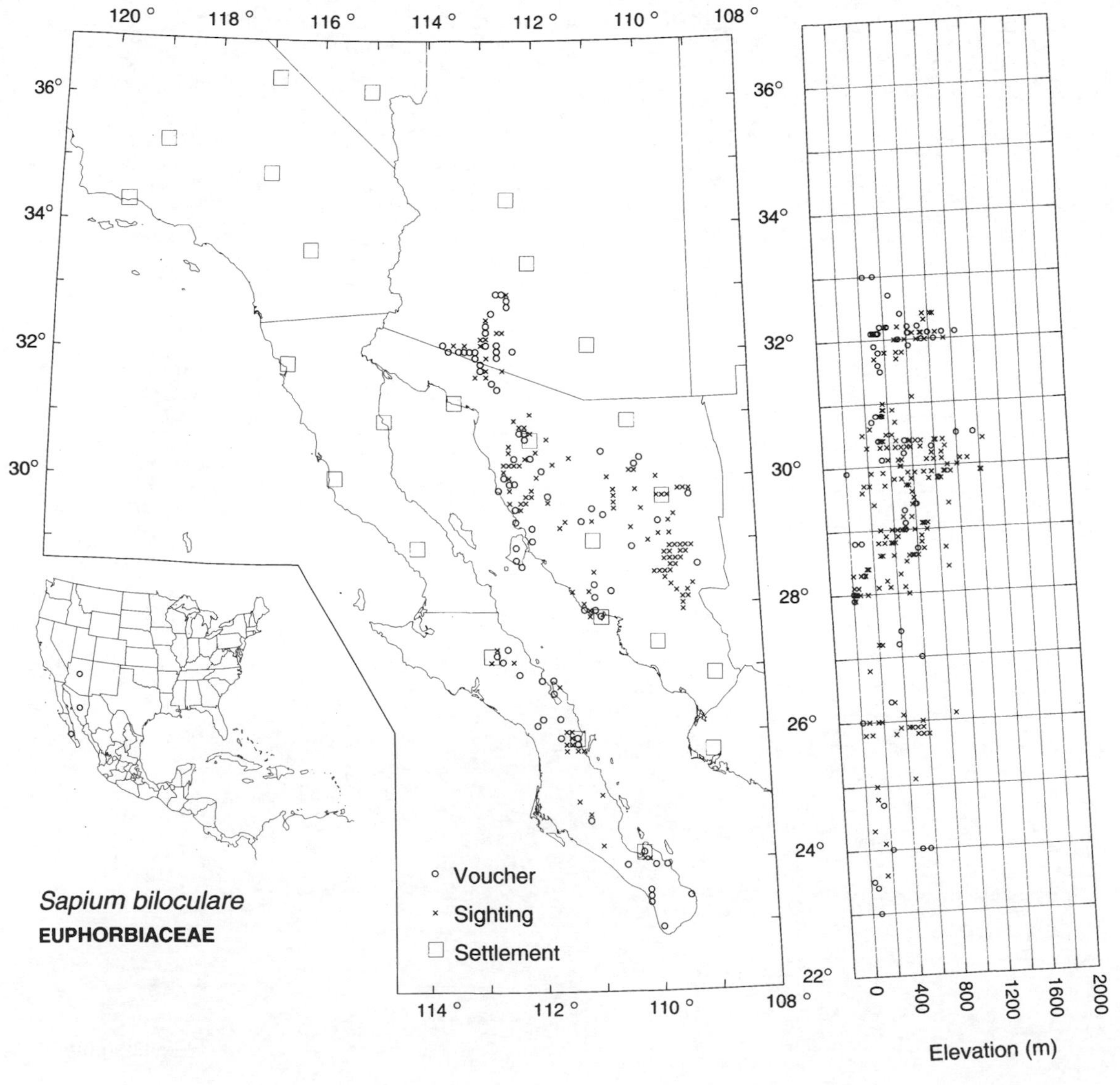

western limit at about 113°W.

Fossil packrat midden assemblages from the Puerto Blanco Mountains in southwestern Arizona suggest that *S. biloculare* arrived at its northern limit only 3,400 years ago, coincident with a shift to warmer, drier conditions (Van Devender 1987).

Flowers appear from March to November (Kearney and Peebles 1960). Leaves are evergreen (Shreve 1964).

The poisonous sap has been used as an arrow poison (Standley 1923; Kearney and Peebles 1960; Felger and Moser 1985). In Baja California Sur, the chopped branches were thrown into streams to stupefy fish (Standley 1923). Smoke from the burning wood is said to cause sore eyes (Standley 1923; Kearney and Peebles 1960). Even gusts of wind apparently can release enough toxins from the foliage to cause stinging eyes and congested lungs (Corbett 1991). Despite the toxins, the leafcutter ant *Atta mexicana* will cut and collect the foliage if annual plants are not available (Mintzer 1979).

Schaefferia cuneifolia A. Gray

Desert yaupon, capul, panalero

Schaefferia cuneifolia is a rigidly branched shrub 1–2 m tall with somewhat spinescent twigs and light gray bark. The obovate to oblanceolate leaves, 1–2.2 cm long and 2–12 mm wide, are glabrous and leathery and are fasciculate on knoblike spur branches. The plants are dioecious, with inconspicuous, solitary or clustered staminate or pistillate flowers. The small, orange or red subsessile drupes contain 2 seeds.

Schaefferia shrevei has puberulent fo-

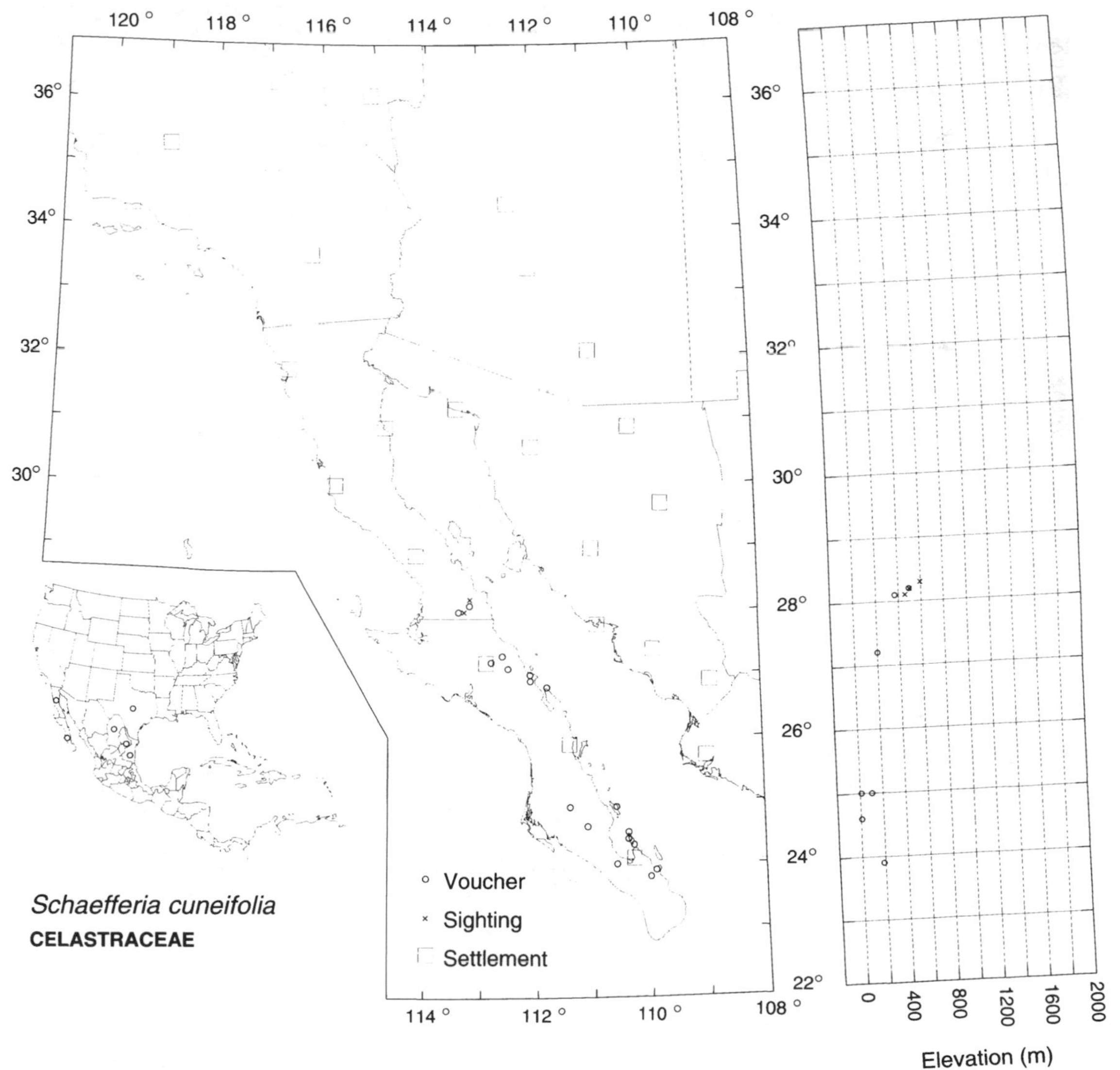

Schaefferia cuneifolia
CELASTRACEAE

liage and distinctly pediceled fruits.

This shrub is rare to locally common on rocky hillsides and mesas, generally below 600 m. The distribution is split between the Baja California peninsula and the northern Tamaulipan thornscrub. Sterile plants of *S. cuneifolia* are difficult to identify, and the species may be more frequent than our map shows. The Sonoran Desert populations occur in maritime tropical arid climates in which rain is usually concentrated in August and September.

Flowers appear from April to July (Wiggins 1964). Ripe fruits can be found as late as November and December. The leaves are evergreen. Growing from a root crown, the stems can regenerate fairly quickly after frost or fire (Flinn et al. 1992).

Standley (1923) reported that the roots were used in treatment of venereal disease. Triterpene quinones isolated from *S. cuneifolia* are pharmaceutically active (Moujir et al. 1990).

Schaefferia shrevei
Lundell

A somewhat rigid shrub up to 4 m tall, *Schaefferia shrevei* has obovate to oblanceolate leaves 1–2.2 cm long on spur branches. Foliage, branchlets, and fruits are puberulent. The plants are dioecious, with inconspicuous flowers in small clusters or single in the leaf axils. The small, 2-seeded drupes are red or orange at maturity and are on pedicels 4–7 mm long.

Schaefferia cuneifolia has glabrous foliage and subsessile fruits.

According to Annetta Carter (personal communication 1988), material from Baja California Sur—in particular Reid Moran's collection from the Sierra San Francisco (27.5°N) and her Sierra Giganta collections—should be assigned to *S. pilosa* Standley, described from Oaxaca. Further study may show that *S. shrevei* merits varietal status under *S. pilosa* (Annetta Carter, personal communication 1988).

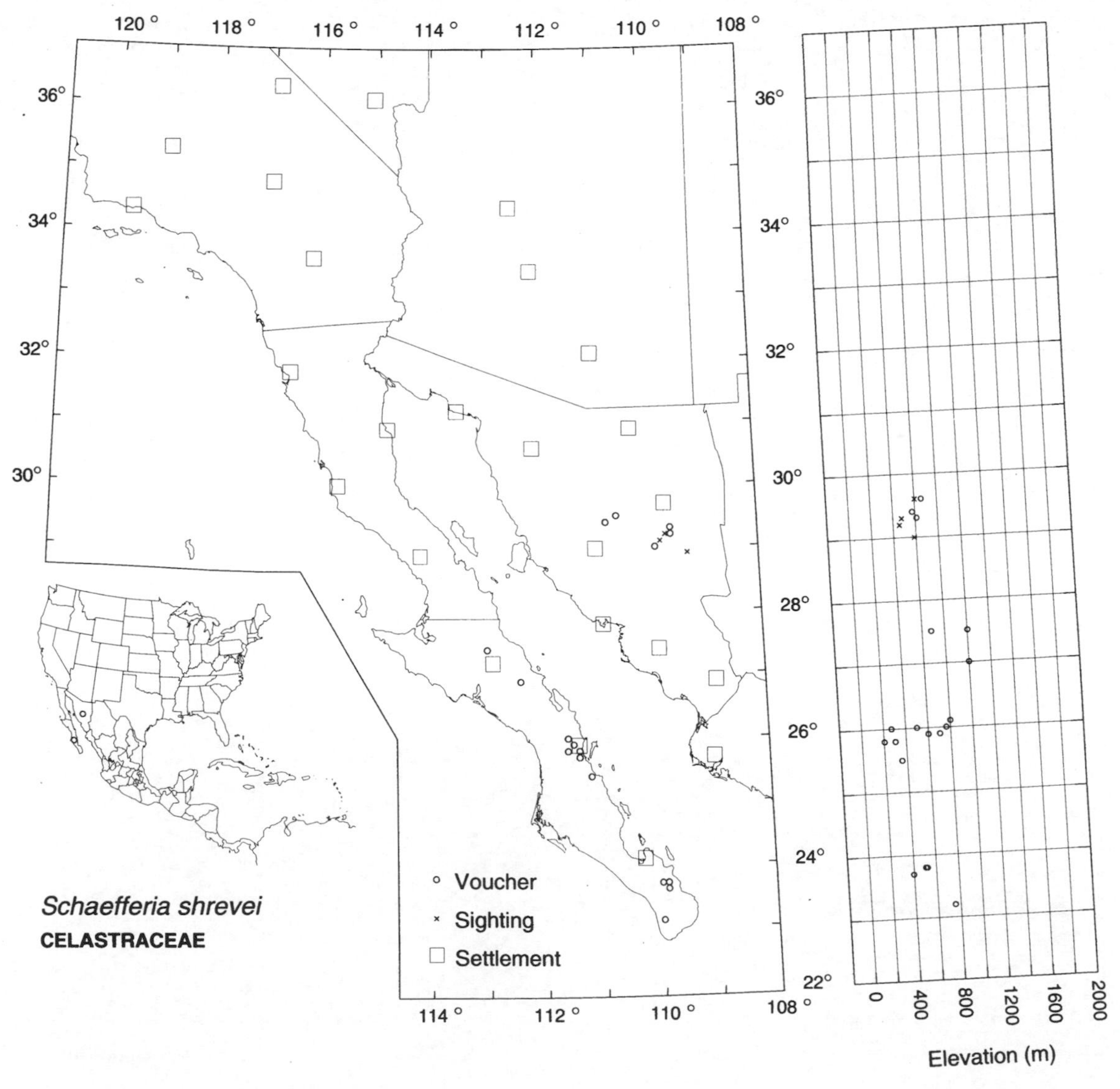

Schaefferia shrevei
CELASTRACEAE

This plant grows along washes, on canyon slopes, and on dry hillsides. The distribution is split between Baja California Sur and central Sonora. At the arid peninsular locations, rainfall is erratic and occurs most often in August and September. At the Sonoran sites, rainfall is biseasonal, with substantial winter moisture and fairly predictable summer rains beginning in July.

Flowers appear in late spring (Wiggins 1964). Mature fruits can be found from September into November.

Schoepfia californica

Brandegee

Schoepfia californica is a rounded shrub or small tree 2–6 m tall with brittle, divaricate branches. The alternate, entire leaves (2–5 mm wide and 10–25 mm long) are oblanceolate to spatulate, glaucous, and leathery. The reddish yellow, urceolate flowers, 5–7 mm long, are single or paired in the axils. Fruits are drupes 6–8 mm long.

Without flowers or fruits, *S. californica* resembles many unrelated desert shrubs and can be difficult to identify.

Most frequent along washes and on alkaline flats, *S. californica* is endemic to the Baja California peninsula. The relatively broad geographical and altitudinal distribution suggests tolerance of summer and winter drought. Over most of the range, rainfall is most predictable during late summer and early fall except in the northwest, where populations are more likely to receive winter moisture.

Flowers appear from October to March (rarely as early as September or as late as May).

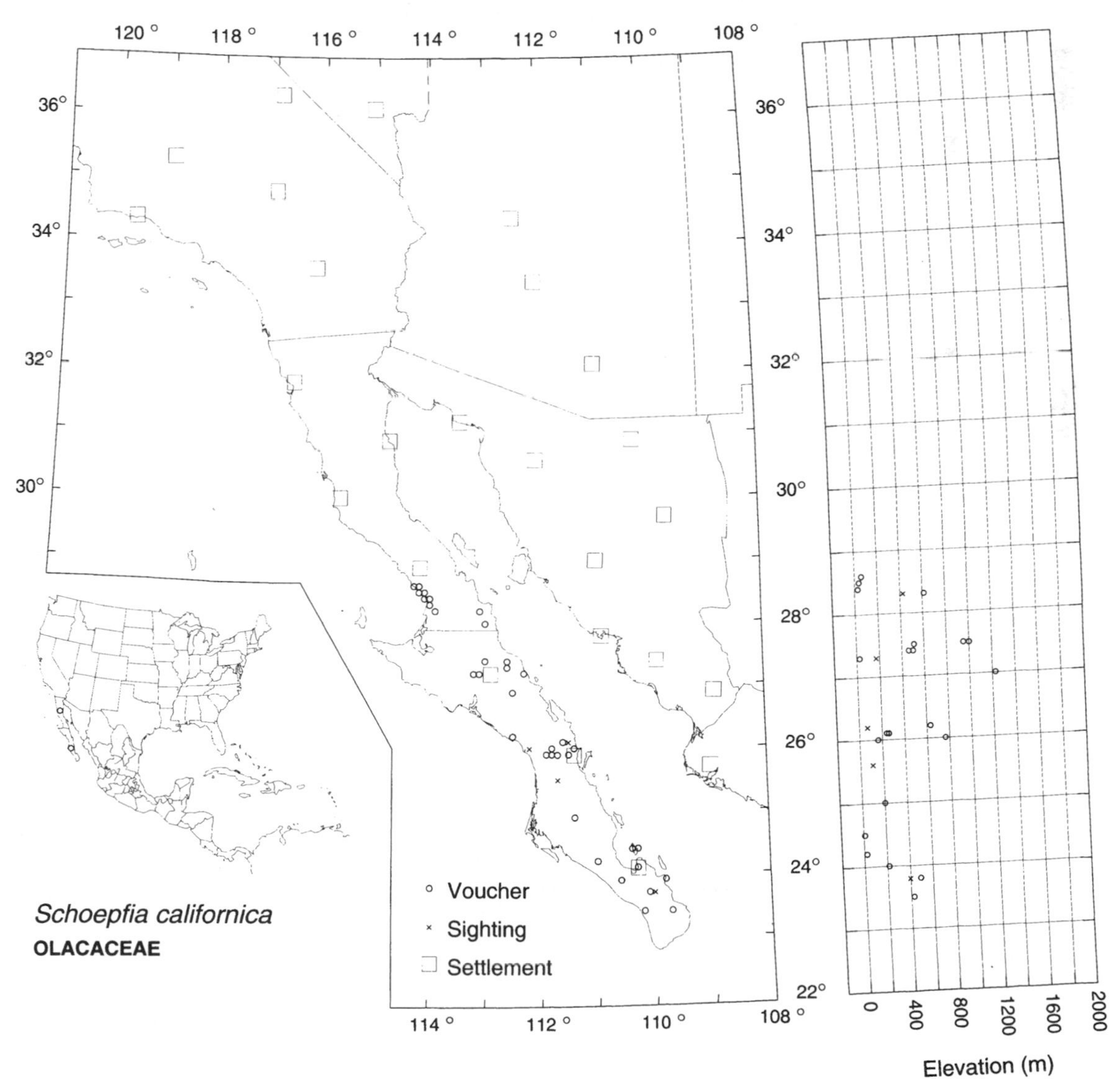

Schoepfia californica
OLACACEAE

Schoepfia shreveana

Wiggins

This shrub 2–6 m tall has rigid, slightly zigzag branches and oblanceolate to elliptic leaves 2–6 mm wide and 12–25 mm long. The leaves are glaucous and leathery. The urceolate flowers, solitary or paired in the axils, are 6–7 mm long. Fruits are single-seeded drupes 6–8 mm long.

Flowers or fruits are necessary for certain identification.

This extremely narrow endemic is known only from alkaline playas near Guaymas, Empalme, and Peón, Sonora. Its restricted distribution is in marked contrast to the wider range of *S. californica*. Perhaps the two taxa were sympatric and conspecific before gulf-floor spreading separated the peninsula from the mainland. Speciation may have occurred once peninsular and mainland populations were sufficiently isolated from one another.

Flowers appear from February to March (Wiggins 1964) and again in August.

Senna armata

(S. Watson) Irwin & Barneby

[=*Cassia armata* S. Watson]

Spiny senna

Senna armata is a green-barked shrub up to 1.5 m high with grooved, tapering stems. The ephemeral, once-pinnate leaves have 2–4 pairs of oblong-ovate leaflets 5 mm long and 4 mm wide. The leaf rachis, 1.5–8 cm long, extends beyond the last pair of leaflets and remains as a weak, green spine after the leaflets drop. The yellow flowers are solitary or paired in the upper leaf axils. The turgid, tan pods

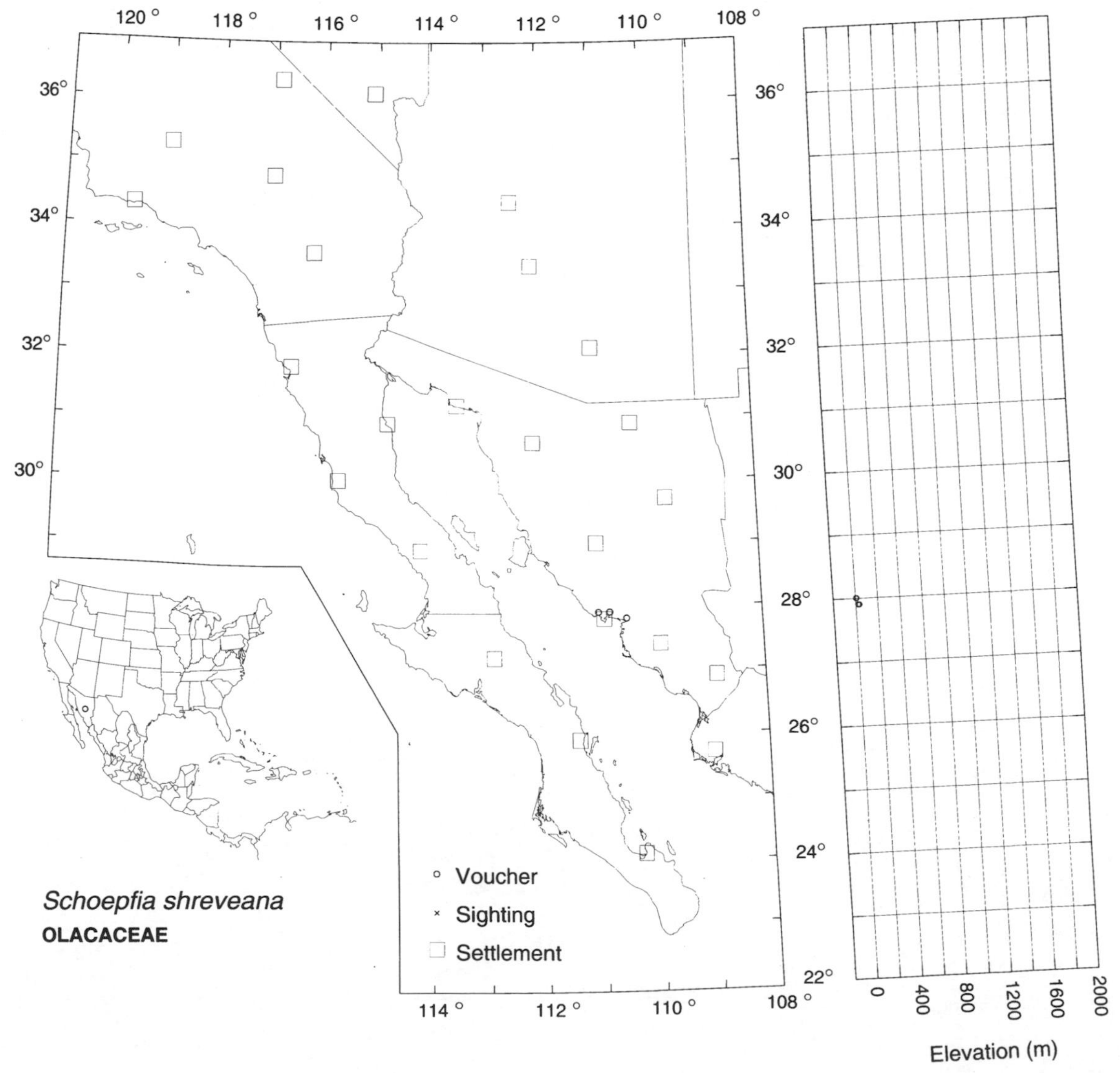

Schoepfia shreveana
OLACACEAE

(4–7 mm broad and 2.5–4.5 cm long) persist for many months. The haploid chromosome number is 14 (Raven et al. 1965).

Asclepias subulata and *A. albicans* have copious milky juice. *Ephedra* species are brushier in habit and lack spines.

Irwin and Barneby (1982) separated the genus *Senna* from *Cassia* on the basis of inflorescence, fruit, and staminal characters.

Taxonomically isolated from other species in the genus, *Senna armata* is the only member of series Armatae (Irwin and Barneby 1982). The rushlike habit is unique among North American *Senna* species, although several South American species (series Aphyllae) also converged upon the leafless, green-barked life-form (Irwin and Barneby 1982).

Locally common in sandy or gravelly washes, *S. armata* also grows on alluvial fans and floodplains throughout much of the Mojave Desert. It enters the Sonoran Desert in northeastern Baja California and southeastern California, where it experiences winter rain and hot, dry summers. The plants are well adapted to their arid environment by the photosynthesizing bark and ephemeral leaves.

Flowers appear from March to May, occasionally in other months in response to rain. The drought-deciduous leaves appear with or somewhat later than the flowers and drop by May or June.

The dimorphic anthers dehisce by terminal pores. Solitary bees, the only pollinators, "milk" them using vibratile pollination. As bees extract food pollen from the four short anthers, fertile pollen from the three long anthers adheres to their bodies (Irwin and Barneby 1982).

The shrubs make attractive, low-water-use ornamentals.

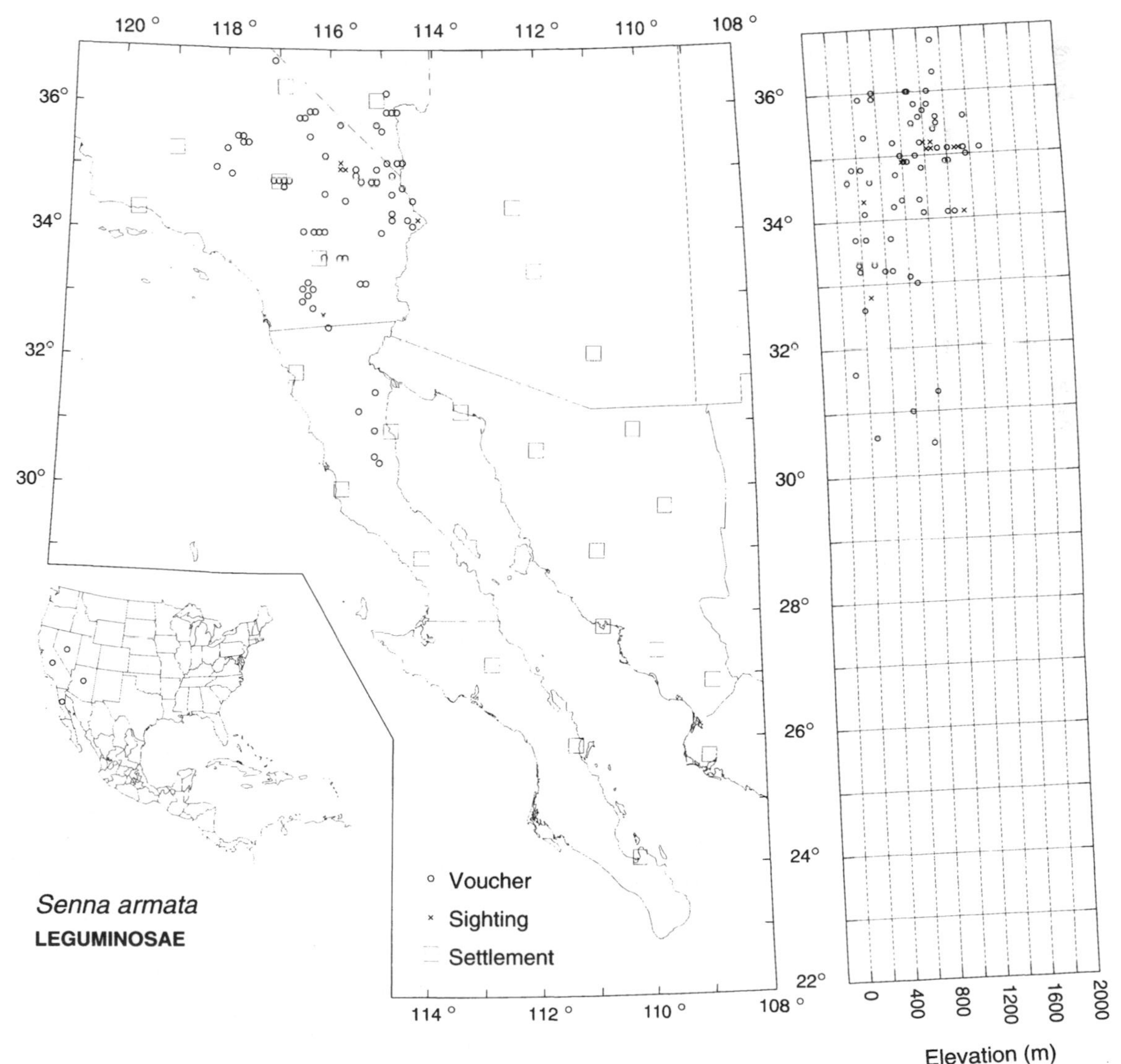

Senna atomaria

(L.) Irwin & Barneby

[=*Cassia emarginata* L.]

Sorillo, alcaparro, chile perro, flor de San José, hediondillo, palo de zorillo, vara de San José, viche

Senna atomaria, a tree up to 20 m tall, has once-pinnate leaves with 2–5 pairs of revolute leaflets as much as 7 cm wide and 15 cm long. The yellow flowers are on peduncles 5–12 cm long. The indehiscent pods, up to 4 dm long and 10–15 mm wide, break transversely along irregular fractures.

This tree occurs throughout most of Mexico and the Caribbean and extends through Central America into northern South America. East of the Sonoran Desert, it grows in Sinaloan thornscrub and Sinaloan deciduous forest. Within the desert, it is confined to watercourses.

Flowers appear in June with the new leaves. The pods are conspicuous in autumn when the trees are leafless.

The crushed foliage has a strong, disagreeable odor (Gentry 1942). *Senna atomaria* shows outstanding promise as a fast-growing wood source in the subtropics (Hughes and Styles 1984; Ngulube 1989). The seeds require scarification with acid (Ngulube 1989).

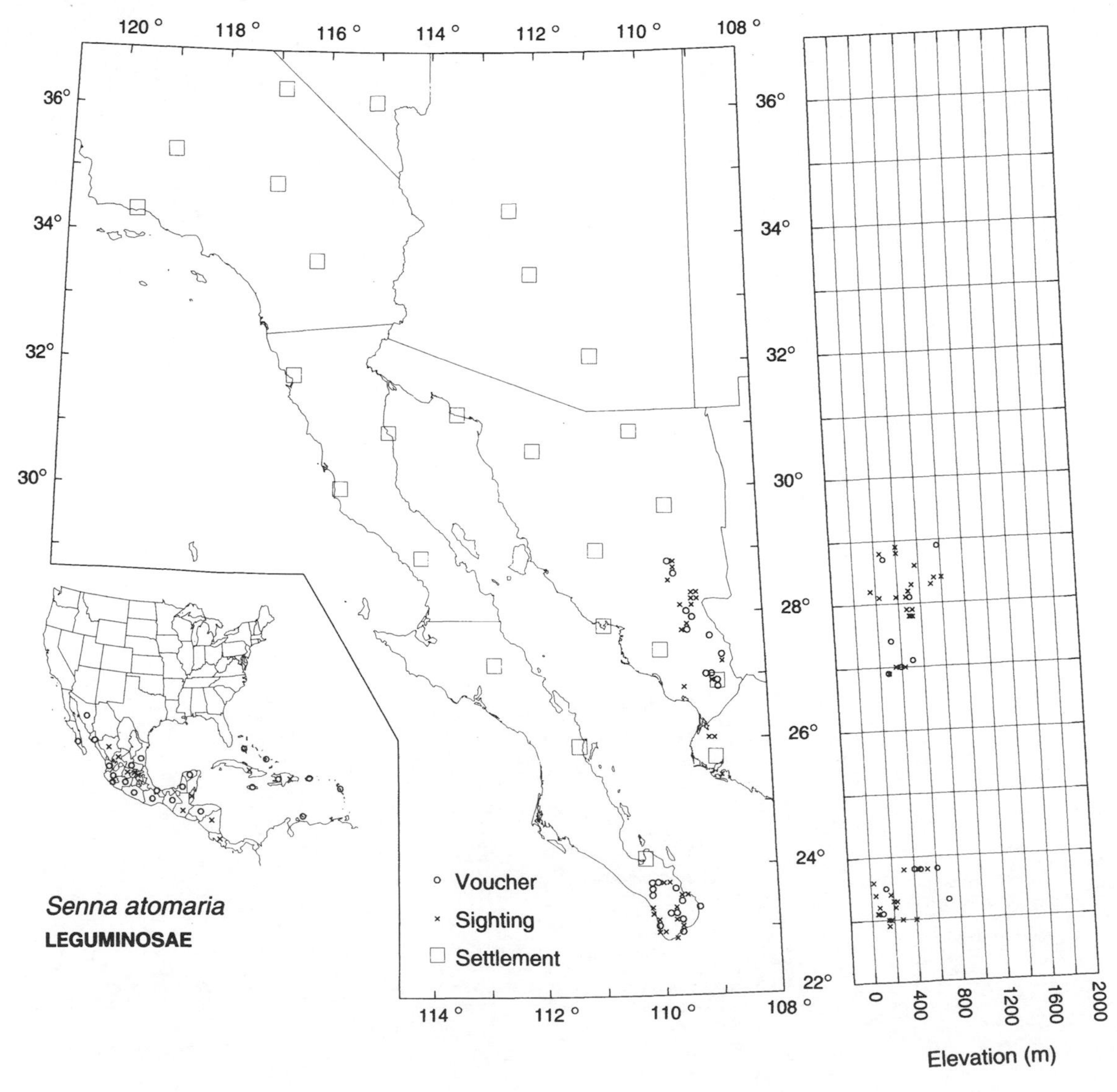

Senna pallida

(Vahl) Irwin & Barneby

[=*Cassia biflora* L.]

Ejotillo de monte, abejón, ronrón, flor de San José, viche

This slender shrub has ascending, decurved branches and thin, glabrous leaflets 1–2 cm long in 4–10 pairs on once-pinnate leaves. Leaflets are prominently veined underneath and have thickened leaf margins. The large, yellow flowers are usually paired on a thin stalk 2 cm long. The dehiscent fruits are 3–4 mm wide and 6–8 cm long. The rectangular seeds are marked with a raised "X" on the surface. The haploid chromosome number is 14 (Bawa 1973).

The 1–3 mm long, yellow-orange nectary (gland) between the lowest pair of leaflets is distinctive.

In the Sonoran Desert, *S. pallida* is largely confined to river and arroyo borders where extra moisture is available. Outside the desert, it often grows in open Sinaloan deciduous forest.

The lengthy flowering period begins in July and reaches a peak by November. If moisture is plentiful, flowering may continue until April (Gentry 1942; Janzen 1983a).

The flowers are typically visited by large bees, which remove the pollen by vibratile pollination (Wille 1963; Buchman and Hurley 1978; Michener et al. 1978). In Veracruz, Mexico, hybridization with other *Senna* species is prevented at least in part by pollinator behavior (Delgado-Salinas and Sousa-Sanchez 1977).

Each plant produces about 20–100 pods containing 6–29 seeds apiece. Several species of bruchid beetles attack the seeds (Silander 1978; Janzen 1980).

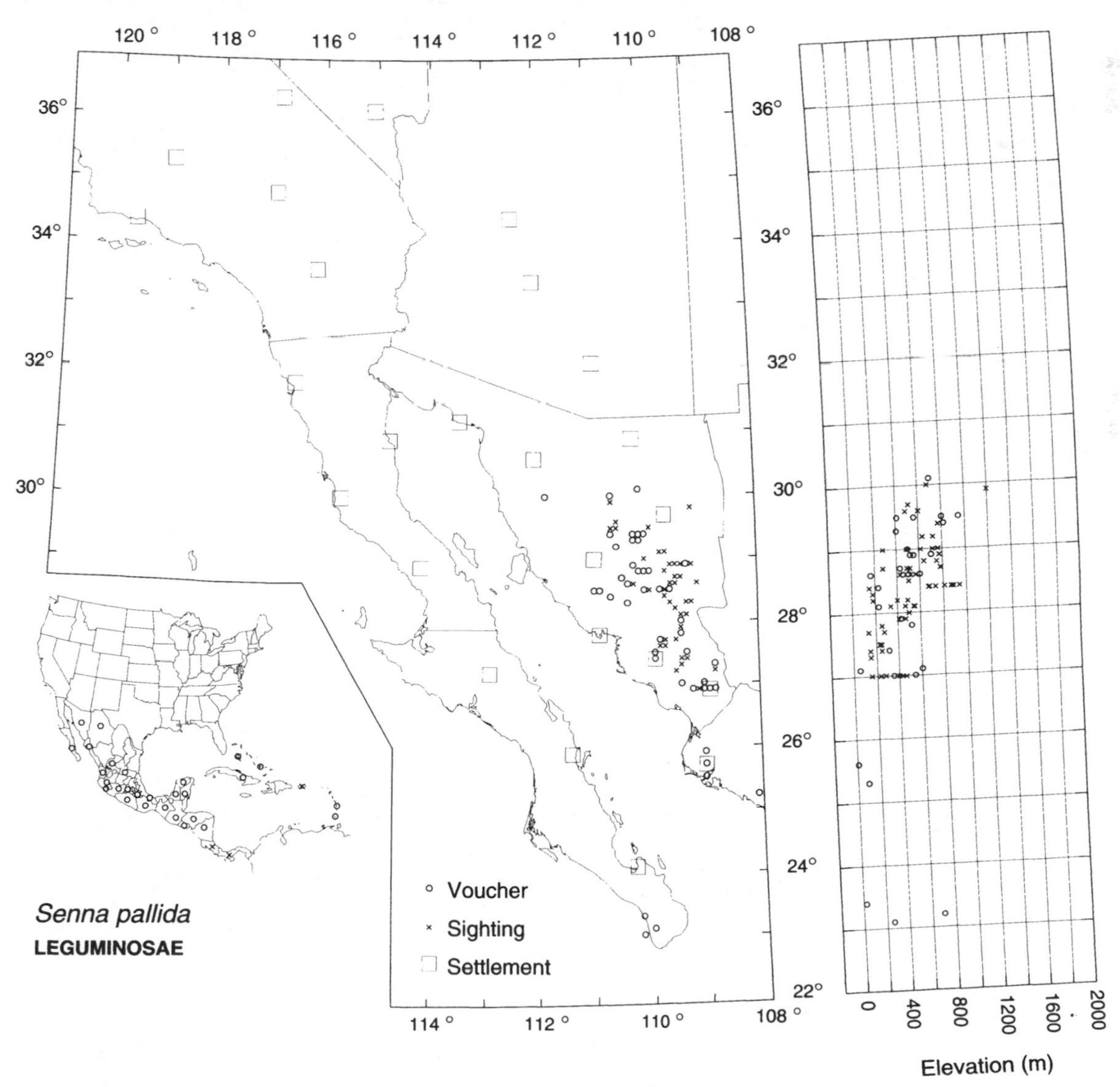

Senna polyantha

(Collad.) Irwin & Barneby

[=*Cassia goldmannii* Rose]

An openly branched tree 3–6 m tall, *Senna polyantha* has once-pinnate leaves with 5–12 pairs of leaflets 3–5 mm wide and 10–16 mm long. The fruits, 2 cm wide and 8–12 cm long, are narrowly winged.

Before 1980 this species was considered endemic to the Sierra de la Giganta in Baja California Sur (Wiggins 1964). Now it is also known from slopes and shaded canyons in the Sierra del Viejo, Sonora (30.3°N and 30.4°N) (Yatskievych and Fischer 1984). The peninsular and Sonoran populations experience biseasonal rain concentrated largely in summer or early fall. Temperatures are more extreme at the Sonoran sites.

Flowers appear from June to October. Fruits persist into the winter. The plant is found in leaf during all seasons.

Senna purpusii

(Brandegee) Irwin & Barneby

[=*Cassia purpusii* Brandegee]

A straggling shrub 1–2 m tall, *Senna purpusii* has gray-black bark and glaucous, fleshy leaves 2–4 cm long. The leaves are once-pinnate with 2–4 pairs of broadly oval leaflets 6–12 mm wide and 9–20 mm long. The flowers, bright yellow veined with purplish brown, are in terminal panicles. The turgid but flat pods are 10–12 mm wide and 3–7 cm long.

No other shrub in its range has once-pinnate leaves that are glaucous and somewhat succulent.

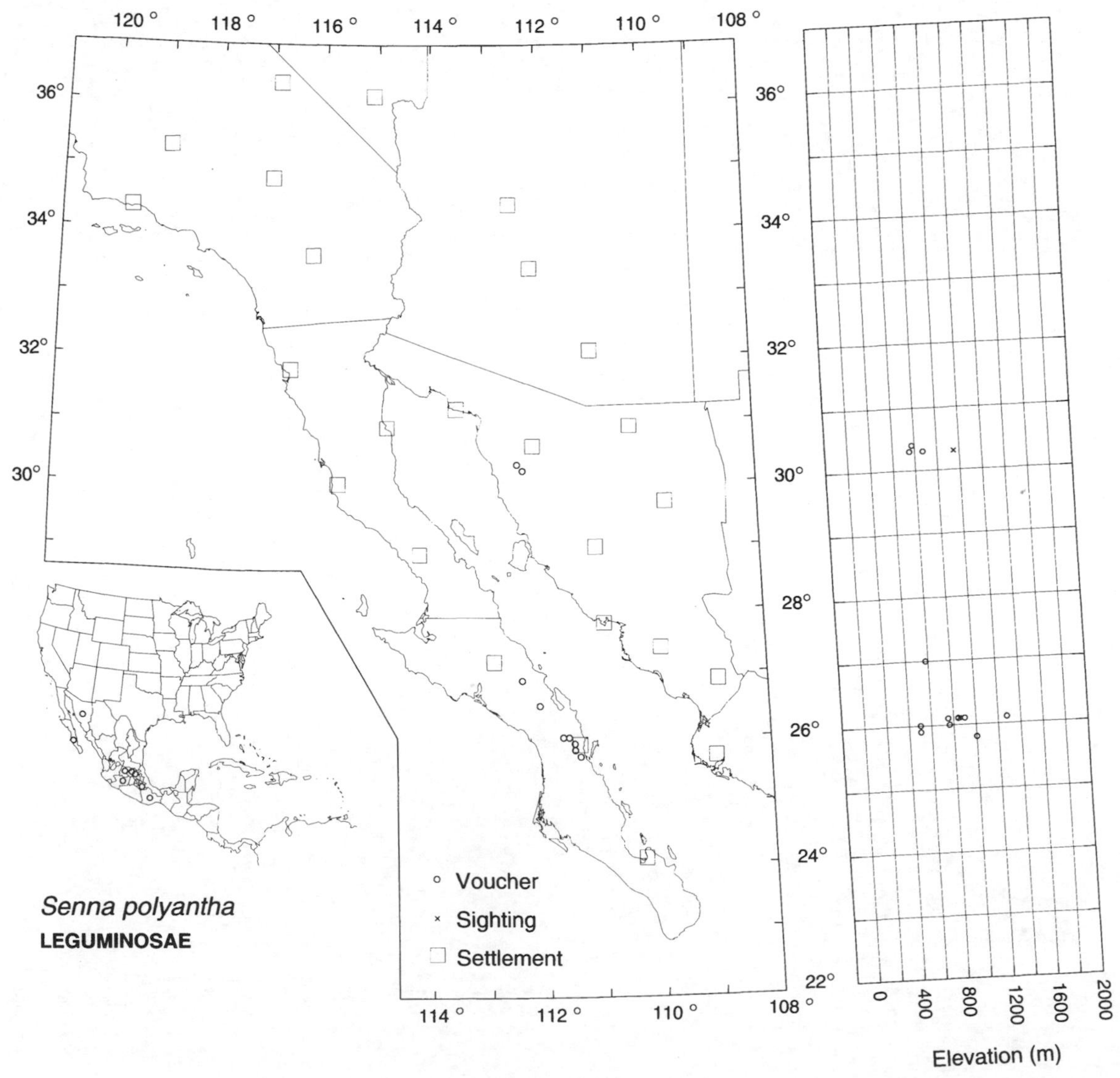

Senna polyantha
LEGUMINOSAE

Irwin and Barneby (1982) separated the genus *Senna* from *Cassia* on the basis of floral characters. *Senna purpusii* is taxonomically isolated and has no obvious relatives in Mexico (Irwin and Barneby 1982). Its closest affinities may lie with South American species in the series Pachycarpae (Irwin and Barneby 1982).

Rare to common on flats, slopes, and washes, *S. purpusii* is narrowly distributed along the Pacific coast of the central Baja California peninsula, usually below 300 m. Most populations occur well within the zone of heavy fog, which may compensate for the low, sporadic rainfall. The range is confined to the maritime arid zone with erratic winter rainfall and cool, dry summers.

Leaves are drought deciduous. Flowers appear mainly from December through April and in other months after significant rainfall.

Senna purpusii is worthy of cultivation in Mediterranean climates. Plants can be grown from seed or stem cuttings. They tend to die during hot, wet weather. They are frost hardy down to –4°C (Johnson 1993).

Senna wislizeni
(A. Gray) Irwin & Barneby

[=*Cassia wislizenii* A. Gray]

Shrubby senna, pinacate, yerba del pinacate, pinacatillo, palo prieto

A shrub up to 3 m tall, *Senna wislizeni* has dark brown or gray bark conspicuously dotted with numerous white lenticels. The leaves, 1.5–3 cm long, are once-pinnate with 2–7 pairs of obovate leaflets 4–10 mm long and 2–6 mm wide. The yellow flowers are 4–5 cm wide. The linear pods, shiny and dark brown or black at maturity, are 6–15 cm long and 5–6 mm wide.

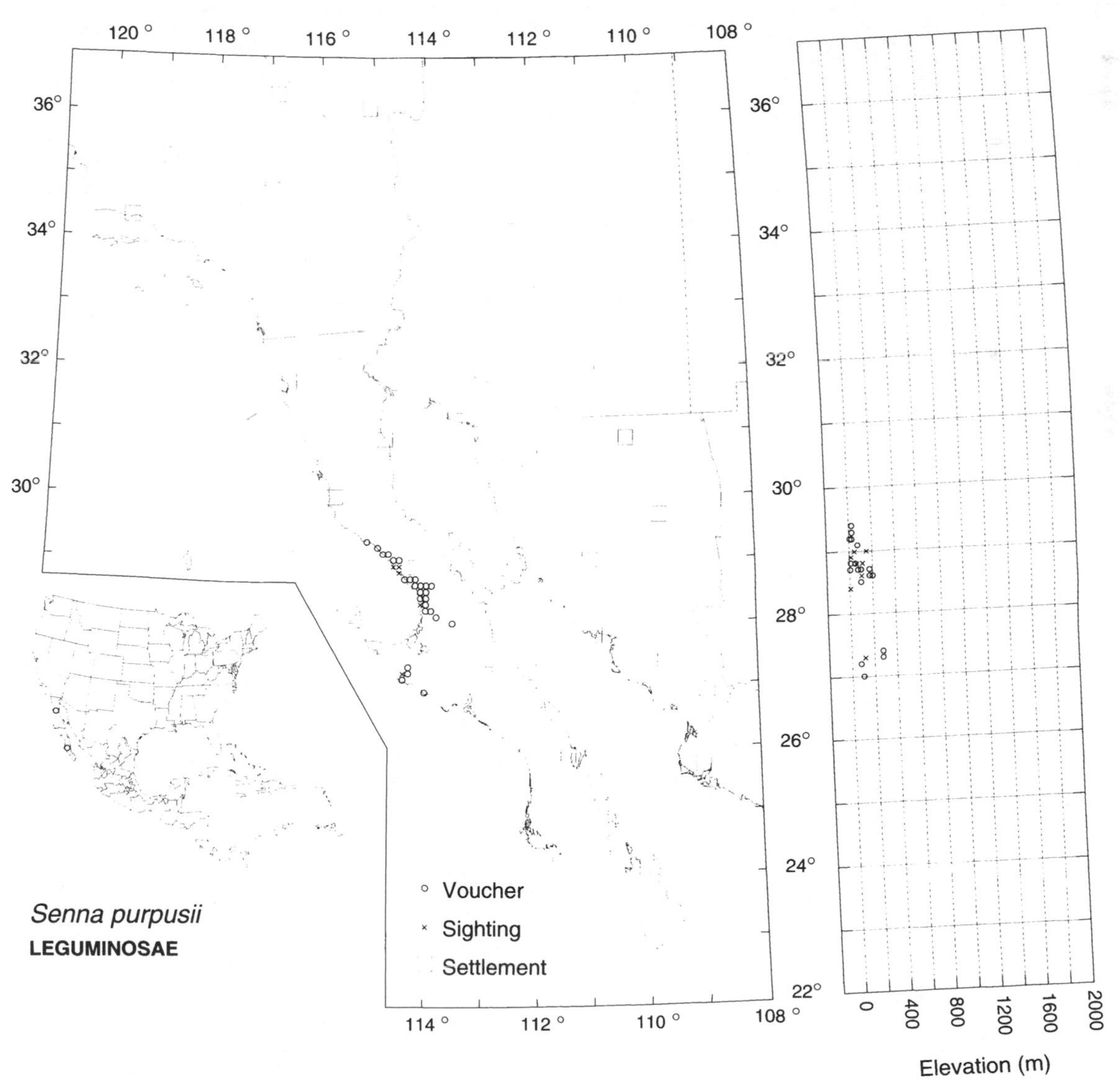

Senna purpusii
LEGUMINOSAE

Senna polyantha has 5–12 pairs of oblong leaflets. *Caesalpinia* species are bipinnate.

Irwin and Barneby (1979, 1982) separated the genus *Senna* from *Cassia* on the basis of inflorescence, fruit, and staminal characters. They defined four varieties in *S. wislizeni*. The southernmost, var. *pringlei*, represents the "primitive core" from which the others evolved as they moved north into progressively colder and drier climates (Irwin and Barneby 1982). Variety *wislizeni*, the northernmost, is the most cold- and drought-tolerant taxon in the group. It is the only variety in the area under study in this atlas (Irwin and Barneby 1982).

Common in the Chihuahuan Desert on rocky limestone hills and canyon slopes, *S. wislizeni* var. *wislizeni* enters the Sonoran Desert in the vicinity of Santa Ana, Sonora, where it was collected by Forrest Shreve in 1933 (30.3°N) and Janice E. Bowers in 1983 (30.2°N). The Sonoran population receives rain from July through September and again in winter. This widely disjunct population may be a relict of Pleistocene interglacials, when relatively high temperatures and decreased rainfall might have allowed the westward movement of this and other Chihuahuan Desert species (Yatskievych and Fischer 1984).

Leaves are drought- and cold-deciduous. Flowers appear from June to October in response to rain. Occasional plants may bloom as early as April. The flowers are pollinated by solitary bees, which use vibratile pollination to extract pollen from the terminal pores (Irwin and Barneby 1982).

Senna wislizeni is grown as an ornamental in southern Arizona and Texas. Fresh, untreated seed germinates readily, and the seedlings grow rapidly (Nokes 1986). Plants can also be grown from semihardwood cuttings taken in late summer. Newly rooted cuttings should be protected from frost (Nokes 1986). Once established, the plants require only occasional irrigation, except where annual rainfall is less than 25 cm (Duffield and Jones 1981). Too little water inhibits blossoming (Duffield and Jones 1981).

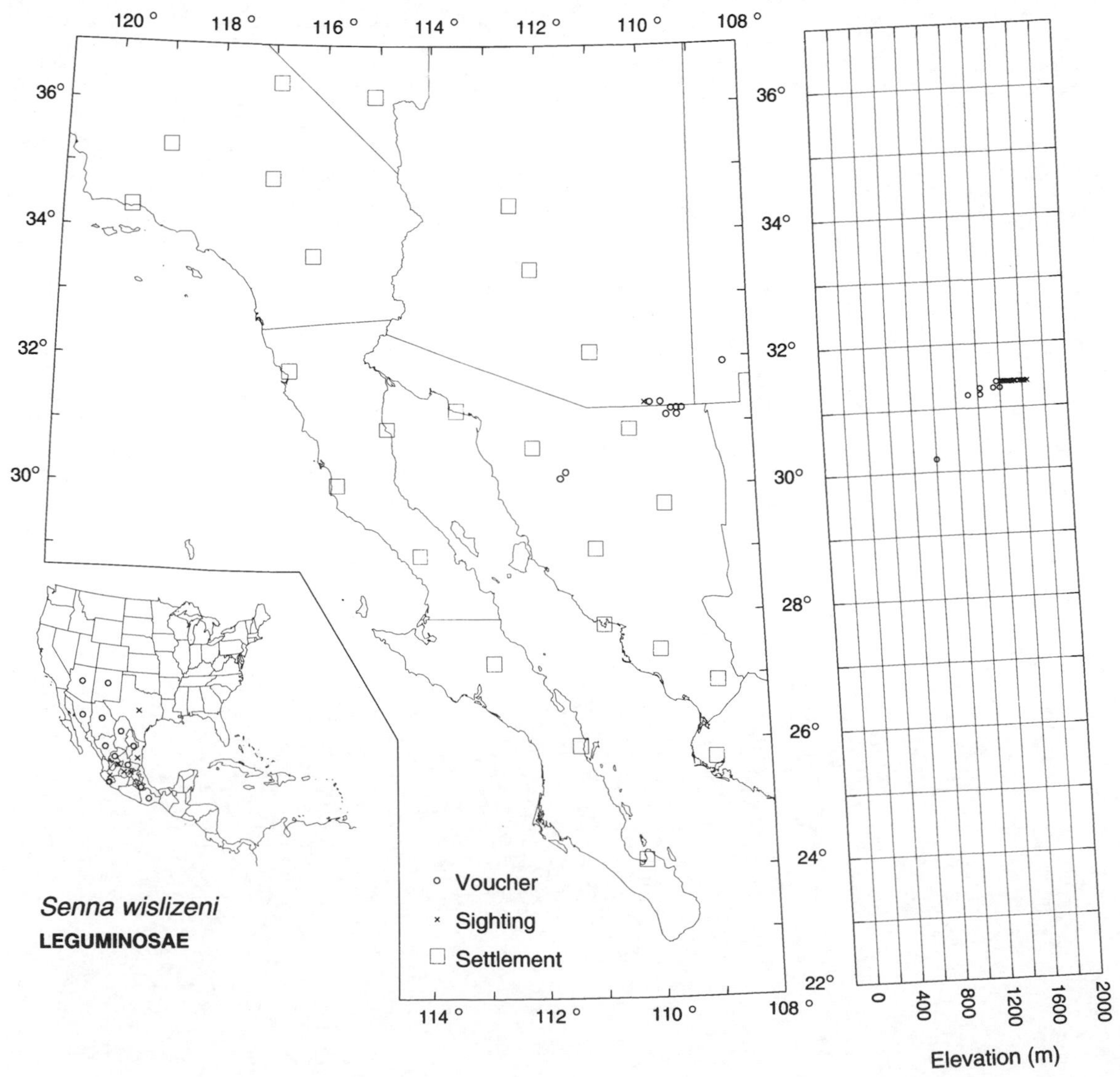

The plants tolerate freezing temperatures as low as −12°C (Johnson 1993).

Sideroxylon leucophyllum

S. Watson

[=*Pouteria leucophylla* (S. Watson) Cronq.]

This shrub or tree grows 3–10 m tall and has densely tomentose oblong, lanceolate, or oblanceolate leaves 1.5–3.5 cm wide and 4–10 cm long. The inconspicuous flowers are densely clustered in the leaf axils. The small, ovoid drupes are 2–2.5 cm long.

Cronquist (1946) limited use of the genus name *Sideroxylon* to several related species in Africa.

Endemic to Baja California, *S. leucophyllum* is rare to locally common along washes and on rocky slopes, typically in canyons. It grows on both Isla Angel de la Guarda and Isla San Esteban (Moran 1983b). The northernmost populations (32.2°N and 32.1°N) are in and around Guadalupe Canyon in the Sierra Juárez. The species continues along the east base of the mountains south to Bahía de los Ángeles (Wiggins 1964). The spotty distribution on our map may reflect a lack of collecting activity on the east side of the Sierra Juárez and Sierra San Pedro Mártir (Geoffrey Levin, personal communication 1991). The range is essentially arid and frost free. Summer temperatures are often very high. Rain is most likely in September or during winter.

Flowers appear from March to June, sometimes again from October through January. The leaves are apparently persistent.

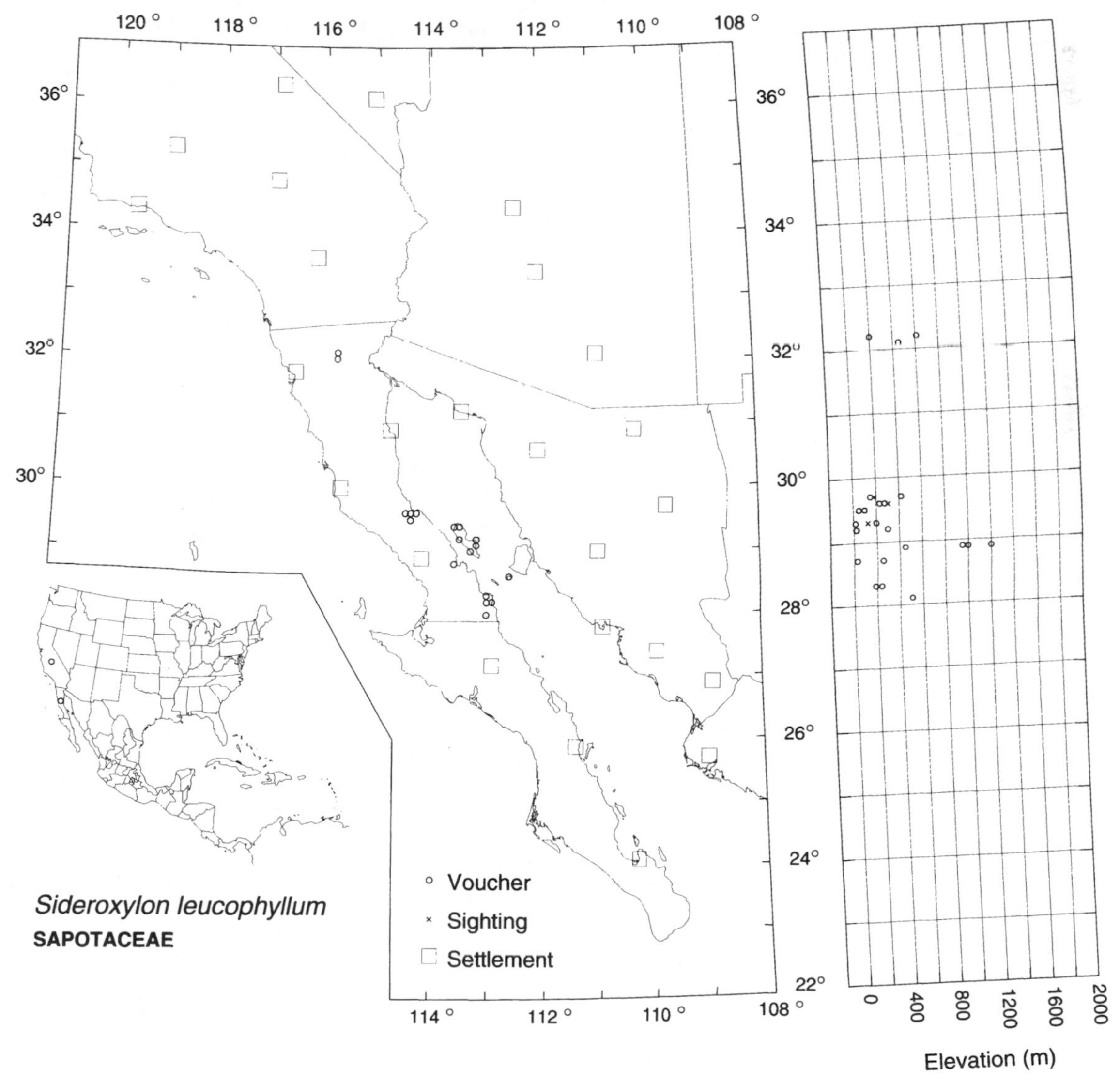

Sideroxylon leucophyllum
SAPOTACEAE

Sideroxylon occidentale
(Hemsl.) Pennington

[=*Bumelia occidentalis* Hemsl.]

Bebelama

Sideroxylon occidentale is a spiny shrub or tree up to 8 (rarely 10.5) m tall with stiff branches and a dense crown. The oblanceolate, elliptic, spatulate, or obovate leaves, 8–30 mm long and 2–20 mm wide, are glabrous or puberulent above, finely gray-puberulent below. The flowers, 4.5–5.2 mm long, are in axillary clusters. In some flowers the stamens are sterile (Pennington 1990). The blue-black drupes are 7–16 mm long.

The leaves of *S. lanuginosum* Michx. are conspicuously reticulate above and often loosely woolly-villous beneath. The leaves of other *Sideroxylon* species in southern Sonora and Sinaloa are more than 5 cm long (Cronquist 1945; Pennington 1990).

Sideroxylon occidentale is most closely related to *S. celastrinum* (Kunth) Pennington of southern Texas, eastern Mexico, and northern Venezuela (Pennington 1990).

This plant grows along arroyos, on alluvial plains, and near water holes. It occurs throughout the state of Sonora from 31.0°N to the vicinity of Navojoa and at scattered localities in Baja California Sur, often along the arid margin of Sinaloan thornscrub. Northern and upper elevational limits appear to be truncated by freezing. The biseasonal rainfall regime includes a substantial proportion of summer precipitation.

Flowers and fruits are produced throughout the year.

The fruit was eaten fresh by the Seri Indians (Felger and Moser 1985).

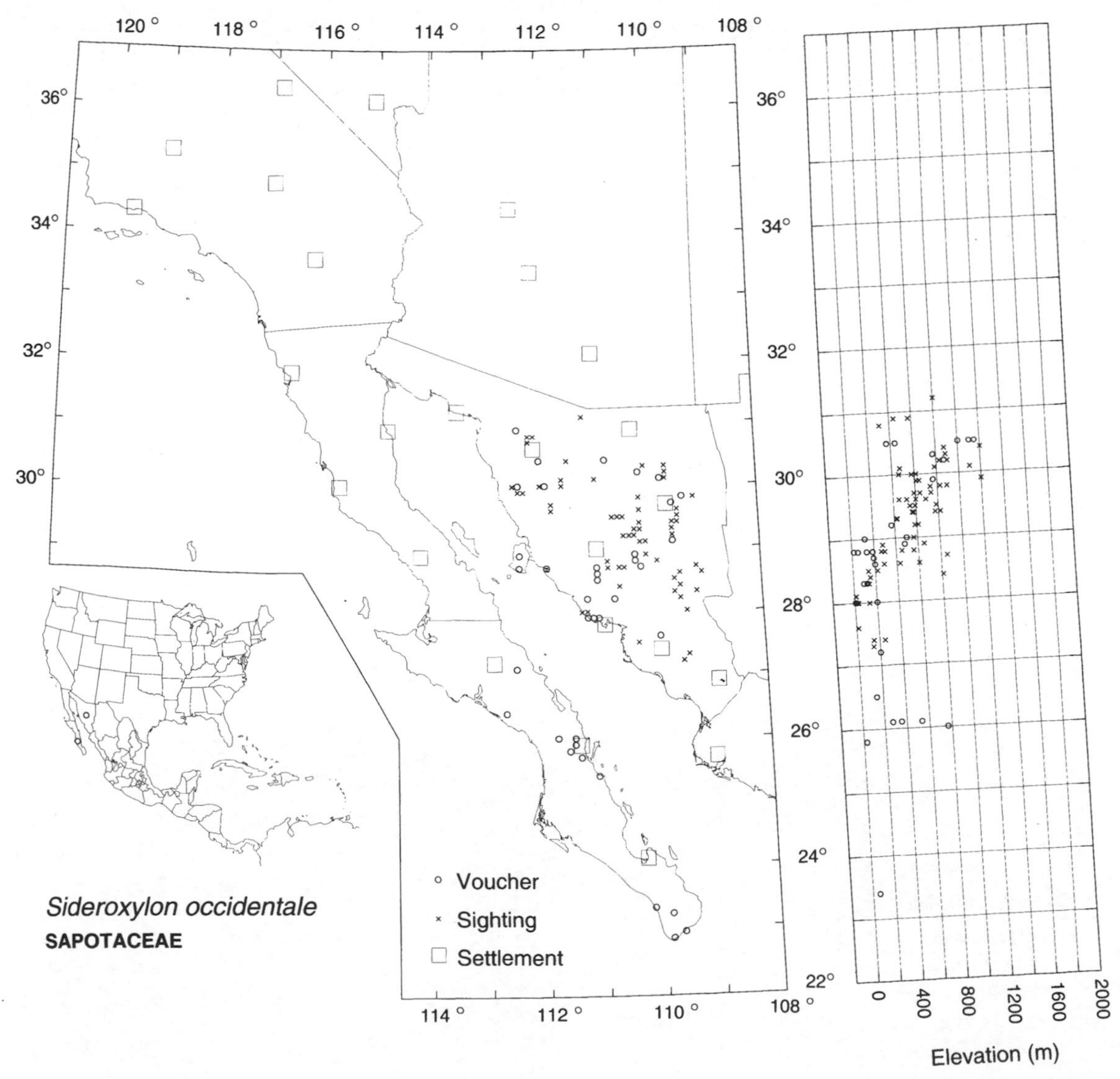

Sideroxylon occidentale
SAPOTACEAE

Sideroxylon peninsulare

(Brandegee) Pennington

[=*Bumelia peninsularis* Brandegee]

This shrub or small tree, 3–5 m tall, has stiff, spine-tipped branchlets. The alternate, leathery leaves, 1–2.5 cm wide and 2.5–4.5 cm long, are ovate to elliptic. They are dull green and glabrous above, pubescent along the midrib beneath. The fascicled, yellow flowers are 6–8 mm long. The ellipsoid fruits, 12–15 mm long, are blue-black at maturity.

Sideroxylon occidentale has smaller, glaucous leaves finely puberulent on both surfaces.

Sideroxylon peninsulare may be conspecific with *S. cartilagineum* (Cronq.) Pennington, known from Sinaloa to Guerrero (Pennington 1990). Their close similarity is the basis for Standley's (1924) report of *S. peninsulare* from the mainland.

This plant is narrowly endemic to the Cape Region, where it grows in canyons and along arroyos. It is not found on any gulf islands (Moran 1983b). This and other Cape Region endemics (among them *Calliandra brandegeei*, *Mimosa xanti*, and *Tephrosia cana* Brandegee) may have evolved after sea-level changes during the Pliocene isolated the Cape Region from the remainder of the peninsula (Gastil et al. 1983).

The flowering season is December to January (Wiggins 1964); sporadic flowering may occur in other months.

Simmondsia chinensis

(Link) Schneid.

Jojoba, goatnut, gray box bush, pignut, sheepnut, wild hazel, quinine plant, coffeeberry

Simmondsia chinensis is an intricately and rigidly branched shrub 1–3 m tall with a dense canopy of leathery, opposite, entire leaves. The leaves, 2–5 cm long and 1–2 cm wide, are ovate to oblong-elliptic. Their

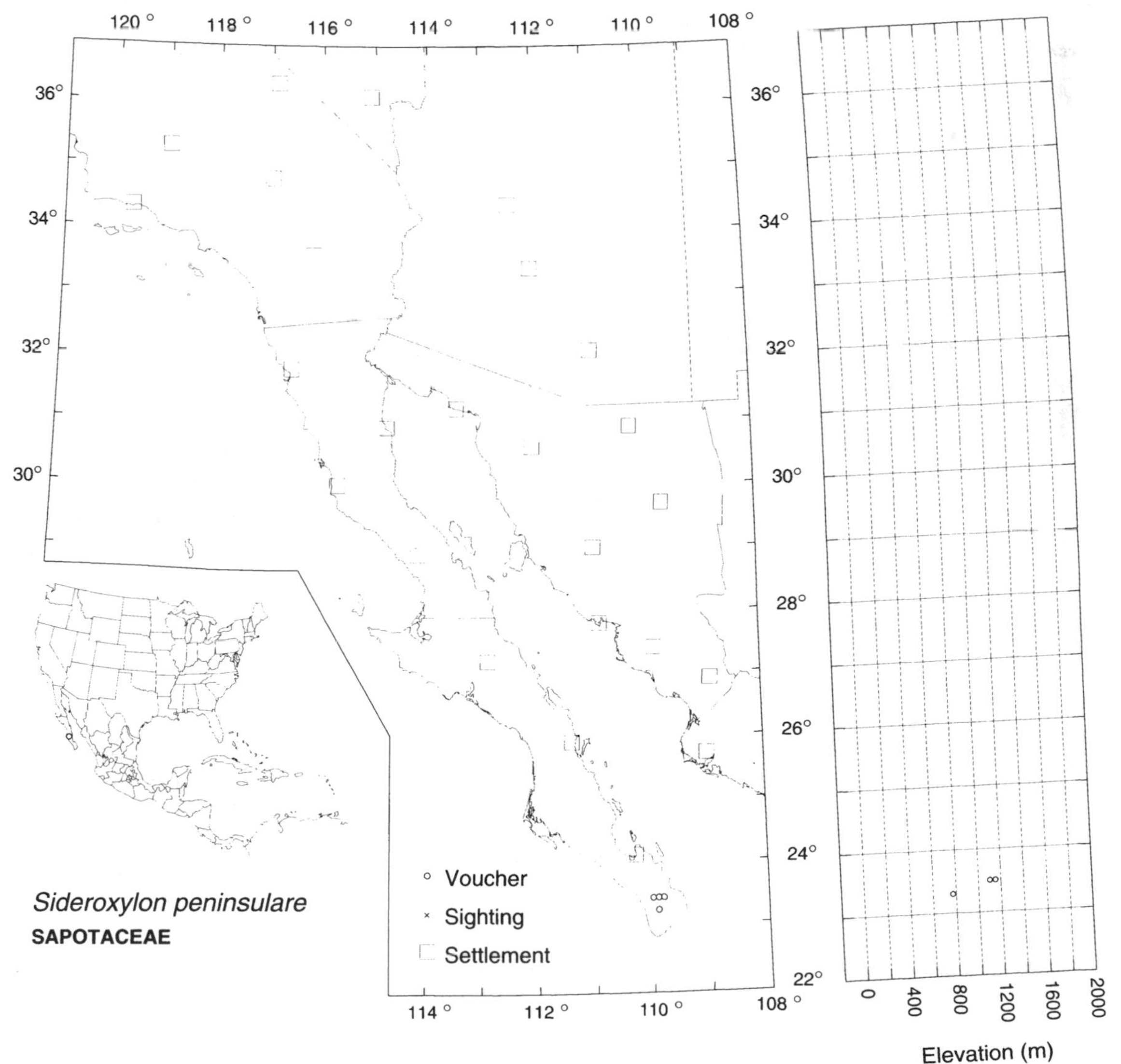

Sideroxylon peninsulare
SAPOTACEAE

upright orientation is distinctive. Plants are dioecious, with either solitary pistillate flowers or axillary clusters of staminate flowers. The ratio of female to male plants in a population is usually 1:1 (Wallace and Rundel 1979). The single-seeded capsules are 1.5–2 cm long. The diploid chromosome number is 52 (Raven et al. 1965; Lee and Sherman 1985). Jauhar (1983) suggested that *S. chinensis* is an allotetraploid; however, given its apparent phylogenetic isolation, the hybridization would be very old, and it seems more likely that the plants are autotetraploids (Geoffrey Levin, personal communication 1991).

The opposite, entire, leathery leaves and large, 1-seeded capsules are distinctive.

Simmondsia, a genus with a single species, has been placed in three different families—Euphorbiaceae, Buxaceae, and Simmondsiaceae—and as many orders (Scogin 1980). Serotaxonomic evidence suggests that it is best treated as Simmondsiaceae, within Euphorbiales (Scogin 1980), as first suggested by Takhtajan (1969). Pollen morphology also supports placement in Simmondsiaceae (Nowicke and Skvarla 1984).

The origin of *Simmondsia* has long been a matter for speculation. Gentry (1958) suggested that it originated in a Mediterranean climate. Kadish (1985) thought it might have arisen 5–15 million years ago as a hybrid between the genera *Buxus* and *Sarcococca.* Stebbins and Major (1965) classified *S. chinensis* as a paleoendemic, but Gail (1964) argued that the large degree of heterozygosity and great variation between populations suggests it is quite young and still expanding its range. Our map furnishes little evidence to support or refute these suggestions. The scattered disjuncts along the eastern edge of the range could perhaps represent an expanding species. Certainly *S. chinensis* is absent from large areas that would seem to provide suitable habitat.

Simmondsia chinensis is locally common on gravelly or rocky hillsides and canyon slopes, sometimes forming dense thickets. In the more arid portions of its range, it is occasional to common along

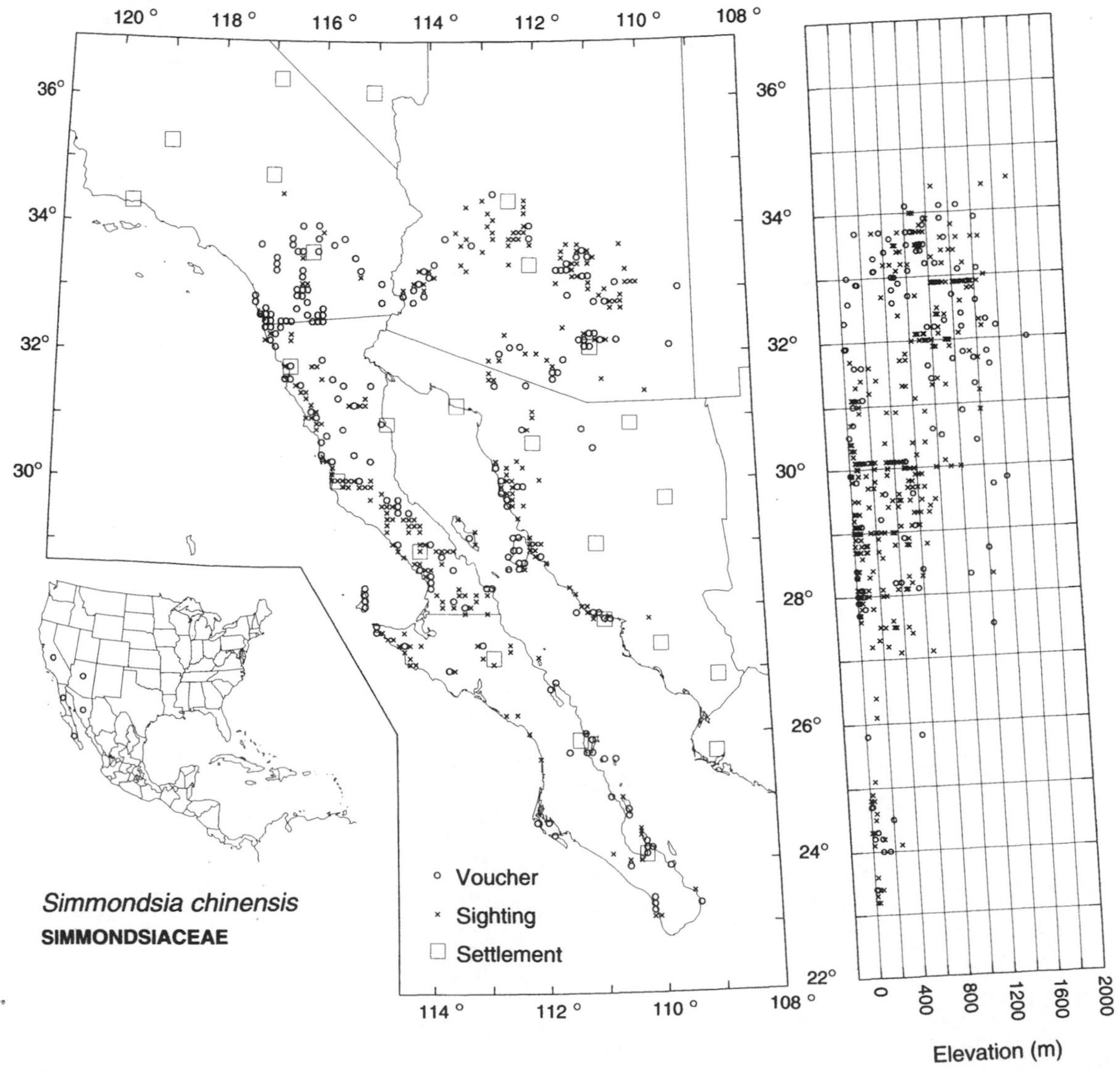

washes and runnels. It grows in nonsaline to slightly saline and near-neutral to alkaline soils (Franco-Vizcaíno and Khattak 1990).

The range encompasses most of the Sonoran Desert except for those areas in central and southern Sonora where average summer rainfall is high and winters are relatively dry. The eastern parts of the range in Arizona, Sonora, and Baja California Sur receive biseasonal rainfall. This pattern shifts from mostly summer rain in the south to equal amounts of summer and winter rain in the north. Western populations receive most of their moisture in winter. Summer storms are occasional except along the Pacific coast. Across the range, winters vary from almost warm-temperate with occasional frosts in the north to essentially tropical in the southern peninsula. The area around the Colorado Delta is apparently too arid for *S. chinensis.*

The evergreen leaves last 2–3 years (Gentry 1958). Although the plants are slow to show drought stress (Humphrey 1975; Nilsen et al. 1984), they may lose many or most of their leaves in severe drought years (Felger 1966). Foliage grows mainly in response to winter-spring rains (Gentry 1958). Summer rains may stimulate stem and leaf growth in early autumn (Humphrey 1975). Flower buds appear in fall (Gentry 1958) or winter (Humphrey 1975). Peak bloom is from February through April. Occasional plants flower as early as December or as late as July. When winter rains have been light, buds might not break dormancy (Gentry 1958). Ripe fruits can be found by June and often remain on the plant into August (Humphrey 1975).

The copious pollen, small pollen grains with almost no surface sculpturing, and enlarged stigmatic surfaces are typical of wind-pollinated plants (Buchmann 1987b). The leaves act as foils in pollen capture (Niklas and Buchmann 1985). Ancestral *Simmondsia* was apparently insect pollinated (Gail 1964; Niklas and Buchmann 1985). If wind pollination in present-day *Simmondsia* is secondarily derived, that would account for the rich concentrations of total crude protein and carbohydrates in the pollen, which are much higher than is normal for a wind-pollinated species (Buchmann 1987b).

The leaves are morphologically and anatomically adapted to endure severe drought stress. The high percentage of dry matter (often more than 40%) makes them rigid and therefore resistant to wilting (Benzioni and Dunstone 1986). The prominent cuticle reduces water loss (Benzioni and Dunstone 1986). A high solute potential may help maintain leaf water content under drought conditions (Wardlaw et al. 1983), as does diurnal osmotic adjustment (Nilsen et al. 1984). The vertical orientation maximizes interception of solar radiation at sunrise and sunset, when evaporative demand is low, and reduces interception at midday, when temperatures are high (Benzioni and Dunstone 1986).

These adaptations for reducing water loss and resisting desiccation also restrict carbon dioxide uptake, resulting in slow growth (Wardlaw et al. 1983; Benzioni and Dunstone 1986). At high temperatures (30°C), *Simmondsia* has a low photosynthetic rate and a limited allocation of photosynthate to new leaves (Wardlaw et al. 1983). The maximum rate of net carbon dioxide exchange in one study, 16 mg CO_2/dm^2/hr, occurred at 19–25°C (Wardlaw et al. 1983). Positive photosynthetic rates are maintained at water potentials as low as −7 MPa and at temperatures as high as 40–47°C (Al-Ani et al. 1972). Because *Simmondsia* continues net photosynthesis even during severe drought, the plants maintain a favorable carbon balance year-round (Al-Ani et al. 1972). Wardlaw and coauthors (1983:299) concluded that "with low rates even under optimal conditions [the strategy] is essentially one of adaptation for survival rather than adaptation for production." The ability to transpire and extract water from soil even when plant water potential is quite low probably allows *Simmondsia* to reduce interspecific competition for water (Roundy and Dobrenz 1989).

Across its range, *Simmondsia* varies substantially in frost susceptibility (Khalafalla and Palzkill 1990), chilling requirements (Ferriere et al. 1989), and other environmental responses (Al-Ani et al. 1972; Benzioni and Dunstone 1986). When plants from desert and coastal populations are subjected to the same water stress, desert plants show significantly less reduction in photosynthetic rates and water loss than coastal plants, which suggests that drought has been a potent selective force in the evolution of ecotypes (Al-Ani et al. 1972).

In desert populations, the female shrubs have larger, thicker leaves with a higher water content than male shrubs. Also, the canopies of females are more open, with fewer secondary branches and less mutual shading (Wallace and Rundel 1979). Allocating a greater proportion of photosynthate to leaves enables female plants to meet the energy requirements of seed bearing. Whereas male shrubs allocate 10–15% of their resources to reproductive tissues, females must allocate 30–40% if all the flowers are to set seed (Wallace and Rundel 1979).

Simmondsia seeds contain 50% liquid wax by weight (Rost et al. 1977). The wax content declines to about 10% within 30 days after germination, which suggests that it is metabolized by the embryo and seedling (Rost et al. 1977; Sherbrooke 1977). During this time, the taproot elongates rapidly, reaching 16 cm in 10–12 days, while the shoot, which emerges 14–18 days after germination, grows at a much slower rate (Benzioni and Dunstone 1986).

Given adequate moisture, seeds germinate over a wide range of temperatures: Sherbrooke (1977) found recently germinated seedlings in November, December, February, July, August, and September. Seedling survival on his study site in the Tucson Mountains, Arizona, was 12% after 12 months. The main causes of death appeared to be drought, freezing, and rodent predation (Sherbrooke 1977). Seedlings in protected sites (associated with rocks or shrubs) had a higher survival rate

(22%) than unprotected seedlings (6%). Survival rates continued to decline over the following nine years (Sherbrooke 1989). At another site, seedling establishment was affected more by vegetative growth and environmental conditions before the onset of spring drought than by predation (Castellanos and Molina 1990).

The seeds contain cyanogenic glucosides that may deter seed predation. Nevertheless, the heteromyid rodent *Perognathus baileyi* eats and stores them. Evidently the rodents are able to detoxify the seeds (Sherbrooke 1976). Other heteromyids in the Sonoran Desert lack this detoxification mechanism, which suggests that *P. baileyi* coevolved with developing seed toxicity (Sherbrooke 1976). Although rodent predation on seeds in the soil is high (up to 98% in one study; see Castellanos and Molina 1990), *P. baileyi* does facilitate seed dispersal and germination (Sherbrooke 1976).

This species is an attractive ornamental in desert gardens. Although young plants are slow growing and cold sensitive, mature plants are heat and drought tolerant and grow fairly rapidly with supplemental irrigation (Duffield and Jones 1981).

More than 200 species of insects have been collected from *S. chinensis* in California and Arizona. Nearly half were plant feeders, and most of those were generalist species (Pinto et al. 1987). *Simmondsia* is browsed by cattle and wildlife throughout its range (Roundy and Dobrenz 1989; Roundy et al. 1989).

Seri Indians made shampoo from the ground fruits and used various preparations of the seeds to treat colds and sores (Felger and Moser 1985). Other tribes ate the fruits parched or raw and brewed a coffeelike beverage from roasted seeds (Standley 1923).

The liquid wax in the seeds resembles sperm whale oil. According to Sherbrooke and Haase (1974), sporadic attempts to cultivate *Simmondsia* have been made since 1933, when the unique nature of the wax was discovered (Greene and Foster 1933), but widespread interest in its commercial potential did not develop until 1972, when importation of sperm whale oil into the United States was banned (Benzioni and Dunstone 1986). Since then, *Simmondsia* has been the focus of intensive cultivation and domestication efforts (Sherbrooke and Haase 1974). Some of the numerous potential uses for *Simmondsia* wax include lubricants, cosmetics, soaps, and pharmaceuticals (Sherbrooke and Haase 1974). Extensive bibliographies emphasizing the economic botany of *Simmondsia* are provided by Sherbrooke and Haase (1974) and Sherbrooke (1978).

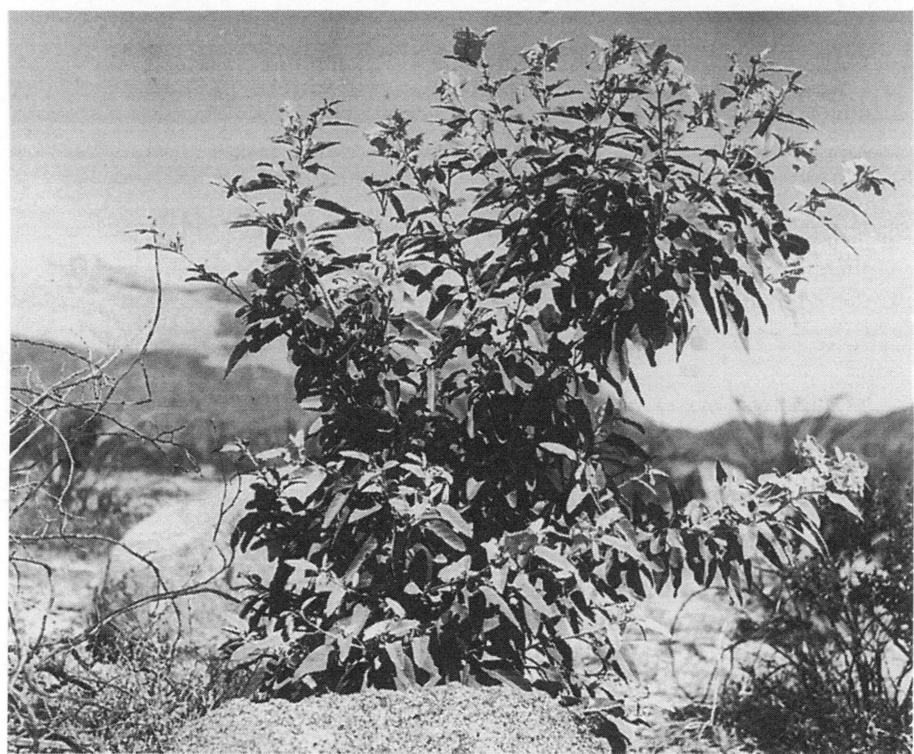

Figure 77. Solanum hindsianum *near Bahía de los Angeles, Baja California. (Photograph by J. R. Hastings.)*

Solanum hindsianum

Bentham

Mariola

Solanum hindsianum (figure 77) is a straggling shrub 1–3 m tall with scattered spines. The entire to weakly undulate leaves, 2–6 cm long and 5–20 mm wide, are ovate to oblong. The foliage is densely stellate-tomentose, often with tawny hairs. The rotate, lavender to purple flowers, 2.5–5 cm wide, are single in the upper leaf axils. The smooth, globose berries 1–2 cm wide are pale green with dark green stripes. The haploid chromosome number is 12 (Averett and Powell 1972).

Solanum amazonium Ker, a shrub or semiwoody perennial up to 1 m tall, has sinuate to lobed leaves up to 10 cm wide and 18 cm long.

Solanum hindsianum is often common on rocky hillsides and bajadas and along arroyos and canyons. It occurs widely throughout the peninsula and gulf islands, more sparingly along the Sonoran coast. Before M. C. Lechner collected it at Organ Pipe Cactus National Monument (32.0°N) in 1986, *S. hindsianum* was known neither from the monument (Bowers 1980) nor from the state of Arizona (Kearney and Peebles 1960). The range covers arid climates where frost is rare or absent and summers are less torrid than in the lower Colorado River Valley. Rainfall seasonality varies from almost entirely winter for the Pacific coast populations to mostly summer near the southern and eastern limits of the species.

Flowers and fruits are produced throughout the year in response to rain. The leaves are evergreen except in unusually dry years and at the arid margins of the range (Humphrey 1975).

The flowers are probably pollinated by native bees that use indirect flight muscles to vibrate pollen out of terminal pores in the anthers (Buchmann et al. 1977).

Seri Indians made a tea from the roots and flowers as a remedy for diarrhea (Felger and Moser 1985).

Stegnosperma halimifolium Bentham

Stegnosperma watsonii D. J. Rogers

Amole, bledo carbonero, ojo de zanate

These spreading shrubs grow 1–5 m tall and have arching branches (figure 78). The oblong to obovate, alternate leaves, 1–5 cm long, are thick, leathery, and glaucous. The white, 5-petaled flowers, 6–8 mm wide, are in axillary or terminal clusters. The red, ovoid capsules are 6–8 mm long. The haploid chromosome number of *S. halimifolium* is 18 (Gilmartin and Neighbours 1982). Horak (1981) described the anatomy of the vascular tissues.

Stegnosperma halimifolium and *S. watsonii* cannot be reliably separated with the keys in Rogers (1949) or Wiggins (1964). According to Gilmartin and Neighbours (1982), *S. watsonii* has axillary inflorescences, stamens 2–3 mm long, and spheroidal seeds, whereas *S. halimifolium* has terminal inflorescences, stamens 4–6 mm long, and ovoid seeds. Certain specimens, especially from the gulf coast of Sonora, do not readily fit this pattern, however, and we suspect that the two taxa are conspecific. We have mapped them together here. Embryological evidence suggests

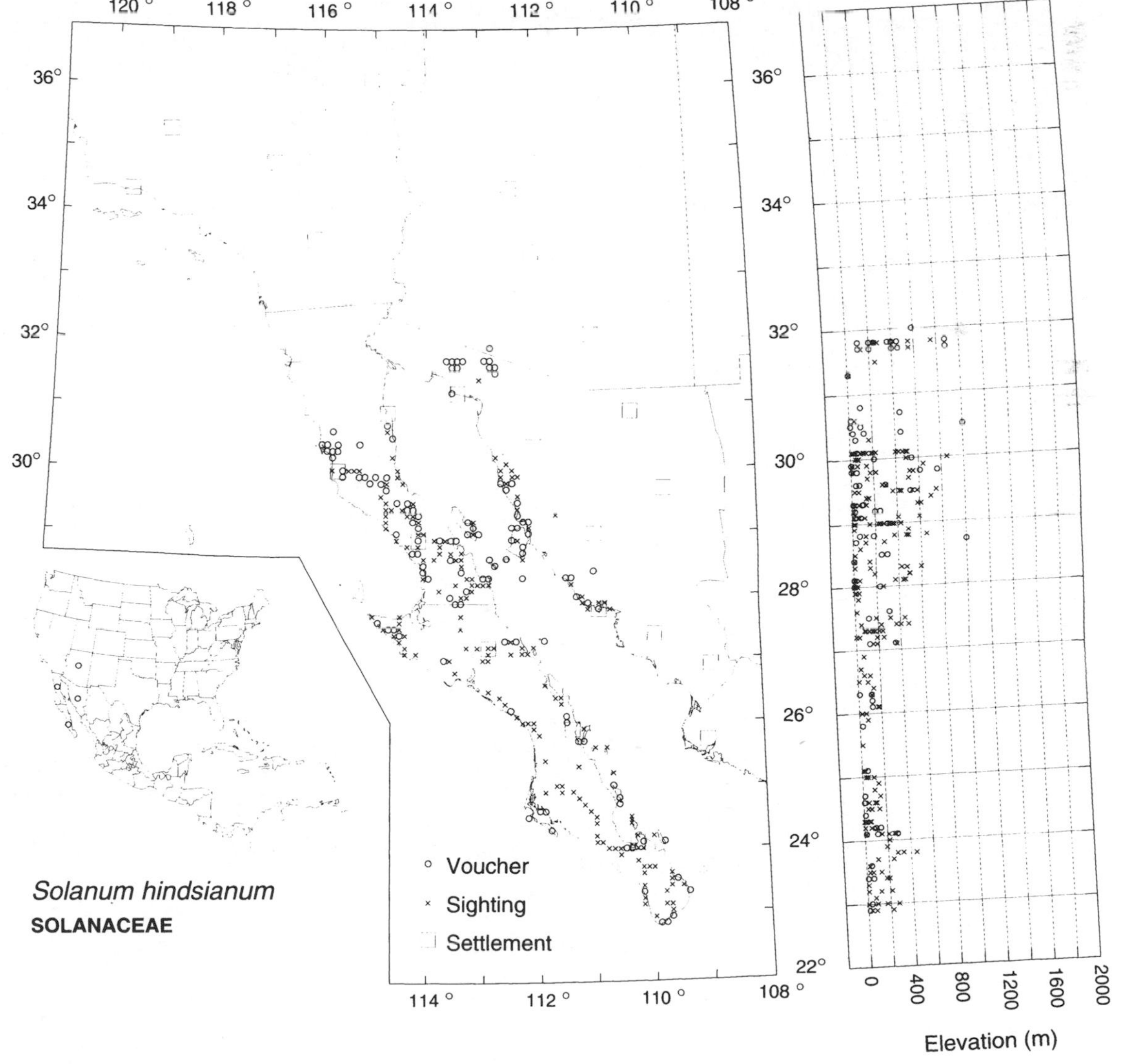

Solanum hindsianum
SOLANACEAE

Figure 78. Stegnosperma *species near Rancho Americano east of Puerto Libertad, Sonora. The upturned leaf tips (above) and the arching branches, some of which reach the ground, are characteristic. (Photograph by R. M. Turner.)*

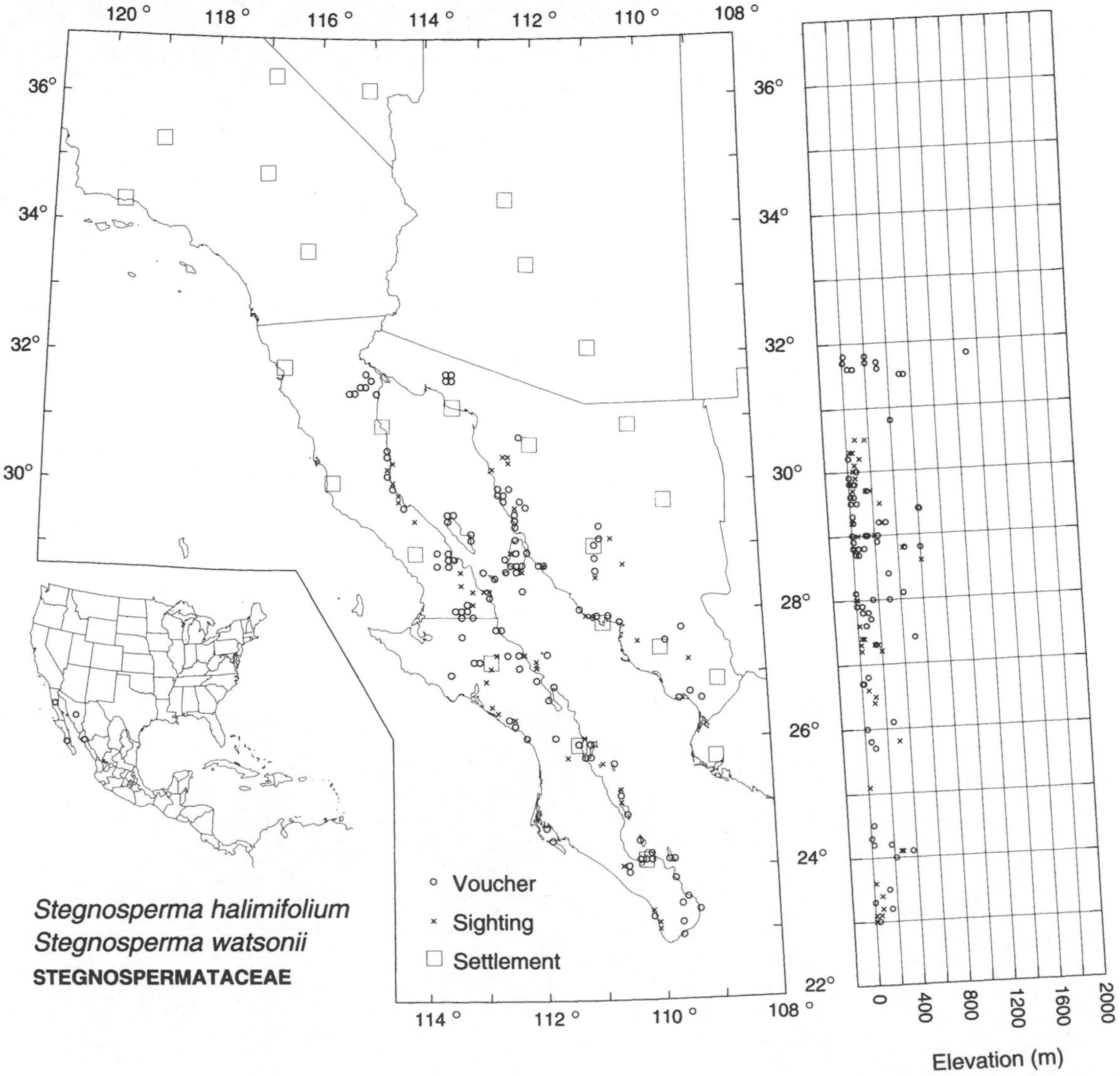

Stegnosperma halimifolium
Stegnosperma watsonii
STEGNOSPERMATACEAE

○ Voucher
× Sighting
□ Settlement

Elevation (m)

that the genus *Stegnosperma* is best placed in its own family, the Stegnospermataceae (Narayana and Narayana 1986).

The plants are locally common on the coastal strand and occasional along inland washes at low elevations. The cluster of four points in northwestern Sonora represent collections by Richard S. Felger from the Gran Desierto.

Given enough rain, the plants produce flowers and fruits throughout the year. The leaves are evergreen. Sheathed in a fleshy, white aril that turns red at maturity (Wiggins 1964), the seeds may be dispersed largely by birds (Gilmartin and Neighbours 1982).

In the Baja California peninsula, the powdered roots are used as a soap substitute (Standley 1922). The Seri used the leaves to treat snakebite and headache (Felger and Moser 1985).

Stenocereus alamosensis

(Coulter) Gibson & Horak

[=*Rathbunia alamosensis* Coulter, *R. sonorensis* (Runge) Britton & Rose]

Sina, cina, nacido, tasajo

Stenocereus alamosensis (figure 79) is a sprawling cactus with many slender, arching stems 6–10 cm in diameter and 1–2 m long. The stems have 5–8 ribs and often form impenetrable thickets as much as 10 m across. The sparse spines are borne in clusters of 11–18 straight, white or gray radial spines and 1–4 stouter central spines up to 5 cm long. The scarlet, tubular flowers are 7–10 cm long and remain open all day. The globular fruits, 3–4.5 cm in diameter, are bright red at maturity. The haploid chromosome number is 11 (Pinkava et al. 1977). Gibson (1988b) described the stem anatomy.

Stenocereus gummosus is a more massive plant with 8–12 radial and 3–6 central spines per areole, and white to purplish nocturnal flowers. *Stenocereus kerberi* (K. Schum.) Gibson & Horak has only 4 ribs per stem.

Gibson and Horak (1978) placed this

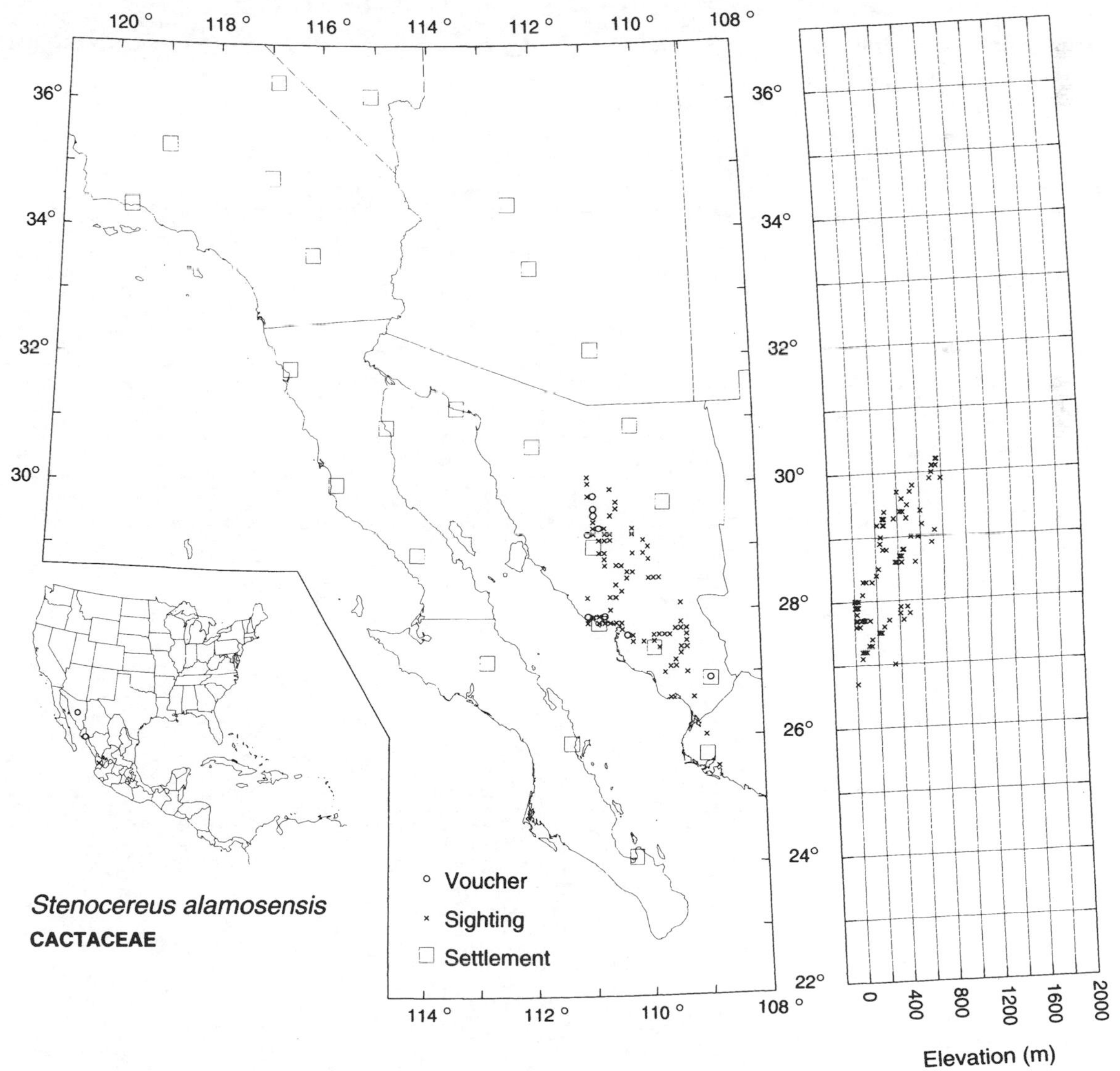

Stenocereus alamosensis
CACTACEAE

species in the genus *Stenocereus* based largely on floral and seed characters. The seeds are an uncommon type in subtribe Stenocereinae (Gibson and Horak 1978; Gibson 1988b). Plants with relatively thick stems (7–10 cm) were formerly recognized as *Rathbunia sonorensis* (Bravo-Hollis 1978).

Stenocereus alamosensis grows in alluvial or sandy soil on plains and bajadas in Sinaloan thornscrub. Near its arid and northern limits it most often grows under tree canopies. To the west, the species is probably limited by inadequate summer rainfall. Some plants sustained severe frost damage during the unusually cold winter of 1937, and the species is most likely limited to the north by low temperatures (Shreve 1933; Turnage and Hinckley 1938).

The high ratio of photosynthetic surface area to volume suggests that light could be limiting for this species (Cody 1984). If, in order to maximize its photosynthetic surface area, *S. alamosensis* has adopted narrow stems and a sprawling habit, this strategy would effectively limit it to open habitats where shading would not be a problem (Cody 1984).

Flowers appear from March to May and again in July and August. Hummingbirds are the main pollinators (Gibson and Horak 1978; Gibson and Nobel 1986), and the spring bloom occurs during the northward migration of several hummingbird species (Johnsgard 1983). Gibson and Nobel (1986) suggested that *S. alamosensis* and *S. kerberi,* another red-flowered, hummingbird-pollinated species, arose from a common ancestor, then diverged along climatic lines, with *S. alamosensis* becoming adapted to the desert margins and *S. kerberi* remaining in the more me-

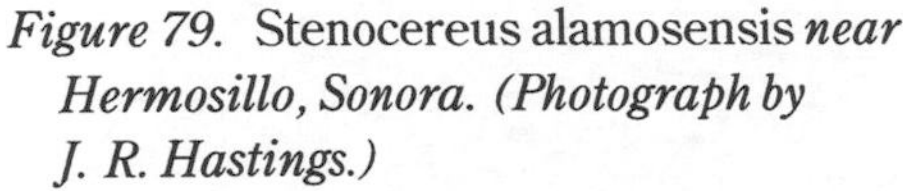

Figure 79. Stenocereus alamosensis *near Hermosillo, Sonora. (Photograph by J. R. Hastings.)*

Figure 80. Stenocereus eruca *on the Llanos de Magdalena, Baja California Sur. (Photograph by J. R. Hastings.)*

sic thornforest habitat.

The seeds of *S. alamosensis,* like those of other fleshy-fruited cacti, are probably dispersed mostly by birds and small mammals. Plants are established by seed, evidently a rare event (Wiggins 1964), and by vegetative propagation as the stems bend to the ground, then root in place. Mature plants form large, colonial rings that eventually die out in the center (Gentry 1942). The juvenile plant is a tangle of prostrate, slender stems that root profusely along the lower side (Wiggins 1964).

In Mexico the stems are occasionally planted as living fences (Gentry 1942). *Stenocereus alamosensis* is occasionally grown as an ornamental in Tucson but requires protection from frost.

Stenocereus eruca

(Brandegee) Gibson & Horak

[=*Machaerocereus eruca* (Brandegee) Britton & Rose]

Creeping devil, chirinola, casa de ratas

Stenocereus eruca (figure 80) has prostrate, creeping stems up to 5 m long and 4–12 cm in diameter (exclusive of spines). The stems turn upward at the tip and are densely armed with stout, gray or white reflexed spines, the longest about 4 cm long. The funnelform flowers, tinged with pink, are 10–12 cm long and 8–10 cm wide. The globose fruits, 3–4 cm in diameter, are scarlet and spiny at maturity.

No other cactus in its range has long, prostrate stems with upraised tips.

Gibson and Horak (1978) transferred this species from *Machaerocereus* to *Stenocereus* based on stem anatomy, triterpene chemistry, and seed structure. The species has been described as a relatively recent derivative of an ancestor that resembled *S. gummosus* (Gibson 1989). Speciation no doubt occurred on sandy flats of the Magdalena region (Gibson 1989).

An extremely narrow endemic, *S. eruca* is known only from the Magdalena Plain of central Baja California Sur where it is locally common on silty or sandy flats not far inland from the Pacific coast. The area has a maritime arid climate. Rain is

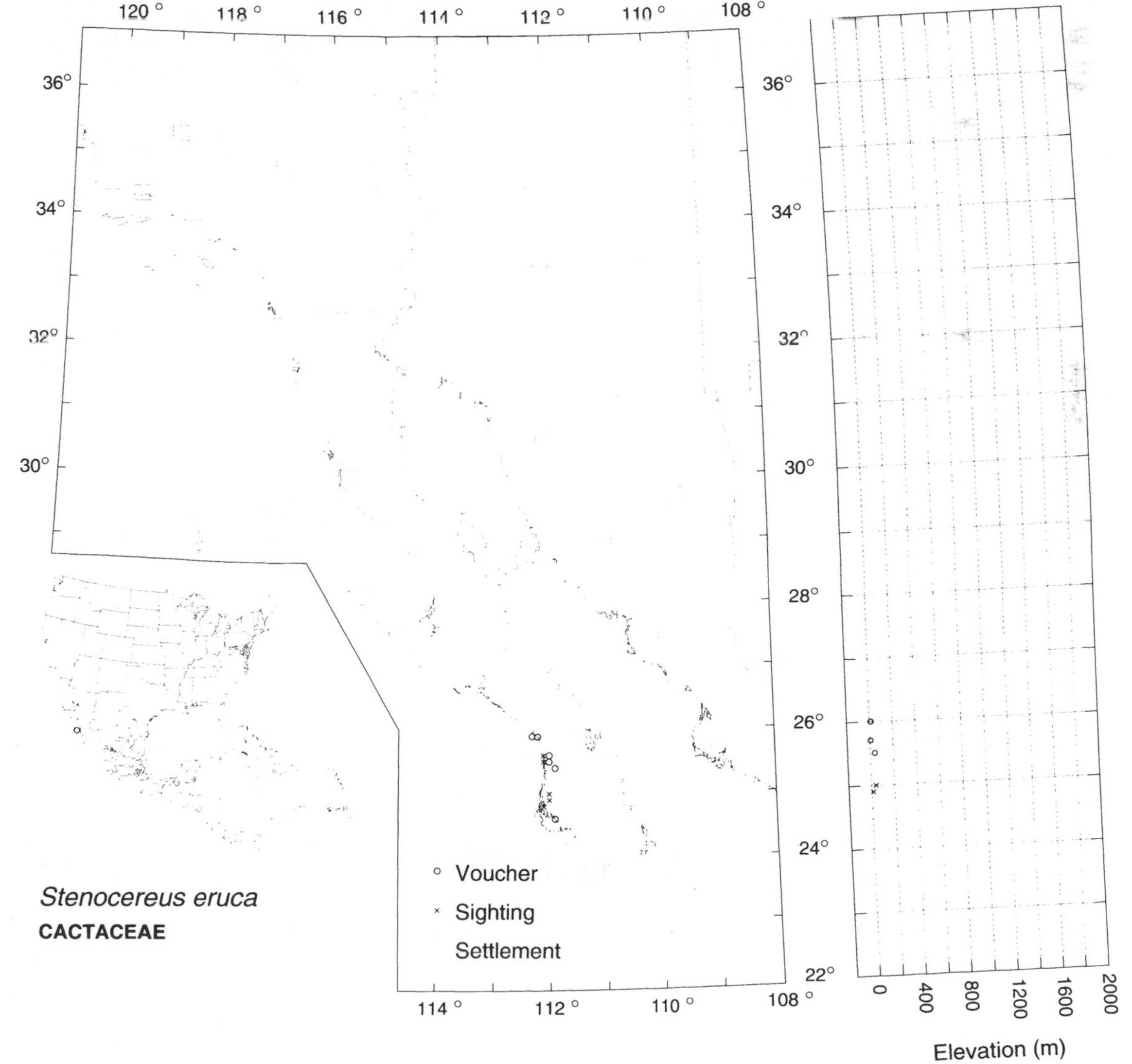

erratic but probably falls most frequently during August and September or in midwinter.

The stems root along the underside where they touch the ground. As the growing tip elongates, the older, basal portion dies (Wiggins 1964; Moran 1975), a habit of growth that limits the plants to loose, friable soils where they can root easily. Seedlings are rare in nature. Most propagation in the wild is evidently vegetative. The existing colonies may be several hundred years old (Gibson 1989).

Blooming specimens have been collected in August, September, October, and February. The nocturnal flowers are most likely pollinated by hawk moths (Moran 1975; Gibson and Horak 1978).

The fruits are edible (Moran 1975). Currently, this species is endangered by agricultural development and highway construction (Wiggins 1980).

Stenocereus gummosus

(Engelm.) Gibson & Horak

[=*Machaerocereus gummosus* (Engelm.) Britton & Rose]

Pitahaya, pitahaya agria

This large, sprawling, multistemmed cactus grows up to 3 (occasionally 5) m tall (figure 81). The leaning, arching, or scrambling stems can form impenetrable thickets 10 m or more across. Ribs number 8 to 9 on stems 5–8 cm thick. Placed about 2 cm apart, the spine clusters are of 8–12 radials and 3–6 centrals, the longest up to 4 cm. The white to purplish flowers are 10–15 cm long and 6–8 cm wide. The globose fruits, 6–8 cm in diameter, are red and nearly spineless when ripe. The haploid chromosome number is 11 (Pinkava et al. 1977).

Stenocereus alamosensis has red, tubular flowers, somewhat thinner stems, and 11–18 radial spines. *Myrtillocactus cochal, Lophocereus schottii,* and *S. thurberi* are all candelabraform rather than sprawling in habit.

Gibson and Horak (1978) subsumed

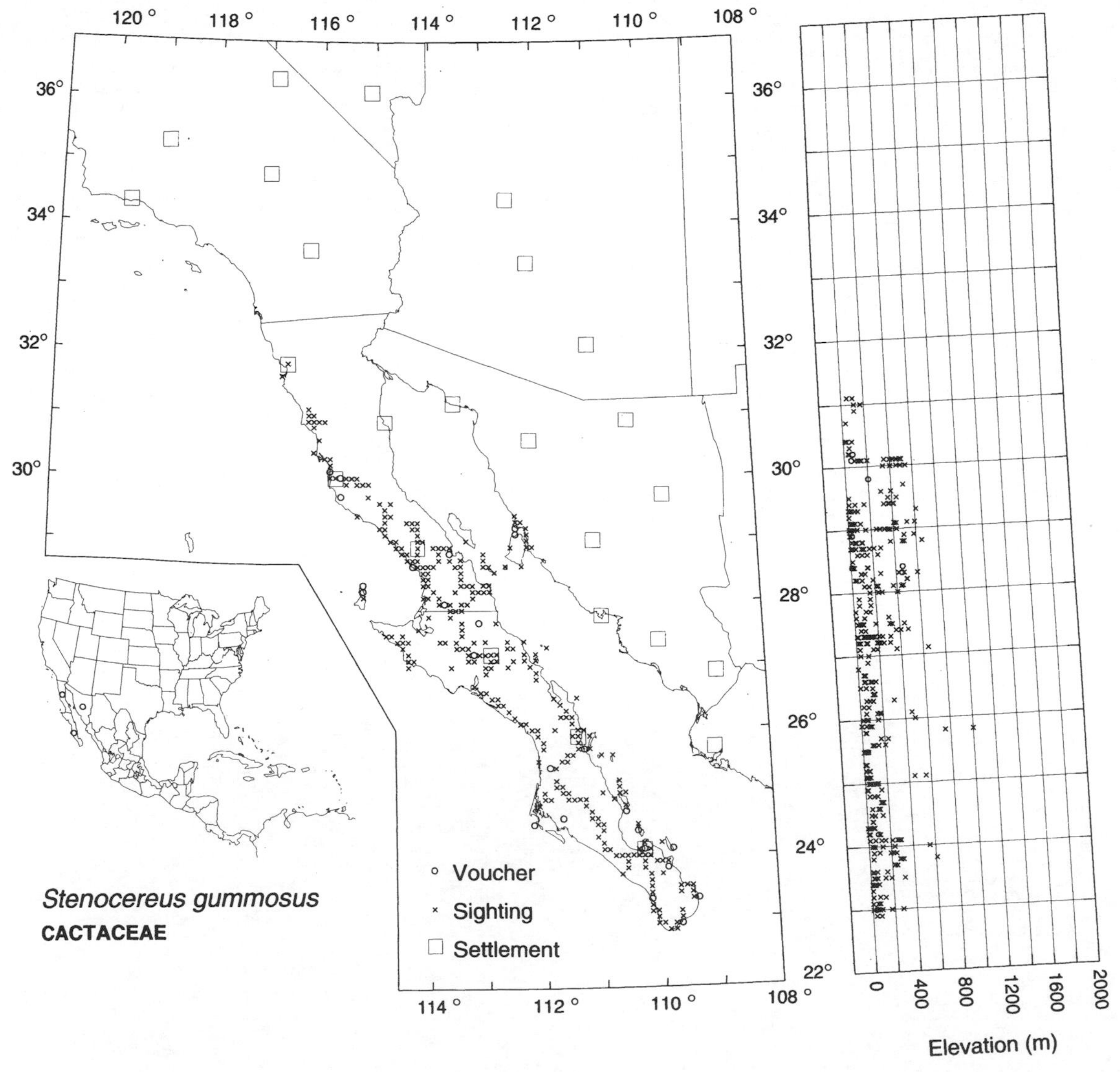

Figure 81. Stenocereus gummosus *in Arroyo El Rosario, Baja California. (Photograph by J. R. Hastings.)*

Machaerocereus under *Stenocereus* on the basis of similarities in triterpene chemistry and stem anatomy, particularly silica bodies in the skin and highly mucilaginous stem cortex.

Stenocereus gummosus is locally abundant on gravelly plains, rocky hillsides, and arroyo margins. The distribution is split between the Baja California peninsula and the Sonoran coast. Throughout the range, frosts are essentially absent. In the northwest, rains are largely limited to winter and spring. Rainfall regimes shift to biseasonal in midpeninsula and along the Sonoran coast. In the Cape Region, Baja California Sur, rainfall is concentrated in late summer and early fall.

Cody and coauthors (1983) suggested that *S. gummosus* may have migrated from the peninsula to the mainland via the midriff islands. The northernmost sightings are Reid Moran's from the vicinity of Ensenada (31.9°N) and the north side of Punta Banda (31.7°N). The sighting on Isla Angel del la Guarda is Ivan M. Johnston's (1924). *Stenocereus gummosus* has not been seen there in recent years (Reid Moran, personal communication 1971). The species does grow along the Sonoran coast despite the doubts expressed by Gates (1957): it has been collected from Punta Sargento (29.3°N) by George Lindsay and on the north shore of Isla Tiburón (29.2°N) by A. F. Whiting.

Stenocereus gummosus is closely related to the *S. stellatus* complex of southern Mexico and may have evolved from that complex either before or after seafloor spreading (Gibson 1989).

The height of *S. gummosus* is positively correlated with the height of the surrounding vegetation. Individuals in coastalscrub communities may be less than 1 m tall, while those in thornscrub often grow to 3–5 m tall (Gibson and Nobel 1986). Where photosynthetically active radiation is high, as in open scrub, plant stature is low because growth is directed laterally, preempting space. Where tree canopies reduce radiation in the understory, *S. gummosus* grows to higher levels where more light is available (Gibson and Nobel 1986).

The plants flower throughout the year, most heavily in July and August (Wiggins 1964). Fruiting may be curtailed in years of lower than normal rainfall (Felger and Moser 1985). Fruits averaged 674 seeds in one study (León de la Luz and Domínguez-Cadena 1991). Stems that bend to the ground often root, forming small, clonal colonies (Gibson 1989). The sweet scent, nocturnal anthesis, trumpet shape, and pale color of the flowers suggest pollination by hawk moths (Gibson and Horak 1978). Bats may also pollinate the blossoms (Hevly 1979).

Mycorrhizae colonize the roots (Rose 1981). A variety of yeasts and fruit flies (*Drosophila* species) inhabit the rotting stems (Starmer et al. 1976; Starmer 1982; Fogleman and Armstrong 1989).

The edible fruits are highly prized by the Seri (Felger and Moser 1985). Fishermen on the peninsula sometimes used the chopped stems as a fish poison (Aschmann 1959).

Stenocereus thurberi

(Engelm.) Buxb.

[=*Lemaireocereus thurberi* (Engelm.) Britton & Rose, *Cereus thurberi* Engelm.]

Organ pipe cactus, pitahaya, pitahaya dulce, marismeña, órgano, mehuelé

Stenocereus thurberi (figure 82) is a columnar cactus 3–8 m high with numerous upright stems from the base or (in the southern part of its range) from a short trunk. The stems are 15–20 cm in diameter with

12–19 ribs. The straight, gray to black spines, clustered in areoles of about 12 radials and 1–3 centrals, are 1.2–2.5 (occasionally 4) cm long. The funnelform, pale pink to lavender flowers are 6–8 (occasionally as few as 4) cm long. The globose fruits, 4–7.5 cm in diameter, contain reddish pulp and numerous black, shining seeds. The haploid chromosome number is 11 (Pinkava and McLeod 1971). Gibson (1990) described the stem anatomy.

Stenocereus martinezii (G. Ort.) Buxb. has a well-defined trunk 1–2 m high with shorter spines (2–10 mm long). *Stenocereus montanus* (Britton & Rose) Buxb., with 7–8 ribs and fewer than 7 spines per areole, has a prominent trunk 1 m or more tall. *Lophocereus schottii* has 5–10 ribs per stem, with dense bristles near the apex of mature stems. *Pachycereus pecten-aboriginum* has a trunk 1.2–5 m tall. Its fruits are large and burlike.

Gibson and Horak (1978) transferred this species from the genus *Lemaireocereus* to *Stenocereus* based on stem anatomy. Benson (1982) treated it as *Cereus thurberi* Engelm. Biologists recognize two varieties: var. *thurberi* occurs throughout the range of the species; var. *littoralis* K. Brandegee is a dwarf (less than 3 m tall), somewhat sprawling form restricted to coastal bluffs near San José del Cabo, Baja California Sur (Gibson and Horak 1978; Gibson 1990).

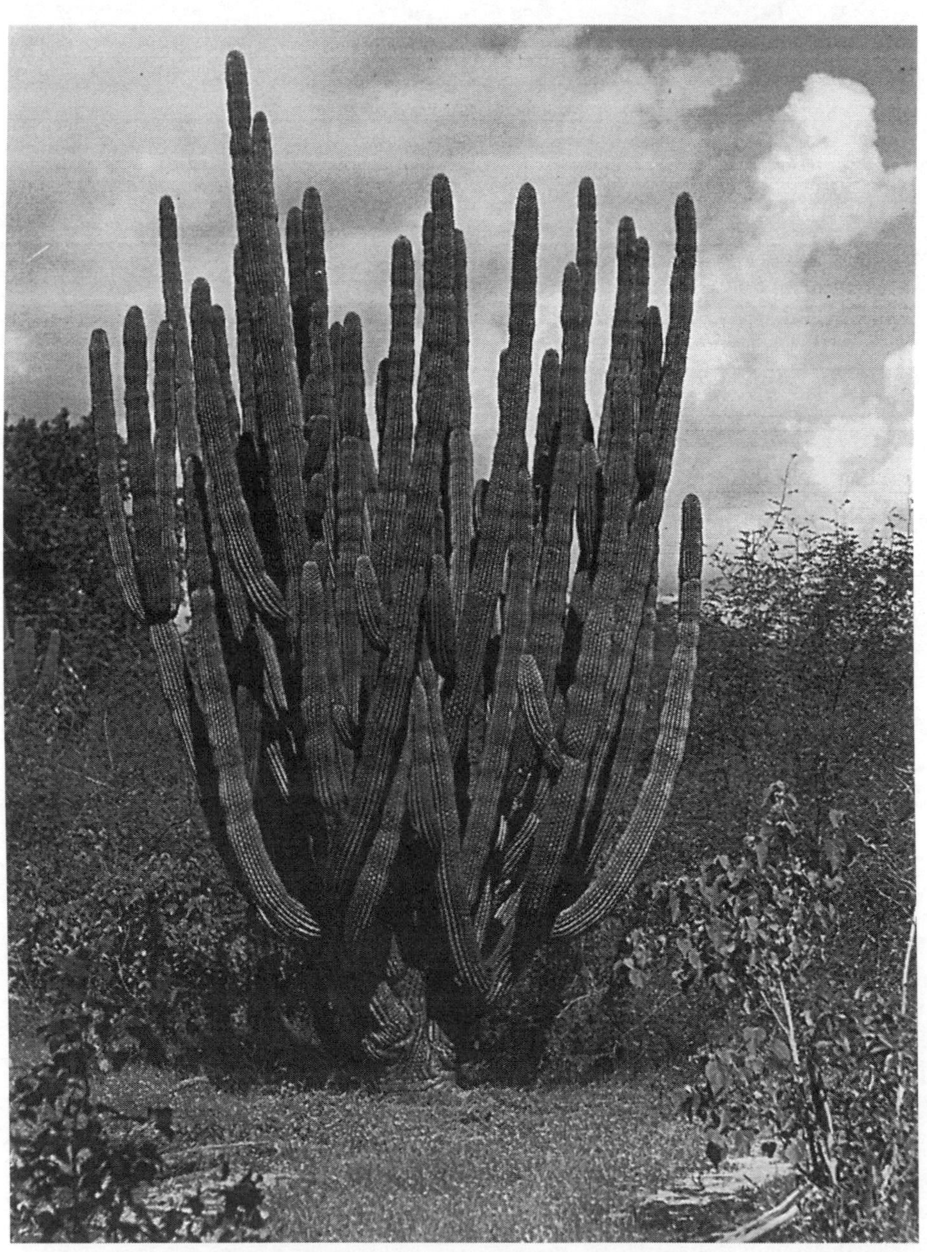

Figure 82. Stenocereus thurberi *near Miraflores, Baja California Sur. (Photograph by J. R. Hastings.)*

According to Gibson (1990), *S. thurberi* belongs to a small group of four columnar cacti characterized by a candelabraform habit, red or brown hairs in the areoles, and conspicuously elevated tubercles. The hypothetical ancestor may have resembled *S. quevedonis* (G. Ortega) Bravo, a small, vase-shaped tree dominant in parts of Michoacán (Gibson 1990). *Stenocereus thurberi* var. *littoralis* might have evolved in a geographically isolated population on coastal bluffs of the Cape Region (Gibson 1990).

Stenocereus thurberi is locally dominant on slopes and plains. At the northern end of their range, the plants are restricted to rocky slopes and cliff ledges where nighttime radiation of heat protects them from frost (Parker 1988a). Even so, frost damage to growing stem tips is not an infrequent occurrence. The plants reach their maximum height on plains where frost is infrequent or unknown. Southward from about 32°N, populations are large and cover many square kilometers, but at the northern limit, they are small and separated by several kilometers. Fossil midden assemblages from the Puerto Blanco Mountains, Arizona, show *S. thurberi* entering the fossil record only 3,500 years B.P. (Van Devender 1987), which suggests that the northernmost populations are advance disjuncts.

The northernmost point represents a sighting from the Slate Mountains. The nearby point denotes a single plant on the south slope of Desert Peak. The plant in question is a large individual with numerous arms. It was damaged by a catastrophic freeze in December 1978. The sighting at 31.5°N, 111.1°W is David E. Brown's from Fresno Canyon near Nogales, Arizona, at an elevation of 1,200 m. This single plant grows against a low, south-facing rock wall above which branch growth is sharply limited by repeated freezes. The southeasternmost point is based on a collection by Paul C. Hutchinson near Guamuchil, Sinaloa. The sighting on Isla Angel de la Guarda (29.0°N, 113.3°W) by Ivan Johnston (1924) is doubtful (Moran 1983a:401).

Flowers appear from May to June (Kearney and Peebles 1960), opening shortly after sunset and closing the next morning. The abundant nectar, nocturnal bloom, and musky odor suggest bat pollination. In an artificial setting, both honeybees and bats pollinated the flowers (Alcorn et al. 1962). The nectar-feeding bats *Choeronycteris mexicana* and *Leptonycteris sanborni* are likely important pollinators (Hevly 1979). Fruits mature throughout the summer. Animal dispersal of seeds is no doubt the primary means of dispersal and may account for disjuncts such as those near Nogales and on Desert Peak.

The range shows dependence on fairly predictable warm-season rain. Plants at the northern end of the range evince frequent frost damage, such as numerous constrictions and calluses near the stem tips (Parker 1988b), and low temperatures probably limit the northward distribution. Cold-hardened plants suffer substantial freezing damage in chlorenchyma cells at –9.0°C. Apical spines shade the stem apex to some extent and help prevent loss of infrared radiation on the coldest nights. Spine-induced shading increases from 41% at lower latitudes to 71% at higher latitudes, perhaps enabling the plants to survive subfreezing temperatures at the northern end of their range (Nobel 1980c). Stem diameter at midheight increases from 10.4 cm at lower latitudes to 15.1 cm at higher latitudes, probably another factor in the survival of the northern populations, since wide stem apices are better buffered against thermal extremes than narrow ones (Nobel 1980c).

Nobel (1980c) pointed out that stems of larger diameter permit more water storage per unit of surface area and suggested that populations in the arid, northwestern part of the range survive by virtue of increased stem diameter. For the species as a whole, in fact, water storage may take

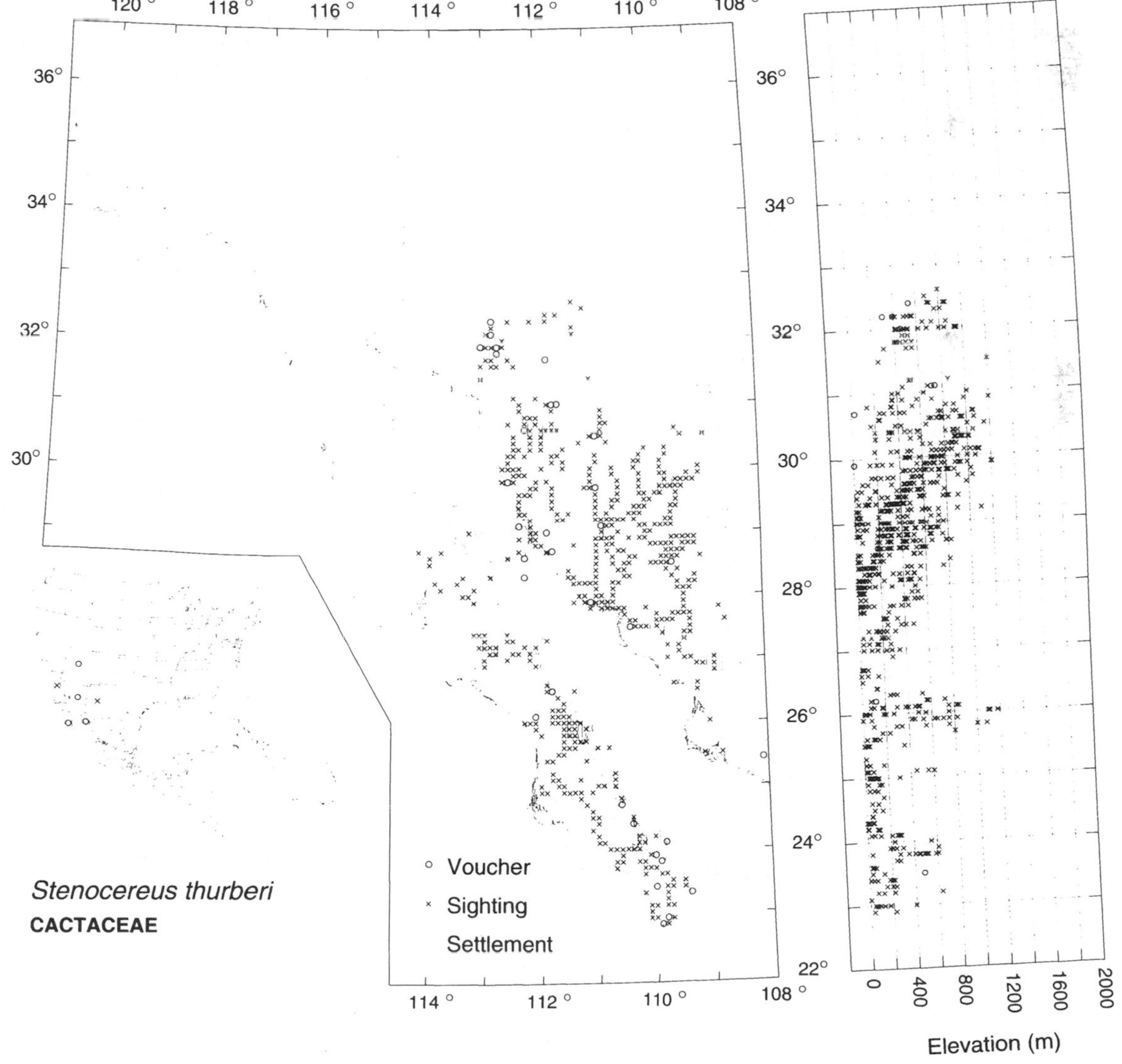

precedence over thermal regulation in determining stem diameter (Cody 1984).

During the growing season, photosynthetically active radiation can be a limiting factor for carbon dioxide uptake for cacti, and self-shading by spines and branches decreases absorption of radiation still further. The profuse branching of *S. thurberi* increases the ratio of surface area to volume of the plants, thus maximizing the absorption of solar radiation and the uptake of carbon dioxide (Geller and Nobel 1986). In open desert vegetation, most branching is at or near ground level. In thornscrub, most branching occurs 4 m above the ground, slightly above the level of the surrounding tree canopy (Cody 1984). These patterns may reflect differing strategies for maximizing capture of photosynthetically active radiation in differently illuminated environments.

Larger plants have greater photosynthetic surface area and water storage capacity than smaller ones and therefore grow more quickly. Plants under 1 m grow 0.07 m/yr, whereas those 5 m or more grow 0.62 m/yr (Parker 1988b). Growth rates tend to be highest after unusually wet winters and lowest after relatively dry ones. A winter of frequent or severe freezing retards growth the following summer (Parker 1988b).

Germination is inhibited as long as the seeds are enmeshed in the pulp (McDonough 1964) but is high (85%) for cleaned seed (Parker 1987a). The seeds require light for germination (McDonough 1964). In one study, germination was somewhat depressed at 20°C and somewhat enhanced at 35°C and 40°C relative to that of *Carnegiea* (McDonough 1964). This difference in thermal response is reflected in the distribution of the two species, with *Carnegiea* showing greater cold tolerance than *S. thurberi*. Germination of *S. thurberi* was initiated at 45°C but was not completed. No germination occurred at 15°C (McDonough 1964).

Most plants at Organ Pipe Cactus National Monument, Arizona, reach reproductive age when 2–2.5 m tall. Typically, they have 4–10 arms by this time (Parker 1987a). Large, mature plants can produce more than 50 fruits in a season for a total of about 98,450 seeds (Parker 1987a).

At the northern end of the range, seedlings thrive best when protected by nurse plants or rocks. At Organ Pipe Cactus National Monument only 15% of plants less than 0.5 m tall grew when unprotected (Parker 1987b). Nurse plants provide shade in summer and heat radiation in winter, ameliorating the stressful effects of seasonal temperature extremes. In the national monument, steep, rocky, south-facing slopes are more favorable for establishment and growth than other sites. On southerly exposures, high solar radiation during the day elevates minimum temperatures at night, and seedlings may escape freeze damage in all but the coldest years (Parker 1987b).

Shreve (1935) estimated that plants 3–4.5 m in height were between 50 and 75 years of age. Parker (1988b) concluded that plants 10 m tall were about 45 years old. A 1914 photograph taken by Forrest Shreve at the Desert Laboratory, Tucson, Arizona, showed a plant 7 dm tall that might have been about 15 years old (Parker 1988b:344). In 1991, this nearly 90-year-old plant was 5 m tall and had roughly 25 stems.

A variety of yeasts and fruit flies (*Drosophila* species) inhabit the rotting stems (Starmer et al. 1976; Fogleman and Starmer 1985; Fogleman and Armstrong 1989). Many different animals eat the fruits.

This species played an important role in the subsistence economy of Native Americans on the central Baja California peninsula. Ripe fruits were eaten immediately, and, in lean times, seeds recovered from excrement provided a second harvest (Aschmann 1959). Indians in Sonora and Arizona ate fresh fruits and dried the pulp for later use (Aschmann 1959; Felger and Moser 1985). Seri Indians pounded the dried stems with animal fat to make pitch for caulking boats. They used the ribs for a variety of purposes, including daub and wattle construction, torches, and tongs (Felger and Moser 1985).

Tabebuia chrysantha

(Jacq.) Nicholson

Amapa amarilla, amapa prieta, roble, verdecillo

This tree grows 7–15 m tall with scaly bark and a rounded, spreading crown. The opposite leaves are palmately compound with 5 obovate to elliptic leaflets up to 18 cm long. The yellow, funnelform flowers, 5–6.5 cm long, are in dense terminal clusters. The cylindrical capsules are 20–30 cm long and contain numerous flat, winged seeds.

Tabebuia palmeri has pink to lavender flowers and smooth bark. *Bombax palmeri* S. Watson has alternate leaves and green bark. *Ceiba acuminata* has conical spines on the trunk.

Apparently sterile hybrids between *T. chrysantha* and *T. palmeri* occur in the Río Mayo region (Gentry 1942). As currently defined, *T. chrysantha* shows great morphological diversity and may actually constitute several species or subspecies (Gentry 1970).

Tabebuia chrysantha is occasional on hillsides and in canyons in dry or moist tropical and subtropical forests. The species grows from South America to Mexico, reaching its northern limit in Sonora in Sinaloan thornscrub.

Flowers typically appear when the canopy is leafless, as early as January to as late as May (Mason and Mason 1987). In the Río Mayo area, peak flowering occurs in February (Gentry 1942). Leaves appear in July once summer rains are underway and last until the autumn (Gentry 1942).

The strong wood is used in cabinetry and construction (Mason and Mason 1987).

Tabebuia palmeri

Rose

Amapa, roble, amapa prieta, amapa rosa, amapa colorada, palo de rosa

An openly branched tree 5–20 m tall, *Tabebuia palmeri* has a rounded to umbrellalike crown and gray, nearly smooth bark. The opposite leaves are palmate with 5 elliptic leaflets to 8 cm wide and 20 cm long. The funnelform, pink to lavender flowers, in crowded terminal cymes, are 6–8.5 cm long. Fruits are smooth, cylindrical capsules 2.5–3.5 cm long and 10–15 mm across. The numerous seeds are thin and winged. The haploid number of chromosomes is 20 (Bawa 1973).

Ipomoea arborescens has alternate, entire, cordate leaves and smooth, white bark. *Ceiba acuminata* bears conical spines on the trunk and has alternate leaves. *Bombax palmeri* S. Watson, also with alternate leaves, has green bark. *Tabebuia chrysantha,* a species of moist or dry subtropical and tropical forests, has scaly bark and yellow flowers.

Apparently sterile hybrids with *T. chrysantha* are known from the Río Mayo region (Gentry 1942).

Tabebuia palmeri can be locally abundant on forested or brushy hillsides (Gentry 1942). It ranges far south into Central America and reaches its northern limit in Sonora, where it occurs in Sinaloan thornscrub.

Flowers appear on the leafless branches in winter and early spring (Krizman 1972). Sometimes they cover the entire crown of the tree (Gentry 1942). With summer rains in July, leaves appear; they last until the late fall or early winter (Krizman 1972). The wind-dispersed seeds are released during the spring and early summer (Gentry 1942).

The wood is hard and strong and is used for ceiling beams, corral posts, railroad ties, and so forth (Standley 1926; Gentry 1942).

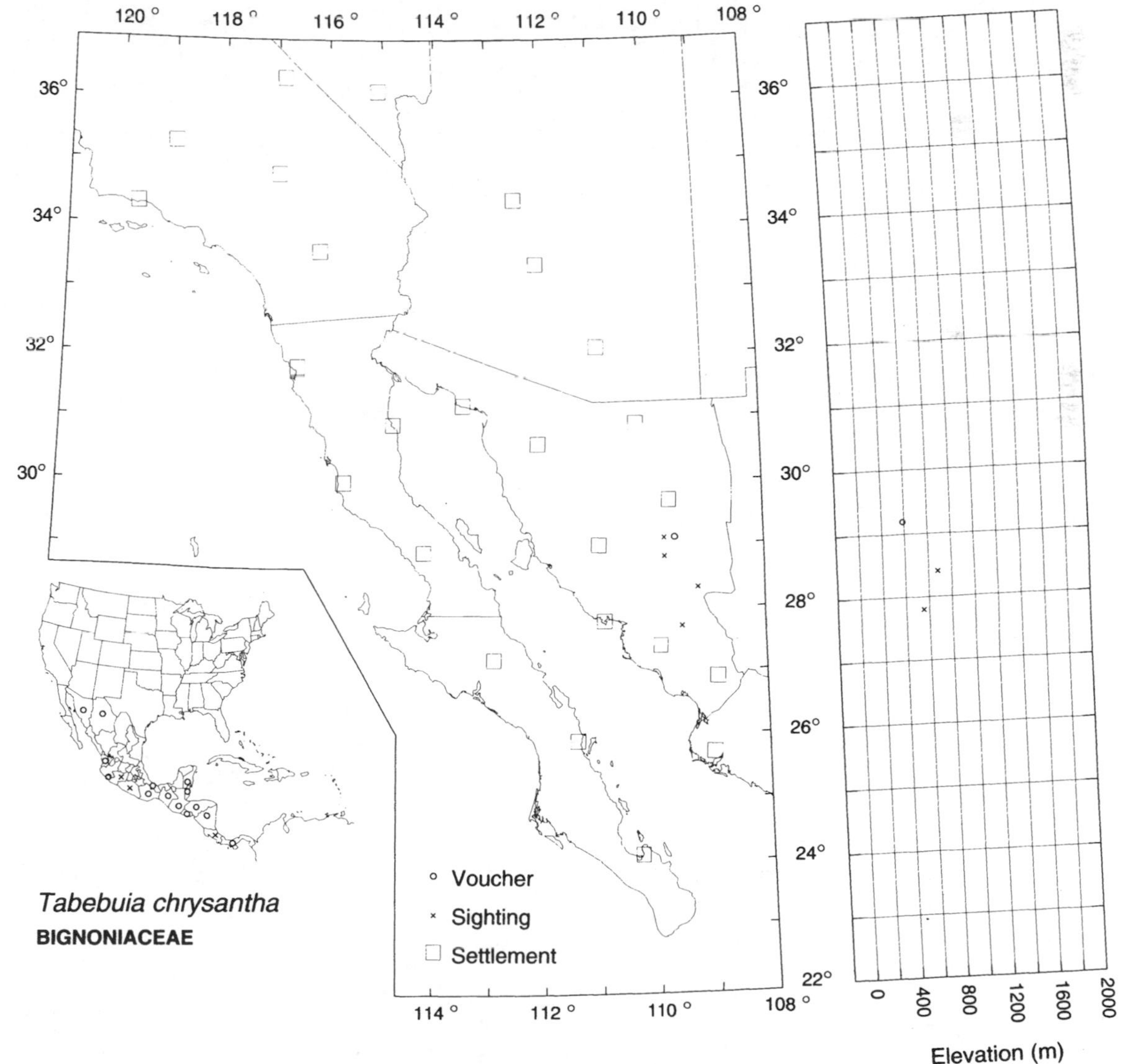

Tabebuia chrysantha
BIGNONIACEAE

Tamarix chinensis
Lour.

Tamarix ramosissima
Ledeb.

[=*T. pentandra* Pall.]

Tamarisk, saltcedar, tamarisco, tamariz, taray, atarfe, talaya

Bushy trees or shrubs up to 5 m tall, *Tamarix chinensis* and *T. ramosissima* can be readily recognized by the abundant scale-like leaves 2–3 mm long and the small, pink or white flowers in racemes or dense panicles. *T. ramosissima* has a diploid chromosome number of 24 (Zhai and Li 1986).

With the exception of *T. aphylla* (L.) Karsten, the various species of *Tamarix* in the Sonoran Desert region can be identified with certainty only on the basis of flowering parts. *Tamarix aphylla* is unique in its apparently leafless stems and clustered, drooping branchlets (Benson and Darrow 1981).

Eight species of *Tamarix* have been introduced into North America from the Old World (Baum 1967). Apparently some were widely cultivated in the United States as ornamentals before 1900 (Horton 1964; Robinson 1965), then gradually escaped from cultivation. Between 1938 and 1955, *T. chinensis* and *T. ramosissima* became thoroughly naturalized along riverbanks and reservoirs throughout the Southwest (Robinson 1965; Turner and Karpiscak 1980).

Tamarix populations on Isla Tiburón (Moran 1983b), Isla Angel de la Guarda (a collection by Raymond M. Turner), and in Volcán Elegante, Sierra del Pinacate (R. M. Turner, unpublished data), are the result of recent dispersal across many kilometers of open water or inhospitable desert.

Because the taxonomic status of introduced and naturalized *Tamarix* species is still unsettled and because it is difficult to accurately determine sterile plants, our map may include sightings of other *Tamarix* species. In Arizona *T. chinensis* and *T. ramosissima* have generally been referred to *T. pentandra* (Baum 1967). *Tamarix pentandra* has also been referred to

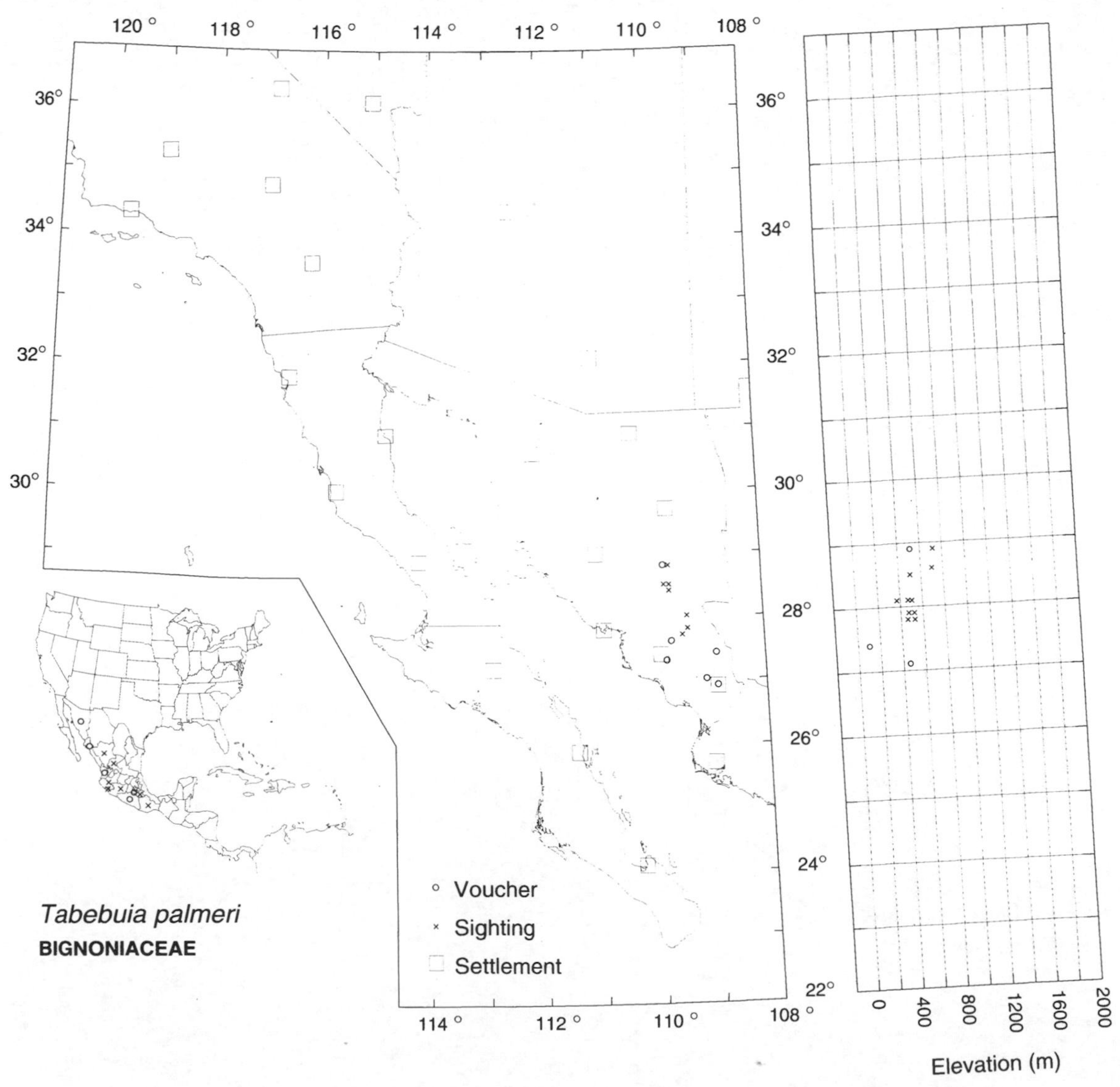

Tabebuia palmeri
BIGNONIACEAE

as *T. gallica* L. (Kearney and Peebles 1960). Horton and Flood (1962) advocated retaining *T. pentandra,* at least for the time being, but Benson and Darrow (1981) considered this name a synonym of *T. ramosissima.*

Often a dominant in permanently moist habitats such as riverbanks, reservoirs, and irrigation ditches, *Tamarix* has replaced native phreatophytes in many areas (Warren and Turner 1975). Its widespread establishment might not have been possible without the altered flow regimes associated with dams (Harris 1966). The genus can be considered highly invasive and, by some, undesirable over much of its range in the Southwest due to its high evaporative demand for water (Robinson 1965; van Hylckama 1974). Its spread has dramatically altered wildlife habitat (Carothers and Brown 1991).

The flowering phenology of *Tamarix* is complex (Horton 1957). In March, racemose flowers appear on old wood of the previous year. Seeds are shed between mid April and early May. Vegetative shoots produced early in the spring give rise to paniculate flowers in April and May, then go to seed. In early August, the central stem of the panicle elongates, producing a second panicle and yet another set of blossoms and seeds (Horton 1957).

Salt deposits often accumulate on the branches (Decker 1961; Campbell and Strong 1964). These salt deposits are excreted by glands located primarily on the abaxial leaf surfaces and on very young stems. Fluid shaken from the twigs contained 41,000 ppm of total solids (Campbell and Strong 1964). Inorganic solutes constitute 5–15% of the leaf dry weight (Hem 1967). Its salt-excreting capacity accounts at least in part for the success of *Tamarix* in moist, alkaline habitats.

Another factor in its success is the adaptability of the root system. Where the water table is far below the surface, plants produce deep taproots with a limited horizontal spread. Where groundwater is close to the surface, secondary roots occupy all zones of the soil profile above the water table (Gary 1963). Root systems of mature plants are able to survive up to 98 days of complete soil saturation (Gary

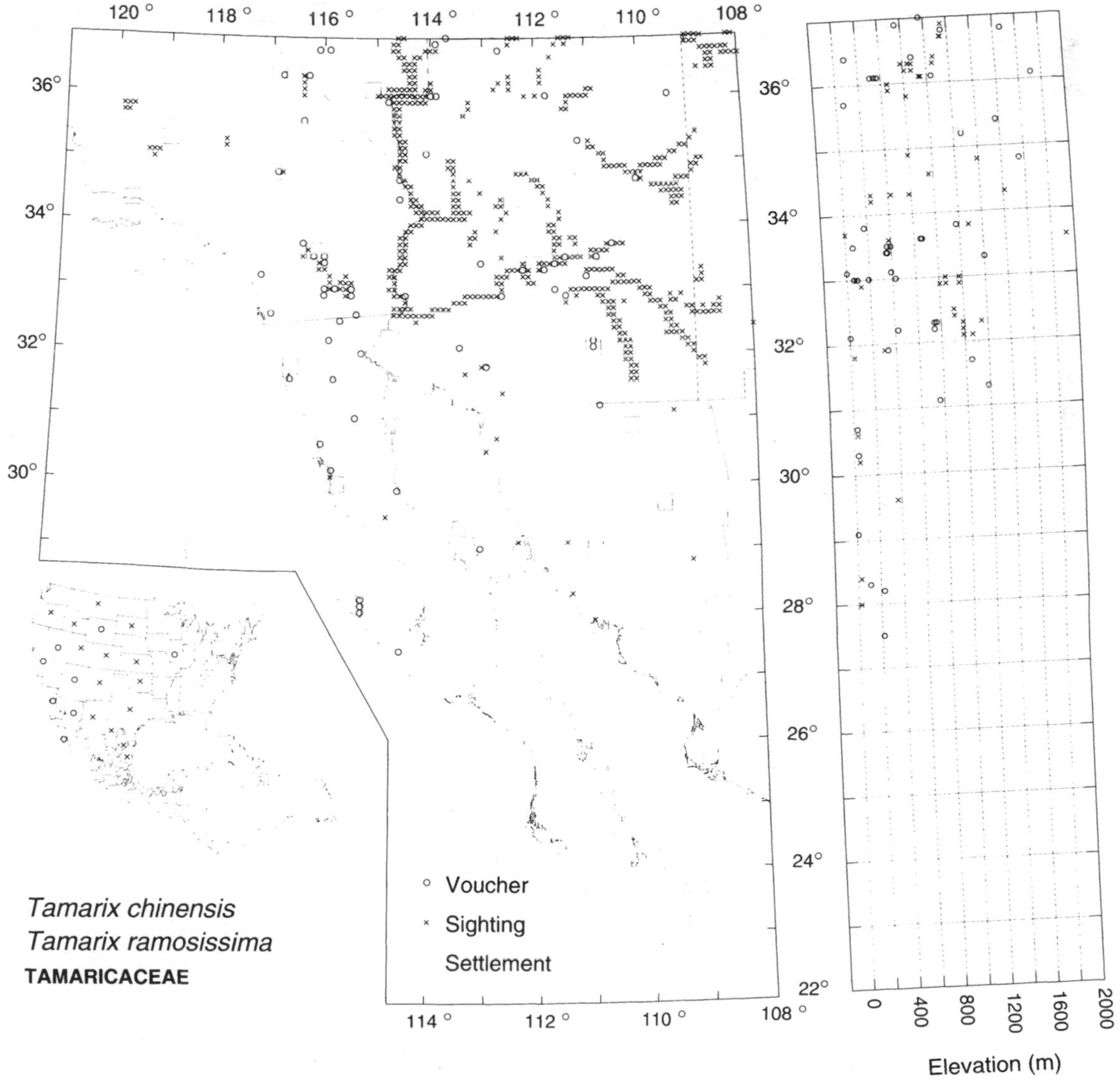

Tamarix chinensis
Tamarix ramosissima
TAMARICACEAE

1963; Warren and Turner 1975).

Water use varies with the depth of the water table and with water salinity (Decker et al. 1962; van Hylckama 1974). Average water use by experimental plants was 215 cm/yr when the water table was 1.5 m below the soil surface but less than half that when the depth to water was 2.7 m (van Hylckama 1974).

Seed production is high: in dense stands of these trees, 15–17 viable seeds fall on each square centimeter of soil each season (Turner 1974; Warren and Turner 1975). The seeds are dispersed by wind and water. Viability is retained for only a few weeks, but the germination percentage of fresh seed is quite high, 76–90% (Horton et al. 1960; Warren and Turner 1975). Seedlings become established along streambanks and lake shores during the warm season. Establishment is highest in May and June. Seedlings can survive submergence for up to 6 weeks; their survival rate increases with age (Horton et al. 1960). Saturated soils are necessary for seedlings to survive the first two to four weeks of growth (Horton et al. 1960). Once established, however, the plants can survive for many years with only the meager rainfall of the Sonoran Desert region (Turner 1974).

Tamarix flowers are a source of nectar for honeybees (Kearney and Peebles 1960). Trees of the genus are occasionally cultivated as an ornamental. The pollen is a powerful allergen to susceptible persons.

Tecoma stans (L.) Juss.

[=*Stenolobium stans* Seem.]

Yellow trumpet flower, yellow elder, retama, tronadora, trompetilla, palo de arco, flor de San Pedro, gloria, esperanza, lluvia de oro

A shrub 1–2 m tall in the northern part of its range, *Tecoma stans* is a small tree up to 8 m tall over much of the area mapped in this atlas and, in the Cape Region, even becomes a liana. The opposite leaves are odd-pinnate with 4–9 (occasionally 13) serrate, lanceolate leaflets 4–12 cm long

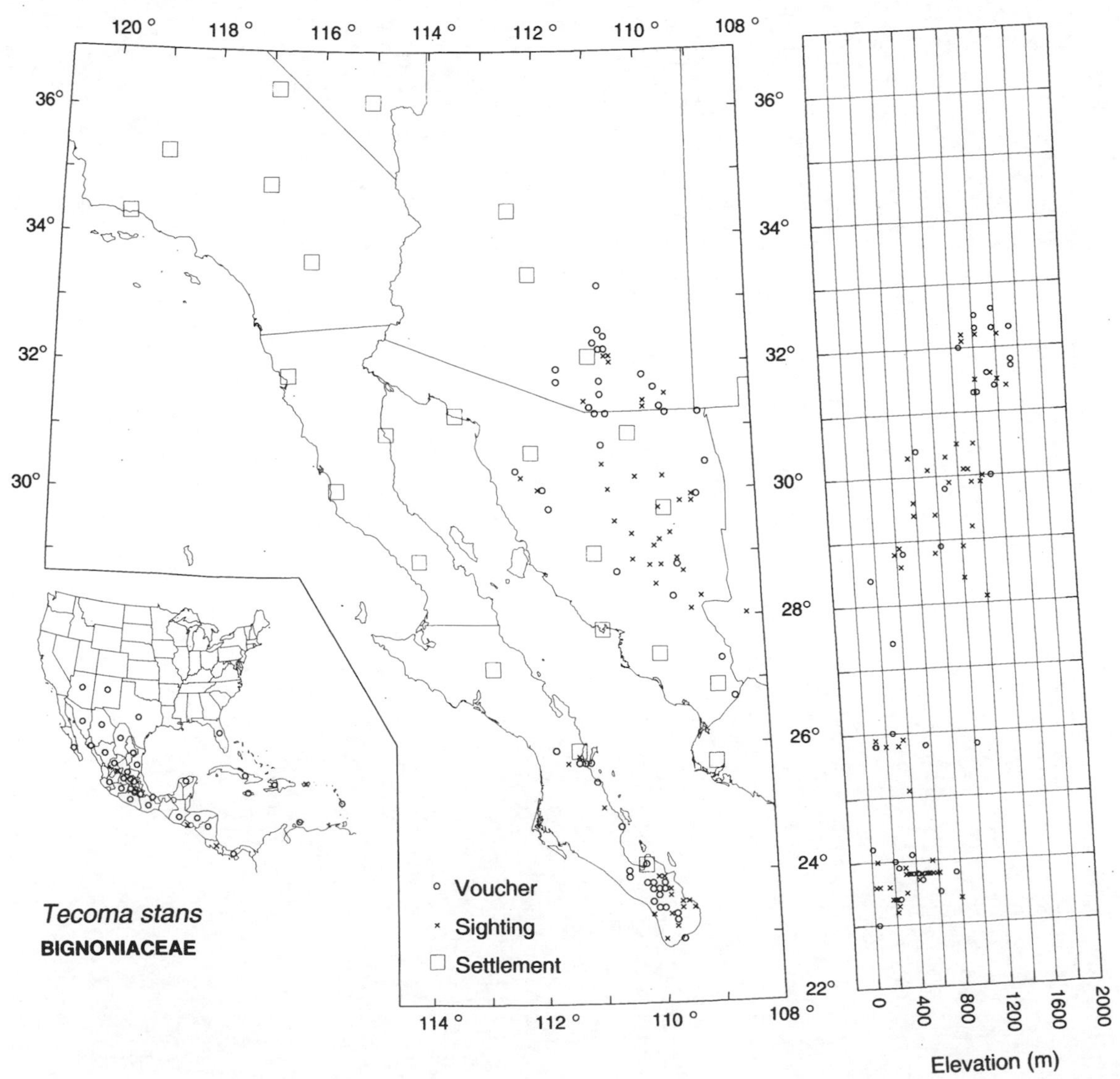

Tecoma stans
BIGNONIACEAE

and 1.5–3 cm wide. The yellow, funnelform flowers, 4–6 cm long, are racemose. Fruits are linear capsules 7–20 cm long containing numerous flat, winged seeds. The diploid chromosome number is 36 (Goldblatt and Gentry 1979). Jain and Singh (1979) and Jain (1978a, 1978b) have studied the floral, shoot, and leaf anatomy.

Sambucus coerulea var. *mexicana* L. Benson, a large shrub or medium-sized tree of desert watercourses, also has opposite, odd-pinnate leaves, but its leaflets are elliptic or ovate rather than lanceolate, and its inflorescences are broad, white-flowered cymes.

Our map combines var. *angustata* Rehd., which occurs from central Sonora northward, and var. *stans,* found from southern Sonora and the Cape Region into South America. The principal difference between them is the narrower (not more than 1 cm wide), more deeply incised leaflets of var. *angustata.*

Tecoma stans is a tropical and subtropical species that reaches its northern limit at the upper margin of the Sonoran Desert. Unprotected plants suffer frost damage at –4.4°C (Kinnison 1979), and the upper elevational limit is likely determined by low temperatures. At the northern end of its range, *T. stans* grows on rocky slopes, often among boulders, where it obtains some protection from frost, and along washes. Farther south it grows on floodplains, flats, and gentle slopes. In southern Arizona, the species typically occupies the ecotone between Sonoran desertscrub and semidesert grassland, where it finds both protection from frost and adequate moisture.

The northernmost location on our map represents an 1887 collection by a Mrs. Aguirre from the Pinal Mountains, Pinal County, Arizona. The cluster of points in western Sonora occurs, for the most part, in isolated desert mountain ranges. Inadequate summer rainfall probably controls the western and lower elevational limits of this species.

In the Cape Region and southern Sonora, plants can be found in flower and leaf almost throughout the year. Farther north, flowering is generally from May to October. Bumblebees (*Bombus* species)

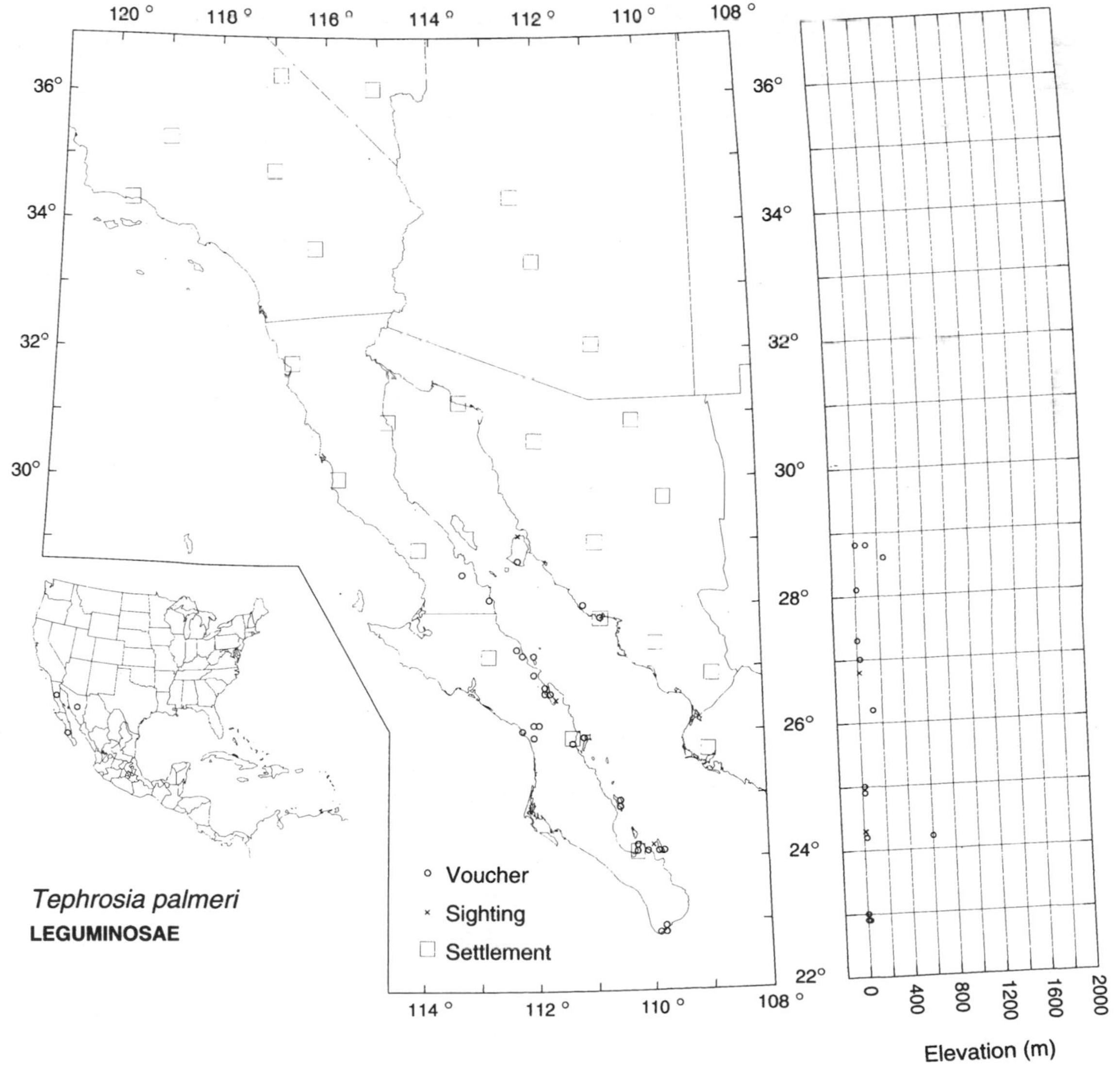

Tephrosia palmeri
LEGUMINOSAE

are probably important pollinators. Carpenter bees (*Xylocopa* species) often take nectar from the flowers.

Tecoma stans is widely cultivated in tropical and subtropical regions of the New and Old World. Stored seed does not germinate well, but fresh, untreated seed germinates within a few weeks of harvest (Nokes 1986). Semihardwood cuttings taken from midsummer through fall root under intermittent mist. They require protection from frost the first winter (Nokes 1986). Below –2°C, mature plants may be killed to the ground, but recovery is rapid (Duffield and Jones 1981). Moderate to ample irrigation improves bloom (Duffield and Jones 1981).

Various parts of the plant are employed as a diuretic and to treat syphilis and diabetes (Standley 1926; Moore 1989). *Tecoma stans* shows significant pharmacological activity (Hammouda and Khalafallah 1971; Lozoya-Meckes and Mellado-Campos 1985).

Tephrosia palmeri

S. Watson

This openly branched shrub grows up to 1.5 m tall and has silvery, appressed hairs on the foliage. The once-pinnate leaves, 8–15 cm long, comprise 7–15 widely spaced, narrow leaflets 1.5 to 5 cm long. The yellow flowers, streaked with purple, are 15–18 mm long. They are borne 1 to 2 at a node. The pods are 4–6 cm long, 4–5 mm wide.

Found both on the Baja California peninsula and in Sonora, *T. palmeri* also grows on several of the gulf islands (Moran 1983b). Most of the occurrences are below 250 m, although one sighting (by Reid Moran) on Isla Cerralvo placed it at 600 m.

Flowering occurs mainly in spring. Some flowers are produced in summer and fall.

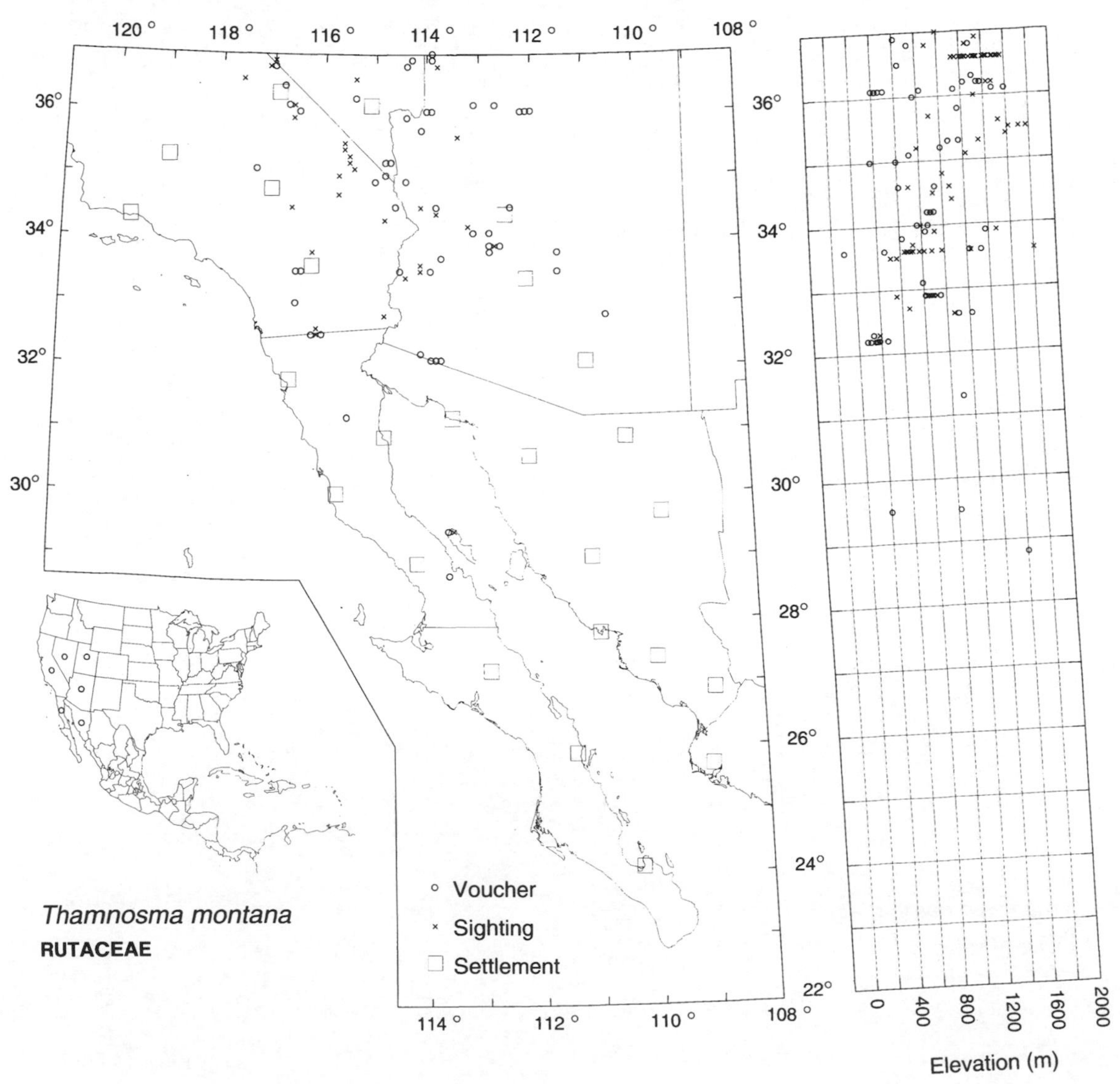

Thamnosma montana
RUTACEAE

The plants grow on rocky or gravelly slopes and mesas, often on limestone. The outliers in Baja California (28.8°N and 29.5°N) represent a sighting and several collections made by Reid Moran. The easternmost collection (32.9°N, 110.6°W) was made by Peter Warren in Aravaipa Creek, Pinal County.

Thamnosma montana generally blooms in spring (February–May), rarely in late summer (August–September). Leaves appear with the flowers and fall about the same time as the flowers fall.

Native Americans reportedly used *T. montana* as a tonic and in the treatment of gonorrhea (Kearney and Peebles 1960).

Tricerma phyllanthoides

(Benth.) Lundell

[=*Maytenus phyllanthoides* Benth.]

Mangle dulce, mangle, aguabola, mangle aguabola, granadilla

Tricerma phyllanthoides, a shrub 2–4 (occasionally 8) m tall, has a rounded crown and smooth, light gray bark. The alternate, obovate leaves, 2–4 cm long and 1–2 cm wide, are both leathery and succulent. The inconspicuous flowers are in few-flowered, axillary clusters. Fruits are dark red, coriaceous capsules 7–10 mm long. A fleshy, red aril encloses the seeds.

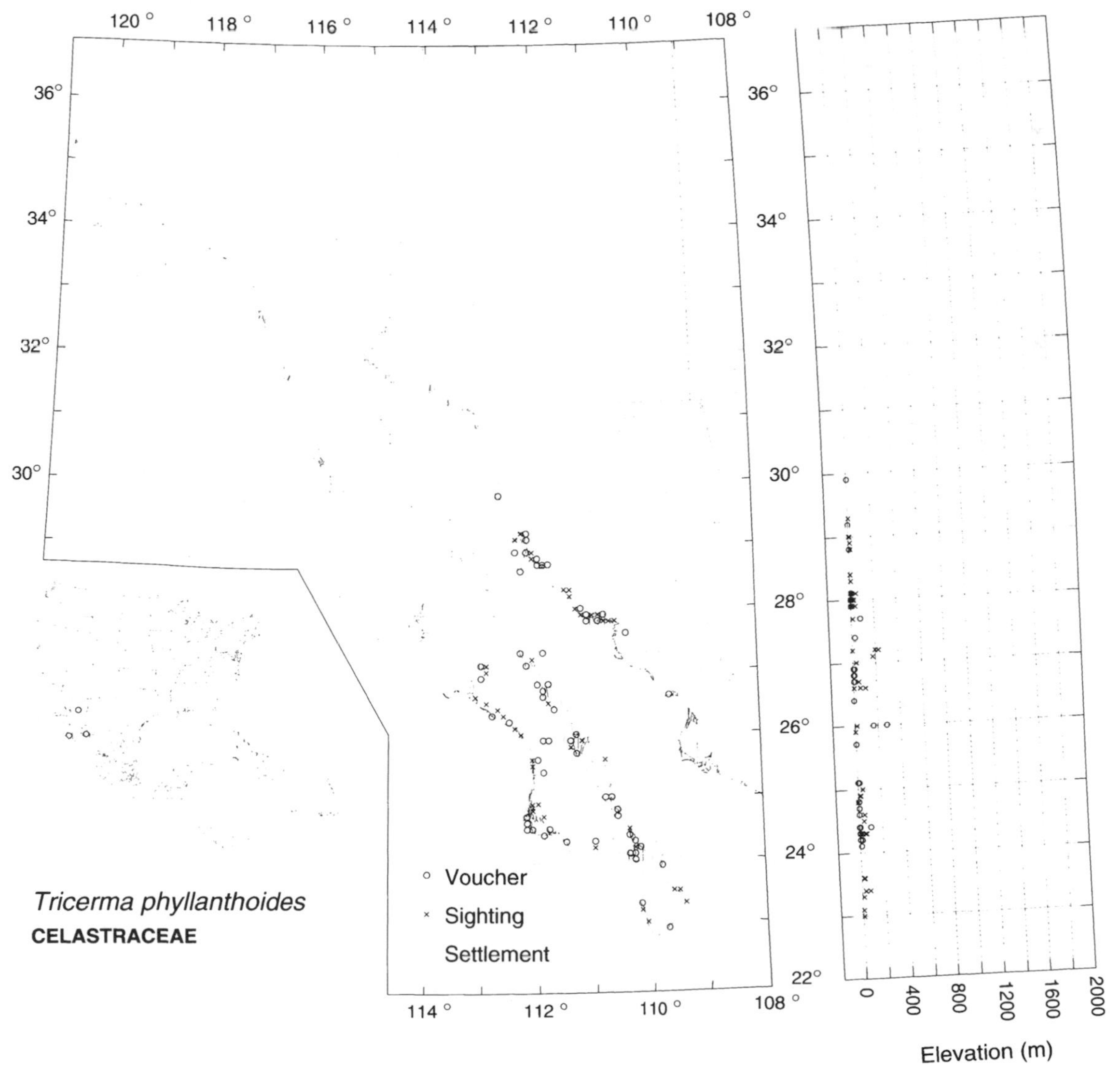

Tricerma phyllanthoides
CELASTRACEAE

In habit and habitat, *T. phyllanthoides* superficially resembles the true mangroves of the gulf coast. *Rhizophora mangle, Avicennia germinans,* and *Laguncularia racemosa,* however, all have opposite leaves. The alternate leaves of *Conocarpus erecta* have two small glands at the juncture of the petiole and blade.

Lundell (1971) transferred this and several closely related taxa from the genus *Maytenus* to *Tricerma,* a small, natural group of species distinguished by thick, fleshy leaves and 3-celled ovaries with one ovule in each cell.

Tricerma phyllanthoides tolerates inundation by seawater and attains its best development on wet, saline soils, as on coastal flats, where it sometimes forms thickets (Felger and Moser 1985). Occasional individuals grow well back from the beach, where inundation is infrequent. This species also occurs inland near saline lakes (Wiggins 1964). The northernmost collection was made at Puerto Libertad in 1935. We have not seen it there.

Unlike *A. germinans, L. racemosa,* and *R. mangle, T. phyllanthoides* rounds the tip of Baja California Sur. Presumably, the rocky coast and rough surf of the cape afford few of the quiet, backwater habitats that *Avicennia, Laguncularia* and *Rhizophora* require, whereas *T. phyllanthoides,* able to thrive away from the tidal zone, can find suitable habitat where the others cannot.

Flowers appear throughout the year and are probably wind pollinated. The leaves are evergreen. The arillate seeds may be dispersed by animals, particularly birds. Here again *T. phyllanthoides* exhibits its greater independence of the tidal habitat than the true mangroves, which depend on ocean currents for seed dispersal.

Seri Indians used the wood for fuel, the fruits for food, the bark as a gargle for sore throats and a tea for dysentery, and the leafy branches for basket linings (Felger and Moser 1985).

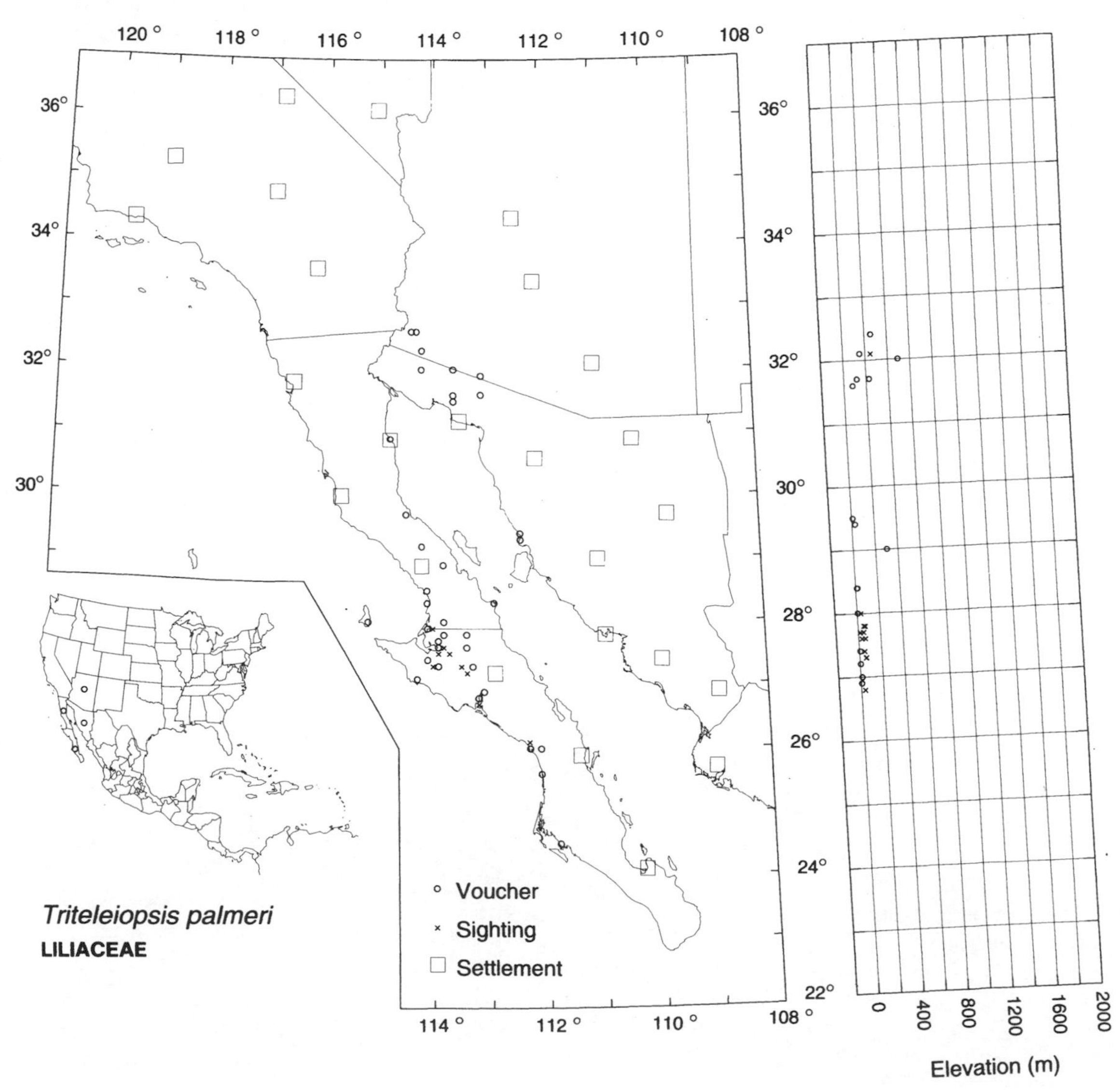

Triteleiopsis palmeri
LILIACEAE

Triteleiopsis palmeri

(S. Watson) Hoover

Blue sand lily

A perennial herb from a fibrous-coated corm, *Triteleiopsis palmeri* has a stout, terete stem 3–10 dm high and 6–15 mm wide at the base. The leaves, 2–5 dm long and clustered at the base of the stem, are fleshy and V-shaped in cross section. The deep blue to purplish blue flowers are in umbels of 30 to 100. The diploid number of chromosomes is 33 [sic] (Lenz 1966).

Dichelostemma pulchellum (Salisb.) Heller, a common geophyte of rocky slopes and plains in the northern Sonoran Desert, has lilac rather than blue flowers and fewer flowers per umbel (2–20). *Hesperocallis undulata* A. Gray, a white-flowered lily of sandy desert areas, also has long, narrow basal leaves, but these are strongly undulate.

Triteleiopsis palmeri is most common on dunes and sand sheets, occasional in rocky or heavy soil. Its low elevational range (sea level to 490 m) may reflect its proclivity for areas of loose sand, which, in the Sonoran Desert, are mostly near the coast. The populations are widely disjunct in southwestern Arizona, northwestern Sonora, the Sonoran coast, and the Vizcaíno Plain of the Baja California peninsula.

Flowers generally appear in the spring (February–April), sometimes in autumn (October–November). Reproduction is by multiplication of corms (Felger and Moser 1985) and by seed.

Seri Indians, especially young children, ate the corms either raw or cooked (Felger and Moser 1985).

Trixis californica

Kellogg

[=*T. peninsularis* S. F. Blake]

This profusely branched shrub, 3–6 dm tall, has brittle, whitish stems. The bright

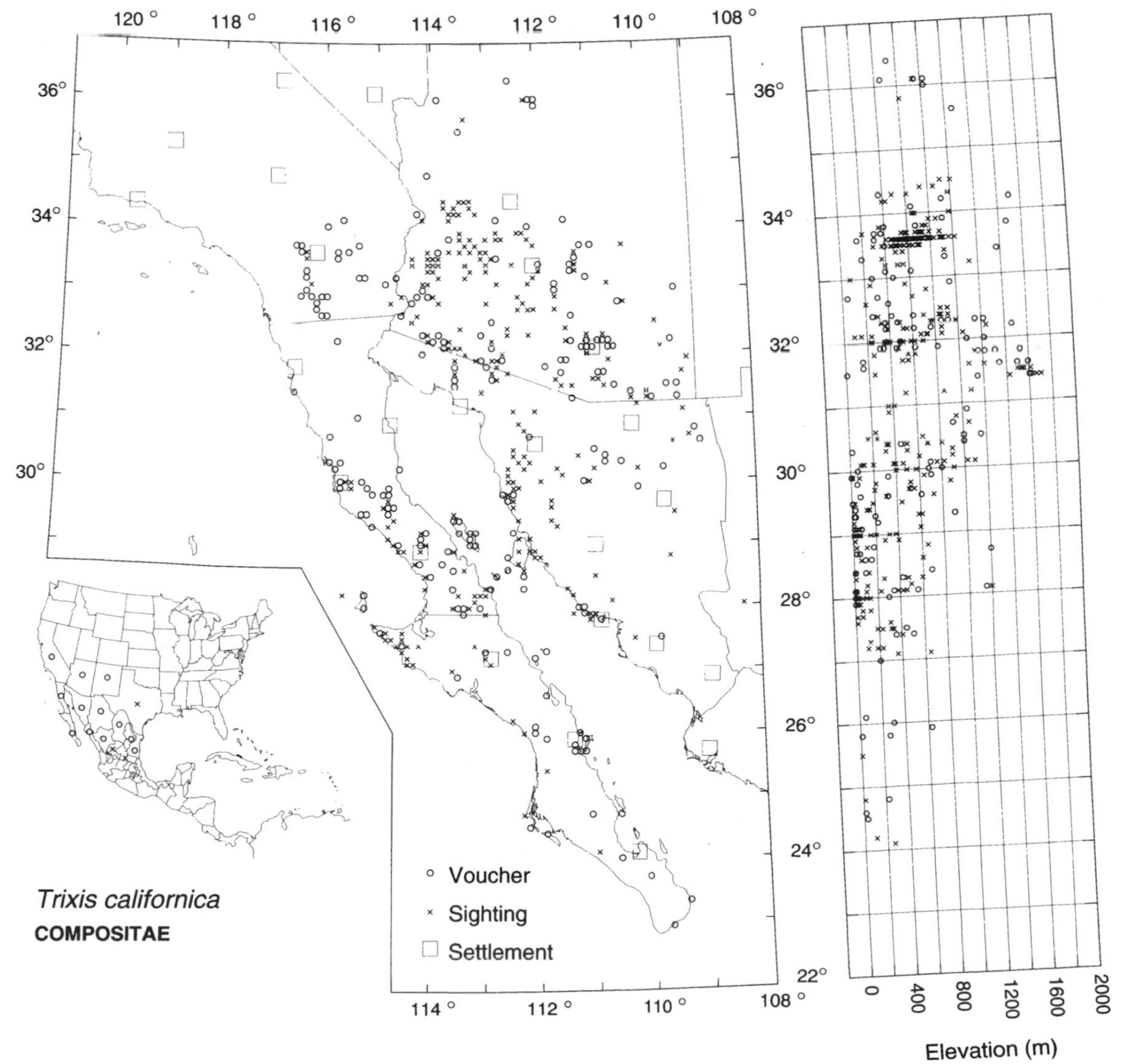

green, lanceolate or linear-lanceolate leaves, 2–10 cm long, are entire to dentate. The yellow-flowered heads, 1.5–2 cm high, are in dense, rather flat panicles. The slender achenes have a tawny or whitish pappus of capillary bristles. The haploid chromosome number is 27 (Turner et al. 1962; Pinkava and Keil 1977).

Anderson (1971) treated *T. peninsularis* as a variety of *T. californica*. Confined to the Cape Region, var. *peninsularis* (S. F. Blake) C. Anderson differs from var. *californica* in its denser leaf pubescence and sparsely pubescent corollas.

Trixis californica is occasional to common along washes and on rocky slopes, often under trees. This widespread species may have several different ecotypes. Populations at 1,700 m in the Mule Mountains, southeastern Arizona, evidently tolerate frequent subfreezing temperatures, whereas those at 820 m on Tumamoc Hill, Tucson, Arizona, are killed nearly to the ground in the coldest winters. The northern limit occurs in the western Grand Canyon, which has apparently served as an eastward distributional corridor.

Spring (February–April) is the main blooming season, but flowers appear in response to rain in other months. Leaves are cold and drought deciduous. The seeds are doubtless wind dispersed. Recruitment is apparently infrequent; during 4 years of observation on a 557 m² plot at Tumamoc Hill, Tucson, Arizona, mass germination occurred only once.

The Seri smoked the leaves as tobacco. Seri women used a tea made from the roots to hasten childbirth (Felger and Moser 1985).

Tumamoca macdougalii Rose

Tumamoc globe-berry

Tumamoca macdougalii is a slender-stemmed vine growing from thickened, tuberlike roots; it has thin, 3-lobed leaves 2–4 cm long. The staminate flowers are in few-flowered racemes. The pale yellow

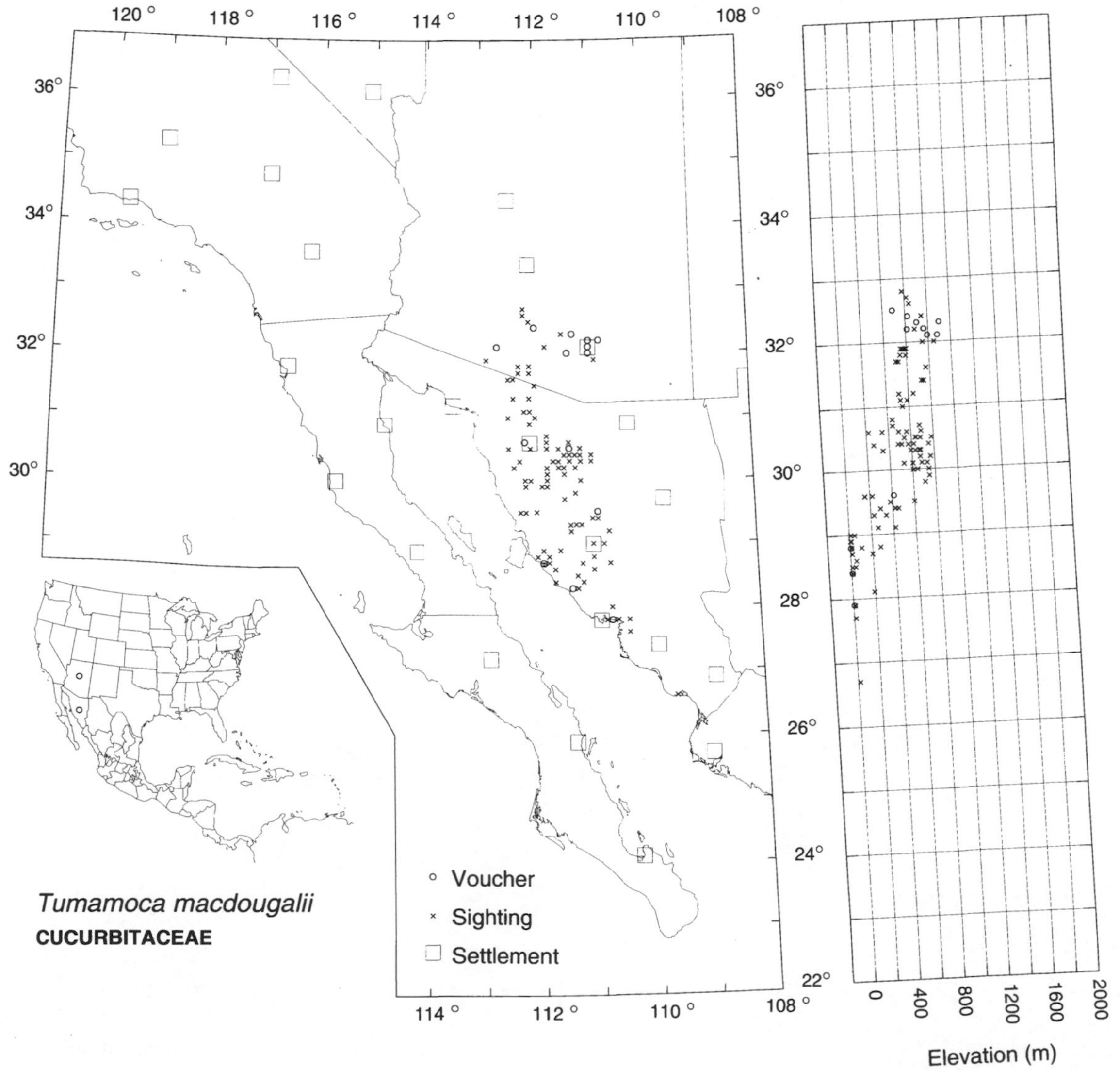

Tumamoca macdougalii
CUCURBITACEAE

pistillate flowers are solitary. The globose, red to yellow berries contain 2 to several seeds.

Within its relatively broad geographic range, *T. macdougalii* occurs on a wide variety of soil types on bajadas, valleys, and coastal plains (Reichenbacher 1990). Along the coast, it is largely restricted to saline soils (Reichenbacher 1990). The vines frequently clamber into shrubs, where they are difficult to detect until the brightly colored fruits appear.

While a candidate for the U. S. Fish and Wildlife Service endangered species list, *T. macdougalii* was the subject of an intensive botanical survey (Reichenbacher 1985b, 1990). The record from the vicinity of Huatabampo (26.7°N) suggests that the species may extend as far south as northern Sinaloa (Reichenbacher 1990).

The tuberous roots are united into a woody crown that produces a short, persistent, woody stem. In late spring, the perennial stem sprouts an annual stem that climbs into shrubs or trees and produces a few leaves and flower buds (Reichenbacher 1985b). With the onset of summer rains in July, the production of leaves, flowers, and fruits dramatically increases. As the soil dries out in early fall, the annual stem dies back. Fruits mature in 4–5 weeks. In the Tucson area, the seeds are dispersed by ants (Reichenbacher 1985b). Scarification is necessary for germination. Seedlings can be found in August (Reichenbacher 1985b).

The dependence of *T. macdougalii* on summer rainfall no doubt defines the western boundary of the species, while low winter minimum temperatures likely limit it to the north and northeast (Reichenbacher 1985b, 1990). The southern limit seems to occur where Sinaloan thornscrub becomes predominant; possibly the closed canopy is too humid and shady for *T. macdougalii* (Reichenbacher 1985b).

Javelina root out and eat the tubers (Reichenbacher 1985b). The species was added to the federal list of endangered species in 1986 (Reichenbacher 1990).

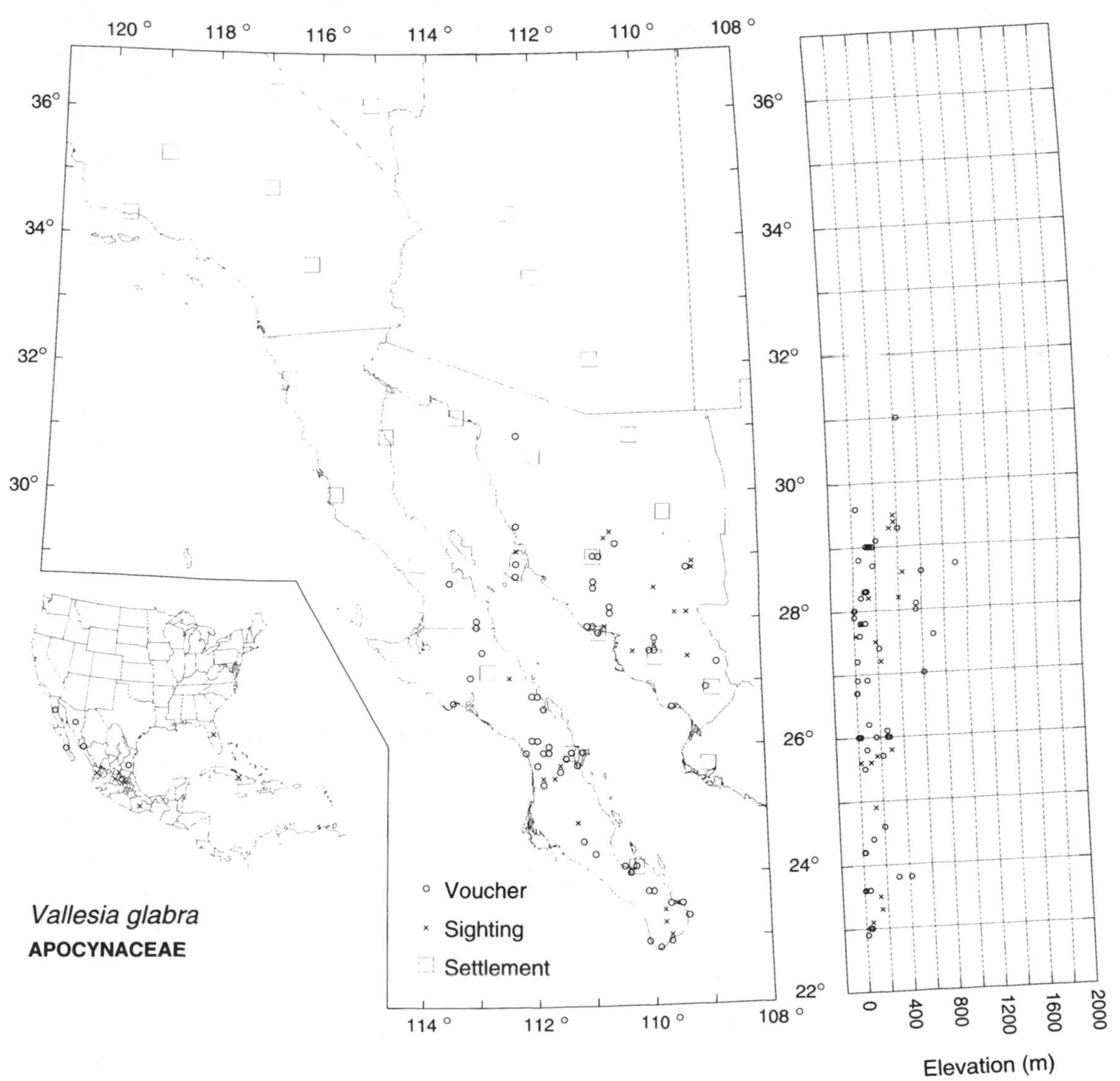

Vallesia glabra
APOCYNACEAE

Vallesia glabra

(Cav.) Link

Cacarahue, otatave, frutilla, huelatave, huevito, sitavaro

Vallesia glabra is a shrub or small tree up to 6 m tall with slender branches. The bright green, leathery leaves, 6–20 mm wide and 2.5–7 cm long, are glabrous and lanceolate. The petioles are 3–5 mm long. The salverform, yellowish or cream-colored flowers, 5–7 mm long, are in axillary cymes opposite the leaves. Fruits are white drupes 8–10 mm long with thin, translucent flesh. The diploid number of chromosomes is 22 (Norman and Roper 1981).

The foliage of *Vallesia laciniata* Brandegee is densely puberulent. *Vallesia baileyana* Woods., known only from the Guaymas area, has larger flowers (10–16 mm long) and longer petioles (7–9 mm long).

Vallesia glabra, widespread in the tropics and subtropics of the New World, reaches its northern limit in the Sonoran Desert region, where it is rare to locally common along washes, in valley bottoms, and on plains. The disjunct at 31.0°N, 112.4°W represents a 1936 collection by Ira L. Wiggins near Tajitos, Sonora.

Leaves are evergreen. Flowering specimens have been collected in every month. A relatively copious fruit crop is produced several times a year (Felger and Moser 1985).

The fruits are edible. Their juice was used to treat eye inflammations, and the burned, powdered leaves were used to alleviate rashes (Standley 1924; Felger and Moser 1985).

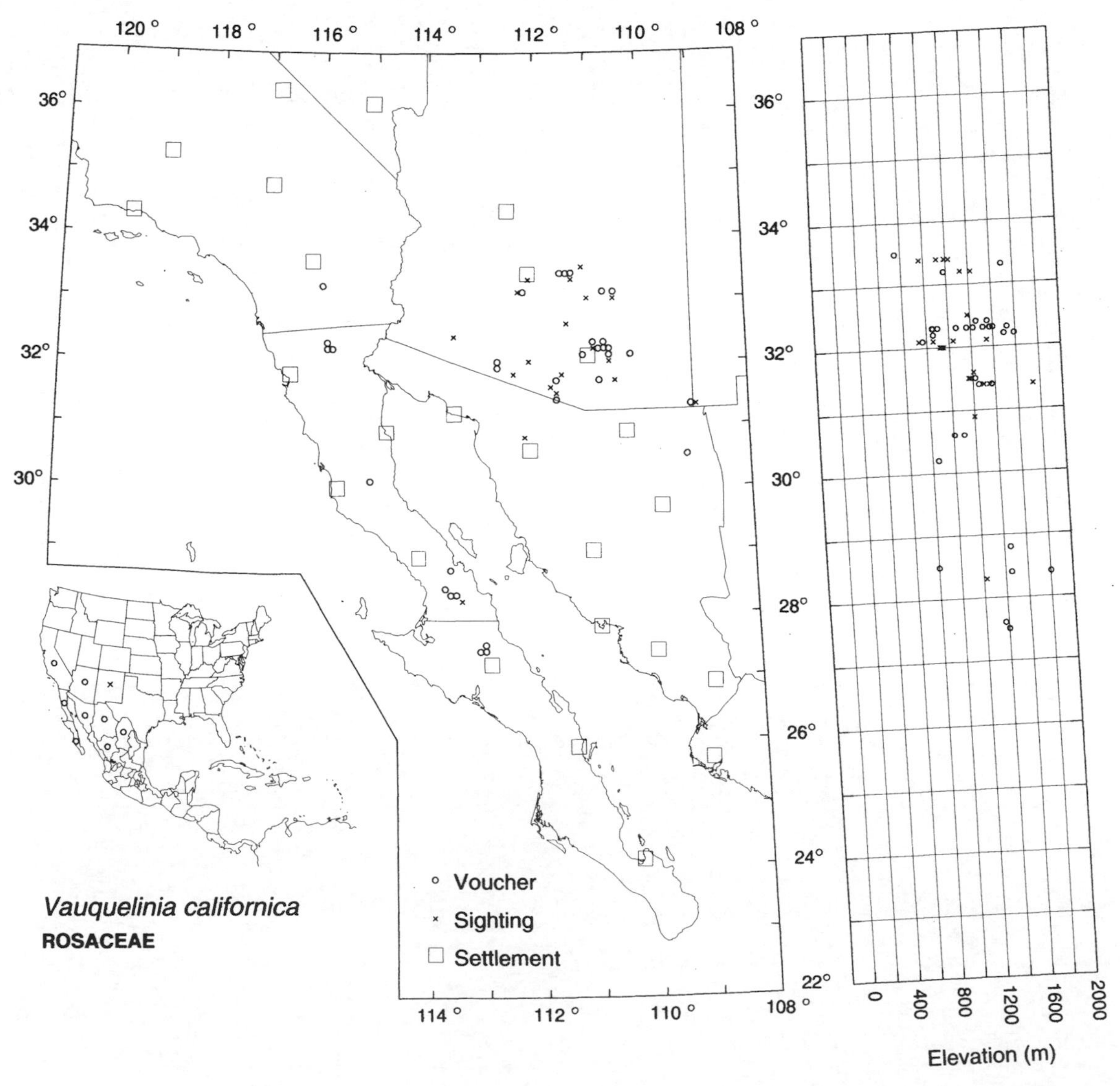

Vauquelinia californica
ROSACEAE

Vauquelinia californica

(Torr.) Sarg.

[=*V. pauciflora* Standley]

Arizona rosewood

Vauquelinia californica is a large shrub or small tree 3–8 m high with a dense, dark green canopy. The alternate leaves, 4–10 cm long and 9–15 mm wide, are leathery and serrate. The small, white flowers, 8–9 mm in diameter, are clustered in flat-topped corymbs 5–8 cm broad. The woody, 5-parted capsules are 6 mm long. The bark is reddish brown and scaly. The haploid chromosome number is 15 (Hess and Henrickson 1987).

Largely on the basis of leaf morphology, Hess and Henrickson (1987) subsumed *Vauquelinia pauciflora* under *V. californica* and recognized four subspecies: subsp. *californica* of central Arizona, Sonora, and the Baja California peninsula; subsp. *pauciflora* (Standley) Hess & Henrickson from southeastern Arizona, Sonora, Chihuahua, Coahuila, and Durango; subsp. *sonorensis* Hess & Henrickson from the Ajo Mountains of southwestern Arizona; and subsp. *retherfordii* (I. M. Johnston) Hess & Henrickson from southwestern Coahuila and adjacent northeastern Durango. The leaves of subsp. *pauciflora* are glabrous. Except in central Arizona, where the leaves are typically glabrate beneath, the leaves of subsp. *californica* and subsp. *sonorensis* are white-tomentose on the underside. Subspecies *sonorensis* has markedly narrower leaves than subsp. *californica* and subsp. *pauciflora* (generally less than 11 mm wide), and its leaves have larger marginal spines (Williams and Bonham 1972; Hess and Henrickson 1987).

Vauquelinia californica grows on rocky slopes of hillsides and canyons on a variety of substrates, including rhyolite, andesite, granite, granitic gneiss, limestone, dolomite, sandstone, and tuff (Williams and Bonham 1972). In southeastern Arizona and elsewhere, subsp. *pauciflora* is apparently restricted to limestone (Wells and Johnson 1964; Hess and Henrickson

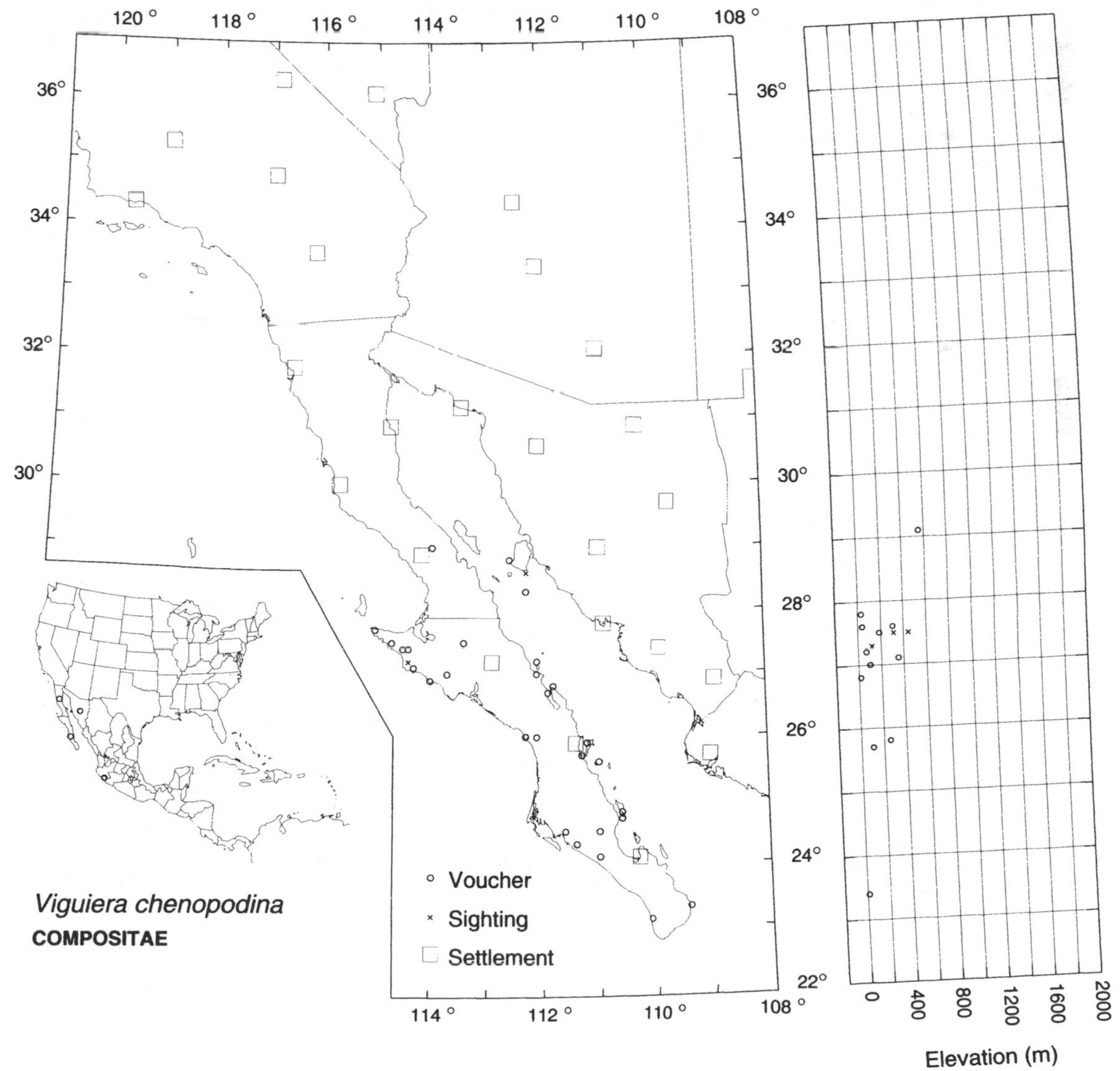

Viguiera chenopodina
COMPOSITAE

1987). Populations are often confined to mesic sites within their generally arid habitat, perhaps an indication that they are relicts of a more widespread distribution during the Holocene (Hess and Henrickson 1987). The New Mexico sighting is Drummond Hadley's from the Peloncillo Mountains.

Vauquelinia californica flowers between May and July (Kearney and Peebles 1960). Its leaves are evergreen.

This attractive plant is occasionally cultivated as an ornamental. Young plants require moderate irrigation until established and need little care thereafter, although growth is most rapid with periodic irrigation (Duffield and Jones 1981).

Viguiera

Golden-eye

There are in the Sonoran Desert several different shrub species in the genus *Viguiera*. We first describe four closely allied species that were formerly regarded as varieties of *Viguiera deltoidea* A. Gray: *V. parishii* Greene (=*V. deltoidea* var. *parishii* [Greene] Vasey & Rose), *V. triangularis* M. E. Jones (=*V. deltoidea* var. *deltoidea*), *V. chenopodina* Greene (=*V. deltoidea* var. *chenopodina* [Greene] S. F. Blake), and *V. deltoidea* (=*V. deltoidea* var. *tastensis* Brandegee).

Viguiera parishii grows to about 0.5 m high. The triangular-ovate leaves are toothed and 1–2 (occasionally 3) cm long and 1–1.5 (occasionally 2) cm wide. The short hairs arising from pustulate bases on the upper sides of the leaves are characteristic of this species. The yellow, radiate heads, 1 cm across the disk, are single (occasionally clustered) on elongate peduncles. *Viguiera triangularis* is taller (1–3 m high) and less densely branched. The dentate leaves are deltoid-ovate, 1.5–6.5 cm long and 0.5–4.5 cm wide; leaf pubescence may be harsh or soft, but the hairs do not arise from pustulate bases. Also, the heads are cymosely arranged rather than solitary on the peduncles. Differing from *V. parishii* and *V. triangularis* in its softer, often tangled, leaf pubescence

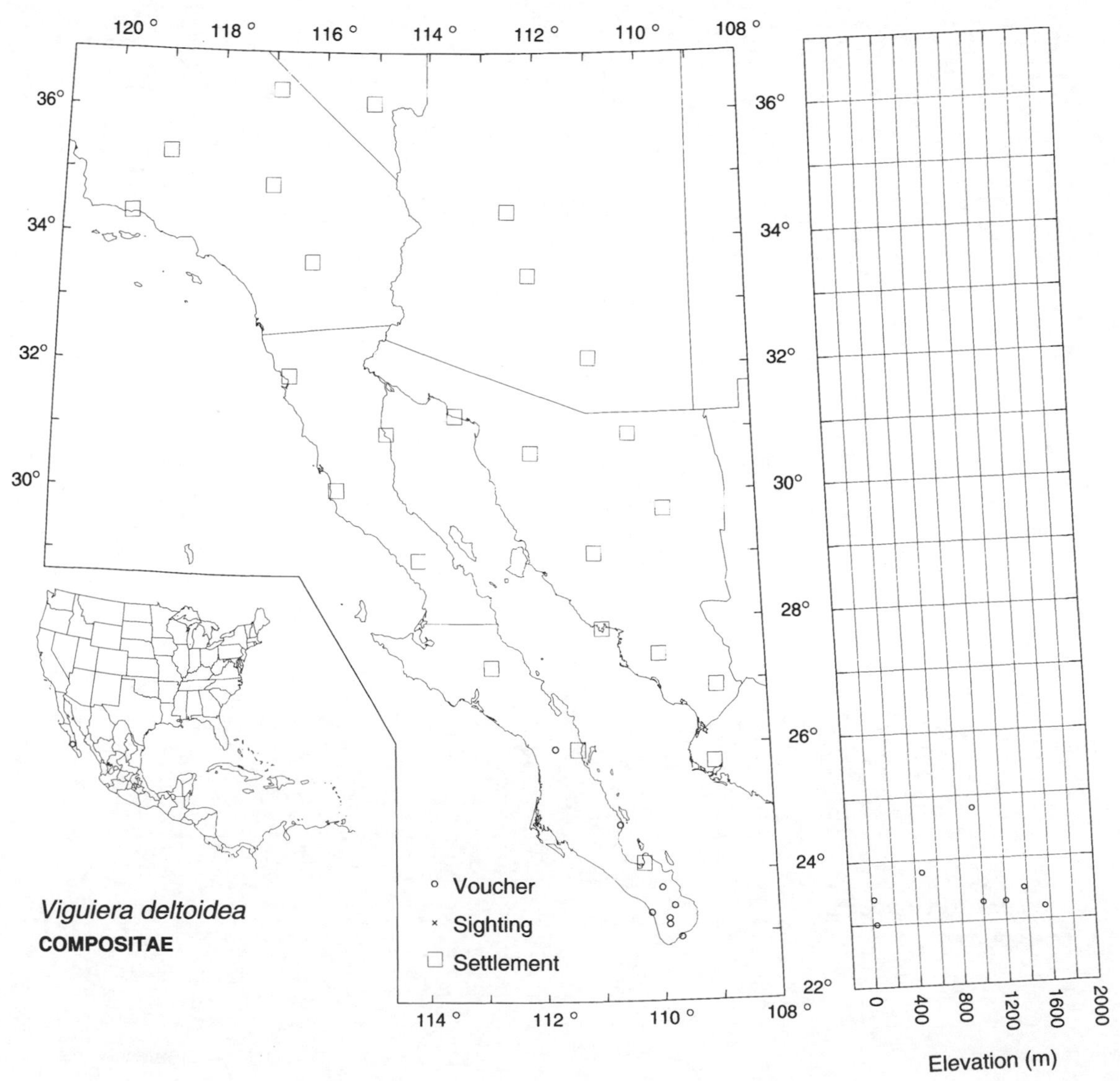

and its usually entire leaves is *Viguiera chenopodina. Viguiera deltoidea* is a much taller shrub than the others, reaching a height of 6 m. The leaves are correspondingly larger: up to 6 cm wide and 5–10 cm long. They are softly and densely pubescent beneath. The flower heads are in corymbose clusters 4.5–5 cm wide. Blake (1918) discerned a south-to-north cline in leaf size and pubescence among *V. deltoidea, V. triangularis,* and *V. parishii.*

The base chromosome number in this group is n=18 (Schilling 1990). Polyploidy occurs in *V. chenopodina, V. deltoidea,* and *V. triangularis; V. parishii* is diploid (Schilling and Schilling 1986).

Viguiera laciniata has green, laciniate leaves. *Viguiera reticulata* S. Watson (a shrub mostly confined to the southern Mojave Desert) has larger leaves than *V. parishii* (3–5 cm long); moreover, the upper leaf surface is silvery white from densely pilose hairs. Leaves of *V. microphylla* are markedly smaller than in any species discussed in this section—only 3–8 mm wide and 5–12 mm long. In the genus *Encelia* the leaves are alternate, whereas in *Viguiera* the lower leaves are opposite.

The species in this group often grow on gravelly or sandy plains, in dry washes, and on rocky hillsides and mesas. Only *V. parishii* is widespread; even so it is absent from areas where favorable habitat appears to exist. *Viguiera deltoidea* is the most localized, occurring mainly in the Cape Region on granitic hillsides and among boulders in arroyos (Schilling 1990). *Viguiera chenopodina* extends from the Vizcaíno region of Baja California to the Cape Region. It is on several gulf islands, including Tiburón and San Pedro Nolasco, and also occurs on Isla Socorro, well south of the peninsula (Schilling 1990).

Viguiera deltoidea, V. chenopodina, and *V. triangularis* apparently flower from March through April (occasionally as late as May) and in October (occasionally as early as August) through January. *Viguiera parishii* flowers year-round, presumably in response to rainfall.

Seri Indians cooked the mashed roots of *V. parishii* in water to make a contracep-

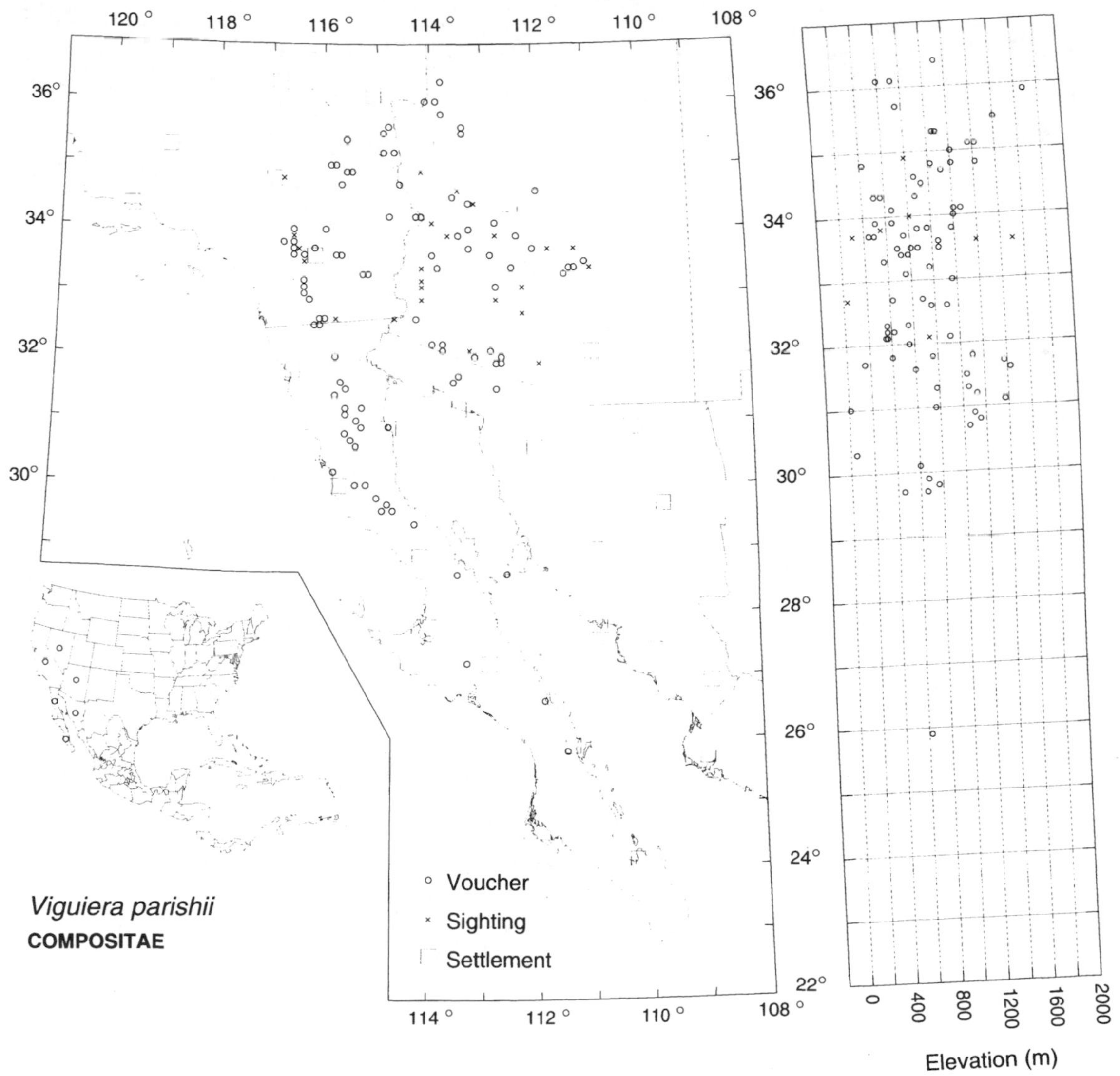

Viguiera parishii
COMPOSITAE

tive tea (Felger and Moser 1985).

In addition to the foregoing *Viguiera* species, we include the three following species.

Viguiera laciniata

A. Gray

Ragged-leaf golden-eye

This straggling shrub grows up to 1.5 m tall and has green, harshly pubescent foliage. The leathery, laciniate leaves, 1.5–4.5 cm long and up to 1.5 cm wide, are lanceolate to ovate. The yellow, radiate heads, 1.5–3 cm across, are in cymose panicles. The black, 4-angled achenes are about 3 mm long, with a pappus of lanceolate scales. The haploid number of chromosomes is 18 (Solbrig et al. 1972).

Other shrubby *Viguiera* species in its range have entire or, at most, serrate leaves. Leaves of ***Encelia ventorum*** Brandegee are glabrous, fleshy and pinnatifid with linear divisions. Foliage of ***Encelia laciniata*** Vasey & Rose is finely and densely puberulent on both surfaces rather than green and hispidulous.

Viguiera laciniata is occasional to common on rocky hillsides and along washes, especially on granitic soils. It grows in Californian coastalscrub and Californian chaparral in the northern part of its range, in Sonoran desertscrub in the southern part. It is also common in the transition between the two (Shreve 1936). The distribution suggests a tolerance for summer drought, especially when moderated by coastal fogs. On Cerro Potrero (29.8°N, 112.6°W), it grows as high as 1,400 m.

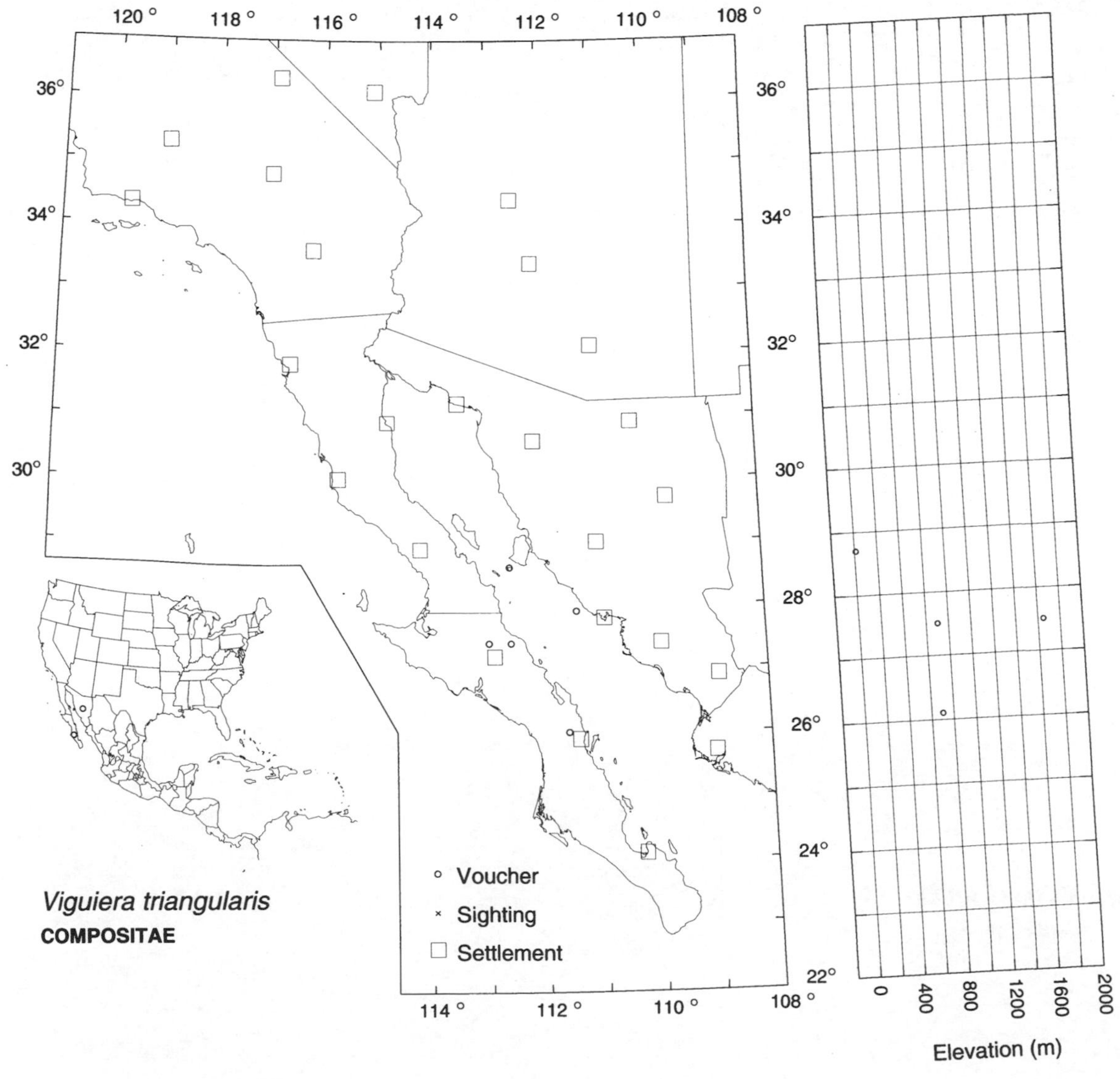

Viguiera triangularis
COMPOSITAE

Cold sensitivity apparently restricts the plants to coastal habitats in the north (Shreve 1936). The upper elevational limit may also be determined by low temperatures.

The sole Sonoran locality—Punta Cirio, a few km south of Puerto Libertad—was first vouchered in 1979 by Janice E. Bowers and Tony L. Burgess. This location has been a favorite haunt of botanists since the 1923 discovery there of *Fouquieria columnaris,* disjunct by 125 km from the Baja peninsula, and it is odd that no early collectors noted the presence of disjunct *V. laciniata.*

Flowering specimens have been collected in every month of the year.

Viguiera microphylla

Vasey & Rose

This slenderly branched shrub up to 1 m tall has ovate, entire leaves 3–8 mm wide and 5–12 mm long. The small flower heads are corymbose or paniculate.

The diminutive leaves and flower heads separate this species from the shrubs discussed in the earlier *Viguiera* section.

Blake (1918) noted that *V. microphylla* is close to *V. chenopodina,* and later chromosome studies (Schilling and Schilling 1986) confirmed his supposition. Indeed, *V. chenopodina* may include one or more genomes from *V. microphylla* (Schilling 1990).

A narrow endemic of the peninsular Vizcaíno region, *V. microphylla* grows on plains and washes, generally in sandy soils. It is locally abundant along roadways (Schilling 1990).

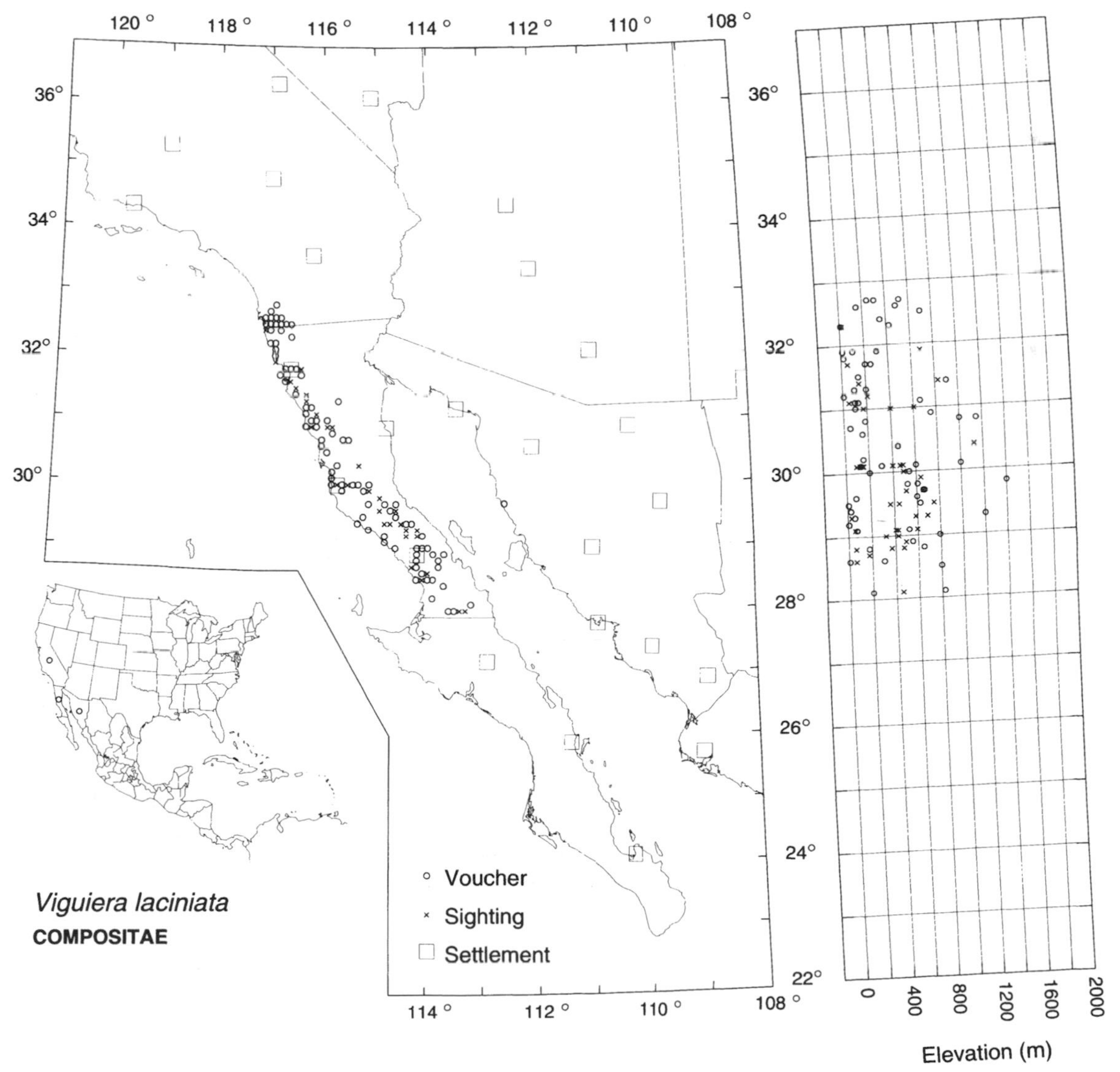

Viguiera laciniata
COMPOSITAE

Viguiera purisimae
Brandegee

Viguiera purisimae, a shrub or suffrutescent perennial up to 12 dm tall, has dark green, obovate to elliptic leaves 1–4 cm wide and 2–6 cm long. They are strongly 3-veined from the base. The large flower heads, in corymbose panicles, have conspicuous dark green to blackish ovate phyllaries.

The dark, rounded phyllaries readily distinguish this species from others in the genus.

This peninsular endemic grows in basaltic uplands, along arroyos, and in bottomlands, often in heavy clay soils. The scattered populations range in elevation from sea level to 1,625 m.

The flowering season is November to March (Wiggins 1964).

Viscainoa geniculata
(Kellogg) Greene

Guayacán

This compact, rounded shrub 2–6 m tall has rigid branches and ashy-puberulent foliage. The alternate, leathery leaves, 1.5–5 cm long and 8–30 mm wide, may be simple or pinnate with 3–5 leaflets. The rather showy, 5-petaled flowers are cream-colored or yellow. Fruits are distinctive spindle-shaped capsules with, usually, 4 angles. The black seeds are embedded in an oily coat. The haploid number of chromosomes is 13 (Porter 1963).

Simmondsia chinensis has opposite

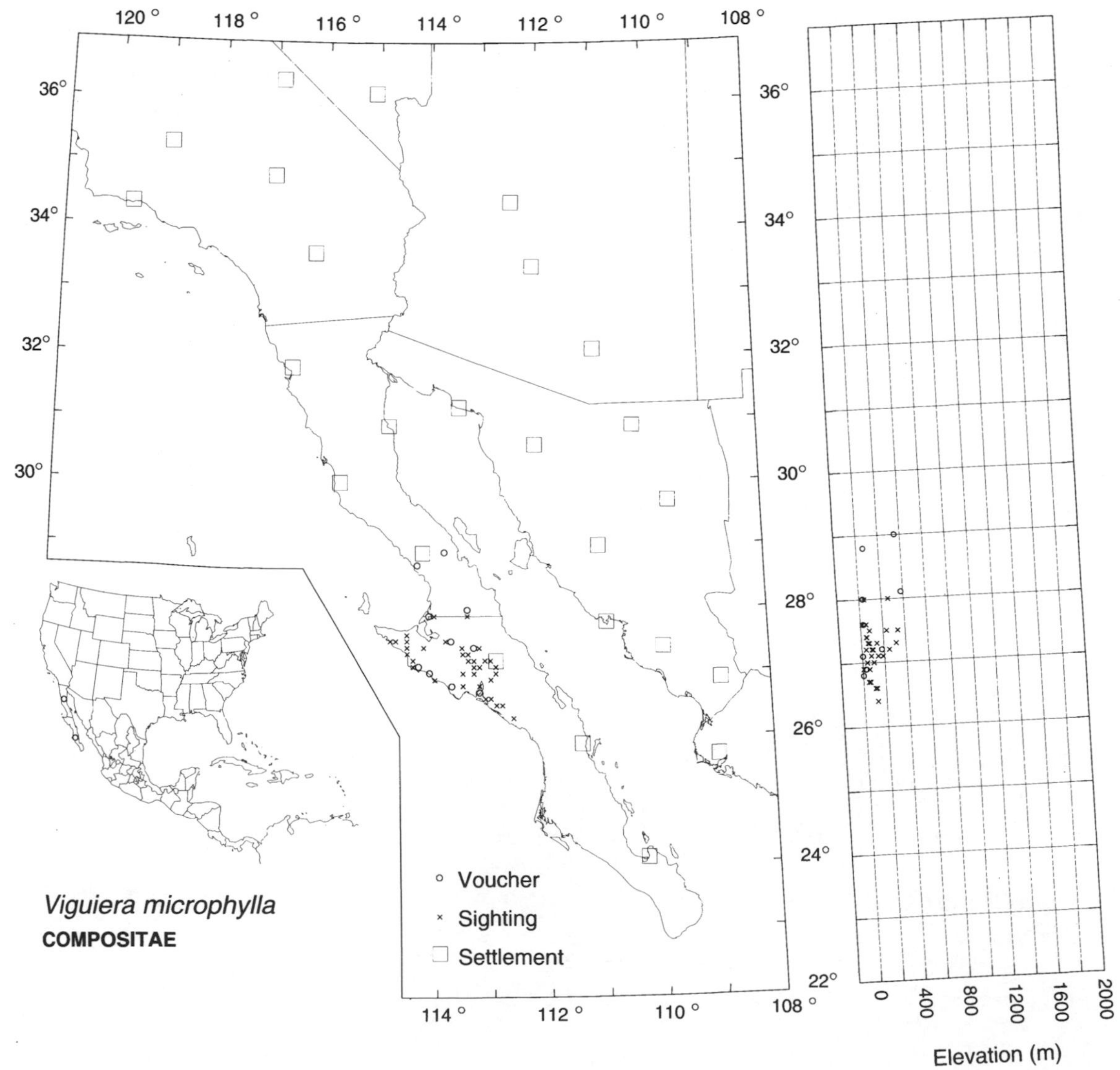

Viguiera microphylla
COMPOSITAE

leaves. *Stegnosperma* species have obovate, somewhat fleshy, glabrous leaves.

Botanists have recognized two varieties of *V. geniculata:* var. *geniculata,* which occurs throughout the range of the species, and var. *pinnata* I. M. Johnston, found on the western side of Baja California Sur near San Juánico (Wiggins 1964). Variety *pinnata* has green foliage and odd-pinnate leaves.

Rare to locally common in washes, on dunes, and on coastal plains, *V. geniculata* is largely restricted to sandy soils. The species reaches its greatest frequency in the Vizcaíno Subdivision of Baja California. From there it extends southward along the Gulf and Pacific coasts. The Sonoran populations also hug the coast, which is where the preferred sandy habitat predominates. The species may have migrated from the peninsula to the mainland via gulf island "stepping stones" (Cody et al. 1983). The oily seed coat around the seeds may function as an aril, a feature often associated with bird-dispersal.

Flowers appear from October to July. The leaves are evergreen (Shreve 1964).

Cattle and goats heavily browse both varieties (Porter 1963).

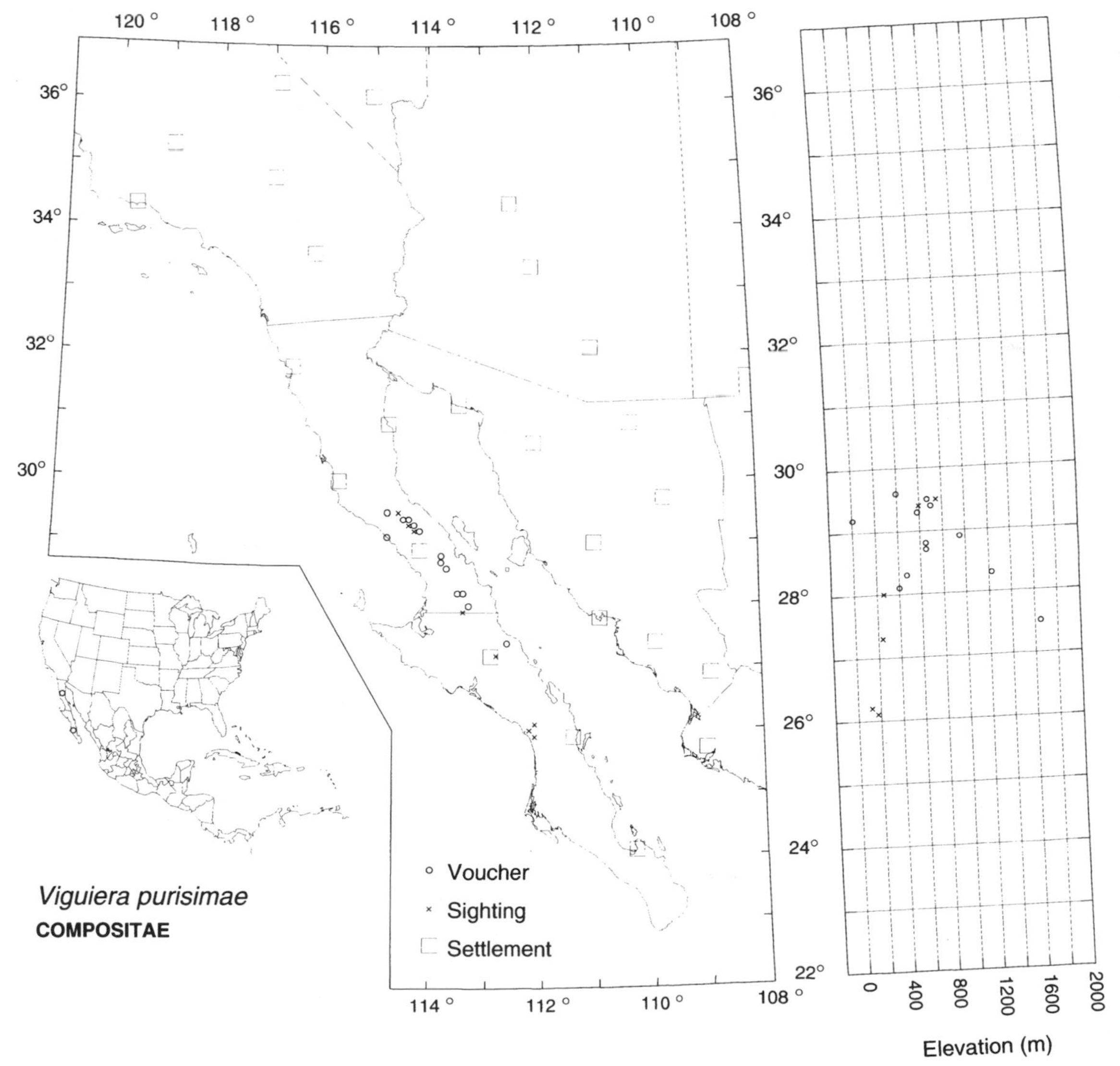

Viguiera purisimae
COMPOSITAE

Washingtonia filifera
(L. Linden) H. Wendl.

California fan palm, Washington fan palm, desert fan palm, petticoat palm, palma de castilla

Washingtonia filifera is a stout tree up to 15 m tall with a single trunk up to 1 m in diameter. Dead leaves accumulate around the trunk in a characteristic shag or thatch. If left undisturbed, the shag can reach 8 feet in diameter and 40–50 feet long (McCurrach 1960); often, however, the shag is burned before it reaches this size (Vogl and McHargue 1966). The gray-green, fan-shaped leaves are 1–2.3 m wide and are split into narrow segments with filamentous margins. The segments usually extend more than halfway to the base of the leaf. The petiole margins are spiny or almost entire on older plants. Inconspicuous flowers are on panicles 3 to 4 m long that often extend beyond the canopy. The wrinkled, dry fruits are 7–10 mm long. The haploid chromosome number is 18 (Uhl and Dransfield 1987).

The trunk of the closely related *Washingtonia robusta* H. Wendl. flares abruptly at the base. *Brahea armata,* another large palm, has markedly blue-glaucous leaves.

There is reason to believe that the type locality is not California, as has been long stated in the literature, but Castle Creek, Arizona (Miller 1983).

Because the plants require a constant source of water, they grow in arroyos or canyons where there is permanent subsurface flow and at seeps and springs. Many southern California populations occur along the San Andreas fault, where exposed strata or other geological structures have produced hillside seeps (Vogl and McHargue 1966). *Washingtonia filifera* is highly tolerant of alkali and often grows in fine, fluffy soils that are covered with salt deposits (Vogl and McHargue 1966; Brown et al. 1976). Plants in canyon bottoms are vulnerable to uprooting by severe floods; those on hillsides are sometimes toppled by soil slumping after heavy rains (Vogl and McHargue 1966).

According to Axelrod (1950), *Washing-*

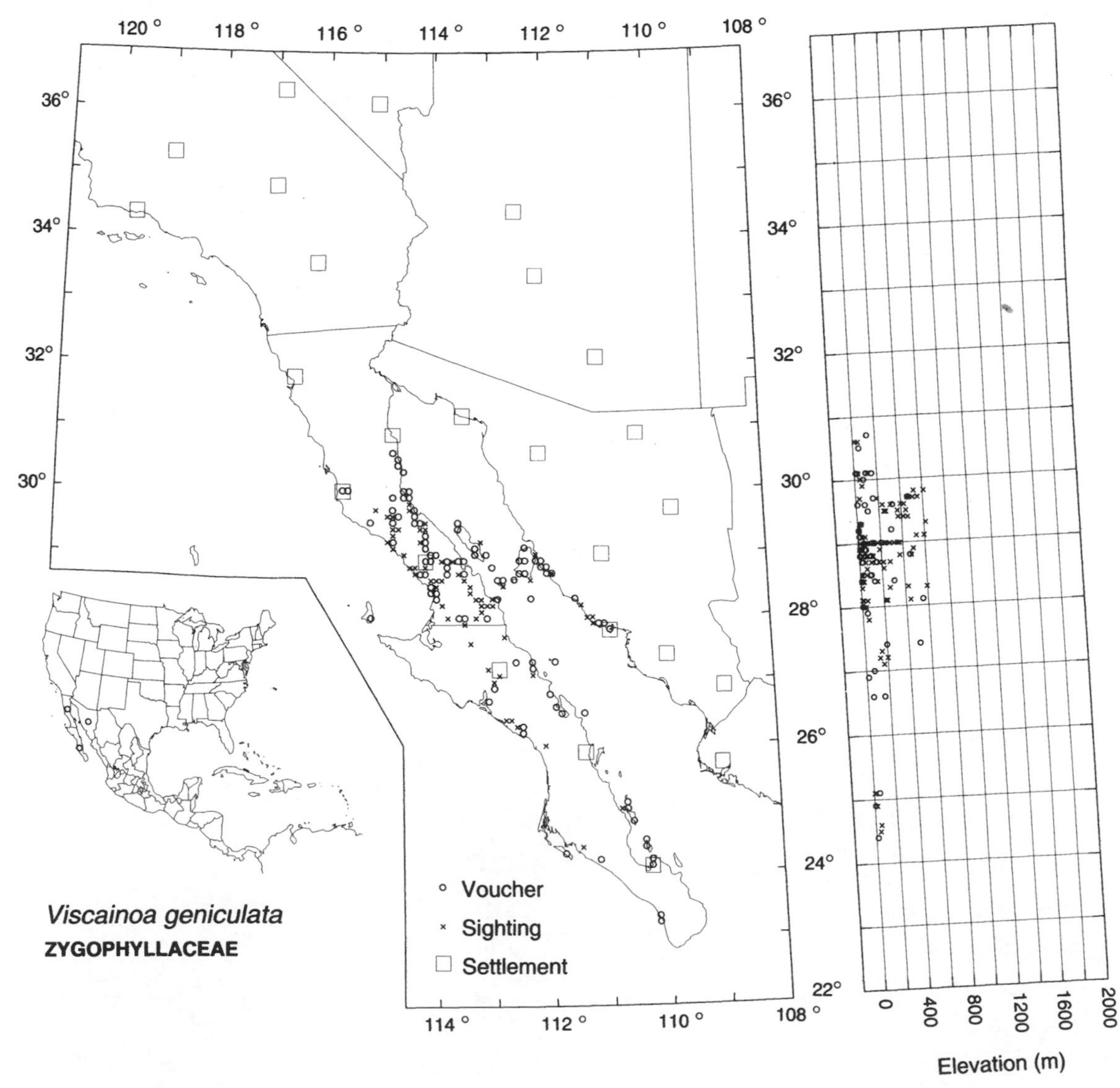

Viscainoa geniculata
ZYGOPHYLLACEAE

tonia is a relic from the Miocene and Pliocene. Once widespread from the Mojave Desert to the Pacific coast, the genus was eliminated across much of its range by climatic and geologic changes and was finally restricted to the more favorable climate of the western Sonoran Desert, particularly along the shores of the former Lake Cahuilla (now the Salton Basin) (Axelrod 1950). Our map shows that the preponderance of *W. filifera* populations are still restricted to this area. Outliers in western Arizona and northwestern Baja California may be relics, or they may have arisen via long-distance dispersal, or they may have been planted by Native Americans (Moran 1979). The population near Castle Creek, Arizona (34.0°N, 112.4°W), seems to have been established before the species was widely cultivated and is evidently native (Brown et al. 1976), although planting by Native Americans cannot be ruled out. The sightings north of 36°N all represent adventive populations. Plants have also naturalized in Kern County, California (Munz 1974), and doubtless at many other locations, as well. Once introduced, *W. filifera* may persist as long as a century at some locations (Cornett 1988).

Natural populations tolerate temperatures down to −11°C and can withstand up to 22 hours of freezing (Cornett 1987). Seeds are even hardier and will germinate after being exposed to −21°C for 36 hours. It seems likely that some factor other than low temperatures accounts for the absence of this species from the eastern Sonoran Desert and parts of the Mojave Desert where suitable habitat would seem to exist (Cornett 1987).

The plants bloom in spring and, occasionally, in summer. Fruits disperse by the following autumn. Birds and coyotes consume much of the crop, and coyotes are believed to be a primary agent of dispersal (Vogl and McHargue 1966). Seeds appear to remain dormant on the ground until leached of germination inhibitors (Vogl and McHargue 1966). Recruitment of seedlings is heavy during exceptionally wet winters, resulting in even-aged stands. Optimum conditions for germination need occur only once every hundred years for most populations in the Salton

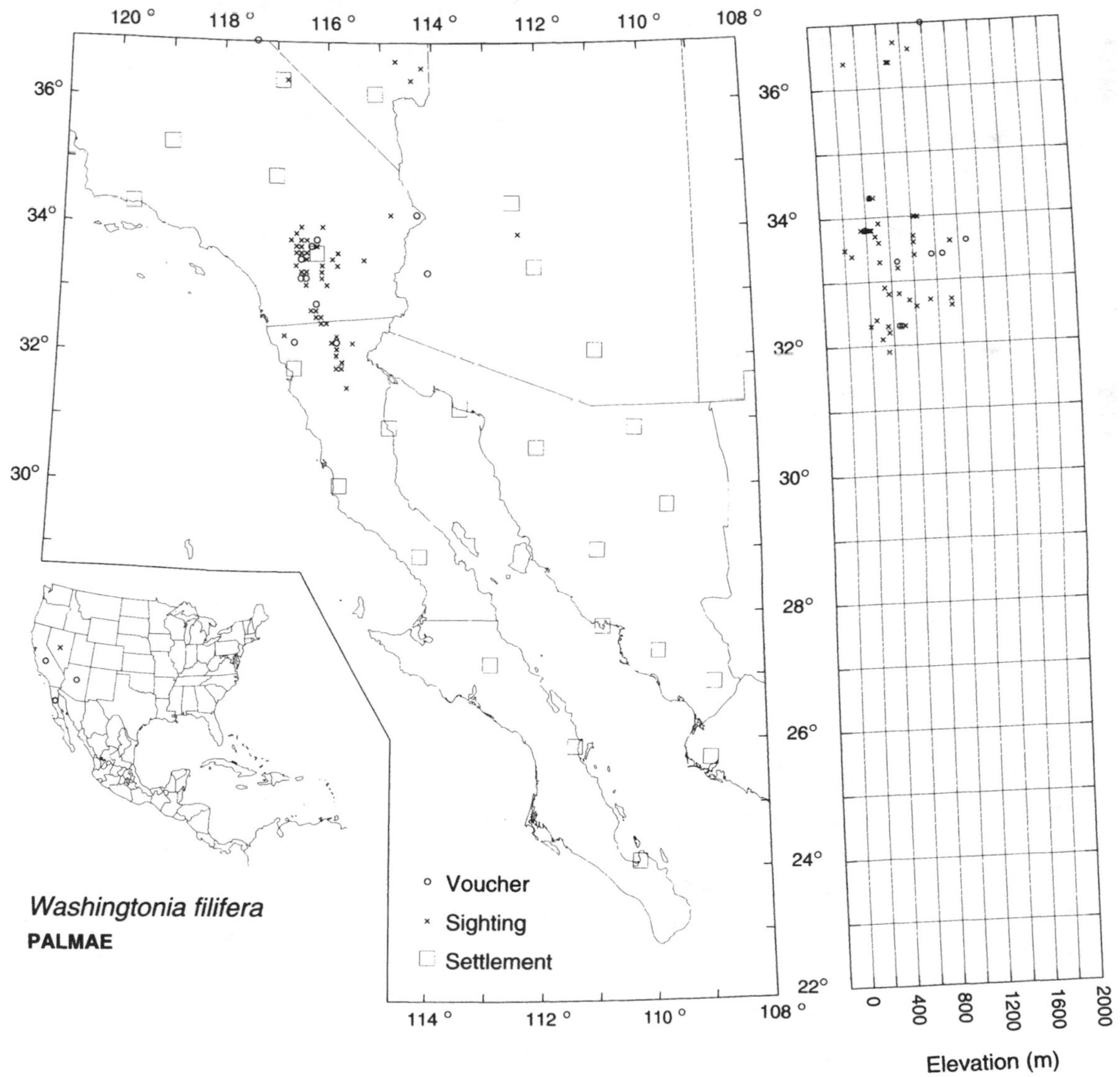

Basin to maintain their present levels (Vogl and McHargue 1966).

Using the fire history of individual palm groves, Vogl and McHargue (1966) estimated a maximum age of 200 years for *W. filifera* in southeastern California. Most plants die before reaching 150 years old. Although evidence of fire is prominent in most *W. filifera* groves, the plants are able to survive repeated burning of the thatch and appear to be adapted to fire (Vogl and McHargue 1966). By suppressing the growth of woody competitors and by liberating nutrients in the litter, fire may actually benefit the palms.

The most widely planted of all palms, *W. filifera* has been known to horticulturists since 1879 (McCurrach 1960). It is a common ornamental in Arizona and southern California. The plants are susceptible to bud rot (Duffield and Jones 1981). Where annual rainfall exceeds 180 mm, no supplemental irrigation is required, although fastest growth is obtained with generous feeding and watering (Duffield and Jones 1981).

The Cahuilla Indians of southeastern California made extensive use of *W. filifera,* employing the leaves as thatch for dwellings and as fiber in rope making and basketry, eating the fresh and preserved fruits, grinding the seeds into a meal, and baking and eating the terminal buds (Parish 1907; Bailey 1936).

Willardia mexicana

(S. Watson) Rose

Nesco, palo piojo, taliste

Willardia mexicana, a small tree 6–8 m tall, has smooth gray bark. Usually unbranched toward the lower half, the trunk branches profusely above to form a dense round crown (Gentry 1942). The once-compound, minutely pubescent leaves are about 1 dm long and are composed of 5–15 leaflets, each 2–4 cm long. Their margins are curled under. The deep blue or lilac flowers, 10–15 mm long, are in axillary racemes. The pods, 3–6 cm long, are constricted between the seeds.

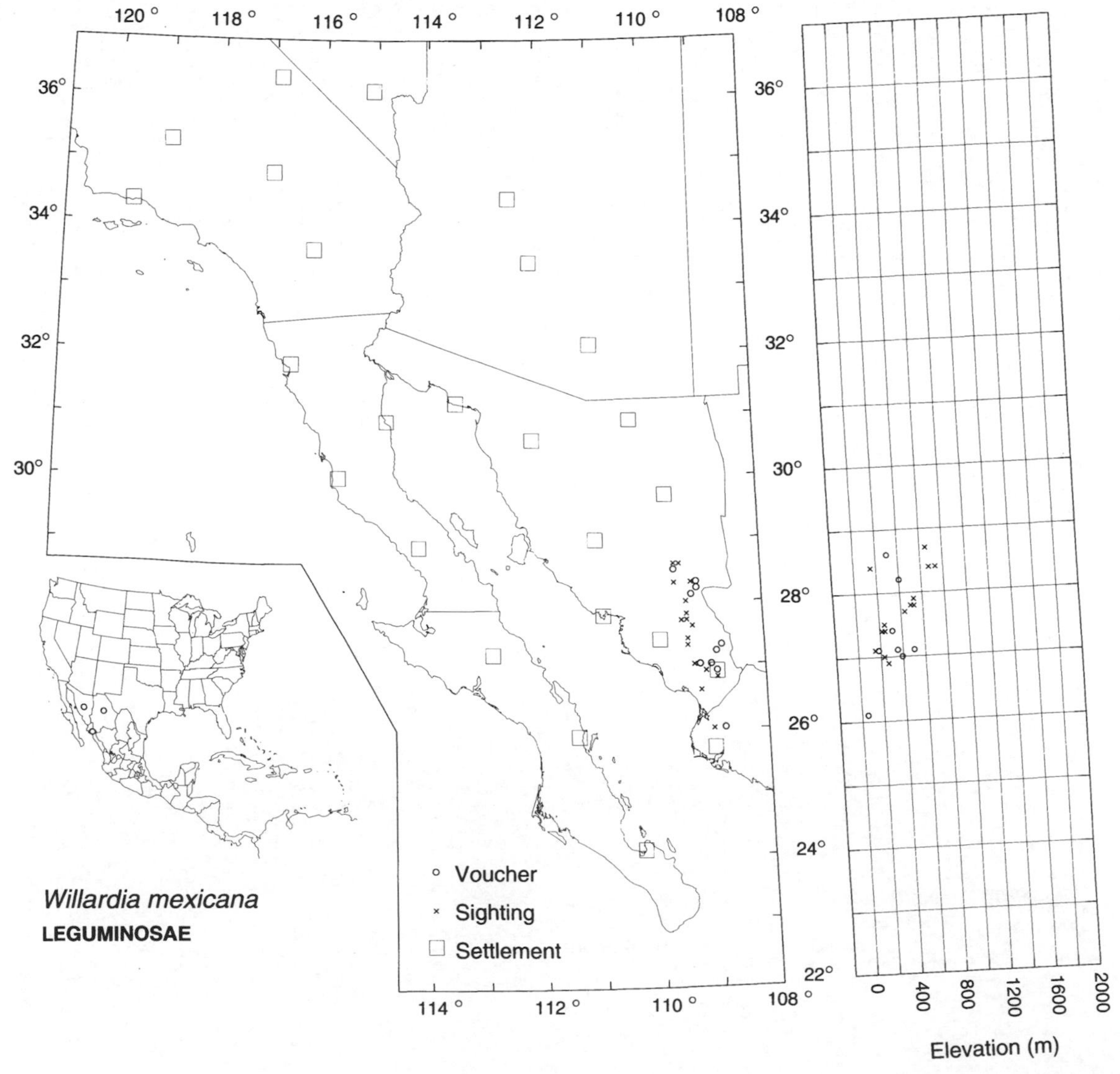

Willardia mexicana
LEGUMINOSAE

This distinctive species is not likely to be confused with other trees in the Sonoran Desert region.

This tree occurs mainly east of the Sonoran Desert, from Tónichi, Sonora, and nearby Chihuahua southward into Sinaloa. It extends into the desert along drainageways. The western limit is probably set by moisture supply, the northern limit by temperature.

The flowers appear in May and June at the height of the dry season.

The wood is used for support beams, and a decoction of the bark is used to kill parasites on livestock (Standley 1922).

Xylonagra arborea

Donn. Smith & Rose

This shrub up to 1.7 m tall has short, ascending branches and alternate leaves 3–10 mm wide and up to 2.5 cm long. The red, tubular flowers, 3–4 cm long, are in terminal racemes. Fruits are dehiscent, 4-celled capsules 10–15 mm long. The haploid number of chromosomes is 7 (Kurabayashi et al. 1962).

Donnell Smith and Rose (1913) took *X. arborea* out of the genus *Hauya* because *Xylonagra* deserved generic status based on its habit and floral characters. At the same time, they retained *Xylonagra* in the tribe Hauyeae. Several decades later, Raven and Lewis (1960), based on anatomical differences and chromosome counts of *Xylonagra* (n=7) and *Hauya* (n=10), argued that the two genera are not closely related and that Watson (1873) was correct in suggesting that the affinities of *Xylonagra* are with *Oenothera* subgenus *Chylismia*. *Xylonagra* is apparently closely related to other Onagraceae that have a basic chromosome number of 7, including the genera *Oenothera, Clarkia, Gayophytum,* and *Gaura* (Raven 1962). Two subspecies of *X. arborea* have been recognized: foliage of subsp. *arborea* has both simple and glandular pubescence, that of subsp. *wigginsii* Munz has nonglandular hairs only (Munz 1960).

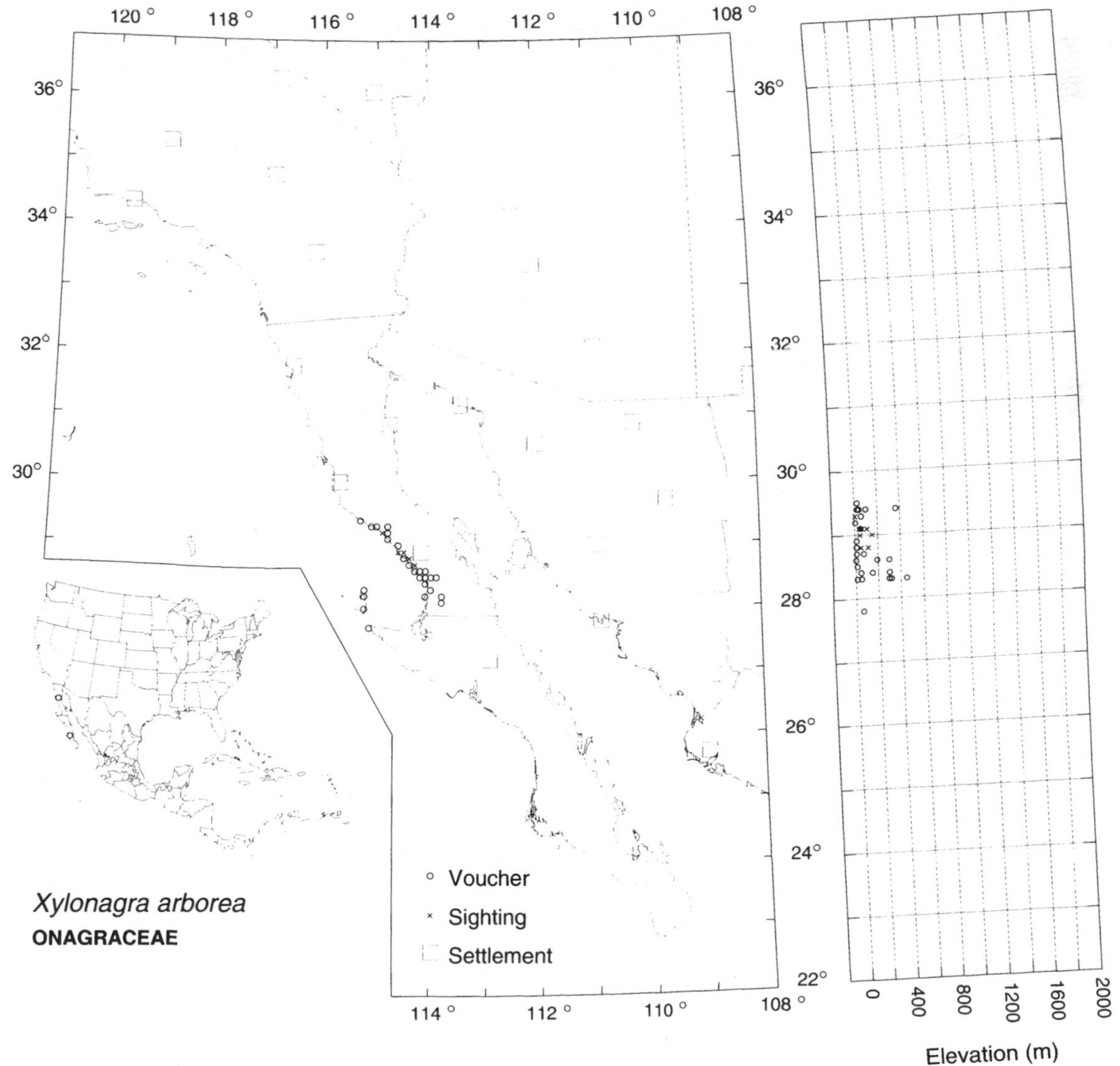

Xylonagra arborea
ONAGRACEAE

Xylonagra arborea looks much like *Zauschneria californica* Presl, another red-flowered Onagraceae, but their ranges do not overlap and they belong to different tribes within the family. *Zauschneria,* which has a basic chromosome number of 15, may belong in *Epilobium* (Lewis and Raven 1961; Raven 1976). The similarities between *Xylonagra* and *Zauschneria* are probably a result of convergent evolution, specifically, selection for and by hummingbird pollinators.

A peninsular endemic, *X. arborea* grows on gravelly mesas, rocky slopes, sandy plains, and arroyos. Subspecies *arborea* occurs only on Isla Cedros. Subspecies *wigginsii* is largely restricted to the mainland coast opposite. Our map combines them. The species is confined to areas where summers are very dry and relatively cool.

Variety *arborea* blooms January through August, var. *wigginsii* February through October (Wiggins 1980). Both are undoubtedly pollinated mainly by hummingbirds. The plants usually flower when leafless, using reserves stored in the tuberous roots.

Xylothamnia diffusa

(Benth.) Nesom

[=*Ericameria diffusa* Benth., *Haplopappus sonoriensis* (A. Gray) S. F. Blake]

This diffusely branched shrub with slender, leafy stems reaches 2.5 m in height. The strongly revolute leaves, 0.5 mm wide and 5–30 mm long, are filiform and nearly round in cross section. The foliage is resinous with punctate glands. The small (4–4.5 mm tall), turbinate heads are in paniculate clusters. Ray flowers are generally lacking. The haploid chromosome num-

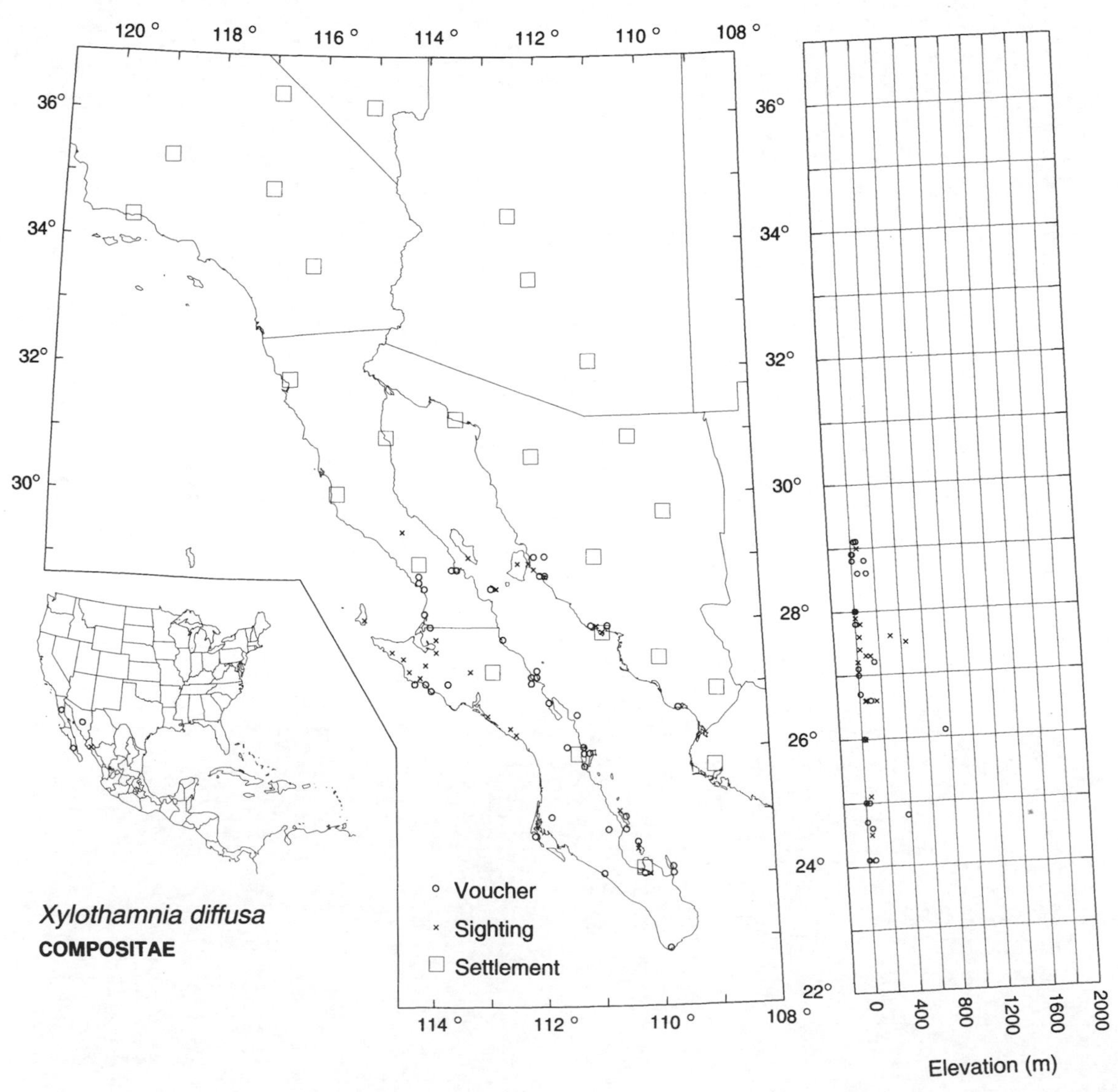

ber is 18 (Pinkava and Keil 1977).

Nesom and coworkers (1990) segregated the genus *Xylothamnia* from *Ericameria* on the basis of floral characters and chloroplast DNA. *Xylothamnia diffusa* is the only species in this small genus found in western Mexico; the others are distributed primarily in the Chihuahuan Desert (Nesom et al. 1990).

A plant of sandy plains, arroyos, and dunes, *X. diffusa* tolerates slightly alkaline soil and is seldom found far from the coast (Hall 1928). Most populations occur below 200 m; the collection at 750 m by Annetta Carter from Valle de los Encinos in the Sierra de la Giganta is a notable exception. Although fairly widespread on the southern peninsula, the species has only a limited distribution on the Sonoran coast. Climates range from those with almost entirely winter rainfall and relatively cool, dry summers in the northwestern part of the range to those with mostly summer rainfall in the south. Frost is essentially absent throughout the range.

The leaves are evergreen. Flowers appear March–May and August–December (Wiggins 1964), apparently in response to rain.

Yucca arizonica

McKelvey

[= *Y. baccata* Torr. var. *brevifolia* (Schott) Benson & Darrow]

Spanish dagger, palma criolla, dátil

This variable species typically grows in large, crowded clumps of stems up to 2.5 m tall (figure 83). The usually unbranched trunk is hidden by the thatch of dead, deflexed leaves. Young plants may appear acaulescent. The yellowish green to bluish green leaves are 3–8 dm long, 12–40 mm wide. Young leaves have fine marginal fibers that are lacking on older ones.

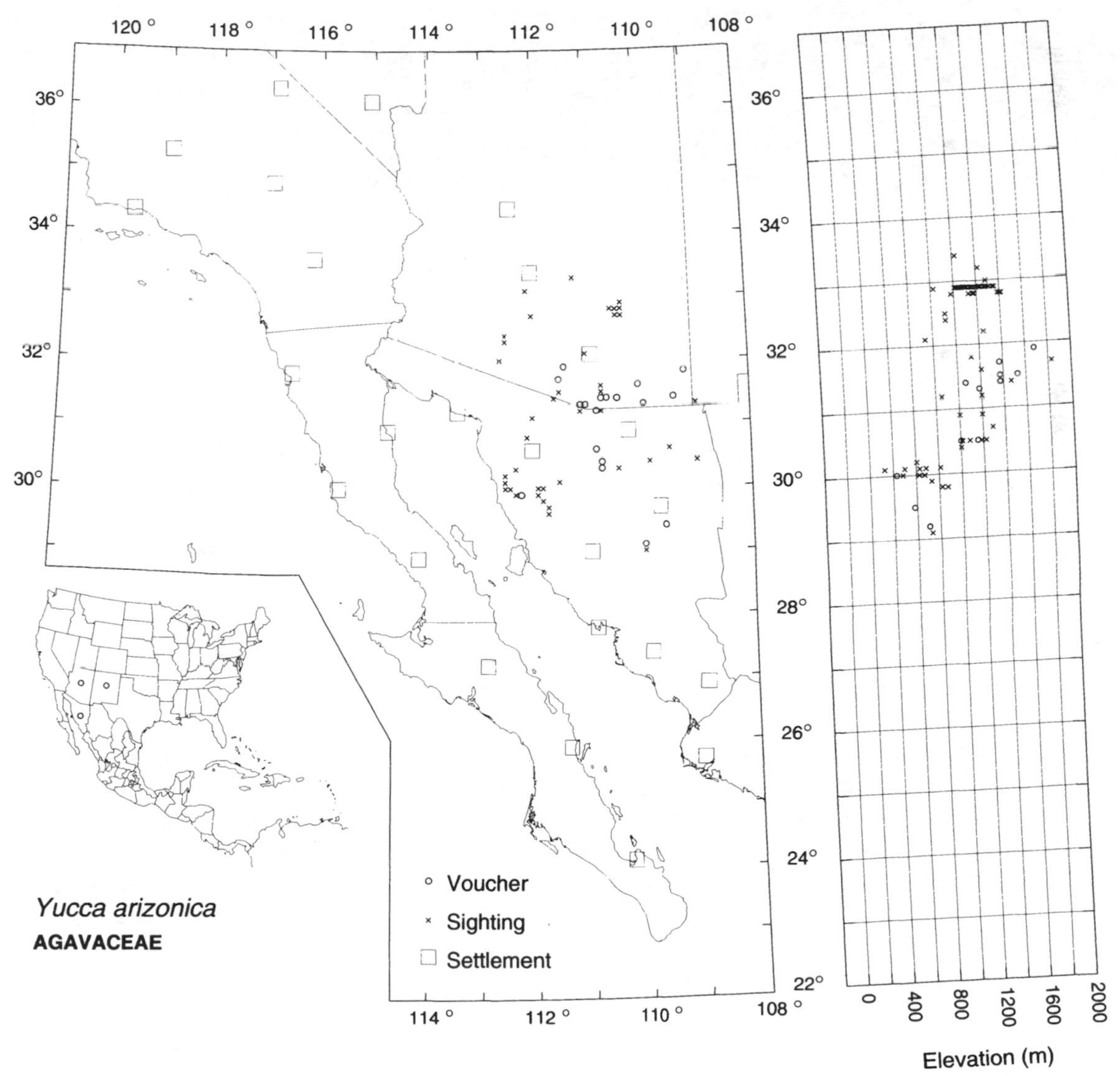

Yucca arizonica
AGAVACEAE

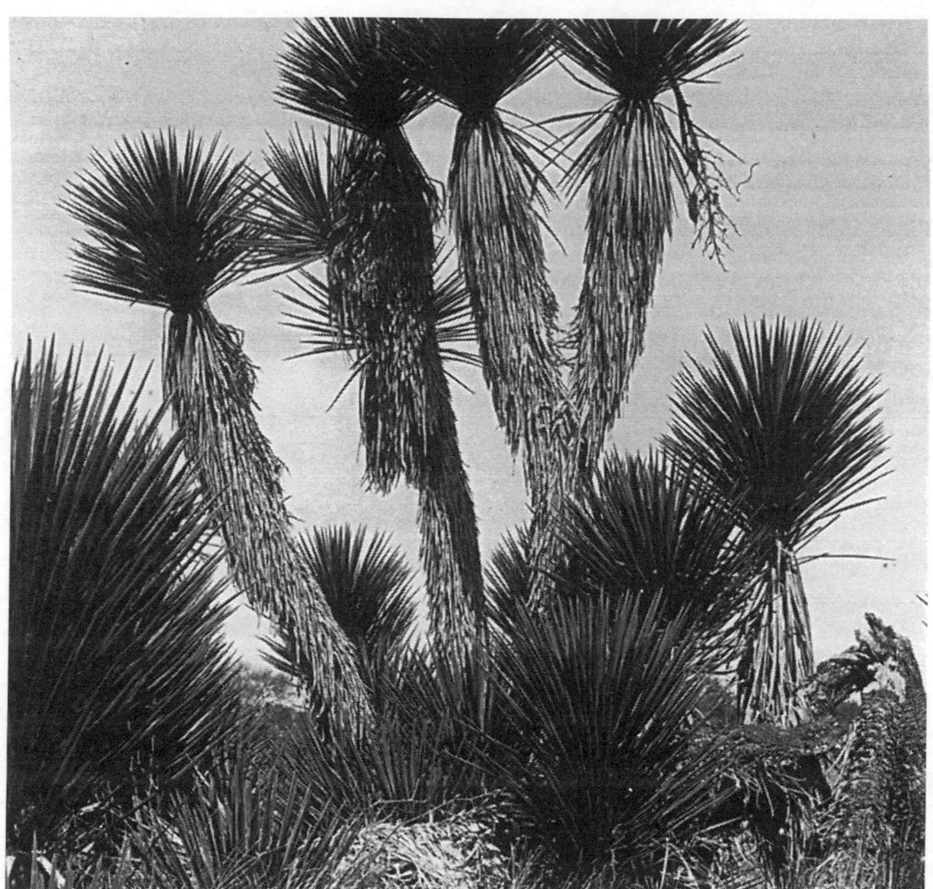

Figure 83. Yucca arizonica *near La Unión, Sonora. (Photograph by H. L. Shantz.)*

All are sharply pointed with a terminal spine 7–15 mm long. The inflorescence, 4–15 dm high, has pendent white flowers 6–13 cm long. The fleshy fruits, 8–20 cm long, are indehiscent.

Yucca baccata is acaulescent or has short, creeping stems. Its leaves are persistently and coarsely fibrous and often blue-green in color. *Yucca thornberi* McKelvey leaves lack marginal fibers when young but become markedly fibrous in old age. Its stems are usually less than 1.5 m high, and its leaves are up to 1.3 m long. *Yucca schottii* Engelm., a plant of oak woodland and pine forest, has more flexible leaves up to 6.5 cm wide. They may have a few marginal fibers or lack them entirely. *Yucca grandiflora* Gentry, a Sonoran plant, branches in the crown, and its leaves are wider (40–50 mm), thinner, and more flexible.

Benson and Darrow (1981) treated *Y. arizonica* (along with *Y. thornberi*) as *Y. baccata* var. *brevifolia.* We follow McKelvey (1938), Webber (1953), Wiggins (1964), and Gentry (1972) in maintaining it as a separate species, while recognizing that hybridization might make it difficult to identify individual specimens with any confidence. Hybridization may account for much of the notorious variability in the genus (Webber 1960). *Yucca arizonica* is known to produce fertile hybrids when artificially crossed with *Y. neomexicana* Woot. & Standl. and *Y. glauca* Nutt. (Webber 1960), and it apparently hybridizes naturally with *Y. grandiflora* (Gentry 1972) and *Y. baccata* (Webber 1953), as well. *Yucca thornberi* may have originated as a hybrid between *Y. baccata* and *Y. arizonica* (Webber 1953). Some of our sightings in the north may represent *Y. thornberi;* the delineation between *Y. arizonica* and *Y. thornberi* is not precise.

Rare to common along washes and on slopes and plains, *Y. arizonica* grows in Sonoran desertscrub at its lower elevational limit and Madrean evergreen woodland at its upper elevational limit. The populations are widely scattered, as McKelvey (1938) noted. As interpreted in this atlas, *Y. arizonica* occurs in an area where rainfall is biseasonal with a pronounced summer maximum. Its western outposts mark the arid limits of fleshy-fruited yuccas east of the Gulf of California. Along its northern limit it is mostly replaced by *Y. thornberi,* which seems to wedge between *Y. arizonica* and *Y. baccata* farther north and higher up. Along its upper elevational limits *Y. arizonica* often abuts *Y. schottii,* which is centered in denser woodland vegetation. Along its southeastern limit, *Y. arizonica* may grade into *Y. grandiflora.* Gentry (1972) noted that in eastern Sonora a yucca population exhibits characters of both species.

As in other species of *Yucca* (Engelmann 1872; Riley 1881, 1892, 1893; Powell 1992), *Tegeticula yuccasella,* a prodoxid moth species complex, is probably the sole pollinator (McKelvey 1947). The female moth collects pollen from anthers on one flower, then rubs the pollen into the stigmas of another flower after depositing her eggs in its ovary. The developing larvae feed on the seeds (their only food source). Seed loss to larvae is highly variable (Addicott 1986); substantial losses have been anecdotally reported for *Y. arizonica* (Gentry 1972).

A given plant does not flower every year (McKelvey 1938; Gentry 1972), a fact that Gentry ascribed to rainfall variability. McKelvey (1938) noted that, in addition, entire populations failed to flower following a year of drought. In the closely related *Y. baccata,* three years elapse between inflorescences, this being the period of time required to build up sufficient carbon and water reserves (Wallen and Ludwig 1978). A similar requirement may also account for sporadic flowering in *Y. arizonica.* Fruit production may be poor even in years when the plants flower abundantly (McKelvey 1938). Since the flowers are obligately outcrossing (Webber 1953), a paucity of pollinators could limit fruit production in some years.

Bell and Castetter (1941) reported use of this species, under the name *Y. baccata,* by the Tohono O'odham (formerly called Papago Indians). The Tohono O'odham dried fresh fruits for later consumption; macerated whole plants for use as shampoo; employed leaf fibers in making baskets, mats, and cordage; and used root fibers in making baskets.

Yucca baccata

Torrey

Banana yucca, blue yucca, Spanish dagger, dátil

Rosettes of this variable species may be single or clumped, acaulescent, or with a short, creeping stem usually less than 1 m long. The leaves, 5–7.5 dm long and 2.5–4 cm wide (5–6 cm when flattened), are often twisted and falcate and may be bluish or yellow-green. Their coarse, flattened marginal fibers are persistent. The flowers, 5–15 cm long, are in dense inflorescences up to 7.5 dm tall. Fruits are fleshy and up to 17 cm in length.

Yucca arizonica reaches 2.5 m in height and typically grows in large, crowded clumps. Leaves of *Y. thornberi* McKelvey are longer and somewhat narrower, and they lack marginal fibers when young. *Yucca schottii* Engelm. leaves usually lack marginal fibers entirely. *Yucca schottii* stems are much longer.

We follow McKelvey (1938) rather than Benson and Darrow (1954) in our treatment of this confusing and often difficult species group. Two rather poorly defined varieties have been recognized: var. *baccata,* which occurs throughout the range of the species and grows in clumps of one to six individuals; and var. *vespertina* McKelvey of northern and central Arizona, which grows in large, caespitose

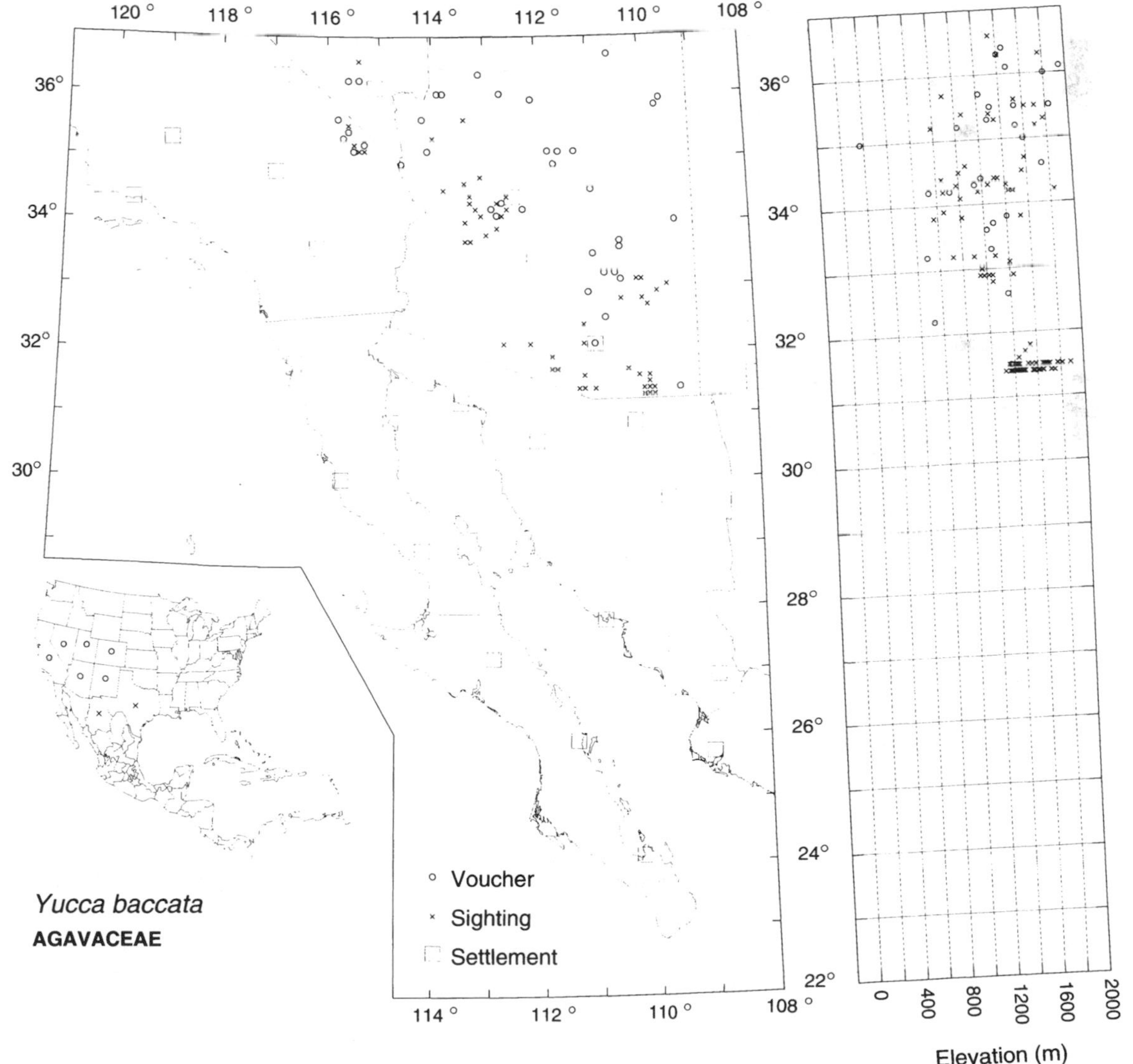

Yucca baccata
AGAVACEAE

clumps from a procumbent stem. *Yucca thornberi* may represent a hybrid between this species and *Y. arizonica* (Webber 1953). Most of our southern sightings probably represent populations of *Y. thornberi*. Additional research is needed to define and delimit ecological entities in this species complex.

Occasional to common on slopes and plains and in washes, *Y. baccata* occurs from 700 to 2,285 m. A collection from an exceptionally low elevation was made at 165 m by John G. Lemmon at Fort Mohave on the Colorado River (35.0°N, 114.6°W). Most typical, perhaps, of pinyon-juniper woodland, this species enters the Sonoran Desert in southern Arizona, where it occurs with *Carnegiea gigantea, Larrea tridentata,* and similar Sonoran Desert plants. We know of no Sonoran locations, nor did Gentry (1972) list *Y. baccata* from that state.

Throughout its range, rainfall is biseasonal, shifting from mainly summer in the east to mainly winter in the west. Yeaton and coworkers (1985) found evidence that its lower elevational limit in California is determined by competitive interactions with *Y. schidigera,* which replaces it where summers are drier. Apparent introgression between these two species occurs near Kingman, Arizona (Hanson et al. 1989). To a significant extent the limits of *Y. baccata* are determined by biological interactions with other fleshy-fruited yuccas.

Given adequate rainfall, *Y. baccata* produces new leaves in spring and fall (Wallen and Ludwig 1978). April to June is the season of peak flowering, although occasional individuals may bloom as early as March or as late as October. Ten weeks elapse between the appearance of flower buds and the production of mature fruit (Wallen and Ludwig 1978). The prodoxid moth *Tegeticula yuccasella,* a species complex (Powell 1992), is the primary pollinator of this and many other *Yucca* species, as described by Engelmann (1872), Riley (1881, 1892, 1893), and others.

Seed loss to larvae (which feed on the developing seeds) varies: one study found that only 10–12% of the seeds in a capsule were destroyed (Keeley et al. 1984); another reported losses of 27% (Wallen and Ludwig 1978). Sometimes no larvae develop in a fruit (Addicott 1986). Although viable seed is produced in abundance, seedlings are rare in the wild due to grazing by herbivores (particularly rodents and lagomorphs) and as a result of sporadic and scanty rainfall (Webber 1953). Nabhan (1989) suggested that the edible, nutritious fruits of the baccate *Yucca* species evolved for dispersal by Pleistocene megafauna. Currently, rodents (especially *Neotoma* species) and lagomorphs are the primary dispersal agents (Wallen and Ludwig 1978). Exposure to high temperature reduces seed viability (Keeley and Meyers 1985), and seedling establishment after fires is probably delayed.

In a given population, the number of reproductive plants varies from year to year, apparently in response to climatic variability (Wallen and Ludwig 1978). Flowering is initiated in spring only when several conditions are met: (1) the carbon storage pool exceeds 10% of the plant dry weight, (2) the water storage pool exceeds 75% of potential, (3) the photoperiod equals or exceeds 11.25 hours, and (4) mean night temperatures exceed 3°C (Wallen and Ludwig 1978). Once an individual plant produces flowers, about three years are required for carbon reserves to build up to the threshold level again (Wallen and Ludwig 1978).

Schaffer and Schaffer (1977a, 1979) contrasted the reproductive strategies of yuccas and agaves, making much of the former's polycarpic growth form (reproducing many times over its life span) versus the latter's monocarpic habit (reproducing once, then dying). This distinction seems to be based upon a misunderstanding of yucca morphology. In *Y. elata* (and presumably in all *Yucca* species except *Y. whipplei*), what appears to be an individual rosette is actually a compound rosette comprising an apical and several lateral meristems and their surrounding leaves (Steven P. McLaughlin, personal communication 1987). Each meristem can bloom only once. Afterwards, it eventually dies. The meristems in a compound rosette seldom bloom simultaneously. Thus, as the meristems produce flower stalks in successive years, the plant gives an impression of polycarpic reproduction, but since every meristem dies after reproducing, the separate meristems are essentially monocarpic. The major difference between yuccas and agaves, therefore, is not one of reproductive habit but one of morphology: the multiple rosettes of *Yucca* species generally occur in clusters on aboveground stems, while those of some *Agave* species are produced individually from underground rootstocks. As mentioned, yucca rosettes persist for some time after flowering whereas agave rosettes die shortly thereafter.

Yucca baccata is a CAM plant, taking up carbon dioxide in the dark and, when well watered, in the light also (Kemp and Gardetto 1982). As in other *Yucca* species, photosynthesis is sensitive to water stress: in one study, after two weeks without water, greenhouse plants showed a leaf water potential of –2.1 MPa and ceased virtually all carbon dioxide exchange (Kemp and Gardetto 1982). This reaction, although not characteristic of the more succulent CAM plants, is not unexpected in *Y. baccata,* which has thin leaves and low water storage capacity. Sensitivity to water stress restricts carbon dioxide uptake to seasons when soil moisture is readily available (Kemp and Gardetto 1982).

Although not often grown as ornamentals, the plants are easily propagated from seed or basal sprouts (Webber 1953). Native Americans throughout the greater Southwest made extensive use of *Y. baccata.* The nutritious, flavorful fruits were prepared in a variety of ways and were often harvested while immature to avoid competition with insects, rodents, birds, and deer (Bell and Castetter 1941). The leaf fibers were used in making sandals, baskets, mats, cords, fabric, and netting. Leaves, whole plants, or roots were macerated and used for shampoo (Bell and Castetter 1941).

Yucca schidigera

Roezl ex Ortgies

[=*Y. mohavensis* Sarg.]

Mojave yucca, Spanish dagger, dátil

This robust, often clumped, shrub or tree grows 1–5 m tall (or occasionally up to 9 m tall [Munz 1973]). The trunks, thatched with dead, reflexed leaves, may be simple or branched above. Yellow green to dark green, the sturdy leaves are 30–60 cm long (occasionally up to 150 cm long) and 3–4 cm wide. A sharp spine terminates each leaf, and the margins separate into coarse fibers. The creamy white or purplish flowers, 2.5–4.5 cm long, are clustered on a sessile or short-stalked panicle 4–8 dm high. The fruits, 5–10 cm long, are fleshy and indehiscent.

Where *Y. schidigera* and *Y. baccata* overlap, the latter is acaulescent and generally has bluish green leaves. *Yucca brevifolia* Engelm. and *Y. valida* are highly branched trees with relatively short leaves. *Yucca whipplei* is acaulescent and has much narrower leaves.

An unnamed, blue-leaved form with narrow, falcate leaves occurs in the southern Mojave Desert (Webber 1953). Apparent hybrids with *Y. baccata* var. *vespertina* are known (Webber 1953).

Locally common on rocky slopes and along sandy washes, *Y. schidigera* occurs with *Larrea tridentata, Atriplex* species, and *Ambrosia dumosa* in the Mojave Desert (Beatley 1976). The species reaches its upper elevational limits in pinyon-juniper woodland and filters westward to the Pacific coast through Californian chaparral communities. It extends southward into Baja California along the west base of the Sierra Juárez, again in Californian chaparral. The species enters the Sonoran Desert in Baja California in the vicinity of La Ramona (29.8°N, 115.5°W), where it grows under a winter rainfall regime with *Fouquieria columnaris, Ambrosia dumosa, Larrea tridentata,* and *Simmondsia chinensis.* Its southern limit is almost contiguous with the northern limit of *Y. valida.* The distribution suggests a species toler-

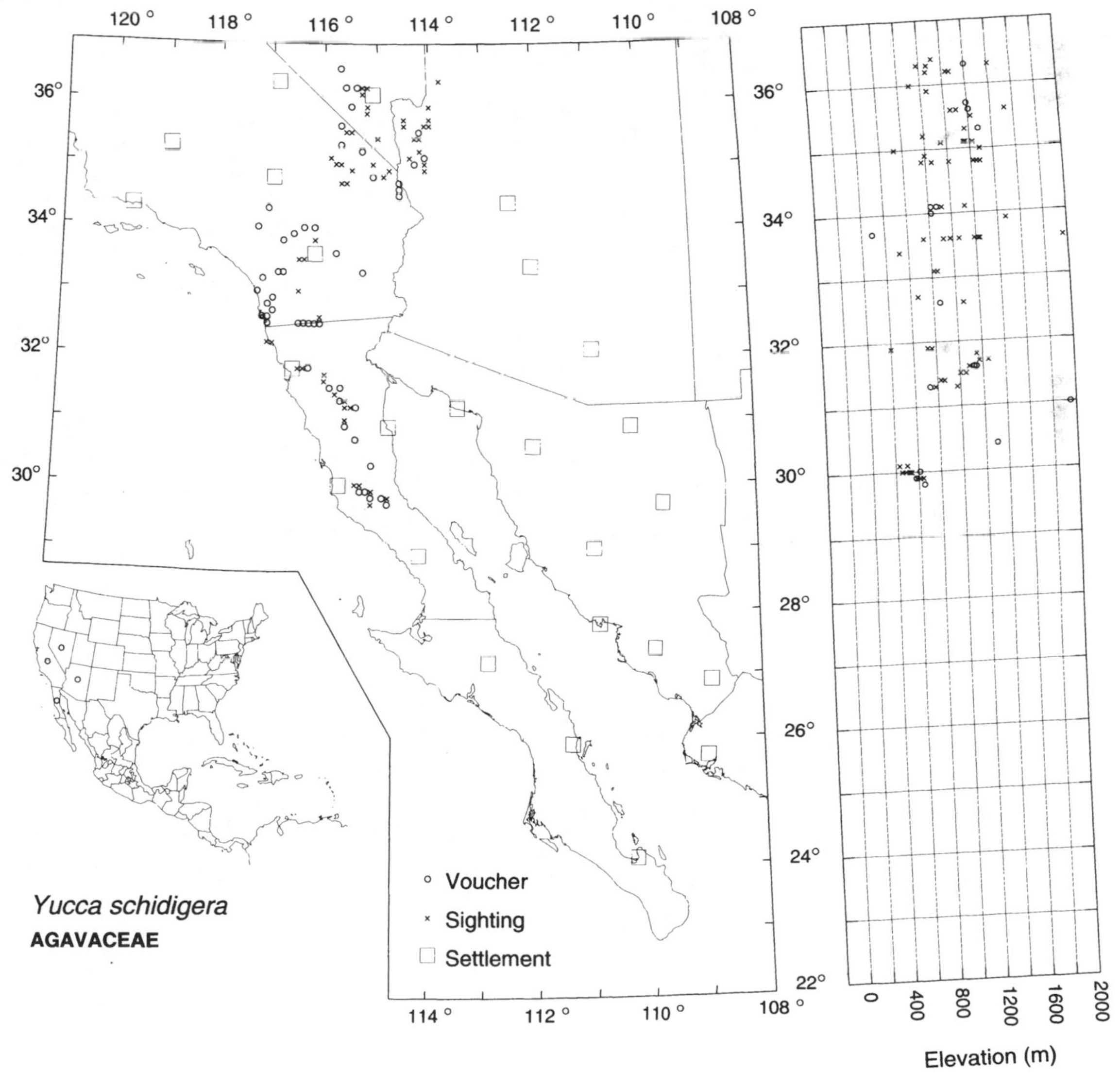

Yucca schidigera
AGAVACEAE

ant of low temperatures and dry summers and also suggests a need for cool-season rain. In the eastern Mojave Desert, the population density of *Y. schidigera* is highest near its upper elevational limits, where it is replaced by *Y. baccata* (Yeaton et al. 1985). McKelvey (1938) noted that these two species are often associated. Apparently the range of *Y. schidigera* is partly determined by interactions with other fleshy-fruited *Yucca* species.

Brief measurements in one study showed a physiology similar to that of *Y. baccata* (Yeaton et al. 1985). Photosynthesis via crassulacean acid metabolism was indicated by high internal carbon dioxide concentrations (Cockburn et al. 1979) and by heavy isotope enrichment of water within the leaves (Cooper and DeNiro 1989).

Plants flower from February to June. Their sole pollinator is the prodoxid moth *Tegeticula yuccasella,* a species complex, which also pollinates many other *Yucca* species in the United States (Powell and Mackie 1966; Powell 1992). The symbiosis between yuccas and *Tegeticula* species has been known for many years (Engelmann 1872; Riley 1881, 1892, 1893). After gathering a ball of pollen from the anthers of one flower, the female moth flies to another plant, deposits eggs in the ovary of a single flower, then rubs pollen into the stigmas of that flower. The developing larvae feed on the seeds, their only food source, but do not destroy the entire crop; in *Y. schidigera,* only 3–5% of the seeds in a capsule are eaten (Keeley et al. 1984).

In most populations, some plants flower every year, but only rarely do the majority flower in the same year (Beatley 1976). *Yucca baccata* needs about three years to build up sufficient food reserves for flowering (Wallen and Ludwig 1978); a similar requirement may also account for sporadic flowering in *Y. schidigera.* Fruit production is also variable from year to year (McKelvey 1938), which may be a result of fluctuations in *T. yuccasella* populations, as described for *Y. whipplei* and its prodoxid pollinator (Aker 1982). Alternatively, since a proportion of the flowers are self-fertile (Webber 1953), resource limitations may account for poor fruiting in some years. Seeds briefly exposed to high temperatures in one study had reduced germinability, indicating that fires probably retard seedling establishment (Keeley and Meyers 1985).

Yucca schidigera makes an attractive ornamental. The plants are easily propagated if the gardener cuts off and roots young sprouts (Webber 1953). Native Americans of the Mojave Desert made extensive use of this species. The fruits were eaten fresh and dry; the stems were cut, pulped, and employed as soap; and the leaf fibers were fashioned into ropes, twine, nets, hats, hairbrushes, shoes, mattresses, and saddle blankets (McKelvey 1947). Saponins extracted from the crushed stems are sold commercially for livestock odor control (Johnston et al. 1981; Headon and Killeen 1993).

Figure 84. Yucca valida *south of Laguna Chapala, Baja California. (Photograph by J. R. Hastings.)*

Yucca valida

Brandegee

Datilillo, dátil cimarrón

A straggling tree 3–7 (10) m tall, *Yucca valida* (figure 84) may branch at the base or at any point along the slender trunk. Older plants develop an enlarged base from which new shoots may arise. The sharply pointed, filiferous leaves, 15–35 cm long and 1.5–2.5 cm wide, become deflexed in age and thatch the stem. The white or purplish flowers, 2.5–4.5 cm long, are in short, crowded panicles 4–6 dm tall. They are reported to smell faintly of dill. The fruits, 5–10 cm long, are indehiscent.

Yucca brevifolia, also an arborescent, much-branched yucca, is limited to the Mojave Desert and the adjacent edge of the Sonoran Desert. *Yucca schidigera* has shorter stems, fewer branches, and somewhat longer leaves.

A dominant on fine-textured soils of wide valleys and gentle slopes, *Y. valida* often forms "forests" with an understory of cacti and low shrubs (Shreve 1964; Humphrey 1974). In the northern Vizcaíno Plain, it is by far the most frequent large perennial (Shreve 1964). Plants in the Cape Region have somewhat larger leaves but are not considered different enough to warrant taxonomic segregation. Over its latitudinal range, climates vary from arid with mainly winter rain to semiarid with mostly warm-season rain. This broad climatic range is unusual for a yucca. In Sonora, an equivalent climatic range has been split among several species (Gentry 1972). Probably its peninsular isolation has allowed *Y. valida* to radiate along a more extensive gradient than if other species had been present. This broad climatic adaptation suggests that the species is restricted to lower elevations because of a lack of suitable substrate at higher elevations.

During the upper Tertiary and Pleistocene, several seaways evidently divided the Baja peninsula into discrete land masses (Durham and Allison 1960; Gentry 1978; Gastil et al. 1983). The present-day populations of *Y. valida* form distinct

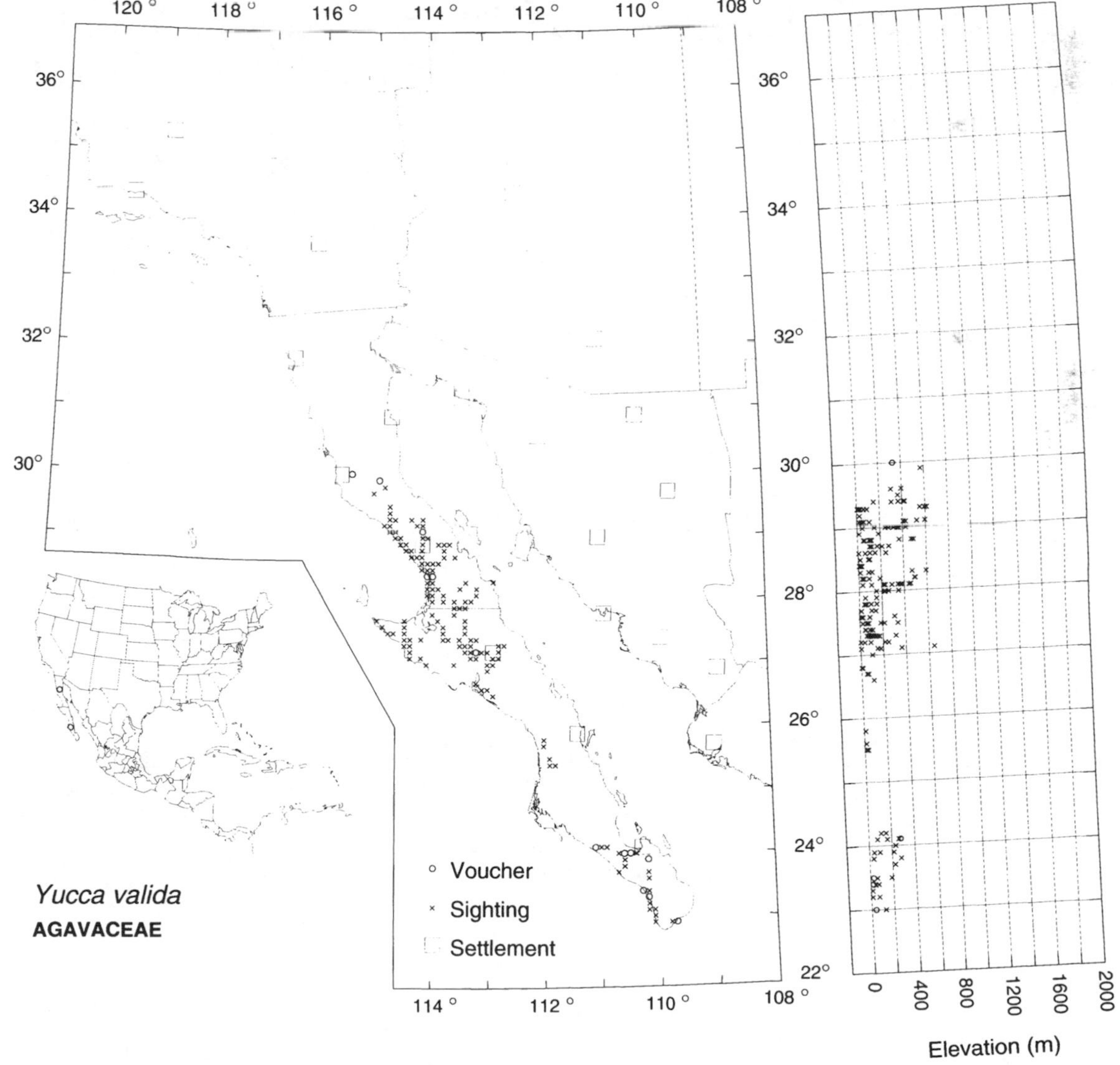

groups that roughly correspond to three of these ancient islands: an extensive range centering north of the Vizcaíno Plain and two smaller aggregations on the Magdalena Plain and in the Cape Region. Possibly these three groups have been separated since the upper Tertiary or early Pleistocene. Alternatively, the geographic pattern may have formed during climatic shifts in the Holocene.

Mycorrhizal fungi in the roots (Rose 1981) may enhance phosphate uptake.

The plants flower in spring (April and May) or fall (September through November), rarely in summer. Fruits of spring-flowering plants mature in September and October. The prodoxid moth *Tegeticula maculata* is presumably the only pollinator.

Native peoples of the peninsula harvested and ate the fruits, which resemble little dates (thus the Spanish common name *datilillo*) (Aschmann 1959). The trunks, although soft and not very durable, are used locally for fence posts and corrals (Humphrey 1974) and are occasionally cut for saponin extraction.

Yucca whipplei

Torrey

[=*Y. newberryi* McKelvey, *Y. peninsularis* McKelvey]

Our Lord's candle, chaparral yucca, quiote, lechuguilla

This rosette plant has silvery-glaucous leaves 2.5–7 dm long and 7–40 mm wide (figure 85). They are sharply spine tipped and denticulate along the margins. The pendent, white flowers, 3.5–5 cm long, are in dense panicles up to 4.5 m tall. Fruits are woody capsules 3–4 cm long and 2.5 cm in diameter. The haploid chromosome

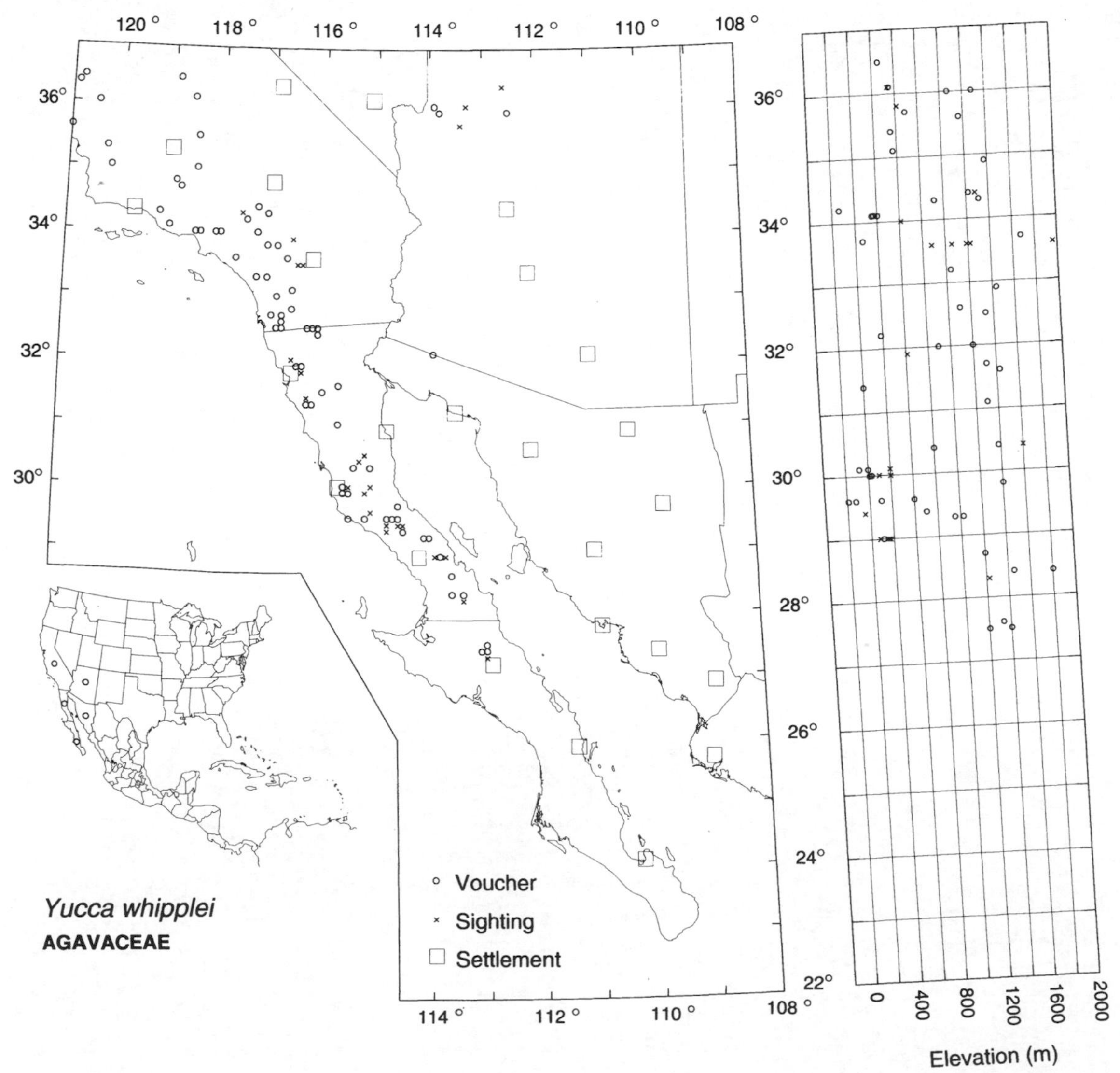

number is 30 (McKelvey and Sax 1933). Diggle and DeMason (1983a, 1983b) described the anatomy of the vegetative stem.

The basal rosette of silvery leaves and the massive flowering stalk are distinctive. Moreover, no other yucca in its range combines an acaulescent habit with leaves that lack marginal fibers.

Haines (1941) recognized five varieties based largely on growth form, whether single rosette, multiple rosettes produced by branching, or multiple rosettes produced by rhizomes. Two of these (vars. *whipplei* and *parishii* [Jones] Haines) are monocarpic, producing a single rosette of leaves that dies after flowering. The rest (vars. *caespitosa* [Jones] Haines, *percursa* Haines, and *intermedia* Haines) are polycarpic: because several rosettes are produced, the plant survives after each flowering rosette dies. Recent work tends to call into question the validity of these varietal distinctions. Wild populations often contain a mixture of monocarpic, caespitose, and rhizomatous plants (Keeley and Tufenkian 1984); even more telling, seeds from a single capsule of var. *percursa* produced all possible growth forms (DeMason 1984). Plants in the Vizcaíno region of Baja California usually produce clumps of rosettes by branching. Vizcaíno populations were segregated as *Y. peninsularis* McKelvey on the basis of their shorter, broader leaves (McKelvey 1947).

Figure 85. Yucca whipplei *east of Bahía de los Angeles, Baja California. (Photograph by J. R. Hastings.)*

Yucca whipplei occurs more or less continuously in mountainous areas from southwestern California to northern Baja California, spanning a range of plant communities from Californian chaparral and Californian coastalscrub to Sonoran desertscrub. Southern populations are scattered and rather restricted. Habitats include cliff faces, rocky slopes, and coastal bluffs, often in locations where the soil is shallow. The distribution suggests reliance on winter rainfall and tolerance of summer drought and, in the north, freezing temperatures. The presence or absence of its sole pollinator, the prodoxid moth *Tegeticula maculata,* may also affect the distribution of *Y. whipplei,* especially in the desert populations, where the moth may be uncommon (Powell and Mackie 1966; Keeley et al. 1986).

Hastings and coworkers (1972) mapped a disjunct point at Yuma, Arizona, based on an 1894 collection by E. A. Mearns. It seems likely that the voucher in question actually came from Jacumba, California (McKelvey 1947), and we have omitted this location from the present map. Genuine disjuncts include the scatter of points in the western Grand Canyon and the lone population of monocarpic rosettes in the Pinacate region of northwestern Sonora (collected by Tony Burgess). (The Grand Canyon plants, originally described as *Y. newberryi* McKelvey, have since been subsumed under *Y. whipplei* var. *caespitosa* [Benson and Darrow 1981].) That these widely dispersed populations are relicts of a more continuous Pleistocene distribution (as Benson and Darrow [1981] suggested) is borne out by fossil evidence: 13,000–19,000 years B.P., *Y. whipplei* was present in Death Valley, California, widely disjunct from its current range (Wells and Woodcock 1985). Packrat midden deposits record its presence at 285 m near Picacho Peak, California, about 13,000 years B.P., and in the Whipple Mountains, California, from 13,800 to about 8,900 years B.P. (Van Devender 1990b). Fourteen thousand years ago, it was in the Puerto Blanco Mountains of southwestern Arizona, far east of its present range (Van Devender 1990b). It is absent from midden deposits at its present outpost in the lower Grand Canyon; hence, the date of arrival in the area is uncertain (Spaulding 1990).

If the taxonomic significance of the different reproductive forms is obscure, their ecological significance is hardly more clear. Keeley and Tufenkian (1984) found little correlation between growth form and germination requirements or between growth form and biomass allocation. In general, seedlings of all subspecies respond to low moisture and high temperatures by allocating more biomass to corms than to leaves (Keeley and Tufenkian 1984). Casual observation suggests that many more flowering stalks are formed in wet years than in dry ones except in the desert, where the flowering of subsp. *caespitosa* does not appear to be correlated with rainfall (Powell and Mackie 1966). Rather, many plants in desert populations flower every year, perhaps an adaptation to erratic rainfall patterns (Powell and Mackie 1966). In deserts, if anywhere, low pollinator abundance is apt to limit seed set (Powell and Mackie 1966; Aker 1982); but, as Powell and Mackie (1966) noted, population stability in this subspecies is accomplished, despite pollinator scarcity and low percentage of seed set, by the plant's perennating habit. Over the years, a steady supply of flowers is produced. The monocarpic habit of the Sonora population is anomalous.

Across the range of the species, plants flower from February to June (rarely as early as January or as late as July). Six to seven years are required before rosettes accumulate enough food reserves to produce an inflorescence (Wolf 1935). After flowering, the rosette dies. A given population blooms for about 75 days, with individual plants flowering for 15–24 days. Each flower lasts 4–5 days (Powell and Mackie 1966).

The prodoxid moth, *Tegeticula maculata,* is the only pollinator; conversely, seeds of *Y. whipplei* are the sole food source for *T. maculata* larvae (Powell and Mackie 1966). The mechanics of pollination, known for many years (Engelmann 1872; Riley 1881, 1892, 1893), have been concisely described by Aker and Udovic (1981:96): "The female moths use a piercing ovipositor to insert their eggs into the ovaries of open yucca flowers and then actively pollinate the same flowers, using pollen which they carry about with them in a ball held against the thorax. The larvae then hatch out inside the fruits and feed on the developing seeds." Powell and Mackie (1966) and Aker and Udovic (1981) provided additional information on the life history of *T. maculata.* Seed loss to larvae ranges from 2.4 to 24.5% of the seeds in a capsule and is generally less than 10% (Keeley et al. 1986). Germination of fresh seed is high (80–90%), but in greenhouse trials, only 48% of germinated seed produced seedlings (Arnott 1962).

Yucca whipplei consistently produces many more flowers than fruits, and fruit set is usually less than 10% (Udovic 1981). Fruit production is closely correlated with plant size, indicating that the limiting factor in fruit set is the amount of resources available to support developing fruits (Udovic 1981; Aker 1982). In most years and in most locations, pollinators are abundant enough that they do not limit fruit set (Udovic 1981; Aker 1982). In some years, however, pollinators are scarce enough that at least some plants in a population are pollinator-limited (Aker 1982). In either case, why are the inflorescences so large when so few fruits can be supported? One hypothesis is that the floral display is the result of selection for pollen dissemination at the expense of fruit set. In other words, large inflorescences increase the male component of fitness at the expense of the female component (Udovic 1981). A second hypothesis is that the large inflorescences are a "bet-hedging" strategy to either increase the probability of adequate pollination when pollinators are unusually rare or to enable plants to support more fruits when resources are unusually abundant (Udovic 1981; Aker 1982). These hypotheses are not mutually exclusive, and all may be valid explanations of inflorescence size (Udovic 1981).

Schaffer and Schaffer (1977a, 1979) suggested a fourth possibility to account for the large number of flowers relative to fruits: pollinator preference for larger inflorescences selected for the monocarpic growth-form. If pollinators were attracted to stalks with numerous flowers (as they might well be for energetic reasons), they would select for increasingly large flower stalks, with the result being a stalk so large that a rosette would necessarily put all of its food reserves into the production of that single stalk (the so-called big bang) rather than expending a proportion of its energy to produce smaller stalks at widely spaced intervals (Schaffer and Schaffer 1977a, 1979). However, if fruit production is determined primarily by available resources rather than pollinator abundance, pollinators would have little influence on inflorescence size or reproductive habit (Udovic 1981; Aker 1982).

Fire is more likely to suppress than to stimulate seed germination (Keeley and Meyers 1985). Fire-killed rosettes do not resprout, but because plants tend to grow in open sites, they often survive chaparral fires (Keeley and Keeley 1981).

Yucca whipplei is protected by law in California; nevertheless, urban development and livestock grazing have made significant inroads on lowland populations (Powell and Mackie 1966). Stems and young inflorescences were roasted and eaten, and the leaves have been used as a source of fiber (McKelvey 1947). *Yucca whipplei* makes an attractive ornamental and deserves wider cultivation in the arid Southwest.

Zanthoxylum arborescens

Rose

Pipima

Zanthoxylum arborescens is a shrub or tree up to 6 m tall with scattered, stout, slightly curved spines on the branches and heavier, conical spines on the trunk. The leaves, 3–7.5 cm long and 1.5–4 cm wide, are pinnate with 3–7 oblong to obovate leaflets velvety pubescent beneath and dotted with pellucid glands. The leaf rachis is winged. Dioecious or polygamodioecious, the small, greenish yellow flowers are paniculate. Fruits are follicles 4–6 mm long.

The pinnate, pellucid-dotted leaves, sparse armature, and winged rachises distinguish *Zanthoxylum* from other genera in its range. *Zanthoxylum sonorense* has smaller leaves (2.5–5 cm long) that are glabrate to hirtellous beneath.

Occasional in rocky canyons and along arroyos, *Z. arborescens* is a Sinaloan thornscrub species known from Sinaloa and the Cape Region. Its presence on the peninsula may predate gulf-floor spreading.

September–October is the flowering season.

Zanthoxylum sonorense Lundell

This densely branched shrub, 2–3 m tall, has slender, nearly straight spines 3–8 mm long. The hirtellous to glabrate leaves, 2.5–5 cm long, are odd-pinnate with 5–9 pellucid-dotted, oblanceolate or obovate leaflets. The rachis is winged. The inconspicuous flowers are in short, axillary racemes. Fruits are follicles 4–5 mm long.

The slender spines, winged rachises, and pinnate, pellucid-dotted leaves distinguish *Z. sonorense* from other species in the Sonoran Desert region. *Zanthoxylum arborescens* has larger, velvety pubescent leaves, and *Z. fagara* (L.) Sarg. is heavily armed with recurved spines. The specimens from Baja California evidently approach *Z. arborescens* in their broader, more leathery, and more densely puberulent leaves (Wiggins 1964).

Rare to common on rocky hillsides, canyon slopes, and ridges, this Sinaloan thornscrub species reaches its northern limit near Arizpe (vouchered by Raymond M. Turner). The northern populations often grow on slopes above the level of nighttime temperature inversion. *Zanthoxylum sonorense* enters the Sonoran Desert proper only near Guaymas, where it was collected by Arthur and Barbara Phillips in 1975 and Martha Ames in 1977.

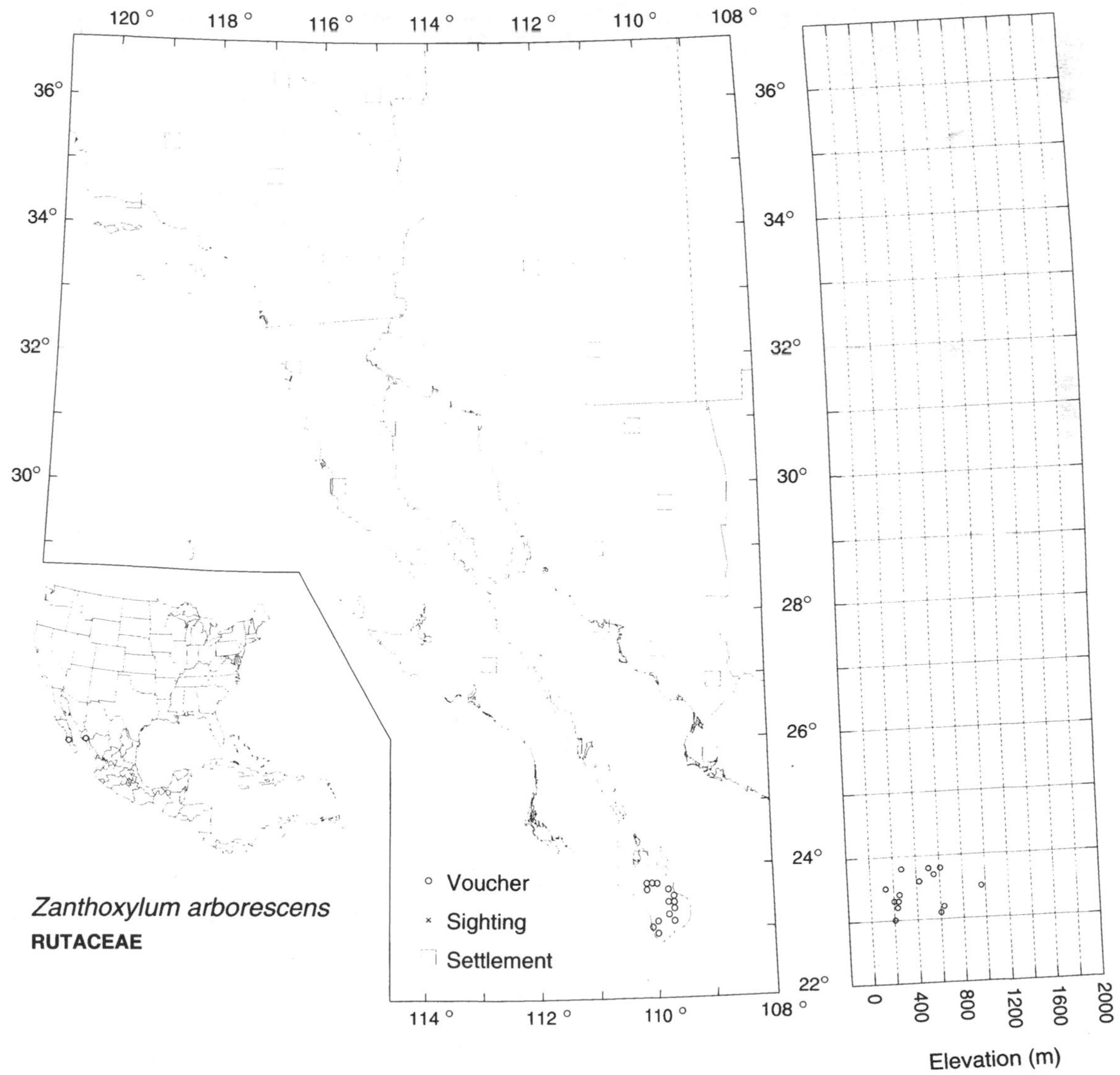

Zanthoxylum arborescens
RUTACEAE

These disjunctions may be relicts of a once widespread thornscrub formation that, under increasing aridity, retreated to relatively mesic sites.

Flowers appear from February to March (Wiggins 1964) and sometimes again in summer.

Zapoteca formosa

(Kunth.) H. Hern.

[=*Calliandra schottii* Torr. ex S. Watson, *Calliandra rosei* Wiggins]

Schott calliandra, huaje de caballo, tosapolo

This unarmed shrub, 1.5 m tall, has thinly puberulent or glabrous foliage and straw-colored stems. The 1–3 pairs of pinnae bear 4–7 pairs of oblong leaflets 3–13 mm wide and up to 20 mm long. The pinnae are 1.5–4 cm long. The whitish or pinkish flowers are in capitate clusters about 15 mm in diameter. The flattened, linear pods are 4–6 cm long and 6–8 mm wide. Pollen is present as permanent, 16-grained units called polyads (Hernández 1989). The diploid chromosome number of subsp. *rosei* is 26 (Hernández 1989).

Leaflets of *Calliandra eriophylla* are minute. The pinnae of *Acacia crinita* are swollen at the base (Turner and Busman 1984). *Caesalpinia pumila* has red-brown or dark gray bark conspicuously dotted with white lenticels. *Mimosa distachya* is sparsely armed with small, curved spines.

Wiggins (1964) treated this taxon separately from *C. rosei*. Felger and Lowe (1970) regarded *C. rosei* as a subspecies of *C. schottii*. We follow Hernández (1989), who elevated *Calliandra* subsection *Zapoteca* to genus *Zapoteca* and created the

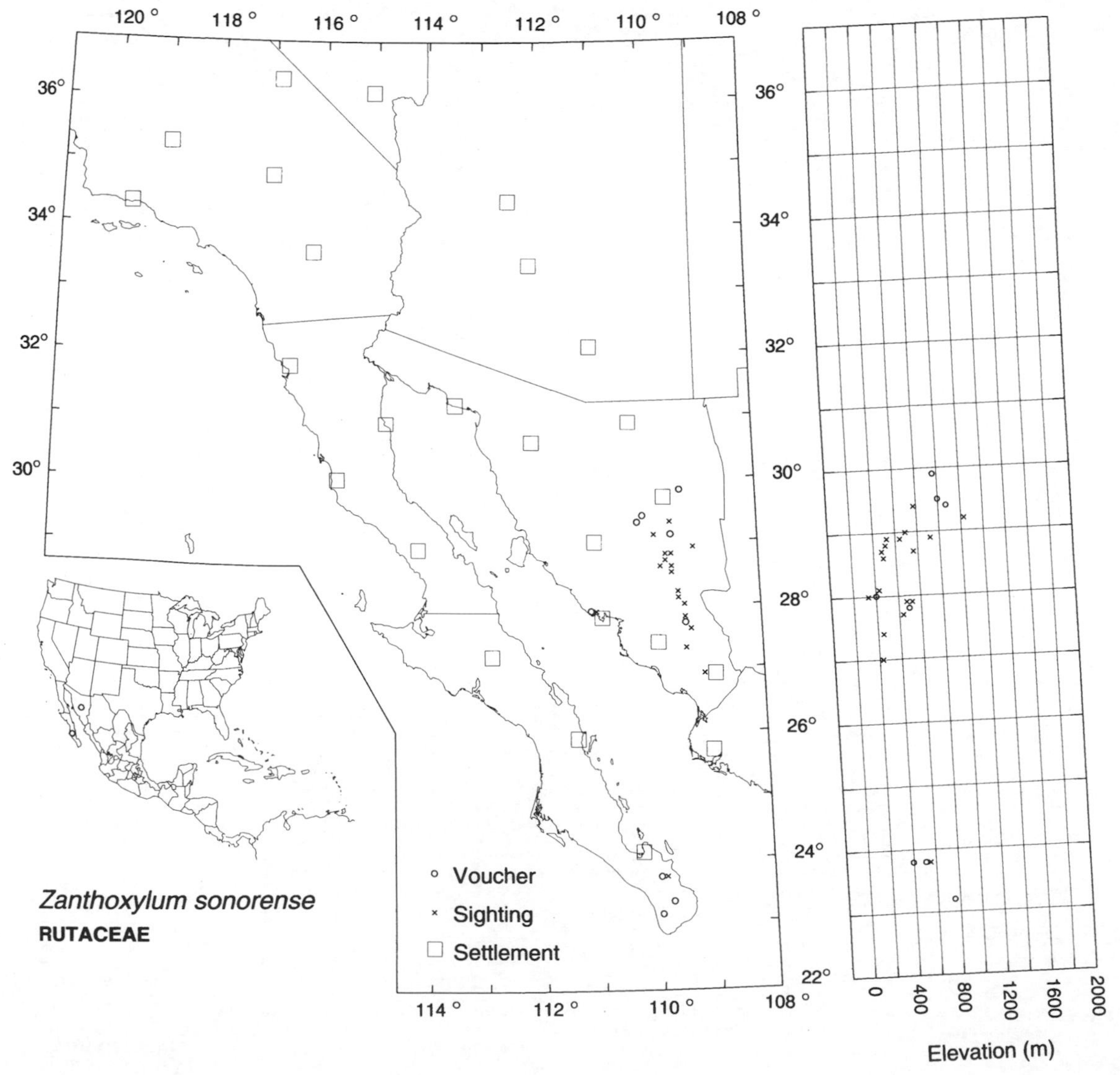

Zanthoxylum sonorense
RUTACEAE

new combination *Z. formosa* for *Calliandra schottii.* Hernández recognized two taxa within *Z. formosa.* Subspecies *schottii,* which occurs in northern Sonora and southern Arizona, has leaflets 2–4 mm wide and 9–14 mm long. Subspecies *rosei* (Wiggins) H. Hern., which is found from Sonora to the Isthmus of Tehuantepec, has leaflets 3–8 mm wide and 21–31 mm long (Hernández 1989). Our map combines the two.

A shrub of rocky slopes and canyon bottoms, *Z. formosa* is widely but spottily distributed in the Sonoran Desert region. The northernmost populations are on the south slope of the Santa Catalina Mountains, Pima County, Arizona (specifically, in Pima, Sabino, and Soldier canyons). The Isla Tiburón sighting is based on Felger and Lowe (1970). The record from Baja California Sur is based upon a voucher cited in Hernández (1989).

At their northern limit, the plants sustain frost damage in the coldest winters.

In the Sonoran Desert region, flowers appear from July to September, occasionally October. The leaves are drought and cold deciduous. Certain characteristics suggest that the inflorescence is well adapted for moth pollination, especially by moths that settle on the flowers instead of hovering before them (for example, moths in the Noctuidae, Pyralidae, and Geometridae): first, anthesis occurs at night; and second, although a low ratio of polyads to ovules suggests that the chances for successful pollination are rather low, settling moths spend a considerable length of time on an individual flower, thus greatly increasing the chance that a polyad will reach a stigma (Hernández 1989).

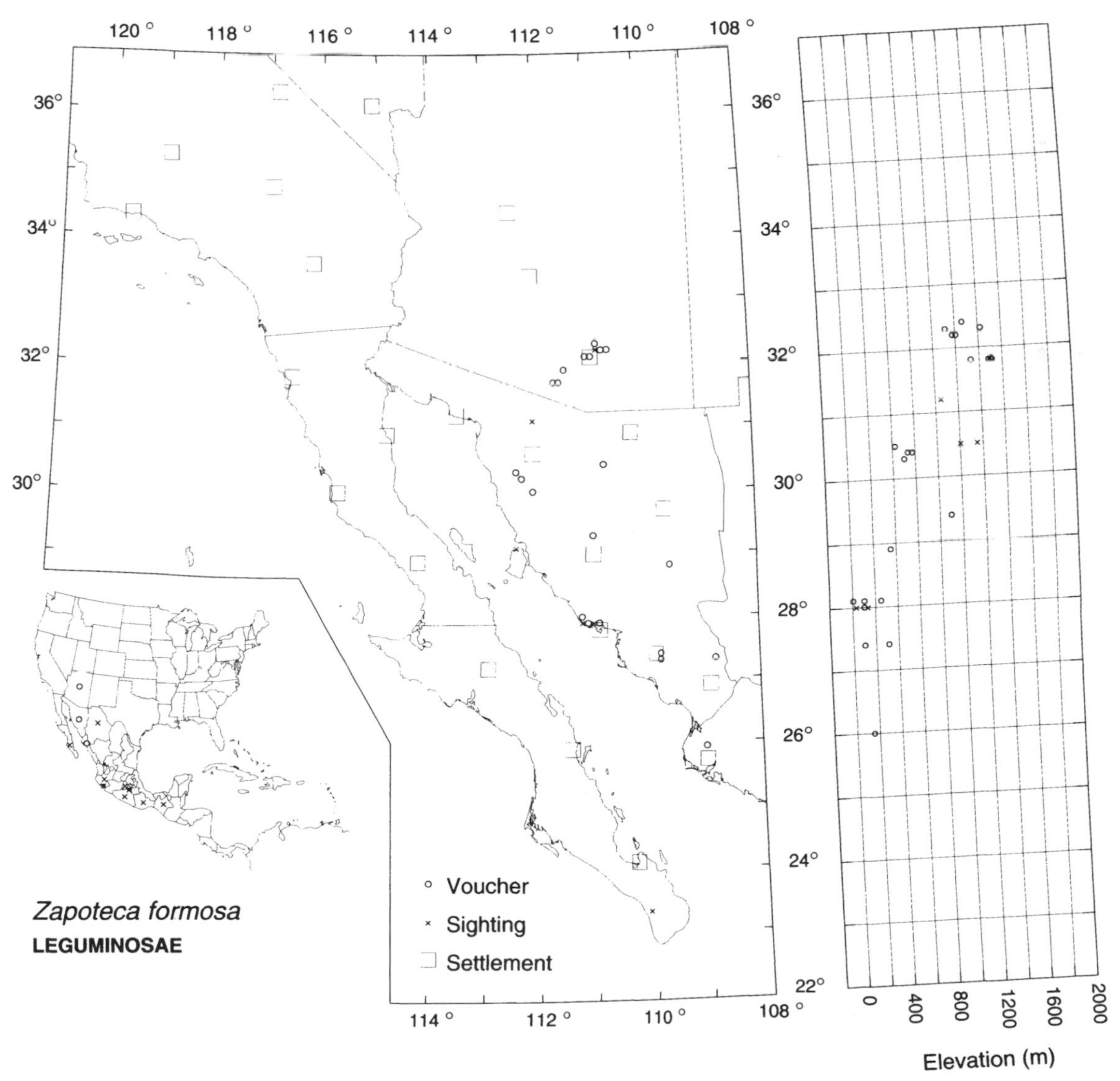

Zapoteca formosa
LEGUMINOSAE

Ziziphus amole

(Sess. & Moc.) M. C. Johnston

[=*Z. sonorensis* S. Watson]

Nanche de la costa, amole dulce, manzanita, brasilillo

This shrub or small tree grows up to 12 m tall and has paired spines, 2–8 mm long, at the nodes. The ovate to orbicular leaves, 3–6 cm long and 2–3.5 cm wide, may be alternate or subopposite, and entire or remotely crenate. The inconspicuous flowers are in axillary cymes. Fruits are hard-seeded, reddish drupes 8–10 mm long.

Only a few simple-leaved species in the Sonoran Desert area have paired spines at the nodes. *Celtis pallida* leaves are smaller than *Z. amole* leaves. *Celtis iguanea* has stout, recurved spines 7–14 mm long and arched, scandent branchlets. *Randia thurberi* has clustered leaves at the tips of short, lateral spur shoots.

Regional floras (e.g., Standley 1923; Wiggins 1964) treat this species as *Z. sonorensis.* Johnston (1963b) pointed out that *Z. amole* has priority, and we follow his treatment here. It is most closely related to *Z. mistol* Griseb. of South America and *Z. obtusifolia* of the southwestern United States and northern Mexico (Johnston 1963b).

A plant of subtropical bottomlands, floodplains, flats, and arroyos, *Z. amole* is widespread in Mexico and reaches its northern limit in the southern part of the area mapped in this atlas. According to Wiggins (1964, 1980), it occurs in the Cape Region of Baja California Sur; however, we have seen no specimens from there.

The flowering period is August to September (Wiggins 1964).

Fruits of *Z. amole* have been used as a substitute for soap (Standley 1923; Johnston 1963b).

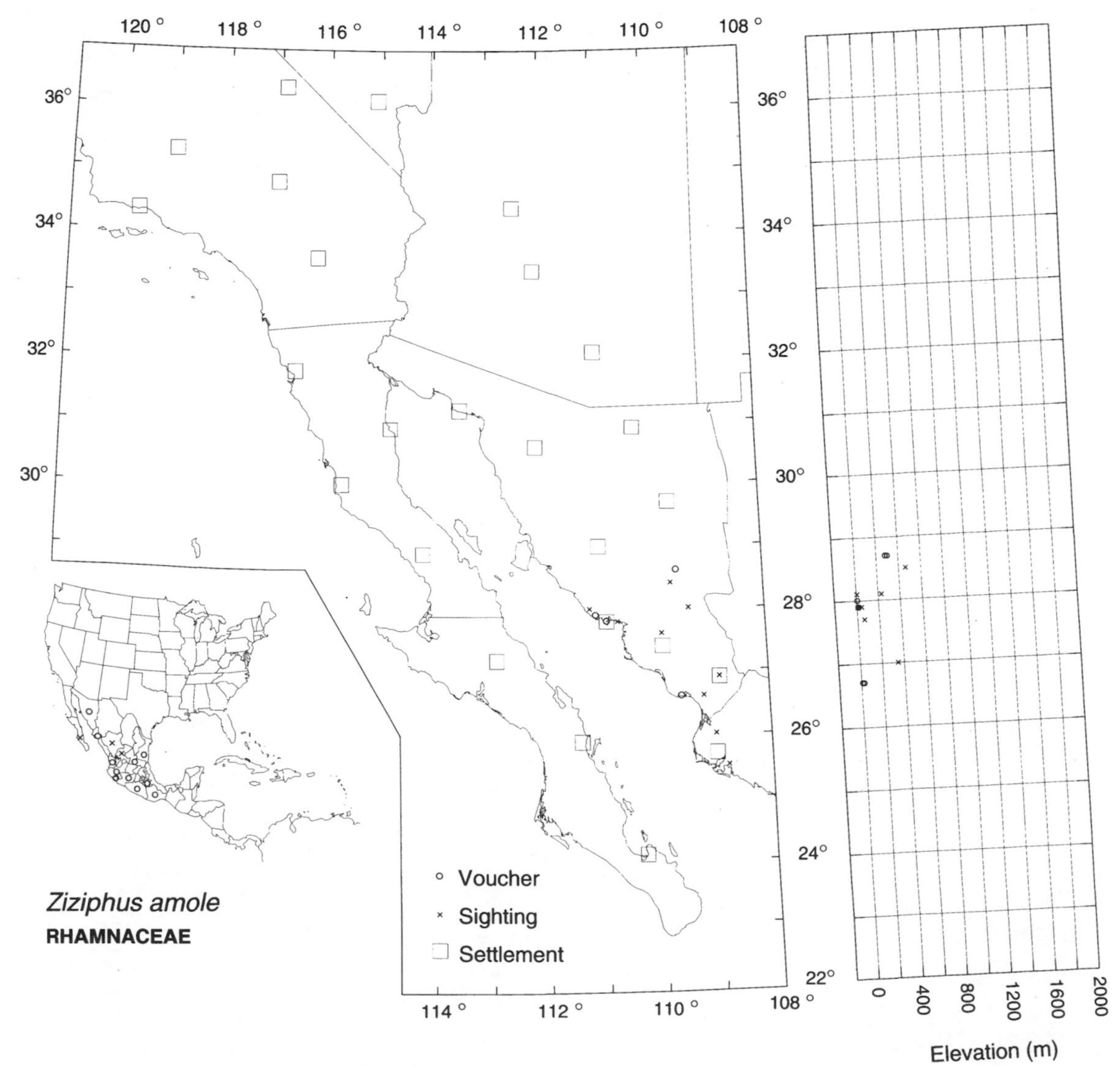

Ziziphus amole
RHAMNACEAE

Ziziphus obtusifolia

(Hook. ex T. & G.) A. Gray

[=*Condaliopsis lycioides* (A. Gray) Suessing., *Condalia lycioides* (A. Gray) Weberb.]

Lotebush, gray-thorn, gumdrop tree, white crucillo, bachata, garrapata, garambullo, barbachatas, palo blanco

A shrub 1–4 m tall, *Ziziphus obtusifolia* has zigzag, gray branches that are persistently tomentose in one variety and soon glabrate in another. Twigs spread at right angles from the branches and end in stout thorns. The elliptic to oblong leaves are 8–19 mm long and 5–9 mm wide, either persistently tomentose or glabrate; occasionally they are serrate. The inconspicuous flowers have 5 petals and are in umbels. Fruits are dark blue or black ellipsoid drupes 6–8 mm long.

Castela emoryi, Koeberlinia spinosa, and *Canotia holacantha* differ in having green-barked twigs and branches that only briefly bear leaves. Except for *Condalia brandegei,* leaves of the *Condalia* species in the area mapped in this atlas are much smaller—half the size or less. The large-leaved *Condalia brandegei* can be distinguished by the lack of petals on the flowers. *Celtis pallida* has markedly zigzag, glabrous stems. Most *Lycium* species also have glabrous stems, and none has persistent white tomentum.

Two varieties have been described: var. *obtusifolia,* which is glabrate, and var. *canescens* (A. Gray) M. C. Johnston, which is tomentose. The former is mainly a plant of the Chihuahuan Desert and adjacent woodlands, the latter of the Sonoran Desert and adjacent semidesert grassland (Johnston 1963b; Benson and Darrow 1981).

Regional manuals generally treat this species as *Condalia lycioides* (Kearney and Peebles 1960) or *Condaliopsis lycioides* (Wiggins 1964; Munz 1974). We follow Johnston (1962a), who argued that *Condaliopsis* could be easily separated from *Condalia* on characters of ovary, disk, and style and, further, that *Condali-*

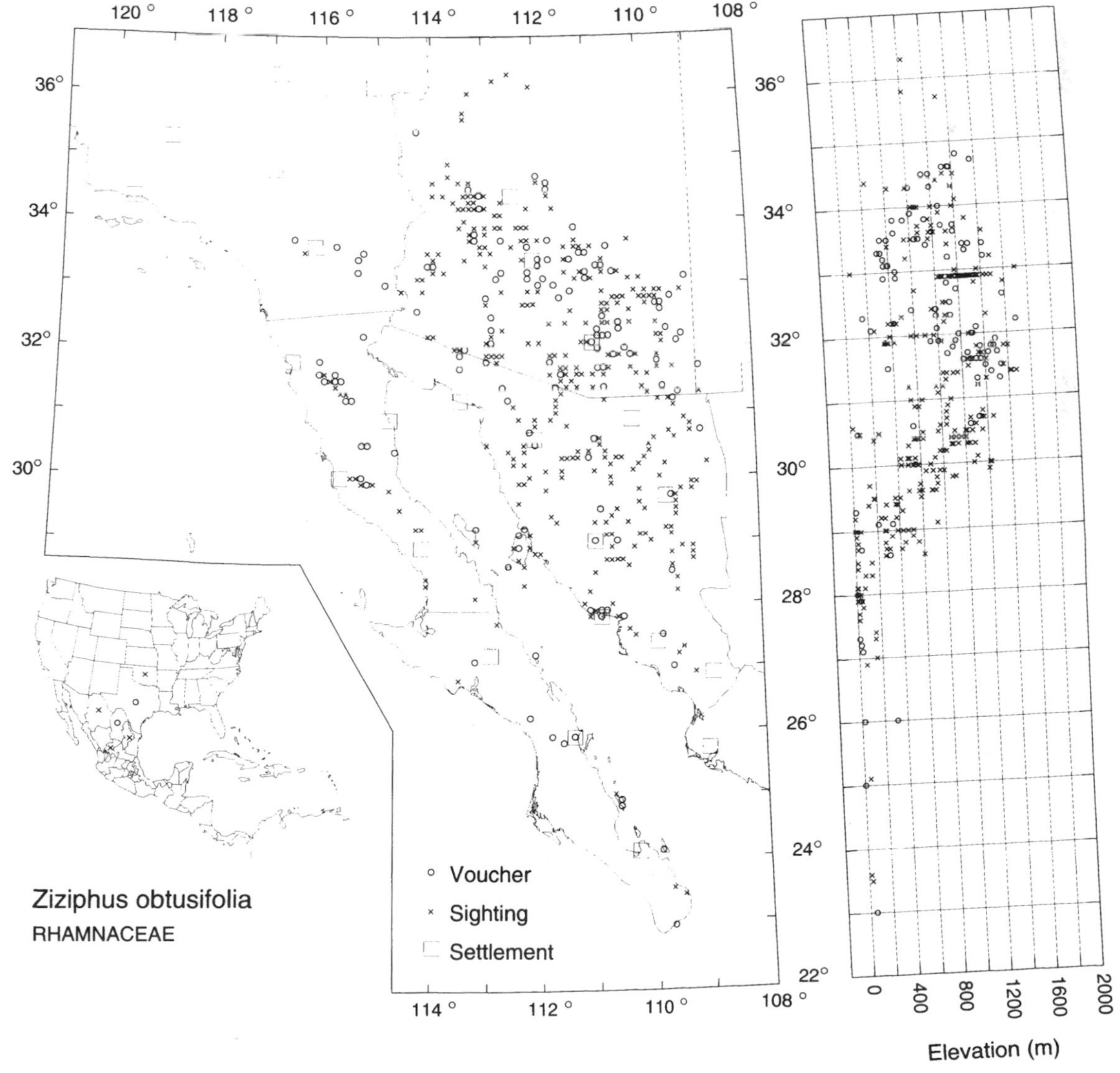

Ziziphus obtusifolia
RHAMNACEAE

opsis was insufficiently distinct from *Ziziphus* to warrant a separate genus.

Occasional to abundant on flats, slopes, mesas, bottomlands, and washes, *Z. obtusifolia* can form impenetrable thickets. Its range of habitats mirrors its broad distribution in the arid and semiarid regions of western North America. Although this plant is absent along the lower Colorado River, the northernmost disjunctions follow the course of that river, which may have served as a distributional corridor. *Ziziphus obtusifolia* is an important component of stream-channel vegetation in Arizona (Campbell and Green 1968), and it is also frequent in mesquite bosques. In California the species is uncommon (Munz 1974). The outlier in that state (33.8°N) is A. F. Gilman's 1922 collection from Andreas Canyon near Palm Springs.

The plants bloom and set fruit intermittently throughout the year in response to rain. Leaves are drought deciduous.

Pima Indians used a decoction of the roots to treat sore eyes (Kearney and Peebles 1960), and the Seri Indians placed powdered roots on skin and scalp sores (Felger and Moser 1985). The roots have also been used as a substitute for soap (Kearney and Peebles 1960; Felger and Moser 1985). Seri Indians ate the fresh fruits (Felger and Moser 1985).

Glossary

abaxial.	On the side away from the axis. (Compare *adaxial*)
abscission.	The normal separation of flowers, fruit, and leaves from plants, usually caused by the development of a special tissue called the abscission layer.
acaulescent.	Literally, without a stem; in common botanical usage, having the leafy part of the stem so short that the leaves are all clustered in a basal rosette. (Compare *caulescent*)
accrescent.	Increasing in size with age.
achene.	A small, dry, hard, one-seeded, indehiscent fruit.
acuminate.	Gradually and concavely tapering to a narrow tip or sharp point. (Compare *acute, attenuate*)
acute.	Sharply pointed, with more or less straight distal margins that form an angle of less than 90° at the tip.
adaxial.	On the side toward the axis, or turned toward the axis. (Compare *abaxial*)
allopatric.	Occupying different geographic regions. (Compare *sympatric*)
allotetraploid.	Same as amphidiploid; a plant having the chromosome numbers of two parental species, owing to the doubling of the chromosomes in a sterile hybrid of the two species.
amphiploid.	A functioning diploid derived by chromosome doubling after interspecific hybridization, with each functioning genome consisting of a full genome from each parent species.
aneuploid.	An organism having a chromosome number that differs from the diploid number by more, or less, than a genome or a single set of chromosomes (e.g., 2n+1, 2n−1, where n is a full genome).
anthesis.	The period during which a flower is fully expanded and functional.
apetalous.	Without petals.
apomictic.	Setting seed without fertilization.
appressed.	Pressed flat against another organ.

areole. A sharply defined space, such as the spine-bearing area in cacti.

aril. Loosely, any appendage or fleshy thickening of the seed coat.

arillate. Provided with an aril.

armed. Possessing spines, thorns, or prickles.

attenuate. Tapering very gradually to a slender tip; more extreme than acute or acuminate.

autochthonous. Formed in the region where found.

autopolyploid. Polyploid, with the genomes all derived from the same species.

autotetraploid. A plant having four haploid sets of chromosomes that are derived from the same species through one of the mechanisms producing chromosome doubling.

auxin. Substance that acts as a growth hormone in plants, where it controls cell enlargement.

axil. The angle between the upper side of a leaf and its supporting stem.

baccate. Berrylike and pulpy.

banner. The upper, usually enlarged petal of members of the legume family having pea-like flowers.

bifid. More or less deeply cleft from the tip into two, usually equal, parts.

binomial. A taxonomic name consisting of a generic term and an adjective.

bipinnate. Twice-pinnate; the pinnae again pinnate.

bipinnatifid. Twice-pinnatifid; the deeply cut divisions themselves deeply cut.

bract. A specialized leaf, from the axil of which a flower or flower stalk arises; more loosely, any more or less reduced or modified leaf associated with a flower or an inflorescence but not a part of the flower itself.

bulbil. A bulblike structure produced by some plants in the axils of leaves or in the place of flowers.

caespitose. Growing in dense, low tufts.

Californian chaparral. A dense scrub vegetation widespread in California mountains below 2,750 m, dominated by species with sclerophyllous leaves, including *Ceanothus* species, *Quercus* species, and *Adenostoma fasciculatum*.

Californian coastalscrub. Low scrub vegetation characterized by *Artemisia californica, Salvia* species, and *Encelia californica* that occurs along the Pacific coast from the northern edge of the Sonoran Desert north past San Francisco Bay.

campanulate. Bell shaped.

CAM plant. CAM is the acronym for crassulacean acid metabolism; this term refers to plants that fix carbon dioxide during the night and store it in the form of an organic acid until daylight.

canescent. Pale or gray, because of a fine, close, whitish pubescence. (Compare *cinereous*)

capillary. With the form of a hair.

capitate. Headlike or in a head.

capsule. A dry, dehiscent fruit composed of more than one carpel.

carpel. Floral organ that contains ovules in flowering plants. May occur singly or as several fused together, forming simple or compound pistils, respectively.

caudex. A short, vertical, thickened, often woody, persistent base of an herbaceous or succulent perennial.

caulescent. With an obvious, leafy stem that has visible internodes. (Compare *acaulescent*)

C_4 plant. A plant in which carbon dioxide is initially fixed as a four-carbon compound that is transported elsewhere in the leaf for reinsertion into the Calvin cycle.

chamaephyte. A low-growing, partly woody perennial plant having renewal buds near the ground (up to 50 cm aboveground).

chlorenchyma. A type of parenchyma; a tissue containing chlorophyll.

chlorophyllous. Containing chlorophyll.

ciliate. With a fringe of marginal hairs.

cinereous. Ashy in color, usually because of a covering of short hairs.

circumscissile. Dehiscing by an encircling transverse line, so that the top comes off as a lid or cap.

clavate. Shaped like a club or a baseball bat, being thicker at the distal end than at the base.

coastal thornscrub. Dense vegetation of woody and stem-succulent shrubs and short trees (up to 2–8 m tall), usually dominated by microphyllous and thorny species, found on coastal plains and foothills that experience long, dry winters.

competitive displacement. A process by which a species is prevented from becoming established in an otherwise suitable area by the presence of another species or group of species. Biogeographically this is expressed by a series of related or ecologically similar species that replace one another along a geographic or environmental gradient. To show that competitive displacement is the responsible mechanism, removal of one or more species must permit another species to expand its range. In the absence of this evidence, the mechanism shaping the biogeographic pattern in question remains circumstantial and conjectural.

connivent. Converging or coming together, but not physically united.

cordate. Shaped like a stylized heart, with the notch at the base.

coriaceous. Leathery in texture.

corm. A short, vertical, underground stem that is thickened as a perennating food-storage organ, without prominently thickened leaves.

corneous. Composed of hornlike material.

corolla. All of the petals of a flower, collectively.

cortex. Stem tissue immediately below the epidermis and outside the vascular tissue.

corymb. A simple, racemose inflorescence that is flat-topped or round-topped because the outer pedicels are progressively longer than the inner.

crassulaccan acid metabolism.	See CAM plant.
crenate.	Provided with rounded teeth.
crenulate.	Diminutive of crenate.
crustose lichen.	Lichens that form a crust closely appressed to the substrate.
C_3 plant.	A plant in which carbon dioxide fixation occurs exclusively through the Calvin cycle.
culm.	The aerial stem of a grass or sedge.
cuneate.	Wedge shaped or triangular with the narrow end at the point of attachment.
cuticle.	The waxy layer covering the epidermis of a leaf or stem.
cyme.	A broad class of inflorescences characterized by having the terminal flower bloom first, commonly also with the terminal flower of each branch blooming before the others on that branch; a determinate inflorescence. (Compare *raceme*)
decurrent.	With a wing or margin extending down the stem or leaf below the point of attachment.
dehiscent.	Opening at maturity, releasing or exposing the contents.
deltoid.	Shaped more or less like an equilateral triangle, with one of the sides as the base.
dendritic.	Of a branching, treelike form.
dentate.	Provided with spreading, pointed teeth; frequently used to describe leaf margins. (Compare *serrate, crenate*)
dichotomous.	Forking more or less regularly into two branches of equal size.
dimorphic.	Of two forms.
dioecious.	Producing male and female flowers on separate individuals. (Compare *monoecious*)
diploid.	Having two full chromosome complements per cell. (Compare *haploid*)
discoid.	Resembling a disc; in the Compositae, with the flowers of a head all tubular (i.e., no strap-shaped flowers) and perfect (or functionally staminate).
divaricate.	Widely spreading from the axis or rachis.
dopamine.	A chemical widely occurring in biological systems. It takes part in the transmission of nerve impulses in animals and is a constituent of healthy tissue in plants such as *Carnegiea gigantea*.
drupe.	Fleshy fruit, such as a plum, containing one or a few seeds, each enclosed in a stony layer that is part of the fruit wall.
echinate.	Provided with prickles.
ecocline.	A gradient zone that is relatively heterogeneous with respect to at least one environmental factor but ecologically stable enough to allow organisms to adapt within its context.
ecotype.	The individuals of a species that are adapted to a particular environment, however defined.
edaphic.	Pertaining to soil.
eglandular.	Without glands.

ellipsoid. Elliptic in outline and circular in cross section (applied only to three-dimensional bodies).

endemic. Confined to a particular geographic area.

endocarp. The innermost of the three layers of a pericarp. (Compare *mesocarp, exocarp*)

entire. Refers to leaf margins that are not toothed or otherwise cut.

epicuticular. Upon the cuticle.

epidermis. A thin layer of cells forming the outer covering of seed plants and ferns.

even-pinnate. Pinnately compound leaves having only paired leaflets without a solitary terminal leaflet. (Compare *odd-pinnate*)

exfoliate. To peel off in layers.

exocarp. The outermost layer of the three layers of a pericarp. (Compare *mesocarp, endocarp*)

exserted. Projecting beyond an envelope, as stamens from a corolla.

facultative. Having the capacity to live under more than one specific set of environmental conditions or possessing more than one contrasting physiologic or genetic trait. (Compare *obligate*)

falcate. Sickle shaped, or curved like a hawk's beak.

farinose. Covered with a meal-like powder.

fascicle. A close bundle or cluster.

fecundity. Reproductive capacity of an organism.

filament. The stalk of a stamen, i.e., the part supporting the anther.

filiferous. Bearing threadlike fibers.

filiform. Slender, threadlike.

flabelliform. Fan shaped.

flexuous. The quality of being pliant, limber, flexible.

floccose. Covered by very fine, long, soft, loosely spreading and more or less tangled hairs; like tomentose, but looser and more open.

follicle. A dry fruit composed of a single carpel, which at maturity splits along the ventral (seed-bearing) suture only. (Compare *legume*)

fugacious. Fleeting; lasting only a short time.

funnelform. Shaped like a funnel.

genome. The haploid chromosome set; the set of chromosomes contributed by one gamete in a basic diploid species.

geophyte. A plant with underground perennating buds.

glabrate. Nearly glabrous, or becoming glabrous.

glabrous. Smooth, without hairs or glands.

gland. A protuberance, appendage, or depression on the surface of an organ (or on the end of a trichome) that produces a sticky or greasy, viscous substance.

glaucous. Covered with a fine, waxy, removable powder that imparts a whitish or bluish cast to a surface, such as a prune or a cabbage leaf.

globose. More or less spherical.

glochid. A hair or hairlike spine with minute retrorse, hooklike projections.

glomerule. A small, compact, headlike cyme; any dense, small cluster.

halophyte. A plant adapted to growth in salty soil.

haploid. Having only one complement of chromosomes per cell.

hastate. Shaped like an arrowhead, but with the basal lobes more divergent.

head. An inflorescence of sessile or subsessile flowers crowded closely together at the tip of a peduncle.

hermaphroditic. With both sexes together in the same flower.

heterostylous. With styles of different (usually two) lengths in flowers of different individuals, some surpassing, others surpassed by the stamens.

hirsute. Pubescent with rather coarse or stiff but not sharp hairs; often bent; coarser than villous but less firm and sharp than hispid.

hirtellous. Diminutive of hirsute.

hispid. Pubescent with coarse and firm, often sharp-pointed hairs. (Compare ***hirsute***)

hispidulous. Diminutive of hispid.

homeostasis. Tendency of a biological system to resist change and to maintain itself in a state of stable equilibrium.

hypocotyl. That part of a stem below cotyledons and above the root.

imbricate. Overlapping in sequence, as tiles or shingles on a roof.

indehiscent. Not dehiscent; remaining closed at maturity.

indoleacetic acid. Chemical that acts as a growth hormone in plants, where it controls cell enlargement.

indument. Collectively, the epidermal appendages of a plant or an organ (hairs, glands, etc.).

indurate. Hard; hardened.

inflorescence. A flower cluster of a plant or, more correctly, the arrangement of the flowers on the axis.

internode. The part of a stem between two adjacent nodes.

introgression. Process by which genes of one species spread into the germ plasm of another as the result of hybridization.

involucre. A set of bracts beneath an inflorescence; more generally, any set of structures that surround the base of another structure.

isozyme. Species of enzyme that exists in two or more structural forms which are easily identified by electrophoretic methods.

karyotype. All of the chromosomes of a nucleus, with reference to related size and shape as well as number.

keel. The two partly united lower petals in flowers of many plants in the legume family.

Kranz-type leaf anatomy. Leaf anatomy characterized by a sheath of enlarged cells surrounding each vascular bundle; usually associated with C_4 photosynthesis.

K selection. Selection producing superior competitive ability in stable, predictable environments so that the population is maintained at or near the carrying capacity of the habitat. (Compare ***r selection***)

laciniate.	Cut into narrow and usually unequal segments.
lanate.	Woolly; covered by soft hairs resembling wool.
lanceolate.	Lance shaped; much longer than wide, widest below the middle, and tapering to both ends; like ovate but narrower.
leaflet.	An ultimate unit of a compound leaf.
legume.	The fruit of a member of the Leguminosae (legume family), composed of a single carpel, typically dry and dehiscing down both sutures.
lenticel.	A slightly raised area in the bark of a stem or root, consisting of loosely arranged, nearly or quite unsuberized cells.
lepidote.	Scaly; covered with small scales.
ligniferous.	Containing lignin and hence somewhat woody.
lignin.	A complex chemical substance found in many plant cell walls. Its function seems to be to cement together and anchor cellulose fibers and to stiffen the cell wall.
limb.	The expanded part of a sympetalous corolla above the throat.
linear.	Line shaped; very long and narrow, with essentially parallel sides.
lunate.	Crescent shaped.
Madrean evergreen woodland.	Vegetation of trees with an open canopy structure; growing at medium and higher elevations of the Sierra Madre; usually dominated by evergreen *Quercus* species, *Juniperus* species, and *Pinus cembroides.*
Madrean montane conifer forest.	Vegetation with a closed canopy of trees found at higher elevations in the Sierra Madre; typical dominants are *Pinus* species, *Pseudotsuga menziesii,* and *Quercus* species.
mammillate.	Having nipples.
melanin.	A dark pigment found in the skin and hair of many animals and in plants under some conditions, e.g., around wounds in stems of *Carnegiea gigantea* where it is derived from dopamine.
meristem.	A tissue composed of actively dividing cells.
-merous.	Greek suffix, referring to the parts in each circle of floral organs, usually with a numerical prefix, as trimerous, pentamerous, etc.
mesic.	Of, pertaining to, or adapted to a habitat with a balanced supply of moisture. (Compare *xeric*)
mesocarp.	The central layer of a pericarp, which is differentiated into three layers. (Compare *endocarp, exocarp*)
mesophyll.	The tissue, other than vascular tissue, between the upper and lower epidermis of a leaf.
mesophyte.	A plant adapted to growth under balanced moisture conditions; intermediate between hydrophyte and xerophyte.
monocarpic.	Blooming just once and then dying.
monoecious.	Condition where separate male and female flowers are borne on the same plant. (Compare *dioecious*)
mycorrhiza(e).	A symbiotic association between a fungus and the root of a vascular plant.
node.	A place on a stem where a leaf is or has been attached.

obcordate. Like cordate, but with the notch at the tip instead of at the base.

oblanceolate. Like lanceolate, but broadest above the middle and tapering toward the base.

obligate. Limited to a single life condition. (Compare *facultative*)

oblong. Much longer than broad, with nearly parallel sides.

obovate. Like ovate, but larger toward the distal end.

odd-pinnate. Pinnately compound leaves having a terminal leaflet so that the number of leaflets is an odd number. (Compare *even-pinnate*)

offset. A short, usually prostrate shoot, primarily propagative in function, originating near the ground level at the base of another shoot.

once-pinnate. A pinnately compound leaf with two rows of leaflets along the rachis. (Compare *twice-pinnate*)

operculum. A little lid; the deciduous cap of a circumscissile fruit or other organ.

orbicular. Essentially circular in outline.

orographic. Pertaining to mountains.

outcross. Pollination between two different plants of the same species.

ovate. Shaped like a long section through a bird's egg, with the broader end toward the base. (Term applied to plane surfaces; compare *ovoid, obovate*)

ovoid. Shaped like a hen's egg. (Term applied to three-dimensional objects; compare *ovate*)

ovule. A young or undeveloped seed.

palmate. With three or more lobes, nerves, leaflets, or pinnae arising from a common point. (Compare *pinnate*)

panicle. A compound racemose inflorescence.

paniculate. Adjectival form for panicle.

papillate. Covered with papillae, i.e., with short, rounded, blunt projections.

pappus. The modified calyx crowning the ovary (and achene) of the members of the sunflower family (Compositae) consisting variously of hairs, scales, bristles, or a mixture of these.

parenchyma. A tissue composed of relatively unspecialized, usually thin-walled cells.

pedicel. The stalk of a single flower within an inflorescence.

peduncle. The stalk of an inflorescence or of a solitary flower.

pellucid. Transparent or translucent.

peltate. Shield shaped, attached at the center of the lower surface instead of at the margin.

pentamerous. With five parts of a kind.

perfect. Refers to a flower with both stamens and pistil(s).

pericarp. The wall of the fruit.

periderm. Protective tissue that replaces the epidermis as the stem or root it covers ages.

petiole. A leaf stalk that attaches the blade or rachis to a stem.

phenology. Pattern of the recurrence of seasonal growth and reproductive activities of a species.

photoperiod. The interval in a 24-hour period during which an organism is exposed to light.

phreatophyte. A plant that absorbs water from the permanent water table.

phyllary. An involucral bract of the Compositae.

phyllode. A dilated petiole serving as a leaf blade.

pilose. With long, straight, rather soft, spreading hairs.

pinna. One of the primary lateral divisions of a pinnately compound leaf.

pinnate. A compound leaf having the leaflets arranged on each side of a rachis; featherlike. (Compare *palmate*)

pinnatifid. More or less deeply cut in pinnate fashion.

pistil. The female organ of the flower, composed of one or more carpels, and ordinarily differentiated into ovary, style, and stigma.

pistillate. Having one or more pistils but no stamens.

platyopuntia. A subgroup in the genus *Opuntia* having flattened stem segments.

-ploid. Suffix used with a numerical prefix to indicate the number of chromosome complements in each cell, as haploid, diploid, triploid, etc.

plumose. Feathery or brushlike.

plumule. The part of the embryo of a seed that gives rise to the shoot.

pneumatophore. Specialized "breathing" root developed in some plant species that grow in waterlogged or strongly compacted soils, e.g., mangroves.

pod. Any kind of dry, dehiscent fruit.

polygamodioecious. Nearly dioecious but some of the flowers bisexual.

polygamous. With intermingled perfect and unisexual flowers.

polyploid. Having a chromosome number that is more than twice the basic or haploid number.

prickle. A sharp outgrowth of the epidermis or bark. (Compare *spine, thorn*)

propagules. Any structure that functions in propagation and dispersal.

puberulent. Minutely pubescent; provided with fine, short, loose, curled hairs.

pubescent. Bearing hairs of any sort.

punctate. Dotted, usually with small pits that may be translucent or glandular.

pustulate. With little blisters or pustules.

raceme. A more less elongate inflorescence with pedicellate flowers arising in acropetal succession (from the bottom upward) from an unbranched axis. (Compare *cyme, spike*)

rachilla. A small or secondary rachis, such as the axis of a pinna in a twice-pinnate leaf.

rachis. A main axis, such as that of a pinnately compound leaf or inflorescence.

radiate. In the sunflower family, having the marginal flowers of the head strap shaped (ligulate) and the central ones tubular (discoid).

radicle. The part of the embryo within a seed that gives rise to the primary root.

ramet. An individual plant of a clone.

reniform. Kidney shaped.

repand. With a shallowly sinuate or slightly wavy margin.

reticulate. Forming a network, as in the veins of a leaf.

retrorse. Directed backward or downward.

rhizobial. Refers to ***Rhizobium,*** a genus of bacteria that brings about the formation of nitrogen-fixing nodules on leguminous plants.

rhizomatous. Bearing rhizomes.

rhizome. A creeping underground stem.

rosette. A cluster of leaves or other organs arranged in a circle or disk, often in basal position.

rotate. Flat and circular in outline, as a sympetalous corolla with spreading lobes and without a tubular basal part; saucer shaped.

r selection. Selection favoring a rapid rate of population increase, typical of species that colonize short-lived environments or of species that undergo large fluctuations in population size. (Compare *K selection*)

ruderal. Weedy, growing in waste places.

rugose. Wrinkled.

rugulose. Diminutive of rugose.

salverform. With a slender tube and an abruptly spreading limb.

samara. A dry, indehiscent, usually one-seeded, winged fruit.

sapogenous. Producing saponins.

saponin. Substance characterized by the ability to foam and form emulsions in water; used as a detergent.

sarcocaulescent. Fleshy stemmed; a term applied to trees and shrubs that have exaggerated stem diameters and relatively soft wood.

scaberulous. Diminutive of scabrous.

scabrous. Rough to the touch as the result of epidermal structure or presence of short stiff hairs.

scandent. Climbing or adapted for climbing.

scarification. Any chemical or physical treatment of a hard seed coat that renders it permeable.

sclerophyll. A firm leaf, with a relatively large amount of strengthening tissue, which retains its firmness (and often also its shape) when physiologically wilted.

scurfy. Beset with small branlike scales.

semidesert grassland. A biotic community located along a moisture gradient between desertscrub and woodland, usually in areas with warm-season rainfall. The vegetation typically consists of an open savanna with scattered low trees (*Prosopis* species), shrubs (*Celtis pallida, Calliandra eriophylla, Ziziphus obtusifolia, Acacia greggii, Opuntia* species) and rosette plants (*Yucca* species, *Dasylirion* species, *Nolina* species) in a lower matrix of grasses (*Hilaria* species, *Digitaria californica, Aristida* species, *Bouteloua* species, especially *B. eriopoda*) and chamaephytes (*Gutierrezia* species, *Isocoma tenuisecta, Eriogonum wrightii*). Also called desert grassland or Apacherian savanna. Extensive areas have lost much of their former grass cover while woody plants have increased over the past century.

senescent. Aging; growing old.

sepal. One of the outermost set of floral leaves, typically green or greenish and more or less leafy in texture.

sericeous. Silky, from the presence of long, slender, soft, more or less appressed hairs.

serrate. Toothed along the margin with sharp, forward-pointing teeth.

serrulate. Diminutive of serrate.

sessile. Attached by the base, as in leaves without petioles.

Sinaloan thornscrub. Vegetation covering much of southern and southeastern Sonora, neighboring Sinaloa, and southern Baja California Sur having a preponderance of drought-deciduous, often thorny, pinnate-leaved, multitrunked trees and shrubs forming an open to dense canopy 2–7 m high. Typical plants include *Fouquieria* species, *Acacia* species, *Cercidium praecox*, and *Jatropha* species. This is the home of the elegant quail, *Lophortyx douglasii*.

sinuate. With a strongly wavy margin.

Sonoran desertscrub. Vegetation occupying the arid region around the head of the Gulf of California, taking in the western half of the state of Sonora, Mexico, as well as large areas in southeastern California, southwestern Arizona, and the Baja California peninsula. Typical plants include *Cercidium microphyllum, Pachycereus pringlei, Carnegiea gigantea, Olneya tesota, Larrea tridentata,* and *Ambrosia* species.

Sonoran savanna grassland. A fairly open vegetation of scattered microphyllous shrubs and short trees with an understory dominated by short-lived warm-season grasses including species of *Bouteloua, Aristida,* and *Cathesticum.* This vegetation is thought to have occurred on deep, fine-textured soils in subtropical semiarid climates near the southeastern margins of the Sonoran Desert, but it has almost completely changed to thornscrub with the advent of livestock ranching.

spatulate. Shaped like a spatula, rounded above and narrowed toward the base.

spike. A more or less elongate inflorescence of the racemose (indeterminate) type, with sessile or subsessile flowers. (Compare *head, raceme*)

spine. A firm, slender, sharp-pointed structure, representing a modified leaf or stipule; more loosely, any structure having the appearance of a true spine. (Compare *prickle, thorn*) spinose. Provided with spines.

staminate flower. A flower with one or more stamens but no pistil.

staminode. A modified stamen that does not produce pollen.

stellate. Star shaped. Stellate hairs have several or many branches connected at their bases.

stigma. The part of the pistil that is receptive to pollen.

stipitate. Borne on a stipe, which broadly speaking is the stalk upon which a structure is borne.

stipule. One of a pair of appendages found at the base of many leaf petioles.

stoma(ta). A special kind of intercellular opening in epidermal tissue, bounded by a pair of guard cells, which, under certain conditions, close off the space by changing shape.

stratification. When referring to seed germination, exposure to cold temperatures to effect germination.

striate. Marked with fine, more or less parallel lines.

sub-. Latin prefix, meaning under, almost, or not quite.

suberin. A waxlike substance present in the walls of certain plant cells.

subulate. Awl shaped: prolonged toward apex with extended point.

suffrutescent. Half-shrubby, or somewhat shrubby, or dying back to a persistent, woody base.

suture. A seam or line of fusion; usually applied to the vertical lines along which a fruit may dehisce.

sympatric. Occupying the same geographic region. (Compare *allopatric*)

sympetalous. With petals attached to each other, at least at the base.

taxon. Any taxonomic entity of whatever rank (e.g., family, genus, subspecies).

tendril. A slender, coiling or twining organ by which a climbing plant grasps its support.

terete. Cylindrical; round in cross section.

thorn. A stiff, woody, modified stem with a sharp point. (Compare *prickle, spine*)

tomentose. Covered with matted or tangled, woolly hairs.

tomentulose. Diminutive of tomentose.

tomentum. A covering of tangled or matted, woolly hairs.

torose. Alternately contracted and expanded.

torulose. Diminutive of torose.

trichome. Any hairlike outgrowth of the epidermis.

trigonous. With three angles.

trimerous. With three parts of a kind.

truncate. With apex or base transversely straight, as if cut off.

tubercle. A small swelling or projection, usually distinct in color or texture from the organ on which it is borne.

tuberculate. Bearing tubercles.

tuffaceous. Pertaining to tuff, a rock consisting of compressed aerially ejected volcanic detritus, usually more or less stratified.

turgor. The rigidity of a plant and its cells and organs resulting from hydrostatic pressure exerted on the cell walls.

twice-pinnate. Doubly pinnately compound leaf with the leaflets attached to paired rachillae, which are themselves attached to a primary rachis; same as bipinnate.

umbel. An inflorescence in which a number of pedicels, nearly equal in length, spread from a common center.

unarmed. Lacking spines, thorns, or prickles.

uncinate. Hooked at the tip.

urceolate. Urn shaped or pitcherlike, contracted at or just below the mouth.

utricle. A small, thin-walled, one-seeded, more or less inflated fruit.

valve. One of the portions of the ovary wall into which a capsule separates at maturity.

vascular. Pertaining to conduction. In plants, the xylem and phloem are vascular tissues.

velutinous. Velvety; covered with a fine, soft, short, spreading pubescence.

vespertine. Pertaining to the evening.

vibratile pollination. A type of pollination in which native bees coax pollen from anthers by shaking them.

villous. Pubescent with long, soft, often bent or curved but not matted hairs.

virgate. Wandlike; slender, straight, and erect.

viscid. Sticky or greasy.

viscin. The sticky substance forming threads that unite pollen grains.

viviparous. Bearing seeds that germinate within the fruit from which they obtain nourishment (e.g., mangroves).

water potential. The difference between the energy of water in the system being considered and of pure, free water at the same temperature. The water potential of pure water is zero, so that of a solution will be negative. If there is a gradient of water potential between two cells, water will diffuse down the gradient until equilibrium is reached.

wing. One of the two lateral petals in flowers of the legume family having pea-like flowers.

xeric. Having very little moisture; tolerating or adapted to dry conditions.

xerophyte. A plant adapted to life in dry places.

Literature Cited

Abe, Y. 1982. Phenology of tetraploid creosotebush, *Larrea tridentata* (DC.) Cov., at the northeastern edge of the Sonoran Desert. Ph.D. dissertation, University of Arizona, Tucson.

Ackerman, T. L. 1979. Germination and survival of perennial plant species in the Mojave Desert. *Southwestern Naturalist* 24: 399–408.

Ackerman, T. L., E. M. Romney, A. Wallace, J. E. Kinnear. 1980. Phenology of desert shrubs in southern Nye County, Nevada. *Great Basin Naturalist Memoirs* 4: 4–23.

Acuna, P. I., N. C. Garwood. 1987. Effect of light and scarification on seeds of five species of secondary tropical trees. *Revista de Biología Tropical* 35: 203–208.

Adams, W. W., S. D. Smith, C. B. Osmond. 1987. Photoinhibition of the CAM succulent *Opuntia basilaris,* growing in Death Valley, California: Evidence from ^{77}K fluorescence and quantum yield. *Oecologia* 71: 221–228.

Addicott, J. F. 1986. Variation in the costs and benefits of mutualism: The interaction between yuccas and yucca moths. *Oecologia* 70: 486–494.

Aker, C. L. 1982. Spatial and temporal dispersion patterns of pollinators and their relationship to the flowering strategy of *Yucca whipplei. Oecologia* 54: 243–252.

Aker, C. L., D. Udovic. 1981. Oviposition and pollination behavior of the yucca moth, *Tegeticula maculata* (Lepidoptera: Prodoxidae), and its relation to the reproductive biology of *Yucca whipplei* (Agavaceae). *Oecologia* 49: 96–101.

Al-Ani, H., B. R. Strain, H. A. Mooney. 1972. The physiological ecology of diverse populations of the desert shrub *Simmondsia chinensis. Journal of Ecology* 60: 41–57.

Albee, B. J., L. M. Shultz, S. Goodrich. 1988. *Atlas of the Vascular Plants of Utah.* Utah Museum of Natural History, Salt Lake City.

Al Charchafchi, F. M. R., M. A. Clor. 1989. Inhibition of germination and seedling development of *Atriplex canescens. Annals of Arid Zone* 28: 113–116.

Alcorn, S. M., E. B. Kurtz. 1959. Some factors affecting the germination of seed of the saguaro cactus (*Carnegiea gigantea*). *American Journal of Botany* 46: 526–529.

Alcorn, S. M., C. May. 1962. Attrition of a saguaro forest. *Plant Disease Reporter* 46: 156–158.

Alcorn, S. M., S. E. McGregor, G. D. Butler, E. B. Kurtz. 1959. Pollination requirements of the saguaro (*Carnegiea gigantea*). *Cactus and Succulent Journal (U.S.)* 31: 39–41.

Alcorn, S. M., S. E. McGregor, G. Olin. 1961. Pollination of saguaro cactus by doves, nectar-feeding bats, and honey bees. *Science* 133: 1594–1595.

Alcorn, S. M., S. E. McGregor, G. Olin. 1962. Pollination requirements of the organ pipe cactus. *Cactus and Succulent Journal (U.S.)* 34: 135–138.

Alfieri, F. J., P. M. Mottola. 1983. Seasonal changes in the phloem of *Ephedra californica* Wats. *Botanical Gazette* 144: 240–246.

Ambasht, R. S. 1963. Ecological studies on *Alhagi camelorum* Fisch. *Tropical Ecology* 4: 72–82.

Anderson, C. 1971. New names and combinations in *Trixis* (Compositae). *Brittonia* 23: 347–353.

Anonymous. 1902. A wonderful plant doomed. *Plant World* 5: 12–13.

Anonymous. 1979. The severe freeze of 1978–1979 in the southwestern United States. *Desert Plants* 1: 37–39.

Archer, S., C. Scifres, C. R. Bassham, R. Maggio. 1988. Autogenic succession in a subtropical savanna: Conversion of grassland to thorn woodland. *Ecological Monographs* 58: 111–127.

Argus, G. W., D. J. White (eds.). 1982. *Atlas of the Rare Vascular Plants of Ontario*. Botany Division, National Museum of Natural Sciences, Ontario.

Armond, P. A., V. Schreiber, O. Bjorkman. 1978. Photosynthetic acclimation to temperature in the desert shrub, *Larrea divaricata*, part 2: Light-harvesting efficiency and electron transport. *Plant Physiology* 61: 411–415.

Arnott, H. J. 1962. The seed, germination, and seedling of *Yucca*. *University of California Publications in Botany* 35: 1–164.

Aronson, J. A., D. Pasternak, A. Danon. 1988. Introduction and first evaluation of 120 halophytes under seawater irrigation. In E. E. Whitehead, C. F. Hutchinson, B. N. Timmermann, R. G. Varady (eds.), *Arid Lands Today and Tomorrow*, 737–746. Westview Press, Boulder, Colo.

Arreguin-Sanchez, M., R. Palacios-Chavez, D. L. Quiroz-Garcia, D. Ramos-Zamora. 1986. Morfología de los granos de polen de *Jacquinia* (Theophrastaceae) de Chamela, Jalisco, México. *Phytologia* 61: 161–163.

Arroyo, M. T. K. 1981. Brongniartieae (Benth.) Hutch. In R. M. Polhill, P. H. Raven (eds.), *Advances in Legume Systematics*, vol. 2, part 1, 387–390. Proceedings of the International Legume Conference, 24–29 July 1978, Kew, England.

Artz, M. C. 1989. Impacts of linear corridors on perennial vegetation in the east Mojave Desert: Implications for environmental management and planning. *Natural Areas Journal* 9: 117–129.

Aschmann, H. 1959. *The Central Desert of Baja California: Demography and Ecology.* University of California Press, Berkeley.

Askham, L. R., D. R. Cornelius. 1971. Influence of desert saltbush saponin on germination. *Journal of Range Management* 24: 439–442.

Askham, L. R., D. R. Cornelius. 1972. A preliminary investigation of water soluble inhibitors in *Atriplex polycarpa* using the fungus *Trichoderma viride*. *Agronomy Journal* 64: 82–83.

Augspurger, C. K. 1986. Morphology and dispersal potential of wind-dispersed diaspores of neotropical trees. *American Journal of Botany* 73: 353–363.

Ault, J. R., W. J. Blackmon. 1987. In vitro propagation of *Ferocactus acanthodes* (Cactaceae). *HortScience* 22: 126–127.

Averett, J. E., A. M. Powell. 1972. Chromosome numbers in *Physalis* and *Solanum* (Solanaceae). *Sida* 5: 3–7.

Axelrod, D. I. 1950. *Evolution of Desert Vegetation in Western North America.* Carnegie Institute of Washington Publication no. 590, pp. 215–306. Washington, D.C.

Axelrod, D. I. 1979. *Age and Origin of Sonoran Desert Vegetation.* Occasional Papers of the California Academy of Sciences no. 132. San Francisco.

Babos, K. 1979. Fiber length frequency of some Cuban tree species. *Acta Botanica, Academiae Scientiarum Hungaricae* 25: 177–186.

Babos, K., A. Borhidi. 1978. Xylotomic study of some woody plant species from Cuba: 2. *Acta Botanica, Academiae Scientiarum Hungaricae* 24: 235–262.

Badan-Dangon, C., J. Koblinsky, T. Baumgartner. 1985. Spring and summer in the Gulf of California: Observations of surface thermal patterns. *Oceanológica Acta* 8: 13–22.

Bailey, L. H. 1936. *Washingtonia. Gentes Herbarum* 3: 53–82.

Bailey, L. H. 1937. *Erythea. Gentes Herbarum* 4: 84–118.

Baker, H. G. 1983. *Ceiba pentandra* (Ceyba, Ceiba, Kapok Tree). In D. H. Janzen (ed.), *Costa Rican Natural History,* 212–215. University of Chicago Press, Chicago.

Baker, H. G., R. W. Cruden, I. Baker. 1971. Minor parasitism in pollination biology and its community function: The case of *Ceiba acuminata. BioScience* 21: 1127–1129.

Baker, M. A., B. D. Parfitt. 1986. Chromosome number reports XCI. *Taxon* 35: 405–406.

Baker, M. A., D. J. Pinkava. 1987. A cytological and morphometric analysis of a triploid apomict, *Opuntia* X *kelvinensis* (subgenus Cylindropuntia, Cactaceae). *Brittonia* 39: 387–401.

Bamberg, S. A., A. T. Vollmer, G. E. Kleinkopf, T. L. Ackerman. 1976. A comparison of seasonal primary production of Mojave Desert shrubs during wet and dry years. *American Midland Naturalist* 95: 398–405.

Banerjee, S., A. K. Sharma. 1988. Structural differences of chromosomes in diploid *Agave. Cytologia* 53: 415–420.

Barajas-Morales, J. 1983. Detalles ultraestructurales de la madera de algunas Boraginaceae de México. *Boletín de la Sociedad Botánica de México* 45: 3–14.

Barbour, M. G. 1968. Germination requirements of the desert shrub *Larrea divaricata. Ecology* 49: 915–923.

Barbour, M. G. 1969. Patterns of genetic similarity between *Larrea divaricata* of North and South America. *American Midland Naturalist* 81: 54–67.

Barbour, M. G., G. Cunningham, W. C. Oechel, S. A. Bamberg. 1977. Growth and development, form and function. In T. J. Mabry, J. H. Hunziker, D. R. Difeo (eds.), *Creosotebush: Biology and Chemistry of Larrea in New World Deserts,* 48–91. Dowden, Hutchinson & Ross, Stroudsburg, Penn.

Barbour, M. G., T. M. De Jong. 1977. Response of West Coast beach taxa to salt spray, seawater inundation, and soil salinity. *Bulletin of the Torrey Botanical Club* 104: 29–34.

Barcikowski, W., P. S. Nobel. 1984. Water relations of cacti during desiccation: Distribution of water in tissues. *Botanical Gazette* 145: 110–115.

Barkley, F. A. 1937. A monographic study of *Rhus* and its immediate allies in North and Central America, including the West Indies. *Annals of the Missouri Botanical Garden* 24: 265–500.

Barneby, R. C. 1977. *Daleae imagines:* An illustrated revision of *Errazurizia* Philippi, *Psorothamnus* Rydberg, *Marina* Liebmann, and *Dalea* Lucanus emend. Barneby, including all species of Leguminosae tribe *Amorpheae* Borissoua ever referred to *Dalea. Memoirs of the New York Botanical Garden* 27: 1–891.

Barneby, R. C. 1989. Fabales. In A. Cronquist, A. H. Holmgren, N. H. Holmgren, J. L. Reveal, P. K. Holmgren, *Intermountain Flora: Vascular Plants of the Inter-*

mountain West, U.S.A., vol. 3, part B, New York Botanical Garden, Bronx, N.Y.

Barneby, R. C. 1991. *Sensitivae censitae:* A description of the genus *Mimosa* Linnaeus (Mimosaceae) in the New World. *Memoirs of the New York Botanical Garden* 65: 1–835.

Barrow, J. R. 1987. The effects of chromosome number on sex expression in *Atriplex canescens. Botanical Gazette* 148: 379–385.

Barth, R. C., J. O. Klemmedson. 1982. Amount and distribution of dry matter, nitrogen, and organic carbon in soil-plant systems of mesquite and palo verde. *Journal of Range Management* 35: 412–418.

Barth, R. C., J. O. Klemmedson. 1986. Seasonal and annual changes in biomass, nitrogen, and carbon of mesquite and palo verde ecosystems. *Journal of Range Management* 39: 108–112.

Bassett, I. J. 1969. IOPB chromosome number reports XXI. *Taxon* 18: 310–315.

Bassett, I. J., C. W. Crompton. 1971. IOPB chromosome number reports XXXIV. *Taxon* 20: 785–797.

Bates, D. M. 1976. Chromosome numbers in the Malvales, part 3: Miscellaneous counts from the Byttneriaceae and Malvaceae. *Gentes Herbarum* 11: 143–150.

Bates, D. M., O. J. Blanchard. 1970. Chromosome numbers in the Malvales, part 2: New or otherwise noteworthy counts relevant to classification in the Malvaceae, tribe Malveae. *American Journal of Botany* 57: 927–934.

Bates, J. 1987. Winter ecology of the gray vireo (*Virea vicinior*) in Sonora, Mexico. M.S. thesis, University of Arizona, Tucson.

Baum, B. R. 1967. Introduced and naturalized tamarisks in the United States and Canada (Tamaricaceae). *Baileya* 15: 19–25.

Bawa, K. S. 1973. Chromosome numbers of tree species of a lowland tropical community. *Journal of the Arnold Arboretum* 54: 422–434.

Beatley, J. C. 1974a. Effects of rainfall and temperature on the distribution and behavior of *Larrea tridentata* (creosote-bush) in the Mojave Desert of Nevada. *Ecology* 55: 245–261.

Beatley, J. C. 1974b. Phenologic events and their environmental triggers in Mohave Desert ecosystems. *Ecology* 55: 856–863.

Beatley, J. C. 1976. *Vascular Plants of the Nevada Test Site and Central-Southern Nevada: Ecologic and Geographic Distributions.* National Technical Information Service, Springfield, Va.

Beauchamp, R. M. 1980. New names in American acacias. *Phytologia* 46: 5–9.

Beauchamp, R. M. 1986. *A Flora of San Diego County, California.* Sweetwater River Press, National City, Calif.

Becker, R. 1983. Nutritional evaluation and chemical composition of seeds from desert ironwood (*Olneya tesota*). *International Tree Crops Journal* 2: 297–312.

Becker, R., O. K. K. Grosjean. 1980. A compositional study of pods of two varieties of mesquite (*Prosopis glandulosa, P. velutina*). *Journal of Agricultural and Food Chemistry* 28: 22–25.

Beckett, R. E., R. S. Stitt. 1935. *The Desert Milkweed (Asclepias subulata) as a Possible Source of Rubber.* USDA Technical Bulletin no. 472. Government Printing Office, Washington, D.C.

Beckett, R. E., R. S. Stitt, E. N. Duncan. 1938. *Rubber Content and Habits of a Second Desert Milkweed (Asclepias erosa) of Southern California and Arizona.* USDA Technical Bulletin no. 604. Government Printing Office, Washington, D.C.

Bell, C. R. 1965. Documented chromosome numbers of plants 65:3. *Sida* 2: 168–170.

Bell, E. A. 1971. Comparative biochemistry of non-protein amino acids. In J. B. Harborne, D. Boulter, B. L. Turner (eds.), *Chemotaxonomy of the Leguminosae,* 179–206. Academic Press, New York.

Bell, W. H., E. F. Castetter. 1937. The utilization of mesquite and screwbean by the aborigines in the American Southwest. Ethnobiological studies in the American Southwest no. 5. *University of New Mexico Bulletin, Biological Series* 5 (2): 1–55.

Bell, W. H., E. F. Castetter. 1941. The utilization of yucca, sotol, and beargrass by the aborigines in the American Southwest. Ethnobiological studies in the American Southwest no. 7. *University of New Mexico Bulletin, Biological Series* 5 (5): 1–74.

Belmares, H., A. Barrera, F. Hernández, L. F. Ramos, E. Castillo, A. B. Motomochi. 1981. Research and development of *Larrea tridentata* as a source of raw materials. In E. Campos Lopez, T. J. Mabry, S. Fernandez Tavizon (eds.), *Larrea,* 247–276. Consejo Nacional de Ciencia y Tecnológia, México, D.F.

Bennert, H. W., B. Schmidt. 1983. Investigations on the salt secretion of *Atriplex hymenelytra* (Chenopodiaceae). *Flora* 174: 341–355.

Bennert, H. W., B. Schmidt. 1984. On the osmoregulation in *Atriplex hymenelytra* (Torr.) Wats. (Chenopodiaceae). *Oecologia* 62: 80–84.

Bennert, W. H., H. A. Mooney. 1979. The water relations of some desert plants in Death Valley, California. *Flora* 168: 405–427.

Benseler, R. W. 1975. Floral biology of California buckeye. *Madroño* 23: 41–53.

Benson, L. 1941. The mesquites and screw-beans of the United States. *American Journal of Botany* 28: 748–754.

Benson, L. 1943. Revisions of status of southwestern trees and shrubs. *American Journal of Botany* 30: 230–240.

Benson, L. 1950. *The Cacti of Arizona.* 2d ed. University of Arizona Press, Tucson.

Benson, L. 1969. *The Cacti of Arizona.* 3d ed. University of Arizona Press, Tucson.

Benson, L. 1982. *The Cacti of the United States and Canada.* Stanford University Press, Stanford, Calif.

Benson, L., R. A. Darrow. 1954. *Trees and Shrubs of the Southwestern Deserts.* University of Arizona Press, Tucson.

Benson, L., R. A. Darrow. 1981. *Trees and Shrubs of the Southwestern Deserts.* 3d ed. University of Arizona Press, Tucson.

Benzioni, A., R. L. Dunstone. 1986. Jojoba: Adaptation to environmental stress and the implications for domestication. *Quarterly Review of Biology* 61: 177–199.

Berry, W. L., P. S. Nobel. 1985. Influence of soil and mineral stresses on cacti. *Journal of Plant Nutrition* 8: 679–696.

Blackwell, W. H., M. J. Powell. 1981. A preliminary note on pollination in the Chenopodiaceae. *Annals of the Missouri Botanical Garden* 68: 524–526.

Blake, S. F. 1913. A revision of *Encelia* and some related genera. *Proceedings of the American Academy of Arts and Sciences* 49: 346–396.

Blake, S. F. 1918. A revision of the genus *Viguiera. Contributions from the Gray Herbarium* 54: 1–197.

Blauer, E. C., A. P. Plummer, E. D. McArthur, R. Stevens, B. C. Giunta. 1976. *Characteristics and Hybridization of Important Intermountain Shrubs,* vol. 2, *Chenopods.* USDA Forest Service Research Paper INT–77. Ogden, Utah.

Blom, P. E. 1987. Host plants of *Moneilema michelbacheri* Linsley in Baja California. *Coleopterists' Bulletin* 41: 358–360.

Blom, P. E., W. H. Clark. 1980. Observations of ants (Hymenoptera: Formicidae) visiting extrafloral nectaries of the barrel cactus *Ferocactus gracilis* Gates (Cactaceae) in Baja California, Mexico. *Southwestern Naturalist* 25: 181–196.

Bloss, H. E. 1988. Symbiotic microflora and their role in the ecology of desert plants. In E. E. Whitehead, C. F. Hutchinson, B. N. Timmermann, R. G. Varady (eds.), *Arid Lands Today and Tomorrow,* 1031–1040. Westview Press, Boulder, Colo.

Bontrager, O. E., C. J. Scifres, D. L. Drawe. 1979. Huisache control by power grubbing. *Journal of Range Management* 32: 185–188.

Booth, D. T. 1985. The role of fourwing saltbush in mined land reclamation: A viewpoint. *Journal of Range Management* 38: 562–565.

Bovey, R. W., J. R. Baur, H. L. Morton. 1970. Control of huisache and associated woody species in South Texas. *Journal of Range Management* 23: 47–50.

Bowers, J. E. 1980. Flora of Organ Pipe Cactus National Monument. *Journal of the Arizona-Nevada Academy of Science* 15: 1–11, 33–47.

Bowers, J. E. 1980–81. Catastrophic freezes in the Sonoran Desert. *Desert Plants* 2: 232–236.

Bowers, J. E. 1982. The plant ecology of inland dunes in western North America. *Journal of Arid Environments* 5: 199–220.

Bowers, J. E. 1984. Plant geography of southwestern sand dunes. *Desert Plants* 6: 31–42, 51–54.

Bowers, J. E. 1986. *Seasons of the Wind: A Naturalist's Look at the Plant Life of Southwestern Sand Dunes.* Northland Press, Flagstaff, Az.

Bowers, J. E. 1994. Natural conditions for seedling emergence of three woody plants in the northern Sonoran Desert. *Madroño* 41:73–84.

Bowers, J. E., M. A. Dimmitt. 1994. Flowering phenology of six woody plants in the northern Sonoran Desert. *Bulletin of the Torrey Botanical Club* 121: 215–29.

Boyce, S. G. 1954. The salt spray community. *Ecological Monographs* 24: 26–67.

Boyd, R. S., G. D. Brum. 1983a. Predispersal reproductive attrition in a Mojave Desert population of *Larrea tridentata* (Zygophyllaceae). *American Midland Naturalist* 110: 14–24.

Boyd, R. S., G. D. Brum. 1983b. Postdispersal reproductive biology of a Mojave Desert population of *Larrea tridentata* (Zygophyllaceae). *American Midland Naturalist* 110: 25–36.

Bradley, W. G. 1966. Populations of two Sonoran Desert plants, and deductions as to factors limiting their northward extension. *Southwestern Naturalist* 11: 395–401.

Brandegee, T. S. 1891. Flora of the Cape Region of Baja California. *Proceedings of the California Academy of Sciences* 3: 218–227.

Bravo-Hollis, H. 1978. *Las Cactaceas de México,* vol. 1. Universidad Nacional Autónoma de México, México, D.F.

Brenan, J. P. M. 1963. Notes on African Caesalpinioideae: The genus *Peltophoropsis* Chiov. and its relationships. *Kew Bulletin of Miscellaneous Information* 17: 203–209.

Briggs, J. A. 1984. Seed production of *Atriplex canescens* (Pursh) Nutt. in southern Arizona. In A. R. Tiedemann, E. D. McArthur, H. C. Stutz, R. Stevens, K. L. Johnson (comps.), *Proceedings—Symposium on the Biology of Atriplex and Related Chenopods, Provo, Utah, May 2–6, 1983,* 187–190. USDA Forest Service General Technical Report INT–172. Ogden, Utah.

Brookbank, G. 1978. *Native Mesquite Trees.* University of Arizona Cooperative Extension Service Series Q–355. Tucson.

Brooks, R. E., R. L. McGregor. 1979. New records and notes on the vascular flora of Kansas for 1978. *State Biological Survey of Kansas Technical Publication No. 8,* 87–92. Lawrence.

Brown, D. E. 1982a. Sinaloan thornscrub. *Desert Plants* 4: 101–105.

Brown, D. E. 1982b. Biotic communities of the American Southwest—United States and Mexico. *Desert Plants* 4: 1–342.

Brown, D. E., N. B. Carmony, C. H. Lowe, R. M. Turner. 1976. A second locality for native California fan palms (*Washingtonia filifera*) in Arizona. *Journal of the Arizona Academy of Science* 11: 37–41.

Brown, D. E., C. H. Lowe. 1980. *Biotic Communities of the Southwest.* USDA Forest Service General Technical Report RM–78. Fort Collins, Colo.

Brown, D. E., R. A. Minnich. 1986. Fire and changes in creosote bush scrub of the western Sonoran Desert, California. *American Midland Naturalist* 116: 411–422.

Brown, J. H. 1984. On the relationship between abundance and distribution of species. *American Naturalist* 124: 255–279.

Brown, J. H., A. Kodric-Brown, T. G. Whitham, H. W. Bond. 1981. Competition between hummingbirds and insects for the nectar of two species of shrubs. *Southwestern Naturalist* 26: 133–146.

Brown, J. H., G. A. Lieberman, W. F. Dengler. 1972. Woodrats and cholla: Dependence of a small mammal population on the density of cacti. *Ecology* 72: 310–313.

Brum, G. D. 1973. Ecology of the saguaro (*Carnegiea gigantea*): Phenology and establishment in marginal populations. *Madroño* 22: 195–204.

Bryson, R. A. 1957. *The Annual March of Precipitation in Arizona, New Mexico, and Northwestern Mexico.* Institute of Atmospheric Physics, University of Arizona Technical Reports on the Meteorology and Climatology of Arid Regions no. 6. Tucson.

Buchmann, S. L. 1987a. The ecology of oil flowers and their bees. *Annual Review of Ecology and Systematics* 18: 343–369.

Buchmann, S. L. 1987b. Floral biology of jojoba (*Simmondsia chinensis*), an anemophilous plant. *Desert Plants* 8: 3, 111–124.

Buchmann, S. L., J. P. Hurley. 1978. A biophysical model for buzz pollination in angiosperms. *Journal of Theoretical Biology* 72: 639–657.

Buchmann, S. L., C. E. Jones, L. J. Colin. 1977. Vibratile pollination of *Solanum douglasii* and *S. xantii* (Solanaceae) in southern California. *Wasmann Journal of Biology* 35: 1–25.

Buchmann, S. L., M. K. O'Rourke, K. J. Niklas. 1989. Aerodynamics of *Ephedra trifurca,* part 3: Selective pollen capture by pollination droplets. *Botanical Gazette* 150: 122–131.

Buehrer, T. F., L. Benson. 1945. *Rubber Content of Native Plants of the Southwestern Desert.* Arizona Agricultural Experiment Station Technical Bulletin no. 108. Tucson.

Buffington, L. C., C. H. Herbel. 1965. Vegetational changes on a semidesert grassland range. *Ecological Monographs* 35: 139–164.

Bullock, A. A. 1936. Notes on the Mexican species of the genus *Bursera. Kew Bulletin of Miscellaneous Information* 1936: 346–387.

Bullock, S. H., R. Ayala, I. Baker, H. G. Baker. 1987. Reproductive biology of the tree *Ipomoea wolcottiana* (Convolvulaceae). *Madroño* 34: 304–314.

Bunting, S. C., H. A. Wright, L. F. Neuenschwander. 1980. Long-term effects of fire on cactus in the southern mixed prairie of Texas. *Journal of Range Management* 33: 85–88.

Burgess, T. L. 1985. *Agave* adaptation to aridity. *Desert Plants* 7: 39–50.

Burgess, T. L. 1988. The relationship between climate and leaf shape in the *Agave cerulata* complex. Ph.D. dissertation, University of Arizona, Tucson.

Burk, J. H., W. A. Dick-Peddie. 1973. Comparative production of *Larrea divaricata* Cav. on three geomorphic surfaces in southern New Mexico. *Ecology* 54: 1094–1102.

Burkart, A. 1964. Leguminosas nuevas o críticas VI. *Darwiniana* 13: 428–448.

Burkart, A. 1976. A monograph of the genus *Prosopis* (Leguminosae, subfam. Mimosoideae). *Journal of the Arnold Arboretum* 57: 219–249, 450–525.

Burkart, A., A. Carter. 1976. Notas en el género *Cercidium* (Caesalpinioideae) en Sud America. *Darwiniana* 20: 305–311.

Bush, J. K., O. W. Van Auken. 1986. Light requirements of *Acacia smallii* and *Celtis laevigata* in relation to secondary succession on floodplains on South Texas, U.S. A. *American Midland Naturalist* 115: 118–122.

Butanda-Cervera, A., A. Vázquez-Yanes, L. Trejo. 1978. La polinazación quiropterofila: Una revisión bibliográfica. *Biótica* 3: 29–35.

Bye, R. A., D. Burgess, A. M. Trias. 1975. Ethnobotany of the western Tarahumara of Chihuahua, Mexico. *Botanical Museum Leaflets, Harvard University* 24: 85–112.

Cable, D. R. 1975. *Range Management in the Chaparral Type and Its Ecological Basis: The Status of Our Knowledge.* USDA Forest Service Research Paper RM–155. Fort Collins, Colo.

Cable, D. R. 1977a. Soil water changes in creosotebush and bur-sage during a dry period in southern Arizona. *Journal of the Arizona Academy of Science* 12: 15–20.

Cable, D. R. 1977b. Seasonal use of soil water by mature velvet mesquite. *Journal of Range Management* 30: 4–11.

Caceres, A., O. Cano, B. Samayoa, L. Aguilar. 1990. Plants used in Guatemala for the treatment of gastrointestinal disorders. 1. Screening of 84 plants against enterobacteria. *Journal of Ethnopharmacology* 30: 55–74.

Cadbury, D. A., J. G. Hawkes, R. C. Readett. 1971. *A Computer-Mapped Flora: A Study of the County of Warwickshire.* Academic Press, New York.

Camilleri, J. C., G. Ribi. 1983. Leaf thickness of mangroves (*Rhizophora mangle*) growing in different salinities. *Biotropica* 15: 139–141.

Campbell, C. J., W. Green. 1968. Perpetual succession of stream-channel vegetation in a semiarid region. *Journal of the Arizona Academy of Science* 5: 86–98.

Campbell, C. J., J. E. Strong. 1964. Salt gland anatomy in *Tamarix pentandra* (Tamaricaceae). *Southwestern Naturalist* 9: 232–238.

Cancino, J., J. L. León de la Luz, R. Coria, H. Romero. 1993. Effect of heat treatment on germination of seeds of cardon (*Pachycereus pringlei* [S. Wats.] Britt. & Rose, Cactaceae). *Journal of the Arizona-Nevada Academy of Science* 27: 47–54.

Cannon, W. A. 1905. On the transpiration of *Fouquieria splendens. Bulletin of the Torrey Botanical Club* 32: 397–414.

Cannon, W. A. 1908. *The Topography of the Chlorophyll Apparatus in Desert Plants.* Carnegie Institution of Washington Publication no. 98. Washington, D.C.

Cannon, W. A. 1911. *The Root Habits of Desert Plants.* Carnegie Institution of Washington Publication no. 131. Washington, D.C.

Cannon, W. A. 1912. Deciduous rootlets of desert plants. *Science* 35: 632–633.

Cannon, W. A. 1916a. Distribution of the cacti with especial reference to soil temperature and soil moisture. *American Naturalist* 50: 435–442.

Cannon, W. A. 1916b. Rate of growth of *Covillea tridentata* in relation to the temperature of the soil. *Carnegie Yearbook* 15: 75–76.

Carlquist, S. 1985. Observations on functional wood histology of vines and lianas: Vessel dimorphism, tracheids, vasicentric tracheids, narrow vessels, and parenchyma. *Aliso* 11: 139–157.

Carlquist, S. 1989a. Wood anatomy of *Cercidium* (Fabaceae), with emphasis on vessel wall sculpture. *Aliso* 12: 235–255.

Carlquist, S. 1989b. Wood and bark anatomy of the New World species of *Ephedra*. *Aliso* 12: 441–483.

Carlquist, S., V. M. Eckhart, D. C. Michener. 1984. Wood anatomy of Polemoniaceae. *Aliso* 10: 547–572.

Carlquist, S., M. A. Hanson. 1991. Wood and stem anatomy of Convolvulaceae, a survey. *Aliso* 13: 51–94.

Carothers, S. W., and B. T. Brown. 1991. The Colorado River through the Grand Canyon: Natural history and human change. The University of Arizona Press, Tucson.

Carpenter, D. E., M. G. Barbour, C. J. Bahre. 1986. Old field succession in Mojave Desert scrub. *Madroño* 33: 111–122.

Carter, A. M. 1974a. Evidence for the hybrid origin of *Cercidium sonorae* (Leguminosae: Caesalpinioideae) of northwestern Mexico. *Madroño* 22: 266–272.

Carter, A. M. 1974b. The genus *Cercidium* (Leguminosae: Caesalpinioideae) in the Sonoran Desert of Mexico and the United States. *Proceedings of the California Academy of Sciences* 40: 17–57.

Carter, A. M., N. C. Rem. 1974. Pollen studies in relation to hybridization in *Cercidium* (Leguminosae: Caesalpinioideae). *Madroño* 22: 303–311.

Carter, A. M., V. E. Rudd. 1981. A new species of *Acacia* (Leguminosae: Mimosoideae) from Baja California Sur, Mexico. *Madroño* 28: 220–225.

Castellanos, A. E., F. E. Molina. 1990. Differential survivorship and establishment in *Simmondsia chinensis* (jojoba). *Journal of Arid Environments* 19: 65–76.

Castetter, E. F., W. H. Bell, A. R. Grove. 1938. The early utilization and the distribution of *Agave* in the American Southwest. Ethnobiological studies in the American Southwest no. 6. *University of New Mexico Bulletin, Biological Series* 5 (4): 1–92.

Cave, M. S. 1964. Cytological observations on some genera of the Agavaceae. *Madroño* 17: 163–170.

Center, T. D., C. D. Johnson. 1974. Coevolution of some seed beetles (Coleoptera: Bruchidae) and their hosts. *Ecology* 55: 1096–1103.

Center, T. D., C. D. Johnson. 1976. Host plants and parasites of some Arizona seed-feeding insects. *Annals of the Entomological Society of America* 69: 195–201.

Chamberland, M. 1991. Biosystematics of the *Echinocactus polycephalus* complex (Cactaceae). M.S. thesis, Arizona State University, Tempe.

Chase, U. C., B. R. Strain. 1966. Propagation of some woody desert perennials by stem cuttings. *Madroño* 18: 240–243.

Chatterton, N. J., J. R. Goodin, C. Duncan. 1971. Nitrogen metabolism in *Atriplex polycarpa* as affected by substrate nitrogen and NaCl salinity. *Agronomy Journal* 63: 271–274.

Chatterton, N. J., C. M. McKell. 1969. *Atriplex polycarpa,* part 1: Germination and growth as affected by sodium chloride in water cultures. *Agronomy Journal* 61: 448–450.

Chatterton, N. J., C. M. McKell, J. R. Goodin, F. T. Bingham. 1969. *Atriplex polycarpa,* part 2: Germination and growth in water cultures containing high levels of boron. *Agronomy Journal* 61: 451–453.

Chatterton, N. J., C. M. McKell, B. R. Strain. 1970. Intraspecific differences in temperature-induced respiratory acclimation of desert saltbush. *Ecology* 51: 545–547.

Chazaro, M. 1977. El huizache, *Acacia pennatula* (Schlecht. & Cham.) Benth., una invasora del centro de Veracruz. *Biótica* 2: 1–18.

Cherubini, C. 1954. Numero de cromosomas de algunas especies del género *Prosopis* (Leguminosae-Mimosoideae). *Darwiniana* 10: 637–643.

Chetti, M. B., P. S. Nobel. 1987. High-temperature sensitivity and its acclimation for photosynthetic electron transport reactions of desert succulents. *Plant Physiology* 84: 1063–1067.

Chew, R. M., A. E. Chew. 1965. The primary productivity of a desert shrub (*Larrea tridentata*) community. *Ecological Monographs* 35: 355–375.

Chew, R. M., W. G. Whitford. 1992. A long-term positive effect of kangaroo rats (*Dipodomys spectabilis*) on creosotebushes (*Larrea tridentata*). *Journal of Arid Environments* 22: 375–386.

Clark, W. H., P. E. Blom. 1982. Una localidad adicional de *Lophocereus schottii* forma *monstrosus* en Baja California. *Cactaceas y Suculentas Mexicanas* 27: 75–81.

Clark, W. H., P. L. Comanor. 1973. A quantitative examination of spring foraging of *Veromessor pergandei* (Mayr) in northern Death Valley, California (Hymenoptera: Formicidae). *American Midland Naturalist* 90: 467–474.

Clarke, H. D., D. S. Seigler, J. E. Ebinger. 1989. *Acacia farnesiana* (Fabaceae: Mimosoideae) and related species from Mexico, the southwestern U.S., and the Caribbean. *Systematic Botany* 14: 549–564.

Cockburn, W., I. P. Ting, L. O. Sternberg. 1979. Relationships between stomatal behavior and internal carbon dioxide concentration in crassulacean acid metabolism plants. *Plant Physiology* 63: 1029–1032.

Cockrum, E. L., and Y. Petryszyn. 1991. The long-nosed bat, *Leptonycteris:* An endangered species in the Southwest? *Occasional Papers of the Museum of Texas Tech University* 142: 1–32.

Cody, M. L. 1984. Branching patterns in columnar cacti. In N. S. Margaris, M. Arianoustou-Farragitako, W. C. Oechel (eds.), *Being Alive on Land,* 201–236. Dr. Junk, The Hague, Netherlands.

Cody, M. L., R. Moran, H. Thompson. 1983. The plants. In T. J. Case, M. L. Cody (eds.), *Island Biogeography in the Sea of Cortez,* 49–97. University of California Press, Berkeley.

Cole, K. L. 1986. The lower Colorado Valley: A Pleistocene desert. *Quaternary Research* 25: 392–400.

Cole, K. L. 1990. Late Quaternary vegetation gradients through the Grand Canyon. In J. L. Betancourt, T. R. Van Devender, P. S. Martin (eds.), *Packrat Middens: The Last 40,000 Years of Biotic Change,* 240–258. University of Arizona Press, Tucson.

Comstock, J. P., T. A. Cooper, J. R. Ehleringer. 1988. Seasonal patterns of canopy development and carbon gain in nineteen warm desert shrub species. *Oecologia* 75: 327–335.

Comstock, J. P., J. R. Ehleringer. 1984. Photosynthetic responses to slowly decreasing leaf water potentials in *Encelia frutescens. Oecologia* 61: 241–248.

Comstock, J. P., J. R. Ehleringer. 1986. Canopy dynamics and carbon gain in response to soil water availability in *Encelia farinosa* Gray, a drought-deciduous shrub. *Oecologia* 68: 271–278.

Comstock, J. P., J. R. Ehleringer. 1988. Contrasting photosynthetic behavior in leaves and twigs of *Hymenoclea salsola,* a green-twigged warm desert shrub. *American Journal of Botany* 75: 1360–1370.

Conde, L. F. 1975. Anatomical comparisons of five species of *Opuntia* (Cactaceae). *Annals of the Missouri Botanical Garden* 62: 425–473.

Conn, J. S., E. K. Snyder-Conn. 1981. The relationship of the rock outcrop microhabitat to germination, water relations, and phenology of *Erythrina flabelliformis* (Fabaceae) in southern Arizona. *Southwestern Naturalist* 25: 443–451.

Cooper, L. W., M. J. DeNiro. 1989. Covariance of oxygen and hydrogen isotopic compositions in plant water: Species effects. *Ecology* 70: 1619–1628.

Corbett, J. 1991. *Goatwalking.* Viking Penguin, New York.

Cornett, J. W. 1987. Cold tolerance in the desert fan palm, *Washingtonia filifera* (Arecaceae). *Madroño* 34: 57–62.

Cornett, J. W. 1988. The occurrence of the desert fan palm, *Washingtonia filifera,* in southern Nevada. *Desert Plants* 8: 169–171.

Correll, D. S., H. B. Correll. 1972. *Aquatic and Wetland Plants of Southwestern United States.* Environmental Protection Agency, Washington, D.C.

Correll, D. S., M. C. Johnston. 1970. *Manual of the Vascular Plants of Texas.* Texas Research Foundation, Renner.

Cottam, W. P. 1937. Has Utah lost claim to the Lower Sonoran zone? *Science* 85: 563–564.

Coyle, J., N. C. Roberts. 1975. *A Field Guide to the Common and Interesting Plants of Baja California.* Natural History Publishing Company, La Jolla, Calif.

Croizat, L. 1945. New or critical Euphorbiaceae from the Americas. *Journal of the Arnold Arboretum* 26: 181–196.

Cronquist, A. 1944. Studies in the Simaroubaceae, part 1: The genus *Castela. Journal of the Arnold Arboretum* 25: 122–128.

Cronquist, A. 1945. Studies in the Sapotaceae, part 3: *Dipholis* and *Bumelia. Journal of the Arnold Arboretum* 26: 435–471.

Cronquist, A. 1946. Studies in the Sapotaceae, part 2: Survey of the North American genera. *Lloydia* 9: 241–292.

Crosswhite, F. S. 1980. The annual saguaro harvest and crop cycle of the Papago, with reference to ecology and symbolism. *Desert Plants* 2: 3–61.

Crosswhite, F. S., C. D. Crosswhite. 1984. A classification of life forms of the Sonoran Desert, with emphasis on the seed plants and their survival strategies. *Desert Plants* 5: 131–161, 186–190.

Cruden, R. W., S. M. Hermann-Parker. 1979. Butterfly pollination of *Caesalpinia pulcherrima,* with observations on a psychophilous syndrome. *Journal of Ecology* 67: 155–168.

Cruden, R. W., K. G. Jensen. 1979. Viscin threads, pollination efficiency, and low pollen-ovule ratios. *American Journal of Botany* 66: 875–879.

Cunningham, G. L., J. H. Burk. 1973. The effect of carbonate deposition layers ("caliche") on the water status of *Larrea divaricata. American Midland Naturalist* 90: 474–480.

Cunningham, G. L., J. F. Reynolds. 1978. A simulation model of primary production and carbon allocation in the creosotebush (*Larrea tridentata* [DC.] Cov.). *Ecology* 59: 37–52.

Cunningham, G. L., B. R. Strain. 1969. An ecological significance of seasonal leaf variability in a desert shrub. *Ecology* 50: 400–408.

Cunningham, G. L., J. P. Syvertsen, J. F. Reynolds, J. M. Wilson. 1979. Some effects of soil moisture availability on aboveground production and reproductive allocation in *Larrea tridentata. Oecologia* 40: 113–124.

Cutler, H. C. 1939. Monograph of the North American species of the genus *Ephedra. Annals of the Missouri Botanical Garden* 26: 373–424.

Daniel, T. F. 1982. *Anisacanthus andersonii* (Acanthaceae), a new species from northwestern Mexico. *Bulletin of the Torrey Botanical Club* 109: 148–151.

Daniel, T. F. 1984. The Acanthaceae of the southwestern United States. *Desert Plants* 5: 162–179.

Daniel, T. F., B. D. Parfitt, M. A. Baker. 1984. Chromosome numbers and their systematic implications in some North American Acanthaceae. *Systematic Botany* 9: 346–355.

Darrow, R. A. 1943. Vegetative and floral growth of *Fouquieria splendens*. *Ecology* 24: 310–322.

Date, R. A. 1991. Nitrogen fixation in *Desmanthus:* Strain specificity of *Rhizobium* and responses to inoculation in acidic and alkaline soil. *Tropical Grasslands* 25: 47–55.

Davis, P. H. 1970. *Flora of Turkey and the Aegean Islands*. Edinburgh University Press, Edinburgh.

Davis, W. S., H. J. Thompson. 1967. A revision of *Petalonyx* (Loasaceae) with a consideration of affinities in subfamily Gronovioideae. *Madroño* 19: 1–18.

Day, A. G., R. Moran. 1986. *Acanthogilia,* a new genus of Polemoniaceae from Baja California, Mexico. *Proceedings of the California Academy of Sciences* 44: 111–126.

Decker, J. P. 1961. Salt secretion by *Tamarix pentandra* Pall. *Forest Science* 7: 214–217.

Decker, J. P., W. G. Gaylor, F. D. Cole. 1962. Measuring transpiration of undisturbed tamarisk shrubs. *Plant Physiology* 37: 393–397.

Dehgan, B., G. L. Webster. 1978. Three new species of *Jatropha* (Euphorbiaceae) from western Mexico. *Madroño* 25: 30–39.

Dehgan, B., G. L. Webster. 1979. Morphology and infrageneric relationships of the genus *Jatropha* (Euphorbiaceae). *University of California Publications in Botany* 74: 1–73.

DeJong, T. M., M. G. Barbour. 1979. Contributions to the biology of *Atriplex leucophylla,* a C_4 California beach plant. *Bulletin of the Torrey Botanical Club* 106: 9–19.

Delgado-Salinas, A. O., M. Sousa-Sanchez. 1977. Flower biology in the genus *Cassia* in the region of Los Tuxtlas, Veracruz, Mexico. *Boletín de la Sociedad Botánica de México* 37: 5–52.

DeMason, D. A. 1984. Offshoot variability in *Yucca whipplei* subsp. *percursa* (Agavaceae). *Madroño* 31: 197–202.

Despain, D. G. 1974. The survival of saguaro (*Carnegiea gigantea*) seedlings on soils of differing albedo and cover. *Journal of the Arizona Academy of Science* 9: 102–107.

Didden-Zopfy, B., P. S. Nobel. 1982. High temperature tolerance and heat acclimation of *Opuntia bigelovii*. *Oecologia* 52: 176–180.

Diggle, P. K., D. A. DeMason. 1983a. The relationship between the primary thickening meristem and the secondary thickening meristem in *Yucca whipplei* Torr, part 1: Histology of the mature vegetative stem. *American Journal of Botany* 70: 1195–1204.

Diggle, P. K., D. A. DeMason. 1983b. The relationship between the primary thickening meristem and the secondary thickening meristem in *Yucca whipplei* Torr, part 2: Ontogenetic relationship within the vegetative stem. *American Journal of Botany* 70: 1205–1216.

Dimmitt, M. A. 1987. The hybrid palo verde "Desert Museum": A new, superior tree for desert landscapes. *Desert Plants* 8: 99–103.

Dodgson, C. L. 1966. *The Hunting of the Snark*. Pantheon Books, New York.

Donnell Smith, J., J. N. Rose. 1913. A monograph of the Hauyeae and Gongylocarpeae, tribes of the Onagraceae. *Contributions from the U.S. National Herbarium* 16: 287–298.

Dorado, O. 1992. A systematic and evolutionary study of the genus *Brongniartia* (Fabaceae). Ph.D. dissertation, Claremont Graduate School, Claremont, Calif.

Dressler, R. L. 1956. *Prosopis globosa* in Baja California. *Madroño* 13: 172–174.

Dressler, R. L. 1957. The genus *Pedilanthus* (Euphorbiaceae). *Contributions from the Gray Herbarium* 182: 1–188.

Duffield, M. R., W. D. Jones. 1981. *Plants for Dry Climates: How to Select, Grow, and Enjoy.* H. P. Books, Tucson, Ariz.

Dunford, M. P. 1984. Cytotype distribution of *Atriplex canescens* (Chenopodiaceae) of southern New Mexico and adjacent Texas. *Southwestern Naturalist* 29: 223–228.

Dunford, M. P. 1985. A statistical analysis of morphological vegetation in cytotypes of *Atriplex canescens* (Chenopodiaceae). *Southwestern Naturalist* 30: 377–384.

DuPree, E., J. A. Ludwig. 1978. Vegetative and reproductive growth patterns in desert willow (*Chilopsis linearis* [Cav.] Sweet). *Southwestern Naturalist* 23: 239–246.

Durham, J. W., E. C. Allison. 1960. The geologic history of Baja California and its marine faunas. *Systematic Zoology* 9: 47–91.

Ehleringer, J. R. 1982. The influence of water stress and temperature on leaf pubescence development in *Encelia farinosa. American Journal of Botany* 69: 670–675.

Ehleringer, J. R. 1983. Characterization of a glabrate *Encelia farinosa:* Mutant morphology, ecophysiology, and field observations. *Oecologia* 57: 303–310.

Ehleringer, J. R. 1984. Intraspecific competitive effects on water relations, growth, and reproduction in *Encelia farinosa. Oecologia* 63: 153–158.

Ehleringer, J. R. 1988. Comparative ecophysiology of *Encelia farinosa* and *Encelia frutescens,* part 1: Energy balance considerations. *Oecologia* 76: 553–561.

Ehleringer, J. R., O. Bjorkman. 1978. Pubescence and leaf spectral characteristics in a desert shrub, *Encelia farinosa. Oecologia* 36: 151–162.

Ehleringer, J. R., O. Bjorkman. 1982. A comparison of photosynthetic characteristics of *Encelia* species possessing glabrous and pubescent leaves. *Plant Physiology* 62: 185–190.

Ehleringer, J. R., O. Bjorkman, H. A. Mooney. 1976. Leaf pubescence: Effects on absorptance and photosynthesis in a desert shrub. *Science* 192: 376–377.

Ehleringer, J. R., C. S. Cook. 1984. Photosynthesis in *Encelia farinosa* Gray in response to decreasing leaf water potential. *Plant Physiology* 75: 688–693.

Ehleringer, J. R., C. S. Cook. 1990. Characteristics of *Encelia* spp. differing in leaf reflectance and transpiration rate under common garden conditions. *Oecologia* 82: 484–489.

Ehleringer, J. R., D. House. 1984. Orientation and slope preference in barrel cactus (*Ferocactus acanthodes*) at its northern distribution limit. *Great Basin Naturalist* 44: 133–139.

Ehleringer, J. R., H. A. Mooney. 1978. Leaf hairs: Effects on physiological activity and adaptive value to a desert shrub. *Oecologia* 37: 183–200.

Eichhorn, A. 1957. Nouvelle contribution a l'étude caryologique des Palmiers. *Revue de Cytologie et de Biologie Vegetales* 18: 139–151.

Eley, T. J., S. W. Harris. 1976. Fall and winter foods of American coots along the lower Colorado River. *California Fish and Game* 62: 225–227.

Ellner, S., A. Shmida. 1981. Why are adaptations for long-range seed dispersal rare in desert plants? *Oecologia* 51: 133–144.

Engelmann, G. 1872. The flower of yucca and its fertilization. *Bulletin of the Torrey Botanical Club* 3: 33.

Engel-Wilson, R. W., R. D. Ohmart. 1979. Floral and attendant faunal changes on the lower Rio Grande between Fort Quitman and Presidio, Texas. *USDA Forest Service General Technical Report* WO–12, 139–147. Washington, D.C.

Eskew, D. L., I. P. Ting. 1978. Nitrogen fixation by legumes and blue-green algal-lichen crusts in a Colorado Desert environment. *American Journal of Botany* 65: 850–856.

Espejel, I. 1986. Coastal dune vegetation of the Yucatan Peninsula, part 2: The Nature Reserve Sian Ka'an, Quintana Roo, Mexico. *Biótica* 11: 7–24.

Ettershank, G., J. Ettershank, M. Bryant, W. G. Whitford. 1978. Effects of nitrogen fertilization on primary production in a Chihuahuan Desert ecosystem. *Journal of Arid Environments* 1: 135–139.

Evans, D. D., T. W. Sammis, D. R. Cable. 1981. Actual evapotranspiration under desert conditions. In D. D. Evans, J. L. Thames (eds.), *Water in Desert Ecosystems,* 195–218. Dowden, Hutchinson & Ross, Stroundsburg, Penn.

Everett, R. L., R. O. Meeuwig, J. H. Robertson. 1978. Propagation of Nevada shrubs by stem cuttings. *Journal of Range Management* 31: 426–429.

Everitt, J. H., M. A. Alaniz. 1981. Nutrient content of cactus and woody plant fruits eaten by birds and mammals in South Texas. *Southwestern Naturalist* 26: 301–306.

Everitt, J. H., C. L. Gonzalez. 1981. Germination of honey mesquite (*Prosopis glandulosa*) seeds after passage through the digestive tracts of peccaries (*Pecari tajacu*). *Southwestern Naturalist* 26: 432.

Ezcurra, E., S. Arizaga, P. L. Valverde, C. Mourelle, A. Flores-Martinez. 1992. Foliole movement and canopy architecture of *Larrea tridentata* (DC.) Cov. in Mexican deserts. *Oecologia* 92: 83–89.

Ezcurra, E., R. S. Felger, A. D. Russell, M. Equihua. 1988. Freshwater islands in a desert sand sea: The hydrology, flora, and phytogeography of the Gran Desierto oases of northwestern Mexico. *Desert Plants* 9: 35–44, 55–63.

Ezcurra, E., J. Rodrigues. 1986. Rainfall patterns in the Gran Desierto, Mexico. *Journal of Arid Environments* 10: 13–28.

Faegri, K., L. van der Pijl. 1979. *The Principles of Pollination Ecology.* 3d ed. Pergamon Press, New York.

Faure, P. 1968. Contribution a l'étude caryo-taxonomique des Myrsinaceae et des Theophrastaceae. *Mémoires Muséum National d'Histoire Naturelle Ser. B* 18: 37–58.

Felger, R. S. 1966. Ecology of the gulf coast and islands of Sonora, Mexico. Ph.D. dissertation, University of Arizona, Tucson.

Felger, R. S. 1977. Mesquite in Indian cultures of southwestern North America. In B. B. Simpson (ed.), *Mesquite: Its Biology in Two Desert Scrub Ecosystems,* 150–176. Dowden, Hutchinson & Ross, Stroudsburg, Penn.

Felger, R. S. 1980. Vegetation and flora of the Gran Desierto, Sonora, Mexico. *Desert Plants* 2: 87–114.

Felger, R. S., C. H. Lowe. 1967. Clinal variation in the surface-volume relationships of the columnar cactus *Lophocereus schottii* in northwestern Mexico. *Ecology* 48: 530–536.

Felger, R. S., C. H. Lowe. 1970. New combinations for plant taxa in northwestern Mexico and southwestern United States. *Journal of the Arizona Academy of Science* 6: 82–84.

Felger, R. S., M. B. Moser. 1971. Seri use of mesquite (*Prosopis glandulosa* var. *torreyana*). *Kiva* 37: 53–60.

Felger, R. S., M. B. Moser. 1985. *People of the Desert and Sea: Ethnobotany of the Seri Indians.* University of Arizona Press, Tucson.

Felker, P., G. H. Cannell, P. R. Clark. 1981. Variation in growth among thirteen *Prosopis* (mesquite) species. *Experimental Agriculture* 17: 209–218.

Felker, P., P. R. Clark. 1980. Nitrogen fixation, acetylene reductions, and cross-inoculation in twelve *Prosopis* mesquite species. *Plant and Soil* 57: 177–186.

Felker, P., P. R. Clark. 1981. Nodulation and nitrogen fixation (acetylene reduction) in desert ironwood (*Olneya tesota*). *Oecologia* 48: 292–293.

Felker, P., P. R. Clark. 1982. Position of mesquite (*Prosopis* spp.) nodulation and nitrogen fixation (acetylene reduction) in 3-m-long phreatophytically simulated soil columns. *Plant and Soil* 64: 297–305.

Felker, P., P. R. Clark, A. E. Laag, P. F. Pratt. 1981. Salinity tolerance of the tree legumes: mesquite (*Prosopis glandulosa* var. *torreyana, P. velutina,* and *P. articulata*), algarroba (*P. chilensis*), kiawe (*P. pallida*), and tamarugo (*P. tamarugo*) grown in sand culture on nitrogen-free media. *Plant and Soil* 61: 311–318.

Felker, P., P. R. Clark, P. Nash, J. F. Osborn, G. H. Cannell. 1982. Screening *Prosopis* (mesquite) for cold tolerance. *Forest Science* 28: 556–562.

Ferguson, C. W., R. A. Wright. 1962. Botanical studies. *Kiva* 28: 108–114.

Fernandez, S., M. A. Ponce, F. Hernández, L. Hurtado, V. Gonzalez. 1979. Creosote bush (*Larrea tridentata*) industrial potential. In J. R. Goodin, D. K. Northington (eds.), *Arid Land Plant Resources, Proceedings of the International Arid Lands Conference on Plant Resources,* 284–293. Texas Tech University, Lubbock.

Ferriere, J., P. L. Milthorpe, R. L. Dunstone. 1989. Variability in chilling requirements for the breaking of flower bud dormancy in jojoba (*Simmondsia chinensis* [Link] Schneider). *Journal of Horticultural Science* 64: 379–387.

Ferrucci, M. S. 1981. Recuentos cromosómicos en Sapindaceas. *Bonplandia* 5: 73–81.

Fick, W. H., R. E. Sosebee. 1981. Translocation and storage of ^{14}C-labeled total nonstructural carbohydrates in honey mesquite (*Prosopis glandulosa* var. *glandulosa*). *Journal of Range Management* 34: 205–208.

Fisher, C. E., C. H. Meadors, R. Behrens, E. D. Robison, P. T. Marion, H. L. Morton. 1959. *Control of Mesquite on Grazing Lands.* Texas Agricultural Experiment Station Bulletin no. 935. College Station.

Flath, R. A., T. R. Mon, G. Lorenz, C. J. Whitten, J. W. Mackley. 1983. Volatile components of *Acacia* spp. blossoms. *Journal of Agricultural and Food Chemistry* 31: 1167–1170.

Flinn, R. C., C. J. Scifres, S. R. Archer. 1992. Variation in basal sprouting in co-occurring shrubs: Implications for stand dynamics. *Journal of Vegetation Science* 3: 125–128.

Fogleman, J. C., L. Armstrong. 1989. Ecological aspects of cactus triterpene glycosides, part 1: Their effect on fitness components of *Drosophila mojavensis. Journal of Chemical Ecology* 15: 663–676.

Fogleman, J. C., W. T. Starmer. 1985. Analysis of the community structure of yeasts associated with the decaying stems of cactus, 3. *Stenocereus thurberi. Microbial Ecology* 11: 165–174.

Fontana, B. L. 1980. Ethnobotany of the saguaro: An annotated bibliography. *Desert Plants* 2: 62–78.

Fonteyn, P. J., B. E. Mahall. 1981. An experimental analysis of structure in a desert plant community. *Journal of Ecology* 69: 883–896.

Fosberg, F. R. 1938. The Lower Sonoran in Utah. *Science* 87: 39–40.

Franco, A. C., P. S. Nobel. 1988. Interactions between seedlings of *Agave deserti* and the nurse plant *Hilaria rigida. Ecology* 69: 1731–1740.

Franco, A. C., P. S. Nobel. 1989. Effect of nurse plants on the microhabitat and growth of cacti. *Journal of Ecology* 77: 870–886.

Franco, A. C., P. S. Nobel. 1990. Influences of root distribution and growth on predicted water uptake and interspecific competition. *Oecologia* 82: 151–157.

Franco-Vizcaíno, E., G. Goldstein, I. P. Ting. 1990. Comparative gas exchange of leaves and bark in three stem succulents of Baja California. *American Journal of Botany* 77: 1272–1278.

Franco-Vizcaíno, E., R. A. Khattak. 1990. Elemental composition of soils and tissues of natural jojoba populations of Baja California, Mexico. *Journal of Arid Environments* 19: 55–64.

Freeman, C. E. 1973. Germination responses of a Texas population of ocotillo (*Fouquieria splendens* Engelm.) to constant temperature, water stress, pH, and salinity. *American Midland Naturalist* 89: 252–256.

Freeman, C. E. 1982. Seasonal variation in leaf nitrogen in creosotebush (*Larrea tridentata* [DC.] Coville: Zygophyllaceae). *Southwestern Naturalist* 27: 354–356.

Freeman, C. E., W. H. Reid, J. E. Becvar. 1983. Nectar sugar composition in some species of *Agave* (Agavaceae). *Madroño* 30: 153–158.

Freeman, C. E., R. S. Tiffany, W. H. Reid. 1977. Germination responses of *Agave lecheguilla, Agave parryi,* and *Fouquieria splendens. Southwestern Naturalist* 22: 195–204.

Freeman, D. C., E. D. McArthur. 1984. The relative influences of mortality, nonflowering, and sex change on the sex ratio of six *Atriplex* species. *Botanical Gazette* 145: 385–394.

Freeman, T. P. 1970. The developmental anatomy of *Opuntia basilaris:* Apical meristem, leaves, areoles, glochids. *American Journal of Botany* 57: 616–622.

Freeman, T. P. 1973. Developmental anatomy of epidermal and mesophyll chloroplasts in *Opuntia basilaris* leaves. *American Journal of Botany* 60: 86–91.

Friedman, W. E. 1990. Sexual reproduction in *Ephedra nevadensis* (Ephedraceae): Further evidence of double fertilization in a nonflowering seed plant. *American Journal of Botany* 77: 1582–1598.

Fryxell, J. E. 1983. A revision of *Abutilon* sect. *Oligocarpae* (Malvaceae), including a new species from Mexico. *Madroño* 30: 84–92.

Fulbright, T. E., K. S. Flenniken, G. L. Waggerman. 1986. Enhancing germination of spiny hackberry seeds. *Journal of Range Management* 39: 552–554.

Gail, P. A. 1964. *Simmondsia chinensis* (Link) Schneider: Anatomy and morphology of flowers. M.S. thesis, Claremont College, Claremont, Calif.

Ganter, P. F., W. T. Starmer, M. A. Lachance, H. J. Phaff. 1986. Yeast communities from host plants and associated *Drosophila* in southern Arizona: New isolations and analysis of the relative importance of hosts and vectors on community composition. *Oecologia* 70: 386–392.

Garcia, E., C. Soto, F. Miranda. 1961. *Larrea* y clima. *Anales del Instituto de Biología* 31: 133–190.

Gary, H. L. 1963. Root distribution of five-stamen tamarisk, seepwillow, and arrowweed. *Forest Science* 9: 313–314.

Garza, A., T. E. Fulbright. 1988. Comparative chemical composition of armed saltbush and fourwing saltbush. *Journal of Range Management* 41: 401–403.

Garza, R., M. del Rosario. 1981. Creosote bush resin: A possible dormancy breaking agent. In E. Campos Lopez, T. J. Mabry, S. Fernandez Tavizon (eds.), *Larrea,* 305–316. Consejo Nacional de Ciéncia y Tecnología, México, D.F.

Gastil, G., J. Minch, R. P. Phillips. 1983. The geology and ages of the islands. In T. J. Case, M. L. Cody (eds.), *Island Biogeography in the Sea of Cortez,* 13–25. University of California Press, Berkeley.

Gates, H. E. 1957. Distribución de las cactaceas de Baja California. *Cactaceas y Suculentes Mexicanas* 2: 69–75.

Geller, G. N., P. S. Nobel. 1984. Cactus ribs: Influences on PAR interception and CO_2 uptake. *Photosynthetica* 18: 482–494.

Geller, G. N., P. S. Nobel. 1986. Branching patterns of columnar cacti: Influences on PAR interception and CO_2 uptake. *American Journal of Botany* 73: 1193–1200.

Geller, G. N., P. S. Nobel. 1987. Comparative cactus architecture and PAR interception. *American Journal of Botany* 74: 998–1005.

Gentry, A. H. 1970. A revision of *Tabebuia* (Bignoniaceae) in Central America. *Brittonia* 22: 246–264.

Gentry, H. S. 1942. *Rio Mayo Plants: A Study of the Flora and Vegetation of the Valley of the Rio Mayo, Sonora.* Carnegie Institution of Washington Publication no. 527. Washington, D.C.

Gentry, H. S. 1949a. *Land Plants Collected by the Velero III, Allan Hancock Pacific Expeditions 1937–1941.* Allan Hancock Pacific Expeditions, vol. 13. University of California Press, Los Angeles.

Gentry, H. S. 1949b. On the hundredth anniversary of *Fouquieria splendens. Desert Plant Life* 21: 11–13.

Gentry, H. S. 1950. Studies in the genus *Dalea. Madroño* 10: 225–250.

Gentry, H. S. 1958. The natural history of jojoba (*Simmondsia chinensis*) and its cultural aspects. *Economic Botany* 12: 261–295.

Gentry, H. S. 1972. *The Agave Family in Sonora.* USDA Handbook no. 399. Government Printing Office, Washington, D.C.

Gentry, H. S. 1978. *The Agaves of Baja California.* Occasional Papers of the California Academy of Sciences no. 130. San Francisco.

Gentry, H. S. 1982. *Agaves of Continental North America.* University of Arizona Press, Tucson.

Ghimpu, V. 1929. Contribution a l'étude chromosomique de *Acacia. Comptes Rendus Hebdomadaire des Séances Académie des Science (Paris)* 187: 1429–1431.

Gibbs, J. G., D. T. Patten. 1970. Plant temperatures and heat flux in a Sonoran Desert ecosystem. *Oecologia* 5: 165–184.

Gibson, A. C. 1979. Anatomy of *Koeberlinia* and *Canotia* revisited. *Madroño* 26: 1–12.

Gibson, A. C. 1981. Vegetative anatomy of *Pachycormus* (Anacardiaceae). *Botanical Journal of the Linnean Society* 83: 273–284.

Gibson, A. C. 1983. Anatomy of photosynthetic old stems of nonsucculent dicotyledons from North American deserts. *Botanical Gazette* 144: 347–362.

Gibson, A. C. 1988a. The systematics and evolution of subtribe Stenocereinae. 3. *Myrtillocactus. Cactus and Succulent Journal (U.S.)* 60: 109–116.

Gibson, A. C. 1988b. The systematics and evolution of subtribe Stenocereinae. 5. Cina and its relatives. *Cactus and Succulent Journal (U.S.)* 60: 283–288.

Gibson, A. C. 1989. The systematics and evolution of subtribe Stenocereinae. 7. The *Machaerocerei* of *Stenocereus. Cactus and Succulent Journal (U.S.)* 61: 104–112.

Gibson, A. C. 1990. The systematics and evolution of subtribe Stenocereinae. 8. Organpipe cactus and its closest relatives. *Cactus and Succulent Journal (U.S.)* 62: 13–24.

Gibson, A. C., K. E. Horak. 1978. Systematic anatomy and phylogeny of Mexican columnar cacti. *Annals of the Missouri Botanical Garden* 65: 999–1057.

Gibson, A. C., P. S. Nobel. 1986. *The Cactus Primer.* Harvard University Press, Cambridge, Mass.

Gibson, D. N. 1972. Studies in American plants, part 3. *Fieldiana: Botany* 34: 57–87.

Gilbertson, R. L., E. R. Canfield. 1972. *Poria carnegiea* and decay of saguaro cactus in Arizona. *Mycologia* 64: 1300–1311.

Gill, A. M., P. B. Tomlinson. 1969. Studies on the growth of red mangrove (*Rhizophora mangle* L.). *Biotropica* 1: 1–9.

Gill, A. M., P. B. Tomlinson. 1971. Studies on the growth of red mangrove (*Rhizophora mangle* L.) 3. Phenology of the shoot. *Biotropica* 3: 109–124.

Gill, A. M., P. B. Tomlinson. 1977. Studies on the growth of red mangrove (*Rhizophora mangle* L.) 4. The adult root system. *Biotropica* 9: 145–155.

Gilmartin, A. J., M. L. Neighbours. 1982. Variability within *Stegnosperma halimifolium* Benth. (Phytolaccaceae). *Southwestern Naturalist* 27: 63–72.

Gilreath, M. E., J. W. Smith. 1988. Natural enemies of *Dactylopius confusus* (Homoptera: Dactylopiidae): Exclusion and subsequent impact on *Opuntia* (Cactaceae). *Environmental Entomology* 17: 730–738.

Glendening, G. E., H. A. Paulson. 1955. *Reproduction and Establishment of Velvet Mesquite as Related to Invasion of Semidesert Grasslands.* USDA Forest Service Technical Bulletin no. 1127. Government Printing Office, Washington, D.C.

Glinski, R. L., D. E. Brown. 1982. Mesquite (*Prosopis juliflora*) response to severe freezing in southeastern Arizona. *Journal of the Arizona-Nevada Academy of Science* 17: 15–18.

Goeden, R. D., D. W. Ricker. 1976a. The phytophagous insect faunas of the ragweeds *Ambrosia chenopodiifolia, Ambrosia eriocentra,* and *Ambrosia ilicifolia* in southern California. *Environmental Entomology* 5: 923–930.

Goeden, R. D., D. W. Ricker. 1976b. Life history of the ragweed plume moth, *Adaina ambrosiae* (Murtfeldt), in southern California. *Pan-Pacific Entomologist* 52: 251–255.

Goeden, R. D., D. W. Ricker. 1986. Phytophagous insect fauna of the desert shrub *Hymenoclea salsola* in southern California. *Annals of the Entomological Society of America* 79: 39–47.

Goen, J. P., B. E. Dahl. 1982. Factors affecting budbreak in honey mesquite in West Texas. *Journal of Range Management* 35: 533–534.

Goldberg, A. 1967. The genus *Melochia* L. (Sterculiaceae). *Contributions from the U.S. National Herbarium* 34: 191–363.

Goldberg, D. E. 1985. Effects of soil pH, competition, and seed predation on the distributions of two tree species. *Ecology* 66: 503–511.

Goldberg, D. E., R. M. Turner. 1986. Vegetation change and plant demography in permanent plots in the Sonoran Desert. *Ecology* 67: 695–712.

Goldblatt, P. 1976. New or noteworthy chromosome records in the Angiosperms. *Annals of the Missouri Botanical Garden* 63: 889–895.

Goldblatt, P., A. H. Gentry. 1979. Cytology of Bignoniaceae. *Botaniska Notiser* 132: 475–482.

Graham-Gadzia, J. S., J. A. Ludwig. 1983. Mesquite (*Prosopis glandulosa* var. *torreyana*) age and size in relation to dunes and artifacts. *Southwestern Naturalist* 28: 89–94.

Grant, V., K. A. Grant. 1971. Dynamics of clonal microspecies in cholla cactus. *Evolution* 25: 144–155.

Grant, V., K. A. Grant. 1979a. Pollination of *Echinocereus fasciculatus* and *Ferocactus wislizenii. Plant Systematics and Evolution* 132: 85–90.

Grant, V., K. A. Grant. 1979b. Systematics of the *Opuntia phaeacantha* group in Texas. *Botanical Gazette* 140: 199–207.

Grant, V., K. A. Grant. 1979c. Pollination of *Opuntia basilaris* and *Opuntia littoralis. Plant Systematics and Evolution* 132: 321–326.

Grant, V., K. A. Grant, P. D. Hurd. 1979. Pollination of *Opuntia lindheimeri* and related species. *Plant Systematics and Evolution* 132: 313–320.

Grant, V., P. D. Hurd. 1979. Pollination of the southwestern *Opuntias. Plant Systematics and Evolution* 133: 15–28.

Grant, W. F. 1955. A cytogenetic study in the Acanthaceae. *Brittonia* 8: 121–149.

Graves, W. L., B. L. Kay, W. A. Williams. 1975. Seed treatment of Mojave Desert shrubs. *Agronomy Journal* 67: 773–777.

Graves, W. L., B. L. Kay, W. A. Williams. 1978. Revegetation of disturbed sites in the Mojave Desert with native shrubs. *California Agriculture* 32: 4–5.

Greene, R. A., E. O. Foster. 1933. The liquid wax of seeds of *Simmondsia californica*. *Botanical Gazette* 94: 826–828.

Grime, J. P. 1979. *Plant Strategies and Vegetation Processes*. John Wiley & Sons, New York.

Guinet, P., J. Vassal. 1978. Hypotheses on the differentiation of the major groups in the genus *Acacia* (Leguminosae). *Kew Bulletin* 32: 509–527.

Gulmon, S. L., A. J. Bloom. 1979. C_3 photosynthesis and high temperature acclimation of CAM in *Opuntia basilaris* Engelm. & Bigel. *Oecologia* 38: 217–222.

Gulmon, S. L., H. A. Mooney. 1977. Spatial and temporal relationships between two desert shrubs, *Atriplex hymenelytra* and *Tidestromia oblongifolia*, in Death Valley, California. *Journal of Ecology* 65: 831–838.

Haas, R. H., R. E. Meyer, C. J. Scifres, J. H. Brock. 1973. Growth and development of mesquite. *Texas Agricultural Experiment Station, Research Monograph No. 1*, 10–19. College Station.

Haiman, R. L. 1974. Human modification of the coastal vegetation of northwestern Baja California. *Journal of the Arizona-Nevada Academy of Science* 9 (Supplement): 26.

Haines, L. 1941. Variation in *Yucca whipplei*. *Madroño* 6: 33–64.

Hall, H. M. 1928. *The Genus Haplopappus: A Phylogenetic Study in the Compositae*. Cai negie Institution of Washington Publication no. 389. Washington, D.C.

Hall, H. M., F. E. Clements. 1923. *The Phylogenetic Method in Taxonomy: The North American Species of Artemisia, Chrysothamnus, and Atriplex*. Carnegie Instit tion of Washington Publication no. 326. Washington, D.C.

Hall, H. M., F. L. Long. 1921. *Rubber Content of North American Plants*. Carnegie Insti tution of Washington Publication no. 313. Washington, D.C.

Hallmark, C. T., B. L. Allen. 1975. The distribution of creosotebush in West Texas and eastern New Mexico as affected by selected soil properties. *Soil Science Soci ety of America Proceedings* 39: 120–124.

Halvorson, W. L., D. T. Patten. 1974. Seasonal water potential changes in Sonoran Desert shrubs in relation to topography. *Ecology* 55: 173–177.

Hammouda, Y., N. Khalafallah. 1971. Stability of tecomine, the major anti-diabetic factor of *Tecoma stans* (Bignoniaceae). *Journal of Pharmaceutical Sciences* 60: 1142–1145.

Hanley, T. A., W. W. Brady. 1977. Feral burro impact on a Sonoran Desert range. *Journal of Range Management* 30: 374–377.

Hanscom, Z., I. P. Ting. 1977. Physiological responses to irrigation in *Opuntia basilaris*. *Botanical Gazette* 138: 159–167.

Hanscom, Z., I. P. Ting. 1978. Irrigation magnifies Crassulacean acid metabolism photosynthesis in *Opuntia basilaris* (Cactaceae). *Oecologia* 33: 1–16.

Hanson, M. A., L. H. Rieseberg, D. M. Arias. 1989. Isozyme analysis of hybridization in *Yucca*. *American Journal of Botany* 76: 243–244.

Harden, M. L., R. Zolfaghari. 1988. Nutritive composition of green and ripe pods of honey mesquite, *Prosopis glandulosa* (Fabaceae). *Economic Botany* 42: 522–532.

Hardin, J. W. 1957a. A revision of the American Hippocastanaceae. *Brittonia* 9: 145–171.

Hardin, J. W. 1957b. A revision of the American Hippocastanaceae, part 2. *Brittonia* 9: 173–195.

Harrington, G. N. 1991. Effects of soil moisture on shrub seedling survival in a semiarid grassland. *Ecology* 72: 1138–1149.

Harris, D. R. 1966. Recent plant invasions in the arid and semi-arid Southwest of the United States. *Annals of the Association of American Geographers* 56: 408–422.

Hartsock, T. L., P. S. Nobel. 1976. Watering converts a CAM plant to daytime CO_2 uptake. *Nature* 262: 574–576.

Hastings, J. R. 1961. Precipitation and saguaro growth. *University of Arizona Arid Lands Colloquia* 1959–60/1960–61: 30–38.

Hastings, J. R. 1964. *Climatological Data for Baja California.* Technical Reports on the Meteorology and Climatology of Arid Regions no. 14. University of Arizona Institute of Atmospheric Physics, Tucson.

Hastings, J. R., S. M. Alcorn. 1961. Physical determination of growth and age in the giant cactus. *Journal of the Arizona Academy of Science* 2: 32–39.

Hastings, J. R., R. R. Humphrey. 1969. *Climatological Data and Statistics for Sonora and Northern Sinaloa.* Technical Reports on the Meteorology and Climatology of Arid Regions no. 19. University of Arizona Institute of Atmospheric Physics, Tucson.

Hastings, J. R., R. M. Turner. 1965a. Seasonal precipitation regimes in Baja California, Mexico. *Geografiska Annaler* 47, Ser. A: 204–223.

Hastings, J. R., R. M. Turner. 1965b. *The Changing Mile: An Ecological Study of Vegetation Change with Time in the Lower Mile of an Arid and Semiarid Region.* University of Arizona Press, Tucson.

Hastings, J. R., R. M. Turner, D. K. Warren. 1972. *An Atlas of Some Plant Distributions in the Sonoran Desert.* Technical Reports on the Meteorology and Climatology of Arid Regions no. 21. University of Arizona Institute of Atmospheric Physics, Tucson.

Headon, D. R., G. Killeen. 1993. Effects of *Yucca schidigera* extract on *Bacillus pasteurii* urease activity. *Animal Production* 56: 459.

Helbsing, T., K. H. Kreeb. 1985. Effect of water stress on the water uptake and CAM of young plants of *Carnegiea gigantea. Verhandlungen der Gesellschaft für Okologie* 18: 645–648.

Hem, J. D. 1967. *Composition of Saline Residues on Leaves and Stems of Saltcedar (Tamarix pentandra Pallas).* U.S. Geological Survey Professional Paper 491-C. Government Printing Office, Washington, D.C.

Hennessy, J. T., R. P. Gibbens, M. Cardenas. 1984. The effect of shade and planting depth on the emergence of fourwing saltbush. *Journal of Range Management* 37: 22–24.

Henrickson, J. 1969. Anatomy of periderm and cortex of Fouquieriaceae. *Aliso* 7: 97–126.

Henrickson, J. 1972. A taxonomic revision of the Fouquieriaceae. *Aliso* 7: 439–537.

Henrickson, J. 1977. Leaf production and flowering in ocotillos. *Cactus and Succulent Journal (U.S.)* 49: 133–137.

Henrickson, J. 1985. A taxonomic revision of *Chilopsis* (Bignoniaceae). *Aliso* 11: 179–197.

Hernández, H. M. 1989. Systematics of *Zapoteca* (Leguminosae). *Annals of the Missouri Botanical Garden* 76: 781–862.

Herrera, C. M. 1989. Vertebrate frugivores and their interaction with invertebrate fruit predators: Supporting evidence from a Costa Rican dry forest. *Oikos* 54: 185–188.

Hess, W. J., J. Henrickson. 1987. A taxonomic revision of *Vauquelinia* (Rosaceae). *Sida* 12: 101–163.

Hevly, R. H. 1979. Dietary habits of two nectar and pollen feeding bats in southern Arizona and northern Mexico. *Journal of the Arizona-Nevada Academy of Science* 14: 13–18.

Hicks, S. 1966. *Desert Plants and People.* Naylor, San Antonio, Tex.

Hilu, K. W., S. Boyd, P. Felker. 1982. Morphological diversity and taxonomy of California mesquites (*Prosopis*, Leguminosae). *Madroño* 29: 237–254.

Hodgkinson, H. S. 1987. Relationship of saltbush species to soil chemical properties. *Journal of Range Management* 40: 23–26.

Hodgkinson, K. C., R. E. Oxley. 1990. Influence of fire and edaphic factors on germination of the arid zone shrubs *Acacia aneura, Cassia nemophila,* and *Dodonaea viscosa. Australian Journal of Botany* 38: 269–279.

Hodgson, W., G. Nabhan, L. Ecker. 1989. Conserving rediscovered *Agave* cultivars. *Agave* 3: 9–11.

Hoffmeister, D. F. 1986. *Mammals of Arizona.* University of Arizona Press, Tucson.

Holland, D. C. 1987. *Prosopis* (Mimosaceae) in the San Joaquin Valley, California: Vanishing relict or recent invader. *Madroño* 34: 324–333.

Holland, D. C., B. Anderson. 1988. Additional support for the recent invasive advent of mesquite (Mimosaceae: *Prosopis*) in the San Joaquin Valley, California. *Madroño* 35: 329.

Holthe, P. A., S. R. Szarek. 1985. Physiological potential for survival of propagules of Crassulacean acid metabolism species. *Plant Physiology* 79: 219–224.

Horak, K. 1981. The three-dimensional structure of vascular tissues in *Stegnosperma* (Phytolaccaceae). *Botanical Gazette* 142: 545–549.

Horton, J. S. 1957. Inflorescence development in *Tamarix pentandra* Pallas (Tamaricaceae). *Southwestern Naturalist* 2: 135–139.

Horton, J. S. 1964. *Notes on the Introduction of Deciduous Tamarisk.* USDA Forest Service Research Note RM–16. Fort Collins, Colo.

Horton, J. S., J. E. Flood. 1962. Taxonomic notes on *Tamarix pentandra* in Arizona. *Southwestern Naturalist* 7: 23–28.

Horton, J. S., F. C. Mounts, J. M. Kraft. 1960. *Seed Germination and Seedling Establishment of Phreatophyte Species.* USDA Forest Service Station Paper RM–48. Fort Collins, Colo.

Howell, D. J. 1972. Physiological adaptations in the syndrome of chiropterophily with emphasis on the bat *Leptonycteris* Lydekker. Ph.D. dissertation, University of Arizona, Tucson.

Howell, D. J. 1979. Flock foraging in nectar-feeding bats. *American Naturalist* 114: 23–49.

Howell, D. J., B. S. Roth. 1981. Sexual reproduction in agaves: The benefits of bats; the cost of semelparous advertising. *Ecology* 62: 1–7.

Howell, J. T., E. McClintock. 1960. Supplement. In T. H. Kearney, R. H. Peebles, *Arizona Flora,* 2d ed., 1033–1085. University of California Press, Berkeley.

Hufford, L. D. 1989. Structure of the inflorescence and flower of *Petalonyx linearis* (Loasaceae). *Plant Systematics and Evolution* 163: 211–226.

Huft, M. J. 1984. A review of *Euphorbia* (Euphorbiaceae) in Baja California. *Annals of the Missouri Botanical Garden* 71: 1021–1027.

Hughes, C. E., B. T. Styles. 1984. Exploration and seed collection of multiple-purpose dry-zone trees in Central America. *International Tree Crops Journal* 3: 1–32.

Hughes, L. E. 1982. A grazing system in the Mojave Desert. *Rangelands* 4: 256–257.

Hull, H. M. 1956. Studies on herbicidal absorption and translocation in velvet mesquite seedlings. *Weeds* 4: 22–42.

Humphrey, R. R. 1931. Thorn formation in *Fouquieria splendens* and *Idria columnaris. Bulletin of the Torrey Botanical Club* 58: 263–264.

Humphrey, R. R. 1935. A study of *Idria columnaris* and *Fouquieria splendens. American Journal of Botany* 22: 184–207.

Humphrey, R. R. 1936. Growth habits of barrel cacti. *Madroño* 3: 348–352.

Humphrey, R. R. 1953. *Forage Production on Arizona Ranges,* vol. 3, *Mohave County.* Arizona Agricultural Experiment Station Bulletin no. 244. Tucson.

Humphrey, R. R. 1958. The desert grassland: A history of vegetation change and an analysis of causes. *Botanical Review* 24: 193–252.

Humphrey, R. R. 1970. The cirio: The tallest tree of the Sonoran Desert? *Cactus and Succulent Journal (U.S.)* 42: 99–101.

Humphrey, R. R. 1974. *The Boojum and Its Home: Idria Columnaris Kellogg and Its Ecological Niche.* University of Arizona Press, Tucson.

Humphrey, R. R. 1975. Phenology of selected Sonoran Desert plants at Punta Cirio, Sonora, Mexico. *Journal of the Arizona Academy of Science* 10: 50–67.

Humphrey, R. R. 1981. A climatological summary for Punta Cirio, Sonora, Mexico. *Desert Plants* 3: 92–97.

Humphrey, R. R. 1991. Montevideo Valley and its tallest recorded cirio. *Cactus and Succulent Journal (U.S.)* 63: 239–240.

Humphrey, R. R., A. B. Humphrey. 1990. *Idria columnaris:* Age as determined by growth rate. *Desert Plants* 10: 51–54.

Humphrey, R. R., D. B. Marx. 1980. Distribution of the boojum tree (*Idria columnaris*) on the coast of Sonora, Mexico, as influenced by climate. *Desert Plants* 2: 183–187.

Humphrey, R. R., F. G. Werner. 1969. Some records of bee visitations to the flowers of *Idria columnaris. Journal of the Arizona Academy of Science* 5: 243–244.

Huning, J. R. 1978. *A Characterization of the Climate of the California Desert.* U.S. Department of the Interior, Bureau of Land Management, Riverside, Calif.

Hunt, C. B. 1966. *Plant Ecology of Death Valley.* U.S. Geological Survey Professional Paper 509. Government Printing Office, Washington, D.C.

Hunt, E. R., P. S. Nobel. 1987. Allometric root-shoot relationships and predicted water uptake for desert succulents. *Annals of Botany* 59: 571–578.

Hunter, R. 1989. Competition between adult and seedling shrubs of *Ambrosia dumosa* in the Mojave Desert, Nevada. *Great Basin Naturalist* 49: 79–84.

Hunter, R. B., E. M. Romney, A. Wallace. 1980. Rodent-denuded areas of the northern Mojave Desert. *Great Basin Naturalist Memoirs* 4: 208–211.

Hunziker, J. H., R. A. Palacios, L. Poggio, C. A. Naranjo, T. W. Yang. 1977. Geographic distribution, morphology, hybridization, cytogenetics, and evolution. In T. J. Mabry, J. H. Hunziker, D. R. Difeo Jr. (eds.), *Creosotebush: Biology and Chemistry of Larrea in New World Deserts,* 10–46. Dowden, Hutchinson & Ross, Stroudsburg, Penn.

Hunziker, J. H., L. Poggio, C. A. Naranjo, R. A. Palacios, A. B. Andrada. 1975. Cytogenetics of some species and natural hybrids in *Prosopis* (Leguminosae). *Canadian Journal of Genetics and Cytology* 17: 253–262.

Hurd, P. D., E. G. Linsley. 1975. *The Principal Larrea Bees of the Southwestern United States (Hymenoptera: Apoidea).* Smithsonian Contributions to Zoology no. 193. Washington, D.C.

Hutto, R. L., J. R. McAuliffe, L. Hogan. 1986. Distributional associates of the saguaro (*Carnegiea gigantea*). *Southwestern Naturalist* 31: 469–476.

Idso, S. B., B. A. Kimball, M. G. Anderson, S. R. Szarek. 1986. Growth response of a succulent plant, *Agave vilmoriniana,* to elevated CO_2. *Plant Physiology* 80: 796–797.

Ingham, J. L., S. Tahara, S. Shibaki, J. Mizutani. 1989. Isoflavonoids from the root bark of *Piscidia erythrina* and a note on the structure of piscidone. *Zeitschrift für Naturforschung, Section C, Biosciences* 44: 905–913.

Irwin, H. S., R. C. Barneby. 1979. New names in *Senna* P. Mill. and *Chamaecrista*

Moench (Leguminosae: Caesalpinioideae) precursory to the Chihuahuan Desert flora. *Phytologia* 44: 499–501.

Irwin, H. S., R. C. Barneby. 1982. The American Cassiinae: A synoptical revision of Leguminosae tribe Cassieae subtribe Cassiinae in the New World. *Memoirs of the New York Botanical Garden* 35: 1–918.

Isely, D. 1969. Legumes of the United States, part 1: Native *Acacia*. *Sida* 3: 365–386.

Isely, D. 1971. Legumes of the United States, part 4: *Mimosa*. *American Midland Naturalist* 85: 410–424.

Isely, D. 1972. Legumes of the United States, part 6: *Calliandra, Pithecellobium, Prosopis*. *Madroño* 21: 273–298.

Isely, D. 1973. Leguminosae of the United States, part 1: Subfamily Mimosoideae. *Memoirs of the New York Botanical Garden* 25 (1): 1–152.

Isely, D. 1975. Leguminosae of the United States, part 2: Subfamily Caesalpinioideae. *Memoirs of the New York Botanical Garden* 25 (2): 1–228.

Jackson, R. C. 1960. Documented chromosome numbers of plants. *Madroño* 15: 219–221.

Jain, D. K. 1978a. Studies in Bignoniaceae, 4. Shoot apex organization. *Acta Botanica Indica* 6 (Supplement): 48–53.

Jain, D. K. 1978b. Studies in Bignoniaceae, part 3: Leaf architecture. *Journal of the Indian Botanical Society* 57: 369–386.

Jain, D. K., V. Singh. 1979. Studies in Bignoniaceae, 6. Floral anatomy. *Proceedings of the Indian Academy of Sciences* 88: 379–390.

Janzen, D. H. 1967. Synchronization of sexual reproduction of trees within the dry season in Central America. *Evolution* 21: 620–637.

Janzen, D. H. 1970. *Jacquinia pungens*, a heliophile from the understory of tropical deciduous forest. *Biotropica* 2: 112–119.

Janzen, D. H. 1974. *Swollen-thorn Acacias of Central America*. Smithsonian Contributions to Botany no. 13. Washington, D.C.

Janzen, D. H. 1975. Intra- and interhabitat variations in *Guazuma ulmifolia* (Sterculiaceae) seed predation by *Amblycerus cistelinus* (Bruchidae) in Costa Rica. *Ecology* 56: 1009–1013.

Janzen, D. H. 1980. Specificity of seed-attacking beetles in a Costa Rican deciduous forest. *Journal of Ecology* 69: 929–952.

Janzen, D. H. 1982. Natural history of guacimo fruits (Sterculiaceae, *Guazuma ulmifolia*) with respect to consumption by large mammals. *American Journal of Botany* 69: 1240–1250.

Janzen, D. H. 1983a. *Cassia biflora* (Abejon). In D. H. Janzen (ed.), *Costa Rican Natural History,* 210–211. University of Chicago Press, Chicago.

Janzen, D. H. 1983b. *Jaquinia pungens* (burriquita, siempre viva, siempre verde, false evergreen needle bush). In D. H. Janzen (ed.), *Costa Rican Natural History,* 265–266. University of Chicago Press, Chicago.

Janzen, D. H., P. S. Martin. 1982. Neotropical anachronisms: The fruits the gomphotheres ate. *Science* 215: 19–27.

Janzen, D. H., D. E. Wilson. 1974. The cost of being dormant in the tropics. *Biotropica* 6: 260–262.

Jauhar, P. P. 1983. Cytogenetic studies in relation to breeding superior cultivars of jojoba. In A. Elias-Cesnik (ed.), *Jojoba and Its Uses through 1982, Proceedings of the Fifth International Conference on Jojoba and Its Uses,* 63–77. University of Arizona, Tucson.

Jenkins, M. B., R. A. Virginia, W. M. Jarrell. 1989. Ecology of fast-growing and slow-growing mesquite-nodulating rhizobia in Chihuahuan and Sonoran desert ecosystems. *Soil Science Society of America Journal* 53: 543–549.

Jeune, B. 1983. Biometric study of the foliar development of *Parkinsonia aculeata* (Leguminosae). *Canadian Journal of Botany* 61: 87–92.

Johnsgard, P. A. 1983. *The Hummingbirds of North America.* Smithsonian Institution Press, Washington, D.C.

Johnson, A. F. 1977. A survey of the strand and dune vegetation along the Pacific and southern gulf coasts of Baja California, Mexico. *Journal of Biogeography* 7: 83–99.

Johnson, A. F. 1982. Dune vegetation along the eastern shore of the Gulf of California. *Journal of Biogeography* 9: 317–330.

Johnson, C. D. 1982. Host preferences of *Stator* in nonhost seeds. *Environmental Entomology* 10: 857–863.

Johnson, D. S. 1918. *The Fruit of Opuntia Fulgida.* Carnegie Institution of Washington Publication no. 269. Washington, D.C.

Johnson, D. S. 1924. The influence of insolation on the distribution and on the developmental sequence of the flowers of the giant cactus of Arizona. *Ecology* 5: 70–82.

Johnson, M. B. 1988. Horticultural characteristics of seven Sonoran Desert legumes with potential for southwestern landscaping. M.S. thesis, University of Arizona, Tucson.

Johnson, M. B. 1991. Sonoran tree catclaw. *Plant Press* 15: 12–13.

Johnson, M. B. 1992. The genus *Bursera* in Sonora, Mexico and Arizona, U.S.A. *Desert Plants* 10: 126–143.

Johnson, M. B. 1993. Woody legumes in southwest desert landscapes. *Desert Plants* 10: 147–175.

Johnston, I. M. 1924. Expedition of the California Academy of Sciences to the Gulf of California in 1921. The botany (the vascular plants). *Proceedings of the California Academy of Sciences* 12: 951–1218.

Johnston, I. M. 1949. Studies in the Boraginaceae, part 17. *Journal of the Arnold Arboretum* 30: 85–110.

Johnston, M. C. 1957. *Celtis spinosa* Sprengel var. *pallida* (Torrey) M. C. Johnston, new combination. *Southwestern Naturalist* 2: 172.

Johnston, M. C. 1962a. Revision of *Condalia* including *Microrhamnus* (Rhamnaceae). *Brittonia* 14: 332–368.

Johnston, M. C. 1962b. The North American mesquites *Prosopis* sect. *Algarobia* (Leguminosae). *Brittonia* 14: 72–90.

Johnston, M. C. 1963a. Novelties in *Colubrina* including *Cormonema* and *Hybosperma. Wrightia* 3: 91–96.

Johnston, M. C. 1963b. The species of *Ziziphus* indigenous to the United States and Mexico. *American Journal of Botany* 50: 1020–1027.

Johnston, M. C. 1966. Systematic studies in the plant genus *Karwinskia* in Mexico and Central America. *Yearbook of the American Philosophical Society* 1966: 351–357.

Johnston, M. C. 1971. Revision of *Colubrina* (Rhamnaceae). *Brittonia* 23: 2–53.

Johnston, N. L., C. L. Quarles, D. J. Fagerberg, D. D. Caveny. 1981. Evaluation of *Yucca schidigera* saponin on broiler performance and ammonia suppression. *Poultry Science* 60: 2289–2292.

Jolad, S. D., R. B. Bates, J. R. Cole, J. J. Hoffmann, T. J. Siahaan, B. N. Timmermann. 1986. Cardenolides and a lignan from *Asclepias subulata. Phytochemistry* 25: 2581–2590.

Jones, C. E. 1978. Pollinator constancy as a pre-pollination isolating mechanism between sympatric species of *Cercidium. Evolution* 32: 189–198.

Jones, C. S. 1984. The effect of axis splitting on xylem pressure potentials and water movement in the desert shrub *Ambrosia dumosa* (Asteraceae). *Botanical Gazette* 145: 125–131.

Jones, W. D. 1979. Effects of 1978 freeze on native plants of Sonora, Mexico. *Desert Plants* 1: 33–36.

Jones, W. D., C. W. Lee, L. Hogan. 1980. Selection and evaluation of plant materials for landscape use in arid regions. *HortScience* 15: 397.

Jordan, P. W., P. S. Nobel. 1979. Infrequent establishment of seedlings of *Agave deserti* (Agavaceae) in the northwestern Sonoran Desert. *American Journal of Botany* 66: 1079–1084.

Jordan, P. W., P. S. Nobel. 1981. Seedling establishment of *Ferocactus acanthodes* in relation to drought. *Ecology* 62: 901–906.

Jordan, P. W., P. S. Nobel. 1982. Height distributions of two species of cacti in relation to rainfall, seedling establishment, and growth. *Botanical Gazette* 143: 511–517.

Jordan, P. W., P. S. Nobel. 1984. Thermal and water relations of roots of desert succulents. *Annals of Botany* 54: 705–717.

Juncosa, A. M., P. B. Tomlinson. 1987. Floral development in mangrove Rhizophoraceae. *American Journal of Botany* 74: 1263–1279.

Kadish, A. 1985. A search for the origin of jojoba. In J. Wisniak, J. Zabicky (eds.), *Proceedings of the Sixth International Conference on Jojoba and Its Uses, 21–26 October 1984,* 287–292. Ben-Gurion University of the Negev, Beer Sheeva, Israel.

Kay, B. L., C. M. Ross, W. L. Graves. 1977. *White Burrobush.* Revegetation Notes no. 8. Department of Agronomy and Range Science, University of California, Davis.

Kearney, R. H., R. H. Peebles. 1960. *Arizona Flora.* 2d ed. Supplement by J. T. Howell, E. McClintock, and others. University of California Press, Berkeley.

Kee, S. C., P. S. Nobel. 1986. Concomitant changes in high temperature tolerance and heat-shock proteins in desert succulents. *Plant Physiology* 80: 596–598.

Keeley, J. E., S. C. Keeley. 1981. Post-fire regeneration of southern California chaparral. *American Journal of Botany* 68: 524–530.

Keeley, J. E., S. C. Keeley, D. A. Ikeda. 1986. Seed predation by yucca moths on semelparous, iteroparous, and vegetatively reproducing subspecies of *Yucca whipplei* (Agavaceae). *American Midland Naturalist* 115: 1–9.

Keeley, J. E., S. C. Keeley, C. C. Swift, J. Lee. 1984. Seed predation due to the *Yucca*-moth symbiosis. *American Midland Naturalist* 112: 187–191.

Keeley, J. E., A. Meyers. 1985. Effect of heat on seed germination of southwestern *Yucca* species. *Southwestern Naturalist* 30: 303–304.

Keeley, J. E., D. A. Tufenkian. 1984. Garden comparison of germination and seedling growth of *Yucca whipplei* subspecies (Agavaceae). *Madroño* 31: 24–29.

Kemp, P. R., P. E. Gardetto. 1982. Photosynthetic pathway types of evergreen rosette plants (Liliaceae) of the Chihuahuan Desert. *Oecologia* 55: 149–156.

Khalafalla, M. S., D. A. Palzkill. 1990. Seasonal patterns of carbohydrates and proline in jojoba clones that differ in frost susceptibility. *HortScience* 25: 103–105.

Khatamian, H., Z. Abuelgasim. 1986. Auxins aid western soapberry cuttings. *American Nurseryman* 164: 65, 72.

Killingbeck, K. T. 1990. Leaf production can be decoupled from root activity in the desert shrub ocotillo (*Fouquieria splendens* [Engelm.]). *American Midland Naturalist* 124: 124–129.

Kingsolver, J. M., C. D. Johnson, S. R. Swier, A. Teran. 1977. *Prosopis* fruits as a resource for invertebrates. In B. B. Simpson (ed.), *Mesquite: Its Biology in Two Desert Scrub Ecosystems,* 108–122. Dowden, Hutchinson and Ross, Stroudsburg, Penn.

Kinnison, W. A. 1979. Preliminary evaluation of cold-hardiness in desert landscaping plants at Central Arizona College. *Desert Plants* 1: 29–32.

Kleinkopf, G. E., T. L. Hartsock, A. Wallace, E. M. Romney. 1980. Photosynthetic strategies of two Mojave Desert shrubs. *Great Basin Naturalist Memoirs* 4: 100–109.

Klikoff, L. G. 1967. Moisture stress in a vegetational continuum in the Sonoran Desert. *American Midland Naturalist* 77: 128–137.

Knipe, D., C. H. Herbel. 1966. Germination and growth of some grassland species treated with aqueous extract from creosote bush. *Ecology* 47: 775–781.

Krizman, R. D. 1972. Environment and season in a tropical deciduous forest in northwestern Mexico. Ph.D. dissertation, University of Arizona, Tucson.

Krukoff, B. A. 1969. Supplementary notes on the American species of *Erythrina,* part 3. *Phytologia* 19: 113–175.

Kurabayashi, M., H. Lewis, P. H. Raven. 1962. A comparative study of mitosis in the Onagraceae. *American Journal of Botany* 49: 1003–1026.

Kyhos, D. W. 1971. Evidence of different adaptations of flower color variants of *Encelia farinosa* (Compositae). *Madroño* 21: 49–61.

Lajtha, K. 1987. Nutrient reabsorption efficiency and the response to phosphorus fertilization in the desert shrub *Larrea tridentata* (DC.) Cov. *Biogeochemistry* 4: 265–276.

Lajtha, K., M. Klein. 1988. The effect of varying nitrogen and phosphorus availability on nutrient use by *Larrea tridentata,* a desert evergreen shrub. *Oecologia* 75: 348–353.

Lajtha, K., J. Weishampel, W. H. Schlesinger. 1987. Phosphorus and pH tolerances in the germination of the desert shrub *Larrea tridentata* (Zygophyllaceae). *Madroño* 34: 63–68.

Lang, J. M., D. Isely. 1982. *Eysenhardtia* (Leguminosae, Papilionoideae). *Iowa State Journal of Research* 56: 393–418.

Lansford, H. H. 1967. The desert's declining king. *The Denver Post, Empire Magazine.* January 15: 32–35.

Lavin, M. 1988. Systematics of *Coursetia* (Leguminosae—Papilionoideae). *Systematic Botany Monographs* 21: 1–167.

Lee, C. W., R. A. Sherman. 1985. Meiosis in jojoba, *Simmondsia chinensis. Israel Journal of Botany* 34: 1–6.

Lee, Y. S., D. S. Seigler, J. E. Ebinger. 1989. *Acacia rigidula* (Fabaceae) and related species in Mexico and Texas. *Systematic Botany* 14: 91–100.

Leenhouts, P. W. 1983. Notes on the extra-Australian species of *Dodonaea* (Sapindaceae). *Blumea* 28: 271–279.

Lenz, L. W. 1966. Chromosome numbers in the Allieae (Liliaceae). *Aliso* 6: 81–82.

León de la Luz, J. L., R. Domínguez-Cadena. 1989. Flora of the Sierra de la Laguna, Baja California Sur, Mexico. *Madroño* 36: 61–83.

León de la Luz, J. L., R. Domínguez-Cadena. 1991. Evaluación de la reproducción por semilla de la pitaya agria (*Stenocereus gummosus*) en Baja California Sur, México. *Acta Botánica Mexicana* 14: 75–87.

Leonard, E. C. 1958. The Acanthaceae of Columbia, part 3. *Contributions from the U.S. National Herbarium* 31: 323–781.

Lersten, N. R., K. A. Carvey. 1974. Leaf anatomy of ocotillo (*Fouquieria splendens,* Fouquieriaceae), especially vein endings and associated veinlet elements. *Canadian Journal of Botany* 52: 2017–2022.

Lessani, H., S. Chariat-Panahi. 1979. IOPB chromosome number reports LXV. *Taxon* 28: 635–636.

Leuenberger, B. E. 1979. Contribution to the succulent flora of Togo, West Africa. *Willdenowia* 9: 71–86.

Lewis, D. A., P. S. Nobel. 1977. Thermal energy exchange model and water loss of a barrel cactus, *Ferocactus acanthodes*. *Plant Physiology* 60: 609–616.

Lewis, H., P. H. Raven. 1961. Phylogeny of the Onagraceae. In *Recent Advances in Botany: From Lectures and Symposia Presented to the IX International Botanical Congress, Montreal, 1959,* 1466–1468. University of Toronto Press, Toronto.

Lightfoot, D. C., W. G. Whitford. 1987. Variation in insect densities on desert creosote bush: Is nitrogen a factor? *Ecology* 68: 547–557.

Lim, G., H. L. Ng. 1977. Root nodules of some tropical legumes in Singapore. *Plant and Soil* 46: 317–327.

Lindsay, G. E. 1955. The taxonomy and ecology of the genus *Ferocactus*. Ph.D. dissertation, Stanford University, Stanford, Calif.

Lindsay, G. E. 1963. The genus *Lophocereus*. *Cactus and Succulent Journal (U.S.)* 35: 176–192.

Lindsey, D. L., S. E. Williams, W. D. Beavis, E. F. Aldon. 1984. Vesicular-arbuscular mycorrhizae associations in *Atriplex canescens* (Pursh) Nutt. and *Ceratoides lanata* (Pursh) J. T. Howell. In A. R. Tiedemann, E. D. McArthur, H. C. Stutz, R. Stevens, K. L. Johnson (comps.), *Proceedings—Symposium on the Biology of Atriplex and Related Chenopods, Provo, Utah, May 2–6, 1983,* 75–79. USDA Forest Service General Technical Report INT–172. Ogden, Utah.

Little, E. L. 1971. *Atlas of United States Trees,* vol. 1, *Conifers and Important Hardwoods.* USDA Miscellaneous Publication no. 1146. Government Printing Office, Washington, D.C.

Little, E. L. 1976. *Atlas of United States Trees,* vol. 3, *Minor Western Hardwoods.* USDA Miscellaneous Publication no. 1314. Government Printing Office, Washington, D.C.

Lloret, F., P. H. Zedler. 1991. Recruitment of *Rhus integrifolia* populations in periods between fire in chaparral. *Journal of Vegetation Science* 2: 217–230.

Lonard, R. I., F. W. Judd. 1991. Comparison of the effects of the severe freezes of 1983 and 1989 on native woody plants in the Lower Rio Grande Valley, Texas. *Southwestern Naturalist* 36: 213–217.

Lopez, F. B., P. S. Nobel. 1991. Root hydraulic conductivity of two cactus species in relation to root age, temperature, and soil water status. *Journal of Experimental Botany* 42: 143–150.

Lopez-Portillo, J., E. Ezcurra. 1989. Response of three mangroves to salinity in two geoforms. *Functional Ecology* 3: 355–361.

Lozoya-Meckes, M., V. Mellado-Campos. 1985. Is the *Tecoma stans* infusion an antidiabetic remedy? *Journal of Ethnopharmacology* 14: 1–9.

Luckow, M. A. 1989. Systematics of *Desmanthus* Willd. (Leguminosae: Mimosoideae). Ph.D. dissertation, University of Texas, Austin.

Luckow, M. A., C. D. Johnson. 1987. New host records of Bruchidae (Coleoptera) from *Desmanthus* (Leguminosae) from Texas and Mexico. *Pan-Pacific Entomologist* 63: 48–49.

Ludwig, J. A., P. Flavill. 1981. Productivity patterns of *Larrea* in the northern Chihuahuan Desert. In E. Campos Lopez, T. J. Mabry, S. Fernandez Tavizon (eds.), *Larrea,* 139–150. Consejo Nacional de Ciéncia y Tecnológia, México, D.F.

Ludwig, J. A., J. F. Reynolds, P. D. Whitson. 1975. Size-biomass relationships of several Chihuahuan Desert shrubs. *American Midland Naturalist* 94: 451–461.

Lugo, A. E., C. P. Zucca. 1977. The impact of low temperature stress on mangrove structure and growth. *Tropical Ecology* 18: 149–161.

Lundell, C. L. 1971. Studies of American plants, part 3. *Wrightia* 4: 153–170.

Lunt, O. R., J. Letey, S. B. Clark. 1973. Oxygen requirements for root growth in three species of desert shrubs. *Ecology* 54: 1356–1362.

Lux, A., P. R. Earl. 1989. The leaf structure of coyotillo, *Karwinskia humboldtiana* (Rhamnaceae). *Publicaciones Biológicas Facultad de Ciéncias, Biológicas Universidad Autónoma de Nuevo Leon* 3: 83–90.

McArthur, E. D. 1977. Environmentally induced changes of sex expression in *Atriplex canescens. Heredity* 38: 97–103.

McArthur, E. D., A. C. Blauer, G. L. Noller. 1984. Propagation of fourwing saltbush (*Atriplex canescens*) by stem cuttings. In A. R. Tiedemann (comp.), *Proceedings—Symposium on the Biology of Atriplex and Related Chenopods, 1983 May 2–6, Provo, UT,* 261–264. USDA Forest Service General Technical Report INT–172. Ogden, Utah.

McArthur, E. D., D. C. Freeman. 1982. Sex expression in *Atriplex canescens:* Genetics and environment. *Botanical Gazette* 143: 476–482.

McArthur, E. D., S. C. Sanderson, D. C. Freeman. 1986. Isozymes of an autopolyploid shrub, *Atriplex canescens* (Chenopodiaceae). *Great Basin Naturalist* 46: 157–160.

MacArthur, R. H., E. O. Wilson. 1967. *The Theory of Island Biogeography.* Princeton University Press, Princeton, N.J.

McAuliffe, J. R. 1984a. Sahuaro-nursetree associations in the Sonoran Desert: Competitive effects of sahuaros. *Oecologia* 64: 319–321.

McAuliffe, J. R. 1984b. Prey refugia and the distributions of two Sonoran Desert cacti. *Oecologia* 65: 82–85.

McAuliffe, J. R. 1986. Herbivore-limited establishment of a Sonoran Desert tree, *Cercidium microphyllum. Ecology* 67: 276–280.

McAuliffe, J. R. 1988. Markovian dynamics of simple and complex desert plant communities. *American Naturalist* 131: 459–490.

McAuliffe, J. R. 1990. Paloverdes, pocket mice, and bruchid beetles: Interrelationships of seeds, dispersers, and seed predators. *Southwestern Naturalist* 35: 329–337.

McAuliffe, J. R. 1991. Demographic shifts and plant succession along a late Holocene soil chronosequence in the Sonoran Desert of Baja California. *Journal of Arid Environments* 20: 165–178.

McAuliffe, J. R., P. Hendricks. 1988. Determinants of the vertical distributions of woodpecker nest cavities in the sahuaro cactus. *Condor* 90: 791–801.

McAuliffe, J. R., F. J. Janzen. 1986. Effects of intraspecific crowding on water uptake, water storage, apical growth, and reproductive potential in the sahuaro cactus, *Carnegiea gigantea. Botanical Gazette* 147: 334–341.

McCurrach, J. C. 1960. *Palms of the World.* Harper & Brothers, New York.

McDonough, W. T. 1964. Germination responses of *Carnegiea gigantea* and *Lemaireocereus thurberi. Ecology* 45: 155–159.

McDonough, W. T. 1965. Pattern changes associated with the decline of a species in a desert habitat. *Vegetatio* 13: 97–101.

MacDougal, D. T. 1912. The water balance of desert plants. *Annals of Botany* 26: 71–93.

MacDougal, D. T. 1915. General course of depletion in starving succulents. *Physiological Researches* 16: 292–298.

MacDougal, D. T. 1924. Dendrographic measurements. In *Growth in Trees and Massive Organs of Plants,* 1–88. Carnegie Institution of Washington Publication no. 350. Washington, D.C.

McGinnies, W. G. 1983. Flowering periods for common desert plants—southwestern

Arizona. Wall poster. Office of Arid Lands Studies, University of Arizona, Tucson.

McGregor, S. E., S. M. Alcorn. 1959. Partial self-sterility of the barrel cactus. *Cactus and Succulent Journal (U.S.)* 31: 88.

McKee, K. L., I. A. Mendelssohn. 1987. Root metabolism in the black mangrove (*Avicennia germinans* [L.] L): Response to hypoxia. *Environmental and Experimental Botany* 27: 147–156.

McKee, K. L., I. A. Mendelssohn, M. W. Hester. 1988. Reexamination of pore water sulfide concentrations and redox potentials near the aerial roots of *Rhizophora mangle* and *Avicennia germinans. American Journal of Botany* 75: 1352–1359.

McKelvey, S. D. 1938. *Yuccas of the Southwestern United States,* part 1. Arnold Arboretum of Harvard University, Jamaica Plain, Mass.

McKelvey, S. D. 1947. *Yuccas of the Southwestern United States,* part 2. Arnold Arboretum of Harvard University, Jamaica Plain, Mass.

McKelvey, S. D., K. Sax. 1933. Taxonomic and cytological relationships of *Yucca* and *Agave. Journal of the Arnold Arboretum* 14: 76–81.

McLaughlin, S. P. 1989. Floristic analysis of the southwestern United States. *Great Basin Naturalist* 46: 46–65.

McLaughlin, S. P. 1992. Are floristic areas hierarchically arranged? *Journal of Biogeography* 19: 21–32.

McLaughlin, S. P., J. E. Bowers. 1982. Effects of fire on a Sonoran desert community. *Ecology* 63: 246–248.

McLaughlin, S. P., J. J. Hoffmann. 1982. Survey of biocrude-producing plants from the Southwest. *Economic Botany* 36: 323–339.

McLeod, M. G. 1974. A cytotaxonomic investigation of the juicy-fruited prickly pears of Arizona (Cactaceae *Opuntia* series *Opuntiae*). *American Journal of Botany* 61 (Supplement): 47.

McLoskey, R. T. 1983. Desert rodent activity response to seed production by two perennial plant species. *Oikos* 41: 233–238.

McMillan, C. 1971. Environmental factors affecting seedling establishment of the black mangrove on the central Texas coast. *Ecology* 52: 927–930.

McMillan, C. 1986. Isozyme patterns among populations of black mangrove, *Avicennia germinans,* from the Gulf of Mexico–Caribbean and Pacific-Panama. *Contributions in Marine Science* 29: 17–26.

McMillan, C., C. L. Sherrod. 1986. The chilling tolerance of black mangrove, *Avicennia germinans,* from the Gulf of Mexico coast of Texas, Louisiana, and Florida. *Contributions in Marine Science* 29: 9–16.

McPherson, G. 1982. Studies in *Ipomoea* (Convolvulaceae), part 1: The *arborescens* group. *Annals of the Missouri Botanical Garden* 68: 527–545.

McPherson, G. R., H. A. Wright, D. B. Wester. 1988. Patterns of shrub invasion in semiarid Texas grasslands. *American Midland Naturalist* 120: 391–397.

McVaugh, R. 1945. The genus *Jatropha* in America: Principal intrageneric groups. *Bulletin of the Torrey Botanical Club* 72: 271–294.

McVaugh, R. 1987. *Flora Novo-Galiciana: A Descriptive Account of the Vascular Plants of Western Mexico,* vol. 5, *Leguminosae.* University of Michigan Press, Ann Arbor.

McVaugh, R., J. Rzedowski. 1965. Synopsis of the genus *Bursera* L. in western Mexico, with notes on the material of *Bursera* collected by Sessé and Moçino. *Kew Bulletin* 18: 317–382.

Maddox, J. C., S. Carlquist. 1985. Wind dispersal in Californian desert plants: Experimental studies and conceptual considerations. *Aliso* 11: 77–96.

Madhusudana-Rao, I., P. M. Swamy, V. S. R. Das. 1979. Some characteristics of Crassulacean acid metabolism in five nonsucculent scrub species under natural semi-arid conditions. *Zeitschrift für Pflanzenphysiologie* 94: 201–210.

Mahall, B. E., R. M. Callaway. 1991. Root communications among desert shrubs. *Proceedings of the National Academy of Sciences* 88: 874–876.

Malik, K. A., R. Bilal, G. Rasul, K. Mahmood, M. I. Sajjad. 1991. Associative nitrogen-fixation in plants growing in saline sodic soils and its relative quantification based on ^{15}nitrogen natural abundance. *Plant and Soil* 137: 67–74.

Mallery, C. H., H. J. Teas. 1984. The mineral ion relations of mangroves, part 1: Root cell compartments in a salt excluder and a salt secreter species at low salinities. *Plant and Cell Physiology* 25: 1123–1131.

Markham, C. G. 1972. Baja California's climate. *Weatherwise* 25: 64–101.

Markley, J. L., C. McMillan, G. A. Thompson. 1982. Latitudinal differentiation in response to chilling temperatures among populations of three mangroves, *Avicennia germinans, Laguncularia racemosa,* and *Rhizophora mangle,* from the western tropical Atlantic and Pacific Panama. *Canadian Journal of Botany* 60: 2704–2715.

Marsh, C. D., A. B. Clawson. 1928. *Coyotillo (Karwinskia humboldtiana) as a Poisonous Plant.* USDA Technical Bulletin no. 29. Washington, D.C.

Martin, S. C. 1948. Mesquite seeds remain viable after 44 years. *Ecology* 29: 393.

Martin, S. C., R. M. Turner. 1977. Vegetation change in the Sonoran Desert region, Arizona and Sonora. *Journal of the Arizona Academy of Science* 12: 59–69.

Martinez, M. 1979. *Catálogo de Nombres Vulgares y Científicos de Plantas Mexicanas.* Fondo de Cultura Económica, México, D.F.

Maruyama, E., T. Yokoyama, K. Migata. 1989. Effect of temperature and pre-heating on germination of *Guazuma crinita* Mart. and *Guazuma ulmifolia* Lam. seeds. *Journal of the Japanese Forestry Society* 71: 65–68.

Mason, C. T., P. B. Mason. 1987. *A Handbook of Mexican Roadside Flora.* University of Arizona Press, Tucson.

Mata, R., J. L. Contreras, D. Crisanto, R. Pereda-Miranda, P. Castaneda, F. del Rio. 1991. Chemical studies on Mexican plants used in traditional medicine, part 18: New secondary metabolites from *Dodonaea viscosa. Journal of Natural Products* 54: 913–917.

Mattson, W. J. 1980. Herbivory in relation to plant nitrogen content. *Annual Review of Ecology and Systematics* 11: 119–161.

Mayeux, H. S., C. J. Scifres, R. E. Meyer. 1979. *Some Factors Affecting the Response of Spiny Aster to Herbicide Sprays.* Texas Agricultural Experiment Station Bulletin no. B–1197. College Station.

Meinzer, F. C., P. W. Rundel, M. R. Sharifi, E. T. Nilsen. 1986. Turgor and osmotic relations of the desert shrub *Larrea tridentata. Plant, Cell, and Environment* 9: 467–476.

Meinzer, F. C., C. S. Wisdom, A. Gonzalez-Coloma, P. W. Rundel, L. M. Shultz. 1990. Effects of leaf resin on stomatal behaviour and gas exchange of *Larrea tridentata. Functional Ecology* 4: 579–584.

Melanson, M. A., S. M. Carter. 1981. Rehabilitation of disturbed arid lands: Long-range planning and the role of mycorrhizae in rehabilitation. In D. H. Graves (ed.), *Proceedings, 1981 Symposium on Surface Mining Hydrology, Sedimentology, and Reclamation,* 79–87. College of Engineering, University of Kentucky, Lexington.

Menke, J. W., M. J. Trlica. 1981. Carbohydrate reserve, phenology, and growth cycles of nine Colorado range species. *Journal of Range Management* 34: 269–277.

Meyer, B. N., Y. A. H. Mohamed, J. L. McLaughlin. 1980. Beta-phenethylamines from the cactus genus *Opuntia. Phytochemistry* 19: 719–720.

Meyer, M. W., W. H. Karasov. 1989. Antiherbivore chemistry of *Larrea tridentata:* Effects of woodrat (*Neotoma lepida*) feeding and nutrition. *Ecology* 70: 953–961.

Meyer, R. E. 1982. Brush response to spacing and individual plant herbicide treatments. *Weed Science* 30: 378–384.

Michener, C. D., M. L. Winston, R. Jander. 1978. Pollen manipulation and related activities and structures in bees of the family Apidae. *University of Kansas Science Bulletin* 51: 575–601.

Miege, J. 1962. Quatrième liste de nombres chromosomiques d'espèces d'Afrique Occidentale. *Revue de Cytologie et de Biologie Vegetales* 24: 149–164.

Mielke, J. L. 1986. Stars of the desert: Shrubs for Southwestern landscapes. *American Nurseryman* 164: 71–76.

Miller, J. S. 1988. A revised treatment of Boraginaceae for Panama. *Annals of the Missouri Botanical Garden* 75: 456–521.

Miller, V. C. 1983. Arizona's own fan palm: *Washingtonia filifera. Desert Plants* 5: 99–104.

Mintzer, A. 1979. Foraging activity of the Mexican leaf cutting ant, *Atta mexicana,* in a Sonoran Desert habitat (Hymenoptera: Formicidae). *Insectes Sociaux* 26: 364–372.

Mitchell, J. D., D. C. Daly. 1991. *Cyrtocarpa* Kunth (Anacardiaceae) in South America. *Annals of the Missouri Botanical Garden* 78: 184–189.

Mitchell, R. 1977. Bruchid beetles and seed packaging by palo verde. *Ecology* 58: 644–651.

Mitscher, L. A., S. R. Gollapudi, D. S. Oburn, S. Drake. 1985. Antimicrobial agents from higher plants: 2-dimethylbenzisochromans from *Karwinskia humboldtiana. Phytochemistry* 24: 1681–1683.

Moldenke, H. N. 1960. Materials toward a monograph of the genus *Avicennia,* part 2. *Phytologia* 7: 179–232.

Monson, R. K., S. D. Smith. 1982. Seasonal water potential components of Sonoran Desert plants. *Ecology* 63: 113–123.

Monson, R. K., S. D. Smith, J. L. Gehring, W. D. Bowman, S. R. Szarek. 1992. Physiological differentiation within an *Encelia farinosa* population along a short topographic gradient in the Sonoran Desert. *Functional Ecology* 6: 751–759.

Mooney, H. A., B. Bartholomew. 1974. Comparative carbon balance and reproductive modes of two California *Aesculus* species. *Botanical Gazette* 135: 306–313.

Mooney, H. A., J. Ehleringer, O. Bjorkman. 1977. The energy balance of leaves of the evergreen desert shrub *Atriplex hymenelytra. Oecologia* 29: 301–310.

Mooney, H. A., W. A. Emboden. 1968. The relationship of terpene composition, morphology, and distribution of populations of *Bursera microphylla* (Burseraceae). *Brittonia* 20: 44–51.

Mooney, H. A., B. B. Simpson, O. T. Solbrig. 1977. Phenology, morphology, physiology. In B. B. Simpson (ed.), *Mesquite: Its Biology in Two Desert Scrub Ecosystems,* 26–43. Dowden, Hutchinson and Ross, Stroudsburg, Penn.

Mooney, H. A., B. R. Strain. 1964. Bark photosynthesis in ocotillo. *Madroño* 17: 230–233.

Moore, M. 1989. *Medicinal Plants of the Desert and Canyon West.* Museum of New Mexico Press, Santa Fe.

Moran, R. 1962a. The unique *Cereus* of Rosario Bay. *Cactus and Succulent Journal (U.S.)* 34: 184–188.

Moran, R. 1962b. *Pachycereus orcuttii*—a puzzle solved. *Cactus and Succulent Journal (U.S.)* 34: 88–94.

Moran, R. 1962c. Visitors to the flowers of *Pachycereus pringlei. National Cactus and Succulent Journal* 17: 21.

Moran, R. 1964. Floración de *Agave goldmaniana* a los 31 anos. *Cactaceas y Suculentas Mexicanas* 9: 87–88.

Moran, R. 1965. Revisión de *Bergerocactus. Cactaceas y Succulentas Mexicanas* 10: 51–59.

Moran, R. 1966. The fruit of *Bergerocactus. National Cactus and Succulent Journal* 21: 30–31.

Moran, R. 1968. Cardón. *Pacific Discovery* 21: 2–9.

Moran, R. 1969. Twelve new dicots from Baja California, Mexico. *Transactions of the San Diego Society of Natural History* 15: 265–295.

Moran, R. 1975. Creeping devil. *Environment Southwest* (February 1975): 14–16.

Moran, R. 1977a. Palms in Baja California. *Environment Southwest* (Summer 1977): 10–14.

Moran, R. 1977b. Plant notes from the Sierra Juarez of Baja California, Mexico. *Phytologia* 35: 205–215.

Moran, R. 1979. Palmas of Valle de las Palmas. *Environment Southwest* (Spring 1979): 16–18.

Moran, R. 1983a. The vascular flora of Isla Angel de la Guarda. Appendix 4.2. In T. J. Case, M. L. Cody (eds.), *Island Biogeography in the Sea of Cortez,* 382–403. University of California Press, Berkeley.

Moran, R. 1983b. Vascular plants of the Gulf Islands. Appendix 4.1. In T. J. Case, M. L. Cody (eds.), *Island Biogeography in the Sea of Cortez,* 348–381. University of California Press, Berkeley.

Moran, R., R. Felger. 1968. *Castela polyandra,* a new species in a new section; union of *Holacantha* with *Castela* (Simaroubaceae). *Transactions of the San Diego Society of Natural History* 15: 31–40.

Moreno-Casasola, P., I. Espejel. 1986. Classification and ordination of coastal sand dune vegetation along the Gulf and Caribbean Sea of Mexico. *Vegetatio* 66: 147–182.

Moujir, L., A. M. Gutierrez-Navarro, A. G. Gonzalez, A. G. Ravelo, J. G. Luis. 1990. The relationship between structure and antimicrobial activity in quinones from the Celastraceae. *Biochemical Systematics and Ecology* 18: 25–28.

Mozafar, A., J. R. Goodin. 1970. Vesiculated hairs: A mechanism for salt tolerance in *Atriplex halimus* L. *Plant Physiology* 45: 62–65.

Muller, C. H. 1953. The association of desert annuals with shrubs. *American Journal of Botany* 40: 53–60.

Muller, W. H., C. H. Muller. 1956. Association patterns involving desert plants that contain toxic products. *American Journal of Botany* 43: 354–361.

Munson, R. H. 1984. Germination of western soapberry as affected by scarification and stratification. *HortScience* 19: 712–713.

Munz, P. A. 1960. The genus *Xylonagra* (Onagraceae). *Aliso* 4: 499–500.

Munz, P. A. 1973. Record of an unusually tall *Yucca schidigera. Aliso* 8: 13–14.

Munz, P. A. 1974. *A Flora of Southern California.* University of California Press, Berkeley.

Munz, P. A., and D. D. Keck. 1959. *A California Flora.* University of California Press, Berkeley.

Musick, H. B. 1977. The physiological basis for calcicoly of *Larrea divaricata.* Ph.D. dissertation, University of Arizona, Tucson.

Mutz, J. L., C. J. Scifres, W. C. Mohr, D. L. Drawe. 1979. *Control of Willow Baccharis and Spiny Aster with Pelleted Herbicides.* Texas Agricultural Experiment Station Bulletin no. B–1194. College Station.

Nabhan, G. P. 1989. *Enduring Seeds: Native American Agriculture and Wild Plant Conservation.* North Point Press, San Francisco.

Nabhan, G. P., R. S. Felger. 1985. Wild desert relatives of crops: Their direct uses as food. In G. E. Wickens, J. R. Goodin, D. V. Field (eds.), *Plants for Arid Lands,* 19–33. George Allen and Unwin, London.

Nair, P. K. K., K. N. Singh. 1974. A study of two honey plants, *Antigonon leptopus* Hook. and *Moringa pterigosperma* Gaertn. *Indian Journal of Horticulture* 31: 375–379.

Narayana, P. S., L. L. Narayana. 1986. The embryology of Stegnospermataceae with a discussion of its status, affinities, and systematic position. *Plant Systematics and Evolution* 154: 139–145.

Nedoff, J. A., I. P. Ting, E. M. Lord. 1985. Structure and function of the green stem tissue in ocotillo (*Fouquieria splendens*). *American Journal of Botany* 72: 143–151.

Neilson, R. P. 1987. On the interface between current ecological studies and the paleobotany of pinyon-juniper woodlands. In R. L. Everett (comp.), *Proceedings—Pinyon Juniper Conference,* 93–98. USDA Forest Service General Technical Report INT–215. Ogden, Utah.

Nesom, G. L., Y. Suh, D. R. Morgan, B. B. Simpson. 1990. *Xylothamnia,* a new genus (Asteraceae: Astereae) related to *Euthamia. Sida* 14: 101–116.

Neufeld, H. S., F. C. Meinzer, C. S. Wisdom, M. R. Sharifi, P. W. Rundel, M. S. Neufeld, Y. Goldring, G. L. Cunningham. 1988. Canopy architecture of *Larrea tridentata* (DC.) Cov., a desert shrub: Foliage orientation and direct beam radiation interception. *Oecologia* 75: 54–60.

Newland, K. C., S. Ives, G. E. Joseph, M. A. Dimmitt, M. Mittleman, K. E. Foster, C. Scannell, W. R. Feldman, F. S. Crosswhite, C. Hansen. 1980. Propagation techniques for desert plants, part 1. *Desert Plants* 2: 205–216.

Ngulube, M. R. 1989. Seed germination, seedling growth, and biomass production of eight Central American multipurpose trees under nursery conditions in Malawi. *Forest Ecology and Management* 27: 21–28.

Nickerson, N. H., F. R. Thibodeau. 1984. Association between pore water sulfide concentrations and the distribution of mangroves. *Biogeochemistry* 1: 183–192.

Niering, W. A., C. H. Lowe. 1984. Vegetation of the Santa Catalina Mountains, Arizona, USA: Community types and dynamics. *Vegetatio* 58: 3–28.

Niering, W. A., R. H. Whittaker. 1965. The saguaro problem and grazing in southwestern national monuments. *National Parks Magazine* 39: 4–9.

Niering, W. A., R. H. Whittaker, C. H. Lowe. 1963. The saguaro: A population in relation to environment. *Science* 142: 15–23.

Niklas, K. J., S. L. Buchmann. 1985. Aerodynamics of wind pollination in *Simmondsia chinensis* (Link) Schneider. *American Journal of Botany* 72: 530–539.

Niklas, K. J., S. L. Buchmann. 1986. Wind pollination in two sympatric species of *Ephedra. American Journal of Botany* 73: 64.

Niklas, K. J., S. L. Buchmann, V. Verchner. 1986. Aerodynamics of *Ephedra trifurca,* part 1: Pollen grain velocity fields around stems bearing ovules. *American Journal of Botany* 73: 966–979.

Nilsen, E. T., F. C. Meinzer, P. W. Rundel. 1989. Stem photosynthesis in *Psorothamnus spinosus* (smoke tree) in the Sonoran Desert of California. *Oecologia* 79: 193–197.

Nilsen, E. T., M. R. Sharifi, P. W. Rundel. 1984. Comparative water relations of phreatophytes in the Sonoran Desert of California. *Ecology* 65: 767–778.

Nilsen, E. T., M. R. Sharifi, P. W. Rundel, I. N. Forseth, J. R. Ehleringer. 1990. Water relations of stem succulent trees in north-central Baja California. *Oecologia* 82: 299–303.

Nilsen, E. T., M. R. Sharifi, P. W. Rundel, W. M. Jarrell, R. A. Virginia. 1983. Diurnal and seasonal water relations of the desert phreatophyte *Prosopis glandulosa* (honey mesquite) in the Sonoran Desert of California. *Ecology* 64: 1381–1393.

Nilsen, E. T., M. R. Sharifi, R. A. Virginia, P. W. Rundel. 1987. Phenology of warm desert phreatophytes: Seasonal growth and herbivory in *Prosopis glandulosa* var. *torreyana* (honey mesquite). *Journal of Arid Environments* 13: 217–229.

Nisbet, R. A., D. T. Patten. 1974. Seasonal temperature acclimation of a prickly-pear cactus in south-central Arizona. *Oecologia* 15: 345–352.

Nobel, P. S. 1976a. Water relations and photosynthesis of a desert CAM plant, *Agave deserti. Plant Physiology* 58: 576–582.

Nobel, P. S. 1976b. Photosynthetic rates of sun versus shade leaves of *Hyptis emoryi. Plant Physiology* 58: 218–223.

Nobel, P. S. 1977a. Water relations of flowering *Agave deserti. Botanical Gazette* 138: 1–6.

Nobel, P. S. 1977b. Water relations and photosynthesis of a barrel cactus, *Ferocactus acanthodes,* in the Colorado Desert. *Oecologia* 27: 117–133.

Nobel, P. S. 1978. Surface temperatures of cacti—influences of environmental and morphological factors. *Ecology* 59: 986–996.

Nobel, P. S. 1980a. Interception of photosynthetically active radiation by cacti of different morphology. *Oecologia* 45: 160–166.

Nobel, P. S. 1980b. Morphology, nurse plants, and minimum apical temperatures for young *Carnegiea gigantea. Botanical Gazette* 141: 188–191.

Nobel, P. S. 1980c. Morphology, surface temperature, and northern limits of columnar cacti in the Sonoran Desert. *Ecology* 61: 1–7.

Nobel, P. S. 1980d. Water vapor conductance and CO_2 uptake for leaves of a C_4 desert grass, *Hilaria rigida. Ecology* 61: 252–258.

Nobel, P. S. 1980e. Influences of minimum stem temperatures on ranges of cacti in the southwestern United States and central Chile. *Oecologia* 47: 10–15.

Nobel, P. S. 1981a. Influences of photosynthetically active radiation on cladode orientation, stem tilting, and height of cacti. *Ecology* 62: 982–990.

Nobel, P. S. 1981b. Spacing and transpiration of various sized clumps of a desert grass, *Hilaria rigida. Journal of Ecology* 69: 735–742.

Nobel, P. S. 1982a. Low-temperature tolerance and cold hardening of cacti. *Ecology* 63: 1650–1656.

Nobel, P. S. 1982b. Orientation of terminal cladodes of platyopuntias. *Botanical Gazette* 143: 219–224.

Nobel, P. S. 1983a. Nutrient levels in cacti: Relation to nocturnal acid accumulation and growth. *American Journal of Botany* 70: 1244–1253.

Nobel, P. S. 1983b. Spine influence on PAR interception, stem temperature, and nocturnal acid accumulation by cacti. *Plant, Cell, and Environment* 6: 153–159.

Nobel, P. S. 1984a. Productivity of *Agave deserti:* Measurement by dry weight and monthly prediction using physiological responses to environmental parameters. *Oecologia* 64: 1–7.

Nobel, P. S. 1984b. Extreme temperatures and thermal tolerances of desert succulents. *Oecologia* 62: 310–317.

Nobel, P. S. 1985. Water relations and carbon dioxide uptake of *Agave deserti*—special adaptations to desert climates. *Desert Plants* 7: 51–56, 70.

Nobel, P. S. 1986. Relation between monthly growth of *Ferocactus acanthodes* and an environmental productivity index. *American Journal of Botany* 73: 541–547.

Nobel, P. S. 1987. Water relations and plant size aspects of flowering for *Agave deserti. Botanical Gazette* 148: 79–84.

Nobel, P. S. 1988. *The Environmental Biology of Agaves and Cacti.* Cambridge University Press, New York.

Nobel, P. S. 1989a. Influence of photoperiod on growth for three desert CAM species. *Botanical Gazette* 150: 9–14.

Nobel, P. S. 1989b. A nutrient index quantifying productivity of agaves and cacti. *Journal of Applied Ecology* 26: 635–646.

Nobel, P. S. 1989c. Temperature, water availability, and nutrient levels at various soil depths: Consequences for shallow-rooted desert succulents including nurse plant effects. *American Journal of Botany* 76: 1486–1492.

Nobel, P. S. 1990. Soil oxygen and carbon dioxide effects on apparent cell viability for roots of desert succulents. *Journal of Experimental Botany* 41: 1031–1038.

Nobel, P. S., J. Cavelier, J. L. Andrade. 1992. Mucilage in cacti: Its apoplastic capacitance, associated solutes, and influence on tissue water relations. *Journal of Experimental Botany* 43: 641–648.

Nobel, P. S., A. C. Franco. 1986. Annual root growth and intraspecific competition for a desert bunchgrass. *Journal of Ecology* 74: 1119–1126.

Nobel, P. S., G. N. Geller, S. C. Kee, A. D. Zimmerman. 1986. Temperatures and thermal tolerances for cacti exposed to high temperatures near the soil surface. *Plant, Cell, and Environment* 9: 279–288.

Nobel, P. S., T. L. Hartsock. 1981. Shifts in the optimal temperature for nocturnal CO_2 uptake caused by changes in growth temperature for cacti and agaves. *Physiologia Plantarum* 53: 523–527.

Nobel, P. S., T. L. Hartsock. 1986a. Environmental influence on the productivity of three desert succulents in the southwestern USA. *Plant, Cell, and Environment* 9: 741–750.

Nobel, P. S., T. L. Hartsock. 1986b. Leaf and stem carbon dioxide uptake in the three subfamilies of the Cactaceae. *Plant Physiology* 80: 913–917.

Nobel, P. S., P. W. Jordan. 1983. Transpiration stream of desert species: Resistances and capacitances for a C_3, a C_4, and a CAM plant. *Journal of Experimental Botany* 34: 1379–1391.

Nobel, P. S., D. J. Longstreth, T. L. Hartsock. 1978. Effect of water stress on the temperature optima of net carbon dioxide exchange for two desert species. *Physiologia Plantarum* 44: 97–101.

Nobel, P. S., R. G. McDaniel. 1988. Low temperature tolerances, nocturnal acid accumulation, and biomass increases for seven species of *Agave. Journal of Arid Environments* 15: 147–156.

Nobel, P. S., J. A. Palta. 1989. Soil oxygen and carbon dioxide effects on root respiration of cacti. *Plant and Soil* 120: 263–272.

Nobel, P. S., C. E. Russell, P. Felker, J. G. Medina, E. Acuna. 1987. Nutrient relations and productivity of prickly pear cacti. *Agronomy Journal* 79: 550–555.

Nobel, P. S., J. Sanderson. 1984. Rectifier-like activities of roots of two desert succulents. *Journal of Experimental Botany* 35: 727–737.

Nobel, P. S., S. D. Smith. 1983. High and low temperature tolerances and their relationships to distribution of agaves. *Plant, Cell, and Environment* 6: 711–719.

Nobs, M. A. 1978. Chromosome numbers in *Atriplex. Carnegie Yearbook* 77: 240–241.

Nokes, J. 1986. *How to Grow Native Plants of Texas and the Southwest.* Texas Monthly Press, Austin.

Nord, E. C., P. F. Hartless, W. D. Nettleton. 1971. Effects of several factors on saltbush establishment in California. *Journal of Range Management* 24: 216–223.

Norem, M. A., A. D. Day, K. L. Ludeke. 1982. An evaluation of shrub and tree species used for revegetating copper mine wastes in the southwestern United States. *Journal of Arid Environments* 5: 299–304.

Norman, E. M., S. Roper. 1981. Chromosome number reports LXXII. *Taxon* 30: 697.

Noster, S., L. Kraus. 1990. In vitro antimalarial activity of *Coutarea latiflora* and *Exostema caribaeum* extracts on *Plasmodium falciparum. Planta Medica* 56: 63–65.

Nowicke, J. W., J. J. Skvarla. 1984. Pollen morphology and the relationships of *Simmondsia chinensis* to the order Euphorbiales. *American Journal of Botany* 71: 210–215.

Nutting, W. L. 1979. Biological notes on a rare dry-wood termite in the Southwest, *Incisitermes banksi* (Kalolermitidae). *Southwestern Entomologist* 4: 308–310.

Odening, W. R., B. R. Strain, W. C. Oechel. 1974. The effect of decreasing water potential on net CO_2 exchange of intact desert shrubs. *Ecology* 55: 1086–1095.

Odum, W. E., C. C. McIvor. 1990. Mangroves. In R. L. Myers, J. J. Ewel (eds.), *Ecosystems of Florida,* 517–548. University of Central Florida Press, Orlando.

Oechel, W. C., B. R. Strain, W. R. Odening. 1972. Tissue water potential, photosynthesis, ^{14}C-labeled photosynthate utilization, and growth in the desert shrub *Larrea divaricata* Cav. *Ecological Monographs* 42: 127–141.

Okunade, A. L., D. F. Wiemer. 1985. Jacquinonic acid, an ant-repellent triterpenoid from *Jacquinia pungens. Phytochemistry* 24: 1203–1205.

O'Leary, J. F., R. A. Minnich. 1981. Postfire recovery of creosote bush scrub vegetation in the western Colorado Desert. *Madroño* 28: 61–66.

Olin, G., S. M. Alcorn, J. M. Alcorn. 1989. Dispersal of viable saguaro seeds by white-winged doves (*Zenaida asiatica*). *Southwestern Naturalist* 34: 281–284.

Opler, P. A., H. G. Baker, G. W. Frankie. 1975. Reproductive biology of some Costa Rican *Cordia* spp. (Boraginaceae). *Biotropica* 7: 234–247.

Opler, P. A., D. H. Janzen. 1983. *Cordia alliodora* (Laurel). In D. H. Janzen (ed.), *Costa Rican Natural History,* 219–221. University of Chicago Press, Chicago.

Osborn, M. M., P. G. Kevan, M. A. Lane. 1988. Pollination biology of *Opuntia polyacantha* and *Opuntia phaeacantha* (Cactaceae) in southern Colorado. *Plant Systematics and Evolution* 159: 85–94.

Osmond, C. B., O. Bjorkman, D. J. Anderson. 1980. *Physiological Processes in Plant Ecology: Toward a Synthesis with Atriplex.* Springer-Verlag, New York.

Ottley, A. M. 1944. The American Loti with special consideration of a proposed new section, Simpeteria. *Brittonia* 5: 81–123.

Palacios, R. A., L. D. Bravo. 1974/75. Morphological study of the seeds of *Prosopis,* part 2: Some North American and tropical species. *Darwiniana (Buenos Aires)* 19: 357–372.

Palta, J. A., P. S. Nobel. 1989a. Root respiration of *Agave deserti:* Influence of temperature, water status, and root age on daily patterns. *Journal of Experimental Botany* 20: 181–186.

Palta, J. A., P. S. Nobel. 1989b. Influences of water status, temperature, and root age on daily patterns of root respiration of two cactus species. *Annals of Botany* 63: 651–662.

Parfitt, B. D., W. D. Clark. 1978. Possible origin of *Opuntia curvospina* (Cactaceae). *Journal of the Arizona-Nevada Academy of Science* 13 (Supplement): 23–24.

Parfitt, B. D., C. H. Pickett. 1980. Insect pollination of prickly pears (*Opuntia:* Cactaceae). *Southwestern Naturalist* 25: 104–106.

Parfitt, B. D., D. J. Pinkava. 1988. Nomenclatural and systematic reassessment of *Opuntia engelmannii* and *O. lindheimeri* (Cactaceae). *Madroño* 35: 342–349.

Parish, S. B. 1907. A contribution toward a knowledge of the genus *Washingtonia. Botanical Gazette* 44: 408–434.

Parker, K. C. 1987a. Seedcrop characteristics and minimum reproductive size of organ pipe cactus (*Stenocereus thurberi*) in southern Arizona. *Madroño* 34: 294–303.

Parker, K. C. 1987b. Site-related demographic patterns of organ pipe cactus populations in southern Arizona. *Bulletin of the Torrey Botanical Club* 114: 149–155.

Parker, K. C. 1988a. Environmental relationships and vegetation associates of columnar cacti in the northern Sonoran Desert. *Vegetatio* 78: 125–140.

Parker, K. C. 1988b. Growth rates of *Stenocereus thurberi* and *Lophocereus schottii* in southern Arizona. *Botanical Gazette* 149: 335–346.

Parker, K. C. 1989. Height structure and reproductive characteristics of senita, *Lophocereus schottii* (Cactaceae), in southern Arizona. *Southwestern Naturalist* 34: 392–401.

Parker, K. C., J. L. Hamrick. 1992. Genetic diversity and clonal structure in a columnar cactus, *Lophocereus schottii. American Journal of Botany* 79: 86–96.

Parker, K. F. 1972. *An Illustrated Guide to Arizona Weeds.* University of Arizona Press, Tucson.

Parker, K. W., S. C. Martin. 1952. *The Mesquite Problem on Arizona Ranges.* USDA Forest Service Circular no. 908. Government Printing Office, Washington, D.C.

Pase, C. P., D. E. Brown. 1982. California coastalscrub. *Desert Plants* 4: 86–94.

Pasternak, D., A. Danon, J. A. Aronson, R. W. Benjamin. 1985. Developing the seawater agriculture concept. *Plant and Soil* 89: 337–348.

Patten, D. T. 1978. Productivity and production efficiency of an Upper Sonoran Desert ephemeral community. *American Journal of Botany* 65: 891–895.

Patten, D. T., G. H. Cave. 1984. Fire temperatures and physical characteristics of a controlled burn in the upper Sonoran Desert. *Journal of Range Management* 37: 277–280.

Paulsen, H. A. 1950. Mortality of velvet mesquite seedlings. *Journal of Range Management* 3: 281–286.

Payne, W. W. 1962. The unique morphology of the spines of an armed ragweed, *Ambrosia bryantii* (Compositae). *Madroño* 116: 233–236.

Payne, W. W. 1963. The morphology of the inflorescence of ragweeds (*Ambrosia-Franseria:* Compositae). *American Journal of Botany* 50: 872–880.

Payne, W. W. 1964. A re-evaluation of the genus *Ambrosia* (Compositae). *Journal of the Arnold Arboretum* 45: 401–438.

Payne, W. W., P. H. Raven, D. W. Kyhos. 1964. Chromosome numbers in Compositae, part 4: Ambrosieae. *American Journal of Botany* 51: 419–424.

Peacock, J. T., C. McMillan. 1965. Ecotypic differentiation in *Prosopis* (mesquite). *Ecology* 46: 35–51.

Peacock, J. T., C. McMillan. 1968. The photoperiodic response of American *Prosopis* and *Acacia* from a broad latitudinal distribution. *American Journal of Botany* 55: 153–159.

Pearcy, R. W. 1976. Temperature response of growth and photosynthetic CO_2 exchange rates in coastal and desert races of *Atriplex lentiformis. Oecologia* 26: 245–255.

Pearcy, R. W. 1977. Acclimation of photosynthetic and respiratory carbon dioxide exchange to growth temperature in *Atriplex lentiformis* (Torr.) Wats. *Plant Physiology* 59: 795–799.

Pearcy, R. W., J. A. Berry, D. C. Fork. 1977. Effects of growth temperature on the thermal stability of the photosynthetic apparatus of *Atriplex lentiformis. Plant Physiology* 59: 873–878.

Pearcy, R. W., A. T. Harrison. 1974. Comparative photosynthetic and respiratory gas exchange characteristics of *Atriplex lentiformis* (Torr.) Wats. in coastal and desert habitats. *Ecology* 55: 1104–1111.

Pearcy, R. W., A. T. Harrison, H. A. Mooney, O. Bjorkman. 1974. Seasonal changes in net photosynthesis of *Atriplex hymenelytra* shrubs growing in Death Valley, California. *Oecologia* 17: 111–121.

Pearl, M. B., R. Kleiman, F. R. Earle. 1973. Acetylenic acids of *Alvaradoa amorphoides* seed oil. *Lipids* 8: 627–630.

Pedley, L. 1975. Revision of the extra-Australian species of *Acacia* subg. *Heterophyllum*. *Contributions from the Queensland Herbarium* 18: 1–24.

Pemberton, R. W. 1988. The abundance of plants bearing extrafloral nectaries in Colorado and Mojave desert communities of southern California. *Madroño* 35: 238–246.

Pennington, C. W. 1963. *The Tarahumar of Mexico: Their Environment and Material Culture*. University of Utah Press, Salt Lake City.

Pennington, T. D. 1990. Sapotaceae. *Flora Neotropica Monograph* 52: 1–770.

Perez, G. R. M., Z. A. Ocegueda, L. J. L. Munoz, A. J. G. Avila, W. W. Morrow. 1984. A study of the hypoglycemic effect of some Mexican plants. *Journal of Ethnopharmacology* 12: 253–262.

Perring, F., S. M. Walters. 1976. *Atlas of the British Flora*. 2d ed. Botanical Society of the British Isles, E. P. Publishing, Wakefield.

Perry, H. M., E. F. Aldon, J. H. Brock. 1987. Reclamation of an asbestos mill waste site in the southwestern United States. *Reclamation and Revegetation Research* 6: 187–196.

Petersen, C., J. H. Brown, A. Kodric-Brown. 1982. An experimental study of floral display and fruit set in *Chilopsis linearis* (Bignoniaceae). *Oecologia* 55: 7–11.

Peterson, K. M., W. W. Payne. 1973. The genus *Hymenoclea* (Compositae: Ambrosieae). *Brittonia* 25: 243–256.

Peterson, K. M., W. W. Payne. 1974. On the correct name for the appressed-winged variety of *Hymenoclea salsola*. *Brittonia* 26: 397.

Pezeshki, S. R., R. D. Delaune, W. H. Patrick. 1990. Differential response of selected mangroves to soil flooding and salinity, gas exchange and biomass partitioning. *Canadian Journal of Forestry Research* 20: 869–874.

Phillips, A. M. 1977. Packrats, plants, and the Pleistocene in the lower Grand Canyon. Ph.D. dissertation, University of Arizona, Tucson.

Phillips, A. M., T. R. Van Devender. 1974. Pleistocene packrat middens from the lower Grand Canyon of Arizona. *Journal of the Arizona Academy of Science* 9: 117–119.

Phillips, W. S. 1963. Depths of roots in soil. *Ecology* 44: 424.

Pinkava, D. J. 1976. A new species of cholla (Cactaceae: *Opuntia anteojoensis* sp. nov.) from Coahuila, Mexico. *Madroño* 23: 292–294.

Pinkava, D. J., M. A. Baker. 1985. Chromosome and hybridization studies of agaves. *Desert Plants* 7: 93–100.

Pinkava, D. J., M. A. Baker, B. D. Parfitt, M. W. Mohlenbrock, R. D. Worthington. 1985. Chromosome numbers in some cacti of western North America, part 5. *Systematic Botany* 10: 471–483.

Pinkava, D. J., D. J. Keil. 1977. Chromosome counts of Compositae from the United States and Mexico. *American Journal of Botany* 64: 680–686.

Pinkava, D. J., L. A. McGill, T. Reeves, M. G. McLeod. 1977. Chromosome numbers in some cacti of western North America, part 3. *Bulletin of the Torrey Botanical Club* 104: 105–110.

Pinkava, D. J., M. G. McLeod. 1971. Chromosome numbers in some cacti of western North America. *Brittonia* 23: 171–176.

Pinkava, D. J., M. G. McLeod, L. A. McGill, R. C. Brown. 1973. Chromosome numbers in some cacti of western North America, part 2. *Brittonia* 25: 2–9.

Pinkava, D. J., B. D. Parfitt. 1982. Chromosome numbers in some cacti of western North America, part 4. *Bulletin of the Torrey Botanical Club* 109: 121–128.

Pinkava, D. J., B. D. Parfitt. 1988. Nomenclatural changes in Chihuahuan Desert *Opuntia* (Cactaceae). *Sida* 13: 125–130.

Pinkava, D. J., B. D. Parfitt, M. A. Baker. 1992. Chromosome numbers in some cacti of western North America—VI, with some nomenclatural changes. *Madroño* 39: 98–113.

Pinto, J. D., S. I. Frommer, S. A. Manweiler. 1987. The insects of jojoba, *Simmondsia chinensis,* in natural stands and plantations in southwestern North America. *Southwestern Entomologist* 12: 287–298.

Poole, D. K., P. C. Miller. 1975. Water relations of selected species of chaparral and coastal sage communities. *Ecology* 56: 1118–1128.

Poole, F. N. 1958. Seed germination requirements of four desert tree species. M.S. thesis, University of Arizona, Tucson.

Popham, R. A. 1947. Developmental anatomy of seedling of *Jatropha cordata. Ohio Journal of Science* 47: 1–20.

Porter, D. M. 1963. The taxonomy and distribution of the Zygophyllaceae of Baja California, Mexico. *Contributions from the Gray Herbarium* 192: 99–135.

Potter, R. L., D. N. Ueckert, J. L. Petersen, M. L. McFarland. 1986. Germination of fourwing saltbush seeds: Interaction of temperature, osmotic potential, and pH. *Journal of Range Management* 39: 43–46.

Powell, J., T. W. Box, C. V. Baker. 1972. Growth rate of sprouts after top removal of huisache (*Acacia farnesiana* [L.] Willd) (Leguminosae) in South Texas, U.S.A. *Southwestern Naturalist* 17: 191–195.

Powell, J. A. 1992. Interrelationships of yuccas and yucca moths. *Trends in Ecology and Evolution* 7: 10–15.

Powell, J. A., R. A. Mackie. 1966. Biological interrelationships of moths and *Yucca whipplei. University of California Publications in Entomology* 42: 1–46.

Proctor, M., P. Yeo. 1972. *The Pollination of Flowers.* Taplinger, New York.

Proksch, P., C. Sternberg, E. Rodriguez. 1981. Epicuticular alkanes from desert plants of Baja California, Mexico. *Biochemical Systematics and Ecology* 9: 205–206.

Prose, D. V., S. K. Metzger, H. G. Wilshire. 1987. Effects of substrate disturbance on secondary plant succession: Mojave Desert, California. *Journal of Applied Ecology* 24: 305–313.

Rabinowitz, D. 1978. Dispersal properties of mangrove propagules. *Biotropica* 10: 47–57.

Rada, F., G. Goldstein, A. Orozco, M. Montilla, O. Zabala, A. Azocar. 1989. Osmotic and turgor relations of three mangrove ecosystem species. *Australian Journal of Plant Physiology* 16: 477–486.

Rao, J. V. S., K. R. Reddy. 1980. Seasonal variation in leaf epicuticular wax of some semiarid shrubs. *Indian Journal of Experimental Biology* 18: 495–499.

Raphael, D. O., P. S. Nobel. 1986. Growth and survivorship of ramets and seedlings of *Agave deserti:* Influences of parent-ramet connections. *Botanical Gazette* 147: 78–83.

Rasmussen, G. A., C. J. Scifres, D. L. Drawe. 1983. Huisache (*Acacia farnesiana*) growth, browse quality, and use following burning. *Journal of Range Management* 36: 337–342.

Raven, P. H. 1962. The systematics of *Oenothera* subgenus *Chylismia. University of California Publications in Botany* 34: 1–122.

Raven, P. H. 1963. Amphitropical relationships in the floras of North and South America. *Quarterly Review of Biology* 38: 151–177.

Raven, P. H. 1967. *Holacantha emoryi* Gray. *Madroño* 19: 134–136.

Raven, P. H. 1976. Generic and sectional delimitation in Onagraceae, tribe Epilobieae. *Annals of the Missouri Botanical Garden* 63: 326–340.

Raven, P. H., D. W. Kyhos, A. J. Hill. 1965. Chromosome numbers of spermatophytes, mostly Californian. *Aliso* 6: 105–113.

Raven, P. H., D. W. Kyhos, D. E. Breedlove, W. W. Payne. 1968. Polyploidy in *Ambrosia dumosa* (Compositae: Ambrosieae). *Brittonia* 20: 205–211.

Raven, P. H., H. Lewis. 1960. Observations on the chromosomes and relationships of *Hauya* and *Xylonagra*. *Aliso* 4: 483–484.

Reddi, C. B., A. J. Bai, E. U. B. Reddi, K. V. R. Raju. 1980. Pollen productivity, release, and dispersal in *Dodonaea viscosa* (Linn.) Jacq. *Proceedings of the Indian National Science Academy, Biological Sciences, Part B* 46: 184–190.

Reeder, J. R. 1977. Chromosome numbers in western grasses. *American Journal of Botany* 64: 102–110.

Reeder, J. R., C. J. Reeder. 1988. *Hilaria annua* (Gramineae), a new species from Mexico. *Madroño* 35: 6–9.

Reich, P. B., R. Borchert. 1984. Water stress and tree phenology in a tropical dry forest in the lowlands of Costa Rica. *Journal of Ecology* 72: 61–74.

Reichenbacher, F. W. 1985a. Conservation of southwestern agaves. *Desert Plants* 7: 88, 103–106.

Reichenbacher, F. W. 1985b. *Status and Distribution of the Tumamoc Globe-berry (Tumamoca macdougalii Rose)*. Department of Biology and Microbiology, Arizona State University, Tempe.

Reichenbacher, F. W. 1990. *Tumamoc Globe-berry Studies in Arizona and Sonora, Mexico.* Bureau of Reclamation, Arizona Projects Office, Phoenix.

Reid, W., R. Lozano, R. Odom. 1983. Non-equilibrium population structure in three Chihuahuan Desert cacti. *Southwestern Naturalist* 28: 115–117.

Reveal, J. L. 1989. A review of the genus *Harfordia* (Polygonaceae: Eriogonoideae). *Phytologia* 66: 221–227.

Reveal, J. L., R. Moran. 1977. Miscellaneous chromosome counts of western American plants, part 4. *Madroño* 24: 227–235.

Rhoades, D. F. 1977. The antiherbivore chemistry of *Larrea*. In T. J. Mabry, J. H. Hunziker, D. R. Difeo (eds.), *Creosotebush: Biology and Chemistry of Larrea in New World Deserts,* 135–175. Dowden, Hutchinson and Ross, Stroudsburg, Penn.

Rhodes, D., P. Felker. 1988. Mass screening of *Prosopis* mesquite seedlings for growth at seawater salinity concentrations. *Forest Ecology and Management* 24: 169–176.

Rico-Gray, V. 1989. The importance of floral and circum-floral nectar to ants inhabiting dry tropical lowlands. *Biological Journal of the Linnean Society* 38: 173–181.

Riley, C. V. 1881. Further notes on the pollination of yucca and on *Pronuba* and *Prodoxus. Proceedings of the American Association for the Advancement of Science* 1880: 617–639.

Riley, C. V. 1892. The yucca moth and yucca pollination. In *Missouri Botanical Garden, Third Annual Report,* 99–158. St. Louis, Mo.

Riley, C. V. 1893. Further notes on yucca insects and yucca pollination. *Proceedings of the Biological Society of Washington* 8: 41–54.

Rindfleisch, J. K. 1979. Beekeeping in Barbados, West Indies. *American Bee Journal* 119: 131–135.

Rivera, E. R., B. N. Smith. 1979. Crystal morphology and $^{13}C/^{12}C$ composition of solid oxalate in cacti. *Plant Physiology* 64: 966–970.

Rivera, R. L., C. E. Freeman. 1979. The effects of some alternating temperatures on germination of creosotebush, *Larrea tridentata* (Zygophyllaceae). *Southwestern Naturalist* 24: 711–714.

Robberecht, R., B. E. Mahall, P. S. Nobel. 1983. Experimental removal of intraspecific competitors—effects on water relations and productivity of a desert bunchgrass, *Hilaria rigida*. *Oecologia* 60: 21–24.

Robinson, M. K. 1973. *Atlas of Monthly Mean Sea Surface and Subsurface Temperatures in the Gulf of California, Mexico.* San Diego Society of Natural History Memoir no. 5.

Robinson, T. W. 1965. *Introduction, Spread, and Areal Extent of Saltcedar (Tamarix) in the Western States.* U.S. Geological Survey Professional Paper 491-A. Government Printing Office, Washington, D.C.

Robinson, W. S. 1904. The spines of *Fouquieria*. *Bulletin of the Torrey Botanical Club* 31: 45–50.

Rodriguez, E. 1977. Ecogeographic distribution of secondary constituents in *Parthenium* (Compositae). *Biochemical Systematics and Ecology* 5: 207–218.

Rogers, D. J. 1949. *Stegnosperma:* A new species and a generic commentary. *Annals of the Missouri Botanical Garden* 36: 475–477.

Rogers, G. F. 1985. Mortality of burned *Cereus giganteus*. *Ecology* 66: 630–632.

Rogers, G. F. 1989. Asymmetrical growth of the crowns of neighboring desert shrubs. *Journal of Arid Environments* 17: 319–326.

Rollins, R. C. 1946. Interspecific hybridization in *Parthenium*, part 2: Crosses involving *P. argentatum*, *P. incanum*, *P. stramonium*, *P. tomentosum*, and *P. hysterophorus*. *American Journal of Botany* 33: 21–30.

Rollins, R. C. 1950. The guayule rubber plant and its relatives. *Contributions from the Gray Herbarium* 172: 1–73.

Romanczuk, M. C., M. A. del Pero de Martinez. 1978. Las espécies del género *Celtis* (Ulmaceae) en la flora argentina. *Darwiniana* 21: 541–578.

Romney, E. M., A. Wallace. 1980. Ecotonal distribution of salt-tolerant shrubs in the northern Mojave Desert. *Great Basin Naturalist Memoirs* 4: 134–139.

Rose, J. N. 1891. List of plants collected by Dr. Edward Palmer in 1890 in western Mexico and Arizona. *Contributions from the U.S. National Herbarium* 1: 87–127.

Rose, S. L. 1981. Vesicular-arbuscular endomycorrhizal associations of some desert plants of Baja California. *Canadian Journal of Botany* 59: 1056–1061.

Ross, R. 1982. Initiation of stamens, carpels, and receptacle in the Cactaceae. *American Journal of Botany* 69: 369–379.

Rost, T. L., A. D. Simper, P. Schell, S. Allen. 1977. Anatomy of jojoba (*Simmondsia chinensis*) seed and the utilization of liquid wax during germination. *Economic Botany* 31: 140–147.

Rotkis, P. T., S. M. Alcorn. 1984. Susceptibility of native plants to three soil borne fungi endemic to the southwestern United States. *Phytopathology* 74: 853–854.

Roundy, B. A., A. K. Dobrenz. 1989. Herbivory and plant water status of jojoba (*Simmondsia chinensis* [Link] Schn.) in the Sonoran Desert in Arizona. *Journal of Arid Environments* 16: 283–291.

Roundy, B. A., G. B. Ruyle, J. Ard. 1989. Estimating production and utilization of jojoba. *Journal of Range Management* 42: 75–78.

Rudd, V. E. 1955. The American species of *Aeschynomene*. *Contributions from the U.S. National Herbarium* 32: 1–172.

Rudd, V. E. 1966. *Acacia cochliacantha* or *Acacia cymbispina* in Mexico? *Leaflets in Western Botany* 10: 257–262.

Rudd, V. E. 1969. A synopsis of the genus *Piscidia* (Leguminosae). *Phytologia* 18: 473–499.

Rudd, V. E. 1975. Supplementary studies in *Aeschynomene*, part 3: Series *Scopariae* in Mexico and Central America. *Phytologia* 31: 431–434.

Rudd, V. E., A. M. Carter. 1983. *Acacia pacensis* (Leguminosae: Mimosoideae), a new species from Baja California Sur, Mexico. *Madroño* 30: 176–180.

Ruffner, G. A., W. D. Clark. 1986. Extrafloral nectar of *Ferocactus acanthodes* (Cactaceae): Composition and its importance to ants. *American Journal of Botany* 73: 185–189.

Runyon, E. H. 1934. The organization of the creosote bush with respect to drought. *Ecology* 15: 128–138.

Rydberg, P. A. 1923. (Rosales) Fabaceae, Indigofereae, Galegeae (pars). *North American Flora* 24: 137–200.

Rzedowski, J. 1988. Análisis de la distribución geográfica del complejo *Prosopis* (Leguminosae, Mimosoideae) en Norteamerica. *Acta Botánica Mexicana* 3: 7–19.

Rzedowski, J., H. Kruse. 1979. Algunas tendéncias evolutivas en *Bursera* (Burseraceae). *Taxon* 28: 103–116.

Rzedowski, J., F. M. Leal. 1958. El limite sur de distribución geográfica de *Larrea tridentata. Acta Cientifica Potosina* 2: 133–147.

Sakakibara, M., D. DiFeo, N. Nakatani, B. Timmermann, T. J. Mabry. 1976. Flavonoid methyl ethers on the external leaf surface of *Larrea tridentata* and *L. divaricata. Phytochemistry* 15: 727–731.

Salunkhe, S. S., B. A. Karadge. 1989. Influence of salinity and water stress on growth and organic constituents of *Dodonaea viscosa* L. *Geobios* 16: 252–258.

Sammis, T. W. 1974. The microenvironment of a desert hackberry plant (*Celtis pallida*). Ph.D. dissertation, University of Arizona, Tucson.

Sammis, T. W., D. L. Weeks. 1977. Variations in soil moisture under natural vegetation. *Hydrology and Water Resources in Arizona and the Southwest* 7: 235–240.

Sanchez-Diaz, M. F., H. A. Mooney. 1979. Resistance to water transfer in desert shrubs. *Physiologia Plantarum* 46: 139–146.

Sankary, M. N., M. G. Barbour. 1972. Autecology of *Atriplex polycarpa* from California. *Ecology* 53: 1155–1162.

Sankary, M. N., J. R. Goodin. 1988. Ecology of the Chenopodiaceae shrubs and their breeding for forage, fuel, wood production, and sand dune fixation in the Arabian Mediterranean-type arid areas. In E. E. Whitehead, C. F. Hutchinson, B. N. Timmermann, R. G. Varady (eds.), *Arid Lands Today and Tomorrow,* 149–164. Westview Press, Boulder, Colo.

Saruwatari, M. W., S. D. Davis. 1989. Tissue water relations of three chaparral shrub species after wildfire. *Oecologia* 80: 303–308.

Schaefer, C. M. 1988. Halophytes and their potential as landscape plants. M.S. thesis, University of Arizona, Tucson.

Schaffer, W. M., M. V. Schaffer. 1977a. The adaptive significance of variations in reproductive habit in the Agavaceae. In B. Stonehouse, C. Perrins (eds.), *Evolutionary Ecology,* 261–276. Macmillan Press, London.

Schaffer, W. M., M. V. Schaffer. 1977b. The reproductive biology of Agavaceae, part 1: Pollen and nectar production in four Arizona agaves. *Southwestern Naturalist* 22: 157–168.

Schaffer, W. M., M. V. Schaffer. 1979. The adaptive significance of variations in reproductive habit in the Agavaceae II: Pollinator foraging behavior and selection for increased reproductive expenditure. *Ecology* 60: 1051–1069.

Schaffer, W. M., D. W. Zeh, S. L. Buchmann, S. Kleinhaus, M. Valentine-Schaffer, J. Antrim. 1983. Competition for nectar between introduced honey bees (*Apis mellifera*) and native North American bees and ants. *Ecology* 64: 564–577.

Schilling, E. E. 1990. Taxonomic revision of *Viguiera* subg. *Bahiopsis* (Asteraceae: Heliantheae). *Madroño* 37: 149–170.

Schilling, E. E., E. M. Schilling. 1986. Chromosome numbers in *Viguiera* series *Dentatae* (Compositae). *Systematic Botany* 11: 51–55.

Schmidt, J. O., S. L. Buchmann. 1986. Floral biology of the saguaro (*Cereus giganteus*), part 1: Pollen harvest by *Apis mellifera*. *Oecologia* 69: 491–498.

Schmidt, J. O., B. E. Johnson. 1984. Pollen feeding preferences of *Apis mellifera*, a polylectic bee. *Southwestern Entomologist* 9: 41–47.

Schmidt, R. H. 1983. Climate and the Chihuahuan Desert. In L. E. Campos, R. J. Anderson (ed.), *Natural Resources and Development in Arid Regions*, 35–52. Westview Press, Boulder, Colo.

Schmidt, R. H. 1989. The arid zones of Mexico: Climatic extremes and conceptualization of the Sonoran Desert. *Journal of Arid Environments* 16: 241–256.

Schnetter, M. L. 1978. The influence of external factors on the structure of the leaf of *Avicennia germinans* under natural conditions. *Beitrage zur Biologie der Pflanzen* 54: 13–28.

Schnetter, M. L. 1985. Investigations on salt secretion of the leaves of *Avicennia germinans*. *Flora* 177: 157–165.

Scholander, P. F., E. D. Bradstreet, H. T. Hammel, E. A. Hemmingsen. 1966. Sap concentrations in halophytes and some other plants. *Plant Physiology* 41: 529–532.

Scholander, P. F., L. van Dam, S. I. Scholander. 1955. Gas exchange in the roots of mangroves. *American Journal of Botany* 42: 92–98.

Schramm, D. R. 1982. *Floristics and Vegetation of the Black Mountains, Death Valley National Monument, California*. CPSU/UNLV Contribution no. 012–13, University of Nevada, Las Vegas.

Schulte, P. J., P. S. Nobel. 1989. Responses of a CAM plant to drought and rainfall: Capacitance and osmotic pressure influences on water movement. *Journal of Experimental Botany* 40: 61–70.

Schulte, P. J., J. A. C. Smith, P. S. Nobel. 1989. Water storage and osmotic pressure influences on the water relations of a dicotyledonous desert succulent. *Plant, Cell, and Environment* 12: 831–842.

Scifres, C. J. 1974. Salient aspects of huisache seed germination. *Southwestern Naturalist* 18: 383–391.

Scifres, C. J., J. H. Brock. 1971. Thermal regulation of water uptake by germinating honey mesquite seeds. *Journal of Range Management* 24: 157–158.

Scifres, C. J., J. H. Brock. 1972. Emergence of honey mesquite seedlings relative to planting depth and soil temperatures. *Journal of Range Management* 25: 217–219.

Scifres, C. J., J. L. Mutz, W. T. Hamilton. 1979. Control of mixed brush with tebuthiuron. *Journal of Range Management* 32: 155–158.

Scogin, R. 1980. Serotaxonomy of *Simmondsia chinensis* (Simmondsiaceae). *Aliso* 9: 555–559.

Scott, F. M. 1932. Some features of the anatomy of *Fouquieria splendens*. *American Journal of Botany* 19: 673–678.

Scott, F. M. 1935a. Contribution to the causal anatomy of the desert willow, *Chilopsis linearis*. *American Journal of Botany* 22: 333–343.

Scott, F. M. 1935b. The anatomy of *Cercidium torreyanum* and *Parkinsonia microphylla*. *Madroño* 3: 33–41.

Scott, F. M. 1938. Root system and root nodules of the seedling smoke tree, *Parosela spinosa*. *Ecology* 19: 166–168.

Seaman, F. C. 1975. *Hymenoclea platyspina* (Asteraceae: Ambrosieae), a new species from Baja California. *Madroño* 23: 111–114.

Seaman, F. C., T. J. Mabry. 1979a. Sesquiterpene lactones and species relationships among the shrubby *Ambrosia* taxa. *Biochemical Systematics and Ecology* 7: 105–114.

Seaman, F. C., T. J. Mabry. 1979b. Sesquiterpene lactones of diploid and tetraploid *Ambrosia camphorata. Biochemical Systematics and Ecology* 7: 3–6.

Seaman, F. C., T. J. Mabry. 1979c. Sesquiterpene lactone patterns in diploid and polyploid *Ambrosia dumosa. Biochemical Systematics and Ecology* 7: 7–12.

Segura, L. J., C. Calzado F. 1981. Possible amoebicidal activity of *Larrea.* In E. Campos Lopez, T. J. Mabry, S. Fernandez Tavizon (eds.), *Larrea,* 317–326. Consejo Nacional de Ciéncia y Tecnología, México, D.F.

Seigler, D. S., J. E. Ebinger. 1988. *Acacia macrantha, A. pennatula,* and *A. cochliacantha* (Fabaceae: Mimosoideae) species complexes in Mexico. *Systematic Botany* 13: 7–15.

Seigler, D. S., S. Seilheimer, J. Keesy, H. F. Huang. 1986. Tannins from four common *Acacia* species of Texas, USA, and northeastern Mexico. *Economic Botany* 40: 220–232.

Sellers, W. D., R. H. Hill. 1974. *Arizona Climate.* University of Arizona Press, Tucson.

Shaanker, R. U., K. N. Ganeshaiah. 1988. Bimodal distribution of seeds per pod in *Caesalpinia pulcherrima:* Parent-offspring conflict. *Evolutionary Trends in Plants* 2: 91–98.

Shah, J. J., Y. S. Dave. 1971. Ontogeny of tendrils in *Antigonon leptopus* H. & Arn. *Annals of Botany* 35: 411–419.

Shantz, H. L., R. L. Piemeisel. 1924. Indicator significance of the natural vegetation of the southwestern desert region. *Journal of Agricultural Research* 28: 721–801.

Sharifi, M. R., E. T. Nilsen, P. W. Rundel. 1982. Biomass and new primary production of *Prosopis glandulosa* var. *torreyana* (Fabaceae) in the Sonoran Desert of California. *American Journal of Botany* 69: 760–767.

Sharifi, M. R., E. T. Nilsen, R. Virginia, P. W. Rundel, W. M. Jarrell. 1983. Phenological patterns of current season's shoots of *Prosopis glandulosa* var. *torreyana* in the Sonoran Desert of southern California. *Flora* 173: 265–277.

Sharma, A. K. 1970. Annual report, 1967–1968. *Research Bulletin of the University of Calcutta (Cytogenetics Laboratory)* 2: 1–50.

Sheps, L. O. 1973. Survival of *Larrea tridentata* S. & M. seedlings in Death Valley National Monument, California. *Israel Journal of Botany* 22: 8–17.

Sherbrooke, W. C. 1976. Differential acceptance of toxic jojoba seed (*Simmondsia chinensis*) by four Sonoran Desert heteromyid rodents. *Ecology* 57: 596–602.

Sherbrooke, W. C. 1977. First-year seedling survival of jojoba (*Simmondsia chinensis*) in the Tucson Mountains, Arizona. *Southwestern Naturalist* 22: 225–234.

Sherbrooke, W. C. 1978. *Jojoba: An Annotated Bibliographic Update.* Supplement to Arid Lands Resources Information Paper no. 5. Office of Arid Lands Studies, University of Arizona, Tucson.

Sherbrooke, W. C. 1989. Seedling survival and growth of a Sonoran Desert shrub, jojoba (*Simmonsdia chinensis*), during the first ten years. *Southwestern Naturalist* 34: 421–424.

Sherbrooke, W. C., E. F. Haase. 1974. *Jojoba: A Wax-Producing Shrub of the Sonoran Desert: Literature Review and Annotated Bibliography.* Arid Lands Resources Information Paper no. 5. Office of Arid Lands Studies, University of Arizona, Tucson.

Sherbrooke, W. C., J. C. Scheerens. 1979. Ant-visited extrafloral (calyx and foliar) nectaries and nectar sugars of *Erythrina flabelliformis* Kearney in Arizona. *Annals of the Missouri Botanical Garden* 66: 472–481.

Sherrod, C. L., C. McMillan. 1981. Black mangrove, *Avicennia germinans,* in Texas: Past and present distribution. *Contributions in Marine Science* 24: 115–131.

Sherrod, C. L., C. McMillan. 1985. The distributional history and ecology of mangrove vegetation along the northern Gulf of Mexico coastal region. *Contributions in Marine Science* 28: 129–140.

Sheth, K., S. Jolad, R. Wiedhopf, J. Cole. 1972. Tumor inhibitory agent from *Hyptis emoryi* (Labiatae). *Journal of Pharmaceutical Sciences* 61: 1819.

Shibata, K. 1962. Estudio citológicos de plantas columbianas silvestres y cultivadas. *Journal of Agricultural Science, Tokyo Nogyo Daigaku* 8: 49–62.

Shillington, D. C., T. W. Yang. 1975. Phenology of Christmas cactus in the Tucson area. *Journal of the Arizona Academy of Science* 10 (Supplement): 19.

Short, H. L. 1979. *Food Habits of Coyotes in a Semidesert Grass-shrub Habitat.* U.S. Forest Service Research Note RM–364. Fort Collins, Colo..

Shreve, E. B. 1914. *The Daily March of Transpiration in a Desert Perennial.* Carnegie Institution of Washington Publication no. 194. Washington, D.C.

Shreve, E. B. 1923. Seasonal changes in the water relations of desert plants. *Ecology* 4: 266–292.

Shreve, E. B. 1924. Factors governing seasonal changes in transpiration of *Encelia farinosa. Botanical Gazette* 77: 432–439.

Shreve, F. 1910. The rate of establishment of the giant cactus. *Plant World* 13: 235–240.

Shreve, F. 1911a. The influence of low temperature on the distribution of giant cactus. *Plant World* 14: 136–146.

Shreve, F. 1911b. Establishment behavior of the palo verde. *Plant World* 14: 289–296.

Shreve, F. 1917. Establishment of desert perennials. *Journal of Ecology* 5: 210–216.

Shreve, F. 1925. Ecological aspects of the deserts of California. *Ecology* 6: 93–103.

Shreve, F. 1933. Desert investigations. *Carnegie Yearbook* 32: 196–201.

Shreve, F. 1935. The longevity of cacti. *Cactus and Succulent Journal (U.S.)* 7: 66–68.

Shreve, F. 1936. The transition from desert to chaparral in Baja California. *Madroño* 3: 257–264.

Shreve, F. 1937a. Plants of the sand. *Carnegie Institution of Washington News Service Bulletin* 10: 91–96.

Shreve, F. 1937b. The vegetation of the Cape Region of Baja California. *Madroño* 4: 105–113.

Shreve, F. 1937c. Desert investigations. *Carnegie Yearbook* 36: 214–221.

Shreve, F. 1940. The edge of the desert. *Yearbook of the Association of Pacific Coast Geographers* 6: 6–11.

Shreve, F. 1942. The desert vegetation of North America. *Botanical Review* 8: 195–246.

Shreve, F. 1943. Desert investigations. *Carnegie Yearbook* 42: 100–103.

Shreve, F. 1951. *Vegetation of the Sonoran Desert.* Carnegie Institution of Washington Publication no. 591. Washington, D.C.

Shreve, F. 1964. Vegetation of the Sonoran Desert. In F. Shreve and I. L. Wiggins, *Vegetation and Flora of the Sonoran Desert,* 2 vols., 1–186. Stanford University Press, Stanford, Calif.

Shreve, F., T. D. Mallery. 1933. The relation of caliche to desert plants. *Soil Science* 35: 99–112.

Siegel, R. S., J. H. Brock. 1990. Germination requirements of key southwestern woody riparian species. *Desert Plants* 10: 3–8.

Siemens, D. H., C. D. Johnson. 1990. Host-associated differences in fitness within and between populations of a seed beetle (Bruchidae): Effects of plant variability. *Oecologia* 82: 408–413.

Silander, J. A. 1978. Density-dependent control of reproductive success in *Cassia biflora. Biotropica* 10: 292–296.

Silverman, J., R. D. Goeden. 1979. Life history of the lacebug *Corythucha morilli* on the ragweed *Ambrosia dumosa* in southern California. *Pan-Pacific Entomologist* 55: 305–308.

Silverman, J., R. D. Goeden. 1980. Life history of a fruit fly, *Procecidochares* sp., on the ragweed *Ambrosia dumosa* (Gray) Payne in southern California (Diptera: Tephritidae). *Pan-Pacific Entomologist* 56: 283–288.

Simmonds, F. J. 1980. Biological control of *Cordia curassavica* (Boraginaceae) in Malaysia. *Entomophaga* 25: 363–364.

Simmons, N. M. 1966. Flora of the Cabeza Prieta Game Range. *Journal of the Arizona Academy of Science* 4: 93–104.

Simpson, B. B. 1977. Breeding systems of dominant perennial plants of two disjunct warm-desert ecosystems. *Oecologia* 27: 203–226.

Simpson, B. B., A. Burkart, N. J. Carman. 1975. *Prosopis palmeri,* a relict of an ancient North American colonization. *Madroño* 23: 220–227.

Simpson, B. B., J. L. Neff. 1987. Pollination ecology in the arid Southwest. *Aliso* 11: 417–440.

Simpson, B. B., J. L. Neff, A. R. Moldenke. 1977a. Reproductive systems of *Larrea.* In T. J. Mabry, J. H. Hunziker, D. R. Difeo (eds.), *Creosotebush: Biology and Chemistry of Larrea in New World Deserts,* 92–114. Dowden, Hutchinson and Ross, Stroudsburg, Penn.

Simpson, B. B., J. L. Neff, A. R. Moldenke. 1977b. *Prosopis* flowers as a resource. In B. B. Simpson (ed.), *Mesquite: Its Biology in Two Desert Scrub Ecosystems,* 84–107. Dowden, Hutchinson and Ross, Stroudsburg, Penn.

Skvarla, J. J., J. W. Nowicke. 1982. Pollen fine structure and relationships of *Achatocarpus* Triana and *Phaulothamnus* A. Gray. *Taxon* 31: 244–249.

Smith, H. N., C. A. Rechenthin. 1964. *Grassland Restoration,* part 1: *The Texas Brush Problem.* USDA Soil Conservation Service Bulletin no. 4–19114. Government Printing Office, Washington, D.C.

Smith, J. A. C., M. Popp, U. Luttge, W. J. Cram, M. Diaz, H. Griffiths, H. S. J. Lee, E. Medina, C. S. Schafer, K.-H. Stimmel, B. Thonke. 1989. Ecophysiology of xerophytic and halophytic vegetation of a coastal alluvial plain in northern Venezuela, part 6: Water relations and gas exchange of mangroves. *New Phytologist* 111: 293–307.

Smith, S. D., B. Didden-Zopfy, P. S. Nobel. 1984. High-temperature responses of North American cacti. *Ecology* 65: 643–651.

Smith, S. D., D. T. Patten, R. K. Monson. 1987. Effects of artificially imposed shade on a Sonoran Desert ecosystem: Microclimate and vegetation. *Journal of Arid Environments* 13: 65–82.

Smith, T. J., H. T. Chan, C. C. McIvor, B. Robblee. 1989. Comparisons of seed predation in tropical tidal forests from three continents. *Ecology* 70: 146–151.

Smith, W. K. 1978. Temperatures of desert plants: Another perspective on the adaptability of leaf size. *Science* 201: 614–616.

Smith, W. K., P. S. Nobel. 1977a. Influences of seasonal changes in leaf morphology on water-use efficiency for three desert broadleaf shrubs. *Ecology* 58: 1033–1043.

Smith, W. K., P. S. Nobel. 1977b. Temperature and water relations for sun and shade leaves of a desert broadleaf, *Hyptis emoryi. Journal of Experimental Botany* 28: 169–183.

Solbrig, O. T., K. Bawa, N. J. Carman, J. H. Hunziker, C. A. Naranjo, R. A. Palacios, L. Poggio, B. B. Simpson. 1977. Patterns of variation. In B. B. Simpson (ed.), *Mesquite: Its Biology in Two Desert Scrub Ecosystems,* 44–60. Dowden, Hutchinson and Ross, Stroudsburg, Penn.

Solbrig, O. T., P. D. Cantino. 1975. Reproductive adaptations in *Prosopis* (Leguminosae, Mimosoideae). *Journal of the Arnold Arboretum* 56: 185–210.

Solbrig, O. T., D. W. Kyhos, M. Powell, P. H. Raven. 1972. Chromosome numbers in Compositae VIII: Heliantheae. *American Journal of Botany* 59: 869–878.

Soto, R. 1988. Geometry, biomass allocation, and leaf life-span of *Avicennia germinans* (L.) L. (Avicenniaceae) along a salinity gradient in Salinas, Puntarenas, Costa Rica. *Revista de Biología Tropical* 36: 309–323.

Soto, R., L. F. Corrales. 1987. Variación de algunas características foliares de *Avicennia germinans* (L.) L. (Avicenniaceae) en un gradiente climática y de salinidad. *Revista de Biología Tropical* 35: 245–256.

Soto de Villatoro, B., F. Giral-Gónzalez, J. Polonsky, Z. Baskevitch-Varon. 1974. Chrysophanic acid, chrysophanein, and chaparrin from *Alvaradoa amorphoides. Phytochemistry (Oxford)* 13: 2018–2019.

Spalding, E. S. 1905. Mechanical adjustment of the sahuaro (*Cereus giganteus*) to varying quantities of stored water. *Bulletin of the Torrey Botanical Club* 32: 57–68.

Spaulding, W. G. 1983. Late Wisconsin macrofossil records of desert vegetation in the American Southwest. *Quaternary Research* 19: 256–264.

Spaulding, W. G. 1990. Vegetational and climatic development of the Mojave Desert: The last glacial maximum to the present. In J. R. Betancourt, T. R. Van Devender, P. S. Martin (eds.), *Packrat Middens: The Last 40,000 Years of Biotic Change,* 166–199. University of Arizona Press, Tucson.

Spaulding, W. G., J. L. Betancourt, L. K. Croft, K. L. Cole. 1990. Packrat middens: Their composition and methods of analysis. In J. L. Betancourt, T. R. Van Devender, P. S. Martin (eds.), *Packrat Middens: The Last 40,000 Years of Biotic Change,* 59–84. University of Arizona Press, Tucson.

Spellenberg, R. W. 1973. IOPB chromosome number reports XLII. *Taxon* 22: 647–654.

Spellenberg, R. W. 1981. Polyploidy in *Dalea formosa* (Fabaceae) in the Chihuahuan Desert. *Brittonia* 33: 309–324.

Springfield, H. W. 1970. *Germination and Establishment of Fourwing Saltbush in the Southwest.* USDA Forest Service Research Paper RM–55.

Stahl, B. 1989. A synopsis of Central American Theophrastaceae. *Nordic Journal of Botany* 9: 15–30.

Standley, P. C. 1920. Trees and shrubs of Mexico. *Contributions from the U.S. National Herbarium,* vol. 23, part 1, 1–170. Smithsonian Institution, Washington, D.C.

Standley, P. C. 1922. Trees and shrubs of Mexico. *Contributions from the U.S. National Herbarium,* vol. 23, part 2, 171–515. Smithsonian Institution, Washington, D.C.

Standley, P. C. 1923. Trees and shrubs of Mexico. *Contributions from the U.S. National Herbarium,* vol. 23, part 3, 517–848. Smithsonian Institution, Washington, D.C.

Standley, P. C. 1924. Trees and shrubs of Mexico. *Contributions from the U.S. National Herbarium,* vol. 23, part 4, 849–1312. Smithsonian Institution, Washington, D.C.

Standley, P. C. 1926. Trees and shrubs of Mexico. *Contributions from the U.S. National Herbarium,* vol. 23, part 5, 1313–1721. Smithsonian Institution, Washington, D.C.

Stark, N., L. D. Love. 1969. Water relations of three warm desert species. *Israel Journal of Botany* 18: 175–190.

Starmer, W. T. 1982. Analysis of the community structure of yeasts associated with the decaying stems of cactus, 1. *Stenocereus gummosus. Microbial Ecology* 8: 71–82.

Starmer, W. T., W. B. Heed, M. Miranda, M. W. Miller, H. J. Phaff. 1976. The ecology of yeast flora associated with cactiphilic *Drosophila* and their host plants in the Sonoran Desert. *Microbial Ecology* 3: 11–30.

Starmer, W. T., H. J. Phaff, J. M. Bowles, M. A. Lachance. 1988. Yeasts vectored by insects feeding on decaying saguaro cactus. *Southwestern Naturalist* 33: 362–363.

Starr, G. 1988. *Dalea*—a genus of horticulturally promising legumes for desert landscapes. *Desert Plants* 9: 3–5, 28–31.

Stebbins, G. L., J. Major. 1965. Endemism and speciation in the California flora. *Ecological Monographs* 35: 1–35.

Steelink, C., M. Yeung, R. L. Caldwell. 1967. Phenolic constituents of healthy and wound tissues in the giant cactus (*Carnegiea gigantea*). *Phytochemistry* 6: 1435–1440.

Steenbergh, W. F. 1972. Lightning-caused destruction in a desert plant community. *Southwestern Naturalist* 16: 419–429.

Steenbergh, W. F., C. H. Lowe. 1969. Critical factors during the first years of life of the saguaro (*Cereus giganteus*) at Saguaro National Monument, Arizona. *Ecology* 50: 825–834.

Steenbergh, W. F., C. H. Lowe. 1976. Ecology of the saguaro, part 1: The role of freezing weather in a warm-desert population. In *Research in the Parks,* 49–92. National Park Service Symposium Series no. 1. Washington, D.C.

Steenbergh, W. F., C. H. Lowe. 1977. *Ecology of the Saguaro,* part 2: *Reproduction, Germination, Establishment, Growth, and Survival of the Young Plant.* National Park Service Scientific Monograph Series no. 8. Washington, D.C.

Steenbergh, W. F., C. H. Lowe. 1983. *Ecology of the Saguaro,* part 3: *Growth and Demography.* National Park Service Scientific Monograph Series no. 17. Washington, D.C.

Stein, R. A., J. A. Ludwig. 1979. Vegetation and soil patterns on a Chihuahuan desert bajada. *American Midland Naturalist* 101: 28–37.

Stern, W. L., G. K. Voit. 1959. Effect of salt concentration on growth of red mangrove in culture. *Botanical Gazette* 121: 36–39.

Sternberg, L. 1976. Growth forms of *Larrea tridentata. Madroño* 23: 408–417.

Sternburg, C., E. Rodriguez. 1982. Hydrocarbons from *Pedilanthus macrocarpus* (Euphorbiaceae) of Baja California and Sonora, Mexico. *American Journal of Botany* 69: 214–218.

Stone, E. C., G. Juhren. 1951. The effect of fire on the germination of the seed of *Rhus ovata. American Journal of Botany* 38: 368–372.

Strain, B. R. 1969. Seasonal adaptations in photosynthesis and respiration in four desert shrubs growing in situ. *Ecology* 50: 511–513.

Strain, B. R., V. C. Chase. 1966. Effect of past and prevailing temperatures on the carbon dioxide exchange capacities of some woody desert perennials. *Ecology* 47: 1043–1045.

Strojan, C. L., F. B. Turner, R. Castetter. 1979. Litterfall from shrubs in the northern Mojave Desert. *Ecology* 60: 891–900.

Stutz, H. C., S. C. Sanderson. 1979. The role of polyploidy in the evolution of *Atriplex canescens.* In J. R. Goodin, D. K. Northington (eds.), *Arid Land Plant Resources, Proceedings of the International Arid Lands Conference on Plant Resources,* 615–621. Texas Tech University, Lubbock.

Sundberg, S. D. 1987. *Aster spinosus* Benth. (Compositae): an *Erigeron* misplaced in *Aster. American Journal of Botany* 74: 757–758.

Sutherland, S. D. 1982. The pollination biology of paniculate agaves: Documenting the importance of male fitness in plants. Ph.D. dissertation, University of Arizona, Tucson.

Sutherland, S. D. 1987. Why hermaphroditic plants produce many more flowers than fruits: Experimental tests with *Agave mckelveyana. Evolution* 41: 750–759.

Sykes, G. G. 1982. The naming of the boojum. *Journal of Arizona History* 23: 351–356.

Syvertson, J. P., G. L. Cunningham, T. V. Feather. 1975. Anomalous diurnal patterns of stem xylem water potentials in *Larrea tridentata. Ecology* 56: 1423–1428.

Szarek, S. R., P. A. Holthe, I. P. Ting. 1987. Minor physiological response to elevated CO_2 by the CAM plant *Agave vilmoriniana. Plant Physiology* 83: 938–940.

Szarek, S. R., H. B. Johnson, I. P. Ting. 1973. Drought adaptation in *Opuntia basilaris. Plant Physiology* 52: 539–541.

Szarek, S. R., I. P. Ting. 1974. Seasonal patterns of acid metabolism and gas exchange in *Opuntia basilaris. Plant Physiology* 54: 76–81.

Szarek, S. R., I. P. Ting. 1975. Physiological responses to rainfall in *Opuntia basilaris* (Cactaceae). *American Journal of Botany* 62: 602–609.

Szarek, S. R., J. H. Troughton. 1976. Carbon isotope ratios in Crassulacean acid metabolism plants: Seasonal patterns from natural stands. *Plant Physiology* 58: 367–370.

Szarek, S. R., R. M. Woodhouse. 1976. Ecophysiological studies of Sonoran Desert plants, part 1: Diurnal photosynthesis patterns of *Ambrosia deltoidea* and *Olneya tesota. Oecologia* 26: 225–234.

Szarek, S. R., R. M. Woodhouse. 1977. Ecophysiological studies of Sonoran Desert plants, part 2: Seasonal photosynthesis patterns and primary production of *Ambrosia deltoidea* and *Olneya tesota. Oecologia* 28: 365–375.

Szarek, S. R., R. M. Woodhouse. 1978a. Ecophysiological studies of Sonoran Desert plants, part 3: Daily course of photosynthesis for *Acacia greggii* and *Cercidium microphyllum. Oecologia* 35: 285–294.

Szarek, S. R., R. M. Woodhouse. 1978b. Ecophysiological studies of Sonoran Desert plants, part 4: Seasonal photosynthetic capacities of *Acacia greggii* and *Cercidium microphyllum. Oecologia* 37: 221–229.

Takhtajan, A. 1969. *Flowering Plants: Origin and Dispersal.* Oliver and Boyd, Edinburgh.

Taylor, N. P. 1979. Notes on *Ferocactus* B. & R. *Cactus and Succulent Journal (Great Britain)* 41: 88–94.

Taylor, N. P. 1984. A review of *Ferocactus* Britton & Rose. *Bradleya* 2: 19–38.

Theimer, T. C., G. C. Bateman. 1992. Patterns of prickly-pear herbivory by collared peccaries. *Journal of Wildlife Management* 56: 234–240.

Thibodeau, F. R., N. H. Nickerson. 1986. Differential oxidation of mangrove substrate by *Avicennia germinans* and *Rhizophora mangle. American Journal of Botany* 73: 512–516.

Thomas, C. M., S. D. Davis. 1989. Recovery patterns of three chaparral shrub species after wildfire. *Oecologia* 80: 309–320.

Thompson, R. L. 1980. A revision of the genus *Lysiloma* (Leguminosae). Ph.D. dissertation, Southern Illinois University, Carbondale.

Thorn, K. A., A. M. Tinsley, C. W. Weber, J. W. Berry. 1983. Antinutritional factors in legumes of the Sonoran Desert. *Ecology of Food and Nutrition* 13: 251–256.

Tiedemann, A. R., J. O. Klemmedson. 1986. Long-term effects of mesquite (*Prosopis juliflora*) removal on soil characteristics, part 1: Nutrients and bulk density. *Soil Science Society of America Journal* 50: 472–475.

Tiedemann, A. R., E. D. McArthur, D. C. Freeman. 1987. Variations in physiological metabolites and chlorophyll in sexual phenotypes of "Rincon" fourwing saltbush. *Journal of Range Management* 40: 151–155.

Timmermann, B. N. 1977. Practical uses of *Larrea.* In T. J. Mabry, J. H. Hunziker, D. R. DiFeo Jr. (eds.), *Creosote Bush: Biology and Chemistry of Larrea in New World Deserts,* 252–256. Dowden, Hutchinson and Ross, Stroudsburg, Penn.

Ting, Y. C., A. E. Kehr, J. C. Miller. 1957. A cytological study of the sweet potato plant

Ipomoea batatas (L.) Lam. and its related species. *American Naturalist* 91: 197–203.

Tipton, J. L. 1983. Hulling enhances creosotebush *Larrea tridentata* germination. *HortScience* 18: 574.

Tipton, J. L. 1985. Light, osmotic stress, and fungicides affect hulled creosotebush mericarp germination. *Journal of the American Society of Horticultural Science* 110: 615–618.

Tipton, J. L. 1990. Vegetative propagation of Mexican redbud, larchleaf goldenweed, littleleaf ash, and evergreen sumac. *HortScience* 25: 196–198.

Tipton, J. L., E. L. McWilliams. 1979. Ornamental potential of the creosotebush (*Larrea tridentata*). In J. R. Goodin, D. K. Northington (eds.), *Arid Land Plant Resources, Proceedings of the International Arids Lands Conference on Plant Resources,* 699–711. Texas Tech University, Lubbock.

Tissue, D. T., P. S. Nobel. 1988. Parent-ramet connections in *Agave deserti:* Influences of carbohydrates in growth. *Oecologia* 75: 266–271.

Tissue, D. T., P. S. Nobel. 1990. Carbon relations of flowering in a semelparous clonal desert perennial. *Ecology* 71: 273–281.

Tixier, P. 1965. Données cytologiques sur quelques Legumineuses cultivées ou spontanées du Vietnam et du Laos. *Revue de Cytologie et de Biologie Vegetales* 28: 133–155.

Toledo, V. M. 1974. Observations on the relationship between hummingbirds and *Erythrina* species. *Lloydia* 37: 482–487.

Tomlinson, P. B. 1986. *The Botany of Mangroves.* Cambridge University Press, New York.

Tomlinson, P. B., R. B. Primack, J. S. Bunt. 1979. Preliminary observations on floral biology in mangrove Rhizophoraceae. *Biotropica* 11: 256–277.

Traveset, A. 1990. *Ctenosaura similis* Gray (Iguanidae) as a seed disperser in a Central American deciduous forest. *American Midland Naturalist* 123: 402–404.

Tschirley, F. H. 1963. A physio-ecological study of jumping cholla (*Opuntia fulgida* Engelm.). Ph.D. dissertation, University of Arizona, Tucson.

Tschirley, F. H., S. C. Martin. 1960. Germination and longevity of velvet mesquite seed in soil. *Journal of Range Management* 13: 94–97.

Tschirley, F. H., R. F. Wagle. 1964. Growth rate and population dynamics of jumping cholla (*Opuntia fulgida* Englem.). *Journal of the Arizona Academy of Science* 3: 67–71.

Tucker, S. C. 1988. Dioecy in *Bauhinia* resulting from organ suppression. *American Journal of Botany* 75: 1584–1597.

Tucker, S. C., O. L. Stein, K. S. Derstine. 1985. Floral development in *Caesalpinia* (Leguminosae). *American Journal of Botany* 72: 1424–1434.

Turnage, W. V., A. L. Hinckley. 1938. Freezing weather in relation to plant distribution in the Sonoran Desert. *Ecological Monographs* 8: 530–550.

Turnage, W. V., T. D. Mallery. 1941. *An Analysis of Rainfall in the Sonoran Desert and Adjacent Territory.* Carnegie Institution of Washington Publication no. 529. Washington, D.C.

Turner, B. L. 1959. *The Legumes of Texas.* University of Texas Press, Austin.

Turner, B. L. 1963. Documented chromosome numbers of plants. *Madroño* 17: 116–117.

Turner, B. L., W. L. Ellison, R. M. King. 1961. Chromosome numbers in the Compositae, part 4: North American species, with phyletic interpretations. *American Journal of Botany* 48: 216–223.

Turner, B. L., O. S. Fearing. 1960. Chromosome numbers in the Leguminosae, part 3:

Species of the southwestern United States and Mexico. *American Journal of Botany* 47: 603–608.

Turner, B. L., D. Flyr. 1966. Chromosome numbers in the Compositae, part 10: North American species. *American Journal of Botany* 53: 24–33.

Turner, B. L., A. M. Powell, R. M. King. 1962. Chromosome numbers in the Compositae, part 6: Additional Mexican and Guatemalan species. *Rhodora* 64: 251–271.

Turner, B. L., A. M. Powell, T. J. Watson. 1973. Chromosome numbers in Asteraceae. *American Journal of Botany* 60: 592–596.

Turner, F. B., D. C. Randall. 1987. The phenology of desert shrubs in southern Nevada, U.S.A. *Journal of Arid Environments* 13: 119–128.

Turner, F. B., D. C. Randall. 1989. Net production by shrubs and winter annuals in southern Nevada. *Journal of Arid Environments* 17: 23–26.

Turner, R. M. 1963. Growth in four species of Sonoran Desert trees. *Ecology* 44: 760–765.

Turner, R. M. 1974. *Quantitative and Historical Evidence of Vegetation Changes along the upper Gila River, Arizona.* U.S. Geological Survey Professional Paper 66-H. U.S. Government Printing Office, Washington, D.C.

Turner, R. M. 1982. Mohave desertscrub. *Desert Plants* 4: 157–168.

Turner, R. M. 1990. Long-term vegetation change at a fully protected Sonoran Desert site. *Ecology* 71: 464–477.

Turner, R. M., S. M. Alcorn, G. Olin. 1969. Mortality of transplanted saguaro seedlings. *Ecology* 50: 835–844.

Turner, R. M., S. M. Alcorn, G. Olin, J. A. Booth. 1966. The influence of shade, soil, and water on saguaro seedling establishment. *Botanical Gazette* 127: 95–102.

Turner, R. M., J. E. Bowers. 1988. Long-term changes in populations of *Carnegiea gigantea,* exotic plant species and *Cercidium floridum* at the Desert Laboratory, Tumamoc Hill, Tucson, Arizona. In E. Whitehead, C. F. Hutchinson, B. N. Timmermann, R. G. Varady (eds.), *Arid Lands Today and Tomorrow,* 445–455. Westview Press, Boulder, Colo.

Turner, R. M., D. E. Brown. 1982. Sonoran desertscrub. *Desert Plants* 4: 181–221.

Turner, R. M., C. L. Busman. 1984. Vegetative key for identification of the woody legumes of the Sonoran Desert region. *Desert Plants* 6: 189–202.

Turner, R. M., M. M. Karpiscak. 1980. *Recent Vegetation Changes along the Colorado River between Glen Canyon Dam and Lake Mead, Arizona.* U.S. Geological Survey Professional Paper 1132. Government Printing Office, Washington, D.C.

Udovic, D. 1981. Determinants of fruit set in *Yucca whipplei:* Reproductive expenditure vs. pollinator availability. *Oecologia* 48: 389–399.

Uhl, N. W., J. Dransfield. 1987. *Genera Palmarum.* The L. H. Bailey Hortorium and the International Palm Society, Lawrence, Kans.

Valentine, K. A., J. B. Gerard. 1968. *Life-History Characteristics of the Creosote-bush, Larrea tridentata.* University of New Mexico Agricultural Experiment Station Bulletin no. 526. Las Cruces.

Vales, M. A., C. Martinez. 1983. Contribution to the anatomical study of the xylem of the Simarubaceae family in Cuba: 1. *Alvaradoa* and *Simarouba. Acta Botanica, Academiae Scientiarum Hungaricae* 29: 231–240.

Van Auken, O. W., J. K. Bush. 1985. Secondary succession on terraces of the San Antonio River. *Bulletin of the Torrey Botanical Club* 112: 158–166.

Van Auken, O. W., E. M. Gese, K. Connors. 1985. Fertilization response of early and late successional species *Acacia smallii* and *Celtis laevigata. Botanical Gazette* 146: 564–570.

van der Maarel, E. 1990. Ecotones and ecoclines are different. *Journal of Vegetation Science* 1: 135–138.

Vandermeer, J. 1980. Saguaros and nurse trees: A new hypothesis to account for population fluctuations. *Southwestern Naturalist* 25: 357–360.

Van Devender, T. R. 1987. Holocene vegetation and climate in the Puerto Blanco Mountains, southwestern Arizona. *Quaternary Research* 27: 51–72.

Ven Devender, T. R. 1990a. Late Quaternary vegetation and climate of the Chihuahuan Desert, United States and Mexico. In J. L. Betancourt, T. R. VanDevender, P. S. Martin (eds.), *Packrat Middens: The Last 40,000 Years of Biotic Change,* 104–133. University of Arizona Press, Tucson.

Van Devender, T. R. 1990b. Late Quaternary vegetation and climate of the Sonoran Desert, United States and Mexico. In J. L. Betancourt, T. R. VanDevender, P. S. Martin (eds.), *Packrat Middens: The Last 40,000 Years of Biotic Change,* 134–163. University of Arizona Press, Tucson.

Van Devender, T. R., T. L. Burgess. 1985. Late Pleistocene woodlands in the Bolsón de Mapimí: A refugium for the Chihuahuan Desert biota? *Quaternary Research* 24: 346–353.

Van Devender, T. R., T. L. Burgess, R. S. Felger, R. M. Turner. 1990. *Holocene Vegetation of the Hornaday Mountains of Northwestern Sonora, Mexico.* Proceedings of the San Diego Society of Natural History no. 2.

Van Devender, T. R., L. J. Toolin. 1983. Late Quaternary vegetation of the San Andres Mountains, Sierra County, New Mexico. In P. L. Eidenbach (ed.), *The Prehistory of Rhodes Canyon: Survey and Mitigation,* 33–54. Human Systems Research, Inc., report to Holloman Air Force Base, Tularosa, N.M.

Van Devender, T. R., L. J. Toolin, T. L. Burgess. 1990. The ecology and paleoecology of grasses in selected Sonoran Desert communities. In J. L. Betancourt, T. R. Van Devender, P. S. Martin (eds.), *Packrat Middens: The Last 40,000 Years of Biotic Change,* 326–349. University of Arizona Press, Tucson.

Van Epps, G. A. 1975. Winter injury to four-wing saltbush. *Journal of Range Management* 28: 157–159.

van Hylckama, T. E. A. 1974. *Water Use by Saltcedar as Measured by the Water Budget Method.* U.S. Geological Survey Professional Paper 491-E. Government Printing Office, Washington, D.C.

Van Kessel, C., J. P. Roskoski, T. Wood, J. Montano. 1983. $^{15}N_2$ fixation and H_2 evolution by six species of tropical leguminous trees. *Plant Physiology* 72: 909–910.

Vasek, F. C. 1980a. Creosote bush: Long-lived clones in the Mojave Desert. *American Journal of Botany* 67: 246–255.

Vasek, F. C. 1980b. Early successional stages in Mojave Desert scrub vegetation. *Israel Journal of Botany* 28: 133–148.

Vasek, F. C. 1983. Plant succession in the Mojave Desert. *Crossosoma* 9: 1–23.

Vasek, F. C., H. B. Johnson, D. H. Eslinger. 1975. Effects of pipeline construction on a creosote bush scrub vegetation of the Mojave Desert. *Madroño* 23: 1–13.

Vassal, J., P. Guinet. 1972. Un *Acacia* Américain a pétiole diaphyllodinisé *A. willardiana* Rose. *Adansonia* 12: 421–428.

Vaughan, T. A., S. T. Schwartz. 1980. Behavioral ecology of an insular woodrat. *Journal of Mammalogy* 61: 205–218.

Vaurie, P. 1971. Review of *Scyphophorus* (Curculionidae: Rhynchophorinae). *Coleopterists Bulletin* 25: 1–8.

Viereck, L. A., E. L. Little. 1975. *Atlas of United States Trees,* vol. 2, *Alaska Trees and Common Shrubs.* USDA Miscellaneous Publication no. 1293. Government Printing Office, Washington, D.C.

Vines, R. A. 1960. *Trees, Shrubs and Woody Vines of the Southwest.* University of Texas Press, Austin.

Vivó, J. A., J. C. Gomez. 1946. *Climatologia de México.* Instituto Panamericano de Geografía e Historía Publicación no. 19.

Vogl, R. J., L. T. McHargue. 1966. Vegetation of California fan palm oases on the San Andreas fault. *Ecology* 47: 532–540.

Vora, R. S. 1989. Seed germination characteristics of selected native plants of the lower Rio Grande Valley, Texas, U.S. A. *Journal of Range Management* 42: 36–40.

Wadsworth, F. H. 1959. Growth and regeneration of white mangrove in Puerto Rico. *Caribbean Forestry* 20: 59–71.

Wagner, H., C. Ludwig, L. Grotjahn, M. S. Y. Khan. 1987. Biologically active saponins from *Dodonaea viscosa. Phytochemistry* 26: 697–701.

Wagner, W. L., E. F. Aldon. 1978. *Manual of the Saltbushes (Atriplex spp.) in New Mexico.* USDA Forest Service General Technical Report RM–57. Fort Collins, Colo.

Walker, P. A., K. D. Cocks. 1991. HABITAT: A procedure for modelling a disjoint environmental envelope for a plant or animal species. *Global Ecology and Biogeography* 1: 108–118.

Wallace, A., S. A. Bamberg, J. W. Cha. 1974. Quantitative studies of roots of perennial plants in the Mojave Desert. *Ecology* 55: 1160–1162.

Wallace, A., J. W. Cha, R. T. Mueller, E. M. Romney. 1980. Retranslocation of tagged carbon in *Ambrosia dumosa. Great Basin Naturalist Memoirs* 4: 168–171.

Wallace, A., E. M. Romney, R. T. Ashcroft. 1970. Soil temperature effects on growth of seedlings of some shrub species which grow in the transitional area between the Mojave and the Great Basin deserts. *BioScience* 20: 1158–1159.

Wallace, A., E. M. Romney, J. W. Cha. 1980. Depth distribution of roots of some perennial plants in the Nevada Test Site area of the northern Mojave Desert. *Great Basin Naturalist Memoirs* 4: 201–207.

Wallace, A., E. M. Romney, V. Q. Hale. 1973. Sodium relations in desert plants: 1. Cation contents of some plant species from the Mojave and Great Basin deserts. *Soil Science* 115: 284–287.

Wallace, A., E. M. Romney, R. T. Mueller. 1982. Sodium relations in desert plants: 7. Effects of sodium chloride on *Atriplex polycarpa* and *Atriplex canescens. Soil Science* 134: 65–68.

Wallace, C. S., P. W. Rundel. 1979. Sexual dimorphism and resource allocation in male and female shrubs of *Simmondsia chinensis. Oecologia* 44: 34–39.

Wallen, D. R., J. A. Ludwig. 1978. Energy dynamics of vegetative and reproductive growth in Spanish bayonet (*Yucca baccata* Torr.). *Southwestern Naturalist* 23: 409–422.

Walsh, G. E. 1979. Mangroves: A review. In R. J. Reimold, W. H. Queen (eds.) *Ecology of Halophytes,* 51–174. Academic Press, New York.

Walter, H. 1979. *Vegetation of the Earth and Ecological Systems of the Geobiosphere.* 2d ed. Springer-Verlag, New York.

Walter, H., E. Stadelmann. 1974. A new approach to the water relations of desert plants. In G. W. Brown (ed.), *Desert Biology,* vol. 2, 213–310. Academic Press, New York.

Walters, J. P., C. E. Freeman. 1983. Growth rates and root-shoot ratios in seedlings of the desert shrub *Larrea tridentata. Southwestern Naturalist* 28: 357–364.

Wan, C., R. E. Sosebee. 1990. Characteristics of photosynthesis and conductance in early and late leaves of honey mesquite. *Botanical Gazette* 151: 14–20.

Ward, D. E. 1984. Chromosome counts from New Mexico and Mexico. *Phytologia* 56: 55–60.

Wardlaw, I. F., J. E. Begg, D. Bagnall, R. L. Dunstone. 1983. Jojoba: Temperature adaptation as expressed in growth and leaf form. *Australian Journal of Plant Physiology* 10: 299–312.

Waring, G. 1986. Creosote bush: The ultimate desert survivor. *Agave* 2: 3–12.

Waring, G. L., P. W. Price. 1990. Plant water stress and gall formation (Cecidomyiidae: *Asphondylia* spp.) on creosote bush. *Ecological Entomology* 15: 87–95.

Waring, G. L., R. L. Smith. 1987. Patterns of faunal succession in *Agave palmeri. Southwestern Naturalist* 32: 489–497.

Warren, D. K. 1979. Precipitation and temperature as climate determinants of the distribution of *Fouquieria columnaris.* Ph.D. dissertation, Walden University, Naples, Florida.

Warren, D. K., R. M. Turner. 1975. Saltcedar (*Tamarix chinensis*) seed production, seedling establishment, and response to inundation. *Journal of the Arizona Academy of Science* 10: 135–144.

Warrick, G. D., P. R. Krausman. 1989. Barrel cacti consumption by desert bighorn sheep. *Southwestern Naturalist* 34: 483–486.

Waser, N. M. 1979. Pollinator availability as a determinant of flowering time in ocotillo (*Fouquieria splendens*). *Oecologia* 39: 107–121.

Watson, M. C., J. W. O'Leary, E. P. Glenn. 1987. Evaluation of *Atriplex lentiformis* (Torr.) S. Wats. and *Atriplex nummularia* Lindl. as irrigated forage crops. *Journal of Arid Environments* 13: 293–303.

Watson, S. 1873. Revision of the extra-tropical North American species of the genus *Oenothera. Proceedings of the American Academy of Arts and Sciences* 8: 573–618.

Watson, S. 1886. Contributions to American botany. 1. List of plants collected by Dr. Edward Palmer in southwestern Chihuahua, Mexico, in 1885. *Proceedings of the American Academy of Arts and Sciences* 21: 414–445.

Webb, R. H., J. W. Steiger, E. B. Newman. 1987. *The Response of Vegetation to Disturbance in Death Valley National Monument, California.* U.S. Geological Survey Bulletin 1793. Government Printing Office, Washington, D.C.

Webb, R. H., H. G. Wilshire. 1980. Recovery of soils and vegetation in a Mojave Desert ghost town, Nevada, U.S.A. *Journal of Arid Environments* 3: 291–303.

Webber, I. E. 1936. Systematic anatomy of the woods of the Simaroubaceae. *American Journal of Botany* 23: 577–587.

Webber, J. M. 1953. *Yuccas of the Southwest.* USDA Agricultural Monograph no. 17. Government Printing Office, Washington, D.C.

Webber, J. M. 1960. Hybridization and instability of *Yucca. Madroño* 15: 187–192.

Weedin, J. R., A. M. Powell. 1978a. Chromosome numbers in Chihuahuan Desert Cactaceae, trans-Pecos Texas. *American Journal of Botany* 65: 531–537.

Weedin, J. R., A. M. Powell. 1978b. IOPB chromosome number reports LX. *Taxon* 27: 223–231.

Wells, P. V. 1961. Succession in desert vegetation on streets of a Nevada ghost town. *Science* 134: 670–671.

Wells, P. V., R. Berger. 1967. Late Pleistocene history of coniferous woodland in the Mohave Desert. *Science* 155: 1640–1647.

Wells, P. V., R. R. Johnson. 1964. *Vauquelinia pauciflora* (Rosaceae) from Guadalupe Canyon, Arizona: A species of tree newly reported for the United States. *Southwestern Naturalist* 9: 151–154.

Wells, P. V., D. Woodcock. 1985. Full-glacial vegetation of Death Valley, California: Juniper woodland opening to *Yucca* semidesert. *Madroño* 32: 11–23.

Welsh, R. G., R. F. Beck. 1976. Some ecological relationships between creosotebush and bush muhly. *Journal of Range Management* 29: 472–475.

Welsh, S. L., N. D. Atwood, L. C. Higgins, S. Goodrich. 1987. *A Utah Flora.* Great Basin Naturalist Memoirs no. 9. Brigham Young University Press, Provo, Utah.

Wendelken, P. W., R. F. Martin. 1987. Avian consumption of *Guaiacum sanctum* fruit in the arid interior of Guatemala. *Biotropica* 19: 116–121.

Went, F. W. 1948. Ecology of desert plants, part 1: Observations on germination in the Joshua Tree National Monument, California. *Ecology* 29: 242–253.

Went, F. W. 1957. *The Experimental Control of Plant Growth.* Chronica Botanica, Waltham, Mass.

Went, F. W., M. Westergaard. 1949. Ecology of desert plants, part 3: Development of plants in the Death Valley National Monument, California. *Ecology* 30: 26–38.

Werk, K. S., J. R. Ehleringer. 1983. Photosynthesis by flowers in *Encelia farinosa* and *Encelia californica* (Asteraceae). *Oecologia* 57: 311–315.

Werner, A., R. Stelzer. 1990. Physiological responses of the mangrove *Rhizophora mangle* grown in the absence and presence of sodium chloride. *Plant, Cell, and Environment* 13: 243–256.

West, J. G. 1984. A revision of *Dodonaea* Miller (Sapindaceae) in Australia. *Brunonia* 7: 1–194.

Westman, W. E. 1981a. Factors influencing the distribution of Californian coastal sage scrub. *Ecology* 62: 439–455.

Westman, W. E. 1981b. Diversity relations and succession in Californian coastal sage scrub. *Ecology* 62: 170–184.

Westman, W. E., J. F. O'Leary, G. P. Malanson. 1981. The effects of fire intensity, aspect, and substrate on post-fire growth of Californian coastal sage scrub. In N. S. Margaris, H. A. Mooney (eds.), *Components of Productivity of Mediterranean-Climate Regions—Basic and Applied Aspects,* 151–180. Dr. W. Junk, The Hague, Netherlands.

Wetten, M., H. C. Ruben-Sutter. 1982. *Verbreitungsatlas der Farn- und Blütenpflanzen der Schweiz.* 2 vols. Birkhäuser Verlag, Basel.

Whalen, M. A. 1987. Systematics of *Frankenia* (Frankeniaceae) in North and South America. *Systematic Botany Monographs* 17: 1–93.

Wheeler, E. A., C. A. LaPasha, R. B. Miller. 1989. Wood anatomy of elm (*Ulmus*) and hackberry (*Celtis*) species native to the United States. *IAWA Bulletin* 10: 5–26.

Whisenant, S. G., D. N. Ueckert. 1981. Germination responses of *Anisacanthus wrightii* (Torr.) Gray (Acanthaceae) to selected environmental variables. *Southwestern Naturalist* 26: 379–384.

White, S. S. 1949. The vegetation and flora of the region of the Rio de Bavispe in northeastern Sonora, Mexico. *Lloydia* 11: 229–302.

Whitham, T. G. 1977. Coevolution of foraging in *Bombus* and nectar dispensing in *Chilopsis:* A last dreg theory. *Science* 197: 593–596.

Wieland, P. A. T., E. F. Frolich, A. Wallace. 1971. Vegetative propagation of woody shrub species from the northern Mojave and southern Great Basin deserts. *Madroño* 21: 149–152.

Wierenga, P. J., J. M. Hendricks, M. H. Nash, L. A. Daugherty. 1987. Variation of soil and vegetation with distance along a transect in the Chihuahuan Desert. *Journal of Arid Environments* 13: 53–63.

Wiggins, I. L. 1940. New and poorly known species from the Sonoran Desert. *Contributions from the Dudley Herbarium* 3: 65–68.

Wiggins, I. L. 1942. *Acacia angustissima* (Mill.) Ktze. and its near relatives. *Contributions from the Dudley Herbarium* 3: 227–239.

Wiggins, I. L. 1964. Flora of the Sonoran Desert. In F. Shreve, I. L. Wiggins, *Vegetation and Flora of the Sonoran Desert,* 2 vols., 187–840. Stanford University Press, Stanford, Calif.

Wiggins, I. L. 1980. *Flora of Baja California.* Stanford University Press, Stanford, Calif.

Wille, A. 1963. Behavioral adaptations of bees for pollen collecting from *Cassia* flowers. *Revista de Biología Tropical* 11: 205–210.

Williams, K. B., C. D. Bonham. 1972. Leaf variation in *Vauquelinia californica* populations of Arizona. *Journal of the Arizona Academy of Science* 7: 47–50.

Williams, S. E., A. G. Wollum, E. F. Aldon. 1974. Growth of *Atriplex canescens* (Pursh) Nutt. improved by formation of vesicular-arbuscular mycorrhizae. *Soil Science Society of America Proceedings* 38: 962–965.

Wilson, R. C. 1976. *Abronia,* part 4: Acanthocarp dispersibility and its ecological implications for nine species of *Abronia. Aliso* 8: 493–506.

Wilson, R. T., D. R. Krieg, B. E. Dahl. 1974. A physiological study of developing pods and leaves of honey mesquite. *Journal of Range Management* 27: 202–203.

Wisdom, C. S. 1985. Use of chemical variation and predation as plant defenses by *Encelia farinosa* against a specialist herbivore. *Journal of Chemical Ecology* 11: 1553–1566.

Wisdom, C. S., E. Rodriguez. 1982. Quantitative variation of the sesquiterpene lactones and chromenes of *Encelia farinosa. Biochemical Systematics and Ecology* 10: 43–48.

Wolf, C. B. 1935. California plant notes, part 1. *Occasional Papers of the Rancho Santa Ana Botanic Garden,* ser. 1, 1: 31–43.

Wollenweber, E., D. Hradetzky, K. Mann, J. N. Roitman, G. Yatskievych, M. Proksch, P. Proksch. 1987. Exudate flavonoids from aerial parts of five *Ambrosia* species. *Journal of Plant Physiology* 131: 37–44.

Woodhouse, R. M., J. G. Williams, P. S. Nobel. 1980. Leaf orientation, radiation interception, and nocturnal acidity increases by the CAM plant *Agave deserti* (Agavaceae). *American Journal of Botany* 67: 1179–1185.

Woodson, R. E. 1945. The North American species of *Asclepias* L. *Annals of the Missouri Botanical Garden* 41: 1–211.

Wright, S. J., H. F. Howe. 1987. Pattern and mortality in Colorado Desert plants. *Oecologia* 73: 543–552.

Wunderlin, R. P. 1973. IOPB chromosome number reports XLII. *Taxon* 22: 647–654.

Wunderlin, R. P. 1983. Revision of the arborescent *Bauhinias* (Fabaceae: Caesalpinioideae: Cercideae) native to Middle America. *Annals of the Missouri Botanical Garden* 70: 95–127.

Yang, T. W. 1967a. Chromosome numbers in populations of creosote bush (*Larrea divaricata*) in the Chihuahuan and Sonoran subdivisions of the North American Desert. *Journal of the Arizona Academy of Science* 4: 183–184.

Yang, T. W. 1967b. Ecotypic variation in *Larrea divaricata. American Journal of Botany* 54: 1041–1044.

Yang, T. W. 1970. Major chromosome races of *Larrea divaricata* in North America. *Journal of the Arizona Academy of Science* 6: 41–45.

Yang, T. W., C. H. Lowe. 1968. Chromosome variation in ecotypes of *Larrea divaricata* in the North American desert. *Madroño* 19: 161–164.

Yatskievych, G., P. C. Fischer. 1984. New plant records from the Sonoran Desert. *Desert Plants* 5: 180–185.

Yeaton, R. I. 1978. A cyclical relationship between *Larrea tridentata* and *Opuntia leptocaulis* in the northern Chihuahuan Desert. *Journal of Ecology* 66: 651–656.

Yeaton, R. I., R. Karban, H. B. Wagner. 1980. Morphological growth patterns of saguaro (*Carnegiea gigantea:* Cactaceae) on flats and slopes in Organ Pipe Cactus National Monument, Arizona. *Southwestern Naturalist* 25: 339–349.

Yeaton, R. I., R. W. Yeaton, J. P. Waggoner, J. E. Horenstein. 1985. The ecology of *Yucca* (Agavaceae) over an environmental gradient in the Mojave Desert, California, USA: Distribution and interspecific interactions. *Journal of Arid Environments* 8: 33–44.

Yensen, N. P., E. P. Glenn, M. R. Fontes. 1983. Biogeographical distribution of salt marsh halophytes on the coasts of the Sonoran Desert. *Desert Plants* 5: 76–81.

Yoshioka, H., K. Kondo, M. Segawa, K. Nehira, S. Maeda. 1984. Karyomorphological studies in five species of mangrove genera in the Rhizophoraceae. *La Kromosomo* ser. 2, 35–36: 1111–1116.

Young, D. A. 1974a. Introgressive hybridization in two southern California species of *Rhus* (Anacardiaceae). *Brittonia* 26: 241–255.

Young, D. A. 1974b. Comparative wood anatomy of *Malosma* and related genera (Anacardiaceae). *Aliso* 8: 133–146.

Young, D. A. 1975. Correction of the geographic distribution of *Rhus microphylla* (Anacardiaceae). *Madroño* 23: 78.

Young, J. A., R. A. Evans, B. L. Kay. 1977. *Ephedra* seed germination. *Agronomy Journal* 69: 209–211.

Young, J. A., B. L. Kay, H. George, R. A. Evans. 1980. Germination of three species of *Atriplex. Agronomy Journal* 72: 705–709.

Yuasa, H., H. Shimizu, S. Kashiwai, N. Kondo. 1974. Chromosome numbers and their bearing on the geographic distribution in the subfamily Opuntioideae (Cactaceae). *Rep. Inst. Breed. Research, Tokyo Univ. Agric.* 4: 1–10.

Zabriskie, J. G. 1979. *Plants of Deep Canyon and the Central Coachella Valley, California.* Philip L. Boyd Deep Canyon Desert Research Center, University of California, Riverside.

Zabriskie, N., J. Zabriskie. 1976. Vegetation. In I. P. Ting, B. Jennings (eds.), *Deep Canyon, A Desert Wilderness for Science,* 117–124. Philip L. Boyd Deep Canyon Research Center, University of California, Riverside.

Zavortink, T. J. 1975. Host plants, behavior, and distribution of the eucerine bees *Idiomelissodes duplocincta* (Cockerell) and *Syntrichalonia exquisita* (Cresson) (Hymenoptera: Anthophoridae). *Pan-Pacific Entomologist* 51: 236–242.

Zedler, P. H. 1981. Vegetation change in chaparral and desert communities in San Diego County, California. In D. C. West, H. H. Shugart, D. B. Botkin (eds.), *Forest Succession,* 406–430. Springer Verlag, New York.

Zhai, S.-h., M.-x. Li. 1986. Chromosome number of *Tamarix* L. *Acta Phytotaxonomica Sinica* 24: 273–274.

Zona, S. 1990. A monograph of *Sabal* (Arecaceae: Coryphoideae). *Aliso* 12: 583–666.

Index

About the Authors

RAYMOND M. TURNER graduated in botany from the University of Utah in 1948 and received his doctorate in plant ecology from Washington State University in 1954. He worked as an ecologist in the U.S. Geological Survey from 1962 until his retirement in 1989. The principal focus of his work has been documentation of landscape changes in the southwestern United States, northwestern Mexico, and Kenya. He is coauthor, with J. R. Hastings, of *The Changing Mile* and has published studies of vegetation change along the Colorado and Gila rivers in Arizona and of changes with time of permanent vegetation plots. Work on the atlas began in October 1963 when Hastings and Turner took their first trip to Baja California with the main purpose of matching old photographs and of establishing permanent plots for the study of apical growth in *Fouquieria columnaris* and *Pachycereus pringlei* and for evaluating changes in the populations of these species through time. At one of their first camps after leaving their base in Tucson, they decided to collect plant distribution data in a systematic fashion during this first traverse of the peninsula and to expand the effort later to take in the rest of the Sonoran Desert. This work has continued through three decades, resulting in publication of the 1972 report by Hastings, Turner, and Douglas K. Warren, *An Atlas of Some Plant Distributions in the Sonoran Desert,* and culminating in the present greatly expanded volume.

JANICE E. BOWERS, a botanist with the U.S. Geological Survey, has been stationed at the Desert Laboratory on Tumamoc Hill, Tucson, Arizona, since 1982. There, she has studied flowering time of the saguaro and other desert plants; seedling emergence of brittlebush, ocotillo, and paloverde; and plant succession on Grand Canyon debris flows. She is author of *A Sense of Place: The Life and Work of Forrest Shreve* (University of Arizona Press, 1988) and several other pieces about the Desert Laboratory and its staff, as well as *The Mountains Next Door* (University of Arizona Press, 1991) and *A Full Life in a Small Place and Other Essays from a Desert Garden* (University of Arizona Press, 1993). She has devoted much of the past 20 years to collecting and identifying southern Arizona plants and has prepared several plant checklists for local areas, including Organ Pipe Cactus National Monument, Tumamoc Hill, the Rincon Mountains, and the northern Santa Rita Mountains. Currently, she is studying flower and fruit production of prickly pears and barrel cacti and, with Steven P. McLaughlin, is preparing a flora for the Huachuca Mountains on the U.S.-Mexico border.

TONY L. BURGESS is an assistant staff scientist with the Desert Laboratory on Tumamoc Hill, Tucson, Arizona. He graduated from the University of Arizona in 1971 with a degree in biology after writing an honors thesis on the mammals of Canyon de Chelly. Following service in Vietnam with the U.S. Army, he entered Texas Tech University, where he studied vegetation and flora of the Guadalupe Mountains, vegetation of the Middle Pecos Valley, and hybridization in agaves. Burgess was awarded his master's degree in botany in 1977 and joined the U.S. Geological Survey in 1978, with the primary task of working on this Sonoran Desert atlas. During the 1980s he began work with Biosphere 2 on the design and assembly of synthetic desert, thornscrub, and savanna communities. He completed his doctorate in ecology in 1988 at the University of Arizona, with a dissertation on the relationship between climate and *Agave* leaf morphology in the Vizcaíno region of Baja California. Recently he has collaborated at the Tumamoc Desert Laboratory in studies of plant growth forms in arid climates and exotic plant invasions, and he continues as a consultant with Biosphere 2 investigating vegetation and soil patterns of Apacherian savannas.